DIGITAL COMMUNICATIONS

DIGITAL COMMUNICATIONS

Mehmet Şafak

This edition first published 2017
© 2017 John Wiley & Sons Ltd

Registered Office
John Wiley & Sons Ltd, The Atrium, Southern Gate, Chichester, West Sussex, PO19 8SQ, United Kingdom

For details of our global editorial offices, for customer services and for information about how to apply for permission to reuse the copyright material in this book please see our website at www.wiley.com.

Library of Congress Cataloging-in-Publication Data

Names: Şafak, Mehmet, 1948– author.
Title: Digital communications / Mehmet Şafak.
Description: Chichester, UK ; Hoboken, NJ : John Wiley & Sons, 2017. |
  Includes bibliographical references and index.
Identifiers: LCCN 2016032956 (print) | LCCN 2016046780 (ebook) | ISBN 9781119091257 (cloth) |
  ISBN 9781119091264 (pdf) | ISBN 9781119091271 (epub)
Subjects: LCSH: Digital communications.
Classification: LCC TK5103.7 .S24 2017 (print) | LCC TK5103.7 (ebook) | DDC 621.382–dc23
LC record available at https://lccn.loc.gov/2016032956

A catalogue record for this book is available from the British Library.

Cover Design: Wiley
Cover Image: KTSDESIGN/Gettyimages

Set in 10/12pt Times by SPi Global, Pondicherry, India

To my children Emre and Ilgın

Contents

Preface

Telecommunications is a rapidly evolving area of electrical engineering, encompassing diverse areas of applications, including RF communications, radar systems, ad-hoc networks, sensor networks, optical communications, radioastronomy, and so on. Therefore, a solid background is needed on numerous topics of electrical engineering, including calculus, antennas, wave propagation, signals and systems, random variables and stochastic processes and digital signal processing. In view of the above, the success in the telecommunications education depends on the background of the student in these topics and how these topics are covered in the curriculum. For example, the Fourier transform may not usually be taught in relation with time- and frequency-response of the systems. Similarly, concepts of probabiliy may not be related to random signals. On the other hand, students studying telecommunications may not be expected to know the details of the Maxwell's equations and wave propagation. However, in view of the fact that wireless communication systems comprise transmit/receive antennas and a propagation medium, it is necessary to have a clear understanding of the radiation by the transmit antenna, propagation of electromagnetic waves in the considered channel and the reception of electromagnetic waves by the receive antenna. Otherwise, the students may face difficulties in understanding the telecommunications process in the physical layer.

The engineering education requires a careful tradeoff between the rigour provided by the theory and the simple exposure of the corresponding physical phenomena and their applications in our daily life. Therefore, the book aims to help the students to understand the basic principles and to apply them. Basic principles and analytical tools are provided for the design of communication systems, illustrated with examples, and supported by graphical illustrations.

The book is designed to meet the needs of electrical engineering students at undergraduate and graduate levels, and those of researchers and practicing engineers. Though the book is on digital communications, many concepts and approaches presented in the book are also applicable for analog communication systems. The students are assumed to have basic knowledge of Maxwell's equations, calculus, matrix theory, probability and stochastic processes, signals and systems and digital signal processing. Mathematical tools required for understanding some topics are incorporated in the relevant chapters or are presented in the appendices. Each chapter contains graphical illustrations, figures, examples, references, and problems for better understanding the exposed concepts.

Chapter 1 Signal Analysis summarizes the time-frequency relationship and basic concepts of Fourier transform for deterministic and random signals used in the linear systems. The aim was to provide a handy reference and to avoid repeating the same basic concepts in the subsequent chapters. *Chapter 2 Antennas* presents the fundamentals of the antenna theory with emphasis on the telecommunication aspects rather than on the Maxwell's equations. *Chapter 3 Channel Modeling* presents the propagation processes following the conversion of the electrical signals in the transmitter into electromagnetic waves by the transmit antenna until they are reconverted into electrical signals by the receive antenna. *Chapter 4 System Noise* is mainly based on the standards for determining the receiver noise of internal and external origin and provides tools for calculating SNR at the receiver output; the SNR is known to be the figure-of-merit of communication systems since it determines the system performance. Chapters 2, 3 and 4 thus relate the wireless interaction between transmitter and receiver in the physical layer. It may be worth mentioning that, unlike many books on wireless communications, covering only VHF and UHF bands, Chapters 2, 3 and 4 extend the coverage of antennas, receiver noise and channel modeling to SHF and EHF bands. A thorough understanding of the materials provided in these chapters is believed to be critical for deeper understanding of the rest of the book. These three chapters are believed to close the gap between the approaches usually followed by books on antennas and RF propagation, based on the Maxwell's equations, and the books on digital communications, based on statistical theory of communications. One of the aims of the book is to help the students to fuse these two complementary approaches.

The following chapters are dedicated to statistical theory of digital communications. *Chapter 4 Pulse Modulation* treats the conversion of analog signals into digital for digital communication systems. Sampling, quantization and encoding tradeoffs are presented, line codes used for pulse transmission are related to the transmission bandwidth. Time division multiplexing (TDM) allows multiple digital signals to be transmitted as a single signal. At the receiver they are reconverted into analog for the end user. PCM and other pulse modulations as well as audio and video coding techniques are also presented. *Chapter 5 Baseband Modulation* focuses on the optimal reception of pulse modulated signals and intersymbol interference (ISI) between pulses, due to filtering so as to limit the transmission bandwidth or to minimize the received noise power. In an AWGN channel, the optimum receiver maximizes the output SNR by matching the receive filter characteristics to those of the transmitter. The optimal choice of pulse shape, for example, Nyquist, raised-cosine, or correlative-level coding (partial-response signaling) is also presented in order to mitigate the ISI. *Chapter 7 Optimum Receiver in AWGN Channels* is focused on the geometric representation of the signals so as to be able to identify the two functionalities (demodulation and detection) of an optimum receiver. Based on this approach, derivation of the bit error probability (BEP) is presented and upper bounds are provided when the BEP can not be obtained exactly. *Chapter 8 Passband Modulation Techniques* starts with the definition of bandwidth and the bandwidth efficiency, followed by the synchronization (in frequency, phase and symbol timing) between transmitted and received symbols. The PSD, bandwidth and power efficiencies and bit/symbol error probabilities are derived for M-ary coherent, differentially coherent and noncoherent modulations, for example, M-ary PSK, M-ary ASK, M-ary FSK, M-ary QAM and M-ary DPSK. This chapter also provides a comparasion of spectrum and power efficiencies of the above-cited passband modulation techniques. *Chapter 9 Error Control Coding* presents the principles of channel coding in order to control (detect and/or correct) Gaussian (random) and burst errors occuring in the channel due to noise, fading, shadowing and other potential sources of interference. Source coding is not addressed in the book. Channel coding usually comes at the expense of increased transmission rate, hence wider transmission bandwidth, due to the inclusion of additional (parity check) bits among the data bits. Use of parity check bits reduces energy per channel bits and hence leads to higher channel BEP. However, a good code is expected to correct more errors than it creates and the overall coded BEP decreases at the expense of increased transmission bandwidth. This tradeoff between the BEP and the transmission bandwidth is well-known in the coding theory. As shown by the Shannon capacity theorem, one can achieve error-free communications as the transmission bandwidth goes to infinity, that is, by using infinitely many parity check bits, as long as the ratio of the energy per bit to noise PSD (E_b/N_0) is higher than $-1.6\,$dB. This chapter addresses block and convolutional codes which are capable of correcting random and burst-errors. Automatic-repeat request (ARQ) techniques based on error-detection codes and hybrid ARQ (HARQ) techniques exploiting codes which can both detect and correct channel bit errors are also presented. *Chapter 10 Broadband Transmission Techniques* is composed of mainly two sections, namely spread-spectrum (SS) and the orthogonal frequency division multiplexing (OFDM). SS and OFDM provide alternative approaches for transmission of multi-user signals over wide transmission bandwidths. In SS, spread multi-user signals are distinguished from each other by orthogonal codes, while, in OFDM, narrowband multi-user signals are transmitted with different orthogonal subcarriers. The chapter is focused on two versions of SS, namely the direct sequence (DS) SS and frequency-hopping (FH) SS. Intercarrier- and intersymbol-interference, channel estimation and synchronization, adaptive modulation and coding, peak-to-average power ratio, and multiple access in up- and down-links of OFDM systems are also presented. *Chapter 11 Fading Channels* accounts for the effects of multipath propagation and shadowing. Fading channels are usually characterized by delay and Doppler spread of the received signals. The fading may be slow or fast, frequency-flat or frequency-selective. If the receiver can not collect coherently all the incoming signal components spread in time and frequency, then the received signal power level will be decreased drastically, hence leading to sigificant performance losses. This chapter is focused on the principal approaches for the channel fading and shadowing, for example, Rayleigh, Rician, Nakagami, and log-normal. The effect of fading and shadowing on the BEP are presented. Resource allocation and scheduling in fading channels is also treated. *Chapter 12 Diversity and Combining Techniques* addresses the approaches to alleviate the degradation caused by fading and shadowing. This is achieved by providing the receiver with multiple, preferably independent, replicas (in time, frequency, space) of the transmitted signal, and combine these signals in various ways, for example, selection, equal-gain, maximal-ratio, square-law. The performance improvement provided by diversity and combining techniques is presented as a function of the correlation and power balance between the diversity branches. Transmit and receive diversity, pre-detection and post-detection combining of diversity branches and channel capacity in fading and shadowing channels are also addressed. In contrast with telecommunication systems with single-transmit and single receive antennas (i.e., the so-called single-input single-output (SISO) systems), Chapter 12 is also concerned with systems using multiple antennas at the receiver or the transmitter. The receive diversity systems with multiple receive antennas are also called as SIMO (single-input multiple-output). Similarly, the transmit diversity systems with multiple antennas at the transmitter are referred to as MISO (multiple-input single-output) systems. *Chapter 13 MIMO* (multiple-input multiple-output) Systems is concerned with telecommunication systems with multiple antennas both at the transmit and the receive sides. A MIMO system, equipped with N_t transmit and N_r receive antennas, can benefit an $N_t N_r$-fold antenna diversity ($N_t N_r$ independent paths between transmitter and receiver). The MIMO channels is usually characterized by Wishart distribution, presented in Appendix E. The eigenvalues of random Wishart matrices determine the dominant characteristics of the MIMO channels, which may suffer correlation between the transmitted and/or received signals. This determines the number and the relative weights of the eigenmodes; water-filling algorithm can be used to equalize the transmit power or the data rate supported by each eignmode. Transmit antenna selection (TAS) implies the selection of one or a few of the multiple transmit antennas with highest instantaneous SNRs. TAS makes good use of the transmit diversity by dividing the transmit power only between the transmit antennas with highest instantaneous SNRs. MIMO systems enjoy full coordination between transmit and receive antennas. Consequently, by adjusting the complex antenna weights at the transmit- and receive-sides, the SNR at the output of a MIMO beamforming system can be maximized, hence minimizing the BEP. *Chapter 14 on Cooperative Communications* is based on dual-hop relaying with amplify-and-forward, detect-and-forward and coded cooperation protocols. The source-relay-destination link is modeled as a single link with an equivalent SNR, the relay with the highest equivalent SNR may be selected amongst a number of relays, and multiple antennas may be used at the source, at the relay and/or the destination. The source-destination link is usually selection- or maximal-ratio-combined with the source-relay-destination link. In coded cooperation, relaying and channel coding are simultaneously used to make better use of the cooperation.

The appendices are believed to provide convenient references, and useful background for better understanding of the relevant concepts. *Appendix A Vector Calculus in Spherical Coordinates* provides tools for conversion between spherical and polar coordinates required for Chapter 2 Antennas. *Appendix B Gaussian Q Function* is useful for determining the BEP of majority of modulation schemes. *Appendix C* presents a list of Fourier Transforms usually encountered in telecommunication applications. *Appendix D Mathematical Tools* presents series, integrals and functions used in the book, minimizing the need to resort to another mathematical handbook. *Appendix E Wishart Distribution* provides the necessary background for the Chapter 13 MIMO Systems. *Appendix F Probability and Random Variables* aims to help students with probabilistic concepts, widely used probability distributions and random processes.

Topics to be taught at undergraduate and graduate levels may be decided according to the priority of the instructor and the course contents. Some sections and/or chapters may be omitted or covered partially depending on the preferences of the instructor. However, it may not be easy to give a unique approach for specifying the curriculum.

During my career, I benefited from numerous excellent books, publications and Internet web pages. I would like to thank the authors of all sources who contributed for the accumulation of the knowledge reflected in this book. I would like to thank all my undergraduate and graduate students who, with their response to my teaching approaches, helped enormously for determining the contents and the coverage of the topics of this book. Valuable cooperation and help from Sandra Grayson, Preethi Belkese and Adalfin Jayasingh from John Wiley and Sons is highly appreciated.

Mehmet Şafak
July 2016

List of Abbreviations

ACK	acknowledgment	COST	European Cooperation for
ADC	analog-to-digital conversion		Scientific and Technical
ADM	adaptive delta modulation		Research
AF	amplify and forward	CP	cyclic prefix
AGC	automatic gain control	CPA	co-polar attenuation
AJ	anti jamming	CRC	cyclic redundancy check
AMR	adaptive multi rate	CSI	channel state information
AOA	angle of arrival	DAC	digital to analog conversion
AOD	angle of departure	DCT	discrete cosine transform
AOF	amount of fading	DF	detect and forward
ARQ	automatic repeat request	DFT	discrete Fourier transform
ASK	amplitude shift keying	DGPS	differential GPS
AT&T	American Telephone &	DM	delta modulation
	Telegraph Company	DMC	discrete memoryless channel
AWGN	additive white Gaussian noise	DPCM	differential PCM
BCH	Bose-Chaudhuri-Hocquenghem	DPSK	differential phase shift keying
	codes	DS	direct sequence
BEP	bit error probability	DSSS	direct sequence spread spectrum
BPSK	binary phase shift keying	E1	European telephone multiplex-
BS	base station		ing hierarchy
BSC	binary symmetric channel	EGC	equal gain combining
C/N	carrier-to-noise ratio	EGNOS	European geostationary
CCITT	International Telegraph and		navigation overlay service
	Telephone Consultative	EHF	extremely high frequencies
	Committee		(30-300 GHz)
CD	compact disc	EIRP	effective isotropic radiative
CDF	cumulative distribution function		power
CDMA	code division multiple access	EP	elliptical polarization
CIR	channel impulse response	ESD	energy spectral density

ETSI	European Telecommunications Standards Institute	ISU	international system of units
FDM	frequency division multiplexing	ITU	International Telecommunications Union
FEC	forward error correction	JPEG	joint photographic experts group
FFT	fast Fourier transform	Ka-band	26.5-40 GHz band
FH	frequency hopping	Ku-band	12.4-18 GHz band
FHSS	frequency hopping spread spectrum	L band	1-2 GHz band
		LAN	local area network
FIR	finite impulse response	LDPC	low-density parity check codes
FOM	figure of merit	LEO	low Earth orbiting
FSK	frequency shift keying	LF	low frequencies (30-300 kHz)
FT	Fourier transform	LHCP	left hand circular polarization
G/T	figure of merit of a receiver (antenna gain to system noise temperature ratio)	LMS	least mean square
		LNA	low noise amplifier
		LORAN-C	radio navigation system by land based beacons
GALILEO	European global navigation satellite system	LOS	line of sight
GBN	go-back-N ARQ	LP	linear polarization
GEO	geostationary	LPF	low pass filter
GLONASS	Russian global navigation satellite system	LPI	low probability of intercept
		LTI	linear time invariant
GNSS	global navigation satellite systems	MAC	multiple access
		MAI	multiple access interference
GPS	global positioning system	MAP	maximum a posteriori
GS	greedy scheduling	MEO	medium Earth orbit
GSC	generalized selection combining	MF	medium frequencies (300-3000 kHz)
GSM	global system for mobile communications		
		MGF	moment generating function
H.264/AVC	advanced video coding	MIMO	multiple-input multiple-output
HARQ	hybrid ARQ	MIP	multipath intensity profile
HDD	hard decision decoding	MISO	multiple-input single-output
HDTV	high definition TV	ML	maximum likelihood
HEVC	high efficiency video coding	MLD	maximum likelihood detection
HF	high frequencies (3-30 MHz)	MPEG	motion photograpic experts group
HPA	high power amplifier		
ICI	inter carrier interference	MRC	maximal ratio combining
IDFT	inverse discrete Fourier transform	MS	mobile station
IEEE	Institute of Electrical and Electronics Engineers	MUD	multiuser detection
		MUI	multiuser interference
IFFT	inverse fast Fourier transform	NACK	negative acknowledgment
IMT-2000	international mobile telephone standard	NAVSTAR	NAVigation Satellite Timing And Ranging (GPS satellite network)
IP	Internet protocol		
ISI	inter symbol interference	NFC	near field communications
ISM	industrial, scientific, and medical frequency band	NRZ	non return to zero
		OC	optimum combining

OFDM	orthogonal frequency division multiplexing	SF	spreading factor
OLC	optical lattice clock	SGT	satellite ground terminal
OOK	on-off keying	SHF	super high frequencies (3-30 gHz)
OVSF	orthogonal variable spreading factor	SIMO	single-input multiple-output
PAL	phase alternating line	SINR	signal-to-interference and noise ratio
PAM	pulse amplitude modulation	SIR	signal-to-interference ratio
PAPR	peak to average power ratio	SISO	single-input single-output
PCM	pulse code modulation	SLC	square-law combining
PDF	probability density function	SNR	signal-to-noise ratio
PDM	pulse duration modulation	SPS	standard positioning system
PFS	proportionally fair scheduling	SR	source-relay link
PLL	phase lock loop	SRD	source-relay-destination link
PN	pseudo noise	SRe	selective repeat ARQ
PPM	pulse position modulation	SS	spread spectrum
PPS	precise positioning system	SSC	switch-and-stay combining
PRS	partial response signaling	SW	stop-and-wait ARQ
PSD	power spectral density	T1	AT&T telephone multiplexing hierarchy
PSK	phase shift keying		
QAM	quadrature-amplitude modulation	TAS	transmit antenna selection
		TDM	time division multiplexing
QPSK	quadrature phase shift keying	TEC	total electron content
RCPC	rate compatible punctured convolutional	TPC	transmit power control
		UHF	ultra high frequencies (300-3000 MHz)
RD	relay destination link		
RFID	radio frequency identification	ULA	uniform linear array
RGB	red green blue	UMTS	universal mobile telecommunications system
RHCP	right hand circular polarization		
RPE-LTP	regular pulse excited long term prediction	UTC	universal coordinated time
		VHF	very-high frequencies (30-300 MHz)
RR	round robin		
RS	Reed-Solomon	WAN	wide area networks
RSC	recursive systematic convolutional	WCDMA	wideband code division multiple access (CDMA)
RZ	return to zero	WiFi	wireless fidelity
SA	selective availability	WiMax	worldwide interoperability for microwave access
SATCOM	satellite communications		
SC	selection combining	X-band	8.2-12.4 GHz band
SC-FDMA	single carrier frequency division multiple access	XPD	cross polar discrimination
		XPI	cross polar isolation
SDTV	standard definition TV		

About the Companion Website

Don't forget to visit the companion website for this book:

www.wiley.com/go/safak/Digital_Communications

There you will find valuable material designed to enhance your learning, including:

- Solutions manual

Scan this QR code to visit the companion website

1

Signal Analysis

In the course of history, human beings communicated with each other using their ears and eyes, by transmitting their messages via voice, sound, light, smoke, signs, paintings, and so on. [1] The invention of writing made written communications also possible. Telecommunications refers to the transmission of messages in the form of voice, image or data by using electrical signals and/or electromagnetic waves. As these messages modulate the amplitude, the phase or the frequency of a sinusoidal carrier, electrical signals are characterized both in time and frequency domains. The behavior of these signals in time and frequency domains are closely related to each other. Therefore, the design of telecommunication systems takes into account both the time- and the frequency-characteristics of the signals.

In the time-domain, modulating the amplitude, the phase and/or the frequency at high rates may become challenging because of the limitations in the switching capability of electronic circuits, clocks, synchronization and receiver performance. On the other hand, the frequency-domain behavior of signals is of critical importance from the viewpoint of the bandwidth they occupy and the interference they cause to signals in the adjacent frequency channels. Frequency-domain analysis provides valuable insight for the system design and efficient usage of the available frequency spectrum, which is a scarce and valuable resource. Distribution of the energy or the power of a transmitted signal with frequency, measured in terms of energy spectral density (ESD) or power spectral density (PSD), is important for the efficient use of the available frequency spectrum. ESD and PSD are determined by the Fourier transform, which relates time- and frequency-domain behaviors of a signal, and the autocorrelation function, which is a measure of the similarity of a signal with a delayed replica of itself in the time domain. Spectrum efficiency provides a measure of data rate transmitted per unit bandwidth at a given transmit power level. It also determines the interference caused to adjacent frequency channels.

Signals are classified based on several parameters. A signal is said to be periodic if it repeats itself with a period, for example, a sinusoidal signal. A signal is said to be aperiodic if it does not repeat itself in time. The

Digital Communications, First Edition. Mehmet Şafak.
© 2017 John Wiley & Sons Ltd. Published 2017 by John Wiley & Sons Ltd.
Companion website: www.wiley.com/go/safak/Digital_Communications

signals may also be classified as being analog or discrete (digital). An analog signal varies continuously with time while a digital signal is defined by a set of discrete values. For example, a digital signal may be defined as a sequence of 1 s and 0 s, which are transmitted by discrete voltage levels, for example, $\pm V$ volts. A signal is said to be deterministic if its behavior is predictable in time- and frequency-domains. However, a random signal, for example, noise, can not be predicted beforehand and is therefore characterized statistically. [2][3][4][5][6][7][8][9]

In this book, we will deal with both baseband and passband signals. The spectrum of a baseband signal is centered around $f = 0$, while the spectrum of a passband signal is located around a sufficiently large carrier frequency f_c, such that the transmission bandwidth remains in the region $f > 0$. The baseband signals, though their direct use is limited, facilitate the analysis and design of the passband systems. A baseband signal may be up-converted to become a passband signal by a frequency-shifting operation, that is, multiplying the baseband signal with a sinusoidal carrier of sufficiently high carrier frequency f_c. Shifting the spectrum of a baseband signal, with spectral components for $f < 0$ and $f > 0$, to around a carrier frequency f_c, implies that the bandwidth of a passband signal is doubled compared to a baseband signal. Most telecommunication systems employ passband signals, that is, the messages to be transmitted modulate carriers with sufficiently large carrier frequencies. Bandpass transmission has numerous advantages, for example, ease of radiation/reception by antennas, noise and interference mitigation, frequency-channel assignment by multiplexing and transmission of multiple message signals using a single carrier. In addition, passband transmission has the cost advantage, since it usually requires smaller, more cost-effective and power-efficient equipments.

This chapter will deal with analog/digital, periodic/aperiodic, deterministic/random and baseband/passband signals. Noting that the fundamental concepts can be explained by analogy to analog systems, this chapter will mostly be focused on analog baseband signals unless otherwise stated. The conversion of an analog signal into digital and the characterization of a digital signal will be treated in the subsequent chapters. Since the characteristics of passband signals can easily be derived from those of the baseband signals, the focus will be on the baseband signals. Telecommunication systems operate usually with random signals due to the presence of the system noise and/or fading and shadowing in wireless channels. [4][5][6][7]

Assuming that the student is familiar with probabilistic concepts, a short introduction is presented on random signals and processes. One may refer to Appendix F, Probability and Random Variables, for further details. Diverse applications of these concepts will be presented in the subsequent chapters.

1.1 Relationship Between Time and Frequency Characteristics of Signals

Fourier series and Fourier transform provide useful tools for characterizing the relationship between time- and frequency-domain behaviors of signals. For spectral analysis, we generally use the Fourier series for periodic signals and the Fourier transform for aperiodic signals. These two will be observed to merge as the period of a periodic signal approaches infinity. [3][9]

1.1.1 Fourier Series

The Fourier series expansion of a periodic function $s_{T_0}(t)$ with period $T_0 = 1/f_0$ is given by

$$s_{T_0}(t) = a_0 + \sum_{n=1}^{\infty} [a_n \cos(nw_0 t) + b_n \sin(nw_0 t)]$$

(1.1)

Using the orthogonality between $\cos(nw_0 t)$ and $\sin(nw_0 t)$, the coefficients a_n and b_n are found as

$$a_0 = E[s_{T_0}(t)] = \frac{1}{T_0} \int_0^{T_0} s_{T_0}(t)\, dt$$

$$a_n = \frac{2}{T_0} \int_0^{T_0} s_{T_0}(t)\, \cos(nw_0 t)\, dt, \quad n = 1, 2, \cdots$$

$$b_n = \frac{2}{T_0} \int_0^{T_0} s_{T_0}(t)\, \sin(nw_0 t)\, dt, \quad n = 1, 2, \cdots$$

$$(1.2)$$

where a_0 denotes the average value of $s_{T_0}(t)$. Using the Euler's identity $e^{\pm jx} = \cos x \pm j \sin x$, the Fourier series expansion given by (1.1) may be rewritten as a complex Fourier series expansion:

$$s_{T_0}(t) = \sum_{n=-\infty}^{\infty} c_n\, e^{jnw_0 t}$$

$$(1.3)$$

$$c_n = \frac{1}{T_0} \int_0^{T_0} s_{T_0}(t)\, e^{-jnw_0 t}\, dt$$

From the equivalence of (1.1) and (1.3), one may easily show that

$$c_0 = a_0$$

$$c_{\pm n} = \frac{1}{2}(a_n \mp jb_n)$$

$$(1.4)$$

According to the Parseval's theorem, the power of a periodic signal may be expressed in terms of the Fourier series coefficients:

$$P = \frac{1}{T_0} \int_0^{T_0} s_{T_0}(t)\, s_{T_0}^*(t)\, dt$$

$$= \frac{1}{T_0} \int_0^{T_0} s_{T_0}(t) \sum_{n=-\infty}^{\infty} c_n^*\, e^{-jnw_0 t}\, dt$$

$$= \sum_{n=-\infty}^{\infty} c_n^* \underbrace{\frac{1}{T_0} \int_0^{T_0} s_{T_0}(t) e^{-jnw_0 t}\, dt}_{c_n}$$

$$= \sum_{n=-\infty}^{\infty} |c_n|^2 = a_0^2 + \frac{1}{2}\sum_{n=1}^{\infty}\left[|a_n|^2 + |b_n|^2\right]$$

$$(1.5)$$

One may also observe from (1.5) that the power of a periodic signal is equal to the sum of the powers $|c_n|^2$ of its spectral components, located at nf_0, and its PSD is discrete with values $|c_n|^2$. Hence, the signal power is the same irrespective of whether it is calculated in time- or frequency-domains.

Example 1.1 Fourier Series of a Rectangular Pulse Train.

Consider the rectangular pulse train shown in Figure 1.1. Using (1.3) we determine the complex Fourier series of $s_{T_0}(t)$:

$$c_n = \frac{1}{T_0} \int_{-T_0/2}^{T_0/2} s_{T_0}(t)\, e^{-jnw_0 t} = \frac{1}{T_0} \int_{-T/2}^{T/2} e^{-jnw_0 t}\, dt$$

$$= \frac{T}{T_0}\, \mathrm{sinc}(nf_0 T)$$

$$(1.6)$$

where $f_0 = 1/T_0$. The sinc function is defined by

$$\mathrm{sinc}(x) = \sin(\pi x)/(\pi x)$$

$$(1.7)$$

Being a damped sinusoid, the zeros of the sinc function are the same as those of the sine function, that is, $x = \pm 1, \pm 2, \ldots$, except for at $x = 0$ where it is equal to unity:

$$\mathrm{sinc}(n) = \delta(n), \quad n = 0, \pm 1, \pm 2, \cdots$$

$$(1.8)$$

Figure 1.1(a) shows the variation of c_n as a function of $nf_0 T$. Note that as T_0 goes to infinity as in Figure 1.1(b), the periodic rectangular pulse train reduces to a single pulse at the origion. Then, the spectral lines merge, that is, $f_0 \to 0$, and the discrete spectra shown in Figure 1.1(a) becomes continuous and is described by $\mathrm{sinc}(fT)$ whose zeros are given by k/T, $k = 1, 2, \ldots$. On the other hand, as $T_0 \to T$ we have $s_{T_0}(t) = 1$ (see Figure 1.1(c)) and the corresponding Fourier coefficients become

$$c_n = \mathrm{sinc}(n) = \delta(n), \quad T_0 \to T$$

$$(1.9)$$

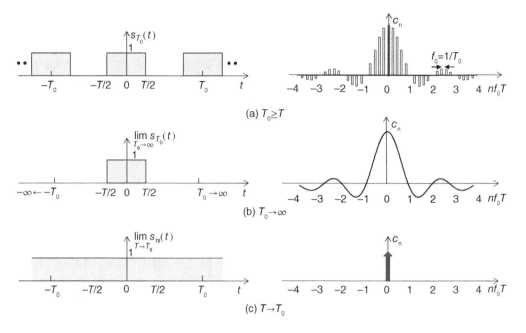

Figure 1.1   Rectangular Pulse Train and the Coefficients of the Complex Fourier Series.

1.1.2   Fourier Transform

As T_0 goes to infinity as shown in Figure 1.1(b), $s_{T_0}(t)$ tends to become an aperiodic signal, which will be shown hereafter as $s(t)$. Then, $f_0 = 1/T_0$ approaches zero, spectral lines at nf_0 merge and form a continuous spectrum. The Fourier transform $S(f)$ of an aperiodic continuous function $s(t)$ is defined by [2][3][9]

$$S(f) = \Im[s(t)] = \int_{-\infty}^{\infty} s(t)\, e^{-jwt}\, dt$$
$$s(t) = \Im^{-1}[S(f)] = \int_{-\infty}^{\infty} S(f)\, e^{jwt}\, df \tag{1.10}$$

This Fourier transform relationship will also be denoted as

$$s(t) \;\Leftrightarrow\; S(f) \tag{1.11}$$

Unless otherwise stated, small-case letters will be used to denote time-functions while capital letters will denote their Fourier transforms. It is evident from (1.10) that the value

of $S(f)$ at the origin gives the mean value of the signal:

$$S(0) = E[s(t)] = \int_{-\infty}^{\infty} s(t)\, dt$$
$$s(0) = E[S(f)] = \int_{-\infty}^{\infty} S(f)\, df \tag{1.12}$$

while $s(0)$ denotes the average value of $S(f)$.

The so-called Rayleigh's energy theorem states that the energy of an aperiodic signal found in time- and frequency-domains are identical to each other:

$$E = \int_{-\infty}^{\infty} |s(t)|^2\, dt$$
$$= \int_{-\infty}^{\infty} s^*(t)\, dt \int_{-\infty}^{\infty} S(f) e^{jwt} df$$
$$= \int_{-\infty}^{\infty} S(f)\, df \underbrace{\int_{-\infty}^{\infty} s^*(t)\, e^{jwt}\, dt}_{S^*(f)} \tag{1.13}$$
$$= \int_{-\infty}^{\infty} |S(f)|^2\, df$$

In view of the integration over $-\infty < f < \infty$ in (1.13), the energy of $s(t)$ is equal to the area under the energy spectral density $\Psi_s(f)$ of $s(t)$, that is, the energy per unit bandwidth:

$$\Psi_s(f) = |S(f)|^2, \quad (J/Hz) \qquad (1.14)$$

One may observe from (1.10) that the Fourier transform, hence the ESD, of a real signal $s(t)$ with even symmetry $s(t) = s(-t)$, has also even symmetry with respect to $f = 0$.

Example 1.2 Fourier Transform of a Rectangular Pulse.
Let $s(t)$ be defined as

$$s(t) = A\Pi(t/T) = A \begin{cases} 1 & |t| < T/2 \\ 0 & |t| > T/2 \end{cases} \qquad (1.15)$$

where A is a constant. Using (1.10), the Fourier transform of (1.15) is found as follows:

$$S(f) = A \int_{-T/2}^{T/2} e^{-jwt} dt \qquad (1.16)$$

$$= AT \ \text{sinc}(fT)$$

where $\text{sinc}(x)$ is defined by (1.7) (see (1.6) and Figure 1.1(b)).

Figure 1.2(a) shows the Fourier transform relationship between the (1.15) and (1.16). The frequency components of the pulse, which is time-limited to $\pm T/2$, extends over $(-\infty, \infty)$ in the frequency domain. In the limiting case where $T \rightarrow \infty$, then the pulse extends uniformly over $(-\infty, \infty)$ in the time domain, hence a dc signal (band-limited but not time-limited). Then, the pulse can be represented by only a single frequency component at $f = 0$, hence by a delta function, in the frequency domain (see Figure 1.2(b)). On the other hand, if we let $A = 1/T$ so that the area under the pulse becomes unity (see Figure 1.2(c)), and let $T \rightarrow 0$, then the time-limited pulse approaches a delta function and its Fourier transform $S(f)$ tends to be flat in the frequency domain. As one may also observe from the Fourier transform relationship in (1.10), a time-limited signal has frequency components over $(-\infty, \infty)$, hence not band-limited, while a band-limited signal can not be time-limited. To have a better feeling about the time-frequency relationship, we observe from Figure 1.2(a) that the bandwidth between dc and the first null of the sinc function is given by $W = 1/T$. If we use this bandwidth as a measure of the spectrum occupancy of the pulse $s(t)$, the product of the pulse duration T and the bandwidth is $WT = 1$.

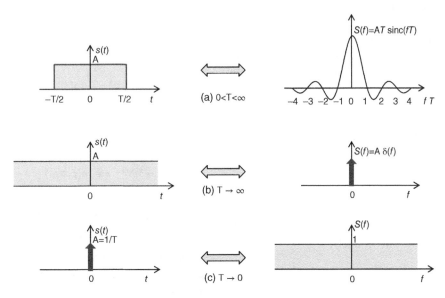

Figure 1.2 Fourier Transform of a Rectangular Pulse with Amplitude A.

In general, the so-called time-bandwidth product, which relates time and frequency behaviors of a signal, is a constant and its value depends on the considered pulse. This implies that bandwidth and the pulse duration are inversely related to each other. Hence, faster changing pulses, that is, pulses with smaller T, occupy wider bandwidths. In other words, higher data rate transmissions require wider transmission bandwidths. Indeed, high data rate transmissions (with minimum pulse durations) in minimum possible transmission bandwidths is one of the challenging issues in telecommunications engineering.

1.1.2.1 Impulses and Transforms in the Limit

Dirac delta function does not exist physically, but is widely used in many areas of engineering. Some functions are used to approximate the Dirac delta function. For example, the rectangular pulse shown in Figure 1.2 approximates Dirac delta function in the time domain as T approaches zero. As the area under a Dirac delta function should be equal to unity, the amplitude of the rectangular pulse is chosen as $A = 1/T$. Figure 1.2(c) also shows that a Dirac delta function in the time domain has a flat spectrum, that is, its PSD is uniform. Similarly, the Fourier transform of a rectangular pulse approximates a Dirac delta function located at $f = 0$ as $T \rightarrow \infty$ (see also Figure 1.2(b)), which implies that the Dirac delta function in the frequency domain implies a time-invariant signal.

Some alternative definitions of the Dirac delta function are listed below: [2][3][9]

$$\delta(t-t_0) = \begin{cases} \infty & t = t_0 \\ 0 & t \neq t_0 \end{cases}$$

$$\delta(t) = \lim_{T \to 0} \frac{1}{T} \Pi\left(\frac{t}{T}\right) \tag{1.17}$$

$$\delta(f) = \lim_{T \to \infty} T \ \mathrm{sinc}(fT)$$

$$\delta(t-t_0) = \int_{-\infty}^{\infty} e^{\pm j2\pi(t-t_0)\lambda} d\lambda$$

Some properties of the Dirac delta function are summarized below:

a. Area under the Dirac delta function is unity:

$$\int_{-\infty}^{\infty} \delta(t-t_0) dt = 1 \tag{1.18}$$

b. Sampling of an ordinary function $s(t)$ which is continuous at t_0:

$$\int_{a}^{b} s(t)\delta(t-t_0) dt = \begin{cases} s(t_0) & a < t_0 < b \\ s(t_0)/2 & t_0 = a, \ t_0 = b \\ 0 & otherwise \end{cases} \tag{1.19}$$

c. Relation with the unit step function

$$u(t-t_0) = \int_{-\infty}^{t} \delta(\tau-t_0) \ d\tau = \begin{cases} 1 & t > t_0 \\ 1/2 & t = t_0 \\ 0 & t < t_0 \end{cases}$$

$$\delta(t-t_0) = \frac{d}{dt} u(t-t_0) \tag{1.20}$$

d. Multiplication and convolution with a continuous function:

$$s(t)\delta(t-t_0) = s(t_0)\delta(t-t_0)$$
$$s(t) \otimes \delta(t-t_0) = s(t-t_0) \tag{1.21}$$

e. Scaling

$$\delta(at) = \frac{1}{|a|}\delta(t) \tag{1.22}$$

f. Fourier transform

$$\delta(t \mp t_0) \Leftrightarrow e^{\mp jwt_0}$$
$$e^{\pm jw_0 t} \Leftrightarrow \delta(f \mp f_0) \tag{1.23}$$

1.1.2.2 Signals with Even and Odd Symmetry and Causality

Assuming that $s(t)$ is a real-valued function, its Fourier transform $S(f)$ may be decomposed into its even and odd components as follows: [2]

$$S(f) = \int_{-\infty}^{\infty} s(t)e^{-jwt}dt = S_e(f) + jS_0(f)$$

$$S_e(f) = \int_{-\infty}^{\infty} s(t)\cos(wt)dt = S_e(-f)$$

$$S_0(f) = -\int_{-\infty}^{\infty} s(t)\sin(wt)dt = -S_0(-f)$$

$$(1.24)$$

It is clear from (1.24) that

$$S^*(f) = S_e(f) - jS_0(f) = S(-f) \qquad (1.25)$$

When the signal has an even symmetry with respect to $t = 0$, that is, $s(t) = s(-t)$, then $S(f)$ becomes a real and even function of frequency:

$$S(f) = S_e(f) = 2\int_0^{\infty} s(t)\cos(wt)dt$$
$$(1.26)$$
$$S_0(f) = 0$$

When the signal has an odd symmetry with respect to $t = 0$, that is, $s(t) = -s(-t)$, then $S(f)$ becomes a purely imaginary and odd function of frequency:

$$S(f) = jS_0(f) = -j2\int_0^{\infty} s(t)\sin(wt)dt$$

$$S_e(f) = 0$$
$$(1.27)$$

A *causal system* is defined as a system where the output at time t_0 depends on only past and current values but not on future values of the input. In other words, the output $y(t_0)$ depends only on the input $x(t)$ for values of $t \leq t_0$. Therefore, the output of a causal system has no time

symmetry and its spectrum has both real and imaginary components.

Example 1.3 Fourier Transform of a Decaying Exponential Function.
Consider a decaying exponential function as shown in Figure 1.3. Its Fourier transform is obtained using (1.10) and (1.20) as follows:

$$s(t) = e^{-\alpha t}u(t),\ \alpha > 0 \ \Leftrightarrow\ S(f) = \frac{1}{2}\delta(f) + \frac{1}{\alpha + jw}$$
$$(1.28)$$

We can use (1.28) to rewrite the unit step function as follows:

$$u(t) = \lim_{\alpha \to 0} s(t) \qquad (1.29)$$

The Fourier transform of the unit step function may then be expressed as follows:

$$U(f) = \Im(u(t)) = \lim_{\alpha \to 0} \Im(s(t))$$

$$= \lim_{\alpha \to 0} \left[\frac{1}{2}\delta(f) + \frac{1}{\alpha + j2\pi f}\right]$$

$$= \frac{1}{2}\delta(f) + \frac{1}{j2\pi f} \qquad (1.30)$$

Based on (1.10) and (1.28), the following Fourier transform pairs are valid for $\alpha > 0$:

$$s(-t) \ \Leftrightarrow\ S(-f) = \frac{1}{2}\delta(f) + \frac{1}{\alpha - jw}$$

$$s(t) - s(-t) \ \Leftrightarrow\ S(f) - S(-f) = -j\frac{2w}{\alpha^2 + w^2}$$
$$(1.31)$$

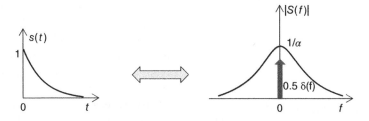

Figure 1.3 A Decaying Exponential and its Fourier Transform.

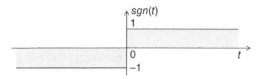

Figure 1.4 Signum function.

The signum function may be defined in terms of the unit step function as follows (see Figure 1.4):

$$\text{sgn}(t) = u(t) - u(-t)$$

$$= \lim_{\alpha \to 0}[s(t) - s(-t)] = \begin{cases} 1 & t > 0 \\ 0 & t = 0 \\ -1 & t < 0 \end{cases} \quad (1.32)$$

The Fourier transform of the signum function, which has an odd symmetry, is found using (1.30) and (1.31):

$$\mathfrak{I}[\text{sgn}(t)] = U(f) - U(-f)$$

$$= \lim_{\alpha \to 0}[S(f) - S(-f)] = \frac{1}{j\pi f} \quad (1.33)$$

which is purely imaginary, as expected (see (1.27)).

1.1.2.3 Fourier Transform Relations

Some properties of the Fourier transform are listed below: [2][3][9]

a. Superposition
    If $s_i(t) \Leftrightarrow S_i(f)$, $i = 1, 2, \cdots$, then using the linearity property of the Fourier transform in (1.10),

$$s(t) = \sum_i \alpha_i s_i(t) \quad \Leftrightarrow \quad S(f) = \sum_i \alpha_i S_i(f)$$

$$(1.34)$$

b. Time and frequency shift
    If $s(t) \Leftrightarrow S(f)$, then

$$s(t \mp \tau) \Leftrightarrow S(f) e^{\mp j w \tau}$$

$$s(t) e^{\pm j w_c t} \Leftrightarrow S(f \mp f_c) \quad (1.35)$$

Proof: The use of variable transformations in the following equations lead to

$$\mathfrak{I}[s(t \mp \tau)] = \int_{-\infty}^{\infty} s(t \mp \tau) e^{-jwt} dt = e^{\mp j w \tau} S(f)$$

$$\mathfrak{I}[s(t) e^{\pm j w_c t}] = \int_{-\infty}^{\infty} s(t) e^{-j(w \mp w_c)t} dt = S(f \mp f_c)$$

$$(1.36)$$

Example 1.4 Frequency Translation and Modulation.
Consider the rectangular pulse and its Fourier transform given by (1.15) and (1.16). From (1.35), one can write

$$\Pi(t/T) \quad \Leftrightarrow \quad T \, \text{sinc}(fT)$$
$$\Pi(t/T) e^{\pm j w_c t} \quad \Leftrightarrow \quad T \, \text{sinc}[(f \mp f_c)T] \quad (1.37)$$

One may easily observe from (1.37) that multiplying a time function by an exponential term $\exp(jw_c t)$ simply shifts the center frequency of the signal spectrum from dc to f_c. One may use linearity property, given by (1.34), to obtain the following from (1.37):

$$\Pi(t/T) \cos(w_c t) \Leftrightarrow \frac{T}{2} \, \text{sinc}[(f + f_c)T]$$
$$+ \frac{T}{2} \, \text{sinc}[(f - f_c)T] \quad (1.38)$$

Figure 1.5 shows the effect of multiplying a rectangular pulse by a sinusoidal carrier with frequency f_c. Multiplying a rectangular pulse with $\cos(w_c t)$ with $-\infty < t < \infty$ makes $\cos(w_c t)$ time-limited to $-T/2 < t < T/2$. This operation, which results in shifting the spectrum of the rectangular pulse from dc to $\pm f_c$, is widely used in transmitters for upconverting baseband signals to higher frequencies for passband transmission. Here, it is important to note that the passband signals occupy twice the transmission bandwidth compared to baseband signals due to the fact that negative frequency components of the baseband signals are also shifted around the carrier frequency (see Figure 1.5(b)). As we shall see in

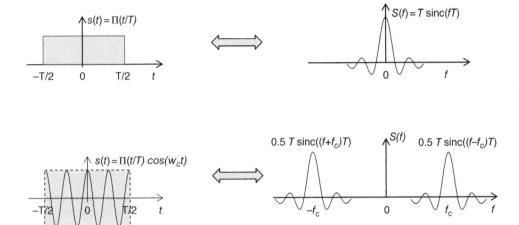

Figure 1.5   Frequency Translation.

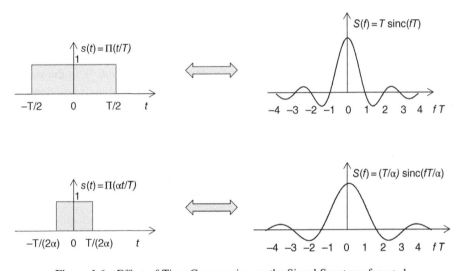

Figure 1.6   Effect of Time Compression on the Signal Spectrum for $\alpha > 1$.

the subsequent chapters, multiplication of the bandpass signal $\Pi(t/T)\cos w_c t$ by a carrier of the same frequency restores the baseband signal at the receiver, since $E[\Pi(t/T)\ \cos^2 w_c t] = (1/2)\Pi(t/T)$. Here, the expectation operation may be implemented by a low-pass filter.

c. Time and frequency scaling
   If $s(t) \Leftrightarrow S(f)$, then, for a real number α,

$$s(\alpha t) \Leftrightarrow |\alpha|^{-1} S(f/\alpha) \qquad (1.39)$$

Proof:

$$\Im[s(\alpha t)] = \int_{-\infty}^{\infty} s(\alpha t)e^{-jwt}\,dt$$

$$= \alpha^{-1}\int_{-\infty}^{\infty} s(\alpha t)e^{-j(w/\alpha)(\alpha t)}\,\alpha dt \qquad (1.40)$$

$$= |\alpha|^{-1} S(f/\alpha)$$

As shown in Figure 1.6, $s(\alpha t)$ is compressed in time compared to $s(t)$ by a factor $\alpha > 1$. However, $S(f/\alpha)$ is expanded in frequency by the same factor. Compression in

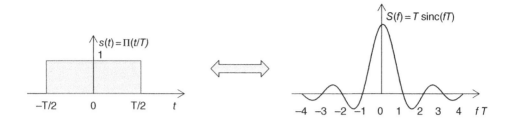

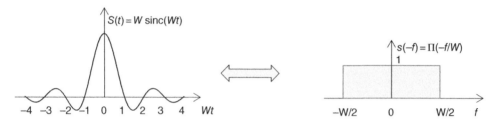

Figure 1.7   Demonstration of the Duality Theorem for the Rectangular Pulse.

the time domain, that is, faster variation of the signal with time, implies an expansion of its Fourier transform in the frequency domain, that is, the signal occupies a larger bandwidth. However, the time bandwidth product remains the same; $(\alpha W) \cdot (T/\alpha) = WT$.

d. Duality
If $s(t) \Leftrightarrow S(f)$, then

$$S(t) \Leftrightarrow s(-f) \qquad (1.41)$$

Proof: First changing the sign of t and interchanging t and f leads to

$$s(-t) = \int_{-\infty}^{\infty} S(f)\, e^{-j2\pi ft}\, df \;\Rightarrow$$

$$s(-f) = \int_{-\infty}^{\infty} S(t)\, e^{-j2\pi ft}\, dt \;\Rightarrow\; S(t) \Leftrightarrow s(-f)$$

$$(1.42)$$

Example 1.5   Duality.
Using the Fourier transform relationship given by (1.15) and (1.16), one can write from (1.41)

$$s(t) = \Pi(t/T) \;\Leftrightarrow\; S(f) = T \; \text{sinc}(fT)$$

$$S(t) = W \; \text{sinc}(Wt) \;\Leftrightarrow\; s(-f) = \Pi(-f/W)$$

$$(1.43)$$

by interchanging T with W in the second line (see Figure 1.7).

e. Conjugation
If $s(t) \Leftrightarrow S(f)$ then

$$s^*(\pm t) \Leftrightarrow S^*(\mp f) \qquad (1.44)$$

Proof:

$$s^*(t) = \int_{-\infty}^{\infty} S^*(f)\, e^{-jwt}\, df$$

$$= \int_{-\infty}^{\infty} S^*(-f)\, e^{jwt}\, df \;\Leftrightarrow\; S^*(-f)$$

$$(1.45)$$

$$s^*(-t) = \left[\int_{-\infty}^{\infty} S(f)\, e^{-jwt}\, df \right]^*$$

$$= \int_{-\infty}^{\infty} S^*(f)\, e^{jwt}\, df \;\Leftrightarrow\; S^*(f)$$

f. Convolution
If $s(t) \Leftrightarrow S(f)$ and $r(t) \Leftrightarrow R(f)$, then

$$r(t)s(t) \Leftrightarrow R(f) \otimes S(f)$$

$$r(t) \otimes s(t) \Leftrightarrow R(f) S(f)$$

$$(1.46)$$

where $\otimes$ denotes the convolution operator:

$$r(t)\otimes s(t)=\int_{-\infty}^{\infty}r(\lambda)s(t-\lambda)d\lambda=\int_{-\infty}^{\infty}s(\lambda)r(t-\lambda)d\lambda \tag{1.47}$$

Proof: Inserting the Fourier transform of $r(t)$ into the first line of the following and interchanging the order of integration completes the proof of the first line of (1.46):

$$\Im[r(t)s(t)]=\int_{-\infty}^{\infty}r(t)s(t)\,e^{-jwt}\,dt$$

$$=\int_{-\infty}^{\infty}s(t)e^{-jwt}dt\int_{-\infty}^{\infty}R(f')e^{jw't}df'$$

$$=\int_{-\infty}^{\infty}R(f')df'\underbrace{\int_{-\infty}^{\infty}s(t)e^{-j(w-w')t}dt}_{S(f-f')}$$

$$=\int_{-\infty}^{\infty}R(f')S(f-f')df'=R(f)\otimes S(f) \tag{1.48}$$

Similarly, we first write the convolution integral in the Fourier transform expression and change the order of integration to complete the proof:

$$\Im[r(t)\otimes s(t)]=\int_{-\infty}^{\infty}\left[\int_{-\infty}^{\infty}s(t')r(t-t')dt'\right]e^{-jwt}\,dt$$

$$=\int_{-\infty}^{\infty}s(t')dt'\underbrace{\int_{-\infty}^{\infty}r(t-t')e^{-jwt}\,dt}_{R(f)e^{-jwt'}}$$

$$=R(f)\int_{-\infty}^{\infty}s(t')\,e^{-jwt'}\,dt'=R(f)\,S(f) \tag{1.49}$$

Example 1.6 Fourier Transform of a Triangular Waveform.

Figure 1.8 shows that the convolution of a pulse of duration T with itself yields a triangular pulse in $(-T, T)$ with an amplitude T:

$$\Pi(t/T)\otimes\Pi(t/T)$$

$$=\begin{cases}0 & t<-T \text{ and } t>T\\[4pt]\int_{-T/2}^{t+T/2}dt=t+T & -T\le t<0\\[4pt]\int_{t-T/2}^{T/2}dt=-t+T & 0\le t<T\end{cases}$$

$$=T\Lambda(t/T) \tag{1.50}$$

where the triangular waveform is defined as:

$$\Lambda(t/\tau)=\begin{cases}1-|t|/\tau & |t|<\tau\\0 & otherwise\end{cases} \tag{1.51}$$

Using (1.15), (1.16) and (1.46), Fourier transform of a triangular waveform is found to be

$$\Pi(t/T)\;\Leftrightarrow\;T\;\mathrm{sinc}(fT)$$

$$\Lambda(t/T)=\frac{1}{T}\Pi(t/T)\otimes\Pi(t/T)\;\Leftrightarrow\;T\;\mathrm{sinc}^2(fT) \tag{1.52}$$

Example 1.7 Evaluation of the Integral $\int_{-\infty}^{\infty}\mathrm{sinc}(fT)\otimes\mathrm{sinc}(\ell fT)\,df$.

From (1.15), (1.16) and the convolution property (1.46) of the Fourier transform, one can write the following for $\ell>1$:

$$v(t)=\frac{1}{T}\Pi\Big(\frac{t}{T}\Big)\frac{1}{\ell T}\Pi\Big(\frac{t}{\ell T}\Big)=\frac{1}{\ell T^2}\Pi\Big(\frac{t}{T}\Big)$$

$$\Leftrightarrow\;V(f)=\mathrm{sinc}(fT)\otimes\mathrm{sinc}(f\ell T) \tag{1.53}$$

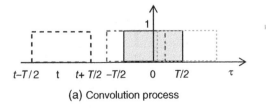

(a) Convolution process

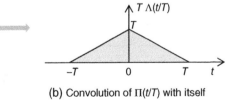

(b) Convolution of $\Pi(t/T)$ with itself

Figure 1.8   Convolution of a Rectangular Pulse with Itself.

From (1.12), the integral of the convolution on the right hand side of (1.53) is equal to $v(0)$:

$$v(0) = \int_{-\infty}^{\infty} V(f)\, df = \int_{-\infty}^{\infty} \text{sinc}(fT) \otimes \text{sinc}(f\ell T)\, df$$

$$= \frac{1}{\ell T^2} \tag{1.54}$$

Using (1.12) and the Fourier transform relationship given by (1.15) and (1.16), one gets

$$\int_{-\infty}^{\infty} \text{sinc}(fT)\, df = \frac{1}{T}\Pi(0) = \frac{1}{T} \tag{1.55}$$

g. Differentiation
   If $s(t) \Leftrightarrow S(f)$ then

$$\frac{d^n s(t)}{dt^n} \Leftrightarrow (j2\pi f)^n S(f)$$

$$\frac{d^n S(f)}{df^n} \Leftrightarrow (-j2\pi t)^n s(t) \tag{1.56}$$

Proof: Noting that differentiation is not with respect to the variable of integration, one can easily find

$$\frac{d^n}{dt^n} s(t) = \frac{d^n}{dt^n}\left[\int_{-\infty}^{\infty} S(f)\, e^{jwt}\, df\right]$$

$$= \int_{-\infty}^{\infty} (j2\pi f)^n S(f)\, e^{jwt}\, df \;\; \Leftrightarrow \;\; (j2\pi f)^n S(f)$$

$$\frac{d^n}{df^n} S(f) = \frac{d^n}{df^n}\left[\int_{-\infty}^{\infty} s(t)\, e^{-jwt}\, dt\right]$$

$$= \int_{-\infty}^{\infty} (-j2\pi t)^n s(t)\, e^{-jwt}\, dt \;\; \Leftrightarrow \;\; (-j2\pi t)^n s(t) \tag{1.57}$$

Note that the differentiation with respect to time enhances the high-frequency components of a signal since $|(j2\pi f)^n S(f)| > |S(f)|$ for $|2\pi f| > 1$ [2].

Example 1.8   Fourier Transform of a Triangular Waveform Using the Differentiation Property.
We use differentiation property to determine the Fourier transform of a triangular waveform defined by (1.51). Using (1.56), we can express the Fourier transform of a triangular waveform as

$$\mathfrak{F}[\Lambda(t/T)] = \frac{1}{(j2\pi f)^2}\, \mathfrak{F}\left[\frac{d^2}{dt^2}\Lambda(t/T)\right] \tag{1.58}$$

The first two time-derivatives of a triangular waveform may be written as (see also Figure 1.9)

$$\frac{d}{dt}\Lambda(t/T) = \begin{cases} 1/T & -T < t < 0 \\ -1/T & 0 \le t < T \\ 0 & |t| \ge T/2 \end{cases}$$

$$\frac{d^2}{dt^2}\Lambda(t/T) = \frac{1}{T}\delta(t+T) - \frac{2}{T}\delta(t) + \frac{1}{T}\delta(t-T) \tag{1.59}$$

The second derivative of a triangular waveform may be expressed in terms of Dirac delta functions, whose Fourier transform is straightforward. Inserting the second derivative of $\Lambda(t/T)$ in (1.59) into (1.58) and using (1.23), one gets

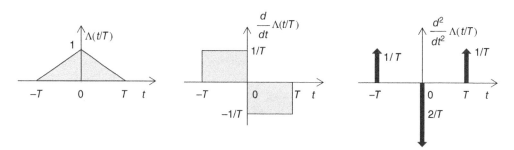

Figure 1.9   First- and Second-Order Derivatives of a Triangular Function.

$$\mathfrak{F}[\Lambda(t/T)] = \frac{1}{(j2\pi f)^2 T}\left[-2 + e^{-jwT} + e^{jwT}\right]$$

$$= T \operatorname{sinc}^2(fT) \tag{1.60}$$

which is identical to (1.52), as expected.

h. Integration
   If $s(t) \Leftrightarrow S(f)$ then

$$\int_{-\infty}^{t} s(t')dt' \;\Leftrightarrow\; \frac{1}{2}S(0)\delta(f) + \frac{S(f)}{j2\pi f} \tag{1.61}$$

Proof: The above integral may be expressed as the convolution of $s(t)$ with the unit step function:

$$\int_{-\infty}^{t} s(t')dt' = \int_{-\infty}^{\infty} s(t')u(t-t')dt' = s(t)\otimes u(t)$$

$$\Leftrightarrow\; S(f)U(f) \tag{1.62}$$

The use of (1.30) and the convolution property given by (1.46) leads to (1.61). Note that the integration of a function suppresses its high-frequency components since $S(f)/(2\pi f)$ $< S(f)$ for $|2\pi f| > 1$ [2].

Example 1.9 Fourier Transforms of the Unit Step and the Dirac Delta Functions Using Integration and Differentiation Properties.
In view of (1.20), (1.23), (1.30) and (1.61), one can write for $t > t_0$

$$u(t-t_0) = \int_{-\infty}^{t} \delta(\tau - t_0)\, d\tau$$

$$\Leftrightarrow\; U(f)\, e^{-j2\pi f\, t_0} = \frac{1}{2}\delta(f) + \frac{e^{-j2\pi f\, t_0}}{j2\pi f} \tag{1.63}$$

Fourier transform of a delta function may be determined using the differentiation property as follows

$$\mathfrak{F}[\delta(t-t_0)] = \mathfrak{F}\left[\frac{d}{dt}u(t-t_0)\right]$$

$$= (j2\pi f)\, U(f)e^{-j2\pi f\, t_0} \tag{1.64}$$

$$= e^{-j2\pi f\, t_0}$$

which is the same as (1.23), as expected.

1.1.3 Fourier Transform of Periodic Functions

Fourier transform of a periodic function, of period T_0, with a complex Fourier series expansion given by (1.3) is given by [4]

$$\mathfrak{F}[s_{T_0}(t)] = \sum_{n=-\infty}^{\infty} c_n \mathfrak{F}\left[e^{j2\pi n f_0 t}\right] = \sum_{n=-\infty}^{\infty} c_n \delta(f-nf_0) \tag{1.65}$$

The Fourier transform (1.65) is hence a discrete function of frequency, that is, is non-zero at only the discrete frequencies nf_0. The PSD may be written as the square of the magnitude of its Fourier transform:

$$G_s(f) = |\mathfrak{F}[s_{T_0}(t)]|^2 = \sum_{n=-\infty}^{\infty} |c_n|^2 \delta(f-nf_0) \tag{1.66}$$

The power of a periodic signal is given by the area under the PSD:

$$P = \int_{-\infty}^{\infty} G_s(f)\,df = \int_{-\infty}^{\infty} \sum_{n=-\infty}^{\infty} |c_n|^2 \delta(f-nf_0)\,df$$

$$= \sum_{n=-\infty}^{\infty} |c_n|^2 \tag{1.67}$$

Similarly, as given by (1.6) and shown in Figure 1.1, the coefficients of the complex Fourier series expansion of a periodic rectangular pulse train, with period $T_0 = 1/f_0$, also exhibit discrete frequency lines. One may observe from (1.6) that the coefficients satisfy $c_n T_0 = S(nf_0)$ where $S(f)$ denotes the Fourier transform of a single rectangular pulse centered at the origin (see (1.15), (1.16) and the pulse train $s_{T_0}(t)$ as $T_0 \to \infty$ in Figure 1.1(b)). The PSD and the power of a periodic signal, given by (1.5), determined by the Fourier series analysis, are in complete in agreement with (1.66) and (1.67), found by the Fourier transform approach.

The PSD of an aperiodic signal $s(t)$ with finite power can not be expressed by a Fourier series expansion. Instead, it may be written as

$$G_s(f) = \lim_{T \to \infty} \frac{1}{T} |S(f)|^2 \qquad (1.68)$$

where $S(f)$ denotes the Fourier transform of $s(t)$ over the time interval $[-T/2, T/2]$.

Example 1.10   Fourier Transform of a Periodic Train of Impulses in Time.
Consider the following periodic discrete-time function, which consists of a periodic train of impulses with period T:

$$s_T(t) = \sum_{m=-\infty}^{\infty} \delta(t - mT) \qquad (1.69)$$

Using (1.3) with $f_0 = 1/T$, the complex Fourier series expansion of (1.69) may be written as follows:

$$s_T(t) = \sum_{m=-\infty}^{\infty} \delta(t - mT) = \sum_{n=-\infty}^{\infty} c_n e^{j\frac{2\pi nt}{T}} \qquad (1.70)$$

where c_n is given by

$$
\begin{aligned}
c_n &= \frac{1}{T} \int_{-T/2}^{T/2} s_T(t) e^{-j\frac{2\pi nt}{T}} dt \\
&= \frac{1}{T} \sum_{m=-\infty}^{\infty} \int_{-T/2}^{T/2} \delta(t - mT) e^{-j\frac{2\pi nt}{T}} dt \\
&= \frac{1}{T} \int_{-T/2}^{T/2} \delta(t) e^{-j\frac{2\pi nt}{T}} dt \\
&= 1/T
\end{aligned} \qquad (1.71)
$$

The Fourier transform of the RHS of (1.70) with $c_n = 1/T$, as given by (1.65), shows that the Fourier transform of a train of impulses with period T in the time domain consists of a train of impulses in the frequency domain separated from each other by $1/T$:

$$\sum_{m=-\infty}^{\infty} \delta(t - mT) \iff \frac{1}{T} \sum_{n=-\infty}^{\infty} \delta\left(f - \frac{n}{T}\right) \qquad (1.72)$$

Example 1.11   Sampling.
In digital communication systems, the first step of converting analog signals into digital, is to sample analog signals. Consider the following discrete-time function which represents the sequence of samples of an analog signal $g(t)$ with a sampling period of T (see Figure 1.10):

$$g_\delta(t) = g(t) \sum_{m=-\infty}^{\infty} \delta(t - mT) \qquad (1.73)$$

Since the sampled analog signal $g_\delta(t)$ may be written as the product of $g(t)$ and the periodic sequence of Dirac delta functions, its Fourier transform $G_\delta(f)$ may be expressed, using (1.72), as the convolution of their Fourier transforms in the frequency domain:

$$
\begin{aligned}
g_\delta(t) &= g(t) \sum_{m=-\infty}^{\infty} \delta(t - mT) \\
&\iff G_\delta(f) = \frac{1}{T} \sum_{n=-\infty}^{\infty} G\left(f - \frac{n}{T}\right)
\end{aligned} \qquad (1.74)
$$

The left-hand side of Figure 1.10 shows the process of sampling of an analog signal $g(t)$, bandlimited to $W/2$, by a periodic train of impulses with period T. The right-hand side of the Figure 1.10 shows the corresponding Fourier transforms. The sampled signal has a Fourier transform consisting of replicas of $G(f)$ located at frequencies defined by n/T, $-\infty < n < \infty$ (see (1.74)); hence, the bandwidth occupied by the sampled signal is infinitely large. In practice, the spectrum of the sampled signal is filtered at the transmitter by a low-pass filter (LPF) with a cut-off frequency of $W/2$ so as to eliminate the interference caused to other systems using the adjacent frequency channels. However, as one may also observe from Figure 1.10, LPF can successfully filter out the replicas $G(f-n/T)$ for $n > 0$ if the condition $1/T > W$ is satisfied. Otherwise, tails of $G(f-n/T)$ will overlap and filtering would distort the transmitted signal, represented by $G(f)$. This topic will be treated in detail in Chapter 5.

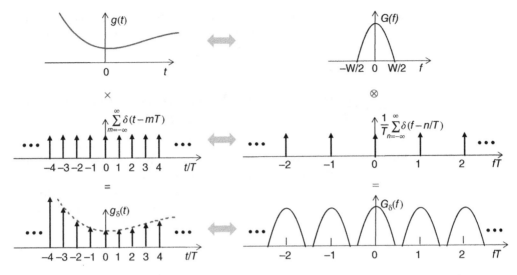

Figure 1.10   Sampling.

1.2   Power Spectal Density (PSD) and Energy Spectral Density (ESD)

1.2.1   Energy Signals Versus Power Signals

The energy of a signal $s(t)$ is defined as

$$E = \int_{-\infty}^{\infty} |s(t)|^2 dt \qquad (1.75)$$

The power of a signal is defined as

$$P = \lim_{T \to \infty} \frac{1}{T} \int_{-T/2}^{T/2} |s(t)|^2 dt \qquad (1.76)$$

As a special case when the power signal is periodic with a period T_0, the signal power defined by (1.76) simplifies to

$$P = \frac{1}{T_0} \int_{-T_0/2}^{T_0/2} |s_{T_0}(t)|^2 dt \qquad (1.77)$$

Based on (1.75)–(1.77), signals may be classified as

a. An energy signal is defined as a signal with finite energy and zero power:

$$0 < E < \infty \quad \Rightarrow \quad P = 0 \qquad (1.78)$$

Aperiodic signals are typically energy signals as they have a finite energy and zero power.

b. A power signal has finite power and infinite energy:

$$0 < P < \infty \quad \Rightarrow \quad E = \infty \qquad (1.79)$$

Periodic signals are typically power signals since they have finite power and infinite energy. [4][5][6][7]

Example 1.12   Energy and Power Signals. Let us first consider the following decaying exponential function:

$$s(t) = A e^{-\alpha t} u(t), \quad \alpha \geq 0 \qquad (1.80)$$

where $u(t)$ denotes the unit step function and α is a positive finite real number. The energy and power of this signal are calculated using (1.75) and (1.76) as follows:

$$E = \int_0^\infty A^2 e^{-2\alpha t}\, dt = \frac{A^2}{2\alpha}$$

$$P = \lim_{T \to \infty} \frac{1}{T}\int_0^T A^2 e^{-2\alpha t}\, dt = \lim_{T \to \infty} \frac{A^2}{2\alpha}\frac{1-e^{-2\alpha T}}{T} \to 0$$

(1.81)

Hence, this is an energy signal. However, if we assume A = 1 and $\alpha = 0$ in (1.80), then

$$u(t) = \lim_{\substack{\alpha \to 0 \\ A \to 1}} s(t) \;\Rightarrow\; E \to \infty,\; P = 1$$

(1.82)

which shows that a unit step function is a power signal.

Now consider the following signal:

$$s(t) = \begin{cases} t^{-1/4} & t \ge t_0 > 0 \\ 0 & elsewhere \end{cases}$$

(1.83)

Its energy and power are found to be

$$E = \int_{t_0}^\infty t^{-1/2}\, dt = 2\sqrt{t}\,\Big|_{t_0}^\infty \to \infty$$

$$P = \lim_{T \to \infty} \frac{1}{T}\int_{t_0}^{t_0+T} t^{-1/2}\, dt$$

(1.84)

$$= \lim_{T \to \infty} 2\frac{\sqrt{T+t_0}-\sqrt{t_0}}{T} \to 0$$

This signal is neither a power signal nor an energy signal.

Finally we consider a cosine function with amplitude A:

$$s_{T_0}(t) = A\cos(w_0 t + \theta)$$

(1.85)

Since this function is periodic with a period $T_0 = 1/f_0$, its power can be determined by integration over one period:

$$P = \frac{1}{T_0}\int_0^{T_0} A^2\cos^2(w_0 t + \theta)\, dt = \frac{A^2}{2}$$

(1.86)

The energy of this signal is given by the product of its power and the number of periods

considered, which is infinite. Therefore, this signal having infinite energy, is classified as a power signal.

1.2.2   Autocorrelation Function and Spectral Density

ESD and PSD are employed to describe the distribution of, respectively, the signal energy and the signal power over the frequency spectrum. ESD/PSD is defined as the Fourier transform of the autocorrelation function of an energy/power signal. [4][5][6][7]

1.2.2.1   Energy Signals

Autocorrelation of a function $s(t)$ provides a measure of the degree of similarity between the signal $s(t)$ and a delayed replica of itself. The autocorrelation function of an energy signal $s(t)$ is defined by

$$R_s(\tau) = \int_{-\infty}^\infty s^*(t)s(t+\tau)\, dt$$

$$= \int_{-\infty}^\infty s(t)s^*(t-\tau)\, dt, \quad -\infty < \tau < \infty$$

(1.87)

Note that τ, that varies in the interval $(-\infty,\infty)$, measures the time shift between $s(t)$ and its delayed replica. The sign of the time shift τ shows whether the delayed version of $s(t)$ leads or lags $s(t)$. If the time shift τ becomes zero, then $s(t)$ will overlap with itself and the resemblence between them will be perfect. Consequently, the value of the autocorrelation function at the origion is equal to the energy of the signal:

$$E = R_s(0) = \int_{-\infty}^\infty |s(t)|^2\, dt$$

(1.88)

It is logical to expect that the resemblence between a signal and its delayed replica to decrease with increasing values of τ. The properties of the autocorrelation function of a real valued energy signal may be listed as follows:

a. Symmetry about $\tau = 0$:

$$R_s(\tau) = R_s(-\tau) \qquad (1.89)$$

b. Maximum value occurs at $\tau = 0$:

$$|R_s(\tau)| \leq |R_s(0)| = E, \quad \forall \tau \qquad (1.90)$$

c. Autocorrelation and ESD are related to each other by the Fourier transform:

$$R_s(\tau) \Leftrightarrow \Psi_s(f) \qquad (1.91)$$

where ESD is defined in (1.14) as the variation of the energy E of $s(t)$ with frequency. One can prove the relationship given by (1.91) by taking the Fourier transform of the autocorrelation function of an energy signal, given by (1.87):

$$\Psi_s(f) = \Im[R_s(\tau)] = \int_{-\infty}^{\infty} R_s(\tau) e^{-j\omega\tau} \, d\tau$$

$$= \int_{-\infty}^{\infty} e^{-j\omega\tau} d\tau \int_{-\infty}^{\infty} s^*(t) \, s(t+\tau) \, dt$$

$$= \int_{-\infty}^{\infty} s^*(t) dt \int_{-\infty}^{\infty} s(t+\tau) e^{-j\omega\tau} d\tau$$

$$= \underbrace{\int_{-\infty}^{\infty} s^*(t) e^{j\omega t} dt}_{S^*(f)} \underbrace{\int_{-\infty}^{\infty} s(v) e^{-j\omega v} dv}_{S(f)}$$

$$= |S(f)|^2 \qquad (1.92)$$

Example   1.13   Convolution   Versus Correlation.
In order to discover the similarity between the correlation and convolution, we first define the cross correlation of two energy signals $r(t)$ and $s(t)$ as follows:

$$R_{rs}(\tau) = \int_{-\infty}^{\infty} r(t) s^*(t-\tau) \, dt = \int_{-\infty}^{\infty} s^*(t) r(t+\tau) dt \qquad (1.93)$$

which reduces to (1.87) for $R_s(\tau) = R_{ss}(\tau)$. Comparison of (1.93) with the expression for

the convolution of $r(t)$ and $s(t)$, given by (1.47), shows that

$$R_{rs}(\tau) = r(t) \otimes s^*(-t)$$
$$R_s(\tau) = R_{ss}(\tau) = s(t) \otimes s^*(-t) \qquad (1.94)$$

For real-valued functions with even symmetry, that is, for $s(t) = s(-t)$, auto and cross correlation functions are the same as the convolution:

$$R_{rs}(\tau) = r(t) \otimes s(t)$$
$$\qquad\qquad\qquad\qquad for \; s(-t) = s(t) \quad (1.95)$$
$$R_s(\tau) = s(t) \otimes s(t)$$

1.2.2.2   Power Signals

The autocorrelation function of a power signal $s(t)$ is defined by

$$R_s(\tau) = \lim_{T \to \infty} \frac{1}{T} \int_{-T/2}^{T/2} s^*(t) \, s(t+\tau) \, dt \quad (1.96)$$

If the power signal is periodic with a period T_0, then the autocorrelation can be computed by integration over a single period:

$$R_s(\tau) = \frac{1}{T_0} \int_{-T_0/2}^{T_0/2} s^*(t) \, s(t+\tau) \, dt \qquad (1.97)$$

The value of the autocorrelation function at the origin is equal to the power of $s(t)$:

$$P = R_s(0) = \lim_{T \to \infty} \frac{1}{T} \int_{-T/2}^{T/2} |s(t)|^2 \, dt \qquad (1.98)$$

The properties of the autocorrelation function of a real-valued power signal $s(t)$ are given by

a. Symmetry about $\tau = 0$:

$$R_s(\tau) = R_s(-\tau) \qquad (1.99)$$

b. Maximum value occurs at $\tau = 0$:

$$|R_s(\tau)| \leq |R_s(0)| = P, \quad \forall \tau \qquad (1.100)$$

c. Autocorrelation and power spectral density (PSD) are related to each other by Fourier transform:

$$R_s(\tau) \Leftrightarrow G_s(f) \qquad (1.101)$$

PSD, the variation of the power P of $s(t)$ with frequency, is given by (1.66) and (1.68) for respectively periodic and aperiodic power signals.

Based on (1.87), (1.92), (1.96) and (1.101), the PSD and ESD of a signal may be obtained either by taking the magnitude-square of its Fourier transform or the Fourier transform of the autocorrelation function (see Figure 1.11):

$$s(t) \Leftrightarrow S(f)$$
$$\Downarrow \qquad \Downarrow$$
$$R_s(\tau) \Leftrightarrow |S(f)|^2$$

Figure 1.11 Two Alternative Approaches for Obtaining the ESD/PSD of a Signal $s(t)$.

Example 1.14 PSD and ESD of a Rectangular Pulse.
To have a better feeling about the two alternative approaches shown in Figure 1.11 to determine the PSD/ESD, consider a rectangular pulse $s(t)$ and its Fourier transform $S(f)$ (see (1.15) and (1.16)):

$$s(t) = A\Pi(t/T) \Leftrightarrow S(f) = AT\mathrm{sin}c(fT)$$
$$(1.102)$$

which is an energy signal, with energy $A^2 T$. The autocorrelation function and the ESD of $s(t)$ are related to each other by (1.50), (1.52) and (1.95):

$$R_s(\tau) = A^2 T \Lambda(\tau/T) \Leftrightarrow \Psi_s(f) = A^2 T^2 \mathrm{sin}c^2(fT)$$
$$(1.103)$$

Noting that $R_s(0) = A^2 T$ denotes the energy of $s(t)$, $\Psi_s(f) = |S(f)|^2$ clearly represents the ESD, since the area under it is equal to $A^2 T$. The corresponding PSD is given by

$$G_s(f) = \frac{\Psi_s(f)}{T} = \frac{|S(f)|^2}{T} = A^2 T \, \mathrm{sin}c^2(fT)$$
$$(1.104)$$

The integration of (1.104) with frequency gives A^2, which denotes the signal power over the finite pulse duration T.

1.3 Random Signals

In a digital communication system, one of the M symbols is transmitted, in a finite symbol duration, using a pre-defined M-ary alphabet. These symbols are unknown to the receiver when they are transmitted. At the receiver, these signals bear a random character in AWGN, fading or shadowing channels. Since a random signal can not be predicted before reception, a receiver processes the received random signal in order to estimate the transmitted symbol. In view of the above, the performance of telecommunication systems can be evaluated by characterizing the random signals statistically.

This section aims to provide a short introduction to random signals and processes encountered in telecommunication systems. The reader may refer to Appendix F for further details about the random signals and their characterization. [3][9]

1.3.1 Random Variables

Given a sample space S and elements $s \in S$, we define a function $X(s)$ whose domain is S and whose range is a set of numbers on the real line (see Figure 1.12). The function $X(s)$ is called a random variable (rv). For example, if we consider coin flipping with outcomes head (H) and tail (T), the sample space is defined

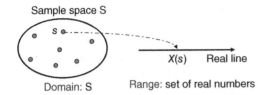

Sample space S

$X(s)$ Real line

Domain: S   Range: set of real numbers

Figure 1.12   Definition of a Random Variable.

by $S = \{H, T\}$ and the rv may be assumed to be $X(s) = 1$ for $s = H$ and -1 for $s = T$. Note that a rv, which is unknown and unpredictable beforehand, is known completely once it occurs. For example, one does not know the outcome before flipping a coin. However, once the coin is flipped, the outcome (head or tail) is known.

A rv is characterized by its probability density function (pdf) or cumulative distribution function (cdf), which are interrelated. The cdf $F_X(x)$ of a rv X is defined by

$$F_X(x) = P(X \le x), \quad -\infty < x < \infty \quad (1.105)$$

which specifies the probability with which the rv X is less than or equal to a real number x. The pdf of a rv X is defined as the derivative of the cdf.

$$f_X(x) = \frac{d}{dx} F_X(x), \quad -\infty < x < \infty \quad (1.106)$$

Conversely, the cdf is defined as the integral of the pdf:

$$F_X(x) = \int_{-\infty}^{x} f_X(u) \, du \quad (1.107)$$

In view of (1.105)–(1.107), a rv is characterized by the following properties:

a. $f_X(x) \ge 0$ and the area under $f_X(x)$ is always equal to unity: $F_X(\infty) = \int_{-\infty}^{\infty} f_X(u) \, du = 1$.

b. $0 \le F_X(x) \le 1$ since $F_X(\infty) = 1$ and $F(-\infty) = \int_{-\infty}^{-\infty} f_X(u) \, du = 0$.

c. The cdf of a continuous rv X is a non-decreasing and smooth function of X. Therefore, for $x_2 \ge x_1$, $P(x_1 < X \le x_2) = \int_{x_1}^{x_2} f_X(u) \, du = F_X(x_2) - F_X(x_1) \ge 0$.

d. When a rv X is discrete or mixed, its cdf is still a non-decreasing function of X but contains discontinuities.

A rv is characterized by its moments. The n-th moment of a rv X is defined as

$$E[X^n] = \int_{-\infty}^{\infty} x^n f_X(x) \, dx \quad (1.108)$$

where $E[.]$ denotes the expectation. The rv's encountered in telecommunication systems are mostly characterized by their first two moments, namely the mean (expected) value m_X and the variance σ_X^2:

$$m_X = E[X] = \int_{-\infty}^{\infty} x f_X(x) \, dx$$

$$\mathrm{var}(X) = \sigma_X^2 = E\left[(X - m_X)^2\right] \quad (1.109)$$

$$= E[X^2] - m_X^2$$

When X is discrete or of a mixed type, the pdf contains impulses at the points of discontinuity of $F_X(x)$. In such cases, the discrete part of $f_X(x)$ may be expressed as

$$f_X(x) = \sum_{i=1}^{N} P(X = x_i) \, \delta(x - x_i) \quad (1.110)$$

where the rv X may assume one of the N values, $x_1, x_2, \ldots, x_N$ at the discontinuities. For example, in case of coin flipping, two outcomes may be represented as x_1 for head and x_2 for tail. Then, $P(X = x_1) = p \le 1$ shows the probability of head, while $P(X = x_2) = 1 - p$ denotes the probability of tail.

Mean value and the variance of a discrete rv is found by inserting (1.110) into (1.109):

$$m_X = E[X] = \sum_{i=1}^{N} x_i P(X = x_i)$$

$$\sigma_X^2 = \sum_{i=1}^{N} x_i^2 P(X = x_i) - m_X^2 \quad (1.111)$$

For the special case of equally-likely probabilities, that is, $P(X=x_i)=1/N$, then (1.111) simplifies to the well-known expression

$$m_X = E[X] = \frac{1}{N}\sum_{i=1}^{N} x_i$$

(1.112)

$$\sigma_X^2 = \frac{1}{N}\sum_{i=1}^{N}(x_i-m_X)^2 = \frac{1}{N}\sum_{i=1}^{N}x_i^2 - m_X^2$$

Now consider two rv's X_1 and X_2, each of which may be continuous, discrete or mixed. The joint cdf is defined as

$$F_{X_1X_2}(x_1,x_2) = P(X_1 \le x_1, X_2 \le x_2)$$

$$= \int_{-\infty}^{x_1}\int_{-\infty}^{x_2} f_{X_1X_2}(u_1,u_2)\, du_2\, du_1$$

$$F_{X_1X_2}(-\infty,-\infty) = 0$$

$$F_{X_1X_2}(\infty,\infty) = 1$$

(1.113)

and is related to the joint pdf as follows:

$$f_{X_1X_2}(x_1,x_2) = \frac{\partial^2}{\partial x_1 \partial x_2} F_{X_1X_2}(x_1,x_2)$$

(1.114)

The marginal pdf's are found as follows:

$$\int_{-\infty}^{\infty} f_{X_1X_2}(x_1,x_2)\, dx_1 = f_{X_2}(x_2)$$

$$\int_{-\infty}^{\infty} f_{X_1X_2}(x_1,x_2)\, dx_2 = f_{X_1}(x_1)$$

(1.115)

The conditional pdf $f(x_i|x_j)$, which gives the pdf of x_i for a given deterministic value of the rv x_j, has the following properties

$$f(x_i|x_j) = f_{X_iX_j}(x_i,x_j)/f_{X_j}(x_j)$$

$$F(-\infty|x_j) = 0, \quad F(\infty|x_j) = 1 \qquad i \ne j, \; i,j = 1,2$$

(1.116)

Generalization of (1.113)-(1.116) to more than two rv's is straightforward. Statistical independence of random events has the

following implications: If the experiments result in mutually exclusive outcomes, then the probability of an outcome in one experiment is independent of an outcome in any other experiment. The joint pdf's (cdf's) may then be written as the product of the pdf's (cdf's) corresponding to each outcome.

Example 1.15   Averaging.
In order to model the wireless channel between base and mobile stations of a mobile radio system, the average signal level received by a mobile station is measured at points evenly distributed on a circle of radius $d_0 = 100$ meters from the base station. These measurement results are rescaled for arbitrary distances in order to develop a reliable channel path-loss model. Therefore, sufficiently many short-range measurements are required in order to take into account all potential propagation effects, for example, climate, topography of the terrain, vegetation. Now consider for simplicity that ten measurements are conducted in dBm and (1.112) is used to determine the average signal level and its standard deviation at $d_0 = 100$ meters:

$$X = [-41.1 \quad -45 \quad -39.6 \quad -42 \quad -40.3 \quad -44$$

$$-48 \quad -43 \quad -43.3 \quad -41.5]$$

$$m_X = \frac{1}{N}\sum_{i=1}^{N} x_i = -42.78 \; dBm$$

$$\sigma_X = \left(\frac{1}{N}\sum_{i=1}^{N} x_i^2 - m_X^2\right)^{\frac{1}{2}} = 2.35 \; dBm \qquad (1.117)$$

Consequently, the average signal level at a distance of $d_0 = 100$ meters may be modeled by $m_X = -42.78 \; dBm$ and a standard deviation $\sigma_X = 2.35 \; dBm$.

1.3.2   Random Processes

Future values of a deterministic signal can be predicted from its past values. For example, $A\cos(wt + \Phi)$ shows a deterministic signal as long as A, w and Φ are deterministic and known; its future values can be determined exactly using its value at a certain time. However, it describes a *random signal* if any of A, w or Φ is random. For example, a noise generator generates a noise with a random amplitude and phase at any instant of time and these values change randomly with time. Similarly, amplitude and phase of a multipath fading signal vary randomly not only at a given instant of time but also with time. Hence, a random signal changes randomly not only at an instant of time but also with time. Therefore, future values of random signals cannot be predicted by using their observed past values. [4][6][7]

A rv may be used to describe a random signal at a given instant of time. *Random process* extends the concept of a rv to include also the time dimension. Such a rv then becomes a function of the possible outcomes s of a random event (experiment) and time t. In other words, every outcome s will be associated with a time (sample) function. The family (ensemble) of all such sample functions is called a *random process* and denoted as $X(s, t)$. [3][4][9]

A random process does not need to be a function of a deterministic argument. For example, the terrain height between transmitter and receiver may be a random process of the distance, since the location and the height of the obstacles may change randomly. The random process may be continuous or discrete either in time or in the value of the rv.

For example, consider L Gaussian noise generators. The time variation of the noise voltages at the output of each of the noise generators is called as a *sample function*, or a *realization* of the process. Thus, each of the L sample functions, corresponding to a specific event s_j, changes randomly with time; noise voltages at any two instants are independent from each other (see Figure 1.13). The totality of sample functions is called an *ensemble*. On the other hand, at an instant of time t_k, the value of the rv $X(s, t_k)$ depends on the event. For a specific event and time t_k, $X(s_j, t_k) = X_j(t_k)$ is a real

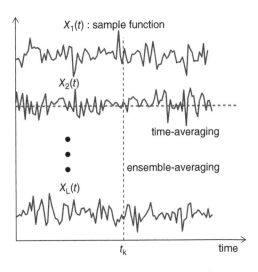

Figure 1.13   Random (Gaussian Noise) Process.

number. The properties of $X(s,t)$ is summarized below: [3]

a. $X(S, t)$ represents a family (ensemble) of sample functions, where $S = \{s_1, s_2, \cdots, s_L\}$.
b. $X(s_j, t)$ represents a sample function for the event s_j (an outcome of S or a realization of the process).
c. $X(S, t_k)$ is a rv at time t_k.
d. $X(s_j, t_k)$ is a non-random number.

For the sake of convenience, we will denote a random process by $X(t)$ where the presence of the event is implicit.

1.3.2.1   Statistical Averages

Empirical determination of the pdf of a random process is neither practical nor easy since it may change with time. For example, the pdf of a Gaussian noise voltage may change with time as the temperature of the noise generator (a resistor) increases; then, the noise variance would be higher. Nevertheless, the mean and the autocorrelation function adequately describe the random processes encountered in telecommunication systems. The mean of a random process $X(t)$ at time t_k is defined as

$$E[X(t_k)] = m_X(t_k) = \int_{-\infty}^{\infty} x f_{X_k}(x)\,dx \quad (1.118)$$

where the pdf of $X(t_k)$ is defined over the ensemble of events at time t_k. Evidently, the mean would be time-invariant if the pdf is time-invariant. On the other hand, identical time variations of the mean and the variance of two random processes do not imply that these process are equivalents. For example, one of these processes may be changing faster than the other, and hence have higher spectral components. Therefore, autocorrelation functions (and corresponding PSDs) of these processes should also be compared with each other. The autocorrelation function of a random process $X(t)$ provides a measure of the degree of similarity between the rv's $X(t_j)$ and $X(t_k)$:

$$R_X(t_j, t_k) = E\left[X(t_j)X^*(t_k)\right] \quad (1.119)$$

A random process $X(t)$ is said to be *stationary in the strict sense*, if none of its statistics changes with time, but they depend only on the time difference $t_k - t_j$. This implies that the joint pdf satisfies the following:

$$f(x_1, x_2, \cdots, x_L; t_1, t_2, \cdots, t_L)$$
$$= f(x_1, x_2, \cdots, x_L; t_1 + \tau, t_2 + \tau, \cdots, t_L + \tau) \quad (1.120)$$

Here, (1.120) should be read as the joint pdf of $x_1, x_2, \cdots, x_L$ at times $t_1, t_2, \cdots, t_L$. For example, a strict sense stationary process satisfies the following for $L = 1$ and 2:

$$f(x; t) = f(x; t + \tau)$$
$$f(x_1, x_2; t_1, t_2)|_{t_2 = t_1 + \tau} = f(x_1, x_2; \tau) \quad (1.121)$$

A random process $X(t)$ is said to be *wide sense stationary* (WSS), if only its mean and autocorrelation function are unaffected by a shift in the time origin. This implies that the

mean is a constant and the autocorrelation function depends only on the time difference $t_k - t_j$:

$$E[X(t)] = m_X, \ \forall t$$
$$R_X(t_j, t_k) = R_X(\tau), \ \tau \triangleq t_j - t_k \quad (1.122)$$

Note that strict-sense stationarity implies wide-sense stationarity, but the reverse is not true. In most applications, the analysis based on WSS assumption provides sufficiently accurate results at least over some time intervals of interest. The autocorrelation function of WSS random processes, which is an even function of τ, provides a measure of the degree of correlation between the random values of a process with τ seconds time shift from each other.

The properties of the autocorrelation function of a real-valued WSS random process are listed below:

a. An even function of τ:

$$R_X(\tau) = R_X(-\tau) \quad (1.123)$$

b. Maximum value occurs at $\tau = 0$:

$$|R_X(\tau)| \leq |R_X(0)|, \ \forall \tau \quad (1.124)$$

c. The value of the autocorrelation function at the origin is equal to the average power of the signal

$$R_X(0) = P = E\left[|X(t)|^2\right] \quad (1.125)$$

d. Autocorrelation and PSD are related to each other by the Fourier transform:

$$R_X(\tau) \ \Leftrightarrow \ G_X(f) \quad (1.126)$$

Example 1.16 Poisson Process.
Poisson process is a discrete random process which describes the number of occurrences of an event as a function of time. The event may represent the number of customers

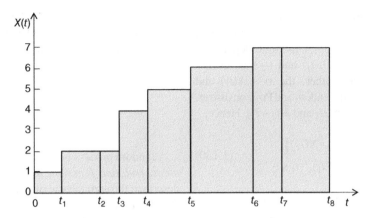

Figure 1.14 Poisson Process.

arriving to a bank or a supermarket, the failure of some components in a system, the number of planes arriving to and departing from an airport, the number of goals scored in a football game, and so on. A single event of the process, which consists of counting the number of occurrences (arrivals) with time, is also random.

Let the rv k denote the number of occurrences (arrivals) during a time interval $\tau = t_2 - t_1$ where t_1 and t_2 are arbitrary times with $t_2 \geq t_1$. The average number of arrivals per unit time is defined by the arrival rate λ in arrivals/s. The probability of k arrivals during τ is given by

$$P_k(\tau) = e^{-\lambda\tau}\frac{(\lambda\tau)^k}{k!}, \quad k = 0, 1, 2, \cdots \quad (1.127)$$

For example, the probability that no customers arrive to a bank during $\tau = 1$ minute is given by $P_0(\tau) = \exp(-\lambda\tau) = 0.905$ if one customer arrives on the average per 10 minutes, that is, $\lambda = 0.1$ arrivals/minutes. The pdf of the number of arrivals during τ may be written as

$$f_k(x) = e^{-\lambda\tau}\sum_{k=0}^{\infty}\frac{(\lambda\tau)^k}{k!}\,\delta(x-k) \quad (1.128)$$

Noting that $\sum_{k=0}^{\infty}P_k(\tau) = 1$ (see (D.1)), the area under the pdf given by (1.128) is unity.

Based on (1.127) and (1.128), we now define a discrete random process $X(t)$, accounting for the number of occurrences during (t_i, t_j), with discontinuities at random time instants t_i and t_j. As shown in Figure 1.14, this process may represent, for example, the number of uncoordinated customers entering into a bank office with random arrival times t_i. Figure 1.14 shows that one customer arrives during $(0, t_1)$, no customers arrive during (t_2, t_3) and two customers arrive during (t_3, t_4). Since the time intervals of any two events do not overlap, the corresponding rv's are independent. [3]

The first two moments of the rv $X(t_k)$ at a specific time t_k are found, with the help of (D.1), as follows:

$$E[X(t_1)] = e^{-\lambda t_1}\sum_{k=0}^{\infty}k\frac{(\lambda t_1)^k}{k!}$$

$$= \lambda t_1 e^{-\lambda t_1}\sum_{k=1}^{\infty}\frac{(\lambda t_1)^{k-1}}{(k-1)!}$$

$$= \lambda t_1$$

$$E[X^2(t_1)] = e^{-\lambda t_1}\sum_{k=0}^{\infty}k^2\frac{(\lambda t_1)^k}{k!} \quad (1.129)$$

$$= \lambda t_1 e^{-\lambda t_1}\sum_{k=1}^{\infty}k\frac{(\lambda t_1)^{k-1}}{(k-1)!}$$

$$= \lambda t_1 e^{-\lambda t_1}\sum_{k'=0}^{\infty}(k'+1)\frac{(\lambda t_1)^{k'}}{k'!}$$

$$= \lambda t_1[1 + \lambda t_1]$$

It is clear from (1.129) that the first two moments of the rv, defined as the number of occurrences during $(0, t_1)$, are time-dependent. Since the intervals $(0, t_1)$ and $(t_2 - t_1)$ do not overlap with each other, the rv's $X(t_1)$ and $X(t_2) - X(t_1)$ are independent and Poisson distributed with parameters λt_1 and $\lambda(t_2 - t_1)$. Hence,

$$E[X(t_1)\{X(t_2)-X(t_1)\}]$$
$$= \lambda t_1 \, \lambda(t_2 - t_1), \quad t_1 \le t_2 \qquad (1.130)$$

The autocorrelation function for $t_2 \ge t_1$ may then be found using (1.129) and (1.130) as follows:

$$R(t_1, t_2) = E[X(t_1)X(t_2)]$$
$$= E\left[X(t_1)\{X(t_2)-X(t_1)\}+X^2(t_1)\right]$$
$$= \lambda t_1 \lambda(t_2 - t_1) + \lambda t_1(1 + \lambda t_1)$$
$$= \lambda^2 t_1 t_2 + \lambda t_1, \quad t_1 \le t_2$$
$$(1.131)$$

Since $R(t_1, t_2) = R(t_2, t_1)$ for a real random process, the autocorrelation function is given by

$$R(t_1, t_2) = \begin{cases} \lambda t_2 + \lambda^2 t_1 t_2 & t_1 \ge t_2 \\ \lambda t_1 + \lambda^2 t_1 t_2 & t_1 \le t_2 \end{cases} \qquad (1.132)$$

Since the mean of the random process and the autocorrelation are time dependent, the Poisson process is not stationary.

1.3.2.2 Time Averaging and Ergodicity

The computation of the mean and the autocorrelation function of a random process by *ensemble averaging* is often not practical since it requires the use of all sample functions. For the so-called *ergodic processes*, ensemble averaging may be replaced by *time averaging*. This means that the mean and the autocorrelation function of an ergodic process may be determined by using a single sample function. [3][4][5][7][8]

A random process is said to be *ergodic in the mean* if its mean can be calculated using a sample function, that is, using time averaging instead of ensemble averaging (see (1.118)):

$$m_X = \lim_{T \to \infty} \frac{1}{T} \int_{-T/2}^{T/2} X(t)\, dt \qquad (1.133)$$

A random process is said to be *ergodic in the autocorrelation function*, if $R_X(x)$ is correctly described through the use of a sample function, that is, time averaging instead of ensemble averaging (see (1.119)):

$$R_X(\tau) = \lim_{T \to \infty} \frac{1}{T} \int_{-T/2}^{T/2} X^*(t)\, X(t+\tau)\, dt \qquad (1.134)$$

It is not easy to test whether a random process is ergodic or not. However, in most applications, it is reasonable to assume that time and ensemble averaging are interchangeable.

Example 1.17 Test for Stationarity. Consider a random process defined by

$$X(t) = A\cos(wt + \Phi) \qquad (1.135)$$

where A and w are constants but Φ is uniformly distributed in $[0, 2\pi]$:

$$f_\Phi(\phi) = \frac{1}{2\pi}, \quad 0 \le \phi \le 2\pi \qquad (1.136)$$

In order to determine whether this process is WSS, we first determine its mean:

$$m_X = E[X(t)] = \int_{-A}^{A} x f_{X(t)}(x)\, dx$$
$$= \int_0^{2\pi} A\cos(wt+\phi) f_\Phi(\phi)\, d\phi = 0 \qquad (1.137)$$

which is obviously independent of time. Note that the evaluation of the integral on the first line requires the knowledge of the pdf of $X(t)$, while the second integral uses the pdf of Φ. The pdf of

$X(t)$ is obtained from the pdf of Φ with a variable transformation (see Example F.6):

$$f_{X(t)}(x) = f_\Phi(\phi)\left|\frac{d\phi}{dx}\right|_{\phi=\phi_1} + f_\Phi(\phi)\left|\frac{d\phi}{dx}\right|_{\phi=\phi_2}$$

$$= \frac{1}{\pi\sqrt{A^2-x^2}}, \quad -A \leq x \leq A \qquad (1.138)$$

where $\phi_1 = \phi_2 - \pi = \pi/2 - wt$ are the two roots of the multi-valued function (1.135). Note that the pdf is independent of w and time t, accounting for the time-invariance of the mean. Also note that the pdf peaks at $x = \pm A$.

The autocorrelation of $X(t)$ is given by

$$R_X(t_1,t_2) = E[X(t_1)X(t_2)]$$

$$= A^2 E[\cos(wt_1 + \phi)\cos(wt_2 + \phi)]$$

$$= \frac{1}{2}A^2 E[\cos(w\{t_1 + t_2\} + 2\phi) + \cos(w\{t_1 - t_2\})]$$

$$= \frac{1}{2}A^2 \cos(w\tau), \quad \tau = t_1 - t_2$$

$$(1.139)$$

Since the mean is constant and the autocorrelation is a function of the time difference, this random process is WSS.

Example 1.18 Complex Random Process. A received signal, described by the following complex random process

$$Z(t) = \sum_{\ell=1}^{L} \alpha_\ell e^{j(wt+\Phi_\ell)} \qquad (1.140)$$

represents the sum of L replicas of a complex carrier signal e^{jwt} scattered from L obstacles with random channel gains α_ℓ and phases Φ_ℓ. They are all assumed to be statistically independent of each other and the pdf of Φ_ℓ is given by (1.136). In view of (1.137), the mean value of (1.140) is identically equal to zero. Its autocorrelation function is given by

$$R_Z(t,t+\tau) = E[Z^*(t)Z(t+\tau)] = R_Z(\tau)$$

$$= E\left[\sum_{\ell=1}^{L}\alpha_\ell e^{-j(wt+\Phi_\ell)}\sum_{k=1}^{L}\alpha_k e^{j(w(t+\tau)+\Phi_k)}\right]$$

$$= e^{jw\tau}\sum_{k=1}^{L}\sum_{\ell=1}^{L}E[\alpha_\ell \alpha_k]\underbrace{E\left[e^{j(\Phi_k-\Phi_\ell)}\right]}_{\delta(k-\ell)}$$

$$= e^{jw\tau}\sum_{\ell=1}^{L}E[\alpha_\ell^2]$$

$$(1.141)$$

The random process $Z(t)$ is evidently WSS. The value of the autocorrelation function at the origin $R_Z(0)$ shows that the received power is given by the sum of the powers of independent channel gains. The pdf of the received power can be determined using the pdf's of α_ℓ's (see for example (F.109)–(F.111) and Example F.14).

1.3.2.3 PSD of a Random Process

Random processes are generally classified as power signals with a PSD as given by (1.68) and with the following properties:

a. The PSD is positive real-valued function of frequency:

$$G_X(f) \geq 0 \qquad (1.142)$$

b. The PSD has even symmetry with frequency:

$$G_X(f) = G_X(-f) \qquad (1.143)$$

c. The PSD and autocorrelation function are Fourier transform pairs:

$$R_X(\tau) \iff G_X(f) \qquad (1.144)$$

d. The average normalized power of a random process is given by the area under the PSD:

$$P_X = R_X(0) = \int_{-\infty}^{\infty} G_X(f)\, df \qquad (1.145)$$

Example 1.19 Autocorrelation and PSD of a Random Binary Sequence.

A baseband signal consisting of a binary sequence can be expressed as follows:

$$s(t) = A \sum_{n=-\infty}^{\infty} b_n p(t-nT) \qquad (1.146)$$

where $b_n \in \{-1, +1\}$ represents the bit 0 and 1 with equal likelihood, $P(b_n = 1) = P(b_n = -1) = 1/2$, $p(t)$ denotes the pulse shape, T is the bit duration and A is the amplitude. Therefore, (1.146) represents a sequence of pulses with amplitudes ± 1. The infinitely long bit sequence, which is a power signal, becomes perfectly random if the bits are independent from each other. The autocorrelation function of a random bit sequence is given by a delta function, since all dibit combinations are equally likely with probability ¼:

$$R_b[k] = E[b_n^* b_{n+k}]$$

$$= \begin{cases} \frac{1}{4}[(1)\times(1)+(1)\times(-1)+(-1)\times(1)+(-1)\times(-1)]=0 & k\neq 0 \\ 1 & k=0 \end{cases}$$

$$R_b[k] = \delta(k) \iff G_b(f) = 1 \qquad (1.147)$$

where the PSD of the infinitely long random bit sequence in the second line is flat and given by (1.23).

The autocorrelation function of $s(t)$ is given by the expectation of $s(t)$ with a delayed replica of itself:

$$R_s(\tau) = E[s^*(t)s(t+\tau)]$$

$$= A^2 \sum_{m=-\infty}^{\infty} \sum_{n=-\infty}^{\infty} \underbrace{E[b_n^* b_m]}_{\delta(n-m)}$$

$$E[p^*(t-nT)p(t+\tau-mT)]$$

$$= A^2 \sum_{n=-\infty}^{\infty} E[p^*(t-nT)p(t+\tau-nT)]$$

$$= A^2 \sum_{n=-\infty}^{\infty} \int_{nT}^{(n+1)T} p^*(t-nT)p(t+\tau-nT)dt$$

$$= A^2 \int_{-\infty}^{\infty} p^*(u)p(u+\tau)dt = A^2 R_p(\tau)$$

$$(1.148)$$

It is clear from (1.148) that the autocorrelation of a random binary sequence is identical to that of the pulse $p(t)$. For example, if we assume $p(t) = \Pi(t/T)$, which is an even function of t, then, in view of (1.50), (1.95) and (1.104), the autocorrelation function is triangular and the PSD is sinc-squared:

$$R_s(\tau) = A^2 R_p(\tau) = A^2 T \Lambda(t/T)$$
$$\iff G_s(f) = A^2 T \ \text{sinc}^2(fT) \qquad (1.149)$$

An alternative approach for determining the PSD of $s(t)$ is based on the observation that (1.146) is the convolution of the random bit sequence and the rectangular pulse $\Pi(t/T)$. Then, in view of the convolution property of the Fourier transform, given by (1.46), the PSD of $s(t)$ may be written as the product of the PSD's of $\Pi(t/T)$ and the infinite bit sequence. The result is evidently identical to (1.149). Noting that the area under $T\text{sinc}^2(fT)$ in (1.149) is equal to unity, the integration of $G_s(f)$ gives the total signal power A^2, as expected.

1.3.2.4 Noise

The thermal noise voltage $n(t)$ that we encounter in telecommunication systems is a random process with zero mean and an autocorrelation function described by Dirac delta function:

$$E[n(t)] = 0$$
$$R_n(\tau) = E[n(t)\,n(t+\tau)] = \frac{N_0}{2}\delta(\tau) \qquad (1.150)$$

The formula (1.150) implies that any two different noise samples are uncorrelated with each other, no matter how close the time difference between the samples is. Using the Fourier transform relationship between the autocorrelation function and the PSD, the two-sided noise PSD is given by

$$G_n(f) = \mathfrak{F}_\tau[R_n(\tau)] = \frac{N_0}{2} \ (W/Hz) \qquad (1.151)$$

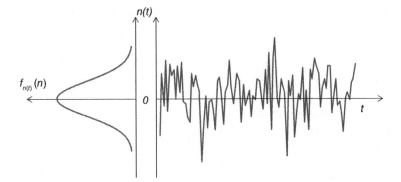

Figure 1.15   Time Variation and the Pdf of Gaussian Noise.

The PSD of the Gaussian noise is flat in the frequency bands of interest for telecommunication applications; hence it is called as *white* noise. White Gaussian noise is characterized by Gaussian pdf with zero mean:

$$f_{n(t)}(n) = \frac{1}{\sqrt{2\pi}\sigma_n} \exp\left(-\frac{n^2}{2\sigma_n^2}\right), \quad -\infty < n < \infty \tag{1.152}$$

where $\sigma_n^2 = R_n(0) = N_0/2$ denotes the noise variance. Figure 1.15 shows a sample function of a Gaussian noise process, which is white with zero mean and unity variance. The pdf of the noise voltage is depicted on the left-hand side.

1.4   Signal Transmission Through Linear Systems

The telecommunication systems are mostly linear. Therefore, we will focus our attention on signal transmission through linear systems. The systems generally consist of cascade-connected subsystems. The signal $y(t)$ at the output of a linear system may be written as the convolution of the input signal $x(t)$ with the impulse response $h(t)$ of the system (see Figure 1.16): [4][5][7][8][9]

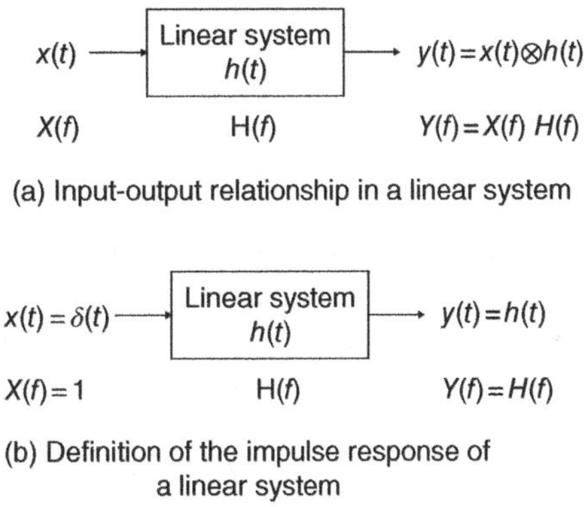

$x(t)$ → | Linear system $h(t)$ | → $y(t) = x(t)\otimes h(t)$

$X(f)$ $H(f)$ $Y(f) = X(f)\, H(f)$

(a) Input-output relationship in a linear system

$x(t) = \delta(t)$ → | Linear system $h(t)$ | → $y(t) = h(t)$

$X(f) = 1$ $H(f)$ $Y(f) = H(f)$

(b) Definition of the impulse response of a linear system

Figure 1.16   Input-Output Relationship and the Definition of the Impulse Response in a Linear System.

$$y(t) = x(t)\otimes h(t)$$

$$= \int_{-\infty}^{\infty} x(\tau)h(t-\tau)\,d\tau \tag{1.153}$$

$$= \int_{-\infty}^{\infty} h(\tau)x(t-\tau)\,d\tau$$

Impulse response represents the response of a linear system to an impulse at its input, that is, $x(t) = \delta(t)$ (see (1.21) and Figure 1.16(b)):

$$y(t) = x(t)\otimes h(t)\big|_{x(t) = \delta(t)} = h(t) \tag{1.154}$$

Using the convolution property of the Fourier transform, the Fourier transform of (1.153) leads to

$$Y(f) = H(f)X(f)$$
$$G_Y(f) = |H(f)|^2 G_X(f) \qquad (1.155)$$

where $H(f) = \Im[h(t)]$ is called as the channel transfer function or the channel frequency response. $H(f)$, which is generally complex, can be expressed in terms of the channel amplitude response and the phase response. Note that the PSD of the output signal is not the same as that of the input signal but modified by the channel frequency response.

For distortionless transmission of signals through a system, the output signal is desired to be the same as the input signal. A close look at (1.153) shows that this is possible only when the channel impulse response (CIR) is a delta function, that is, $h(t) = \delta(t)$. However, this is not physical since the response of a system to an input signal at $t = 0$ is delayed by τ seconds by electronic equipments/processors and the propagation time in a channel. Then, to account for the delay, the CIR may be written in the form of a delayed delta function with channel amplitude gain α and channel phase θ:

$$h(t) = \alpha e^{-j\theta} \delta(t - \tau) \iff H(f) = \alpha e^{-j\theta} e^{-jw\tau}$$
$$(1.156)$$

Since $|H(f)| = \alpha$, we conclude from (1.155) and (1.156) that, apart from a constant amplitude change due to α, such a channel does not cause any distortion on the PSD of the input signal. However, the channel induces a frequency- and delay-dependent change in the phase of the output signal. A CIR with a flat frequency response as in (1.156) is not physically realizable since any physical channel has a non-flat frequency response with a finite bandwidth. For example, electronic devices do not respond identically to all frequency components of the input signals. Similarly, both wired and wireless channels behave differently at different frequency bands. The analysis of (1.153) becomes even more complicated for channels where the channel gain α, the phase θ and the delay τ vary with time. Moreover, the channel may provide

multiple replicas of the same signal with different gains, phases and delays due to scattering from different obstacles in wireless channels and reflections in discontinuities in wired channels. These issues will be discussed in detail in the subsequent chapters.

The definition of the bandwidth is not unique. Nevertheless, the PSD $G_Y(f)$ of the output signal determines the bandwidth occupied by the signal in the channel. As one may observe from (1.155), the *signal bandwidth* is determined by $G_X(f)$, while $|H(f)|^2$ is responsible for the *channel (coherence) bandwidth*. A channel with a bandwidth narrower than that of the signal distorts the signal transmitted through this channel. In other words, the channel bandwidth is required to be sufficiently larger than the signal bandwidth for signal transmission with minimal distortion. On the other hand, bearing in mind that the frequency spectrum is a scarce and valuable resource, the signals at the transmitter output are usually filtered so as the minimize the bandwidth requirements at the expense of some acceptable distortion. Hence, the *transmission bandwidth* of a signal is determined by the signal bandwidth, the channel bandwidth and filtering at the transmitter output.

Noting that the PSD of a real-valued random process is even and has zero-mean, the following definition of the rms baseband bandwidth is commonly used:

$$W_{rms,bb} = \left(\frac{\displaystyle\int_{-\infty}^{\infty} f^2 G_Y(f)\, df}{\displaystyle\int_{-\infty}^{\infty} G_Y(f)\, df} \right)^{1/2} \qquad (1.157)$$

The spectrum of a passband signal is obtained by shifting its equivalent baseband spectrum to around the carrier frequency f_c along the frequency axis. Shifting a baseband spectrum to a passband frequency brings the negative frequency portions of the baseband spectrum to positive frequencies around f_c, hence leads to doubling the bandwidth (see Figure 1.5). Therefore, the passband rms bandwidth is simply the twice the baseband rms bandwidth:

$$W_{rms,pb} = 2W_{rms,bb} \qquad (1.158)$$

For the sake of simplicity, assume $G_Y(f) = \Pi(f/W)$. The rms baseband bandwidth is found using (1.157) as

$$W_{rms,bb} = \left(\frac{\int_0^\infty f^2 G_Y(f) df}{\int_0^\infty G_Y(f) df} \right)^{1/2} = \left(\frac{\int_0^{W/2} f^2 df}{W/2} \right)^{1/2} = \frac{W}{2\sqrt{3}} \qquad (1.159)$$

Example 1.20 Convolution of Triangular and Window Functions.

In this example, we evaluate the convolution of triangular and rectangular functions:

$$s(t) = \Lambda(t/T) \otimes \Pi(t/T) \qquad (1.160)$$

As shown in Figure 1.17(a), we keep the triangular function fixed and slide the rectangular function along the time axis:

$$s(t) = \int_{t-T/2}^{t+T/2} \Lambda(\tau/T) d\tau, \quad -3T/2 < t < 3T/2$$

$$= \begin{cases} (2t+3T)^2/(8T) & -3T/2 < t < -T/2 \\ 3T/4 - t^2/T & -T/2 < t < T/2 \\ (2t-3T)^2/(8T) & T/2 < t < 3T/2 \end{cases} \qquad (1.161)$$

One may observe from Figure 1.17(a) that $s(t) = 0$ for $|t| > 3T/2$ and $s(0) = 3T/4$. From

(1.46), the Fourier transform of $s(t)$ is given by the product of the Fourier transforms of $\Lambda(t/T)$ and $\Pi(t/T)$, which is $T^2 \text{sinc}^3(fT)$ (see (1.52)).

Figure 1.17(b) clearly shows that the convolution causes a time dispersion of the signal waveforms. This means that $s(t)$ is dispersed over the time axis in the interval $-3\,T/2 < t < 3\,T/2$ compared to $\Lambda(t/T)$, which is non-zero over $-T < t < T$. A close look at Figure 1.17(a) shows that the dispersion will be lower as the rectangular function gets narrower. In the limiting case where the rectangular function approaches a delta function, the triangular function does not suffer any dispersion since the result of convolution of any function with a delta function is the function itself (see (1.21)).

Example 1.21 Impulse Response of a LPF.
We consider a LPF with the following transfer function

$$H(f) = \Pi(f/W) e^{-jw\tau} \qquad (1.162)$$

where the exponential term represents the delay caused by the ideal LPF. The frequency response of the considered LPF is flat and equal to unity for $-W/2 < f < W/2$. The impulse response of the considered LPF is found by the inverse Fourier transform of (1.162):

$$h(t) = \int_{-\infty}^{\infty} \Pi(f/W) e^{-jw\tau} e^{jwt} df$$

$$= \int_{-W/2}^{W/2} e^{j2\pi f(t-\tau)} df \qquad (1.163)$$

$$= W \text{sinc}(W(t-\tau))$$

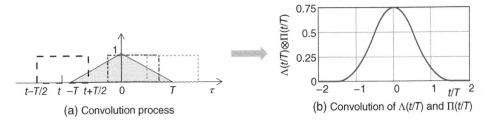

(a) Convolution process

(b) Convolution of $\Lambda(t/T)$ and $\Pi(t/T)$

Figure 1.17 Convolution of Triangular and Rectangular Functions in the Time Domain for $T = 1$.

The sinc function centered at $t = \tau$ is evidently not a delta function but can approximate the delta function $\delta(t-\tau)$ as $W \to \infty$ (see (1.17)).

Example 1.22   Transmission of a Pulse and White Noise Through an Ideal LPF.

Now consider that a pulse of duration T, centered at the origin, and an additive white Gaussian noise $w(t)$ are at the input of a LPF, described by (1.162) and (1.163):

$$s(t) = \Pi(t/T) + w(t) \qquad (1.164)$$

The signal at the LPF output may be written as

$$
\begin{aligned}
r(t) &= s(t) \otimes h(t) = y(t) + n(t) \\
&= \Pi(t/T) \otimes h(t) + w(t) \otimes h(t)
\end{aligned}
\qquad (1.165)
$$

The Fourier transforms of $\Pi(t/T)$ and $h(t)$ are given by (1.16) and (1.162), respectively. The PSD of the signal component $y(t)$ at the LPF output is found using (1.104), (1.155) and (1.162):

$$G_Y(f) = T \, \operatorname{sinc}^2(fT) \, \Pi(f/W) \qquad (1.166)$$

It is clear from (1.166) that the LPF filters out the frequency components of the sinc-squared

signal spectrum falling outside its bandwidth. Figure 1.18(a) shows the effect of low-pass filtering with a bandwidth $W = 1/T$, $2/T$ and $5/T$. Apparently, the signal spectrum will be distorted more as the filter bandwidth gets narrower. Since filtering suppresses high frequency components of the pulse, this corresponds to smoothing the sharp corners of the pulse in the time-domain (see Figure 1.18(b)). However, this also leads to time-spreading of the pulse, as in the convolution process. Then, the tails of the filtered pulse will interfere with preceeding and following pulses, leading to the so-called intersymbol interference (ISI). The distortion and ISI will be less for larger values of the time-bandwidth product WT, as expected. Since the frequency bandwidth is a scarce and expensive resource, tradeoff is usually sought between a tolerable distortion level and a minimum acceptable bandwidth.

Now we consider low-pass filtering the white noise with $G_w(f) = N_0/2$. Using (1.155), the PSD of the noise component at the LPF output is found to be:

$$G_n(f) = |H(f)|^2 G_w(f) = \frac{N_0}{2} \Pi(f/W) \qquad (1.167)$$

The noise power at the LPF output is found from (1.167) as $WN_0/2$. Hence, the use of a LPF

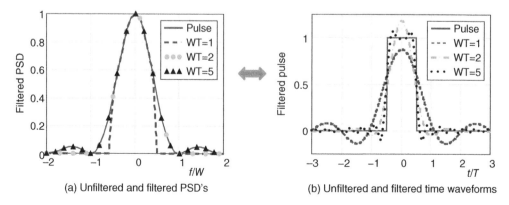

(a) Unfiltered and filtered PSD's   (b) Unfiltered and filtered time waveforms

Figure 1.18   Effect of Low-Pass Filtering of a Rectangular Pulse.

with a larger bandwidth reduces the signal distortion at the expense of increased noise power. The autocorrelation function of the filtered noise is found by the inverse Fourier transform of (1.167)

$$R_n(\tau) = \frac{N_0}{2} W \; \text{sinc}(W\tau) \qquad (1.168)$$

As suggested by (1.150), the autocorrelation function of the unfiltered white noise is a delta function, implying that the noise samples are uncorrelated with each other and it has a flat frequency spectrum (white noise). However, (1.168) clearly shows that the noise samples become correlated with each other when they are filtered, unless $W\tau$ is a non-zero integer. Therefore, the white noise is said to be colored after filtering, implying that its PSD is not any more white. As $W \to \infty$, the sinc function approaches a delta function, that is, the correlation between noise samples will decrease, and the PSD will look more flat.

References

[1] P. Chakravarthi, *The history of communications from cave drawings to mail messages*, IEEE AES Magazine, pp. 30–35, April 1992.
[2] A. B. Carlson, P. B. Crilly and J. C. Rutledge, *Communication Systems* (4th ed.), McGraw Hill: Boston, 2002.
[3] A. Papoulis and S. U. Pillai, *Probability, Random Variables and Stochastic Processes* (4th ed.), McGraw Hill: Boston, 2002.
[4] B. Sklar, *Digital Communications: Fundamentals and Applications* (2nd ed.), Prentice Hall PTR: New Jersey, 2003.
[5] S. Haykin, *Communication Systems (3rd ed.)*, J. Wiley: New York, 1994.
[6] A. Goldsmith, *Wireless Communications*, Cambridge University Press: Cambridge, 2005.
[7] J. G. Proakis, *Digital Communications* (3rd ed.), McGraw Hill: New York, 1995.
[8] P. Beckmann, *Probability in Communication Engineering*, Harcourt, Brace and World, Inc.: New York, 1967.
[9] P. Z. Peebles, Jr., *Probability, Random Variables, and Random Signal Principles* (3rd ed.), McGraw Hill: New York, 1993.

Problems

1. Determine the complex exponential Fourier series for

$$s(t) = \sin w_0 t + 2\cos^2(2w_0 t) - \cos(3w_0 t)$$

2. Determine the Fourier series expansion of

$$s(t) = |\cos t|$$

3. Using Parseval's energy theorem and the properties of the Fourier transform, determine

$$\int_{-\infty}^{\infty} \text{sinc}^n(fT) \; df, \quad n = 1, 2, 3, 4, 6$$

4. Using (1.166) and (1.167), determine the signal power and the SNR at the LPF output.

5. Determine the autocorrelation function of $s(t) + w(t)$, where $s(t)$ denotes the signal and $w(t)$ Gaussian white noise.

6. Show that the cross-correlation function between $x(t)$ and $y(t)$, which is defined by

$$R_{xy}(\tau) = E[x(t)y^*(t-\tau)] = E[x(t+\tau)y^*(t)]$$

satisfies $R_{xy}(\tau) = R_{yx}^*(-\tau)$.

7. Prove (1.98) for a power signal, for example, $\cos w_c t$.

8. Determine the Fourier transform of the function

$$s(t) = \begin{cases} 0 & t < -1/2 \\ t + 1/4 & -1/2 \le t < 3/4 \\ 1 & t \ge 3/4 \end{cases}$$

by using the properties of the Fourier transform only.

9. Determine the Fourier transform of a triangular pulse $\Lambda(t/T)$ by using

$$\Pi\left(\frac{t+T/2}{T}\right)-\Pi\left(\frac{t-T/2}{T}\right)$$

and the properties of the Fourier transform only.

10. Evaluate

$$\int_{-\infty}^{\infty}\text{sinc}(2\tau)\ \text{sinc}(t-\tau)d\tau$$

11. Determine the convolution $\Pi(t/T)\otimes\Pi(t/2\ T)$.

12. Determine the Fourier transform of

$$h(t)=\delta(t)-\delta(t-T)$$

Determine the time waveform and its Fourier transform if a unit step function $u(t+t_0)$ is at the input of a system described by the above impulse response.

13. Draw the block diagram for and determine the Fourier transform of

$$h(t)=\int_{-\infty}^{t}[\delta(\tau)-\delta(\tau-T)]\,d\tau$$

14. Determine the Fourier transform of the following:

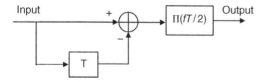

15. Determine the impulse response and the Fourier transform of the following:

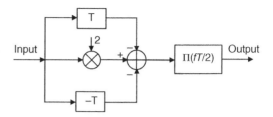

16. A raised-cosine pulse, which is defined by the following

$$P(f)=\frac{1}{4W}\left[1+\cos\left(\frac{\pi f}{2W}\right)\right]\Pi\left(\frac{f}{4W}\right)$$

is commonly used for intersymbol interference (ISI) mitigation, instead of the so-called Nyquist filter:

$$P_N(f)=\frac{1}{2W}\Pi\left(\frac{f}{2W}\right)$$

Determine the impulse response of $P(f)$ and compare with that of $P_N(f)$.

17. Using the Fourier transform of $s(t)$ and $s(t)\pm s(-t)$ from (1.28) and (1.31) and the Rayleigh energy theorem, show that

$$\int_0^{\infty}\frac{1}{a^2+x^2}dx=\frac{\pi}{2a}$$

$$\int_0^{\infty}\frac{1}{\left(a^2+x^2\right)^2}dx=\frac{\pi}{4a^3}$$

$$\int_0^{\infty}\frac{x^2}{\left(a^2+x^2\right)^2}dx=\frac{\pi}{4a}$$

2

Antennas

Antennas are important components of wireless systems, such as communications, radars, remote-sensing, and so on. Antenna is a transducer which converts electrical signals (voltage and current) into electromagnetic (EM) waves and vice versa (see Figure 2.1). Therefore, antennas are used for transmitting electrical signals via EM waves in the direction of intended remote receivers, and collecting incident EM waves carrying electrical signals transmitted by distant transmitters. In duplex communication systems, the same antenna can be used both for transmission and reception. By virtue of the reciprocity theorem, an antenna has the same performance in transmission and reception.

Theoretically, any structure can radiate and receive EM waves but not all structures can do it efficiently. Antennas are conducting or dielectric structures that allow efficient radiation and reception of EM waves. We know from Maxwell's equations that a conducting structure radiates if the current flowing through it changes sufficiently fast with time and its size is comparable to the wavelenth. An antenna is desired to have low ohmic losses, so that the largest percentage of the power fed at its

terminals is radiated. Impedance matching between antenna and transceiver minimizes the return losses. Similarly, polarization matching between transmit and receive antennas is required for efficient coupling (power transfer) between them. Antennas are usually desired to focus their radiation in some desired directions, for example, in the direction of the intended receiver. However, in broadcasting applications, for example, radio and TV transmissions, they are usually required to radiate uniformly on the Earth's surface so that all receivers in the coverage area can equally receive these signals. Antennas can also be designed so as not to receive radiations from certain directions, for example, that of a jammer or another transmitter, or to null their radiation for interference avoidance.

Depending on the platform used, the coverage area, and the frequency and bandwidth of operation, antennas are required to have different shapes and sizes. An antenna has potentially better performance as its electrical size (maximum antenna dimension measured in wavelength) increases. This implies that antennas have potentially better performance at

Digital Communications, First Edition. Mehmet Şafak.
© 2017 John Wiley & Sons Ltd. Published 2017 by John Wiley & Sons Ltd.
Companion website: www.wiley.com/go/safak/Digital_Communications

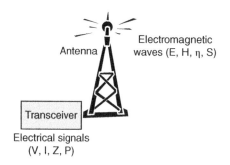

Figure 2.1 An Antenna as a Transducer Between the Transceiver (Including Feeder Cables/Waveguides) and the Propagation Medium.

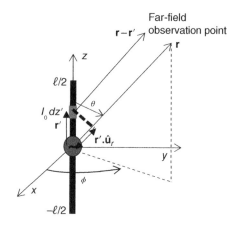

Figure 2.2 Radiation By an Infinitesmall Thin Current Element.

higher frequencies. Special attention is required for the design of broadband antennas, that is, antennas with wider bandwidths measured in percentage of the carrier frequency. The shape of an antenna usually determines its coverage area. Antennas can be erected on the surface of earth, at the top of a building or a tower, on board of a satellite or a space vehicle. They may as well be located in an indoor environment or even integrated to the receiver. Antennas may be operated in the fixed, transportable or mobile modes.

Antenna analysis and design is a complex and interesting domain of research and plays a critical role in the performance of wireless systems. This chapter does not present a detailed analysis of antennas, which requires in-depth treatment of Maxwell's equations, but provides some fundamental concepts on antennas. The aim is to help the reader to establish the link between antennas and propagation concepts of RF engineering and the statistical theory of communications.

2.1 Hertz Dipole

Assume that a linear thin wire of length $\ell < < \lambda$ is directed along the z-axis as shown in Figure 2.2. Consider a Hertz dipole which is defined as an infinitesimal current element $I_0 \, dz'$, which implies a constant current I_0 flowing along a conductor dz' long. Such a current element does not

exist in real life in isolation but it forms the basic building block of a practical antenna, since a dipole may be assumed to be made up of these current elements connected together end to end, each with a different current I_0.

Assume that the $\exp(j2\pi f t)$ factor describes the time-harmonic variation of the current with time, where f denotes the frequency of operation, hence the time-rate of change of the antenna current. The current is assumed constant along the wire:

$$\mathbf{I}(z) = I_0 \, \hat{\mathbf{u}}_z \qquad (2.1)$$

where $\hat{\mathbf{u}}_z$ denotes the unit vector along z-axis. The EM fields radiated by this current element may be determined using the Maxwell's equations. The magnetic vector potential is defined by [1]

$$\mathbf{A}(\mathbf{r}) = \mu \int_{-\ell/2}^{\ell/2} \mathbf{I}(z') \frac{\exp(-jk|\mathbf{r}-\mathbf{r}'|)}{4\pi|\mathbf{r}-\mathbf{r}'|} dz' \quad (2.2)$$

where $k = 2\pi/\lambda$ (λ is the wavelength) and the magnetic permeability of free space is given by $\mu = 4\pi 10^{-7}$ H/m. Here, $\mathbf{r}$ and $\mathbf{r}'$ denote the spherical coordinates $r = (r,\theta,\phi)$ of observation and source points, respectively. At observation points which are sufficiently far from the antenna, we may assume that $|\mathbf{r}-\mathbf{r}'| = |\mathbf{r}-z'\hat{\mathbf{u}}_z| \simeq r$ (see Figure 2.2) since $\ell < < \lambda$.

Using (A.4), the magnetic vector potential may be written, as

$$A(r) = \frac{\mu I_0 \ell}{4\pi} \frac{e^{-jkr}}{r} \hat{u}_z$$

$$= \frac{\mu I_0 \ell}{4\pi} \frac{e^{-jkr}}{r} (\cos\theta \hat{u}_r - \sin\theta \hat{u}_\theta) \quad (2.3)$$

Example 2.1 Far-Field Region.

Here it may be appropriate to question the limits of validity of the approximation $|r-r'| \simeq r$ in (2.3). Consider an observation point described by $r = (x,y,z) = (r\sin\theta\cos\phi \ \ r\sin\theta \ \sin\phi \ \ r\cos\theta)$ and the source coordinates along infinitely thin wire antenna by $r' = z'\hat{u}_z$, as shown in Figure 2.2. Then, noting that $r^2 = |r|^2 = x^2 + y^2 + z^2$, one may simply write that

$$|r-r'| = \sqrt{x^2 + y^2 + (z-z')^2}$$

$$= r\sqrt{1 - 2(z'/r)\cos\theta + (z'/r)^2} \quad (2.4)$$

Inserting the first four terms of the binomial expansion (see (D.9))

$$\sqrt{1+x} = 1 + x/2 - x^2/8 + x^3/16$$
$$- 5x^4/128 + \cdots, \quad x \to 0 \quad (2.5)$$

into (2.4), one gets

$$|r-r'| = r - z'\cos\theta + \frac{(z')^2}{2r}\sin^2\theta$$
$$- \frac{(z')^3}{2r^2}\cos\theta\sin^2\theta + \cdots \quad (2.6)$$

where terms $(z')^n$ with $n \geq 4$ are ignored. The vectors r and $r-r'$ are considered to be parallel at observation points when the terms with $(z')^n$, $n \geq 2$ may be neglected in (2.6). The usual assumption is that the third term may be neglected if its maximum value satisfies the following [1]

$$\max\left\{ \frac{(z')^2}{2r}\sin^2\theta \right\} \leq \frac{\lambda}{16} \quad (2.7)$$

Since the maximum value of z' is $\ell/2$, (2.7) reduces to

$$r \geq \frac{2\ell^2}{\lambda} \quad (2.8)$$

Hence, the distance beyond which the two vectors may be considered parallel is determined by the maximum antenna size and the wavelength. This distance is the border of the so-called far-field region. In summary, in the far-field region, (2.6) may be approximated as

$$|r-r'| \simeq r - z'\cos\theta = r - r'.\hat{u}_r \quad (2.9)$$

In the derivation of (2.3), the second term in (2.9) is also neglected since the antenna size is assumed to be very short compared to the wavelength.

One may easily observe from (2.3) and (A.1) that the magnetic field induced by an Hertzian dipole has only ϕ-component:

$$H(r) = \frac{1}{\mu}\nabla \times A(r) = \frac{1}{\mu r}\left[\frac{\partial}{\partial r}(rA_\theta) - \frac{\partial A_r}{\partial r}\right]\hat{u}_\phi \ \left(\frac{A}{m}\right)$$

$$H_\phi = j\frac{kI_0\ell\sin\theta}{4\pi r}\left[1 + \frac{1}{jkr}\right]e^{-jkr} \ \left(\frac{A}{m}\right)$$
$$(2.10)$$

The corresponding electric field is found from the magnetic field and (A.1) as follows:

$$E(r) = \frac{1}{jw\varepsilon}\nabla \times H(r)$$

$$= \frac{1}{jw\varepsilon r\sin\theta}\frac{\partial}{\partial\theta}(H_\phi\sin\theta)\hat{u}_r$$

$$- \frac{1}{jw\varepsilon r}\frac{\partial}{\partial r}(rH_\phi)\hat{u}_\theta$$

$$E_r = \eta\frac{I_0\ell\cos\theta}{2\pi r^2}\left[1 + \frac{1}{jkr}\right]e^{-jkr} \ (V/m)$$

$$E_\theta = j\eta\frac{kI_0\ell\sin\theta}{4\pi r}\left[1 + \frac{1}{jkr} + \frac{1}{(jkr)^2}\right]e^{-jkr}(V/m)$$

$$E_\phi = 0 \ (V/m) \quad (2.11)$$

where $\varepsilon = 1/(c^2\mu) = 10^{-9}/(36\pi)\ (F/m)$ is the permittivity of free-space and $\eta = \sqrt{\mu/\varepsilon} = 120\pi\ \Omega$ denotes the intrinsic impedance of free-space.

The complex Poynting vector, which measures the direction and the intensity of the power flowing from the antenna, is defined by

$$S = \frac{1}{2}\mathbf{E} \times \mathbf{H}^* = \underbrace{\frac{1}{2}E_\theta H_\phi^* \hat{u}_r}_{radiated\ field \approx 1/r^2}$$

$$- \underbrace{\frac{1}{2}E_r H_\phi^* \hat{u}_\theta}_{no\ radiation \approx 1/r^3} \quad (W/m^2) \qquad (2.12)$$

where the factor ½ accounts for the fact that $\mathbf{E}$ and $\mathbf{H}$ denote the peak values of the EM fields. It is important to note that the power flow (radiation) from the antenna takes place in two directions. The r-component of the Poynting vector represents the radiated field, that is, the power flow away from the antenna, while its θ-component represents the power which circulates/remains around the antenna, hence no radiation in the far-field region. The components of the complex Poynting vector are given by

$$S_r = \frac{1}{2}E_\theta\ H_\phi^*$$

$$= \frac{\eta}{8}\left|\frac{I_0\ell}{\lambda}\right|^2 \frac{\sin^2\theta}{r^2}\left[1 - j\frac{1}{(kr)^3}\right]\left(\frac{W}{m^2}\right)$$

$$S_\theta = -\frac{1}{2}E_r\ H_\phi^*$$

$$= j\eta\frac{k|I_0\ell|^2\sin\theta\cos\theta}{16\pi^2 r^3}\left[1 + \frac{1}{(kr)^2}\right]\left(\frac{W}{m^2}\right)$$

$$\qquad (2.13)$$

The real component of S_r, proportional to $1/r^2$, becomes the dominant term in the far-field region of the transmit antenna. However, the imaginary components of S_r and S_θ, which are proportional to $1/r^3$ and $1/r^5$, are stronger in the close vicinity of the antenna but fade away rapidly with increasing values of the distance r from the transmit antenna. The reactive components of the Poynting vector

represent the power stored in the vicinity of the antenna and have no contribution to radiation. On the other hand, as can easily be observed from (2.13), the Hertzian dipole does not equally radiate in all directions. For example, there is no radiation along the antenna axis ($\theta = 0, \pi$) but radiation is maximum in the boresight direction ($\theta = \pi/2$), that is, in the direction perpendicular to the antenna axis.

The complex radiated power is defined as the power flowing outward through a closed spherical surface of radius r:

$$P = \int_0^{2\pi}\int_0^\pi S.\hat{u}_r\,r^2\sin\theta\,d\theta\,d\phi = \frac{\pi}{3}\eta\left|\frac{I_0\ell}{\lambda}\right|^2\left[1 - j\frac{1}{(kr)^3}\right]$$

$$\qquad (2.14)$$

The total radiated power P_{tr} is the real component of (2.14), since only this component is available in the far-field region and the other component, which is imaginary, attenuates as $1/r^3$:

$$P_{tr} = \text{Re}[P] = \frac{\pi}{3}\eta\left|\frac{I_0\ell}{\lambda}\right|^2 \qquad (2.15)$$

The total radiated power by a Hertzian dipole may be considered as it is dissipated in a so-called radiation resistance R_r:

$$P_{tr} = \frac{1}{2}|I_0|^2 R_r \quad \Rightarrow \quad R_r = 80\pi^2(\ell/\lambda)^2 \quad \ell << \lambda$$

$$\qquad (2.16)$$

One may observe from (2.16) that total radiated power by a Hertzian dipole is proportional to the square of the current (to the transmit power), and the radiation resistance. Hence, the power radiated by a Hertzian dipole, like any other antenna, is determined by the power at its input and its electrical length ℓ/λ, which increases with frequency.

In the far-field region, reactive components of the EM fields fade away and the electric field intensity incident at the receiving antenna has constant-phase surfaces. Their wavefronts may then be approximated as planes; hence they are called as plane-waves. Using (2.10), (2.11) and (2.13), the EM waves and corresponding Poynting vector intensity in the far-field region may simply be written as

$$E_\theta = j\eta \frac{kI_0\ell \sin\theta}{4\pi r} e^{-jkr} \quad \left(\frac{V}{m}\right)$$

$$H_\phi = \frac{E_\theta}{\eta} = j\frac{kI_0\ell \sin\theta}{4\pi r} e^{-jkr} \quad \left(\frac{A}{m}\right) \quad (2.17)$$

$$S_r = \frac{1}{2}E_\theta \, H_\phi^* = \frac{\eta}{8}\left|\frac{I_0\ell}{\lambda}\right|^2 \frac{\sin^2\theta}{r^2} \quad \left(\frac{W}{m^2}\right)$$

2.1.1   Near- and Far-Field Regions

The so-called near- and far-field regions of an antenna are determined by the interaction between charge and/or current elements carried by an antenna and the observed EM fields, as a function of the distance between them. Note for example that the EM field components of a Hertzian dipole, given by (2.10) and (2.11), are proportional to 1, $1/(jkr)$ and $1/(jkr)^2$. The relative magnitudes of these components are equal to each other at $kr = 1$. However, for $kr < 1$, the components proportional to $1/(jkr)$ and $1/(jkr)^2$ are larger than those proportional to 1. For $kr \ll 1$, in the *reactive near-field region*, the reactive field components, which represent the reactive power stored in the close vicinity of the antenna, dominate the EM fields. With increasing distances from antenna, the radiative components tend to emerge as the dominant term in the EM fields. *Radiative near-field (Fresnel) region* is defined as the region where radiative fields dominate and the electric and magnetic field distributions are dependent on the distance from the antenna. This region is located between reactive near-field and the far-field regions. The border between radiative near-field (Fresnel) and reactive near-field regions is determined by the assumption that the phase of the maximum value of the fourth term of (2.6) is less than $\pi/8$:

$$\max\left\{\frac{k(z')^3}{2r^2}\cos\theta\sin^2\theta\right\} \le \pi/8 \quad (2.18)$$

Inserting into (2.18) $\max(z') = \ell/2$ and $\theta = \tan^{-1}(\sqrt{2})$, which maximizes the term

$\cos\theta \, \sin^2\theta$, the border between the Fresnel and reactive-near field regions is found to be

$$r \ge 0.62\sqrt{\ell^3/\lambda} \quad (2.19)$$

The border with far-field region, given by (2.8), corresponds to the assumption that third term in the binomial expansion (2.6) of the distance between source and observation points is less than or equal to $\lambda/16$. This term represents the distance component which influences the angular (spatial) distribution of the EM field in the Fresnel region.

In the *radiative near-field (Fresnel) region*, radiative fields dominate the EM fields whose angular distribution is dependent on the distance from the antenna. The near-field radiative components are dominant in this region but decay rapidly with distance. Far-field components decay more slowly and become dominant in the far-field region. Hence, Fresnel region represents a transition region between near- and far-field regions where both types of EM field behavior may be observed.

In the *far-field (Fraunhofer) region*, the angular field distribution is essentially independent of the distance from the antenna; electric and magnetic field magnitudes are inversely proportional to the distance from the antenna, and the ratio of electric to magnetic field intensities is equal to free-space intrinsic impedance $\eta = 120\pi$ (Ω).

In most applications, receiving antenna is located in the far-field region of a transmit antenna so that communication between them takes place without the interference of reactive near-field components. In this region, the real component of the Poynting vector intensity, which is proportional to $1/r^2$, contributes to the far-field radiation, that is, carries the transmitted signal information to the intended receiver with minimum attenuation. However, the imaginary components of S_r and S_θ, proportional to $1/r^3$ and $1/r^5$, represent the power stored around the antenna and have no contribution to the far-field radiation (see (2.13)). In brief, these components are not useful for communications between antennas located in far-field regions of each other.

It is commonly assumed that the far-field region starts at distances larger than $2L^2/\lambda$ (L denotes the largest antenna dimension) from the transmitting antenna (see (2.8)) where the EM field behaves as a plane wave. As shown in Figure 2.3, the region described by the distance range, $0.62\sqrt{L^3/\lambda} < r < 2L^2/\lambda$, from the transmitting antenna is called the radiative

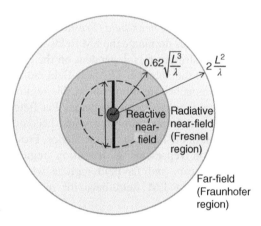

Figure 2.3 Near- and Far-Field Regions of an Antenna.

near field (Fresnel) region. In this region, the distances from an antenna are comparable with antenna size L and EM fields have spherical wave fronts (see Figure 2.4). Note that near- and far-field distances defined above are meaningful for antenna sizes exceeding the wavelength ($L > \lambda$). For electrically small antennas with $L < \lambda/2$, the Fresnel region is usually assumed to be bounded by the range $\lambda/(2\pi) < r < \lambda$. [1]

Example 2.2 Near-Field Communications and Radio-Frequency Identification (RFID) System.

Near-field communications refers to communications in the Fresnel region, where electric and magnetic fields do not have plane wavefronts and the ratio of electric to magnetic field intensities is not equal to the free-space intrinsic impedance, $\eta = 120\pi$ (Ω). The radiative near-field components, which dominate the reactive near-field field components, are used for communications.

The success of near-field communications is critically dependent on coupling between transmit and receive antennas which are located in each other's Fresnel regions. Depending on the frequency of operation, the propagation

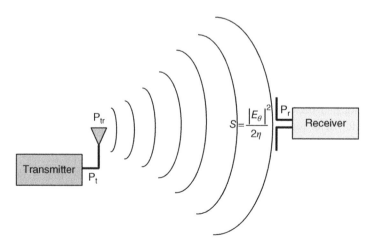

Figure 2.4 EM Waves Have Spherical Wave-Fronts In the Near Field Region but They Behave as Plane Waves In the Far-Field Region.

medium and the types of antennas used, it might be more suitable to use electric or magnetic coupling between antennas. The electrical characteristics (permittivity, magnetic permeability and conductivity) of the medium between transmit and receive antennas plays a critical role in choosing the type of coupling. For example, magnetic coupling is preferred in a lossy propagation medium, which is characterized by high conductivity. Since electric field suffers high attenuation in such a medium, electrical coupling between transmit and receive antennas would be much weaker compared to magnetic coupling.

One of the most notable examples of near-field communications by magnetic inductive coupling between transmitting and receiving antennas is radio-frequency identification (RFID) systems. RFID is a technology for identifying objects using radio signals. For example, it may be used for inventory control via distant reading of codes, for automatically identifying or tracking a person, an animal, a product or an item. Figure 2.5 shows a simplistic view of a RFID system. The magnetic field radiated by modulated time-varying currents in the 'RFID reader' loop antenna charges the capacitor of the 'RFID tag'. When the capacitor buids up enough

energy, it releases this energy to induce an open circuit voltage at the terminals of the loop antenna of the 'RFID tag'. The induced voltage triggers the RFID tag to transmit back the information required by the RFID reader over a short distance. The range is chosen so as to support data rates in the order of several hundred kbps. For example, some smart cards are simply RFID tags that use the ISO15693 communication protocol and operate at 13.56 MHz.

We use the duality principle in order to determine the open-circuit voltage at the terminals of a RFID-tag. The magnetic field radiated by a current I_ℓ flowing through a small loop antenna of radius a may be obtained by the following transformation in (2.11):

$$
\begin{aligned}
E_r &\rightarrow H_r \\
E_\theta &\rightarrow H_\theta \\
I_0 \ell &\rightarrow jw\varepsilon(\pi a^2)I_\ell
\end{aligned}
\tag{2.20}
$$

The coupling between transmitter and receiver is determined by the magnetic flux due to RFID reader's loop antenna that induces an open circuit voltage at the terminals of the RFID tag's loop antenna. This requires an integral of the magnetic fields found by (2.20) over

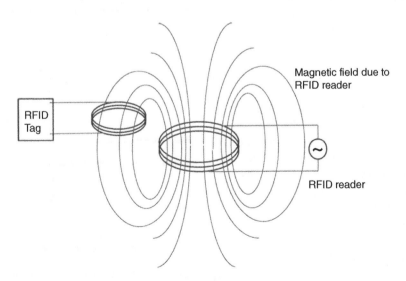

Figure 2.5   RFID Principle.

the surface of the RFID tag's loop-antenna. Near-field coupling between antennas is highly dependent on the types, electrical sizes, radiation patterns, relative orientations and impedance matching of the antennas employed and the distance between them. [2][3] Some tags can be read from several meters away and in the presence of obstructions between the reader and the tag.

The same architecture may also be used for distant charging of a battery connected to the terminals of the receive loop antenna, that is for RF energy harvesting. The RF energy received by the RFID tag may be converted into chemical energy in a battery for future use.

2.2 Linear Dipole Antenna

In the previous section on Hertzian dipole, the current flowing along an infinitesmall linear antenna is assumed to be uniform. However, the current distribution along a linear antenna can not be uniform and is required for calculating the radiated fields. Transmission line approximation provides a first order and heuristic approach for determining the current distribution along a linear antenna. We know from the transmission line theory that the current flowing in $\pm z$-directions along a two-wire transmission line, as shown in Figure 2.6a, may be written as

$$I(z) = I_0' \, e^{-jkz} + \rho I_0' \, e^{+jkz} \qquad (2.21)$$

where ρ denotes the reflection coefficient at the end of the line and I_0' shows the peak current. Since the current at the end of an open-circuited transmission line vanishes, $I(\ell/2) = 0$; therefore the reflection coefficient is given by $\rho = -\exp(-jk\ell)$. Consequently, the current distribution along the transmission line may be written as

$$I(z) = I_0' \left(e^{-jkz} - e^{-k\ell} e^{+jkz} \right)$$

$$= I_0 \sin k \left(\frac{\ell}{2} - z \right), \quad 0 \le z \le \ell/2 \qquad (2.22)$$

where $I_0 = 2I_0' \, \exp[-j(k\ell/2 - \pi/2)]$. The interference between currents flowing in $\pm z$-directions, forms a standing wave along the line; the distance between two consecutive maxima/minima is equal to $\lambda/2$. Apart from a phase factor, the maximum value of the current I_0 corresponds to the constructive interference between currents flowing in $\pm z$-directions.

As the transmission line is bent 90° as shown in Figure 2.6b, as a first-order approximation, the current distribution may be assumed to be unchanged. Then, the current distribution of an infinitely thin linear antenna located at the origin and directed along the z-axis may be approximated as

$$\mathbf{I}(z') = I_0 \, \sin \left[k \left(\frac{\ell}{2} - |z'| \right) \right] \hat{\mathbf{u}}_z, \quad -\ell/2 \le z' \le \ell/2 \qquad (2.23)$$

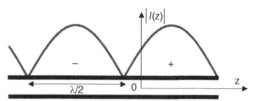

(a) Standing wave current distribution along a transmission line

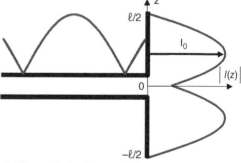

(b) Transmission line model for current distribution along a linear dipole antenna

Figure 2.6   Current Distribution Along a Lossless Two-Wire Transmission Line and a Linear Dipole, Based on the Transmission-Line Theory.

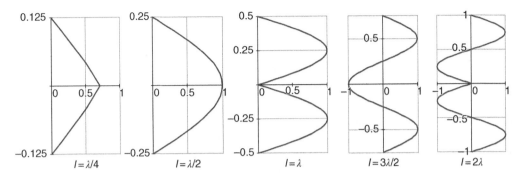

Figure 2.7   Current Distributions Along a Dipole Antenna of Length ℓ.

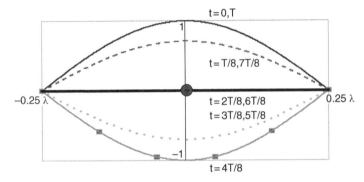

Figure 2.8   Normalized Current Distribution Along a $\lambda/2$ Dipole Antenna at Different Fractions of the Period T, Where the Frequency of Operation is Given by $f = 1/T$.

where the antenna is assumed to be center-fed. In contrast with the Hertzian dipole, this current distribution is not uniform along the antenna, vanishes at the edges and has a maximum value of I_0. Figure 2.7 shows the current distribution of a linear antenna for various values of the electrical antenna size ℓ/λ.

Figure 2.7 shows the current distributions frozen in time. In order that radiation takes place from an antenna, the current should change sufficiently fast with time. The variation with time of the current distribution, given by (2.23), may be expressed as follows:

$$\mathbf{I}(z',t) = \mathrm{Re}\{\mathbf{I}(z')e^{j2\pi ft}\} = I_0 \sin\left(k\left(\frac{\ell}{2} - |z'|\right)\right)\cos(2\pi ft) \tag{2.24}$$

Time variation of the current along a half-wave dipole is shown in Figure 2.8 at $t = 0$, $T/8$, $T/4$, $3T/8$, $T/2$, $5T/8$, $3T/4$, $7T/8$ and T.

Using (2.17), the far-field electrical field radiated by a current element of length dz' may be written as

$$dE_\theta = j\eta k I_z(z')\sin\theta\frac{e^{-jk|\mathbf{r}-\mathbf{r}'|}}{4\pi|\mathbf{r}-\mathbf{r}'|}dz' \quad \left(\frac{V}{m}\right) \tag{2.25}$$

where $|\mathbf{r}-\mathbf{r}'|$ denotes the distance between current element and the far-field observation point, where $\mathbf{r}-\mathbf{r}'$ and $\mathbf{r}$ may be assumed to be parallel to each other (see (2.9)):

$$|\mathbf{r}-\mathbf{r}'| = \begin{cases} r - z'\cos\theta & \text{for phase} \\ r & \text{for amplitude} \end{cases} \tag{2.26}$$

Inserting (2.26) into (2.25) and summing all contributions from infinitesmall current elements leads to the following integration for the radiated electrical field in the far-field region:

$$E_\theta = j\eta k \frac{e^{-jkr}}{4\pi r} \sin\theta \int_{-\ell/2}^{\ell/2} I_z(z') e^{jkz'\cos\theta} dz' \quad \left(\frac{V}{m}\right)$$

(2.27)

(2.27) suggests that the far-field electric field radiated by a linear antenna is proportional to the inverse Fourier transform of the current distribution. Hence, the current distribution plays an important role in determining the radiation from an antenna. Substituting the current distribution given by (2.23) into (2.27) and using the symmetry of the current distribution with respect to z', one can rewrite (2.27) as follows

$$E_\theta = j\eta k \frac{e^{-jkr}}{4\pi r} \sin\theta \int_0^{\ell/2} I_z(z')\, 2\cos(kz'\cos\theta)\, dz'$$

$$= j\eta k I_0 \frac{e^{-jkr}}{4\pi r} \sin\theta \int_0^{\ell/2} \left[\sin\left(\frac{k\ell}{2} - k(1+\cos\theta)z'\right) \right.$$

$$\left. + \sin\left(\frac{k\ell}{2} - k(1-\cos\theta)z'\right) \right] dz'$$

(2.28)

After integration, one gets the radiated EM fields in the far-field region as

$$E_\theta = \eta H_\phi = j\eta I_0 \frac{e^{-jkr}}{2\pi r} \left[\frac{\cos\left(\frac{k\ell}{2}\cos\theta\right) - \cos\left(\frac{k\ell}{2}\right)}{\sin\theta} \right] \left(\frac{V}{m}\right)$$

(2.29)

The Poynting vector intensity in the far-field region is given by

$$S_r = \frac{1}{2\eta}|E_\theta|^2 = \eta \frac{|I_0|^2}{8\pi^2 r^2} \left[\frac{\cos\left(\frac{k\ell}{2}\cos\theta\right) - \cos\left(\frac{k\ell}{2}\right)}{\sin\theta} \right]^2 \left(\frac{W}{m^2}\right)$$

(2.30)

which is evidently independent of the azimuth and depends on the square of the maximum current intensity (hence the transmitted power), the distance from the transmit antenna, elevation angle θ and the electrical antenna length ℓ/λ.

The input impedance Z_m of a dipole antenna of physical length ℓ, referred to the current maximum I_0, is given by [1]

$$Z_m = R_m + jX_m$$

$$R_m = R_r = \frac{2}{|I_0|^2} \int_0^{2\pi}\int_0^\pi S_r r^2 \sin\theta\, d\theta\, d\phi$$

$$= \frac{\eta}{2\pi} \left[\begin{array}{l} C + \ln(k\ell) - C_i(k\ell) + \frac{1}{2}\sin(k\ell)\{S_i(2k\ell) - 2S_i(k\ell)\} + \\ \frac{1}{2}\cos(k\ell)\{C + \ln(k\ell/2) + C_i(2k\ell) - 2C_i(k\ell)\} \end{array} \right]$$

$$X_m = \frac{\eta}{4\pi} \left\{ \begin{array}{l} 2S_i(k\ell) + \cos(k\ell)[2S_i(k\ell) - S_i(2k\ell)] \\ -\sin(k\ell)\left[2C_i(k\ell) - C_i(2k\ell) - C_i\left(\frac{2ka^2}{\ell}\right)\right] \end{array} \right\}$$

(2.31)

where $C = 0.5772$ is the Euler's constant, a is the radius of the wire, and

$$C_i(x) = -\int_x^\infty \frac{\cos t}{t} dt$$

$$S_i(x) = \int_0^x \frac{\sin t}{t} dt$$

(2.32)

Figure 2.9 shows the variation of the radiation resistance and the input reactance of a dipole antenna both referred to the current maximum. The radiation resistance is determined only by the electrical length ℓ/λ of the dipole antenna while the input reactance is a function of the antenna radius and ℓ/λ (see (2.31)). The non-monotonic increase of the radiation resistance (total radiated power) with ℓ/λ implies that each increase in the antenna length does not always lead to increased radiation from the antenna. This may be attributed to constructive/destructive interferences between the radiated EM fields due to infinitesmall current elements along the antenna. As for the total radiated power, it may be considered as the power dissipated when a current I_0 flows in a fictitious resistance R_r, the so-called radiation resistance. The reactance of a dipole antenna also shows an oscillatory behavior and depends strongly on a/λ. As the thickness of the antenna wire increases, the reactance varies more slowly with the frequency of operation and the antenna becomes less frequency selective. Impedance matching circuits are generally used to minimize impedance mismatches between the feeder line and the antenna and hence to

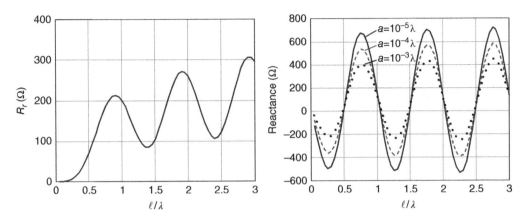

Figure 2.9 Radiation Resistance and the Reactance (Referred to the Current Maximum) of a Linear Dipole Antenna.

maximize the power transfer. On the other hand, the reactance vanishes for ℓ/λ slightly less than 0.5, Therefore, the use of thicker dipoles with electrical lengths slightly less than 0.5 is preferred for broadband operation and for impedance matching purposes.

It is important to distinguish between the radiation resistance, which is equal to the input resistance referred to the current maximum, and the input resistance, which is referred to the feeding point. Since the antenna is center-fed, the current at the feeding point is given by (2.23):

$$I(0) = I_0 \sin(k\ell/2) \qquad (2.33)$$

The input impedance Z_{in} of a linear dipole antenna at the feeding point is related to Z_m as follows:

$$\frac{1}{2}Z_{in} |I(0)|^2 = \frac{1}{2}Z_m |I_0|^2$$

$$\Rightarrow Z_{in} = \frac{|I_0|^2}{|I(0)|^2}Z_m = \frac{Z_m}{\sin^2(k\ell/2)} \qquad (2.34)$$

For a center-fed half-wave dipole ($\ell = \lambda/2$), the current maximum occurs at the feeding point, that is, $I(0) = I_0$. Consequently, the input resistance is equal to the radiation resistance; $R_r = R_{in} = 73\,\Omega$. The input reactance of a

half-wave dipole is equal to $42.5\,\Omega$ (see Figure 2.9). Hence, input impedance of a half-wave dipole is given by $Z_{in} = R_{in} + jX_{in} = 73 + j42.5\,\Omega$.

2.3 Aperture Antennas

Aperture antennas are used in diverse application areas. Reflector, lens, slot and horn antennas are typical examples of aperture antennas. The EM waves radiated by such antennas may be considered as being emanating from their physical apertures. Instead of finding the far-field radiation by integration over physical sources, the radiated fields may be directly determined by the integration of electric and/or magnetic fields over the aperture, using field equivalence principle. Field equivalence principle simply states that the EM fields induced by a primary source over an aperture radiate as secondary sources and the radiated EM fields determined using primary and secondary sources are identical to each other. For example, consider a parabolic reflector antenna which consists of a feed located at its focus, acting as a primary source, and a parabolic reflector. Reflection, scattering and diffraction of the EM fields due to feed by the parabolic reflector are responsible for the

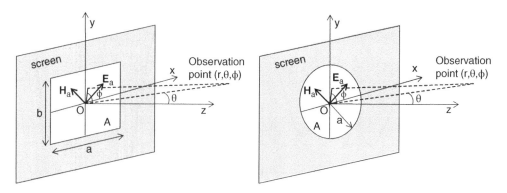

Figure 2.10   Radiation From Rectangular and Circular Apertures.

far-field radiated fields. According to the field equivalence principle, far-field radiated fields can also be determined by using the EM fields over the reflector aperture.

In order to emphasize physical concepts and to simplify calculations, we will be concerned with the far-field radiation. Figure 2.10 shows rectangular and circular apertures on infinitely large screens in $z=0$ plane (aperture plane). The electric and magnetic fields are assumed to be known over the aperture plane and their tangential components (polarized in the xy-plane) are denoted by $\mathbf{E}_a$ and $\mathbf{H}_a$. The far-field radiated fields may be uniquely determined for $z>0$ if tangential electric and/or magnetic fields are known over the aperture plane. However, in practice, electric and/or magnetic fields may not be known exactly over the aperture plane. If the screen is a perfect electrical conductor, then the far-field radiation can be determined using the tangential electric field $\mathbf{E}_a$ over the aperture surface A (see Figure 2.10), since the tangential electric field vanishes on the perfect electrical conductor screen. As for the case of a perfect magnetic conductor screen, the far-field radiation may be determined using only the tangential magnetic field $\mathbf{H}_a$ over the aperture surface A since $\mathbf{H}_a$ vanishes on the perfect magnetic conductor screen. The image theory can be exploited to remove the conducting plane and calculate the radiated fields for $z>0$ by doubling the current EM field intensities over

A and assuming free-space radiation environment.[4]

We will study below the two alternative forms of the field equivalence principle when only the electric or the magnetic field is known over the aperture A. Using the image theory for a perfect electric conductor screen, the equivalent current may be written as

$$\mathbf{J}_{ms} = \begin{cases} -2\hat{\mathbf{u}}_n \times \mathbf{E}_a & over\ A \\ 0 & on\ the\ screen \end{cases} \quad (2.35)$$

where $\hat{\mathbf{u}}_n$ denotes the unit normal vector to the aperture surface A. For a perfect magnetic conductor screen, the equivalent current is given by

$$\mathbf{J}_s = \begin{cases} 2\hat{\mathbf{u}}_n \times \mathbf{H}_a & over\ A \\ 0 & on\ the\ screen \end{cases} \quad (2.36)$$

Perfect electric or magnetic conductor screens are not used in practice. However, for physical apertures sufficiently large compared to the wavelength, EM field may be assumed to be negligibly small over the aperture plane except for over the physical aperture A. Hence far-field radiation may be found from the tangential electric and/or magnetic field over the physical aperture A only.

The magnetic vector potential is given by

$$\mathbf{A}(\mathbf{r}) = \mu \iint_A \mathbf{J}_s(\mathbf{r}') \frac{e^{-jk|\mathbf{r}-\mathbf{r}'|}}{4\pi|\mathbf{r}-\mathbf{r}'|} dS' \quad (2.37)$$

where A denotes the aperture in the xy-plane (see Figure 2.10) over which the current density is nonzero. The coordinates of observation and source points are denoted by $\mathbf{r}$ and $\mathbf{r}'$, respectively, and $\hat{\mathbf{u}}_r$ is the unit vector in the direction of $\mathbf{r}$. Inserting the far-field approximation

$$|\mathbf{r}-\mathbf{r}'| = \begin{cases} r - \mathbf{r}'.\hat{\mathbf{u}}_r & \text{for phase} \\ r & \text{for amplitude} \end{cases} \quad (2.38)$$

into (2.37) one gets

$$\mathbf{A}(\mathbf{r}) = \frac{\mu e^{-jkr}}{4\pi r}\iint_A \mathbf{J}_s(\mathbf{r}') e^{jk\hat{\mathbf{u}}_r.\mathbf{r}'} dS' = \frac{\mu e^{-jkr}}{2\pi r}\hat{\mathbf{u}}_z \times \mathbf{h}$$

$$\mathbf{h} = \iint_A \mathbf{H}_a(\mathbf{r}') e^{jk\hat{\mathbf{u}}_r.\mathbf{r}'} dS'$$

$$(2.39)$$

If the magnetic field is uniformly distributed over the aperture, then one gets

$$\mathbf{h} = Af(\theta,\phi)\mathbf{H}_a$$
$$f(\theta,\phi) = \frac{1}{A}\iint_A e^{jk\hat{\mathbf{u}}_r.\mathbf{r}'} dS' \quad (2.40)$$

Using (A.4), the electric field in the far-field region may be expressed as follows:

$$\mathbf{E}(\mathbf{r}) = -jw\mathbf{A}(\mathbf{r}) = -\frac{jk\eta}{2\pi r}e^{-jkr}\hat{\mathbf{u}}_z \times \mathbf{h}$$

$$E_\theta = \frac{-jk\eta}{2\pi r}e^{-jkr}\cos\theta(h_x\sin\phi - h_y\cos\phi)$$

$$E_\phi = \frac{-jk\eta}{2\pi r}e^{-jkr}(h_x\cos\phi + h_y\sin\phi)$$

$$(2.41)$$

Similarly, the electric vector potential in the far-field region is found as

$$\mathbf{A}_e(\mathbf{r}) = \frac{\varepsilon e^{-jkr}}{4\pi r}\iint_A \mathbf{J}_{ms}(\mathbf{r}') e^{jk\hat{\mathbf{u}}_r.\mathbf{r}'} dS'$$

$$= -\frac{\varepsilon e^{-jkr}}{2\pi r}\hat{\mathbf{u}}_z \times \mathbf{e} \quad (2.42)$$

$$\mathbf{e} = \iint_A \mathbf{E}_a(\mathbf{r}') e^{jk\hat{\mathbf{u}}_r.\mathbf{r}'} dS'$$

In case where the electric field is uniformly distributed over the aperture A, one gets

$$\mathbf{e} = Af(\theta,\phi)\mathbf{E}_a \quad (2.43)$$

Using the far-field approximation $\nabla \cong -jk\hat{\mathbf{u}}_r$ from (A.2), (A.3) and the vector identity, $\mathbf{A} \times (\mathbf{B} \times \mathbf{C}) = \mathbf{B}(\mathbf{A}.\mathbf{C}) - \mathbf{C}(\mathbf{A}.\mathbf{B})$, one can express the electric field in the far-field region as

$$\mathbf{E}(\mathbf{r}) = \frac{-1}{\varepsilon}\nabla \times \mathbf{A}_e(\mathbf{r}) = jw\eta\hat{\mathbf{u}}_r \times \mathbf{A}_e(\mathbf{r})$$

$$= -\frac{jke^{-jkr}}{2\pi r}\hat{\mathbf{u}}_r \times (\hat{\mathbf{u}}_z \times \mathbf{e})$$

$$E_\theta = \frac{jk}{2\pi r}e^{-jkr}(e_x\cos\phi + e_y\sin\phi)$$

$$E_\phi = \frac{jk}{2\pi r}e^{-jkr}\cos\theta(-e_x\sin\phi + e_y\cos\phi)$$

$$(2.44)$$

Example 2.3 Use of Aperture Electric and Magnetic to Determine the Far-Field Radiation. An alternative approach in determining far-field radiation by aperture fields consists of using both electric and magnetic fields over the aperture A. This leads to the average of the EM fields determined by the two approaches summarized above. Hence, if the knowledge of both tangential electric and magnetic fields over the aperture A is used, then the electric field radiated in the far-field region may be written as, by averaging E_θ and E_ϕ given by (2.41) and (2.44):

$$E_\theta = \frac{jk}{4\pi r}e^{-jkr}[(e_x + \eta\cos\theta h_y)\cos\phi$$
$$+ (e_y - \eta\cos\theta h_x)\sin\phi]$$

$$E_\phi = \frac{jk}{4\pi r}e^{-jkr}[(-e_x\cos\theta - \eta h_y)\sin\phi$$
$$+ (e_y\cos\theta - \eta h_x)\cos\phi] \quad (2.45)$$

When the aperture fields $\mathbf{E}_a$ and $\mathbf{H}_a$ are not in the far-field region of the primary source, they do not behave as plane waves and are not related to each other by free-space intrinsic impedance. However, if the aperture fields E_a, H_a are in the far-field region of the source, then they behave as Huygens' source and are related to each other by

$$H_a = \frac{1}{\eta}\hat{u}_z \times E_a \quad \Rightarrow \quad h_y = e_x/\eta, \quad h_x = -e_y/\eta$$

$$(2.46)$$

Inserting the values of h_x and h_y from (2.46) into (2.45), the electric field in the far-field region may be written in terms of the Cartesian components of (2.43) as follows:

$$E_\theta = \frac{jk}{4\pi r}e^{-jkr}(1+\cos\theta)(e_x\cos\phi + e_y\sin\phi)$$

$$E_\phi = \frac{jk}{4\pi r}e^{-jkr}(1+\cos\theta)(-e_x\sin\phi + e_y\cos\phi)$$

$$(2.47)$$

The far-field magnetic field intensities corresponding to (2.41), (2.44) and (2.47) may be expressed in terms of the electric field components as follows:

$$H_\theta = -\frac{1}{\eta}E_\phi, \quad H_\phi = \frac{1}{\eta}E_\theta \qquad (2.48)$$

The Poynting vector intensity in the far-field is given by

$$S(r,\theta,\phi) = \frac{1}{2\eta}\left(|E_\theta|^2 + |E_\phi|^2\right) \qquad (2.49)$$

In view of (2.40), the ability of an aperture to direct its radiations is determined by $f(\theta,\phi)$, based on the assumption that the EM fields E_a and H_a are uniformly distributed over the aperture. For a circular aperture of radius a, $f(\theta,\phi)$ is easily found to be

$$f(\theta,\phi) = \frac{1}{\pi a^2}\int_0^a\int_0^{2\pi} e^{jkr'\sin\theta\cos(\phi-\phi')}d\phi'r'dr'$$

$$= \frac{2J_1(ka\sin\theta)}{ka\sin\theta}$$

$$(2.50)$$

where the integration with respect to ϕ' gives the Bessel function of the first kind and zero-th order. The second integral with respect to r' is evaluated by the following (see (D.63) and (D.92))

$$J_0(x) = \frac{1}{2\pi}\int_0^{2\pi} e^{jx\cos\theta}d\theta$$

$$\int_0^1 J_0(xr)r\,dr = J_1(x)/x$$

$$r'.\hat{u}_r = (r'\cos\phi'\hat{u}_x + r'\sin\phi'\hat{u}_y).$$

$$(\sin\theta\cos\phi\hat{u}_x + \sin\theta\sin\phi\hat{u}_y + \cos\theta\hat{u}_z)$$

$$= r'\sin\theta\cos(\phi-\phi')$$

$$(2.51)$$

Note that $f(\theta,\phi)$ is normalized to unity since $J_1(x)\simeq x/2$ as x approaches zero (see (D.94)). Figure 2.11 shows the variation of (2.50), the pattern $f(\theta,\phi)$ in dB, of a circular aperture of diameter D_a. The main lobe gets narrower and the number of sidelobes increases in proportion with aperture diameter but the sidelobe envelope remains the same irrespective of the aperture size. The first sidelobe level is 17.6 dB below the main lobe.

Consider now a uniformly illuminated rectangular aperture of dimensions a and b along x-and y-axes respectively. Using (A.3) we find $r'.\hat{u}_r = (x'\hat{u}_x + y'\hat{u}_y).\hat{u}_r = \sin\theta(x'\cos\phi + y'\sin\phi)$ and

$$f(\theta,\phi) = \frac{1}{ab}\int_{-a/2}^{a/2}\int_{-b/2}^{b/2} e^{jk\sin\theta(x'\cos\phi + y'\sin\phi)}dx'dy'$$

$$= \frac{\sin\left(\frac{ka}{2}\sin\theta\cos\phi\right)}{\frac{ka}{2}\sin\theta\cos\phi}\frac{\sin\left(\frac{kb}{2}\sin\theta\sin\phi\right)}{\frac{kb}{2}\sin\theta\sin\phi}$$

$$(2.52)$$

The above formulations for aperture radiation are based on geometrical optics and are capable of providing the antenna patterns for $\theta \le \pi/2$. Alternative approach based on the geometrical theory of diffraction [5][6] provides more accurate description of the radiation for

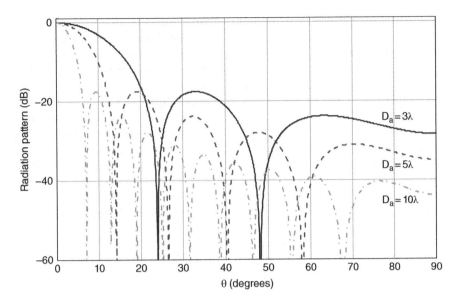

Figure 2.11   Radiation Pattern of a Circular Aperture of Diameter D_a.

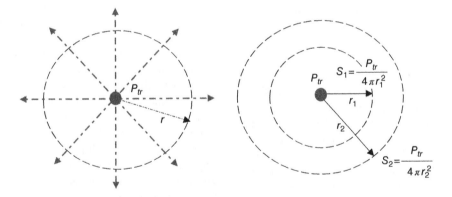

Figure 2.12   Radiation By an Isotropic Antenna.

$0 \leq \theta \leq \pi$, except for the close vicinity of the boresight direction ($\theta = 0$).

2.4   Isotropic and Omnidirectional Antennas

An *isotropic antenna* is assumed to radiate equally in all directions (in three dimensions). Isotropic antenna is a hypothetical antenna and is used as a reference. In practice, antennas always have directive properties, hence radiated fields in certain directions are higher compared to other directions.

As shown in Figure 2.12, the Poynting vector intensity due to an isotropic antenna is independent of $\{\theta, \phi\}$:

$$S_{isotropic} = \frac{P_{tr}}{4 \pi r^2} \quad \left(\frac{W}{m^2}\right) \qquad (2.53)$$

Since the total power radiated by an antenna P_{tr} is conserved, the power flux flowing out of a spherical surface is constant:

$$S_1 = \frac{P_{tr}}{4\pi\, r_1^2} \quad S_2 = \frac{P_{tr}}{4\pi\, r_2^2} \qquad (2.54)$$
$$\Rightarrow \quad P_{tr} = 4\pi r_1^2 S_1 = 4\pi r_2^2 S_2$$

An *omnidirectional antenna* acts as an isotropic antenna in one plane but is directive in the others. For example, Hertzian and linear dipole antennas aligned along the z-axis radiate isotropically in the so-called equatorial plane, defined by $\theta = \pi/2$, but are directional in elevation plane ($\phi =$ constant). This implies that, in the equatorial plane, the radiated fields and the Poynting vector intensity are independent of the azimuth angle ϕ. Such antennas are widely used in radio and TV broadcasting.

Example 2.4   Solar Radiation.
Due to its nearly spherical shape and based on the (realistic?) assumption that solar activities are independent of the direction, the sun may be assumed to be an isotropic radiator. Sun is known to radiate in a very large frequency spectrum. The mean intensity of the sun's radiation is estimated to be 1.36 kW/m^2; 50% of this radiation is in the infrared band, 40% in the visible, and about 10% in the ultraviolet at the top of the atmosphere. Depending on the climatic conditions, the radiation level is approximately 30% less intense on the earth's surface.

Photovoltaic systems are used to generate electrical power by converting solar radiation into direct current using solar panels, which are composed of solar cells containing semiconductor material that exhibit photovoltaic effect. The conversion efficiency of photovoltaic systems is typically less than 20%. Because sunshine is not always available, photovoltaic systems also incorporate energy storage devices for future use of the generated energy. Satellites use solar panels of tens of square meter areas in order to collect several kWs of electrical power needed for proper operation of the satellite.

Solar energy is widely used for heating. As the required temperature increases, different forms of conversion techniques are used. For example, in solar thermal energy plants, solar radiation is concentrated by mirrors or lenses to obtain higher temperatures. The conversion efficiency of the solar energy for heating purposes is on the order of 40%. On the other hand, the storage of heat is much cheaper and more efficient than the storage of electricity.

2.5   Antenna Parameters

Antennas may be categorized in various ways, depending on the frequency of operation, the bandwidth, the ability to direct radiation, the coverage area, size, polarization, and so on. For example, omnidirectional antennas are widely used in radio and TV broadcasting and in mobile terminals of cellular radio systems. Yagi-Uda antennas are commonly used for receiving TV transmissions which cover a wide frequency band. A multi-band antenna connected to a receiver may be used for simultaneous reception of 3G, 4G, Wi-Fi and Bluetooth signals. Horn, aperture and reflector antennas are used to achieve higher gains in long range and/or high data-rate communications in UHF and higher frequency bands. Similarly, multiple-input multiple-output (MIMO) antenna systems help to realize more reliable, long range and higher data rate communications.

In this section, we will study some parameters which determine the antenna performance. A good understanding of these parameters will help choosing an appropriate antenna for the systems under consideration.

2.5.1   Polarization

Polarization of an antenna in a given direction is defined as the polarization of its radiated electric field. Antenna polarization is therefore a function of direction but is usually referred to that in the direction of maximum radiation.

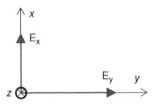

Figure 2.13   x-and y-Components of an Electric Field Propagating in +z-Direction.

Polarization of an electric field is defined by the curve traced by the end-point of the arrow representing the instantaneous value of the electric field vector, as it propagates away from us. The polarization of an electric field may not be the same as it propagates away from us or towards us. For example, a linear dipole antenna directed along the z-axis has an electric field polarized along the unit vector $\hat{\mathbf{u}}_\theta$ in spherical coordinates; hence, polarization is perpendicular to $\hat{\mathbf{u}}_z$ at $\theta = 0$ but is parallel to $\hat{\mathbf{u}}_z$ at $\theta = \pi/2$. Because of the reciprocity theorem, antenna polarization is the same at transmission and reception.

Consider an electric field propagating along +z-axis, as shown in Figure 2.13. The elecric field is assumed to have x- and y-components in the plane perpendicular to the direction of propagation:

$$\mathbf{E} = E_x\,\hat{\mathbf{u}}_x + E_y\,\hat{\mathbf{u}}_y$$
$$= E_1\,\sin(wt - kz)\,\hat{\mathbf{u}}_x + E_2\,\sin(wt - kz + \delta)\,\hat{\mathbf{u}}_y \tag{2.55}$$

where δ shows the phase angle by which E_y leads E_x. The so-called axial ratio (AR) is defined by the ratio of the electric field intensities in y- and x-directions, $AR = E_2/E_1$.

An antenna is said to be *linearly polarized* (LP) if the tip of the radiated electric field traces a line as it propagates away from us. If this line is vertical to the earth's surface it is referred to as vertical polarization (VP), while horizontal polarization (HP) implies that the electric field is polarized parallel to the earth's surface. Assuming that the yz-plane in Figure 2.13 coincides with the earth's surface, an antenna

directed along the x-axis would have its electric field polarized in x-direction in the direction of maximum radiation; hence it is said to be vertically polarized. However, the electric field of an antenna directed along y-axis is polarized in y-direction in the direction of maximum radiation; this antenna is said to be horizontally polarized. A linearly polarized wave is described by (2.55) if $\delta = 0$, that is, when there is no phase difference between x-and y-components of the electric field. Depending on the relative amplitudes of E_1 and E_2, the direction of polarization may shift from x-axis to y-axis. If $E_1 = 0$, then polarization is along y-axis and $AR = \infty$. When $E_2 = 0$, then polarization is along x-axis and $AR = 0$ (see Figure 2.14).

An antenna is said to be *elliptically polarized* (EP) if the tip of the radiated electric field traces an ellipse as it propagates away from us. If the ellipse is traced in the clockwise direction, the antenna is said to be right-hand elliptically polarized (RHEP). If the elipse is traced in the counter-clockwise then the antenna is left-hand elliptically polarized (LHEP). The phase difference between the two unequal orthogonal components determines whether the antenna polarization is RHEP or LHEP. If $\delta > 0$ then the polarization is LHEP, but it is RHEP when $\delta < 0$. As long as $\delta \neq 0$, the polarization is elliptical even if $E_1 = E_2$ (see Figure 2.14).

In case of *circular polarization* (CP), $E_1 = E_2$ (hence $AR = 1$), and the phase difference between them is equal to $\delta = \pm\pi/2$. If $\delta = \pi/2$, the antenna has left-hand circular polarization (LHCP), but is right-hand circularly polarized (RHCP) if $\delta = -\pi/2$. Note that linear and circular polarizations are simply special cases of the elliptical polarization (see Figure 2.14).

The polarizations of transmitting and receiving antennas should be matched for maximum power transfer to the receiving antenna from an incident EM field. Otherwise, there will some reduction in the received signal power. Polarization loss efficiency, which is defined as

$$\eta_{pol} = |\hat{\mathbf{e}}_i \cdot \hat{\mathbf{e}}_r|^2 \tag{2.56}$$

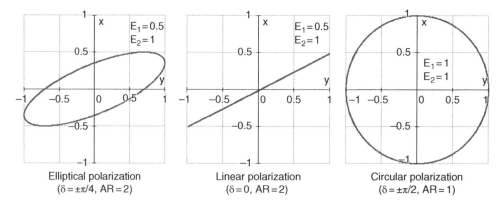

Elliptical polarization
$(\delta = \pm\pi/4,\ AR = 2)$

Linear polarization
$(\delta = 0,\ AR = 2)$

Circular polarization
$(\delta = \pm\pi/2,\ AR = 1)$

Figure 2.14 Elliptical, Circular and Linear Polarizations. The waves are assumed to be propagating (along z-axis) away from us. Axial ratio is defined by $AR = E_2/E_1$. Positive values of δ correspond to left-hand polarizations (anti-clockwise rotation) while negative values of δ implies right-hand polarization (clockwise rotation).

provides a measure of the efficiency of the polarization matching between transmitting and receiving antennas. Here $\hat{\mathbf{e}}_i$ is a unit vector representing the polarization of the electric field of the incident plane wave, that is,

$$\mathbf{E}_i = \hat{\mathbf{e}}_i\, E_i\, e^{-jkr} \qquad (2.57)$$

and $\hat{\mathbf{e}}_r$ denotes a unit vector representing the polarization of the electric field of the receive antenna when it operates as a transmit antenna, that is,

$$\mathbf{E}_r = \hat{\mathbf{e}}_r\, E_r\, e^{-jkr} \qquad (2.58)$$

For example, consider a CP incident wave received by a LP antenna in y-direction:

$$\hat{\mathbf{e}}_i = \left(\hat{\mathbf{u}}_x \pm j\hat{\mathbf{u}}_y\right)/\sqrt{2},\quad \hat{\mathbf{e}}_r = \hat{\mathbf{u}}_y$$
$$\Rightarrow \eta_{pol} = |\hat{\mathbf{e}}_i.\hat{\mathbf{e}}_r|^2 = 0.5\ (-3\ dB) \qquad (2.59)$$

The polarization loss efficiency is 0.5, which implies that a LP receive antenna can absorb only half of the power carried by a CP incident wave.

Circular polarization simplifies the antenna installation on the ground because the antenna feed does not need to be rotated after the antenna is aligned in azimuth and elevation. A receiver system will experience a 3 dB loss due to polarization mismatch when a LP-to-CP or CP-to-LP connection is attempted. Polarization matching is more difficult for LP, but the system design is simpler and polarization losses due to rain is less serious.

Example 2.5 Polarization of a Rectangular Aperture Antenna.

Consider a rectangular aperture in an infinitely large perfectly conducting screen, in the xy-plane, over which only the tangential electric field is known (see Figure 2.10). According to the boundary conditions, the electric field vanishes over the perfect electrical conductor screen. Assuming that the electric field, polarized along the x-axis, is uniformly distributed over the aperture with intensity E_{ax}, using (2.44), one may write the electric field in the far-field region as

$$\mathbf{E} = \frac{jke^{-jkr}}{2\pi r} A f(\theta,\phi) E_{ax} \left(\cos\phi\,\hat{\mathbf{u}}_\theta - \cos\theta\sin\phi\,\hat{\mathbf{u}}_\phi\right)$$

$$(2.60)$$

where $A = ab$ and $f(\theta, \phi)$ is given by (2.52). The polarization vector is given by

$$\hat{e}_i = \frac{\hat{u}_\theta \cos\phi - \hat{u}_\phi \sin\phi \cos\theta}{\sqrt{\cos^2\phi + \sin^2\phi \cos^2\theta}} \qquad (2.61)$$

Assuming that the receiving antenna is linearly polarized along the x-axis, the polarization loss efficiency is found using (A.3):

$$|\hat{e}_i . \hat{e}_r|^2 = |\hat{e}_i . \hat{u}_x|^2$$

$$= \left| \frac{(\hat{u}_\theta . \hat{u}_x)\cos\phi - (\hat{u}_\phi . \hat{u}_x)\sin\phi \cos\theta}{\sqrt{\cos^2\phi + \sin^2\phi \cos^2\theta}} \right|^2$$

$$= \frac{\cos^2\theta}{\cos^2\phi + \sin^2\phi \cos^2\theta}$$

$$(2.62)$$

It is evident from (2.62) that the polarization loss efficiency becomes equal to unity in the direction $\theta = 0$, implying a perfect match between the two polarizations. However, polarization loss efficiency is equal to zero in the direction defined by $\theta = \pi/2$ where the two polarization become orthogonal to each other. In general, the polarization loss efficiency will be less than unity implying some loss in the received power due to polarization mismatch. This may be due to the depolarization of the incident wave in the channel, the misalignment of transmit and receive antenna polarizations and/or the pointing errors. The depolarized component is in the z-direction since

$$|\hat{e}_i . \hat{u}_y|^2 = \left| \frac{(\hat{u}_\theta . \hat{u}_y)\cos\phi - (\hat{u}_\phi . \hat{u}_y)\sin\phi \cos\theta}{\sqrt{\cos^2\phi + \sin^2\phi \cos^2\theta}} \right|^2 = 0$$

$$|\hat{e}_i . \hat{u}_z|^2 = \left| \frac{(\hat{u}_\theta . \hat{u}_z)\cos\phi - (\hat{u}_\phi . \hat{u}_z)\sin\phi \cos\theta}{\sqrt{\cos^2\phi + \sin^2\phi \cos^2\theta}} \right|^2$$

$$= \frac{\sin^2\theta \cos^2\phi}{\cos^2\phi + \sin^2\phi \cos^2\theta}$$

$$(2.63)$$

It is worth noting that

$$|\hat{e}_i . \hat{u}_x|^2 + |\hat{e}_i . \hat{u}_y|^2 + |\hat{e}_i . \hat{u}_z|^2 \equiv 1 \qquad (2.64)$$

which implies the received signal power is conserved in the polarization domain. However,

there is a leak of received signal power into polarizations orthogonal to the nominal polarization. This is the scientific basis of the so-called polarization diversity, which aims to collect all the incoming signal power dispersed in orthogonal polarizations due to depolarization of the EM waves due to channel and/or transmitting and receiving antennas.

2.5.2 Radiation Pattern

Radiation pattern provides a graphical representation of directional properties, that is, the ability to focus the radiated energy, of an antenna as a function of $\{\theta, \phi\}$ in the far-field region. The maximum value of the radiation pattern is normalized to unity. A graphical representation in $\{\theta, \phi\}$ requires a three dimensional plot. However, the radiation pattern is usually depicted in two-dimensional (constant-θ and constant-ϕ) planes. The pattern in a constant-ϕ plane is usually referred to as *elevation pattern*, while the pattern in the equatorial plane, for example, $\theta = \pi/2$ in Figure 2.2 (*azimuthal pattern*), usually coincides with the direction of maximum radiation. As its name implies, the radiation pattern depicts the radiation characteristics of a transmit antenna. However, according to the reciprocity theorem, the radiation pattern of a receiving antenna is the same as that of a transmitting antenna and shows its ability for receiving incident EM fields as a function of $\{\theta, \phi\}$. Either the electric field (field pattern) intensity or the Poynting vector intensity (power pattern) is used as the radiation pattern. Since electric and magnetic fields are related by $E_\theta = \eta H_\phi$ in the far-field region, power pattern is proportional to magnitude-square of the field pattern. If the radiation pattern is plotted in dB scale, then field- and power-patterns are the same.

From (2.11), the elevation field-pattern of a Hertzian dipole is simply given by $\sin\theta$; the radiation is maximized at $\theta = \pi/2$ but there is no radiation along the antenna axis, hence E_θ is equal to zero at $\theta = 0, \pi$. Hertzian dipole is an *omnidirectional* antenna since it radiates iso-tropically in the equatorial plane but directive in the elevation plane. It has a linear polariza-tion along the antenna axis for maximum radiation.

From (2.29), the field-pattern of a linear dipole-antenna in the elevation plane is propor-tional to

$$|E_\theta| \propto \left| \frac{\cos\left(\frac{k\ell}{2}\cos\theta\right) - \cos\left(\frac{k\ell}{2}\right)}{\sin\theta} \right| \qquad (2.65)$$

The proportionality constant is a function of $k\ell$ and normalizes (2.65) to unity. This antenna does not radiate along its axis ($\theta = 0,\pi$), but the radiation is maximized in a direction, which depends on $k\ell$. For antenna lengths $\ell/\lambda < 1.5$, the direction of maximum radiation coincides with $\theta = \pi/2$ and the normalization constant is $1 - \cos(k\ell/2)$. For example, the normalized field pattern of a half-wave dipole in the eleva-tion plane is simply

$$\left| \cos\left(\frac{\pi}{2}\cos\theta\right)/\sin\theta \right|, \quad \ell = \lambda/2 \qquad (2.66)$$

The direction of maximum radiation is $\theta = \pi/2$, that is, in the equatorial plane. Similar to a Hertzian dipole, a linear dipole antenna is also *omnidirectional* since its pattern given by (2.65) is independent of the azimuth angle ϕ and is linearly polarized along the antenna axis. The radiation pattern of a linear dipole antenna given by (2.65) has a rotational symmetry, that is, independent of ϕ. However, the radiation pattern of the rectangular aperture antenna con-sidered in Example 2.5 does not have rotational symmetry and is determined by $f(\theta, 0)$ in the $\phi = 0$ plane and by $\cos\theta f(\theta, \pi/2)$ in the $\phi = \pi/2$ plane (see (2.60)).

Figure 2.15 shows the elevation field-pattern of a linear dipole antenna in polar coordinates. For antenna sizes $\ell \leq \lambda$, the antenna has a single lobe with a beamwidth which becomes nar-rower as its electrical length increases. How-ever, for antenna sizes $\ell > \lambda$, sidelobes begin

to emerge due to constructive and destructive interferences of the current elements which change sign along the antenna (see Figure 2.7). As the sidelobe levels become comparable to that of the main lobe, the antenna loses its ability to direct (focus) its radiation in the direction of the intended receiver. This explains the decrease in the radiation resistance for antenna lengths $1 < \ell/\lambda < 1.5$, in Figure 2.9. For $\ell/\lambda > 1.5$, the direction of maximum radi-ation is shifted from $\theta = \pi/2$ due to the above reasons.

Beamwidth and sidelobe levels are important parameters that characterize a radiation pattern (see Figure 2.16). The beamwidth is a measure of the angular width of the main beam where most of the antenna power is radiated. There is no unique definition of the beamwidth. How-ever, we generally use the half-power (3-dB) beamwidth, which is defined as the angular width between the directions where the radiated power density falls to half of its maximum value. As the beamwidth gets narrower, the ability of a transmitting antenna to focus its

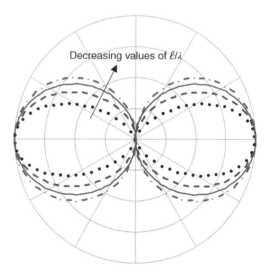

Figure 2.15 Elevation-Plane Field Patterns for a Thin Linear Dipole with Sinusoidal Current Distribution for $\ell/\lambda = 0.1$, 0.5, 0.75, 1, 1.25. Note that the pattern becomes more directive as electrical length of the antenna increases.

radiation is improved and a receiving antenna becomes more selective in receiving incident EM waves.

Sidelobe levels describe local maxima of radiations in a pattern. The directions and the levels of sidelobes may be important since they may cause interference to unintended receivers when transmitting and receive interferences from unwanted transmitters. Antenna sidelobes are particularly important for radar antennas and in electronic warfare (anti-jamming (AJ) and low-probability–of-intercept (LPI)) applications. In addition, sidelobes may also increase the antenna noise, as we shall discuss in chapter 4. Therefore, the minimization, if

not suppression, of the sidelobes is strongly desired. The existence of sidelobes in a radiation pattern is the consequence of the Fourier transform relation, given by (2.27), between the current distribution along a linear dipole-antenna and the radiated electric field intensity. Directions and levels of the sidelobes can hence be controlled by the electrical antenna length and the current distribution. Similar arguments apply for aperture antennas, where Hankel transform determines the correspondence between the aperture field distribution and the far-field radiated fields (see (250) and (2.52)). The pattern synthesis which aims to find the current distribution so as to form a desired radiation pattern is an important topic in the antenna theory. [4][7][8]

Figure 2.17 shows a typical radiation pattern in Cartesian coordinates in linear and dB scales where the 3-dB beamwidth and sidelobes are also shown. Note that the dB scale allows more detailed description of the sidelobes, which are a few tens of dB lower than the main lobe level.

2.5.3 Directivity and Beamwidth

In many applications, antennas are expected to operate in point-to-point links. Therefore, they are required to focus their radiation in the direction of intended receivers. Reciprocally,

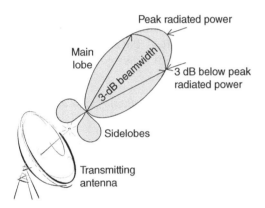

Figure 2.16 Focusing the Radiated Power Within the Half-Power Beamwidth.

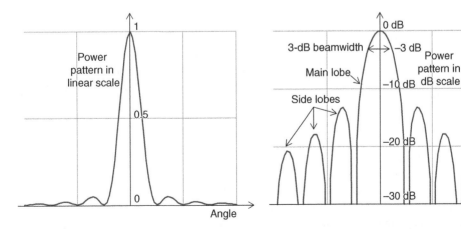

Figure 2.17 Radiation Pattern Shown in Cartesian Coordinates in Linear and dB Scales.

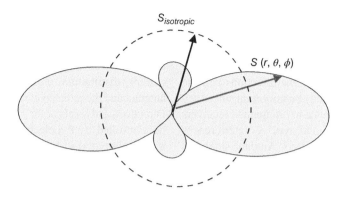

Figure 2.18 Definition of Directivity.

receiving antennas should be selective in receiving signals of the intended transmitters compared to unintended transmissions, usually coming from the directions of sidelobes. Therefore, one needs to quantify the ability of an antenna to focus its radiation when transmitting and its spatial selectivity for receiving incident signals.

Antenna directivity in a particular direction is defined as the ratio of the Poynting vector intensity in that direction, $S(r,\theta,\phi)$ (W/m²), to that of a lossless isotropic antenna (see Figure 2.18):

$$D(\theta,\phi) = \frac{S(r,\theta,\phi)}{S_{isotropic}} = \frac{S(r,\theta,\phi)}{P_{tr}/(4\pi r^2)} \qquad (2.67)$$

which is a unitless quantity. Antenna directivity, being a function of direction $\{\theta,\phi\}$, provides a measure of its spatial selectivity when transmitting and/or receiving. Higher directivity implies improved ability to focus radiation. The directivity of an isotropic antenna is, by definition, equal to unity. The maximum directivity evidently takes place in the direction of the maximum radiation. The directivity of an antenna is usually expressed in dBi, where i indicates that isotropic antenna is taken as reference. Hence, the directivity of an antenna in dBi shows the Poynting vector intensity in dB in a particular direction compared to the

Poynting vector intensity due to an isotropic antenna.

Since the total radiated power is equal to the flux of the far-field Poynting vector over a spherical surface of radius r,

$$P_{tr} = \int_{\varphi=0}^{2\pi} \int_{\theta=0}^{\pi} S(r,\theta,\varphi)\, r^2 \sin\theta\, d\theta\, d\varphi$$

$$(2.68)$$

the directivity expression in (2.67) may be rewritten as

$$D(\theta,\phi) = \frac{S(r,\theta,\phi)}{\dfrac{1}{4\pi} \displaystyle\int_{\varphi=0}^{2\pi} \int_{\theta=0}^{\pi} S(r,\theta,\varphi)\, \sin\theta\, d\theta\, d\varphi}$$

$$(2.69)$$

Since $S = |E_\theta|^2/2\eta$ is proportional to $1/r^2$ in the far-field, the total radiated power and the directivity are independent of r, the distance between transmitting and receiving antennas. It is clear from (2.69) that the directivity must satisfy the following:

$$\frac{1}{4\pi} \int_{\varphi=0}^{2\pi} \int_{\theta=0}^{\pi} D(\theta,\varphi)\, \sin\theta\, d\theta\, d\varphi = 1 \quad (2.70)$$

This implies that the total radiated power is conserved and an antenna merely distributes

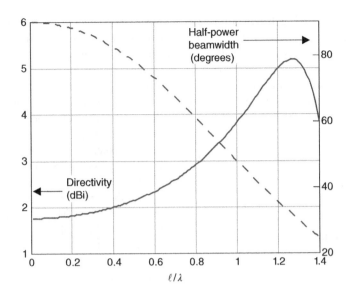

Figure 2.19 Directivity and Half-Power Beamwidth of a Linear Antenna as a Function of its Electrical Length.

the total radiated power into different directions.

Example 2.6 Directivity of a Hertzian Dipole.
Inserting the Poynting vector intensity (2.17) and the total radiated power (2.15) into (2.67), the directivity of a Hertzian dipole is found to be

$$D(\theta) = 4\pi r^2 \frac{S(r,\theta)}{P_{tr}} = 4\pi r^2 \frac{\frac{\eta}{8}\left|\frac{I_0\ell}{\lambda}\right|^2 \frac{\sin^2\theta}{r^2}}{\eta\frac{\pi}{3}\left|\frac{I_0\ell}{\lambda}\right|^2} = \frac{3}{2}\sin^2\theta$$

(2.71)

The maximum directivity of a Hertz-dipole is thus 1.5 (1.77 dBi). One may observe from (2.71) that the power pattern of a Hertzian dipole is given by $\sin^2\theta$, which becomes equal to ½ at $\pi/4$ and $3\pi/4$; hence a 3-dB beamwidth of 90°.

Example 2.7 Directivity and Half-Power Beamwidth of a Linear Dipole Antenna.
Inserting the expressions for the Poynting vector intensity, given by (2.30), and the

radiation resistance, given by (2.31), into (2.67), one may express the maximum directivity of a linear dipole antenna of length ℓ as follows:

$$D_{max} = \frac{4\pi r^2 \max\{S_r(r,\theta,\phi)\}}{\frac{1}{2}|I_0|^2 R_r}$$

$$= \frac{\eta}{\pi R_r}\max\left\{\frac{\cos\left(\frac{k\ell}{2}\cos\theta\right) - \cos\left(\frac{k\ell}{2}\right)}{\sin\theta}\right\}^2$$

(2.72)

Hence, the directivity is a function of direction and the electrical length of the antenna. Figure 2.19 shows the variation of the maximum directivity and 3-dB beamwidth of a linear dipole antenna as a function of its electrical length ℓ/λ. For $\ell \le 1.4\lambda$, the direction of maximum directivity coincides with boresight $(\theta = \pi/2)$, and the terms in paranthesis in (2.72) simplifies to $4\sin^4(k\ell/4)$. The decrease in directivity for $\ell \ge 1.3\lambda$ may be explained by the destructive interference between positive and negative current elements along the antenna (see Figure 2.7). The maximum directivity is equal to 1.64 (2.15 dBi) for a half-wave dipole compared to 1.5 (1.76 dBi) for a Hertzian dipole. The beamwidth of a half-wave dipole is found to be 78° in comparison with 90° for the Hertzian dipole. Figure 2.19 clearly shows the inverse relationship between the directivity and the beamwidth.

Example 2.8   Beamwidth and Directivity of a Horn Antenna.

Horn antennas may be considered as open circular/rectangular waveguides with enlarged mouths so as to have larger apertures. Among other applications, they are also used as feeds for parabolic, Cassegrain and Gregorian type reflector antennas. Consider a horn antenna with a power pattern given by

$$P(\theta,\phi) = \cos^n\theta, \quad 0 < \theta < \pi/2, \quad 0 < \phi < 2\pi$$

$$(2.73)$$

where $\theta = 0$ is the direction of maximum radiation which coincides with the center of the reflector antenna as shown in Figure 2.20. As implied by (2.73), this antenna does not radiate for $\theta > \pi/2$, since its role is simply to illuminate the parabolic reflector. On the other hand, n, a positive integer, is used to adjust the beamwidth of the horn antenna so as to control the taper of the illumination at the reflector edge.

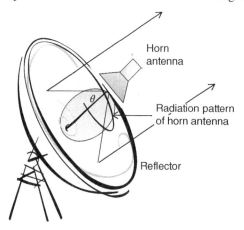

Figure 2.20   Parabolic Reflector Antenna with a Horn Feeder.

Since $\cos\theta$ decreases monotonously as θ increases from 0 to $\pi/2$, the field distribution over the reflector aperture will not be uniform and the level of tapering is controlled by the value of n.

The rays radiated by the horn antenna are all reflected by the reflector surface parallel to each other along the boresight direction. The distances taken by these rays are all equal to each other over the reflector aperture. Hence the reflector aperture forms an equi-phase surface and these rays are added coherently in the boresight direction. This provides the improved focusing ability of a parabolic reflector antenna, which is also called as a parabolic dish or simply a dish antenna.

3-dB beamwidth of the horn antenna for $n = 4$ is found as

$$\theta_{3-dB} = 2 \cos^{-1}\left(1/2^{1/n}\right) = 65.52 \text{ deg.}$$

$$(2.74)$$

The directivity is given by (2.69):

$$D(\theta,\phi) = \frac{\cos^n\theta}{\dfrac{1}{4\pi}\displaystyle\int_0^{2\pi}\int_0^{\pi/2}\cos^n\theta\,\sin\theta\,d\theta\,d\phi}$$

$$(2.75)$$

$$= \frac{\cos^n\theta}{\dfrac{1}{2}\displaystyle\int_0^1 x^n dx} = 2(n+1)\cos^n\theta$$

The maximum directivity is given by $D_{max} = 2(n+1)$; it is equal to 10 dBi for $n = 4$.

Example   2.9   Directivity   of   Aperture Antennas.

Using (2.44) and (2.67), the maximum directivity of an aperture antenna may be expressed as

$$D_{max} = \frac{\max S(r,\theta,\phi)}{P_{tr}/(4\pi r^2)} = 4\pi r^2 \frac{\left(|E_\theta|^2 + |E_\phi|^2\right)|_{\theta=0}/2\eta}{P_{tr}}$$

$$(2.76)$$

From (2.44) and (2.78) we can write

$$\left(|E_\theta|^2 + |E_\phi|^2\right)|_{\theta=0} = \left(\frac{k}{2\pi r}\right)^2 \left(|e_x|^2 + |e_y|^2\right)$$

$$= \left(\frac{k}{2\pi r}\right)^2 \left|\iint_A E_a(\mathbf{r}')\,dS'\right|^2 \tag{2.77}$$

Noting that the term $e^{jk\hat{u}_r\cdot\mathbf{r}'}$ in (2.42) does not appear in the integral in (2.77) since it becomes equal to unity in the direction of maximum radiation, that is, at $\theta=0$. It is not easy to calculate the power radiated by an aperture antenna since it requires integration of the Poynting vector over a closed surface. Instead, the total radiated power may be determined by integrating the flux of the Poynting vector over the aperture, assuming that the aperture field vanishes over the screen (see Figure 2.10). Assuming that only the electric field is known over the aperture, total radiated power is simply given by

$$P_{tr} = \frac{1}{2\eta}\iint_A |E_a(\mathbf{r}')|^2\,dS' \tag{2.78}$$

Inserting (2.77) and (2.78) into (2.76), the maximum directivity is obtained as follows:

$$D_{\max} = \frac{4\pi}{\lambda^2}\eta_a A \tag{2.79}$$

The so-called aperture efficiency η_a, which is defined by

$$\eta_a = \frac{\left|\iint_A E_a(\mathbf{r}')\,dS'\right|^2}{A\iint_A |E_a(\mathbf{r}')|^2\,dS'} \le 1 \tag{2.80}$$

is a measure of the uniformity of the field distribution over the aperture. The aperture efficiency is equal to unity for a uniformly illuminated aperture but becomes smaller as the distribution becomes tapered towards the aperture edge.

Consider a circularly symmetric aperture distribution

$$E_a(\mathbf{r}') = E_{ax}\,e^{-bT(r'/a)^2}\,\hat{u}_x \tag{2.81}$$

where $b=1/(20\log e)$ and the taper which is defined as $T = 20\log(|E_a(0)|/|E_a(a)|)$ shows the value of the electric field intensity with respect to its value at the center of the aperture. Inserting (2.81) into (2.80), one can easily find the aperture efficiency as

$$\eta_a = \frac{2}{bT}\frac{\left(1-e^{-bT}\right)^2}{1-e^{-2bT}} \tag{2.82}$$

Figure 2.21 shows circular aperture distributions with different tapers and the variation of

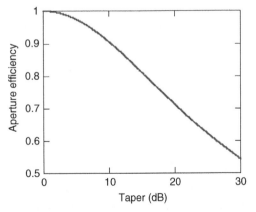

Figure 2.21 Circular Aperture Distributions for Several Values of the Taper T and the Corresponding Aperture Efficiency for $E_{ax}=1$.

the aperture efficiency with taper in dB. The aperture efficiency, hence the antenna gain, decreases with increasing values of the taper. However, lower values of the taper increases the electric field intensity at the aperture edge, which in turn increases diffracted fields and hence the sidelobe levels. Therefore, keeping the aperture taper below some specified level is strongly desired for low-noise applications and interference mitigation purposes.

Example 2.10 Half-Power Beamwidth of Aperture Antennas.
The field pattern of a circular aperture antenna of diameter $D_a = 2a$ is given by (2.50) for uniform aperture distribution. The first sidelobe level is 17.6 dB below the main lobe (see Figure 2.11). Noting that $2J_1(x)/x = 1/\sqrt{2}$ for $x = ka \sin(\theta) = 1.616$, the half-power beamwidth of a circular aperture with uniform distribution is found from (2.50) as

$$\theta_{3-dB} = 1.03 \frac{\lambda}{D_a} \ (rad.)$$
$$= 59^0 \frac{\lambda}{D_a} \ (deg.), \quad uniform \ distribution$$

(2.83)

Uniform aperture distribution, which yields the narrowest 3-dB beamwidth, is hard to realize in practice. Instead, tapered distributions are used. Tapered distributions do not use the antenna aperture as efficiently as uniform distribution since the beamwidth is relatively widened, and directivity is decreased. For an aperture distribution reasonably tapered at the aperture edge (for an aperture efficiency of $\eta_a \approx 0.65$), the 3-dB (half-power) beamwidth is approximately given by

$$\theta_{3-dB} = 1.2 \frac{\lambda}{D_a} (rad.) = 72 \frac{\lambda}{D_a} (deg.) \quad (2.84)$$

For example, half-power beamwidth of a dish antenna with $D_a = 0.6$ m mounted on roofs for satellite TV reception at 12 GHz, with an aperture efficiency of $\eta_a = 0.65$, is found to be

$$\theta_{3-dB} = 72 \frac{\lambda}{D_a} = 72 \times \frac{0.025}{0.6} = 3 \ deg. \quad (2.85)$$

which is sufficiently large so as to avoid pointing errors due to wind load and installation. Such an antenna can receive radiations from satellites operating at 12 GHz at an orbital height of 40000 km from the Earth's surface as long as the satellite is located along an arc of $R\,\theta_{3-dB} \approx 2094 \ km$ around the direction of maximum reception.

Noting that $\sin(x)/x = 1/\sqrt{2}$ for $x = ka \sin\theta/2 = 1.393$ in (2.52), the 3-dB beamwidths of a uniformly illuminated rectangular aperture in two orthogonal planes are given by

$$\theta_{3-dB,x} = 50.76 \frac{\lambda}{a} (deg.), \quad \theta_{3-dB,y} = 50.76 \frac{\lambda}{b} (deg.)$$

(2.86)

Example 2.11 Effects of Nonuniform Aperture Distributions on the Radiation Pattern.
In practice, it may not be easy or desirable to achieve uniform distributions over an aperture. For a linear antenna, far-field radiation pattern is proportional to the Fourier transform of the current distribution. For a circular aperture antenna, the directional properties of the far-field radiation is directly related to the Hankel transform of the aperture field. As for a rectangular aperture antenna, far-field pattern is obtained by the multiplication of the Fourier transforms of aperture distributions in two orthogonal planes. For non-uniform current/aperture distributions, we know from the properties of Fourier and Hankel transforms that the far-field radiations will be less focused in space.

The pattern $f(\theta, \phi)$ for circular and rectangular apertures, respectively given by (2.50) and (2.52), are based on the assumption that the aperture distribution is uniform. Here, we will study the effect of non-uniform aperture distributions on $f(\theta, \phi)$. Assume that the aperture electric field has a triangular distribution along the x-axis but is uniformly distributed along the y-axis:

$$\mathbf{E}_a(x',y') = \left(1 - \frac{|x'|}{a/2}\right)\mathbf{E}_a, \quad \begin{array}{c} -a/2 \le x' \le a/2 \\ -b/2 \le y' \le b/2 \end{array}$$

(2.87)

where $\mathbf{E}_a$ is a constant vector. Using (2.42), (2.43), and $\mathbf{u}_r.\mathbf{r}' = x'\sin\theta\cos\phi + y'\sin\theta\sin\phi$ from (A.3), we get

$$f(\theta,\phi) = \frac{1}{ab}\int\limits_{-a/2}^{a/2}\int\limits_{-b/2}^{b/2}\left(1 - \frac{|x'|}{a/2}\right)e^{jk\sin\theta(x'\cos\phi + y'\sin\phi)}dx'dy'$$

$$= \frac{1}{2}\left[\frac{\sin\left(\dfrac{ka}{4}\sin\theta\cos\phi\right)}{\dfrac{ka}{4}\sin\theta\cos\phi}\right]^2 \frac{\sin\left(\dfrac{kb}{2}\sin\theta\sin\phi\right)}{\dfrac{kb}{2}\sin\theta\sin\phi}$$

(2.88)

The triangular field distribution along the x-axis may be considered as the convolution of uniform distribution with itself. The above result easily follows from the convolution property of the Fourier transform that convolution in the aperture domain corresponds multiplication in the angular domain. Note that the factor ½ denotes the efficiency of the triangular distribution as compared to the uniform distribution (see (2.52)). Note the patterns corresponding to aperture distributions along x- and y-axes are multiplied in (2.88) where field pattern corresponding to the distribution along y-axis is given by the term, $\sin(Y)/Y$, $Y = (kb/2)\sin\theta\sin\phi$.

Now let us consider that the electric field has a cosine distribution along x-axis and is uniformly distributed along the y-axis, as usually encountered at the aperture of rectangular waveguides operating in TE_{10} mode:

$$\mathbf{E}_a(x',y') = \cos\left(\frac{\pi x'}{a}\right)\mathbf{E}_a, \quad \begin{array}{c} -a/2 \le x' \le a/2 \\ -b/2 \le y' \le b/2 \end{array}$$

(2.89)

The corresponding pattern is given by

$$f(\theta,\phi) = \frac{1}{ab}\int\limits_{-a/2}^{a/2}\int\limits_{-b/2}^{b/2}\cos\left(\frac{\pi x'}{a}\right)e^{jk\sin\theta(x'\cos\phi + y'\sin\phi)}dx'dy'$$

$$= \frac{2}{\pi}\frac{(\pi/2)^2\cos\left(\dfrac{ka}{2}\sin\theta\cos\phi\right)}{\left(\dfrac{\pi}{2}\right)^2 - \left(\dfrac{ka}{2}\sin\theta\cos\phi\right)^2}\frac{\sin\left(\dfrac{kb}{2}\sin\theta\sin\phi\right)}{\dfrac{kb}{2}\sin\theta\sin\phi}$$

(2.90)

Note that the factor $2/\pi$ in (2.90) may be attributed to the lower aperture efficiency of the cosine distribution compared to the uniform distribution. Hence, the mean electric field intensity in the direction of maximum radiation is decreased by a factor $2/\pi$ compared to the uniform distribution.

Figure 2.22 shows the variation of $f(\theta,0)$ with its peak value normalized to unity; hence the pattern for triangular distribution is multiplied by 2 and the pattern for cosine distribution by $\pi/2$. Half-power beamwidths, first sidelobe levels and aperture efficiencies of these distributions are compared in Table 2.1. Figure 2.22 and Table 2.1 show that the half-power beamwidth, sidelobe levels and directivity of an antenna are strongly affected by its

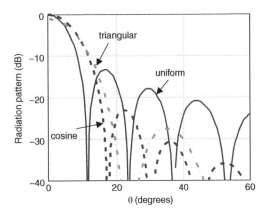

Figure 2.22   Radiation Pattern in the $\phi = 0$ Plane of a Rectangular Aperture of Size 5λ with Different Aperture Distributions.

Table 2.1 Effect of Aperture Distribution on Half-Power Beamwidth, First Sidelobe
Level and the Aperture Efficiency of a Rectangular Aperture Antenna Defined by
$-a/2 \leq x \leq a/2, -b/2 \leq y \leq b/2$. The aperture distribution along y-axis is assumed to be uniform.

Distribution	3–dB beamwidth	First SLL (dB)	Aperture efficiency, η_a
Uniform	0.88 λ/a (rad.) 50.42 λ/a (deg.)	−13.2	1 (0 dB)
Cosine	1.19 λ/a (rad.) 68.2 λ/a (deg.)	−23.0	$8/\pi^2 = 0.81$ (−0.92 dB)
Triangular	1.28 λ/a (rad.) 73.3 λ/a (deg.)	−26.4	$\frac{3}{4} = 0.75$ (−1.23 dB)

aperture distribution. The beamwidth gets wider (hence directivity decreases) and the sidelobe levels decrease (sidelobes' envelope decays faster) as the aperture distribution becomes more tapered. For example, aperture efficiency given by (2.82) shows that the directivity of a rectangular aperture with cosine distribution in one direction and uniform in the other is 0.92 dB lower compared to the uniform distribution.

Example 2.12 Beam Scanning.
This example aims to demonstrate that the direction of maximum radiation can be controlled by the phase distribution of the aperture field. Assume that the magnitude of the electric field is distributed uniformly over a rectangular aperture but its phase has a linear shift along the x-axis as follows:

$$\mathbf{E}_a \, e^{-jkx' \sin\theta_0}, \quad -a/2 \leq x' \leq a/2 \quad (2.91)$$

The factor $f(\theta, \phi)$ corresponding to (2.91) may be found from the Fourier transform property, stating that a phase shift in the aperture domain corresponds to a shift in the direction of maximum radiation. Hence, the corresponding field pattern for uniform illumination is given by

$$f(\theta,\phi) = \frac{1}{a} \int_{-a/2}^{a/2} e^{jkx'(\sin\theta - \sin\theta_0)} dx'$$

$$= \frac{\sin\left(\frac{ka}{2}(\sin\theta - \sin\theta_0)\right)}{\frac{ka}{2}(\sin\theta - \sin\theta_0)} \quad (2.92)$$

where the direction of maximum radiation is shifted from $\theta = 0$ to $\theta = \theta_0$. Linear phase variation over the aperture broadens the half-power beamwidth by a factor $(\cos\theta_0)^{-1}$ in off-boresight directions. For example, the half-power beamwidth of (2.92) is given by 0.88 $\lambda/(a \cos\theta_0)$ (see Table 2.1).

Beam scanning is widely used in radar and space applications. For example radar systems with beam scanning sweep 360° by rotating its antenna mechanically to detect signals transmitted by enemy systems or to detect presence of enemy planes. Beam scanning by controlling the phase distribution electronically over the aperture is easier than rotating the antenna mechanically. Beam scanning can be achieved with a single antenna and by antenna arrays.

2.5.4 Gain

The total power supplied to an antenna, P_t, and total power radiated by an antenna, P_{tr}, are related by

$$P_t = P_{tr} + P_{loss} \quad (2.93)$$

where P_{loss} accounts for the power lost (ohmic losses) in the antenna. The power loss efficiency η_L, which is defined by

$$\eta_L = \frac{P_{tr}}{P_t} = \frac{P_{tr}}{P_{tr} + P_{loss}} \leq 1 \quad (2.94)$$

is a measure of how efficient an antenna is in transmitting the power at its input. The power loss efficiency of a receiving antenna is defined as the ratio of the power transferred P_{tr} to the receiver by the antenna to the power P_t at its input. The fraction of the power lost in the antenna is given by $1 - \eta_L$. The power loss efficiency is equal to unity for a lossless antenna.

The directivity is defined with respect to a lossless antenna and ignores the ohmic losses in the antenna. In practice, an antenna radiates part of the power at its input and the rest is lost as Joule heat. Antenna gain, $G(\theta, \phi)$, is defined by

$$G(\theta,\phi) = \frac{S(r,\theta,\phi)}{P_t/(4\pi r^2)} = \eta_L \, D(\theta,\phi) \qquad (2.95)$$

where the reference isotropic antenna is not any more assumed to be lossless. The gain is equal to the directivity of a lossless antenna and is less than or equal to the directivity. Directivity and gain are often used interchangeably for low-loss antennas.

2.5.5 Effective Receiving Area

A receiving antenna extracts energy from incident EM waves and feeds it through a feeder line into the receiver. The amount of extracted energy depends on the Poynting vector intensity, maximum gain of the receiving antenna, polarization matching between the incident wave and the receiving antenna,

impedance matching between the antenna and the receiver and the direction of arrival of the incident waves.

Assuming perfect matching in polarisation and impedance, the received power P_r may be written as

$$\begin{aligned} P_r &= S(r,\theta,\phi)A_e(\theta,\phi) \\ &= \frac{1}{2\eta}|E_\theta(r,\theta,\phi)|^2 A_e(\theta,\phi) \quad (W) \end{aligned} \qquad (2.96)$$

where $S(r, \theta, \phi)$ denotes the Poynting vector intensity in W/m^2 incident on the receiving antenna aperture (see Figure 2.23). The effective area in a direction (θ,ϕ) of a receiving antenna, $A_e(\theta,\phi)$, is hence defined by

$$A_e(\theta,\phi) \triangleq \frac{P_r}{S(r,\theta,\phi)} \, (m^2) \qquad (2.97)$$

According to the reciprocity principle, stating that the antenna performance is the same for reception and transmission, antenna parameters for transmission and reception are directly related to each other. Generalizing (2.79) for a lossy antenna, the effective receiving area of an antenna is related to its gain by

$$A_e(\theta,\phi) = \frac{\lambda^2}{4\pi} G(\theta,\phi) \qquad (2.98)$$

Because of the direct proportionality between them, effective receiving area and the gain have the same spatial behavior. This implies that an antenna receives highest power

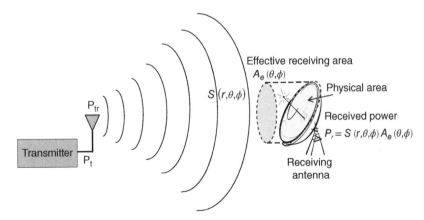

Figure 2.23   Definition of the Effective Receiving Area of an Antenna.

from the direction of its maximum transmission. Maximum effective receiving area of an antenna is related to its physical area A by (see (2.79))

$$A_e(\theta,\phi)_{max} = \eta_a\,\eta_L A = \frac{\lambda^2}{4\pi}G(\theta,\phi)_{max} \quad (2.99)$$

Effective receiving area of an antenna is related but not equal to its physical area. As defined by (2.82), the aperture efficiency, $\eta_a \le 1$, shows that effective receiving area is less than or equal to its physical aperture area and is a measure of how efficiently the antenna aperture is used for radiation and reception.

The maximum value of the effective receiving area of an isotropic antenna is obtained using (2.99):

$$A_{e,max} = \frac{\lambda^2}{4\pi}G_{isotropic} = \frac{\lambda^2}{4\pi}\eta_L \quad (2.100)$$

For aperture antennas, the gain is approximately equal to the directivity since the losses are often negligibly small. The gain of antennas with rectangular and circular aperture is given by

$$G_{max} = \frac{4\pi}{\lambda^2}\eta_a A$$

$$= \begin{cases} \dfrac{4\pi}{\lambda^2}\eta_a ab & \text{rectangular aperture} \\[2mm] \eta_a\left(\dfrac{\pi D_a}{\lambda}\right)^2 & \text{circular aperture} \end{cases}$$

$$(2.101)$$

For apertures with unequal sides, the radiation pattern is asymmetric and the beamwidths in two principal planes are not equal to each other. The maximum gain of a rectangular aperture with uniform field distribution may be written as

$$G = \frac{32383}{\theta_{3-dB,x}\,\theta_{3-dB,y}}, \quad \text{rectangular aperture}$$

$$(2.102)$$

As for a circular aperture with uniform distribution, the gain is related to its 3-dB beamwidth by

$$G = \left(\frac{59\pi}{\theta_{3-dB}(\text{deg.})}\right)^2, \quad \text{circular aperture}$$

$$(2.103)$$

Figure 2.24 shows the variation of the gain and the half-power beamwidth of a circular

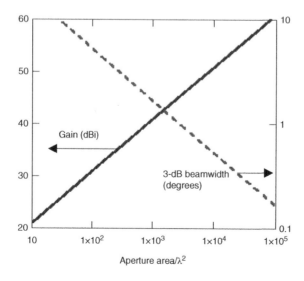

Figure 2.24 Inverse Relationship Between the Gain (2.101) and 3-dB Beamwidth (2.83) of a Uniformly Illuminated Circular Aperture Antenna.

Table 2.2 Directivity and 3-dB Beamwidth of Various Antenna Types.

Antenna type	Directivity	3-dB beamwidth (degrees)
Hertz dipole	1.5 (1.76 dBi)	90
Half-wave dipole	1.64 (2.15 dBi)	78
Circular aperture	$(\pi D_a/\lambda)^2$ D_a: aperture diameter	$59\lambda/D_a$ ($\eta_a = 1$)
Square aperture	$4\pi(a/\lambda)^2$ a,b: aperture width	$50.8\ \lambda/a$ ($\eta_a = 1$)

aperture with electrical aperture size, A/λ^2 for uniform illumination. We observe from (2.101) that the gain is directly proportional to the electrical aperture size. However, (2.83) suggests that the half-power beamwidth is inversely proportional to the square root of the electrical aperture size. The inverse relation between gain and beamwidth implies that the beamwidth gets narrower as the antenna gain increases, as expected.

Table 2.2 provides a summary of the directivity and half-power beamwidths of dipole and uniformly illuminated aperture antennas.

Example 2.13 Effect of Surface Irregularities and Pointing Errors on the Gain of a Parabolic Dish Antenna.

The maximum gain of a parabolic dish antenna cannot easily be realized especially at high frequencies when the wavelength of operation becomes comparable with surface irregularities (errors) mainly due to manufacturing. In this case, EM waves impinging on the reflector surface are not perfectly reflected but undergo scattering. This degrades the focusing ability of the antenna and leads losses in the antenna gain. On the other hand, misalignment between transmit and receive antennas prevents the realization of the maximum achievable gain. This means that antennas may not look at each other at their maximum gain values. For example, a dish on the roof may see a TV broadcasting satellite with an angular error, either due to misalignment of the antenna and/or the wind load.

Gain of a parabolic dish antenna in the presence of pointing errors and surface irregularities may be written as [9]

$$G(\theta,\phi) = \eta_a \left(\frac{\pi D_a}{\lambda}\right)^2 e^{-\left(\frac{4\pi\sigma}{\lambda}\right)^2} e^{-2.76\left(\frac{\theta}{\theta_{3-dB}}\right)^2}$$

(2.104)

where σ shows rms surface roughness (error) of the reflector, and θ_{3-dB} shows the half-power beamwidth of the antenna. The gain is assumed to be independent of the azimuth angle ϕ, and θ is measured from the boresight direction. Since gain is maximum at $\theta = 0$, a pointing error of θ will cause a reduction in the antenna gain as given by the last term in (2.104); the antenna gain would be reduced to ½ of its maximum value when $\theta = \theta_{3-dB}/2$.

Let us now determine the gain of a parabolic dish of $D_a = 0.6$ m diameter and an aperture efficiency of $\eta_a = 0.65$. This receiving antenna is assumed to directed 1° off the DVB-S transmitter operating at $f = 12$ GHz. The maximum gain of this antenna is found to be

$$G_{max} = \eta_a \left(\frac{\pi D_a}{\lambda}\right)^2 = 0.65\left(\frac{\pi 0.6}{0.025}\right)^2$$

$$= 3.7 \times 10^3 \ (35.68\ dBi)$$

(2.105)

Its 3-dB beamwidth is given by

$$\theta_{3-dB} = \frac{72\lambda}{D_a} = \frac{72 \times 0.025}{0.6} = 3 \text{ deg.}$$

(2.106)

The loss due to pointing error is given by

$$L_p = \exp\left(-2.76\left(\frac{\theta}{\theta_{3-dB}}\right)^2\right)$$

$$= \exp\left(-\frac{2.76}{9}\right) = 0.736 \quad (-1.33 \ dB)$$

$$(2.107)$$

Assuming a rms surface error of $\sigma = 1$ mm, the loss due to surface roughness:

$$L_s = \exp\left(-\left(\frac{4\pi\sigma}{\lambda}\right)^2\right) = \exp\left(-\left(\frac{4\pi}{25}\right)^2\right)$$

$$= 0.777 \quad (-1.1 \ dB)$$

$$(2.108)$$

Hence, the antenna gain in the presence of surface irregularities and pointing errors is given by

$$G_{\max} = 35.68 - 1.33 - 1.1 = 33.25 \ dBi \quad (2.109)$$

This implies a 2.34 dB loss in gain due surface roughness and pointing inaccuracy.

Example 2.14 Maximizing Antenna Gain in the Presence of Surface Irregularities.
We know from (2.101) that gain of a reflector antenna increases in proportion to $(D_a/\lambda)^2$. However, the reflector surface gradually loses its ability of coherently reflecting the incident rays due to scattering as the rms surface roughness becomes comparable to the wavelength. The degradation in antenna gain becomes more noticeable with increasing values of $(\sigma/\lambda)^2$, which is typical for antennas operating at mm wavelengths. For very large reflector antennas, as in radio astronomy and satellite communications, it may not be easy to reduce surface roughness σ below a certain level due to gravitational distortion of the reflector surface or adjustment errors of the panels which make up the reflector surface. Consequently, increasing the frequency and/or the antenna diameter does not always lead to increased gain since it begins to decrease if a critical value of σ/D_a is exceeded. Here, we will determine the maximum achievable antenna gain for a given value of σ/D_a, assuming no pointing errors (see Figure 2.25).
In the presence of only the surface errors, using (12.104) the antenna gain can be expressed as

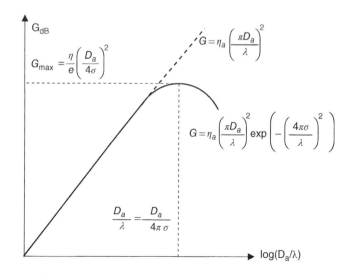

Figure 2.25 Degradation of Antenna Gain with Surface Irregularities.

$$G = \eta_a \left(\frac{\pi D_a}{\lambda}\right)^2 \exp\left[-\left(\frac{4\pi\sigma D_a}{D_a \; \lambda}\right)^2\right] \quad (2.110)$$

Considering that the surface errors increase with the antenna diameter, we assume the ratio σ/D_a to be a constant. The maximum value of the gain is obtained by taking the derivative of gain with respect to D_a/λ and equating to zero. The result is that the antenna gain is maximized when the following equation is satisfied:

$$\frac{D_a}{\lambda} = \frac{D_a}{4\pi\sigma} \quad (2.111)$$

This implies that the wavelength of operation should be equal to $\lambda = 4\pi\sigma$ for maximizing the gain. Inserting (2.111) into (2.110), one gets the maximum gain of a reflector antenna with an rms surface error of σ:

$$G_{max} = \frac{\eta_a}{e}\left(\frac{\pi D_a}{\lambda}\right)^2 = \frac{\eta_a}{e}\left(\frac{D_a}{4\sigma}\right)^2 \quad (2.112)$$

which corresponds to 1/2.718 of the maximum gain without surface error.

For $D_a = 0.6$ m, $\sigma = 1$ mm and $\eta_a = 0.65$, the antenna gain at 12 GHz is $G = 35.68 - 1.1 = 34.58$ dBi. For the same parameters as above, the maximum antenna gain is found from (2.112) as $G_{max} = 37.3$ dBi if $D_a/\lambda = 47.75$ is chosen; this corresponds to operation at 23.8 GHz ($\lambda = D_a/47.75 = 1.26$ cm) with $D_a = 0.6$ m or the choice of an antenna diameter of $D_a = 47.75\lambda \simeq 1.2 m$ at 12 GHz.

2.5.5.1   Friis Transmission Formula

Consider a transmitting antenna with gain $G_t(\theta, \phi)$ in the direction $\{\theta, \phi\}$. Using (2.95), the Poynting vector intensity (W/m^2) at a distance r in free space due to this antenna of transmit power P_t may be written as:

$$S(r,\theta,\phi) = \frac{P_t}{4\pi r^2}G_t(\theta,\phi) \quad (W/m^2) \quad (2.113)$$

From (2.96), (2.98) and (2.113), the power received P_r by a receiving antenna, with perfect polarization and impedance matching, may be written as

$$P_r = S(r,\theta,\phi)A_e(\theta,\phi) = \frac{P_t}{4\pi r^2}G_t(\theta,\phi)\frac{\lambda^2}{4\pi}G_r(\theta,\phi)$$
$$= \frac{P_t G_t(\theta,\phi) G_r(\theta,\phi)}{L_{fs}}$$

$$(2.114)$$

where P_t denotes the total power fed to the transmit antenna, and G_t (G_r) is the gain of the transmit (receive) antenna. Note that (2.114), which is known as the Friis transmission formula, relates the power received by a receiver to the transmitted power under free-space propagation conditions. Free space loss accounts for the spatial spreading of EM energy in all directions:

$$L_{fs} = (4\pi r/\lambda)^2$$
$$L_{fs,dB} = 32.44 + 20\log_{10}(f_{MHz}) + 20\log_{10}(r_{km}) \quad (dB)$$

$$(2.115)$$

The capability of the transmitter is usually measured by its Effective Isotropic Radiated Power (EIRP):

$$EIRP = P_t G_{t,max} = P_{tr} D_{t,max} \quad (W) \quad (2.116)$$

For example, for $P_{tr} = 100$ W and $D_{t,max} = 1.64$, $EIRP = 100 \times 1.64 = 164$ (22.2 dBW).

In practice, the received power is reduced by a factor $\eta_{loss} < 1$, which accounts for losses due to implementation, atmospheric absorption, rain as well as impedance and polarisation mismatches:

$$P_r = \eta_{loss}\frac{P_t G_t G_r}{(4\pi r/\lambda)^2} \quad (2.117)$$

From (2.96) and (2.113), the rms electric field intensity at a distance r from the transmitter may be expressed in terms of the EIRP of the transmitter:

$$S(r,\ \theta,\phi) = \frac{|E(\theta,\phi)|_{rms}^2}{\eta} = \frac{P_t\ G_t(\theta,\phi)}{4\pi r^2}\ \left(\frac{W}{m^2}\right)$$

$$(2.118)$$

$$|E(\theta,\phi)|_{rms} = \frac{\sqrt{30P_t G_t(\theta,\phi)}}{r} \left(\frac{V}{m}\right) \quad (2.119)$$

The power delivered to the terminals of a (polarization and impedance) matched receiver is given by

$$P_r = \frac{|E|_{rms}^2}{\eta} A_e(\theta,\phi) = \frac{|E|_{rms}^2}{120\pi} \frac{\lambda^2}{4\pi} G_r(\theta,\phi)$$

$$= \left(\frac{|E|_{rms}\lambda}{2\pi}\right)^2 \frac{G_r(\theta,\phi)}{120}$$

$$(2.120)$$

which provides an alternative formula for the Friis equation given by (2.114). This formula is commonly used in LF, MF and HF bands where linear antennas are used for signal reception and the received signal level is commonly measured in terms of the rms electric field intensity.

Example 2.15   Power Budget of an Analog Repeater.
A TV broadcasting station is required to extend its coverage area to a region where line-of-sight transmission is not possible due to the Earth's curvature. Therefore, a repeater (relay) with an overall gain of G_{rep} (including antennas and amplifier) is used in the middle of the path.

Assume that free-space propagation conditions apply and losses due to polarization and impedance mismatches are ignored. In the absence of a repeater the received power at a distance of R is given by

$$P_r = P_t G_t G_r/L \quad (2.121)$$

where $L = (4\pi R/\lambda)^2$ denotes the free-space propagation loss. If a repeater, with an overall gain of G_{rep}, is used in the mid-way between transmitter and receiver, the received signal power will be

$$P_{r,rep} = \frac{P_t G_t G_{r,rep}}{L_1} G_A \frac{G_{t,rep} G_r}{L_1} = \frac{16 P_t G_t G_{rep} G_r}{L^2}$$

$$(2.122)$$

where $L_1 = L/4$ since the repeater is located at mid-way and $G_{rep} = G_{r,rep} G_A G_{t,rep}$ denotes the overall repeater gain including the gains of receiving and transmitting antennas and the amplifier.

From (2.121) and (2.122), the ratio of the received power with and without repeater may be written as

$$\frac{P_{r,rep}}{P_r} = \frac{16 G_{rep}}{L} \quad (2.123)$$

In order to have the same received signal power with and without repeater, we need $G_{rep} = L/16$. Assuming a frequency of operation of 500 MHz and $R = 50$ km, the required repeater gain will be $G_{rep} = 108.4$ dB. If an amplifier of, for example, 60 dB gain is used at the repeater, then one needs receiving and transmitting antennas of 3.9 m diameter and 65% aperture efficiency. Since the repeater system amplifies not only the received signal power but also the system noise power, the link budget should take into account the system noise as well.

Example 2.16   Health Hazard by a 2G Base Station.
The output power of a 2G BS is approximately 16 W (12 dBW). The antenna gains are typically 12 dBi in urban areas and 17 dBi in rural areas. Table 2.3 shows typical EIRP levels of a BS transmitter in urban, suburban and rural regions.

Widespread use of wireless communication and broadcasting systems is the source of serious concerns about potential health hazards caused by EM waves to living tissues. EM

Table 2.3   Typical EIRP Levels of a 900 MHz GSM BS.

Region	EIRP (dBW)	EIRP
Urban and tourist areas	23.5	224
Suburban	26.5	447
Rural areas and roads	30.5	1127

waves may potentially cause health hazards to living tissues by heating effects and other adverse effects on cells. Concerning the heating effects, there are international standards [10] [11] limiting the maximum power levels or electric field intensities a transmitter can induce at a living tissue. The standards limit the maximum allowed rms electric field intensity E_{rms} for living tissues to 41 V/m, 58 V/m, and 67 V/m at 900 MHz, 1800 MHz and 2400 MHz, respectively. The corresponding values for the Poynting vector intensity is determined by (2.96). The power received by a lossless isotropic receiving antenna (as a reference) is calculated from (2.96) and (2.100). Table 2.4 presents the maximum allowed electric field

intensity, the Poynting vector intensity and the received power level at 900, 1800 and 2400 MHz bands. E_{rms}, denotes rms value of the maximum electric field intensity not exceeded, as set by the standards. S_{max} is the power density in W/m^2 corresponding to E_{rms}, given by (2.96). Free space propagation between transmitter and receiver is assumed.

The safe range R_{safe} denotes the distance where the electric field intensity received by the mobile user is less than or equal to E_{rms}, that is $R_{safe} \geq \sqrt{30 EIRP}/E_{rms}$ (see (2.119)). Figure 2.26 shows the variation of the distance beyond which radiations from a typical BS are not harmful to living tissues as a function of the BS EIRP. According to Table 2.4 and Figure 2.26, a BS transmitter does not cause a health hazard if you are located more than 2 meters away from a BS operating at 900 MHz in an urban environment.

Table 2.4 Maximum Allowed Electric Field Intensity and Received Power Levels.

Frequency (MHz)	E_{rms} (V/m)	S_{max} (mW/cm^2)	Received power P_r
900	41	0.45	40 mW (16 dBm)
1800	58	0.9	20 mW (13 dBm)
2400	67	1.2	15 mW (12 dBm)

Example 2.17 Radar Equation.
Assume a scenario as shown in Figure 2.27 where a transmitter, with transmit power P_t and antenna gain G_t, transmit signals in the direction of an unknown target and the signals

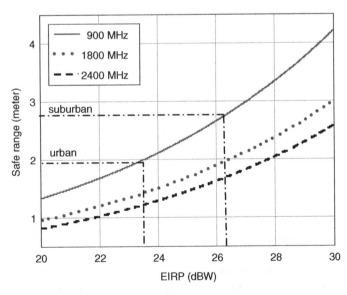

Figure 2.26 The Variation of the Safe Range, Beyond Which BS Radiation Does Not Cause a Health Hazard, as a Function of the EIRP for Operation at 900 MHz, 1800 MHz and 2400 MHz.

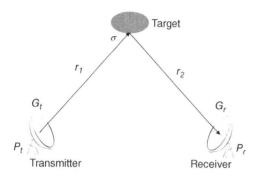

Figure 2.27   Block Diagram of a Bistatic Radar.

Table 2.5   Typical Values for Radar Cross-Section of Some Targets for a Centimeter Wave Radar. [12][13]

Target	RCS (m^2)
Large commercial airplane	100
Large combat aircraft	5–6
Small combat aircraft	2–3
Man	1
Small bird	0.01
F-117 fighter, Surface-to-air missile	0.1
B-2 bomber	0.01

scattered by the target are detected by a receiver, which might either be co-located with the transmitter or be based elsewhere. The signal power 'received' by the target may be written as

$$P_\sigma = \frac{P_t G_t}{4\pi\, r_1^2}\sigma \quad (W) \qquad (2.124)$$

where σ (m^2) denotes the so-called radar cross-section of the target. We assume that the signal power given by (2.124) is reradiated isotropically by the target. Hence, the received power by an antenna of gain G_r may be written as

$$P_r = \frac{P_\sigma}{4\pi\, r_2^2}\frac{\lambda^2}{4\pi}G_r = \frac{P_t G_t G_r \sigma}{(4\pi)^3}\left(\frac{\lambda}{r_1 r_2}\right)^2 \quad (W)$$
$$(2.125)$$

where λ denotes the wavelength and r_1 and r_2 show the distances from the target to the transmitter and to the receiver, respectively.

A radar is said to be monostatic when transmitter and receiver are co-located ($r_1 = r_2$), and bi-static when their geographic locations are separated from each other. It is important to note that, unlike point-to-point communications where the receiver signal power is proportional to r^{-2}, the received signal power is proportional to $(r_1 r_2)^{-2}$ for bistatic and to r^{-4} for monostatic radar. This implies that radar signals attenuate much more rapidly with distance and that more sensitive receivers are needed in comparison with communication systems.

Radar cross-section (RCS) is a measure of the ability of a target to scatter the EM energy incident upon it (see Table 2.5). Since a larger RCS indicates a larger object, which scatters more energy, RCS shows how detectable an object is by radar. A number of factors determine the RCS, including the material of which the target is made, electrical size of the target, the angles of incidence and scattering, and the polarization of transmitted and the received radiation in respect to the orientation of the target. Since the RCS is a property of the target, it does not depend on the distances. RCS is used in various areas, including military applications and remote sensing. For example, military aircraft and ballistic missiles are usually desired to have low RCS for detection avoidance.

2.5.6   Effective Antenna Height and Polarization Matching

The concept of effective receiving area of an antenna, defined by (2.97), is more appealing for determining the signal power received by aperture antennas. However, it may not be

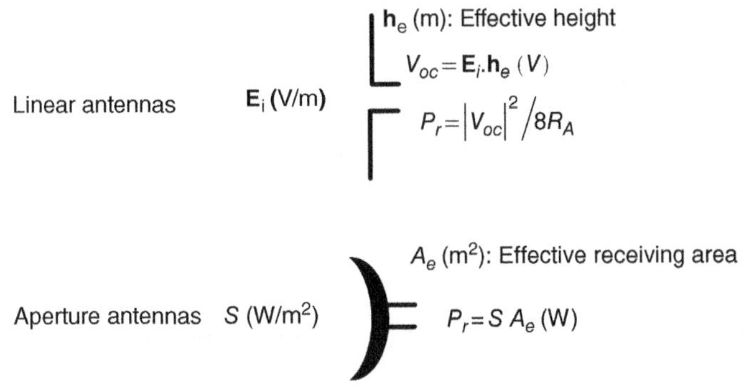

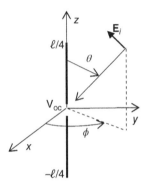

Figure 2.29 shows. Related figure at top:

h_e (m): Effective height

$$V_{oc} = E_i.h_e \ (V)$$

Linear antennas E_i (V/m)

$$P_r = |V_{oc}|^2 \big/ 8R_A$$

A_e (m²): Effective receiving area

Aperture antennas S (W/m²)

$$P_r = S\,A_e \ (W)$$

Figure 2.28   Correspondence Between Effective Height and Effective Aperture of a Receiving Antenna (R_A Denotes Antenna Input Resistance Which Is Matched To the Receiver Input Impedance).

easy to visualize the effective receiving area for linear antennas. Instead, the concept of effective antenna height is introduced to determine the open-circuit voltage, hence the power received by linear antennas. Similar to the concept of effective receiving area, the effective height of an antenna is a measure of its receiving capability; higher the effective height, higher the received open-circuit voltage. For linear antennas, the received signal level is measured by the electric field intensity and the receiving antenna is characterized by its effective height. The received open-circuit voltage is found by the dot product of these vector quantities. We know from the network theory that the received signal power is proportional to the square of the open-circuit voltage. Figure 2.28 shows the correspondence between effective antenna height and effective receiving area.

Consider a linear dipole antenna directed along the z-axis and an electric field $E_i = |E_i|\hat{e}_i$ is incident on this antenna at an angle θ from the z-axis, as shown in Figure 2.29. The open-circuit voltage at the terminals of the receiving dipole antenna may be written as

$$V_{oc} = h_e.E_i \ (V) \qquad (2.126)$$

where h_e (m) is a vector quantity; its magnitude denotes the effective antenna height. The direction of h_e shows the polarization of the

Figure 2.29   An Electric Field E_i Incident On a Half-Wave Dipole.

receiving antenna in the direction of the incident signal:

$$h_e = |h_e|\hat{e}_r \qquad (2.127)$$

Similar to the effective receiving area, effective antenna height is proportional to its physical height but is not identical to it. Using (2.126) and (2.127), the received open-circuit voltage may be written as

$$V_{oc} = h_e.E_i = |h_e||E_i|(\hat{e}_i.\hat{e}_r) = |h_e||E_i|\cos\beta \qquad (2.128)$$

where β is the angle between the direction of polarization of the incident field and that of the receiving antenna. The open-circuit voltage

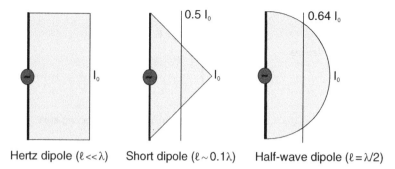

Hertz dipole ($\ell \ll \lambda$)   Short dipole ($\ell \sim 0.1\lambda$)   Half-wave dipole ($\ell = \lambda/2$)

Figure 2.30   Average Current for a Hertz Dipole with Uniform Current Distribution, a Short Dipole with Triangular Current Distribution and a Half-Wave Dipole ($\ell = \lambda/2$) with Sinusoidal Current Distribution.

is maximized when the polarizations of the incident field and the receiving antenna are matched, that is, for $\cos\beta = 1$. In case of polarization mismatch, the received signal power is reduced by the polarization loss efficiency $\eta_{pol} = \cos^2\beta = |\hat{\mathbf{e}}_i \cdot \hat{\mathbf{e}}_r|^2 \leq 1$ as defined by (2.56).

Now, let us calculate the effective height of a $\lambda/2$ linear dipole antenna. The current distribution on this antenna with physical height $\ell = \lambda/2$ is nearly sinusoidal and is given by (2.23)

$$I(z') = I_0 \cos\left(\frac{2\pi}{\lambda}z'\right), \quad -\frac{\lambda}{4} \leq z' \leq \frac{\lambda}{4} \quad (2.129)$$

The effective height for this half-wavelength linear dipole, which is given by

$$h_e = \frac{1}{I_0}\int_{-\ell/2}^{\ell/2} I(z')\,dz' = \int_{-\lambda/4}^{\lambda/4} \cos\left(\frac{2\pi}{\lambda}z'\right) dz'$$

$$= \frac{\lambda}{\pi} = \frac{2\lambda}{\pi 2} = \frac{2}{\pi}\ell = 0.64\ell$$

$$(2.130)$$

is smaller than its physical length by a factor of $2/\pi = 0.64$. Comparison with (2.90) clearly shows that a linear dipole antenna may be considered as a one-dimensional rectangular aperture antenna. Inserting $\theta = \pi/2$ into (2.27), one may show that the resulting electric field will be reduced by the same factor in the direction

of maximum radiation. Assuming perfect polarization matching, (2.128) shows that the corresponding open-circuit voltage is $V_{oc} = h_e E_i = 0.64\ell E_i$. If the current distribution were uniform, then the effective height would be maximized and become equal to the antenna physical height, that is, $h_e = \ell$ (see Figure 2.30). If the same antenna is used at longer wavelengths so that $\ell \ll \lambda$, then the current distribution would be approximately triangular. The effective height and the corresponding open-circuit voltage are then given by $h_e = 0.5\ell$ (see (2.88)) and $V_{oc} = h_e E_i = 0.5\ell E$, respectively. It is clear from the above discussion that, as the current distribution goes to uniform, the antenna effective height approaches its physical length and the resulting electric field would reach its maximum value; the beamwidth would then become narrower and directivity increases (see Table 2.1).

2.5.7   Impedance Matching

When an EM wave impinges on a receiving antenna, it induces an open-circuit voltage V_{oc} at the antenna terminals when antenna is not connected to the receiver. Therefore, Thevenin equivalent of an open-circuited receiving antenna consists of the open-circuit voltage induced by the incident EM field and the antenna input impedance Z_A. The input impedance of the antenna is given by

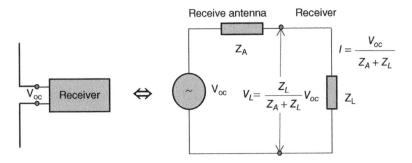

Figure 2.31 Block Diagram of a Receive.

$Z_A = R_A + R_\ell + jX_A$ where R_A is the input resistance, X_A is the input reactance, as defined by (2.31) and (2.34), and R_ℓ is the loss resistance, accounting for the Joule (ohmic) losses in the antenna. The Thevenin equivalent circuit of a receiving antenna coupled to a receiver with input impedance $Z_L = R_L + jX_L$ is shown in Figure 2.31.

Complex power delivered to the load impedance, Z_L, connected to the terminals of an antenna:

$$P_L = \frac{1}{2}V_L I^* = \frac{|V_{oc}|^2 Z_L}{2|Z_L + Z_A|^2} = \frac{|V_{oc}|^2 (R_L + jX_L)}{2|Z_L + Z_A|^2} \tag{2.131}$$

Power delivered to the load resistance, R_L, connected to the terminals of an antenna is given by

$$P_r = \frac{1}{2}\mathrm{Re}[V_L I^*] = \frac{|V_{oc}|^2 R_L}{2|Z_L + Z_A|^2}$$
$$= \frac{|V_{oc}|^2}{8R_A}\frac{4R_A R_L}{|Z_L + Z_A|^2} = \eta_{imp} P_{r,max} \tag{2.132}$$

$P_{r,max}$ denotes the maximum received power when antenna and load impedances are matched:

$$P_{r,max} = \frac{|V_{oc}|^2}{8R_A}, \quad if \; Z_L = Z_A^* \tag{2.133}$$

The impedance matching (between the antenna and the receiver) efficiency $\eta_{imp} \le 1$ is defined by

$$\eta_{imp} = 1 - |\Gamma_{load}|^2 \tag{2.134}$$

Here, the standing wave ratio is given by

$$\Gamma_{load} = \frac{Z_L - Z_A^*}{Z_L + Z_A} \tag{2.135}$$

Impedance matching efficiency shows the fraction of maximum power transferred to the receiver. For perfect impedance matching, that is $\eta_{imp} = 1$, maximum power is transferred to the receiver. In case of impedance mismatch, the antenna will not be able to maximize the transfer of power it receives from incident EM waves. In practice, impedance matching circuits are generally used to maximize power transfer to the receiver in the frequency band of operation.

Using (2.96) and (2.133), the effective height may be expressed in terms of antenna gain and radiation resistance as follows:

$$P_{r,max} = \frac{|V_{oc}|^2}{8R_A} = \frac{|E_i|^2}{2\eta}\frac{\lambda^2}{4\pi}G_r(\theta,\phi) \Rightarrow$$

$$|h_e(\theta,\phi)| = \left|\frac{V_{oc}}{E_i}\right| = \frac{\lambda}{\pi}\sqrt{\frac{G_r(\theta,\phi)R_A}{120}} \; (m) \tag{2.136}$$

where perfect impedance and polarization matching is assumed. For a $\lambda/2$ dipole, inserting $G_r = 1.64$, $R_A = R_r + R_\ell = 73.2\Omega$, one finds $h_e = \lambda/\pi$, as expected. For maximum power transfer to the load, perfect impedance matching implies $R_A = R_L = R_r + R_\ell = 73.2\Omega$ and

$$P_{r,\max} = \frac{|V_{oc}|^2}{8R_A} = \frac{(\mathbf{E}_i.\mathbf{h}_e)^2}{8R_A} = \frac{|\mathbf{E}_i|^2|\mathbf{h}_e|^2\eta_{pol}}{8R_A}\,(W)$$

$$\text{(2.137)}$$

Effective antenna height and effective receiving area may be related to each other as follows

$$P_{r,\max} = S\,A_e = \frac{|\mathbf{E}_i|^2}{2\eta}A_e = \frac{|\mathbf{E}_i|^2|\mathbf{h}_e|^2}{8R_A} \implies$$

$$A_e = \frac{30\pi}{R_A}|\mathbf{h}_e|^2$$

$$\text{(2.138)}$$

For a Hertz dipole one has $R_l = 0$, $h_e = \ell$, $R_r = 20(2\pi\ell/\lambda)^2$ and $A_e = 3\lambda^2/(8\pi)\,(m^2)$.

Assuming perfect impedance matching, and using (2.119), (2.128) and (2.136), the rms open-circuit voltage may be expressed as

$$|V_{oc}|_{rms} = |\mathbf{E}|_{rms}|\mathbf{h}_e|\sqrt{\eta_{pol}}$$

$$= \frac{\lambda}{2\pi r}\sqrt{P_t G_t G_r R_A \eta_{pol}}\,(V)$$

$$\text{(2.139)}$$

Example 2.18 Directivity, Effective Receiving Area and Effective Height of a Half-Wave Dipole.
Using the Poynting vector intensity given by (2.30) and $R_r = 73\Omega$, the directivity of a $\lambda/2$-dipole antenna is found to be

$$D(\theta) = \frac{S(r,\theta)}{P_{tr}/(4\pi r^2)} = 4\pi r^2 \frac{S(r,\theta)}{R_r\,I_0^2/2}$$

$$\text{(2.140)}$$

$$= 1.64\left[\cos\left(\frac{\pi}{2}\cos\theta\right)/\sin\theta\right]^2$$

The effective receiving area is given by

$$A_e(\theta) = \frac{\lambda^2}{4\pi}G(\theta) = \frac{\lambda^2}{4\pi}\eta_L\,1.64\left[\cos\left(\frac{\pi}{2}\cos\theta\right)/\sin\theta\right]^2\,(m^2)$$

$$\text{(2.141)}$$

where η_L denotes the loss efficiency. The effective height of a $\lambda/2$-dipole antenna is found from (2.138) and (2.141) as

$$\mathbf{h}_e(\theta) = \frac{\lambda}{\pi}\frac{\cos\left(\frac{\pi}{2}\cos\theta\right)}{\sin\theta}\sqrt{\frac{1.64\eta_L(R_r+R_l)}{120}}\,\hat{\mathbf{u}}_\theta$$

$$= \frac{2\lambda}{\pi 2}\frac{\cos\left(\frac{\pi}{2}\cos\theta\right)}{\sin\theta}\sqrt{\eta_L}\,\hat{\mathbf{u}}_\theta\,(m)$$

$$\text{(2.142)}$$

where the factor $2/\pi$ accounts for the decrease in effective height due to cosine distribution of the antenna current (see (2.90)).

Example 2.19 A Transmitting Half-Wave Dipole Antenna [1].
Thevenin equivalent circuit of a transmit antenna connected to a transmitter, represented by a voltage source of V_S and an internal impedance Z_S is shown in Figure 2.32 (compare with Figure 2.31 for a receive antenna). The transmit

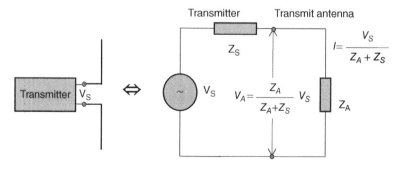

Figure 2.32   Block Diagram of a Transmit Antenna.

antenna is represented by an antenna impedance of Z_A. The antenna impedance may be written as $Z_A = R_A + jX_A$ where X_A is the antenna reactance and $R_A = R_r + R_\ell$ is the sum of the radiation resistance R_r and the loss resistance R_ℓ.

Now consider that a $\lambda/2$ transmitting dipole antenna, with radiation resistance $R_r = 73\ \Omega$, loss resistance $R_\ell = 1\ \Omega$ and $X_A = 0$, is connected to a source (transmitter) with $V_S = 1$ V (peak), $R_S = 50\ \Omega$ and $X_S = 0$. We will find the power supplied by the source, power radiated by the antenna, and power dissipated by the antenna:

Peak current:

$$I = \frac{V_S}{R_r + R_\ell + R_S} = \frac{1}{73 + 1 + 50} = 8.065\ mA \tag{2.143}$$

Power supplied by the source:

$$P_{supplied} = \frac{1}{2} V_S I^* = \frac{1}{2} \times 1 \times 8.065 = 4.032\ mW \tag{2.144}$$

Power dissipated by the source resistance:

$$P_S = \frac{1}{2} I^2 R_S = \frac{1}{2} \times (8.065)^2 \times 50 = 1.626\ mW \tag{2.145}$$

The net power supplied by the source:

$$P_{supplied} - P_S = 4.032\ mW - 1.625\ mW = 2.406\ mW \tag{2.146}$$

Power radiated and dissipated by the antenna:

$$P_{tr} = \frac{1}{2} I^2 R_r = \frac{1}{2} \times (8.065)^2 \times 73 = 2.374\ mW$$

$$P_{loss} = \frac{1}{2} I^2 R_\ell = \frac{1}{2} \times (8.065)^2 \times 1 = 0.03252\ mW \tag{2.147}$$

The total power transferred to the transmitting antenna:

$$P_{tr} + P_{loss} = 2.374\ mW + 0.033\ mW = 2.406\ mW \tag{2.148}$$

Antenna loss efficiency:

$$\eta_L = \frac{P_{tr}}{P_t} = \frac{P_{tr}}{P_{tr} + P_{loss}} = \frac{R_r}{R_r + R_\ell} = \frac{73}{73 + 1} = 98.6\% \tag{2.149}$$

Antenna mismatch loss:

$$10\log\left(1 - \left|\frac{R_A - R_S}{R_A + R_S}\right|^2\right) = -0.166\ dB \tag{2.150}$$

Example 2.20   A Receiving Hertz Dipole. A short dipole antenna, made of copper, of length $\ell = 1$ m and diameter $d = 0.6$ cm operates at 1 MHz ($\lambda = 300$ m). Since $\ell = \lambda/300$, the current distribution is assumed uniform and the effective antenna height is $h_e = \ell = 1$ m. Received electric field strength at the antenna input is assumed to be $E_{rms} = 1$ mV/m. Radiation resistance is given by (2.16):

$$R_r = 80\pi^2 (\ell/\lambda)^2 = 8.77\ m\Omega \tag{2.151}$$

Loss resistance is related to the skin depth δ_s of the conductor wire as

$$R_\ell = \frac{\ell}{\pi d \delta_s \sigma} = \frac{\ell \sqrt{\pi f \mu_0 \sigma}}{\pi d \sigma} = \frac{\ell}{d} \sqrt{\frac{f \mu_0}{\pi \sigma}} = 6.9\ m\Omega \tag{2.152}$$

where $\mu = 4\ \pi\ 10^{-7}$ (H/m) and conductivity of the copper is given by $\sigma = 5.96 \times 10^7$ (S/m). Antenna gain is given by $G = \eta_L\ 1.5$, where the loss efficiency is given by

$$\eta_L = \frac{R_r}{R_r + R_\ell} = \frac{8.77}{8.77 + 6.9} = 0.56 \tag{2.153}$$

Open circuit voltage (rms value):

$$V_{oc,rms} = E_{rms} h_e = 1\ (mV/m) \times 1\ (m) = 1\ (mV) \tag{2.154}$$

Maximum power delivered to a matched receiver:

$$P_{r,max} = \frac{V_{oc,rms}^2}{4(R_r + R_\ell)} = 16\ \mu W \qquad (2.155)$$

Example 2.21 FM Signal Reception.
An FM broadcast station with $P_t = 1$ kW transmit power and an omnidirectional antenna of 16 (12 dBi) gain is operating at 100 MHz ($\lambda = 3$ m). We will determine the maximum power delivered to a FM radio receiver with a $\lambda/2$ dipole antenna, located at 40 km from the transmitter. Free-space propagation is assumed.

Solution 1: Using (2.114) and (2.115), we get

$$L_{fs,dB} = 32.44 + 20\log(f_{MHz}) + 20\log(r_{km})$$

$$= 104.44\ dB$$

$$P_r(dB) = P_{t,dB} + G_{t,dB} + G_{r,dB} - L_{fs,dB}$$

$$= 30dBW + 12 + 2.15 - 104.4$$

$$= -60.29\ dBW\ (0.94\ \mu W)$$

$$(2.156)$$

Solution 2: Using (2.119), (2.128) and (2.133), we get

$$E_{rms} = \frac{\sqrt{30P_t G_t}}{r} = \frac{\sqrt{30 \times 10^3 \times 16}}{4 \times 10^4} = 17.3\ \left(\frac{mV}{m}\right)$$

$$V_{oc,rms} = |E|_{rms} h_e = |E|\frac{\lambda}{\pi} = 17.3 \times 10^{-3} \times \frac{3}{\pi}$$

$$= 16.5\ (mV)$$

$$P_r = \frac{V_{oc,rms}^2}{4R_r} = \frac{(16.5 \times 10^{-3})^2}{4 \times 73} = 0.94\ (\mu W)$$

$$(2.157)$$

Solution 3: One gets directly from (2.157) and (2.120)

$$P_r = \left(\frac{E_{rms}\lambda}{2\pi}\right)^2 \frac{G_r}{120} = \left(\frac{17.3 \times 10^{-3} \times 3}{2\pi}\right)^2 \frac{1.64}{120}$$

$$= 0.94\ (\mu W)$$

$$(2.158)$$

Example 2.22 Mutual Impedance.
When linear antennas are located close to each other, EM fields radiated from one of these antennas induce currents in the others. In so-called array systems, where multiple antennas operate as a single antenna, array elements may be actively fed, or one array element may be active but the others are loaded with impedances. The latter is called as a parasitic array. Yagi-Uda antenna widely used for TV reception constitutes a well-known example of parasitic array.

Currents and voltages across the terminals of a two-element parasitic array may be modeled as a two-port system as shown in Figure 2.33, where one antenna is fed and the other is loaded with an impedance Z_{L1}. [14]

The current-voltage relations at the two terminals may be written as follows:

$$V_0 = I_0 Z_{00} + I_1 Z_{01}$$
$$V_1 = I_0 Z_{01} + I_1 Z_{11}$$

$$(2.159)$$

where Z_{00} and Z_{11} denote the self impedances of 0th and 1st antennas, respectively. Z_{01} is the mutual impedance between 0th and 1st

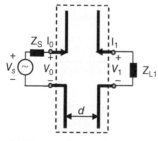

(a) Two parallel dipole antennas

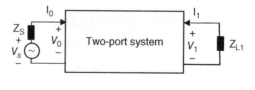

(b) Two-port system model

Figure 2.33 Mutual Impedance Between Two Dipole Antennas.

antennas and, according to the reciprocity theorem, is identical to Z_{10}.

Now assume that the 0th antenna is active and is fed by a voltage source with V_S and impedance Z_S and the 1st antenna is loaded by an impedance of Z_{L1}. Then, the following can be written from Figure 2.33: [15]

$$V_0 = V_S - I_0 Z_S, \quad V_1 = -I_1 Z_{L1} \qquad (2.160)$$

Inserting (2.160) into (2.159)

$$\begin{aligned} V_S - I_0 Z_S &= I_0 Z_{00} + I_1 Z_{01} \\ -I_1 Z_{L1} &= I_0 Z_{01} + I_1 Z_{11} \end{aligned} \qquad (2.161)$$

one may write the currents on active and parasitic antennas as

$$\begin{aligned} \begin{pmatrix} I_0 \\ I_1 \end{pmatrix} &= \begin{pmatrix} Z_{00} + Z_S & Z_{01} \\ Z_{01} & Z_{11} + Z_{L1} \end{pmatrix}^{-1} \begin{pmatrix} V_S \\ 0 \end{pmatrix} \\ &= \frac{V_S}{(Z_{00} + Z_S)(Z_{11} + Z_{L1}) - Z_{01}^2} \begin{pmatrix} Z_{11} + Z_{L1} & -Z_{01} \\ -Z_{01} & Z_{00} + Z_S \end{pmatrix} \begin{pmatrix} 1 \\ 0 \end{pmatrix} \\ &= \begin{cases} \dfrac{V_S (Z_{11} + Z_{L1})}{(Z_{00} + Z_S)(Z_{11} + Z_{L1}) - Z_{01}^2} \\[2ex] \dfrac{-V_S Z_{01}}{(Z_{00} + Z_S)(Z_{11} + Z_{L1}) - Z_{01}^2} \end{cases} \end{aligned} \qquad (2.162)$$

The input impedance of the active antenna is given by (2.159) and (2.162):

$$Z_{in} = V_0/I_0 = Z_{00} + Z_{01}(I_1/I_0) = Z_{00} - Z_{01}^2/(Z_{11} + Z_{L1}) \qquad (2.163)$$

The fraction of transmitter power that is radiated by the active antenna (impedance mismatch loss) is given by

$$\eta_{imp} = 1 - \left| \frac{Z_S - Z_{in}}{Z_S + Z_{in}} \right|^2 \qquad (2.164)$$

The input impedances Z_{00} and Z_{11} are already given by (2.31) and (2.34). The mutual impedance Z_{01} between parallel dipoles of length ℓ and separated by a distance d from each other is given by: [1]

$$\begin{aligned} Z_{01} &= R_{01} + jX_{01} \\ R_{01} &= 30[2 \cdot Ci(kd) - Ci(\mu_1) - Ci(\mu_2)] \\ X_{01} &= -30[2 \cdot Si(kd) - Si(\mu_1) - Si(\mu_2)] \\ \mu_1 &= k\left(\sqrt{d^2 + \ell^2} + \ell\right), \quad \mu_2 = k\left(\sqrt{d^2 + \ell^2} - \ell\right) \end{aligned} \qquad (2.165)$$

where $k = 2\pi/\lambda$ and $Ci(x)$ and $Si(x)$ are defined by (2.32). Figure 2.34 shows the variation of

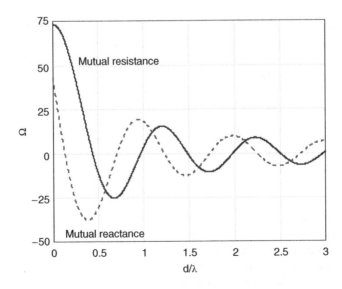

Figure 2.34 Mutual Impedance Between Two Parallel Half-Wave Dipoles.

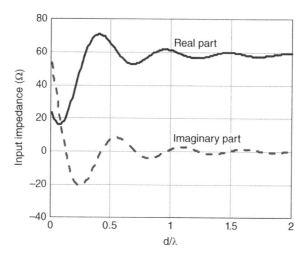

Figure 2.35 Input Impedance of the Active Dipole. Both dipoles are assumed to have a length of $\ell = 0.464\lambda$, $Z_{L1} = 50\Omega$, $R_r = 58\Omega$ and $X_{in} = 0$.

the mutual impedance between two parallel half-wave dipoles. Note that the two dipoles coincide with each other at $d = 0$, leading to $R_m = 73\,\Omega$ and $X_m = 42.5\,\Omega$. As the distance between them increases, the interaction between the two dipoles becomes weaker and consequently R_m and X_m both tend to disappear.

Figure 2.35 shows the variation of the real and imaginary parts of (2.163), which gives the input impedance of the active dipole. Both dipoles are assumed to be slightly shorter than half-wave, $\ell = 0.464\lambda$, implying that input impedances are real, that is, $Z_{00} = Z_{11} = 58\Omega$. The load impedance is also assumed to be real $Z_{L1} = 50\Omega$. The presence of the parasitic dipole was observed to cause oscillations around the input impedance of the active dipole for $d < \lambda$, but it converges rapidly to Z_{00} for $d > \lambda$. In practice, impedance matching circuits are used to match the impedance of the active dipole with that of the transceiver to stabilize the transmit/receive power level. The parasitic dipole, which is loaded by a load impedance Z_{L1} radiates as a secondary source by using the current I_1 induced by the active dipole. The intensity of radiation evidently depends of the magnitude of I_1. The variation of real

and imaginary parts as well as the magnitude of I_1/I_0 is shown in Figure 2.36. The intensity of the currents induced on the parasitic dipole decreases with electrical distance d/λ, as expected. Since no transmit power is allocated to the parasitic dipole, the radiation by parasitic dipole comes free. One may adjust phase and amplitude of I_1 via the load impedance Z_{L1} in order to maximize radiation/reception in a desired direction. The direction of maximum radiation and the amount of increase in the radiated signal level depend on the phase and magnitude of I_1 with respect to I_0 and d/λ. This issue will be discussed in detail in Chapter 13 on parasitic MIMO systems.

The radiated electric field by active and parasitic dipoles, both directed along z-axis and separated from each other by a distance d along the y-axis, may be written as

$$E_\theta(r,\theta,\phi) = E_{\theta,active}(r,\theta,\phi) + E_{\theta,parasitic}(r,\theta,\phi)$$
$$= E_{\theta,active}(r,\theta,\phi)\left(1 + \frac{I_1}{I_0}e^{jkd\hat{u}_y.\hat{u}_r}\right)$$
$$= E_{\theta,active}(r,\theta,\phi)\left(1 + \frac{I_1}{I_0}e^{jkd\sin\theta\sin\phi}\right)$$

$$(2.166)$$

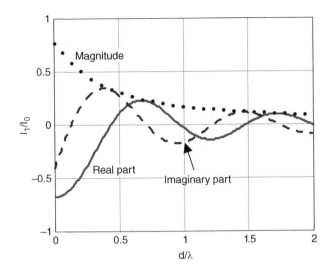

Figure 2.36 The Current of the Parasitic Dipole Normalized with Respect To That of the Active Dipole. Both dipoles are assumed to have a length of $\ell = 0.464\lambda$, $Z_{LI} = 50\Omega$, $R_r = 58\Omega$ and $X_{in} = 0$.

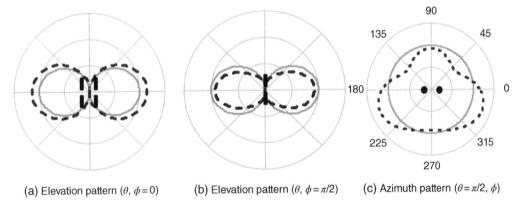

(a) Elevation pattern (θ, $\phi = 0$)   (b) Elevation pattern (θ, $\phi = \pi/2$)   (c) Azimuth pattern ($\theta = \pi/2$, ϕ)

Figure 2.37 Elevation and Azimuth Patterns of the Parasitic Array (Dashed Lines) for $d = \lambda/2$ in Comparison of a Single Dipole Antenna (Continuous Line).

where the electric field intensity of the active dipole is given by (2.29) and the path differences is accounted for by (2.38). Note that the presence of the parasitic dipole changes the electric field intensity of the active dipole antenna by the so-called array factor shown in parenthesis in (2.166). Figure 2.37 shows the field pattern of this parasitic dipole system in elevation (θ, $\phi = 0$ and $\pi/2$) and

equatorial (azimuth) planes ($\theta = \pi/2$, ϕ) planes for $d = \lambda/2$. A parasitic dipole can evidently be exploited to change the radiation pattern of a single dipole antenna so as to enhance the radiation in some directions. In this system, no increase in the transmit power is required but changes in the input impedance due to the mutual impedance should be accounted for.

References

[1] C. A. Balanis, *Antenna Theory: Analysis and Design* (2nd ed.), John Wiley: New York, 1997.

[2] R. Want, *An introduction to RFID technology*, IEEE Pervasive Computing, Jan.-March 2006, pp. 25–33.

[3] R. Weinstein, RFID: A Technical overview and its application to the enterprise, IT Pro, pp. 27–33, May-June 2005.

[4] R. E. Collin and F. J. Zucker, *Antenna Theory, Part 1*, McGraw Hill: New York, 1969.

[5] M. Şafak, Complete radiation pattern of a focus fed off-set paraboloidal reflector, *IEEE Trans. Antennas Propagation*, vol. **AP-33**, no. 5, May 1985, pp. 566–570.

[6] M. Şafak, Radiation pattern of an offset hyperbolic reflector, IEEE Trans. Antennas Propagation, vol. **AP-37**, Feb. 1989, no. 2, pp. 251–253.

[7] C. L. Dolph, A current distribution for broadside arrays which optimizes the relationship between beamwidth beamwidth and sidelobe level, *Proc. IRE*, vol. **34**, no. 6, pp. 335–348, 1946.

[8] T. T. Taylor, Design of line source antennas for narrow beamwidth and low sidelobes, *IRE Trans. Antennas Propagation*, vol. **3**, pp. 16–28, Jan. 1955.

[9] M. Şafak, Limitations on the reflector antenna gain by random surface errors, pointing errors and the angle-of-arrival jitter, *IEEE Trans. Antennas Propagation*, vol. **AP-38**, no. 1, Jan. 1990, pp. 117–121.

[10] International Commission on Non-Ionizing Radiation Protection (ICNIRP). *Statement on the* Guidelines for limiting exposure to time-varying electric, magnetic and electromagetic fields (up to 300 GHz), 2009.

[11] IEEE standard for safety levels with respect to human exposure to radio frequency electromagnetic fields, 3 kHz to 300 GHz, IEEE Std C95.1, 2005.

[12] M. C. Rezende et al., Radar cross section measurements (8–12 GHz) of magnetic and dielectric microwave absorbing thin sheets, *Revista de Fisica Applicada e Instrumentaçao*, vol. **15**, no. 1, Dec. 2002, pp. 24–29.

[13] R. A. Stonier, Stealth aircraft and technology from World War II to the Gulf. Part II: Applications and design. *SAMPLE Journal*, vol. **27**, no. 5, Spt/Oct. 1991.

[14] A. Mohammadi and F. M. Ghannouchi, Single RF front-end MIMO receivers, IEEE Commun. Magazine, pp. 104–109, Dec. 2011.

[15] O. N. Alrabadi, C. B. Papadias, A. Kalis, N. Marchetti, ve R. Prasad, "MIMO transmission and reception techniques using three-element ESPAR antennas", *IEEE Communications Letters*, vol. **13**, no. 4, pp. 236–238, April 2009.

Problems

1. Consider a half-wave dipole with rms current $I_0 = 1$A.

    a. Use (2.29) to determine the electric field intensity at a distance of 10 km in the direction perpendicular to the axis of a half-wave dipole.

    b. If the radiation resistance of the dipole is 73 Ω, calculate the directivity in dB of this dipole, compared with that of an antenna, which radiates uniformly the same transmit power.

2. Assume that two isotropic antennas, which are aligned along the z-axis as shown below, receive transmissions from a far-field source in LOS:

    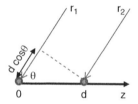

    a. Determine the phase difference $\Delta\phi$ between the signals received by two antennas.

    b. Determine the angle-of-arrival (AOA) θ of the wave in terms of the phase difference between the received signals by the two receive antennas.

    c. Assuming that the phase difference is measured as $\Delta\phi = 45$ degrees with ±5 degrees measurement error between the antenna signals and antennas are separated from each other by half-wavelength $d/\lambda = 0.5$, determine the AOA of the signal, and the AOA prediction error.

    d. Assume that, due to measurement errors and/or multipath effects, the AOA randomly changes and is described by a Gaussian distribution with a mean value corresponding to $\Delta\phi$ and a standard deviation, σ_θ. Determine the probability that the AOA error is larger than $k\sigma_\theta$, where k is a real number.

3. Verify the results given in Table 2.1.

4. A satellite transmits at 1.2 GHz (L-band) using a spot-beam aperture antenna to

illuminate the Earth. Assume an antenna aperture efficiency of 65%.

a. Assuming a lower earth orbiting (LEO) satellite at 1000 km height, determine the required antenna diameter to cover a footprint of 1000 km². What is the resulting antenna gain?

b. Repeat (a) for a medium earth orbiting (MEO) satellite at 20000 km height and a desired ground footprint of 2 000 000 km².

5. Consider the radiation by a rectangular aperture uniformly illuminated by an electric field polarized along x-axis. The radiated electric field is given by (2.60). Assuming that a receiving antenna in the far-field is linearly polarized along the x-axis, find the polarization loss factor.

6. A circularly polarized electric field of intensity E_0, $\mathbf{E}_i = E_0 \left(\hat{\mathbf{u}}_\theta - j\hat{\mathbf{u}}_\phi\right)/\sqrt{2}$ is incident on a half-wave dipole antenna, making an angle of θ with the antenna axis along z-direction, as shown in Figure 2.2. The dipole is connected to a load resistance of R_L. The dipole radiation resistance is $R_r = 73\Omega$ and the loss resistance is $R_\ell = 1\Omega$. Assume that $E_0 = 1\mu V/m$, $\theta = 60^0$ and $\lambda = 1m$.

a. Write down the effective height of the receiving antenna.
b. Determine the polarization loss efficiency.
c. Draw the equivalent Thevenin circuit and determine the open circuit voltage.
d. Determine the power delivered to R_L, when it is impedance matched to the antenna.
e. Determine the impedance mismatch loss when antenna is connected to a 50Ω coaxial cable.

7. Widespread use of wireless communication and broadcasting systems caused serious concerns about the hazards caused by the electromagnetic waves to human beings. Electromagnetic waves may potentially cause health hazards to living tissues by heating effects and by distorting the cellular structure. Concerning the heating effects, there are international standards limiting the maximum power levels or electric field strengths a transmitter can induce at a living tissue, as shown in Table P 2.1:

Table P 2.1

Frequency (MHz)	E_{max} (rms) (V/m)	Power density (mW/cm²)
900	41	
2400	67	

On the other hand, typical transmit power, transmitting antenna gain and the operational frequency of some of the considered communication systems are listed in Table P 2.2:

Table P 2.2

Transmitter	Frequency	Transmit power	Tx antenna gain
GSM Base Station	900 MHz	42 dBm	17 dBi (rural areas)
GSM Mobile Station	900 MHz	30 dBm	2 dBi
Wireless LAN	2400 MHz	15 dBm	2 dBi

a. Fill in the blanks in Table P 2.1.
b. Express the rms received electric field strength in terms of the transmit EIRP and the range.
c. Using the data given in Table P 2.2, determine the minimum distance in the broadside direction of the antenna of the considered transmitters beyond which the received electromagnetic radiation is considered to be safe.
d. According to the results obtained in part (c), which of the following transmitters are potentially more harmful to health of human beings.

8. A direct broadcast satellite (DVB-S) operating at 12 GHz uses a spot-beam antenna to transmit TV signals over a region on earth. The minimum and maximum angles subtended by this region at the satellite are $1°$ and $2.5°$, respectively. The maximum Poynting vector intensity on the surface of Earth 40000 km away from the satellite is limited to -110 dBW/m². The satellite has a transmitting power of 60 W. The losses in the system are specified to be 5 dB.

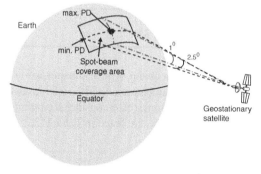

a. Determine the required transmitting antenna diameter at the satellite, ignoring surface and pointing errors. Assume 65% antenna efficiency.

b. Determine maximum and minimum values of the PD on the earth's surface.

c. Assuming an rms reflector surface error of 1 mm and a pointing error loss of 1 dB, determine the diameter of the satellite transmitter antenna to achieve the required Poynting vector intensity -110 dBW/m² on the Earth's surface.

d. If maximum and minimum values of the Poynting vector intensity differ by 5 dB and 1 meter dish is sufficient to receive the radiations with maximum Poynting vector intensity, determine the required antenna diameter for reception of signals with minimum Poynting vector intensity.

9. Consider a communication system operating at 600 MHz. The transmitter has an output power of 1 mW. The receiver antenna is assumed to be matched to the receiver with 50Ω internal impedance.

a. Determine the rms electric field intensity at the receiver.

b. Determine the effective height of the receiving antenna.

c. Assuming that transmit and receive antenna gains are 2 dBi and 0 dBi respectively, determine the maximum distance where a peak signal voltage of 1 μV is received across 50Ω.

10. The power radiated by a lossless aperture antenna is 1W. The Poynting vector intensity (W/m²) due to this antenna is described in spherical coordinates (r, θ, ϕ) by the following expression:

$$S_r(r,\theta,\phi) = S_0 \cos\theta \sin\phi / r^2, \quad \begin{matrix} 0 \le \theta \le \pi/2 \\ 0 \le \phi \le \pi \end{matrix}$$

a. Determine S_0?

b. Determine the direction of maximum radiation.

c. Determine the maximum directivity of the antenna.

d. Calculate the 3-dB beamwidth of this antenna

e. Determine the maximum Poynting vector intensity in W/m² at a distance of 1 km.

11. Assume free-space propagation between a 100 mW transmitter and a receiver located at a distance of 5 km. The frequency of operation is 900 MHz. The antenna gains are $G_t = 20$ and $G_r = 1$. The receiver antenna has a purely real impedance of 36.5 ohm and feeds a receiver with 50 ohm input impedance. The receive antenna is linearly polarized but transmit antenna transmits right hand circularly polarized (RHCP) waves. Determine

a. the received power.
b. the magnitude of the E-field at the receiver antenna input.
c. the magnitude of the open-circuit voltage.
d. the effective height of the receive antenna.

12. Consider that a 60 cm diameter dish antenna with 65% aperture efficiency is used for the reception of digital DVB-S signals at 12 GHz from a geostationary satellite.

a. Determine the maximum gain of the antenna.
b. Determine the 3-dB beamwidth θ_{3-dB} of the antenna.
c. Assuming that the power radiation pattern of the antenna can be approximated as $\exp\left(-2.76(\theta/\theta_{3-dB})^2\right)$ around the mean beam, where θ denotes angle with respect to the boresight direction, write down the expression for the antenna gain as a function of θ.
d. If the antenna is directed 1° off the satellite transmitter, calculate the gain loss due to pointing error.

13. Consider an antenna onboard a geostationary orbit with global coverage. This means that the radiation pattern of the antenna is designed so as to see the edge of the Earth with 3-dB loss when it is directed to the center of the Earth. The Earth's radius is 6371 km and a geostationary satellite is located at 35768 km from the surface of the Earth. Antenna aperture efficiency is assumed to be 65%.

a. Determine the half-power beamwidth of the satellite antenna.
b. What diameter of an Earth coverage antenna should have at 2 GHz, 12 GHz, 18 GHz and 40 GHz.
c. Determine the gain of an Earth-coverage antenna at 2 GHz, 12 GHz, 18 GHz and 40 GHz.

14. Consider a monostatic radar operating 9 GHz with an antenna gain of 30 dBi and a sensitivity level of −150 dBW. If this radar is required to detect aircrafts with radar cross section areas as small as 0.01 m^2 at 10 km range, what transmit power levels must use?

3

Channel Modeling

The SNR at the detector input of a receiver is an important measure of telecommunication systems. If the SNR is 'sufficiently' high, that is the received signal power level is sufficiently higher than the received noise power level, the detector can decode the received symbols with minimum bit errors. Otherwise, reliable transmission of information from the transmitter to the receiver may not be possible.

The receiver noise will be studied in the next chapter. In this chapter, we will focus on the received signal power level that was calculated in Chapter 2 for free-space propagation conditions in a wireless channel. In such channels, the received signal power level is determined by the transmit signal power level, transmit and receive antenna gains, range, and frequency (see (2.114)). However, in most applications, free-space propagation may be hindered by the Earth's curvature and/or the presence of obstacles, such as hills, buildings, and trees in the propagation path. Then, replicas of the transmitted signals may reach the receiver antenna via multi-path propagation due to reflection, diffraction and scattering from the obstacles and refraction of electromagnetic waves in the troposhere.

Electromagnetic waves are reflected from obstacles with dimensions comparable to the wavelength; reflected field strength is then a fraction of the incident field strength. Diffraction occurs when the propagation path is obstructed by objects with sharp edges; diffracted rays are received even in the regions with no LOS with the transmitter. The scattering is caused by rough obstacles that have irregularities much smaller than the wavelength. Consequently, electromagnetic wave propagation in a wireless channel largely depends on the frequency of operation and the characteristics of the propagation medium (cable, air, sea, ground, indoor/outdoor, and so on).

In this chapter, we will briefly mention about wave propagation in low-frequency (LF: 30–300 kHz), medium-frequency (MF: 300–3000 kHz) and high-frequency (HF:3–30 MHz) bands. However, more emphasis will be given to wave propagation in very high frequency (VHF:30–300 MHz) and ultra high frequency (UHF:300 MHz–3 GHz) bands, allocated to FM radio, TV broadcasting, cellular radio, satellite communications (SATCOM), and so on. VHF and UHF bands provide a

Digital Communications, First Edition. Mehmet Şafak.
© 2017 John Wiley & Sons Ltd. Published 2017 by John Wiley & Sons Ltd.
Companion website: www.wiley.com/go/safak/Digital_Communications

low-noise spectral window for electromagnetic wave propagation (thus leading to lower receiver noise power). In addition, they allow electromagnetic waves to penetrate through walls with relatively low signal attenuation; this helps the provision of radio/TV broadcasting services and mobile radio signals to indoor users. Super-high frequency (SHF) (3–30 GHz) and extremely high-frequency (EHF) (30–300 GHz) bands (wavelengths $\lambda < 10$ cm) enable the use of physically small and yet high-gain (narrow-beamwidth) antennas. Such antennas can easily be installed on high grounds or towers to free the propagation path from obstacles so as to create free-space propagation conditions. At frequencies higher than 10 GHz, atmospheric absorption and precipitation, such as snow and rain, cause additional signal attenuation and increase in the receiver noise. SHF and EHF bands are used for point-to-point LOS communications, satellite communications, radioastronomy, remote sensing and radar applications.

3.1 Wave Propagation in Low- and Medium-Frequency Bands (Surface Waves)

In LF and MF bands, the waves are guided by the Earth's surface. Hence the so-called *surface waves* are influenced by physical and electrical characteristics of the Earth's surface. Bearing in mind that the wavelength in LF and MF bands changes between 10 m–1 km and 1000 m–100 m respectively, transmit and receive antennas are electrically short monopoles and are erected over the surface of the Earth. Since the ground is highly conductive at these frequencies, electromagnetic waves radiated by horizontal antennas are attenuated much more rapidly compared to those radiated by vertical antennas. Radiated electric fields with horizontal polarization attenuate rapidly with distance by inducing currents on the Earth's surface, instead of carrying the transmitted information to long distances. Therefore, the use of vertical antennas above the

Earth's surface lead to surface waves with vertical polarization. The received electric field in the far-field is formed by the complex interaction of direct (LOS) and Earth-reflected signals. In LF and MF bands, electromagnetic waves follow the contour of the Earth. Transmit monopole antennas with directivities typically less than 2 dBi and transmit power levels exceeding 10 kW are commonly used.

Received electric field intensity of the surface wave with vertical polarization may be written, in terms of the LOS electric field strength, E_{LOS}, as [1]

$$E_s \approx 2A_s E_{LOS} \tag{3.1}$$

assuming that the Earth's surface is flat, that is, the range is less than $80/f_{MHz}^{1/3}$, and that the relative dielectric constant ε_r of the Earth is larger than 10. As shown by (3.1), the presence of the Earth modifies the received LOS electric field strength by a factor of $2A_s$. Since $A_s \leq 1$, the received field strength is less than $2E_{LOS}$ which corresponds to the case where direct and Earth-reflected fields arrive in phase to the receiver. For flat-Earth assumption, the surface wave attenuation factor A_s is given by [1]

$$A_s = \begin{cases} \exp(-0.43p + 0.01p^2) & p < 4.5 \\ 1/(2p - 3.7) & p > 4.5 \end{cases}$$

$$p \cong \frac{\pi r/\lambda}{\sqrt{\varepsilon_r^2 + x^2}} \tag{3.2}$$

$$x = \frac{\sigma}{w\varepsilon_0} = \frac{18000\sigma}{f_{MHz}}$$

The Earth surface is electrically characterized by its conductivity σ (S/m) and relative dielectric constant ε_r. Some typical values are presented in Table 3.1.

Since the sea water behaves like a perfect conductor, LF waves propagate over sea surface much larger distances (with relatively small attenuation) compared to free space propagation. Therefore, spherical Earth propagation conditions are often needed. Attenuation

Table 3.1 Typical Values of Conductivity and Relative Dielectric Constant of the Earth's Surface.

Earth's surface	Conductivity σ (S/m)	Relative dielectric constant ε_r
Poor ground	0.001	4–7
Average ground	0.005	15
Good ground	0.02	25–30
Sea water	5	81
Fresh water	0.01	81

of surface waves over spherical Earth is more rapid compared to the flat-Earth.

Surface waves are guided by the Earth surface; they are attenuated with increasing height from the surface of the Earth and cannot penetrate through the surface of the Earth. The penetration depth d_p of the electric field into the ground/sea is determined by its skin depth:

$$d_p = \exp(-z/\delta_s)$$
$$\delta_s = 1\sqrt{\pi f \mu_0 \sigma}$$
(3.3)

where z denotes the distance from the Earth's surface in the direction of the center of the Earth. The penetration depth in the sea water is found to be $\delta_s = 0.25$ m at 1 MHz.

In LF, surface waves are used for long distance communications and navigation, while MF band (frequencies lower than 2 MHz) is mostly exploited for AM broadcasting due to larger available bandwidth. The range of surface waves can typically be several hundred kilometers at broadcast frequencies, dropping to 20 km or so at 100 MHz, but much higher over the sea. At broadcast frequencies, the noise level in urban areas is so high that field strength of 1–10 mV/m at the receiving antenna is required for acceptable reception. In rural areas signal levels of an order of magnitude less are satisfactory, due to lower external noise

coupled into the receive antenna. During the night-time, slow fading is observed with fade durations of tens of seconds due to interference between surface and ionospheric waves (3–30 MHz) at the upper part of the MF band.

Example 3.1 Surface Wave Propagation at 1 MHz.

In this example, we will calculate the surface-wave attenuation for vertical polarization at 1 MHz. Assuming flat-Earth propagation over average ground ($\varepsilon_r = 15$, $\sigma = 5 \ 10^{-3}$ S/m) at a range $r = 60$ km, we find

$$x = \frac{18000\sigma}{f_{MHz}} = 90$$

$$p = \frac{\pi r/\lambda}{\sqrt{\varepsilon_r^2 + x^2}} = 6.9 \qquad (3.4)$$

$$2A_s = \frac{2}{2p - 3.7} = 0.199 \ (-14 dB)$$

Hence, the received signal power level due to the surface wave is 14 dB below the level for free space propagation. The corresponding skin depth is found from (3.3) as 7.1 m.

3.2 Wave Propagation in the HF Band (Sky Waves)

The HF band covers the frequencies between 3–30 MHz; the corresponding wavelength interval is 10–1 m. The wave propagation in the HF band may be considered as guided by the parallel-plate waveguide formed by the Earth surface and the layers of the ionosphere, between 40–400 km heights over the Earth's surface. The so-called *sky waves* are returned down to Earth from the ionosphere due to the refraction phenomenon. Ability of ionosphere to refract (return) electromagnetic waves toward the Earth depends on frequency, angle of transmission and the ion/electron density, which is a function of time of the day. This ability is improved with the increased ionization density which is higher during the day than at night, in summer than in winter and during periods with higher solar activity than in periods with quiet sun. As the frequency of operation increases in the HF band, we observe an increase in the ionospheric height at which ionospheric waves are returned. Electromagnetic waves at frequencies higher than the HF band go through the ionosphere since ionization density is not sufficiently high so as to refract (return) them back to the Earth. At higher end of the MF and lower end of the HF bands, interference between ionospheric and surface waves may be observed.

Sky waves can travel a number of hops, back and forth between ionosphere and Earth's surface. The refracting/reflecting process between the ionosphere and the ground is called skipping. Skip distance is in the order of several hundred km. This allows long distance communications and broadcasting over thousands of km. HF signals undergo strong fading due to time-varying and nonhomogeneous character of the ionization density, which varies with geographical location, time of day, and season.

The ionosphere is usually considered to be stratified in terms of the height over the Earth's surface into D-layer (40–90 km), E-layer (90–140 km) and F-layer (140–400 km). D-layer has the ability to refract signals of lower frequencies due to lower electron concentration. High frequencies pass through it with partial attenuation. Being affected by the solar activity, this layer disappears after sunset, due to rapid rate of recombination which is almost complete by midnight. Consequently, signals normally refracted by D-layer are refracted at night by higher layers, resulting in longer skip distances. E-layer refracts signals of frequencies less than 30 MHz but higher compared to frequencies that can be refracted by the D-layer. F-layer separates into F_1 and F_2 layers during daylight hours. Ionization level is very high and varies widely during the day and recombination occurs slowly after sunset. A fairly constant ionization layer is present at all times but F_2 is most affected by solar activity. E- and F-layers are permanent but their heights differ.

HF band is mainly used for short-wave radio, amateur radio and military communications. Since the antenna noise is very high and multipath propagation is inherent, sky waves undergo strong fading; the SNR at the receiver is generally low and undergo fast random fluctuations. Consequently, the available 27 MHz bandwidth in the HF band is troublesome for high data rate and high quality communications.

The following ionospheric effects are usually accounted for in terms of the total electron content (TEC) along the propagation path: [2]

a. The so-called Faraday rotation accounts for the progressive rotation of the plane of polarization of a linearly polarized wave, propagating through the ionosphere. The amount of rotation is a function of the TEC. In many applications, TEC is estimated and induced polarization rotation is corrected.

b. Dispersion of ionospheric signals results in time delay differences across the bandwidth of the transmitted signal;

c. Inhomogeneities in the TEC lead to amplitude fluctuations called as scintillations. The resulting phase and amplitude scintillations cause focusing or defocusing of radio waves. Scintillations decrease with increasing frequency and depend upon path geometry, location, season, solar activity and local time.

3.3 Wave Propagation in VHF and UHF Bands

VHF and UHF bands cover 30–300 MHz and 300–3000 MHz bands, respectively. The corresponding wavelengths are 10–1 m for VHF and 1–0.1 m for UHF. Therefore, the required physical size of an antenna for a given antenna electrical size is relatively small. Consequently, antennas can be mounted to masts or on the roofs of high buildings to enlarge the coverage area, which is restricted mainly to LOS and influenced by the presence of Earth and other obstacles. Coupling between transmit and receive antennas is hence accomplished via direct, Earth-reflected and scattered rays. Consequently, the receiver proces multiple copies of the same signal with varying amplitude, phase and delay for signal detection. Constructive and destructive interferences between direct (LOS) and Earth-reflected rays result in oscillations around the received LOS signal power level which attenuates with distance as r^{-2}. As the range exceeds a crital distance, the received signal power level fades away in proportion to r^{-4}.

VHF and UHF bands are allocated mainly to FM and TV broadcasting, radio-link, mobile radio, SATCOM, and so on. This allocation is justified by the fact that the waves in these bands can penetrate walls with relatively small loss. Since the antenna noise decreases with increasing frequencies, the systems operating in VHF and UHF bands have the potential of being less noisy. On the other hand, atmospheric absorption and rain attenuation are negligibly small in these bands. The effects of troposheric refraction, that is, bending of radio waves as they propagate through the troposhere, are accounted for by assuming an equivalent Earth radius of 8500 km which is 4/3 times the actual Earth radius. The problems related to digital signaling such as frequency dependence of attenuation and delay as well as intersymbol interference (ISI), due to bandwidth limitation by the channel and/or multipath propagation, will be studied in detail in the following chapters.

3.3.1 Free-Space Propagation

The wave propagation is said to take place in free space, if there is no obstacle, including the Earth, and if atmospheric absorption and precipitation do not have any effect on wave propagation between transmit and receive antennas. The power received by the receive antenna under free space propagation conditions is given by (2.117).

Free-space communications, which comprises a single propagation path, may be modelled by an AWGN channel. In this channel, the low-pass equivalent received signal $r_\ell(t)$ may be written in terms of the low-pass equivalent transmitted signal $s_\ell(t)$ as

$$r_\ell(t) = \alpha e^{-j\phi} s_\ell(t - \tau) + w(t)$$

$$\alpha \propto \sqrt{G_t G_r / L_{fs}}, \tag{3.5}$$

$$\phi = kr + \theta_{channel}, \tau = r/c$$

where α (see (2.114)) and ϕ denote respectively the attenuation constant and the phase shift due to the channel and are assumed to be deterministic. The delay τ is defined by the ratio of the range r to the velocity of light. One may rewrite (3.5) in the classical form of an AWGN channel by dividing both sides of it with $\alpha e^{-j\phi}$:

$$r_\ell'(t) = \frac{r_\ell(t)}{\alpha e^{-j\phi}} = s_\ell(t - \tau) + w'(t)$$

$$w'(t) = \frac{w(t)}{\alpha e^{-j\phi}} \tag{3.6}$$

Note that rescaling the received signal level by $\alpha e^{-j\phi}$ does not affect SNR since the SNR given by (3.5) and (3.6) are the same.

3.3.2 Line-Of-Sight (LOS) Propagation

In practice, the propagation environment between transmit and receive antennas can not be modelled as free space due to the presence of buidings, trees, hills and other obstacles causing scattering of electromagnetic waves. Even if transmit and receive antennas are in line-of-sight (LOS) of each other, in addition to the direct ray, some rays arrive to the receive antenna terminal via scattering from these obstacles.

The low-pass equivalent signal $r_\ell(t)$ received in a multipath channel may be characterized in terms of the low-pass equivalent transmit signal $s_\ell(t)$ and the impulse response $h(t)$ by

$$r_\ell(t) = s_\ell(t) \otimes h(t) + w(t)$$

$$= \sum_{i=1}^{L} \alpha_i e^{-j\phi_i} s_\ell(t - \tau_i) + w(t) \qquad (3.7)$$

$$h(t) = \sum_{i=1}^{L} \alpha_i e^{-j\phi_i} \delta(t - \tau_i)$$

where L replicas of the transmitted low-pass equivalent signal $s_\ell(t)$ with different amplitudes, phases and delays reach the receiver. If the amplitudes α_i of some multi-path components are small compared to the dominant one, then they may be ignored. The channel induces ISI if $\tau_{max} = \max\{\tau_i\}, i = 1, 2, \ldots, L$ is larger than a significant fraction of the signal duration. Then signal components, which are scattered from obstacles far from the LOS, arrive late and interfere with subsequently transmitted signals. This causes ISI and leads to errors in signal detection. However, if τ_{max} is small fraction of the signal duration, then the signaling becomes ISI free.

Here, the fundamental problem is to decide about the number of scatterers which contribute significantly to the received signal. This decision is closely related to the electrical size of an obstacle and its distance to the LOS path. Noting that larger obstacle sizes are needed for significantly high scattered signal power levels at lower frequencies, the number L of scatterers is expected to be larger at higher frequencies in a propagation path. Then, the received signal given by (3.7) is better described by integration rather than summation. For small L, one may

trace the rays between transmitter and receiver via scatterers which induce sufficiently large signal power levels at the receiver. This may be a convenient approach in VHF and in the lower part of the UHF band but, at higher frequencies, ray tracing becomes practically impossible since L is potentially very large. The multipath channel is then characterized statistically. This case will be studied in detail in the forthcoming chapters.

3.3.3 Fresnel Zones

A LOS path is said to be *clear* when there are no scatterers in the close vicinity of the direct path between the transmitter and the receiver. As shown in Figure 3.1, the clearance of the LOS path is affected by heights of transmit and receive antennas, terrain profile (hills, buildings), terrain cover (vegetation), Earth's curvature, troposheric refraction, and so on. The concept of *Fresnel zone* provides a practical measure of the clearance of the LOS path.

An ellipse is defined as the locus of points on a plane so that sum of the distances to the foci (two fixed focal points) is constant. If an ellipse is rotated around the axis which connects the two focal points, then the three-dimensional curve thus formed becomes an ellipsoid. If transmit and receive antennas are located at the focal points, then the path lengths of signals reflected, scattered or diffracted from obstacles located on the surface of the same ellipsoid are the same (see Figure 3.2). This also implies that

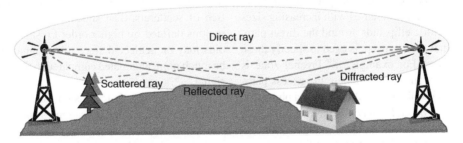

Figure 3.1   Propagation Path with Reflection, Diffraction and Scattering.

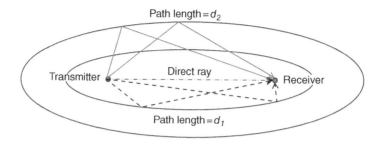

Figure 3.2 Signals Arrive at the Receiver Antenna at the Same Time if They are Scattered by Obstacles That are Located on the Surface of the Same Ellipsoid.

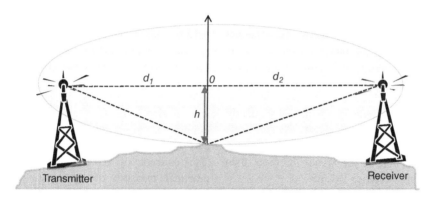

Figure 3.3 Fresnel Zone Height.

these signals also suffer the same propagation delay and the same phase shift at the receiver input. The line connecting the two focal points represents the direct path between transmit and receive antennas. Thus, the excess path length, that is, the path length difference between the direct ray and the rays scattered from the surface of an ellipsoid, is also a constant. Thus, phase and delay differences are also constants. It should also be evident that path length/phase/ delay difference increases with increasing sizes of concentric ellipsoids around the direct path.

Fresnel zone concept is used to specify these differences. For example, n^{th} Fresnel zone is defined by an ellipsoid where the excess path-length is less than $n\lambda/2$; this implies a phase difference of $n\pi$ and a delay difference of $n\lambda/(2c)$ where c denotes the velocity of light. The first Fresnel zone is defined as an ellipsoid where

the excess path length is less than or equal to $\lambda/2$. Hence, the phase difference between rays scattered from obstacles located in the first Fresnel zone is always less than or equal to 180 degrees. Objects within a series of concentric ellipsoids around the LOS between transceivers have therefore constructive/destructive interference effects on the received signal. In practice, the communications is assumed to take place in free space conditions if the first Fresnel zone is free of scatterers. The signals scattered from regions defined by higher order Fresnel ellipsoids may be neglected since these signals will be much more attenuated compared to the direct signal.

Now consider that transmit and receive antennas of a communication system are located at the foci of an elllipsoid as shown in Figure 3.3. The reflection point on the ground, which remains on the surface of an ellipsoid with distance h to the LOS at that point, is assumed be d_1 (m) away from the transmitter. The excess path length between reflected and direct rays is found from Figure 3.3:

$$\Delta l = \sqrt{d_1^2 + h^2} + \sqrt{d_2^2 + h^2} - d_1 - d_2$$

$$\approx d_1 \left(1 + \frac{h^2}{2d_1^2}\right) + d_2 \left(1 + \frac{h^2}{2d_2^2}\right) - d_1 - d_2$$

$$= \frac{h^2}{2} \left(\frac{1}{d_1} + \frac{1}{d_2}\right)$$

$$(3.8)$$

where we used the binomial approximation $\sqrt{1+x} \cong 1 + x/2$ as $x \to 0$. This approximation is valid when the height h is much less than d_1 and d_2, which is reasonable in practical applications. The loci of points at which the excess path length is equal to $n\lambda/2$ is given by h_n, which denotes the height of nth Fresnel zone at distance d_1 from the transmitter:

$$\Delta l = n\frac{\lambda}{2} = \frac{h_n^2}{2} \left(\frac{1}{d_1} + \frac{1}{d_2}\right) \Rightarrow h_n = \sqrt{\frac{n\lambda d_1 d_2}{d_1 + d_2}}$$

$$(3.9)$$

The corresponding phase difference is given by

$$\Delta\phi = \frac{2\pi}{\lambda} \Delta l = \frac{2\pi h^2}{\lambda 2} \left(\frac{1}{d_1} + \frac{1}{d_2}\right) \quad (3.10)$$

The phase difference between the direct ray and the rays scattered by scatterers on the surface of the nth Fresnel ellipsoid is given by

$$\Delta\phi = \frac{2\pi}{\lambda} \Delta l = \frac{2\pi}{\lambda} n\frac{\lambda}{2} = n\pi \quad (3.11)$$

Contributions to the received signal from successive Fresnel zones tend to be in phase opposition and interfere with each other

destructively. The contributions to the received signal level from scatterers in the nth Fresnel zone would be less than those in the $(n-1)^{th}$ Fresnel zone.

Received signal level is considered to be significant only from scatterers which are located in the first Fresnel zone. Antenna heights in point-to-point links are usually selected so that the first Fresnel zone is clear. We know from geometrical considerations that the maximum value of h_n occurs at the mid-point of the direct path $d_1 = d_2$. In terrestrial links, one should check whether the point of reflection remains in the first Fresnel zone. However, in Earth-space links, the scatterers in close vicinity of the ground station antenna may remain in the first Fresnel zone. In VHF and UHF bands, the Earth is usually located in the first Fresnel zone and free space propagation conditions do not exist. Therefore, signal propagation takes place via direct, Earth-reflected and scattered rays due to obstacles located in the first Fresnel zone. For example, consider a communication system with $d_1 = d_2 = 10$ km. The height of the first Fresnel zone is found from (3.9) as $h_1 = \sqrt{\lambda d_1/2}$, which is equal to 122.5 m for a FM radio channel at 100 MHz and 40.8 m for a GSM system operating at 900 MHz. In both cases, the Earth surface will be located in the first Fresnel zone in most operational scenarios. On the other hand, the height of a Fresnel ellipsoid is proportional to $f^{-1/2}$. This means that a Fresnel zone becomes more concentrated along the direct path with increasing frequencies. Consequently, it is easier to clear the first Fresnel zone at higher frequency bands, where antennas are physically smaller and can easily be mounted on masts and/or roof tops.

We will hereafter assume that free-space propagation conditions apply when the first Fresnel zone is clear. If not, in addition to the direct (LOS) signal, the received signal will be determined by taking into account of the contributions of all obstacles located in the first Fresnel zone. These can reflect, diffract or scatter the incident signals. The reflection is often from the Earth's surface which may be smooth to cause specular reflection, or rough so as to cause diffuse reflection or scattering of the incident rays. Besides, hills with sharp edges, roof tops of buildings and other metallic objects may diffract electromagnetic waves.

Below we will first consider the diffraction of electromagnetic waves from obstacles located in the first Fresnel zone and then study reflection from Earth's surface with flat and spherical Earth assumptions as well as for the cases of smooth and rough surfaces.

3.3.4   Knife-Edge Diffraction

Here we will study the diffraction of electromagnetic waves from sharp edges of conducting obstacles located in the first Fresnel zone. As shown in Figure 3.4, a receiver may be located either in the illuminated or the shadow region, which are separated from each other by the *shadow boundary*. Illuminated (shadow) region is defined as the region where the direct (LOS) ray is (not) available. When a receiver located in the *illuminated region*, it can receive both the direct (LOS) signal and the signal diffracted by the edge. However, a receiver located in the *shadow region* can receive only

the diffracted ray, since the direct ray is blocked.

Using (3.10) the phase difference between direct and diffracted rays may be related to the so-called Fresnel-Kirchhoff diffraction parameter v as

$$\Delta\phi = \frac{2\pi h^2}{\lambda \, 2}\left(\frac{1}{d_1}+\frac{1}{d_2}\right)=\frac{\pi}{2}v^2$$

$$\Rightarrow v = h\sqrt{\frac{2}{\lambda}\left(\frac{1}{d_1}+\frac{1}{d_2}\right)}$$

(3.12)

As shown in Figure 3.5, h denotes the height of the diffracting screen above (below) the LOS, if the LOS is (not) blocked by the screen. If the diffracting edge just touches the LOS between transmitter and receiver, that is, in the shadow boundary, then $h = v = 0$. Note that, $h, v > 0$ when LOS is blocked by the edge and $h, v < 0$ in the presence of LOS between transmitter and receiver. Hence, being directly proportional to h, the Fresnel-Kirchhoff diffraction parameter v provides a measure of the closeness of the diffracted ray to the LOS.

According to Huygens-Fresnel principle, the electric field intensity E_d at the observation point may be written in terms of the electric field intensity E_i incident at the point of diffraction as [1]

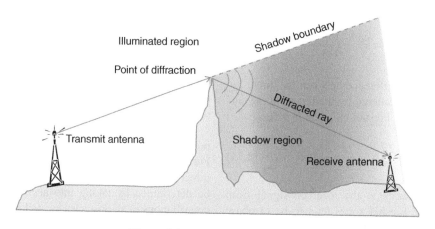

Figure 3.4   Knife-Edge Diffraction.

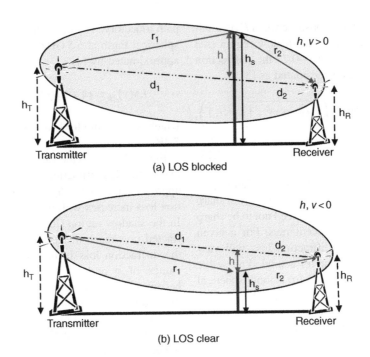

Figure 3.5 Knife-Edge Diffraction Scenarios With and Without LOS.

$$E_d = E_i \left(\frac{1}{2} - \frac{\exp(j\pi/4)}{\sqrt{2}} F(v) \right) e^{-jkr_2}$$

$$F(v) = \int_0^v \exp\left(-j\pi t^2/2\right) dt = C(v) - jS(v)$$

$$C(v) = \int_0^v \cos\left(\pi t^2/2\right) dt$$

$$S(v) = \int_0^v \sin\left(\pi t^2/2\right) dt$$

$$(3.13)$$

The Fresnel integrals $C(v)$ and $S(v)$ vanish and the received electric field intensity becomes equal to $E_d = E_i/2$ at $v = 0$, that is, when the receiver located on the shadow boundary. Physically, this means that half of the incident electric field intensity ($E_i/2$) goes directly to the receiver, whereas the other half is reflected back by the surface of the screen. For large positive values of v (in the shadow region), the Fresnel integrals $C(v)$ and $S(v)$ approach ½ and the received electric field

intensity E_d vanishes (see Figure D.6). On the other hand, as $v \to -\infty$ (in the illuminated region), one may use $C(-v) = -C(v)$ and $S(-v) = -S(v)$ to show that the received electric field intensity E_d oscillates around and eventually approaches the direct ray E_i.

The so-called *diffraction loss* represents the signal power loss between diffraction and observation points. Using (3.13), the diffraction loss may be expressed as follows:

$$L(v)_{dB} = 10\log \left| \frac{E_i}{E_d} \right|^2$$

$$= 20\log \left| \frac{2}{1 - C(v) - S(v) + j(S(v) - C(v))} \right|$$

$$= 10\log \left(\frac{2}{(0.5 - C(v))^2 + (0.5 - S(v))^2} \right)$$

$$= 20\log(\pi v), \quad v \to \infty$$

$$(3.14)$$

which is defined to be positive when E_d is smaller than E_i, that is, in the shadow region (see Example D.1). For $v > -0.7$, the diffraction loss $L(v)$ may be approximated as [3]

$$LA(v)_{dB} = 6.9 + 20 \log\left(\sqrt{(v-0.1)^2 + 1} + v - 0.1\right),$$
$$v > -0.7$$

$$(3.15)$$

The knife-edge diffraction formulation presented above is valid for diffraction from sharp edges. However, the edges need not to be sharp for diffracting the incident rays. For a given path clearance, the diffraction loss will vary from a minimum value for a single knife-edge diffraction to a maximum for smooth spherical Earth, which may be more applicable in rural areas. Empirical diffraction loss for average terrain is approximated by [4]

$$LE(v)_{dB} \approx 10 + 10\sqrt{2}v \ (dB) \qquad (3.16)$$

which is valid for losses greater than about 15 dB and h is defined as the height difference between most significant path blockage and path trajectory. The diffraction loss for smooth spherical Earth at 6.5 GHz and $k = 4/3$ Earth is approximated by [4]

$$LS(v)_{dB} \approx 14 + 14\sqrt{2}v \ (dB) \qquad (3.17)$$

which is also valid for losses greater than about 5 dB.

Figure 3.6 shows the variation of the theoretical knife-edge diffraction loss by (3.14) and its approximation by (3.15). Note that the diffraction loss increases with increasing values of v. In the shadow region (for $v > 0$), only the diffracted rays reach the receiver. Consequently, the diffraction loss increases with increasing values of v and h, that is, as shadowing becomes more intense. However, for $v < 0$ (in the illuminated region), E_d given by (3.13), which represents the phasor sum of direct and diffracted rays, oscillates around E_i (represented by 0 dB level in Figure 3.6). Consequently, the negative values of the diffraction loss imply constructive interference between direct and diffracted rays; hence, signal levels is higher than the direct signal level E_i. Figure 3.6 also shows empirical diffraction loss

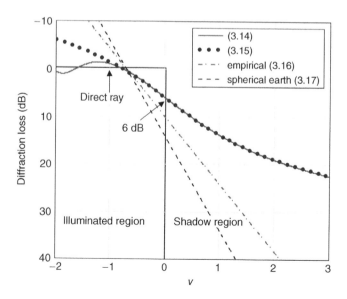

Figure 3.6 Diffraction Loss.

and smooth Earth diffraction loss in comparison with the theoretical knife-edge diffraction
loss. Diffraction loss was observed to increase
as the diffracting edge becomes smoother. For
$v = 2$, the diffraction losses for theoretical knife
edge, empirical and smooth Earth are found
using (3.15)–(3.17) to be 19.1 dB, 38.4 dB
and 53.8 dB, respectively.

In order to determine the electric field intensity at the observation point, we first express
the rms value of the incident electric field intensity at the point of diffraction, as given
by (2.119),

$$E_i = \frac{e^{-jkr_1}}{r_1}\sqrt{30P_tG_t} \ (V/m) \qquad (3.18)$$

where P_t denotes the transmit power, G_t is the
transmit antenna gain in the direction of the
diffracting edge and r_1 denotes the distance
between the transmitter and the point of diffraction. From (3.14) and (3.18), the rms value of the
diffracted electric field intensity is found to be

$$|E_d|^2 = \frac{|E_i|^2}{L(v)} = \frac{30P_tG_t}{r_1^2 L(v)} \qquad (3.19)$$

The received diffracted signal power at the
observation point is found using (2.120)
and (3.19):

$$P_d = \frac{|E_d|^2}{120\pi}A_e = \frac{|E_d|^2}{120\pi}\frac{\lambda^2}{4\pi}G_r$$

$$= \frac{1}{120\pi}\frac{30P_tG_t}{L(v)}\frac{\lambda^2}{r_1^2}\frac{1}{4\pi}G_r \qquad (3.20)$$

$$= \frac{P_tG_tG_r}{L_{fs_1}L(v)}$$

where $L_{fs_1} = (4\pi r_1/\lambda)^2$ denotes the free space
loss between the transmitter and the point of
diffraction.

When the separation between transmitter and
receiver is much longer compared to the height
of the diffracting screen, one might assume

$d_i \approx r_i, i = 1,2$ in Figure 3.5. Then, the ratio of
the received diffracted signal power and free-
space signal powers at the observation point
may be written as

$$\frac{P_d}{P_{r,fs}} = \frac{\dfrac{P_tG_tG_r}{L(v)}\left(\dfrac{\lambda}{4\pi r_1}\right)^2}{P_tG_tG_r\left(\dfrac{\lambda}{4\pi(r_1+r_2)}\right)^2} = \frac{1}{L(v)}\left(\frac{r_1+r_2}{r_1}\right)^2$$

$$(3.21)$$

This ratio is mostly less than unity, implying
that received signal power due to diffraction is
lower than that for free-space propagation.

Example 3.2 Calculation of the Diffraction Loss.
Assume $d_1 = d_2 = 3$ km, $f = 2$ GHz ($\lambda = 0.15$ m)
and $h = 3$ m, and a sharp diffracting edge blocks
LOS between transmitter and receiver. The diffraction loss is found by (3.12) and (3.15)

$$v = h\sqrt{\frac{2(d_1+d_2)}{\lambda d_1 d_2}} = 3\sqrt{\frac{2(3000+3000)}{0.15 \times 3000^2}} = 0.283$$

$$LA(v)_{dB} = 6.9$$

$$+ 20\log\left(\sqrt{(v-0.1)^2+1}+v-0.1\right)\Big|_{v=0.283}$$

$$= 8.48 \, dB$$

$$(3.22)$$

From (3.21), the ratio of the received signal
powers due to diffraction and direct ray is
found as

$$\frac{P_d}{P_{r,fs}} = \frac{1}{LA(v)}\left(\frac{d_1+d_2}{d_1}\right)^2$$

$$\Rightarrow -8.48dB + 10\log(4) = -2.48dB$$

$$(3.23)$$

shows that the diffracted signal power is 2.48 dB
lower than that for free space propagation.

Example 3.3 Double Knife-Edge Diffraction.
Consider a propagation scenario, as shown in
Figure 3.7, where there are two diffracting

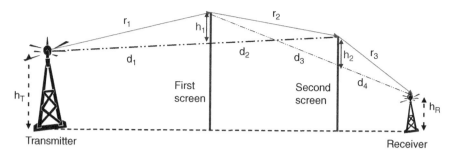

Figure 3.7   Geometry of Double Knife-Edge Diffraction.

screens between transmitter and receiver. If $E_{i,1}$ and $E_{i,2}$ denote respectively the incident electric field intensities at the first and the second diffracting edges, the incident field at the diffracting point of the second screen may be written as

$$E_{i,2} = \frac{E_{i,1}}{\sqrt{L(v_1)}} e^{-jkr_2} = \frac{\sqrt{30 P_t G_t}}{r_1 \sqrt{L(v_1)}} e^{-jk(r_1 + r_2)}$$

$$(3.24)$$

The diffracted field at the observation point is given by

$$E_d = \frac{E_{i,2}}{\sqrt{L(v_2)}} e^{-jkr_3} = \frac{\sqrt{30 P_t G_t}}{r_1 \sqrt{L(v_1) L(v_2)}} e^{-jk(r_1 + r_2 + r_3)}$$

$$(3.25)$$

where Fresnel-Kirchoff diffraction parameters for the first and the second diffraction points are given by [5]

$$v_1 = h_1 \sqrt{\frac{2(d_1 + d_2)}{\lambda d_1 d_2}}, \quad v_2 = h_2 \sqrt{\frac{2(d_3 + d_4)}{\lambda d_3 d_4}}$$

$$(3.26)$$

Using (3.25), the received diffracted signal power at the observation point may be written as

$$P_d = \frac{|E_d|^2}{120\pi} \frac{\lambda^2}{4\pi} G_r = \frac{P_t G_t G_r}{L_{fs_1} L(v_1) L(v_2)}$$

$$(3.27)$$

The decrease in received diffracted signal power compared to the LOS signal power is given by

$$\frac{P_d}{P_{r,fs}} = \frac{L_{fs}}{L_{fs_1} L(v_1) L(v_2)}$$

$$\cong \frac{1}{L(v_1) L(v_2)} \left(\frac{d_1 + d_2 + d_4}{d_1} \right)^2$$

$$(3.28)$$

where L_{fs_1} and L_{fs} respectively denote the free space propagation loss along r_1 and the direct path between transmitter and transmitter.

Example 3.4   Knife-Edge Diffraction From an Obstacle in the First Fresnel Zone.
Transmitting and receiving antenna heights of a communication link operating at 300 MHz at a range of $d = 20\,\text{km}$ are 100 m and 200 m, respectively (see Figure 3.8). The Earth is assumed to be flat along the propagation path.

Maximum height of first Fresnel zone is located at the mid-point between transmitter and receiver, $d_1 = d_2 = d/2$:

$$h_{1,\max} = \sqrt{\frac{\lambda d_1 d_2}{d_1 + d_2}} = \sqrt{\frac{\lambda d}{4}} = 70.7\,\text{m}$$

$$(3.29)$$

Since $h_{1,max}$ is less than the antenna heights, the First Fresnel zone is free of the Earth surface. Now assume that a sharp diffracting edge of 80 m high is located at a distance of 8 km from the transmit antenna. From the right triangle formed by transmitter, receiver, and point

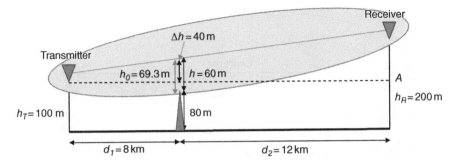

Figure 3.8 A Knife-Edge Diffracting Obstacle in the First Fresnel Zone.

A in Figure 3.8, we find that the LOS line at $d_1 = 8$ km from the transmitter is 140 m above the Earth's surface. The distance betwen the diffracting edge and LOS line is therefore $h = -60$ m. Since the height of the First Fresnel zone at this point is $h_0 = 69.3$ m, the diffracting edge is located in the first Fresnel zone and its contribution to the received signal should be taken into account. Using $d_1 = 8$ km, and $d_2 = 12$ km, the Fresnel-Kirchhoff diffraction parameter is found to be

$$v = h\sqrt{\frac{2}{\lambda}\left(\frac{1}{d_1}+\frac{1}{d_2}\right)}$$

$$= -60\sqrt{\frac{2}{1}\left(\frac{1}{8000}+\frac{1}{12000}\right)} = -1.22$$

$$(3.30)$$

Inserting (3.30) into (3.15), the diffraction loss is found to be

$$LA_{dB}(-0.02)$$

$$= 6.9 + 20\log\left(\sqrt{1+(-0.02-0.1)^2}+v-0.1\right)$$

$$= -2.57\,dB$$

$$(3.31)$$

It is clear from Figure 3.8 that the receive antenna receives both the LOS and the diffracted signal and the received signal power level is 2.57 dB higher than that of the LOS signal.

3.3.5 Propagation Over the Earth Surface

In VHF and UHF bands, free-space propagation conditions apply only under restricted cases. Effects of obstructions on the ground or the turbulent sea should be considered, since the first Fresnel zone usually comprises the Earth's surface. Therefore, VHF and UHF wave propagation in urban, suburban, and rural areas and over sea show significant differences.

Two issues are of primary interest in propagation over the Earth surface in VHF and UHF bands. Firstly, the Earth surface may be assumed to be flat when the range r satisfies

$$r_{km} < 80/f_{MHz}^{1/3} \qquad (3.32)$$

For example, at $f = 1$ MHz the Earth may be assumed to be flat for ranges not exceeding 80 km, but the range reduces to 8.3 km at $f = 900$ MHz. Therefore, transmit and receive antennas should be raised more to compensate for the curvature of the Earth, which limits the LOS distance and hence the communication range. The second issue stems from the roughness of the Earth surface, which is measured in terms of the standard deviation of surface irregularities compared to the local mean level

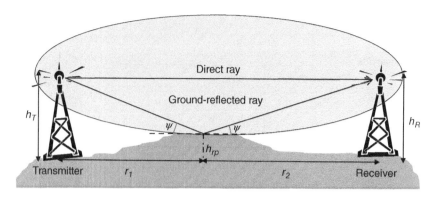

Figure 3.9   Signal Transmission Via Direct and Ground-Reflected Rays.

at the point of reflection/scattering. Depending on surface roughness, the electrical parameters of the ground, frequency of operation and the angle of incidence ψ (see Figure 3.9), electromagnetic waves undergo specular/diffuse reflection or scattering. Consequently, the roughness of the Earth surface strongly affects the intensity of the surface-reflected signals at the receiver.

We will first determine the reflected electric field intensity at the receiver, as shown in Figure 3.9, based on the assumption that the Earth surface is flat and smooth at the reflection point. It is also assumed that the height of the reflection point h_{rp} coincides with the line connecting transmit and receive antennas. Otherwise, transmit and receive antenna heights should be adjusted accordingly, that is, as h_T-h_{rp} and h_R-h_{rp} (see Figure 3.9). [4] Based on these assumptions, the point of reflection is determined by the law that the angle of incidence is equal to the angle of reflection:

$$\frac{h_T}{r_1} = \frac{h_R}{r_2} \;\Rightarrow\; r_1 = \frac{h_T}{h_T + h_R} r \qquad (3.33)$$

Here $r = r_1 + r_2$ denotes the distance between transmitter and receiver where r_1 (r_2) is the distance between the point of reflection and the transmitter (receiver). The point of reflection moves away from the transmitter with increasing values of h_T. For the special case

when $h_T = h_R$, one gets $r_1 = r_2 = r/2$. Similarly, if $h_T = 10$ m, $h_R = 20$ m and $r = 30$ km, then $r_1 = 10$ km and $r_2 = 20$ km.

Reflection coefficient, which is defined as the ratio of the reflected field intensity to the incident field intensity at the point of reflection, depends on the carrier frequency, dielectric constant and conductivity of the Earth, the angle of incidence and the polarization of the incident wave. We assume that the Earth is flat and smooth at the point of reflection. Furthermore, the point of reflection is assumed to be in the far-field of the transmitter and receiver so that the fields have planar wave-fronts. For horizontal and vertical polarization, the reflection coefficient over a lossy dielectric flat Earth surface is given by: [4][6]

$$\rho_h = |\rho_h|e^{j\theta_h} = \frac{\sin\psi - \sqrt{(\varepsilon_r - jx) - \cos^2\psi}}{\sin\psi + \sqrt{(\varepsilon_r - jx) - \cos^2\psi}}$$

$$\rho_v = |\rho_v|e^{j\theta_v} = \frac{(\varepsilon_r - jx)\sin\psi - \sqrt{(\varepsilon_r - jx) - \cos^2\psi}}{(\varepsilon_r - jx)\sin\psi + \sqrt{(\varepsilon_r - jx) - \cos^2\psi}}$$

$$x = \frac{\sigma}{w\varepsilon_0} = 18 \times 10^3 \frac{\sigma}{f_{MHz}}$$

$$(3.34)$$

where ψ denotes the angle of incidence with respect to the horizon. Similarly, ε_r and σ denote respectively the relative dielectric

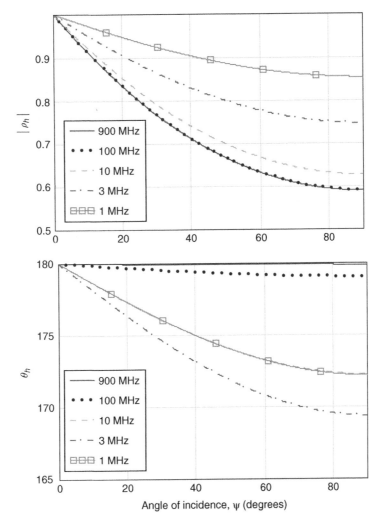

Figure 3.10 Magnitude and Phase of the Reflection Coefficient for Horizontal Polarisation Over Average Ground ($\sigma = 5 \times 10^{-3}$ and $\varepsilon_r = 15$).

constant and the conductivity of the Earth's surface at the point of reflection. Figures 3.10 and 3.11 show the variation of the complex reflection coefficients over average Earth ($\sigma = 5 \times 10^{-3}$ and $\varepsilon_r = 15$ from Table 3.1) for horizontal and vertical polarizations, respectively. Horizontal polarization refers to the case where the incident electric field is polarized parallel to the surface of the Earth, that is, perpendicular to the surface of the paper in Figure 3.9. In vertical polarization, the incident electric field is

polarized parallel to the surface of the paper and perpendicular to the direction of propagation. The parameter x denotes the ratio of the conduction current to the displacement current at the point of reflection at a given frequency. The magnitudes of the reflection coefficients for both horizontal and vertical polarization increase with increasing values of x, that is, decreasing values of the frequency. This implies that the Earth's surface behaves as more conductive and yields higher reflection

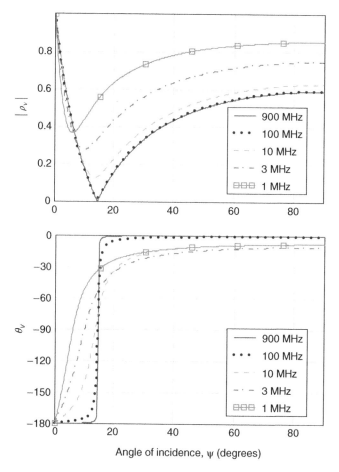

Figure 3.11   Magnitude and Phase of the Reflection Coefficient for Vertical Polarisation Over Average Ground ($\sigma = 5 \times 10^{-3}$ and $\varepsilon_r = 15$).

coefficients at lower frequencies. In case of vertical polarization, the reflection coefficient has a minimum at the so-called pseudo-Brewster angle, which is a function of x.

In many cases of interest, the height of transmit and receive antennas are much smaller compared to r_1 and r_2. The angle of incidence then becomes very small so that $sin\psi \approx 0$; then, the reflection coefficients for horizontal and vertical polarizations tend to be -1:

$$\rho_h = |\rho_h|e^{j\theta_h} \rightarrow 1e^{j\pi} = -1 \quad as \quad \psi \rightarrow 0$$
$$\rho_v = |\rho_v|e^{j\theta_v} \rightarrow 1e^{j\pi} = -1 \quad as \quad \psi \rightarrow 0 \tag{3.35}$$

Typical values for the conductivity and the relative dielectric constant of the Earth's surface are presented in Table 3.1.

3.3.5.1   Propagation Over Flat Earth Surface

As shown in Figure 3.12 transmit and receive antennas are located above the Earth surface which is assumed to be flat and smooth. Thus direct- and ground-reflected rays contribute to the received electric field intensity:

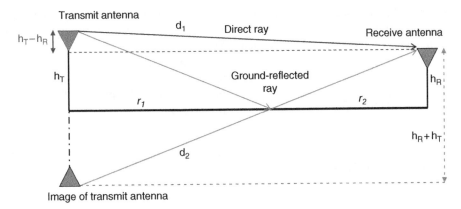

Figure 3.12   Direct and Reflected Rays Between Transmitter and Receiver.

$$E_r = E_{LOS} + E_{REF} = E_0 \frac{e^{-jkd_1}}{d_1}$$
$$+ \rho_v e^{j\theta_v} E_0 \frac{e^{-jkd_2}}{d_2} \qquad (3.36)$$

where d_1 and d_2 denote respectively the path lengths of direct and reflected rays. One may rewrite (3.36)

$$E_r = E_0 \underbrace{\frac{e^{-jkd_1}}{d_1}}_{E_{LOS}} \underbrace{\left(1 + \rho_v e^{j\theta_v} \frac{d_1}{d_2} e^{-jk(d_2-d_1)}\right)}_{f_a}$$

$$(3.37)$$

where the received electric field intensity is expressed as the product of the direct ray and the so-called array factor f_a, which represents the contribution of the Earth-reflected ray to the received signal strength. The excess path length d_2-d_1 for large $r = r_1 + r_2$ and $r \gg h_T, h_R$, is found from Figure 3.12:

$$d_1^2 = r^2 + (h_R - h_T)^2$$
$$d_2^2 = r^2 + (h_R + h_T)^2 \qquad (3.38)$$
$$d_2^2 - d_1^2 = 4h_T h_R$$
$$d_2 - d_1 = 4h_T h_R / (d_1 + d_2) \cong 2h_T h_R / r$$

For sufficiently large separations r between transmitter and receiver, we can assume

$\rho_v e^{j\theta_v} \approx -1$ (see 3.35). Then the array factor reduces to

$$|f_a| = \left|1 - e^{-j2\pi \frac{(d_2-d_1)}{\lambda}}\right| = \left|1 - e^{-j\frac{4\pi h_T h_R}{r}}\right|$$

$$= 2 \left|\sin\left(\frac{2\pi h_T h_R}{\lambda r}\right)\right| \qquad (3.39)$$

It is evident from (3.37) and (3.39) that the Earth-reflected ray interferes with the direct ray; the received electric field intensity is equal to twice that of the direct ray when the phase difference between them is an even multiple of π and vanishes for odd multiples of π. Hence, the maxima of the array factor given by (3.39) occur at

$$|f_a| = 2 \Rightarrow \frac{2\pi h_T h_R}{\lambda r} = (2n-1)\frac{\pi}{2} \quad n = 1,2\ldots$$

$$(3.40)$$

Similarly, the minima occur at

$$|f_a| = 0 \Rightarrow \frac{2\pi h_T h_R}{\lambda r} = n\pi \quad n = 1,2,\ldots \quad (3.41)$$

Figure 3.13 shows the variation of the array factor given by (3.39) with the electrical height of the receive antenna h_R/λ for $r/h_T = 10$. Free space signal propagation condition is represented by $|f_a| = 1$. The optimum receiver height corresponds to the minimum value of h_R that maximizes the array factor, that is, $n = 1$ in (3.40). The optimal receiver height is then given by

$$h_{R,opt} = \frac{\lambda r}{4h_T} \qquad (3.42)$$

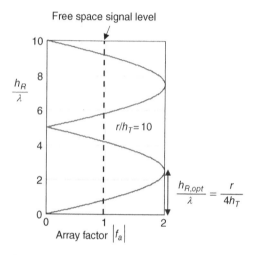

Figure 3.13 Variation of the Array Factor with h_R/λ for $r/h_T = 10$.

For $\lambda = 0.1$ m ($f = 3$ GHz), $h_T = 10$ m and $r = 1$ km, the optimum receiver height is found to be $h_{R,opt} = 2.5$ m.

In the presence of only direct and Earth-reflected rays, the received signal power P_r may be written as the product of the LOS signal power and $|f_a|^2$:

$$P_r = P_{LOS}|f_a|^2 = P_t G_t G_r \left(\frac{\lambda}{4\pi r}\right)^2 4\sin^2\left(\frac{2\pi h_T h_R}{\lambda r}\right)$$

$$(3.43)$$

Figure 3.14 shows the variation of (3.43) with distance r for $h_T = 10$ m, $h_R = 1.6$ m, $\lambda = 1/3$ m ($f = 900$ MHz) and $P_t G_t G_r = 1$. For small values of r, one observes strong oscillations in the received signal power due to constructive/destructive interferences between direct and reflected rays. Consequently, the received signal power might be 6 dB higher than the free-space value at points determined by (3.40) and vanish at points determined by (3.41). For large values of r, the reflection coefficient approaches -1 since the elevation angle approaches zero and the frequency of oscillations decreases.

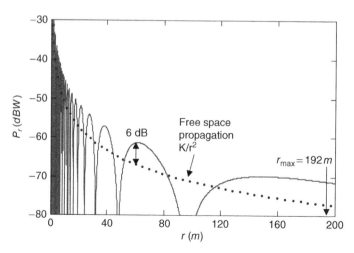

Figure 3.14 Variation of the Received Signal Power with Range for $h_T = 10$ m, $h_R = 1.6$ m, $\lambda = 1/3$ m ($f = 900$ MHz) and $P_t G_t G_r = 1$.

The furthest signal maximum from the transmitter corresponds to the largest value of r that maximizes the array factor, that is, $n = 1$ in (3.40):

$$r_{max} = 4h_T h_R / \lambda \ (m) \qquad (3.44)$$

Note from (3.38) that r_{max} represents the largest distance at which the path difference d_2-d_1 between direct and reflected rays is equal to $\lambda/2$. This implies that the point determined by r_{max} is located on the surface of the first Fresnel ellipsoid; ignoring the reflection coefficient, which is equal to −1 two paths interfere destructively with each other since the phase difference between them is equal to π radians. At distances shorter than r_{max}, the signal power decreases in proportion to $1/r^2$, since the array factor varies over the range −2 and 2. However, at distances $r > r_{max}$, the argument of sine in (3.39) becomes much smaller than $\pi/2$ and one can use the approximation $\sin(x) \cong x$ in (3.43) to get

$$P_r \cong P_t G_t G_r \left(\frac{h_T h_R}{r^2} \right)^2$$
$$= \frac{P_t G_t G_r}{L_{pe}} \propto \frac{1}{r^4} (W) \ for \ r > r_{max} \qquad (3.45)$$

where L_{pe} denotes the so-called plane-Earth path-loss:

$$L_{pe} = \left(\frac{r^2}{h_T h_R} \right)^2 \qquad (3.46)$$

Beyond r_{max} the signal power decreases monotonically as $1/r^4$.

Example 3.5 Aeronautical Channel.
Consider a transmitter with antenna height of $h_T = 10$ m, and an air-borne receiver with antenna height $h_R = 1000$ m. The frequency of operation is $f = 127.5$ MHz. The Earth's surface is characterized by a relative dielectric constant of $\varepsilon_r = 15$, and conductivity $\sigma = 0.012$ S/m. The communication takes place with direct and Earth-reflected rays. The corresponding path loss between the transmitter and the air-borne receiver is presented in Figure 3.15 as a function of the horizontal distance between them.

For small values of the horizontal distance, direct and reflected rays were observed to interfere constructively so that the received signal power is twice that of the free space value. However, at larger horizontal separations, the interference between them leads to alternating flares and fades in the received signal power.

3.3.5.2 Propagation Over Spherical Earth Surface

When the distance between transmitter and receiver does not satisfy the condition given by (3.32), the Earth's surface can not be assumed flat. We then need to account for the effect of the Earth's curvature on the electric field reflected from the Earth's surface. In this case, the Earth is assumed to be spherical with an effective radius of $r_e = (4/3)a = 8500$ km, where $a = 6371$ km denotes the physical Earth radius. As it will be explained in the following sections, the factor 4/3 accounts for the refraction (bending) of electromagnetic waves in the troposphere. Therefore, the electromagnetic waves are implicitly assumed to propagate along straight lines over the surface of the Earth with an equivalent radius of 8500 km.

Unlike on the flat Earth-surface, one may observe from Figure 3.16a that the 'effective' antenna height observed on spherical Earth-surface is lower than its physical height. From Figure 3.16b, one may find the effective antenna heights as

$$r_1^2 = [r_e + (h_T - h'_T)]^2 - r_e^2 \cong 2r_e (h_T - h'_T)$$
$$\Rightarrow h'_T = h_T - \frac{r_1^2}{2r_e}$$

$$r_2^2 = [r_e + (h_R - h'_R)]^2 - r_e^2 \cong 2r_e (h_R - h'_R)$$
$$\Rightarrow h'_R = h_R - \frac{r_2^2}{2r_e} \qquad (3.47)$$

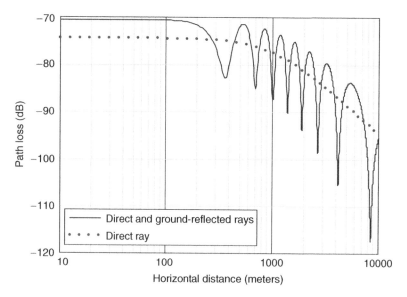

Figure 3.15 Path Loss as a Function of the Horizontal Distance Between a Ground-Based Transmitter and an Air-Borne Receiver. Heights of transmit and receive antennas are given by $h_T = 10$ m and $h_R = 1000$ m, respectively.

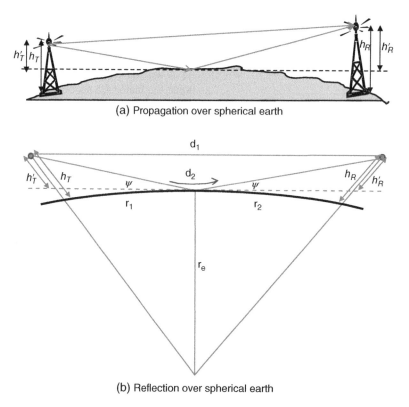

Figure 3.16 Effective Antenna Heights for Reflection Over Spherical Earth.

For field calculations over spherical Earth, $\{h_T, h_R\}$ appearing in (3.33)–(3.46) for propagation over flat-Earth should be replaced by $\{h'_T, h'_R\}$.

On the other hand, a bundle of parallel rays incident on the surface of flat Earth are reflected parallel to each other. However, they diverge after reflection from the surface of spherical Earth, thus leading to a decrease in the reflection coefficient (see Figure 3.17). The rays will evidently diverge more for decreasing values of the angle of incidence, which is directly proportional to the antenna heights and inversely proportional to the range. The divergence factor D, which accounts for the divergence of electromagnetic waves as they are reflected over spherical Earth-surface, is given by: [4]

$$D = \sqrt{\frac{1 - m(1 + b^2)}{1 + m(1 - 3b^2)}}$$

$$m = \frac{r^2}{4r_e(h'_T + h'_R)}$$

$$b = 2\sqrt{\frac{m+1}{3m}}\cos\left[\frac{\pi}{3}\right.$$
$$\left. + \frac{1}{3}\cos^{-1}\left(\frac{3}{2}\, h'_T - h'_R\, h'_T + h'_R\sqrt{\frac{3m}{(m+1)^3}}\right)\right]$$

(3.48)

Note that $D = 1$ for flat surface but $D < 1$ for the spherical Earth. Figure 3.18 shows the variation of the divergence factor as a function of

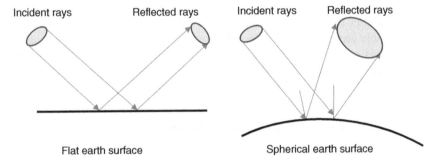

Flat earth surface   Spherical earth surface

Figure 3.17   Reflection Over Flat and Spherical Earth Surface.

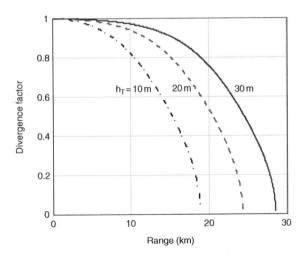

Figure 3.18   Divergence Factor Versus Range in km for $h_R = 2$ m and Various Values of h_T.

range for $h_R = 2$ m and various values of h_T. Note that small values of the divergence factor at large distances from the transmitter may help reducing the fades due to interference between direct and reflected rays.

3.3.5.3 Effect of Surface Roughness on Reflection

So far we discussed the reflection of electromagnetic waves from smooth surfaces only; then, the reflection is in the specular direction, which is determined by the law that the angle of incidence equals the angle of reflection. For specular reflection, smooth Earth reflection coefficients, given by (3.34), are used to determine the reflected field strength (see Figure 3.19a). As shown in Figure 3.19b, incident rays reflected by a rough surface in off-specular directions have a nonnegligible part of the total reflected signal energy. In this so-called diffuse reflection process, the energy of the reflected field in the specular direction is reduced at the expense of

the energy reflected in other directions. Thus, the reflection coefficient becomes smaller and the direction of reflection broadens. In the extreme case of reflection from rough surfaces as shown in Figure 3.19c, the process is described as scattering rather than reflection. In this process, the smooth Earth reflection coefficient can not be used since incident rays are scattered randomly in all directions and only a small fraction of the scattered energy is received by the receiver. This implies that the scattered signals have negligible contribution to the received electric field intensity.

The extent of scattering depends on the angle of incidence and the surface roughness in comparison to the wavelength. The apparent roughness of a surface is reduced as the angle of incidence becomes closer to grazing incidence, and/or the wavelength becomes larger. Here we will define the surface roughness to make a quantitative distinction between smooth and rough surfaces. Consider the reflection of two rays from a rough surface, as shown in Figure 3.20, one from the local mean level and the other from an obstacle

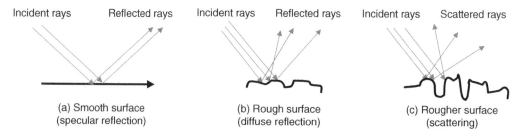

(a) Smooth surface (specular reflection)

(b) Rough surface (diffuse reflection)

(c) Rougher surface (scattering)

Figure 3.19   Effect of Surface Roughness on the Reflection.

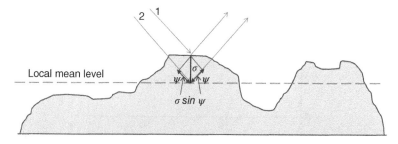

Figure 3.20   Definition of the Surface Roughness.

of height σ, which represents the standard deviation of the surface irregularities with respect to the local mean height. The phase difference between the two rays may be written as

$$\Delta\phi = \frac{2\pi}{\lambda}\Delta\ell = \frac{2\pi}{\lambda}2\sigma\sin\psi = \frac{4\pi\sigma}{\lambda}\sin\psi \quad (3.49)$$

where $\Delta\ell$ shows the excess path length, that is, the path length difference between rays 1 and 2. The reflection is said to be specular if $\sigma \ll \lambda$ (reflecting surface is smooth), that is, when $\Delta\phi$ is sufficiently small. On the other hand, a rough surface corresponds to a phase difference of $\Delta\phi \approx \pi$ where direct and reflected rays cancel each other. The dividing line between rough and smooth surfaces is taken as $\Delta\phi = \pi/2$. Hence, a reflecting surface is said to be smooth if

$$\Delta\phi < \frac{\pi}{2} \implies \sigma < \frac{\lambda}{8\sin\psi} \quad (3.50)$$

where one may use (3.33) to express ψ in terms of h_T, h_R, and r as

$$\sin\psi = \frac{h_T + h_R}{\sqrt{r^2 + (h_T + h_R)^2}} \quad (3.51)$$

The Rayleigh roughness criterion specifies the maximum allowable surface deviation (relative to a perfectly flat surface) in order for a surface to be considered smooth. The variation of the standard deviation of the surface irregularities is plotted in Figure 3.21 as function of the angle of incidence to for a surface to be considered smooth; a surface standard deviation of $(\sigma/\lambda \leq 0.72)$ is sufficient to make a surface smooth for $\psi \geq 10$ degrees. The standard deviation of surface irregularities will then be $\sigma \leq 0.24$ m and 2.16 m at 900 MHz and 100 MHz, respectively. Therefore, surface roughness can mostly be ignored in the VHF/UHF bands at sufficiently far ranges.

When the surface is rough, the reflection coeffient given by (3.34) for a smooth surface should be multiplied by the surface roughness factor $f_R(\sigma)$ given by [4]

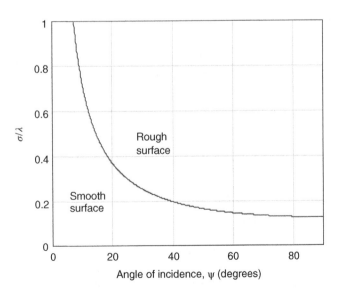

Figure 3.21 Rayleigh Criterion for Surface Roughness. Reflection from rough surfaces can not be accurately modeled by Fresnel reflection coefficients.

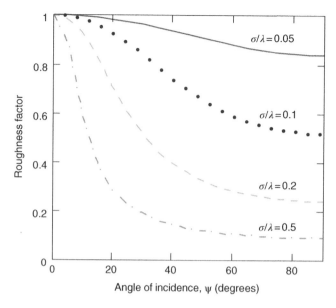

Figure 3.22   Roughness Factor for Several Values of σ/λ.

$$f_R(\sigma) = \sqrt{\frac{1+z}{1+2.35z+2\pi z^2}}$$

(3.52)

$$z = \frac{1}{2}\left(\frac{4\pi\sigma\sin\psi}{\lambda}\right)^2$$

The surface roughness factor, which is plotted in Figure 3.22 decreases with increasing surface irregularities and angle of incidence, as expected.

Hence, the effective reflection coefficient may be written as

$$\rho_{eff} = \rho D f_R(\sigma)$$

(3.53)

Here ρ is given by (3.34) and denotes the reflection coefficient for smooth Earth for horizontal or vertical polarization. The divergence factor D accounts for the decrease in the smooth Earth reflection coefficient due to Earth's curvature and $f_R(\sigma)$ provides a measure of the decrease in the smooth Earth reflection coefficient due to surface roughness at the point of reflection.

3.4   Wave Propagation in SHF and EHF Bands

In SHF (3–30 GHz) and EHF (30–300 GHz) bands, the wavelength is shorter than 10 cm. Therefore, transmit and receive antennas can easily be mounted on masts and roof-tops. Desired antenna gains can be achieved with smaller physical antenna sizes. Moreover, very narrow first Fresnel regions help clearing the propagation paths so as to allow free space propagation. In these bands, signal attenuation by atmospheric gases and hydrometeors including rain, fog, hail, snow and clouds and the corresponding increase in the system noise constitute the two major sources of impairment.

Atmospheric losses are caused by the absorption of the energy of electromagnetic waves by the gaseous constituents (dry air and water vapour) of the atmosphere. Thus, electromagnetic waves transfer part of their energy to heat the air in the propagation medium. Attenuation by atmospheric gases changes with frequency, elevation angle, pressure, temperature, elevation angle, altitude above sea level and humidity. Atmospheric absorption loss, which is minimized in zenith direction, increases with decreasing values of elevation angles, due to the increased path length in the atmosphere. For elevation angles above 10°, the absorption losses do not exceed 2 dB at frequencies below 22 GHz and can be neglected at frequencies below 10 GHz. [7] Peak of water vapour absorption is at 22 GHz, whereas the peak of oxygen absorption is located at 60 GHz. Low atmospheric absorption windows for communications are located in 28–42 GHz and 75–95 GHz bands.

For moderate rain rates, the rain attenuation becomes significant (higher than ~0.1 dB/km) only for frequencies higher than several GHz. Significant rain intensity occurs only for small percentages of time and does not generally cover the whole propagation path. Therefore, the rain attenuation is characterized statistically. Zenith attenuation due to rain is estimated not to exceed 0.15 dB for about 1% of the time for frequencies below 20 GHz. Meanwhile cloud and fog attenuations in the zenith path do not exceed approximately 0.4 dB for 5% of the time. [8] Nevertheless, cloud attenuation can be significant at frequencies above 10 GHz and low elevation angles, where more than 2 dB of attenuation can occur. [7]

The mechanisms and the sources of the propagation losses largely depend on the frequency, geographic location and whether the link is terrestrial or Earth-space. Terrestrial links use the troposphere, which is the non-ionized part of the atmosphere with height less than 15 km above Earth's surface. The troposphere is responsible for most of the weather effects due to clouds, rainfall, and snow, as well as for the tropospheric refraction. Earth-space links are affected by the troposphere, and the ionosphere, the ionized part of the atmosphere with height between 30 km and up to 1000 km. Troposphere and ionosphere can give rise to the following significant signal impairments in terrestrial and Earth-space links, whenever applicable: [2][4][7]

a. absorption losses by atmospheric gases;

b. absorption, scattering and depolarization by hydrometeors (raindrops, ice crystals, clouds, etc.), sand, and dust;

c. signal depolarization by Faraday rotation in the ionosphere;

d. noise emission due to absorption of electromagnetic waves (important at frequencies above 10 GHz);

e. loss of signal and decrease in effective receive antenna gain, due to phase decorrelation across the antenna aperture, caused refraction and relatively slow fading;

f. scintillation (rapid fluctuations) in the received signal's amplitude, phase and signal-of-arrival due to small-scale refractivity variations, especially at low elevation angles, and ionization density fluctuations in the ionosphere;

g. bandwidth limitations due to multiple scattering or multipath, dispersion and delay jitter, especially in high-capacity digital systems;

h. signal attenuation caused by the local environment of the receiver terminal (buildings, trees, etc.);

i. short-term variations of the ratio of attenuations at the up- and down-link frequencies, which may affect the accuracy of adaptive power control;

j. elevation angle variations in non-geostationary satellite systems.

Tropospheric effects on Earth-space paths become significant only for low elevation angles (<3°) or at frequencies above 10 GHz. Ionospheric impairments are dominant on Earth-space links only for lower frequencies (<1GHz). However, scintillations can be observed up to around 6 GHz at high latitudes or within ±20° of the geomagnetic equator. [7] For Earth-space links with elevation angles above 10°, attenuations may be significant only from rain, atmospheric absorption and possibly scintillation. However, in certain climatic zones, snow and ice accumulations on the surfaces of antenna reflectors and feeds can produce severe attenuation. In certain other climatic zones, attenuation by sand and dust storms may be significant. [2]

3.4.1   Atmospheric Absorption Losses

Attenuation due to absorption by dry air and water vapour is always present, and should be included in the calculation of total propagation loss at frequencies above 10 GHz. At a given frequency, the contribution of the dry-air is relatively constant, while both the density and the vertical profile of the water vapour are quite variable. Typically, the maximum

gaseous attenuation occurs during the season of maximum rainfall. [9]

For a terrestrial LOS path of length r (km), or for slightly inclined paths close to the ground, the path attenuation, A, due to atmospheric absorption may be written as: [4]

$$A = \gamma r = (\gamma_0 + \gamma_w) r \quad (dB) \qquad (3.54)$$

where r (km) is path length and γ (dB/km) denotes the sum of the specific attenuation of dry air γ_0 (dB/km) and of the water vapour γ_w (dB/km). The specific attenuations γ, γ_0 and γ_w (dB/km) are shown in Figure 3.23 as

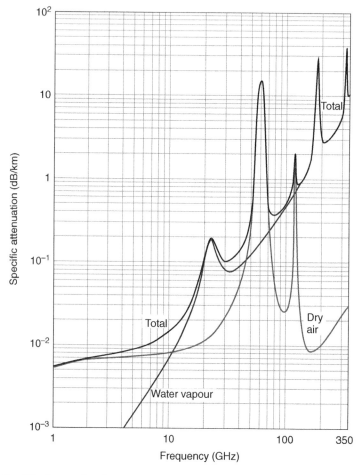

Figure 3.23   Specific Attenuation Due to Atmospheric Gases for Atmospheric Pressure $p = 1013$ hPa, Temperature $T = 15°C$ and Water Vapor Density $e = 7.5$ g/m^3. [9]

a function of frequency for atmospheric pressure $p = 1013$ hPa, temperature $T = 15°C$ and water vapor density $e = 7.5$ g/m^3. [9] The peak of the specific attenuation due to water vapor absorption occurs at 22 GHz and reaches a level ~0.2 dB/km. The peak of the oxygen absorption is located at 60 GHz with a specific attenuation ~15 dB/km; this implies 1.5 dB loss at 100 m for a LAN operating at 60 GHz. Figure 3.24 shows the total zenith attenuation at sea level, as well as the attenuation due to dry air and water vapour, using the mean annual global reference atmosphere ($p = 1013$ hPa, $T = 15°C$ and $e = 7.5$ g/m^3) [10].

Based on surface meteorological data, the path attenuation for Earth-space paths may be approximated using the cosecant law for elevation angles θ between 5° and 90°:

$$A = A_{zenith}/\sin\theta \ \ (dB) \quad\quad (3.55)$$

where A_{zenith} denotes the zenith attenuation. A more accurate approximation to the path attenuation may be obtained by integrating the water vapour content along the slant-path. [9]

Figure 3.25 shows that the atmospheric absorption losses given by (3.55) become more effective at higher frequencies and lower elevation angles, where the length of the slant path in the atmosphere becomes much larger.

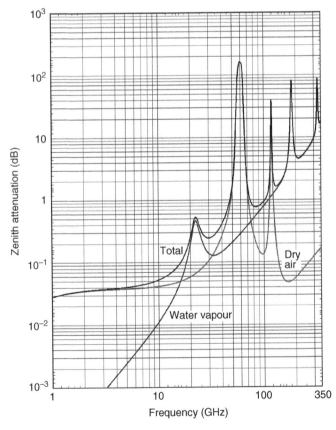

Figure 3.24 Total, Dry Air and Water-Vapour Attenuation at Sea Level for $p = 1013$ hPa, $T = 15°C$ and $e = 7.5$ g/m^3. [9]

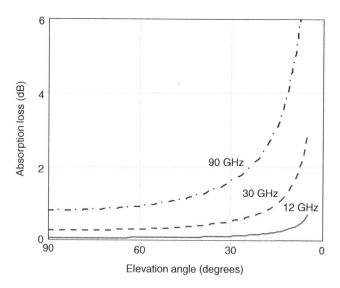

Figure 3.25 Atmospheric Absorption Loss Versus Elevation Angle at 12 GHz, 30 GHz and 90 GHz.

3.4.2   Rain Attenuation

3.4.2.1   Specific Attenuation

The specific attenuation γ_R (dB/km) due to rain is related to the rain rate R (mm/h) by

$$\gamma_R = k R^\alpha \quad (dB/km) \qquad (3.56)$$

One can determine the specific attenuation using the nomogram provided in Figure 3.26. For linear and circular polarizations, and for all path geometries, the coefficients k and α in (3.56) may also be determined at frequency f (GHz) using the following and Table 3.2: [11]

$$k = \frac{1}{2}\left[k_H + k_V + (k_H - k_V)\cos^2(\theta)\cos(2\tau)\right]$$

$$\alpha = \frac{1}{2k}\left[k_H\alpha_H + k_V\alpha_V + (k_H\alpha_H - k_V\alpha_V)\cos^2(\theta)\cos(2\tau)\right]$$

$$(3.57)$$

where θ denotes the elevation angle and τ is the polarization tilt angle relative to the horizontal. Hence, $\tau = 0°$ for horizontal polarization, $45°$ for circular polarization, and $90°$ for vertical polarization.

3.4.2.2   Long-Term Statistics of Rain Attenuation in Terrestrial LOS Paths

Absorption and scattering of electromagnetic waves by rain, snow, hail and fog lead to signal attenuation. The rain attenuation cannot be ignored at frequencies above 5 GHz. On paths at high latitudes or high altitude paths at lower latitudes, wet snow can cause significant attenuation over a larger range of frequencies. The following procedure, which is proposed in [4] for estimating the long-term statistics of rain attenuation, is considered to be valid in all parts of the world at least for frequencies up to 100 GHz and path lengths up to 60 km:

Step 1: Obtain the rainfall rate $R_{0.01}$ exceeded for 0.01% of the time (with an integration time of 1 min) from local sources (they generally have long-term measurements). Otherwise, an estimate can be obtained from Figure 3.27 for Europe and similar curves presented in [12], for other regions of the World. For example, Figure 3.27 shows the rain rate (mm/hr) exceeded for 0.01% of the average year over Europe.

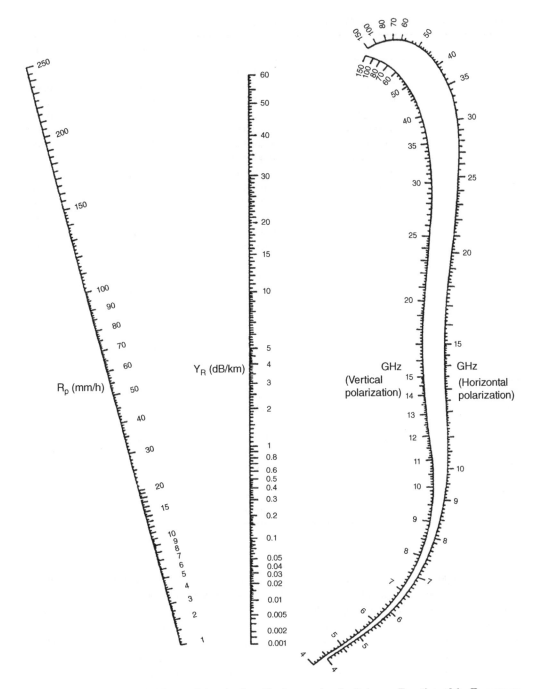

Figure 3.26 Nomogram for Determining the Specific Attenuation for Rain as a Function of the Frequency (GHz) and Rain rRate (mm/hr). [7]

Table 3.2 Frequency Dependent Coefficients for Estimating Specific Attenuation. [11]

Freq.(GHz)	k_H	k_V	α_H	α_V
1	0.0000387	0.0000352	0.9122	0.8801
1.5	0.0000868	0.0000784	0.9341	0.8905
2	0.0001543	0.0001388	0.9629	0.9230
2.5	0.0002416	0.0002169	0.9873	0.9594
3	0.0003504	0.0003145	1.0185	0.9927
4	0.0006479	0.0005807	1.1212	1.0749
5	0.001103	0.0009829	1.2338	1.1805
6	0.001813	0.001603	1.3068	1.2662
7	0.002915	0.002560	1.3334	1.3086
8	0.004567	0.003996	1.3275	1.3129
9	0.006916	0.006056	1.3044	1.2937
10	0.01006	0.008853	1.2747	1.2636
12	0.01882	0.01680	1.2168	1.1994
15	0.03689	0.03362	1.1549	1.1275
20	0.07504	0.06898	1.0995	1.0663
25	0.1237	0.1125	1.0604	1.0308
30	0.1864	0.1673	1.0202	0.9974
35	0.2632	0.2341	0.9789	0.9630
40	0.3504	0.3104	0.9394	0.9293
45	0.4426	0.3922	0.9040	0.8981
50	0.5346	0.4755	0.8735	0.8705
60	0.7039	0.6347	0.8266	0.8263
70	0.8440	0.7735	0.7943	0.7948
80	0.9552	0.8888	0.7719	0.7723
90	1.0432	0.9832	0.7557	0.7558
100	1.1142	1.0603	0.7434	0.7434
120	1.2218	1.1766	0.7255	0.7257
150	1.3293	1.2886	0.7080	0.7091
200	1.4126	1.3764	0.6930	0.6948
300	1.3737	1.3665	0.6862	0.6869
400	1.3163	1.3059	0.6840	0.6849

Step 2: Compute the specific attenuation γ_R (dB/km) for the frequency, polarization and rain rate of interest using (3.56), (3.57) and Table 3.2 or directly from Figure 3.26.

Step 3: Compute the effective path length, $r_{eff} = \mu\, r$, of the link by multiplying the actual path length r (km), by a distance factor μ. An estimate of this factor is given by

$$\mu = \frac{1}{0.477\, r^{0.663}\, R_{0.01}^{0.073}\, \alpha f^{0.123} - 10.579(1 - e^{-0.024\, r})} \tag{3.58}$$

where f (GHz) is the frequency, α is the exponent in the specific attenuation model in (3.56) and $R_{0.01}$ denotes the point rainfall rate for the location for 0.01% of an average year (mm/h). Maximum recommended μ is 2.5, so if the denominator of equation (3.58) is less than 0.4, use $\mu = 2.5$.

Step 4: Path attenuation exceeded for 0.01% of the time is estimated by

$$A_{0.01} = \gamma_R r_{eff} = \gamma_R \mu\, r \quad (dB) \tag{3.59}$$

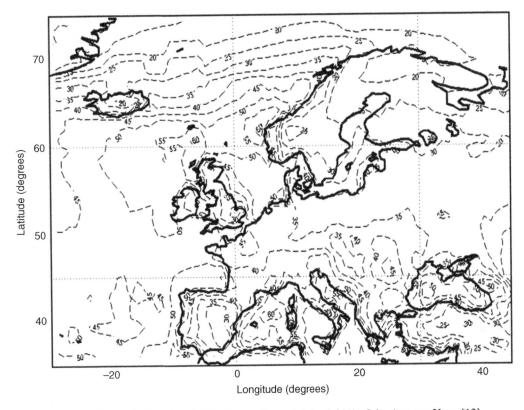

Figure 3.27 Rain Rate (mm/hr) in Europe Exceeded for 0.01% of the Average Year. [12]

Step 5: The attenuation exceeded for other percentages of time in the range $0.001\% \leq p \leq 1\%$ is determined from the following: [4]

$$A_p/A_{0.01} = C_1 p^{-(C_2 + C_3 \log_{10} p)} \qquad (3.60)$$

where

$$C_1 = 0.07^{C_0} 0.12^{(1-C_0)}$$

$$C_2 = 0.855 C_0 + 0.546(1 - C_0)$$

$$C_3 = 0.139 C_0 + 0.043(1 - C_0)$$

$$C_0 = \begin{cases} 0.12 + 0.4 \log_{10}(f/10)^{0.8} & f \geq 10 \text{ GHz} \\ 0.12 & f < 10 \text{ GHz} \end{cases}$$

$$(3.61)$$

3.4.2.3 Frequency Scaling of Rain Attenuation

If reliable long-term attenuation statistics are available at an elevation angle and a frequency different from those for which prediction is needed, the average attenuation statistics may be predicted accurately by scaling the available data to the elevation angle and the frequency. Frequency scaling is used to predict the rain statistics at one frequency using the statistics available for a different frequency. The ratio between the rain attenuation at two frequencies can vary during a rain event, and the variability of the ratio generally increases as the rain attenuation increases.

First of the two methods predicts the statistics of the rain attenuation at frequency f_2 conditioned on the rain attenuation at frequency f_1. This method requires the cumulative distributions of rain attenuation at both frequencies. A second method predicts the equiprobable rain attenuation at frequency f_2 conditioned on the rain attenuation at frequency f_1. This simple method does not require the cumulative distribution of the rain attenuation at either frequency.

When reliable long-term statistics of rain attenuation are available at one frequency f_1, attenuation statistics in the same climatic region for another frequency f_2 may be estimated in the range 7 to 50 GHz as follows: [4]

$$A_2 = A_1 \left(\Phi(f_2) / \Phi(f_1) \right)^{1-H(f_1,f_2,A_1)}$$

$$\Phi(f) = \frac{f^2}{1 + 10^{-4} f^2}$$

$$H(f_1, f_2, A_1) = 1.12 \times 10^{-3} \left(\Phi(f_2) / \Phi(f_1) \right)^{0.5} \left(\Phi(f_1) A_1 \right)^{0.55}$$

$$(3.62)$$

Here, A_1 and A_2 denote the equiprobable values of the rain attenuation at frequencies f_1 and f_2 (GHz), respectively.

These prediction methods may be applicable to uplink power control and adaptive coding and modulation (ACM). Uplink power control is used to adaptively change the transmit power so as to compensate for time-variations in the uplink attenuation. ACM adaptively changes the modulation alphabet and the code rate so as to make best use of the SNR variations. For example, the first method predicts the instantaneous uplink rain attenuation at frequency f_2 based on the measured instantaneous downlink rain attenuation at frequency f_1 for a p % probability that the actual uplink rain attenuation will exceed the predicted value. The second method predicts the uplink rain attenuation at frequency f_2 based on knowledge of the downlink rain attenuation at frequency f_1 at the same probability of exceedance. [2]

3.4.2.4 Depolarization Due to Rain

In addition to causing attenuation and increasing the receiver noise, rain also causes the depolarization of electromagnetic waves. Raindrops flatten and look like oblate spheroids with their major axis nearly horizontal as they fall. The components of electromagnetic waves polarized along minor and major axes

of raindrops are not attenuated by the same amount. Consequently, the relative intensities in two orthogonal directions would be changed after propagation in a rainy region. This would lead to depolarization of electromagnetic waves. Attenuation statistics is already predicted in the previous section. However, depolarization and scattering of electromagnetic waves depend on the amount of raindrops in the propagation path, their size, shape and orientation distribution. [13] Now assume that transmitted electric field is polarized along x-direction. In addition to a sufficiently strong co-polar (CP) component in the x-direction, the received electric field will also have a cross-polar (XP) component along the y-direction. Since the receiver antenna will be designed to receive co-polar component, the XP component will represent a loss in the link budget. In dual-polarized systems, that is, systems using two orthogonal polarizations for communications, the XP component will will interfere with the CP channel.

Cross-polar discrimination (XPD) is defined as the ratio of the powers of the CP signals to the XP signals. In the absence of rain depolarization of the received electromagnetic waves may be due to transmit and/or receive antennas and electrical characteristics, size, shape and orientation of the scatterers in the channel. Nevertheless, at frequencies above 10 GHz, where rain attenuation may be nonnegligible, free-space communication conditions exist and scatterers may not play a significant role in the depolarization of electromagnetic waves. Design of reflector antennas above 10 GHz allows XPD levels higher than 25 dB. Therefore, depolarization of electromagnetic waves and accompanying degradation of XPD due to intense rain is a serious issue in terrestrial links, even if it occurs for small percentages of time.

A rough estimate of the distribution of XPD can be obtained from the distribution of the co-polar attenuation for rain using the equiprobability relation: [4]

$$XPD_p = U(f) - V(f)\log_{10}(A_p) \quad (dB) \quad (3.63)$$

where the co-polar attenuation (CPA) A_p (expressed in dB) exceeded for p % of the time is predicted by (3.59) and (3.60). For LOS paths with small elevation angles and horizontal or vertical polarization, the coefficients $U(f)$ and $V(f)$ may be approximated by:

$$U(f) = U_0 + 30\log(f_{GHz})$$

$$V(f) = \begin{cases} 12.8 f_{GHz}^{0.19} & 8 \le f \le 20\,\text{GHz} \\ 22.6 & 20 < f \le 35\,\text{GHz} \end{cases} \quad (3.64)$$

An average value of U_0 of about 15 dB, with a lower bound of 9 dB for all measurements, has been obtained for attenuations greater than 15 dB. The difference between the CPA values for vertical and horizontal polarizations is reported to be insignificant when evaluating XPD. The user is advised to use the value of CPA for circular polarization when working with (3.63). [4]

Long-term XPD statistics obtained at one frequency can be scaled to another frequency using the following:

$$XPD_2 = XPD_1 - 20\log(f_2/f_1), \quad 4 \le f_1, f_2 \le 30\,\text{GHz} \quad (3.65)$$

where XPD_1 and XPD_2 are the XPD values not exceeded for the same percentages of time at frequencies f_1 and f_2. $V(f)$ is least accurate for large differences between the respective frequencies. It is most accurate if XPD_1 and XPD_2 correspond to the same polarization (horizontal or vertical).

Example 3.6 Rain Attenuation and XPD in Terrestrial Links.

Long-term rain statistics at a given location show that the rain does not exceed 32.5 mm in one hour. This corresponds to a rain rate of

$R = 32.5$ mm/hr for $1/(24 \times 365) \approx 0.01\%$ of the year. Consider now a LOS radio-link system with vertical polarization operating over a horizontal path of range $r = 30$ km. The factors α_V and k_V and the corresponding specific attenuation for vertical polarization are approximated from the curves provided in [11] as follows:

$$k_V = 1.82 - 0.555\log_{10}(f_{GHz})$$

$$\alpha_V = 10^{-5.075 + 2.962\log_{10}(f_{GHz})}$$

The rain attenuation exceeded for 0.01% of the time is given by (3.59)

$$A_{0.01} = k_V R_{0.01}^{\alpha_V} \mu r$$

where the distance factor μ is given by (3.58). The rain attenuation exceeded for 0.1% of the time is obtained using (3.59). Figure 3.28a clearly shows that the rain attenuation rapidly increases with frequency. The rain attenuation that will be exceeded for 0.01 and 0.1% of the time at 12 GHz is found to be 9.3 dB and 3.5 dB, respectively. This implies that the rain attenuation at 12 GHz will be less than 9.3 dB for 99.99% of the time and 3.5 dB for 99.9% of the time. This radio-link system must employ 9.3 dB rain margin in order to limit the rain outage duration to less than one hour per year. The required 9.3 dB rain margin (higher EIRP) for providing 99.99% availability against rain attenuation may not be cost-effective. Instead, telecommunication systems use automatic power control in real time for mitigating rain attenuation.

The XPD curves corresponding to rain attenuations exceeded for 0.01% and 0.1% of time are shown in Figure 3.28b. Here, the depolarization due to transmit and receive antennas is ignored. Increase in the rain attenuation with frequency clearly deteriorates the XPD and this becomes a serious source of concern.

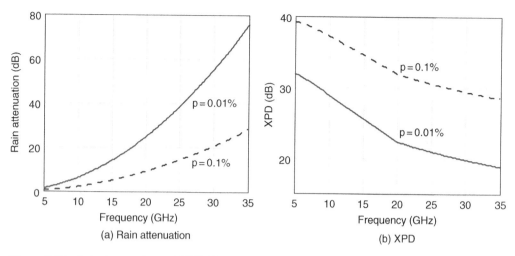

Figure 3.28 Rain Attenuation and XPD Versus Frequency for a Rain Rate of 32.5 mm/hr for 0.01% of the Year.

3.4.2.5 Rain Attenuation in Earth-Space Paths

The following procedure by [2] provides estimates of the long-term statistics of the slant-path rain attenuation using point rainfall rate at a given location for frequencies up to 55 GHz. The following parameters are required (see Figure 3.29):

h_s: Earth station height above mean sea level (km)

θ: elevation angle of the Earth station (degrees)

θ_E: latitude of the Earth station (+ for north, – for south) (degrees)

ϕ_S, ϕ_E: longitude (+ for east, – for west) of the satellite and the Earth station, respectively.

f: frequency (GHz)

r_e: effective radius of the Earth, accounting for tropospheric refraction ($r_e = 8500$ km for $h_s \leq 1$ km).

Step 1: For regions where no specific information is available, the rain height, h_R, may be determined using the mean annual 0° C isotherm height above mean sea level that is provided in [14], where h_0 is also available.

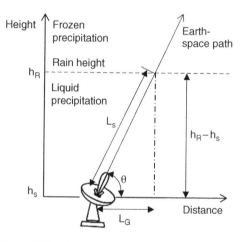

Figure 3.29 Parameters For the Earth-Space Path Used for the Attenuation Prediction Process.

The mean annual rain height above mean sea level, h_R, is given by [7].

$$h_R = h_0 + 0.36 \quad (km)$$

$$= \begin{cases} 5 - 0.075(\theta_E - 23) & \theta_E > 23° & \text{Northern hemisphere} \\ 5 & 0 \leq \theta_E \leq 23° & \text{Northern hemisphere} \\ 5 & -21° \leq \theta_E \leq 0 & \text{Southern hemisphere} \\ 5 + 0.1(\theta_E + 21) & -71° \leq \theta_E \leq -21° & \text{Southern hemisphere} \\ 0 & \theta_E \leq -71° & \text{Southern hemisphere} \end{cases}$$

$$(3.66)$$

Step 2: The elevation angle θ and the distance d_S to a geostationary satellite located at a longitude of ϕ_S are given by [13]

$$\theta = \cos^{-1}\left[\frac{\sin\gamma}{(1.02274 - 0.301596\cos\gamma)^{1/2}}\right]$$

$$d_S = 42242(1.02274 - 0.301596\cos\gamma)^{1/2} \quad (km)$$

$$\gamma = \cos^{-1}[\cos\theta_E\cos(\phi_E - \phi_S)]$$

$$(3.67)$$

Step 3: The slant-path length, L_s (km), below the rain height (see Figure 3.29) is given by

$$L_s = \begin{cases} (h_R - h_s)/\sin\theta & \theta \geq 5° \\ \dfrac{2(h_R - h_s)}{\left(\sin^2\theta + 2(h_R - h_s)/r_e\right)^{1/2} + \sin\theta} & \theta < 5° \end{cases} \quad (km)$$

$$(3.68)$$

If $h_R - h_s$ is less than or equal to zero, the predicted rain attenuation for any time percentage is zero and the following steps are not required.

Step 4: Obtain the rainfall rate, $R_{0.01}$, exceeded for 0.01% of an average year (with an integration time of 1 min) using the same procedure followed for terrestrial LOS links. If $R_{0.01}$ is equal to zero, the predicted rain attenuation is zero for any time percentage and the following steps are not required.

Step 5: Specific attenuation, γ_R, is obtained inserting α and k given by (3.57) and Table 3.2 and the rainfall rate, $R_{0.01}$ into (3.56):

$$\gamma_R = k R_{0.01}^\alpha \quad (dB/km) \qquad (3.69)$$

One may also use the nomogram shown in Figure 3.26 for determining the specific attenuation.

Step 6: The effective path length is given by

$$L_E = \begin{cases} L_s\mu_{0.01}v_{0.01} & \zeta > \theta \\ v_{0.01}(h_R - h_s)/\sin\theta & otherwise \end{cases} \quad (km)$$

$$\zeta = \tan^{-1}\left(\frac{h_R - h_s}{L_s\cos\theta\mu_{0.01}}\right) \quad (degrees)$$

$$(3.70)$$

where $\mu_{0.01}$ and $v_{0.01}$ denote respectively horizontal and vertical reduction factors for 0.01% of the time:

$$\mu_{0.01} = \left[1 + 0.78\sqrt{L_s\gamma_R\cos\theta/f_{GHz}} \right.$$

$$\left. -0.38\left(1 - e^{-2L_s\cos\theta}\right)\right]^{-1}$$

$$(3.71)$$

$$v_{0.01} = \left[1 + 31\left(1 - e^{-\theta/(1+\chi)}\right)\sqrt{L_R\,\gamma_R\sin\theta}/f_{GHz}^2 \right.$$

$$\left. -0.45\sqrt{\sin\theta}\right]^{-1}$$

$$(3.72)$$

$$|\chi| = \begin{cases} 36 - |\theta_E| & |\theta_E| < 36° \\ 0 & otherwise \end{cases} \quad (deg.)$$

$$(3.73)$$

$$L_R = \begin{cases} L_s\mu_{0.01} & \varsigma > \theta \\ (h_R - h_s)/\sin\theta & otherwise \end{cases}$$

Step 7: The predicted attenuation exceeded for 0.01% of an average year is obtained from:

$$A_{0.01} = \gamma_R L_E \quad (dB) \qquad (3.74)$$

where L_E denotes the effective path length.

Step 8: The estimated attenuation to be exceeded for other percentages of an average year, in the range $0.001\% \leq p \leq 5\%$, is determined from the attenuation to be exceeded for 0.01% for an average year:

$$A_p = A_{0.01}\left(\frac{p}{0.01}\right)^{-(0.665 + 0.033\ln p - 0.045\ln A_{0.01} - \beta(1-p)\sin\theta)} \quad (dB)$$

$$(3.75)$$

where

$$\beta = \begin{cases} 0 & p \geq 1\% \, or \, |\theta_E| \geq 36° \\ -0.005(|\theta_E| - 36) & p < 1\%, \, |\theta_E| \geq 36°, \, \theta \geq 25° \\ -0.005(|\theta_E| - 36) + 1.8 - 4.25\sin\theta & otherwise \end{cases}$$

$$(3.76)$$

In [7], a simpler approach is proposed to estimate the attenuation to be exceeded for 0.01% for an average year:

$$A_{0.01} = \frac{\gamma_R L_s}{1 + L_s \cos\theta/L_0} \quad (dB)$$

$$L_0 = \begin{cases} 35 e^{-0.015 R_{0.01}} & R_{0.01} \leq 100 mm/hr \\ 35 e^{-1.5} & R_{0.01} > 100 mm/hr \end{cases}$$

$$(3.77)$$

The attenuation to be exceeded for other percentages A_p of the year in the range 0.001% to 1% is estimated from the attenuation to be exceeded for 0.01% for an average year by using:

$$A_p = 0.12 A_{0.01} p^{-(0.546 + 0.043 \log_{10} p)} \quad (3.78)$$

Note that (3.78) yields $A_p/A_{0.01} = 0.12$, 0.38, 1.0 and 2.14 for 1%, 0.1%, 0.01% and 0.001% respectively.

When comparing measured statistics with the above prediction, based on long-term statistics of rain attenuation, allowance should be made for large year-to-year variability in the rainfall rate statistics. [15][16]

Example 3.7 Rain Attenuation and Rain Margin in SATCOM Links.

Consider that an Earth-based TV receiver, with latitude $\theta_E = 40°$, longitude $\phi_E = 33°E$ and an altitude of $h_s \approx 0.9$ km above the sea level, receives circularly polarized digital TV signals at 12 GHz, broadcasted by a geostationary satellite located a longitude of $\phi_S = 7°E$. Long-term rain statistics at the receiver site show that the rain does not exceed 32.5 mm in one hour. This corresponds to a rain rate of $R = 32.5$ mm/hr for $1/(24 \times 365) \approx 0.01\%$ of the year. The factors α and k and the corresponding specific attenuation, exceeded for 0.01% of the time, are found using (3.56), (3.57) and Table 3.2 as follows:

$$k = (k_H + k_V)/2 = 0.01781$$
$$\alpha = (\alpha_H k_H + \alpha_V k_V)/(2k) = 1.209$$
$$\gamma_R = k R^\alpha = 1.197 \, dB/km$$

Using (3.67), the elevation angle to the geostationary satellite is found to be $\theta \approx 37°$. Assuming $h_R = 3.6$ km,[14] and using (3.66)–(3.74), we find $\mu_{0.01} = 0.935$, $\nu_{0.01} = 1.218$, $L_E = 4.71 km$ and $A_{0.01} = 5.63 dB$. Hence, the rain attenuation to be exceeded for 0.01% of the year is estimated to be 5.63 dB. This DVB-S system limits outages to 0.01% of the time or, in other words, ensures an availability of 99.99% of the time if it employs a rain margin of 5.63 dB. This value of the rain attenuation is evidently much lower than that for terrestrial systems. Similarly, the reduction in XPD in SATCOM links due to rain attenuation is not as serious as in terrestrial links. SATCOM systems usually use real-time adaptive power control to mitigate signal attenuation due to atmospheric precipitation, for example, rain, snow and fog.

3.5   Tropospheric Refraction

Troposphere is part of the atmosphere of 5–6 km thickness above the Earth's surface and refracts (bends) electromagnetic waves as they propagate through it. Refraction is caused by changes in the refractive index n of the troposphere with altitude as a function of temperature T in K, pressure P in mb, and water vapour partial pressure e in mb. Since changes in the refractive index n with these parameters is very small, we use the refractivity N instead of the refractive index n:

$$N = (n-1) \times 10^6 = 77.6 \frac{P}{T} + 3.73 \times 10^5 \frac{e}{T^2}$$

$$(3.79)$$

Refractivity is dimensionless, but 'N units' is used. For example, for $P = 1000$ mb (sea level), $e = 11$ mb, and $T = 290$ K (17 C), we find the refractivity on the surface of the Earth as $N_s =$ 316. As shown in Figure 3.30, the variation of the pressure P, the temperature T, and the water vapour partial pressure e with altitude results in an exponential decrease of N, with height h above the Earth: [17]

$$N(h) = N_s \exp(-h/h_0) \quad (N \text{ units}) \quad (3.80)$$

where $N_s = 316$ is the surface refractivity and $h_0 = 7.35$ km denotes the scale height (see Figure 3.30). Over the first km above Earth,

$N(h)$, for the so-called standard atmosphere, is approximately linear and has a slope

$$\Delta N = \frac{dN}{dh}\bigg|_{h \leq 1 \, km} = N(1) - N(0) = -40(N \, units/km) \quad (3.81)$$

Refraction of electromagnetic waves may be visualized in a stratified troposhere as shown in Figure 3.31. Assume that the troposphere is stratified into layers, each with a refractive index decreasing with altitude $n_0 > n_1 > n_2 > \ldots$. At the

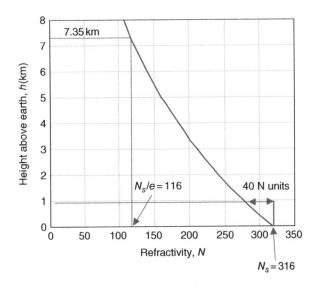

Figure 3.30 Standard Refractivity Versus Height Above Earth.

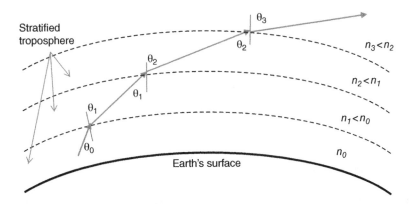

Figure 3.31 Refraction of Electromagnetic Waves in Stratified Troposphere.

interface between i^{th} and $(i+1)^{\text{th}}$ layers, the waves are incident with an angle θ_i and are refracted with an angle θ_{i+1}, measured from the zenith. According to the Snell's law, one may write

$$n_0 \sin\theta_0 = n_1 \sin\theta_1 = n_2 \sin\theta_2 = \cdots \quad (3.82)$$

Since the refraction index decreases with height above Earth surface, the waves refract (bend) towards the horizontal direction as they propagate upward in the troposhere:

$$n_{i+1} < n_i \;\Rightarrow\; \sin\theta_{i+1} > \sin\theta_i \;\Rightarrow\; \theta_{i+1} > \theta_i \quad (3.83)$$

Hence, the slow decrease of the refractive index with height causes waves to propagate, not in straight lines, but along circular arcs with radius of curvature ρ [17]

$$\frac{1}{\rho} = -\frac{1}{n}\frac{dn}{dh}\cos\psi \quad (3.84)$$

where ψ denotes the incidence angle of ray to the local horizontal plane. Using (3.79) and (3.81), one gets

$$\frac{dn}{dh} = 10^{-6}\frac{dN}{dh} = -40 \times 10^{-6} \;\; (units/km) \quad (3.85)$$

For terrestrial links in standard atmosphere we can assume $\psi \cong 0$ degrees and $n \approx 1$. Inserting (3.85) into (3.84), the radius of curvature of the refracted rays in the troposhere is found to be

$$\rho \cong \frac{1}{-dn/dh} = \frac{10^6}{40} = 25000\,km \quad (3.86)$$

As shown in Figure 3.32a, electromagnetic waves propagate along a curved path in the troposhere with a radius of curvature of 25000 km. It is evidently not easy to account for the propagation of electromagnetic waves along a curved path with a radius of curvature of 25000 km above the spherical Earth surface with 6371 km radius. The propagation of electromagnetic waves along a curved path due to troposheric refraction is usually accounted for using the so-called straight line model. In this model for the standard atmosphere $(dN/dh = -40)$, electromagnetic waves propagate along straight lines over the surface of the Earth $(h < 1\,km)$

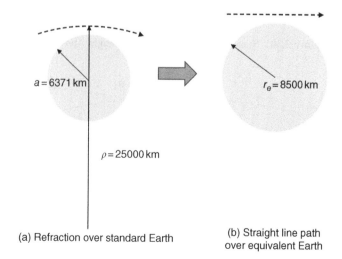

$a = 6371\ km$

$\rho = 25000\ km$

$r_e = 8500\ km$

(a) Refraction over standard Earth

(b) Straight line path over equivalent Earth

Figure 3.32 Straight Line Ray Path Model Over Spherical Earth Surface with an Equivalent Radius of $r_e = 8500$ km to Account for Troposheric Refraction.

with an equivalent Earth radius r_e (see Figure 3.32b) [17]

$$\frac{1}{r_e} = \frac{1}{a} + \frac{dn}{dh} \Rightarrow$$

$$k = \frac{r_e}{a} = \frac{1}{1 + a(dn/dh)} = \frac{1}{1 - 6371 \times 40 \times 10^{-6}} \cong \frac{4}{3}$$

$$r_e = ka = \frac{4}{3} \times 6371\,km = 8500\,km$$

(3.87)

In the absence of the tropospheric refraction, electromagnetic waves would propagate along straight lines on the surface of the Earth with physical radius of 6371 km ($k = 1$). However, tropospheric refraction forces electromagnetic waves to propagate along straight lines over the Earth's surface with an effective radius of $r_e = 8500$ km. Atmospheric conditions may sometimes affect the slope of the refraction index to be more or less than that for the standard atmosphere ($dN/dh = -40$, $k = 4/3$). As shown in Figure 3.33, for $dN/dh > -40$ (sub-refraction), electromagnetic waves propagate along a upward-curved path over the Earth's surface of radius r_e. For $dN/dh = -157$, the effective radius of the Earth would be equal to $r_e = \infty$, implying that electromagnetic waves would propagate along straight lines on the surface of the Earth having an equivalent radius of infinity. For $-157 < dN/dh < -40$ (super-refraction), electromagnetic waves propagate along a downward-curved path over the Earth's surface with an effective radius of r_e. In case

when $dN/dh < -157$ ($k < 0$), the refraction leads to a phenomenon called ducting, where electromagnetic waves are guided in a parallel-plate waveguide formed by the Earth's surface and a certain height of the troposphere. Propagation in ducts leads to signal reception at unusually long ranges due to very low signal attenuation. Tropospheric refraction can therefore be accounted for by assuming that electromagnetic waves propagate over the Earth's surface with an equivalent radius determined by (3.87) as a function of the refractive index gradient.

3.5.1 Ducting

In certain geographical regions, the index of refraction may occasionally have a rate of decrease with a slope less than $dN/dh < -157$ N/km ($k = \infty$) over short distances. This leads to the formation of a duct which may potentially trap electromagnetic waves between the Earth's surface and a specified height of the troposphere (see Figure 3.34). Nevertheless, the presence of a duct does not necessarily imply efficient coupling of the signal energy. Ducting is possible if both transmit and receive antennas are located within the duct for effective coupling, which requires sufficiently low elevation angles (typically a small fraction of a degree), sufficiently high frequency and a sufficiently thick ducting layer. Efficiently coupled waves propagate in

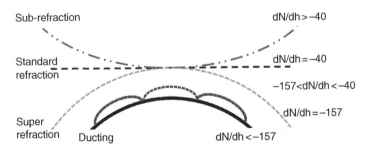

Figure 3.33   Ray Trajectories Over the Earth's Surface With an Equivalent Radius of $r_e = 8500$ km.

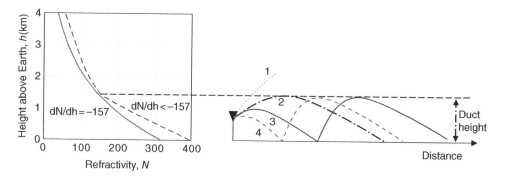

Figure 3.34 Trapping of Electromagnetic Waves in a Surface Duct.

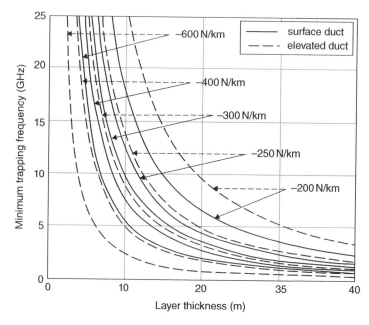

Figure 3.35 Minimum Frequency for Trapping in Ducts of Constant Refractivity Gradients. [17]

ducts over long distances with much less attenuation compared to free space propagation. Figure 3.35 shows the variation of the minimum trapping frequency as a function of the duct thickness for several values of the refraction index gradient in surface and elevated ducts. Below minimum trapping frequency, increasing amounts of signal energy will leak through the duct boundaries. Minimum trapping frequency strongly depends

and is inversely proportional to the duct thickness and magnitude of the refractive index gradient. For example, for $dN/dh = -200$ N/km, minimum trapping frequencies of about 18 GHz and 2.5 GHz are required for duct thicknesses of 10 m and 40 m, respectively. The corresponding minimum trapping frequencies for −400 N/km are approximately 8GHz and 1GHz (see Figure 3.35). For terrestrial systems operating typically in 8–16 GHz, a ducting layer of about 5–15 m minimum thickness is required. [17]

Ducts are formed primarily by fast changes with altitude of the water vapour content of the troposphere due to its stronger influence on the index of refraction. Therefore, ducts usually occur over large bodies of water. Ducts can be near Earth's surface (surface ducts) and/or at some altitudes (elevated ducts). Ground ducts are produced by a mass of warm air arriving over a cold ground or sea, night frosts, and high humidity in lower troposphere. Elevated ducts are formed by subsistence of an air mass in a high pressure area. They mainly occur above clouds and interfere with communications between aircraft and ground. Ducts are also potential sources of interference to other services and may cause multipath interference. Surface ducts often cause TV transmissions to be received at locations several hundreds of km away.

3.5.2 Radio Horizon

Radio horizon is defined as the maximum LOS range over smooth spherical Earth between transmit and receive antennas of heights h_T and h_R, respectively. The radio horizon may be determined from Figure 3.16 when the LOS line is tangent to the Earth surface with an equivalent radius of $r_e = 8500$ km:

$$r_1^2 = (h_T + r_e)^2 - r_e^2 = h_T^2 + 2h_T r_e \cong 2h_T r_e$$

$$r_2^2 = (h_R + r_e)^2 - r_e^2 = h_R^2 + 2h_R r_e \cong 2h_R r_e$$

$$r = r_1 + r_2 = \sqrt{2r_e}\left(\sqrt{h_T} + \sqrt{h_R}\right)$$

$$(3.88)$$

Inserting $r_e = 8500$ km into (3.88), the radio horizon (maximum LOS distance) may be rewritten as

$$r_{km} = \sqrt{17h_T} + \sqrt{17h_R} = 4.12\sqrt{h_T} + 4.12\sqrt{h_R}$$

$$(3.89)$$

where h_T and h_R are expressed in meters. For example, $r_{km} = 18.2$ for $h_T = 10$ m and $h_R = 1.6$ m. Elevated antennas with $h_T = h_R = 50$ m increase the radio horizon distance to $r_{km} = 58$.

3.6 Outdoor Path-Loss Models

Determination of the received signal power in VHF and UHF bands requires firstly the identification of the obstacles in the first Fresnel zone. Secondly, ray-tracing is accomplished between transmit and receive antennas to identify possible propagation paths via these obstacles. All these depend on transmit and receive antenna heights, frequency of operation, curvature, roughness and vegetative cover of the Earth's surface, as well as the type and density of obstacles. This time-consuming and expensive process is deterministic and should be repeated for each propagation path depending on the topography of the Earth's surface at that specific site.

Instead of using the deterministic ray-tracing techniques mentioned above, radio propagation models are derived using a combination of analytical and empirical methods for predicting the received signal power in a generic propagation environment. The empirical approach is based on curve fitting or developing analytic expressions to interpolate a set of measured data. Hence, all propagation factors are implicitly taken into account through actual measurements. The validity of an empirical model may be limited to operating frequencies or propagation environments used to derive the model. Path-loss models, still evolving in time, are used to predict indoor and outdoor coverage for mobile communication systems in terms of the received signal level as a function of distance, antenna heights, frequency of operation, and the density/height of obstacles in the propagation environment.

Received signal power in free space, given by (2.114), may be rewritten as

$$P(r) = P_t G_t G_r \left(\frac{\lambda}{4\pi r_0}\right)^2 \left(\frac{r_0}{r}\right)^2$$

$$= P(r_0) \left(\frac{r_0}{r}\right)^2 \propto \frac{1}{r^2} \tag{3.90}$$

where $P(r_0)$ denotes the received signal power at a reference distance r_0. In free space propagation, the received signal is proportional to $1/r^2$. When only direct and Earth-reflected rays reach the receiver, (3.43) and (3.45) give the received signal power:

$$P(r) = \frac{P_t G_t G_r}{(4\pi r/\lambda)^2} 4 \sin^2 \left(\frac{2\pi h_T h_R}{\lambda r}\right)$$

$$\propto \begin{cases} 1/r^2 & r < r_{max} \\ 1/r^4 & r > r_{max} \end{cases} \tag{3.91}$$

where $r_{max} = 4 h_T h_R / \lambda$ as given by (3.44). Hence, the received signal power decreases as $1/r^2$ for distances shorter than r_{max} and as $1/r^4$ at longer distances.

In view of the above, one may adopt a simple model for received signal power in a multipath propagation environment:

$$P(r) = P(r_0) (r_0/r)^n$$

$$P(r)_{dB} = P(r_0)_{dB} + 10 n \log(r_0/r) \tag{3.92}$$

where

$P(r)$, $P(r_0)$: ensemble average of all possible received signal powers at distances r and r_0, respectively.

r_0: the close-in reference distance which is determined from measurements close to but in the far-field of the transmitter. $r_0 = 1$ km is commonly used for macro-cellular systems and $r_0 = 100$ m–1 m in microcellular systems.

n: the path-loss exponent which describes the rate at which the path-loss increases with distance. The path-loss, which has a slope of $-10n$ dB/decade, depends on the

Table 3.3 Typical Values of the Path-Loss Exponent for Various Propagation Environment.

Environment	Path-loss exponent, n
Free space	2
Urban cellular radio	2.7–3.5
Shadowed urban cellular	3–5
Indoor (LOS)	1.6–1.8
Indoor (obstructed)	4–6
In factory (obstructed)	2–3

considered propagation environment (see Table 3.3).

3.6.1 Hata Model

When the assumption of free space propagation does not hold, a realistic path-loss model is needed which would accurately predict the path loss for the propagation scenarios usually encountered in practice. For estimating the path loss in wireless communication systems, one must consider the effects of range, the frequency, transmit and receive antenna heights, and the irregular terrain, that is, the Earth's curvature, its electrical characteristics, the terrain profile (roughness), and the presence of buildings, trees and other obstacles. Therefore, models which are more accurate than those given by (3.90)–(3.92) are needed. Since the terrain irregularities may change considerably in urban, suburban and rural areas, the path-loss models should distinguish between them. On the other hand, the path-loss is also affected by atmospheric losses and attenuation due to precipitation (rain, snow, etc.), which show variability with location. The losses are not considered in the path-loss models but are determined separately and added to those predicted by the path-loss models outlined below.

The widely used empirical model by Hata is originally developed for 2G cellular radio systems. It provides a simple and practical path loss prediction for narrowband cellular mobile systems. Hata model is valid for the following ranges of the considered parameters:

Carrier frequency f_c in MHz:	150–1500 MHz
Transmitter antenna height h_T in meters:	30–200 m
Receiver antenna height h_R in meters:	1–10 m
Range r in km:	1–20 km

In this model the median path loss is used since it is easier to determine using the measured data compared to the mean value. The median value can be determined by ordering all the measurement values from the highest to lowest and picking the middle one. Hence, the median value separates the higher half of the measured data from the lower half. If there is an even number of observations, then the median value is usually defined to be the mean of the two middle values. Median path loss (exceeded for 50 % of the time) in *urban area* is predicted by

$$L_{50}(urban) = 69.55 + 26.16 \log f_c - 13.82 \log h_T$$
$$- a(h_R) + (44.9 - 6.55 \log h_T) \log r$$
$$(3.93)$$

Here $a(h_R)$ denotes the correction factor for mobile antenna height. For a *small to medium sized city*, it is given by

$$a(h_R)_{dB} = (1.1 \log f_c - 0.7) h_R - (1.56 \log f_c - 0.8)$$
$$(3.94)$$

For a *large city*:

$$a(h_R)_{dB} = \begin{cases} 8.29(\log(1.54 h_R))^2 - 1.1 & f_c \le 200\,MHz \\ 3.2(\log(11.75 h_R))^2 - 4.97 & f_c > 200\,MHz \end{cases}$$
$$(3.95)$$

The median path loss in a *suburban area* is predicted by

$$L_{50}(suburban) = L_{50}(urban)$$
$$- 2[\log(f_c/28)]^2 - 5.4\ dB$$
$$(3.96)$$

The median path loss in *open rural area* is found as

$$L_{50}(rural) = L_{50}(urban) - 4.78(\log f_c)^2$$
$$+ 18.33 \log f_c - 40.94\ (dB)$$
$$(3.97)$$

Predictions by the Hata model compare very closely with the original Okumura model, as long as the distance, r, exceeds 1 km. Due to its slow response to rapid changes in terrain, the model is good in urban and suburban areas, but may not be as good in rural areas. Note that the received signal power is found from (2.114) by replacing the free-space propagation loss with the path loss as found above.

Figure 3.36 shows the path loss in urban, suburban and rural areas at 900 MHz and 1800 MHz for $h_T = 30$ m and $h_R = 1.7$ m. The path loss difference between 900 MHz and 1800 MHz remains nearly constant at 6 dB as in free space propagation. The path loss is higher in urban areas compared to others, as expected. Similarly, the path loss in urban environment is at least 50 dB higher than for free space propagation conditions at distances longer than 5 km.

Figure 3.37 shows the effect of BS antenna height on the median path. Based on this model, it is clear that increasing the BS antenna height beyond some reasonable value (e.g., 30–50 m) does not cause significant decrease in the path loss, though it might help increasing the LOS distance.

3.6.2 COST 231 Extension to Hata Model

European Cooperation for Scientific and Technical Research (COST) 231 proposed the following extension to Hata model for frequencies up to 2 GHz, which covers also the 1900 MHz cellular radio band. The median path loss in *urban areas* is predicted to be

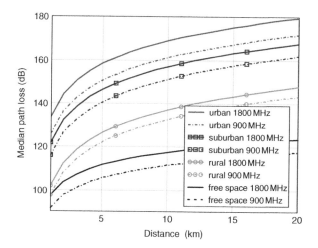

Figure 3.36 Prediction of Downlink Path Loss by Hata Model for a Small/Medium City in Urban, Suburban and Rural Areas at $f_c = 900$ and 1800 MHz. BS antenna height is $h_T = 30$ m and MS antenna height is $h_R = 1.7$ m. Free space propagation loss is also included as a reference.

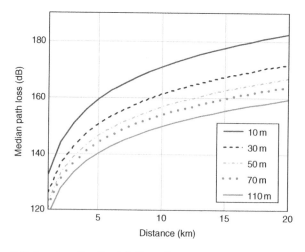

Figure 3.37 Median Path Loss in a Small to Medium City in Urban Environment at 900 MHz for Various Values of the BS Antenna Height h_T. MS antenna height is assumed to be $h_R = 1.7$ m.

$$L_{50} = 46.3 + 33.9 \log f_c - 13.82 \log h_T - a(h_R)$$
$$+ (44.9 - 6.55 \log h_T) \log r + C_M \ (dB)$$

$$C_M = \begin{cases} 0 \, dB & \text{medium sized city \& suburban areas} \\ 3 \, dB & \text{metropolitan centers} \end{cases}$$

$$(3.98)$$

where $a(h_R)$ is defined as in Hata model. The other parameters are defined as:

Carrier frequency f_c in MHz:	150–2000 MHz
Transmit antenna height h_T in meters:	30–200 m
Receive antenna height h_R in meters:	1–10 m
Range r in km:	1–20 km

Figure 3.38 shows a comparison between Hata and COST 231 models for $h_T = 30$ m and

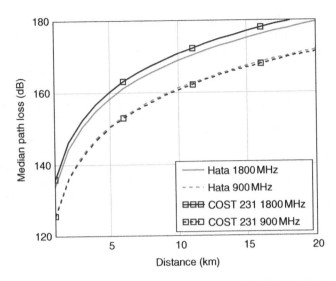

Figure 3.38   Comparison Between Hata and COST 231 Models for $h_T = 30$ m and $h_R = 1.7$ m for a Small to Medium City in Urban Area.

$h_R = 1.7$ m for a small to medium city in urban area. Hata and COST 231 models show perfect agreement at 900 MHz but the slight difference between them at 1800 MHz is believed to be insignificant since both of these models need to be calibrated with the data measured at a specific site anyhow. In both Hata and COST 231 models, propagation losses increase with frequency, range and in built-up areas but they decrease with increasing antenna heights, as expected.

Example 3.8   Link Budget for an Isolated Cell. Consider a single isolated cell of a cellular communication system in a suburban area. We will determine the BS average transmit power required to close the link by achieving a sensitivity level of −107 dBm at a mobile station (MS) that is 5 km away from the BS. Assume that the BS and MS antennas are omnidirectional, with gains equal to 12 dBi and 2 dBi, respectively. The BS and MS antenna heights are 25 meters and 1.7 meters respectively and the carrier frequency is 900 MHz. The propagation is assumed to be modelled by the COST 231 model.

In a suburban area, $C_M = 0$ dB and the correction factor for the mobile antenna height is found to be $a(1.7) = 0.526$ using (3.94). The median loss is then found from (3.98) as $L_{50} = 151.6 dB$. The transmit power required to achieve a received power level of −107 dBm at a distance of 5 km is given by (2.114)

$$P_{t,dBm} = P_{r,dBm} - G_{t,dB} - G_{r,dB} + L_{50} = 30.6 dBm$$
$$(3.99)$$

If the coverage area is desired to be extended to a radius of 10 km from 5 km, then the median pah loss becomes $L_{50} = 162.35 dB$ and the required transmit power increases from 30.6 dBm to 41.35 dBm.

Example 3.9   Comparison of Path-Loss Models.

Assuming that the received signal power is $P_0(r_0) = 1\ \mu W$ at $r_0 = 1$ km, the received signal power at a distance of $r = 10$ km will be determined and compared by using the following models: The model given by (3.92) with $n = 2$, 3, 4, (3.91) based on reflection from flat-Earth and (3.98), the COST 231 extension to Hata model for metropolitan centers. Assuming $f_c =$ 1800 MHz, $h_T = 25$ m, and $h_R = 1.7$ m, (3.92) yields:

$$P_r(r) = P_0(r_0)\left(\frac{r_0}{r}\right)^n = 10^{-6}\left(\frac{1}{10}\right)^n$$

$$= \begin{cases} -50\,dBm & n=2 \\ -60\,dBm & n=3 \\ -70\,dBm & n=4 \end{cases} \qquad (3.100)$$

The signal power received at $r = 10$ km via direct and reflected rays over flat Earth is found using (3.91):

$$P_r(r) = -50dBm + 6 + 10\log\left\{\sin^2\left(\frac{2\pi h_T h_R}{\lambda r}\right)\right\}$$

$$= -59.94\,dBm$$

$$(3.101)$$

The received power level based on the COST231 extension to Hata model (3.98) is found as follows:

$$a(h_R) = 3.2(\log_{10}11.75h_R)^2 - 4.97 = 0.44\,dB$$

$$L_{50}(1km) = 139.89\,dB$$

$$L_{50}(10km) = 183.29\,dB$$

$$P_r = P_0(dBm) - [L_{50}(10km) - L_{50}(1km)]$$

$$= -74.4\,dBm$$

$$(3.102)$$

where $C_M = 3$ dB is taken.

Note from the above results that the received power levels predicted by various models vary considerably from each other. Therefore, one should 'calibrate' the particular model used with the measurement data. [18]

3.6.3   Erceg Model

Hata model is unsuitable for broadband wireless systems working at higher frequencies with fixed MS and with lower BS heights for operation in hilly terrain and terrain with moderate-to-heavy tree density. Erceg model, which is

valid for fixed wireless broadband systems in suburban environments, is based on measurements in 95 existing macrocells in the US at 1.9 GHz with omnidirectional MS antennas at the height of 2 m.

The model distinguishes between three different terrain categories:

Terrain category A:	hilly with moderate-to-heavy tree density (maximum path loss)
Terrain category B:	hilly with light tree density or flat with moderate-to-heavy tree density (middle path loss)
Terrain category C:	flat with light tree density (minimum path loss)

For omnidirectional MS antennas with 2 m height and at 1.9 GHz, the median path loss is predicted by (3.92) with the inclusion of a log-normal shadowing parameter X[19]

$$L_s(dB) = L_{fs}(r_0) + 10n\log_{10}(r/r_0) + X, \quad r \ge r_0$$

$$L_{fs}(r_0) = 20\log(4\pi r_0/\lambda) = -27.56$$

$$+ 20\log(r_0) + 20\log(f_{MHz})$$

$$(3.103)$$

r:	the distance from the BS in meters (100–8000 m)
r_0:	the close-in reference distance in meters ($r_0 = 100$ m)
f_{MHz}:	the frequency in MHz
X:	the shadow fading component in dB

Path-loss exponent n is a Gaussian random variable over a population of macrocells within each terrain category:

$$n = a - bh_b + c/h_b + x\sigma_n, \quad 10m \ge h_b \ge 80m$$

$$(3.104)$$

h_b:	height of BS antenna in meters (10–80 m)
x:	a zero-mean Gaussian variable of unit standard deviation
σ_n:	standard deviation of the path-loss exponent n

a, b and c are consistent units listed in Table 3.4 for the terrain categories A, B and C.

Table 3.4 The Values of the Parameters in Erceg Model for Terrain Categories A, B and C. [19]

Model parameter	Terrain category, A	Terrain category, B	Terrain category, C
a	4.6	4.0	3.6
b (1/m)	0.0075	0.0065	0.0050
c (m)	12.6	17.1	20.0
σ_n	0.57	0.75	0.59
μ_σ	10.6	9.6	8.2
σ_σ	2.3	3.0	1.6

Shadow fading component X (expressed in dB) varies randomly from one terminal location to another within any macrocell and is a zero-mean Gaussian variable:

$$X = \sigma y = (\mu_\sigma + z\sigma_\sigma)y \qquad (3.105)$$

y, z:	zero-mean Gaussian variables of unit standard deviation
σ	standard deviation of X, which is itself a Gaussian variable over the population of macrocells within each terrain category, with mean μ_σ and standard deviation σ_σ.
μ_σ:	the mean of σ
σ_σ:	the standard deviation of σ

Including the correction terms for frequencies different from 1.9 GHz, and MS antenna heights other than 2 meters, the median path loss may be rewritten as

$$L_s(dB) = 20\ \log(4\pi r_0/\lambda) + 10n\ \log(r/r_0) + X$$
$$+ 6\ \log(f_{MHz}/1900) - 10.8\ \log(h_s/2) \qquad (3.106)$$

f_{MHz}:	frequency in MHz (450 MHz–11.2 GHz)
h_s:	height of the omnidirectional MS antenna in meters (2–10 m)

The parameters which are used in the Erceg model for terrain categories A, B, and C are tabulated in Table 3.4.

3.7 Indoor Propagation Models

Wireless local area networks (LANs) are implemented in indoor environment in order to reduce the cost and inconvenience of wired network, for example, in houses, libraries, shopping malls and airport lounges. IEEE 802.11 high-speed wireless LAN standard for indoor environments in 2.4, 3.6, 5 and 60 GHz bands aim to meet the user requirements for higher transmission rates. On the other hand, reception/transmission of mobile radio signals from/to outdoor BSs is also an important issue for indoor users.

Indoor propagation differs from outdoor propagation because of the shorter distances involved and the variability of the propagation environment in much shorter distances. Indoor radio propagation is dominated by reflection, diffraction and scattering as in outdoor channels. In addition, transmission loss through walls, floors and other obstacles, tunneling of energy, especially in corridors and mobility of persons and objects are special features of indoor propagation. Buildings have a wide variety of partitions and features that form the internal and external structure. Hard partitions, formed as part of the building structure, are denser and thicker compared to soft partitions, which may be moved and do not often cause high signal attenuations. Therefore, indoor propagation is strongly influenced by building type, its layout, construction materials, walls and floors as well as the presence of portable materials (furniture) and human beings. In addition to the factors cited above, signal levels vary greatly depending on antenna locations. Consequently, indoor propagation suffers impairments such as path loss, temporal and spatial variation of path loss, multipath effects from reflected and diffracted components, and polarization mismatch due to depolarization of electromagnetic waves due to multiple reflections and diffractions.

Similar to outdoor systems, indoor systems aim to ensure efficient coverage of the required area, and to mitigate inter- and intra-system interference. The indoor coverage area is defined by the geometry of the building, and the presence of the building itself will affect the propagation. Frequency reuse on the same floor and/or between floors of the same building creates further interference issues. Especially in the millimetre wave frequencies, small changes in the propagation path may have substantial effects on the channel characteristics. Nevertheless, for initial system planning, it is necessary to estimate the number of BSs to serve the distributed MSs within the coverage area and to estimate potential interference from/to other services and systems.

where

$L_m(r)$:	mean path-loss at a distance r from the transmitter
n:	path-loss exponent;
f_{MHz}:	frequency in MHz;
r:	separation distance (m) between the BS and MS (where $r > r_0 = 1$ m);
$L_f(N)$:	floor penetration loss factor (dB);
N:	number of floors between BS and MS ($N \geq 1$).
X:	zero-mean log-normal random variable (in dB) with a standard deviation of σ dB.

3.7.1 Site-General Indoor Path Loss Models

Indoor path-loss model considered here assumes that the BS and MSs are located inside the same building. The path loss between BS and MSs can be estimated with either site-independent or site-specific models. In this section, we will be interested in site-independent 'generic' models, which require little path or site-specific information.

Several indoor path loss models account for the signal attenuation through multiple walls and/or multiple floors. Site-specific models usually account for the loss due to each wall and floor explicitely. The site-general path-loss model described here implicitly accounts for transmission through walls and scattering from obstacles, and other loss mechanisms likely to be encountered within a single floor. However, the losses through floors are explicitly shown in the formulation.

The considered indoor path-loss model, which is characterized as the sum of an average path loss and a random shadow fading component, is based on (3.92): [20]

$$L(r) = L_m(r) + X \quad (dB)$$
$$L_m(r) = 20 \log f_{MHz} + 10 n \log r + L_f(N) - 28$$
$$\text{(3.107)}$$

Based on the measurement results, the path-loss exponent is given in Table 3.5 and the floor penetration loss is listed in Table 3.6 as a function of frequency for residential, office and commercial indoor environments.

Table 3.5 Path-Loss Exponent n for Indoor Transmission Loss Calculation. [20]

Frequency	Residential	Office	Commercial
900 MHz	– [3]	3.3	2
1.2–1.3 GHz	– [3]	3.2	2.2
1.8–2 GHz	2.8	3.0	2.2
2.4 GHz	2.8	3.0	
3.5 GHz	– [3]	2.7	
4 GHz	– [3]	2.8	2.2
5.2 GHz	3.0 (apartment) 2.8 (house) [2]	3.1	–
5.8 GHz	– [3]	2.4	
60 GHz[1]	– [3]	2.2	1.7
70 GHz[1]	– [3]	2.2	–

[1] 60 GHz and 70 GHz values assume propagation within a single room or space. Gaseous absorption around 60 GHz should also be considered for distances greater than about 100 m. [9]

[2] *Apartment*: Single or double storey dwellings for several households. In general most walls separating rooms are concrete walls. *House*: Single or double storey dwellings with wooden walls.

[3] For the frequency bands where the path-loss exponent is not stated for residential buildings, the value given for office buildings could be used.

Table 3.6   Floor Penetration Loss Factors, $L_f(N)$ dB with $N \geq 1$ Being the Number of Floors Penetrated, for Indoor Transmission Loss Calculation. [20]

Frequency	Residential	Office	Commercial
900 MHz	–	9 (1 floor) 19 (2 floors) 24 (3 floors)	–
1.8–2 GHz	$4N$	$15 + 4(N-1)$	$6 + 3(N-1)$
2.4 GHz	$10^{(1)}$ (apartment) 5 (house)	14	
3.5 GHz		18 (1 floor) 26 (2 floors)	
5.2 GHz	$13^{(1)}$ (apartment) $7^{(2)}$ (house)	16 (1 floor)	–
5.8 GHz		22 (1 floor) 28 (2 floors)	

$^{(1)}$ Per concrete wall, $^{(2)}$ Wooden mortar.

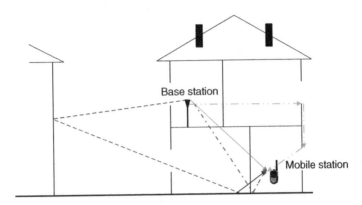

Figure 3.39   Alternative Propagation Paths Between Floors.

The losses between floors are determined by external dimensions, construction material, the type of construction used to build the floors, external surroundings and the number of windows. Note that some paths other than those through the floors may help establishing the link between transmitter and receiver with lower losses (see Figure 3.39). The richness of paths between transmitter and receiver makes the observed path loss to increase insignificantly after about five or six floor separations. When the external paths are excluded, measurements at 5.2 GHz have shown that at normal incidence the mean additional loss due to a typical reinforced concrete floor with a suspended false ceiling is 20 dB, with a standard deviation of 1.5 dB. Lighting fixtures increase the mean loss to 30 dB, with a standard deviation of 3 dB, and air ducts under the floor increase the mean loss to 36 dB, with a standard deviation of 5 dB. The accuracy of the existing models will improve as more data, experience and test environments will be available. [21]

The received signal power may then be written as

$$P_r(r)_{dB} = P_{r,m}(r)_{dB} - X \quad (dB)$$
$$P_{r,m}(r)_{dB} = P_{t,dB} + G_{t,dB} + G_{t,dB} - L_m(r) \tag{3.108}$$

where P_t, G_t, and G_r denote, respectively, the transmit power, transmit antenna gain and receive antenna gain all expressed in dB. The received power at distance r is a hence Gaussian random variable with mean $P_{r,m}(r)$ and standard deviation σ. The standard deviation (dB) for log-normal shadowing in indoor

Table 3.7 Standard Deviation (dB) for Log-Normal Shadowing in Indoor Environment. [20]

Frequency (GHz)	Residential	Office	Commercial
1.8–2	8	10	10
3.5		8	
5.2	–	12	–
5.8		17	

environment is listed in Table 3.7. The probability that the received signal power exceeds a specified threshold level of P_{th} (dB) (also called as system availability) may then be written as

$$P\left(P_r(r)_{dB} > P_{th}\right)$$

$$= \frac{1}{\sqrt{2\pi}\sigma} \int_{P_{th}}^{\infty} \exp\left(-\frac{\left(x - P_{r,m}(r)_{dB}\right)^2}{2\sigma^2}\right) dx$$

$$= Q\left(\frac{P_{th} - P_{r,m}(r)_{dB}}{\sigma}\right)$$

$$(3.109)$$

where $Q(x) = (2\pi)^{-1/2} \int_x^{\infty} \exp(-t^2/2)dt$ denotes the Gaussian Q function (see Appendix B). Note that (3.109) is synonymous to the probability that the path loss is less than a threshold loss level.

3.7.2 Signal Penetration Into Buildings

Signal strength received inside a building due to an outdoor transmitter is an important issue for wireless communication and broadcasting systems. Even though limited data is available for accurate modeling of such propagation environments, the signal strength penetrated inside a building from an outdoor transmitter increases with the height of the receiving terminal. At lower floors, multipath fading and shadowing induces greater attenuation and reduces the level of penetration. However, stronger incident signals are observed at the exterior walls at higher floors and hence higher penetration levels are expected. As a rule of thumb, the received signal power was observed to increase approximately 2 dB per floor as one goes up. Penetration loss is lower through windows compared to walls which cause higher attenuations to the signals. Signal attenuation through walls depends on the construction material and loss increases with the wall thickness in wavelengths. Stronger penetrated signal levels with increasing frequency may therefore be explained by the increased electrical aperture surface of the windows. Penetration loss was also observed to strongly depend on the angle of incidence and the elevation pattern of the outdoor transmit antenna (see Figure 3.40).

As a measure of the excess loss due to the presence of a building wall (including windows and other features), the penetration loss is useful for evaluating radio coverage and interference calculations between indoor and outdoor systems. Penetration loss is a function of the angle of incidence, the wall thickness, the materials used in the wall, and the frequency of operation but independent of the height. If the building is in the far-field of a transmitter, path loss between transmitter and wall should also be taken into account. If the building is in the near-field of the transmitter, then near-field effects should be considered. At 5.2 GHz, the penetration loss was measured to have a 12 dB mean and 5 dB standard deviation through an external building wall made of brick and concrete with glass windows. The wall thickness was 60 cm and the window-to-wall ratio was about 2:1. Table 3.8 shows the measurements results at 5.2 GHz through an external wall made of stone blocks, for incidence angles between 0° and 75°. The wall was 40 cm thick, with two layers of 10 cm thick blocks and loose fill between. Particularly at larger incident angles, the wall attenuation was extremely

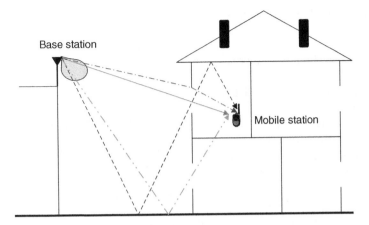

Figure 3.40   Penetration of Electromagnetic Waves Into an Indoor Environment.

Table 3.8   Building Entry Loss Due to Stone Block Wall, of 60 cm Thickness, at Various Incident Angles at $f = 5.2\,\text{GHz}$. [21]

Angle of incidence (degrees)	0	15	30	45	60	75
Wall attenuation factor (dB)	28	32	32	38	45	50
Standard deviation (dB)	4	3	3	5	6	5

sensitive to the position of the receiver, as evidenced by the large standard deviation.

Example 3.10   System Availability with Log-Normal Shadowing.

A propagation channel, undergoing log-normal shadowing with standard deviation σ, can be modelled by (3.92) and (3.108). Average received power levels of 0 dBm and $-35\,\text{dBm}$ were measured at, respectively, 1 m and 10 m from the transmitter.

If we let $r_0 = 1\,\text{m}$ in (3.92), then the mean received signal power at $r = 10\,\text{m}$ is given by

$$P_{r,m}(r)_{dB} = P(r_0)_{dB} + 10 n \log(r_0/r) \qquad (3.110)$$
$$= -10 n \; \log 10 = -35\,dBm$$

implying that the path-loss exponent is $n = 3.5$. If 10% of the measurements at $r = 10\,\text{m}$ were observed to be higher than $P_{th} = -25\,\text{dBm}$, then

the probability that the received power level will be higher than P_{th} is given by (3.109):

$$Prob\left(P(r)_{dB} > P_{th}\right) = Q\left(\frac{P_{th} - P_{r,m}(r)_{dB}}{\sigma}\right) = 0.1 \qquad (3.111)$$

From Table B.1, we find $Q(1.28) \approx 0.1$. Then, the standard deviation due to shadowing is found to be

$$\frac{P_{th} - P_{r,m}(r)_{dB}}{\sigma} \cong 1.28 \Rightarrow \sigma = \frac{-25 + 35}{1.28} = 7.8 dB \qquad (3.112)$$

An outage probability of 2% implies

$$P_{outage} = Prob\left(P(r)_{dB} < P_{th}\right)$$
$$= 1 - Q\left(\frac{P_{th} - P_{r,m}(r)_{dB}}{\sigma}\right) = 0.02 \qquad (3.113)$$

Using $Q(x) = 1 - Q(-x)$, the threshold level is found to be 16 dB below the mean received signal level (see Table B.1):

$$\frac{P_{r,m}(r)_{dB} - P_{th}}{\sigma} \cong 2.05 \Rightarrow P_{r,m}(r)_{dB} - P_{th}$$
$$= 2.05\sigma = 16 dB \qquad (3.114)$$

The corresponding threshold level is therefore given by $P_{th} = P_{r,m}(r)_{dB} - 16 = -51 \, dBm$

Example 3.11 An Intuitive Approach for Indoor Channel Modeling.

An indoor Wi-Fi transmitter at 2.4 GHz is separated from an indoor receiver located at a distance of 20 m. The signals must go through a wall at a distance of $r_1 = 16$ m from the transmitter and then propagate through the office for $r_2 = 4$ m before reaching the receiver. The wall causes 5-dB attenuation to the signal. The path loss exponent is estimated to be $n_1 = 3$ in the transmitter side but the office environment is characterized by $n_2 = 2.8$.

An intuitive approach to determine the received signal power behind the wall could consist of the following steps:

a. calculate the Poynting vector intensity at the wall input at distance r_1, using (3.92) with $r_0 = 1$ m and the path-loss exponent n_1;
b. atenuate the signal by the loss $L_W \geq 1$ due to wall;
c. using the signal at the wall ouput as the reference signal level in (3.92), determine the received signal power and the channel loss at $r = r_1 + r_2$:

$$P_r(r_1 + r_2) = \frac{P_t G_t}{4\pi \, r_0^2} \left(\frac{r_0}{r_1}\right)^{n_1} \frac{1}{L_W} \left(\frac{r_1}{r_1 + r_2}\right)^{n_2} \frac{\lambda^2}{4\pi} G_r$$

$$= \frac{P_t G_t G_r}{(4\pi r_0/\lambda)^2} \left(\frac{r_0}{r_1}\right)^{n_1} \frac{1}{L_W} \left(\frac{r_1}{r_1 + r_2}\right)^{n_2}$$

$$L = \left(\frac{4\pi r_0}{\lambda}\right)^2 \left(\frac{r_1}{r_0}\right)^{n_1} L_W \left(\frac{r_1 + r_2}{r_2}\right)^{n_2}$$

$$L_{dB} = 20 \log f_{MHz} + 20 \log r_0 + 10 n_1 \log(r_1/r_0)$$
$$+ 10 n_2 \log(1 + r_2/r_1) + L_{w,dB} - 27.56$$

$$(3.115)$$

Inserting the given parameters into (3.115) one gets $L_{dB} = 83.9$ dB. The path-loss predicted by (3.107) is found to be 78.6 dB for $r = r_1 + r_2$ and $n = 3$.

3.8 Propagation in Vegetation

Trees and bushes located in the first Fresnel zone can potentially lead to multipath propagation via diffraction and scattering. Hence, considerably high signal attenuation levels were observed. Multipath effects, scattering and depolarization of waves are highly dependent on type, density and water content of vegetation as well as wind and seasonal changes.

Consider a communication scenario as shown in Figure 3.41 where the receiver is located at a depth r of vegatation. The attenuation in excess of both free-space and diffraction loss, A_{ev}, due to the presence of the vegetation is given by [22]

$$A_{ev} = A_{max}[1 - \exp(-r\gamma/A_{max})](dB) \quad (3.116)$$

where

r :	length of path within woodland (m)
γ :	specific attenuation for vegetative paths (dB/m)
A_{max} :	maximum attenuation within a specific type and depth of vegetation (dB).

Figure 3.42 shows the specific attenuation as a function of frequency between 30 MHz–30 GHz for vertical and horizontal polarizations. However, attenuation due to vegetation varies widely due to irregular nature of the medium and the wide range of species, densities, and

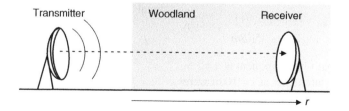

Figure 3.41 A Radio Path in Woodland.

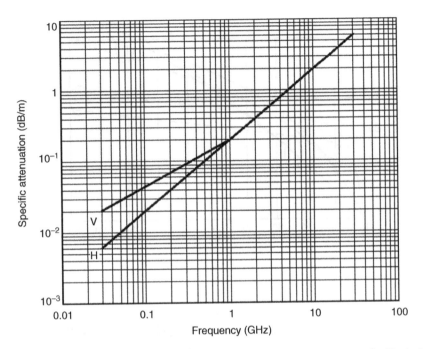

Figure 3.42 Specific Attenuation Due to Woodland as a Function of the Frequency for Vertical (V) and Horizontal (H) Polarizations. [22]

water content. At frequencies of the order of 1 GHz, the specific attenuation through trees in leaf appears to be about 20 % greater (dB/m) than for leafless trees. There can be variations of attenuation due to the movement of foliage, for example, due to wind.

The maximum attenuation A_{max} is limited by scattering and depends on the species and density of the vegetation, the antenna pattern of the terminal within the vegetation, and the vertical distance between the antenna and the top of the vegetation. Frequency dependence of A_{max} is described by

$$A_{max} = A_1 f_{MHz}^{\alpha} \ (dB) \qquad (3.117)$$

Measurements carried out in the frequency range 900–1800 MHz in a park with tropical trees in Rio de Janeiro (Brazil) with a mean tree height of 15 m have yielded $A_1 = 0.18$ dB and $\alpha = 0.752$. The receiving antenna height was 2.4 m.

Note that (3.116) does not apply for a radio path obstructed by a single vegetative obstruction where both terminals are outside the vegetative medium, such as a path passing through the canopy of a single tree. For frequencies below 3 GHz, the total excess loss through the canopy of a single tree is upper-limited by

$$A_{et} = \gamma r \ (dB) \qquad (3.118)$$

where A_{et} is lower than or equal to the lowest excess attenuation for other paths (dB).

One may refer to [22] for an empirical model of propagation through vegetation for frequencies above 5 GHz. Figure 3.43 shows the excess loss due to the presence of a volume of foliage which attenuates the signals passing through it. In practical situations, the signal beyond such a volume will receive contributions due to propagation both through the vegetation and via diffraction around it. The dominant propagation mechanism will then limit the total vegetation loss. The excess loss is shown in Figure 3.43 as a function of the vegetation depth for various frequencies and illumination areas with in- and out-of-leaf.

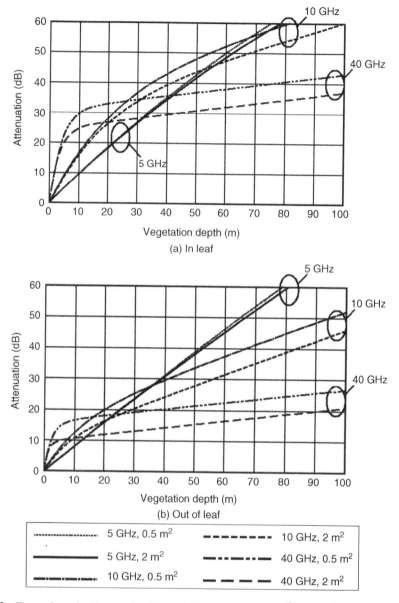

Figure 3.43 Excess Loss for Propagation Through Vegetation for 0.5 m² and 2 m² Illumination Area a) in Leaf, and b) Out of Leaf for Frequencies 5 GHz, 10 GHz and 40 GHz. [22]

When the vegetation is out-of-leaf, the attenuation was observed to decrease with increasing frequency and illumination area. However, the behavior of the vegetation in-leaf becomes more complicated though attenuation increases with increased vegetation thickness.

Measurements at 38 GHz for large vegetation depths suggest that depolarization through vegetation may be so large that co- and cross-polar signals may reach similar orders of magnitudes. Very high attenuation levels could even cause both components to stay below the receiver dynamic range. [22]

References

[1] E. C. Jordan and K. G. Balmain, *Electromagnetic waves and radiating systems* (2nd ed.), Prentice Hall: New Jersey, 1968.

[2] Recommendation ITU-R P.618–11 (09/2013), Propagation data and prediction methods required for the design of Earth-space telecommunication systems.

[3] Recommendation, ITU-R P.526.10, 2007, Propagation by diffraction.

[4] Recommendation ITU-R P.530–15 (09/2013), Propagation data and prediction methods required for the design of terrestrial line-of-sight systems.

[5] R. Fraile et al., Multiple diffraction shadowing simulation, *IEEE Com. Letters*, vol. **11**, no. 4, April 207, pp. 319–321.

[6] C. A. Balanis, *Antenna Theory: Analysis and Design* (2nd ed.), J. Wiley: New York, 1997.

[7] ITU-R Handbook (on Satellite Communications), Radio Propagation Information for Predictions for Earth-to-Space Path Communications (3rd ed.)

[8] C. Ho, A. Kantak, S. Slobin and D. Moratibo, Link analysis of a telecommunication system on Earth, in geostationary orbit and at the moon: Atmospheric attenuation and noise temperature effects, IPN Progress Report 42–168, 15 Feb. 2007.

[9] Recommendation ITU-R P.676–10 (09/2013), Attenuation by atmospheric gases.

[10] Recommendation ITU-R P.835–5 (02/2012) Reference standard atmospheres.

[11] Recommendation ITU-R P.838–3 (2005), Specific attenuation model for rain for use in prediction methods.

[12] Recommendation ITU-R P.837–6 (02/2012), Characteristics of precipitation for propagation modelling.

[13] T. Pratt and C.W. Bostian, *Satellite Communications*, J. Wiley: New York, 1986.

[14] Recommendation ITU-R P.839–4 (09/2013), Rain height model for prediction methods.

[15] Recommendation ITU-R P.678–2 (09/2013), Characterization of the variability of propagation phenomena and estimation of the risk associated with propagation margin.

[16] Recommendation ITU-R P.842–3 (04/2003), Conversion of annual statistics to worst-month statistics.

[17] Recommendation ITU-R P.834–6 (01/2007), Effects of troposhperic refraction on radiowave propagation.

[18] T. S. Rappaport, *Wireless Communications: Principles and Practice* (2nd edition), Prentice Hall: New Jersey, 2002.

[19] V. Erceg, et al., An empirically based path loss model for wireless channels in suburban environments, *IEEE J. SAC*, vol. **17**, no. 7, pp. 1205–1211, July 1999.

[20] Recommendation ITU-R P.1238–7 (02/2012), Propagation data and prediction methods for the planning of indoor radiocommunication systems and radio local area networks in the frequency range 900 MHz to 100 GHz.

[21] Recommendation ITU-R P.1411–1 (2001), Propagation data and prediction models for the planning of short range outdoor radiocommunication systems and radio LANs in the frequency range 300 MHz to 100 GHz.

[22] Recommendation ITU-R P.833–4, 2003, Attenuation in vegetation.

[23] M. Support (2010), Overview of 3GPP *Release* 9, http://www.3gpp.org).

[24] D. Astley et al., LTE: The Evolution of Mobile Broadband, *IEEE Communications Magazine*, vol. **47**, no. 4, pp. 44–51, 2009.

Problems

1. In LF (30–300 kHz) and MF (0.3–3 MHz) bands, electromagnetic waves propagate following the Earth's surface with relatively small losses. However, they attenuate rapidly as you go above and below the surface of the earth. Consider a surface wave at 1.5 MHz propagating over a ground with $\varepsilon_r = 10$ $\mu_r = 1$ and $\sigma = 0.002$ and over the sea water with $\varepsilon_r = 81$, $\mu_r = 1$ and $\sigma = 5$. The electric field E_S due to the surface wave is given by (3.2) as a function of the distance in the far-field region.

   a. Determine the distances over ground and sea surfaces at which the surface wave intensity E_S is 13 dB below E_{LOS}, which denotes the received electric field strength under free-space propagation conditions.

   b. Compare the attenuations for ground and sea water at a depth of 10 m below the earth/sea surface assuming that the electric field at d meters below the Earth's surface is given by

$$E_S(d) = E_S \, e^{-d/d_s} \; (V/m)$$

$$d_s = \frac{1}{\sqrt{\pi f \mu_0 \sigma}} \; (m)$$

where d_s denotes the skin depth, f is the frequency and the permeability of free space is given by $\mu = 4\pi\ 10^{-7}$.

2. Consider a communication link operating at 50 MHz in an environment where the ground constants are $\varepsilon_r = 10$ and $\sigma = 5\ 10^{-3}$ S/m. Transmit and receive antennas have 0dBi gains, the transmit power is 4 W, and the received noise power in the 5 kHz receiver bandwidth is $N = 7.7 \times 10^{-16}(W)$.

a. Determine the SNR under free space propagation conditions as a function of distance.
b. Determine the SNR assuming surface-wave propagation conditions and compare it with (a).

3. A portable AM radio receiver, having a ferrite-core antenna of gain $1.5\eta_L$ ($\eta_L = 10^{-6}$) receiving an AM broadcasting station operating at 1.5 MHz carrier frequency, which has a dipole antenna with a transmit power of 20 kW. If the receiver noise is 2×10^{-13} (W) in 10 kHz receiver bandwidth, determine the SNR at a distance of 80 km. Electrical parameters of the ground are assumed to be $\sigma = 10^{-2}$ S/m and $\varepsilon_r = 12$.

4. Using Example D.1, determine the asymptotic value of the diffraction loss given by (3.14) as $|v| \to \infty$.

5. Use (3.14) to plot the variation of the diffraction loss with frequency (1 MHz $< f <$ 10 GHz) for $d_1 = d_2 = 1000$m and $h = 5$, 10, 20, 40, 80 m.

6. Consider a knife-edge obstacle lying between the transmitter and receiver of a wireless communication system. Suppose $d_1 = d_2 = 500$m and signal carrier frequency is $f_c = 1$ GHz. If the system design

allows for a maximum diffraction loss of 15 dB, then

a. Determine the maximum permitted diffraction height.
b. Determine the power level of the diffracted field at the receiver in comparison with free-space propagation between transmitter and receiver.

7. Consider a downlink, from base station (BS) to mobile station (MS), transmission at a range of 15 km in a cellular radio system operating at 900 MHz. The heights of the MS and BS are assumed to be 10 meters each. There is an obstacle, which can be modeled as a knife-edge diffracting obstacle of height 30 meters, at a distance of 5 km from the transmitter. This obstacle prevents the LOS between the transmitter and the receiver. There is also a large building, in the first Fresnel zone, at a horizontal distance of D = 30 meters from the mid-point of the line connecting transmitter and receiver. This building reflects the signals with a reflection coefficient of $\rho = -0.8$.

a. Express analytically the received signal power levels for reflected and diffracted rays in terms of the system parameters.
b. Determine ratio of received power levels of reflected to diffracted rays in dB.

8. A BS is transmitting P_t W at 900 MHz with an antenna height of $h_T = 30$ m and 10 dBi antenna gain. The receiver with 0 dBi antenna gain is located on board of a baloon which carries tourists in the Cappadocea region and moves vertically in the time frame of interest. There is a mountain of 1000 m height between the BS and the baloon at a distance of 4 km from the BS transmitter as shown below. We can use

the knife-edge diffraction model for diffraction from the mountain top.

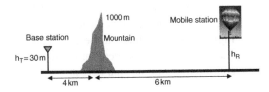

1000 m
Base station
Mountain
Mobile station
$h_T = 30$ m
h_R
4 km
6 km

a. Compute the transmit power P_t such that a mobile receiver at a distance of 10 km can operate adequately with a sensitivity level of -100 dBm at a balloon height of $h_R = 30$ m.

b. At which height of the balloon, the diffraction loss is 6 dB? Determine the corresponding value of the received signal power at this height.

c. The balloon can go up to the maximum height of 3 km. Determine the corresponding value of the diffraction loss and the received signal level.

d. Repeat part (a) for $f = 1800$ MHz and comment on the effect of frequency on the diffracted field strength.

9. Consider the diffraction from a knife-edge of height $h = 10$m, which is located at midpoint between transmitter and receiver, that is, $d_1 = d_2 = 500$ m.

a. Determine the diffraction loss at 0.3 GHz and 3 GHz and compare them.

b. What is the value of h at 3 GHz that gives the same diffraction loss for $h = 10$ m and at 0.3 GHz?

c. Now consider the case where the source is located at far distances from the point of diffraction, that is, $d_1 \gg d_2$. Calculate the diffraction loss and compare with the results found in (a).

d. Repeat (a) for $d_1 = 100$ m, $d_2 = 900$ m and $d_1 = 900$ m, $d_2 = 100$ m. Compare the results with those found in (a) and comment about the differences between them.

e. Assuming that $d_1 + d_2$ is constant, what combination of (d_1, d_2) minimizes the diffraction loss?

10. A BS of height $h_t = 30$m is transmitting 20 W at 900 MHz with an antenna gain of 17 dBi. The MS is located at 20 km from the BS, it has $h_r = 2$m antenna height and has 0 dBi gain.

a. Determine the received power in dBm at the MS assuming free-space propagation model.

b. Is the first Fresnel zone free? If not, determine the received power in dBm at the MS by taking into account direct and reflected rays.

c. Determine the required received power level in dB m to keep the coverage radius unchanged if a 60 m height building, with a sharp roof acting as a diffracting edge, is built at a distance of 1 km far from the BS.

d. Repeat (c) when the 60 m height building is 1 km far from the MS.

11. A link is described by a 2-ray ground reflection model, where $r \gg h_t$, h_r and $P_t = 50$ W, $f_c = 1800$ MHz, $G_t = 20$ dBi and $G_t = 3$ dBi.

a. If the received LOS signal power was measured to be $P_r = 1\mu$W at a range of $r = 1$ km, determine the path loss and the corresponding path-loss exponent.

b. Consider the 2-ray ground reflection model with the following system values: $h_t = 40$ m, $h_r = 3$ m and $G_r = 3$ dBi. Calculate the received power level at $r = 2$ km.

c. Consider the 2-ray ground reflection model of (b). Suppose the receiver antenna height hr is adjustable (but all other system values are fixed). Find the value of h_r which maximizes the received power.

12. Assume that the antenna gains of BS and MS are 6 dBi and 2 dBi, respectively. Corresponding antenna heights are 10 m and 1.5 m, respectively and they operate in an environment where the ground plane can be treated as perfectly conducting. The BS transmits with a maximum of 40W and the MS with a power of 0.1W. The center frequency of the links (duplex) are both at 900 MHz, even if uplink and downlink frequencies differ from each other.

a. Using plane-earth propagation assumption, calculate how much received power is available at the output of the BS and MS receive antennas, as a function of distance d.
b. Assuming that the receiver sensitivity of both BS and MS is -140 dBW, what would be the range of the downlink and the uplink ?
c. In order that the range of the uplink and the downlink be the same, what would be the relation between the sensitivities of the BS and MS receivers?

13. Consider a link where the plane-earth propagation conditions apply. The received signal power at a distance of d_1 is satisfactory when the transmitting BS antenna height $h_t = 50$ m and a mobile antenna height is h_r.

a. If the BS antenna height is lowered to 10 m, what will be the effect on reception in terms of distance?
b. If the distance remains the same, that is, d_1, and the BS antenna height is lowered to 10 m, how high must the mobile antenna be raised to ensure satisfactory reception?

14. In a 2-ray ground reflected model, where the received electric field intensity is given by (3.37) and (3.39). The carrier frequency is 900 MHz and the receiver antenna height is 2 m. If the angle of incidence, shown in Figure 3.9, is required to be less than 5° and $2\pi h_T h_R/(\lambda r) \leq \pi$, determine the range and the transmit antenna height.

15. Consider a communication system operating at 3 GHz along a path of 15 km over sea water with relative dielectric constant of 81 and conductivity of 5 S/m. Transmitting antenna is located on a tower of 20 m high. The receive antenna height h_R is required to be higher than 16 m.

a. Explain if these antennas are in LOS of each other, assuming worst-case conditions?
b. Can the sea water be considered as a good conductor at this frequency?
c. Determine the angle of arrival with respect to horizon.
d. Determine the reflection coefficient.
e. Can earth curvature be neglected at this frequency?
f. Determine the optimal height for the receiving antenna in order to maximize the received signal level.
g. Determine and compare the signal level for free space transmission with the level found in part (f).
h. Using the above-determined receiving antenna height, determine the signal level when the sea level is decreased by 2 meters due to tide and compare this value with the optimal value found in part (f).

16. Transmit power of a BS is limited to 20 W. Assume that the receiver sensitivity is -100 dBm and receiver antenna height is 2 m. Also assume that receiver and transmitter antennas are omnidirectional. Assuming that the plane-earth propagation

model applies, determine transmit antenna height required for a coverage area radius of 2 km and 10 km.

17. Consider a channel with impulse response $h(t) = \alpha_1 \delta(t) + \alpha_2 \delta(t-\tau_2)$ based on a two-ray model operating at 900 MHz in the VHF/UHF band (see (3.8) and Figure 3.3). Assume that transmit and receive antennas have the same height $h = 20$ m above the ground. Also assume that the transmit power is $P_t = 1$ W. and the transmit antenna has a gain $G_t = 100$.

a. Show that the distance d between transmitter and receiver, as well as α_1 and α_2, can be determined, if τ_2 can be measured. Assume that the point reflection is in the far-field of both transmitter and receiver, and that the reflection coefficient is equal to -1. Determine d, α_1 and α_2 for $\tau_2 = 1$ns.
b. Does the path loss vary as $1/d^2$ or $1/d^4$ as a function of distance?

18. A VHF system operating at 200 MHz has a transmitting antenna height of 40 m.

a. Determine the height of the receiver antenna located at a distance of 2 km from the transmitter for maximum received signal.
b. Can the same signal level as in part (a) be obtained by raising the receiving antenna higher than in part (a)?
c. Determine the horizontal distance to get the peak signal level closest to that found in part (a).

19. A terrestrial LOS link operating at 12 GHz over a range of 30 km. Transmit and receive antenna gains are 18 dBi, transmit power is 0 dBW and the received noise power is −150 dBW.

a. Determine the average SNR with no rain.
b. Estimate the rain attenuations exceeded for 0.1% and 0.01% of the time and the corresponding effective SNRs based on the point rainfall rates 35mm/hr and 50mm/hr exceeded for 0.01% of the time.

20. For an earth station, 0.8km above the sea-level, located at 40° North latitude and 33° E longitude, the measured rain rates are shown below:

Time Interval		Maximum Rain rate	
in time	in % of time	mm	mm/hr
5 minutes	0.00095	12	144
10 minutes	0.0019	16	96
15 minutes	0.00285	18	72
30 minutes	0.0057	24.5	49
1 hour	0.0114	32.5	32.5
2 hours	0.0228	44.5	22.25
24 hours	0.274	69.8	2.9
1 month	8.3	121.9	0.169
1 year	100	612.6	0.07

Assuming that the earth station receives circularly-polarized TV signals from a DVB-S satellite located at 42°E longitude at 12 GHz, determine

a. the rain attenuation exceeded for 0.01% and 0.1% of the year.
b. the downlink XPD that would be associated with the attenuations calculated in part (a) for 0.01% and 0.1% of the time.
c. Repeat (a) and (b) for 18 GHz.

21. Site diversity is used to mitigate the rain attenuation. If two satellite ground terminal antennas are located sufficiently far from each other, then the probability of simultaneously having heavy rain at

two locations is expected to be very low. The receiver selects the signal received by the ground terminal experiencing lower rain attenuation, i.e., $A = \min(A_1, A_2)$. If the rain rate distributions along the two paths of a site diversity system are statistically independent, then the probability that the attenuation A exceeds some threshold value A_{th} is given by the product of the probabilities that the attenuations of site 1 and site 2 exceed A_{th}:

$$P(A \geq A_{th}) = P(A_1 \geq A_{th})P(A_2 \geq A_{th})$$

Assume that there are two diversity sites with identical distributions and the probability that

$$P(A_i > 12dB) = 0.1\%, \quad P(A_i > 24dB)$$
$$= 0.01\%, \quad i = 1, 2$$

Calculate the *diversity gain* that would be achieved at $p = 0.01\%$ if the rain rate distributions on the two paths were statistically independent. *Diversity gain* is defined as the difference, in dB, between the average single-site attenuation exceeded for $p\%$ of the time and the joint attenuation exceeded $p\%$ of the time.

22. We consider a single cell of a cellular communication system in a suburban area. BS and MS antennas are assumed to be omnidirectional with gains 10 dBi and 0 dBi, respectively. Antenna heights of BS and MS are 30 m, and 2 m, respectively. The carrier frequency is assumed to be 900 MHz and log-normal shadowing is ignored. Using the COST 231 model, determine the average BS transmit power required to achieve a minimum received signal power of −107 dBm at a MS that is 10 km from the BS.

23. We consider the following path loss model defined by (3.92) where $P(r_0)$ denotes the received power at a reference distance r_0

and n is a random variable. Based on the measurement results, the pdf of n may be written as

$$f_n(n) = 0.25\ \delta(n-2) + 0.3\ \delta(n-2.5)$$
$$+ 0.35\ \delta(n-3) + 0.1\ \delta(n-4)$$

Determine the pdf and the mean value of the received SNR at a distance r if the received noise power is N (W).

24. The path-loss exponent is usually determined experimentally as the best fit to the measured signal power. Using the model (3.92) where $P(r_0) = 0\ dBm$ at $r_0 = 10$ m, measured and predicted signal levels (by the model) at distances of 10 m, 20 m, 100 m and 500 m from the transmitter are shown below:

Distance from transmitter	Measured power level (dBm)	Predicted power level by the model (dBm)
10 m	0	0
20 m	−17	−3 n
100 m	−30	−10 n
500 m	−55	−17 n

a. Determine the minimum mean square estimate of the path-loss exponent
b. Determine the standard deviation around the mean value
c. Estimate the received power at $r = 300$ m

25. Consider a GSM uplink operating at 1800 MHz. The MS has 100 mW transmit power and the receiver sensitivity is −105 dBm. The distance between the BS and MS is 1km. The propagation model is given by (3.92) with $r_0 = 50$ m and $n = 4$. Transmit and receiving antenna gains are 0 dBi and 12 dBi, respectively. Compute the available margin.

26. Consider a 10 W transmitter communicating at 400 MHz with a mobile receiver

having a sensitivity of −100 dBm. Assume that the receiver antenna height is 2 m, and the transmitter and receiver antenna gains are 1 dB.

a. Using the plane-earth path-loss model, determine the height of the BS antenna so as to provide a coverage area of 10 km radius.
b. Solve the same problem by using Hata model for open rural area. How can you explain the difference between the results obtained in (a) and (b).
c. If the transmit power is restricted to be 10 W or less, how can you increase the coverage area?

27. Consider a cellular radio system operates at 900 MHz band with 0 dBi transmit and receive antenna gains and has a 1W transmit power. The received SNR is required to be higher than 25 dB. The received noise power over $B = 30$ kHz bandwidth is −150 dBW. The propagation model is given by (3.92) with $r_0 = 1$ km. Determine maximum range for $n = 2$ and $n = 4$.

28. A police patrol car driving at a speed of 120 km/h on the highway has an X-band Doppler radar operating at 9.9 GHz. Tracking starts when police observes a car exceeding the speed limitations at a distance of 100m ahead of the police car.

a. If the police radar receiver measures a Doppler frequency shift of −759 Hz, determine relative and absolute speeds of the tracked car.
b. Assuming that both cars have constant speeds all the time and that the channel model is given by (3.92) with $r_0 = 100$ m, $P(r_0) = -60$ dBm and $n = 2.1$, when would the police radar lose connection to the tracked car if the radar receiver sensitivity is −100 dBm.

c. How long it takes to catch tracked car if the speed of the police car is increased to 210 km/s?

29. A wireless channel in an urban cellular radio system can be modelled by (3.92), where the received signal power at $r_0 = 1$ m is 1 mW and the average path loss exponent is $n = 3$. Determine the cell radius normalized to the distance between the centres of the considered and the single interfering cell if the signal-to-interference ratio (SIR) is required to exceed 20 dB.

30. Consider a receiver in a cellular radio system which detects 1 mW signal power at a distance of 1 meter from the BS transmitter and −60 dBm at 100 meter. The receiver system noise power is −122 dBW and the channel model is given by (3.92).

a. Determine the path loss exponent n.
b. Determine the cell radius if the minimum required SNR at the mobile receiver is 13 dB.
c. The system is required to operate in the presence of co-channel interference from another BS as long as signal-to-interference ratio (SIR) is larger than 20 dB. Determine the minimum separation between the two BSs.

31. Consider an 802.11b Wi-Fi BS operating at $f = 2.5$ GHz with a transmit power of $P_t = 20$ dBm, and $G_t = 5$ dBi transmit antenna gain. The Wi-Fi transmitter is located in an outdoor area, where the channel loss is modelled as $L_{dB} = 20\log_{10}(f) + 16.9\log_{10}(d) + 32.8$ [23], [24]. In an office, a user is receiving Wi-Fi signals at 20 m from the transmitter, behind a wall at 16 m from the transmitter and 4 m from the receiver. Assuming that the wall attenuates the signal by 5 dB, and the indoor channel loss is estimated by (3.92) with $n = 2.3$, determine the received signal level.

Transmit and receive antennas are assumed to be omnidirectional and the receive antenna gain is 2.0 dBi. Determine the link margin if the receiver has a sensitivity of −50 dBm.

32. A minimum SNR of 16 dB is required for indoor reception of the signals due to an outdoor transmitter. The transmit power is 20 W, transmit antenna gain is 7 dBi, the receive antenna gain is 0 dBi, the frequency of operation is 900 MHz. BS and MS antenna heights are assumed to 30 m and 2 m, respectively. Assuming that the receiver noise power is −100 dBm, determine the maximum building penetration loss (including the loss due to indoor propagation) that is acceptable for a BS with a coverage radius of 10 km, if the following path loss models are used.

a. Free space path-loss model.
b. Two-ray path-loss model and compare with the result of part (a).

33. The average power received at a MS, which is located 100 m from a BS is 0 dBm. Lognormal shadowing with 8 dB standard deviation is experienced at that distance.

a. Determine the probability that the received power at the MS will exceed 0 dBm.
b. Determine the probability that the MS will have a signal level higher than 10dBm.
c. Determine the outage probability when the received power level falls 10 dB below the mean received power level.

34. The signal quality for a particular mobile communication system operating at 1800 MHz is acceptable when the received power at the MS is −95 dBm.

a. Find the maximum acceptable propagation loss and the corresponding range for the system in free-space propagation conditions, given that the BS transmit power is 2W, system implementation losses are 10 dB, the BS antenna gain is 9 dBi, and the MS antenna gain is 0 dBi.
b. In a shadowing environment with 6 dB standard deviation, determine the probability of receiving signal power levels lower than −115 dBm.

35. Assume a propagation model given by (3.92) with $n = 2.7$, $r_0 = 10$ m and $P(r_0) = 0$ dBm. Standard deviation due to shadowing is $\sigma = 6$ dB and the receiver sensitivity is −100 dBm. Transmit and receive antennas are assumed to be omnidirectional. The system has 10 dB shadowing margin for 99% availability (1% outage).

a. Determine the range if outage is limited to less than 1% .
b. Determine the range for $n = 3$ and compare with the range found in part (a).
c. For $n = 3$, determine the increase in the transmit power so as to maintain the same coverage area as for (a)?

36. Consider a 900 MHz cellular transmitter with an EIRP of 20 dBW. The noise power received by the 0 dBi gain receive antenna over $B = 30$ kHz bandwidth is −140 dBW. The SNR is required to be higher than 25 dB. The propagation model is given by (3.92) with $r_0 = 1$ km, $n = 3.5$, and $\sigma = 6$ dB due to shadowing. Determine the probability that the desired SNR is higher than 25 dB at a distance of 10 km from the transmitter.

4

Receiver System Noise

In Chapters 2 and 3, we studied the prediction of the signal power received under free-space propagation conditions and in the presence of the Earth, that is, obstacles, mountains, hills, buildings, and so on. Consequently, we predicted the received signal power due to reflection, refraction, diffraction and scattering of electromagnetic waves by the obstacles in the propagation path. However, the SNR, the principal figure-of-merit of a communication system, is determined not only by the received signal level but also the received noise and interference levels. Noise is defined as a random time-varying electromagnetic phenomenon which may be superimposed on, or combined with, a wanted signal. Noise varies randomly with time and does not carry information. [1] There are many sources of noise in the RF band that are internal and external to the receiving system. Internal noise is mostly of thermal origin and is generated by the receiver itself. Antenna losses, feeder lines connecting the antenna to the receiver and the receiver itself are the major contributors of the internal noise. The external noise is collected by the antenna and may be due to: [1][2][3]

a. Emissions from atmospheric gases and hydrometeors (water vapour, oxygen, nitrogen, rain, snow);
b. Radiation from extra-terrestrial sources such as galaxies, sun, moon and stars;
c. The Earth's surface;
d. Atmospheric noise due to lightning discharges (atmospheric noise due to lightning);
e. Man-made noise due to electrical machinery, power transmission lines, electronic equipment, engine ignition systems, and so on.

A receiver may also receive co-channel transmissions whether intentionally or unintentionally radiated or spurious emissions by radars, radio/TV stations, multi-user systems and jammers. Unlike noise, such signals carry information depending on their source and distance; hence interfere with the received signals.

A communication receiver should operate successfully in the presence of noise and interference since it can not distinguish between wanted and unwanted (noise and interference) signals impinging at the receive antenna in

---

Digital Communications, First Edition. Mehmet Şafak.
© 2017 John Wiley & Sons Ltd. Published 2017 by John Wiley & Sons Ltd.
Companion website: www.wiley.com/go/safak/Digital_Communications

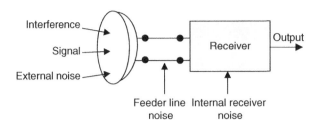

Figure 4.1   Sources of Receiver Noise.

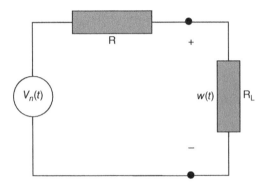

Figure 4.2   Physical Model of Thermal Noise.

the frequency band of operation (Figure 4.1). The amount of noise and interference power coupled into a receiver depends not only on the antenna radiation pattern but also on the spatial and spectral distribution of the sources of noise and interference. The noise perform- ance of a receiver is hence determined by the noise collected by its antenna and its internal noise.

In this chapter, we will be interested in the statistical characterization and the effects of internal and external noise in the SNR perform- ance of a receiver.

4.1   Thermal Noise

Every object in the universe, with a temperature above 0 K, generates thermal noise which is caused by motion of electrons. The physical model for the thermal noise may be described, as shown in Figure 4.2, by considering a resis- tor R at temperature T above 0 K. The motions of the electrons in their orbits and occasional collisions between them give rise to a noise

voltage $v_n(t)$ which has a zero mean $E[v_n]=0$; this implies that the polarity of the noise voltage has equal probabilities of being positive and negative. However, the mean-square noise voltage (the noise variance), $\bar{v}_n^2 = E\left[\left|v_n\right|^2\right]$, gen- erated at the terminals of a resistor R at tempera- ture T (K) in a bandwidth Δf (Hz) is non-zero. If a load resistor R_L is connected to the terminals of R, the power delivered to R_L is maximized when $R_L = R$ and is given by Planck's law: [4]

$$P_{n,\,\text{max}} = \frac{\bar{v}_n^2}{4R} = \frac{hf}{\exp(hf/kT)-1}\Delta f \ \ (W) \ \ (4.1)$$

where $k = 1.38 \ 10^{-23}$(J/K) denotes the Boltz- mann's constant and $h = 6.6254 \ 10^{-34}$(J.s) is the Planck's constant. At radio frequencies where $hf \ll kT$, one may use the binomial expansion $e^x \cong 1+x, \ x \to 0$ (see (D.1)) to express the single-sided power spectral density (PSD) of the thermal noise as

$$N_0 = P_{n,\,\text{max}}/\Delta f \cong kT(W/Hz) \ \ \text{for} \ \ hf \ll kT$$
$$(4.2)$$

Figure 4.3 shows the variation of the noise PSD, given by (4.1), with frequency for various values of T. One may observe that the PSD of thermal noise is white (flat) in the RF frequen- cies of interest and may be approximated by (4.2) for $f < 10^{12}$ Hz at temperatures of practical interest. Hence, the double-sided noise PSD, which applies for both positive and negative frequencies, is $N_0/2$.

Based on the above considerations, we define the random noise voltage $w(t)$ across a 1Ω load resistor which is matched to the source resistor, hence $w(t) = v_n(t)/2$ (see Figure 4.2).

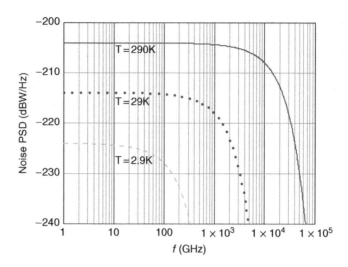

Figure 4.3   Thermal Noise Power Spectral Density (PSD).

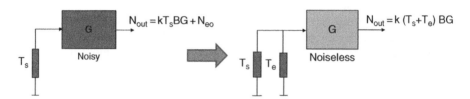

Figure 4.4   Definition of Equivalent Noise Temperature.

The random noise voltage $w(t)$ is described by a Gaussian probability density function (pdf) with zero mean and a variance equal to the double sided PSD $N_0/2$:

$$f_{w(t)}(x) = \frac{1}{\sqrt{2\pi\sigma^2}} \exp\left(-\frac{x^2}{2\sigma^2}\right)$$

$$E[w(t)] = 0$$

$$\sigma^2 = E\left[|w(t)|^2\right] = E\left[\frac{1}{T}\int_0^T\int_0^T w(t)w^*(t')\, dt\, dt'\right]$$

$$= \frac{1}{T}\int_0^T\int_0^T R_w(t-t')\, dt\, dt' = \frac{N_0}{2}$$

$$(4.3)$$

where $R_w(\tau)$ denotes the autocorrelation function of $w(t)$. The white approximation for the PSD $S_w(f)$ of the thermal noise defined by (4.2) implies that any two samples of $w(t)$ are uncorrelated from each other unless they occur

at exactly the same time; hence the autocorrelation function of $w(t)$ is represented by a delta function in time:

$$S_w(f) = \frac{N_0}{2} \quad \Leftrightarrow$$

$$R_w(\tau) = E[w(t)w^*(t+\tau)] = \frac{N_0}{2}\delta(\tau)$$

$$(4.4)$$

4.2   Equivalent Noise Temperature

The physical model described above for the thermal noise generated by a resistance at temperature T may be generalized for modelling the thermal noise generated by any thermal noise source. Consider a narrowband noisy device with gain G and bandwidth B. The noise power kT_sB due to a resistance of temperature T_s at the input of the noisy device is amplified by the device of gain G so that the output noise power due to this resistance is given by kT_sBG (see Figure 4.4). On the other hand, the internal noise of the device also induces a noise power, say N_{eo}, at the output. Let T_e (K) denotes the temperature of an *equivalent resistance* which delivers the same average output noise power N_{eo} as the considered device in a given bandwidth B. Then, the total equivalent noise power at the device input is given by

$$N_i = k(T_s + T_e)B \quad (W) \qquad (4.5)$$

The total equivalent noise power at the device output may be written as

$$N_{out} = kT_s GB + N_{eo} = k(T_s + T_e)GB \quad (W) \qquad (4.6)$$

where the device is now assumed to be noiseless. Note that T_e is not equal to the physical temperature of the noisy device but is related to it; it simply provides a measure of the noise power generated by the device. For example, let $G = 20$ dB, $T_s = 290$ K, $B = 1$ MHz and $T_e = 580$ K. The total equivalent input noise power $N_i = k(T_s + T_e)B$ is equal to -139.2 dBW, and total output noise power $N_{out} = N_i$ G is -119.2 dBW.

Example 4.1 Measurement of the Equivalent Noise Temperature.
Equivalent noise temperature of a device may be measured by driving the considered device with a known noise source and measuring the increase in the output noise power. Because the thermal noise power generated by a resistor is too small to be measured, active noise sources, for example, a diode or an electron tube, are used to drive the device under consideration. Y-factor method is a reliable technique to measure the generated low noise levels. [5]

As shown by Figure 4.5, the device under test is connected to one of two different matched loads at temperatures T_1 and T_2, which are not close to each other. Let T_1 and N_1 be respectively the noise temperature and the output noise power of the load with higher noise. Similarly, let T_2 and N_2 denote respectively the equivalent noise temperature of the load with lower noise. T_e designates the equivalent noise temperature of the device under test.

Since the source noise is uncorrelated with the noise of the device under test, the total output noise power is additive. From the measurements using T_1 and T_2, one can write

$$N_1 = Kk(T_1 + T_e)BG$$
$$N_2 = Kk(T_2 + T_e)BG \qquad (4.7)$$

where K is a constant different from unity accounting for the possibility that a relative power meter at the amplifier output cannot measure the true (absolute) output noise power. The ratio of the measured output noise powers using T_1 and T_2 is given by

$$Y = \frac{N_1}{N_2} = \frac{T_1 + T_e}{T_2 + T_e} \qquad (4.8)$$

Using (4.8), the equivalent noise temperature of the device under test is found in terms of Y, and the temperatures T_1 and T_2 of the calibrated noise sources:

$$T_e = \frac{T_1 - YT_2}{Y - 1} \qquad (4.9)$$

It is evident from (4.8) and (4.9) that this method yields accurate results when the two

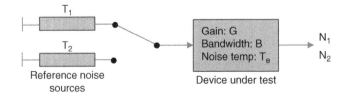

Figure 4.5 The Y-Factor Method for Measuring the Equivalent Noise Temperature.

source temperatures, T_1 and T_2, are not too close to each other so that Y differs significantly from unity.

The following variant of the above measurement procedure requires a calibrated source of white noise, such as a diode noise generator which is impedance-matched to the input of the amplifier: [6]

a. Set the source noise temperature at $T_1 = T_0$, and record the output meter reading N_1 as given by (4.7).
b. Increase the source temperature to $T_2 = T_0 + T_X$ such that the meter reading has doubled, that is, the meter reading is $N_2 = 2 N_1$.
c. From the ratio $Y = N_1/N_2 = 1/2$, one obtains the equivalent noise temperature and the noise figure of the amplifier (to be defined in section 4.3) as $T_e = T_X - T_0$ and $F_e = T_X/T_0$ without knowing G, B_n and K.

4.2.1 Equivalent Noise Temperature of Cascaded Subsystems

A communication receiver consists of cascaded devices, for example, antenna, RF amplifier, frequency convertor, demodulator, detector and audio amplifier, as shown in Figure 4.6. These devices may be active or passive but are all noisy with different noise levels. The internal noise of a receiver system of cascaded components may be determined in terms of the equivalent noise temperatures and gains of the individual components. Let G_i and T_{ei}, $i = 1, 2, \ldots, n$ denote, respectively, power gains and equivalent noise temperatures of these components. Output noise power may be written as

$$N_{out} = (kT_{e1}B)G_1 G_2 \cdots G_n + (kT_{e2}B)G_2 G_3 \cdots G_n + \cdots (kT_{en}B)G_n$$

(4.10)

If T_e denotes the equivalent noise temperature of the overall system, which has a power gain of $G = G_1 G_2 \cdots G_n$, then the output noise power is given by

$$N_{out} = (kT_e B)G = (kT_e B)G_1 G_2 \cdots G_n \quad (4.11)$$

Equating (4.10) and (4.11) to each other, the equivalent noise temperature of the system is found to be

$$T_e = T_{e1} + \frac{T_{e2}}{G_1} + \frac{T_{e3}}{G_1 G_2} + \cdots + \frac{T_{en}}{G_1 G_2 \cdots G_{n-1}}$$

(4.12)

One may observe from (4.12) that the equivalent noise temperature T_e of the receiver is dominated by T_{e1} and G_1, which usually correspond to the feeder line connecting the antenna to the receiver. In practice, low-noise amplifiers (LNAs) are used between the receiving antenna and the feeder line in order to decrease the receiver noise, since low T_{e1} and high G_1 decrease T_e significantly. One can also infer from

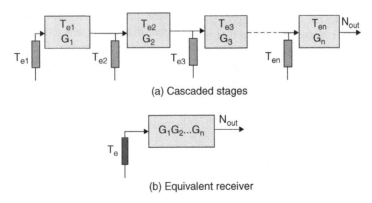

(a) Cascaded stages

(b) Equivalent receiver

Figure 4.6 Equivalent Noise Temperature of a Cascaded System.

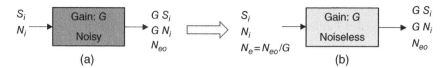

Figure 4.7 Definition of Noise Figure.

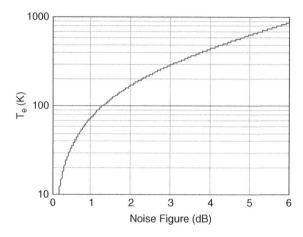

Figure 4.8 Variation of Equivalent Noise Temperature $T_e = (F-1)T_0$, $T_0 = 290$ K Versus Noise Figure F (dB).

(4.12) that the noise temperatures of the devices towards the end of the receiver do not have significant contribution to the receiver noise.

4.3 Noise Figure

Noise figure provides a measure of SNR degradation caused by a noisy device compared to a reference noise source at its input. Consider a noisy device with gain G as shown in Figure 4.7a. This device amplifies signal and noise powers S_i and N_i at its input so that the output signal and noise powers at the device output are given by GS_i and GN_i, respectively. In addition, the internal noise of the device also induces a noise power of N_{eo} at its output, which may be referred to its input as $N_e = N_{eo}/G$, as shown in Figure 4.7b.

The noise figure is defined as the ratio of the input SNR to the output SNR, for an input noise source at reference temperature $T_0 = 290$ K, that is, $N_i = kT_0B$, which implies an input noise

PSD of $N_0 = N_i/B = kT_0 = 4 \times 10^{-21}$ (W/Hz) $(-204 \ dBW/Hz)$:

$$F = \frac{SNR_i}{SNR_{out}} = \frac{S_i/N_i}{\dfrac{GS_i}{G(N_i + N_e)}} = 1 + \frac{N_e}{N_i} > 1 \ \ for \ \ N_i = kT_0B$$

$$(4.13)$$

For a noiseless device, one has $T_e = 0K$, $N_e = kT_eB = 0$ and $F = 1$ (0 dB). Since each device has $T_e \geq 0$ K, noise figure is always higher than 0 dB; hence, higher the noise figure, noisier the device. Inserting the internal noise power $N_e = kT_eB$ of the considered device when referred to its input into (4.13), the noise figure and the equivalent noise temperature may be related to each other as follows:

$$F = 1 + \frac{N_e}{N_i} = 1 + \frac{kT_eB}{kT_0B} \quad \Rightarrow \quad T_e = (F-1) \ T_0$$

$$(4.14)$$

Figure 4.8 shows the variation of the equivalent noise temperature T_e with noise figure F, usually expressed in dB. Note that T_e is less than 100 K for $F < 1.5$ dB; this is typical for a low-noise amplifier which may require cooling. A typical LNA has a noise figure of ~3 dB, which implies an equivalent noise temperature of 290 K.

Choice of $N_i/B = kT_0B$, as a reference input noise PSD, is meaningful for systems operating at around $T_0 = 290$ (K), that is, for terrestrial applications. However, $N_i/B = kT_0$ is not a good reference for space applications, where the ambient temperature may be highly different from $T_0 = 290$ K. For a reference temperature T_A different from T_0, the corresponding noise figure is called as *operating noise figure F_A*:

$$T_e = (F-1)T_0 = (F_A - 1)T_A$$
$$\Rightarrow F_A = 1 + (F-1)\frac{T_0}{T_A} \qquad (4.15)$$

It is clear from (4.15) that the value of noise figure depends on the assumption about the reference source temperature, but the value of the equivalent noise temperature T_e is independent of the reference source temperature.

Using (4.12) and (4.14), the equivalent noise figure F_e of cascaded subsystems may easily be written as

$$F_e = 1 + \frac{T_e}{T_0} = F_1 + \frac{F_2 - 1}{G_1} + \frac{F_3 - 1}{G_1 G_2} + \cdots$$
$$+ \frac{F_n - 1}{G_1 G_2 \cdots G_{n-1}} \qquad (4.16)$$

Example 4.2 Operating Noise Figure.
Consider a narrowband receiver system with multiple stages, each having the same bandwidth. The noise figure of the first stage is $F_1 = 10$ dB and it has a loss of $L_1 = -G_1 = 8$ dB. The following stage has a noise figure of $F_2 = 3$ dB and higher gain. The input noise temperature and the input SNR are given by 273 K (0 C) and $SNR_{in} = 50$ dB, respectively. This receiver system gives satisfactory results when $SNR_{out} \geq 40$ dB.

Ignoring the effects of stages other than the first two, and noting that the noise figures are

defined for input noise temperature of $T_0 = 290$ K, the equivalent noise figure of the receiver system when referred to $T_0 = 290$ K may be written from (4.16) as

$$F_{290} = F_1 + L_1(F_2 - 1) = 10 + 10^{0.8}(2-1) = 16.31 \qquad (4.17)$$

Using (4.15), the equivalent noise figure when referred to $T = 273$ K is given by

$$F_{273} = 1 + (F_{290} - 1)\frac{290}{273} = 17.26 \qquad (4.18)$$

The output SNR is found from (4.13) as

$$SNR_{out} = \frac{SNR_{in}}{F_{273}} = \frac{10^5}{17.26} \qquad (4.19)$$
$$\cong 5.79 \times 10^3 \ (37.6 \ dB)$$

Since $SNR_{out} < 40$ dB in the present configuration, one may use a pre-amplifier at the receiver input. In order to achieve $SNR_{out} > 40$ dB, it is evident from (4.19) that the overall noise figure should be less than 10 dB. The use of a pre-amplifier with F and G leads to

$$F_{290} = F + \frac{F_1 - 1 + L_1(F_2 - 1)}{G} = F + \frac{9 + 6.31}{G}$$
$$= F + \frac{15.31}{G} \qquad (4.20)$$

By inspection, when $F = 8$ and $G = 10$, one gets $F_{290} = 9.53$ and $F_{273} = 1 + (9.53 - 1)290/273 \cong 10$. Hence, the requirement for $SNR_{out} > 40$ dB is met if $F \leq 8$ and $G \geq 10$.

Example 4.3 Noise Figure of Broadband Devices.
The noise figure formulations presented until now are based on the assumption that the considered noisy devices are narrowband. However, the narrowband assumption may not be valid for devices used in broadband communication systems, for example, when the

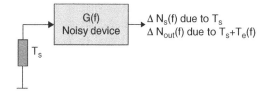

Figure 4.9   Broadband Noisy Device.

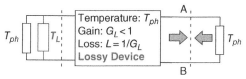

Figure 4.10   Noise Figure of a Lossy Device.

bandwidth exceeds several percent of the carrier frequency. Consider now a broadband noisy device with gain $G(f)$, as shown in Figure 4.9. For a source noise temperature T_s, which is assumed to be frequency-independent, the output noise power in a narrow bandwidth Δf around the frequency f may be written as [7]

$$\Delta N_s(f) = kT_s G(f)\, \Delta f$$
$$\Rightarrow\ N_s = kT_s \int_0^\infty G(f)\, df \tag{4.21}$$

The operating noise figure at frequency f and the corresponding output noise power are given by

$$F_{op}(f) = \Delta N_{out}(f)/\Delta N_s(f)$$
$$\Rightarrow\ N_{out} = kT_s \int_0^\infty F_{op}(f)\, G(f)\, df \tag{4.22}$$

where $\Delta N_{out}(f)$ denotes the total output noise power due to the source and the internal noise of the device in a narrow bandwidth Δf around the frequency f.

Using (4.13), (4.21) and (4.22), the average operating noise figure is found to be

$$\overline{F}_{op} = \frac{N_{out}}{N_s} = \frac{\displaystyle\int_0^\infty F_{op}(f)\, G(f)\, df}{\displaystyle\int_0^\infty G(f)\, df} \tag{4.23}$$

From (4.14), $T_e(f)$ and $\overline{T}_e$ correspond to $F_{op}(f)$ and $\overline{F}_{op}$, respectively:

$$F_{op}(f) = 1 + T_e(f)/T_s \ \Rightarrow\ \overline{F}_{op} = 1 + \overline{T}_e/T_s \tag{4.24}$$

Using (4.23) and (4.24), one may express the average equivalent noise temperature in terms of the equivalent noise temperature $T_e(f)$ at frequency f as

$$\overline{T}_e = \frac{\displaystyle\int_0^\infty T_e(f)\, G(f)\, df}{\displaystyle\int_0^\infty G(f)\, df} \tag{4.25}$$

If the equivalent noise temperature is independent of frequency, $T_e(f) = T_e$, then (4.25) leads to $\overline{T}_e = T_e$, as expected.

4.3.1   Noise Figure of a Lossy Device

Consider a lossy device with gain $G_L < 1$ and loss $L = 1/G_L > 1$, as shown in Figure 4.10. Let T_L and T_{ph} denote, respectively, equivalent noise temperature and physical temperature of this device. The device is assumed to be in thermal equilibrium, that is, the device and the two resistances connected to its input and output have the same physical temperatures T_{ph}. Since there can be no net flow of noise power at either end of this system, one may equate the noise powers flowing in opposite directions at the interface defined by AB in Figure 4.10:

$$k(T_{ph} + T_L)BG_L = kT_{ph}B \ \ at\ AB \tag{4.26}$$

From (4.26), equivalent noise temperature and the noise figure of this lossy device may be written as

$$T_L = (1/G_L - 1)T_{ph} = (L-1)T_{ph}$$
$$F = 1 + \frac{T_L}{T_0} = 1 + (L-1)\frac{T_{ph}}{T_0} \tag{4.27}$$
$$F = L \ \ for\ \ T_{ph} = T_0$$

In view of (4.27), the noise figure increases with increasing values of loss L. Therefore, the attenuation caused by a lossy device not only decreases the received signal level but also generates noise. Since signal attenuation and the noise figure are proportional to the feeder-cable length, the use of short (low-loss) feeder cables is strongly advised. In contrast with lossy devices, the SNR degradation in active devices (amplifiers) was already observed to result only from the injection of the amplifier internal noise into the link.

Example 4.4   Cascaded Lossy Devices.
In this example we will determine the equivalent noise temperature and the equivalent noise figure of two cascaded lossy devices; one with loss L_1 at physical temperature T_1 and the other with loss L_2 at physical temperature T_2. Using (4.12), the equivalent noise temperature and the corresponding noise figure are given by

$$T_e = T_{e1} + T_{e2}/G_1 = (L_1 - 1)T_1 + L_1(L_2 - 1)T_2$$
$$F_e = 1 + T_e/T_0$$
$$= 1 + (L_1 - 1)T_1/T_0 + L_1(L_2 - 1)T_2/T_0$$
$$(4.28)$$

For the special case of $T_1 = T_2 = T_0$, the equivalent noise figure reduces to $F_e = L_1 L_2$ and the corresponding noise temperature becomes $T_e = (L_1 L_2 - 1)T_0$.

If the cascade-connected lossy devices above represent two feeder cables and replace the lossy device shown in Figure 4.10 with $T_{ph} = T_0$, then the output SNR may be related to the input SNR as follows:

$$SNR_{out} = \frac{S_{out}}{N_{out}} = \frac{S_i/(L_1 L_2)}{N_i} = \frac{1}{L_1 L_2} SNR_i$$
$$SNR_i = \frac{S_i}{N_i} = \frac{S_i}{kT_0 B}$$
$$(4.29)$$

where $N_{out} = N_i = kT_0 B$ since the system is in thermal equilibrium.

4.4 External Noise and Antenna Noise Temperature

Antenna noise temperature provides a measure of the noise power received by an antenna from external noise sources. It depends on the radiation pattern, the frequency of operation, the system bandwidth and spatial and spectral distribution of the noise sources surrounding an antenna. The external noise sources include the troposphere, the ionosphere, the Earth's surface, the moon, the planets, the sun and the other stars and galaxies. Man-made noise also contributes the antenna noise temperature.

The contribution of the atmospheric noise increases in proportion to the thickness of the atmosphere seen by the antenna; hence, it is minimum at zenith and increases with decreasing elevation angle. The clear sky has a background noise temperature of 3 K, which is attributed to the big bang. Contribution of planets, stars and galaxies depend on whether the antenna is looking towards them, the frequency of operation and the brightness of the source. In summary, the antenna noise temperature is determined by the received noise power from external noise sources, weighted by the antenna radiation pattern in frequency and space coordinates.

In telecommunication systems, transmitters are usually considered as point sources when they are located in the far-field of the receiver. The Poynting vector intensity radiated by a point source decreases with the square of the distance as signal propagates away from the transmitter (see for example (2.13)). The power received by a receiver is proportional to the intercepted flux density in (W/m^2) times the effective aperture of the receiver (see (2.96) and Figure 4.11). The noise sources may behave as point sources if their size is relatively small or their distance to the receiver is very high. However, in majority of cases, the noise sources are extended in space; hence they are analysed differently.

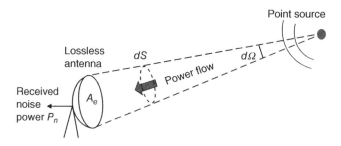

Figure 4.11   Reception From a Point Noise Source Through a Solid Angle $d\Omega$.

4.4.1   Point Noise Sources

if a noise source may be assumed as a point source, then it is meaningful to describe the received signals by the Poynting vector intensity (flux density) in W/m². Power flux density from a point source, which is assumed to be isotropic, is given by the Poynting vector intensity (see also (2.53))

$$S_n(r) = \frac{P_{tn}}{4\pi r^2} \quad \left(\frac{W}{m^2}\right) \qquad (4.30)$$

where P_{tn} denotes the total radiated power by the noise source. When the source is not isotropic, the Poynting vector intensity changes with direction $S_n(r, \theta, \phi)$. Then, the power flux through a solid angle $d\Omega$ (steradian) is given by (see Figure 4.11)

$$d\Phi(r) = S_n(r,\theta,\phi)\ dS = S_n(r,\theta,\phi)r^2 d\Omega \quad (W)$$
$$(4.31)$$

In spherical coordinates, the surface element dS is given by

$$dS = r^2 d\Omega = r^2 \sin\theta d\theta d\phi \qquad (4.32)$$

The power received by an antenna located at distance r may be written as

$$P_n = \iint_{A_A} S_n(r,\theta,\phi)\ dS \quad (W) \qquad (4.33)$$

where integration is over the antenna aperture A_A. Figure 4.11, (4.31) and (4.33) show that, in order to have a constant flux through a solid

angle $d\Omega$ (steradian), the antenna size should be increased in proportion with r^2. If antenna size is kept constant, then the power received by the antenna (hence the power flux) will decrease in proportion with r^2. If the Poynting vector intensity may be assumed to be constant over the antenna aperture, then (4.33) simply reduces to (2.96), which states that the received power is equal to the product of the Poynting vector intensity and the effective receiving area of the antenna.

4.4.2   Extended Noise Sources and Brightness Temperature

In many cases, the sources contributing to the antenna noise can not be described as point sources since they are extended in space coordinates. Consider an extended noise source as shown in Figure 4.12 and a receiving antenna at distance r collects the noise power from all the extended source region.

The concept of blackbody radiation is suitable for characterizing radiation from extended noise sources. Although no perfect blackbody exists in nature, many objects (e.g., the sun, stars) behave approximately like blackbody over at least a range of frequencies. A blackbody is an idealized body which absorbs the electromagnetic energy at all frequencies impinging upon it. A blackbody in thermal equilibrium has the same rate of energy emission and absorption. According to the Planck's Law, the brightness, that is, the power

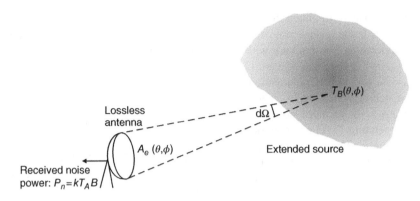

Figure 4.12   Noise Power Reception From an Extended Noise Source.

spectral density (PSD) radiated isotropically from a unit area of blackbody at temperature T (K) per solid angle, is given by [4]

$$B = \frac{2hf^3}{c^2} \frac{1}{e^{hf/kT}-1} \quad \left(\frac{W/m^2}{Hz \times sr}\right) \quad (4.34)$$

where $k = 1.38 \ 10^{-23}$(J/K) is the Boltzmann's constant and $h = 6.6254 \ 10^{-34}$(J.s) denotes the Planck's constant. At RF frequencies, where $hf << kT$, one can use $e^x \approx 1 + x$ in (4.34) to get the Rayleigh-Jeans approximation:

$$B = \frac{2kT}{\lambda^2} \quad \left(\frac{W/m^2}{Hz \times sr}\right), \quad for \ hf << kT \quad (4.35)$$

which is valid for all RF frequencies of interest, $f << T \times 20 \ GHz$, (e.g., for $f << 6 \ THz$ for $T = 300$ K).

We now determine the power received from an extended noise source as shown in Figure 4.12 by a receiving antenna whose beam is confined to a solid angle $d\Omega$. The antenna beam intercepts an extended noise source with brightness $B(\theta, \phi)$, which implies that T in (4.34) and (4.35) is replaced by $T_B(\theta, \phi)$, which is called the *brightness temperature* of the noise source in the direction $\{\theta, \phi\}$. If an antenna with effective receiving area $A_e(\theta, \phi)$ is pointed in the direction (θ, ϕ), the power received per bandwidth Δf from an angular region defined by $d\Omega$ may be written as

$$P_n = \Delta f \frac{1}{2} \int\int B(\theta, \phi) \, A_e(\theta, \phi) \, d\Omega \quad (W)$$

$$= \Delta f \frac{\lambda^2}{8\pi} \int\int B(\theta, \phi) \, G(\theta, \phi) \, d\Omega \quad (W)$$

$$= \Delta f \frac{1}{4\pi} \int\int \frac{hf}{\exp[hf/kT_B(\theta,\phi)]-1} \, G(\theta, \phi) \, d\Omega \quad (W)$$

$$(4.36)$$

where (2.98) is used to express the effective area of the receive antenna in terms of its gain. The factor ½ in front of the first expression in (4.36) accounts for the fact that the randomly polarized noise power is equally divided between two orthogonal polarizations and an antenna can extract noise power only from one of the two orthogonal polarizations.

In the special case where Rayleigh-Jeans approximation is valid, that is, the noise PSD is flat in the frequency range of interest, then (4.36) for a lossless antenna simplifies to

$$P_n = kT_A B$$

$$T_A = \frac{1}{4\pi} \int_0^{2\pi} \int_0^{\pi} D(\theta, \phi) T_B(\theta, \phi) \, \sin\theta d\theta d\phi$$

$$(4.37)$$

Note that Δf is replaced by B, the bandwidth, since the PSD is flat. On the other hand, the gain $G(\theta, \phi)$ is replaced by the directivity $D(\theta, \phi)$ for the lossless antenna. It is clear from (4.37) that the brightness temperature of the extended noise source is weighted by the antenna radiation pattern to obtain the antenna noise temperature. In the special case where the brightness temperature does not vary with direction, $T_B(\theta,\phi) = T_B$, then the use of (2.70) leads to $T_A = T_B$ for perfect polarization and impedance matching. Then, (4.36) reduces to (4.1), which gives the received power from a resistor at temperature T. As one may observe from (4.37), a highly directive receive antenna (with very narrow beamwidth), whose directivity $D(\theta,\phi)$ approximated as a delta function in (θ,ϕ), may be used to map $T_B(\theta,\phi)$. The brightness temperature of a noise source is not identical but is related to its physical temperature as given by

$$T_B(\theta,\phi) = \varepsilon(\theta,\phi)T_{ph}(\theta,\phi) + \rho T_{atm} \quad (4.38)$$

where $\varepsilon(\theta,\phi)$ denotes the emissivity of the source, $T_{ph}(\theta,\phi)$ is its physical temperature, ρ is the reflection coefficient and T_{atm} denotes the weighted average of the sky brightness temperature. The term ρT_{atm} is more suited to describe the brightness temperature of the Earth's surface and accounts for the reflection of the atmospheric noise. Up to about 100 GHz, but particularly below 10 GHz, the reflection coefficient ρ from the Earth's surface is generally high and the emissivity is low. [1]

4.4.3   Antenna Noise Figure

The received noise power due to external sources is expressed in (4.37) in terms of the equivalent noise temperature of a lossless antenna. Similarly, the received noise power may also be measured by the equivalent noise figure F_A of a lossless antenna, which is defined by the ratio of the received noise power P_n to the reference noise power $kT_0 B$: [1]

$$F_A = \frac{P_n}{kT_0 B} \quad (4.39)$$

where $k = 1.38 \times 10^{-23}$ J/K is the Boltzmann's constant and $T_0 = 290$ K is the reference ambient noise temperature. The noise figure F_A is related to the equivalent noise temperature T_A of a lossless antenna as [1]

$$F_A = \frac{P_n}{kT_0 B} = \frac{kT_A B}{kT_0 B} = \frac{T_A}{T_0} \quad (4.40)$$

In view of (4.37), the value of T_A is determined by the parameters of the noise source and the antenna radiation pattern. Using (4.39), the available noise power P_n from an equivalent lossless antenna may be written as

$$P_{n,dB} = 10 \log F_A + 10 \log B - 204 \quad (dBW) \quad (4.41)$$

In some low-frequencies applications, dipole antennas are used for signal reception. In these cases, the received signal and noise levels are determined by the received rms electric field intensity. Using $P_n/B = |E_n|^2 \lambda^2 G/(480\pi^2)$ given by (2.120) the received rms field intensity due to noise may be expressed as

$$E_{n,dB} = 10 \log F_A + 20 \log f_{MHz} \\ -96.8 - G_{dB} \quad dB(\mu V/m) \quad (4.42)$$

where G_{dBi} denotes the gain of the receiving dipole antenna. It is evident from (4.41) and (4.42) that the antenna noise figure F_A, hence T_A, may be determined if the received noise power or the rms electric field intensity is known.

4.4.4   Effects of Lossy Propagation Medium on the Observed Brightness Temperature

The brightness temperature of external noise sources is observed through a propagation

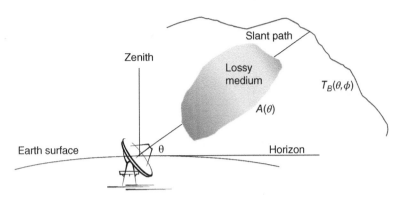

Figure 4.13 Geometry of the Slant Path.

medium which attenuates signals and also generates noise; this affects the noise power received by the antenna. At frequencies higher than 10 GHz, rain attenuation and atmospheric absorption may cause significant degradation in the system performance due to increases both in system noise and in signal attenuation. For example, the propagation medium between the sky, of brightness temperature $T_B(\theta, \phi)$, and the receive antenna attenuates signals and also changes the observed brightness temperature (see Figure 4.13). The observed brightness temperature $T_B'(\theta, \phi)$ may be expressed as

$$T_B'(\theta, \phi) = \frac{T_B(\theta, \phi)}{A(\theta)} + \left(1 - \frac{1}{A(\theta)}\right) T_{medium}$$

(4.43)

The medium temperature T_{medium} is usually related to the earth-surface temperature $T_{surface}$ as [8][9]

$$T_{medium} = 1.12 T_{surface} - 50 \ (K)$$ (4.44)

Here $A(\theta)$ denotes the signal attenuation at elevation angle θ due to absorption of electromagnetic waves by the gaseous constituents (oxygen, hydrogen, water vapour, and so on)

of the atmosphere and/or attenuation due to precipitation, for example, rain, snow. If the loss in the direction of zenith is denoted by A_z, the loss at elevation angle θ is approximated by (3.55). Similarly, the attenuation due to rain is predicted by (3.74) or (3.77). Note that (4.43) accounts for the noise caused by a lossy propagation medium, for example, atmospheric absorption, rain, snow and water vapour. Since T_{medium} (~270 K) is 'hotter' than the brightness temperature of the sky (~3 K), a lossy propagation medium always increases the observed brightness temperature.

Figure 4.14 shows the variation of $T_B'(\theta)$ in (4.43) as a function of the the elevation angle θ. The loss is modeled as $A(\theta) = A_z / \sin \theta$ (dB) where A_z denotes the zenith attenuation of the lossy propagation medium. The physical temperature of the medium is given by (4.44) with $T_{surface} = 290°K$ and $T_B(\theta) = T_{B,z} / \sin \theta$, where the zenith brightness temperature is assumed to be $T_{B,z} = 3K$. Note from (4.43) that if $A(\theta) = 1$ (lossless medium), then $T_B'(\theta) = T_B(\theta)$. However, in a propagation medium with high losses, one has $A(\theta) \gg 1$ and $T_B'(\theta) = T_{medium}$. This implies that an antenna looking towards a region in the sky with high losses, for example, heavy rain intensity, suffers not only from heavy signal losses but also from the increased noise power received by the antenna.

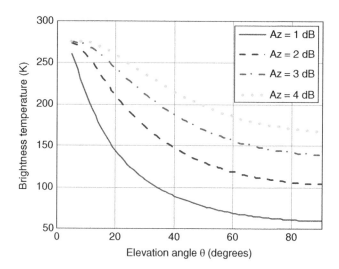

Figure 4.14 Brightness Temperature as a Function of the Elevation Angle Based on the Values of Zenith Attenuation $A_z = 1$–4 dB. The background brightness temperature and the path-loss are assumed to have a 1/sinθ dependence. The sky brightness temperature in the direction of zenith is assumed to be $T_{B,z} = 3$ K. The physical temperature of the medium is assumed to be $T_{surface} = 290$ K.

Example 4.11 Calculation of Antenna Noise Temperature.

As shown in Figure 4.15, we consider a satellite digital TV (DVB-S) receiving system operating at 12 GHz. The dish antenna of diameter $D_a = 0.6$ m $= 24\lambda$ of the receiver looks at the satellite with an elevation angle of θ_{el} through a lossy medium. For uniform illumination, this antenna has 2.46 degrees half-power beamwidth (see (2.83)) and 37.55 dBi gain (see (2.101)).

The radiation pattern of this uniformly illuminated dish antenna is given by (2.50) (see Figure 2.11)

$$P(\theta,\phi) = \begin{cases} |f(\theta)|^2 & 0 \le \theta \le \pi/2 \\ |f(\pi/2)|^2 & \theta > \pi/2 \end{cases} \quad (4.45)$$

where $f(\theta) = 2J_1(u)/(u)$, $u = \pi(D_a/\lambda)\sin\theta$ and θ is defined with respect to the boresight direction (direction of maximum radiation of the antenna). The sky brightness temperature of $T_B(\theta,\theta_{el}) = 4K$ is observed by the antenna through atmosphere which also generates

noise for $\theta \le \theta_{el}$. For $\theta > \theta_{el}$, the brightness temperature seen by the antenna is assumed to be the ground noise temperature T_0:

$$T_B(\theta,\theta_{el}) = \begin{cases} T_B'(\theta,\theta_{el}) & \theta < \theta_{el} \\ T_0 & \theta \ge \theta_{el} \end{cases}$$

$$T_B'(\theta,\theta_{el}) = \frac{T_B(\theta,\theta_{el})}{A(\theta,\theta_{el})} + \left[1 - \frac{1}{A(\theta,\theta_{el})}\right] T_{medium}$$

$$(4.46)$$

where $A(\theta,\theta_{el}) = A_z/\sin(\theta_{el}-\theta)$ denotes the direction-dependent attenuation caused by the propagation medium and A_z is the zenith attenuation. T_{medium} is determined using (4.44) with $T_{surface} = T_0$.

Since $P(\theta,\phi)$ is proportional to the Poynting vector intensity, using (2.69), the antenna directivity may be written as

$$D(\theta,\phi) = \frac{4\pi P(\theta,\phi)}{\displaystyle\int_{\phi=0}^{2\pi}\int_{\theta=0}^{\pi} P(\theta,\phi)\sin\theta \, d\theta \, d\phi} \quad (4.47)$$

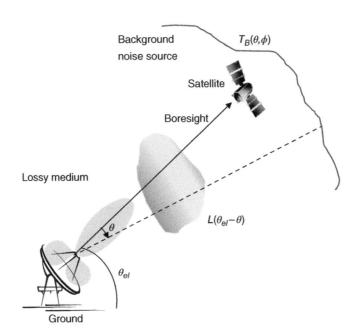

Figure 4.15 An Antenna Looking at the Sky with an Elevation Angle θ_{el} Through a Lossy Medium.

Inserting (4.47) into (4.37), the noise temperature for a lossless antenna ($\eta_L = 1$) may be written as

$$T_A = \frac{\displaystyle\int\limits_{\phi=0}^{2\pi}\int\limits_{\theta=0}^{\pi} T_B(\theta,\theta_{el})P(\theta,\phi)\sin\theta \, d\theta \, d\phi}{\displaystyle\int\limits_{\phi=0}^{2\pi}\int\limits_{\theta=0}^{\pi} P(\theta,\phi)\sin\theta \, d\theta \, d\phi}$$

(4.48)

where $T_B'(\theta,\theta_{el})$ is defined by (4.46). Figure 4.16 shows the variation of (4.48) and the corresponding value of G/T_A as a function of the elevation angle for a typical zenith attenuation of $A_z = 0.2$ dB due to atmospheric absorption. For low elevation angles, the atmospheric absorption losses are high due to increased atmospheric thickness. This leads to higher antenna noise temperatures and lower G/T_A values. With increased elevation, antenna noise temperature decreases and G/T_A becomes higher. Increased absorption losses and antenna noise makes communications impractical at low

elevation angles. In practice, elevation angle is usually chosen to be higher than 10 degrees.

Assume that this lossless antenna is connected to a receiver of noise figure of $F = 10$ dB by first a low-noise amplifier with $F_{PA} = 3$ dB noise figure and $G_{PA} = 20$ dB gain using a cable of $L = 1$ dB loss. The effective noise figure and noise temperature of the LNA-cable–receiver combination are found to be

$$F_R = F_{PA} + \frac{(FL-1)}{G_{PA}} = 2 + \frac{(10\times 10^{0.1} - 1)}{100} = 2.1$$

$$T_R = (F_R - 1)T_0 = 323.6K$$

(4.49)

The system noise temperature is given by the sum of T_R and T_A. Correspondingly, the figure of merit G/T_S of the receiver system will be less than shown in Figure 4.16 since $T_S > T_A$.

In 5 MHz bandwidth, the noise power received by this system at an elevation angle of $\theta_{el} = 60$ degrees is found as

$$N_{dBW} = 10 \log[k(T_A + T_R)B] = -136.3 \; dBW$$

(4.50)

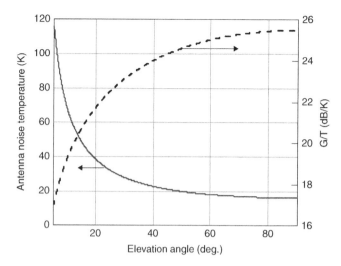

Figure 4.16   The Variation of G/T_A and the Antenna Noise Temperature T_A Versus Elevation Angle for a Uniformly Illuminated Dish Antenna Operating at 12 GHz with Diameter $D_a = 0.6$ m $(D_a = 24\lambda)$. Antenna radiation pattern and brightness temperature surrounding the antenna are respectively specified by (4.46) and (4.47). Zenith attenuation is assumed to be $A_z = 0.2$ dB.

where $T_A = 17.9$ K. If an SNR of 15 dB is required for proper operation of the system, the receiver can detect the received signals as long as their power level is higher than

$$S_{dB} \geq N_{dBW} + 15 \ dB = -121.3 \ dBW \quad (4.51)$$

which is referred to as the *receiver sensitivity*.

4.4.5   Brightness Temperature of Some Extended Noise Sources

Any object in the universe radiates noise as long as its temperature is higher than 0°K. An antenna receives thermal radiation emanating from sources in its environment by weighting them with its radiation pattern. Since antennas are usually directed with an elevation angle, they see the sky with their main beam and some near-in sidelobes. Thermal radiation from Earth's surface contributes to the noise of a ground-based antenna through its far-out sidelobes. Compared with background 'cold' sky, with a brightness temperature of

approximately 3 K, the 'hot' surface of the earth with 290 K brightness temperature provides the largest contribution to the antenna noise. Therefore, the suppression of antenna sidelobes is of great interest, especially in low-noise receiver systems used in the area of telecommunications, radar, radioastronomy and remote sensing. Special attention should be paid for not directing the main lobe of the antenna to external noise sources like stars, sun and moon since this causes excessive increases in the antenna noise temperature.

Figure 4.17 shows the emissivity and the brightness temperature of a smooth water surface for vertical and horizontal polarizations and for two angles of incidence. The curves for fresh and sea water are indistinguishable for frequencies higher than 5 GHz. The emissivities (and hence the brightness temperatures) of land surfaces are higher than those of water surfaces due to smaller dielectric constant of land. Figure 4.18 shows the brightness temperature of a smooth ground surface for different moisture contents and for vertical,

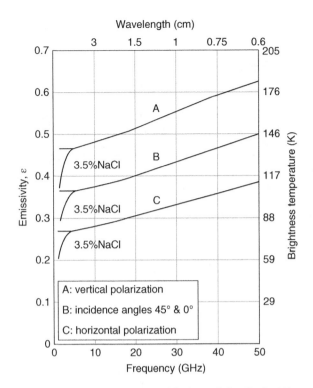

Figure 4.17 Emissivity and Brightness Temperature of the Smooth Sea Surface Versus Frequency for 3.5% Salinity. [1]

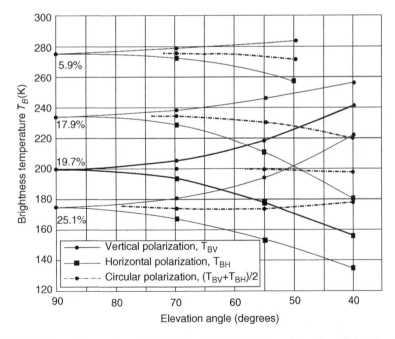

Figure 4.18 Brightness Temperature at 1.43 GHz of the Ground as a Function of the Elevation Angle. Moisture content from 5.9% to 25.1% for a bare smooth ground. [1]

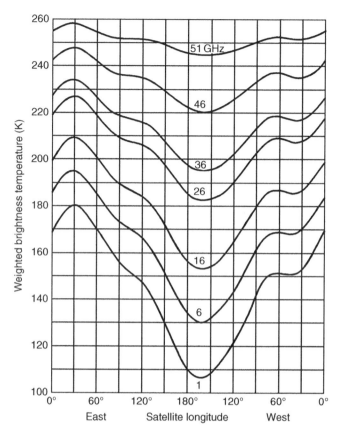

Figure 4.19 Weighted Brightness Temperature of the Earth as a Function of Longitude Viewed From Geostationary Orbit by an Earth-Coverage Beam at Frequencies Between 1 GHz and 51 GHz. [1]

horizontal and circular polarizations. If the moisture content increases, the brightness temperature decreases. The brightness temperature increases as the surface becomes rougher. [1]

We now consider an Earth-coverage antenna onboard a geostationary satellite that looks at the Earth. The radiation pattern of the Earth-coverage antenna pattern is assumed to be

$$G(\theta) = -3(\theta/8.715)^2 \ (dB) \ \ 0 \le \theta \le 8.715 \tag{4.52}$$

where θ denotes the angle off boresight and $\theta_{3-dB} = 2 \times 8.715 = 17.43$ (deg) is the half-power antenna beamwidth. Noting that the Earth, of 6370 km radius, is subtended by

17.43 degrees at the geostationary satellite, located at 35870 km from the surface of the Earth, the Earth fills the main beam of the Earth-coverage satellite antenna between 3 dB points. As the satellite moves around its orbit, the receiving antenna sees the African (hot) land mass at 30^0 East longitude and the (cold) Pacific at 180^0 West to 150^0 West longitude. Therefore, the antenna noise temperature provides a weighted average of the Earth's brightness temperature. Brightness temperature increases with frequency largely due to atmospheric absorption. Curves shown in Figure 4.19 are based on the assumption that water vapour density is 2.5 g/m^3 and 50% cloud coverage.

We now consider the case of a ground-based antenna looking at the sky. Unless directed to a

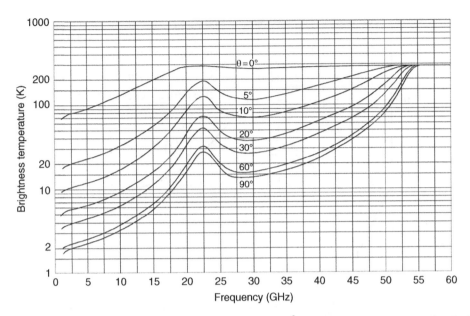

Figure 4.20   Brightness Temperature for Clear Sky for 7.5 g/m^3 Surface Water Vapour Density, Surface Temperature of 288 K, and a Scale Height of 2 km for Several Values of the Elevation Angle. [1]

noise source like sun, moon, stars or a galaxy, the antenna sees the empty sky with brightness temperature of 3 K through the troposphere of $T_{medium} = 274$ K (see (4.44)). In view of (4.43), the cold troposphere masks the sky radiation and the brightness temperature observed by the receiver antenna is that of the troposphere. Figure 4.20 shows the brightness temperature of a ground-based receiver excluding the sky noise contribution of 3 K or other extra-terrestrial sources for frequencies between 1 and 60 GHz. The curves are calculated for several elevation angles and an average atmosphere (7.5 g/m^3 surface water vapour density, surface temperature of 288 K, and a scale height of 2 km for water vapour). Note that the variation of the brightness temperature is similar to that of the atmospheric absorption loss shown in Figure 3.23. Similarly, the variation of the brightness temperature with elevation angle follows the cosecant law for the atmospheric path attenuation in (3.55). This implies that the background sky noise temperature of 3 K is masked by the

'hot' troposphere and the noise temperature of an antenna on the Earth's surface is determined by that of the atmospheric gases.

Figure 4.21 shows the brightness temperature of some extra-terrestrial noise sources in the frequency range 0.1 to 100 GHz. For frequencies up to about 100 MHz, the median brightness temperature for galactic noise for a vertical antenna, neglecting ionospheric shielding, is given by. [1]

$$T_A = 45.96 \times 10^6 / f_{MHz}^{2.3} \quad (K) \qquad (4.53)$$

Below 2 GHz, one needs to be concerned with the Sun and the galaxy (the Milky Way), which appears as a broad belt of strong emission. Above 2 GHz, one needs to consider only the Sun and a few very strong non-thermal sources such as Cassiopeia A, Cygnus A and X and the Crab nebula since the cosmic background contributes 3 K only and the Milky Way appears as a narrow zone of somewhat enhanced intensity. [1]

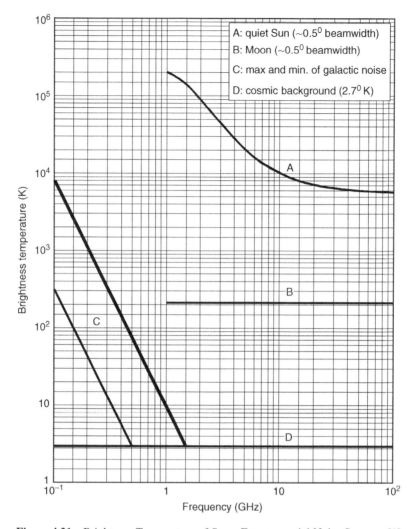

Figure 4.21   Brightness Temperature of Some Extraterrestrial Noise Sources. [1]

The noise due to active sun (with solar flares) is broadband and is much higher compared to those due to planets. Active sun refers to intense solar activity resulting in anomalously large solar flares for three-year durations with a periodicity of 11 years. Figure 4.21 shows that the brightness temperature of the quiet sun is $2 \times 10^5 K$, $10^4 K$, and $6 \times 10^3 K$ at 1 GHz, 10 GHz and 100 GHz, respectively. The brightness temperature of the Moon is ~210 K between 1–100 GHz when a 0.5 deg. beamwidth antenna points to the Moon. [10][11]

4.4.5.1   Frequency Dependence of the Antenna Noise Temperature

Figure 4.22 shows the maximum and minimum hourly median values of T_A in the frequency range 0.1 Hz to 10 kHz, based on measurements on all seasons and times of day on the entire Earth's surface. In this frequency range, seasonal, diurnal or geographic variations are not significant. Larger variability in T_A in the 100–10000 Hz range is due to the variability of the cutoff frequency of the Earth-ionosphere waveguide. Figure 4.23 covers the frequency range

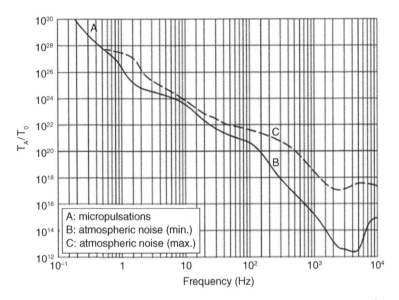

Figure 4.22 Minimum and Maximum Values of T_A Versus Frequency (0.1 to 10^4 Hz). [1]

10 kHz–100 MHz. Minimum and maximum values of the hourly median atmospheric noise correspond to values exceeded 99.5% and 0.5% of the hours, respectively. Atmospheric noise curves are based on measurements on all seasons and times of day on the entire Earth's surface. In the frequency range up to 30 MHz (LF, MF and HF), the antenna noise and hence the receiver performance is dominated by the atmospheric noise. The noise level is so high that it is not feasible to invest for a low-noise receiver.

In the VHF band (30–300 MHz), galactic noise is the dominant contributor. The antenna noise temperature may become 10 times higher when it looks towards the galactic center compared to the galactic pole. Average value of the galactic noise (over the entire sky) is given by the solid curves labelled galactic noise in Figures 4.23 and 4.24. Measurements indicate a ±2 dB variation about this curve, neglecting ionospheric shielding. The minimum galactic noise (narrow beam antenna towards galactic pole) is 3 dB below the solid galactic noise curve shown on Figure 4.24 and the maximum galactic noise for narrow beam antennas is shown via a dashed curve.

The UHF band (0.3–3 GHz) provides a low-noise window for RF communications, since the antenna (external) noise is the lowest. In this band, absorption of the electromagnetic waves by the atmosphere and the water vapor are the major contributors to the antenna noise. Losses due to atmospheric absorption increase with the thickness of the atmosphere seen by the antenna main lobe and the elevation angle. In view of (4.43), this also implies increased antenna noise. Hence, a communication system with an antenna main beam directed close to the horizon has a higher antenna noise temperature compared to when it looks towards the zenith.

The antenna temperature, T_A, is the weighted average of the brightness temperature of the sky and ground by the antenna pattern (see (4.37)). For antennas whose patterns encompass a single source, the antenna temperature and brightness temperature are the same (curves C, D and E of Figure 4.24, for example). The majority of the results shown in Figures 4.22–4.24 are for omnidirectional antennas except unless otherwise noted. For very narrow-beam HF antennas, however, studies have indicated that atmospheric noise due to lightning can vary as much as 10 dB (5 dB around the average

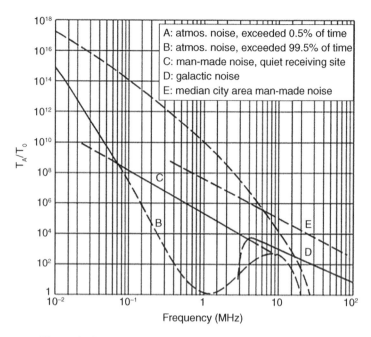

Figure 4.23 T_A Versus Frequency (10 kHz to 100 MHz). [1]

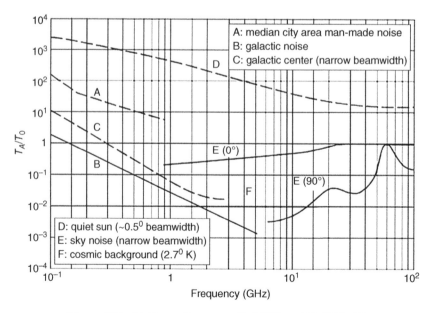

Figure 4.24 T_A Versus Frequency (100 MHz to 100 GHz). [1]

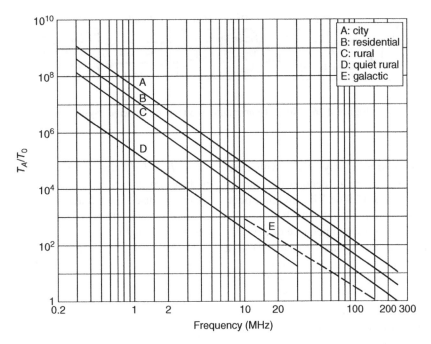

Figure 4.25 Median Values of Man-Made Noise Power received by a Short Vertical Lossless Grounded Monopole Antenna. [1]

Table 4.1 Values of the Constants c and d. [1]

Environmental category	c	d
City (curve A)	7.68	2.77
Residential (curve B)	7.25	2.77
Rural (curve C)	6.72	2.77
Quiet rural (curve D)	5.36	2.86
Galactic noise (curve E)	5.20	2.30

T_A value shown) depending on antenna pointing direction, frequency and geographical location.

4.4.6 Man-Made Noise

Median values of man-made noise power for several environments is shown in Figure 4.25 which also includes a curve for galactic noise for comparison purposes. In all cases, the median value of T_A varies with frequency as

$$T_A = T_0 \, 10^c / f_{MHz}^d \qquad (4.54)$$

where c and d take the values given in Table 4.1. Note that (4.54) is valid in the range 0.3 to 250 MHz for all the considered environments except those of curves D and E in Figure 4.25. The man-made noise considered here has a Gaussian distribution. Man-made noise often has an impulsive component and this may have an important effect on the performance of radio systems and networks.

4.5 System Noise Temperature

Besides the noise generated within the receiver system itself (including the noise due to antenna losses), an antenna collects the external noise by weighting with its radiation pattern. Thermal noise generated by a lossy antenna will also contribute to the antenna noise temperature. As shown in Figure 4.26, a lossy antenna is usually represented as an ideal

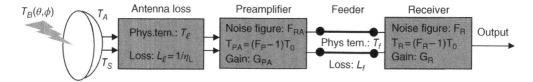

Figure 4.26   Noise Temperature of a Generic Receiver System.

antenna followed by a lossy device, with a loss factor $L_\ell = 1/\eta_L$, where η_L denotes the antenna loss efficiency (see (2.94)). The reference point for the overall noise level, which is usually measured by the noise temperature of a receiving system, is the terminals (front-end) of a noise-free receiving antenna. The system noise temperature T_S of a receiving terminal, shown in Figure 4.26, accounts for the receiver internal noise, the noise due to lossy feeder line, internal noise of a pre-amplifier (optional), and the antenna noise (see (4.12)):

$$T_S = T_A + T_{aL} + L_\ell T_{PA} + \frac{L_\ell}{G_{PA}} T_F + \frac{L_\ell L_f}{G_{PA}} T_R$$

$$= T_A + (L_\ell - 1) T_\ell + L_\ell (F_{PA} - 1) T_0$$

$$+ \frac{L_\ell}{G_{PA}} (L_f - 1) T_f + \frac{L_\ell L_f}{G_{PA}} (F_R - 1) T_0$$

$$(4.55)$$

where $\{T_{aL}, T_\ell, L_\ell, F_\ell\}$ denote the equivalent noise temperature, the physical temperature, attenuation and the noise figure of the antenna losses. Similarly, $\{T_F, T_f, L_f, F_f\}$ denote the equivalent noise temperature, the physical temperature, attenuation and the noise figure of the feeder line. They are related to each other by (4.27):

$$T_{aL} = (L_\ell - 1) T_\ell = (F_\ell - 1) T_0$$
$$T_F = (L_f - 1) T_f = (F_f - 1) T_0$$

$$(4.56)$$

If $T_\ell = T_f = T_0$, then (4.55) simplifies to

$$\frac{T_S}{T_0} = F_S - 1 = F_A - 1 + L_\ell F_{PA} + \frac{L_\ell}{G_{PA}} (L_f F_R - 1)$$

$$(4.57)$$

where $F_A = T_A/T_0$ from (4.40). The noise figure of the receiver system F_S, which is related to the system noise temperature T_S by (4.14), accounts for the effects of both both the internal and the external noise.

For the special case when there is no preamplifier, that is, $F_{PA} = G_{PA} = 1$, then (4.57) reduces to

$$\frac{T_S}{T_0} = F_S - 1 = F_A + L_\ell L_f F_R - 1 \qquad (4.58)$$

where the system noise figure is given by $F_S = F_A + L_\ell L_f F_R$. Furthermore, if the antenna noise temperature can be approximated as $T_A \cong T_0$ in the 1–10 GHz band (see Figure 4.24), then insertion of $F_A = T_A/T_0 \cong 1$ into (4.58) leads to

$$\frac{T_S}{T_0} = F_S - 1 = L_\ell L_f F_R \qquad (4.59)$$

where the system noise figure is determined by the product of antenna losses, feeder cable losses and the receiver noise figure. This evidently shows the importance of keeping antenna and feeder line losses as low as possible.

Now consider a simplified baseline receiver system, as shown in Figure 4.27, by shifting the pre-amplifier in Figure 4.26 to the receiver input and combining with the receiver and combining the antenna losses with the feeder-line losses. Then, the considered receiver system may simply be described by a receiving antenna, of noise temperature T_A, connected to the receiver of noise figure F and gain G:

$$F = F_{PA} + (F_R - 1)/G_{PA}$$
$$G = G_{PA} G_R$$

$$(4.60)$$

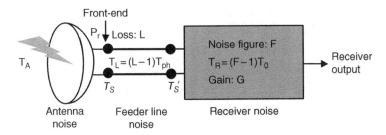

Figure 4.27   Definition of Input and Output SNR's In a Baseline Communication Receiver.

(see (4.16)) via a feeder line of physical temperature T_{ph} and loss L (see Example 4.4 and (4.58)):

$$L = L_\ell L_f \tag{4.61}$$

The effective gain and the effective noise figure of the feeder line-receiver combination are given by G/L and FL, respectively (see (4.58)). The effective noise temperature and the noise figure of this simplified baseline receiving system, shown in Figure 4.27, are given by

$$T_S = (F_S - 1)T_0 = T_A + T_L + LT_R$$
$$= T_A + (L-1)T_{ph} + L(F-1)T_0$$
$$= \begin{cases} T_A + (LF-1)\,T_0 & T_{ph} = T_0 \\ \\ LFT_0 & T_A = T_{ph} = T_0 \end{cases} \tag{4.62}$$

The term LT_R shows the equivalent noise temperature of the receiver referred to the input of the feeder line. The noise figure of the feeder-receiver referred combination is given by LF for $T_{ph} = T_0$. If LF is high, then T_S is dominated by $(LF-1)T_0$ by masking the antenna noise temperature T_A. On the other hand, if LF is sufficiently low, then T_S approaches T_A. In this case, the noise of the line-receiver combination becomes negligible.

The system noise temperature T_S may be written as when it is referred to the receiver input

$$T_S' = \frac{T_S}{L} = \frac{1}{L}[T_A + T_L + LT_R]$$
$$= \frac{T_A}{L} + \left(1 - \frac{1}{L}\right)T_{ph} + T_R \tag{4.63}$$

This implies that the antenna noise temperature is seen at the receiver input just as the way the sky brightness temperature is seen at the receiver antenna through a lossy propagation medium (see (4.43)).

The SNR at the receiver output may be written as

$$SNR = \frac{P_r}{kT_S B} \tag{4.64}$$

where P_r denotes the received signal power and T_S is the noise temperature of the receiver system, both referred to the receiver front-end. Note that (4.64) would be the same irrespective of whether it is referred to the front-end, the receiver input or the receiver output:

$$SNR = \frac{P_r}{kT_S B} = \frac{P_r/L}{kT_S B/L} = \frac{P_r G/L}{kT_S BG/L} \tag{4.65}$$

since the signal and the noise would be amplified by the same factors at each point. On the other hand, the presence of the feeder lines and the receiver decrease the output SNR by a factor

$$\frac{SNR_i}{SNR} = \frac{P_r/(kT_A B)}{P_r/(kT_S B)} = \frac{T_S}{T_A}$$
$$= \frac{T_A + (L-1)T_{ph} + LT_R}{T_A}$$
$$= \begin{cases} 1 + (F-1)T_0/T_A & L = 1 \\ \\ LF & T_A = T_{ph} = T_0 \end{cases} \tag{4.66}$$

When transmitter and receiver are in LOS of each other, the output SNR given by (4.65) may be written as

$$
\begin{aligned}
SNR &= \frac{P_r}{kT_SB} = \eta_{loss}\frac{P_tG_tG_r}{kT_SBL_{fs}} \\
&= \underbrace{EIRP}_{transmitter}\ \underbrace{\frac{1}{kBL_{fs}}}_{channel}\ \underbrace{\eta_{loss}\frac{G_r}{T_S}}_{receiver} \qquad (4.67)
\end{aligned}
$$

where the received power P_r is given by (2.117). Here, η_{loss} denotes the overall receiver efficiency accounting for polarization and impedance matching as well as implementation losses. A careful look at (4.67) shows that the SNR is determined by the EIRP, which is a transmitter parameter defined by (2.116), by the channel parameters, namely the bandwidth B and free-space loss L_{fs}, depending on the range and the frequency, and the receiver parameters $\eta_{loss}G_r/T_S$, depending on the antenna size, frequency, receiver system noise and the receiver efficiency parameters. Here, G_r/T_S, usually expressed in dBi/K, is called the receiver figure of merit (FOM) and used as a quality measure for the receiver system (see Figure 4.16). Therefore, (4.67) clearly shows that the SNR at the receiver output requires a trade-off between many system parameters related to the transmitter, the receiver and the channel.

Example 4.5 Design Tradeoffs for Uplinks and Downlinks of Satellite Communication Systems.

Physical size of an antenna onboard a satellite and the available transmit power empose major constraints in the design of satellite communication systems. Antenna size onboard a satellite is limited by launching costs and space limitations. Moreover, receive antennas in satellite uplinks looking at the 'hot' Earth (see Figure 4.19) have higher noise temperatures compared to downlink receiving antennas on the Earth surface, which see the 'cold' sky (see Figure 4.20). Therefore, G_r/T_S

ratio of the satellite receive antenna limits the performance of satellite uplinks (see (4.67)).

One way to increase the G_r/T_S ratio of a satellite receive antenna is to choose higher frequency in the uplinks so that higher antenna gains can be achieved onboard the satellite with smaller physical sizes. The use of different frequencies in up- and down-links prevents coupling of transmitted signals to the receiver on the Earth-surface and onboard the satellite. Otherwise, isolation between uplink and downlink frequencies would be technically more challenging and costly. Especially, the high gain of the satellite transponder would lead to positive feedback of transmitted signals. The choice of large frequency difference between uplink and downlink frequencies, for example, 6/4 GHz (C-band), 14/11 GHz (Ku-band), 30/20 GHz (Ka-band), facilitates the isolation between up- and down-links. Intersatellite links operate at 60 GHz at which atmospheric absorption losses exceed 15 dB/km (see Figure 3.23). Therefore, intersatellite links could neither cause interference to or intercepted from the Earth-based systems. In addition, intersatellite links would not suffer absorption losses since there is no atmosphere at the geostationary orbit (at a height of 35870 km from the Earth's surface).

Relatively low-efficiency of generating electric power from solar radiation and decreasing efficiency of solar panels with aging dictates economic use of the available electrical power. On the other hand, a very large percent of the electrical power generated by the solar panels is served for domestic purposes, for example, for heating the satellite during the night, cooling during the day, sensors and control channel for keeping the satellite in orbit. Therefore, the EIRP is the major constraint in the design of the downlinks, that is, SATCOM downlinks are limited by the transmitter parameters (see (4.67)).

A typical bent-pipe (non-regenerative) satellite transponder (**trans**mitter + res**ponder**) consists of a limiter (to limit the maximum received signal level for AJ purposes), a band pass filter, a low-noise amplifier (LNA), designed to amplify the received signals, a frequency convertor used to convert the uplink frequency of the received signal to the downlink frequency of the relayed signal, a band pass filter and a high-power amplifier. Such satellites, with multiple wide-band transponders, act as repeater stations in orbit with no on-board processing; they simply amplify, frequency-convert and retransmit the received signals. On the other hand, the so-called regenerative satellites have onboard processing capability to demodulate, decode, re-encode and remodulate. On-board processing has many advantages, but is more complex, more vulnerable and usually requires more on-board power for signal processing.

Since there is practically no possibility of repairing a failed satellite module, most robust and tested technologies are used in the satellite design. In the nuclear radiation environment of the orbit due to sun and other stars, large temperature differences between day and night, scarcity of the electric power, and the necessity for orbital corrections makes the satellite design with a lifetime of, for example 15 years, with 99.99% reliability is definitely a challenging and multidisciplinary task.

Since the gravitational force (GMm/r^2 where $GM = 3.986 \ 10^{14}$ (m^3/s^2), G is the gravitational constant, M is the mass of the Earth and m is the mass of the satellite) acting on a satellite at distance r from the center of the Earth is balanced by the centripetal force (mv^2/r, v is the tangential velocity of the satellite), a satellite has to move with higher tangential velocities at lower altitudes. Therefore, the satellite period, that is, the time required to make a complete revolution around the Earth, becomes shorter at lower altitudes. For example, a satellite with an orbital height of 1500 km over the surface of the Earth, makes 12 revolutions per day, that is, one revolution per two hours. Consequently, lower-orbiting satellites are observed only during a limited time-interval at a particular location on the surface of the Earth during each revolution. In systems with lower-orbiting satellites, Doppler shift due to satellite motion is an issue to be resolved. At the geostationary orbit

(35870 km orbital height), one revolution lasts 24 hours, that is, the satellite has the same angular velocity as the Earth. Consequently, geostationary satellites seem to be 'stationary' when looked from the surface of the Earth. Therefore, signals relayed by geostationary satellites do not undergo Doppler shifts. However, a time delay of ¼ seconds is required for the signals to propagate from the transmitter to the receiver via a geostationary satellite. A geostationary satellite provides continuous coverage of approximately 1/3 of the globe with a global-coverage antenna, of 17.4 degrees half-power beamwidth, and three geostationary satellites spaced by 120^0 from each other are capable of providing global coverage. Nevertheless, elevation angles for satellite ground terminals at high latitudes are very low, requiring much higher G_r/T_s ratios to compensate for increased signal attenuation, scintillations and higher noise levels due to atmospheric absorption. Satellite systems are not generally used for elevation angles lower than 5^0.

Satellites may be classified according to their orbital height. Depending on the requirements, satellites might be low-Earth orbiting (LEO), with orbital heights ~500–1500 km, medium-Earth orbiting (MEO), with ~6000–20000 km orbital heights, or geostationary (GEO) at 35870 km. Satellites may also be classified according to their orbital type. The center of a circular orbit coincides with Earth's center while the Earth's center is located at one foci of an elliptical orbit. A satellite may revolve around the Earth in different orbital planes in equatorial orbit above Earth's equator, polar orbit passing over both poles or in an inclined orbit, which is located in a plane making a certain (inclination) angle with the equator.

Satellites became part of our daily life, with the launch of the first satellite SPUTNIK, based on the idea by A. C. Clarke in 1945. [12] Nowadays, satellites have diverse application areas in telecommunications, radio/TV broadcasting, military, radio-astronomy, remote

sensing (including meteorology) and navigation and localization, for example, GPS.

Example 4.6 TV Receiver System.
Receiving antennas for satellite TV (DVB-S)
are usually mounted on roof-tops and transmission lines are used to carry the received signals
down to the TV receiver. The antenna noise temperature of such a TV receiver system is
assumed to be $T_A = 1000$ K, the transmission
line (with $T_{ph} = T_0$) loss is $L = 2$ dB and the noise
figure of the TV receiver is estimated to be $F =$
13 dB. The SNR at the receiver output of this
configuration is measured to be $SNR = 30$ dB.

a. Equivalent noise temperature of the transmission line-TV receiver combination may
   be written from (4.62) as

$$T_a = (FL - 1)T_0 = (20 \times 10^{0.2} - 1)T_0 = 8902K$$
(4.68)

The input SNR SNR_i corresponding to a
given output SNR is given by

$$SNR_a = \frac{S_i}{k(T_A + T_a)B} = \frac{S_i}{kT_A B} \frac{T_A}{T_A + T_a}$$

$$= SNR_i \frac{T_A}{T_A + T_a}$$

$$SNR_i = SNR_a (1 + T_a/T_A) = 9902 \ (39.96 \ dB)$$
(4.69)

where $SNR_a = SNR = 30 dB$. This implies
that an input SNR of $SNR_i = 39.96$ dB is
required in order to achieve an output
SNR of 30 dB.

b. Now consider that a preamplifier with
   $F_{PA} = 6$ dB noise figure and $G_{PA} = 20$ dB
   gain is connected between the transmission
   line and the input to the TV receiver. The
   equivalent noise figure and temperature of
   this triple-combination is given by

$$F_b = L\left(F_{PA} + \frac{F-1}{G_{PA}}\right) = 10^{0.2}\left(4 + \frac{20-1}{100}\right)$$

$$= 6.64 \ (8.2 \ dB)$$

$$T_b = (F_b - 1)T_0 = 1636K$$
(4.70)

The output SNR then becomes

$$SNR_b = \frac{S_i}{k(T_A + T_b)B} = SNR_a \frac{T_A + T_a}{T_A + T_b}$$
(4.71)

$$= 3756 \ (35.75 \ dB)$$

This implies a 5.75 dB increase in the output
SNR compared to the case when no pre-
amplifier is used.

c. In this case, the preamplifier is moved to the
   antenna site and connected between the
   antenna and the transmission line. Then,
   the effective noise figure is obtained by cascade connection of the pre-amplifier with
   the combination of the feeder-line and
   receiver with an effective noise figure of FL:

$$F_c = F_{PA} + \frac{FL-1}{G_{PA}} = 4 + \frac{20 \times 10^{0.2} - 1}{100}$$

$$= 4.31 \ (6.35 \ dB)$$
(4.72)

$$T_c = (F_c - 1)T_0 = 960K$$

In this case, the noise figure is 8.2–6.35 =
1.85 dB lower compared to (4.70) and corresponding output SNR is given by

$$SNR_c = \frac{S_i}{k(T_A + T_c)B} = SNR_a \frac{T_a + T_A}{T_c + T_A}$$
(4.73)

$$= 5052 \ (37 \ dB)$$

which implies a 7 dB increase compared to
the case where no pre-amplifier is used, and
1.25 dB improvement compared to the case
where the pre-amplifier is inserted between
the transmission line and the TV receiver.

Example 4.7 Effect of Feeder Loss on SNR.
An antenna with an equivalent noise temperature of T_A is connected to a receiver of noise figure $F = 3$ dB via a transmission line of loss L
and physical temperature $T_{ph} = T_0$. Using
(4.62), the system noise temperature when the
feeder line is lossless ($L = 0$ dB) and has a 3
dB loss is found as

$$T_S = T_A + (LF - 1)T_0 = \begin{cases} T_A + 290K & L=0 \ dB \\ T_A + 870K & L=3 \ dB \end{cases}$$
(4.74)

From (4.64), the reduction in output SNR due to a lossy line with $L = 3$ dB is given by

$$10 \log \left(\frac{T_A + 870}{T_A + 290} \right) = \begin{cases} 4.3 \ dB & T_A = 50K \\ 3 \ dB & T_A = 290K \end{cases}$$

(4.75)

Hence, (4.75) shows that a 3 dB lossy feeder line decreases the SNR by 4.3 dB for $T_A = 50$ K and 3 dB for $T_A = 290$ K.

Example 4.8 Repeater System.
Consider a wired communication system which operates in AWGN channel over a transmission bandwidth of B. Total line attenuation between transmitter and receiver is assumed to be L (see Figure 4.28). When the total loss in the channel is very high, repeaters are used to boost the signal level, and thus, to offset the effect of signal attenuation in transmission lines.

Let us first consider a repeaterless system shown in Figure 4.28a, as a reference. In view of (4.62), the combination of a feeder-line loss L and a receiver with noise figure F may be considered as a single component with overall gain G/L and noise figure LF. Hence, the SNRs at the transmitter and receiver outputs of a repeaterless system may be written as

$$SNR_i = \frac{P_t}{N_i B}$$

$$SNR_0 = \frac{1}{LF} SNR_i = \frac{P_t}{LFN_i B}$$

(4.76)

where P_t denotes the transmit power.

We now assume that $N-1$ nonregenerative (amplify and forward) repeaters with identical gains G_r and noise figures F_r are inserted in the wireline channel, as shown in Figure 4.28b. The system with $N-1$ repeaters consists of one transmitter and N repeaters, including the receiver, each with gain $G_r = G/L_s$ and equivalent noise figure $F_r = FL_s$, where $L_s = L/N$ denotes the loss per segment:

$$G_r = G_1 = G_2 = \cdots = G_N = G/L_s$$
$$F_r = F_1 = F_2 = \cdots = F_N = FL_s$$

(4.77)

Using (4.16) and (4.77), the equivalent noise figure of this repeater system is found to be

$$F_e = F_1 + \frac{F_2 - 1}{G_1} + \frac{F_3 - 1}{G_1 G_2} + \cdots + \frac{F_N - 1}{G_1 G_2 \cdots G_{N-1}}$$

$$= 1 + (F_r - 1) \frac{1 - G_r^{-N}}{1 - G_r^{-1}}$$

(4.78)

where the identity $\sum_{n=0}^{N-1} x^n = (1 - x^N)/(1 - x)$ (see (D.12)) is used. If the relay gains are chosen so as to compensate the segment loss, that is, $G_r = G/L_s = 1$, then the first expression in (4.78) reduces to

$$F_e = 1 + (FL_s - 1)N \cong NL_s F \ \text{ for } \ G_r = G/L_s = 1$$

(4.79)

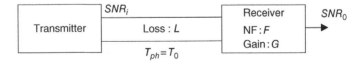

(a) System with no repeater

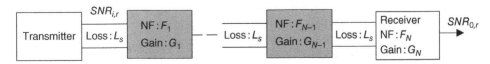

(b) System with repeaters

Figure 4.28   Block Diagram of the Wired Transmission System.

which is approximately the same as that for a repeaterless system with a receiver of noise figure F and a total path loss L, as given by (4.76).

The output SNR in this repeater system is

$$SNR_{0,r} = \frac{1}{F_e} SNR_{i,r} = \frac{1}{F_e} \frac{P_{t,r}}{N_i B} \qquad (4.80)$$

where $SNR_{i,r}$ denotes the SNR at the transmitter output and $P_{t,r}$ is the transmit power in this repeater system.

In the special case where (4.79) applies, (4.80) may be rewritten as

$$SNR_{0,r} = \frac{1}{N L_s F N_i B} P_{t,r} = \frac{1}{N} SNR_{0,1} \qquad (4.81)$$

where $SNR_{0,1}$ denotes the SNR at the output of the first relay. It is clear from (4.81) that the accumulation of noise in this amplify-and-forward relay system leads to a gradual decrease of the output SNR as the number N of relays increases.

The ratio of output SNRs with and without repeaters is found by (4.76) and (4.81):

$$\frac{SNR_{o,r}}{SNR_0} = \frac{P_{t,r}}{P_t} \frac{LF}{F_e} \cong \frac{P_{t,r}}{P_t} \frac{L}{N L_s} \qquad (4.82)$$

If the systems with and without repeaters are required to have the same output SNR's, then using $L_{s,dB} = L_{dB}/N$ and (4.82), one may write

$$P_{t,r,dB} = P_{t,dB} + 10 \log N + L_{s,dB} - L_{dB}$$

$$= P_{t,dB} + 10 \log N - \left(1 - \frac{1}{N}\right) L_{dB}$$
$$(4.83)$$

For example, we consider a repeater system with $L_{dB} = 100 dB$ with $N = 10$ segments. In this system, each repeater will have $G = 10$ dB gain so as the compensate the segment loss $L_{s,dB} = 10 dB$. Then, (4.83) shows that the transmit power of the system with repeaters will be 80 dB lower than that for the repeaterless system; which explains the advantage of using repeaters in systems with high path losses [13].

4.6 Additive White Gaussian Noise Channel

The receiver system noise, which is described by T_S at the receiver front-end (see Figure 4.27), accounts for the antenna noise, the feeder line noise and the receiver internal noise. Since these noise terms are uncorrelated with each other, their powers are additive. Hence, the receiver noise may be represented by an additive white Gaussian noise (AWGN), $w(t)$, at the receiver input. In view of (4.3) and (4.4), the PSD of AWGN is white and it is described by a Gaussian pdf with zero mean and variance $N_0/2$. The AWGN is added to the signal at the receiver input. This channel, which is described by Figure 4.29, is called as AWGN channel under free space propagation conditions, when the signal at the receiver input is only an attenuated, delayed and phase shifted version of the transmitted signal. These parameters are generally incorporated into the AWGN term; the delay and phase shift due to channel do not affect the statistical characteristics of the Gaussian noise (see (4.3)) and signal attenuation may be accounted for by scaling the AWGN term so as to generate the desired SNR in the channel (see (3.5) and (3.6)). Therefore, the signal at the receiver input is qualified as 'transmitted signal' and 'the received signal' represents the sum of the 'transmitted signal' and the AWGN term, as shown in Figure 4.29. The received signal is hence described by a Gaussian pdf with its mean equal to the transmitted signal level and variance $N_0/2$, where $N_0 = kT_S$ is determined by the system noise temperature.

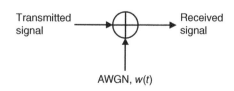

Figure 4.29 Additive White Gaussian Noise (AWGN) Channel.

Note that the noise in some channels may be non-Gaussian. The noise of non-thermal nature also exists, for example, shot noise. However, the noise in the majority of communication channels of practical interest has thermal origin and can be modelled as AWGN.

References

[1] Recommendation ITU-R P.372–10 (10/2009) Radio noise.

[2] Recommendation ITU-R P.618–11 (09/2013) Propagation data and prediction methods required for the design of Earth-space telecommunication systems.

[3] Recommendation ITU-R P.530–15 (09/2013) Propagation data and prediction methods required for the design of terrestrial line-of-sight systems.

[4] J. D. Kraus, *Radio Astronomy*, McGraw Hill: New York, 1966.

[5] D. M. Pozar, *Microwave and RF Design of Wireless Systems*, Wiley: New York, 2001.

[6] A. B. Carlson, P.B. Crilly, and J. C. Rutledge, *Communication systems*, McGraw Hill: Boston, 2002.

[7] P. Z. Peebles, Jr., *Probabiity, Random Variables, and Random Signal Processing* (3rd ed.), Mc.Graw Hill: New York, 1993.

[8] R.L. Freeman, *Telecommunication transmission Handbook* (3rd ed.), John Wiley: New York, 1996.

[9] I. A. Glover and P. M. Grant, *Digital Communications* (2nd ed.), Pearson-Prentice Hall: Harlow, 2004.

[10] C. Ho, A. Kantak, S. Slobin, and D. Morabito, Link analysis of a telecommunication system on earth, in geostationary orbit, and at the moon: Atmospheric attenuation and noise temperature effects, IPN Progress Report 42–168, Feb. 15, 2007.

[11] C. Ho, S. Slobin, A. Kantak, and S. Asmar, Solar brightness temperature and corresponding antenna noise temperature at microwave frequencies, IPN Progress Report 42–175, Nov. 15, 2008.

[12] A. C. Clark, Extra terrestrial relays-can rocket stations give worldwide coverage, Wireless World, October 1945.

[13] J. G. Proakis and M. Salehi, *Communication Systems Engineering* (2nd ed.), Prentice Hall: New Jersey, 2002.

Problems

1. Consider the block-diagram of a receiver as shown below.

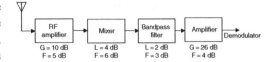

$G = 10\ dB$ $L = 4\ dB$ $L = 2\ dB$ $G = 26\ dB$
$F = 5\ dB$ $F = 6\ dB$ $F = 3\ dB$ $F = 4\ dB$

The received signal power at the input to the first amplifier is –90 dBm, and the received noise power spectral-density is –170 dBm/Hz. The bandpass filter has a bandwidth of 10 MHz. Determine the SNR at the input to the demodulator.

2. An antenna with a gain of 40 dB is connected to a preamplifier (impedance-matched) with a gain of 30 dB and a noise figure of 6 dB. The preamplifier output is connected to a receiver having a noise figure of 10 dB by means of a transmission line having a length of 20 m and an attenuation of 0.5dB/m. The antenna noise temperature is 50 K; the system bandwidth is 10 MHz and the wavelength of operation is 10 cm. Find the following parameters for the system:

  a. The system noise temperature T_s.
  b. What is the required incident power per unit area in order to give a signal-to-noise ratio of 30 dB at the receiver output.
  c. Find the corresponding signal-to-noise ratio at the receiver input.
  d. Solve the above problem again when the transmission line is inserted between the antenna and the preamplifier. Comment on the results obtained.

3. Assuming that the equivalent noise temperature and the gain of a broadband device are characterized by

$$T_e(f) = Kf^n$$

$$G(f) = \frac{G_0}{1 + (f/f_0)^2}, \quad f_1 < f < f_2$$

where K is an arbitrary constant. Use (4.25) and the formulas from Appendix D to show that the mean equivalent noise temperature of this device is given by

$$\overline{T}_e = \begin{cases} Kf_0\dfrac{0.5\ln\dfrac{1+(f_2/f_0)^2}{1+(f_1/f_0)^2}}{\tan^{-1}(f_2/f_0)-\tan^{-1}(f_1/f_0)} & n=1 \\ Kf_0^2\left[-1+\dfrac{f_2/f_0-f_1/f_0}{\tan^{-1}(f_2/f_0)-\tan^{-1}(f_1/f_0)}\right] & n=2 \end{cases}$$

The percentage bandwidth of a device is defined as the percent of the ratio of the bandwidth $B = f_2-f_1$ to the carrier frequency f_0. The bandwidth of a narrowband device is in the order of a several percent of the carrier frequency. Assuming that $f_2/f_0 = f_0/f_1$, plot the above formulas for the mean equivalent noise temperature as a function of $1 < f_2/f_0 < 1.3$. Show that the narrowband assumption does not lead to significant errors in estimating the equivalent noise temperature of this device.

4. Using (D.56), show that the total brightness of a blackbody defined by (4.34) is proportional to the fourth-power of it noise temperature:

$$B = \int_0^\infty \frac{2hf^3}{c^2}\frac{1}{e^{hf/kT}-1}df = \frac{2h}{c^2}\left(\frac{kT}{h}\right)^4\frac{\pi^4}{15}$$

$$= CT^4 \quad \left(\frac{W/m^2}{sr}\right)$$

where $C = 1.8\times 10^{-8}$.

5. Transfer functions of devices vary with frequency. Consider that a source of PSD $|S(f)|^2$ (W/Hz) is connected to the input of a baseband device with frequency-dependent transfer function $H(f)$.

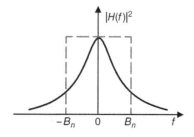

a. Express the PSD of the signal at the output of the device in terms of S(f) and H(f).
b. Determine the total power available at the output of the device.
c. Determine the total power available at the output of the device, P_o, if the source has an equivalent noise temperature of T_e.
d. It is convenient to replace the actual frequency-dependent gain characteristics by a filter with the *noise bandwidth* B_n determined by the condition of equal noise power output for the two cases: $P_0 = kT_eB_nG$, where G denotes the maximum gain of the device under consideration. Determine the noise bandwidth of a device if $|H(f)|^2 = |H(0)|^2/\left[1+(f/f_0)^2\right]$ where $f_0 = 1$ MHz.

6. An antenna with noise temperature of $T_A = 300$K is connected to a receiver system which is formed by the cascade connection of a preamplifier $(G_1, F_1 = 2, B > 10$ MHz$)$, mixer $(G_2 = 1/2, F_2 = 6, B > 10$ MHz$)$ and amplifier $(G_3 = 100, F_3 = 7, B = 10$ MHz$)$.

a. What should be the gain G_1 of the preamplifier so as to keep the overall noise figure of the receiver less than or equal to 3.
b. Find the system noise temperature when the preamplifier is fed by the antenna by using $G_1 = 15$ dB and the value of G_1 found in part (a).
c. Find the noise power at the amplifier output for the two cases of part (b).
d. Repeat part (b) if a feeder line with 2 dB loss connects the antenna to the preamplifier.

7. An antenna with 50K noise temperature feeds a cascade of two impedance-matched attenuators. The first attenuator at a physical temperature of $T_1 = 100$ K and with a loss of $L_1 = 2$ is connected to the antenna. The second attenuator, at a physical temperature of $T_2 = 290$ K and loss $L_2 = 1.4$, drives the receiver of noise temperature $T_R = 1000$K, 3-dB noise bandwidth of $B = 1$ MHz and $G = 60$ dB gain.

a. Determine the effective noise temperatures of the two attenuators.
b. Determine the effective noise temperature of the cascade.
c. Determine the system noise temperature?
d. Determine the noise figure and the noise temperature of this system having the antenna as its source.
e. Determine the average noise power which is available at the system output.

8. An amplifier has $B = 1$ MHz bandwidth and standard (with a noise temperature of 290 K at its input) noise figure $F = 4$. The amplifier's available output noise power is 0.1 μW when its input is connected to a radio receiving antenna with $T_A = 100$ K noise temperature. Determine the amplifier's input effective noise temperature T_e, its operating noise figure F_{op} with an antenna of $T_A = 100$ K at its input, and its power gain.

9. Consider a system with 0.1 mW transmit power, 0dBi gain for transmit and receive antennas, operating at 500 MHz carrier frequency with 100 kHz bandwidth. The system operates in a suburban environment. Assume the receiver noise figure is 6 dB and the noise temperature of the receive antenna is 200K, independent of frequency.

a. Determine the SNR received at a distance of 1km, assuming free space propagation?
b. How does the SNR change when the carrier frequency is increased to 5 GHz?
c. Explain why does the 5 GHz system show a significantly lower SNR? Is the receiver antenna noise temperature independent of frequency (see Figures 4.20 to 4.24)? For a given physical size, are the gains of transmit and receive antennas independent of frequency? In view of the above, what can you say about the frequency dependence of SNR?

10. A 15m diameter antenna, with 65 % aperture efficiency, is used to receive transmissions from a geostationary satellite operating at 10 GHz with an elevation angle of $\theta = 10^0$. When an antenna is pointed at zenith, the measured noise temperature at the feed output is 20K. Use Figure 4.20 for the brightness temperature of the sky. Ignore the noise temperature contribution from the antenna sidelobes at all angles.

a. Assume that the atmospheric absorption loss can be modelled as $A(\theta) = A_z / \sin(\theta)$ where A_z denotes the atmospheric absorption loss at zenith. Determine A_z.
b. Estimate the antenna noise temperature at an elevation angle of 10^0
c. Find the earth station G/T in clear sky in zenith direction and at 10^0 elevation angle.

11. Consider a 12 GHz terrestrial link operating in free space propagation conditions. The signal power at the receiving antenna terminals is measured to be −42 dBm. The overall noise figure of the receiver is 6 dB and the noise bandwidth is 10 MHz. The clear sky antenna noise temperature at 12 GHz is given as 130 K.

a. Estimate the clear sky SNR at the terminals of the receiving antenna assuming that the antenna has a loss efficiency of 97 % and is at a physical temperature of 300 K.
b. Determine the effective SNR during a rainfall in the vicinity of the receiver, causing a signal attenuation of 3 dB. The physical temperature of the rain cloud is assumed to be 273 K.

12. The rain attenuation exceeded for 0.1% and 0.01% of the time is given by 1.1 dB and 7.5 dB, respectively, in a SAT-COM link operating at 12 GHz. We assume that the brightness temperature

of the clear sky is 3K and the physical temperature of the rain cloud is estimated by (4.44).

a. Determine the corresponding increases in the receiver system noise temperature.
b. Repeat (a) for 18 GHz.

13. The effect of rain on the signal attenuation and system noise temperature of a satellite ground terminal operating in 14/11 GHz band is given as follows:

Time parameter exceeded (hours/year)	Rain attenuation (dB)	System noise temperature (K)
–	0	250
12	2	350
6.5	4	420
3.0	6	460
1.8	8	480
1.0	10	495
0.5	12	510

a. Determine the decrease in SNR for each time parameter with respect to the case of 0 dB rain attenuation.
b. Determine the margin, which would ensure the SNR would be the same as in the case with no rain attenuation, for 99.99 % of the time.

14. The effective temperature of an antenna looking toward zenith is approximately T_A and is connected to a receiver by using a transmission line of loss L. The physical temperature of the transmission line (waveguide) is T_{ph}. The received signal power at the antenna terminals is S_i (W).

a. Express the effective noise temperature, noise power and SNR at antenna and receiver terminals.
b. Assuming that $L = 0.1$ dB/m, $T_A = 5$ K and $T_{ph} = 300$ K, determine the effective temperature at the receiver terminals when the length of the transmission line

is 1 m and 30 m. Compare the values obtained with the receiver noise temperature of 100 K.

15. Consider a satellite at an altitude of 900 km transmitting at a carrier frequency of $f = 1.5$ GHz to a mobile user with a half-wave dipole antenna. The satellite transmit power is $P_t = 100$ W, and the antenna gain is $G_t = 30$ dBi. Total system losses are assumed to be 4 dB.

a. Find the received signal power at the mobile receiver antenna terminals
b. Determine the system noise power if the receiver noise figure is 6 dB, antenna noise temperature 60 K and the receiver bandwidth is 1 MHz.
c. Determine the SNR.
d. Determine the decrease in received SNR in dB if there is 3 dB rain attenuation and the medium temperature is 290 K.

16. Antenna noise temperature of a mobile terminal is $T_A = 290°$K. The first amplifier stage of the radio provides a 20 dB gain and has a noise figure of 5. The second amplifier stage has 10 dB gain and a noise temperature of 3000°K. The contribution of other stages on the system noise is ignored.

a. Determine the equivalent noise temperature of the first two stages of the radio and compare with system noise temperature.
b. If the baseband processing of the radio requires an SNR of 10 dB in 10 kHz, what is the receiver sensitivity?

17. A satellite ground station sees the satellite through a cloud which induces a 1 dB loss to electromagnetic waves propagating through it. Assume that the brightness temperature due to background radiation is 30 K and the cloud has a physical temperature of 275 K.

a. Determine the antenna noise temperature with and without the cloud. How much the antena noise temperature is increased due to the presence of the cloud?

b. Determine the ratio of input SNRs with and without cloud. Compare the signal attenuation, the increase in the system noise temperature and the decrease in SNR due to the presence of the cloud.

c. Repeat (b) assuming that the receiver has a noise temperature of 60 K. Compare the results obtained in (a) and (b).

18. A preamplifier with gain = 13 dB and noise figure = 1 dB is inserted between the antenna with $T_A = 290$ K and the receiver with 50 dB amplifier gain, 500 MHz bandwidth, 7 dB noise figure and 8×10^{-11} W input signal power.

a. Determine the equivalent noise figure of the whole receiver. How much the noise figure is improved due to the use of the preamplifier?

b. Determine the output SNR improvement due to the use of preamplifier.

c. Repeat part (b) for $T_A = 29$ K and $T_A = 2900$ K. Determine the SNR improvement in dB.

d. Discuss how the receiver noise figure with regard to the antenna noise temperature affects the SNR.

19. The effective noise temperature of an antenna looking toward empty sky is approximately 4 K. The antenna is connected to a receiver, which has a noise temperature of about 54 K, with a transmission line of physical temperature of 300 K. The transmission line has an attenuation coefficient 0.1 dB/m and its length is 10 m.

a. Find the effective noise temperature at the receiver terminals and calculate the increase in the noise power in dB at the receiver input due to the use of the lossy cable.

b. Repeat part (a) when there is a cloud in the boresight direction of the antenna which has a temperature of 250 K and causing a loss of 2 dB at the frequency of operation.

20. When a directional antenna, of gain 30 dBi, loss efficiency of $\eta_L = 0.9$ and physical temperature of 290°K, is pointing toward empty sky, the noise temperature falls to about 3°K at frequencies between 1 GHz and 10 GHz.

a. If the antenna is connected directly to a low noise amplifier with a gain of 40 dB and a noise figure of 1.4, what is the system noise temperature?

b. If a cable of 3 dB loss is inserted between the antenna and the receiver, determine the system noise temperature.

c. Calculate the increase in noise power due to the use of the cable of 3 dB loss.

21. A microwave receiver used in satellite communications has an antenna with $T_A = 20$ K cascaded with an LNA with $T_e = 6$ K, $G = 30$ dB, an amplifier with $F = 6$ dB, $G = 20$ dB and a mixer + IF combination with $F = 13$ dB and $G = 10$ dB.

a. Determine the equivalent noise figure and the noise temperature of the receiver (excluding the antenna). Compare the equivalent noise temperature of the receiver with that of the first amplifier. Calculate how many dB's the noise power is increased compared with that of the first amplifier.

b. Each stage of the receiver introduces some band limiting. However, it may be convenient to consider that all the stages of the system are of unlimited bandwidth and band limiting is done in a single IF filter at the output end of the IF amplifier. Under these circumstances, the noise input to this IF filter would have a white (uniform) power spectral density. Determine the noise power spectral density (W/Hz) present at the demodulator input.

c. Explain whether the noise introduced by the demodulator or any succeeding decoder will further degrade the signal-to-noise ratio of the signal at the output of the receiver.

d. Consider that the received signal has a baseband bandwidth of 4 MHz. If we want the signal to noise ratio at the demodulator input to be 20 dB, then find the signal power at the demodulator input? Determine the corresponding signal power at the receiver input (output of the receive antenna).

22. The noise figure of a cell phone receiver is specified as 16 dB. Since the internal noise is very high, the external antenna noise can be ignored. This receiver requires an SNR of 13 dB for reliable detection of a signal with 200 kHz bandwidth. Determine the receiver sensitivity.

23. A satellite is located at 10^0 west longitude in the geosynchronous orbit. The transmitting ground station is located at 10^0 east longitude and 30^0 north latitude, and the receiving ground station is located at 33^0 east longitude and 40^0 north latitude. The uplink and downlink frequencies are 12GHz and 10GHz, respectively. All antennas are 3 m in diameter with 65% aperture efficiencies.

Ignoring atmospheric absorption and noting that the transmission bandwidth in the downlink and the uplink are the same, the overall C/N ratio may be expressed in terms of the uplink and downlink carrier-to-noise ratios as

$$\frac{C}{N} = \frac{1}{(C/N)_u^{-1} + (C/N)_d^{-1}}$$

a. Use (3.67) to determine the uplink and downlink ranges to the satellite.

b. Find the uplink and downlink carrier-to-noise ratios so as to give an overall C/N ratio of 20 dB in a transmission bandwidth of 10 MHz. The G/T of satellite and ground-station receive antennas are 4 dB and 7 dB, respectively.

c. Using the C/N values for the uplink and downlink of part (a), determine the transmit powers for the ground station and the satellite.

24. In a SATCOM system, a ground receiver antenna has 64 dBi gain and 65% aperture efficiency. The relevant noise temperatures are given as follows:

T_S at 10^0 elevation angle:	234 K (clear-sky conditions)
Sky noise temperature at 10^0 elevation:	17 K (see Figure 4.20)
Receiver noise temperature:	150 K

During heavy rain, the slant path attenuation reaches 8 dB for 0.01 % of the year. Calculate the G/T_S for this percentage of the time and the corresponding reduction in the SNR.

25. Consider a mobile radio system operating at 900 MHz with 25 kHz bandwidth. Transmit and receive antenna gains are 8 dBi and 0 dBi, respectively and the antenna noise temperature is $T_A = 300$ K. Losses in cables, combiners, and so on at the transmitter are 3 dB. The voltage reflection coefficients at the transmit and receive side are given as 0.25. The received power in the channel is modelled by (3.92) with $r_0 = 10$ m and $n = 4$. The operating SNR at the receiver, with a 7 dB noise figure, is required to be 18 dB. Assuming a fading margin of 7 dB, determine the minimum transmit power for a coverage radius of 2 km.

26. A wireless microwave link, operating in the 10 GHz band with $B = 10$ MHz bandwidth, is used to relay TV signals at a range of 10 km. Transmitting and receiving antennas both have 15 dBi gains. For acceptable performance 30 dB SNR is required. Loss due to polarization mismatch is limited to 3 dB. Perfect impedance matching and free space propagation conditions are assumed.

   a. For $T_A = 40$ K and $F_R = 5$, determine the minimum required received power at the receiver antenna terminals.
   b. Determine the required transmit power level.

27. A transmitter and a receiver both on the surface of the earth establish a link between them by using the moon as a relay. Transmit and receive antenna gains are given by G_t and G_r, respectively. Transmitter and the receiver are located at distances d_1 and d_2 from the moon and are given by $d_1 = d_2 \cong 3.8 \times 10^8$ (m). The system operates with a transmit power P_t at $f = 4$ GHz and has a noise bandwidth of B Hz.

   The radar cross section (RCS) σ of the moon is reported to be 0.070 times the physical cross section. Since the diameter of the moon is 3.474×10^6 (m), its physical cross section is 9.5×10^{12} (m²), and its RCS is approximately equal to $\sigma = 0.07 \times 9.5 \times 10^{12} = 6.65 \times 10^{11}$ (m²) in the frequency band of operation. Since the receive antenna 'looks' at the moon, the mean antenna noise temperature may be approximated by 210 K (see Figure 4.21). Differences in the lunar surface temperature may decrease its brightness temperature by a factor of two during the night and increases by the same factor during the day.

   a. Using (2.125), express the SNR at the output of the receiver with noise figure F and antenna noise temperature

   T_A in terms of the relevant system parameters.
   b. For $F = 3$, $P_t = 1$ MW, $B = 30$ kHz and transmit antennas with 65% aperture efficiencies and 3 m diameter antennas, determine the value of the SNR in dB.
   c. Calculate the variations with respect to the mean SNR due to night-time and day-time variations in the antenna noise temperature.

28. A police traffic-control radar operating at 9.9 GHz has a transmit power of 0.1 W and uses dish antennas with 20 cm diameter and 65% aperture efficiency. The minimum SNR level required for reliable detection is 10 dB with a receiver system noise temperature of 2000 K. The system has a noise bandwidth of $B = 1$ kHz.

   a. Determine the maximum range at which cars with radar cross sections larger than 0.2 m² can be detected.
   b. A driver has a radar receiver which scans the radar frequency channels for detecting the presence of the police radar. At what range can the police radar be detected by this receiver with an antenna gain of 16 dBi and a signal sensitivity of –60 dBm.

29. Consider a LOS communication system operating at 9 GHz requires 10 dB SNR. The EIRP of the transmitter is 0 dBW. The receiving antenna diameter is 13 cm (with an aperture efficiency of 0.65), and the antenna noise temperature is 800 K. The distance between the transmitter and receiver is 12 km. The transmission bandwidth is $B = 10$ MHz.

   a. Determine the maximum allowable receiver noise figure
   b. Determine the value of the noise figure in part (a), when the transmission bandwidth is doubled.

c. Determine the value of the noise figure in part (a), when the antenna diameter is doubled.

30. In order that a satellite keeps revolving in an orbit, the gravitational force should be balanced by the centripetal force

$$\frac{GMm}{r^2} = \frac{mv^2}{r}$$

where $GM = 3.986 \ 10^{14} \ (m^3/s^2)$, G is the gravitational constant, M is the Earth's mass, m is the mass of the satellite, v is the tangential velocity of the satellite and r denotes the distance between the Earth's center and the satellite. Inserting $v = wr = 2\pi r/T$ into the above equation, one can write the orbital period T, the time required for one complete revolution around earth in terms of h, the height above Earth, is as follows:

$$T = 2\pi \sqrt{\frac{(h+R)^3}{GM}}$$

where $r = h + R$ and $R = 6371$ km denotes the radius of the Earth.

a. Plot the variation of T as a function of h, the height of the satellite above Earth's surface and show that T increases with increasing height and becomes equal to 24 hours at a height of $h = 35870$ km. What is the period of a satellite at an orbital height of 900 km?

b. Plot the variation of the tangential velocity v as a function of h and show that the velocity decreases with increasing orbital height.

31. A narrowband signal is to be transmitted a distance of 100 km over a wireline channel that has an attenuation of 1 dB/km. Assume that noise equivalent bandwidth

of each repeater is $B = 4$ kHz and that $N_0 = kT_0 = 4 \ 10^{-21}$ W/Hz.

a. Using (4.76), determine the transmitted power P_t required to achieve an $SNR_0 = 30$ dB at the output of a receiver with a 5 dB noise figure, if no relays are used.

b. Repeat the calculation when a repeater is inserted every 10 km in the wireline channel, where the repeater has a gain of 10 dB and a 5 dB noise figure.

32. Consider a 802.11b WiFi uplink with the following parameters: $f = 2.5$ GHz, BS antenna gain of 5 dBi, and MS antenna gain of 0 dBi. Uplink transmit power is assumed to be 0.1 mW. The base station receiver has a noise figure of 7 dB, $T_A = 290$ K antenna noise temperature and a 4 MHz noise bandwidth. The indoor path loss is modelled by

$$L(d) = 20\log_{10}(f_{GHz}) + 16.9\log_{10}(d) + 32.8 + X \quad (dB)$$

where $X = 8$ dB denotes the link margin in dB due to shadowing.

a. For a distance of 20 m between the MS and the BS, calculate the SNR at the detector input of the BS receiver.

b. How can the range be increased?

33. Specific attenuation due to rain can also be calculated by $\gamma_R = kR^\alpha (dB/km)$ where R is the rain rate in mm/hr exceeded for p% of the time, and k and α depend on the rain drop size and frequency. Consider a terrestrial radio link system with a range of 20 km and operating at 12 GHz, for which $k = 0.0178$ and $\alpha = 1.209$ (see Table 3.2). Consider a 30 mm/hr rain rate for 0.01% of the time.

a. Determine the rain attenuation exceeded for 0.01% of the time.

b. Ignoring the sky noise temperature of 3 K in (4.43) and (4.44), the increase in antenna noise temperature due to rain may be estimated as

$$\Delta T = (1 - 1/A)T_{medium} \quad (K)$$

where A denotes the rain attenuation found in part (a). Determine the increase in received noise power, in dB, if the receiver system noise temperature is 200 K.

c. Determine the degradation in signal to noise power ratio, SNR.

34. A satellite ground station is transmitting signals to a geostationary satellite located at a distance of 39000 km. For proper system operation, a power density of -120 dBW/m^2 is required at the satellite. Assume 2 dB loss for antenna pointing errors, polarization mismatch, and so on and 5 dB loss due to rain attenuation. Operating frequency is 14 GHz.

a. Determine the required ground station EIRP and the transmitter power if the transmitting antenna has a gain of 40 dB.

b. Determine the figure-of-merit of the satellite receiver antenna to achieve a C/N$_0$ = 95 dB.Hz. What is the SNR in a bandwidth of 40 MHz?

c. If E$_b$/N$_0$ = 20 dB is required to achieve a certain probability of error, what is the maximum data rate which can be supported by this link?

35. The path loss model in 3GPP Release 9 (2010) is given as follows:

$$L_{dB} = 20\log_{10}(f)$$
$$+ \begin{cases} 20\log(d) + 32.44 & \textit{Free Space} \\ 16.9\log_{10}(d) + 32.8 & \textit{WiFi} \\ 43.3\log_{10}(d) + 11.5 & \textit{Femtocell} \\ 22\log_{10}(d) + 28 & \textit{Macrocell} \end{cases}$$

where d is in meters and f denotes the frequency in GHz. An LTE macrocell BS has the following parameters: $f = 2$ GHz, $P_t = 46$ dBm, $G_t = 15$ dBi, the length of the grid range $a = 500$ meters. For LTE Femtocell BS one has $f = 2$ GHz, $P_t = 21$ dBm, $G_t = 5$ dBi, the length of the grid range $a = 20$ meters. Similarly, for 802.11b WiFi BS, $f = 2.5$ GHz, $P_t = 20$ dBm, $G_t = 5$ dBi, the length of the grid range $a = 20$ meters. Receive antenna gains for all users are assumed to be 2.3 dBi.

Consider a WiFi receiver at a distance of d meters from the BS and receives at a data rate of $R = 11$ Mbps.

a. Determine the bit energy $E_b = P_r/R$ at the receiver located at a distance of $d = 50$ m from the WiFi BS.

b. A WiFi link is considered to be good if the received $E_b/N_0 \geq 40$ dB. Determine the requirement on the noise temperature of the WiFi receiver system. Comment if this is a tight requirement.

5

Pulse Modulation

Telecommunication systems aim to provide a faithful replica of transmitted signals to be successfully decoded by distant receivers. Transmitted signals may carry analog or digital information. For example, a speech or an analog TV signal is analog but a computer file is digital. A communication system may be designed for transmission of analog and/or digital signals. Analog (AM/FM) transmission of signals is continuous in time and amplitude. Accordingly, the demodulated received signals are also continuous in time and amplitude. However, in digital transmission, signals are discrete in time and amplitude. Therefore, for digital transmission of analog signals, they should be converted into digital by using analog-to-digital convertors (ADC). Similarly, received digital signals should be converted back into analog by digital-to-analog convertors (DAC) at the receiver. Digital communications is preferred mainly because it offers much improved performance compared to analog communications.

Figure 5.1 shows the role of ADC and DAC in a simple block diagram of a baseband communication system for digital transmission of analog signals. The modulator modulates the amplitude, the duration or the position of the information pulses to make best use of the baseband channel. Evidently, there is no need for ADC and DAC blocks for transmitting digital information.

This chapter will be concerned with digital transmission of analog signals and will present the fundamentals and tradeoffs in this area. Firstly, the basics of ADC, comprising sampling, quantization and encoding, will be presented. At the receiver side, DAC performance will be evaluated. Time-division multiplexing (TDM) will be used to time-interleave multiple digital signals for transmission in the baseband channel. The performance analysis of pulse code modulation (PCM) systems will be followed by the study of differential quantization techniques and their applications in digital communications. Finally, audio and video encoding principles will be presented.

Digital Communications, First Edition. Mehmet Şafak.
© 2017 John Wiley & Sons Ltd. Published 2017 by John Wiley & Sons Ltd.
Companion website: www.wiley.com/go/safak/Digital_Communications

Figure 5.1   Block Diagram of a Baseband Communication System.

5.1   Analog-to-Digital Conversion

ADC is accomplished in three steps, namely sampling, quantizing and encoding. *Sampling* is a basic operation in digital signal processing and digital communications. Sampling consists of taking samples of an analog signal at time intervals sufficiently short so that the original signal does not change appreciably from one sample to the next and hence can be uniquely reconstructed from these samples. This implies that the sampling results in a discrete-time signal and sampling interval, T_s, decreases as the time rate of change of the analog signal increases. Decreasing sampling interval is synonymous to increasing the sampling frequency (rate), $f_s = 1/T_s$. This implies that a rapidly changing signal should be sampled more frequently compared for a slowly-changing signal.

The next step in ADC is the *quantization*, that is, the conversion of analog signal samples with continuously varying amplitudes into discrete-amplitude samples. Instead of transmitting the original signal amplitudes as they are (in analog form), the peak-to-peak voltage range of the analog signal is divided into L intervals, where L is chosen as a power of 2, that is, $L = 2^R$, where R denotes the number of bits per sample. Then, the sampled signal levels within one of these L intervals are represented by a single voltage level. The task of a receiver is hence facilitated since it will simply estimate which of the L levels an incoming signals has. It is evident that quantization results in an irrecoverable loss of information. However, by chosing L to be sufficiently large, one may decrease the so-called quantization error, which is defined as the difference between the original level of the signal sample and its quantized value. However, large values of L implies the representation and transmission of a particular signal sample with larger number of bits. This in turn

implies that information transmission with higher accuracy results in higher transmission rates (higher transmission bandwidth).

The last step in ADC is *encoding*. Instead of transmitting the quantized sample voltage as it is, the quantized sample voltage is encoded by R bits, with constant amplitudes. Consequently, an analog signal sampled at a frequency of f_s (samples/s) and quantized with $L = 2^R$ levels (R bits/sample) is transmitted at a rate of $f_s R$ bits per second.

Figure 5.2 shows the ADC process of an analog signal of amplitude varying between (−4 V, +4 V). Since the signal is linearly quantized into $L = 4$ levels, samples signal voltages between 2–4 V are represented by 3 V, 0-2 V interval by 1 V and so on. Each sample is encoded with $R = 2$ bits; hence, the −3 V quantization level is represented by dibits **00**, −1 V by dibits **01** and so on. The assignment of bits to quantization levels is made so that the neighboring dibits differ from each other by only a single bit. The receiver will most likely be mistaken about the level of the received sample by choosing one-level lower or higher. Therefore, with this assignment scheme, only one of the two bits will be received in error. If the sample duration is assumed to be infinitely short, then each sample may be approximated by a delta function, hence instantaneous sampling. Each sample is then (binary) encoded with bit duration T_b. $RT_b \leq T_s$. In this particular example, the dibit **10** will be transmitted to inform the receiver about the +3 V amplitude of the quantized sample.

Example 5.1   ADC Fundamentals.

Consider a signal, whose amplitude is confined to ±1 V. Assume that this signal is sampled with 1000 samples of per second, implying a $T_s = 1$ ms sampling interval. Also assume that each sample is represented by 6 bits. This

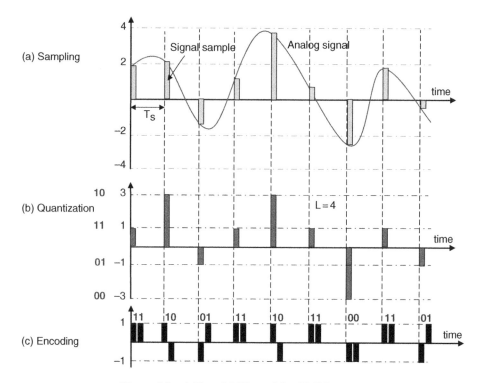

Figure 5.2   A Pictorial View of the ADC Process.

implies that we divide peak-to-peak signal level of $V_{pp} = 2$ volts into $L = 2^R = 64$ levels, thus having an amplitude interval of $\Delta = V_{pp}/L = 31.25$ mV. Any analog signal level falling in one of these 64 levels will be represented by a unique representation level, typically at the middle of the corresponding 31.25 mV interval. The maximum amplitude difference (quantization error) between the original analog signal and its representation level will then be limited to $\Delta/2 = 15.6$ mV.

It is clear that L is required to be increased with increasing values of the V_{pp} to limit the quantization error to tolerable levels. Similarly, higher sampling rates are required for converting rapidly changing analog signals into digital. The following sections will present the scientific background for correctly choosing f_s and L.

Since we take 1000 samples/s and represent each sample by $R = 6$ bits, the bit rate will be $f_s R = 1000(\text{samples/s}) \times 6(\text{bits/sample}) = 6000$

(bits/s). This implies that we use whole sampling interval T_s to transmit R bits. However, one can use only part of the sampling interval T_s to transmit at rate $f_s R$ by choosing $T_b \leq T_s/R$ (see Figure 5.2). In that case, the transmission bandwidth will increase since it is inversely proportional to T_b. The remaining time interval, $T_s - RT_b$, may be used for transmitting other signals by time-multiplexing with the original signal. For example, a bit duration of $T_b = 0.25$ ms implies that 0.5 ms is used for transmitting $R = 2$ bits for the original signal and the remaining 0.5 ms may be used for transmitting a second signal in the same channel by time-division multiplexing (TDM).

5.1.1   Sampling

5.1.1.1   Instantaneous Sampling

By sampling, an analog signal is converted into a sequence of samples that are usually spaced uniformly in time with a sampling interval of T_s. The choice of the sampling rate (frequency) $f_s = 1/T_s$, that is, the number of samples per second, is a critical issue. In order that the samples can provide a faithful representation of the sampled analog signal, the sampling rate should increase in proportion with the rate at which the signal varies in time, which determines the signal bandwidth W.

Consider that a baseband signal $g(t)$, band-limited to $0 < f < W$ Hz, is sampled at uniform time intervals T_s. Sampling duration is assumed to very short so that the samples may be represented by delta functions located at sampling times nT_s, with $-\infty < n < \infty$. The sampled signal $g_\delta(t)$ may then be expressed as

$$g_\delta(t) = \sum_{n=-\infty}^{\infty} g(nT_s)\, \delta(t - nT_s)$$

$$= g(t) \sum_{n=-\infty}^{\infty} \delta(t - nT_s) \tag{5.1}$$

which follows from (1.21). It is logical that the sampling interval should get shorter as the band-with occupied by the signal becomes larger, since this implies more rapid time variations in the signal and closer samples are required to reconstruct a faithful replica of the analog signal from its samples.

Since (5.1) may be written as the multiplication of $g(t)$ and an infinite sequence of delta functions, with the following Fourier transform pair (see (1.72)),

$$h_\delta(t) = \sum_{n=-\infty}^{\infty} \delta(t - nT_s)$$

$$\Leftrightarrow H_\delta(f) = \frac{1}{T_s} \sum_{k=-\infty}^{\infty} \delta\left(f - \frac{k}{T_s}\right) \tag{5.2}$$

the Fourier transform of (5.1), $G_\delta(f)$, may be expressed as the convolution of $G(f)$ and $H_\delta(f)$ (see (1.46)):

$$G_\delta(f) = G(f) \otimes H_\delta(f) = f_s \sum_{k=-\infty}^{\infty} G(f - kf_s) \tag{5.3}$$

The spectrum of a band-limited baseband analog signal g(t) sampled at a sampling rate f_s, is thus the superposition of its spectrum frequency shifted at multiples of f_s (see Figure 1.10). Figure 5.3 shows the effect of the sampling rate on the spectrum of the uniformly sampled signal $g_\delta(t)$. When the sampling rate is less than 2 W, the signal is said to be under-sampled. Then, the sampling interval is longer than it should be for recovering the original signal, since the spectra of $G(f + nf_s)$ can not be separated from each other. To solve this problem one needs to shorten the sampling interval and take more samples per unit time, that is, to increase the sampling rate. When the sampling rate is equal to twice the signal bandwidth, $f_s = 2 W$, then the spectrum $G(f + nf_s)$ does not overlap with neighboring spectra for $n - 1$ and $n + 1$, and a low-pass filter with infinitely sharp cut-ff may be used to recover the original spectrum $G(f)$ (see Figure 5.4). This sampling rate, $f_s = 2 W$, that is referred to as Nyquist sampling rate, is considered to be the minimum sampling rate that allows the recovery of the original signal from its samples. The signal is said to be over-sampled when sampling rate is larger than 2 W. Then, the guard band between the spectra, as shown in Figure 5.3, facilitates the recovery of the original signal without distortion by using a LPF (see Figure 5.4). In practice, the sampling rate is usually taken 10–20% higher than the Shannon's sampling rate in order to avoid aliasing. For example, a stereo audio system with each of the two channels bandlimited to 20 kHz, is sampled at 44.1 ksamples/s instead of the Nyquist rate 40 ksamples/s.

In summary, an analog signal $g(t)$ bandlimited to W may be recovered from its time samples using a LPF with a passband $0 < f < W$, as long as the sampling rate satisfies $f_s \geq 2 W$:

If $G(f) = 0$ for $|f| \geq W$ and $f_s \geq 2 W$, then

$$G(f) = \frac{1}{f_s} G_\delta(f), \quad -W < f < W \tag{5.4}$$

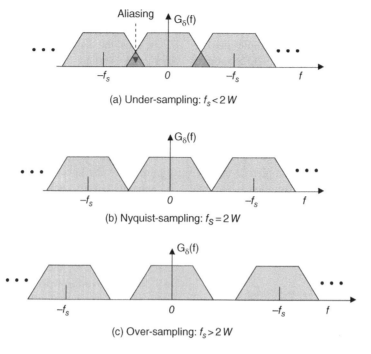

(a) Under-sampling: $f_s < 2W$

(b) Nyquist-sampling: $f_s = 2W$

(c) Over-sampling: $f_s > 2W$

Figure 5.3 Effect of the Sampling Rate on the Spectrum of the Sampled Signal $g_\delta(t)$.

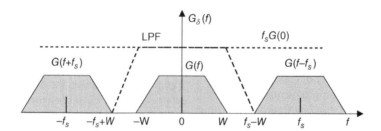

Figure 5.4 The Recovery of an Over-Sampled Signal by a LPF.

Note that $G_\delta(f)$ may also be expressed in terms of its time samples at the receiver:

$$G_\delta(f) = \Im[g_\delta(t)] = \sum_{n=-\infty}^{\infty} g(nT_s)\, \Im[\delta(t-nT_s)]$$

$$= \sum_{n=-\infty}^{\infty} g(nT_s)\, \exp(-j2\pi f nT_s) \tag{5.5}$$

where $\Im[.]$ denotes the Fourier transform and, consequently, (5.5) is simply the discrete-time Fourier transform of $g_\delta(t)$. Noting the original signal $g(t)$ can be uniquely determined from $G_\delta(f)$ (see (5.4)), the time-samples $\{g(nT_s)\}$ in (5.5) have all the information contained in $g(t)$ as long as $f_s = 1/T_s \geq 2W$. Hence, a transmitter does not need to transmit an analog signal at all times. Transmission of time-samples at time intervals $T_s \leq 1/(2W)$ is sufficient for

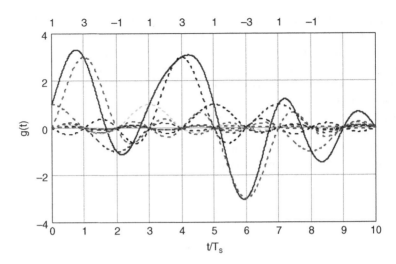

Figure 5.5 Reconstruction of the Original Analog Signal $g(t)$ (Continuous Line) From its Quantized Samples Shown in Figure 5.2b by Using Interpolation Function in (5.6) for $T_s = 1/(2\,W) = 1$ s.

exact recovery by the receiver of the original signal $g(t)$ from these samples.

Using (5.4), the original signal $g(t)$ can be reconstructed from the sequence of sample values $\{g(n/2\,W)\}$:

$$g(t) = \int_{-W}^{W} \frac{1}{2W} G_\delta(f)\, \exp(j2\pi f t)\, df$$

$$= \int_{-W}^{W} \frac{1}{2W} \sum_{n=-\infty}^{\infty} g\left(\frac{n}{2W}\right) \exp\left(-\frac{j\pi nf}{W}\right) \exp(j2\pi f t)\, df$$

$$= \sum_{n=-\infty}^{\infty} g\left(\frac{n}{2W}\right) \frac{1}{2W} \int_{-W}^{W} \exp\left[j2\pi f\left(t - \frac{n}{2W}\right)\right] df$$

$$= \sum_{n=-\infty}^{\infty} g\left(\frac{n}{2W}\right) \underbrace{\mathrm{sinc}(2Wt - n)}_{\textit{interpolation function}} \quad -\infty < t < \infty$$

(5.6)

where $\mathrm{sinc}(2Wt\text{-}n)$ is the interpolation function used for reconstructing. Figure 5.5 shows the recovery of an analog signal $g(t)$ from the received quantized samples $\{1, 3, -1, 1, 3, 1, -3, 1, -1\}$ shown in Figure 5.2b. Interpolation function for each quantized sample is shown in broken line while the recovered analog signal is shown by continuous line. The recovered signal is very similar to the originally transmitted signal shown in Figure 5.2a.

Example 5.2 Music CDs.

Music CDs employ uniform quantization, which implies that L quantization intervals have the same width. Two analog (left and right) channels of stereo music are each bandlimited to $W = 20$ kHz to include high frequency components created by musical instruments. Each channel is oversampled at $f_s = 44100$ samples/s (10% higher than the Nyquist rate), to avoid aliasing. Each sample is then quantized using $R = 16$ bits/ samples. The encoding rate for music on CDs is thus $2 \times f_s \times R = 44.1$ ksamples/s $\times 16$ quantizing bits/sample $\times 2$ channels for stereo $=$ 1.411 Mbits/s. CDs can thus store up to 74 minutes of music and audio, that is, (74×60) s $\times$ 1.411 Mbits/s $= 6.26484$ Gbits.

5.1.1.2 Flat-Top Sampling

In instantaneous sampling, a message signal $g(t)$ is sampled by a periodic train of impulses with sampling period $T_s = 1/f_s$, according to the Shannon's sampling theorem. The duration of each sample is assumed to be infinitely short so that the samples may be represented by impulses. Instantaneous sampling is not realistic since each sample, represented by an impulse, requires infinite bandwidth. However, in practice, a signal is sampled as a finite duration pulse and the resulting sampled signal has a finite channel bandwidth.

In flat-top sampling, a sample-and-hold circuit (see Example 5.3) is used to sample an analog message signal $g(t)$ to obtain a constant sampled signal level during the sampling time. Hence, the analog signal $g(t)$ is not multiplied by a periodic train of impulses but by a periodic train of rectangular pulses $h(t)$:

$$h(t) = \begin{cases} 1 & 0 < t < T \\ 0 & otherwise \end{cases} \tag{5.7}$$

Lengthening the duration of each sample to a given value T avoids the use of excessive channel bandwidth, which is inversely proportional to the pulse duration. As shown in Figure 5.2, the duration T of each rectangular pulse $h(t)$ satisfies $T/T_s \ll 1$. However, the sampling frequency for flat-top sampling is still $f_s = 1/T_s$ as in instantaneous sampling.

The expression for a flat-top sampled signal $g_\Pi(t)$ may be written as

$$g_\Pi(t) = h(t) \otimes g_\delta(t) = \sum_{n=-\infty}^{\infty} g(nT_s)h(t-nT_s) \tag{5.8}$$

which implies that the impulses in (5.1) are replaced by h(t)'s. Using (5.3) and the convolution property of the Fourier transform, the Fourier transform of (5.8) is found as

$$G_\Pi(f) = H(f)\, G_\delta(f) = f_s\, H(f) \sum_{k=-\infty}^{\infty} G(f - kf_s)$$

$$= f_s\, G(f)H(f) + f_s\, H(f) \sum_{\substack{k=-\infty \\ k \neq 0}}^{\infty} G(f - kf_s)$$

$$(5.9)$$

Noting that the spectrum of the rectangular pulse $h(t)$, which is given by (see (1.16))

$$H(f) = \Im[h(t)] = T\operatorname{sinc}(fT)\exp(-j\pi fT)$$

$$(5.10)$$

is not flat for $f < W$ (see Figure 5.6), low-pass filtering of a flat-top sampled signal spectrum (5.9) at the receiver results in a distortion of the original signal spectrum. However, since $T \ll T_s$ and the first null of (5.10) is located at $f = 1/T$, one can write $1/T \gg 1/T_s = f_s = 2\,W$. Therefore, the spectral components of $H(f)$ overlapping with $G(f)$ are practically flat and the distortion caused by $H(f)$ in the recovery of G(f) may be negligibly small. For example,

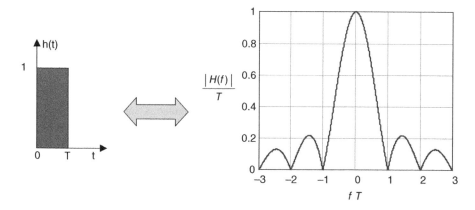

Figure 5.6   Rectangular Pulse h(t) and the Magnitude of its Fourier Transform.

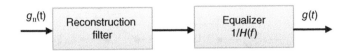

Figure 5.7   Recovery of Flat-Top Sampled Signals.

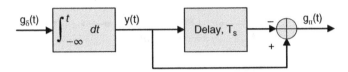

Figure 5.8   A Simple Hold Circuit Used for Flat-Top Sampling.

for $T/T_s = 0.1$ we have $WT = 0.05$. The magnitude of (5.10) at the edge of the signal bandwidth W is then equal to $0.996\,T$, which is 0.035 dB below its peak value and hence negligible. Nevertheless, modern communication systems employ an equalizer, in cascade with the low-pass reconstruction filter, which multiplies the incoming signal with $1/H(f)$, in order to equalize the first term in (5.9) and to minimize the distortion (see Figure 5.7).

Example 5.3   Zero-Order Hold Circuit for Flat-Top Sampling.
A simple zero-order hold circuit, shown in Figure 5.8, may be used to convert instantaneously sampled signals to flat-top sampled ones.

Note from (5.1) and (1.20) that

$$y(t) = \int_{-\infty}^{t} g_\delta(t)\, dt = \sum_{n=-\infty}^{\infty} g(nT_s)\, u(t-nT_s)$$

$$(5.11)$$

where $u(t)$ denotes the unit step function:

$$u(t) = \begin{cases} 0 & t < 0 \\ 1 & t > 0 \end{cases} \qquad (5.12)$$

From Figure 5.8, one can show that the output is the flat-top sampled signal given by (5.8):

$$g_\Pi(t) = y(t) - y(t - T_s) = \sum_{n=-\infty}^{\infty} g(nT_s)\, h(t-nT_s)$$

$$h(t-nT_s) = u(t-nT_s) - u(t-nT_s-T_s)$$

$$(5.13)$$

Example 5.4   Sampling Band-Pass Signals.
In certain applications, need may arise to sample bandpass signals. Consider a narrowband bandpass signal $s(t)$ at carrier frequency f_c and bandlimited to $B = 2\,W \ll f_c$. Direct sampling a bandpass signal requires sampling rates much higher than those required for baseband signals. Instead, a bandpass signal can be first down-converted to the baseband and then inphase and quadrature components at the baseband may be sampled separately.

A bandpass signal $s(t)$ may be expressed in terms of its complex envelope and its low-pass in-phase and quadrature components, $\tilde{s}(t) = \tilde{s}_R(t) + j\tilde{s}_I(t)$, as follows:

$$s(t) = \mathrm{Re}[\tilde{s}(t)\, e^{-jw_c t}] = \mathrm{Re}\{[\tilde{s}_R(t) + j\tilde{s}_I(t)]e^{-jw_c t}\}$$
$$= \tilde{s}_R(t)\cos w_c t + \tilde{s}_I(t)\sin w_c t$$

$$(5.14)$$

In-phase and quadrature baseband components, $\tilde{s}_R(t)$ and $\tilde{s}_I(t)$, are each bandlimited to W. Sampling these components at $f_s = 1/T_s = 2\,W$ allows the recovery of the original signal,

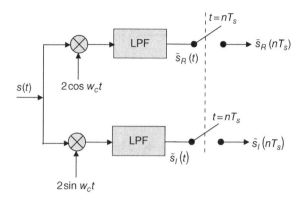

Figure 5.9   Sampling of Bandpass Signals.

s(t). The resource (the number of samples) required for sampling a bandpass signal is therefore twice that required for sampling a baseband signal. Figure 5.9 shows the block diagram of a circuit for sampling narrowband band-pass signals.

Example 5.5   ADC of Video Signals.
A given scene is recorded as three separate analog signals by a color video camera, in three colors, namely red, green, and blue (RGB). In digital video broadcasting (DVB) systems, these analog video signals are sampled, quantized, compressed, encoded, and modulated before transmission. The sampling process of a video signal also takes place in RGB. However, the transmission is in YUV, defined as

Luminance:	Y = 0.30 R + 0.59 G + 0.11 B.
Red:	U = 0.88 (R-Y)
Blue:	V = 0.49 (B-Y)

The conversion of RGB signals to YUV was historically carried out to provide backward compatibility with black-white TV transmissions since a black-white TV receiver uses only the luminance signal.

The sampling rate of a video signal is determined by the Nyquist sampling theorem. In analog video broadcasting, the luminance (black & white) channel of standard definition

TV (SDTV) is assigned a bandwidth of approximately 5 MHz. On the other hand, the color resolution of the human eye is less than its resolution for the luminance components. This sensitivity difference is exploited to deliver the color difference channels U and V in a significantly reduced bandwidth of ~2 MHz each. Consequently, a 8 MHz bandwidth is allocated to an analog TV channel in the phase alternating line (PAL) system.

In the moving picture experts group (MPEG) system, the sampling rate of the luminance component is therefore higher than 10 MHz. By taking advantage of the color sensitivity difference, sampling rates of the color difference components are lower than that for the luminance component.

In the early days of digital video, 4 times the color subcarrier frequency was considered as a convenient rate for sampling the luminance signal, thus a sampling rate of 4×4.434 MHz = 17.7 Msamples/s for the PAL system. The international standards for digital video now specify 13.5 Msamples/s for sampling the luminance component in all systems, with the colour difference sampling rates adjusted accordingly. For color difference signals, U and V, the same rate (4 times) was proposed for internal studio use, but half that rate (2 times) is used for transmissions outside the studio.

These two schemes became known as 4:4:4 and 4:2:2 sampling (Y,U,V components, respectively. A third scheme, known as 4:2:0, was proposed for cases where lower video quality could be tolerated and the available bandwidth might be limited. 4:2:0 involves sampling the color difference signal on only every second scan line, and filling in the missing color information by extrapolation between lines. 4:2:0 or 4:2:2 format is preferred in most applications. In 4:x:x, 4 indicates 13.5 Msamples/s for standard definition video, but may be much higher in high definition (HD) applications. A 4:2:2 picture at 8 bits/sample requires a raw transmission rate of 13.5 Msamples/s × 16 bits/sample = 216 Mbps ignoring overheads, so the need for compression is obvious.

A frame of a standard definition TV (SDTV) signal (excluding overheads) consists of 720 × 576 pixels (aspect ratio:4/3) while it is 1920 × 1080 pixels (aspect ratio: 16/9) for high definition (HD) TV. If we use $Y = U = V = 8$ bits/pixel to represent each color component of a given pixel (4:4:4), then the bit rate for the uncompressed SDTV signal (24 bits/pixel and 25 frames/s) is $24 \times 720 \times 576 \times 25 = 248.832$ Mbps for 4:4:4. For 4:2:2, 8 bits/pixel are used for Y and 4 bits/pixel for U and V components (16 bits/pixel); this implies $16 \times 720 \times 576 \times 25 = 165.888$ Mbps.

MPEG-2 is capable of compressing (by source coding) the bit rate of SDTV for 4:2:0 down to about 3-6 Mbit/s. At lower bit rates in this range, the impairments introduced by the MPEG-2 coding and decoding process may become intolerable. For MPEG-4, bit rates as low as 2 Mbp/s is thought to provide a good compromise between picture quality and transmission bandwidth efficiency.

5.1.2   Quantization

Analog signals and their samples vary in a continuous range of amplitudes. Instead of transmitting the amplitudes of the samples as they are, discrete (quantized) amplitude levels may be transmitted. This offers certain advantages in digital communications at the expense of an irrecoverable loss of signal information. However,

this loss is controlled by adjusting the spacings between the discrete (quantized) amplitude levels. If the quantization levels are chosen with sufficiently close spacings, the loss will be reduced to tolerable limits and the original signal could be recovered with minimum error.

Amplitude quantization is defined as the process of transforming the sample amplitude $g(nT_s)$ of a message signal $g(t)$ at time $t = nT_s$ into a discrete amplitude $v(nT_s)$ taken from a finite set of possible amplitude levels. The mapping $v(nT_s) = q(g(nT_s))$ specifies the quantizer characteristics. Quantization process is assumed to be memoryless and instantaneous. This means that the mapping at time $t = nT_s$ is not affected by earlier or later samples of the message signal. Quantizers are symmetric about the origin, that is, for positive and negative values of $g(t)$. All sample amplitude levels $g(nT_s)$ which are located in one of the partition cells are quantized by the corresponding representation level, as shown by Figure 5.10. A quantizer is said to be uniform if the step size is constant and nonuniform if it is not.

5.1.2.1   Uniform Quantization

In a uniform quantizer, decision and representation levels are uniformly spaced, that is, the step size Δ is constant. The number of representation levels L is chosen as a power of 2, that is, $L = 2^R$, where R is an integer and denotes the number of bits used to represent a quantized sample amplitude. Figure 5.11 shows two different uniform quantizer characteristics, which are of mid-tread or mid-rise type.

Let a quantizer maps the input sample $g(nT_s)$, which is a zero-mean random variable G with continuous amplitude in the range $\{-g_{max}, g_{max}\}$, into a discrete random variable V. The peak-to-peak signal amplitude $2g_{max}$ is divided into L partition cells, each with step size Δ:

$$\Delta = \frac{2g_{max}}{L} = \frac{2g_{max}}{2^R} \qquad (5.15)$$

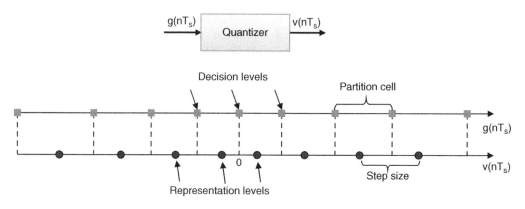

Figure 5.10   Decision and Representation Levels in a Quantizer.

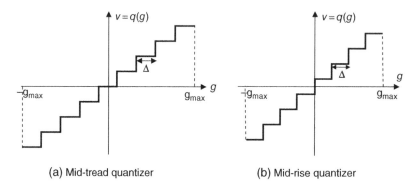

(a) Mid-tread quantizer   (b) Mid-rise quantizer

Figure 5.11   Midtread and Midrise Type Uniform Quantizers.

Noting that R can assume only positive integer values, L can be 2, 4, 8,... The quantization error is denoted by the random variable Q of sample value $q(t)$ and is characterized by the step size Δ and the number of partition cells L:

$$q(t) = g(t) - v(t) \qquad (5.16)$$

G having zero-mean and the quantizer assumed to be symmetric, V and Q will also have zero-mean. Assuming that the representation levels are located at the middle of the partition cells, the quantization error Q will be bounded by $-\Delta/2 \leq q \leq \Delta/2$, since the maximum difference between the input signal sample and its quantized value (representation level) cannot exceed $\Delta/2$ (see Figure 5.10).

If Δ is sufficiently small, then Q may be assumed to a uniformly distributed:

$$f_Q(q) = \begin{cases} 1/\Delta & -\Delta/2 < q \leq \Delta/2 \\ 0 & otherwise \end{cases} \qquad (5.17)$$

Using the pdf given by (5.17), the variance of the quantization error Q is found to be

$$\sigma_Q^2 = E\left[Q^2\right] = \int_{-\infty}^{\infty} q^2 f_Q(q)\, dq = \frac{1}{\Delta} \int_{-\Delta/2}^{\Delta/2} q^2\, dq$$

$$= \frac{\Delta^2}{12} = \frac{g_{max}^2}{3 \times 2^{2R}}$$

$$(5.18)$$

Assuming that the thermal noise power is much lower compared to the variance of the quantization error, the output signal to quantization noise ratio, SNR_Q, of a uniform quantizer is defined as the ratio of the average signal power P to the variance of the quantization error:

$$SNR_Q = \frac{P}{\sigma_Q^2} = \frac{3 \times 2^{2R}}{g_{max}^2/P} = \frac{3 \times 2^{2R}}{\alpha} \quad (5.19)$$

$$\alpha = g_{max}^2/P = P_{peak}/P$$

where α denotes peak to average power ratio (PAPR). Output signal to quantization noise ratio, SNR_Q, is usually expressed in dB:

$$SNR_Q = 4.77 + 6\,R - \alpha_{dB} \quad (dB) \quad (5.20)$$

Note that $\alpha = 2$ (3 dB) for a sinusoidal signal $A\cos w_c t$ since $P_{peak} = A^2$ and $P = A^2/2$. For an audio signal $\alpha = 10$ dB is usually assumed.

Example 5.6 Signal/Quantization Noise Ratio Versus the Number of Bits/Sample.
Consider a signal with 1 W peak power and an average power of 1 mW; thus $\alpha = 30$ dB. If we employ a 5-bit ADC, the SNR_Q is given by (5.20)

$$SNR_Q = 4.77 + 6 \times 5 - 30 = 4.77\ dB \quad (5.21)$$

The corresponding step size is $\Delta = 2g_{max}/2^R = 63$ mV. However, if a 6-bit ADC is used, one gets an $SNR_Q = 10.77$ dB with a step size of $\Delta = 31$ mV.

Example 5.7 Uniform Quantization of a Random Signal With Laplacian Pdf.
Consider that a random signal X described by the Laplacian pdf,

$$f_X(x) = K e^{-b|x|}, \quad -4 \le x \le 4 \quad (5.22)$$

is constrained to the range [−4, +4] by a limiter before uniform quantization. The limiting operation serves the purpose of clipping

infrequent signal amplitudes in order to achieve a smaller Δ, and hence higher SNR_Q.

a. We first determine the normalization constant K and the PAPR. Since the area under the pdf curve is equal to unity:

$$\int_{-4}^{4} K e^{-b|x|} dx = 2K \int_{0}^{4} e^{-b\,x} dx = \frac{2K}{b}\left(1 - e^{-4b}\right) = 1$$

$$\Rightarrow K = \frac{b}{2\left(1 - e^{-4b}\right)}$$

$$(5.23)$$

which becomes $K = 0.5093$ for $b = 1$. The average signal power is given by

$$P = \int_{-4}^{4} x^2 f_X(x)\ dx$$

$$= \frac{2}{b^2\left(1 - e^{-4b}\right)}\left[1 - e^{-4b}\left(1 + 4b + 8b^2\right)\right]$$

$$(5.24)$$

For $b = 1$, the PAPR is found as

$$\alpha = \frac{P_{peak}}{P} = \frac{(4)^2}{1.552} = 10.3 \quad (10.1\ dB) \quad (5.25)$$

b. Figure 5.12 shows the uniform quantizer characteristics, assuming that the number of quantization levels is $L = 4$ and the corresponding step size is $\Delta = 2$.

c. The pdf of the discrete random variable V, for b=1, at the quantizer output:

$$Pr(-2 < x < 0) = Pr(0 < x < 2)$$

$$= K \int_{0}^{2} e^{-x} dx = K\left(1 - e^{-2}\right) = 0.44$$

$$Pr(-4 < x < -2) = Pr(2 < x < 4)$$

$$= K \int_{2}^{4} e^{-x} dx = K\left(e^{-2} - e^{-4}\right) = 0.06$$

$$f_V(v) = Pr(-4 < x < -2)\ \delta(v + 3) \quad (5.26)$$
$$+ Pr(-2 < x < 0)\ \delta(v + 1)$$
$$+ Pr(0 < x < 2)\ \delta(v - 1)$$
$$+ Pr(2 < x < 4)\ \delta(v - 3)$$
$$= 0.06\ \delta(v + 3) + 0.44\ \delta(v + 1)$$
$$+ 0.44\ \delta(v - 1) + 0.06\ \delta(v - 3)$$

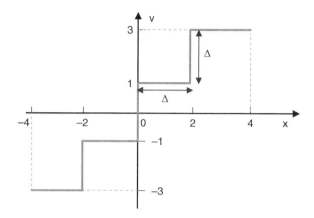

Figure 5.12 Four-Level Uniform Quantizer for [−4 V,+4 V].

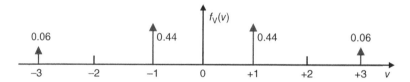

Figure 5.13 Pdf of the Quantizer Output Shown in Figure 5.12 For an Input Random Signal x with Laplacian pdf, $f_X(x) = Ke^{-b|x|}$, $-4 \le x \le 4$, $b = 1$.

The pdf of the quantizer output is shown in Figure 5.13:

d. Noting that X is a Laplacian random variable at the quantizer input and $\Delta = 2$, using the symmetry with respect to the origin, the variance of the quantization error for $b = 1$ is given by

$$\sigma_Q^2 = 2\int_0^2 (x-1)^2 Ke^{-x}dx + 2\int_2^4 (x-3)^2 Ke^{-x}dx$$

$$= 2K(e^{-1} + e^{-3})\int_{-1}^1 u^2 e^{-u}du$$

$$= 2K(e^{-1} + e^{-3})(e - 5e^{-1})$$

$$= 0.374$$

(5.27)

The signal to quantization noise ratio is given by

$$SNR_Q = \frac{P}{\sigma_Q^2} = \frac{1.552}{0.374} = 4.15 \quad (6.18 \ dB)$$

(5.28)

It is important to note that uniform quantization is optimal for input analog signals with uniform pdf's at the quantizer input. However, for $b \ge 1$, the pdf of the signal at the input and the output of the quantizer will be more concentrated around the origin. Then, it may be wise to reduce the step sizes near the origin in order to reduce the quantization noise for the samples that occur more frequently at the expense of increased quantization noise at larger values of the signal amplitudes, which occur less frequently.

For the special case for $b = 0$, the signal X has a uniform pdf, $f_X(x) = 1/8$, $-4 < x < 4$. The variance of the quantization error is then given by

$$\sigma_Q^2 = 2\int_0^2 (x-1)^2 f_X(x) \ dx + 2\int_2^4 (x-3)^2 f_X(x) \ dx$$

$$= \frac{1}{4}\int_0^2 (x-1)^2 dx + \frac{1}{4}\int_2^4 (x-3)^2 dx$$

$$= \frac{1}{2}\int_{-1}^1 u^2 du = \frac{1}{3}$$

(5.29)

which is identical to $\Delta^2/12$, as expected. The corresponding value of the signal to quantization noise ratio is now improved to be 6.68 dB. Note that, for $b=0$, the discrete pdf at the quantizer output, given by (5.26) and shown in Figure 5.13, will be uniform, like the pdf of X at the quantizer input.

e. When X is allowed to change in the range $(-\infty, \infty)$ at the quantizer input instead of limiting its values between -4 and $+4$, the normalizing factor becomes $K = 0.5$ and the probability of saturation is given by

$$P_{sat} = \Pr(|x|>4) = 2\int_{4}^{\infty} f_X(x)\,dx = e^{-4} = 0.018$$

(5.30)

5.1.2.2 Non-Uniform Quantization and Companding

Some signals have very large PAPRs, for example, PAPR > 10 dB for voice. For example, voice signals, which may be described by Laplacian pdf, assume lower voltage levels more frequently. Therefore, the quantization noise power becomes higher if a uniform quantizer is used for quantizing voice signals. However, if the step size gradually increases with increasing amplitudes, then quantization errors will be higher in infrequent large amplitude ranges but smaller at more frequent amplitudes. This will lead to a drastic decrease in the quantization noise. Therefore, non-uniform quantization will improve the speech quality. The increased quantization errors associated with large amplitude ranges may still be relatively small when viewed in percentage of the signal's amplitude. Their impact will not be significant on the perceived speech quality, since ear perceives volume changes logarithmically rather than linearly.

Instead of employing non-uniform quantization directly, a nonuniform quantizer first compresses an analog signal and then digitizes the compressed signal using uniform quantization. The inverse operation at the receiver consists of converting the quantized signal back into compressed form and then decompressing (expanding) the resulting signal by an expander (see Figure 5.14). Expander used at the receiver restores the signal samples to their correct relative levels, since it performs the inverse of the compression at the receiver. Thus, expander output provides a piecewise linear approximation to the (desired) signal at the compressor input. The combination of compressor and an expander is called a compander.

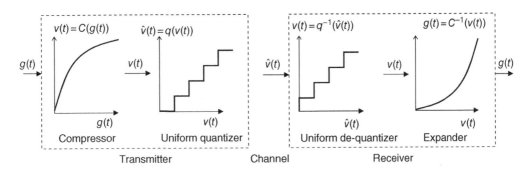

Figure 5.14 Block Diagram for Nonuniform Quantization.

There are mainly two compression standards for non-uniform quantization. The so-called µ-law compression is commonly used in the United States:

$$v = v_{max} \frac{\ln(1 + \mu |g/g_{max}|)}{\ln(1 + \mu)} \operatorname{sgn}(g)$$

$$g = \frac{g_{max}}{\mu} \left[e^{\ln(1+\mu)^{|v|/v_{max}}} - 1 \right] \operatorname{sgn}(v) \qquad (5.31)$$

$$= \frac{g_{max}}{\mu} \left[(1+\mu)^{|v|/v_{max}} - 1 \right] \operatorname{sgn}(v)$$

where $\mu \geq 0$. On the other side, the A-law compression is widely used in Europe:

$$v = \begin{cases} v_{max} \dfrac{A|g|/g_{max}}{1 + \ln A} \operatorname{sgn}(g) & 0 < |g|/g_{max} \leq 1/A \\[2mm] v_{max} \dfrac{1 + \ln(A|g|/g_{max})}{1 + \ln A} \operatorname{sgn}(g) & 1/A < |g|/g_{max} < 1 \end{cases}$$

$$(5.32)$$

where $A \geq 1$. Figure 5.15 shows the input-output characteristics of the µ-law compressor. As a result of compression, for $\mu > 0$, the SNR for low-level signals increases at the expense of the SNR for high level signals. A compromise is usually made in choosing the values of μ.

Typical values used in practice are $\mu = 255$ and $A = 87.6$. Note that $\mu = 0$ and $A = 1$ imply no compression.

Example 5.8 Uniform Versus Nonuniform Quantization.

We observed in Example 5.7 that when a random signal with Laplacian pdf is uniformly quantized, the quantization noise power is higher than that for a uniformly distributed input signal. In this example, we will consider nonuniform quantization of the signal considered in Example 5.7 when µ-law companding is used.

Let us first consider the pdf of the signal at the output of a µ-law compressor. One may easily show that the pdf of the signal at the compressor output is given by

$$f_V(v) = f_G(g) \left| \frac{dg}{dv} \right| = \frac{b \ln(1+\mu)}{2\mu(1 - e^{-4b})} (1+\mu)^{|v/v_{max}|}$$

$$\times \exp \left[-\frac{b \, g_{max}}{\mu} \left\{ (1+\mu)^{|v/v_{max}|} - 1 \right\} \right]$$

$$(5.33)$$

where $f_G(g) = K e^{-b|g|}$, $-4 \leq g \leq 4$.

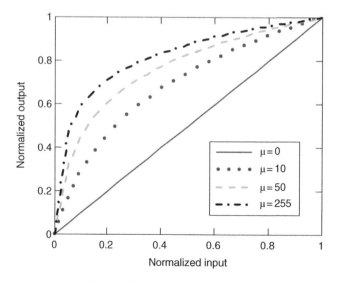

Figure 5.15   µ-law compression

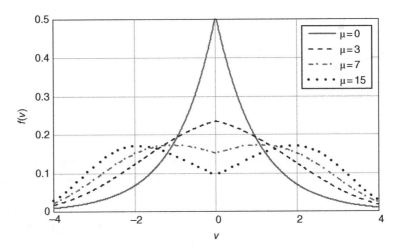

Figure 5.16 The Pdf of the Signal at the μ-Law Compressor Output for Various Values of μ. The signal at the compressor input is given by $f_G(g) = K e^{-b|g|}$, $-4 \le g \le 4$.

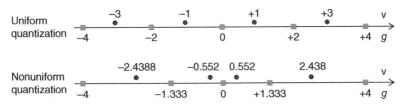

Figure 5.17 Decision (Square) and Representation (Circle) Levels for Uniform and Nonuniform Quantization with $\mu = 3$.

Figure 5.16 shows the variation of the pdf (5.33) at the output of a μ-law compressor for $b = 1$ and various values of μ. One may easily observe that the pdf at the compressor output is more uniformly distributed than the input pdf, given by the curve for $\mu = 0$. Consequently, nonuniform quantization of the signal, that is, uniform quantization of the compressed signals (with $\mu > 0$), is expected to provide lower quantization noise power compared to non-compressed signal (with $\mu = 0$).

Noting that uniform quantization of the compressor output is synonymous to nonuniform quantization of the quantizer input, let us first find the decision and representation levels, for the special case of $\mu = 3$. Inserting $v = 0$, 1, 2, 3 and 4 into (5.31), we get the corresponding values of g as 0, 0.552, 1.333, 2.438 and 4.

Figure 5.17 shows how decision and representation levels change when we use a μ-law compander with $\mu = 3$. Note that the step size becomes smaller near the origin where the signal is more likely to occur. This may easily be observed from the probabilities that the signal amplitude remain in the corresponding partition cells:

$$\Pr(0 < |v| < 2) = \Pr(0 < |g| < 1.333)$$

$$= K \int_0^{1.333} e^{-x} dx = K(1 - e^{-1.333}) = 0.375$$

$$\Pr(2 < |v| < 4) = \Pr(1.333 < |g| < 4)$$

$$= K \int_{1.333}^4 e^{-x} dx = K(e^{-1.333} - e^{-4}) = 0.125$$

$$(5.34)$$

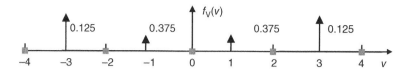

Figure 5.18   Pdf of the Output of a Nonuniform Quantizer with $\mu = 3$.

The pdf of the nonuniform quantizer output V given by

$$f_V(v) = \Pr(-4 < v < -2)\ \delta(v+3)$$

$$+ \Pr(-2 < v < 0)\ \delta(v+1)$$

$$+ \Pr(0 < v < 2)\ \delta(v-1)$$

$$+ \Pr(2 < v < 4)\ \delta(v-3)$$

$$= 0.125\ \delta(v+3) + 0.375\ \delta(v+1)$$

$$+ 0.375\ \delta(v-1) + 0.125\ \delta(v-3) \tag{5.35}$$

is shown in Figure 5.18. Compared with Figure 5.13, the pdf of the quantized signal is more uniform.

The quantization error variance for nonuniform quantization with $\mu = 3$ is given by

$$\sigma_Q^2 = 2\int_0^{1.333} (x-0.552)^2 Ke^{-x}dx$$

$$+ 2\int_{1.333}^{4} (x-2.438)^2 Ke^{-x}dx = 0.233 \tag{5.36}$$

It is clear from (5.36) that non-uniform quantization with $\mu = 3$ of a signal with Laplacian pdf has a quantization error variance of $\sigma_Q^2 = 0.233$, compared with $\sigma_Q^2 = 0.374$ (see (5.27)) for uniform quantization of the same signal. Hence, for a random signal with Laplacian pdf ($b = 1$), non-uniform quantization with $\mu = 3$ offers 10log (0.374/0.233)=2 dB decrease in quantization noise power compared to uniform quantization.

It may be useful to note that the quantization noise power can be minimized (maximizing

SNR_Q) if the compression characteristics (here μ) is matched to the pdf of the signal to be quantized. Figure 5.19 and Figure 5.20 show respectively the variation of the quantization noise power, and the signal to quantization noise ratio as a function of μ for various values of b. Note that $b = 0$ corresponds to uniform pdf of the input signal. As b increases, the pdf of the input signal becomes more concentrated around the origin and uniform quantization ($\mu = 0$) leads to increased variance of the quantization noise. However, nonuniform quantization ($\mu > 0$) prevents the signal power concentration around the origin and the variance of the quantization noise reduces and reaches its minimum at a specific value of μ. This implies that there is an optimum compression ratio μ which minimizes the variance of the quantization noise. As one may easily observe from Figure 5.20, the optimum value of the compression ratio also maximizes the signal to quantization noise power ratio.

5.1.2.3   Signal-to-Quantization Noise Ratio with Companding

The variance of the quantization noise may be written as

$$\sigma_Q^2 = \sum_{i=1}^{L} \int_{a_{i-1}}^{a_i} (x - \hat{x}_i)^2 f_X(x)\, dx$$

$$\cong \sum_{i=1}^{L} f_X(a_{i-1}) \int_{a_{i-1}}^{a_i} (x - \hat{x}_i)^2\, dx \tag{5.37}$$

$$= \sum_{i=1}^{L} f_X(a_{i-1})\ \frac{1}{12}\Delta_i^3$$

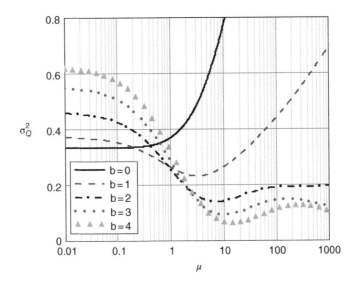

Figure 5.19 Variation of σ_Q^2 with μ for a Nonuniformly Quantized Signal with Laplacian Pdf $f_X(x) = 0.5\ e^{-x}$, $-4 < x < 4$ for $L = 4$. Note that $\sigma_Q^2 = 0.374$ for $b = 1$ and $\mu = 0$, which corresponds to uniform quantization. Similarly, $\sigma_Q^2 = 0.333$ for $b = 0$ and $\mu = 0$, corresponding to uniform quantization of a signal with uniform pdf. Optimum value of μ should be chosen for the type of signals to be quantized.

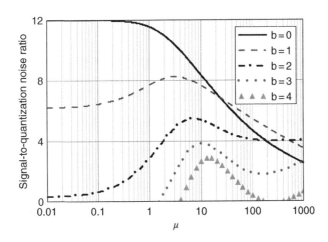

Figure 5.20 Signal to Quantization Noise Power Ratio Versus μ For Various Values of b.

where a_i, $i = 0,1,\ldots,L$ denote the borders of the decision levels, $\hat{x}_i = (a_{i-1} + a_i)/2$, and $\Delta_i = a_i - a_{i-1}$. Noting that the input-output curve of the nonuniform quantizer has a slope (see Figure 5.21)

$$y'(x) = \frac{\Delta}{\Delta_i} = \frac{y_{max}/2^{R-1}}{\Delta_i} \qquad (5.38)$$

where Δ_i and $\Delta = 2y_{max}/2^R$ denote respectively the step sizes before and after companding (see

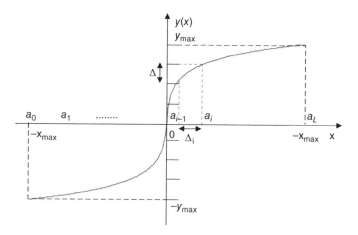

Figure 5.21   Input-Output Characteristics of a Compander.

Figure 5.21). Inserting Δ_i^2 from (5.38) into (5.37), one gets

$$\sigma_Q^2 = \sum_{i=1}^{L} f_X(a_{i-1}) \frac{1}{12} \left(\frac{\Delta}{y'(x)} \right)^2 \Delta_i$$

$$= \frac{y_{max}^2}{3 \times 2^{2R}} \sum_{i=1}^{L} f_X(a_{i-1}) \frac{1}{[y'(x)]^2} \Delta_i \quad (5.39)$$

$$= \frac{y_{max}^2}{3 \times 2^{2R}} \int_{-\infty}^{\infty} \frac{f_X(x)}{[y'(x)]^2} dx$$

where the integral expression follows from the previous expression for an infinitely large number quantization levels. The variance of the quantization error for a given compresser characteristics $y(x)$ can be minimized to obtain the optimal compresser [2][3][4][5]:

$$y(x) = y_{max} \left[\frac{2 \int_{-\infty}^{x} [f_X(t)]^{1/3} dt}{\int_{-\infty}^{\infty} [f_X(t)]^{1/3} dt} - 1 \right] \quad (5.40)$$

Inserting the derivative of (5.40) into (5.39), the minimum value of the quantization noise variance is found to be

$$\sigma_Q^2 = \frac{1}{12 \times 2^{2R}} \left[\int_{-\infty}^{\infty} [f_X(t)]^{1/3} dt \right]^3 \quad (5.41)$$

Using (5.41), the minimum value of the variance of the quantization noise is found to be 0.229 for the Laplacian pdf (5.22) for b=1.

Taking the derivative of y from (5.31) and inserting into (5.39), one can determine the variance of the quantization noise for μ-law companding as follows:

$$\sigma_Q^2 = \frac{x_{max}^2}{3 \times 2^{2R} \mu^2} [\ln(1+\mu)]^2 \int_{-\infty}^{\infty} [1 + \mu |x/x_{max}|]^2 f_X(x) \, dx$$

$$= \frac{x_{max}^2}{3 \times 2^{2R} \mu^2} [\ln(1+\mu)]^2 \int_{-\infty}^{\infty} [1 + \mu |\lambda|]^2 f_\Lambda(\lambda) \, d\lambda$$

$$= \frac{x_{max}^2}{3 \times 2^{2R} \mu^2} [\ln(1+\mu)]^2 \left[1 + 2\mu E[\Lambda] + \mu^2 E[\Lambda^2] \right]$$

(5.42)

where $\Lambda = X/x_{max}$ has a variance equal to the inverse of the PAPR:

$$E[\Lambda^2] = \frac{E[X^2]}{x_{max}^2} = \frac{P}{P_{peak}} = \frac{1}{PAPR} \quad (5.43)$$

Signal-to-quantization noise ratio may then be expressed as

$$SNR_Q = \frac{E[X^2]}{\sigma_Q^2} = \frac{3 \times 2^{2R}}{[\ln(1+\mu)]^2 \mu^2 E[\Lambda^2] + 2\mu E[\Lambda] + 1}$$

$$= SNR_{Q,uniform} \frac{\mu^2}{[\ln(1+\mu)]^2 [\mu^2 E[\Lambda^2] + 2\mu E[\Lambda] + 1]}$$

(5.44)

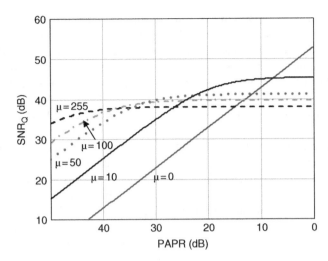

Figure 5.22 Signal-to-Quantization Noise Power Ratio for Nonuniform Quantization with μ-Law Companding. $R = 8$ is assumed.

Figure 5.22 shows the variation of the signal-to-quantization noise ratio as a function of PAPR for various values of μ. $E[\Lambda] = 0$ since the pdf $f_X(x)$ is symmetric around $x = 0$. For μ = 0, which corresponds to uniform quantization, the SNR_Q decreases linearly with increasing values of the PAPR in log-log scale. This implies that the system performance does not remain the same over the dynamic range of the signal and rapidly deteriorates with increasing values of PAPR. For example, SNR_Q varies 40 dB for speech signals which typically have 40 dB dynamic range. However, SNR_Q becomes less sensitive to PAPR variations with increasing values of μ and improves at high values of PAPR at the expense of some deterioration at lower PAPR values. For μ = 255, SNR_Q is almost flat for 0 dB < PAPR < 40 dB; hence, equal quality performance is guaranteed for very low and very high volumes of the speech signals.

5.1.3 Encoding

Following sampling and quantization, we obtain discrete-time and discrete-amplitude signal samples. The encoding process translates the discrete set of quantized signal samples into a more appropriate form for transmission. The

next step in the ADC process is to represent these samples by voltage levels with equal amplitude and opposite polarity, corresponding to binary digits **1** and **0**.

5.1.3.1 Differential Encoding

A binary stream may be encoded in an absolute manner or differentially. In *absolute encoding*, each bit is encoded independently of the others. For example, a binary **1** corresponds to a representation level of +1 V and a binary **0** to −1 V or vice versa. However, *differential encoding* is not absolute and is based on the transitions in the bit stream to be encoded. For example, consider a bit stream $\{b_k\}$ to be differentially encoded according to the rule:

$$d_k = \overline{d_{k-1} \oplus b_k} \qquad (5.45)$$

where b_k and d_k denote respectively the kth input and transmitted bits (see Table 5.1). The bar over the binary sum denotes the complement. According to the rule, d_k undergoes a transition (d_k changes from 0 to 1 or from 1 to 0) when $b_k = 0$, but is unchanged when the input bit $b_k = 1$. Hence, a transition is used to designate **0** in the incoming binary bit stream $\{b_k\}$ and a no-transition is used to designate **1**.

Differential encoding requires the use of a reference bit d_0 for initiating the encoding process. Use of **0** or **1** as the reference bit simply inverts the encoded bit stream. Inverting a differentially encoded stream does not affect its interpretation.

Now assume that the differentially encoded bit stream $\{d_k\}$ in Table 5.1 is received with its third bit in error and an incorrect reference bit is assumed at the receiver. The received bit stream and the incorrect reference bit are shown in bold in Table 5.2. The transmitted bit stream is estimated, in the last line of Table 5.2, by simply observing whether or not a transition has occurred in the polarity of adjacent bits of the received bit stream $\{d_k\}$. The use of an incorrect reference bit was observed to alter only the first decoded bit (shown in bold in the last line). On the other hand, the decoded bit stream also shows that a bit received in error causes erroneous decoding of two bits in the decoded bit stream. For differential encoding, each bit error in the received stram $\{d_k\}$ causes two bit errors

in the decoded bit stream $\{b_k\}$; hence, the bit error probability is doubled compared to absolute encoding. Table 5.3 shows that the original received bit stream and its inverted version result in the same decoded bit stream. Therefore, differential encoding is immune to the inversion of the polarity of the received bit sequence, that is, to the use of 1 or 0 as the reference bit.

Despite its degraded bit error performance, differential encoding has the advantage of being immune to polarity inversions. Another advantage of differential encoding is in robust decoding performance. In decoding symbols with absolute encoding, a receiver needs a stable phase reference. However, in some propagation channels, it may not be easy to obtain such a stable phase reference; this may considerably increase the decoding errors. In differential encoding, a receiver does not need a phase reference since transmitted symbols can be estimated by simply comparing the phase of a received symbol with that of the previous one.

5.1.4   Pulse Modulation Schemes

In ADC, sampled and quantized signal samples are encoded by R bits/sample, consisting of 1's and 0's. This is the basis of binary pulse code modulation (PCM), to be explored further in the following sections. In M-ary PCM, k bits are grouped to form a symbol and each symbol is assigned to a different amplitude level. This scheme is called as pulse amplitude modulation (M-ary PAM). For M=2, M-ary PAM reduces to binary PCM. In addition to the PCM, ADC output can be transmitted by three other pulse modulation techniques, namely the pulse amplitude modulation (PAM), the pulse position modulation (PPM) and the pulse duration modulation (PDM) (see Figure 5.23).

Table 5.1   Differential Encoding According to the Rule $d_k = \overline{d_{k-1} + b_k}$ with $d_0 = 1$ as the reference bit.

k	0	1	2	3	4	5	6	7	8
b_k		0	1	1	0	1	0	0	1
d_k	1	0	0	0	1	1	0	1	1

Table 5.2   Differential Decoding the Differentially Encoded Bit Stream in Table 5.1.

k	0	1	2	3	4	5	6	7	8
d_k	**0**	0	**1**	0	1	1	0	1	1
b_k		**1**	**0**	0	0	1	0	0	1

Table 5.3   Immunity of Differential Decoding to the Polarity Inversion of the Received Bit Sequence.

k	0	1	2	3	4	5	6	7	8
d_k (original)	**1**	0	0	0	1	1	0	1	1
d_k (inverted)	**0**	1	1	1	0	0	1	0	0
b_k *(decoded)*		0	1	1	0	1	0	0	1

In *pulse amplitude modulation* (PAM), k bits in the encoded bit stream are grouped to form one of $M = 2^k$ amplitude levels. For example, each of M amplitude levels in M-ary PAM can be transmitted by using pulses with amplitudes ± 1 V, ± 3 V.... For $M = 4$, the dibits 00, 01, 11 and 10 may be transmitted by the voltage levels, -3V, -1 V, $+1$ V, and $+3$V, respectively. Compared with binary PAM, M-ary PAM requires higher average transmit power but reduced transmission rate $(R = R_b/\log_2 M)$, where R and R_b denote, respectively, the transmission rates for M-ary and binary PAM. For $M = 4$, the transmission rate and, accordingly, the required channel bandwidth is halved compared to binary PAM. Considering that the transmitted amplitude levels are now ± 1 and ± 3 compared to ± 1 in the binary case, this occurs at the expense of 5-fold, $(1^2 + 3^2)/2 = 5$, increase in average transmitted power. The receiver estimates the transmitted quantized amplitude level from the received pulse amplitude (see Figure 5.23).

In *pulse duration modulation* (PDM), the amplitude levels are encoded into the duration of the transmitted pulse. For example, the information of different amplitude levels is transmitted in the channel with pulses of the same amplitude but different pulse durations. The receiver estimates the transmitted voltage level by looking at the duration of the received pulse (see Figure 5.23).

In *pulse position modulation* (PPM), the amplitude levels are encoded into the position (time of transmission) of transmitted pulses of equal amplitudes and durations. For a 4-level quantized transmit signal, ± 3 V may be transmitted with some shift before/after the reference time of transmission, while ± 1 V is transmitted with smaller shift before/after the reference time of transmission. The receiver extracts the information about the amplitude of the transmitted pulse by estimating the time-shift of the received pulse with respect to the reference transmission time (see Figure 5.23).

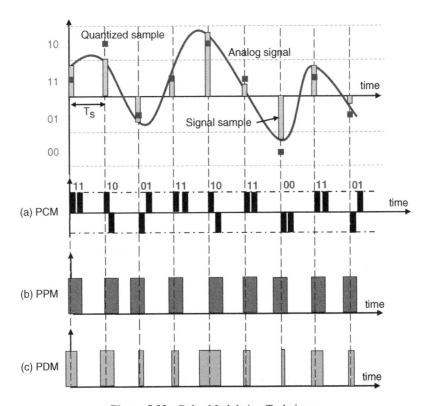

Figure 5.23   Pulse Modulation Techniques.

5.1.4.1 Binary Versus Gray Encoding

Consider that the bit stream {10010011} is obtained as a result of ADC process. In binary encoding, +1 V is transmitted during T_b (the bit duration) to represent the bit **1** and −1 V for the bit **0**, or vice versa. Figure 5.24 shows two different assignments of the above bit sequence where two bits represent a single amplitude level, that is, 4-ary PAM. Thus, each of the four amplitude levels, ±1 V and ±3 V, is transmitted during $2T_b$ period for two bits.

In natural encoding shown in Figure 5.24 transmitted amplitude levels are determined by the arithmetic value of the symbol (*binary encoding*), that is, **00**, **01**, **10** and **11** correspond to −3 V, −1 V, +1 V, and +3 V, respectively. If the channel noise causes an erroneous reception of the transmitted voltage level, then the error will most likely be between the neighboring voltage levels. In binary encoding, if +1 V is received in error when −1 V is transmitted, the receiver estimates **10** as the transmitted symbol while **01** was actually sent; hence, the receiver suffers two bit errors. *Gray encoding* assigns voltage levels −3, −1, +1, and +3 volts to **00**, **01**, **11**, and **10**, respectively, so that an error in the neighboring received voltage levels causes only a single bit error.

5.1.4.2 Line Codes and the Transmission Bandwidth

Above, we implicitly assumed to use finite-amplitude pulses (rectangular pulse) to transmit the information carried by binary or M-ary symbols in the transmission channel. Similarly, binary symbols are assumed to be represented by + V volts and −V volts respectively. However, various other pulses, so-called line codes, of

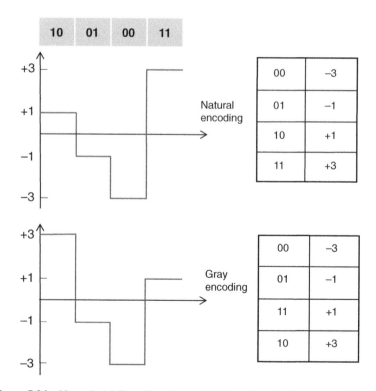

Figure 5.24 Natural and Gray Encoding of Di-Bits of the Bit Sequence 10010011.

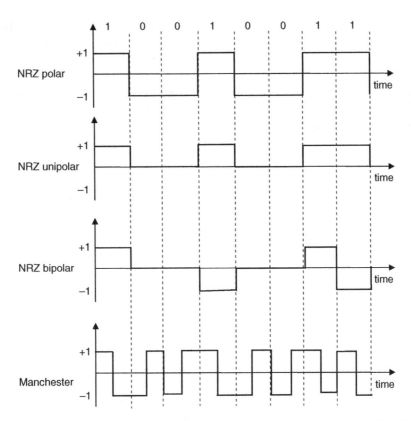

Figure 5.25 Some Commonly Used Line Codes to Represent the Bit Sequence 10010011. NRZ denotes non-return to zero.

various shapes, polarity and duration may be used for electrical representation of a bit stream in the transmission channel. Similarly, the bits may be represented in numerous ways other than ± V volts. Figure 5.25 shows several line codes to represent the binary data stream {10010011}, where the bits 0 and 1 are assumed to be equiprobable. Since the PSD of a time-pulse is proportional to the square of its Fourier transform, the choice of a line code is crucial in determing the transmission bandwidth and intersymbol interference (ISI).

Consider a bit sequence $\{b_n\}$ encoded by a line code $g(t)$ as follows:

$$x(t) = \sum_k b_k \, g(t - kT_b) \qquad (5.46)$$

where $B = \{b_k\}$ is an random binary random stream of equiprobable bits. Noting that (5.46) is in the form of discrete convolution, one may express the PSD of the signal $x(t)$ as the product of the PSD's of the bit sequence $B = \{b_k\}$ and $g(t)$ (see Example 1.19):

$$G_X(f) = G_B(f) \, G_{g(t)}(f) \qquad (5.47)$$

The autocorrelation function and the PSD of the NRZ polar code are given by (1.147)–(1.49). Here we will first derive the auto-correlation funcion of the NRZ unipolar code, where binary 1 and 0 are equiprobable and are represented by 1 V and 0 V, respectively. The PSD of $B = \{b_k\}$ will be found by taking the Fourier transform of the autocorrelation function (see Figure 1.11). Since the bits are i.i.d., the autocorrelation of B, which is a power signal, may be written as

$$R_B[n] = E[b_k \, b_{k+n}]$$

$$= \begin{cases} (1 \times 1)/2 + (0 \times 0)/2 & n = 0 \\ (1 \times 1 + 0 \times 1 + 1 \times 0 + 0 \times 0)/4 & n \neq 0 \end{cases}$$

$$= \begin{cases} 1/2 & n = 0 \\ 1/4 & n \neq 0 \end{cases} \tag{5.48}$$

The PSD of the power signal B, $G_B(f)$, may easily be found by taking the Fourier transform of $R_B[n]$:

$$G_B(f) = \sum_{n=-\infty}^{\infty} R_B[n] e^{-j2\pi f n T_b}$$

$$= \frac{1}{2} + \frac{1}{4} \sum_{\substack{n=-\infty \\ n \neq 0}}^{\infty} e^{-j2\pi f n T_b}$$

$$= \frac{1}{4} + \frac{1}{4} \sum_{n=-\infty}^{\infty} e^{-j2\pi f n T_b}$$

$$= \frac{1}{4} + \frac{1}{4T_b} \sum_{n=-\infty}^{\infty} \delta\left(f - \frac{n}{T_b}\right) \tag{5.49}$$

where the last expression in (5.49) follows from (1.70) or the Poisson sum formula given by (D.78). Assuming that $g(t)$ is a rectangular pulse of amplitude A and duration T_b, then the PSD of $g(t)$ is given by (1.104). Hence inserting (1.104) and (5.49) into (5.47), one gets

$$G_X(f) = A^2 T_b \sin c^2(fT_b) \left[\frac{1}{4} + \frac{1}{4T_b} \sum_{n=-\infty}^{\infty} \delta\left(f - \frac{n}{T_b}\right) \right]$$

$$= \frac{A^2}{4} \delta(f) + \frac{A^2}{4} T_b \sin c^2(fT_b) \tag{5.50}$$

The null-to-null bandwidth of $G_x(t)$ is $2/T_b$ since the first null of (5.50) is located at $fT_b = \pm 1$. Table 5.4 shows the autocorrelation functions and the PSD's of some commonly used line codes. The PSDs of the considered line codes are depicted in Figure 5.26. The considered line codes all require infinite transmission bandwidth, even though the PSD of each code is different. For example, Figure 5.26 shows that Manchester code requires a null-to-null transmission bandwidth twice as large as the other line codes. One can intuitively justify this since the Manchester code changes twice per bit duration T_b, while the others change only once (see Figure 5.25). In practice,

Table 5.4 Autocorrelation Function the PSD of Some Line Codes (*A* Denotes the Pulse Amplitude).

Line code	$R_B[n]$	$G_X(f)$
NRZ Unipolar	$\begin{cases} 1/2 & n = 0 \\ 1/4 & n \neq 0 \end{cases}$	$\frac{A^2}{4} \delta(f) + \frac{A^2}{4} T_b \sin c^2(fT_b)$
NRZ Polar	$\begin{cases} 1 & n = 0 \\ 0 & n \neq 0 \end{cases}$	$A^2 T_b \sin c^2(fT_b)$
NRZ Bipolar	$\begin{cases} 1/2 & n = 0 \\ -1/4 & n = \pm 1 \\ 0 & \text{otherwise} \end{cases}$	$A^2 T_b \sin c^2(fT_b) \sin^2(\pi fT_b)$
Manchester	$\begin{cases} +2 & n = 0 \\ -1 & n = \pm 1/2 \\ 0 & \text{otherwise} \end{cases}$	$A^2 T_b \sin c^2(fT_b/2) \sin^2(\pi fT_b/2)$

for economic use of the available bandwidth, the spectral components of the line codes with sufficiently low energy content are filtered out, for example, at 3-dB points. Bearing in mind that the distortion increases as the transmission bandwidth becomes narrower, the transmission bandwidth is usually determined as a result of a tradeoff between the distortion due to filtering and the cost and the availability of the bandwidth.

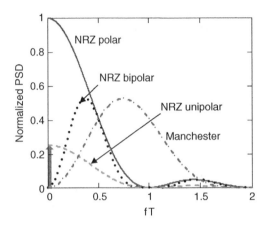

Figure 5.26 PSD of Some Commonly Used Line Codes.

5.2 Time-Division Multiplexing

5.2.1 Time Division Multiplexing

As shown in Figure 5.27a, the message samples occupy the channel for only a fraction of the sampling interval on a periodic basis. Therefore, the time axis may be used more economically if the time interval between adjacent samples is used by other messages on a time-shared basis. Thus, time division multiplexing (TDM) enables the joint utilization of a communication channel by several message sources without interference between them. TDM consists of time interleaving of samples from several sources so that the information from these sources can be transmitted serially

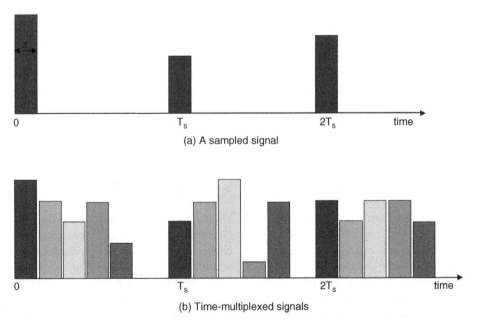

Figure 5.27 A Sampled Signal (at the Top) and Five Signals Time-Division-Multiplexed Into the Same Channel (at the Bottom).

over a single communication channel (see Figure 5.27b).

A time-division multiplexer uses a selector switch at the transmitter which sequentially takes samples from each of the signals to be multiplexed and then interleaves these signal samples in the time-domain as a single high-speed data stream. At the receiver, this high-speed data stream is de-interleaved into low-speed individual components, by a switch synchronized to the switch at the transmitter, and then delivered to their respective destinations.

Since TDM interleaves N samples, each bandlimited to W, into a time slot equal to one sampling interval $T_s \sim 1/2W$ (Shannon sampling theorem), it introduces a N-fold bandwidth expansion, compared to the bandwidth required by a single signal. Consequently, the time duration which may be allocated to each sample must satisfy $NT \leq T_s$. The corresponding transmission bandwidth is proportional to $k/T \geq N(k/T_s) = N\ (k2W)$, where k is a proportionality constant. Since frequency-division multiplexing (FDM) of N signals also requires an N-fold increase in the transmission bandwidth, TDM and FDM have comparable performances as far as their transmission bandwidth requirements are concerned.

Timing operations at the receiver must be synchronized with the corresponding operations at the transmitter. Then, the received multiplexed signals can be properly deinterleaved. Synchronization is therefore essential for satisfactory operation of TDM systems. Its implementation depends on the method of pulse modulation used to transmit multiplexed sequence of samples. The multiplexed signal must include some form of framing so that its individual components can be identified at the receiver. A synchronization word of a given format is usually transmitted within each frame for achieving synchronization.

TDM is highly sensitive to channel dispersion since frequency components of the multiplexed narrow pulses suffer differently from the channel conditions. Therefore, equalization may be required. However, TDM signals are

immune to channel nonlinearities because, at a certain time, the channel is accessed only by a single signal sample.

In TDM, analog signals are sampled sequentially at a common sampling rate and then multiplexed over a common channel. By multiplexing digital signals at different bit rates, several digital signals can be combined, for example, computer outputs, digitized voice/facsimile/TV signals, into a single data stream at considerably higher bit rates than any of the inputs.

5.2.1.1 Bit Stuffing

A multiplexer has to accommodate small variations in the bit rates of incoming digital signals. Therefore, bit stuffing is used for tailoring the requirements of synchronization and rate adjustment. Incoming digital signals are stuffed with a number of stuffing bits in order to increase the bit rate to be equal to that of a locally generated clock. Outgoing bit rate of a multiplexer is made slightly higher than the sum of the maximum expected bit rate of the input channel and some additional (stuffed) non-information carrying pulses. Elastic store is used for bit stuffing. A bit stream is stored in an elastic store so that the stream may be read out at a rate different from the rate at which it is read in. At the de-multiplexer, the stuffed bits are removed from the multiplexed signal. A method is required to identify the stuffed bits.

5.2.2 TDM Hierarchies

Digital multiplexers are commonly used for data transmission services provided by telecommunication operators. A digital hierarchy is used for time-division multiplexing low-rate bit streams into much higher-rate bit streams. By avoiding the use of separate carriers for each signal to be transmitted, TDM provides a cost-effective utilization of a common communication channel by several independent

Table 5.5 Digitial Hierarchy Standards for the United States (T1) and Europe (E1).

TDM hierarchy	AT&T (T1 system) American standard		CCIT (E1 system) European standard	
	Number of inputs	Output rate, Mbps	Number of inputs	Output rate, Mbps
DS0	1	0.064	1	0.064
DS1	24	1.544	30	2.048
DS2	4	6.312	4	8.448
DS3	7	44.736	4	34.368
DS4	6	274.176	4	139.264
DS2	2	560.160	4	565.148

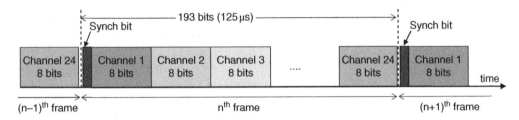

Figure 5.28 DS1 Frame Structure of the T1 Multiplexing System Combines 24 DS0 Streams to Obtain a DS1 Signal at 193 bits/125 μs = 1.544 Mbps.

message sources without mutual interference among them.

Standard PCM representation of a single voice signal, bandlimited to 3.4 kHz, at 64 kbps (8000 samples/s × 8 bits/sample) is called the DS0 Signal. In AT&T's T1 system, which is the adopted standard in the USA and Canada, *first-level multiplexer* combines 24 DS0 streams to obtain a digital signal one (DS1) at 1.544 Mbps, which is higher than $24 \times 64 = 1.536$ Mbps to account for the overhead and synchronization bits (see Table 5.5). This bit stream is called the primary rate, because it is the lowest bit rate that exists outside a digital switch. *The second-level multiplexer* combines four DS1 bit streams to obtain a DS2 at 6.312 Mbps. *The third-level multiplexer* combines seven DS2 bit streams to obtain a DS3 at 44.736 Mbps. *The fourth-level multiplexer* combines six DS3 bit streams to obtain a DS4 at 274.176 Mbps. *The fifth-level multiplexer* combines two

DS4 bit streams to obtain a DS5 at 560.160 Mbps. Higher-level multiplexers are used as the carried traffic increases. The hierarchy of the European E1 system is also shown in Table 5.5.

Example 5.9 Frame Structure of T1 TDM System. [1]

As shown in Figure 5.28, in the T1 system, 24 voice signals sampled and then time-division multiplexed to form a DS1 signal. The sampling operation uses flat-top samples with T seconds duration. The multiplexing operation includes provision for synchronisation by adding an extra pulse of sufficient amplitude and also T seconds duration. By using a LPF, the highest frequency component of each voice signal is limited to 3.4 kHz. Sampling rate is $f_s = 8$ kHz, each sample is represented by 7 bits and an additional bit is used for signalling. The sampling interval is $T_s = 1/8$ kHz = 125 μs. There are 24 channels, each having 8 bits, and 1 synch (signalling) bit, so total number of bits in a frame of 125 μs duration is $24 \times 8 + 1 = 193$ bits. The transmission rate of 24 multiplexed channels is 193 bits/125 μs = 1.544 Mbps and the bit duration is 0.648 μs. Since each of the 24 multiplexed channels sends one sample encoded by 8 bits in a 125 μs frame, each channel has a transmit rate of 8 bits/125 μs = 64 kbps. The received signal samples can be used to recover the transmitted analog signal in virtue of the Shannon's sampling theorem. Noting that each of the 24 multiplexed channels has a transmission rate of 64 kbps, the net transmission rate of 24×64 kbps = 1536 kbps; the difference 1544 kbps−1536 kbps = 8 kbps accounts for the synchronization bit, that is, 1 bit/125 μs = 8 kbps.

5.2.3 Statistical Time-Division Multiplexing

TDM allows multiple users to transmit and receive data simultaneously by assigning the same, fixed number of time slots to each user, even if they have different data rate requirements. TDM works well and is suitable for homogeneous traffic where all users have similar requirements, but does not perform well for different and/or time varying data rate requirements of the users. For example, a busy laser printer shared by many users might need to operate at much higher transmission rates than a personal computer attached to the same line.

Statistical time division multiplexing (STDM) is a method for multiplexing several types of data, with different rates and/or priorities, into a single transmission line. The STDM scheme first analyzes statistics related to the typical workload of each user/device (printer, fax, computer), based on its past and current transmission needs, and then determines in real-time the number of time slots to allocate to each user/device for transmission. The statistics used in STDM include user/device's peak data rate and the duty factor, that is, the

percentage of time the user/device typically spends either transmitting or receiving data. Hence, the STDM uses the total available transmission bandwidth more efficiently than TDM.

5.3 Pulse-Code Modulation (PCM) Systems

PCM is a digital transmission system and consists of transmitting analog message signals by a sequence of encoded pulses obtained by sampling, quantizing and encoding using an ADC at the transmitter and reconstructing the transmitted signals with a DAC at the receiver (see Figure 5.1).

The data rate of PCM signals satisfies the condition $f_s R \geq 2WR$, which implies an R-fold increase in the channel bandwidth compared to analog transmission of the same signal. However, this increase can be compensated by TDM of multiple signals. Since the transmission bandwidth is proportional to R for binary-PCM, (5.19) manifests an exponential increase of SNR_Q with bandwidth. Therefore, in spite of the noise power introduced by quantization, the PCM requires much less transmit signal power due to the exponential increase of SNR_Q with bandwidth. Figure 5.29 presents

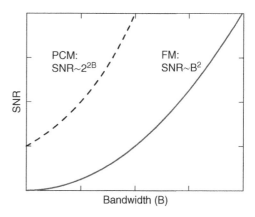

Figure 5.29 SNR-Bandwidth Tradeoff in PCM and FM.

a comparison of the SNRs for PCM and FM, where SNR is proportional to the square of the bandwidth. For a given value of the SNR, FM requires a higher bandwidth. [1] Alternatively, for a given bandwidth, FM offers a lower SNR. This is the fundamental reason of the rapid migration of analog systems into digital.

5.3.1 PCM Transmitter

In practice, a low-pass (anti-aliasing) filter is used at the front-end of the sampler to filter out signal components at frequencies higher than W, which are assumed to have negligible energy. Following the anti-aliasing filter, the message signal is sampled with a stream of narrow rectangular pulses (flat-top sampling). To ensure perfect reconstruction of the message signal at the receiver, the sampling frequency must satisfy $f_s > 2W$. Following amplitude quantization, a stream of discrete-time and discrete-amplitude samples is obtained. The quantized sample amplitudes are then encoded by assigning R bits to each sample. Following M-ary PAM modulation, each symbol, consisting of $k = \log_2 M$ bits, is transmitted using a

suitable line code from Table 5.5 at a symbol rate $R_s = R_b/k$ symbols/s (see Figure 5.30).

To alleviate the effects of attenuation, distortion and noise in long ranges, regenerative repeaters may be used in the transmission path. Regenerative repeaters, located at sufficiently close spacing along the transmission path, receive, recover and retransmit the PCM signals. Following a number of repeaters, a PCM receiver, which is also a regenerative repeater, carries out the final regeneration of impaired signals, decoding, regeneration of the stream of quantized samples of the transmitted signal by using a DAC.

Since we already studied the ADC process in detail and a PCM transmitter, which is basically an ADC, we will consider hereafter only the regenerative repeaters and the PCM receiver.

5.3.2 Regenerative Repeater

The repeaters, in analog transmission systems, amplify-and-forward not only the input signal but also the input noise (see Figure 5.31). Therefore, effects of noise and distortion from individual links accumulate with increasing ranges. This leads to a gradual decrease of the SNR with the number hops (see

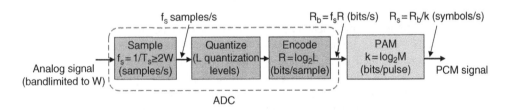

Figure 5.30 Block Diagram of a PCM Transmitter.

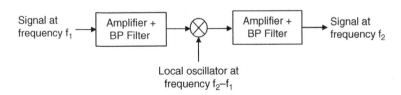

Figure 5.31 Block Diagram of an Analog Repeater.

Figure 5.32   Block Diagram of a Multi-Hop Regenerative PCM Repeater System.

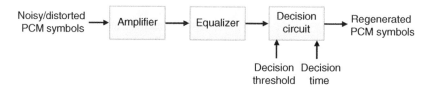

Figure 5.33   Block Diagram of a Regenerative Repeater.

Example 4.8). However, regenerative repeaters in PCM receive, and decode the incoming bit stream and retransmit after amplification. Hence, in PCM systems, accumulation of distortion and noise in a cascade of regenerative repeaters is removed (see Figure 5.32). Consequently, except for the delay and some occasional bit errors due to channel noise, the regenerated data stream is exactly the same as the originally transmitted signal. If the repeaters are located sufficiently close to each other, the attenuation and distortion of the received pulses do not reach to a level which cannot be compensated. Repeater spacing can be optimized under the constraints on cost and the received SNR, so that each repeater can successfully detect, remodulate and retransmit the signals at their input. Therefore, PCM systems are used for reliable transmission of digital information over long ranges. Synchronization is of crucial importance for proper operation of PCM systems. Synchronization issues become more serious with increasing data rates due to shorter bit durations and less tolerance to timing errors.

As shown in Figure 5.33, the basic functions performed by a regenerative repeater are equalization, timing and decision making. Equalizer shapes the received pulses so as to compensate for the effects of amplitude and phase distortions. Timing circuitry determines the starting

and sampling times of the received pulses as well as their durations so as to generate a clean periodic pulse train at the output. Each sample is compared to a predetermined threshold, at specific time instants (usually in the middle of each bit period), in the decision-making device so as to decide whether the received symbol is a 0 or 1.

The signal regenerated by the repeater may depart from the original signal for two reasons. Firstly, the channel noise and the interference force the repeater to make incorrect decisions, thus causing bit errors in the regenerated signal. Secondly, if the starting times of the received pulses are strongly affected by noise and/or delay introduced by the channel, a jitter is introduced into the regenerated pulse position, thereby causing distortion.

In a PCM system, the signal can be regenerated as often as necessary. Thus, amplitude, phase and distortions in one hop (between two repeaters) have no effect on the regenerated input signal to the next hop. In that sense, transmission requirements in PCM are independent of range.

5.3.3   PCM Receiver

The receiver, as the last regenerative repeater, amplifies, reshapes and cleans up the effects of noise. The cleaned pulses are regrouped into

R-bit code words and decoded into quantized PAM signal samples. The decoding process consists of regenerating signal samples with amplitudes determined by the weighted sum of the *R*-bits in the code word. The weight of each of the *R*-bits is determined by its place value $(2^0, 2^1,..., 2^R)$ in the code word. The message signal is recovered at the receiver by passing the decoder output through a low-pass reconstruction filter whose cut-off frequency is equal to the message bandwidth *W*.

Regenerative repeaters may suffer channel noise and distortion introduced by the quantization process at the transmitter. If the transmission path in one hop is error free, then the recovered signal is not affected by the noise, but the distortion due to quantization process always remains. The quantization noise, introduced at the transmitter, is irrecoverable, since it is signal dependent and disappears only when the message signal is switched off. Quantization noise can be reduced by increasing $L = 2^R$ and selecting a compander matched to the signal characteristics.

5.3.3.1   Receiver Noise

As already studied in Chapter 4, the receiver suffers from internal and external noise, which is always present as long as the system is switched on. Hence, unlike the quantization noise, the receiver noise is independent of the presence of the signal. The noise, which is usually assumed to be AWGN, introduces bit errors depending on the received signal SNR. The bit error performance of a PCM system may hence be improved by using closer repeater spacing, increasing repeater transmit power and/or using low-noise receivers. Then, the PCM noise performance will be limited by the quantization noise only.

The channel noise may cause bits errors in regenerative repeaters and/or the PCM receiver. Assuming that bits 1 and 0 are transmitted by $+A$ volt and $-A$ volt, respectively, a bit error occurs when the receiver decodes a received bit as 0 when 1 was transmitted or vice versa. When $+A$ volt is transmitted for sending bit 1, the received signal at the receiver may be written as $\sqrt{E_b} + w(t)$, where $w(t)$ denotes the Gaussian noise voltage at time t, $E_b = A^2 T_b$ is the bit energy and T_b is the bit duration (This will be studied in detail in Chapter 6). Hence, the pdf of the received signal is Gaussian with mean $\sqrt{E_b}$ and variance σ^2. A bit error occurs when the received voltage level $\sqrt{E_b} + w(t) < 0$; then the receiver decodes the received bit as 0. Assuming that the probabilities of transmitting 0 and 1 are equal to each other, that is, $p_0 = p_1 = 1/2$, the probability of bit error may be written as the area of the tail of the Gaussian pdf from $(-\infty, 0)$:

$$P_b = p_0 \text{Prob}(1|0) + p_1 \text{Prob}(0|1) = \text{Prob}(0|1)$$

$$= \frac{1}{\sqrt{2\pi}\sigma} \int_{-\infty}^{0} e^{-\frac{\left(x - \sqrt{E_b}\right)^2}{2\sigma^2}} dx = Q\left(\sqrt{\frac{2E_b}{N_0}}\right)$$

$$(5.51)$$

where $\text{Prob}(i|j)$, $i, j = 0, 1$ denotes the probability of receiving bit i when bit j is transmitted. From the channel symmetry, $\text{Prob}(1|0) = \text{Prob}(0|1)$. The probability of bit error depends only on the received $\gamma_b = E_b/N_0 > 0$ where E_b denotes the bit energy and $N_0 = 2\sigma^2$ is the one-sided PSD of the AWGN. The so-called Gaussian Q function is defined by (see (B.1))

$$Q(x) = \frac{1}{\sqrt{2\pi}} \int_x^{\infty} e^{-t^2/2} dt \qquad (5.52)$$

Figure 5.34 shows the variation of the error probability versus E_b/N_0 in dB. Note that $Q(0) = 1/2$ implies that the bit error probability is upper-bounded by ½. This corresponds to the case where $E_b/N_0 = 0$ and the likelihood of receiving 1 or 0 at the receiver is 50% irrespective of the transmitted symbol. Then there is no need for communications and throwing coin is more economical. The bit error probability

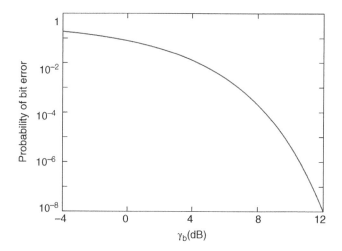

Figure 5.34   Bit Error Probability Versus $\gamma_b = E_b/N_0$ For Binary PCM.

decreases with increasing values of E_b/N_0 since $Q(x)$ vanishes as x goes to infinity. If the error probability is required to be lower than 10^{-8}, then the E_b/N_0 should be larger than 12 dB. This corresponds to a single bit error on the average per 10^8 received bits and also implies that, at a transmission rate of 1 Mbps, one suffers a single bit error in every 100 seconds. To improve the error performance, one evidently needs to increase E_b/N_0, that is, increase the average bit energy, E_b, or decrease the noise PSD, N_0. Even a single bit error per 100 seconds may not be acceptable in some communication systems and techniques other than increasing E_b/N_0 are needed to reduce the bit error probability; these techniques will be studied in the following chapters.

Example 5.10   System Design with Regenerative Repeaters.
Consider a wired PCM system where transmitter and repeater are connected via regenerative repeaters. The attenuation caused by coaxial cables to the signals increases with increasing carrier frequencies. Let us assume that the cable attenuation is $L_0 = 15$ dB/km at 100 MHz, 20 dB/km at 200 MHz, 30 dB/km at 400 MHz, and 60 dB/km at 1 GHz. The transmit power of

a repeater is assumed to be $P_t = 1$ mW and a repeater receiver has a noise figure of $F = 10$ dB. We desire to transmit digital signals at $R_b = 1$ Gb/s.

Note that the received power can be expressed in terms of the bit energy E_b and the bit rate R_b as

$$P_r = E_b \left(\frac{Joule}{bit}\right) R_b \left(\frac{bits}{s}\right) = E_b \, R_b \, (W)$$

$$(5.53)$$

Using (4.76), the E_b/N_0 at the input of a repeater receiver is found to be

$$\frac{E_b}{N_0} = \frac{P_t}{L R_b k (F-1) T_0}$$

$$(5.54)$$

Here, $T_0 = 290$ K, $k = 1.38 \; 10^{-23}$ J/K is the Boltzmann constant and $L = L_0 \times d$ is the total attenuation per hop with d denoting the hop distance.

If the bit error probability is required to be less than 10^{-8}, then $E_b/N_0 > 13$ dB including 1 dB implementation loss (see Figure 5.34). From (5.54), total attenuation per hop should be lower than 61.4 dB. This implies hop distances of 4.1 km, 3.1 km, 2 km and 1 km at

100 MHz, 200 MHz, 400 MHz and 1GHz, respectively. In practice, high data rate traffic between repeaters is carried by fiber cables, which have typical attenuation constants 0.3 dB/km. Therefore, fiber systems may operate over very long distances without repeaters.

5.3.3.2 SNR at the Decoder Output

The signal-dependent quantization noise, due to the quantization process at the transmitter, and the receiver noise, which is omnipresent, both contribute to the errors in the decoding process at the receiver. To study the effects of decoding errors, let us assume that a zero-mean signal with uniform pdf is uniformly quantized, with L quantization levels and step size Δ. The pdf of the allowed signal levels at the decoder input may be written as

$$f_V(v) = \sum_{\substack{k=-L \\ k\ \text{odd}}}^{L} \frac{1}{L} \delta\left(v - k\frac{\Delta}{2}\right) \tag{5.55}$$

which is shown in Figure 5.35.

Using (D.32), the mean-square signal power at the decoder output is found to be

$$P_d = \int_{-\infty}^{\infty} v^2 f_V(v)\, dv$$

$$= \frac{2}{L}\left(\frac{\Delta}{2}\right)^2 \left[1^2 + 3^2 + 5^2 + \dots + (L-1)^2\right]$$

$$= \frac{2}{L}\left(\frac{\Delta}{2}\right)^2 \frac{L(L^2-1)}{6} = \sigma_Q^2(L^2-1)$$

$$\tag{5.56}$$

where $\sigma_Q^2 = \Delta^2/12$ denotes the quantization noise power, as given by (5.18). Mean decoded

signal to quantization noise power ratio at the decoder output for a uniformly quantized signal with uniform pdf may be written as

$$SNR_d = \frac{P_d}{\sigma_Q^2} = L^2 - 1 \cong L^2 = 2^{2R} \tag{5.57}$$

The peak decoded signal power level and the corresponding peak decoded signal to quantization noise (power) ratio is given by

$$P_{d,peak} = \left((L-1)\Delta/2\right)^2$$

$$SNR_{d,peak} = \frac{P_{d,peak}}{\sigma_Q^2} = \frac{\left((L-1)\Delta/2\right)^2}{\Delta^2/12} = 3(L-1)^2$$

$$\tag{5.58}$$

which is 3 times (4.78 dB) higher than its mean value.

5.3.3.3 Output SNR of a PCM Receiver

A decoder groups the received bits into R-bit codewords and reconstructs the quantized samples. These samples are low-pass filtered and expanded in the case of nonuniform quantization. All bits in a received R-bit codeword have the same bit error probability P_b which is uniquely determined by the E_b/N_0 as given by (5.51).

In reconstructing the analog waveform of the original message signal, different bit errors are weighted differently since an error in the most significant bit (MSB) in a codeword is more harmful than the one in the least significant bit (LSB). To consider the effects of unequal weighting of the bit errors, we assume that natural binary coding is employed for the binary representation

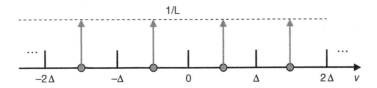

Figure 5.35   The Pdf of the Quantized Sample Amplitudes at the Input of the PCM Decoder.

Table 5.6 The Effects of Bit Errors for Decoding the Codewords of $R = 3$ Bits for Binary-Coded PCM Systems.

Representation levels	Correctly decoded bits	Error in LSB (3rd bit)	Error in 2nd bit	Error in MSB (1st bit)
$+7\Delta/2$	111	111	111	111
$+5\Delta/2$	110	110	110	110
$+3\Delta/2$	101	101	**111**	101
$+1\Delta/2$	100	**101**	100	100
$-1\Delta/2$	011	011	**001**	011
$-3\Delta/2$	010	010	010	010
$-5\Delta/2$	001	**000**	001	**101**
$-7\Delta/2$	000	000	000	**100**

of each of the L quantized values, that is, the lowest quantized level is mapped into a sequence of all zeros; the largest level is mapped into a sequence of all ones; and all other levels are mapped according to their relative values.

Table 5.6 shows in bold some bit errors in the third (least significant) bit, second bit and the first (most significant) bit for a PCM system with $L = 8$ quantization levels. One may easily observe from Table 5.6 that if an error occurs in the least significant bit, this leads to an error of $2^0\Delta$ in the representation level. However, if an error occurs in the second bit, its effect on the representation level is $2^1\Delta$. Finally, an error in the most significant bit causes an error of $2^{R-1}\Delta$ in the representation level.

The possible errors in the decoded representation levels may then be written as

$$\varepsilon_i = 2^{i-1}\Delta, \quad i = 1, 2, \ ....,R \qquad (5.59)$$

The mean-square decoding error may then be expressed in terms of the errors in the representation levels as follows:

$$\sigma_d^2 = \underbrace{\sum_{i=1}^{R} \varepsilon_i^2 \, \text{Prob}\left(error \ in \ i^{th} \ bit\right)}_{P_b}$$

$$= P_b \sum_{i=1}^{R} \left(2^{i-1}\Delta\right)^2 = \Delta^2 P_b \frac{4^R - 1}{3} \qquad (5.60)$$

$$= 4\sigma_Q^2 P_b SNR_d$$

where the series in (5.60) is evaluated using (D.33). Note that, P_b, σ_Q^2 and SNR_d are given by (5.51), (5.18) and (5.57), respectively. Since decoding and quantization errors are statistically independent of each other, the two can be added on a power basis to calculate the output SNR:

$$SNR_{out} = \frac{P_d}{\sigma_Q^2 + \sigma_d^2} = \frac{SNR_d}{1 + \sigma_d^2/\sigma_Q^2}$$

$$= \frac{SNR_d}{1 + 4 P_b SNR_d} = \begin{cases} 1/(4P_b) & P_b \ \text{high} \\ \\ SNR_d & P_b \ \text{low} \end{cases}$$

$$(5.61)$$

Note that quantization noise, with variance σ_Q^2, is due to quantization of the sample amplitudes at the transmitter, P_b arises from the bit transitions due to the receiver noise and the decoding error, with variance σ_d^2, is due to weighting of the bit errors in the decoding process.

Figure 5.36 shows the variation of SNR_{out} versus the probability of bit error P_b. Note from Figure 5.34 that $P_b = 10^{-2}, 10^{-4}, 10^{-6}$ and 10^{-8} correspond to $E_b/N_0 = 4.3$ dB, 8.4 dB, 10.6 dB and 12 dB, respectively. Hence, for small values of the E_b/N_0, P_b is high and the output SNR_{out} is significantly greater than the E_b/N_0. This is because, at very low input E_b/N_0 values,

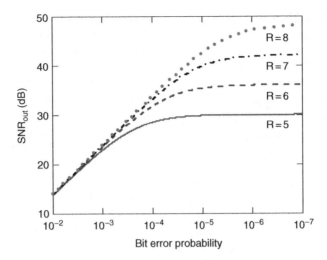

Figure 5.36   SNR$_{out}$ Versus P$_b$ for a Uniformly Quantized Signal with Uniform Pdf.

the noise amplitude is comparable to that of the PCM pulses, and the interpretation of the PCM codeword becomes unreliable. Even a single error in a PCM codeword can change its numerical value by a large amount, the SNR$_{out}$ in this range decreases rapidly. However, if the E_b/N_0 is sufficiently large, then P_b is low and the total noise is dominated by the quantization process. Then, the output SNR$_{out}$, which is proportional to $SNR_d \sim 2^{2R}$, improves with R (hence bandwidth) and bandwidth can be exchanged for SNR.

Example 5.11   SNR at the PCM Receiver Output.

Consider a digital communication system which is to carry a single voice signal using uniformly quantized PCM. The signal is assumed to have a peak to average power ratio of 10 dB. An anti-aliasing filter with a cutoff frequency of 3.4 kHz is used at the transmitter and the signal to quantization noise ratio is required to be higher than 50 dB.

The Nyquist sampling rate is

$$f_s = 2W = 2 \times 3.4 \ kHz = 6800 \ samples/s$$
$$(5.62)$$

On the other hand, from (5.20)

$$SNR_Q = 4.77 + 6R - \alpha_{dB}$$
$$= 4.77 + 6R - 10 > 50 \ dB \qquad (5.63)$$

the condition above implies $R \geq 10$ bits/sample. The corresponding PCM bit rate:

$$R_b = f_s \times R = 6800 \ samples/s \times 10 = 68 \ kbps$$
$$(5.64)$$

The actual signal to quantization ratio for uniform quantization:

$$SNR_Q = 4.77 + 6 \times 10 - 10 = 54.77 \ dB \quad (3 \times 10^5)$$
$$(5.65)$$

Mean decoded signal to quantization ratio:

$$SNR_d = L^2 - 1 \cong 2^{2R} = 2^{20} = 10^6 \quad (60 \ dB)$$
$$(5.66)$$

The overall SNR for the reconstructed analog voice if receiver noise induces an error probability of 10^{-6} is

$$SNR_{out} = \frac{SNR_d}{1 + 4P_b SNR_d} = 2 \times 10^5 (53 \ dB)$$
$$(5.67)$$

5.4 Differential Quantization Techniques

5.4.1 Fundamentals of Differential Quantization

In absolute quantization, each signal sample $x[n]$ is quantized as $x_q[n]$ independently of the other samples and is forwarded to the encoder (see Figure 5.37a). However, in differential quantization, instead of quantizing each sample separately, only the differences between successive time samples are quantized. Since the difference between successive samples is always smaller than the magnitudes of individual samples, the peak-to-peak voltage variation and the time rate of change of the difference signal will be smaller. In view of $V_{pp} = L\Delta$, this implies smaller quantization error ($\Delta/2$) for a given L, or a smaller L for a given value of the quantization error.

The block diagram of a differential quantizer is depicted in Figure 5.37 in comparison with an absolute quantizer. Here, the error signal, which is defined as the difference between a signal sample $x[n]$ and its estimate $\hat{x}[n]$, is quantized and then forwarded to the encoder. The estimate $\hat{x}[n]$ is obtained by a predictor which estimates the current signal sample using the knowledge of the previous signal samples. The receiver reconstructs the original signal samples using the quantized and encoded error signal.

Figure 5.38 shows samples of a sinusoidal signal, $x(t) = \cos(2\pi f_c t)$, sampled at $f_s = 20 f_c$, and the error signal $e[k] = x[k] - \hat{x}[k]$ where $\hat{x}[k] = x[k-1]$ is assumed for simplicity. The peak to peak voltage of $x(t)$ is 2 V while the corresponding value of the error signal is 0.31 V; the average power of the error signal is reduced by $20\log(1/0.31) = 10.2$ dB compared to $x(t)$. Therefore, for a given value of L, the step size (quantization error) will be smaller for the error signal. Conversely, for a given value of the step size, we can use fewer bits (R), hence narrower transmission bandwidth, to represent the quantized samples. Since the signal $x(t)$ is oversampled, that is, much faster than the Nyquist rate, the resulting encoded signal contains redundant information and adjacent samples are highly correlated.

In summary, differential quantization removes the redundancy due to correlation between adjacent samples. The predictors, which use higher numbers of past samples, may predict the current signal sample $\hat{x}[n]$ more accurately and hence may reduce the peak-to-peak variation of the

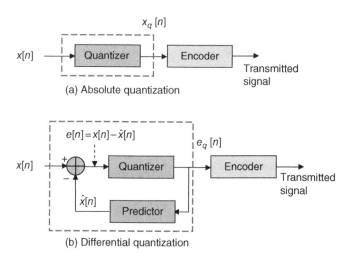

(a) Absolute quantization

(b) Differential quantization

Figure 5.37 Absolute Versus Differential Quantization.

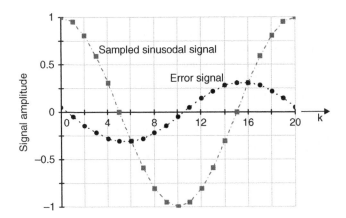

Figure 5.38   A Sinusoidal Signal x(t) Sampled at a Sampling Rate of $f_s = 20 f_c$ (Square) $x[k] = \cos(2\pi k/20)$ and the Error Signal, $e[k] = x[k] - \hat{x}[k]$ Where $\hat{x}[k] = x[k-1]$ (Circles).

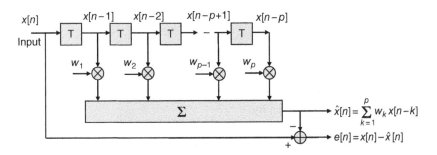

Figure 5.39   The Block Diagram of a FIR Filter Used as a Predictor with p Taps.

error signal; this allows the use of fewer quantization bits and leads to the reduction of the transmission bandwidth.

5.4.2   Linear Prediction

Depending on the application of interest, the prediction (adaptive filter output) or the prediction-error may be used as the system output. A prediction filter (predictor) is a finite-duration impulse response (FIR) filter that provides the best prediction (in some sense) of the present signal sample x[n] of a random signal *x(t)* using previous samples of the same signal. The prediction represents the desired filter response;

previous signal samples supply input to the filter.

Figure 5.39 shows the block diagram of an adaptive FIR filter with a prediction order of *p*, that is, using *p* previous time samples of the input signal, *x*[n-p],..., x[n-2], x[n-1], to provide a linear prediction of the input signal x[n]:

$$\hat{x}[n] = \sum_{k=1}^{p} w_k x[n-k] \qquad (5.68)$$

The predicted signal, $\hat{x}[n]$, is desired to be as close as possible to the input signal x[n]. Hence, the prediction error

$$e[n] = x[n] - \hat{x}[n] \qquad (5.69)$$

is desired to be minimum in some sense. Here, we will determine the effect of the number of taps of the prediction filter in minimizing the prediction error in (5.69) in the mean-square sense. Ignoring the noise, we first write the variance of the prediction error as

$$\sigma_e^2 = E[e^2[n]] = E\left[\left(x[n] - \sum_{k=1}^{p} w_k x[n-k]\right)^2\right]$$

$$= E\left[x^2[n] - 2x[n]\sum_{k=1}^{p} w_k x[n-k]\right.$$

$$\left. + \sum_{k=1}^{p}\sum_{j=1}^{p} w_k w_j x[n-k]x[n-j]\right]$$

$$= \sigma_X^2 - 2\sum_{k=1}^{p} w_k R_X[k]$$

$$+ \sum_{k=1}^{p}\sum_{j=1}^{p} w_k w_j R_X[k-j]$$

(5.70)

The power of nth sample $x[n]$ (average input power to the prediction filter) is given by

$$P = \sigma_X^2 = E[x^2[n]] - (E[x[n]])^2 = E[x^2[n]]$$

(5.71)

where the mean value of x[n] is equal to zero. The autocorrelation of x[n] is defined by

$$R_X(kT_s) = R_X[k] = E[x[n]x[n-k]] \quad (5.72)$$

The variance of the prediction error may be minimized (the so-called minimum mean square error, mmse) by differentiating σ_e^2 in (5.70) with respect to w_k and equating to zero:

$$\sum_{j=1}^{p} w_j R_X[k-j] = R_X[k] \quad k = 1,2,...,p \quad (5.73)$$

The above is the so-called Wiener-Hopf equation for linear prediction. The p unknown filter coefficients w_1, w_2,..., w_p can easily be found from p equations in (5.73). For this

purpose, we write the Wiener-Hopf equation in matrix form as

$$\mathbf{R}_X \mathbf{w} = \mathbf{r}_X \quad (5.74)$$

where

$$\mathbf{w} = [w_1 \ w_2 \ \ ..... \ w_p]^T$$

$$\mathbf{r}_X = [R_X[1] \ R_X[2] \ \ ..... \ R_X[p]]^T$$

$$\mathbf{R}_X = \begin{bmatrix} R_X[0] & R_X[1] & \cdots & R_X[p-1] \\ R_X[1] & R_X[0] & \cdots & R_X[p-2] \\ \vdots & \vdots & \ddots & \vdots \\ R_X[p-1] & R_X[p-2] & \cdots & R_X[0] \end{bmatrix}$$

(5.75)

Optimum weight vector which minimizes the mean-square error is given by

$$\mathbf{w}_0 = \mathbf{R}_X^{-1} \mathbf{r}_X \quad (5.76)$$

To find the minimum prediction error, we insert (5.73) into (5.70) and get

$$\sigma_{e,min}^2 = \sigma_X^2 - \sum_{k=1}^{p} w_k R_X[k] = \sigma_X^2 - \mathbf{w}_0^T \mathbf{r}_X$$

$$= \sigma_X^2 - \left(\mathbf{R}_X^{-1}\mathbf{r}_X\right)^T \mathbf{r}_X = \sigma_X^2 - \mathbf{r}_X^T \mathbf{R}_X^{-1} \mathbf{r}_X$$

(5.77)

Example 5.12   Linear Prediction Filter. Let us consider a linear prediction filter with a sinusoidal signal with amplitude A and frequency f_c, $x(t) = A \cos(w_c t + \phi)$, at its input. Here, ϕ is uniformly distributed in $[0,2\pi]$. The sampling frequency is assumed to be $f_s = 10f_c$. We will determine the input and output powers (variance of the prediction error) for single- and two-taps. We will also determine the optimum values of the tap weights and the predictor output.

Autocorrelation function of $x(t)$ may be written as

$$R(t_1 - t_2) = E[x(t_1)x(t_2)]$$

$$= A^2 \, E[\cos(w_c t_1 + \phi) \, \cos(w_c t_2 + \phi)]$$

$$= 0.5A^2 E[\cos\{w_c(t_1 - t_2)\}]$$

$$+ \cos\{w_c(t_1 + t_2) + 2\phi\}]$$

$$= 0.5A^2 \, \cos\{w_c(t_1 - t_2)\}$$

$$R(nT_s) = R[n] = 0.5A^2 \, \cos(w_c nT_s)$$

$$= 0.5A^2 \, \cos\left(\frac{2\pi n}{f_s/f_c}\right)$$

$$(5.78)$$

Inserting $f_s/f_c = 10$ into (5.78), one gets

$$R[0] = \sigma_x^2 = 0.5A^2$$
$$R[1] = 0.5A^2 \, \cos(2\pi/10) = 0.405A^2 \quad (5.79)$$
$$R[2] = 0.5A^2 \, \cos(4\pi/10) = 0.155A^2$$

For single-tap prediction, using (5.68) and (5.76), optimum tap weights and the predictor output are found as

$$\mathbf{r}_X = [0.405], \quad \mathbf{R}_X = [0.5]$$
$$\mathbf{w}_0 = \mathbf{R}_X^{-1} \, \mathbf{r}_X \;\Rightarrow\; w_1 = R_X[1]/R_X[0] = 0.81$$

$$\hat{x}[n] = \sum_{k=1}^{p} w_k \, x[n-k] = w_1 \, x[n-1] = 0.81 \, x[n-1]$$

$$(5.80)$$

Minimum value of the prediction error variance for a single-tap predictor is found using (5.77)

$$\sigma_{e,\min}^2/\sigma_x^2 = 1 - \mathbf{r}_X^T \mathbf{R}_X^{-1} \mathbf{r}_X = 1 - (R[1]/R[0])^2$$

$$= 1 - (0.405/0.5)^2 = 0.346$$

$$(5.81)$$

Ratio of the power at the input of the predictor to the power at the output of the predictor

$$\frac{P}{P_{out}} = \frac{\sigma_x^2}{\sigma_{e,\min}^2} = \frac{1}{0.346} = 2.89 \; (4.61 \; dB)$$

$$(5.82)$$

Hence, for single-tap prediction, the variance of the prediction error is 4.61 dB lower than the original signal power. In other words, differential quantization reduces the peak-to-peak voltage variation by 4.61 dB compared to absolute quantization.

For two-taps prediction, one has from (5.76) and (5.79)

$$\mathbf{r}_X = [0.405 \quad 0.155]^T$$

$$\mathbf{R}_X = \begin{bmatrix} 0.5 & 0.405 \\ 0.405 & 0.5 \end{bmatrix}$$

$$\mathbf{w}_0 = \mathbf{R}_X^{-1} \mathbf{r}_X = \begin{bmatrix} 0.5 & 0.405 \\ 0.405 & 0.5 \end{bmatrix}^{-1} \begin{bmatrix} 0.405 \\ 0.155 \end{bmatrix}$$

$$= \begin{bmatrix} 1.625 \\ -1.006 \end{bmatrix}$$

$$(5.83)$$

The predictor output is given by (5.68):

$$\hat{x}[n] = \sum_{k=1}^{2} w_k \, x[n-k]$$

$$= 1.625 \, x[n-1] - 1.006 \, x[n-2]$$

$$(5.84)$$

The minimum value of the variance of the prediction error is found from (5.77) as

$$\sigma_{e,\min}^2/\sigma_x^2 = 1 - \mathbf{r}_X^T \mathbf{R}_X^{-1} \mathbf{r}_X = 1 - \mathbf{r}_X^T \mathbf{w}_0$$

$$= 1 - [0.405 \quad 0.155] \begin{bmatrix} 1.625 \\ -1.006 \end{bmatrix}$$

$$= 0.498$$

$$(5.85)$$

For two-taps prediction, the ratio of the input and output powers is found as

$$\frac{P}{P_{out}} = \frac{\sigma_x^2}{\sigma_{e,min}^2} = \frac{1}{0.498} = 2.01 \ (3 \ dB) \quad (5.86)$$

Comparison of (5.74) and (5.86) shows that two-tap prediction performs worse than single-tap prediction in this particular example. However, better prediction results are anticipated with increasing values of the number of taps.

Example 5.13 Equalization of a Deterministic Channel.
Linear prediction filters are also used channel equalization, that is, for minimizing the channel distortion. Consider that a signal at the receiver input consists of two multipath components with path gains h_0 and h_1. For the sake of simplicity, the channel is assumed to be linear time-invariant and deterministic; hence, the path gains are time-invariant. The k-th signal sample may be written as

$$x_k = h_0 a_k + h_1 a_{k-1} + w_k \quad (5.87)$$

where data symbols $a_k = \pm 1$ are equiprobable and uncorrelated. This implies that k-th received signal sample x_k consists of the k-th bit a_k with a path gain h_0 and the intersymbol interference (ISI) represented by (k-1)-th bit weighted by h_1. The noise samples w_k are independent, zero-mean samples of noise with variance σ^2. A linear equalizer with 3-tap coefficients is used for removing the ISI due to the channel. The k-th equalizer output may be written as the convolution of the input signal samples with tap coefficients c_n of the equalizer (see (5.68)):

$$y_k = \sum_{n=0}^{2} c_n x_{k-n} \quad (5.88)$$

We will determine the tap gains $\{c_n\}$ and the mean square error (mse) between the kth equalizer output sample y_k and the kth data symbol,

a_k, for $h_0 = 1$, $h_1 = 0.46$ and $\sigma^2 = 0.001$. Note that (5.88) may be represented by Figure 5.39 with $p = 3$.

From (5.87) and (5.88), the k-th equalizer output may be written as

$$y_k = \sum_{n=0}^{2} c_n x_{k-n} = c_0 x_k + c_1 x_{k-1} + c_2 x_{k-2}$$

$$= c_0 [h_0 a_k + h_1 a_{k-1} + w_k]$$

$$+ c_1 [h_0 a_{k-1} + h_1 a_{k-2} + w_{k-1}]$$

$$+ c_2 [h_0 a_{k-2} + h_1 a_{k-3} + w_{k-2}]$$

$$= \underbrace{c_0 h_0}_{1} a_k + \underbrace{(c_0 h_1 + c_1 h_0)}_{0} a_{k-1}$$

$$+ \underbrace{(c_1 h_1 + c_2 h_0)}_{0} a_{k-2} + c_2 h_1 a_{k-3}$$

$$+ \underbrace{(c_0 w_k + c_1 w_{k-1} + c_2 w_{k-2})}_{w}$$

$$(5.89)$$

Ideally, y_k is desired to be equal to a_k; then, the ISI in (5.87) will be completely eliminated. The channel is said to be equalized when only a_k is received at time kT at the equalizer output. This can be achieved by forcing the coefficients of a_{k-1} and a_{k-2} to zero, and the coefficient of a_k to unity. Hence, the three tap coefficients of this zero-forcing equalizer are found as

$$c_0 = 1/h_0$$
$$c_1 = -c_0 h_1/h_0 = -h_1/h_0^2 \quad (5.90)$$
$$c_2 = -c_1 h_1/h_0 = h_1^2/h_0^3$$

The output of this zero-forcing equalizer at time $t = kT$ is then given by

$$y_k = a_k + (h_1/h_0)^3 a_{k-3} + w \quad (5.91)$$
$$w = c_0 w_k + c_1 w_{k-1} + c_2 w_{k-2}$$

The mean square error between the equalizer output and the input bit is then given by

$$\sigma_e^2 = E\left[|y_k - a_k|^2\right] = E\left[\left|(h_1/h_0)^3 a_{k-3} + w\right|^2\right]$$

$$= (h_1/h_0)^6 + E\left[|w|^2\right]$$

$$= (h_1/h_0)^6 + \left(c_0^2 + c_1^2 + c_2^2\right)\sigma^2$$

$$= (h_1/h_0)^6 + \frac{1}{h_0^2}\sigma^2 + \left(\frac{h_1}{h_0^2}\right)^2 \sigma^2 + \left(\frac{h_1^2}{h_0^3}\right)^2 \sigma^2$$

$$(5.92)$$

The mean square error, σ_e^2 between y_k and a_k is found as $\sigma_e^2 = 0.011$ by inserting $h_0 = 1$, $h_1 = 0.46$ and the noise variance $\sigma^2 = 0.001$ into (5.92), where the contribution of ISI is higher compared to that of the noise. Note that the above analysis based on zero-forcing equalization does not take into consideration of the noise effects.

The channel gains h_0 and h_1, which are assumed to be deterministic and time-invariant, usually vary randomly with time depending on the wireless channel conditions. Then, one needs an equalizer that adaptively changes the tap coefficients to respond the random variations in the channel.

5.4.2.1 Linear Adaptive Prediction

Linear prediction requires the knowledge of the statistics of signal samples in terms of the auto-correlation function R[k], $k = 0, 1,... p$, where p is the prediction order. The accurate determination of R[k], $k = 0, 1, ..., p$ requires the statistics of sufficiently large number of bits. This is often accomplished, in time varying channels, by using a training sequence at periodic time intervals; this brings an overhead and decreases the effective data rate. Hence, in time varying channels, the values of R[k], $k = 0, 1,... p$ need to be updated at time intervals shorter than the channel coherence time, that is defined as the maximum time interval during which the channel gain does not change appreciably. It might therefore be attractive to use an adaptive prediction algorithm, where the tap weights w_k,

$k = 1, 2,.. p$ are computed in a recursive manner, starting from some arbitrary initial values, without needing the statistics of R[k] but using only the instantaneously available data.

The adaptive prediction algorithm aims to minimize σ_e^2, which corresponds to the minimum point of the bowl-shaped error surface that describes the dependence of σ_e^2 on the tap weights. Successive (iterative) adjustments to the tap-weights are made in the direction of the steepest descent of the error surface, that is, in the opposite direction of the gradient, where σ_e^2 shows fastest decrease. The gradient of σ_e^2 in the directions of w_k may be written as

$$\nabla \sigma_e^2 = [g_1 \ g_2 \ \cdots \ g_p], \quad g_k = \frac{\partial \sigma_e^2}{\partial w_k} \quad (5.93)$$

By using σ_e^2, given by (5.70), one may express g_k as follows:

$$g_k = \partial \sigma_e^2/\partial w_k = -2R_X[k] + 2\sum_{j=1}^{p} w_j R_X[k-j]$$

$$= -2E[x[n]x[n-k]] + 2\sum_{j=1}^{p} w_j E[x[n-j]x[n-k]]$$

$$= -2E\left[x[n-k]\left(x[n] - \sum_{j=1}^{p} w_j \ x[n-j]\right)\right]$$

$$(5.94)$$

In view of (5.73), the Wiener-Hopf equation for linear prediction, (5.94) vanishes. However, the use of instantaneous values as estimates of the autocorrelation function facilitates the adaptive process:

$$\hat{g}_k[n] = -2x[n-k]e[n] \quad (5.95)$$

where the prediction error at n-th iteration is defined as

$$e[n] = x[n] - \hat{x}[n]$$

$$\hat{x}[n] = \sum_{j=1}^{p} \hat{w}_j[n]x[n-j] \quad (5.96)$$

Here $\hat{w}_j[n]$ denotes the estimate of the j-th tap weight at n-th iteration. In view of the above, after sufficiently large number of iterations (5.95) and (5.96) are expected to be minimized. This implies that the tap weights are determined with sufficient accuracy so that the error signal vanishes.

The kth tap weight at $(n+1)^{th}$ iteration may be expressed in terms of its value at n^{th} iteration $w_k[n]$, the step size μ and the prediction error as follows:

$$\hat{w}_k[n+1] = \hat{w}_k[n] - \frac{1}{2}\mu\, \hat{g}_k[n]$$

$$= \hat{w}_k[n] + \mu\, x[n-k]\, e[n], k = 1,2,\ldots,p$$

$$(5.97)$$

The iterative approach described above to determine the tap weights decreases the mean square prediction error in the direction of the steepest descent. The step-size μ controls the speed of adaptation. Small μ implies slow convergence and poor tracking performance but more accuracy, while large μ implies rapid convergence but large variations around the optimum values of the tap weights, hence larger prediction error. The iteration stops either after a certain number of iterations or when the change in a tap weight between n^{th} and $(n+1)^{th}$ iteration becomes smaller than a predetermined value. Then, the variance of the prediction error remains in the near vicinity of its minimum value. The speed of convergence is required to be faster than the time rate of change of the channel gain so that the predictor can follow the random time variations in the channel. Depending on the channel coherence time, this may require fast DSP chips and algorithms. Periodic updates of the tap weights enable an adaptive predictor to correctly predict the desired signal sample with a reasonable number of iterations.

The block diagram of a linear adaptive predictor using least-mean-square (LMS) algorithm is shown in Figure 5.40. The input sample is predicted by a predictor using p previous samples of the same signal. The error signal, determined by

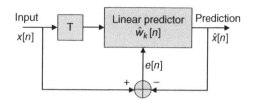

Figure 5.40   Block Diagram of a Linear Adaptive Prediction Process.

difference between the current signal sample and its prediction, is then multiplied with x[n-k] to determine the next iteration of the k-th tap weight. Repeating this for all taps in all iterations minimizes the error signal in the mean square sense and predicts the current signal sample to be used for differential encoding in Figure 5.37.

5.4.3   Differential PCM (DPCM)

DPCM consists of transmitting the prediction error, difference between the actual signal sample and its prediction, as given by (5.69), after quantization (see Figure 5.41). The variance of the error signal is known to be smaller than the variance of the input sample itself (see Figure 5.38 and Example 5.12). Consequently, lower number of quantization levels L, hence fewer bits per sample, may be used to transmit the same signal. Therefore, DPCM allows lower transmission rates and bandwidths.[6]

Figure 5.41 shows the block diagram of a DPCM transceiver. At the transmitter, the signal at the quantizer input is the prediction error, $e[n] = x[n] - \hat{x}[n]$, that is, the difference between the input sample $x[n]$ and its prediction $\hat{x}[n]$ provided by the prediction filter. The prediction error, which is quantized and encoded before transmission, may be reduced by using a predictor with higher prediction order. If prediction is sufficiently good, the variance σ_e^2 of $e[n]$ will be less than the variance σ_X^2 of $x[n]$, which is simply the input signal power. Note that the simplest prediction filter is a single delay element

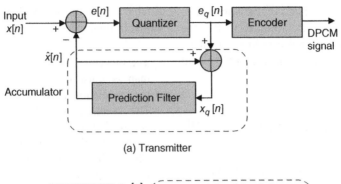

(a) Transmitter

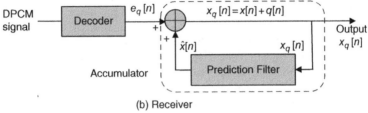

(b) Receiver

Figure 5.41   Block Diagrams of DPCM Transmitter and Receiver.

(a single-tap filter, $p = 1$), predicting the present sample as the previous one.

The quantizer output may be written as the sum of the error signal e[n] and the quantization error q[n]:

$$e_q[n] = e[n] + q[n] \qquad (5.98)$$

The quantization error q[n] will be decreased as the number of quantization levels, L, increases. Multilevel quantization of the error signal provides better information for message reconstruction at the receiver. However, with increasing L, the number of bits per sample, $R = \log_2 L$, and the required transmission rate, $R_b = f_s \log_2 L$, will increase. Note that the predictor provides a prediction $\hat{x}[n]$ for the n-th signal sample $x[n]$ by using $x_q[n]$, the quantized version of the nth signal sample, and the quantized error signal:

$$x_q[n] = \hat{x}[n] + e_q[n]$$

$$= \hat{x}[n] + e[n] + q[n] \qquad (5.99)$$

$$= x[n] + q[n]$$

It is evident from (5.99) that $x_q[n]$ differs from x[n] only by the irrecoverable quantization error q[n] of the error signal. At the receiver, the decoded quantized error signal is added to the prediction filter output to regenerate the quantized signal sample. The same prediction filter as in the transmitter is used to predict a signal sample from its past values. In the absence of the channel noise, the signal at receiver output x_q [n] differs from the signal sample x[n] only by the quantization error.

A quantizer can be designed to produce a quantization error with variance, σ_Q^2, that is smaller than in the standard PCM by choosing a suitable value for the number of quantization levels L. Output SNR of DPCM may be expressed in terms of the variance of the prediction error as

$$SNR_{DPCM} = \frac{P}{\sigma_Q^2} = \frac{P}{\sigma_e^2} \frac{\sigma_e^2}{\sigma_Q^2} = G_p SNR_Q \quad (5.100)$$

where $P = \sigma_X^2$ denotes the variance of the input sample x[n], σ_Q^2 is the variance of the quantization error $q[n]$ and σ_e^2 the variance of the prediction error. According to (5.19), the signal-to-quantization noise ratio is given by $SNR_Q = \sigma_e^2/\sigma_Q^2$. The processing gain G_p due to differential quantization

$$G_p = \frac{P}{\sigma_{e,\min}^2} = \frac{R_X[0]}{R_X[0] - \mathbf{r}_X^T \mathbf{R}_X^{-1} \mathbf{r}_X} \quad (5.101)$$

provides an SNR advantage to DPCM over standard PCM in the order of 5–10 dB for voice signals. [6]

The error $e[n] = x[n] - \hat{x}[n]$ is proportional to the derivative of the input signal. Since $\hat{x}[n]$ changes as much as $\pm(L-1)\Delta$ from sample to sample (see Table 5.6), the maximum rate of increase of the error signal $(L-1)\Delta f_s$ should be at least as fast as the input sequence of samples $\{x[n]\}$ in a region of maximum slope. Hence, the DPCM signal should satisfy the following slope tracking condition:

$$f_s(L-1)\Delta \geq \left| \frac{dx(t)}{dt} \right|_{\max} \quad (5.102)$$

where $f_s = 1/T_s$ is the sampling frequency and Δ is the step size. Otherwise, Δ may be too small for the error signal to follow a steep segment of the input signal $x(t)$; then we suffer the so-called slope overload distortion. On the other hand, when the step size Δ is too large relative to local slope of input signal $x(t)$, granular noise causes oscillations around relatively flat segments of signal samples. Since small values of the step size causes slope overload distortion, while large values of it is the source of granular noise, one should seek for an optimum step size which provides a trade-off between the slope overload distortion and the granular noise.

5.4.4   Delta Modulation

Delta modulation (DM) is basically a special case of DPCM with three fundamental differences. In the block diagram of DPCM shown in Figure 5.41:

a. The message signal is oversampled (higher than Nyquist rate) to purposely increase the correlation between adjacent samples. The rationale behind oversampling is to ensure that adjacent samples do not differ from each other significantly.

b. The predictor has a prediction order $p = 1$ (single-tap) in DM but $p \geq 2$ for DPCM (multi-tap). Since we have $\hat{x}[n] = x_q[n-1]$ at the output of the single-tap predictor, the error signal defined by (5.69) may be rewritten as

$$e[n] = x[n] - \hat{x}[n] = x[n] - x_q[n-1] \quad (5.103)$$

The quantized sample $x_q[n]$ in DM differs from the original sample x[n] by the quantization error $q[n]$ as defined by (5.99).

c. The error signal e[n] in (5.103) is quantized with one-bit; the quantization level is $L = 2$ for DM compared to multilevel $(L > 2)$ quantization in DPCM. As also shown in Figure 5.42, depending on positive (representing bit 1) and negative signs (representing bit 0) of e[n], the quantized error signal assumes one of two levels, $\pm\Delta$:

$$e_q[n] = \Delta \operatorname{sgn}(e[n]) = \begin{cases} +\Delta & x[n] > x_q[n-1] \\ -\Delta & x[n] < x_q[n-1] \end{cases}$$
$$(5.104)$$

DM provides a staircase approximation to the oversampled version of the message signal and achieves digital transmission of analog signals using a much simpler hardware than PCM. Since each sample is quantized with one bit, transmission rate in DM is equal to the sampling rate $f_s = 1/T_s$, which is chosen to be much higher than the Nyquist rate, $f_s = 2W$. Table 5.7 shows the process of DM of the following

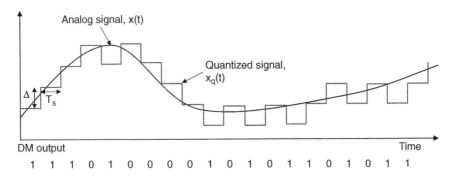

Figure 5.42   Delta Modulation.

Table 5.7   The Steps for Delta Modulating the Sample Sequence {x[1] = 0.9, x[3] = 0.3, x[4] = 1.2, x[5] = −0.2, x[6] = −0.8}.

n	$x[n]$	$x_q[n-1]$	$e[n] = x[n] - x_q[n-1]$	$e_q[n]$	$x_q[n] = e_q[n] + x_q[n-1]$
1	0.9	0 (ref.)	+0.9	+1	1
2	0.3	1	−0.7	−1	0
3	1.2	0	+1.2	+1	1
4	−0.2	1	−1.2	−1	0
5	−0.8	0	−0.8	−1	−1

sample sequence: x[1] = 0.9, x[3] = 0.3, x[4] = 1.2, x[5] = −0.2, x[6] = −0.8.

In a DM receiver, the quantized error signal is added to the one-tap predictor output successively so that the n-th sample of the quantized signal is found as follows:

$$x_q[n] = x_q[n-1] + e_q[n]$$
$$= \Delta \sum_{i=1}^{n} \text{sgn}(e[i]) = \sum_{i=1}^{n} e_q[i] \qquad (5.105)$$

Thus, at the n-th sampling instant, the accumulator increments the predictor output by Δ in positive or negative direction depending on the sign of $e_q[n]$. The staircase approximation $x_q(t)$ is reconstructed by passing the sequence of positive/negative pulses, produced at the decoder output, through an accumulator in a manner similar to that used at the transmitter. Note that (5.105) is the digital equivalent of integration, since it represents the accumulation

of positive and negative increments of the step size Δ. Out-of-band quantization noise in $x_q(t)$ is rejected by passing it through a LPF, of a bandwidth equal to the message bandwidth.

5.4.4.1   Slope-Overload Distortion

In view of (5.99) and (5.103), the error signal at the quantizer input may also be written as

$$e[n] = x[n] - x_q[n-1] = x[n] - x[n-1] - q[n-1] \qquad (5.106)$$

Ignoring the quantization error q[n-1], the error signal is proportional to the derivative of the input signal. Since the maximum rate of increase of the error signal is Δ/T_s and occurs with a sequence of Δ's of the same polarity (see Figure 5.43), the sequence {$x_q[n]$} can increase faster than the input sequence of samples {$x[n]$} in a region of maximum slope if

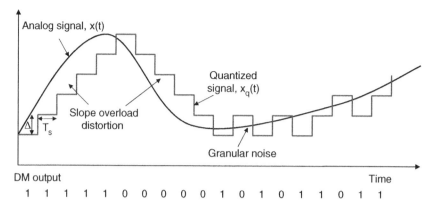

Figure 5.43   Slope Overload Distortion and Granular Noise in Delta Modulation.

$$\frac{\Delta}{T_s} \geq \max\left|\frac{dx(t)}{dt}\right| \qquad (5.107)$$

Otherwise, the slope Δ/T_s of $x_q(t)$ cannot follow a steep segment of the input signal $x(t)$; this leads to slope overload distortion as shown in Figure 5.43.

Example 5.14   Slope Overload Distortion in DM.
Let us find the step size Δ required to prevent slope overload distortion for a sinusoidal input signal $x(t) = A\sin(w_c t)$. From (5.107),

$$\frac{\Delta}{T_s} \geq \max\left|\frac{dm(t)}{dt}\right| = \max|Aw_c \cos w_c t| = Aw_c$$
$$(5.108)$$

For no slope overload distortion, the following requirement should be satisfied:

$$\Delta \geq Aw_c T_s = 2\pi Af_c/f_s \qquad (5.109)$$

For $A = 1$ and $f_s/f_c = 10$, we get $\Delta \geq \pi/5$.

5.4.4.2   Granular Noise

If the signal does not change too rapidly from sample to sample, $x_q[n]$ remains within $\pm\Delta$ of the input sample $x[n]$. If the step size Δ is chosen much larger than the local slope of input

signal $x(t)$, then staircase approximation $x_q(t)$ induces oscillations around relatively flat segments of $x(t)$. Therefore, as in DPCM, an optimum step size is needed so as to compromise the slope overload distortion and the granular noise. The optimum value of the step size maximizes the output SNR of a delta modulator. On the other hand, choice of very small values of T_s (large sampling frequency, f_s) for preventing slope overload distortion is undesirable since it increases the transmission rate. Therefore, one may change the step size adaptively depending on the local slope of the signal.

5.4.4.3   Adaptive Delta Modulation (ADM)

ADM comprises an additional hardware, compared to DM, designed to provide variable step size, thereby reducing slope overload effects without increasing the granular noise. Slope overload appears in $e_q[n]$ as a sequence of pulses having the same polarity, whereas granular noise occurs when the polarity tends to alternate when $x_q(t)$ tracks $x(t)$.

As shown in Figure 5.44a, which shows the block diagram of an ADM, the step-size controller adjusts the variable gain $g(n)$, so that (5.105) is modified as

$$x_q[n] = x_q[n-1] + g[n]\, e_q[n] \qquad (5.110)$$

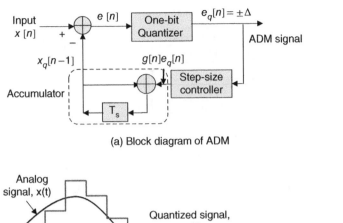

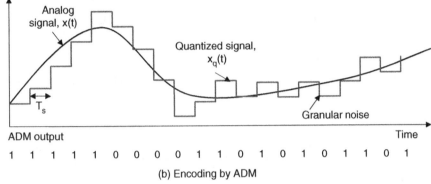

(a) Block diagram of ADM

(b) Encoding by ADM

Figure 5.44 Adaptive Delta Modulation (ADM).

where the step size is adaptively determined as follows:

$$g[n] = \begin{cases} g[n-1] \, K & e_q[n] = e_q[n-1] \\ g[n-1]/K & e_q[n] \neq e_q[n-1] \end{cases} \quad (5.111)$$

Here, K is a constant in the range $1 < K < 2$. In view of (5.110) and (5.111), the effective step size increases in discrete steps by successive powers of K during slope overload conditions. Variable step size yields a wider dynamic range. The SNR of ADM is typically 8–14 dB better than ordinary DM and the transmission bandwidth for voice is typically 24–32 kHz [6]. A scheme known as continuously variable slope delta modulation (CVSD) provides a continuous range of step-size adjustments instead of a set of discrete values as in (5.111). Consequently, CVSD has better performance than DM and ADM.

5.4.4.4 Delta-Sigma Modulation (DΣM)

As can be observed from (5.106), the error signal at the quantizer input of a DM is proportional to the derivative of the message signal. This leads to accumulation of noise power in the demodulated signal due to the presence of accumulator (integrator) in the DM receiver (see Figure 5.41). This drawback can be overcome by integrating the message signal prior to DM at the transmitter as shown in Figure 5.45. By integration at the transmitter, the low-frequency content of the input signal is pre-emphasized and the correlation between adjacent samples of the DM input is increased. The overall system performance is thus improved by reducing the variance of the error signal at the quantizer input and, at the same time, the receiver design is simplified.

The so-called delta-sigma modulation (DΣM) is a version of DM that incorporates integration at its input. Since integration at

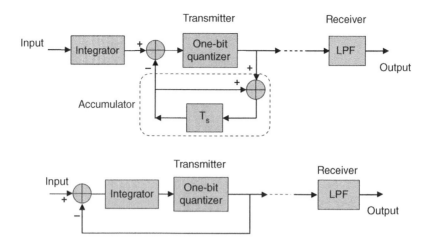

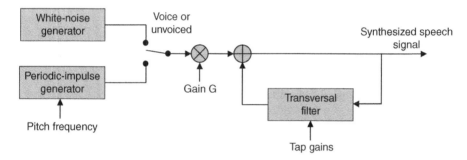

Figure 5.45   Two Equivalent Block Diagrams of the Transceiver of a Delta Sigma Modulation System.

Figure 5.46   Block Diagram of a Speech Synthesizer.

the transmitter input should be compensated for by differentiation at the receiver, this need is satisfied by simply removing the integrator in the conventional DM receiver. Thus, as shown in Figure 5.45, the receiver of a DΣM consists only of a LPF.

Simplicity of implementation of the DΣM transceiver is achieved at the expense of sampling rates far in excess of that needed for PCM, hence increased transmission bandwidth. Then, one may use, for example, a multi-tap prediction filter as in DPCM in order to reduce the transmission bandwidth.

5.4.5   Audio Coding

A voice signal, bandlimited to 3.4 kHz, sampled at 8 ksamples/s and quantizated with

8-bits, is transmitted at 64 kbps. This rate is unacceptably high for some applications that require much lower data rates.

The speech process starts with the generation of a tone in the vocal chords. The throat, mouth, tongue, lips, and nose, and so on, acting as filters, modulate the generated tone. The speech signal consists of active periods separated by unvoiced time intervals. Unvoiced speech is modelled by white noise. Voiced speech may be modelled by periodic sequence of impulses with period which is equal to the reciprocal of the pitch frequency. Figure 5.46 shows a model for the generation of speech signals.

N samples of a speech signal are predicted as

$$x_n = \sum_{i=1}^{p} a_i x_{n-i} + G\, w_n, \quad 1 \le n \le N \qquad (5.112)$$

where w_n denotes the input sequence (white noise or impulses), G is the amplifier gain, $\{a_i\}$ are the tap gains and p denotes the prediction order. The first part in (5.112) represents the output of a prediction filter with p taps as shown in Figure 5.37. However the term Gw_n does not depend on the previous speech samples. Linear prediction is used to determine the tap gains $\{a_i\}$, as outlined in Section 5.4.2. For the optimal choice of the predictor coefficients, the mean square value of the prediction error may be written as

$$\sigma_e^2 = \frac{1}{N}\sum_{n=1}^{N}\left[x_n - \sum_{i=1}^{p} a_i x_{n-i}\right]^2 = G^2 \frac{1}{N}\sum_{n=1}^{N} w_n^2$$

$$(5.113)$$

When the excitation coefficients are normalized to unity, then the gain G is given by [2]

$$\frac{1}{N}\sum_{n=1}^{N} w_n^2 = 1 \;\Rightarrow\; G = \sigma_e \qquad (5.114)$$

Linear predictive coding (LPC) provides an alternative approach to digital transmission of analog voice signals compared to the classical approach, which is based on ADC of the analog voice signals. LPC is based on extracting the parameters of a voice signal, for every 10-25 ms, and transmitting this information with a prespecified number of bits. A transversal filter and some auxiliary components are used to extract the required signal parameters, for example, volume and the duration of the tone, pitch frequency, and so on, and the tap weights of the transversal filter. [2][6] As shown in Figure 5.47, voice samples are analyzed to determine the parameters for the synthesizer. The error signal, defined as the difference between the input and synthesizer output, is encoded along with these parameters for transmission. These parameters and the quantized error signal are used by the receiver to regenerate the original speech signal.

A complete LPC codeword consists of typically 80 bits accounting for pitch frequency, amplifier gain, tap weights of the transversal filter, and so on. Updating the parameters every 10-25 ms is equivalent to sampling at 40-100 Hz. So, LPC requires very modest bit rates in the range of 3.2 kbps (80 bits/25 ms)-8 kbps

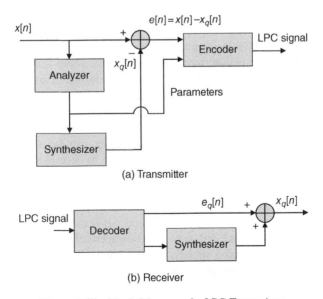

(a) Transmitter

(b) Receiver

Figure 5.47   Block Diagram of a LPC Transceiver.

(80 bits/ 10 ms). [6] Therefore, LPC is very effective for digital voice transmission and is currently employed in many modern communication systems.

The GSM system uses *regular pulse excited long-term prediction* (RPE-LTP) coding, which is based on PLC. ADC of speech signals at 8 ksamples/s and 13 bits/sample leads to a bit rate 104 kbps at the RTP-LTP encoder input. Based on the observation that the speech signals are approximately constant during 20 ms intervals, the filter parameters characterizing a speech signal are updated with 20 ms intervals. This corresponds to sampling speech signal at 1/20 ms = 50 samples/s; such low sampling rates provide a compromise between the speech quality and the bit rate. In the RPE-LPT algorithm, features of each 20 ms speech block are extracted, and encoded into 260 bits with a net bit rate of 260 bits/20 ms = 13 kbits/s. Hence, a RTP-LTP encoder reduces the input bit rate of 104 kbps to 13 kbps. Speech signals are then protected against channel errors with the addition of 196 redundancy bits, so that a total of 456 bits are transmitted during 20 ms interval with a channel rate of 456 bits/ 20 ms = 22.8 kbits/s. Note that this rate is approximately ~22% of 104 kbps obtained using ADC.

The *adaptive multi-rate* (AMR) is a multi-rate narrowband audio compression format optimized for speech coding. It is based on 160 bits for 20 ms frames, and narrowband (200–3400 Hz) signals are encoded at variable bit rates ranging from 4.75 to 12.2 kbit/s. AMR was adopted as the standard speech codec by 3GPP in October 1999 and is now widely used in GSM and UMTS. It adaptively selects one of eight different bit rates (12.2, 10.2, 7.95, 7.40, 6.70, 5.90, 5.15 or 4.75 kbit/s) depending on the channel conditions. [7][8] Although AMR is basically a speech format and may not give ideal results for other audio signals, AMR codec is also used for storing short audio recordings in mobile telephone handsets. The AMR codec can also be used for

applications like random access or synchronization with video.

5.4.6   Video Coding

Multimedia communications in many applications require transmission of audio, video and data signals at low data rates. We know from Example 5.5 that high data rates are needed to transmit digital video services unless video compression is used. As an integral part of video encoding, high compression ratios of video signals are strongly desired in various applications including TV broadcasting, video-conferencing, digital storage, internet streaming, and wireless communications.

Data compression drastically reduces the transmission rate and enables efficient storage and transmission of multimedia signals. Lossless data compression (data compaction) is reversible and operates as ZIP codes used to compress document files. The objective of lossless data compression is to map the original bit stream into a shorter stream, so that the original bit stream can be recovered exactly from the compressed stream. Lossy data compression is irreversible, and aims to compress the original data at the expense of some degradation.

Video compression is possible due to the fact that adjacent parts of a single video frame as well as successive frames are similar to each other. Therefore, transmitting the differences between scenes within the frame and between adjacent frames is much more economical than transmission of all the frames. The interframe redundancy can be significantly reduced to produce a more efficiently compressed video signal. This reduction is achieved through the use of prediction of each frame from its neighbors; the resulting prediction error (differential encoding) is transmitted for motion estimation and compensation. In brief, video compression algorithms exploit two types of redundancy in the signal: temporal (changes from one frame to the next) and spatial (changes in the pixels within one frame). Still image coding algorithms like Joint Photographic Experts Group (JPEG) only exploit the spatial redundancy.

Video compression techniques include

a. Discrete cosine transform (DCT), which is a frequency transform technique similar to Fourier Transform, is used to reduce the spatial redundancy in video frames.
b. Quantization is a technique for deliberately losing some information (lossy compression) from the pictures.
c. Huffman coding, run length coding, and variable length coding are lossless compression techniques and use code tables based on the data statistics to encode more frequently encountered data with shorter codewords.
d. Motion compensated predictive coding aims to exploit temporal redundancy. The differences (changes) between current and preceeding frames are calculated and encoded.
e. Bi-directional prediction where some images are predicted from the pictures immediately preceding and following the image.

The first three of the above compression techniques are also used by the *JPEG* standard, which was developed for compressing full-color or greyscale images of natural, real-world scenes by exploiting known limitations of the human visual system. Low resolution pictures such as computer generated images can be transmitted or stored at data rates up to 1.5 Mbps.

Motion Photographic Experts Group (*MPEG-1*) standard was initially designed to primarily compress video-CD (non-interlaced video) images of size 352×288 at 25 frames/s at a transmission rate of 1.5 Mbps. In contrast with JPEG, MPEG-1 exploits the interframe (temporal) redundancy and provides motion-compensated compression. [9]

MPEG-2 standard, published in 1995, was designed for coding standard-definition (SD) 4:2:0 video at bit rates 3-6 Mbit/s, with a drastic reduction from ~144 Mbps. MPEG-2, used for digital TV broadcast and DVD, has many features in common with MPEG-1 and achieve compressing ratios up to 50:1.

MPEG-3, launched mainly for the HDTV applications, refers to a group of audio and video coding standards designed to handle HDTV signals at 1080p (*p* stands for progressive scanning) in the range of 20 to 40 Mbps. However, as MPEG-2 was observed to have the ability to accommodate HDTV services as well, HDTV was included in 1992 as a separate profile in the MPEG-2 standard.

MPEG-4, though initially introduced in 1998 for low bit-rate video communications, is efficient across a variety of bit-rates ranging from a few kbps to tens of Mbps. MPEG-4 provides improved coding efficiency over MPEG-2, ability to encode multimedia data (video, audio, speech), and robustness against errors. For MPEG-4, bit rates as low as 2 Mbp/s is considered to provide a good compromise between picture quality and transmission bandwidth efficiency. MPEG-4 addresses speech and video synthesis, fractal geometry, computer visualisation, and an artificial intelligence approach for reconstructing images.

ITU-T H.264/MPEG-4 Advanced video coding (AVC) standard, which is published in 2003, allows broadcasting over cable, satellite, cable modem, terrestrial, and so on. Using H.264/AVC, it is possible to transmit video signals at about 1 Mbps with PAL quality, which enables streaming over xDSL connections. H.264 standard contains a set of video coding tools but not all of them are needed for every application. A H.264 decoder may choose to implement any number of them depending on the needs of the application. The adoption of H.264 standard led to doubling the number of TV programmes per satellite in comparison to the use of H.262 (MPEG-2). H.262 also allows

a. Interactive or serial storage multimedia on DVD, optical and magnetic devices,
b. Videoconferencing and videophone,
c. Video-on-demand and multimedia streaming services,

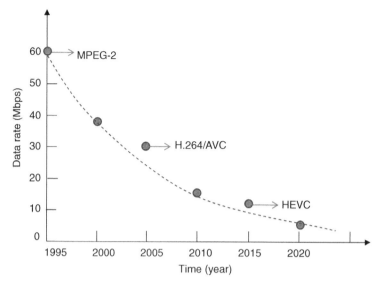

Figure 5.48   Evolution of Video Coding Standards with Time for HD (1080p) TV. [9]

d. Multimedia messaging services, and
e. Multimedia services over packet services, such as multimedia mailing.

High efficiency video coding (HEVC) standard was published in 2013. It aims to achieve a factor of two improvement in compression efficiency compared to the H.264/AVC standard. HEVC is a strong candidate for new generation digital TV services using 1080p.

Figure 5.48 shows the milestones in the evolution of the bit rate required for the transmission of HD(1080p) TV signals. One may easily observe drastic decrease of the required bit rates in this rapidly evolving area.

References

[1] S. Haykin, *Communication Systems* (3rd ed.), J. Wiley: New York, 1978.

[2] J. G. Proakis and M. Salahi, *Communication Systems Engineering* (2nd ed.), Prentice Hall: New Jersey, 2002.

[3] N. Judell, and L. Scharf, A simple derivation of Lloyd's classical result for the optimum scalar quantizer, *IEEE Trans. Information Theory*, vol. **32**, no. 2, pp. 326–328, 1986.

[4] N. S. Jayant, and P. Noll, Digital Coding of Waveforms, Principles and Applications to Speech and Video. Prentice Hall, New Jersey, 1984, pp. 115–251.

[5] Z. Peric, and J. Nikolic, An effective method for initialization of Lloyd-Max's algorithm of optimal scalar quantization for Laplace source, Informatica, vol. **18**, no. 2, pp. 279–288, 2007.

[6] A. B. Carlson, P. B. Crilly, and J. C. Rutledge, *Communication Systems* (4th ed.), McGraw Hill, Boston, 2002.

[7] 3GPP TS 26.071 Mandatory speech CODEC speech processing functions; AMR speech Codec; General description.

[8] 3GPP TS 26.090 Mandatory Speech Codec speech processing functions; Adaptive Multi-Rate (AMR) speech codec; Transcoding functions.

[9] K. McCann, Future Codes, Progress Towards High Efficiency Video Coding, *DVB Scene* **39** March 2012, p. 5, https://www.dvb.org/

Problems

1. The zero-order hold circuit, shown in Figure 5.8, is not only used for flat-top sampling in ADCs but also in DACs to reconstruct signal $x(t)$ from its samples with a sampling interval of T_s.

a. Find the impulse response $h(t)$ of this circuit.

b. Find the transfer function $H(f)$.

c. Show that when a sampled signal
$$x_\delta(t) = \sum_n x(nT_s)\, \delta(t - nT_s) \text{ is applied}$$
at the input of this circuit, the output is a staircase approximation of $x(t)$.

2. A band-limited signal having bandwidth W can be reconstructed by its samples as described by (5.6). Using the samples $g_{-1} = -1$, $g_0 = 2$, $g_1 = 1$ and $g_n = 0$ for $n \neq -1$, 0, 1, plot the received signal as a function of time.

3. The information in an analog voltage waveform, bandlimited to 3.4 kHz, is to be transmitted using 16-PCM. The quantization error is specified not to exceed 0.1 % of the peak-to-peak analog signal.

   a. Determine the minimum number of bits per sample that meets the distortion requirements.
   b. Determine the signal to quantization noise ratio for PAPR = 6 dB.
   c. Determine the minimum required sampling rate, and the resulting bit rate.
   d. Determine the symbol rate for binary PCM.
   e. What is the transmission rate if 16-ary PAM transmission is used instead of the binary PCM?

4. The quantization error in a PCM system is required not to exceed 1% of the peak-to-peak analog signal voltage.

   a. Determine the bandwidth efficiency (bits/s/Hz) of a speech signal bandlimited to 3.4 kHz and sampled at the Nyquist rate.
   b. Repeat (a) for an (high fidelity) audio signal bandlimited to 20 kHz.

5. A compact disc (CD) recording system samples each of two stereo signals with a 16-bit ADC at 44.1 kilosamples/s.

   a. Determine the average output signal-to-quantization noise ratio for PAPR = 1.
   b. If the recorded music is designed to have a PAPR = 20, determine the average output signal-to-quantization noise ratio.
   c. The bit stream at the ADC output is augmented by error-correcting bits, bits for clock extraction and display and control. These additional bits represent 100% overhead, the bit rate at ADC output is doubled. Determine the output bit rate of this CD recorder system.
   d. If the CD can record 75 minutes of music, determine the number of bits recorded on a CD.
   e. Consider a book of 1500 pages, 100 lines/page, 15 words/line, 6 letters/word, and 6 bits/letter. Determine the number of bits required to store the book and estimate the number of comparable books that can be stored on a CD.

6. A music signal has a bandwidth of 20 kHz and a PAPR of 20 dB. Calculate the bit rate required to transmit this as a linearly quantised PCM signal with a $SNR_Q = 50$ dB. Assuming that the minimum ISI-free PCM bandwidth is given ½ of the transmission rate, determine the minimum required transmission bandwidth.

7. An overall SNR of 53 dB is required for the reconstructed analog signal to achieve a bit error rate of 10^{-6} in a uniformly quantized PCM system. The analog signal is modeled by a stationary random process whose average power is 16 W.

   a. Determine the corresponding signal-to-quantization noise ratio, SNR_Q, and the number of quantization levels?
   b. Assuming that the signal is quantized to satisfy the condition of part (a) and assuming the approximate bandwidth of the signal is $W = 4$ kHz, determine the bit rate of a binary PCM signal based on this quantization scheme?

8. A signal varying between ±4V is sampled 10 ksamples/s and is uniformly quantized using 8 bits.

    a. Determine the maximum quantization error.
    b. Determine the average quantization noise power.
    c. Determine the number of bits/sample needed to reduce the maximum quantization error to 10 mV.
    d. Discuss the costs associated with the quantization error reduction achieved in part (c).

9. Consider that an analog signal is sampled at 8 ksamples/s, quantized with 3 bits/sample, and Gray encoded as shown below.

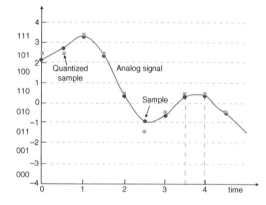

    a. Determine the sampling interval and calibrate the x-axis in the figure above.

    b. Determine the data sequence if uniform quantization is used.
    c. Determine the data sequence if µ-law companding is used with µ = 50. Estimate the original values of the samples by inspecting the waveform.
    d. Compared with uniform quantization, what are the advantages and disadvantages of nonuniform quantization?

10. Consider that an audio signal bandlimited to 4 kHz is sampled at 8 k samples/s.

Assuming that the ratio of peak signal power to average quantization noise power at the output does not exceed 40 dB:

    a. Determine the minimum number of bits/sample needed for uniform quantization.
    b. Determine the bit rate.
    c. Determine the null-to-null bandwidth of this PCM signal using polar NRZ line code.

11. The signal $x(t) = 8\cos(20\pi t)$ is sampled at a rate 50 samples/s, uniformly quantized and encoded with 4 bits.

    a. Determine the values of the 16 representation levels and establish the correspondence between representation levels and Gray-encoded 4-bit symbols.
    b. Insert the uniformly quantized values and the corresponding 4-bit symbols in the table below for the non-companded signal.

Sample Number	$x[n]$	$X_q[n]$	4-bit symbol
0			
1			
2			
3			
4			
5			

    c. Determine the normalized RMS quantization error averaged over one cycle, by using the following formula, where N is the number of samples in one cycle of the signal.

$$\Delta = \frac{1}{N}\left\{\sum_{n=0}^{N-1}\left(\frac{x[n]-x_q[n]}{x_{max}}\right)^2\right\}^{1/2}$$

    d. Now consider that a compander described by $y[n] = 4(x[n])^{1/3}$ is used for non-uniformly quantizing the signal

$x(t)$. Determine the non-uniformly quantized values and insert them and the corresponding 4-bit symbols into the table below.

Sample Number	$x[n]$	$y[n]$	$y_q[n]$	4-bit symbol
0				
1				
2				
3				
4				
5				

e. Determine the normalized RMS quantization error averaged over one cycle for the companded signal.
f. Compare the results obtained in (c) and (e).
g. Show that, without quantization, we get the original signal if we use an expander, as described by $x[n] = \frac{1}{64}\{y[n]\}^3$ where y[n] denotes the input to the expander.

12. A random variable X with pdf $f_X(x) = e^{-|x|}/2$ is quantized using a 3-level quantizer with output y specified as

$$y = \begin{cases} -a & x < -\Delta \\ 0 & -\Delta \leq x \leq \Delta \\ +a & x > \Delta \end{cases}$$

where $0 \leq \Delta \leq a$.

a. Derive an expression for the resulting mean–squared error (MSE) $\sigma_e^2 = E\left[(x-y)^2\right]$ in terms of a and b.
b. Determine the optimal values of a and Δ that minimize the MSE and the minimum value of σ_e^2.
c. Determine the average signal power to quantization noise power ratio.

13. Repeat Problem 5.12 for a 4-level quantizer with output y specified as

$$y = \begin{cases} -b & x < -\Delta \\ -a & -\Delta \leq x < 0 \\ a & 0 \leq x < \Delta \\ b & x \geq \Delta \end{cases}$$

where $0 < a < \Delta < b$.

a. Derive an expression for the resulting mean–squared error (MSE) $\sigma_e^2 = E\left[(x-y)^2\right]$ in terms of a, b and Δ.
b. Find the optimal values of a, b and c such that MSE is minimized. Determine the minimum value of σ_e^2.
c. Determine the average signal power to quantization noise power ratio. Compare your results with those of Example 5.8.

14. A scanner is used to scan images of A4 page (size 210 mm 297 mm) into a data file using 8 bits/pixel (pixel denotes picture element). It takes 100 ms to transmit the data file created by scanning at a transmission rate of 1 Mbps. Determine the resolution of the scanner, that is, number of pixels used per cm^2?

15. Consider a uniform quantizer as shown in Figure 5.12. Assume that a Gaussian-distributed random variable with zero mean and unit variance is applied to this quantizer input.

a. Find the probability density function of the discrete random variable at the quantizer output.
b. Determine the quantization noise power and the signal-to-quantization noise power ratio, and compare your results with that obtained for uniform and Laplace distributions (see Example 5.7).

16. A PCM system uses Nyquist sampling, uniform quantization and 8-bit encoding. The bit rate of the system is equal to 20 Mbps.

   a. What is the maximum message bandwidth for which the system operates satisfactorily when the message signal is sampled at the Nyquist rate?
   b. Determine the output signal-to-quantizing noise ratio for an input sinusoidal modulating wave.

17. A 8-bit ADC with uniform quantizer is designed to operate over ±4 V.

   a. Determine the step size.
   b. Determine the signal-to-quantization noise ratio for a uniformly distributed random signal.
   c. Determine the signal-to-quantization noise ratio for a Gaussian signal with variance σ^2, where the signal voltage is limited to ±4σ.
   d. Determine the probability of signal saturation in part (c).

18. The dynamic range of a signal x(t), whose amplitude is limited by ±x_{max}, is defined as

$$D = 10 \log \left(\frac{P_{peak}}{Average\ noise\ power} \right)$$

   This definition is meaningful since minimum signal power level may approach zero, but the noise is always present. In PCM, the average noise power level is determined by the quantization noise power since it dominates the Gaussian noise. Determine the number of bits required for an ADC to encode music with a dynamic range higher than 120 dB.

19. A message signal with ±1V peak voltage is transmitted using binary PCM with

uniform quantization. The average message power equals 20 mW over a 1 Ohm resistance. If the signal-to-quantization-noise ratio (SNR_Q) is required to be at least 46 dB:

   a. Determine the minimum number of bits per sample and the corresponding number of quantization levels.
   b. Determine the SNR_Q actually obtained with this quantizer.
   c. If the bit error probability of the channel is 10^{-6}, then calculate the SNR in dB at the decoder output.

20. A TV signal, m(t), bandlimited to 4.5 MHz is to be transmitted by binary PCM. The receiver output signal-to-quantization-noise ratio is required to be at least 55 dB.

   a. If all brightness levels are assumed to be equally likely, that is, amplitude of m(t) is uniformly distributed in the range (−m_{max}, m_{max}), determine the minimum required number of quantization levels L, and the corresponding bit rate.
   b. For this value of L, determine the signal-to-quantization noise power ratio SNR_Q.
   c. If the output SNR_Q is required to be increased by 6 dB, what is the new value of L and the corresponding transmission bandwidth?
   d. Repeat b) and c) for 8-level PCM

21. Oversampling implies sampling rates higher than the Nyquist rate, $f_N = 2W$ of a signal bandlimited to W. Oversampling ratio (OSR) is defined as the real sampling rate to the Nyquist rate: $OSR = f_s/f_N > 1$. Oversampled signal evidently occupies a bandwidth larger than W.

   a. Write down the PSD of a quantization noise, bearing in mind that the PSD means the mean power per unit bandwidth.

b. Write down the total amount of quantization noise power in the signal bandwidth W in terms of the OSR.

c. If Nyquist-sampled and oversampled signals have the same signal-to-quantization-noise power ratios, determine the increase in the resolution of an R-bit convertor due to oversampling. Calculate this increase for OSR = 1.2.

22. Consider a bandpass signal which has a spectrum as shown below with a bandwidth $B = f_U\text{-}f_L$ Hz, where f_U and f_L denote upper and lower frequencies, respectively. According to the bandpass sampling theorem, a bandpass signal can be recovered from its samples by bandpass filtering if the sampling frequency is selected as $f_s = 2f_u/K$ where K is the largest integer not exceeding f_u/B. All higher sampling rates are not necessarily usable unless they exceed $2f_u$.

Assuming $B = 10$ KHz and $f_u = 25$ kHz:

a. Plot the spectrum of the sampled signal for the sampling frequencies 25 kHz, 40 kHz, and 50 kHz,

b. Verify the bandpass sampling theorem by observing whether and how the sampled signal can be recovered for these three sampling frequencies,

c. Compare your results with those of Example 5.4, where a bandpass signal is down-converted to baseband before baseband sampling.

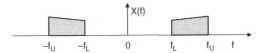

23. Derive the autocorrelation function and the PSD of NRZ bipolar codes in Table 5.4.

24. Determine and plot the power spectrum of a sequence

$$s(t) = \sum_{k=-\infty}^{\infty} b_k\, p(t-kT)$$

of independent and equiprobable message bits $b_k = \pm 1$.

a. Obtain the PSD using the following definition of the Manchester line code

$$p(t) = \Pi\left(\frac{t+T/4}{T/2}\right) - \Pi\left(\frac{t-T/4}{T/2}\right),$$
$$-T/2 < t < T/2$$

where T denotes the bit duration and

$$\Pi\left(\frac{t}{T}\right) = \begin{cases} 1 & 0 < t < T \\ 0 & otherwise \end{cases}$$

b. Derive the autocorrelation function given in Table 5.4 and then obtain the PSD.

25. Draw the pulse stream for the bit stream {11000101} line codes for unipolar NRZ, bipolar NRZ, and Manchester code.

26. Consider that a line code given by

$$g(t) = \begin{cases} \cos(\pi t/T) & |t| < T/2 \\ 0 & |t| > T/2 \end{cases}$$

is used to transmit a bit sequence $\{b_k\}$ as follows:

$$s(t) = \sum_{k=-\infty}^{\infty} b_k\, g(t-kT)$$

where $b_k = \pm 1$ have equal probabilities of transmission and T denotes the bit duration. Determine the power spectral density (PSD) of $s(t)$ by clearly showing all the necessary steps.

27. Consider the unipolar RZ and bipolar RZ binary signaling with bit duration T, as shown below for the binary data stream 110101. Determine the power spectral density (PSD) of these line codes by clearly showing all the necessary steps.

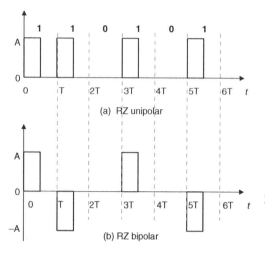

(a) RZ unipolar

(b) RZ bipolar

28. Several audio channels with $W = 20$ kHz are to be transmitted via binary PCM with $R = 12$ bits/sample.

   a. Determine how many of the PCM signals can be accommodated by the first level of the E1 multiplexing hierarchy.
   b. Calculate the corresponding bandwidth efficiency, which is defined as the ratio of transmission bandwidth of the PCM signal to that for the analog transmission (frequency division multiplexing) with single-sideband (SSB). Assume that the minimum ISI-free PCM bandwidth is given ½ of the transmission rate.

29. A TDM carrier supports a frame structure which consists of 30 voice channels, bandlimited to 4 kHz, sampled at Nyquist rate and encoded with 6-bits/samples. Assuming that the frame uses 1 synchronization bit per channel and 1 synchronization bit per frame, determine the required data and frame rates.

30. Ten message signals, each having 2 kHz bandwidth, are time-division multiplexed using uniformly quantized binary PCM. The sampling rate is 20 % higher than the Nyquist rate. The quantization error cannot be exceed 1% of the peak amplitude.

   a. Determine the number of quantization levels and the number of bits per sample.
   b. Determine the sampling rate.
   c. Determine the overall bit rate for the multiplexed signals.

31. Consider a TDM system with four signals, one of them bandlimited to 4 kHz and the remaining three signals are bandlimited to 2 kHz.

   a. Draw the block diagram of a time-division multiplexer for multiplexing these signals, each sampled at their Nyquist rate.
   b. If the output of the multiplexer switch is quantized with 1024 levels and the result is binary-coded, determine the output rate.

32. A TDMA system multiplexes 100 user signals each at 10 Mbps at the multiplexer input. Each user packer consists of 100 bits. A guard time of 1 bit duration is inserted between slots and a preamble of 250 bits used for address information, equalization, and so on.

   a. Determine the frame duration
   b. Determine the throughput efficiency of this multiplexer

c. Determine the effective data rate at the TDM output.

33. Consider the first level E1 multiplexing hierarchy, where 30 voice signals each bandlimited to 4 kHz are sampled at Nyquist rate and encoded by 8 bits/sample. The multiplexed signals are transmitted at a rate of 2.048 Mbps.

   a. Determine the bit duration of the multiplexed signal and that of one of the 30 signals before multiplexing.
   b. If the bit amplitudes are ±1 Volt before multiplexing, determine the bit amplitude of multiplexed signal if the bit energy remains the same before and after multiplexing. Determine the corresponding increase in the transmit power of the multiplexed signal in dB.
   c. Determine the increase in the bandwidth of the multiplexed signal compared to that of a signal before multiplexing. Determine the percentage contribution of overheads to the bandwidth of the multiplexed signals.
   d. If the E_b/N_0 of the multiplexed signal is chosen so as to achieve a bit error probability of 10^{-6}, determine the average number of bit errors that may be observed in 10 seconds?

34. In an adaptive PCM system for speech coding, the input speech signal is sampled 160 times per 20 ms, and each sample is represented by 8 bits. The quantizer step size is recomputed every 10 ms, and it is encoded for transmission using 5 bits. In addition, 1 byte is added at the beginning of each packet (containing 160 voice samples) as header.

   a. Draw the packet structure showing the packet duration and the lengths in bits of the packet constituents.
   b. Compute the effective transmission bit rate of such a speech coder assuming

that information and overhead bits are transmitted in 20 ms duration in real-time applications.
   c. Determine the percentage increase in transmission rate due to overhead bits.

35. Show that $P(0|1) = P(1|0)$ and verify the bit error probability expression in (5.51).

36. Consider a uniformly quantised signal with 8 bits over a noisy channel. What probability of bit error can be tolerated if the reconstructed signal is to have an SNR of 50 dB?

37. A stationary random process has an autocorrelation function given by $R_X(\tau) = A^2 e^{-|\tau|} \cos(2\pi f_0 \tau)$ where A is limited to ±4.

   a. When a μ-law compander is used, show that the output signal-to-quantization noise ratio becomes

$$SNR_Q = \frac{3 \times 2^{2R}}{[\ln(1+\mu)]^2}, \quad \mu^2 \gg PAPR.$$

   where R denotes the number of bits per sample and P denotes the average power of the message signal, whose amplitude is limited to ±A.
   b. Assuming $A = 4$, how many quantization levels are required to guarantee a signal to quantization noise power ratio of at least 50 dB?
   c. Assuming a companded PCM system with $R = 8$ bit and $\mu = 255$, calculate the SNR_Q in dB for very large and very small signal amplitudes and compare with the corresponding uniformly quantized PCM system.
   d. Repeat (c) for $R = 8$ bit and $\mu = 50$.

38. The ramp signal $x(t) = kt$ is applied to a delta modulator that operates with a sampling period T and step size Δ.

a. How should T and Δ be chosen so that slope-overload distortion does not occur.

b. Sketch the DM modulator output for the following three values of step size: $\Delta = 0.75\ kT$,

c. $\Delta = kT$, and $\Delta = 1.25\ kT$.

d. Discuss the implications of choosing larger/smaller values of T and Δ as long as the ratio Δ/T remains constant.

39. The signal waveform shown below is to be transmitted using DM at 10 kbps using a slope of ±5 volts/sec.

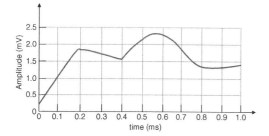

a. Determine the value of Δ.

b. Determine the DM bit stream.

c. Draw the waveform that will appear at the output of the DM decoder prior to low-pass filtering.

40. Consider a delta-modulation system with a first-order predictor, defined as $\hat{x}[n] = w_1 x[n-1]$ where $x[n]$ denotes the n-th sample of a stationary signal of zero mean and the tap gain w_1 is a constant. Assume that the energy of an antipodal signal is described by $\sigma^2 = E\left[|x[n]|^2\right]$.

a. Determine the optimal value of w_1 which minimizes the prediction error in the mmse sense $\sigma_e^2 = E[|x[n] - \hat{x}[n]|^2]$ and find its minimum value.

b. Find the optimal prediction gain which is defined as the ratio of the average signal power to the minimum variance of the prediction error.

c. If the variance of the prediction error is to be less than or equal to σ^2, determine the condition that the channel tap must satisfy.

41. Solve the three-tap equalization problem in Example 5.13 using MSE approach, where the three tap coefficients are required to minimize the mean square value of the error (MSE) between the kth equalizer output sample y_k and the kth data symbol, a_k, that is, $\sigma_E^2 = E\left[|y_k - a_k|^2\right]$. Determine the tap coefficients and the minimum MSE for $h_0 = 1$ and $h_1 = 0.3$, and $\sigma^2 = 0.01$. Compare your results with those found in Example 5.13.

42. Consider a linear predictor with single tap.

a. For optimal single-tap prediction, determine the tap weight if the signal-to-prediction noise power ratio is 28 dB.

b. If the signal-to-quantization noise ratio should be larger than 46 dB and peak-to-average power ratio is 10 dB, determine the required number of bits per sample.

43. Assuming that a voice signal can be approximated by a sinusoid of frequency $f = 800$ Hz, determine the sampling frequency required by DM for 256 quantization levels to meet the slope overload condition. Compare the sampling frequency you found with that required for PCM. Can DM provide an alternative to PCM?

44. Determine the tap gains, the minimum error variance, and the processing gain in dB for DPCM with a single-tap and two-tap prediction filter when the input signal is characterized by the autocorrelation $R(0) = 1$, $R(1) = 0.9$ and $R(2) = 0.6$.

6

Baseband Transmission

In Chapter 5 we considered the conversion of analog signals into digital by an ADC and baseband modulation techniques (see Figure 5.1). The line codes used for symbol transmission require infinite transmission bandwidth, as shown in Figure 5.26. Since frequency spectrum is a scarce resource, its economic use is of paramount importance. Therefore, the frequency components with negligible power of the transmitted signals are filtered out at the transmitter output.

On the other hand, in addition to the intended copy of the transmitted signal, its echos with different amplitude, phase and delay also arrive to the receiver. Hence, the original signal may be spread in time and frequency-domains at the receiver input. On the other hand, the receiver filters out signals and noise outside its bandwidth for mitigating interference and noise. Therefore, the receiver should be able to cope with all these inconveniences to estimate the received pulses with minimum error.

This chapter will address the issues related to the reception of pulses at the receiver after transmission through a linear baseband channel.

6.1 The Channel

Let us consider the transmission of a pulse $g(t)$ through a linear channel, as shown by Figure 6.1, characterized by its impulse response h(t) and the corresponding channel transfer function $H(f)$. The signal $x(t)$ at the channel ouput may be expressed as the convolution of the input pulse $g(t)$ and the channel impulse response $h(t)$. The output frequency response $X(f)$ is the given by the product of the frequency responses $G(f)$ and $H(f)$. [1][2][3][4][5]

Now let us assume that the channel impulse response is described by

$$h(t) = \alpha \delta(t - \tau) \qquad (6.1)$$

where α denotes the channel gain and the time-delay τ refers to the difference between the time the signal is injected into the channel and the

Digital Communications, First Edition. Mehmet Şafak.
© 2017 John Wiley & Sons Ltd. Published 2017 by John Wiley & Sons Ltd.
Companion website: www.wiley.com/go/safak/Digital_Communications

time it is observed at the channel output. The signal at the channel ouput is then given by

$$x(t) = h(t) \otimes g(t) = \alpha g(t-\tau) \quad (6.2)$$

The signal $x(t)$ at the channel output, is the same as the transmitted pulse except for a finite time delay, which corresponds to the time required by the signal to go through the channel, and the channel gain, which accounts for the signal attenuation in the channel. Hence, $x(t)$ has complete information of the transmitted pulse $g(t)$ since its characteristics are preserved at the channel output. This channel is said to be ideal distortionless. In addition, the channel is

said to be linear time-invariant (LTI) if the channel gain α and the time-delay τ do not vary with time. Taking the Fourier transform of both sides of (6.2) leads to

$$X(f) = H(f)G(f) = \alpha G(f) e^{-j2\pi f \tau} \quad (6.3)$$

It is clear from (6.3) that the output spectrum $X(f)$ is the same as that of the transmitted pulse except for a frequency-dependent phase shift due to delay τ and a constant channel gain α, which shows that all input frequency components are amplified/attenuated equally. Since time-limited signals are not bandlimited, a channel providing ideal distortionless transmission requires infinite bandwidth (see (6.1) and Figure 6.2).

Now let us consider signal transmission through a low-pass channel (with finite bandwidth). A low-pass channel may be approximated by an ideal low-pass filter (LPF) with a phase $\exp(-jw\tau)$, accounting for the time delay in the channel. However, as shown in Figure 6.3 (see Figure 1.7), an ideal LPF

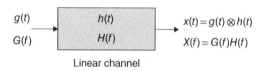

Figure 6.1 Transmission of a Pulse Through a Linear Channel.

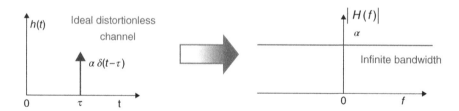

Figure 6.2 Infinite Bandwidth Requirement of an Ideal Distortionless Channel.

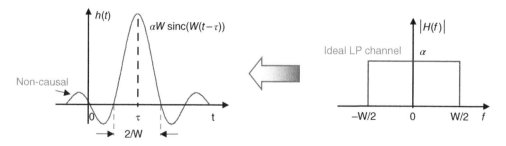

Figure 6.3 Non-Causality in Ideal Low-Pass Channel.

channel provides a finite bandwidth but it is not causal. The non-causality results from the fact that the bandlimited low-pass channel starts responding before a signal is being applied at its input, that is, for $t < 0$. Therefore an ideal low-pass channel is not realistic.

A channel whose frequency response deviates from that of an ideal LPF is dispersive. This implies that different frequency components of a pulse propagating through a dispersive channel are neither amplified/attenuated equally nor arrive with identical time delays. A dispersive channel causes intersymbol interference (ISI), that is, the interference between neighboring pulses. As we shall study in the following Sections, several techniques, for example, pulse shaping and equalization, are used to mitigate the ISI.

It should be clear from the above discussions that there is a trade-off between the bandwidth and the duration of the impulse response (see Figure 6.3). Therefore, the non-causality issue can be largely overcome in a low-pass channel with a relatively flat bandwidth larger than W so as to allow undistorted transmission of the essential frequency components of a pulse. This corresponds to a channel impulse response which fades away more quickly in the time-domain. Hence, pulse transmission through a low-pass channel is affected not only by the channel bandwidth but also by the pulse bandwidth, $W \cong 1/T$, where T denotes the pulse duration. Therefore, low-pass channels with sufficiently large and relatively flat passband and decaying rapidly for $|f| > W/2$ provide a good compromise between ISI and the bandwidth requirements.

Note that a channel may also be dispersive, that is, an ISI channel, when the bandwidth of the pulse $g(t)$ is wider than that of the low-pass channel. Then, essential frequency components of the transmitted pulse remaining outside the channel bandwidth are filtered-out by the channel. Hence, the pulse at the channel output becomes spread in time and causes ISI to preceding and following pulses. To have a better idea about the ISI imposed by the channel bandwidth,

consider a pulse $g(t) = 1$ for $0 < t < T$ is applied at the channel input. The Fourier transform of this pulse $G(f)$, given by (5.10), is obviously not bandlimited. If this pulse goes through a channel, which is modeled by an ideal LPF with bandwidth $W, -W/2 \leq f \leq W/2$, then the low-pass channel will filter out the frequency components of the pulse outside the filter bandwidth. Consequently, the signal at the output is found by low-pass filtering $X(f)$ given by (6.3) with $G(f)$ from (5.10):

$$x_{filtered}(t) = \int_{-W/2}^{W/2} X(f) \, e^{j2\pi f t} \, df$$

$$= \alpha \int_{-W/2}^{W/2} \underbrace{T\,\mathrm{sinc}(fT)\,e^{-j\pi fT}}_{G(f)} \, e^{-j2\pi f \tau} e^{j2\pi f t} \, df$$

$$= \alpha \int_{-WT/2}^{WT/2} \mathrm{sinc}(x) \, \exp\left(-j\pi\left(1 - 2\frac{t-\tau}{T}\right)x\right) dx$$

$$(6.4)$$

Figure 6.4 shows the original pulse $g(t) = 1$ for $0 < t < T$ and the filtered pulse, given by (6.4), for various values of the WT product for $\alpha = 1$, $\tau = 0$ and $T = 1$. $WT = 1$ implies that the filter (channel) bandwidth is equal to null-to-null bandwidth of the transmitted pulse and the original pulse corresponds to $WT = \infty$, hence infinite channel bandwidth. Note that the filtered pulse is spread in time and causes ISI even for large bandwidths of the low-pass channel. Therefore, ISI is a serious problem in digital transmission systems.

A multipath channel may also lead to dispersion of the signals in the time-domain. A multipath channel refers to channel where multiple echos of the transmitted signal arrives to the receiver. The impulse response $h(t)$ of a L-path channel and the signal, $x(t)$, at the channel output may be written as

$$h(t) = \sum_{k=1}^{L} \alpha_k \, \delta(t - \tau_k)$$

$$(6.5)$$

$$x(t) = h(t) \otimes g(t) = \sum_{k=1}^{L} \alpha_k \, g(t - \tau_k)$$

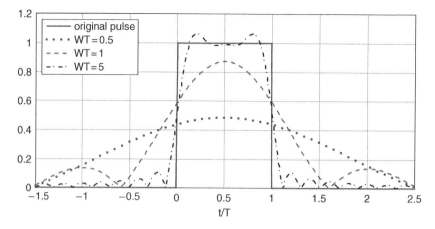

Figure 6.4   Time-Spreading of a Pulse Due to Band Limitation.

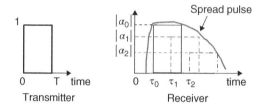

Figure 6.5   Time Spreading In a Multi-Path Channel with Three-Paths.

Note that $x(t)$ consists of weighted sum of replicas of the transmitted pulse $g(t)$ where α_k and τ_k denote, respectively, the complex gain and the delay of the kth path. Comparison of (6.1) and (6.5) shows that an ideal distortionless channel corresponds to signal propagation via a single path. As shown in Figure 6.5, the received pulse duration is spread in time and is longer than the original pulse duration. Consequently, the time-spread received pulses interfere with each other, hence causing ISI. As the value of the maximum delay $\max\{\tau_k\}$ increases in comparison with the pulse duration, the interference between neighboring pulses in time, becomes more serious. Unless the channel is LTI, α_k and τ_k vary with time due to changes in the channel characteristics. The receiver suffers from destructive/constructive interference of multipath components as the time rate of change of

the path gains increases in comparison with the pulse rate.

6.1.1   Additive White Gaussian Noise (AWGN) Channel

In the so-called AWGN channel, the signal at the channel output may be written as in (6.2):

$$x(t) = \alpha g(t - \tau) + w(t) \qquad (6.6)$$

where $w(t)$ denotes white Gaussian noise with zero-mean and single-sided PSD N_0 and accounts for the receiver noise of usually thermal origin. In this ideal distortionless channel, the time delay τ can be ignored if the time that the signal is received by the receiver $t = \tau$ is taken as the reference. Similarly, the channel

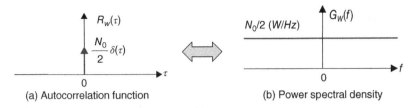

(a) Autocorrelation function   (b) Power spectral density

Figure 6.6 Autocorrelation Function and PSD of AWGN.

gain α can be omitted if the AWGN is scaled accordingly, so that the received signal-to-noise power ratio (SNR) remains the same. Consequently, a binary signal $x(t)$ at the receiver input in an AWGN channel will simply be written hereafter as

$$x(t) = Ag(t) + w(t), \quad 0 \le t \le T \quad (6.7)$$

In binary transmission, the sign A of the pulse amplitude is changed according to whether binary 1 or 0 is transmitted, for example, $A = +1$ for binary 1 and $A = -1$ for binary 0. Then, the receiver has the task of estimating the transmitted bit (1 or 0) corrupted by AWGN. The ability of the receiver to estimate the transmitted bit improves with increasing values of E/N_0, which is defined by the ratio of the pulse energy to the noise PSD N_0.

The autocorrelation function of AWGN is given by

$$R_W(\tau) \triangleq E[w(t)w^*(t+\tau)]$$

$$= E \left[\lim_{T \to \infty} \frac{1}{T} \int_{-T/2}^{T/2} w(t)w^*(t+\tau)\, dt \right]$$

$$= \lim_{T \to \infty} \frac{1}{T} \int_{-T/2}^{T/2} E[w(t)w^*(t+\tau)]\, dt = \frac{N_0}{2}\delta(\tau)$$

$$(6.8)$$

which implies that any two noise samples are uncorrelated unless they are sampled exactly at the same time. It is easy to show by Fourier transform relationship that the PSD of AWGN is flat, hence it is *white* (see Figure 6.6):

$$R_W(\tau) = \frac{N_0}{2}\delta(\tau) \quad \Leftrightarrow \quad G_W(f) = \frac{N_0}{2} \quad (6.9)$$

The AWGN is characterised by Gaussian pdf with zero mean and variance $N_0/2$:

$$f_W(n) = \frac{1}{\sqrt{2\pi}\sigma}\exp\left(-\frac{w^2}{2\sigma^2}\right)$$

$$(6.10)$$

$$E[w(t)] = 0$$

$$\sigma^2 = E[w^2(t)] = R_W(0) = N_0/2$$

Figure 6.7 shows pictorially how the Gaussian pdf characterizes the random time-variations in the AWGN.

In this chapter, the channel is characterized by AWGN, and dispersive, hence causing ISI. We will study the implications and devise methods to mitigate noise and ISI.

6.2 Matched Filter

Consider the detection by receiver of a pulse $g(t)$ transmitted over an AWGN channel. The receiver is modeled as LTI filter of impulse response $c(t)$. The filter input, $x(t)$, is the output of the AWGN channel

$$x(t) = g(t) + w(t), \quad 0 \le t \le T \quad (6.11)$$

where T denotes the pulse duration. Here, the transmitted pulse, $g(t)$, which is an energy signal,

$$E = \int_0^T |g(t)|^2 dt = \int_{-\infty}^{\infty} |G(f)|^2 df \quad (6.12)$$

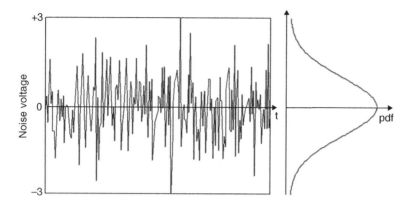

Figure 6.7   Pdf of the AWGN with Zero Mean and Unity Variance.

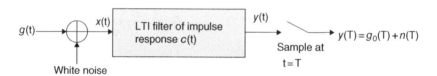

Figure 6.8   Matched Filter Receiver.

is corrupted by $w(t)$, which denotes the sample function of a white noise process of zero mean and double-sided PSD $N_0/2$. The receiver is assumed to have the knowledge of the signal pulse shape $g(t)$. In this Section, we will focus on the design of a receiver filter of impulse response $c(t)$ which detects the pulse, $g(t)$, in an optimum manner in an AWGN channel (see Figure 6.8). Here, optimum manner implies the maximization of the SNR at the receiver output at the sampling time $t = T$.

The signal at the matched-filter receiver output is the convolution of $x(t)$ and $c(t)$:

$$y(t) = x(t) \otimes c(t) = [g(t) + w(t)] \otimes c(t)$$

$$= \underbrace{g(t) \otimes c(t)}_{g_0(t)} + \underbrace{w(t) \otimes c(t)}_{n(t)} \qquad (6.13)$$

The signal at the receiver output is sampled at time $t = T$, to estimate the transmitted signal. For a successful detection, we desire to maximize the received signal-to-noise ratio:

$$SNR = \frac{|g_0(T)|^2}{E[n^2(T)]} \qquad (6.14)$$

where $|g_0(T)|^2$ denotes the output signal power, and $E[n^2(T)]$ is the average output noise power at the sampling time $t = T$. Hence, the problem reduces to specifying $c(t)$ which maximizes the output SNR.

Because $g_0(t)$ is the convolution of $g(t)$ and $c(t)$, the output signal power at $t = T$ may be written in terms of the Fourier transforms of $c(t)$ and $g(t)$ as:

$$|g_0(T)|^2 = \left| \int_{-\infty}^{\infty} C(f)\, G(f) \exp(j2\pi fT)\, df \right|^2 \qquad (6.15)$$

On the other hand, the noise PSD and the average noise power at the matched filter output may be written as

$$E[n^2(T)] = \frac{N_0}{2} \int_{-\infty}^{\infty} |C(f)|^2\, df \qquad (6.16)$$

Inserting (6.15) and (6.16) into (6.14), the SNR may be expressed as:

$$SNR = \frac{\left| \int_{-\infty}^{\infty} C(f)\, G(f)\, \exp(j2\pi fT)\, df \right|^2}{\frac{N_0}{2} \int_{-\infty}^{\infty} |C(f)|^2\, df}$$

(6.17)

For a given a pulse shape $g(t)$, hence $G(f)$, the SNR in (6.17) can be maximized by determining the optimum frequency response $C_{opt}(f)$ of the matched filter. In order to maximize SNR in (6.17), we use Schwartz's inequality for two finite-energy signals $u_1(t)$ and $u_2(t)$:

$$\left| \int_{-\infty}^{\infty} u_1(x)\, u_2(x)\, dx \right|^2 \le \int_{-\infty}^{\infty} |u_1(x)|^2 dx \int_{-\infty}^{\infty} |u_2(x)|^2 dx$$

(6.18)

Equality holds if and only if

$$u_1(x) = k u_2^*(x)$$

(6.19)

where k is an arbitrary constant. Applying Schwartz's inequality to the numerator of (6.17), the SNR may be written as

$$SNR = \frac{2}{N_0} \frac{\left| \int_{-\infty}^{\infty} C(f)\, G(f)\, e^{j2\pi fT}\, df \right|^2}{\int_{-\infty}^{\infty} |C(f)|^2 df} \le \frac{2E}{N_0}$$

(6.20)

where E, the energy of the pulse $g(t)$, is defined by (6.12). The equality in (6.20) is satisfied if $C(f)$ is chosen to be optimum:

$$C_{opt}(f) = k G^*(f) \exp(-j2\pi fT) \;\Leftrightarrow$$
$$c_{opt}(t) = k g(T-t)$$

(6.21)

Then, the SNR is maximized:

$$SNR_{max} = \frac{2}{N_0} \int_{-\infty}^{\infty} |G(f)|^2 df = \frac{2E}{N_0}$$

(6.22)

Note from (6.22) that the maximum SNR SNR_{max} at the matched filter output depends only on the energy E of $g(t)$ but not on its shape. This implies that all pulse waveforms having the same energy yield the same maximum SNR irrespective of their shapes. The pulse shape $g(t)$ is chosen based on other considerations.

For example, the optimal impulse response $c_{opt}(t)$ for a receiver filter matched to a pulse $g(t)$ is shown in Figure 6.9. If $c(t)$ is chosen to be optimum as in (6.21), then the optimum filter is said to be matched to the input signal, hence the matched-filter.

The signal at the matched filter output may then be written as

$$G_0(f) = C_{opt}(f)\, G(f) = k|G(f)|^2 \exp(-j2\pi fT)$$

$$g_0(T) = \int_{-\infty}^{\infty} G_0(f)\, \exp(j2\pi fT)\, df$$

$$= k \int_{-\infty}^{\infty} |G(f)|^2 df = kE$$

(6.23)

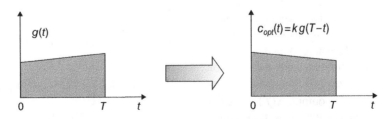

Figure 6.9   Matching of the Receiver Filter to the Input Signal g(t).

Similarly, the average output noise power at the matched filter output is given by (6.16):

$$E\left[n^2(t)\right] = \frac{k^2 N_0}{2} \int_{-\infty}^{\infty} |G(f)|^2 df = k^2 E \frac{N_0}{2}$$

(6.24)

The energy of the matched filter receiver is usually normalized to unity:

$$\int_{-\infty}^{\infty} |C_{opt}(f)|^2 df = k^2 \int_{-\infty}^{\infty} |G(f)|^2 df$$

$$= k^2 E = 1 \quad \Rightarrow \quad k = 1/\sqrt{E}$$

(6.25)

This implies that the matched filter output at $t = T$ may be rewritten as

$$y(T) = g_0(T) + n(T)$$

$$g_0(T) = kE = \sqrt{E}$$

(6.26)

$$\sigma_n^2 = E[n^2(T)] = k^2 E \frac{N_0}{2} = \frac{N_0}{2}$$

One may infer from (6.26) that the matched filter output is a Gaussian random variable with mean value $\sqrt{E}$ and variance $N_0/2$.

Matched-filter reception of a single pulse $g(t)$ can be extended to receive binary 1 and 0 in a digital communication system. For example, signal waveforms $s_1(t) = g(t)$ and $s_2(t) = -g(t)$ may be used to transmit binary 1 and 0, respectively. The signals satisfying $s_2(t) = -s_1(t)$ are said to be antipodal. For antipodal signals, the impulse response of the matched-filter receiver may simply be taken as $s_1(T-t)$. The reception of binary signals may then be analyzed as above by simply noting that $E = A^2 T$, where A and T denote, respectively, the amplitude and the duration of the pulse. Consequently, the positive sign of the matched filter output, $y(T) = \pm \sqrt{E} + w(T)$, simply denotes that 1 is transmitted, and zero otherwise.

Example 6.1   Matched Filter for a Rectangular Pulse.

In the special case where $g(t)$ is a rectangular pulse, then $g_0(t)$ may be written as the convolution integral:

$$g_0(t) = g(t) \otimes c_{opt}(t) = g(t) \otimes k g(T-t)$$

$$= \begin{cases} k \int_0^t g(\tau) g^*(T-t+\tau) \, d\tau = kA^2 t & t \le T \\ k \int_{t-T}^T g(\tau) g^*(T-t+\tau) \, d\tau = kA^2(2T-t) & t \ge T \end{cases}$$

$$g_0(T) = k \int_{-\infty}^{\infty} |g(\tau)|^2 \, d\tau = kE = \sqrt{E}$$

(6.27)

The plot of (6.27) in Figure 6.10c clearly shows that a simple integrator acts as a matched filter receiver if $g(t)$ is a rectangular pulse, since $g_0(t)$ is simply given by the integral of $g(t)$ for $t \le T$. The matched filter output is sampled at $t = T$, and the rest is dumped as shown in Figure 6.10d. Therefore, the matched filter receiver for a rectangular pulse may be realized by an integrate-and-dump circuit (see Figure 6.10a).

6.2.1   Matched Filter Versus Correlation Receiver

Consider a correlation receiver which correlates the input signal with a normalized replica of the pulse, as shown in Figure 6.11. Here, as well, we choose $k = 1/\sqrt{E}$ as in (6.25) so that the input signal is correlated with a unit-energy reference signal. The signal and the noise variance at the output of a correlation receiver are given by, at time $t = T$:

$$g_0(T) = k \int_0^T g^2(\tau) d\tau = kE = \sqrt{E}$$

$$E\left[|n(T)|^2\right] = E\left[\left|k \int_0^T w(t) g(t) dt\right|^2\right]$$

(6.28)

$$= k^2 E \frac{N_0}{2} = \frac{N_0}{2}$$

(a) Integrate-and-dump circuit

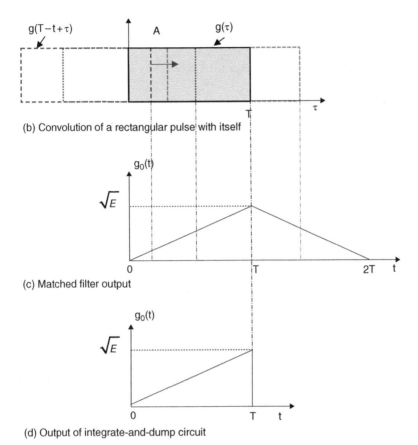

(b) Convolution of a rectangular pulse with itself

(c) Matched filter output

(d) Output of integrate-and-dump circuit

Figure 6.10   Matched Filter for a Rectangular Input Pulse.

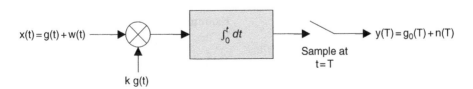

Figure 6.11   Correlation Receiver.

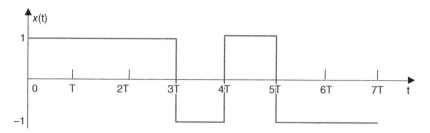

Figure 6.12   Data Sequence to be Received by a Matched Filter.

Comparison of (6.26) and (6.28) shows that correlation receiver output is identical to matched filter output at the sampling time $t = T$.

Example 6.2   Matched Filter Reception of a Data Sequence.
Consider a signal $x(t)$, which consists of a sequence of seven bits, at the input of a matched filter receiver. The signal $x(t)$ shown in Figure 6.12 may be expressed in terms of the transmit pulse $g(t)$ as follows:

$$x(t) = \sum_{k=0}^{6} b_k\, g(t-kT) + w(t),$$

$$b_0 = b_1 = b_2 = b_4 = 1, \quad b_3 = b_5 = b_6 = -1$$

$$(6.29)$$

where $w(t)$ denotes AWGN with zero-mean and variance $N_0/2$. The transmit pulse and its energy are given by

$$g(t) = \begin{cases} 1 & 0 \leq t \leq T \\ 0 & otherwise \end{cases} \qquad (6.30)$$

$$E = \int_0^T |g(t)|^2 dt = T$$

Impulse response of the matched filter receiver is given by

$$c(t) = \frac{g(T-t)}{\sqrt{E}} \qquad (6.31)$$

The matched filter output is given by

$$x(t) \otimes h(t) = \frac{1}{\sqrt{E}} \sum_{k=0}^{6} b_k\, g(t-kT) \otimes g(T-t+kT)$$

$$+ \frac{1}{\sqrt{E}} w(t) \otimes g(T-t+kT)$$

$$(6.32)$$

The matched filter output when sampled at $t = kT$, is equal to the square root of the energy, $E = T$, of the rectangular pulse $g(t)$. Hence, the sampled matched filter output may be written as

$$x(t) \otimes h(t)\big|_{t=kT} = \sqrt{E} \sum_{k=0}^{6} b_k\, \delta(t-kT) + n(T)$$

$$(6.33)$$

where $n(T)$ is shown to be Gaussian with zero-mean and variance $N_0/2$. Therefore, the samples of the matched filter output are Gaussian random variables with mean $\sqrt{E}$ and variance $N_0/2$. The matched filter output for this data sequence is shown in Figure 6.13, where the noise is ignored and the impulses shown at $t = kT$, $k = 0, 1, \ldots, 7$, represent the mean values of the matched filter output.

Example 6.3   Discrete-Time Matched Filter.
In this example, we consider the reception of a discrete-time signal which consists of $L+1$ uniform samples, $x_k = g_k + w_k$, $k = 0, 1, 2, \ldots, L$, at the input of a receiver; each of these $L+1$ samples last for T seconds. A finite-impulse response (FIR) filter with $L + 1$ taps with tap gains c_k, $k = 0, 1, 2, \cdots, L$ is used for the reception of this discrete-time

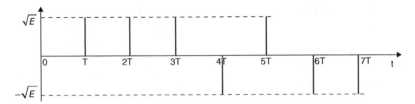

Figure 6.13   Matched Filter Output, Ignoring Noise, For the Data Sequence Shown in Figure 6.12.

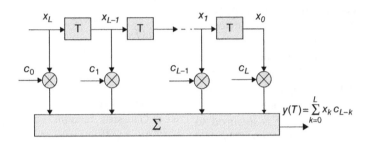

$$y(T) = \sum_{k=0}^{L} x_k c_{L-k}$$

Figure 6.14   A FIR Filter Receiver.

signal (see Figure 6.14). The signal at the FIR filter output may be written as the convolution of the input signal and the impulse response of the FIR filter. Hence, the output of the FIR filter at the sampling instant $t = T$, is given by the discrete convolution of x_k and c_k:

$$y(T) = \sum_{k=0}^{L} x_k c_{L-k} = \sum_{k=0}^{L} (g_k + w_k) c_{L-k} \quad (6.34)$$

From (6.34), the SNR at the matched filter ouput may simply be written as

$$SNR = \frac{\left| \sum_{k=0}^{L} c_k g_{L-k} \right|^2}{E\left[\left| \sum_{k=0}^{L} c_k w_{L-k} \right|^2 \right]} \quad (6.35)$$

$$\leq \frac{\sum_{k=0}^{L} |c_k|^2 \sum_{k=0}^{L} |g_k|^2}{\sum_{k=0}^{L} \sum_{\ell=0}^{L} c_k c_\ell^* \underbrace{E\left[w_{L-k} w_{L-\ell}^* \right]}_{N_0 \delta(k-\ell)/2}}$$

$$= \frac{2}{N_0} \sum_{k=0}^{L} |g_k|^2 = \frac{2E}{N_0}$$

In (6.35) discrete version of the Schwartz's inequality given by (6.18) is employed. The SNR is maximized and the optimum impulse response of the matched filter is obtained for $c_k = g_{L-k}^*$, which is the discrete-time version of the optimality condition, $c(t) = g^*(T-t)$. This implies that the samples of the channel impulse response are matched to the corresponding input samples. The matched filter then collects all the energy $E = \sum_{k=0}^{L} |g_k|^2$ contained in $L+1$ samples of the input signal.

6.2.2   Error Probability For Matched-Filtering in AWGN Channel.

Consider binary antipodal signaling where a binary symbol m_1 is represented by $+g(t)$, and a binary symbol m_0 by $-g(t)$. Note that m_1 and m_0 may represent binary symbols 1 and 0. It is assumed that the pulse $g(t)$ has a duration T_b and an arbitrary shape, which is known to both transmitter and receiver. In an AWGN channel, the signal at the matched-filter receiver input may be written as

$$x(t) = \begin{cases} +g(t) + w(t) & \text{if } m_1 \text{ is sent} \\ -g(t) + w(t) & \text{if } m_0 \text{ is sent} \end{cases} \quad 0 \le t \le T_b$$

(6.36)

where the noise $w(t)$ has zero mean and a PSD of $N_0/2$. The receiver consists of a matched filter, a sampler and a decision device as shown in Figure 6.15. Perfect synchronization is assumed between the transmitter and the receiver. This implies that the receiver is assumed to know the starting and ending times of transmitted pulses and the pulse shape, but not the polarity. The receiver is assumed to be matched to the transmit pulse $g(t)$. Matched filter output is sampled at the end of each signalling interval. Given the noisy signal $x(t)$ at

its input, the receiver is required to make a decision, at the end of each signalling interval as to whether the transmitted symbol is m_1 or m_0. [1][3][4]

The sampled signal and the noise variance at the output of the matched filter is given by (6.26)

$$y(T_b) = g_0(T_b) + n(T_b)$$

$$g_0(T_b) = \begin{cases} +\sqrt{E_b} & \text{if } m_1 \text{ is sent} \\ -\sqrt{E_b} & \text{if } m_0 \text{ is sent} \end{cases}$$

(6.37)

$$\sigma_n^2 = E[n^2(T_b)] = N_0/2$$

In view of (6.37), the matched filter output is a Gaussian random variable with mean $g_0(T_b) = \pm \sqrt{E_b}$, where E_b denotes the bit energy and the noise variance is $\sigma_n^2 = N_0/2$ (see Figure 6.16). Hence, the conditional pdf's

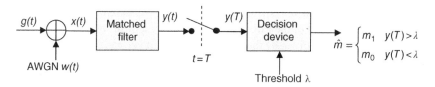

Figure 6.15   Receiver for Baseband Binary Antipodal Signals.

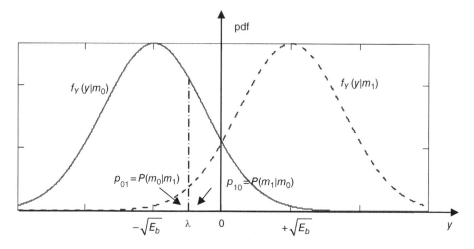

Figure 6.16   Pdf of the Matched Filter Output When m_1 or m_0 is Transmitted.

of the matched-filter output $y(T_b)$, when binary m_1 and binary m_0 are sent, are given by

$$f_Y(y|m_1) = \frac{1}{\sqrt{2\pi}\,\sigma_n} \exp\left(-\frac{(y-\sqrt{E_b})^2}{2\,\sigma_n^2}\right)$$

$$f_Y(y|m_0) = \frac{1}{\sqrt{2\pi}\,\sigma_n} \exp\left(-\frac{(y+\sqrt{E_b})^2}{2\,\sigma_n^2}\right)$$

$$(6.38)$$

Given the matched-filter output (6.37), the decision device makes a decision as to whether m_1 or m_0 was transmitted, in the presence of the AWGN. The decision device compares the sample value $y(T_b)$ to a preset threshold level λ:

$$y(T_b) > \lambda \quad m_1 \text{ is sent}$$
$$(6.39)$$
$$y(T_b) < \lambda \quad m_0 \text{ is sent}$$

Note that if there is no noise in the channel, then the pdfs in (6.38) reduce to delta functions:

$$f_Y(y|m_1) = \delta(y - \sqrt{E_b})$$
$$\text{as } \sigma_n^2 \to 0 \quad (6.40)$$
$$f_Y(y|m_2) = \delta(y + \sqrt{E_b})$$

Since the matched filter output can then assume only one of two values, $\pm\sqrt{E_b}$, transmission is error-free and the decision device makes no errors, as to whether binary symbols m_1 or m_0 was sent. However, in the presence of AWGN, the noise sample at the sampling instant, $n(T_b)$, may change the sign of $y(T_b)$ in (6.37) so that the decision device may make an erroneous decision. The decision error can be one of two kinds; the symbol m_1 is chosen when m_0 is transmitted, which is denoted by the probability p_{10}, or symbol m_0 is chosen when m_1 is transmitted with probability p_{01} (see Figure 6.16).

Conditional probability of error p_{10}, when symbol 0 was transmitted, is found as follows:

$$p_{10} = P(y > \lambda|m_0) = \int_\lambda^\infty f_Y(y|m_0)\,dy$$

$$= \frac{1}{\sqrt{2\pi}\,\sigma_n} \int_\lambda^\infty \exp\left(-\frac{(y+\sqrt{E_b})^2}{2\,\sigma_n^2}\right)dy$$

$$(6.41)$$

$$= \frac{1}{\sqrt{2\pi}} \int_{\frac{\lambda+\sqrt{E_b}}{\sigma_n}}^\infty \exp\left(-\frac{t^2}{2}\right)dt$$

$$= Q\left(\sqrt{\frac{2E_b}{N_0}}\left(1 + \frac{\lambda}{\sqrt{E_b}}\right)\right)$$

where Q is the so-called Gaussian Q function (see (B.1)):

$$Q(x) \triangleq \frac{1}{\sqrt{2\pi}} \int_x^\infty \exp\left(-\frac{t^2}{2}\right)dt \quad (6.42)$$

Similarly, the conditional probability of error p_{01} when symbol m_1 was transmitted, is given by

$$p_{01} = P(y < \lambda|m_1) = \int_{-\infty}^\lambda f_Y(y|m_1)\,dy$$

$$(6.43)$$

$$= Q\left(\sqrt{\frac{2E_b}{N_0}}\left(1 - \frac{\lambda}{\sqrt{E_b}}\right)\right)$$

Since there are two possible kinds of mutually exclusive errors, the average probability of bit error is given by

$$P_2 = p_0 p_{10} + p_1 p_{01}$$

$$= p_0\, Q\left(\sqrt{\frac{2E_b}{N_0}}\left(1 + \frac{\lambda}{\sqrt{E_b}}\right)\right) \quad (6.44)$$

$$+ p_1\, Q\left(\sqrt{\frac{2E_b}{N_0}}\left(1 - \frac{\lambda}{\sqrt{E_b}}\right)\right)$$

where p_0 and $p_1 = 1 - p_0$ denote respectively the a priori probability of transmitting m_0 and m_1. In most applications, a priory probabilities of transmitting 1 and 0 are equal to each other, that is, $p_0 = p_1 = 0.5$.

We now focus on determining the optimum value of the threshold λ in (6.44) which minimizes the average bit error probability. To achieve

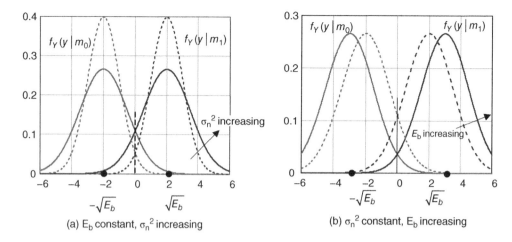

(a) E_b constant, σ_n^2 increasing   (b) σ_n^2 constant, E_b increasing

Figure 6.17 Effect of Increasing Noise Variance σ_n^2 and the Energy Per Bit E_b on the Average Bit Error Probability for a Matched Filter Receiver.

this, we differentiate P_2 with respect to λ and equate to zero. Using the derivative of the Gaussian Q function

$$\frac{d}{du}Q(u) = \frac{d}{du}\left\{\frac{1}{\sqrt{2\pi}}\int_u^\infty \exp\left(-z^2/2\right)\,dz\right\}$$

$$= -\frac{1}{\sqrt{2\pi}}\exp\left(-u^2/2\right)$$

(6.45)

the optimum value of the threshold is found to be

$$\frac{\partial}{\partial\lambda}P_2(\lambda) = 0 \;\Rightarrow\; \lambda_{opt} = \frac{N_0}{4\sqrt{E_b}}\ln\left(\frac{p_0}{p_1}\right) \quad (6.46)$$

When symbols m_1 and m_0 are equiprobable, $p_0 = p_1 = 1/2$ (the so-called binary symmetric channel, BSC), then $\lambda_{opt} = 0$. This implies that the optimum threshold is located at the midpoint between $-\sqrt{E_b}$ and $\sqrt{E_b}$, representing binary symbols m_2 and m_1. Then $p_{01} = p_{10}$, and the average probability of bit error reduces to

$$P_2 = Q\left(\sqrt{\frac{2E_b}{N_0}}\right) \quad (6.47)$$

which depends only on $\gamma_b = E_b/N_0$. Figure 6.17 shows that decreasing the noise variance and increasing the bit energy have similar effects in decreasing p_{10} and p_{01}. Therefore, the average bit error probability depends only on γ_b, which is defined as the ratio of the bit energy to the noise variance. As the noise variance decreases and/or bit energy increases, the overlap between the tails of the two Gaussian pdf's become less; this leads to lower bit error probabilities.

Figure 6.18 shows the variation of the bit error probability with $\gamma_b = E_b/N_0$. The Gaussian Q function, which is defined by (6.42), represents the area under the tail (between x and ∞) of a zero-mean and unit variance Gaussian pdf (see Appendix B). $Q(x)$ is a monotonically decreasing function of x with $x(-\infty) = 1$, $x(0) = 0.5$, and $Q(\infty) = 0$. Therefore the receiver exhibits an exponential improvement in P_2 with increase in γ_b:

$$P_2 = Q\left(\sqrt{2\gamma_b}\right) \cong \frac{1}{\sqrt{2\pi(1+2\gamma_b)}}\exp(-\gamma_b)$$

(6.48)

which implies that even a slight increase in γ_b leads to significant reduction in the bit error

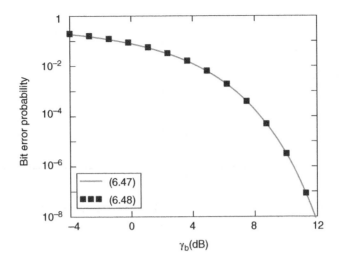

Figure 6.18 Gaussian Q Function, Defined by (6.47) and its Approximation Given by (6.48).

probability. For example, $P_2 = 1.91 \ 10^{-4}$ for $\gamma_b = 8$ dB but reduces to $3.4 \ 10^{-5}$ for $\gamma_b = 9$ dB.

Example 6.4 MAP Versus ML Detection
The decision rule given by (6.39) can be formulated in a more elegant form from a different perspective. Based on the matched-filter output $y(T_b)$ at $t = T_b$, the decision circuit decides that

$$m_1 \text{ is sent if } P(\hat{m} = m_1|y(T_b)) > P(\hat{m} = m_0|y(T_b))$$
$$m_0 \text{ is sent if } P(\hat{m} = m_1|y(T_b)) < P(\hat{m} = m_0|y(T_b))$$

$$(6.49)$$

where m_1 and m_0 denote the binary symbols 1 and 0 and $\hat{m}$ is the estimation of the transmitted symbol at the receiver output. For the sake of brevity, we use y for $y(T_b)$ hereafter. The so-called MAP (maximum a posteriori) decision rule given by (6.49) may also be expressed as follows: [4]

$$L(m|y) = \frac{P(\hat{m} = m_1|y)}{P(\hat{m} = m_0|y)} \overset{\hat{m} = m_1}{\underset{\hat{m} = m_0}{\gtrless}} 1 \qquad (6.50)$$

where the likelihood ratio $L(m|y)$ can be used by the decision device as a test for estimating whether m_1 or m_2 is transmitted.

Given the a priori probabilities $p_1 = p(m_1)$ and $p_0 = p(m_2)$, and using the Bayes theorem, $f_Y(y|m_i)p(m_i) = P(\hat{m} = m_i|y)f_Y(y)$, $i = 0,1$, the MAP decoding rule may be rewritten as follows:

$$L(m|y) = \frac{P(\hat{m} = m_1|y)}{P(\hat{m} = m_0|y)} = \frac{p_1 \ f_Y(y|m_1)}{p_0 \ f_Y(y|m_0)} \overset{\hat{m} = m_1}{\underset{\hat{m} = m_0}{\gtrless}} 1$$

$$(6.51)$$

Here, $f_Y(y|m_i)$, $i = 0,1$ denotes the pdf of the signal $y(T_b)$ at the matched filter output, as given by (6.38), when m_i is transmitted. By inserting the pdf's given in (6.38), into the above equation, it is easy to show that the likelihood ratio test reduces to

$$y(T_b) \overset{m_1}{\underset{m_0}{\gtrless}} \lambda_{opt}, \quad \lambda_{opt} = \frac{N_0}{4\sqrt{E_b}} \ln\left(\frac{p_0}{p_1}\right) \quad (6.52)$$

which is identical to (6.46).

For $p_1 = p_0$, Figure 6.19 shows that, the likelihood of m_1 or m_0 being transmitted at any value of $y(T_b)$ at the matched-filter output may be determined from the ratio

$$L(m|y) = L(y|m) = \frac{f_Y(y|m_1)}{f_Y(y|m_0)}, \quad p_1 = p_0 \quad (6.53)$$

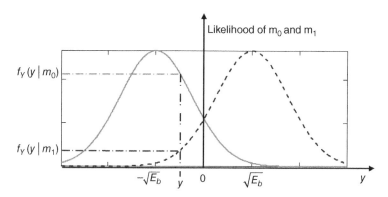

Figure 6.19 Likelihood of m_1 and m_0 at the Matched-Filter Output for Antipodal Signaling with Equiprobable Symbols with Bit Energies $\pm E_b$ in an AWGN Channel.

Note that if $f_Y(y|m_1) > f_Y(y|m_0)$, the likelihood of m_1 being transmitted is higher than that of m_0. Similarly, if $f_Y(y|m_1) < f_Y(y|m_0)$, the likelihood of m_0 being transmitted is higher than that of m_1. For $p_1 = p_0$, the optimum threshold level is 0, the mid-way between the mean values of the two pdf's. It is obvious from Figure 6.19 that the likelihood ratio is larger than unity if $y > 0$ and is smaller than unity if $y < 0$.

Example 6.5 Data Rate Versus E_b/N_0.
Consider antipodal binary signaling with a rectangular pulse of $A = \pm 1$ V amplitude and bit duration T_b. The system operates in an AWGN channel with one-sided PSD of $N_0 = 10^{-5}$ W/Hz. If the received signal is detected with a matched filter, we will determine the maximum bit rate that can be transmitted with a bit error probability of $P_2 < 10^{-3}$:

$$P_2 = Q\left(\sqrt{2E_b/N_0}\right) < 10^{-3} \;\Rightarrow$$

$$\frac{E_b}{N_0} > \frac{1}{2}(3.09)^2 = 4.78 \;\; (6.8\ dB)$$

$$\frac{E_b}{N_0} = \frac{A^2 T_b}{N_0} = \frac{1}{R_b\, N_0} > 4.78 \;\Rightarrow$$

$$R_b < \frac{1}{4.78 \times N_0} = 20.92\ kbps$$

(6.54)

We now determine the value of N_0 to support $R_{b1} = 100$ kbps, instead of 20.92 kbps, with $P_2 < 10^{-3}$. Since we still need to have $E_b/N_0 > 4.78$ for $P_2 < 10^{-3}$, the noise PSD is given by

$$N_0 < \frac{1}{4.78 R_{b1}} = 0.21 \times 10^{-5} \left(\frac{W}{Hz}\right) \quad (6.55)$$

To increase the bit rate from $R_b = 20.92$ kbps to $R_{b1} = 100$ kbps $(100/20.92 = 4.78$ fold-increase$)$ for a given E_b/N_0, the noise PSD should be decreased by the same factor. On the other hand, if we keep $N_0 = 10^{-5}$, $R_{b1} = 100$ kbps, and $P_b = 10^{-3}$, then the value of A^2 (proportional to transmit and receive powers) should be increased by the following:

$$\frac{A^2 T_b}{N_0} = \frac{A_1^2 T_{b1}}{N_0} \;\Rightarrow$$

$$\frac{A_1^2}{A^2} = \frac{T_b}{T_{b1}} = \frac{R_{b1}}{R_b} = \frac{100}{20.92} = 4.78 \;(6.8dB)$$

(6.56)

Example 6.6 Matched Filter Receiver for Correlated Signals.
Consider two baseband signals $s_1(t)$ and $s_0(t)$ which are used to transmit equiprobable symbols at a symbol rate of $1/T$ in an AWGN channel with one-sided noise PSD N_0. The

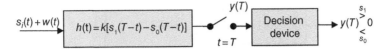

Figure 6.20   Matched Filter Receiver for Correlated Binary Input Signals.

correlation coefficient between the signals, with energies E, is defined by

$$\rho = \frac{1}{E} \int_0^T s_1(t) s_0(t)\, dt \qquad (6.57)$$

The block diagram of a matched filter receiver for these two correlated signals is shown in Figure 6.20. Because of the correlation between $s_1(t)$ and $s_0(t)$, the output can not be found simply by taking the difference between the outputs of two parallel filters matched to $s_1(t)$ and $s_0(t)$. The constant k which normalizes the energy of the matched filter is found as

$$k^2 \int_0^T [s_1(t) - s_0(t)]^2 dt$$

$$= k^2 \int_0^T \left[s_1^2(t) + s_0^2(t) - 2s_1(t)s_0(t) \right] dt$$

$$= 2k^2 E(1-\rho) = 1$$

$$k = \frac{1}{\sqrt{2E(1-\rho)}}$$

$$(6.58)$$

Assuming that $s_1(t)$ is transmitted, the matched-filter output is given by

$$y(t) = y_0(t) + n(t)$$

$$y_0(t) = s_1(t) \otimes k[s_1(T-t) - s_0(T-t)] \quad (6.59)$$

$$n(t) = w(t) \otimes k[s_1(T-t) - s_0(T-t)]$$

The matched-filter output at $t = T$ when $s_1(t)$ is transmitted:

$$y_0(T) = k \int_0^T s_1(\tau)[s_1(\tau) - s_0(\tau)]$$

$$= kE(1-\rho) = \sqrt{E(1-\rho)/2}$$

$$\sigma_n^2 = E\left[|n(T)|^2 \right] \qquad (6.60)$$

$$= k^2 \frac{N_0}{2} \int_0^T [s_1(T-t) - s_0(T-t)]^2 dt$$

$$= k^2 E(1-\rho)N_0 = N_0/2$$

Similarly, the matched-filter output at $t = T$ when $s_0(t)$ is transmitted:

$$y_0(T) = k \int_0^T s_0(t)[s_1(T-t) - s_0(T-t)]$$

$$= -\sqrt{E(1-\rho)/2} \qquad (6.61)$$

$$\sigma_n^2 = E\left[|n(T)|^2 \right] = \frac{N_0}{2}$$

Since $s_1(t)$ and $s_0(t)$ are equiprobable, $P(s_1) = P(s_0) = 0.5$, the optimum decision threshold is $\lambda_{opt} = 0$. The probability of bit error at the matched filter receiver output is given by

$$P_2 = P(y(T) < 0|s_1)P(s_1) + P(y(T) > 0|s_0)P(s_0)$$

$$= P(y(T) < 0|s_1) = P(y(T) > 0|s_0)$$

$$= P\left(n(T) > \sqrt{E(1-\rho)/2}\,|s_0 \right)$$

$$= \frac{1}{\sqrt{\pi N_0}} \int_{\sqrt{E(1-\rho)/2}}^{\infty} \exp\left(-n^2/N_0\right)\, dn$$

$$= Q\left(\sqrt{E(1-\rho)/N_0} \right)$$

$$(6.62)$$

Special case 1: Antipodal signaling
If $s_1(t)$ and $s_0(t)$ are antipodal, that is, $s_0(t) = -s_1(t)$, then the correlation coefficient between them is equal to $\rho = -1$ (see (6.57)). The impulse response of the matched filter then becomes $h(t) = 2k\, s_1(T-t) = s_1(T-t)/\sqrt{E}$ since $k = 1/(2\sqrt{E})$. This implies that a single filter matched to $s_1(t)$ is enough for reception of antipodal signals, as shown in Figure 6.15. In this case, the matched filter output at $t = T$ given by (6.60) and (6.61) reduces to (6.37). Correspondingly, the average bit error probability reduces to $Q\left(\sqrt{2E/N_0} \right)$ as in (6.47).

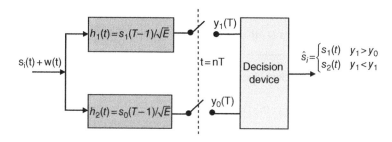

Figure 6.21   Matched Filter Receiver for Two Orthogonal Signals $s_1(t)$ and $s_0(t)$.

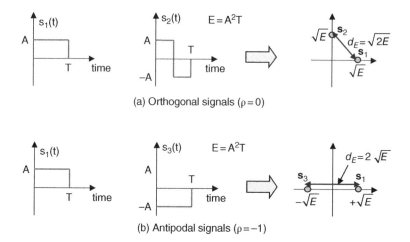

Figure 6.22   Antipodal and Orthogonal Signals and Their Geometric Representations.

Special case 2: Orthogonal signaling

The correlation coefficient between orthogonal signals is equal to $\rho = 0$. Similar to $s_1(t)$ and $s_0(t)$, the matched filter outputs are also independent of each other. Therefore, the matched filter receiver consists of two separate matched filters with impulse responses $s_1(T-t)/\sqrt{E}$ and $s_0(T-t)/\sqrt{E}$ as shown in Figure 6.21. From (6.62), the average bit error probability then reduces to $Q\left(\sqrt{E/N_0}\right)$. This implies that the bit error performance of orthogonal signaling is 3 dB worse compared to antipodal signaling.

Figure 6.22 shows typical examples of antipodal and orthogonal signals and their geometric representations at the matched filter output. The 3-dB performance advantage of antipodal

signaling over orthogonal signaling may be explained in terms of the Euclidean distances between the signals $s_1(t)$ and $s_0(t)$ in these two cases: [2]

$$d_E^2 = \int_0^T |s_1(t) - s_0(t)|^2 dt = \begin{cases} 4E & antipodal \\ 2E & orthogonal \end{cases}$$

(6.63)

Since the Euclidean distance between antipodal signals is $\sqrt{2}$ times larger than that for orthogonal signals, for a given value of signal energy E, the noise variance in antipodal signaling has to be twice as large as that for orthogonal signaling so as to cause the same bit error probability. In other words, orthogonal

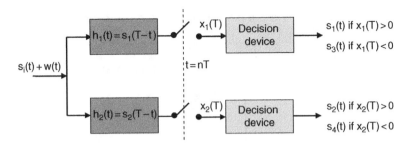

Figure 6.23 Matched Filter Receiver for Bi-Orthogonal Signals.

signaling is more susceptible to bit errors since the Euclidean distance between symbols is shorter compared to antipodal signaling. For the same noise variance, the bit energy for orthogonal signaling should be twice that for antipodal signaling in order to compensate for the shorter Euclidean distance between symbols.

In solving problems related to matched-filtering, one must first determine the degree of correlation between the input signals. If they are uncorrelated ($\rho = 0$), the signaling is said to be orthogonal; then, two parallel matched-filter receiver structure shown in Figure 6.21 is used. In case where the input binary signals are perfectly correlated with each other ($\rho = -1$), the signaling is said to be antipodal, and a single matched-filter receiver structure shown in Figure 6.15 is sufficient. If the input signals have a finite correlation between each other, then the matched filter receiver structure shown in Figure 6.20 is used. Note, however, that the matched-filter structure shown in Figure 6.20 may be used for the reception of both orthogonal and antipodal signals since it reduces to those shown in Figure 6.15 and Figure 6.21 respectively for antipodal and orthogonal signaling. Nevertheless, it is important to note that, in view of the dependence of k in (6.58) on the correlation coefficient, the receiver structure with two parallel matched-filters shown in Figure 6.21 gives incorrect results for correlated signals.

Homework 6.1 Repeat Example 6.6 when $P(s_1)$ and $P(s_0)$ are not equal to each other.

Homework 6.2 Consider a 4-ary signaling system with $s_3(t) = -s_1(t)$ and $s_4(t) = -s_2(t)$, where $s_1(t)$ and $s_2(t)$ are as shown in Figure 6.22. Show that the optimum receiver for this signaling scheme is given by Figure 6.23. Write down the expressions for $x_1(T)$ and $x_2(T)$, determine the noise variance and the output SNRs at the matched-filter output at $t = T$, when the transmitted signal is $s_i(t)$. Determine also the symbol error probability.

6.3 Baseband M-ary PAM Transmission

In binary PAM system ($M = 2$), equally likely and statistically independent bits 1 and 0 of duration T_b are represented by, for example, the amplitude levels $+1$ V and -1 V. The transmission rate is $R_b = 1/T_b$. The average transmit power level is 1 W over 1Ω resistance.

In quaternary PAM system ($M = 4$), the symbols, identified by dibits 00, 01, 11 and 10, are represented by one of the four amplitude levels, for example, -3 V, -1 V, $+1$ V and $+3$ V. Since the symbol duration is $T_s = 2T_b$, the symbol rate $R_s = R_b/2$ becomes one half of that for the binary PAM. However, the average transmit power level over 1Ω resistance is now $(1^2 + 3^2)/2 = 5$ W compared to 1 W for binary PAM, hence, 7 dB higher. Note that the selection of the 2 V difference between the amplitude levels in both 2-PAM and 4-PAM is

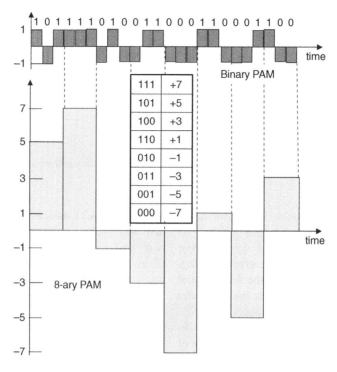

111	+7
101	+5
100	+3
110	+1
010	−1
011	−3
001	−5
000	−7

Figure 6.24   M-ary PAM with Gray Encoding.

not arbitrary. This selection ensures that, the minimum Euclidean distance and hence the error probability between neighboring amplitude levels are the same.

In M-ary PAM, $\log_2 M$ bits are first grouped and the groups of $\log_2 M$ bits are converted into an M-level PAM pulse train by using binary or Gray encoding. In M-ary PAM, the symbol duration and symbol rate are given by $T_s = T_b \log_2 M$ and $R_s = R_b/\log_2 M$, respectively. Hence, the symbol duration increases by a factor of $\log_2 M$ with respect to the binary PAM; the symbol rate (and the transmission bandwidth) decreases by the same factor.

Figure 6.24 depicts the representation of a binary bit sequence in binary and 8-ary PAM. In 2-PAM, binary 1's and 0's are represented by amplitude levels by +1 V and −1 V, respectively. In 8-PAM, Gray encoding is used to transmit the same bit sequence, grouped in 3-bit symbols, 101 111 010 011 000 110 001 100 by sending +5 V, +7 V, −1 V, −3 V, −7 V, +1

V, −5 V and +3 V, respectively. The average transmit power of this signal is $(5^2 + 7^2 + 1^2 + 3^2 + 7^2 + 1^2 + 5^2 + 3^2)/8 = 21$ W   (13.2   dBW). This implies that, in 8-PAM, the three-fold decrease in the symbol rate (transmission bandwidth) compared to the binary PAM is achieved at the expense of 13.2 dB increase in the transmit power.

M-ary PAM signal is corrupted by noise and distortion over an AWGN channel. The received signal is passed through a matched filter and sampled in synchronism with the transmitter. Each sample is compared with appropriate thresholds so as to decide which symbol was transmitted.

Let us first consider a 4-ary PAM system where the dibits 00, 01, 11 and 11 are represented by $\{a_m\} = -3$ V, −1 V, +1 V and +3 V, respectively, using Gray encoding. The signal transmitted by a 4-ary PAM system may then be written as

$$s_m(t) = a_m\, g(t), \quad a_m = \pm 1, \pm 3, \quad 0 \le t \le T_s$$

$$(6.64)$$

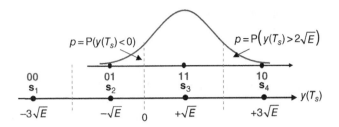

Figure 6.25 Decision Variable for the 4-PAM Signal Constellation with Gray Encoding.

where $T_s = 2T_b$ and the pulse $g(t)$ has energy E and duration T_s. The sampled output of a matched filter receiver may be written as (see (6.26))

$$y(T_s) = s_m + n(T_s)$$

$$= a_m\sqrt{E} + n(T_s), \quad a_m = \pm 1, \ \pm 3$$

$$(6.65)$$

where $n(T_s)$ denotes AWGN noise with variance $N_0/2$. Hence, the matched filter output, (6.65), is a Gaussian random variable with mean $a_m\sqrt{E}$ and variance $N_0/2$.

Figure 6.25 shows the decision variable given by (6.65) for the considered 4-PAM constellation and the decision thresholds. The pdf of the decision variable is also shown when s_3 is transmmitted. The symbol s_3 will be erroneously received when $y(T_s) < 0$ or $y(T_s) > 2\sqrt{E}$. Because of the symmetry of the Gaussian pdf, one has

$$p = P(y(T_s) < 0 | s_3) = P\left(y(T_s) > 2\sqrt{E}|s_3\right)$$

$$= \frac{1}{\sqrt{\pi N_0}} \int_{2\sqrt{E}}^{\infty} \exp\left(-\frac{\left(n - \sqrt{E}\right)^2}{N_0}\right) dn$$

$$= Q\left(\sqrt{\frac{2E}{N_0}}\right)$$

$$(6.66)$$

It is obvious from Figure 6.25 that the probability of error for symbols s_2 and s_3 is equal to

$2p$, but it is equal to p for the symbols s_1 and s_4. Therefore, the average symbol error probability for 4-ary PAM is simply $6p$ averaged over four symbols:

$$P_4 = \frac{1}{4}(6p) = \frac{3}{2}Q\left(\sqrt{\frac{2E}{N_0}}\right)$$

$$(6.67)$$

In M-ary PAM, when either one of the two outside levels $\pm(M-1)$ is transmitted, an error occurs in one direction only. For other amplitudes, the error occurs in two directions. Therefore, the average symbol error probability for M-ary PAM is given by

$$P_M = \frac{1}{M}[p + (M-2)2p + p]$$

$$= \frac{2(M-1)}{M}Q\left(\sqrt{\frac{2E}{N_0}}\right)$$

$$(6.68)$$

Average symbol energy for M-ary PAM may be found as, by using (D.32) and the symmetry of the amplitude levels around 0:

$$E_{s,av} = \frac{2}{M}\sum_{m=1}^{M/2}\left[a_m\sqrt{E}\right]^2$$

$$= \frac{2E}{M}\sum_{m=1}^{M/2}(2m-1)^2 = \frac{M^2-1}{3}E \qquad (6.69)$$

$$E_{b,av} = \frac{E_{s,av}}{\log_2 M} = \frac{M^2-1}{3\log_2 M}E$$

Using (6.68) and (6.69), one may express the symbol error probability for M-ary PAM in terms of average bit SNR:

$$P_M = \frac{2(M-1)}{M} Q\left(\sqrt{\frac{6\log_2 M}{M^2-1} \frac{E_{b,av}}{N_0}}\right) \quad (6.70)$$

Symbol error probability for M-ary PAM is shown in Figure 6.26 for $M = 2, 4, 8$ and 16. The error performance for binary PAM is identical to that for matched-filter reception of antipodal signals (see Figure 5.34, (5.51) and (6.47)). The bit error probability for M-PAM with Gray encoding may be found by dividing (6.70) by the number of bits $\log_2 M$ per PAM symbol.

The effect of the alphabet size M on the symbol error performance is usually studied in terms of the required E_b/N_0 for a given value of the symbol error probability. This implies that the argument of the Q function should remain the same for all values of M, ignoring the terms in front of the Q function in (6.70). Therefore, for achieving the same symbol error probability, the bit energy of M-ary PAM should be increased by

$$\frac{(E_{b,av})_M}{(E_{b,av})_{M=2}} = \frac{E_{b,av}}{E} = \frac{M^2-1}{3\log_2 M} \quad (6.71)$$

compared to the binary PAM. As shown in Table 6.1, for a given symbol error probability, the required average bit energy should be increased by about 4 dB when M is doubled. For a given symbol error probability and bit rate, the 8-PAM requires 8.45 dB higher average bit energy but one-third of the transmission bandwidth compared to binary PAM. Conversely, for a given channel bandwidth

Table 6.1   For a Given Symbol Error Probability, the Increase in the Required Average Bit Energy and the Corresponding Decrease in Symbol Rate with the Number of Amplitude levels M.

M	$(E_{b,av})_M/(E_{b,av})_{M=2}$	R_s/R_b
2	1 (0 dB)	1
4	2.5 (4 dB)	1/2
8	7 (8.45 dB)	1/3
16	21.25 (13.27 dB)	1/4
32	68.2 (18.34 dB)	1/5

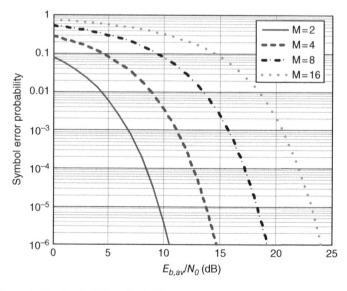

Figure 6.26   Symbol Error Probability of M-ary PAM Versus $E_{b,av}/N_0$ (dB).

(i.e., a given symbol rate R_s), the transmitted bit rate by M-ary PAM is $R_b/R_s = \log_2 M$ faster than the corresponding binary PAM. For example, the bit rate increases by a factor of 3 for 8-ary PAM compared to binary PAM.

The required average bit energy to achieve a given symbol error probability in PAM systems increases as the alphabet size M increases. Therefore, higher peak transmit powers are needed with increasing M values. For M-ary PAM, the required peak transmit power is given by

$$P_{peak} = (M-1)^2 \frac{E}{T_s} \qquad (6.72)$$

where T_s denotes the symbol duration. The average transmit power is readily available from (6.69)

$$P_{av} = \frac{M^2-1}{3} \frac{E}{T_s} \qquad (6.73)$$

The peak-to-average power ratio (PAPR) for M-ary PAM is thus given by

$$PAPR = 3\frac{(M-1)^2}{M^2-1} \qquad (6.74)$$

which changes between 1 and 3 as M goes from 2 to infinity. This means that in M-ary PAM systems, the peak power can be at most 3 times as large as the average transmit power. Also note from (6.72) and (6.73) that the average and peak powers do not depend on the transmission rate.

Example 6.7 Performance Comparison of 4-PAM and 16-PAM.
A 4-ary PAM system has a symbol duration of $T_s = 100\,\mu s$. Hence, the symbol rate is $R_s = 1/T_s = 10$ kbps. The null-to-null transmission bandwidth is given by $B = 2/T_s = R_s = 20$ kHz. Since each symbol comprises two bits in a 4-ary PAM, the corresponding bit rate is $R_b = 2R_s = 20$ kbps.

In order to double the bit rate, we must double the number of bits in each symbol duration of $T_s = 100\,\mu s$. Hence, we need 4 bits/symbol and $M = 2^4 = 16$ amplitude levels, that is, 16-PAM. However, 16-PAM requires $13.27\,dB - 4\,dB = 9.27\,dB$ higher average bit energy compared to 4-PAM (see Table 6.1).

Example 6.8 Limitation of Peak Versus Average Transmit Power on the Performance of 4-ary PAM.
Assume that a 4-PAM transmitter can supply an average transmit power ≤ 1 mW, a peak transmit power ≤ 10 mW, and a maximum symbol rate of 1 M symbols/sec ($T_s = 1\,\mu s$). Noise PSD is assumed to be 5×10^{-11} W/Hz.

Assuming that all $M = 4$ transmission levels are equally likely, we will first determine the constraints resulting from peak and average transmit power levels on the largest voltage spacing between symbols, which is equal to $2\sqrt{E}$ (see Figure 6.25) From (6.72), the constraint on the peak power leads to

$$P_{peak} = \frac{9E}{T_s} \leq 10 mW \Rightarrow$$

$$\qquad (6.75)$$

$$2\sqrt{E} \leq 2\sqrt{\frac{10^{-2}T_s}{9}} = 66.7\,\mu V$$

From (6.73), the constraint on the average energy imposes the following limitation:

$$P_{av} = \frac{M^2-1}{3T_s}E = 5\frac{E}{T_s} \leq 1mW$$

$$\Rightarrow 2\sqrt{E} \leq 2\sqrt{\frac{10^{-3}T_s}{5}} = 28.2\,\mu V$$

$$\qquad (6.76)$$

Limitation comes from the average power, since when the constraint on the average power is satisfied, the constraint on the peak power is also satisfied.

From (6.70), the symbol error probability for 4-PAM:

$$P_4 = \frac{2(M-1)}{M} Q\left(\sqrt{\frac{2E}{N_0}}\right)$$

$$= \frac{3}{2} Q\left(\sqrt{\frac{2(14.1 \times 10^{-6})^2}{5 \times 10^{-11}}}\right) \qquad (6.77)$$

$$= \frac{3}{2} Q\left(\sqrt{8}\right) \cong 2.3 \times 10^{-3}$$

The corresponding bit error probability for Gray encoding is given by

$$P_b = \frac{P_M}{\log_2 M} = 1.12 \times 10^{-3} \qquad (6.78)$$

6.4   Intersymbol Interference

At the beginning of this Chapter, we discussed the physical mechanisms, for example, bandwidth limitation and multipath propagation, by which ISI occurs in communication systems. Here, we will study how ISI degrades the system performance and how this degradation can be mitigated.

Consider a baseband binary PCM system, which consists of a transmitter, AWGN channel and a receiver, as shown in Figure 6.27. The input binary sequence $\{b_k\}$ consists of equally-likely binary digits 1 and 0, each of duration T_b. Pulse amplitude modulator converts the input pulse sequence into a new sequence $\{a_k\}$ of electrical symbols as follows:

$$a_k = \begin{cases} +1\,V & b_k = 1 \\ -1\,V & b_k = 0 \end{cases} \qquad (6.79)$$

Sequence of pulses so produced is applied to a transmit filter of impulse response $g(t)$ so that the signal at the channel input may be written as

$$s(t) = \sum_{k=-\infty}^{\infty} a_k g(t - kT) \qquad (6.80)$$

Signal $s(t)$ undergoes attenuation, phase changes and distortion as it goes through the channel of impulse response $h(t)$. The signal $x(t)$ at the receiver input is simply given by the convolution of the input signal with the channel transfer function:

$$x(t) = s(t) \otimes h(t) + w(t) \qquad (6.81)$$

where $w(t)$ denotes the AWGN with zero-mean and two-sided PSD $N_0/2$. The signal at the receive filter output is given by the convolution of the noisy signal $x(t)$ with the impulse response $c(t)$ of the receive filter:

$$y(t) = c(t) \otimes x(t)$$

$$= c(t) \otimes [h(t) \otimes s(t)] + c(t) \otimes w(t)$$

$$= c(t) \otimes h(t) \otimes \left[\sum_{k=-\infty}^{\infty} a_k g(t - kT)\right] + n(t)$$

$$= \mu \sum_{k=-\infty}^{\infty} a_k p(t - kT) + n(t)$$

$$(6.82)$$

where

$$n(t) = w(t) \otimes c(t)$$
$$\mu p(t) = g(t) \otimes h(t) \otimes c(t) \qquad (6.83)$$

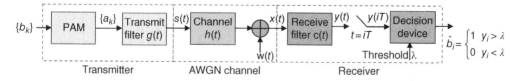

Figure 6.27   Block Diagram of a Binary PAM System.

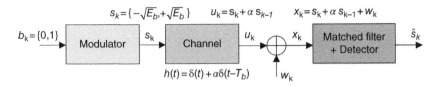

Figure 6.28 A Simplified Block Diagram of a Communication System with a Multipath Channel Leading to ISI.

Here, $n(t)$ denotes the AWGN filtered by the receiver filter with impulse response $c(t)$. On the other hand, the pulse $p(t)$ is proportional to the time convolution of the impulse responses of the transmit filter, the channel and the receive filter. Note that μ is a scaling factor since $p(t)$ is normalized to unity at $t=0$

$$p(t) = \int_{-\infty}^{\infty} P(f)\, e^{j2\pi ft}\, df \Rightarrow$$

$$p(0) = \int_{-\infty}^{\infty} P(f)\, df = 1 \qquad (6.84)$$

Ideally, apart from the noise term, the received signal in (6.82) is desired to be the same as the transmitted signal given by (6.80). One can observe from (6.83) that this possible only if $h(t)$ and $c(t)$ can be represented as delta functions. However, this is not realistic since it requires infinite channel bandwidth. Comparison of (6.80) and (6.82) shows that, due to the convolution process defined by (6.83), $p(t) \neq g(t)$. Therefore, $p(t)$ is a time-spread version of $g(t)$, hence undergoes time dispersion and ISI.

The transmitted data sequence is estimated by the decision device shown in Figure 6.27. Amplitude of each sample is compared to a threshold λ; if $y(iT) > \lambda$, then a decision is made in favor of 1, otherwise, the decision is made in favor of 0.

Example 6.9 Effect of ISI on the Bit Error Probability.
In this example, we will consider ISI due to multipath propagation in a communication system as shown in Figure 6.28, where the bits 0 and 1 are represented by $\mp\sqrt{E}$ by the modulator. Assume the transmission of equi-probable binary antipodal signals, $P(b_k=0) = P(b_k=1) = 1/2$,

over a channel with an impulse response $h(t) = \delta(t) + \alpha\, \delta(t-T_b)$ where α is a real number $0 < \alpha < 1$ and T_b is the bit duration. The channel impulse response implies that a signal and its replica of relative amplitude $\alpha < 1$ arrive at the receiver at times $t=0$ and $t=T_b$, respectively. The signal is also corrupted by the AWGN with double-sided PSD $N_0/2$. The samples at the matched filter output are used by a detector to estimate the transmitted symbol according to the following decision rule:

$$\hat{s}_k = \begin{cases} +\sqrt{E_b} & x_k > 0 \\ -\sqrt{E_b} & x_k < 0 \end{cases} \qquad (6.85)$$

The average bit error probability, P_2, is found in terms of E_b/N_0 and α:

$$P_2 = P(x_k > 0 | b_k = 0) P(b_k = 0)$$
$$\quad + P(x_k < 0 | b_k = 1) P(b_k = 1)$$
$$= P(x_k > 0 | b_k = 0)$$
$$= P(s_k + \alpha s_{k-1} + w_k > 0 | b_k = 0)$$
$$= P\left(-\sqrt{E_b} + \alpha s_{k-1} + w_k > 0\right)$$
$$= P\left(-\sqrt{E_b} + \alpha s_{k-1} + w_k > 0 | b_{k-1} = 0\right) P(b_{k-1} = 0)$$
$$\quad + P\left(-\sqrt{E_b} + \alpha s_{k-1} + w_k > 0 | b_{k-1} = 1\right) P(b_{k-1} = 1)$$
$$= \frac{1}{2} P\left(-\sqrt{E_b}(1+\alpha) + w_k > 0\right)$$
$$\quad + \frac{1}{2} P\left(-\sqrt{E_b}(1-\alpha) + w_k > 0\right)$$
$$= \frac{1}{2} P\left(w_k > \sqrt{E_b}(1+\alpha)\right) + \frac{1}{2} P\left(w_k > \sqrt{E_b}(1-\alpha)\right)$$
$$= \frac{1}{2} Q\left((1+\alpha)\sqrt{\frac{2E_b}{N_0}}\right) + \frac{1}{2} Q\left((1-\alpha)\sqrt{\frac{2E_b}{N_0}}\right)$$
$$\qquad (6.86)$$

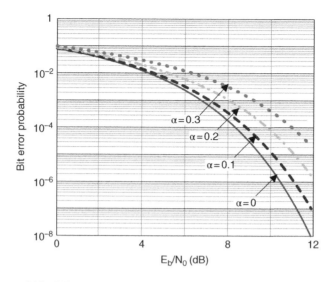

Figure 6.29   Effect of Two-Path Transmission on the Bit Error Probability.

Figure 9.29 shows the effect of α, the gain of the second path relative to the first path, on the bit error probability. The bit error probability performance was observed to degrade as gain of the second path increases. Its effect becomes more noticeable at higher values of the E_b/N_0.

In the absence of ISI, the bit error probability is found for $E_b/N_0 = 10$ by inserting $\alpha = 0$ into (6.86):

$$P_2(\alpha=0) = \frac{1}{2}\, Q\left(\sqrt{2E_b/N_0}\right)$$

$$= \frac{1}{2}\, Q\left(\sqrt{2\times 10}\right) = 3.87\times 10^{-6}$$

$$(6.87)$$

In the presence of ISI, the bit error probability becomes for $\alpha = 0.3$ and $E_b/N_0 = 10$:

$$P_2(\alpha=0.3) = \frac{1}{2}Q\left((1+0.3)\sqrt{2\times 10}\right)$$

$$+ \frac{1}{2}Q\left((1-0.3)\sqrt{2\times 10}\right)$$

$$= \frac{1}{2}Q(5.8138) + \frac{1}{2}Q(3.1305)$$

$$= 4.36\times 10^{-4}$$

$$(6.88)$$

For $E_b/N_0 = 10$, the performance degradation due to ISI is given by the ratio of (6.88) to (6.87):

$$\frac{P_2(\alpha=0.3)}{P_2(\alpha=0)} = \frac{4.36\times 10^{-4}}{3.87\times 10^{-6}} \approx 113 \qquad (6.89)$$

corresponds to a 113-fold increase in the bit error probability. For $P_b = 10^{-5}$, the performance degradation for $\alpha = 0.3$ compared to the case with no ISI ($\alpha = 0$) is $12.35 - 9.58 = 2.77$ dB In the above analysis, we assumed that α is a constant, that is, it does not change with time. In some channels it may change randomly. In that case an averaging over α is necessary.

6.4.1   Optimum Transmit and Receive Filters in an Equalized Channel

Channel equalization is one of the alternative techniques for ISI mitigation. In the equalization process, the coherent receiver determines frequency response $H_{eq}(f)$ of the equalization filter from the estimate, $\hat{H}(f)$, of the channel frequency response and neutralizes the effects of non-ideal channel frequency response, as shown in Figure 6.30. For equalizing the channel, the coherent receiver should be equipped

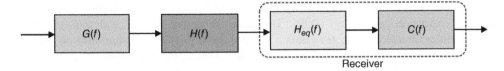

Receiver

Figure 6.30   Effective Frequency Response $\mu P(f) = G(f)H(f)H_{eq}(f)C(f)$ of the Equalized Channel.

with a device that predicts the channel impulse response (hence, the transfer function).

In the presence of the equalization filter, (6.83) and its Fourier transform may be written as

$$\mu p(t) = g(t) \otimes h(t) \otimes h_{eq}(t) \otimes c(t)$$
$$\mu P(f) = G(f)H(f)H_{eq}(f)C(f) \tag{6.90}$$

The transfer function of the equalization filter is chosen as

$$H_{eq}(f) \approx \frac{1}{\hat{H}(f)} \quad \Leftrightarrow \quad \hat{h}(t) \otimes h_{eq}(t) \approx \delta(t) \tag{6.91}$$

When the channel is perfectly equalized, that is, $\hat{H}(f) = H(f)$, (6.90) reduces to

$$\mu p(t) = g(t)\, c(t)$$
$$\mu P(f) = G(f)\, C(f) \tag{6.92}$$

The optimum receiver is then matched to the transmit pulse, $c_{opt}(t) = kg^*(T-t)$ as in (6.21), so as to maximize the output SNR. In view of (6.92), this also implies that the transfer functions of transmit and receive filters may be chosen to be proportional to that of a particular $P(f)$:

$$C_{opt}(f) = kG^*(f)\exp(-j2\pi fT) \quad \Rightarrow$$
$$\mu P(f) = k|G(f)|^2 e^{-j2\pi fT}$$
$$|G(f)|,\ |C_{opt}(f)| \propto \sqrt{|P(f)|} \tag{6.93}$$

where $\propto$ denotes proportionality. One may observe from (6.93) that transmit and receive filters should have the same (frequency and time)

characteristics. This is obviously a desired result for duplex transmission. Nevertheless, we still need an answer as to the determination of $P(f)$. A suitable choice of the line code described by $g(t)$, and a receive filter $c(t)$ matched to $g(t)$ maximizes the output SNR in an equalized channel. We know from match-filter reception that there is no limitation about the shape of $g(t)$ for SNR maximization. Therefore, it is logical to ask if ISI can be mitigated by a clever choice of $g(t)$, hence $p(t)$. This question will be addressed in the next Section.

Example 6.10   Equalization for the Multipath Channel.

Let us now consider the equalization of the multipath channel depicted in Example 6.9, where we assumed that the relative delay suffered by the echo signal is equal to T_b seconds. We will derive an expression for the impulse response of a *zero-forcing* equalization filter, $h_{eq}(t)$, at the receiver so as to mitigate the effects of ISI due to the channel.

Noting the following Fourier transform relation

$$h(t) = \delta(t) + \alpha\, \delta(t - T_b) \Leftrightarrow$$
$$H(f) = 1 + \alpha\, e^{-jwT_b} \tag{6.94}$$

the frequency response of a perfect equalization filter, $H_{eq}(f)$, must satisfy the following:

$$H_{eq}(f) = \frac{1}{H(f)} = \frac{1}{1 + \alpha\, \exp(-jwT_b)}$$

$$= 1 - \alpha\, \exp(-jwT_b) + \alpha^2\, \exp(-j2wT_b)$$

$$- \alpha^3\, \exp(-j3wT_b) + \cdots \tag{6.95}$$

The second expression is obtained by the binomial expansion given by (D.8). The inverse Fourier transform of $H_{eq}(f)$ yields

$$h_{eq}(t) = \delta(t) - \alpha\, \delta(t - T_b) + \alpha^2\, \delta(t - 2T_b)$$
$$- \alpha^3\, \delta(t - 3T_b) + \cdots$$

$$(6.96)$$

The impulse response of the equalizer, given by (6.96), may be realized by a FIR filter as shown in Figure 6.14 with tap weights $\begin{bmatrix} c_0 & c_1 & c_2 & c_3 & \cdots \end{bmatrix} = \begin{bmatrix} 1 & -\alpha & \alpha^2 & -\alpha^3 & \cdots \end{bmatrix}$. Since $\alpha < 1$, several taps/terms may be sufficient for equalization.

From (6.91), for a FIR equalization filter with p taps, one has

$$h(t) \otimes h_{eq}(t) = [\delta(t) + \alpha\delta(t - T_b)] \otimes h_{eq}(t)$$
$$= h_{eq}(t) + \alpha h_{eq}(t - T_b)$$
$$= \delta(t) - (-\alpha)^p \delta(t - pT_b)$$

$$(6.97)$$

Hence, for a FIR equalization filter with $p = 4$ taps, the impulse response of the equalized channel differs from the ideal (perfectly equalized) channel by the residual ISI term $-\alpha^4 \delta(t - 4T_b)$ in (6.97). Since $\alpha < 1$, the degradation caused by the residual ISI term becomes negligibly small.

6.5 Nyquist Criterion for Distortionless Baseband Binary Transmission In a ISI Channel

In the previous section, we considered the equalization of a dispersive channel causing ISI. In this section, we also consider a memory-less dispersive channel, with transfer function $H(f) \neq 1$ $(h(t) \neq \delta(t))$, but with no equalization. Hence, for a dispersive and unequalized channel, we will determine optimum impulse response of $p(t)$ (also those of transmit and receive filters $g(t)$ and $c(t)$) in order to eliminate the ISI and to minimize the bit error probability (see Figure 6.31). Note that each symbol is decoded separately in a memoryless channel since the neighboring symbols in time are independent from each other. Minimization of the bit error probability implies the use matched filter at the receiver. On the other hand, based on the assumption of duplex transmission, each side acts both as transmitter and receiver. Therefore, impulse responses $g(t)$ and $c(t)$ for transmitter and the receiver should be the same.

When the synchronization is perfect between transmitter and receiver, the receive filter output y(t) in Figure 6.27 is sampled synchronously with the transmitter. The receive filter output (6.82), sampled at $t_i = iT$, may be written as

$$y[iT] = \mu \sum_{k=-\infty}^{\infty} a_k p[(i-k)T] + n[iT]$$
$$= \underbrace{\mu a_i}_{desired\ signal} + \underbrace{\mu \sum_{k=-\infty,\ k \neq i}^{\infty} a_k p[(i-k)T]}_{ISI}$$
$$+ \underbrace{n[iT]}_{noise}$$

$$(6.98)$$

where μa_i denotes the contribution of the ith transmitted bit and represents the received bit

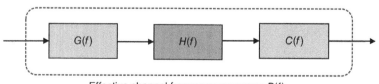

Effective channel frequency response: $P(f)$

Figure 6.31 An Unequalized Channel.

in the absence of noise and ISI. The second term in (6.98) accounts for the ISI due to time dispersion. Note that the term $p[(i-k)T]$, which denotes the value of kth pulse (to transmit a_k) at the sampling time $t_i = iT$, accounts for the ISI caused to a_i by the kth pulse. It is thus evident that not only the AWGN but also the ISI contributes to the bit error probability.

The ISI at the receiver output vanishes if $p[(i-k)T]$ in (6.98) becomes zero for all non-zero values of i-k:

$$p[(i-k)T] = \begin{cases} 1 & i = k \\ 0 & i \neq k \end{cases} \quad (6.99)$$

where T denotes the symbol duration. Then, the received signal becomes ISI-free and is affected only by the AWGN

$$y[iT] = \mu a_i + n[iT] \quad -\infty < i < \infty \quad (6.100)$$

As the matched-filter receiver $c(t)$ will be matched to the effective pulse at the receiver input, $g(t) \otimes h(t)$, that is, $C(f) = kG^*(f)H^*(f) \exp(-j2\pi fT)$. Then, using (1.12) and (6.83) μ may be shown to be equal to the square root of the energy of the ith pulse, and the variance of $n[iT]$ is given by $N_0/2$ (see (6.26)):

$$\mu = k \int_{-\infty}^{\infty} |G(f)|^2 |H(f)|^2 df = kE = \sqrt{E}$$

$$\sigma_n^2 = \frac{N_0}{2} \int_{-\infty}^{\infty} |C(f)|^2 df$$

$$= \frac{N_0}{2} k^2 \int_{-\infty}^{\infty} |G(f)|^2 |H(f)|^2 df = \frac{N_0}{2} \quad (6.101)$$

Hence, the SNR at the receiver output is $SNR = \left(\sqrt{E}\right)^2 / \sigma_n^2 = 2E/N_0$, as expected.

To determine the implication of (6.100) on the frequency domain characterization of $p(t)$, we rewrite, using (5.1)–(5.3), the expressions for the instantaneously sampled version of $p(t)$ and its Fourier transform:

$$p_\delta(t) = \sum_{i=-\infty}^{\infty} p(iT)\delta(t-iT) = p(t)\sum_{i=-\infty}^{\infty} \delta(t-iT)$$

$$P_\delta(f) = P(f) \otimes \frac{1}{T}\sum_{n=-\infty}^{\infty} \delta\left(t - \frac{n}{T}\right)$$

$$= \frac{1}{T}\sum_{n=-\infty}^{\infty} P\left(f - \frac{n}{T}\right)$$

$$(6.102)$$

Using (6.99) and taking the Fourier transform of $p_\delta(t)$ in (6.102), one gets

$$P_\delta(f) = \int_{-\infty}^{\infty} \sum_{i=-\infty}^{\infty} [p(iT)\,\delta(t-iT)]\exp(-j2\pi ft)\,dt$$

$$= \int_{-\infty}^{\infty} p(0)\,\delta(t)\,dt = p(0) = 1$$

$$(6.103)$$

Hence, ISI is eliminated at sampling times provided that $P(f)$ satisfies the following:

$$P_\delta(f) = 1 \quad \Rightarrow \quad \sum_{n=-\infty}^{\infty} P(f-n/T) = T$$

$$(6.104)$$

As shown in Figure 6.31, $P(f)$ denotes the transfer function of the overall system and hence incorporates the transmit filter, the channel and the receive filter. On the other hand, (6.104) implies ISI-free transmission if the sum of the copies of $P(f)$ separated by $1/T$ in the frequency domain is equal to T, independently of the frequency. This is called the *Nyquist criterion*, which is shown pictorially in Figure 6.32 for two different $P(f)$'s. In summary, the ISI will be eliminated and the output SNR will be maximized in a memoryless dispersive channel, if $P(f)$ is chosen so as to satisfy (6.104), and $C(f)$ and $G(f)$ are related to $P(f)$ as in (6.93). Below, we will investigate several alternative $P(f)$'s that satisfy (6.104).

6.5.1 Ideal Nyquist Filter

One of the the simplest ways to satisfy the zero-ISI condition, given by (6.104), is to choose $P(f)$ as

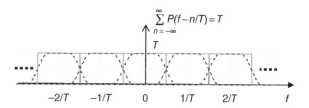

Figure 6.32   Nyquist Criterion for Eliminating ISI.

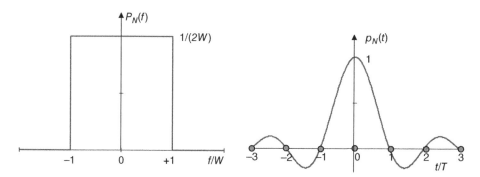

Figure 6.33   Ideal Nyquist Filter and Ideal Nyquist Pulse Where $W = 1/(2\,T) = R_s/2$. Sampling times are shown by circles.

$$P_N(f) = \frac{1}{2W} rect\left(\frac{f}{2W}\right) \Leftrightarrow$$

$$p_N(t) = \sin c(2Wt) \qquad (6.105)$$

which is shown in Figure 6.33 and Figure 6.32 with continuous lines. Since $T = 1/(2W)$, the bandwidth W of the so-called *ideal Nyquist channel* can support a symbol rate $R_s = 1/T$:

$$W = \frac{R_s}{2} = \frac{1}{2T} \qquad (6.106)$$

Ideal Nyquist channel allows a transmission rate of $R_s = f_s = 2\,W$, which is equal to the sampling rate f_s and twice the channel bandwidth W, also known as the Nyquist bandwidth. The resulting spectrum efficiency $R_s/W = 2$ represents the highest available spectrum efficiency and hence the maximum transmission rate provided by all possible $P(f)$'s. In other words, Nyquist channel allows for zero ISI transmission with minimum channel bandwidth.

As shown in Figure 6.33, the corresponding time pulse $p_N(t)$, defined by (6.105), is equal to unity at $t = 0$ and vanishes at $t = iT$, $i \neq 0$; hence it satisfies (6.99). However, $P_N(f)$ being band-limited to W, $p_N(t)$ is non-causal since it is not time-limited. Non-causality exhibits itself by the requirement for starting to send the pulse $p_N(t)$ at $t = -\infty$ instead of at $t = 0$. On the other hand, $p_N(t)$ has a slow decay rate with time and decreases as $1/|t|$ for large t. Consequently, $p_N(t)$ may have unacceptably large values even at times largely separated from the transmission time, $t = 0$. This causes nonnegligible ISI when sampling times deviate from iT due to synchronization errors.

Figure 6.34 shows the time variation of the signal for transmitting the bit sequence 1001011 by using ideal Nyquist pulse. The binary 1 is transmitted by +1 V amplitude while −1 V amplitude is used for transmitting a binary 0. Because $p_N(0) = 1$ and $p_N(iT) = 0$ for $i \neq 0$, the pulses used for transmitting the bit sequence do

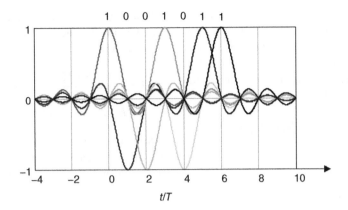

Figure 6.34   Transmission of a Binary Sequence Using Ideal Nyquist Pulse.

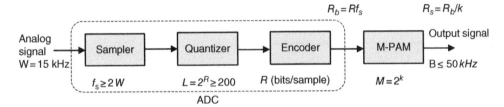

Figure 6.35   The Block Diagram of the M-ary PCM System Considered.

not interfere with each other at $t = iT_b$, hence an ISI-free transmission. Since the envelope of the sinc pulse given by (6.105) decays as $1/t$ with time, ISI occurs if the receiver, with timing errors, does not sample the incoming signal at exactly iT. Hence, an ISI-free transmission requires a receiver which is perfectly time-synchronized with the incoming signal sequence and an ISI-free pulse shape, defined by (6.104), which decays rapidly with time. On the other hand, Figure 6.34 clearly shows that one should-start transmitting the pulse $p_N(t-iT)$ well ahead of the transmission time $t = iT$ of the bit that it represents; hence, $p_N(t)$ is not causal. From the frequency domain perspective, the ideal Nyquist channel requires an ideal low-pass channel, that is, flat frequency response in [−W to W] and zero elsewhere, which is physic-ally unrealizable.

Example 6.11   M-ary PCM Transmission in Ideal Nyquist Channel.
As shown in Figure 6.35, an analog signal ban-dlimited to W = 15 kHz is quantized to $L \geq 200$ levels and transmitted as a M-ary PAM signal in the ideal Nyquist channel. We will determine the sampling rate f_s, the maximum number of bits R per sample, and the corresponding value of k for M-ary PAM when the available trans-mission bandwidth is $B = 50$ kHz.

An ADC samples the analog signal at a given sampling rate, quantizes each analog sample and represents each sample by R bits. Hence, the transmission rate at the ADC output is $R_b = R\, f_s$. The encoder assigns one of the $M = 2^k$ combinations of k bits at its input to one of M amplitude levels. Consequently, the transmission rate at the output of the M-ary encoder is given by $R_s = R_b/k = Rf_s/k$.

The limitation on the number of quantization levels, $L = 2^R \geq 200$, implies that $R > 7.64$. Hence, the actual number of bits/sample and the quantization levels are $R = 8$ and $L = 2^R = 256$, respectively. Then, the bit rate at the ADC output should satisfy $R_b = Rf_s \geq 2RW = 240\ kbps$.

From (6.106) for the ideal Nyquist channel, one gets

$$B = \frac{R_s}{2} = \frac{Rf_s}{2k} \geq \frac{RW}{k} \Rightarrow$$

$$k \geq \frac{RW}{B} = \frac{8 \times 15\ kHz}{50\ kHz} = 2.4 \Rightarrow$$

$$k = 3$$

$$(6.107)$$

Hence $M = 2^k = 8$, and the actual transmission bandwidth and the spectrum efficiency are given by

$$B = \frac{RW}{k} = \frac{8 \times 15\ kHz}{3} = 40\ kHz \qquad (6.108)$$

$$R_b/B = 2k = 6\ bits/Hz$$

Note that the use of 8-ary PAM increases the spectrum efficiency by a factor $\log_2 8$ at the cost of 8.45 dB increase in the required average bit energy (see Table 6.1).

6.5.2   Raised-Cosine Filter

We have shown above that the ideal Nyquist filter, defined by (6.105), satisfies the zero-ISI condition, given by (6.104). It is easy to show that the raised cosine spectrum, defined by

$$P_{RC}(f) = \begin{cases} 1/(2W) & 0 \leq |f| \leq (1-\alpha)W \\ \dfrac{1}{4W}\left[1 + \cos\left(\dfrac{\pi}{2\alpha W}(|f| - (1-\alpha)W)\right)\right] & (1-\alpha)W \leq |f| \leq (1+\alpha)W \\ 0 & |f| > (1+\alpha)W \end{cases} \qquad (6.109)$$

also satisfies the zero-ISI condition, as shown in Figure 6.36 and Figure 6.32 in dash-lines. The frequency spectrum of the raised-cosine pulse and the corresponding pulse waveform are shown in Figure 6.36. As its name implies the

frequency spectrum of the raised-cosine filter consists of a cosine curve raised so that it always assumes positive values. The so-called roll-off factor α changes between 0 and 1 and is used as a measure of the bandwidth in excess of the

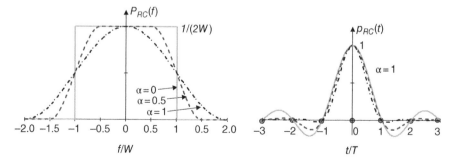

Figure 6.36   Raised-Cosine Filter and Raised-Cosine Pulse Where $W = 1/(2\ T) = R_s/2$. circles show the sampling times.

bandwidth of the ideal Nyquist filter ($W = R_s/2$). The transmission bandwidth of a raised-cosine filter is given by

$$B = W(1+\alpha), \quad 0 \le \alpha \le 1, \quad W = \frac{1}{2T} = \frac{R_s}{2}$$

$$(6.110)$$

The transmission bandwidth B changes between W and $2W$ by adjusting the roll-off factor between 0 and 1.

The time pulse of raised-cosine filter is found by the inverse Fourier transform of $P_{RC}(f)$:

$$p_{RC}(t) = \text{sinc}(2Wt) \frac{\cos(2\pi\alpha Wt)}{1 - 16\alpha^2 W^2 t^2} \quad (6.111)$$

where the factor $\text{sinc}(2Wt)$ characterizes the ideal Nyquist pulse and ensures zero crossings of $p(t)$ at the desired sampling instants of time $t = iT$. The second factor decreases as $1/t^2$ and ensures that the decay rate of the raised-cosine pulse is faster than that of the ideal Nyquist pulse. Hence, raised-cosine pulse is less susceptible to timing errors. On the other hand, ISI due to timing errors decreases as α increases from 0 ($B = W$) to 1 ($B = 2W$), as expected. Hence, raised-cosine filtering eliminates ISI more effectively than Nyquist filtering at the expense of a controlled increase in the transmission bandwidth.

For $\alpha = 1$, (6.109) simplifies to

$$P_{RC}(f) = \begin{cases} \frac{1}{4W}\left[1 + \cos\left(\frac{\pi f}{2W}\right)\right] & 0 < |f| < 2W \\ 0 & |f| \ge 2W \end{cases}$$

$$(6.112)$$

and the corresponding pulse $p_{RC}(t)$ reduces to

$$p_{RC}(t) = \frac{\text{sinc}(4Wt)}{1 - 16W^2 t^2}, \quad (6.113)$$

At $t = \pm T/2 = \pm 1/4\,W$, we have $p_{RC}(t) = 0.5$, that is, pulse width at half amplitude is equal to bit duration T. There are zero crossings at $t = \pm 3\,T/2, \pm 5\,T/2$ in addition to usual crossings at $t = \pm T, \pm 2\,T...$

Example 6.12   Root-Raised Cosine Filter. In view of (6.93), and due to the requirements on duplex transmission in a communication system, it may be wise to choose the frequency responses of transmit and receive filters as root-raised cosine filters so as to realize $P_{RC}(f)$ as the overall frequency response of the communication link. The (root) raised-cosine filter is widely used in communication systems since it allows the removal of the ISI and the control of the transmission bandwidth via the roll-off factor.

Consider a communication system as shown in Figure 6.31, where $g(t)$ is the transmitter pulse shaping filter, $h(t)$ is the impulse response of the channel and $c(t)$ denotes the impulse response of the receiver filter matched to $g(t)$. We use raised-cosine filtering with roll-off factor α for ISI mitigation in this channel. Then, transmit and receive filters are characterized by

$$\mu P_{RC}(f) = G(f)H(f)C(f)$$

$$|G(f)|, \; |C(f)| \propto \sqrt{|P_{RC}(f)/H(f)|}$$

$$(6.114)$$

where $P_{RC}(f)$ denotes the transfer function of the raised-cosine filter. Hence, we use root-raised-cosine filters with roll-off factor α for pulse shaping at transmitter and receiver. In this approach, channel equalization is integrated with the design of transmit and receive filters but requires the estimation of the channel transfer function.

Assuming that 8-PAM is used in a channel band-limited to 10 MHz, the symbol rate should satisfy for $\alpha = 0.25$ (see (6.110))

$$\frac{R_s}{2}(1+\alpha) \le B \implies R_s \le \frac{2B}{1+\alpha} = 16\;Msymbols/s$$

$$(6.115)$$

The corresponding bit rate is given by

$$R_b = (\log_2 M)\,R_s = 3R_s = 48\;Mbps \quad (6.116)$$

Higher bit rates evidently require wider transmission bandwidths and the signal experiences a channel frequency response which is not flat beyond 10 MHz. Consequently, ISI will occur.

6.6 Correlative-Level Coding (Partial-Response Signalling)

In addition to the ideal Nyquist and raised-cosine filters, correlative level coding may also be used to remove the ISI. In this method, which is also called as the partial-response signalling (PRS), ISI is introduced to the transmitted signal in a controlled manner. Using this information, the ISI induced by the channel is removed at the receiver. Signalling rate achieved by PRS is equal to the Nyquist rate (2 W symbols per second) in a channel of bandwidth W Hz.

Consider an input binary sequence $\{b_k\}$ of uncorrelated binary symbols 1 and 0, each of duration T. This sequence is applied to a binary pulse-amplitude modulator, producing

$$a_k = \begin{cases} +1 & b_k = 1 \\ -1 & b_k = 0 \end{cases} \qquad (6.117)$$

Two-level sequence $\{a_k\}$ is first passed through a simple filter involving a single delay element and a summer, as shown in Figure 6.37. The output of the so-called duobinary encoder,

$$c_k = a_k + a_{k-1} \qquad (6.118)$$

has three levels, ±2 and 0, depending on the combinations of a_k and a_{k-1}. Input sequence $\{a_k\}$ of uncorrelated two-level pulses is thus transformed into a sequence $\{c_k\}$ of correlated three-level pulses. The correlation between c_k's, introduced under designer's control, is exploited by the receiver to remove the ISI caused by the channel.

As shown in dashed lines in Figure 6.37, the transfer function of the duobinary encoder in cascade with the ideal Nyquist filter may be written as

$$P_D(f) = P_N(f)\ [1 + \exp(-j2\pi\,fT)]$$
$$= 2\,P_N(f)\ \cos(\pi fT)\exp(-j\pi fT)$$
$$(6.119)$$

Using the frequency response of ideal Nyquist channel of bandwidth $W = 1/2\,T$ given by (6.105), the overall frequency response of duobinary signalling may simply be written as

$$P_D(f) = \begin{cases} 2T\,\cos(\pi fT)\exp(-j\pi fT) & |f| \le 1/2T \\ 0 & otherwise \end{cases}$$
$$(6.120)$$

Hence, unlike the frequency response of ideal Nyquist filtering, the PSD of duobinary signaling is not flat (see Figure 6.38) but the required bandwidth is the same as that for ideal Nyquist filter, that is, $W = 1/(2\,T) = R_s/2$.

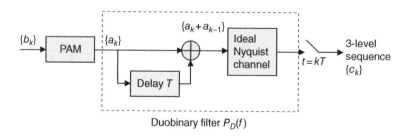

Figure 6.37   Block Diagram of a Duobinary Encoder.

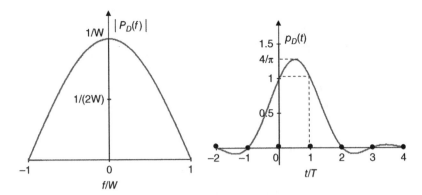

Figure 6.38   Duobinary Filter and Duobinary Pulse Where W = 1/(2 T). Sampling points are shown by circles.

From Figure 6.37 and (6.119), the impulse response for duobinary signaling is found as

$$p_D(t) = [\delta(t) + \delta(t - T_b)] \otimes p_N(t)$$

$$= p_N(t) + p_N(t - T)$$

$$= \frac{\sin(\pi t / T)}{\pi t / T} + \frac{\sin(\pi(t - T)/T)}{\pi(t - T)/T} \quad (6.121)$$

$$= \frac{T^2 \sin(\pi t / T)}{\pi t (T - t)}$$

Note that the response to an input pulse at $t = 0$ is partial (see Figure 6.38), that is, the response is spread over two signalling intervals, at $t = 0$ and $t = T$, hence partial response signaling. On the other hand, the impulse response, given by (6.121), which is the Fourier transform of the cosine-type transfer function, decays as $1/t^2$, that is, faster than the ideal Nyquist pulse.

The original two-level sequence $\{a_k\}$ may be detected from the duobinary-coded sequence $\{c_k\}$ using (6.118). If $\hat{a}_k$ represents the estimate of the pulse a_k by the receiver at time $t = kT$, then subtracting the previous estimate $\hat{a}_{k-1}$ from c_k, one gets $\hat{a}_k = c_k - \hat{a}_{k-1}$. If c_k is received correctly and the previous estimate $\hat{a}_{k-1}$ is correct, then the current estimate $\hat{a}_k$ is also correct. Otherwise, $\hat{a}_k$ will be incorrect and, once an error is made, it propagates through the output.

Hence, the error propagation is the major drawback of this detection procedure.

Propagation of errors can be avoided by using precoding before duobinary encoder as shown in Figure 6.39. Precoding (differential encoding) converts the binary data sequence $\{b_k\}$ into another binary sequence $\{d_k\}$ by, for example, $d_k = b_k \oplus d_{k-1}$. Precoded data sequence is thus given by

$$d_k = b_k + d_{k-1} \implies d_k = \begin{cases} d_{k-1} & b_k = 0 \\ \bar{d}_{k-1} & b_k = 1 \end{cases}$$
$$(6.122)$$

Precoded binary sequence $\{d_k\}$ is applied to a pulse-amplitude modulator, producing $\{a_k\}$:

$$a_k = \begin{cases} +1 & d_k = 1 \\ -1 & d_k = 0 \end{cases} \quad (6.123)$$

Unlike the linear operation of duobinary coding, precoding is nonlinear. Duobinary encoded sequence then becomes

$$c_k = a_k + a_{k-1} = \begin{cases} 0 & b_k = 1 \\ \pm 2 & b_k = 0 \end{cases} \quad (6.124)$$

The receiver for decoding the duobinary encoded sequence is simple. In view of (6.124),

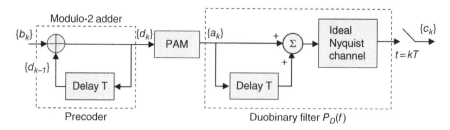

Figure 6.39 Block Diagram of a Precoded Duobinary Encoder.

Table 6.2 Process of Duobinary Encoding of the Binary Sequence {0010110} with Precoding.

Binary sequence $\{b_k\}$	Ref.	0	0	1	0	1	1	0	
Precoded sequence $\{d_k\}$	1	1	1	0	0	1	0	0	see (6.122)
Two-level sequence $\{a_k\}$	+1	+1	+1	−1	−1	+1	−1	−1	see (6.123)
Duobinary encoder output $\{c_k\}$		+2	+2	0	−2	0	0	−2	see (6.124)
Decoded sequence $\{\hat{b}_k\}$		0	0	1	0	1	1	0	see (6.125)

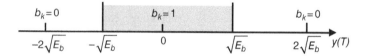

Figure 6.40 Decision Regions for Duobinary Signaling for Equiprobable 1's and 0's.

the receiver for duobinary signalling consists of a rectifier, which takes the absolute value of the received signal sequence $\{c_k\}$ and a detector, which compares the rectifier output to a threshold level of 1 to estimate the transmitted bit:

$$\hat{b}_k = \begin{cases} 1 & \text{if } |c_k| < 1 \\ 0 & \text{if } |c_k| > 1 \end{cases} \qquad (6.125)$$

This detector requires no knowledge of any input sample other than the present one. Hence, neither error propagation nor ISI can occur in this detector.

Example 6.13 Duobinary Signalling.
This example aims to demonstrate the steps in duobinary encoding and decoding processes of a binary sequence $\{b_k\}$ with precoding, as shown in Table 6.2. [1]

6.6.1 Probability of Error in Duobinary Signaling

In an AWGN channel, matched filter output of the received duobinary-encoded bit sequence may be written from (6.124) as

$$c_k = \begin{cases} w_k & \text{if } b_k = 1 \\ \pm 2\sqrt{E_b} + w_k & \text{if } b_k = 0 \end{cases} \qquad (6.126)$$

where w_k denotes the sample of a zero mean Gaussian random variable with variance σ^2 during the period of the kth symbol. From the decision rule given by (6.125) and the decision regions shown in Figure 6.40, one infers that an error occurs if $|w_k| > \sqrt{E_b}$ when $b_k = 1$ is transmitted and if $|\pm 2\sqrt{E_b} + w_k| < \sqrt{E_b}$ when $b_k = 0$ is transmitted:

$$P(error|b_k = 1) = P(|w_k| > \sqrt{E_b})$$

$$= 2P(w_k > \sqrt{E_b})$$

$$P(error|b_k = 0) = \frac{1}{2}P(2\sqrt{E_b} + w_k < \sqrt{E_b})$$

$$+ \frac{1}{2}P(-2\sqrt{E_b} + w_k > -\sqrt{E_b})$$

$$= P(w_k > \sqrt{E_b})$$

$$(6.127)$$

Assuming that $P(b_k = 1) = P(b_k = 0) = 1/2$, the average probability of error for duobinary signalling is found to be

$$P_2 = P(b_k = 1)P(error|b_k = 1)$$

$$+ P(b_k = 0)P(error|b_k = 0)$$

$$= \frac{3}{2}P(w_k > \sqrt{E_b})$$

$$= \frac{3}{2}\int_{\sqrt{E_b}}^{\infty} \frac{1}{\sqrt{2\pi}\sigma} \exp\left(-\frac{x^2}{2\sigma^2}\right) dx$$

$$(6.128)$$

$$= \frac{3}{2}Q\left(\sqrt{\frac{E_b}{\sigma^2}}\right)$$

The average bit energy and the noise variance at the output of the transmitting filter $G(f)$ are given by

$$E_b = E[a_k^2] \int_{-W}^{W} |G(f)|^2 df = E[a_k^2] \int_{-W}^{W} |P_D(f)| df$$

$$\sigma_n^2 = \frac{N_0}{2}\int_{-W}^{W} |G(f)|^2 df = \frac{N_0}{2}\int_{-W}^{W} |P_D(f)| df$$

$$(6.129)$$

where $|P_D(f)| = |G(f)|^2$ is assumed (see (6.93)). The SNR for dubinary signalling may then be written as follows:

$$\left(\frac{E_b}{\sigma^2}\right)_D = \frac{E[a_k^2]}{\sigma_n^2} = \frac{2E_b/N_0}{\left[\int_{-W}^{W} |P_D(f)| df\right]^2}$$

$$= \frac{2E_b/N_0}{(4/\pi)^2}$$

$$(6.130)$$

Ignoring the 3/2 factor multiplying the Q function in P_2 expression for duobinary signaling in (6.129), the degradation in SNR compared to binary signalling is 2.1 dB. However, the 2.1 dB SNR degradation in duobinary signaling is paid for removing the ISI and transmitting at the Nyquist rate.

Example 6.14 PSD of Duobinary Signaling. In this example we present an alternative derivation of the PSD of duobinary signaling to (6.120). For this purpose, the PSD of duobinary signals will be determined through the autocorrelation function of $c_k = a_k + a_{k-1}$. Since $\{a_k\}$ are uncorrelated and can assume values ± 1 in a symbol duration T, the autocorrelation function of $\{c_k\}$ is found as

$$R_c[n] = E[c_k c_{k+n}]$$

$$= E[(a_k + a_{k-1})(a_{k+n} + a_{k+n-1})]$$

$$= \underbrace{E[a_k a_{k+n} + a_{k-1} a_{k+n-1}]}_{2\delta(n)}$$

$$+ \underbrace{E[a_k a_{k+n-1}]}_{\delta(n-1)} + \underbrace{E[a_{k-1} a_{k+n}]}_{\delta(n+1)}$$

$$= 2\delta(n) + \delta(n-1) + \delta(n+1)$$

$$(6.131)$$

By taking the discrete Fourier transform of (6.131), one gets the PSD of $\{c_k\}$ as

$$S_c(f) = \sum_{n=-\infty}^{\infty} R_c[n] e^{-j2\pi fnT}$$

$$= 2 + e^{-j2\pi fT} + e^{j2\pi fT}$$

$$(6.132)$$

$$= 4\cos^2(\pi fT)$$

As shown in Figure 6.37, the PSD of duobinary encoder in cascade with ideal Nyquist filter is given by (see (6.105))

$$|P_D(f)|^2 = S_c(f)|P_N(f)|^2$$

$$= 4T^2\cos^2(\pi fT), \quad |f| \leq 1/(2T)$$

$$(6.133)$$

which is equal to the square of the absolute value of (6.120), as expected.

6.6.2 Generalized Form of Partial Response Signaling (PRS)

As described by (6.118), duobinary signaling uses the two consecutive bits to deliberately insert correlation into the transmitted bit sequence, which is exploited by the receiver to remove the ISI introduced by the channel. This concept can be generalized to other schemes known as correlative level coding or PRS by taking correlation spans more than two binary digits:

$$c_k = \sum_{n=0}^{N-1} w_n a_{k-n} \qquad (6.134)$$

where $\{w_n\}$ denote the weight coefficients. Block diagram of a generalized correlative level encoder is shown in Figure 6.41.

The use of (6.134) leads to different classes of partial response signalling schemes with pulses which may be expressed as a weighted linear combination of N ideal Nyquist pulses (see (6.121)):

$$p_{PRS}(t) = \sum_{n=0}^{N-1} w_n \ \mathrm{sinc}\left(\frac{t-nT}{T}\right) \qquad (6.135)$$

In view of (6.118) and (6.134) (also compare (6.121) and (6.135)), duobinary signalling is described by $N = 2$, and the coefficients $w_0 = w_1 = +1$.

With PRS, binary transmission is possible over baseband channels at the Nyquist rate. Different spectral shapes with gradual cut-off characteristics can be produced. However, larger SNR's are needed for the same error probability compared to corresponding binary PAM systems.

Table 6.3 provides a summary of several pulses used for ISI mitigation in comparison with the commonly used NRZ pulse. Note that Nyquist and duobinary pulses operate at minimum transmission bandwidth $W = 1/(2T)$ and induce no ISI. However, duobinary signalling requires 2.1 dB higher SNR compared to the binary signalling. On the other hand, the transmission bandwidth $B = W(1 + \alpha)$ of the

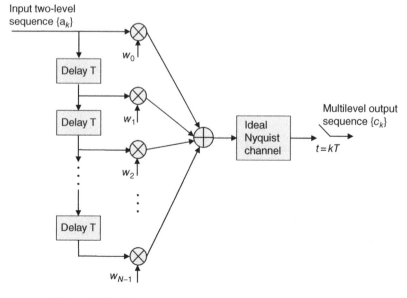

Figure 6.41   Generalized Form of Partial Response Signaling.

Table 6.3 Comparison of Transmission Bandwidth of Various Signaling Pulses.

Pulse type	Time pulse, $p(t)$	Frequency transfer function, $P(f)$		Bandwidth, W	ISI
Nyquist	$\text{sinc}(t/T)$	$T\,\text{rect}(fT)$		$1/(2T)$	No
Raised cosine	$\text{sinc}\left(\dfrac{t}{T}\right)\dfrac{\cos(\pi\alpha t/T)}{1-(2\alpha t/T)^2}$	$\begin{aligned}&T\\&T\cos^2\left(\dfrac{\pi T}{2\alpha}\left(|f|-\dfrac{1-\alpha}{2T}\right)\right)\\&0\end{aligned}$	$\begin{aligned}&0\le|f|\le(1-\alpha)/2T\\&(1-\alpha)/2T\le|f|\le(1+\alpha)/2T\\&|f|>(1+\alpha)/2T\end{aligned}$	$(1+\alpha)/(2T)$	No
Duobinary	$\dfrac{T^2\sin(\pi t/T)}{\pi t(T-t)}$	$\begin{aligned}&2\cos(\pi fT)e^{-j\pi fT}\quad|f|\le 1/2T\\&0\qquad\qquad\qquad otherwise\end{aligned}$		$1/(2T)$	No
NRZ	$\text{rect}(t/T)$	$T\sin c(fT)$		null-null: $2/T$ $B_{3dB}=0.89/T$	Yes

raised-cosine pulse is determined by the roll-off parameter α, $0<\alpha<1$. The increase in the transmission bandwidth is paid by higher robustness to timing/synchronization errors. The raised-cosine pulse reduces to the Nyquist pulse for $\alpha=0$.

Homework 6.3 Modified duobinary signaling Modified duobinary signaling is described by $N=3$, $w_0=1$, $w_1=0$, and $w_2=-1$. Determine anaytical expressions for the time pulse and the frequency response for this signaling. Determine the required transmission bandwidth. Also formulate the decision rule at the receiver output.

6.7 Equalization in Digital Transmission Systems

In an *AWGN channel*, a matched filter receiver is known to be optimal in the sense that it maximizes the output SNR, and minimizes the bit error probability. The impulse response of a matched filter is a time reversed and delayed version of the input pulse, hence matched to the input pulse.

In a *dispersive channel*, transmitted pulses are spread in time, leading to ISI and increased bit errors at the receiver. Equalization is a technique to combat ISI in dispersive channels. An equalizer is actually an inverse filter of the channel, which is inserted at the receiver input to compensate for the fluctuations in the channel amplitude, phase and delay characteristics. In the so-called *frequency selective channels*, where the channel transfer function is frequency-dependent. An equalizer enhances (attenuates) the frequency components with small (large) amplitudes in the received signal in order to provide a flat received frequency response and a linear phase response. For *time-selective channels*, an adaptive equalizer is designed to track the channel variations so as to satisfy the above requirements.

Among the most commonly known equalizers is a zero-forcing equalizer which can be implemented as a FIR filter. The taps of a zero-forcing equalizer are adjusted such that the equalizer output is forced to be zero at a finite number of sample points on each side of the received pulse, ignoring the channel noise. Consequently, the performance of a zero-forcing equalizer degrades as the channel becomes more noisy. However, the taps of the so-called minimum mean-square error (MMSE) equalizer are adjusted such that the mean-square error (MSE) between the equalizer output and the desired signal is minimized. The MMSE equalizer outperforms zero-forcing equalizer since it takes into account the effects of both ISI and channel noise.

Note, however, that channel equalization only cannot remove the ISI caused by the transceivers.

Table 6.4   Equalization Strategies in ISI Only and ISI and AWGN Channels.

Interference	Solution	Characteristics
Noise only	Matched filter receiver	$SNR = 2E_b/N_0$
ISI only	Zero-forcing receiver	Nyquist: $B = W$
		RC: $B = W(1 + \alpha)$
		PRS: $B = W$, higher Tx power
Noise + ISI	Matched filter + equalizer optimized in MMSE sense	Adaptive for time-variant chanels

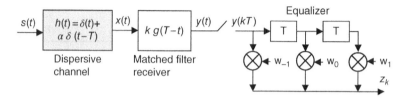

Figure 6.42   Equalizer and Matched Filter in a Multipath Channel.

Therefore, pulse shaping may be required, In a channel with ISI acting only (i.e., ignoring the channel noise), the choices of Nyquist, raised-cosine or partial-response signaling (PRS) would force the resulting ISI to be zero at sampling instants nT, $n > 0$, except for at $n = 0$ where the symbol of interest occurs. A linear receiver implemented as a zero-forcing equalizer is followed by a decision-making device. Symbol-by-symbol detection is then optimal.

In practice, the channel noise and the ISI act together on a data transmission system. Then, design of a linear receiver should be optimized for the general case of a linear channel that is both dispersive and noisy. The optimum linear receiver then consists of the cascade connection of a matched filter, and a transversal (tapped-delay-line) equalizer, with tap spacing equal to the bit duration. Table 6.4 shows a summary of the channels considered, the approaches for removing ISI and the required transmission bandwidth.

Example 6.15   Receiver with Match Filter and Equalizer in a Multipath ISI Channel. Consider an i.i.d. bit sequence $\{b_k\}$ transmitted using a transmit pulse $g(t)$ as given by (6.80).

We assume a multipath channel with ISI and the transmit pulse $g(t)$ has a symbol duration T. The receiver, shown in Figure 6.42, maximizes the output SNR, using a matched filter, and eliminates ISI, using a zero-forcing equalizer.

The channel is assumed to be dispersive and consists of two-paths, with the following channel impulse response

$$h(t) = \delta(t) + \alpha\delta(t - T) \qquad (6.136)$$

where $\alpha < 1$. The signal $x(t)$ at the input of the matched filter is given by the convolution of $x(t)$ and $h(t)$:

$$x(t) = s(t) \otimes h(t) = \sum_{n = -\infty}^{\infty} b_n\, g(t - nT)$$

$$+ \alpha \sum_{n = -\infty}^{\infty} b_n\, g(t - (n + 1)T) \qquad (6.137)$$

We can express the signal $y(t)$ at the output of the matched-filter, whose impulse response is matched to the transmit filter, $c(t) = Kg(T - t)$: [2]

$$y(t) = x(t) \otimes K g(T-t)$$

$$= K \int_{-\infty}^{\infty} x(\tau) \, g(T-t+\tau) d\tau$$

$$= K \sum_{n=-\infty}^{\infty} b_n \int_{-\infty}^{\infty} g(\tau-nT) \, g(T-t+\tau) d\tau$$

$$+ \alpha K \sum_{n=-\infty}^{\infty} b_n \int_{-\infty}^{\infty} g(\tau-(n+1)T) \, g(T-t+\tau) d\tau$$

$$= K \sum_{n=-\infty}^{\infty} b_n \, R_g(t-(n+1)T)$$

$$+ \alpha K \sum_{n=-\infty}^{\infty} b_n \, R_g(t-(n+2)T)$$

$$(6.138)$$

where the autocorrelation function of the transmit pulse $g(t)$ is defined by

$$R_g(t_1-t_2) = R_g(t_2-t_1)$$

$$= \int_{-\infty}^{\infty} g(\tau-t_1) \, g(\tau-t_2) d\tau$$

$$(6.139)$$

E denotes the energy of the pulse $g(t)$ and $K = 1/\sqrt{E}$. When $y(t)$ is sampled at $t = kT$, one gets $y_k = y(kT)$:

$$y(kT) = y_k = k \sum_{n=-\infty}^{\infty} b_n \, R_g((k-n-1)T)$$

$$+ \alpha k \sum_{n=-\infty}^{\infty} b_n \, R_g((k-n-2)T)$$

$$= k \sum_{n=-\infty}^{\infty} b_n \, R_g[k-n-1]$$

$$+ \alpha k \sum_{n=-\infty}^{\infty} b_n \, R_g[k-n-2]$$

$$= \sqrt{E}(b_{k-1} + \alpha \, b_{k-2})$$

$$(6.140)$$

where the autocorrelation function of i.i.d. bits sequence is given by

$$R_g[n] = E\delta(n) \qquad (6.141)$$

Note that if the receive filter is matched to $g(-t)$ instead of to $g(T-t)$, then the output given by (6.140) becomes:

$$y_k = \sqrt{E}(b_k + \alpha \, b_{k-1}) \qquad (6.142)$$

The matched filter output at $t = kT$ contains an ISI term, as shown in (6.140) and (6.142), and a noise term n_k with zero-mean and variance σ^2. Using (6.140) with noise n_k at the input of a zero-forcing equalizer with $p = 3$ taps, the equalizer output may be written as:

$$z_k = \sum_{n=0}^{2} w_n \, (y_{k-n} + n_{k-n})$$

$$= \sqrt{E} \sum_{n=0}^{2} w_n \, (b_{k-n-1} + \alpha b_{k-n-2}) + \sum_{n=0}^{2} w_n \, n_{k-n}$$

$$= w_0 \sqrt{E}(b_{k-1} + \alpha b_{k-2}) + w_1 \sqrt{E}(b_{k-2} + \alpha b_{k-3})$$

$$+ w_2 \sqrt{E}(b_{k-3} + \alpha b_{k-4}) + N_k$$

$$= \underbrace{w_0}_{1} \sqrt{E} b_{k-1} + \underbrace{(w_0 \alpha + w_1)}_{0} \sqrt{E} b_{k-2}$$

$$+ \underbrace{(w_1 \alpha + w_2)}_{0} \sqrt{E} b_{k-3} + w_2 \alpha \sqrt{E} b_{k-4} + N_k$$

$$(6.143)$$

To minimize the ISI, the coefficient of b_{k-1} is forced to be equal to unity, but the coefficients of b_{k-2} and b_{k-3} are forced to be zero. Then, the tap weights are chosen as follows: $w_0 = 1$, $w_1 = -\alpha$, and $w_2 = \alpha^2$. Consequently the signal at the output of the equalizer at $t = kT$ may be written as

$$z_k = \sqrt{E} b_{k-1} + \underbrace{\alpha^3 \sqrt{E} b_{k-4}}_{\text{residual ISI}} + N_k \qquad (6.144)$$

Since $\alpha < 1$, the residual ISI term is usually negligible compared to the first term; hence, the ISI is suppressed. To achieve better ISI suppression, one should evidently increase the number of taps in the equalizer.

The variance of the noise term in (6.143) is given by

$$E\left[|N_k|^2\right] = E\left[\left|\sum_{n=0}^{2} w_n\, n_{k-n}\right|^2\right]$$

$$= E\left[\left|n_k - \alpha n_{k-1} + \alpha^2 n_{k-2}\right|^2\right]$$

$$= \left(1 + \alpha^2 + \alpha^4\right)\sigma^2$$

$$(6.145)$$

Using (6.144), the SNR at the equalizer output may be expressed as follows:

$$SNR = \dfrac{E}{E\left[\left|\displaystyle\sum_{n=0}^{2} w_n\, n_{k-n} + \alpha^3 \sqrt{E}\, b_{k-3}\right|^2\right]}$$

$$= \dfrac{E}{\left(1 + \alpha^2 + \alpha^4\right)\sigma^2 + \alpha^6 E}$$

$$= \dfrac{1}{\left(1 + \alpha^2 + \alpha^4\right)\dfrac{1}{E/\sigma^2} + \alpha^6}$$

$$(6.146)$$

Figure 6.43 shows the variation of (6.146) with E/σ^2, that corresponds to the SNR with $\alpha = 0$. Hence, one may observe the degradation in the output SNR caused by the multipath of channel gain α. The degradation becomes larger at higher values of α and E/σ^2. In order to demonstrate the effect of ISI in (6.146), the SNR with and without ISI term (by neglecting the term α^6 in (6.146) and shown in the curves labelled "no ISI") are compared in Figure 6.44. When the ISI is neglected the output SNR is decreased with respect to E/σ^2 by a factor $\sum_{i=0}^{p-1} \alpha^{2i}$; this simply shifts the SNR curve for $\alpha = 0$ to the right by that amount. One may observe that the degradation for $\alpha > 0.5$ at high values of E/σ^2 is dominated by ISI; so, there is a need more taps in zero-forcing equalization. Performance of zero-forcing equalization may be improved considerably by using equalization in the MMSE sense.

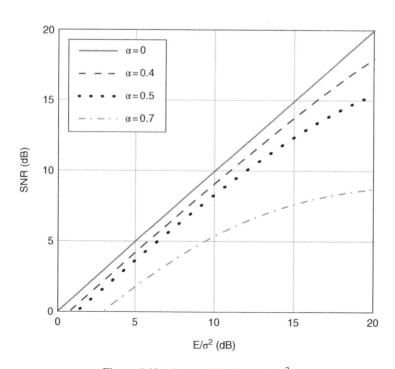

Figure 6.43   Output SNR Versus E/σ^2.

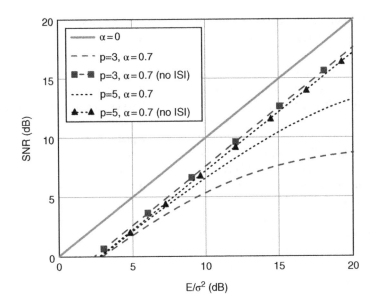

Figure 6.44   Effect of the Residual ISI Term in the Output SNR of a Zero-Forcing Equalizer.

References

[1] S. Haykin, *Communication Systems* (3rd ed.), John Wiley: New York, 1994.

[2] J. G. Proakis and M. Salahi, *Communication Systems Engineering* (2nd ed.), Prentice Hall: New Jersey, 2002.

[3] A. B. Carlson, P. B. Crilly, and J. C. Rutledge, *Communication Systems* (4th ed.), McGraw Hill, Boston, 2002.

[4] B. Sklar, *Digital communications, Fundamentals and Applications* (2nd ed.), Prentice Hall: New Jersey, 2003.

[5] I. A. Glover, and P. M. Grant, *Digital Communications* (2nd ed.), Pearson Prentice Hall: Harlow, 2004.

Problems

1. Consider the following binary signalling schemes with equiprobable signals:

i. $s_1(t) = -s_0(t) = \begin{cases} A & 0 < t < T/2 \\ -A & T/2 < t < T \end{cases}$

ii. $s_1(t) = -s_0(t) = A, \quad 0 < t < T$

iii. $s_1(t) = \begin{cases} A & 0 < t < T \\ 0 & otherwise \end{cases}$,

$s_0(t) = \begin{cases} A & 0 < t < 3T/4 \\ -A & 3T/4 < t < T \\ 0 & otherwise \end{cases}$

iv. $s_0(t) = 0, \quad 0 < t < T \quad s_1(t) = A, \quad 0 < t < T$

v. $s_0(t) = 0, \quad 0 \le t \le T$,

$s_1(t) = \begin{cases} A & 0 \le t \le T/2 \\ -A & T/2 \le t \le T \end{cases}$

where A is a constant.

For each of the above each of the above signalling schemes:

a. Determine the correlation coefficient between the two signals.
b. Give the structure of the matched-filter.
c. Draw the signal-space diagram related to the matched-filter output.
d. Determine the optimal decision threshold level.
e. Determine the probability of error.

2. Binary data is transmitted by using a pulse $g(t)$ for **0** and a pulse $Ag(t)$ for **1**. The

energy of the pulse $g(t)$ is equal to E and A can be positive or negative number.

a. Draw the block diagram of a matched filter receiver used for optimal detection of this signalling scheme,
b. Determine the optimum impulse response of the matched filter,
c. Determine the optimum detection threshold if 0 and 1 are not equally likely.
d. Determine the probability of error if 0 and 1 are not equally likely.
e. Is the choice of the pulses $g(t)$ and $Ag(t)$ a good one? Explain why. Discuss the implications of this choice on the total transmit power of the transmitter. How can the probability of error be minimized in this system?

3. A matched filter is used for the reception of the pulse $g(t) = e^{-\alpha t},\ 0 \le t \le T$ in an AWGN channel with zero mean and PSD $N_0/2$ W/Hz.

a. Find and sketch the impulse response of the filter which is matched to the pulse g(t).
b. Derive an expression for the maximum output SNR in terms of the parameters α, T and N_0.
c. Consider the signalling scheme characterized by $s_0(t) = -g(t)$ and $s_1(t) = g(t)$ for $0 \le t \le T$, with equal probability, in a AWGN channel with PSD $N_0/2$ watts/Hz. Sketch the pdf of the decision variable corresponding to $s_0(t)$ and $s_1(t)$, and derive the probability of error.
d. Repeat (a)–(c) for $g(t) = 2\,t/T,\ 0 \le t \le T$.

4. An antipodal binary signalling scheme uses equiprobable pulses with amplitudes $+A$ and $-A$ during the interval $(0, T)$ in an AWGN channel with single-sided

PSD of N_0. Assume that the received signal is detected with a matched filter:

a. Write down the expression for the P_b at the matched filter output.
b. Express the maximum data rate that can be transmitted with a P_b less than or equal to 10^{-6} in terms of A, T and N_0.
c. Determine the SNR at the matched filter output and the corresponding bit rate for $A = 1\,\text{mV}$ and $N_0 = 10^{-12}$ W/Hz.

5. In an antipodal system, $-A$ is sent for bit **0** and A for bit **1**. The probability of transmitting a bit **0** is p_0 and p_1 for **1**. The matched filter output is characterized by $x_{1,2} = \pm\sqrt{E} + n$ where E denotes the bit energy and n is the noise sample with an exponential distribution:

$$p(n) = \frac{1}{\sigma\sqrt{2}}\exp\left(-\frac{\sqrt{2}|n|}{\sigma}\right)$$

Here, σ denotes the rms-value of the noise.

a. Using the equality $\left|x + \sqrt{E}\right| - \left|x - \sqrt{E}\right| = 2x$, for $-\sqrt{E} \le x \le \sqrt{E}$, determine the optimum decision threshold.
b. Determine the average error probability.
c. Calculate the SNR in dB required for the error probability 10^{-6}, assuming $p_0 = p_1$.
d. Repeat (c) when the noise is Gaussian instead of being exponentially distributed. Discuss whether Gaussian noise or the exponentially distributed noise is more favorable for the system performance.

6. Consider matched-filter reception of the two signals $s_1(t)$ and $s_2(t)$ as given below:

$$s_1(t) = \begin{cases} A & 0 < t < T \\ 0 & otherwise \end{cases},$$

$$s_2(t) = \begin{cases} A & 0 < t < 3T/4 \\ -A & 3T/4 < t < T \\ 0 & otherwise \end{cases}$$

where A is a constant and T denotes the symbol duration.

a. Determine the transmitted signal energies for binary symbols **1** and **0**.
b. Are these signals correlated? If so, determine the correlation coefficient between these signals.
c. Draw the block diagram of a matched-filter receiver for these two signals, using matched filters with unit energy. Write down the analytic expressions at appropriate points when the input signal is $s_1(t) + w(t)$, where $w(t)$ is AWGN with zero mean and variance $N_0/2$.
d. Determine the average SNR at the output of the matched filter receiver.
e. Determine the average bit error rate when $s_1(t)$ is sent for binary **1** and $s_2(t)$ is sent for binary **0**.

7. A binary digital communication system uses the following signalling scheme for information transmission:

$$s_0(t) = 0 \qquad 0 \le t \le T$$
$$s_1(t) = \begin{cases} A & 0 \le t \le T/2 \\ -A & T/2 \le t \le T \end{cases}$$

a. Draw the block diagram of the optimum receiver for an AWGN channel with one-sided PSD N_0, assuming that the probability of transmitting **0** and **1** is $p_0 = p_1 = 0.5$. Write down the output of the optimum receiver when **1** and **0** are sent.
b. Derive analytically the optimum threshold, assuming that the probability of

transmitting **0** and **1** are $p_0 = 0.4$ and $p_1 = 0.6$.
c. Derive analytically the probability of error for the special case of $p_0 = p_1 = 0.5$. Compare the BER performance of this system with antipodal signaling.

8. Consider a binary orthogonal signalling system where E_b denotes the energy for each of the waveforms. The two signals are thus defined as

$$s_1 = \left(\sqrt{E_b} \;\; 0\right) \qquad s_2 = \left(0 \;\; \sqrt{E_b}\right)$$

a. Assuming that s_1 was transmitted in an AWGN channel with zero mean and noise variance $N_0/2$, write down the observation vector, $\mathbf{x} = (x_1 \; x_2)$, at the output of the demodulator.
b. Write down the pdf's of x_1 and x_2.
c. Determine the probability of error when s_1 and s_2 was transmitted. Determine the total probability of error, assuming that s_1 and s_2 are equiprobable.
d. Determine the total probability of error when s_1 and s_2 are defined as below.

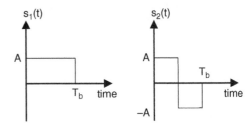

9. Consider that Fourier transform of a time waveform is given by $H(f) = T \operatorname{sinc}^2(fT)$.

a. Determine the impulse response $h(t)$ corresponding to $H(f)$.
b. Determine the impulse response of the filter matched to h(t).
c. Determine the output of the matched filter at the sampling instant $t = T$.

10. A pulse shaping circuit is defined by the following impulse response:

$$h(t) = \frac{1}{4}\delta(t-T) + \frac{1}{2}\delta(t) + \frac{1}{4}\delta(t+T)$$

   a. Determine and plot the frequency response corresponding to $h(t)$.

   b. Determine and plot the frequency response corresponding to $h(t) \otimes p_N(t)$, where $p_N(t)$ denotes the time waveform of the Nyquist pulse. Determine time pulse and explain if the resulting pulse can be categorized as a PRS pulse.

   c. Determine and plot the frequency response corresponding to $h(t) \otimes p_0(t)$, where $p_0(t)$ denotes the time waveform corresponding to $P_0(f) = 1/\text{sinc}(fT)$. Discuss relative advantages and disadvantages of the waveforms obtained in (a), (b) and (c).

11. The figure below shows a signalling scheme with signals $s_1(t)$ and $s_0(t)$ in an AWGN channel of zero-mean and PSD $N_0/2$.

   a. Design a matched-filter receiver that decides in favour of signals $s_1(t)$ and $s_0(t)$ over the observation interval $0 \leq t \leq 3T$, assuming that these two signals are equiprobable. Determine the corresponding error probability at the receiver output.

   b. Now assuming that $s_1(t)$ and $s_0(t)$ both carry three bits over the observation interval $0 \leq t \leq 3T$, design a matched-filter receiver that decides in favour of each of the equiprobable bits carried by $s_1(t)$ and $s_0(t)$ over $(i-1)T \leq t \leq iT$. Determine the corresponding error probability at the receiver output.

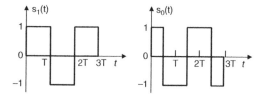

12. A binary signaling has the conditional probability density functions $f_Y(y|m_i)$ at the matched filter output as shown below:

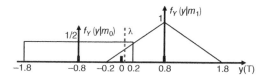

Assume that the probability of transmitting the signal m_i is p_i, $i = 1,2$.

   a. For a given decision threshold level of λ, determine the conditional probability of error when m_i, $i = 0,1$ is transmitted. Determine the average probability of error.

   b. Determine the optimum decision threshold level and the corresponding average probability of error.

   c. Assuming that $p_1 = 0.6$ and $p_2 = 0.4$, plot the average probability of error for $-0.2 < \lambda < 0.2$ and compare with that obtained for the optimal case.

   d. Determine the average error probability for $p_1 = p_2 = 0.5$ and compare with the result you obtained in part (b).

13. A ternary signal configuration at the matched-filter output in an AWGN channel is shown below:

   a. If $p(s_0) = p(s_1) = p(s_{-1}) = 1/3$, determine the optimum decision regions and the corresponding error probability of the optimum receiver.

   b. Repeat part (a) for $p(s_0) = 0.5$, $p(s_1) = 0.2$, and $p(s_{-1}) = 0.3$.

   c. Repeat part (a) for $p(s_0) = 0.5$, $p(s_1) = 0.25$, and $p(s_{-1}) = 0.25$. Compare your results with those in part (a) and (b).

14. The detector following a matched-filter demodulator faces the following hypothesis-testing problem:

$$H_1 : z(T) = \sqrt{E_b} + w(T)$$
$$H_0 : z(T) = -\sqrt{E_b} + w(T)$$

where H_1 and H_0 correspond to the transmission of binary symbols m_1 and m_0, with $p(H_1) = p$, and $p(H_0) = 1-p$. E_b denotes the bit energy and $w(T)$ is the sampled Gaussian random variable with zero mean and variance σ^2. The maximum a posteriori (MAP) rule for deciding whether H_1 or H_0 is transmitted may be written from (6.51) as

$$f_Z(z|H_0)p(H_0) \underset{H_1}{\overset{H_0}{\gtrless}} f_Z(z|H_1)p(H_1) \quad \text{where}$$

$f_Z(z|H_1)$ and $f_Z(z|H_0)$ denote the conditional pdf's.

a. Formulate the MAP test to determine whether H_1 or H_0 is transmitted.
b. Determine an expression for the probability of error for $p = 1/2$.
c. Repeat (a) for $p = 1/2$ when the MAP decision is based on the transmission of the same symbol L times (repetition coding), instead of only once. Compare the performance of this test to the one in (a).

15. In some applications, the noise at the demodulator input may not have a flat PSD. The noise with non-flat PSD is called as colored noise (not being white). In such cases, a whitening filter is inserted prior to the receiver in order to whiten the colored noise. Consider that the noise at the receiver input is Gaussian but with a non-white PSD, $G_n(f)$. The transfer function $H_w(f)$ of the whitening filter is chosen so that the noise $w(t)$ at its output is white with double-sided PSD $N_0/2$:

$$G_w(f) = G_n(f)|H_w(f)|^2 = \frac{N_0}{2}$$

where $n(t)$ is the zero-mean colored noise with the PSD given by $G_n(f)$.

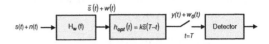

Assume that the whitening filter is a real-valued filter and $\tilde{s}(t)$ represent the output of the whitening filter when the input pulse is s(t). Finally, y(t) and $w_0(t)$ represent the responses of the optimum matched filter $h_{opt}(t)$ to $\tilde{s}(t)$ and w(t), respectively.

a. Determine the frequency response of the whitening filter $H_w(f)$.
b. If h(t) is matched to $\tilde{s}(t)$, determine the noise power, signal power and the SNR at the detector input, that is, $SNR = (y(T))^2/\sigma_{w_0}^2$.

16. Consider that a baseband binary signal $s_i(t)$, $i = 0,1$ and additive colored noise is received by a matched filter. Due to the use of the whitening filter as in Problem 6.15, the rest of the receiver must be matched to the distorted signaling waveforms $\tilde{s}_i(t)$ at the output of the whitening filter, that is, $h_{opt}(t) = \tilde{s}_1(T-t) - \tilde{s}_0(T-t)$.

a. Assuming that $s_1(t)$ and $s_0(t)$ are correlated, show that the signal energy at the matched filter receiver output may be written as follows:

$$E = \int_0^T [\tilde{s}_1(t) - \tilde{s}_0(t)]^2 dt$$
$$= \frac{N_0}{2} \int_{-\infty}^{\infty} \frac{|S_1(f) - S_0(f)|^2}{G_n(f)} df$$

where $S_i(f) = \mathfrak{I}[s_i(t)]$, $i = 0, 1$ and T denotes the symbol duration.

b. Show that the SNR at the matched filter output is given by

$$\frac{2E}{N_0} = \int_{-\infty}^{\infty} \frac{|S_1(f) - S_2(f)|^2}{G_n(f)} \, df$$

Hint: If $s_i(t)$, $i = 0, 1$ is real, then

$$\int_{-\infty}^{\infty} s_1(t)s_2(t)\,dt = \int_{-\infty}^{\infty} S_1(f)S_0^*(f)\,df = \int_{-\infty}^{\infty} S_1^*(f)S_0(f)\,df$$

17. The binary data stream $\{1011100101\}$ is applied to the input of a duabinary encoder.

   a. Give the duobinary encoder output and corresponding receiver outputs, without using a precoder.
   b. Assuming that the fourth bit is erroneously received as 0, determine the new receiver output.
   c. Repeat (a) and (b) when a precoder is used.

18. Consider the following demodulated (modified) duobinary data stream: $\{2\ 2\ 0\ 0 -2\ 0\ 2\ 0\ 0\ 0 -2\ 0\}$

   a. Decode the received stream, assuming it is duobinary. State if there is any error in the demodulated stream. Can you estimate the correct transmitted digit sequence? Is there a single correct data stream?
   b. Repeat (a) assuming that the above stream is modified duobinary.

19. Consider that a pulse $+g(t)$ is sent for binary 1 and $-g(t)$ for 0. The rectangular pulse $g(t)$ is defined by

$$g(t) = \begin{cases} 1 & 0 \le t \le T_b \\ 0 & otherwise \end{cases}$$

The channel impulse response is given by

$$h(t) = \begin{cases} 1 & 0 \le t \le T_m \\ 0 & otherwise \end{cases}$$

where $T_m > T_b$.

   a. Plot the signal at the channel output when the input is a binary 1.
   b. Illustrate the effect of ISI when the input bit sequence is 011.
   c. What is the impulse response of the optimim matched filter for this signal.
   d. Plot the time variation of the matched-filter output for the input bit sequence 011, and when it is sampled at $t = kT_b$.

20. Consider the baseband communication system, where a signal $x(t) = A \sum_{n=-\infty}^{\infty} a_n$ $g(t - nT)$ where $a_i = \pm 1$ with equal probability, is received by a LTI filter. The response of the LTI filter, with impulse response $h(t)$, is given by

$$y(t) = x(t) \otimes h(t) = \sqrt{E_b} \sum_{n=-\infty}^{\infty} a_n\, p(t - nT)$$

When sampled at $t = kT$, the sampled output can be written as

$$y_k = \sqrt{E_b} \sum_{n=-\infty}^{\infty} a_n\, p_{k-n}$$

Assuming that the pulse amplitudes are given by $p_{-1} = 0.1$, $p_0 = 1$, $p_1 = 0.2$, and $p_n = 0$ for all integers n for which $|n| > 1$, we consider the transmission of only three bits, namely a_{-1}, a_0 and a_1. The noise variance at the filter output is $\sigma^2 = N_0/2$. Assuming that $a_0 = 1$ without loss of generality,

   a. Determine the sequence of pulses at $t = 0$ and the corresponding filter output, that is, the desired response and the ISI terms. Determine the maximum and minimum values of the filter output.

b. Determine the corresponding expressions for the maximum, the minimum and the average probability of error in terms of E_b/N_0, if an infinite sequence of pulses is transmitted over the channel.

c. Explain why the same error patterns are observed if $a_0 = -1$.

21. A zero-forcing equalizer (ZFE) forces the ISI to be zero at all instants $t = iT$ at which the channel output, $y(t)$, is sampled except for at $i = 0$ where the wanted symbol occurs. The output of a zero-forcing equalizer can be written as

$$y_i = \sum_{k=-N}^{N} w_k p_{i-k}$$

$$= \begin{cases} 1 & i = 0 \\ 0 & i \neq 0 \end{cases}, \quad i = 0, \pm 1, \pm 2, \cdots, \pm N$$

Therefore, zero-forcing equalization requires the solution of $2N+1$ simultaneous equation in order to determine $2N+1$ tap coefficients w_k.

a. Express the zero-ISI condition in the matrix form as $\mathbf{P} \mathbf{w} = \mathbf{y}$, where $\mathbf{P}$ denotes a $(2N+1) \times (2N+1)$ matrix, the tap vector is defined by $\mathbf{w} = [w_{-N} \; w_{-N+1} \; \ldots w_N]^T$ where T denotes the transpose of a matrix and $\mathbf{y} = [000..1....000]^T$ where all $2N+1$ elements are zero except for the one with $i = 0$.

b. Solve the matrix equation to obtain the tap vector.

22. The samples of a received signal $p_k = p(kT)$ are given by $p_{-1} = 0.2$, $p_0 = 1$, $p_1 = 0.3$ and $p_k = 0$, $|k| > 1$. This signal passes through a 3-tap FIR filter equalizer with tap weights w_{-1}, w_0, and w_1:

$$y_i = y(iT) = \sum_{k=-1}^{1} w_k p_{i-k}$$

a. Determine the tap coefficients for this three-tap zero forcing equalizer that yield zero-ISI.

b. Determine the analytic expression for the equalized pulse y_i and determine its values for $i = 0$, $i = \pm 1$, and $i = \pm 2$. Plot the equalized pulse and compare with the received pulse p_i.

23. Sampled matched filter output of an AWGN channel with ISI is given by $y_k = a_k + i_k + w_k$, where $a_k = \pm \sqrt{E_b}$ and the ISI term is defined by the pdf $f_I(i_k) = 1/3 \; \delta(i_k + \alpha) + 1/3 \; \delta(i_k) + 1/3 \; \delta(i_k - \alpha)$. AWGN channel is characterized by zero-mean and variance $\sigma^2 = N_0/2$.

a. Determine the probability of bit error P_b in terms of E_b/N_0 and α.

b. Plot the variation of the P_b with ISI as a function $0 \leq \alpha \leq 1$ for $E_b/N_0 = 10$, 20, 30 dB.

c. Evaluate P_b for $E_b/N_0 = 10$ dB and $\alpha = 0.1\sqrt{E_b}$, and compare your result with P_b without ISI.

24. Consider a communication channel which is characterized by two signal paths, with the following channel gain α and time delay τ: $(\alpha_1, \tau_1) = (1, 0 \; \mu s)$, and $(\alpha_2, \tau_2) = (0.2, 1 \; \mu s)$.

a. Determine the impulse response of this channel?

b. Determine the path length difference in meters between the two paths? Also determine the ratio of signal powers in dB of the two multipath signals.

c. Express analytically the received signal in terms of the transmitted signal s(t).

d. If the channel transmission rate is 500 kbps, explain whether this channel is distortionless?

e. Determine the tap-weights of a 5-tap FIR filter to remove the ISI induced by this channel. Determine the amplitude of the residual ISI signal?

25. A wireline AWGN channel, acting as an ideal LPF with maximum frequency of 4 kHz, is used to transmit data using M-ary PAM.

a. For binary-PAM, determine the highest bit rate that can be transmitted without ISI?

b. Determine the required E_b/N_0 to achieve a $P_b = 10^{-6}$.

c. Repeat (a) and (b) for 4-ary PAM.

26. A speech signal bandlimited to 4 kHz is sampled at 8000 samples/s.

a. For binary PAM transmission, determine the symbol rate and the minimum system bandwidth with Nyquist filter and a raised-cosine filter wth a roll-off factor of $\alpha = 1$.

b. Repeat (a) for 8-ary PAM transmission.

c. Assuming that both systems have the same pdf, compare binary and 8-ary transmissions from the viewpoint of their requirements for transmission bandwidth and the transmit power.

27. A baseband system transmits at a symbol rate of 160 kbps.

a. Determine the bandwidth required for transmitting this data sequence as a binary PAM signal. Determine the roll-off factor α of a raised cosine filter if the required bandwidth is 100 kHz.

b. Determine M for M-ary PAM transmission over a channel bandwidth of B = 40 kHz.

28. Consider a multipath channel which consists of a direct path and a path delayed by T seconds:

$$h(t) = \delta(t) + \alpha\delta(t-\tau)$$

where $\alpha < 1$ denotes the relative attenuation of the second path. An input signal $g(t)$, whose spectrum is bandlimited to W Hz, is passed through the channel.

a. Determine and plot the magnitude of the channel transfer function assuming that the bandwidth is limited by $W = 1/(2\tau)$, where $\tau = 0.5$ μs.

b. Show that the channel output is given by

$$y(t) = g(t) \otimes h(t) = g(t) + \alpha g(t-\tau)$$

c. Determine the output of the demodulator at $t = kT$, $k = 0$, ±1, ±2, ... that employs a filter matched to the pulse $g(t)$ of duration T at the channel input. Identify the ISI term and explain the mechanism of the occurrence of the ISI.

d. Express the matched filter output at $t = kT$ if the input signal is given by $x(t) = \sum_{k=-\infty}^{\infty} a_k\, g(t-kT)$, where a_k denotes the input bit sequence.

e. Determine the probability of error for $\tau = T$ if the transmitted signal is binary antipodal, $a_k = \pm 1$.

29. A raised-cosine pulse is used for transmitting digital information over a bandlimited channel.

a. Since the bandlimited channel distorts the signal pulses, the following sampled values of the filtered received pulse are obtained:

$$p_{-2} = 0.2, p_{-1} = -0.3, p_0 = 1,$$
$$p_1 = -0.1, p_2 = 0.3.$$

Assuming $a_0 = 1$ for the desired bit, list all possible sequences of +1 s and −1 s and the corresponding value of the maximum ISI.

b. Assuming that all binary digits are equally probable and independent, determine the probability of occurrence of the maximum ISI in (a).

30. Consider a pulse $p(t)$ with the Fourier transform $P(f)$ as given by

$$P(f) = \begin{cases} \dfrac{1}{W}\left(1-\dfrac{|f|}{W}\right) & -W \le f \le W \\ 0 & \text{otherwise} \end{cases}$$

a. Explain whether $P(f)$ satisfies the Nyquist criterion for ISI-free transmission?
b. Determine $p(t)$ and verify your result in part (a). What is the decay rate of this pulse?
c. If the pulse satisfies the Nyquist criterion, determine the transmission rate in bps and the roll-off factor for $W = 10^6$.
d. Compare relative advantages and disadvantages of this waveform compared to raised-cosine waveform.

31. Multipath channels result in dispersion in the time domain and hence lead to ISI. Consider a channel with the following impulse response:

$$h(t) = \delta(t) + 0.2\,\delta(t-T) + 0.2\,\delta(t+T)$$

where T denotes the symbol duration.

a. If $s(t)$ denotes the signal at the channel input, determine the output signal. How can you interpret the signal at the channel output?
b. Determine the channel frequency response.
c. Assuming that the channel frequency response is limited by $|f| \le B$ where $B = 4\,\text{kHz}$, determine the frequency response

characteristics of the optimum root raised-cosine transmit and receive filters with $\alpha = 1$ that yield zero ISI at a rate of $R_s = 1/T$ symbols/s in the AWGN channel with two-sided PSD $N_0/2$. Plot the frequency response of transmit and receive filters.

d. Determine the symbol rate R_s, symbol duration and the bandwidth.

32. We know from Figure 6.4 that low-pass filtering leads to ISI. The magnitude of the frequency response of an RC low-pass filter is given by

$$|H(f)| = \frac{1}{\sqrt{1 + (f/B_{3-dB})^2}}, \quad |f| \le B_{3-dB}$$

where $B_{3-dB} = 1/(2\pi RC)$ denotes the half-power bandwidth of the filter. This filter with $B_{3-dB} = 4\,\text{kHz}$ is used to filter out binary transmission at a data rate of 4 kbps in an AWGN channel.

a. Determine the roll-off factor of a raised-cosine filter that eliminates the ISI in this channel.
b. Determine and plot the frequency transfer functions of optimum transmitting and receiving filters.

33. A signal pulse g(t), defined as

$$g(t) = \cos\frac{\pi}{T}t, \quad -T/2 \le t \le T/2,$$

is transmitted through a ideal low-pass channel with frequency response characteristic as described by

$$H(f) = \begin{cases} 1 & -W/2 < f < W/2 \\ 0 & \text{otherwise} \end{cases}$$

The matched filter receiver channel output corrupted by AWGN with PSD $N_0/2$.

a. Determine and plot the frequency response of the pulse $g(t)$.

b. Determine and plot the frequency response of the output of the ideal low-pass channel.

c. Determine the signal energy and the noise variance at the output of the filter matched to the received signal.

d. What value of WT leads to the maximum signal energy, and the maximum SNR, at the output of the matched filter. Determine the corresponding received pulse energy.

34. A NRZ 8-ary PAM signal plus Gaussian noise with PSD $N_0 = 2 \times 10^{-6}$ W/Hz are applied to a detector. Assume that the bit rate of the signal is $R_b = 150$ kbps; the spacing between transmitted symbol levels is 2 V, that is, ± 1 V, ± 3 V, ± 5 V, ± 7 V; and no inter-symbol interference (ISI) is present at the sampling instants at the output of a raised cosine filter with a roll-off factor of $\alpha = 0.4$.

a. Show the values of the signal points at the matched-filter output,

b. Calculate the signal-to-noise power ratio at the output of the matched-filter receiver at the sampling instant,

c. Calculate the symbol error probability assuming Gray encoding.

d. Determine the transmission bandwidth.

35. We want to transmit at maximum bit rate of 300 kbps in a bandwidth of 100 kHz with $P_b \le 10^{-6}$ using M-ary PAM with Gray encoding in an AWGN channel. In order to mitigate the resulting ISI, raised-cosine pulse shaping is used. In order to minimize the transmit power, we desire to use the minimum possible value of M. Determine M, the symbol rate, the roll-off factor and the E_b/N_0.

36. A communication system, using 4-PAM signalling at a bit rate of 1 Mbps is designed using a raised-cosine filter with a roll-off factor of $\alpha = 1$.

a. Determine the bandwidth of this system

b. Determine the minimum channel bandwidth permitted to avoid ISI.

c. Determine the roll-off factor if the system bandwidth is limited to 400 kHz.

d. Discuss the effects of the roll-off factor in the channel bandwidth and the ISI.

37. A data stream is transmitted at 10 k symbols/s in a 4-PAM system.

a. What is the minimum bandwidth required for transmission, assuming a root-raised cosine filter is used with $\alpha = 0.25$?

b. If it is required to transmit the information in half the time, what would be the value of M for the same transmission bandwidth?

38. Consider a baseband antipodal signaling system where a binary 1 is transmitted as +1 V and binary 0 is transmitted as −1 V. The transmission rate is given by $R_b = 1/T_b$.

a. Plot the PSD of a periodic 1010 … data stream,

b. Plot the PSD of an equiprobable random data stream.

c. Plot the PSD of the transmit Nyquist filter.

39. An analog voice signal, bandlimited to $W = 4000$ Hz, is to be transmitted using M-ary PCM. The signal to quantization noise ratio is required to be higher than 50 dB.

a. Draw the block diagram of the transmitter section of this system

b. What is the minimum number of bits per sample and bits per PCM word that should be used in the PAM? What is the actual value of the signal to quantization noise ratio.

c. What is the minimum required sampling rate, and what is the resulting bit rate?

d. If polar NRZ signalling is used with a pulse duration of 62.5 μs, determine the required transmission bandwidth and the symbol rate. The transmission bandwidth is defined between $f = 0$ to the first null of the signal PSD.

e. Determine the value of M.

f. Determine the spectrum efficiency, which is defined as the number of bits transmitted per unit bandwidth.

40. Consider the unipolar NRZ and bipolar NRZ binary signaling with bit duration T, as shown in Figure 5.25.

a. Determine the null-to-null bandwidths of the two line codes.

b. Compare peak and average powers required for each case.

c. Determine the pdf when the equi-probable unipolar and bipolar NRZ signaling systems are received by a matched filter.

d. Repeat (a)–(c) for unipolar and bipolar RZ binary signaling with bit duration T.

41. Consider 16-PAM transmission of a data stream of $R_b = 9600$ bps. Determine the transmission bandwidth if we use a raised-cosine pulse shape having a roll-off factor of 0.25.

42. A telephone channel is assumed to be flat over 3.4 kHz bandwidth.

a. Select a symbol rate and a power efficient constellation size to achieve 9600 bits/s signal transmission.

b. If a raised cosine pulse is used for the transmission, select the roll-off factor.

43. An music signal with 20 kHz bandwidth is sampled at a rate of 48 kHz and the samples are quantized into 256 levels and coded into M-ary PAM raised-cosine pulses with a roll-off factor $\alpha = 0.5$. If a 60 kHz bandwidth is available to transmit the data, then determine the smallest acceptable value of M, the symbol rate and the transmission bandwidth.

44. A message source outputs one of four possible messages at a rate of 10^6 messages/s. Using these messages at its input, the modulator produces one of eight possible signals, each T seconds, for transmission.

a. Give a table associating each possible message at the modulator input with each of the eight symbols at the modulator output.

b. Determine the symbol rate at the modulator output assuming that there are no overhead bits are inserted by the modulator.

c. Determine the equivalent bit rate at the modulator output.

7

Optimum Receiver in AWGN Channel

In Chapter 6 we considered the matched filter reception of baseband pulses in an AWGN channel, and the ISI due to band-limiting and multipath. We also considered pulse shaping and other techniques for ISI mitigation. In this chapter, we will consider the details of the receiver design for the reception of sequences of M-ary symbols and the resulting error probability in a baseband channel in the presence of AWGN.

7.1   Introduction

Consider a digital communication system with a block diagram as shown in Figure 7.1, where a message source emits every T seconds one symbol from an alphabet of M symbols, m_1, $m_2,\ldots,$ m_M with a priory probabilities p_1, $p_2,\ldots,p_M$. In most cases of interest, M symbols of the alphabet are equally likely:

$$p_i = P(m_i) = \frac{1}{M} \quad i = 1, 2, \ldots, M \qquad (7.1)$$

Transmitter converts the message source output m_i into a distinct electrical signal $s_i(t)$, suitable for transmission over the channel. Hence the transmission of digital information is accomplished by using one of the M signals every T seconds:

$$\{s_i(t), 0 \leq t \leq T, i = 1, 2, \cdots, M\} \qquad (7.2)$$

The energy of the real-valued energy signal $s_i(t)$ over the symbol duration T:

$$E_i = \int_0^T s_i^2(t)\, dt, \ i = 1, 2, \ldots, M \qquad (7.3)$$

Depending on the application of interest, a priori probabilities and the symbol energies of the M signals may be the same. However, we assume that they are not the same. As an example of M-ary transmission, 8-PAM is shown in Figure 6.24.

In this so-called *M-ary transmission*, $s_i(t)$ is used to transmit the information carried by the symbol m_i, consisting of $k = \log_2 M$ bits, in

Digital Communications, First Edition. Mehmet Şafak.
© 2017 John Wiley & Sons Ltd. Published 2017 by John Wiley & Sons Ltd.
Companion website: www.wiley.com/go/safak/Digital_Communications

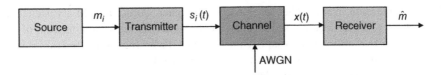

Figure 7.1 Block Diagram of a Baseband Digital Communication System.

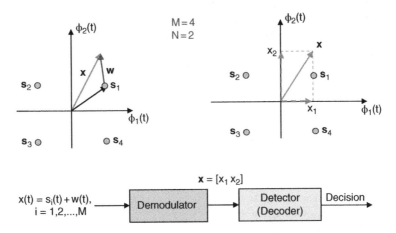

Figure 7.2 Signal Constellation at the Receiver Input and the Observation Vector at the Demodulator Output for a 4-ary Signal of Dimension $N = 2$.

the symbol duration T. Noting that a symbol m_i is $k = \log_2 M$ bits long, the symbol duration is $T = kT_b$, where T_b denotes the bit duration, and the symbol energy is given by $E = kE_b$. The symbol duration and the symbol rate are related to the bit duration and the bit rate as follows:

$$T = kT_b \quad \Rightarrow \quad R = R_b/k \qquad (7.4)$$

where $R_b = 1/T_b$ denotes the bit rate in bits/s and $R = 1/T$ is the symbol rate in symbols/s.

The channel depicted in Figure 7.1 is assumed to be linear and to have sufficiently wide bandwidth for transmission of $s_i(t)$ so as to cause minimum distortion and ISI. In the AWGN channel, where the noise $w(t)$ has zero mean and a one-sided PSD of N_0, the received signal may be written as

$$x(t) = s_i(t) + w(t), \quad 0 \le t \le T, \quad i = 1, 2, \ldots, M \qquad (7.5)$$

A receiver may be subdivided into two functionalities; a signal demodulator and a decoder (detector). The *demodulator* converts the received signal x(t) into an N-dimensional observation vector $\mathbf{x} = [x_1 \ x_2 \ \ldots \ x_N]$ with orthogonal components, where N denotes the dimension of the transmitted signal. Two realizations of the signal demodulator, as a correlator or a matched filter, perform identically at the sampling instants as already demonstrated in Chapter 6. Based on the observation vector **x** and some preset threshold levels, the *detector* estimates the transmitted symbol. Hence, given the received noisy signal (7.5), a receiver makes a decision as to which of the M signals, hence symbols, is transmitted in the considered symbol duration. A receiver is desired to be optimum in the sense that it minimizes the probability of making an error in estimating the transmitted symbol m_i. For example, Figure 7.2 shows the transmission of a 4-ary signal with dimension $N = 2$. The received signal given by (7.5) is decomposed by the demodulator into two orthogonal components $x_1(t)$ and $x_2(t)$ along the orthonormal basis functions $\phi_1(t)$ and $\phi_2(t)$ to enable the detector to make the best estimate of the transmitted symbol. [1][2][3][4]

In this chapter, we will consider the design and the performance evaluation of optimum receivers in an AWGN channel based on the assumption that no ISI or distortion is introduced by the channel.

7.2 Geometric Representation of Signals

Each signal in $\{s_i(t), i = 1, 2, .., M\}$ is defined by the signal vectors $\{s_i, i = 1, 2, ..., M\}$ in an N-dimensional Euclidean signal space, with N mutually perpendicular axes, $\phi_1, \phi_2 ..., \phi_N$:

$$s_i = [s_{i1} \quad s_{i2} \quad \cdots \quad s_{iN}]^T, \quad i = 1, 2, ..., M \quad (7.6)$$

In other words, any M-ary signal may be expressed as a linear combination of N orthonormal basis functions, $\{\phi_i(t), i = 1, 2, ..., N\}$, where $N \leq M$: [1][3]

$$s_i(t) = \sum_{j=1}^{N} s_{ij} \, \phi_j(t), \quad 0 \leq t \leq T, \quad i = 1, 2, ..., M$$

$$s_{ij} = \int_0^T s_i(t) \, \phi_j(t) \, dt, \quad j = 1, 2, \cdots N, \quad i = 1, 2, ..., M$$

$$\int_0^T \phi_i(t) \, \phi_j(t) \, dt = \delta(i-j)$$

$$(7.7)$$

The implementation of a synthesizer for generating $s_i(t)$ in terms of its orthogonal components and of an analyser for determining the orthogonal components of $s_i(t)$ are shown in Figure 7.3.

Length (norm, absolute value) of a signal vector s_i may be expressed as

$$\|s_i\|^2 = s_i^T s_i = \sum_{j=1}^{N} s_{ij}^2, \quad i = 1, 2, ..., M \quad (7.8)$$

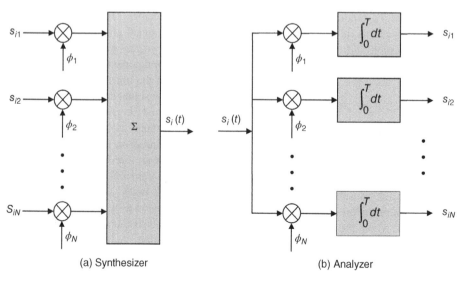

(a) Synthesizer   (b) Analyzer

Figure 7.3   Geometric Representation of the Signal $s_i(t)$. *a)* synthesizer for generating the signal $s_i(t)$, and b) analyzer for obtaining the orthogonal signal components.

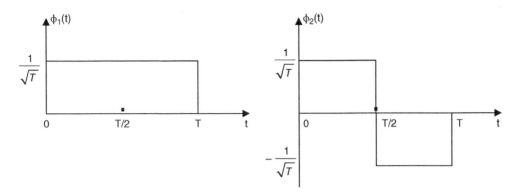

Figure 7.4   Two Orthonormal Basis Functions.

The energy of a signal defined by (7.3) can be shown to be equal to the square of its norm:

$$E_i = \int_0^T s_i^2(t)\,dt$$

$$= \int_0^T \left(\sum_{j=1}^N s_{ij}\phi_j(t)\right)\left(\sum_{k=1}^N s_{ik}\phi_k(t)\right)dt$$

$$= \sum_{j=1}^N \sum_{k=1}^N s_{ij}s_{ik} \underbrace{\int_0^T \phi_j(t)\,\phi_k(t)\,dt}_{\delta(j-k)}$$

$$= \sum_{j=1}^N s_{ij}^2 = \|s_i\|^2$$

(7.9)

The square of the Euclidean distance between the signals $s_i(t)$ and $s_k(t)$ is given by

$$\int_0^T [s_i(t) - s_k(t)]^2\,dt = \int_0^T \left[\sum_{j=1}^N (s_{ij} - s_{kj})\,\phi_j(t)\right]^2 dt$$

$$= \sum_{j=1}^N \sum_{n=1}^N (s_{ij} - s_{kj})(s_{in} - s_{kn})\underbrace{\int_0^T \phi_j(t)\phi_n(t)\,dt}_{\delta(j-n)}$$

$$= \sum_{j=1}^N (s_{ij} - s_{kj})^2 = \|s_i - s_k\|^2$$

(7.10)

Example 7.1   Orthonormal Basis Functions. Consider the two basis functions $\phi_1(t)$ and $\phi_2(t)$ shown in Figure 7.4.

The basis functions are orthonormal:

$$\int_0^T \phi_i(t)\,\phi_j(t)\,dt = \delta(i-j),\ \ i,j = 1,2 \quad (7.11)$$

The square of the Euclidean distance between the two basis functions is found as

$$\phi_1 = \frac{1}{\sqrt{T}}[+1\ \ +1]$$

$$\phi_2 = \frac{1}{\sqrt{T}}[+1\ \ -1]$$

$$\|\phi_1 - \phi_2\|^2 = \frac{1}{T}(1-1)^2\frac{T}{2} + \frac{1}{T}(1+1)^2\frac{T}{2} = 2$$

(7.12)

One may easily show that the Euclidean distance between ϕ_i and $-\phi_i$ (antipodal with each other) is $\sqrt{2}$ times larger than the one between ϕ_1 and ϕ_2, which are orthogonal to each other.

For example, a signal waveform

$$x(t) = \begin{cases} 1 & 0 \le t < T/2 \\ -2 & T/2 \le t < T \end{cases} \quad (7.13)$$

may be expressed as a linear combination of the basis functions:

$$x_1 = \int_0^T x(t)\, \phi_1(t)\, dt = -\sqrt{T}/2$$

$$x_2 = \int_0^T x(t)\, \phi_2(t)\, dt = 3\sqrt{T}/2 \qquad (7.14)$$

$$x(t) = -\frac{\sqrt{T}}{2}\, \phi_1(t) + \frac{3\sqrt{T}}{2}\, \phi_2(t)$$

7.3 Coherent Demodulation in AWGN Channels

As shown in Figure 7.3b, a demodulator decomposes the received signal $x(t)$ given by (7.5) into N orthogonal components. Using the geometric signal representation given by (7.7), the received signal $x(t)$ is applied to the input of the demodulator which may be implemented as a bank of correlators/matched filters:

$$x(t) = s_i(t) + w(t) = \sum_{j=1}^N s_{ij}\, \phi_j(t) + w(t) \quad (7.15)$$

The signal at the output of jth correlator may be written as

$$x_j = \int_0^T x(t)\, \phi_j(t)\, dt$$

$$= \int_0^T [s_i(t) + w(t)]\, \phi_j(t)\, dt = s_{ij} + w_j$$

$$w_j = \int_0^T w(t)\, \phi_j(t)\, dt$$

$$E[w_j] = 0$$

$$\mathrm{var}[w_j] = \int_0^T \int_0^T w(t)\, w^*(t')\underbrace{E\left[\phi_j(t)\phi_j(t')\right]}_{\delta(t-t')}\, dt\, dt'$$

$$= N_0/2$$

$$(7.16)$$

where s_{ij} is given by (7.7). Hence, the observation vector consisting of N orthogonal components at the demodulator output is given by

$$\mathbf{x} = [x_1\ x_2\ \cdots\ x_N]^T \qquad (7.17)$$

where x_i is shown in (7.16) to be a Gaussian random variable with mean s_{ij} and variance $N_0/2$ (see Figure 7.5).

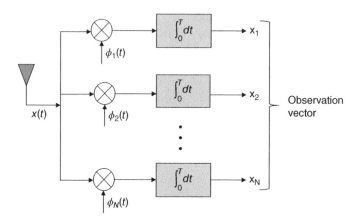

Figure 7.5 Block Diagram of a Demodulator Implemented as a Bank of Correlators.

Noting that the signal at demodulator output is given by

$$x(t) = \sum_{j=1}^{N} x_j \, \phi_j(t) = \sum_{j=1}^{N} \left(s_{ij} + w_j\right) \phi_j(t) \quad (7.18)$$

The signal at demodulator input (7.15) may be written in terms of the signal at the demodulator output as

$$x(t) = s_i(t) + w(t) = \sum_{j=1}^{N} s_{ij} \phi_j(t) + w(t)$$

$$= \sum_{j=1}^{N} \left(x_j - w_j\right)\phi_j(t) + w(t)$$

$$= \underbrace{\sum_{j=1}^{N} x_j \, \phi_j(t)}_{\text{demodulator output}} + \underbrace{w(t) - \sum_{j=1}^{N} w_j \phi_j(t)}_{w'(t):\ \text{remainder noise}}$$

$$(7.19)$$

The last term in (7.19) is called the remainder noise and represents the noise which is orthogonal to the space spanned by $\{\phi_i(t), i = 1,2,\dots N\}$. Note that the set of basis vectors $\{\phi_i(t), i = 1,2,\dots N\}$ span the signal space but not the noise space. According to the so-called *theorem of irrelevance*, for signal demodulation in an AWGN channel, only the projections of the noise onto the basis functions of the signal set affect the sufficient statistics for the detection

problem; the remainder noise is irrelevant. This implies that the decoder does not need the knowledge of the remainder noise to make an estimate as to which signal is transmitted.

Example 7.2 The Theorem of Irrelevance. Consider a matched filter receiver for binary antipodal signals $s_1(t)$ and $s_2(t)$ of equal energies E. In the signal constellation shown in Figure 7.6, these two signals are located at $\pm\sqrt{E}$ along the basis functions $\phi_1(t)$. Therefore, the components $x_1(t) = \pm\sqrt{E} + w_1(t)$ and $x_2(t) = w_2(t)$ of the observation vector $\mathbf{x} = [x_1(t)\ x_2(t)]$ clearly shows that the $x_2(t)$ is determined only by the projection $w_2(t)$ of the circularly symmetric noise $w(t)$ onto $\phi_2(t)$. Since $x_2(t)$ and $w_2(t)$ change the observation vector only along $\phi_2(t)$, they have no effect on changing the decision region and the decoder decision whether $s_1(t)$ or $s_2(t)$ is transmitted. Consequently, the transmitted symbol is estimated based only on $x_1(t) = \pm\sqrt{E} + w_1(t)$. Hence $w_1(t)$, which denotes the projection of the circularly symmetric AWGN input noise $w(t)$ onto the basis function $\phi_1(t)$, provides sufficient statistics for the detector. The noise component $w_2(t)$, which is called as the remainder noise, is irrelevant on the decoder decision. [2]

Example 7.3 Matched Filter Versus Correlation Demodulators. Consider the reception of an arbitrary signal $x(t)$ by a LTI filter with impulse response

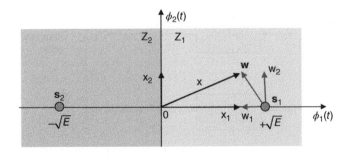

Figure 7.6 Decision Regions Z_1 and Z_2 for Matched Filter Reception of a Binary PAM Signal.

$h_i(t)$. Filter output $x_i(t)$ for the input signal $x(t)$ may be written as the convolution integral:

$$x_i(t) = \int_{-\infty}^{\infty} x(\tau) h_i(t-\tau) \, d\tau \qquad (7.20)$$

If impulse response $h_i(t)$ of a LTI filter is matched to a basis function $\phi_i(t)$ of the input signal $x(t)$, then

$$h_i(t) = \phi_i(T-t) \ \Rightarrow \ h_i(t-\tau) = \phi_i(T-t+\tau) \qquad (7.21)$$

The filter output may then be written as

$$x_i(t) = \int_{-\infty}^{\infty} x(\tau) \phi_i(T-t+\tau) \, d\tau \qquad (7.22)$$

Sampling the matched-filter output at time $t = T$ leads to

$$x_i(T) = \int_0^T x(\tau) \phi_i(\tau) \, d\tau \qquad (7.23)$$

where the integration limits above apply since $\phi_i(t)$ is zero outside the interval $0 \le t \le T$. Note that $x_i(T)$ also represents the correlation of the input signal $x(t)$ with $\phi_i(t)$. Exploiting the equivalence of matched-filter and correlator implementations, as shown in Figure 7.7, one may implement the correlation demodulator shown in Figure 7.5 equivalently as a bank of matched-filters as shown in Figure 7.8.

Irrespective of whether it is implemented as a bank of correlators or matched-filters, the

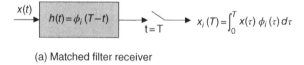

(a) Matched filter receiver

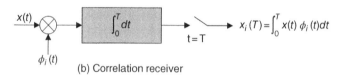

(b) Correlation receiver

Figure 7.7 Equivalence of matched-filter and correlator implementations. Correlator and matched filter outputs are identical at the sampling time $t = T$.

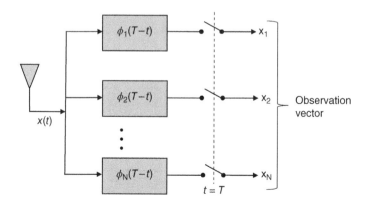

Figure 7.8 Block Diagram of a Demodulator Implemented as a Bank of Matched Filters.

demodulator output in a memoryless AWGN channel consists of an observation vector $\mathbf{x}$ of N independent Gaussian random variables, each with mean s_{ij} and variance $N_0/2$. Hence, the conditional pdf of the vector $\mathbf{x}$, given that the signal $s_i(t)$, that is, m_i was transmitted, may be written as

$$f(\mathbf{x}|m_i) = \prod_{j=1}^{N} f(x_j|m_i), \quad i = 1, 2, ..., M \quad (7.24)$$

where

$$f(x_j|m_i) = \frac{1}{\sqrt{\pi N_0}} \exp\left[-\frac{1}{N_0}(x_j - s_{ij})^2\right],$$

$$\begin{cases} j = 1, 2, ..., N \\ i = 1, 2, ..., M \end{cases}$$
$$(7.25)$$

Inserting (7.25) into (7.24) one gets the joint pdf of the observation vector when m_i is transmitted: [2]

$$f(\mathbf{x}|m_i) = \prod_{j=1}^{N} f(x_j|m_i)$$

$$= (\pi N_0)^{-N/2} \exp\left[-\frac{1}{N_0}\sum_{j=1}^{N}(x_j - s_{ij})^2\right]$$

$$= (\pi N_0)^{-N/2} \exp\left[-\frac{1}{N_0}\|\mathbf{x} - \mathbf{s}_i\|^2\right], \quad i = 1, 2, ...M$$
$$(7.26)$$

where $\|\mathbf{x} - \mathbf{s}_i\|^2$ shows the square of the Euclidean distance between the observation vector and the ith signal vector in the signal-space.

7.3.1 Coherent Detection of Signals in AWGN Channels

The observation vector at the demodulator output, which may be written as $\mathbf{x} = \mathbf{s}_i + \mathbf{w}$, is perturbed by the presence of the additive

noise vector $\mathbf{w}$. The noise vector $\mathbf{w} = [w_1 \ w_2 \ \cdots \ w_N]^T$, consisting of the projections of $w(t)$ along the basis functions, does not contain the remainder noise and yet provides sufficient statistics for the detection process. Since noise components are circularly symmetric Gaussian random variables with zero mean and variance $N_0/2$, the orientation of $\mathbf{w}$ is completely random. The signal vector $\mathbf{s}_i = [s_{i1} \ s_{i2} \ \cdots \ s_{iN}]^T$ may be represented as the transmitted signal point in a Euclidean space of dimension $N \le M$. The set of signal points $\{\mathbf{s}_i\}$, $i = 1, 2, .., M$ are called as the signal constellation. Similarly, the received signal points are represented by the observation vector $\mathbf{x}$ in the same Euclidean space. A received signal point wanders around the message point randomly, lying anywhere inside a Gaussian distributed noise cloud, centred on the message point as shown pictorially in Figure 7.9. As the noise variance increases (E_b/N_0 decreases), the cloud of 100 ksamples of noise becomes larger and the Euclidean distance between the observation and signal vectors increases. This can force the detector to make an erroneous estimate of the transmitted symbol.

Given the observation vector $\mathbf{x}$, the decoder performs a mapping from $\mathbf{x}$ to an estimate $\hat{m}$ of the transmitted symbol, m_i, in a way that would minimize the probability of error in the decision-making process. Given the observation vector $\mathbf{x}$, the probability of error, when m_i is transmitted, may be written as

$$P_e = \sum_{i=1}^{M} p_i P_e(m_i|\mathbf{x})$$
$$(7.27)$$
$$P_e(m_i|\mathbf{x}) = P(\hat{m} \ne m_i|\mathbf{x})$$
$$= 1 - P(\hat{m} = m_i|\mathbf{x})$$

where $p_i = P(m_i)$ denotes the a priori probability of transmitting the symbol m_i. On the other hand, $P(\hat{m} = m_i|\mathbf{x})$ shows the probability that, based on a given observation vector at the decoder input, the decoder estimates as m_i being transmitted.

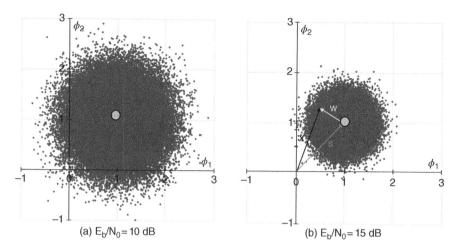

Figure 7.9 Effect of Noise Perturbation on the Location of the Observation Vector in the Signal-Space for $E_b/N_0 = 10$ dB and 15 dB. Simulation is carried out with 100 ksamples of complex Gaussian noise.

7.3.1.1 Optimum Decision Rule

Having formulated by (7.27), the probability of error in the detection process, we now consider minimizing the error probability. The *maximum a posteriori probability (MAP)* rule is an optimum decision rule which minimizes the probability of error in mapping each given observation vector $\mathbf{x}$ into a decision. Minimizing the error probability is equivalent to maximizing $P(\hat{m} = m_i|\mathbf{x})$ in (7.27). Hence, the detector estimates the transmitted symbol as follows: [2]

Decide $\hat{m} = m_i$ if

$$P(\hat{m} = m_i|\mathbf{x}) \geq P(\hat{m} = m_k|\mathbf{x}),$$
$$k = 1, 2, \cdots, M, \quad k \neq i \tag{7.28}$$

Using the Bayes rule

$$P(\hat{m} = m_k|\mathbf{x}) \ f(\mathbf{x}) = f(\mathbf{x}|m_k)P(m_k), \tag{7.29}$$

one may rewrite the MAP rule in (7.28) as follows:

Decide $\hat{m} = m_i$ if

$$P(\hat{m} = m_k|\mathbf{x}) = \frac{f(\mathbf{x}|m_k)\,P(m_k)}{f(\mathbf{x})} \tag{7.30}$$

is maximum for $k = i$. Note that $f(\mathbf{x}|m_k)$ denotes the conditional pdf of $\mathbf{x}$, when m_k is sent by the transmitter. Also note that $f(\mathbf{x})$, the pdf of $\mathbf{x}$,

$$f(\mathbf{x}) = \sum_{k=1}^{M} f(\mathbf{x}|m_k)\,P(m_k), \tag{7.31}$$

is independent of the transmitted symbol m_i.

For binary signaling with given a priori probabilities $p(m_1)$ and $p(m_2)$, the MAP decoding rule, which was already introduced in Example 6.4, may be based on the likelihood (or log-likelihood) ratio test in order to estimate the transmitted symbol: [3]

$$\ell(m|\mathbf{x}) = \frac{P(\hat{m} = m_1|\mathbf{x})}{P(\hat{m} = m_2|\mathbf{x})} \overset{\hat{m}=m_1}{\underset{\hat{m}=m_2}{\gtrless}} 1$$

$$\tag{7.32}$$

$$L(m|\mathbf{x}) = \ln\left[\frac{f(\hat{m} = m_1|\mathbf{x})}{f(\hat{m} = m_2|\mathbf{x})}\right] \overset{\hat{m}=m_1}{\underset{\hat{m}=m_2}{\gtrless}} 0$$

Using the Bayes rule in (7.29), the log-likelihood ratio in (7.32) may be expressed as:

$$\ell(m|\mathbf{x}) = \frac{P(\hat{m} = m_1|\mathbf{x})}{P(\hat{m} = m_2|\mathbf{x})} = \frac{f(\mathbf{x}|m_1)}{f(\mathbf{x}|m_2)} \frac{p(m_1)}{p(m_2)}$$

$$L(m|\mathbf{x}) = \ln\left[\frac{f(\mathbf{x}|m_1)}{f(\mathbf{x}|m_2)}\right] + \ln\left[\frac{p(m_1)}{p(m_2)}\right]$$

$$= L(\mathbf{x}|m) + L(m)$$

$$(7.33)$$

If the likelihood of transmitting m_1 is higher than for m_2, then the log-likelihood ratio (7.32) will be positive and the detector will decide that $\hat{m} = m_1$ is transmitted. Otherwise, the detector will declare that m_2 is sent. Therefore, $L(m|\mathbf{x})$ has a soft value and its sign provides the hard decision as to m_1 or m_2 is transmitted:

$$\hat{m} = m_1 \quad \text{if } L(m|\mathbf{x}) > 0$$

$$(7.34)$$

$$\hat{m} = m_2 \quad \text{if } L(m|\mathbf{x}) < 0$$

The magnitude of $L(m|\mathbf{x})$ represents the reliability of the decision. For example, the decoder is more confident to declare $\hat{m} = m_1$ if $L(m|\mathbf{x})$ is equal to 0.8 compared to 0.3.

If the source symbols are equiprobable that is, $p_k = P(m_k) = 1/M$, $k = 1,2\cdots M$, then the MAP decoding, which is based on $P(\hat{m} = m_k|\mathbf{x})$, may be accomplished by using only $f(\mathbf{x}|m_k)$ since

$$P(\hat{m} = m_k|\mathbf{x}) = \underbrace{\frac{1}{M f(\mathbf{x})}}_{\text{independent of } m_k} f(\mathbf{x}|m_k) \quad (7.35)$$

Therefore, for the special case of equal symbol probabilities, the MAP decoding rule given by (7.30) simply reduces to

Decide $\hat{m} = m_i$ if $f(\mathbf{x}|m_k)$ is maximum for $k = i$.

$$(7.36)$$

This is referred to as the *maximum likelihood* (ML) *decoding* rule and is implemented as a ML decoder. Note that the knowledge of

$f(\mathbf{x}|m_k)$ is already available at the demodulator output, as given by (7.26). A ML decoder calculates the likelihood functions $f(\mathbf{x}|m_k)$, $k = 1,2,\cdots,M$ for all possible message points, compares them, and then decides in favour of the one with maximum likelihood.

For binary signalling, the probabilities of transmitting m_1 and m_2 are the same, that is, $p(m_1) = p(m_2)$. Then, $L(m) = 0$ in (7.33) and the MAP decoding rule reduces to the ML decoding rule:

$$L(m|\mathbf{x}) = L(\mathbf{x}|m)$$

$$= \ln\left[\frac{f(\mathbf{x}|m_1)}{f(\mathbf{x}|m_2)}\right] \overset{\hat{m} = m_1}{\underset{\hat{m} = m_2}{\gtrless}} 0, \quad \text{if } p(m_1) = p(m_2)$$

$$(7.37)$$

For example, consider that antipodal signals s_1 and s_0 of equal energies E_b are used for the transmission of the binary symbols m_1 and m_0 with unequal a priori transmission probabilities $p_1 = P(m_1)$, $p_0 = P(m_0)$ in an AWGN channel. The observation vector at the matched-filter output is given by

$$m_1: \quad x = +\sqrt{E_b} + n(T)$$

$$(7.38)$$

$$m_0: \quad x = -\sqrt{E_b} + n(T)$$

where $n(T)$ has zero mean and variance $\sigma^2 = N_0/2$. Noting that the observation vector $\mathbf{x}$ has a dimension of $N = 1$ in antipodal signaling, the log-likelihood ratio reduces to

$$L(x|m) = \ln\frac{(2\pi\sigma^2)^{-1/2} e^{-\left(x - \sqrt{E_b}\right)^2/2\sigma^2}}{(2\pi\sigma^2)^{-1/2} e^{-\left(x + \sqrt{E_b}\right)^2/2\sigma^2}}$$

$$= \frac{\left(x + \sqrt{E_b}\right)^2 - \left(x - \sqrt{E_b}\right)^2}{2\sigma^2} = \frac{2\sqrt{E_b}}{\sigma^2} x$$

$$L(m|x) = L(x|m) + L(m) = \frac{2\sqrt{E_b}}{\sigma^2} x + \ln\left(\frac{p_1}{p_0}\right)$$

$$(7.39)$$

Applying the log-likelihood test, we get

$$L(m|x) \overset{\hat{m}=m_1}{\underset{\hat{m}=m_0}{\overset{>}{<}}} 0 \;\Rightarrow\; x \overset{\hat{m}=m_1}{\underset{\hat{m}=m_0}{\overset{>}{<}}} \lambda_{opt} \qquad (7.40)$$

$$\lambda_{opt} = \frac{\sigma^2}{2\sqrt{E_b}} \ln\left(\frac{p_0}{p_1}\right)$$

The optimum threshold level in (7.40) is identical to (6.46) and minimizes the probability of error. The minimum probability of error may then be written as (see (6.41))

$$P_e = p_1\, P(x < \lambda_{opt}|m_1) + p_0\, P(x > \lambda_{opt}|m_0)$$

$$= p_1 \frac{1}{\sqrt{2\pi}\sigma} \int_{-\infty}^{\lambda_{opt}} e^{-\frac{(x-\sqrt{E_b})^2}{2\sigma^2}}\, dx$$

$$+ p_0 \frac{1}{\sqrt{2\pi}\sigma} \int_{\lambda_{opt}}^{\infty} e^{-\frac{(x+\sqrt{E_b})^2}{2\sigma^2}}\, dx$$

$$= p_1 Q\left(\frac{\sqrt{E_b}-\lambda_{opt}}{\sigma}\right) + p_0\, Q\left(\frac{\sqrt{E_b}+\lambda_{opt}}{\sigma}\right)$$

$$(7.41)$$

For the special case where $p_1 = p_0 = 0.5$, the optimum threshold becomes $\lambda_{opt} = 0$ and the corresponding error probability reduces to

$$P_e = Q\left(\sqrt{E_b}/\sigma\right) = Q\left(\sqrt{2E_b/N_0}\right) \qquad (7.42)$$

For example, let the received signal level be $x = 0.02$ (see (7.38)), $\sigma^2 = 1$, $E_b = 10$ and $p_1 = 0.3$. From (7.39),

$$L(m|x) = \frac{2\sqrt{10}}{1} 0.02 + \ln\frac{0.3}{0.7} = -0.721,$$

$$(7.43)$$

the negative sign of $L(m|x)$ implies that MAP decoder decides $\hat{m} = m_0$ (hard decision). The absolute value $|L(m|x)| = 0.721$ shows the confidence level in estimating the transmitted symbol. This is in agreement with the fact that the received signal level $x = 0.02$ is lower than the optimum threshold level

$$\lambda_{opt} = \frac{\sigma^2}{2\sqrt{E_b}} \ln\left(\frac{p_0}{p_1}\right) = 0.134 \qquad (7.44)$$

The corresponding error probability is found from (7.41) to be $P_e = 7.12 \times 10^{-4}$.

Example 7.4 MAP Decoding in a Binary Symmetric Channel.
Consider a binary symmetric channel (BSC) with a transition probability of $p = 0.2$ as shown in Figure 7.10. The probability of transmitting symbol 1 is given by $p_1 = P(m_1) = 0.7$.

Because the channel is symmetric, one has

$$P(x_1|m_2) = P(x_2|m_1) = p = 0.2$$
$$P(x_1|m_1) = P(x_2|m_2) = 1-p = 0.8$$
$$(7.45)$$

A priori (transmit) probabilities of the two symbols are given by

$$P(m_1) = p_1 = 0.7$$
$$P(m_2) = p_2 = 1 - p_1 = 0.3$$
$$(7.46)$$

The probabilities of receiving x_1 and x_2 are found as follows:

$$P(x_1) = P(x_1|m_1)p_1 + P(x_1|m_2)p_2$$
$$= p_1(1-p) + p_2 p$$
$$= 0.7 \times 0.8 + 0.3 \times 0.2 = 0.62$$
$$(7.47)$$
$$P(x_2) = P(x_2|m_1)p_1 + P(x_2|m_2)p_2$$
$$= p_1 p + p_2(1-p)$$
$$= 0.7 \times 0.2 + 0.3 \times 0.8 = 0.38$$

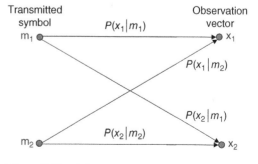

Figure 7.10 A Binary Symmetric Channel (BSC).

Note the difference between transmit and receive probabilities of the two symbols. For the special case of equal transmit probabilities of the two symbols, one also has equal probabilities of receiving x_1 and x_2.

$$p_1 = p_2 = 0.5 \quad \Rightarrow \quad P(x_1) = P(x_2) = 0.5 \quad (7.48)$$

Using the likelihood ratio given by (7.33), the decoder makes the following decisions

$$\ell(m|x_1) = \frac{P(x_1|m_1)\, p_1}{P(x_1|m_2)\, p_2} = \frac{1-p}{p}\frac{p_1}{p_2}$$

$$= \frac{0.8 \, 0.7}{0.2 \, 0.3} = \frac{0.56}{0.06} = 9.33 \quad \Rightarrow \quad \hat{m} = m_1$$

$$\ell(m|x_2) = \frac{P(x_2|m_1)\, p_1}{P(x_2|m_2)\, p_2} = \frac{p}{1-p}\frac{p_1}{p_2}$$

$$= \frac{0.2 \, 0.7}{0.8 \, 0.3} = \frac{0.14}{0.24} = 0.58 \quad \Rightarrow \quad \hat{m} = m_2$$

$$(7.49)$$

Probability of decoding error depends on the transition probability but not on the a priori probabilities of m_1 and m_2:

$$P_e = P(x_1|m_2)\, p_2 + P(x_2|m_1)\, p_1 \qquad (7.50)$$

$$= p\, p_2 + p\, p_1 = p = 0.2$$

Similarly, the probability of correct decoding is given by:

$$P_c = 1 - P_e = P(x_1|m_1)\, p_1 + P(x_2|m_2)\, p_2$$

$$= (1-p)\, p_1 + (1-p)\, p_2 = 1 - p = 0.8 \qquad (7.51)$$

Example 7.5 MAP Hypothesis Testing for Gaussian Signals with Unequal Variances.
A radar receiver aims to discrimate two Gaussian signals $s_0(t)$ and $s_1(t)$ each with zero mean but unequal variances σ_0^2 and σ_1^2. MAP hypothesis testing is used for deciding whether $s_0(t)$ or $s_1(t)$ is sent. The a priori probabilities of $s_0(t)$

and $s_1(t)$ are assumed to be given by p_0 and p_1, respectively.

The MAP decision rule is given by

$$\frac{p_0\, f_0(r|\sigma_0)}{p_1\, f_1(r|\sigma_1)} \underset{s_1}{\overset{s_0}{\gtrless}} 1$$

$$\Rightarrow \quad \frac{p_0 \dfrac{1}{\sqrt{2\pi}\sigma_0} \exp\left(-\dfrac{r^2}{2\sigma_0^2}\right)}{p_1 \dfrac{1}{\sqrt{2\pi}\sigma_1} \exp\left(-\dfrac{r^2}{2\sigma_1^2}\right)} \underset{s_1}{\overset{s_0}{\gtrless}} 1$$

$$(7.52)$$

Taking the logarithms of both sides of the equation on the right hand side, we get the decision rule:

$$|r| \underset{s_1}{\overset{s_0}{\gtrless}} \lambda_{opt}$$

$$(7.53)$$

$$\lambda_{opt} = \left[\frac{2\sigma_0^2\sigma_1^2}{(\sigma_0^2 - \sigma_1^2)} \ln\left(\frac{p_1\sigma_0}{p_0\sigma_1}\right)\right]^{1/2}$$

Note that these two Gaussian signals can be discriminated theoretically as long as their variances are not very close to each other.

Example 7.6 MAP Hypothesis Testing in a Radar System.
Radar system are designed to detect the presence of a target, and to extract useful information about the target. Based on the fact that either there is no target to be detected (hypothesis H_0) or a target is present (hypothesis H_1), the signal received by a radar may be written as

$$x(t) = \begin{cases} s(t) + w(t) & H_1 \\ w(t) & H_0 \end{cases} \qquad (7.54)$$

where $w(t)$ is white Gaussian noise with power spectral density $N_{0/2}$, and $s(t)$ denotes an echo produced as the result of the scattering of radar signals by the target. Assume that $s(t)$ has an energy E, and the two hypotheses are equally likely.

Asssuming that the radar uses a matched filter demodulator, the observation vector may be written as

$$x_1 = \begin{cases} \sqrt{E} + w_1(T) & H_1 \\ w_1(T) & H_0 \end{cases} \tag{7.55}$$

where $w_1(T)$ denotes the Gaussian noise at $t = T$ with zero-mean and PSD $\sigma^2 = N_0/2$. The corresponding pdf's are given by

$$f_{x_1}(r|H_0) = \frac{1}{\sqrt{2\pi}\sigma} \exp\left(-\frac{r^2}{2\sigma^2}\right)$$

$$f_{x_1}(r|H_1) = \frac{1}{\sqrt{2\pi}\sigma} \exp\left(-\frac{(r-\sqrt{E})^2}{2\sigma^2}\right) \tag{7.56}$$

The *probability of false alarm* is defined as the probability that the receiver decides a target is present when it is not:

$$P_{fa} = P(H_1|H_0) = \mathrm{Prob}(x_1 > \lambda|H_0)$$

$$= \int_\lambda^\infty f_{x_1}(r|H_0)\, dr$$

$$= \int_\lambda^\infty \frac{1}{\sqrt{2\pi}\sigma} \exp\left(-\frac{r^2}{2\sigma^2}\right) dr \tag{7.57}$$

$$= Q\left(\frac{\lambda}{\sigma}\right)$$

where λ denotes the threshold level used by the radar detector for decision making. The *probability of detection* is defined as the probability that the receiver decides that the target is present when it is:

$$P_d = P(H_1|H_1) = \mathrm{Prob}(x_1 > \lambda|H_1)$$

$$= \int_\lambda^\infty f_{x_1}(r|H_1)\, dr$$

$$= \int_\lambda^\infty \frac{1}{\sqrt{2\pi}\sigma} \exp\left(-\frac{(r-\sqrt{E})^2}{2\sigma^2}\right) dr \tag{7.58}$$

$$= Q\left(\frac{\lambda-\sqrt{E}}{\sigma}\right) = 1 - Q\left(\sqrt{\frac{2E}{N_0}} - \frac{\lambda}{\sigma}\right)$$

The *probability of miss* is defined as the probability that the receiver decides a target is not present when it is:

$$P_{miss} = P(H_0|H_1) = 1 - P_d = Q\left(\sqrt{\frac{2E}{N_0}} - \frac{\lambda}{\sigma}\right) \tag{7.59}$$

In radar systems, the threshold level used by the detector is usually chosen to limit the probability of false alarm to a certain level. Assume that this radar system limits the false alarm probability to less than 10% and the probability of detection to higher than 90%. Then, using (7.57) and (7.58), one gets $\lambda/\sigma = 1.28$ and $E/N_0 = 3.3$ (5.2 dB). The corresponding miss probability is 10%. The desired E/N_0 level is determined by the transmit power and the antenna gain and the noise of the radar, frequency of operation, distance to the target, and radar cross-section area of the target (see (2.125)).

7.3.1.2 Geometrical Interpretation of Signal Space for MLD

In case where the a priori probabilities of m_k, $k = 1, 2, \cdots, M$ are not equal to each other, the MAP decoding rule leads to minimum error probability if the decision regions are defined by the optimum threshold levels given by (7.40). For equiprobable signals, the MAP decoding reduces to MLD, where the decision regions are defined by the threshold levels that bisect the Euclidean distances between the signal points. To have a better feeling about the MLD, let Z denote the N-dimensional observation space of all possible observation vectors **x** and the observation space is partitioned into M-decision regions, Z_1, Z_2,..., Z_M, corresponding to M symbols likely to be transmitted. Since the threshold levels bisect the Euclidean distances between the signal points, the message points and the observation space are equally distributed between them.

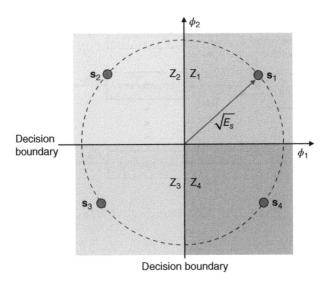

Figure 7.11 The Observation Space with Signal Points and Decision Regions for the Case When Dimensionality is $N = 2$ and the Number of Signal Points is $M = 4$. M transmitted symbols are assumed to be equally likely and have equal energies E_s.

As an example, Figure 7.11 shows the signal points and the observation space for $M = 4$ equal energy signals with identical a priori transmission probabilities. According to MLD decision rule, the observation vector $\mathbf{x}$ lies in region Z_i if the likelihood function $f(\mathbf{x}|m_k)$ is maximum for $k = i$ (see (7.26)). In other words, Z_i is chosen such that $\|\mathbf{x} - \mathbf{s}_k\|$, $k = 1, 2, \ldots, M$ is minimized for $k = i$. MLD decision rule then estimates the signal with minimum Euclidean distance to the observation point as the most likely transmitted signal.

To elaborate this, let us expand the square-Euclidean distance given by (7.26):

$$\|\mathbf{x} - \mathbf{s}_k\|^2 = \sum_{j=1}^{N} (x_j - s_{kj})^2$$

$$= \sum_{j=1}^{N} x_j^2 + \sum_{j=1}^{N} s_{kj}^2 - 2 \sum_{j=1}^{N} x_j s_{kj} \tag{7.60}$$

The first term in (7.60) may be ignored since, being independent of the transmitted signal,

plays no role in estimating the transmitted signal. Noting that $E_k = \sum_{j=1}^{N} s_{kj}^2$ denotes the energy of the signal s_k, the MLD decision rule (7.60) may be rephrased as: Observation vector $\mathbf{x}$ lies in region Z_i (the ML decoder chooses $\hat{m} = m_i$) if

$$\sum_{j=1}^{N} x_j s_{kj} - \frac{E_k}{2} = \mathbf{x}^T \mathbf{s}_k - \frac{E_k}{2} \tag{7.61}$$

is maximum for $k = i$.

Figure 7.12 shows the implementation diagram of an optimum detector defined by (7.61). The accumulators are digital integrators and compute $\mathbf{x}^T \mathbf{s}_i$, $i = 1, 2, \cdots, M$. If signals have equal energies, then the term $E_k/2$ may be ignored.

7.4 Probability of Error

Using the geometrical interpretation of the signal space in MLD, the average probability of symbol error P_e may be written as

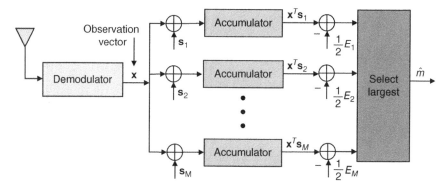

Figure 7.12   Block Diagram of a ML Decoder.

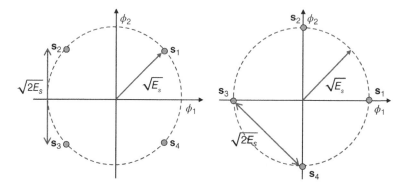

Figure 7.13   Demonstration of the Principle of Rotational Invariance. Rotation of the constellation does not change the error probability.

$$P_e = \frac{1}{M}\sum_{i=1}^{M} P(\mathbf{x} \ does \ not \ lie \ in \ Z_i | m_i \ sent)$$

$$= 1 - \frac{1}{M}\sum_{i=1}^{M} P(\mathbf{x} \ lies \ in \ Z_i | m_i \ sent)$$

$$= 1 - \frac{1}{M}\sum_{i=1}^{M} \int_{Z_i} f_{\mathbf{x}}(\mathbf{x}|m_i) \ d\mathbf{x}$$

$$(7.62)$$

Note that the complex Gaussian noise is circularly symmetric in all directions in the signal space and the probability of error depends only on the Euclidean distances between the message points in the constellation diagram. This leads us

to *the principle of rotational invariance* which states that if a signal constellation is rotated by an orthonormal transformation, then the P_e incurred in MLD in an AWGN channel is unchanged. Figure 7.13 shows an example to the principle of rotational invariance for a signal defined by $N = 2$ and $M = 4$. [2]

If a signal constellation is translated by a constant vector, then the probability of symbol error P_e incurred in ML signal detection over an AWGN channel is unchanged. This so-called principle of *translational invariance* can be exploited so as to minimize the average symbol energy without changing its error probability performance. Given a signal constellation s_k, $k = 1,..,M$, the corresponding

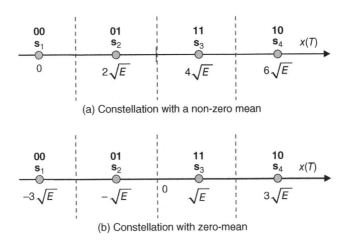

(a) Constellation with a non-zero mean

(b) Constellation with zero-mean

Figure 7.14 Application of the Principle of Translational Invariance to a 4-PAM Signal Constellation.

signal constellation with minimum average energy is obtained by using the translated constellation

$$s_k - E[\mathbf{s}] = \mathbf{s}_k - \sum_{i=1}^{M} \mathbf{s}_i p_i, \quad k = 1,\ 2, \ldots, M,$$

$$(7.63)$$

where $E[\mathbf{s}]$ denotes the mean vector of the original signal constellation.

Figure 7.14 shows a demonstration of the principle of translational invariance for 4-PAM signals. Note that the probability of symbol error is the same in both constellations since the Euclidean distance between symbol points are unchanged. However, the constellation which is centred around the origin achieves this error performance with average symbol energy of $5E$ (see 6.69) while the other one has average symbol energy of $14E$, which is 4.47 dB higher.

7.4.1   Union Bound on Error Probability

In certain cases, it may not be easy to determine the decision regions, or to evaluate (7.62) for determining the error probability. Then, one

may resort to simulation tools or numerical calculations. Another alternative is to find a union bound to the error probability, which provides an analytical error probability which is not exceeded by the exact error probability. The upper-bound is said to be 'loose' if it is not sufficiently close to the exact error probability; then its usefulness is questionable. However, when the union bound is 'tight', then it becomes a valuable tool in predicting the error probability. Therefore, one should be cautious with the upper-bound calculations of the error probability.

The union bound on the probability of symbol error may be written as

$$P_e = \sum_{i=1}^{M} P(m_i) P_e(m_i)$$

$$P_e(m_i) \le \sum_{\substack{k=1 \\ k \ne i}}^{M} P(s_i, s_k), \quad i = 1, 2, \ldots, M \tag{7.64}$$

where $P_e(m_i)$ denotes the probability of error when m_i is sent. The so-called *pairwise error probability* $P(s_i, s_k)$ is defined as the probability that the receiver decides s_k when s_i is sent:

$$P(s_i, s_k) \triangleq Prob(s_k\ received | s_i\ sent), \quad (7.65)$$

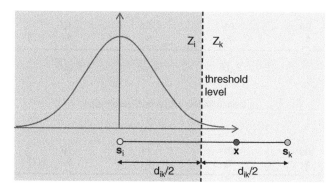

Figure 7.15   Pairwise Error Probability Between Two Equiprobable Signal Points s_i and s_k.

Hence, the union bound in (7.64) is simply the sum of the pair-wise error probabilities that $m_k \neq m_i$, $k = 1,2,...,M$ is received when m_i is sent. Since these probabilities are not mutually exclusive, the union bound is larger than the exact probability.

The decision boundary between equiprobable messages s_k for s_i is determined by the bisector joining the message points (see Figure 7.15). Then, the receiver decision will be in favour of the signal closest to the observation point. This implies that the channel noise will cause an error when its amplitude becomes larger than half the Euclidean distance between s_i and s_k, $d_{ik} = \|s_i - s_k\|$. The pairwise error probability may then be written as

$$P(s_i, s_k) = P(\mathbf{x} \text{ is closer to } s_k \text{ than } s_i | s_i \text{ is sent})$$

$$= P(\|\mathbf{x} - s_k\| < \|\mathbf{x} - s_i\| \,|\, s_i)$$

$$= \int_{d_{ik}/2}^{\infty} \frac{1}{\sqrt{\pi N_0}} \exp\left(-\frac{v^2}{N_0}\right) dv = Q\left(\frac{d_{ik}}{\sqrt{2N_0}}\right)$$

$$(7.66)$$

Inserting (7.66) into (7.64), the union bound for pairwise error probability, when m_i is sent, reduces to

$$P_e(m_i) \leq \sum_{\substack{k=1 \\ k \neq i}}^{M} Q\left(\frac{d_{ik}}{\sqrt{2N_0}}\right), \quad i = 1,2,...,M$$

$$(7.67)$$

Probability of symbol error, averaged over M symbols, is upper-bounded by

$$P_e = \sum_{i=1}^{M} P(m_i)\, P_e(m_i)$$

$$\leq \frac{1}{M} \sum_{i=1}^{M} \sum_{\substack{k=1 \\ k \neq i}}^{M} Q\left(\frac{d_{ik}}{\sqrt{2N_0}}\right) \qquad (7.68)$$

When M is large and/or the calculation of $M(M-1)/2$ Euclidean distances between M signal points is cumbersome, then one may replace all the Euclidean distances with the minimum Euclidean distance d_{min} of the signal constellation. The minimum Euclidean distance is defined as the smallest Euclidean distance between any two transmitted signal points in the constellation:

$$d_{min} = \underbrace{\min}_{\substack{i,k=1,2,...,M \\ i \neq k}} d_{ik} \qquad (7.69)$$

Replacing d_{ik} by d_{min} in (7.68), the average symbol error probability may be expressed as a function of the minimum Euclidean distance:

$$P_e \leq (M-1)Q\left(\frac{d_{min}}{\sqrt{2N_0}}\right) \qquad (7.70)$$

which is evidently looser than (7.68). To overcome this inconvenience, one may replace $(M-1)$ in (7.70) by $M_{closest}$ which denotes the number of signal points closest to the signal point in the constellation, that is, the number of signal points which dominate the symbol errors.

Example 7.7 Symbol Error Probability: Exact Versus Union Bound.

Consider the so-called QPSK signaling, which will be treated in detail in the next chapter. Figure 7.16 shows $M=4$ signal points, the decision regions and boundaries and the observation point. Each of the four signals is assumed to have a symbol energy of E_s. The signal points are described by $\left(\pm\sqrt{E_s/2}, \pm\sqrt{E_s/2}\right)$ in the Cartesian coordinates formed by the basis functions (ϕ_1, ϕ_2); hence $N=2$. Noting that the Euclidean distance between the signal points s_1 and s_2 is $d_{12} = \|s_1 - s_2\| = \sqrt{2E_s}$, the pairwise error probability is determined using (7.66):

$$P(s_1, s_2) = Q\left(\sqrt{E_s/N_0}\right) \qquad (7.71)$$

An error occurs when the AWGN noise brings the observation point to region Z_2 when s_1 is sent, that is, an error in the direction of the basis function ϕ_1. Similarly, an error occurs when the component of the complex Gaussian noise along the basis function ϕ_2 brings the observation point to region Z_4. The probability of the probability of error when s_i is sent is given by

$$P(\hat{m} \neq m_i|m_i) = 1 - P(\hat{m} = m_i|m_i)$$
$$= 1 - \left(1 - Q\left(\sqrt{E_s/N_0}\right)\right)^2$$
$$= 2Q\left(\sqrt{E_s/N_0}\right)$$
$$- Q^2\left(\sqrt{E_s/N_0}\right), \quad i = 1, 2, \cdots, 4 \qquad (7.72)$$

This expression denotes the probability that the observation point lies in regions outside Z_1 when s_1 is sent. The exact expression for

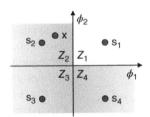

(a) Exact symbol error probability

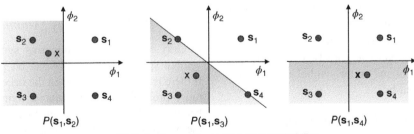

(b) Union bound for symbol error probability

Figure 7.16 Geometric Illustration of the Exact Symbol Error Probability and the Union Bound.

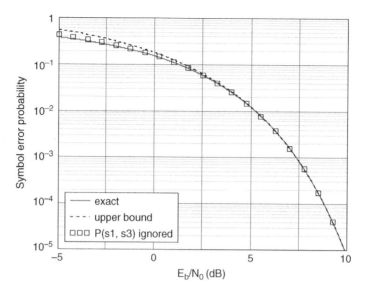

Figure 7.17 Variation of the Exact Symbol Error Probability and the Union Bounds for QPSK Signaling as a Function of E_b/N_0 (dB).

the average probability of error is found by averaging over all symbols:

$$P_e = \sum_{i=1}^{4} p(m_i)\, P(\hat{m} \neq m_i | m_i)$$

$$= 2Q\left(\sqrt{E_s/N_0}\right) - Q^2\left(\sqrt{E_s/N_0}\right) \qquad (7.73)$$

Now let us turn our attention to the calculation of the union bound on symbol error probability. The union bound on the probability of error when s_1 is sent is given by

$$P(\hat{m} \neq m_1 | m_1) \leq P(s_1, s_2) + P(s_1, s_3) + P(s_1, s_4)$$

$$= 2Q\left(\sqrt{E_s/N_0}\right) + Q\left(\sqrt{2E_s/N_0}\right) \qquad (7.74)$$

where the last term accounts for the pairwise error probability between s_1 and s_3, which are separated by an Euclidean distance of $d_{13} = 2\sqrt{E_s}$. Since all signals are equi-probable, the union bound on the symbol error probability is given by

$$P_e = \sum_{i=1}^{4} p(m_i)\, P(\hat{m} \neq m_i | m_i) < 2Q\left(\sqrt{E_s/N_0}\right)$$

$$+ Q\left(\sqrt{2E_s/N_0}\right) \qquad (7.75)$$

Figure 7.17 shows the exact expression (7.73) and union bound (7.75) to the symbol error probability as a function of the E_b/N_0 where $E_b = E_s/2$. One may easily observe from Figure 7.16 that the pair-wise error probabilities are not mutually exclusive and they overlap with each other. Therefore union bound for the symbol error probability based on pair-wise error concept will be higher than the exact expressions. For example, pairwise error probabilities $P(s_1,s_2)$ and $P(s_1,s_4)$ already include for $P(s_1,s_3)$ (see Figure 7.16). Therefore, when $P(s_1,s_3)$ is ignored, the union bound for the symbol error probability becomes tighter.

7.4.2 Bit Error Versus Symbol Error

In Gray coding, the neighbouring symbols in symbol space differ in only one bit position. If there is a symbol error, then the decoder will most probably estimate nearest symbols as being sent, thus causing a single bit error. Then, the bit error probability P_b is approximated by

$$P_b \cong \frac{P_e}{\log_2 M} \qquad (7.76)$$

In binary coding, all symbol errors are assumed equally likely. Therefore, the P_b is related to P_e by

$$P_b = \frac{M/2}{M-1} P_e \qquad (7.77)$$

Example 7.8 Union Bound on the Error Probability.

Three equiprobable symbols m_1, m_2 and m_3 are to be transmitted over an AWGN channel with two-sided noise PSD $N_0/2$. The three signals corresponding to these symbols are given by (see Figure 7.4)

$$s_1(t) = \begin{cases} A & 0 \le t \le T \\ 0 & otherwise \end{cases},$$

$$s_2(t) = -s_3(t) = \begin{cases} A & 0 \le t \le T/2 \quad (7.78) \\ -A & T/2 < t \le T \\ 0 & otherwise \end{cases}$$

All signals have the same energy $E = A^2 T$. Noting that $s_2(t)$ and $s_3(t)$ are antipodal, and $s_1(t)$ is orthogonal to both $s_2(t)$ and $s_3(t)$, one may express the three signal points as $\mathbf{s}_1 = [1\ 1]$, $\mathbf{s}_2 = -\mathbf{s}_3 = [1\ -1]$ in the signal space with $N = 2$. The basis functions $\Phi_1(t)$ and $\Phi_2(t)$ may be written as

$$\Phi_1(t) = \begin{cases} \sqrt{2/T} & 0 < t \le T/2 \\ 0 & otherwise \end{cases}$$
$$\qquad (7.79)$$
$$\Phi_2(t) = \begin{cases} \sqrt{2/T} & T/2 < t \le T \\ 0 & otherwise \end{cases}$$

The three signals may then be expressed in terms of the basis functions as:

$$s_1(t) = \sqrt{E/2}[\Phi_1(t) + \Phi_2(t)],$$
$$\qquad (7.80)$$
$$s_2(t) = -s_3(t) = \sqrt{E/2}[\Phi_1(t) - \Phi_2(t)],$$

Since $s_1(t)$ is orthogonal to both $s_2(t)$ and $s_3(t)$, one may also choose one of the two basis functions to be parallel to $s_1(t)$ and the second one parallel to $s_2(t)$ and $s_3(t)$, as in Figure 7.4:

$$\phi_1(t) = \frac{s_1(t)}{\sqrt{E}} = \begin{cases} 1/\sqrt{T} & 0 \le t \le T \\ 0 & otherwise \end{cases}$$

$$\phi_2(t) = \frac{s_2(t)}{\sqrt{E}} = \begin{cases} 1/\sqrt{T} & 0 \le t \le T/2 \\ -1/\sqrt{T} & T/2 \le t \le T \\ 0 & otherwise \end{cases}$$
$$\qquad (7.81)$$

It is easy to observe that the basis functions $\phi_1(t)$ and $\phi_2(t)$ may be obtained by 45-degrees rotation of $\Phi_1(t)$ and $\Phi_2(t)$ in the clockwise direction, hence rotational invariance:

$$\begin{pmatrix} \phi_1 \\ \phi_2 \end{pmatrix} = \begin{bmatrix} 1/\sqrt{2} & 1/\sqrt{2} \\ 1/\sqrt{2} & -1/\sqrt{2} \end{bmatrix} \begin{pmatrix} \Phi_1 \\ \Phi_2 \end{pmatrix} \qquad (7.82)$$

Figure 7.18 shows a matched filter implementation of an optimum receiver for this signalling system based on the orthonormal basis functions $\phi_1(t)$ and $\phi_2(t)$. Exploiting the orthogonality between basis functions, the receiver first distinguishes s_1 from s_2 and s_3 by simply comparing the matched filter outputs

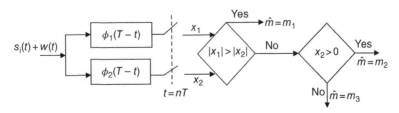

Figure 7.18 Block Diagram of a Matched-Filter Receiver for the Signaling System Described by (7.78).

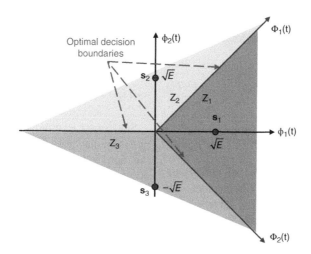

Figure 7.19 Constellation diagram and optimal decision boundaries/regions for the considered signaling scheme in Example 7.8.

$x_1(T)$ and $x_2(t)$, given by $x_k(T) = [s_i(t) + w(t)] \otimes \phi_k(T-t)|_{t=T}$, $i=1, 2$, $k=1, 2$. In case when $|x_2| > |x_1|$, the sign x_2 determines whether s_2 or s_3 is transmitted.

The constellation diagram and optimal decision regions for this signaling scheme is shown in Figure 7.19. In view of the asymmetric decision boundaries, it may not be easy to find an exact analytical expression for the symbol error probability. Instead, we find an union bound on the error probability.

The pairwise error probabilities are given by

$$P(s_1, s_i) = Q\left(\frac{d_{1i}}{\sqrt{2N_0}}\right) = Q\left(\sqrt{\frac{E}{N_0}}\right) = p_0, \quad i=2,3$$

$$P(s_2, s_3) = Q\left(\frac{d_{23}}{\sqrt{2N_0}}\right) = Q\left(\sqrt{\frac{2E}{N_0}}\right) = p_1,$$

$$(7.83)$$

Conditional error probabilities, when s_1, s_2 or s_3 is transmitted, are upper-bounded by

$$P(e|s_1) < P(s_1, s_2) + P(s_1, s_3) = 2p_0$$
$$P(e|s_2) < P(s_2, s_1) + P(s_2, s_3) = p_0 + p_1 \quad (7.84)$$
$$P(e|s_3) < P(s_3, s_1) + P(s_3, s_2) = p_0 + p_1$$

Due to the symmetry in the constellation diagram, the signals $s_2(t)$ and $s_3(t)$ are equally

vulnerable to error. However, the error probability is expected to be higher when the signal $s_1(t)$ is transmitted since its decision region is narrower and the Euclidean distances d_{12} and d_{13} are shorter than d_{23}:

$$d_{1i} = |s_1 - s_i| = \sqrt{2E}, \quad i=2, 3$$
$$d_{23} = |s_2 - s_3| = 2\sqrt{E} \quad (7.85)$$

This is in agreement with (7.83) and (7.84) since $p_0 > p_1$ implies $P(e|s_1) > P(e|s_i)$, $i=2,3$. Union bound on the error probability is found by applying the Bayes rule

$$P_e = P(s_1)P(e|s_1) + P(s_2)P(e|s_2) + P(s_3)P(e|s_3)$$

$$< \frac{1}{3}P(e|s_1) + \frac{1}{3}P(e|s_2) + \frac{1}{3}P(e|s_3)$$

$$= \frac{1}{3}2p_0 + \frac{2}{3}(p_0 + p_1) = \frac{4}{3}p_0 + \frac{2}{3}p_1$$

$$= \frac{4}{3}Q\left(\sqrt{\frac{E}{N_0}}\right) + \frac{2}{3}Q\left(\sqrt{\frac{2E}{N_0}}\right)$$

$$(7.86)$$

For $E/N_0 = 10$ dB, the union bound is found to be

$$P_e < \frac{4}{3}Q\left(\sqrt{\frac{E}{N_0}}\right) = 1.05 \times 10^{-3} \quad (7.87)$$

References

[1] J. G. Proakis and M. Salahi, *Communication Systems Engineering* (2nd ed.), Prentice Hall: New Jersey, 2002.

[2] S. Haykin, *Communication Systems* (3rd ed.), John Wiley: New York, 1994.

[3] B. Sklar, *Digital communications, Fundamentals and Applications* (2nd ed.), Prentice Hall: New Jersey, 2003.

[4] A. B. Carlson, P. B. Crilly, and J. C. Rutledge, *Communication Systems* (4th ed.), McGraw Hill, Boston, 2002.

Problems

1. Consider the signal constellation shown below.

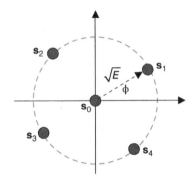

   a. What is the dimension of this constellation?
   b. Determine the average symbol energy in terms of E.
   c. Express the signals in terms of the orthonormal basis functions that you choose.
   d. Show the optimum decision regions in the signal space shown above.
   e. Determine the error probability of the optimum receiver and compare with that of the QPSK.

2. Consider the quaternary signaling schemes shown below:

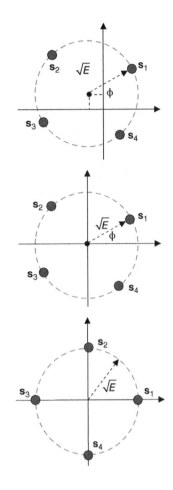

All the signals are equiprobable and the channel noise is white with PSD N_0.

   a. Show the optimum decision regions in these constellations.
   b. Which constellation requires minimum energy.
   c. Determine the error probability of the optimum receiver for these consellations.

3. Consider QPSK signalling defined by the constellation points

$$s_1 = \left(\sqrt{E} \ \ 0\right),\ s_2 = \left(0 \ \ \sqrt{E}\right),$$
$$s_3 = \left(-\sqrt{E} \ \ 0\right),\ s_4 = \left(0 \ \ -\sqrt{E}\right)$$

Assume that the symbols are not equally likely, and have the a priori probabilities $p(s_1) = p(s_3) = 0.3$ and $p(s_2) = p(s_4) = 0.2$.

a. Determine the optimum threshold levels and the decision regions that minimize error probability.
b. Show the decision regions in the constellation diagram.
c. Derive the pair-wise error probability and determine an upper bound for the symbol error probability.

4. Consider the 8-ary amplitude-phase modulated constellation shown below where all symbols are assumed to be equally probable.

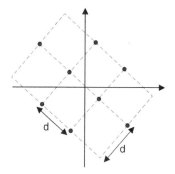

a. Determine the average symbol energy in terms of d and calculate peak-to-average power ratio.
b. Determine the transition probability in terms of E_b/N_0 between two symbols separated by d.
c. Show the decision regions.
d. Use Gray encoding to assign three-bits to each of the constellation points. Express the average symbol and bit error probabilities in terms of E_b/N_0.
e. Determine the nearest neighbour union bound on the bit error probability and compare with the result you found in (d).

5. Consider three equi-probable signals of average energy E. The signals are separated from each other by $120°$ as shown below. The channel is assumed to be AWGN with the double-sided noise PSD $N_0/2$.

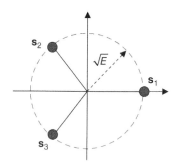

a. Determine a set of orthonormal basis functions for the signal space. Write the expressions for the signals in terms of the basis functions and in vector form.
b. Draw the constellation diagram showing the signal vectors.
c. Determine the Euclidean distances between the signals and show the optimum decision boundaries.
d. Determine an upper bound for the average symbol error probability as a function of E/N_0.

6. Consider the following signal set.

$$s_1(t) = \begin{cases} 1 & 0 \le t < T \\ 0 & otherwise \end{cases}$$

$$s_2(t) = \begin{cases} 2 & 0 \le t < 2T \\ 0 & otherwise \end{cases}$$

$$s_3(t) = \begin{cases} -1 & 0 \le t < T \\ 1 & T \le t < 2T \\ 3 & 2T \le t < 3T \end{cases}$$

a. Determine the dimensionality and the corresponding orthonormal basis functions for this signal set.

b. Express the signals in terms of the orthormal basis functions.

c. Determine the Euclidean distance between the signal pairs and find the minimum distance.

d. Determine the projection of $s_1(t)$ over $s_3(t)$ and compute the angle between $s_1(t)$ and $s_3(t)$.

7. Consider a binary baseband communication system operating in an AWGN channel with zero mean and PSD $N_0/2$. The received signal is $\pm A$ during each symbol duration T. The probability of transmitting $+A$ is 0.4 and the probability of transmitting $-A$ is 0.6. An integrate-and-dump filter is used for demodulation and the detector operates with a threshold level λ.

a. Use MAP rule to formulate the decision rule in terms of the likelihood functions and the threshold level, λ.

b. Determine the optimum threshold level λ_{opt} that minimizes the average error probability.

c. Derive an expression for the minimum average error probability.

8. Consider an asymmetric signal constellation with equally likely symbols:

$$s_1 = \left(\sqrt{E}\ \ 0\right),\ s_2 = \left(0\ \ 2\sqrt{E}\right),$$
$$s_3 = \left(-2\sqrt{E}\ \ 0\right),\ s_4 = \left(0\ \ -\sqrt{E}\right)$$

a. Determine the Euclidean distances between the symbols. What is the minimum Euclidean distance?

b. Compute the average symbol error probability using the two bounds given by

$$P_{e1} \le \frac{1}{M}\sum_{i=1}^{M}\sum_{\substack{k=1 \\ k \ne i}}^{M} Q\left(\frac{d_{ik}}{\sqrt{2N_0}}\right)$$

$$P_{e2} \le (M-1)Q\left(\frac{d_{min}}{\sqrt{2N_0}}\right)$$

c. Which pair-wise error probability dominates the upper bound on the average symbol error probability? Explain.

d. Compute the upper bound when all the Euclidean distances found in (a) are assumed equal to the minimum Euclidean distance. Compare the result you found with the one determined in (b).

9. Consider the four basis functions $\{\phi_1(t), \phi_2(t), \phi_3(t), \phi_4(t)\}$ described by the Hadamard sequences: $\{1111, 1010, 1100, 1001\}$, where the bit 0 is represented by $-1V$ and the bit 1 by $+1V$.

a. Show that these waveforms are orthonormal

b. Determine the Euclidean distances between all the vector pairs.

c. Express the waveform $x(t)$ as a linear combination of $\{\phi_i(t)\}$, and find the coefficients, where $x(t)$ is given by

$$x(t) = \begin{cases} 2 & 0 \le t < 2 \\ -3 & 2 \le t < 3 \\ -1 & 3 \le t < 4 \end{cases}$$

10. Repeat Question 7.9 for the four basis functions $\{\phi_1(t), \phi_2(t), \phi_3(t), \phi_4(t)\}$ described by $\{1000, 0100, 0010, 0001\}$.

11. Consider a quarternary $(M = 4)$ phase shift keying (PSK) system, where the four symbols are located at $\left(\pm\sqrt{E}, 0\right)$, $\left(0, \pm\sqrt{E}\right)$, where E is the symbol energy. Determine the upper bound on the probability of symbol error, assuming that the probability of transmitting each symbol is the

same, and compare with the exact expression for the probability of symbol error.

12. Given a modulation system with $M = 4$ and $N = 2$ with four equiprobable signals defined as, apart from a constant factor $\sqrt{E_0}$, $s_1 = (1,1)$, $s_2 = (-1,1)$, $s_3 = (-2,-1)$, $s_4 = (2,-1)$.

    a. Sketch the signal space (constellation diagram)

    b. Write down the orthonormal basis functions for this signal space.

    c. Write down the corresponding time-domain signals in terms of the orthonormal basis functions.

    d. Determine the average symbol energy.

    e. How can you assign the di-bits to each symbol so as to minmimize the probability of bit error?

    f. Show the optimum decision regions and explain how you obtained them.

    g. Determine the pairwise error probabilities and the upper bound on the average signal error probability in an AWGN channel with zero mean and a variance $N_0/2$. Explain which Euclidean distances dominate the average symbol error probability.

    h. Find the signal most likely to have been sent if all the signals are equally likely and the observation vector is $x = (0.25, -0.25)$.

8

Passband Modulation Techniques

In baseband pulse transmission, a data stream, represented as a sequence of a discrete PAM signals, is transmitted directly over a low-pass channel. However, in digital passband transmission, the input data stream is modulated onto a carrier, which occupies a transmission bandwidth in proportion with the data rate. This is achieved by multiplying the baseband signal $x(t)$ with a carrier of frequency f_c, generated by a local oscillator at the transmitter. The spectrum of the signal thus obtained is shifted to around f_c as shown in Figure 8.1. The energy of the carrier signal generated by the local oscillator is assumed to be unity so that the energy of the baseband signal is not changed with frequency conversion. However, the bandwidth of the passband signal becomes twice as large as that of the baseband signal (see Figure 8.1).

The signal bandwidth, that is, the bandwidth occupied by the signal, should be narrower than the channel bandwidth, that is, the bandwidth allowed by the transmission channel. Otherwise, intersymbol interference (ISI) and some other distortions may arise in the signal at the receiver input. In this context, one is faced with the dilemma between the requirements for higher data rates and lower transmission bandwidths. The transmission bandwidth may be decreased by using suitable line codes (see Chapter 5), higher alphabet sizes, and/or pulse shaping (Nyquist and raised-cosine type filters or partial response signaling) in order to minimize the dispersion, maximize noise immunity, and minimize the resultant ISI (see Chapter 6).

The baseband transmission has a very short range, thus not suitable for long-distance communications. Long-distance communication is made possible by modulation, and, hence, by passband transmission; information-bearing signals modulated by a carrier can be transmitted in a passband channel with much lower attenuation and interference. Passband transmission is more cost-effective due to several reasons, including equipment costs, reliability and capacity (in terms of transmission bandwidth and data rates that can be supported).

The transmitted signal is said to have memory when the signals transmitted in successive symbol intervals are interdependent, as in partial response signaling studied in Chapter 6. When the transmitted signal has memory, the decisions

Digital Communications, First Edition. Mehmet Şafak.
© 2017 John Wiley & Sons Ltd. Published 2017 by John Wiley & Sons Ltd.
Companion website: www.wiley.com/go/safak/Digital_Communications

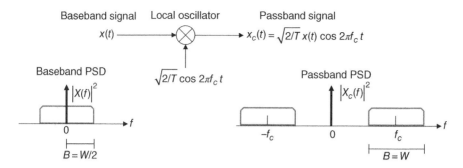

Figure 8.1   Passband Signal. X(*f*) denotes the Fourier transform of the baseband signal *x(t)*.

of the optimum receiver are based on the observation of a sequence of received signals over successive signal intervals. When the signal has no memory, the symbol-by-symbol detector is optimum in the sense of minimizing the symbol error probability. In memoryless channels, we use coherent demodulation when the channel state information (frequency, phase and the delay of the received signal) is available at the receiver, otherwise noncoherent demodulation is employed. In noncoherent demodulation, the receiver does not need the knowledge of the phase and/or the frequency of the carrier wave and treat them as random variables. It basically consists of filtering the signal energy on the allocated spectra and using envelope detectors to estimate the transmitted signal. Noncoherent demodulation techniques are resorted whenever it is impractical to maintain carrier phase and/or frequency synchronization. Noncoherent demodulation performs worse than coherent demodulation, the extent of which depends on the operating E_b/N_0 level.

Modulation consists of switching (keying) the amplitude, the frequency or the phase of a sinusoidal carrier in some fashion in accordance with the incoming data. Hence, similar to analog modulation methods amplitude-modulation (AM), frequency-modulation (FM) and phase-modulation (PM), basically three passband digital modulation schemes are possible; amplitude-shift keying (ASK), frequency-shift keying (FSK), and phase-shift keying (PSK) (see Figure 8.2). Note that PAM is the baseband version of ASK. [1][2][3]

For passband data transmission, PSK and FSK signals have constant envelope and are preferred over nonlinear channels, where the amplitude of the received signal is corrupted by the channel. Since the receiver does not need the amplitude information of received signals modulated by PSK or FSK, these modulation types can perform successfully in propagation channels causing amplitude fluctuations. However, ASK does not have a constant envelope and is hence more successful in linear channels such as cables, with no or minimum amplitude fluctuations. Each modulation scheme has its advantages and weaknesses; a choice is made between them depending on the specific application under consideration.

In this chapter, we will consider the performance characteristics of optimum receivers for various modulation methods in memoryless AWGN channels. Goal is to design an optimum receiver so as to minimize the average probability of symbol error in a passband channel with AWGN. Based on Chapter 7, exact expressions for the symbol error probability are derived for some modulations while union bounds are provided for the others.

8.1   PSD of Passband Signals

A passband signal $s(t)$ may be written in terms of its complex-envelope $\tilde{s}(t)$ as

$$s(t) = A \ \text{Re}[\tilde{s}(t) \ e^{jw_c t}]$$

$$\tilde{s}(t) = \sum_{k=-\infty}^{\infty} b_k \ g(t-kT) \qquad (8.1)$$

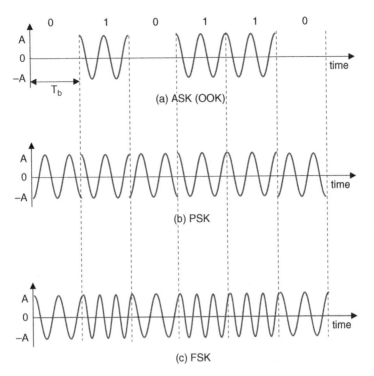

Figure 8.2   Waveforms For the Three Basic Forms of Binary Signaling: a) ASK, b) PSK, and c) FSK.

where A denotes the amplitude, $g(t-kT)$ denotes the pulse over $kT < t < (k+1)T$, T is the symbol duration and b_k is the sequence of independent, equally likely random variables, for example, ± 1 for antipodal signaling.

Assuming that $\tilde{s}(t)$ and $g(t)$ are real, the auto-correlation function of the passband signal $s(t)$ may be written as

$$R_s(\tau) = E[s(t)s(t+\tau)]$$

$$= A^2 E[\tilde{s}(t)\tilde{s}(t+\tau)]\ E[\cos(w_c t)\cos w_c(t+\tau)]$$

$$= \frac{A^2}{2} R_{\tilde{s}}(\tau)\ \cos w_c \tau \qquad (8.2)$$

Power spectral density, PSD, of $s(t)$ is found, by using the Fourier transform relationship between the autocorrelation function and the PSD, as

$$R_{\tilde{s}}(\tau)\ \Leftrightarrow\ G_{\tilde{s}}(f)$$

$$\cos w_c \tau \Leftrightarrow\ \frac{1}{2}\delta(f-f_c) + \frac{1}{2}\delta(f+f_c)$$

$$R_s(\tau)\ \Leftrightarrow\ G_s(f) = \frac{A^2}{4} G_{\tilde{s}}(f-f_c) + \frac{A^2}{4} G_{\tilde{s}}(f+f_c)$$

$$(8.3)$$

8.1.1   Bandwidth

We know from (1.147)–(1.149) that $G_{\tilde{s}}(f) = G_g(f)$ where $G_g(f)$ denotes the PSD of $g(t)$. If $g(t)$ is chosen as NRZ pulse of amplitude A and duration T, then the PSD of $s(t)$ $G_g(f) = T\,sinc^2(fT)$, a squared sinc function with first nulls at $\pm 1/T$ around the carrier frequency (see (1.104)). Then, for a random sequence of NRZ pulses, each with a duration of T, the PSD of the bandpass signal $s(t)$ may be written as

$$G_s(f) = \frac{E}{2}\ sinc^2[(f-f_c)T] + \frac{E}{2}\ sinc^2[(f+f_c)T]\quad (W/Hz)$$

$$(8.4)$$

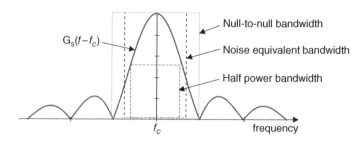

Figure 8.3   Various Definitions of the Bandwidth.

where $E = A^2 T/2$ denotes the energy of $s(t)$. This implies that the choices of the line code and the symbol duration uniquely determine the bandwidth of $s(t)$. However, irrespective of the choice of the line code, PSD of the pass-band signal occupies a very wide bandwidth, though a large percentage of the transmitted signal power is concentrated around the carrier frequency. Therefore, for the economic use of the available frequency spectrum, it may be suitable to filter out the components with neg-ligibly small power of the transmitted signal spectrum at the expense of some acceptable ISI.

There is no unique definition of the band-width. [1][2][3] As shown in Figure 8.3, the *3-dB (half-power) bandwidth* is defined as the spectral width between the frequencies at which the PSD has dropped to half of (3 dB below) its peak value, which occurs at the car-rier frequency. *Noise equivalent bandwidth* is defined by

$$B_{noise\ eq.} = \frac{1}{G_s(0)} \int_{-\infty}^{\infty} G_s(f - f_c)\, df \quad (Hz) \quad (8.5)$$

where $G_S(f - f_c)$ denotes the PSD centered at the carrier frequency. The noise-equivalent band-width refers to that of an ideal rectangular filter, that would pass the same noise power as the actual system. *Null-to-null bandwidth* refers the bandwidth between first nulls on either side of the main spectral lobe, which contains most of the signal power. This is a popular way of describing the bandwidth but may be difficult

to determine in cases where the nulls of the PSD can not be easily identified.

8.1.2   Bandwidth Efficiency

In practice, transmission bandwidth and trans-mitted power are two primary resources in communication systems. Because of the ever-increasing demand for multimedia communica-tions, communication systems are required to operate at highest possible data rates with min-imum error probability by using minimum transmission bandwidths. However, increasing data rates require higher transmission band-widths. Therefore, efficient utilization of these resources calls for a search of spectrally effi-cient transmission and modulation schemes. As a measure of the efficiency of spectrum usage, we define the bandwidth efficiency as the ratio of the data rate in bits/s to the effect-ively utilized transmission bandwidth for a given value of the E_b/N_0:

$$\eta = R_b/B \quad (bps/Hz) \quad (8.6)$$

Note that (8.6) may as well be interpreted as the data rate in bits/s that can be transmitted per Hz (unit bandwidth). Here, R_b denotes the data rate for a given E_b/N_0, and B is the transmission bandwidth. Hence, the spectrum efficiency, which depends on the E_b/N_0, the employed modulation and the line code, may be improved by M-ary power-efficient modulation and spec-tral (pulse) shaping.

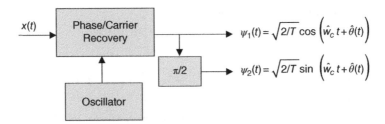

Figure 8.4 Carrier Phase Estimation by PLL.

8.2 Synchronization

In coherent demodulation, the receiver is equipped with a phase-recovery circuit. The phase-recovery circuit ensures that the oscillator supplying the locally generated carrier at the receiver is synchronized in both frequency and phase to the oscillator supplying the carrier at the transmitter. Correlation receiver or matched-filter may be used as demodulator.

Consider a transmitted binary signal with energy E_i and symbol duration of T:

$$s_i(t) = \sqrt{\frac{2E_i}{T}} \cos(w_c t + \phi_i), \quad \begin{array}{l} i = 1,2 \\ 0 < t < T \end{array} \quad (8.7)$$

where a binary 1 is transmitted by $\phi_i = 0$, and a binary 0 is transmiited by $\phi_i = \pi$. The signal at the receiver input is corrupted by a random phase $\theta(t)$ in the AWGN channel:

$$x(t) = \sqrt{\frac{2E_i}{T}} \cos(w_c t + \phi_i + \theta(t)) + w(t), \quad \begin{array}{l} i = 1,2 \\ 0 < t < T \end{array} \quad (8.8)$$

where $w(t)$ denotes the AWGN. If the receiver can estimate $\theta(t)$ accurately, then it can coherently demodulate the transmitted phase ϕ_i, which carries the information whether binary 1 or 0 is transmitted.

Circuits such as phase-locked-loop (PLL), as shown in Figure 8.4, are implemented at the receiver for carrier phase estimation ($\hat{\theta}(t) \approx \theta(t)$). Such a circuit estimates the carrier frequency $\hat{w}_c$ and the phase incurred by the channel $\hat{\theta}(t)$ by using the input signal given by (8.8) and generates in-phase and quadrature components of the local oscillator output with the estimates for the carrier frequency and the phase of the input signal:

$$\psi_1(t) = \sqrt{2/T} \cos(\hat{w}_c t + \hat{\theta}(t))$$
$$\psi_2(t) = \sqrt{2/T} \sin(\hat{w}_c t + \hat{\theta}(t)) \quad (8.9)$$

The so-called orthonormal basis functions in (8.9) have unity energy, as will be shown in Example 8.1.

Based on the assumption that the channel phase changes sufficiently slowly, the process of estimation of the channel phase is adaptive in the sense that estimates are updated by following the fluctuations in the carrier frequency and the phase induced by the time-varying channel.

The performance of a receiver in interpreting an incoming signal is determined by the E_b/N_0, data rate, and the channel bandwidth. The bit error probability is inversely proportional to E_b/N_0 since the negative effects of noise will be less at higher values of E_b/N_0. The bit error probability is directly proportional to the data rate due to the fact that increasing the data rate decreases the energy per bit E_b. On the other hand, higher data rates require wider transmission bandwidths.

Example 8.1 Energy of a Sinusoidal Signal. The amplitude A of a sinusoidal signal $s(t) = A \cos(w_c t + \varphi)$, $0 \le t \le T$, where φ is an arbitrary phase term, may be written in terms of its energy, E:

$$E = \int_0^T s^2(t)\,dt = \int_0^T A^2\cos^2\left(w_c t + \varphi\right)\,dt$$

$$= \frac{A^2}{2}\int_0^T [1 + \cos\left(2w_c t + 2\varphi\right)]dt \tag{8.10}$$

$$= \frac{A^2}{2}T \quad if \ f_c T = integer \Rightarrow$$

$$A = \sqrt{\frac{2E}{T}} = \sqrt{2ER} = \sqrt{2P}$$

where $R = 1/T$ denotes the symbol rate in symbols/s and $P = ER$ stands for the signal power in Watt. Note that the symbol duration is always chosen to contain an integral number of cycles of the carrier, that is, $f_c T = $ integer.

Example 8.2 Error on Carrier Frequency and Phase.

Consider using a correlation receiver for binary transmission in an AWGN channel. The signal at the receiver input is assumed to be given by (8.8). In practice, the correlation demodulator may not have a perfect estimate of the carrier frequency and phase since they are extracted by a recovery circuit from the received signal by performing some nonlinear operations (see Figure 8.5).

In the presence of phase error $\theta(t)$ in the carrier phase, the output of the correlation receiver at the sampling time $t = T$ is given by

$$x_1(T) = \int_0^T x(t)\psi_1(t)\,dt$$

$$= \frac{2}{T}\sqrt{E_i}\int_0^T \cos(w_c t + \phi_i + \theta(t))\cos\left(\hat{w}_c t + \hat{\theta}(t)\right)dt + w_1$$

$$= \sqrt{E_i}\cos(\phi_i + \Delta wt + \Delta\theta(t)) + w_1$$

$$= \pm\sqrt{E_i}\cos(\Delta wt + \Delta\theta(t)) + w_1 \tag{8.11}$$

where $\Delta\theta(t) = \theta(t) - \hat{\theta}(t)$ and $\Delta w = w_c - \hat{w}_c$ denote, respectively, the estimation errors in the channel phase and the carrier frequency. In practice, frequency and phase errors vary slowly and they can be considered constant during a symbol duration, as assumed in the integration in (8.11). The noise component at the sampling instant w_1 is AWGN with a single-sided PSD N_0:

$$w_i = \int_0^T w(t)\,\psi_i(t)\,dt, \quad i = 1, 2$$

$$E[w_i] = \int_0^T E[w(t)]\psi_i(t)\,dt = 0$$

$$\sigma_{w_i}^2 = E\left[|w_i|^2\right]$$

$$= \int_0^T\int_0^T \underbrace{E[w(t)\,w^*(t')]}_{\delta(t-t')N_0/2}\,\psi_i(t)\,\psi_i(t')dt\,dt'$$

$$= N_0/2 \tag{8.12}$$

If the receiver can perfectly estimate the carrier frequency and the channel phase, then

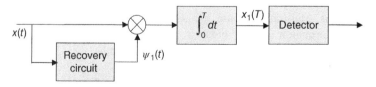

Figure 8.5 Block Diagram of the Correlation Receiver Considered in Example 8.2. The Block Diagram of the Recovery Circuit is given by Figure 8.4.

Table 8.1 The Degradation Caused by the Estimation Errors of the Channel Phase on the Received E/N_0 and the Error Probability.

Phase error, $\Delta\theta$ (degrees)	Degradation in E/N_0 ($\cos^2\Delta\theta$)	Error probability
0	1	0 dB	$1.00 \; 10^{-6}$
5	0.992	-0.03 dB	$1.19 \; 10^{-6}$
10	0.970	-0.13 dB	$1.54 \; 10^{-6}$
15	0.933	-0.30 dB	$2.38 \; 10^{-6}$
20	0.883	-0.54 dB	$4.27 \; 10^{-6}$
45	0.500	-3.00 dB	$4.05 \; 10^{-4}$

inserting $\Delta\theta(t)=0$ and $\Delta w=0$ into (8.11), the signal at the demodulator output simplifies to

$$x_1 = \sqrt{E_i}\cos\phi_i + w_1 = \pm\sqrt{E_i} + w_1, \quad i=1,2 \tag{8.13}$$

as in (6.37). Comparison of (8.11) and (8.13) shows that the estimation errors will decrease the signal level by the factor $\cos(\Delta\theta(t) + \Delta wt)$ at the demodulator output. Assuming $E_1 = E_2 = E$, the received E/N_0 will be decreased by a factor $\cos^2(\Delta\theta(t) + \Delta wt)$. Using (6.47), the resulting error probability may then be written as

$$P_e = Q\left(\sqrt{\frac{2E}{N_0}}\cos(\Delta wt + \Delta\theta(t))\right) \tag{8.14}$$

Based on the assumption that the carrier frequency is perfectly estimated, Table 8.1 presents the degradation caused by the estimation errors of the channel phase on the received E/N_0 and the error probability. Note that the degradation is negligibly small when the phase estimation errors are less than 15 degrees or so.

Example 8.3 Symbol Timing Errors.
Let us now assume that the signal given by (8.8) arrives at the receiver at time $t=0$. However, the demodulator estimates the signal arrival time with an error εT, where ε ($0 \le \varepsilon \le 1$) is a fraction of the symbol time. In other words, the detection of a symbol starts late and completes late by an amount εT. The received and estimated signal timings are shown in

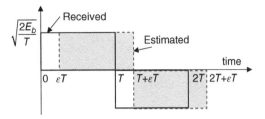

Figure 8.6 Timing Error in Symbol Synchronization.

Figure 8.6 where the transmitted bit sequence is **1 0**. The detection of the symbol sequence starts late and concludes late by an amount εT. [3]

Assuming equally likely antipodal signaling with equal energies $E_1 = E_2 = E$ and perfect frequency and phase synchronization, from (8.8), the received signal may be rewritten as

$$x(t) = s_i(t) + w(t) = \sqrt{\frac{2E}{T}}\cos(w_c t + \phi_i) + w(t), \quad \begin{matrix} i=1,2 \\ 0 < t < T \end{matrix} \tag{8.15}$$

where $s_i(t)$ is defined by (8.7). The signal at the output of the correlation demodulator is given by

$$x_i(T) = \int_{\varepsilon T}^{T+\varepsilon T} x(t)\,\psi_1(t)\,dt$$
$$= \int_{\varepsilon T}^{T+\varepsilon T} s_i(t)\,\psi_1(t)\,dt + w_1 \tag{8.16}$$

where w_1 has zero-mean and single-sided PSD N_0 as given by (8.12).

Assuming that transmitted signal sequence is $\{s_1(t)\ s_2(t)\}$, (8.16) reduces to

$$x_1(T) = \left[\int_{\varepsilon T}^{T} s_1(t)\psi_1(t)\,dt + \int_{T}^{T+\varepsilon T} s_2(t)\psi_1(t)\,dt\right] + w_1$$
$$= \frac{2}{T}\sqrt{E}\left[\int_{\varepsilon T}^{T}\cos^2 w_c t\,dt + \int_{T}^{T+\varepsilon T}(-\cos^2 w_c t)\,dt\right] + w_1$$
$$\cong \frac{1}{T}\sqrt{E}[T - \varepsilon T - (T + \varepsilon T - T)] + w_1$$
$$= \sqrt{E}(1 - 2\varepsilon) + w_1 \tag{8.17}$$

For the sequence $\{s_1(t)\ s_1(t)\}$, we have:

$$x_1(T) = \frac{2}{T}\sqrt{E}\left[\int_{\varepsilon T}^{T} \cos^2 w_c t\ dt + \int_{T}^{T+\varepsilon T} \cos^2 w_c t\ dt\right] + w_1$$

$$= \frac{1}{T}\sqrt{E}[T - \varepsilon T + (T + \varepsilon T - T)] + w_1$$

$$= \sqrt{E} + w_1$$

$$(8.18)$$

The output of the correlation demodulator for the sequences $\{s_2(t)\ s_1(t)\}$ and $\{s_2(t)\ s_2(t)\}$ will be the same as (8.17) and (8.18), except for a sign difference of the signal component. From (8.17) and (8.18) and (6.47), the average error probability is found to be

$$P_e = \frac{1}{2}Q\left(\sqrt{\frac{2E}{N_0}}\right) + \frac{1}{2}Q\left(\sqrt{\frac{2E}{N_0}}(1-2\varepsilon)\right)$$

$$(8.19)$$

Figure 8.7 shows the variation of the bit error probability as a function of E/N_o in dB for various values of ε. One may observe that the error probability increases with symbol timing error. For example, for a given error probability level

of 10^{-4}, $E/N_0 = 8.4$ dB when there is no timing error, that is, $\varepsilon = 0$, while $E/N_0 = 9.92$ dB is needed for $\varepsilon = 0.1$. Hence, the presence of a time offset $\varepsilon = 0.1$ leads to a degradation of $9.92 - 8.4 = 1.52$ dB in the error probability performance at $P_e = 10^{-4}$.

8.2.1 Time and Frequency Standards

Precise time and frequency information is essential for telecommunications, navigation, broadcasting, military, and other scientific applications. For example, 1 ms timing error between transmitter and receiver clocks implies missing 1000 bits at 1 Mbps transmission. Similarly, 1 ms time uncertainty implies an error of 300 km in distance measurements.

Time and frequency standards are interdependent since time and frequency are derived from the same clock. However, the basic unit is the "time". Until 1960s, the most-precise time information, the second, was obtained by synchronizing the best quartz clocks to Earth's revolution around the sun. This was achieved by periodical measurement of the movement

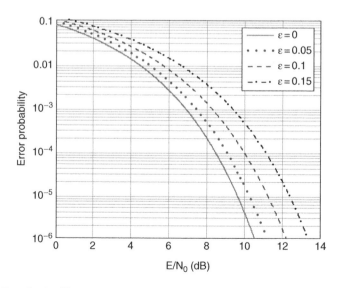

Figure 8.7 Bit Error Probability as a Function of E/N_0 for Various Values of the Symbol Timing Error for Antipodal Signaling.

of stars accross the sky. Measurement uncertainties and the changes in the speed of the Earth limit the time-precision provided by this approach.

Atomic clocks, that provided a quantum leap in the precision of time measurements, use microwave radiation to excite electrons to change orbits. We know from quantum mechanics that, since an electron can occupy only one of a discrete number of orbits around the nucleus of an atom, it can only have a discrete number of energy levels, that is, the energy of an electron is quantized. Therefore, an electron can jump from one orbit to another by absorbing or emitting energy in the form of electromagnetic radiation. Because energy is conserved, the absorption or emission will happen only if the energy corresponding to the frequency of radiation matches the energy difference in the orbital transition. Atomic clocks, including cesium clocks, are based on this principle. The atoms of a clock are hit with a specific frequency of electromagnetic radiation to cause a jump up to a higher energy orbit (level). The closer the frequency of external radiation is to the actual frequency of the clock transition, the higher the probability that the transition will occur. Having adjusted the frequency of the external radiation so as to maximize the orbital transition, the next step is to convert this frequency reference into a clock that provides precise time information. Generally a counter is used to convert the frequency reference into a clock that can count off ticks to indicate the time.

Being one of the most stable frequency generators, the frequency of Cesium atom is used for the purpose of establishing the time standard, the Universal Coordinated Time (UTC). The electron transition of a cesium atom lies in the microwave band. If the clock is kept at absolute zero temperature, and is unperturbed, the transition will occur at 9 192 631 770 Hz. In International System of Units, one second is defined as the time it takes for 9,192, 631,770 cycles of the standard Cesium-133 transition. Cesium clocks moving at relativistic speeds in the sky need correction of the

Doppler shifts that they undergo. The cesium is the primary time/frequency standard, maintained by atomic stations around the world for keeping the UTC. Cesium clocks are widely used in many areas of modern life, including in GPS system, Internet, mobile radio and high-speed communication systems. Accuracy of cesium clock has improved since 1955 increasing by a factor or so every decade. [4] For example, a cesium clock operating at 9.2 GHz with an accuracy of 2×10^{-16}, would suffer a timing error of 0.017 ns per day. Such a clock would therefore make a timing error of one second in 161 million years. The same cesium clock would correspond to an uncertainty of 1.84 µHz in the frequency. [4]

Commercially available rubidium clocks exhibit high accuracy and improved ageing characteristics. Crystal clocks provide lower stability, and are comparatively cheap. Quartz is primarily used for making crystal oscillator clocks.

A new generation of atomic clocks, that use laser light instead of microwaves, can divide time more precisely. Single-ion versions of these optical clocks (OC), made of aluminium or mercury, provide an order of magnitude better timing accuracy. In contrast with single-ion clock, which is based on a single measurement of frequency at a time, optical-lattice clock (OLC) simultaneously measure thousands of atoms brought together by a laser beam, thereby minimize the measurement uncertainty. OLCs are considered to be very promising and may potentially limit the timing error of second over 13.8 billion years, the age of the universe. [4]

The accuracy of cesium clock, which was about 10^{-10} in 1950s, is nowadays in the order of 10^{-16}. Accuracy of OC starting with 10^{-11} in mid-1980s, surpassed cesium clock in 2000s and is nowadays in the order of 10^{-17}. As for OLC, it already has a very impressive accuracy better than 10^{-17}. [4]

Important parameters that characterize a frequency/time standard include frequency accuracy (degree of conformity to a specified frequency), ageing (a process during which the frequency changes irrecoverably), phase noise (random frequency fluctuation of the signal around the nominal frequency, a measure of the purity of the generated frequency), warm-up time and temperature/frequency stability.

There are some terrestrial and space-based navigation systems which are used to disseminate timing information with local or global coverage. Commonly used terrestrial navigation systems are LORAN-C, widely used in the Western world, and its Russian counterpart CHAYKA. These systems enable ships and aircraft to determine time and position information from low frequency radio signals transmitted by fixed land based radio beacons. For example, by using very accurate Cesium clocks, LORAN-C disseminate 5 MHz and pulse per second (pps) signals from high power transmitters to allow navigation receivers to synchronize with typically ±100 ns accuracy.

The so-called global navigation satellite systems (GNSSs) are mainly employed for outdoor applications, provide precise time and positioning information with global coverage. The code division multiple access (CDMA) signals transmitted by these systems enable receivers to determine time and their position (altitude, longitude and latitude). The GPS timing accuracy is typically 10 nanoseconds. However, a typical GPS receiver with a pps output can provide an accuracy of 100 nanoseconds to 1 microsecond. Most commonly used GNSSs are Global Positioning Satellites (GPS) system which has two variants: stand-alone GPS or assisted-GPS (A-GPS). In stand-alone GPS a user determines time and position information by using its own signals, while A-GPS needs additional signals from some reference GPS receivers. The European

GNSS, Galileo, is expected to be fully operational by 2020 and be compatible with the GPS system.

8.3   Coherently Detected Passband Modulations

Passband transmission model is basically a frequency shifted version of the model for baseband transmission (see Figure 8.8). A message source emits one symbol every T seconds, with symbols belonging to an alphabet of M symbols, $m_1, m_2,...., m_M$ with a priory probabilities $p_1, p_2,...., p_M$ specifying the message source output. If M symbols of the alphabet are equally likely, then

$$p_i = P(m_i) = \frac{1}{M} \quad i = 1, 2, ..., M \qquad (8.20)$$

The transmitter assigns the message source output m_i into a distinct signal $s_i(t)$, suitable for transmission over the channel in the symbol duration T. The signal $s_i(t)$ is a real-valued energy signal:

$$E_i = \int_0^T s_i^2(t)\, dt, \quad i = 1, 2, ..., M \qquad (8.21)$$

In binary signalling, we send either one of the two possible symbols $s_1(t)$ or $s_2(t)$, representing m_1 (for 1) and m_2 (for 0), during each signalling interval (bit duration) T_b. In M-ary signalling, one of M possible signals $s_1(t)$, $s_2(t),..., s_M(t)$, is sent during each signalling interval of duration T, to represent one of the M combinations of $\log_2 M$ bits. Thus, the symbol duration is $T = (\log_2 M)\, T_b$.

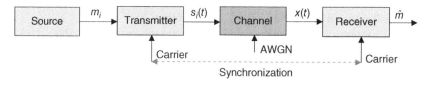

Figure 8.8   Block Diagram of a Passband Communication System.

A sinusoidal carrier is modulated in phase, frequency or amplitude in accordance with the signal $s_i(t)$ to be transmitted. Principal passband modulation techniques, M-ary ASK, M-ary FSK and M-ary PSK, modulate the amplitude, the frequency or the phase of a carrier, respectively. M-ary quadrature-amplitude modulation (QAM) is a combination of M-ary ASK and M-ary PSK. Because of their superior performance coherent systems are preferred. However, noncoherent systems are employed if the receiver does not have information about the carrier phase. Differential phase-shift keying (DPSK) is a noncoherent version of PSK. M-ary FSK and M-ary DPSK are commonly used in noncoherent systems.

The passband channel is assumed to be linear and has a sufficiently wide bandwidth to allow for the transmission of the modulated signal with negligible ISI. The channel noise is characterized by AWGN with zero-mean and PSD $N_0/2$.

The receiver, which usually consists of a correlation demodulator followed by a detector, basically reverses the operations performed at the transmitter. This means that the received passband signal are firstly down converted, filtered, demodulated and detected, as was described in Chapter 7. Using the received signal, the optimum receiver aims to minimize the effect of channel noise on the estimate $\hat{m}$ of the transmitted symbol m_i, hence minimizes the error probability.

The synchronization is accomplished mainly in two phases; acquisition and tracking. In the acquisition process, delay, phase and frequency of the received signal is predicted by the coherent receiver and, subsequently, the random fluctuations in these parameters are tracked so as to preserve synchronization.

8.3.1 Amplitude Shift Keying (ASK)

In ASK, which is the passband version of PAM, the information is transmitted by modulating the amplitude of the carrier signal.

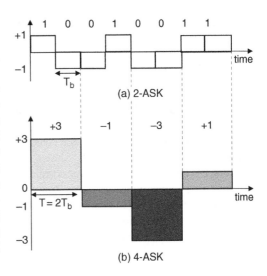

Figure 8.9 Bit Versus Symbol Durations For 2-ASK and 4-ASK.

Because of power-efficiency considerations, M-ary ASK signaling is usually accomplished with symmetric amplitude levels around dc, except for on-off keying (OOK), where for example, a binary 1 is transmitted by turning the transmitter on and a binary 0 is transmitted by turning it off (see Figure 8.2). In ASK, the Euclidean distances between amplitude levels are chosen to be the same in order to prevent any symbol to be more vulnerable to channel noise. Since different amplitude levels in M-ary ASK lead to unequal symbol energies, one can talk only about an average symbol energy and average transmit power (see Figure 8.9).

Let us consider an M-ary ASK signal with symbol duration T:

$$s_i(t) = (2i - M - 1)\sqrt{E_0}\,\psi_1(t), \quad \begin{array}{l} 0 \le t \le T \\ i = 1, 2, \cdots, M \end{array}$$

(8.22)

where E_0 denotes the energy of the symbol with the smallest energy and $\psi_1(t)$ is defined by (8.9). The dimensionality of ASK signaling is $N = 1$; hence a single basis function is sufficient to describe the signal constellation. The M-ary

ASK signal received in an AWGN channel may be written as

$$x(t) = s_i(t) + w(t), \quad i = 1, 2, ...M, \quad 0 \le t \le T$$

(8.23)

where $w(t)$ stands for AWGN with zero mean and double-sided PSD $N_0/2$.

M-ASK signal is coherently detected by using a correlation receiver as shown in Figure 8.10. The correlation receiver correlates the incoming signal with the basis function $\psi_1(t)$ and samples at $t = T$ to obtain the observation vecor with dimension $N = 1$. Hence, the decision variable at the correlator output becomes

$$x_1 = \int_0^T x(t)\psi_1(t)dt = s_i + w_1$$

$$s_i = (2i - M - 1)\sqrt{E_0}, \quad i = 1, 2, \cdots M$$

(8.24)

where s_i denotes the one-dimensional signal vector in the signal-space, shown in Figure 8.11. It is already shown in (8.12) that w_1 has zero mean and a variance $N_0/2$.

A symbol error occurs if the noise, $w_1 = x_1 - s_i$, exceeds one-half of the Euclidean distance between adjacent symbols. For symbols at the edges of the constellation (i.e., for

$m = 1$ and $m = M$), error can occur only in one direction. However, the symbols in the middle are vulnerable to errors in both directions:

$$P_e(s_i) = \begin{cases} P(w_1 = x_1 - s_i > +\sqrt{E_0}) & i = 1 \\ P(|w_1| = |x_1 - s_i| > \sqrt{E_0}) & 2 \le i \le M-1 \\ P(w_1 = x_1 - s_i < -\sqrt{E_0}) & i = M \end{cases}$$

(8.25)

Due to the symmetry of the Gaussian probability density function (pdf) of w_1, the probability of error on both sides of a symbol is the same and is equal to

$$p = P(w_1 < -\sqrt{E_0}) = P(w_1 > \sqrt{E_0})$$

$$= \int_{\sqrt{E_0}}^{\infty} \frac{1}{\sqrt{\pi N_0}} e^{-n^2/N_0} dn$$

(8.26)

$$= Q\left(\sqrt{2E_0/N_0}\right)$$

The average symbol error probability for coherently detected M-ary ASK signaling may then be written as follows:

$$P_M = \frac{1}{M}\sum_{i=1}^{M} P_e(s_i)$$

$$= \frac{1}{M}\left[\underbrace{P(w_1 > \sqrt{E_0})}_{P_e(s_1)} + (M-2)\underbrace{P(|w_1| > \sqrt{E_0})}_{P_e(s_i),\, m=2,3,...,M-1} + \underbrace{P(w_1 < -\sqrt{E_0})}_{P_e(s_M)}\right]$$

$$= \frac{1}{M}[p + (M-2)2p + p]$$

$$= \frac{2(M-1)}{M}Q\left(\sqrt{\frac{2E_0}{N_0}}\right)$$

$$= \frac{2(M-1)}{M}Q\left(\sqrt{\frac{6\log_2 M}{M^2-1}\frac{E_{b,av}}{N_0}}\right)$$

(8.27)

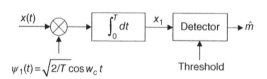

$x(t)$ ⊗ → $\int_0^T dt$ → x_1 → Detector → $\hat{m}$

$\psi_1(t) = \sqrt{2/T}\cos w_c t$

Threshold

Figure 8.10 Correlation Receiver For M-ary ASK.

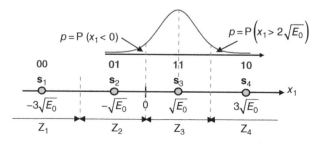

$p = P(x_1 < 0)$

$p = P\left(x_1 > 2\sqrt{E_0}\right)$

| 00 | 01 | 11 | 10 |
| s_1 | s_2 | s_3 | s_4 |
| $-3\sqrt{E_0}$ | $-\sqrt{E_0}$ 0 | $\sqrt{E_0}$ | $3\sqrt{E_0}$ | → x_1
| Z_1 | Z_2 | Z_3 | Z_4 |

Figure 8.11 Occurrence of Error in 4-ASK When Symbol s_3 is Transmitted.

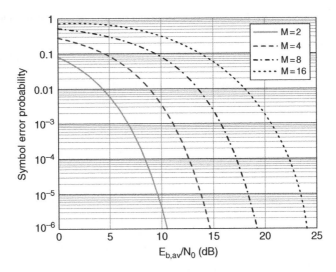

Figure 8.12 Symbol Error Probability for M-ary ASK.

In the last expression above, E_0 is expressed in terms M and the average bit energy $E_{b,av}$ by using the following:

$$E_{s,av} = (\log_2 M)\, E_{b,av}$$

$$= \frac{E_0}{M} \sum_{i=1}^{M} (2i - M - 1)^2 = \frac{E_0}{M/2} \sum_{i=1}^{M/2} (2i-1)^2$$

$$= \frac{(M^2 - 1)E_0}{3}$$

(8.28)

where (D.32) is used to evaluate the sum. Note that mean symbol energy increases with increasing values of M and E_0. Figure 8.12 shows the variation of the symbol error probability with $E_{b,av}/N_0$ for various values of M. For a given symbol error probability, the required $E_{b,av}/N_0$ increases by about 4.5 dB as M goes from 2 to 4 and from 4 to 8. This value saturates at about 5 dB at higher M values. This can be explained by the fact that the minimum Euclidean distance $d_{min} = 2\sqrt{E_0}$ decreases with M for a given $E_{b,av}$, hence making symbols more vulnerable to the channel noise. For a given probability of symbol error, ignoring the factor in front of the Q function in (8.27), the required average bit energy would increase with M by

$$\frac{E_{b,av}}{E_0} = \frac{E_{b,av}}{d_{min}^2/4} = \frac{M^2 - 1}{3\log_2 M}$$

$$d_{min} = 2\sqrt{E_0} = 2\sqrt{\frac{3\log_2 M}{M^2 - 1} E_{b,av}}$$

(8.29)

compared to binary ASK. The variation of this factor with M is given in Table 8.2 together with the decrease in null-to-null bandwidth B_M of M-ary ASK compared to that for binary ASK. Table 8.2 shows the well-known trade-off that, for a given symbol error probability, the required average bit energy increases with increasing values of M, while the corresponding bandwidth decreases by the factor $1/\log_2 M$.

8.3.1.1 PSD of M-ary ASK

An M-ary ASK signal may be written as

$$s(t) = \sqrt{\frac{2}{T}}\, x(t)\, \cos w_c t$$

$$x(t) = \sum_{k=-\infty}^{\infty} b_k\, \Pi\left(\frac{t - kT}{T}\right) \qquad 0 < t \le T \quad (8.30)$$

where $b_k = \pm 1,\ \pm 3,\dots$ The autocorrelation function of $s(t)$ is simply

Table 8.2 Trade-Off Between the Required Bit Energy and the Transmission Bandwidth For a Symbol Error Probability of 10^{-5} for M-ASK.

				$(E_{b,av}/N_0)_M/(E_{b,av}/N_0)_2$ (dB)	
M	B_M/B_2 (8.34)	η (8.35)	$E_{b,av}/N_0$ (dB)	Figure 8.12	(8.29)
2	1	1	9.59	0	0
4	1/2	2	13.75	4.16	4
8	1/3	3	18.29	8.7	8.45
16	1/4	4	23.14	13.55	13.27
32	1/5	5	28.21	18.63	18.34

$$R_s(\tau) = E[s(t)\,s(t+\tau)] = \frac{1}{T}R_x(\tau)\cos w_c\tau$$

$$(8.31)$$

The autocorrelation function and the PSD of an infinite sequence of M-ary ASK pulses of amplitudes $\pm 1,\ \pm 3,\dots$ are given by (1.104) and (1.147)–(1.149):

$$R_x(\tau) = \frac{M^2-1}{3}E_0\Lambda(\tau/T) \quad \Leftrightarrow$$

$$(8.32)$$

$$G_x(f) = \frac{M^2-1}{3}E_0\,T\,\mathrm{sinc}^2(fT)$$

where $E[b_k^2] = (M^2-1)E_0/3$ is found using (8.24) and (8.28). Using the Fourier transform of $\cos w_c\tau$ from (8.3 and (8.31)–(8.32), the PSD of $s(t)$ is found as follows (see also (8.4)):

$$G_s(f) = \frac{1}{T}G_x(f)\otimes\frac{1}{2}[\delta(f-f_c)+\delta(f+f_c)]$$

$$= \frac{M^2-1}{6}E_0\,\mathrm{sinc}^2[(f-f_c)T]$$

$$+ \frac{M^2-1}{6}E_0\,\mathrm{sinc}^2[(f+f_c)T]$$

$$= \frac{E_{s,av}}{2}\,\mathrm{sinc}^2[(f-f_c)T_b\log_2M]$$

$$+ \frac{E_{s,av}}{2}\,\mathrm{sinc}^2[(f+f_c)T_b\log_2M]$$

$$(8.33)$$

The null-to-null bandwidth of M-ary ASK signals is hence given by $2/T$ and the area under $G_s(f)$ gives the average signal power. For larger values of M, the signal power concentrates more around the carrier frequency.

M-ary ASK signalling provides a trade-off between bandwidth efficiency and the increase in transmitted power. For example, the bandwidth for M-ary ASK using NRZ rectangular pulse of duration T, is approximately given by

$$B_M = \frac{1}{T} = \frac{1}{T_b\log_2M} = \frac{B_2}{\log_2M}\ \ (Hz)\ \ (8.34)$$

The spectrum efficiency for M-ary ASK is found from (8.34) as

$$\eta = \frac{R_b}{B_M} = \log_2M\ \ \left(\frac{bps}{Hz}\right)\qquad(8.35)$$

This bandwidth is narrower by a factor of $\log_2M$ than that for binary signalling, $B_2 = 1/T_b = R_b$. However, this does not come free since M-ary ASK requires higher transmit power than the binary PSK. As one may observe from Figure 8.9, the average symbol energy for 4-ASK is given by $(1^2+3^2)/2 = 5$ (7 dB) while it is unity for 2-ASK. Therefore, the bandwidth efficiency of 4-ASK, which is twice that of 2-ASK, comes at the expense of 7 dB (4 dB) increase in the symbol (bit) energy (see Table 8.2).

Example 8.4 On-Off Keying.
On-off keying is a special case of ASK modulation, where a binary 0 is represented by no transmission and a binary 1 is transmitted by a signal of energy E:

$$s(t) = \begin{cases} 0 & for\ 0 \\ \sqrt{2E/T_b}\,\cos w_c t & for\ 1 \end{cases},\ \ 0\le t<T_b$$

$$(8.36)$$

Since the Euclidean distance between the signal points in the signal space is equal to $\sqrt{E}$, the optimum threshold for ML detection is located at the mid-point between the two signal points $\lambda_{opt} = \sqrt{E}/2$. Consequently, the bit error probability is given by

$$P_2 = \frac{1}{2}P(x_1 < \lambda_{opt}|1) + \frac{1}{2}P(x_1 > \lambda_{opt}|0)$$

$$= P(x_1 > \lambda_{opt}|0) = P(w_1 > \lambda_{opt}|0)$$

$$= \frac{1}{\sqrt{2\pi}\,\sigma}\int_{\lambda_{opt}}^{\infty} \exp\left(-\frac{x^2}{2\sigma^2}\right) dx$$

$$= Q\left(\frac{\lambda_{opt}}{\sigma}\right) = Q\left(\sqrt{\frac{E}{2N_0}}\right) = Q\left(\sqrt{\frac{E_b}{N_0}}\right)$$

$$(8.37)$$

where x_1 denotes the observation vector at the demodulator output and w_1 has zero mean and variance $\sigma^2 = N_0/2$. Noting that the average bit energy is $E_b = E/2$, on-off keying performs 3 dB worse than binary ASK for a given value of the P_2.

Example 8.5   3-ASK.
Consider a coherent 3-ary ASK signal at the receiver input as given by

$$x(t) = (i-2)\sqrt{\frac{2E_0}{T}}\ \cos(w_c t + \theta) + w(t),$$

$$0 \le t \le T,\ i = 1,2,3$$

$$(8.38)$$

where $w(t)$ is AWGN with variance $\sigma^2 = N_0/2$. Since symbols have unequal amplitudes, the average symbol energy is given by

$$E_{s,av} = \frac{1}{3}(E_0 + 0 + E_0) = \frac{2E_0}{3} \Rightarrow E_0 = \frac{3E_{s,av}}{2}$$

$$(8.39)$$

As shown in Figure 8.10, the receiver correlates the input signal with the basis function $\psi_1(t) = \sqrt{2/T}\ \cos w_c t$. This implies that the receiver predicts the channel-induced phase

with an error, θ. The observation vector at the demodulator output will be

$$x_i = \int_0^T x(t)\ \psi_1(t)\ dt = (i-2)\sqrt{E_0}\ \cos\theta + w_1, \quad i = 1,2,3$$

$$(8.40)$$

Note that the erroneous prediction of the channel phase reduces the signal power by a factor $\cos^2\theta$. The noise term w_1, which is characterized by (8.12), has zero-mean and variance $\sigma_{w_1}^2 = N_0/2$. The demodulator output will be similar to Figure 8.11. However, the three signals will be located at $0,\ \pm\sqrt{E_0}\cos\theta$ along the basis function $\psi_1(t)$ with the following pdf's:

$$f_{x_i}(x|s_i) = \frac{1}{\sqrt{2\pi}\,\sigma_{w_1}}\exp\left(-\frac{(x \pm \sqrt{E_0}\cos\theta)^2}{2\,\sigma_{w_1}^2}\right), \quad i = 1,3$$

$$f_{x_2}(x|s_2) = \frac{1}{\sqrt{2\pi}\,\sigma_{w_1}}\exp\left(-\frac{x^2}{2\,\sigma_{w_1}^2}\right)$$

$$(8.41)$$

Now let us assume that the a priori probabilities of the three symbols are denoted by p_i, $i = 1,2,3$. The optimum decision threshold between $s_2(t)$ and $s_3(t)$ is determined by the MAP decision rule used by the detector:

$$\frac{p_2 f_{x_2}(x|s_2)}{p_3\ f_{x_3}(x|s_3)} \underset{s_3}{\overset{s_2}{\gtrless}} 1$$

$$x \underset{s_2}{\overset{s_3}{\gtrless}} \lambda_{opt} = \frac{1}{2}\sqrt{E_0}\ \cos\theta - \frac{\sigma_{w_1}^2}{\sqrt{E_0}\ \cos\theta}\ln\left(\frac{p_3}{p_2}\right)$$

$$(8.42)$$

For equiprobable signalling, $p_i = 1/3$, $i = 1,2,3$, the optimum decision threshold will be equal to $\lambda_{opt} = 0.5\ \sqrt{E_0}\cos\theta$, as expected. For $p_1 = p_3$, the optimum threshold level between $s_1(t)$ and $s_2(t)$ will be equal to $-\lambda_{opt}$. Then, the average symbol error probability is given by

$$P_3 = p_1\ P(e|s_1) + p_2\ P(e|s_2) + p_3\ P(e|s_3)$$

$$= p_2\ P(e|s_2) + (1 - p_2\ )P(e|s_3)$$

$$(8.43)$$

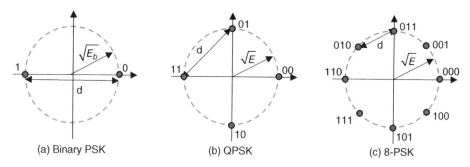

Figure 8.13   Minimum Euclidean Distance in M-ary PSK with Gray-Mapping.

where

$$P(e|s_2) = 2 \int_{\lambda_{opt}}^{\infty} f(x|s_2) \, dx = 2Q\left(\frac{\lambda_{opt}}{\sigma_{w_1}}\right)$$

$$P(e|s_3) = \int_{-\infty}^{\lambda_{opt}} f(x|s_3) \, dx = Q\left(\frac{\sqrt{E_0}\cos\theta - \lambda_{opt}}{\sigma_{w_1}}\right)$$

$$(8.44)$$

8.3.2   Phase Shift Keying (PSK)

PSK is a constant-envelope modulation, that is, with constant symbol energy, which can be demodulated coherently or noncoherently. Since the constant envelope of PSK signals do not carry information, the performance of PSK modulated signals is not adversely affected by the envelope fluctuations due to the channel. Each of the M symbols of M-ary PSK is evenly mapped onto the carrier phase (see Figure 8.13). Each of the M symbols is assigned to a phase, which has an angular separation of $2\pi/M$ rad. from nearby symbols. As M increases, the orbit of the M-ary PSK constellation becomes more crowded and the Euclidean distance between the signal points decreases, increasing the symbol error probability. In order to maintain a given symbol error probability, the Euclidean distance between signal points should not change. Therefore, the constellation radius, which is proportional to the square root of the symbol energy, must increase as M increases. The increase in the symbol energy with M is paid

off by the decrease in the transmission bandwidth as in M-ary ASK (see (8.34)).

8.3.2.1   Binary Phase Shift Keying (BPSK)

In BPSK, the carrier phase is modulated by the phases assigned to binary 1 and 0, which are assumed to equally likely. In the following, the BPSK signals $s_1(t)$ and $s_2(t)$ are used to transmit binary symbols 1 and 0, respectively. These signals are assumed to have equal energies E_b. We assume that binary 1 and binary 0 are transmitted for $0 \le t \le T_b$ by shifting the carrier phase by $\phi_1 = 0$ degrees and $\phi_2 = 180$ degrees, respectively:

$$s_i(t) = \sqrt{2E_b/T_b}\cos(2\pi f_c t + \phi_i) \quad i = 1, 2, 0 \le t \le T_b$$

$$(8.45)$$

As can be observed from (8.45) and Figure 8.13, the BPSK signal is symmetrical with respect to the origin and is the same as binary ASK; such signals are called as antipodal.

The block diagram of a BPSK transmitter shown in Figure 8.14 shows that the input binary sequence of 1 s and 0 s, is encoded by a polar non return-to-zero (NRZ) level encoder with constant amplitude levels $\pm\sqrt{E_b}$. The resulting binary signal is multiplied by the in-phase component $\psi_1(t)$ of the orthormal basis functions defined by (8.9):

$$\begin{aligned}\psi_1(t) &= \sqrt{2/T_b}\cos(2\pi f_c t) \\ \psi_2(t) &= \sqrt{2/T_b}\sin(2\pi f_c t)\end{aligned}, \quad 0 \le t < T_b \quad (8.46)$$

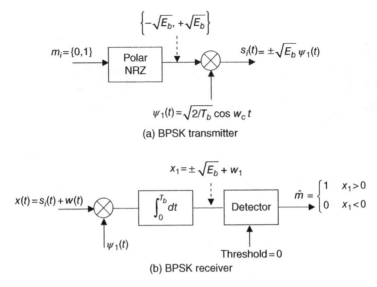

Figure 8.14   Block Diagrams of BPSK Transmitter and Coherent BPSK Receiver.

where $f_c T_b$ = integer. This implies that the carrier frequency, carrier phase and timing pulses are extracted perfectly by the coherent receiver.

Since binary PSK has a one-dimensional constellation ($N = 1$), the transmitted signal may be expressed in terms of the basis function as follows:

$$s_i(t) = \begin{cases} +\sqrt{E_b}\,\psi_1(t) & i=1 \\ -\sqrt{E_b}\,\psi_1(t) & i=2 \end{cases}, \quad 0 \le t < T_b$$

(8.47)

The noisy PSK signal at the receiver input, $x(t) = s_i(t) + w(t)$, is applied to a correlation receiver, which is also supplied with a locally generated coherent reference signal $\psi_1(t)$, where the channel phase is assumed to be estimated perfectly. The observation vector $\mathbf{x} = [x_1]$ at the correlator output is given by

$$x_1 = \int_0^{T_b} [s_i(t) + w(t)]\,\psi_1(t)\,dt = \pm\sqrt{E_b} + w_1$$

(8.48)

where w_1 is AWGN with zero-mean and variance $N_0/2$ (see (8.12)). In this equiprobable signalling, a ML detector compares x_1 with a threshold of zero volts: If $x_1 > 0$, receiver decides in favour of 1, otherwise it decides in favour of 0. Due to circularly symmetric nature of the AWGN noise, only w_1, the noise component along $\phi_1(t)$, affects the detector decision and the noise component w_2, which is the projection of the noise w(t) along $\psi_2(t)$, is irrelevant (see Example 7.2).

One may infer from (8.48) that, in the absence of noise, the signal at the demodulator output (detector input) is $\pm\sqrt{E_b}$ depending on whether $s_1(t)$ or $s_2(t)$ is transmitted. Since the optimal threshold for equiprobable signaling is located at mid-way between the two constellation points (at the origion in this particular example), the presence of noise does not lead the ML detector to erroneous decision as long as $\sqrt{E_b} + w_1 > 0$ when $s_1(t)$ is sent and $-\sqrt{E_b} + w_1 < 0$ when $s_2(t)$ is sent. This is equivalent to say that the detector decides that the signal $s_1(t)$ (i.e., binary symbol 1) was transmitted if $x_1 > 0$ and decides that the signal $s_2(t)$ (i.e., binary symbol 0) was transmitted if $x_1 < 0$. Alternatively, the decision mechanism may be interpreted as follows: The detector decides that binary 1 was transmitted if x_1 lies in decision region Z_1 and binary 0 was transmitted if x_1 lies

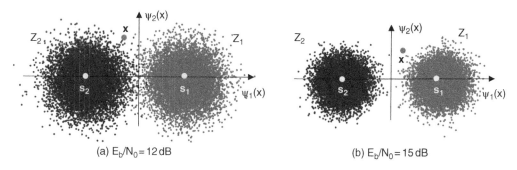

Figure 8.15 Scatter Plot For BPSK Showing the Effect of Noise on the Observation Vector for $E_b/N_0 = 12$ dB and 15 dB with 10000 Noise Samples. **x** denotes the observation vector.

in decision region Z_2 (see Figure 8.15). ML decision of the detector may also be interpreted in terms of the Euclidean distances between the observation point and the signal points. ML detector decides as follows:

$$\hat{m} = \begin{cases} 0 & \|\mathbf{x}-\mathbf{s}_1\| > \|\mathbf{x}-\mathbf{s}_2\| \\ 1 & \|\mathbf{x}-\mathbf{s}_1\| < \|\mathbf{x}-\mathbf{s}_2\| \end{cases} \quad (8.49)$$

The channel noise forces the ML detector to erroneous decision if $\sqrt{E_b} + w_1 < 0$ when $s_1(t)$ is transmitted and $-\sqrt{E_b} + w_1 > 0$ when $s_2(t)$ is transmitted. Therefore, the average bit error probability may be written as

$$P_b = p_1\, P\left(\sqrt{E_b} + w_1 < 0 | 1\right) + p_0\, P\left(-\sqrt{E_b} + w_1 > 0 | 0\right) \quad (8.50)$$

where p_1 and p_0 denote, respectively, a priori probabilities of sending binary 1 and binary 0 and are equal to each other, $p_1 = p_0 = 1/2$, for equi-probable signaling. Using (8.26), the probability of bit error for BPSK given by (8.50) simplifies to

$$P_b = Q\left(\sqrt{\frac{2E_b}{N_0}}\right) \quad (8.51)$$

The variation of the bit error probability with E_b/N_0 is presented in Figure 5.34, where the bit error probability decreases with increasing values of the E_b/N_0.

Example 8.6   Bandwith and Bit Error Probability Calculations for BPSK.
A coherent binary PSK system using a correlation receiver is transmitting at a bit rate of $R_b = 20$ Mbps. Noise power spectral density at the receiver is $N_0 = 5 \times 10^{-11}$ Watts/Hz. The amplitude of the received carrier signal is 100 mV and the carrier frequency is 2.4 GHz. The system uses a raised-cosine pulse with a rolloff factor $\alpha = 0.5$.

The raised cosine bandwidth required for the transmitted passband signal is given by twice the value of (6.110) valid for baseband signals:

$$W = 2\frac{1}{2T_b}(1+\alpha) = R_b(1+\alpha) = 30\text{MHz} \quad (8.52)$$

Hence, the system bandwidth is limited to [2385 MHz 2415 MHz]. The probability of bit error is found to be

$$E_b = A^2 T_b/2 = A^2/(2R_b) = 0.25 \times 10^{-9}J$$
$$P_b = Q\left(\sqrt{2E_b/N_0}\right) = Q\left(\sqrt{10}\right) = 7.83 \times 10^{-4} \quad (8.53)$$

which corresponds to $E_b/N_0 = 5$ (7 dB). If we want to decrease the bit error probability to 10^{-7} or less, then we need $E_b/N_0 = 11.3$ dB. This requires $11.3 - 7 = 4.3$ dB increase in the transmit power.

8.3.2.2 Quadrature Phase Shift Keying (QPSK)

QPSK uses bandwidh more efficiently compared to BPSK due to quadrature-carrier multiplexing. The carrier phase takes one of the four equally spaced values, such as $\pi/4$, $3\pi/4$, $5\pi/4$, and $7\pi/4$, to transmit the information carried by one of the four combinations of dibits, 00, 01, 10 and 11. Since each symbol comprises two bits, the symbol duration is twice the bit duration, $T = 2T_b$, the symbol rate is one-half the bit rate, $R_s = 1/T = 1/(2T_b) = R_b/2$, and the symbol energy is equal to twice the bit energy, $E = 2E_b$.

A QPSK signal may be written, in terms of the four phases $(2i-1)\pi/4$, $i = 1,2,3,4$ assigned to the dibits, as

$$s_i(t) = \begin{cases} \sqrt{2E/T}\cos[2\pi f_c t + (2i-1)\pi/4] & 0 \le t \le T \\ 0 & elsewhere \end{cases}, \quad i = 1,2,3,4$$

(8.54)

By using the trigonometric identity, $\cos(A + B) = \cos A \cos B - \sin A \sin B$, (8.54)

may be expressed in terms of basis functions as follows

$$s_i(t) = \sqrt{E}\cos\left[(2i-1)\frac{\pi}{4}\right]\psi_1(t) - \sqrt{E}\sin\left[(2i-1)\frac{\pi}{4}\right]\psi_2(t)$$

$$= s_{i1}\,\psi_1(t) - s_{i2}\,\psi_2(t), \quad i = 1,2,3,4$$

(8.55)

where the orthonormal basis functions are given by (8.46) if T_b is replaced by T. Message points and associated signal vectors are then given by (see Figure 8.16)

$$\mathbf{s}_i = \begin{bmatrix} s_{i1} \\ s_{i2} \end{bmatrix} = \begin{bmatrix} \sqrt{E}\cos((2i-1)\pi/4) \\ \sqrt{E}\sin((2i-1)\pi/4) \end{bmatrix}, \quad i = 1,2,3,4$$

(8.56)

As shown in Figure 8.17, the input binary data sequence is first transformed into polar form by NRZ encoder, where symbols 1 and 0 are represented by $+\sqrt{E_b}$ and $-\sqrt{E_b}$. Odd- and even-numbered input bits are forwarded to in-phase and quadrature channels by

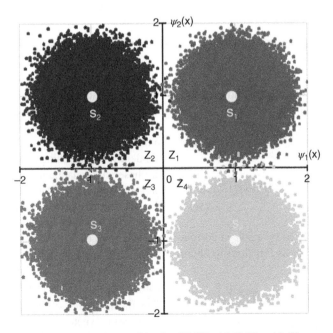

Figure 8.16   Scatter Plot for QPSK with $E_s/N_0 = 13$ dB.

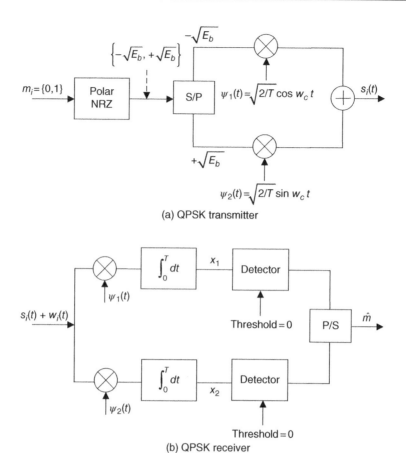

(a) QPSK transmitter

(b) QPSK receiver

Figure 8.17   Block Diagram of QPSK Transmitter and Receiver.

serial-to-parallel (S/P) converter, and modulate quadrature carriers $\psi_1(t)$ and $\psi_2(t)$.

As shown in Figure 8.16, the QPSK signal at the input of a coherent QPSK receiver is corrupted by AWGN:

$$x(t) = s_i(t) + w(t), \quad 0 \le t \le T, \quad i = 1, 2, 3, 4$$
$$(8.57)$$

This signal is correlated by a pair of correlators with orthonormal basis functions to obtain the observation vector:

$$\mathbf{x} = [x_1 \; x_2] = [s_{i1} + w_1 \; s_{i2} + w_2]$$

$$s_{ij} = \int_0^T s_i(t) \, \psi_j(t) dt, \quad i = 1, 2, 3, 4, \; j = 1, 2$$
$$(8.58)$$

where w_1 and w_2 are defined by (8.12). The two binary PSK signals, x_1 and x_2, the components of the observation vector along two orthogonal branches are detected by two ML detectors independently. The resulting two binary sequences at in-phase and quadrature channel outputs are parallel-to-series (P/S) converted to reproduce the transmitted binary sequence with minimum BER in an AWGN channel.

Note from (8.56) and (8.58) that the components of the observation vector are sample values of independent Gaussian random variables with mean $\pm \sqrt{E/2} = \pm \sqrt{E_b}$ and variance $N_0/2$. The mean values of the components of the observation vector evidently define the four signal vectors, $\mathbf{s}_i$, $i = 1, 2, 3, 4$ as given by (8.56). The noise forms a clutter around the signal vectors and may lead the ML detector to erroneous decisions (see Figure 8.16).

It is clear from (8.58) that one of the dibits for each symbol is determined by x_1 and the other by x_2. In other words, a QPSK signal carries one bit of each symbol by the in-phase and the other by the quadrature carrier. Hence, a QPSK system may be characterized by two constituent orthogonal binary PSK systems with decision variables, $x_i = \pm \sqrt{E_b} + w_i$, $i = 1$, 2. Bit errors in in-phase and quadrature channels are statistically independent since x_1 and x_2 are independent of each other. A QPSK symbol is therefore correctly detected if detection is successful in both in-phase and quadrature channels, which implies the correct detection of both bits of a dibit. Hence, the average probability of a symbol error is given by

$$P_4 = 1 - P_c = 1 - (1 - P_b)^2 = 2P_b - P_b^2$$

$$= 2Q\left(\sqrt{\frac{2E_b}{N_0}}\right) - Q^2\left(\sqrt{\frac{2E_b}{N_0}}\right) \qquad (8.59)$$

where P_c denotes the average probability of correct detection of a symbol. P_b shows the average bit error probability of a BPSK signal using a decision variable $x_i = \pm \sqrt{E_b} + w_i$, $i = 1$, 2 and is given by (8.51). The average bit error probability for QPSK with Gray encoding is given by

$$P_b = \frac{P_e}{2} = Q\left(\sqrt{\frac{2E_b}{N_0}}\right) - \frac{1}{2}Q^2\left(\sqrt{\frac{2E_b}{N_0}}\right) \qquad (8.60)$$

For sufficiently large values of E_b/N_0, the second terms in (8.59) and (8.60) are usually neglected since they are much smaller than the corresponding first terms. Hence, average bit error probability for BPSK and QPSK are nearly the same for large values of the E_b/N_0.

Let us now try to clarify the trade-off between the bit rate, the bandwidth and the transmit power in QPSK signaling. Assume a single user transmitting at a bit rate R_b bps using QPSK modulation. The dibits are serial-to-parallel converted to modulate the quadrature carriers. Symbols formed by the dibits are transmitted during each symbol duration $T = 2T_b$ by equally dividing the signal transmit power P in two carriers. The received power by each channel is $P/2$. The bit duration in QPSK channels is doubled, since even and odd bits are transmitted at the same time in in-phase and quadrature channels. Thus, energy per bit in in-phase and quadrature channels, $E_b = (P/2)2T_b = PT_b$, is the same as in BPSK. This justifies the fact that the average bit error probability for BPSK and QPSK is the same. However, QPSK needs only half the bandwidth compared to BPSK since the transmission rate is half that for BPSK: $1/T = 1/(2T_b) = R_b/2$. QPSK is usually preferred to BPSK since it uses channel bandwidth more efficiently than the BPSK.

Let us now consider the case where each QPSK channel carries the same bit rate as BPSK, that is, supports two users each transmitting at rate R_b. In this case, each quadrature carrier transmits one bit during T_b with the same the transmit power as the BPSK system, that is the same $E_b = PT_b$ in order to achieve the same average bit error probability. Thus, bit rate is twice that for BPSK at the expense of doubling the transmit power but using the same bandwidth as BPSK.

Example 8.7 Alternative Derivation/Interpretation of the Symbol Error Probability for QPSK.

Let m_i denote the set of dibits, which are transmitted with equal probabilities:

$$m_i = \{00, 01, 10, 11\}$$
$$P(m_i) = 1/4, \quad i = 1, 2, 3, 4 \qquad (8.61)$$

Due to the rotational symmetry of the QPSK constellation, the average probability of symbol error reduces to

$$P_4 = \sum_{i=1}^{4} P(\hat{m} \neq m_i | m_i) P(m_i)$$
$$= P(\hat{m} \neq m_1 | m_1) \qquad (8.62)$$
$$= 1 - P(\hat{m} = m_1 | m_1)$$

The correct detection of a symbol implies correct detection of its two bits in two quadrature channels. Due to the independence of x_1 and x_2 in (8.58), the bit errors in quadrature channels are independent of each other. From the decision variable for BPSK in (8.48), correct reception of a QPSK symbol requires that the noise components in in-phase and quadrature channels should be both larger than $-\sqrt{E_b}$. Hence, the probability of correct detection of a symbol is given by

$$P(\hat{m}=m_1|m_1)=\text{Prob}(\sqrt{E_b}+w_1>0,\ \sqrt{E_b}+w_2>0)$$

$$=\text{Prob}(\sqrt{E_b}+w_1>0)\times\text{Prob}(\sqrt{E_b}+w_2>0)$$

$$=(1-P_b)^2$$

$$(8.63)$$

where the last expression follows from (8.26). The average symbol error probability obtained using (8.62) and (8.63) is identical to that given by (8.59). Symbol error probability for QPSK may be decomposed into its error components in in-phase (I) and quadrature (Q) channels as follows:

$$P_4=1-(1-P_b)^2$$

$$=2P_b-P_b^2$$

$$=\underbrace{\{error\ I\}}_{P_b}\underbrace{\{error\ q\}}_{P_b}+\underbrace{\{error\ I\}}_{P_b}\underbrace{\{no\ error\ q\}}_{1-P_b}$$

$$+\underbrace{\{no\ error\ I\}}_{1-P_b}\underbrace{\{error\ q\}}_{P_b}$$

$$(8.64)$$

Hence, a symbol error can occur when the first bit, the second bit or both bits are in error. Since each term in the last line of (8.64) defines a different decision region, (8.64) may also be interpreted such that, given that the symbol m_1 is transmitted, a symbol error occurs when the observation vector lies in the decision region Z_2, Z_3 or Z_4.

Offset QPSK

In view of (8.55), the carrier phase changes by 180 degrees whenever both in-phase and quadrature components of QPSK signal changes sign. The carrier phase changes by ±90 degrees when only the in-phase or the quadrature component changes sign. However, it is unchanged when neither the in-phase nor the quadrature components changes sign. Phase changes, particularly 180 degree changes, are of particular concern when the QPSK signal is filtered before transmission due to bandwidth limitations. These shifts in the carrier phase can result in envelope changes in a QPSK signal, thereby causing additional symbol errors.

Offset QPSK is designed mainly to reduce the extent of envelope fluctuations exhibited by QPSK. For this purpose, the basis functions for offset QPSK are chosen as $\psi_1(t)$ and $\psi_2(t-T/2)$ (see (8.46)). Figure 8.18 shows the S/P conversion of a bit stream in QPSK and offset QPSK. In offset QPSK, the transmission of the quadrature carrier starts with a delay of $T/2$, half symbol duration, compared to the in-phase carrier. This implies that the phase transitions in in-phase and quadrature carriers do not occur at the same time. Therefore, in contrast with QPSK, the phase transitions in offset QPSK are limited to a maximum of ±90 degrees, though they occur twice as frequently (see Figure 8.19). Hence, the signal envelope of an offset QPSK signal is better preserved. Nevertheless, the symbol error probability for offset QPSK still remains the same as for QPSK, since the orthogonality between the quadrature carriers is preserved. [1][2][3]

8.3.2.3 M-ary PSK

In M-ary PSK, during each signalling interval T, the carrier phase takes on one of M possible values, namely, $\phi_i=(i-1)$ $2\pi/M,\ i=1,2,\cdots M$. Transmitted M-ary PSK signal may be written as

$$s_i(t)=\sqrt{\frac{2E}{T}}\cos\left(2\pi f_c t+(i-1)\frac{2\pi}{M}\right),\begin{cases}i=1,2,...,M\\0\le t\le T\end{cases}$$

$$(8.65)$$

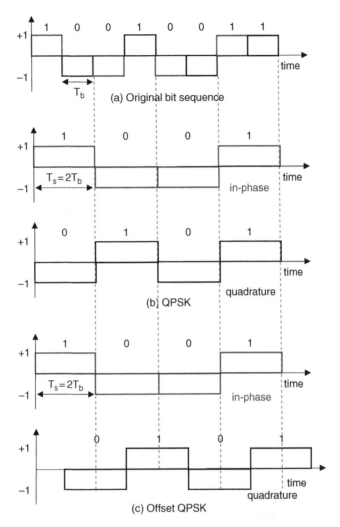

Figure 8.18 In-Phase and Quadrature Components in QPSK and offset QPSK.

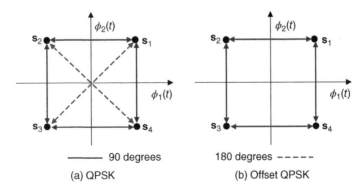

Figure 8.19 Phase Transitions in QPSK and Offset-QPSK.

where $E = \log_2 M \, E_b$ denotes the symbol energy, $T = \log_2 M \, T_b$ the symbol duration and $f_c T = $ integer. As in QPSK signalling, (8.65) may be decomposed into in-phase and quadrature components

$$s_i(t) = \sqrt{E} \cos\left[(2i-1)\frac{\pi}{M}\right]\psi_1(t)$$

$$- \sqrt{E} \sin\left[(2i-1)\frac{\pi}{M}\right]\psi_2(t)$$

$$= s_{i1}\,\psi_1(t) - s_{i2}\,\psi_2(t), \quad i = 1, 2, \cdots, M$$

$$(8.66)$$

where the orthonormal basis functions are defined by (8.46) and the signal vector is given by

$$\mathbf{s}_i = \begin{bmatrix} s_{i1} \\ s_{i2} \end{bmatrix} = \begin{bmatrix} \sqrt{E}\cos((2i-1)\pi/M) \\ \sqrt{E}\sin((2i-1)\pi/M) \end{bmatrix}, \quad i = 1, 2, \cdots, M$$

$$(8.67)$$

The signal constellation of M-ary PSK for $M > 2$ is therefore two-dimensional, that is, $N = 2$ and the M message points are equally spaced on a circle of radius $\sqrt{E}$ with the centre at the origin (see Figure 8.13 and Figure 8.20). Hence, all symbols of M-ary PSK have equal energies.

For coherent demodulation of M-ary PSK symbols, the receiver correlates the received signal corrupted by AWGN with the two basis functions and samples the output at $t = T$. Orthonormal basis functions, defined by (8.46), are in Cartesian coordinates and are more suitable for detecting constellations which can described in Cartesian coordinates,

that is, for $M \leq 4$, For $M > 4$, it may be more suitable to use polar coordinates, since the decision regions are identified by angular regions. Hence, there is a need for transforming the observation vector $\mathbf{x} = [x_1 \ x_2]$ into polar coordinates for $M > 4$. The ML detector for a M-ary PSK signal needs only the phase of the observation vector, defined as $\varphi = \tan^{-1}(x_2/x_1)$, to locate the observation vector in one of the M-1 decision regions.

To derive the pdf of the phase of the observation vector, the phase of the transmitted signal phase is assumed to be $\phi_1 = 0$, which implies the transmission of the signal $s_1(t)$ in Figure 8.20. As shown in Figure 8.21, the observation vector at the output of the quadrature correlator is given by

$$\mathbf{x} = [x_1 \ x_2] = \left[\sqrt{E} + w_1 \ w_2\right] \qquad (8.68)$$

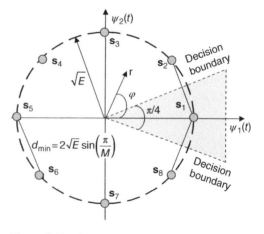

Figure 8.20   Signal-Space Diagram for 8-ary PSK.

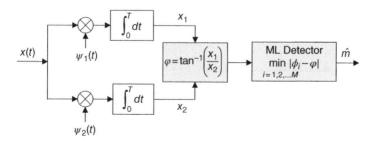

Figure 8.21   Block Diagram of a M-ary PSK Receiver.

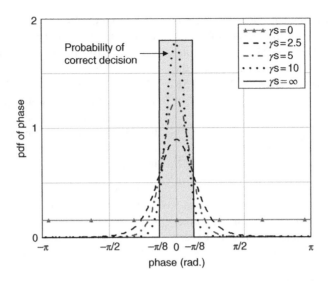

Figure 8.22   Pdf of the Phase For M-ary PSK When the Symbol s_1 is Transmitted.

Because x_1 and x_2 are independent Gaussian random variables, where w_1 and w_2 have zero mean and variance $\sigma^2 = N_0/2$ (see (8.12)), and the decision is based on the phase φ, we transform the joint pdf of x_1 and x_2:

$$f(x_1,x_2) = \frac{1}{2\pi\sigma^2}\exp\left[-\frac{(x_1-\sqrt{E})^2 + x_2^2}{2\sigma^2}\right]$$

$$(8.69)$$

into polar coordinates r and φ, using the following transformation between Cartesian and polar coordinates:

$$\begin{aligned} r = \sqrt{x_1^2 + x_2^2} \\ \varphi = \tan^{-1}(x_2/x_1) \end{aligned} \quad\leftrightarrow\quad \begin{aligned} x_1 = r\cos\varphi \\ x_2 = r\sin\varphi \end{aligned}$$

$$(8.70)$$

Details of the derivation of the pdf of φ by using (8.69) and the transformation given by (8.70) is presented in Example F.7. The joint pdf of R and Φ is given by (F.61)

$$f_{R\Phi}(r,\varphi) = r f(x_1,x_2)\bigg|_{\substack{x_1 = r\cos\varphi \\ x_2 = r\sin\varphi}}$$

$$= \frac{r}{2\pi\sigma^2}\exp\left(-\frac{r^2 + E - 2\sqrt{E}r\cos\varphi}{2\sigma^2}\right)$$

$$(8.71)$$

The pdf of φ is found by integrating (8.71) with respect to r (see (F.61)–(F.64)):

$$f_\Phi(\varphi) = \int_0^\infty f_{a,\Phi}(a,\varphi)d\varphi$$

$$= \frac{1}{2\pi}\exp(-\gamma_s)$$

$$\left[1 + \sqrt{4\pi\gamma_s}\cos\varphi\, e^{\gamma_s\cos^2\varphi}\{1 - Q(\sqrt{2\gamma_s}\cos\varphi)\}\right]$$

$$= \begin{cases} \dfrac{1}{2\pi} & \gamma_s = 0 \\[2mm] \sqrt{\dfrac{\gamma_s}{\pi}}\cos\varphi\, e^{-\gamma_s\sin^2\varphi} & \sqrt{\gamma_s}\cos\varphi \gg 1 \end{cases}$$

$$(8.72)$$

where $\gamma_s = E/N_0$. As shown in Figure 8.22, the pdf of the phase for signal $s_1(t)$ is uniformly distributed for $\gamma_s = 0$, but peaks around $\varphi = 0$ as γ_s increases. If the phase lies in the angular region $-\pi/M \le \varphi \le \pi/M$, then the symbol $s_1(t)$ is correctly detected, otherwise a symbol error occurs. This implies that as γ_s increases, the probability that the phase falls outside the angular region $-\pi/M \le \varphi \le \pi/M$ decreases; this leads to lower symbol error probabilities. The neighboring symbols s_2 or s_8 are more likely to be decoded as being transmitted when a symbol error is made. Therefore, Gray coding helps minimizing bit error probability since a symbol error causes only a single bit error.

When $s_1(t)$ is transmitted, a detection error is made if the noise causes the phase to fall outside the region $-\pi/M \leq \theta \leq \pi/M$. Using the rotational symmetry of the signal space, the probability of a symbol error for M-ary PSK may be written as

$$P_M = 1 - \frac{1}{M}\sum_{m=1}^{M} P_c(s_m)$$

$$= 1 - P_c(s_1) = 1 - \int_{-\pi/M}^{\pi/M} f(\varphi)\, d\varphi \tag{8.73}$$

where $P_c(s_m)$ denotes the probability of correct reception of the transmitted symbol s_m. Except for $M = 2$ and $M = 4$, the above integral can not be evaluated in closed form and requires numerical evaluation. Inserting (8.72) into (8.73), one finds the approximate symbol error probability for large values of γ_s as

$$P_M \approx 2Q\left(\sqrt{\frac{2E}{N_0}}\sin\left(\frac{\pi}{M}\right)\right), \quad E = (\log_2 M)E_b \tag{8.74}$$

which is plotted in Figure 8.23 as a function of E_b/N_0 (dB). Symbol error probabilities for

BPSK and QPSK are nearly the same for large values of E_b/N_0. On the other hand, for a given symbol error probability, higher values of E_b/N_0 (higher transmit power) are required as M increases. Since higher M implies longer symbol duration and narrower transmission bandwidth, this corresponds to the well-known tradeoff between the transmit power and the transmission bandwidth, as already observed in M-ary ASK.

Example 8.8   Upper Bound on Symbol Error Probability For M-ary PSK.
Exploiting the rotational invariance of the error probability, the upper bound on the average symbol error probability for equi-probable symbols may be written as

$$P_M < \sum_{i=1}^{M}\sum_{\substack{k=1 \\ k \neq i}}^{M} p_i Q\left(\frac{d_{ik}}{\sqrt{2N_0}}\right) = \sum_{k=2}^{M} Q\left(\frac{d_{1k}}{\sqrt{2N_0}}\right) \tag{8.75}$$

The Euclidean distance between any two signal points is greater than or equal to the minimum Euclidean distance, that is, $d_{1k} \geq d_{min}$, $k = 2, 3, \cdots M$. Noting that neigboring

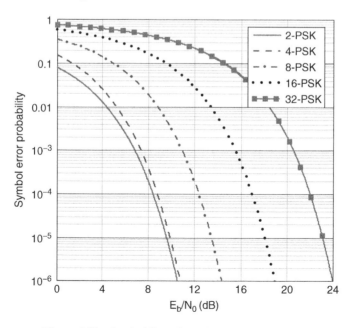

Figure 8.23   Symbol Error Probability For M-ary PSK.

signal points in M-ary PSK are separated from each other by $2\pi/M$ rad., the minimum Euclidean distance in M-ary PSK is given by (see Figure 8.20)

$$d_{min} = 2\sqrt{E}\sin\left(\frac{\pi}{M}\right) \qquad (8.76)$$

Inserting (8.76) into (8.75), the upper bound on the symbol error probability reduces to

$$P_M < (M-1)\, Q\left(\frac{d_{min}}{\sqrt{2N_0}}\right)$$
$$= (M-1)\, Q\left(\sqrt{\frac{2E}{N_0}}\sin\left(\frac{\pi}{M}\right)\right) \qquad (8.77)$$

When compared with (8.74), (8.77) provides a tight bound for the average symbol error probability for coherent M-ary PSK. The bound becomes tighter for larger values of E/N_0 and $M \geq 4$.

8.3.2.4 PSD of M-ary PSK Signals

We assume a random sequence of equiprobable bits, and that the symbols are statistically independent. The M-ary PSK signal given by (8.65) with a polar NRZ transmit pulse waveform suggests an autocorrelation function as given by (see Figure 5.25)

$$s(t) = \sqrt{2E/T}\,\Pi(t/T)\cos(w_c t + (i-1)2\pi/M) \Rightarrow$$
$$R_s(\tau) = \frac{2E}{T}\,T\Lambda(\tau/R)\frac{1}{2}\cos(w_c\tau)$$
$$\qquad (8.78)$$

Note that $\Pi(t/T)$ is an energy signal; hence the value of its autocorrelation function $T\Lambda(\tau/T)$ at the origin denotes its energy. Using (1.103) and (1.104), the baseband PSD of M-ary PSK signals is obtained ignoring the term $(1/2)\cos(w_c\tau)$ as follows:

$$G_s(f) = 2E\,\mathrm{sinc}^2(fT) \qquad (8.79)$$

where $T = T_b\log_2 M$ is the symbol duration, T_b is the bit duration, $E = E_b\log_2 M$ is the symbol

energy, and M shows alphabet size. The passband PSD is found by convolving (8.79) with Fourier transform of $\cos w_c t$ from (8.3):

$$G_s(f) = 2E\,\mathrm{sinc}^2(fT)\otimes\frac{1}{4}[\delta(f-f_c)+\delta(f+f_c)]$$
$$= \frac{E}{2}\,\mathrm{sinc}^2[(f-f_c)T_b\log_2 M]$$
$$+ \frac{E}{2}\,\mathrm{sinc}^2[(f+f_c)T_b\log_2 M]$$
$$\qquad (8.80)$$

Figure 8.24 shows the concentration of the PSD of M-ary PSK signals around the carrier frequency as M increases. Since the first null of the PSD is proportional to $1/T$, increase in symbol duration with M leads to a corresponding decrease in the bandwidth. Comparison of (8.33) and (8.80) shows that, except for amplitude differences, the PSD of M-ary ASK and M-ary PSK signals are the same.

The bandwidth of the passband M-ary PSK (twice that of baseband M-PSK) is approximately given by

$$B_M = \frac{1}{T} = \frac{1}{T_b\log_2 M} = \frac{R_b}{\log_2 M} \quad (Hz) \quad (8.81)$$

Comparison between (8.34) and (8.81) shows that the bandwidth of M-ary ASK and M-ary PSK have the same dependence on M. For example, the null-to-null bandwidth for QPSK is simply half that for BPSK. Using (8.81), the bandwidth efficiency is found to be

$$\eta = \frac{R_b}{B_M} = \log_2 M\left(\frac{bps}{Hz}\right) \qquad (8.82)$$

which improves as M increases. However, this improvement is at the expense of increased transmit power. From Figure 8.23, one may easily determine the increase in E_b/N_0 required to achieve a certain symbol error probability. The increase in E_b/N_0 required to achieve a given symbol error probability is closely related to the minimum Euclidean distance between signal points in the constellation. For a given symbol error probability, the minimum

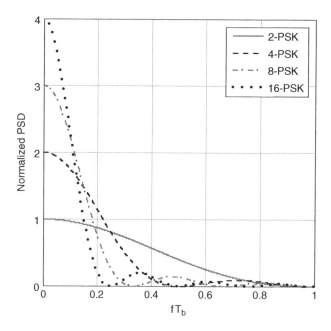

Figure 8.24 Normalized Power Spectral Density of M-PSK Shifted to Baseband.

Table 8.3 Power-Bandwidth Trade-Off in M-PSK For a Symbol Error Probability of 10^{-5}.

M	B_M/B_2 (8.81)	η (8.82)	E_b/N_0 (dB)	$(E_b/N_0)_M/(E_b/N_0)_2$ (dB) Figure 8.23	(8.83)
2	1	1	9.59	0	0
4	1/2	2	9.89	0.31	0
8	1/3	3	13.46	3.88	3.57
16	1/4	4	18.07	8.48	8.18
32	1/5	5	23.08	13.49	13.18

Euclidean distance should remain the same irrespective of M. Therefore, equating the minimum Euclidean distance for arbitrary value of M, given by (8.76), to that for $M = 2$, the increase in E_b/N_0 for any M compared to $M = 2$, required to achieve the same probability of symbol error is found as:

$$\frac{(E_b/N_0)_M}{(E_b/N_0)_2} = \frac{1}{\log_2 M \, \sin^2(\pi/M)} \qquad (8.83)$$

Note that (8.83) represents the asymptotic value and hence is independent of the symbol error probability. Power-bandwidth trade-off

for M-ary PSK for a given symbol error probability of 10^{-5} is presented in Table 8.3. Comparison of Table 8.2 and 8.3 clearly shows that M-PSK is more power-efficient than M-ary ASK, since, for a given value of M, it requires lower E_b/N_0 to achieve a given bit error probability.

8.3.2.5 Differentially Encoded PSK

In M-ary PSK, the signal phase is determined only by the symbol to be transmitted during that particular symbol interval. For example, the signal phases 0 and π may be chosen in BPSK to represent binary 1 and binary 0, respectively.

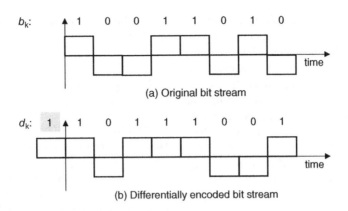

(a) Original bit stream

(b) Differentially encoded bit stream

Figure 8.25 Original and Differentially Encoded ($d_k = \overline{d_{k-1} + b_k}$) Bit Streams, Assuming Reference Bit 1 (in highlighted), and Associated Waveforms Using Polar NRZ Signaling.

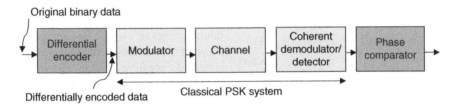

Figure 8.26 Block Diagram of the DEPSK Transceiver.

Similarly, $\pm\pi/4$ and $\pm3\pi/4$ may be used in QPSK represent 00, 01, 10 and 11.

However, as already presented in Chapter 5, in differential encoding, transmitted symbols are encoded based on signal transitions. For example, a transition between two consective symbols may be used to designate symbol 0, while no-transition may be used to designate symbol 1. In differentially encoded BPSK, the bit **0** is transmitted by shifting the phase of the carrier by 180^0 relative to the previous carrier phase (transition), whereas the bit **1** is transmitted by a 0^0 phase shift relative to the phase in the previous signaling interval (no-transition). In differentially encoded QPSK, the relative phase shifts between successive intervals may be chosen as 0^0, 90^0, 180^0, and -90^0, corresponding to the transmission of the dibits 00, 01, 11, and 10, respectively.

Differential encoding requires the use of $\log_2 M$ reference bits to initiate the encoding

process. Use of 0 or 1 as the reference bit in differentially-encoded BPSK inverts the encoded signal. However, a differentially encoded signal is insensitive to inversion, since its interpretation is not affected by inversion.

To have a better insight in differential encoding, let us consider the transmission of the binary sequence 10011010 as shown in Figure 8.25. Differentially encoded bits d_k are computed from the original data bits b_k as $d_k = \overline{d_{k-1} + b_k}$ by a differential encoder preceeding the modulator. The rest of the transmitter of the differentially encoded PSK system is identical to the transmitter of the PSK (see Figure 8.26). The modulator modulates the carrier with differentially encoded bit stream at its input. The original binary information is recovered at the receiver simply by comparing the polarity of adjacent binary symbols to establish whether or not a transition has occurred: $b_k = \overline{d_{k-1} + d_k}$ (see Figure 8.25).

Use of incorrect reference bit alters only the first decoded bit. Demodulation of the differentially encoded PSK is performed by ignoring the phase ambiguities, hence is the same as that for PSK signals. Coherent demodulation is achieved by a correlation receiver. At the end of each signaling interval, the detector estimates the transmitted phase. Then, a phase comparator estimates the transmitted symbol by comparing the phases of demodulated signals over two consecutive intervals.

Noting from $d_k = \overline{d_{k-1} + b_k}$ that $d_k = d_{k-1}$ when a binary 1 is sent ($b_k = 1$), $d_k \neq d_{k-1}$ when a binary 0 is sent ($b_k = 0$), the total probability of bit error may be written as

$$P_2 = \frac{1}{2}P(\hat{d}_k \neq \hat{d}_{k-1}|b_k = 1) + \frac{1}{2}P(\hat{d}_k = \hat{d}_{k-1}|b_k = 0)$$

$$= P(\hat{d}_k = \hat{d}_{k-1}|b_k = 0) = P(\hat{d}_k \neq \hat{d}_{k-1}|b_k = 1)$$

(8.84)

where, using the independence of error events in each signaling intervals,

$$P(\hat{d}_k \neq \hat{d}_{k-1}|b_k = 1)$$
$$= P(\hat{d}_k = d_k, \hat{d}_{k-1} \neq d_{k-1}|b_k = 1)$$
$$+ P(\hat{d}_k \neq d_k, \hat{d}_{k-1} = d_{k-1}|b_k = 1)$$
$$= (1-p)p + p(1-p) = 2p(1-p)$$

(8.85)

Here, $p = P(\hat{d}_k \neq d_k) = Q(\sqrt{2E_b/N_0})$ denotes the probability of making a bit error. For $E_b/N_0 \gg 1$, (8.85) reduces to

$$P_2 = 2p(1-p) \approx 2p = 2Q(\sqrt{2E_b/N_0}) \quad (8.86)$$

In differentially encoded PSK, an error in the demodulated phase of the signal in any given interval will usually result in decoding errors over two consecutive signaling intervals. Therefore, as can be observed from (8.86), coherent demodulation of differentially encoded M-ary PSK results in approximately twice the probability of error for M-ary PSK with absolute phase encoding (see Table 5.2). However, this factor-of-2 increase in the error

probability translates into a relatively small loss in SNR.

8.3.3 Quadrature Amplitude Modulation (QAM)

M-ary PSK is a constant-envelope modulation since its in-phase and quadrature components have the same magnitude and hence the same energy. The message points are located uniformly in a circular constellation with a radius equal to square-root of the symbol energy. However, in M-ary quadrature amplitude modulation (QAM), in-phase and quadrature components may have different amplitudes. Therefore, all QAM symbols do not have the same energy, and we can talk only about an average symbol/bit energy or average transmit power. Consequently, in contrast with PSK signals, QAM signals do not have constant envelope and the QAM carrier experiences both amplitude modulation (AM) and phase modulation (PM).

The QAM is basically a two-dimensional version of ASK; therefore, we need two basis functions, given by (8.46), to characterize the QAM signals. A M-ary QAM signal for k-th symbol may be written as

$$s_k(t) = \sqrt{\frac{2E_0}{T}}a_k\cos(2\pi f_c t) - \sqrt{\frac{2E_0}{T}}b_k\sin(2\pi f_c t)$$

$$= \sqrt{E_0}\,a_k\psi_1(t) - \sqrt{E_0}\,b_k\psi_2(t),$$

$$\begin{cases} \quad\quad 0 \leq t \leq T \\ a_k, b_k = \pm 1, \pm 3, \pm 5, \cdots \end{cases}$$

(8.87)

The signal $s_k(t)$ consists of two quadrature carriers each of which is modulated by a set of discrete amplitudes. In the signal-constellation diagram, the coordinates of the message points are denoted by $(a_k\sqrt{E_0},$ $b_k\sqrt{E_0})$, for $k = 1, 2, 3,..., M$. Hence, $(\pm\sqrt{E_0}, \pm\sqrt{E_0})$ denote the coordinates of the four signals with lowest energy. The minimum Euclidean distance between symbols is given

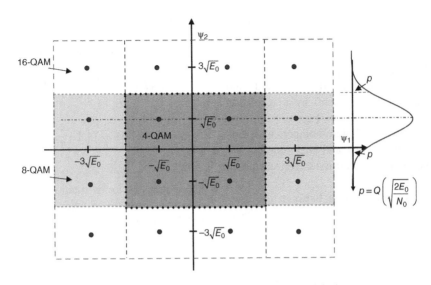

Figure 8.27 Signal Constellation for M-ary QAM.

by $d_{min} = 2\sqrt{E_0}$ (see Figure 8.27). The QAM constellation may have different forms such as hybrids of ASK and PSK as well as square or rectangular forms. For example, the constellation shown in Figure 8.27 is square for M = 4 and 16 but rectangular for M = 8. Note that the number of bits assigned to each symbol is even in a square constellation, whereas in rectangular constellations such as for M = 8 and 32, each symbol carries an odd number of bits. Gray bit-to-symbol mapping is optimal in the sense that a symbol error (involving two adjacent symbols in the QAM signal constellation) results only in a single bit error. Figure 8.28 shows Gray mapping for 16-QAM constellation.

Rectangular/square QAM signal constellations have the distinct advantage of being easily generated as two ASK signals with quadrature carriers. They are also easy to demodulate because it is easy to define the decision boundaries. Although they are not the most power-efficient M-ary QAM signal constellations for $M \geq 16$, the average symbol energy required to achieve minimum Euclidean distance is only slightly greater than that required for the best M-ary QAM constellation. For these reasons, rectangular M-ary QAM signals are most frequently used.

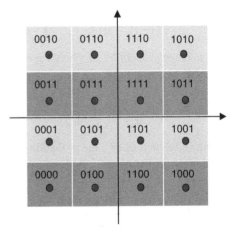

Figure 8.28 Gray Mapping for 16-QAM.

In a square constellation, $M = 2^k$ where the number of bits per symbol k is even. The *even* number of bits assigned to each symbol may be considered to be divided into two and each transmitted by one of the quadrature carriers. Hence, M-ary QAM with square constellation can be viewed as the Cartesian product of a one-dimensional L-ary ASK ($M = L \times L$) constellation with itself. For rectangular constellations in which $M = 2^k$ (k is odd), the M-ary QAM signal constellation may be decomposed into two ASK signals on quadrature carriers, one L-ary ASK and the other T-ary ASK so that $M = L \times T$ (see Table 8.4).

QAM signals with square/rectangular constellations can be perfectly separated into phase-quadrature components by a correlation demodulator as shown in Figure 8.29. The detector for each quadrature-component detects the corresponding bits and their outputs are interleaved in order to estimate the transmitted symbol.

Using the orthogonality between the basis functions, the error probability is easily determined from the probability of error for the corresponding ASK constellation with L and T points. The symbol error probability for M-ary QAM may then be written in terms of P_L^{ASK}, the symbol error probability for L-ary ASK, and P_T^{ASK} that for T-ary ASK:

$$P_M^{QAM} = 1 - \left(1 - P_L^{ASK}\right)\left(1 - P_T^{ASK}\right) \quad (8.88)$$

The average probability of symbol error for L-ary ASK is given by (8.27):

$$P_L^{ASK} = 2\left(1 - \frac{1}{L}\right)Q\left(\sqrt{\frac{2E_0}{N_0}}\right) \quad (8.89)$$

Using (8.88) and (8.89), the probability of a symbol error for the M-ary QAM reduces to

Table 8.4 Decomposing M-ary QAM Into Orthogonal L-ary ASK and T-ary ASK.

M	4	8	16	32	64
L	2	4	4	8	8
T	2	2	4	4	8

$$P_M^{QAM} = 1 - \left(1 - P_L^{ASK}\right)\left(1 - P_T^{ASK}\right)$$

$$= P_L^{ASK} + P_T^{ASK} - P_L^{ASK} \times P_T^{ASK}$$

$$= 2\left(2 - \frac{1}{L} - \frac{1}{T}\right)Q\left(\sqrt{\frac{2E_0}{N_0}}\right)$$

$$- 4\left(1 - \frac{1}{L}\right)\left(1 - \frac{1}{T}\right)Q^2\left(\sqrt{\frac{2E_0}{N_0}}\right)$$

$$(8.90)$$

Note that due to the rotational symmetry of the QAM constellation, interchanging L and T in (8.90) simply rotates the constellation by 90 degrees and hence does not affect the symbol error probability. From (8.87) and Figure 8.27, E_0/N_0 may easily be expressed in terms of the average bit SNR as follows ($M = L \times T$):

$$\frac{E_0}{N_0} = \frac{3}{L^2 + T^2 - 2}\frac{E_{s,av}}{N_0} = \frac{3\log_2(L \times T)}{L^2 + T^2 - 2}\frac{E_{b,av}}{N_0}$$

$$(8.91)$$

Noting that QAM is an extension of ASK to two dimensions; inserting $T = 1$ into (8.90) and (8.91) one gets (8.27) and (8.28), respectively. For square constellations with $M = 2^k$ where k is even, one has $L = T = \sqrt{M}$ and the symbol error probability (8.90) reduces to the well-known expression:

$$P_M^{QAM} = 1 - \left(1 - P_L^{ASK}\right)^2$$

$$= 2P_L^{ASK} - \left[P_L^{ASK}\right]^2 \cong 2P_L^{ASK}$$

$$\cong 4\left(1 - \frac{1}{\sqrt{M}}\right)Q\left(\sqrt{\frac{3\log_2 M}{M - 1}\frac{E_{b,av}}{N_0}}\right)$$

$$(8.92)$$

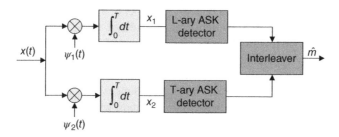

Figure 8.29 Coherent M-ary QAM Receiver For Rectangular Constellations.

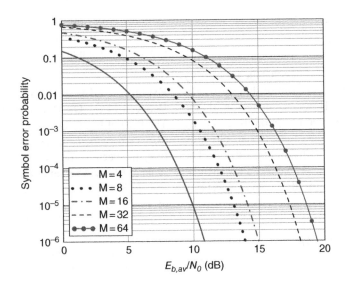

Figure 8.30   Symbol Error Probability For M-ary QAM.

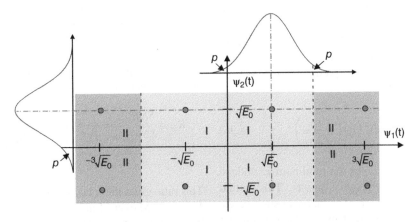

Figure 8.31   Symbol Error Probability Calculation For 8-QAM.

where (8.91) is used to express E_0 in terms of $E_{b,av}$. Figure 8.30 shows the variation of symbol error probability versus $E_{b,av}/N_0$ for 4-QAM to 64-QAM. Note the increase in $E_{b,av}/N_0$ required to achieve a given symbol error probability and the relatively small performance difference between 8-QAM and 16-QAM as well as between 32-QAM and 64-QAM.

Example 8.9   Symbol Error Probability For 8-QAM.
Consider the 8-ary QAM signal constellation as shown in Figure 8.31. An error occurs if the absolute value of the noise amplitude in in-phase and/or quadrature directions exceeds $\sqrt{E_0}$. Noting from (8.91) that $E_{s,av} = 3E_{b,av} = 6E_0$, the error probability p, shown in Figure 8.31, is given by (8.26)

$$p = Q\left(\sqrt{\frac{2E_0}{N_0}}\right) = Q\left(\sqrt{\frac{E_{b,av}}{N_0}}\right) \qquad (8.93)$$

From Figure 8.31, the probabilities of correct reception $P_c(\mathrm{I})$ and $P_c(\mathrm{II})$ in regions I and II, respectively, may be written as

$$P_c(I) = (1-2p)(1-p)$$
$$P_c(II) = (1-p)(1-p)$$
(8.94)

where $1-2p$ in region I denotes the probability of correct reception of a symbol along $\psi_1(t)$ and $1-p$ along $\psi_2(t)$. However, in region II, the probability of correct reception is given by $1-p$, since error occurs only in one direction both for in-phase and quadrature carriers. The average symbol error probability may then be expressed in terms of the correct reception probabilities as follows:

$$P_8 = 1 - \frac{1}{8}[4P_c(I) + 4P_c(II)]$$
$$= 1 - (1-p)\left(1 - \frac{3}{2}p\right) = \frac{5}{2}p - \frac{3}{2}p^2$$
(8.95)

which is identical to (8.90) for $(L, T) = (2, 4)$ or $(4, 2)$.

For nonrectangular M-ary QAM signal constellations, the following union upper bound is generally used for predicting the symbol error probability:

$$P_M < (M-1)Q\left(\frac{d_{min}}{\sqrt{2N_0}}\right)$$
(8.96)

Here, d_{min} denotes the minimum Euclidean distance between signal points. This bound may be loose when M is large. In such cases, $M-1$ may be replaced by the largest number of neighboring points that are at a distance d_{min} from any signal point. For 8-ary QAM shown in Figure 8.31, the minimum Euclidean distance is equal to $d_{min} = 2\sqrt{E_0} = \sqrt{2E_{b,av}}$. For $E_{b,av}/N_0 = 10$ dB, the upper bound on the

symbol error probability is found to be 5.48×10^{-3}, while the exact symbol error probability is equal to 1.96×10^{-3}.

Example 8.10 Performance Comparison Between M-ary QAM and M-ary PSK.
The PSD of M-ary QAM signals is the same as (8.33) where the $E_{s,av}$ is given by (8.91). Therefore, the spectral efficiency of M-ary QAM is the same as M-ary ASK and M-ary PSK (see (8.35) and (8.82)). However, the power-efficiency of M-ary QAM is much better than that for M-ary ASK and M-ary PSK. Noting from Table 8.2 and Table 8.3 that the power efficiency of M-ary PSK is higher than that for M-ary ASK, we compare the symbol error performances of M-ary QAM and M-ary PSK. The symbol error probabilities for M-ary PSK and M-ary QAM, given by (8.74) and (8.90), respectively, are dominated by the argument of the Q function. We therefore ignore the factors in front of (8.74) and (8.90). For the same symbol error probability, we equate arguments of the Q function in (8.74) and (8.90):

$$R_M = \frac{SNR_{PSK}}{SNR_{QAM}} = \frac{3}{(L^2 + T^2 - 2)\sin^2\left(\frac{\pi}{LT}\right)}$$
(8.97)

This ratio gives the relative SNR advantage for M-ary QAM over M-ary PSK, for a given symbol error probability. The first line in Table 8.5 shows the $E_{b,av}/N_0$ required to achieve a symbol error probability of 10^{-5} for M-ary QAM. The second line shows the SNR advantage of M-ary QAM over M-ary PSK, based on (8.97) and on the exact result obtained from (8.74)

Table 8.5   Comparison of M-ary PSK with M-ary QAM. M-QAM is assumed to have a constant symbol error probability of $P_s = 10^{-5}$.

M	4	8	16	32	64
$E_{b,av}/N_0$ (dB) (8.90)	9.89	12.99	14.04	17.25	18.58
R_M (dB) (8.97)	0	0.56	4.20	6.02	9.95
R_M (dB) (8.74) and (8.90)		(0.47)	(4.0)	(5.82)	(9.72)

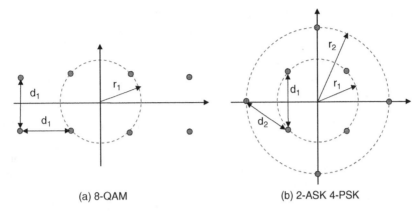

(a) 8-QAM   (b) 2-ASK 4-PSK

Figure 8.32   Signal-Space Diagrams For 8-ary QAM and 2-ASK 4-PSK.

and (8.90) shown in paranthesis. For example, for $E_{b,av}/N_0 = 14.04$ dB, 16-QAM has a symbol error probability of 10^{-5}, whereas 16-PSK needs 4.20 dB (based on (8.97)) higher $E_{b,av}/N_0$ to achieve the same symbol error probability. M-ary QAM clearly outperforms the M-ary PSK, especially as M increases. This can be explained by the fact that QAM utilizes the signal-space more efficiently compared to M-ary PSK, since the signal points in M-QAM are more evenly distributed in the signal space.

Example 8.11   Power Efficiency Comparison Between 8-QAM and Corresponding 2-ASK 4-PSK.

In this example, we will compare the average symbol energies required for the two constellations shown in Figure 8.32 to have the same symbol error probability. [5] The constellation shown in Figure 8.32b is obtained by moving the four highest energy symbols in the constellation diagram in Figure 8.32a to along the two basis functions. It is easy to observe that 8-QAM constellation does not use the signal-space as efficiently as the 2-ASK 4-PSK constellation, in the sense that some signal points use unnecessarily large symbol energies. To compare the two constellations, we assume that the symbol error performances of both constellations are to be the same. This implies that the minimum Euclidean distance in both

constellations should be the same. The signal points in 8-QAM are separated from each other by a Euclidean distance of

$$d_1 = 2r_1 \cos(\pi/4) = \sqrt{2}\, r_1 \qquad (8.98)$$

In 2-ASK 4-PSK constellation shown in Figure 8.32b, d_1 and d_2 are forced to be the same in order to have the same symbol error probability as for 8-ary QAM. The optimum value of the ratio of radii of the two rings $\rho = r_2/r_1$ is found to be

$$d_2^2 = r_1^2\left(1 + \chi^2 - \sqrt{2}\,\chi\right),$$
$$d_1 = \sqrt{2}\, r_1$$
$$d_1^2 = d_2^2 \;\Rightarrow\; \chi = r_2/r_1 = \left(1 + \sqrt{3}\right)/\sqrt{2}$$
$$\qquad (8.99)$$

The average symbol energies for 8-ary QAM and the corresponding PSK-ASK are given by

$$E_{s,av}^{QAM} = 3r_1^2$$
$$E_{s,av}^{PSK-ASK} = \frac{1}{2}\left(r_1^2 + r_2^2\right) = \frac{r_1^2}{2}\left(1 + \chi^2\right) \qquad (8.100)$$

From (8.100), the SNR advantage of 2-PSK 4-ASK over 8-QAM is given by

$$\frac{E_{s,av}^{QAM}}{E_{s,av}^{PSK-ASK}} = \frac{6}{1 + \chi^2} = 1.27 \; (1.03 \; dB) \quad (8.101)$$

Using (8.76) which gives the minimum Euclidean between symbol points in M-ary PSK, the same approach may be followed for computing the SNR advantage of QAM over PSK: equating the minimum Euclidean distances in the two constellations, we get

$$2r_0 \sin(\pi/M) = \sqrt{2}\, r_1 \qquad (8.102)$$

where $r_0 = \sqrt{E}$ for M-ary PSK. Using (8.100) and (8.102), the SNR advantage of 8-QAM over 8-PSK is given by

$$\frac{E_{s,av}^{PSK}}{E_{s,av}^{QAM}} = \frac{r_0^2}{3r_1^2} = \frac{1}{6\sin^2(\pi/8)} = 1.138\ (0.562\ dB) \qquad (8.103)$$

which is in perfect agreement with (8.97). Similarly, the SNR advantage of 2-ASK 4-PSK over 8-PSK is given by

$$\frac{E_{s,av}^{PSK}}{E_{s,av}^{PSK-ASK}} = \frac{r_0^2}{r_1^2(1+\chi^2)/2}$$

$$= \frac{1}{(1+\chi^2)\sin^2(\pi/8)} = 1.443\ (1.59\ dB) \qquad (8.104)$$

where the ratio r_0/r_1 is given by (8.102). In spite of the SNR advantages of hybrid ASK-PSK constellations, rectangular/square QAM constellations are preferred because of facility of decoding with decision boundaries in grid form.

8.3.4   Coherent Orthogonal Frequency Shift Keying (FSK)

Frequency-shift keying (FSK), which is a nonlinear modulation technique, is based on transmitting one of M different frequency tones f_i, $i = 1,2,...,$ M for the message symbols m_i, $i = 1,2,...,$ M. In continuous phase FSK (CPFSK), phase continuity between symbol phases at switching times, from one frequency to the other, is maintained. In orthogonal FSK,

the carrier frequencies are chosen so that the carriers are orthogonal to each other; this facilitates the decoding process. The FSK signals may be demodulated coherently or noncoherently depending on the channel conditions and the ability of the receiver to estimate the frequency and phase of the received tones. In coherent reception, the demodulator correlates the received signal with locally-generated carriers. In noncoherent reception, the phase and/or frequency shift induced by the channel is not available at the receiver. Consequently, the demodulator determines in-phase and quadrature components of the received signal and they are square-combined so as to enable the decoder to estimate the received signal without knowing the channel-induced phase/frequency shifts.

Let us now consider the orthogonality of carriers in M-ary FSK. Two sinusoids with different frequencies and a phase difference of θ between them are orthogonal to each other over the interval $(0, T)$ if the following condition is satisfied: [3]

$$2\int_0^T \cos(w_i t + \theta)\cos(w_j t)\, dt$$

$$= \cos\theta \int_0^T \cos(w_i t)\cos(w_j t)\, dt$$

$$- \sin\theta \int_0^T \sin(w_i t)\cos(w_j t)\, dt$$

$$= \cos\theta \left[\frac{\sin(w_i + w_j)T}{w_i + w_j} + \frac{\sin(w_i - w_j)T}{w_i - w_j}\right]$$

$$+ \sin\theta \left[\frac{\cos(w_i + w_j)T - 1}{w_i + w_j} + \frac{\cos(w_i - w_j)T - 1}{w_i - w_j}\right] \equiv 0 \qquad (8.105)$$

The frequencies f_i and f_j are chosen so as to satisfy $f_i T = n_i$, $i = 1,2,\cdots,$ M where n_i is a positive integer. Inserting $f_i = k_i/T = n_i/(2T)$, $i = 1,2,\cdots,M$ $(n_i = 2k_i)$ into (8.105) leads to

$$\sin(w_i + w_j)T = \sin\pi(n_i + n_j) \equiv 0$$

$$\cos(w_i + w_j)T - 1 = \cos 2\pi(k_i + k_j) - 1 \equiv 0 \qquad (8.106)$$

For $\theta = 0$, the term with $\sin\theta$ in (8.105) vanishes and the orthogonality implies $\sin(w_i - w_j)T \equiv 0$:

$$\sin(w_i - w_j)T \equiv 0 \;\Rightarrow\; f_i - f_j = n\frac{1}{2T} = n\Delta f,$$

$$n = \pm 1, \pm 2, \pm 3, \cdots$$

$$\Delta f = f_i - f_{i-1} = \frac{1}{2T}$$

(8.107)

For $\theta \neq 0$, the orthogonality requires the satisfaction of two conditions simultaneously:

$$\left. \begin{array}{l} \sin(w_i - w_j)T = 0 \Rightarrow f_i - f_j = \dfrac{n}{2T} \\[2mm] \cos(w_i - w_j)T = 1 \Rightarrow f_i - f_j = \dfrac{k}{T} \end{array} \right\} \Delta f = f_i - f_{i-1} = \frac{1}{T}$$

(8.108)

where $k = \pm 1, \pm 2, \pm 3 \ldots$ The minimum frequency separation required for orthogonality of FSK signals is $\Delta f = 1/T$ for $\theta \neq 0$ (noncoherent demodulation) and $\Delta f = 1/(2T)$ for $\theta = 0$ (coherent demodulation). Therefore, for the same symbol rate, coherently demodulated FSK occupies less bandwidth than noncoherently demodulated FSK and still retains orthogonality. Coherent FSK is hence twice more bandwidth-efficient than noncoherent FSK.

The passband frequencies used to transmit M symbols in M-ary FSK may then be written as

$$f_i = f_c + (2i - M - 1)\frac{\Delta f}{2}, \quad i = 1, 2, \cdots, M$$

$$= f_c \pm (2i - 1)\frac{\Delta f}{2}, \quad i = 1, 2, \cdots, M/2$$

(8.109)

where Δf is given by (8.107) and (8.108) for coherently- and noncoherently-detected FSK, respectively. For example, the two orthogonal tones used to transmit $s_1(t)$ and $s_2(t)$ are $f_{1,2} = f_c \pm \Delta f/2$, where Δf is given by (8.107). for coherently demodulated BFSK, and by (8.108) for noncoherent BFSK.

8.3.4.1 Coherent Orthogonal Binary FSK

Binary CPFSK is an orthogonal modulation with $N = 2$ and $M = 2$:

$$s_i(t) = \sqrt{E_b}\,\Phi_i(t), \quad 0 \leq t \leq T_b, \quad i = 1, 2 \quad (8.110)$$

where $s_1(t)$ represents symbol 1, and $s_2(t)$ represents symbol 0. The basis functions, defined as

$$\Phi_i(t) = \sqrt{2/T}\,\cos(2\pi f_i t), \, 0 \leq t \leq T, \, i = 1, 2, \cdots, M$$

(8.111)

are orthonormal over $(0, T_b)$ when T is replaced by T_b:

$$\int_0^{T_b} \Phi_i(t)\Phi_j(t)\,dt = \delta(i - j) \qquad (8.112)$$

Since symbol timing and the carrier frequencies are derived from the same clock, the frequencies f_1 and f_2 are chosen so as to satisfy (8.107).

The invertor, in the block diagram of a binary FSK transmitter shown in Figure 8.33a, switches on the upper oscillator and switches off the lower oscillator if the input bit 1, and vice versa if the input bit is 0. Depending on the input bit, either $s_1(t)$ or $s_2(t)$ is transmitted during $(0, T_b)$.

In coherent binary FSK, the phase induced by the channel on the received signal $s_i(t)$ is assumed to be perfectly estimated by the receiver. Therefore, there is no phase difference between the received signal $s_i(t)$ and the corresponding locally generated basis function in (8.111). Therefore, the minimum frequency difference between neigboring tones can be as narrow as $\Delta f = 1/(2T_b)$ (see (8.107)). In view of the above, the coherent binary FSK receiver shown in Figure 8.33b correlates the received signal $x(t) = s_i(t) + w(t)$, where $s_i(t)$ is given by (8.110), with orthonormal basis functions in (8.111). Since the correlation receiver yields the same output as the corresponding matched-filter receiver, the orthogonality between $s_1(t)$

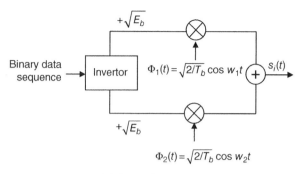

(a) Binary CPFSK transmitter

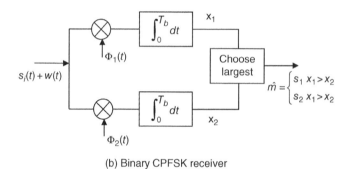

(b) Binary CPFSK receiver

Figure 8.33   Block Diagram of Transmitter and Receiver For Orthogonal BFSK.

and $s_2(t)$ permits the message points $\mathbf{s}_1$ and $\mathbf{s}_2$ to be represented in the signal space diagram at the correlator output as follows (see also Figure 6.22):

$$s_1 = \begin{bmatrix} \sqrt{E_b} & 0 \end{bmatrix} \quad s_2 = \begin{bmatrix} 0 & \sqrt{E_b} \end{bmatrix} \qquad (8.113)$$

The Euclidean distance between $\mathbf{s}_1$ and $\mathbf{s}_2$ is given by $\|\mathbf{s}_1 - \mathbf{s}_2\| = \sqrt{2E_b}$ (see Figure 8.34). The observation vector at the detector input is the noisy version of (8.113), and may be written as

$$\mathbf{x} = \begin{cases} \begin{bmatrix} \sqrt{E_b} + w_1 & w_2 \end{bmatrix} & s_1 \text{ is sent} \\ \begin{bmatrix} w_1 & \sqrt{E_b} + w_2 \end{bmatrix} & s_2 \text{ is sent} \end{cases} \qquad (8.114)$$

where w_i, $i = 1,2$ denotes the projection of the input noise $w(t)$ along $\Phi_i(t)$ with zero mean and variance $N_0/2$ (see (8.12)). If the a-priori probabilities of the two signals are the same, the ML detector chooses the decision threshold at mid-way between the two signals as shown

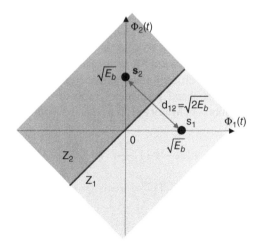

Figure 8.34   Signal-Space Diagram of Orthogonal Binary Phase FSK.

in Figure 8.34. The detector compares the output of the two orthogonal branches and decides in favour of the branch with higher signal level.

In other words, the detector decides in favour of $s_1(t)$ if the observation vector remains in decision region Z_1. Otherwise, the detector decides in favour of $s_2(t)$.

Bit error probability for equiprobable signals may then be written as

$$P_e(s_1) = \text{Prob}\left(\sqrt{E_b} + w_1 < w_2\right)$$

$$= \text{Prob}\left(w_2 - w_1 > \sqrt{E_b}\right) = Q\left(\sqrt{E_b/N_0}\right)$$

$$P_e(s_2) = \text{Prob}\left(\sqrt{E_b} + w_2 < w_1\right)$$

$$= \text{Prob}\left(w_1 - w_2 > \sqrt{E_b}\right) = Q\left(\sqrt{E_b/N_0}\right)$$

$$P_2 = \frac{1}{2}P_e(s_1) + \frac{1}{2}P_e(s_2) = Q\left(\sqrt{E_b/N_0}\right)$$

$$(8.115)$$

which follows from (8.26) bearing in mind that $w_1 - w_2$ is an AWGN noise with zero mean and variance N_0. Comparison of (8.115) with (8.51) shows that the bit error probability performance of coherent binary FSK is 3-dB worse than that for coherent binary PSK. This may also be explained in terms of the differences in Euclidean distances, since for the same E_b, the Euclidean distance between antipodal signal points (as in binary PSK) $d_{min} = 2\sqrt{E_b}$ is $\sqrt{2}$ times larger than that for binary FSK $d_{min} = \sqrt{2E_b}$ (see Figure 6.22).

Example 8.12 Correlation Between Carriers in Coherent BFSK.
Using (8.109), the signals $s_1(t)$ and $s_2(t)$ in (8.110) may as well be written as

$$s_{1,2}(t) = \sqrt{2E_b/T_b}\,\cos[2\pi(f_c \pm \Delta f/2)t], \quad 0 \le t \le T_b$$

$$(8.116)$$

Assuming that $f_c T_b$ to be an integer, the correlation coefficient between $s_1(t)$ and $s_2(t)$ is given by

$$\rho = \frac{1}{E_b}\int_0^{T_b} s_1(t)\,s_2(t)\,dt = \text{sinc}(2\Delta f\,T_b)$$

$$(8.117)$$

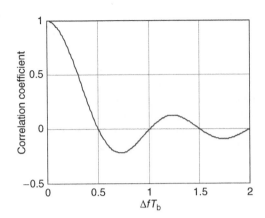

Figure 8.35 Correlation Coefficient Between Two Carriers in Binary Coherent FSK.

where $f_c \gg \Delta f$ is assumed. Figure 8.35 shows that $s_1(t)$ and $s_2(t)$ are orthogonal when the correlation coefficient ρ becomes equal to zero. This implies $\Delta f = k/(2T_b)$, where k is a positive integer, $k = 1,2,\dots$ For other values of Δf, there is a correlation between the carriers. In this case, the average probability of bit error for coherent reception of correlated binary signals is given by (6.62):

$$P_2 = Q\left(\sqrt{\frac{E_b}{N_0}(1-\rho)}\right)$$

$$(8.118)$$

which reduces to (8.115) for $\rho = 0$.

The bit error probability given by (8.118) is minimized for the maximum value of the argument of the Q function. From Figure 8.35, this corresponds to the minimum value of the correlation coefficient $\rho_{min} = -0.217$, which is located at approximately $\Delta f = 0.72/T_b$. Hence, the minimum value of the bit error probability for coherent BFSK is given by

$$P_{2,min} = Q\left(\sqrt{\frac{E_b}{N_0}1.217}\right)$$

$$(8.119)$$

Compared with (8.115) for a coherent BFSK system with $\rho = 0$ and $\Delta f = 1/(2T_b)$, required

E_b/N_0 in (8.119) is decreased by a factor of 1.217 (0.853 dB) for the same bit error probability. The return for 0.853 dB decrease in the required E_b/N_0 is a $0.72/0.5 = 1.44$ fold increase in Δf, and, hence, in the required bandwidth. The receiver shown in Figure 8.33 should then be changed so as to have a single correlator with impulse response, $\Phi_1(t) - \Phi_2(t)$, similar to matched-filter counterpart shown in Figure 6.20.

8.3.4.2 Coherent Orthogonal M-ary FSK (CFSK)

Coherent M-ary FSK signals are defined by

$$s_i(t) = \sqrt{\frac{2E}{T}}\cos(2\pi f_i t) = \sqrt{E}\,\Phi_i(t), \begin{cases} 0 \le t \le T \\ i = 1, 2, \cdots, M \end{cases}$$

(8.120)

M-ary FSK signals, of duration T and equal energies E, are orthogonal since signal frequencies are separated by a multiple of $\Delta f = f_i - f_{i-1} = 1/(2T)$, $i = 2, 3, \cdots M$. The frequencies used to transmit M symbols are given by (8.109).

M-ary FSK has an M-dimensional signal-space diagram, defined by the coefficient s_{ij}:

$$s_{ij} = \int_0^T s_i(t)\,\Phi_j(t)\,dt = \sqrt{E}\,\delta(i-j) \quad (8.121)$$

The signal points are defined by the following vectors in the signal-state diagram with M orthogonal axes along the basis functions $\Phi_i(t)$, $i = 1, 2, \cdots, M$:

$$s_1 = \begin{bmatrix} \sqrt{E} & 0 & \cdots & 0 \end{bmatrix}$$
$$s_2 = \begin{bmatrix} 0 & \sqrt{E} & \cdots & 0 \end{bmatrix}$$
$$\vdots$$
$$s_M = \begin{bmatrix} 0 & 0 & \cdots & \sqrt{E} \end{bmatrix}$$

(8.122)

A coherent M-ary FSK receiver consists of M correlation demodulators, followed by a ML detector, that chooses the largest component of the observation vector. Assuming that the symbol $s_1(t)$ is transmitted, the observation vector at the output of a coherent correlation demodulator, shown in Figure 8.36, may be written as

$$\mathbf{x} = \begin{bmatrix} x_1 & x_2 & \cdots & x_M \end{bmatrix} = \begin{bmatrix} \sqrt{E} + w_1 & w_2 & \cdots & w_M \end{bmatrix}$$
$$x_j = \int_0^T [s_i(t) + w(t)]\,\Phi_j(t)\,dt = \sqrt{E}\delta(i-j) + w_j$$

(8.123)

The noise components w_i denote the projections of the input noise $w(t)$ with zero mean and variance $N_0/2$ along $\Phi_i(t)$ (see (8.12)).

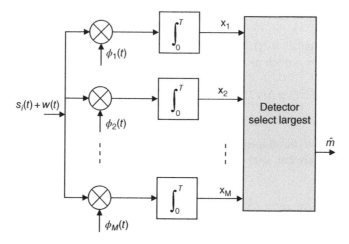

Figure 8.36   Receiver Block Diagram For Coherently Detected of M-ary Orthogonal FSK.

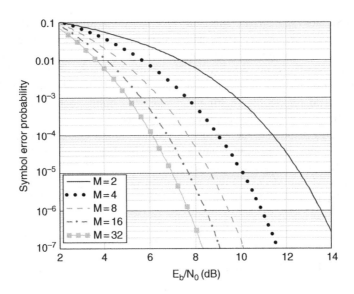

Figure 8.37 Symbol Error Probability For Coherent M-FSK.

Therefore, the components x_m, m = 1,2,3,…,M of the observation vector are statistically independent and identically distributed random variables with equal variances $N_{0/2}$. When $s_1(t)$ is transmitted, the probability of correct decision is simply the joint probability that $x_m < x_1$, m = 2, 3,…, M:

$$P_c(\mathbf{s}_1) = P(x_2 < x_1, x_3 < x_1, \cdots, x_M < x_1)$$
$$= [P(x_2 < x_1)]^{M-1}$$

(8.124)

From (8.115) we get

$$P(x_2 < x_1) = 1 - \text{Prob}(x_2 > x_1)$$

$$= 1 - \text{Prob}\left(w_2 - w_1 > \sqrt{E}\right)$$

$$= 1 - Q\left(\sqrt{E/N_0}\right)$$

(8.125)

The probability of symbol error for coherently detected orthogonal M-ary FSK may then be written as

$$P_M = 1 - \frac{1}{M}\sum_{m=1}^{M} P_c(\mathbf{s}_m)$$

$$= 1 - P_c(\mathbf{s}_1) = 1 - \left[1 - Q\left(\sqrt{E/N_0}\right)\right]^{M-1}$$

(8.126)

For sufficiently large values of E/N_0, using the binomial expansion $(1-x)^n \geq 1 - nx, x \rightarrow 0$ (see (D.7)), (8.126) simplifies to the well-known upper bound for the symbol error probability

$$P_M \leq (M-1)Q\left(\sqrt{\frac{E}{N_0}}\right) = (M-1)Q\left(\sqrt{\frac{(\log_2 M)E_b}{N_0}}\right)$$

(8.127)

The symbol error probability for coherent M-ary orthogonal FSK is shown in Figure 8.37 for several values of M. One may observe that the E_b/N_0 required for a given symbol error probability level decreases as M increases. This can explained by noting from (8.122) that the Euclidean distance between any two signal points of orthogonal M-ary FSK modulation

$$d = \sqrt{2E} = \sqrt{2(\log_2 M)\ E_b}$$

(8.128)

increases with M. This implies that the E_b/N_0 required for a given symbol error probability decreases in proportion with the increase in the Euclidean distance with M. For example, for a given symbol error probability, the E_b/N_0 for 4-ary FSK will be $\log_2 4 = 2$ (3 dB) lower than for binary FSK (see Figure 8.37). This is in contrast with M-ary ASK, M-ary PSK and M-ary QAM, where the minumum Euclidean distance decreases, and the E_b/N_0 required for a given symbol error probability increases, with increasing values of M (see (8.28), (8.76) and (8.91)). As will be seen below, this improvement in the required E_b/N_0 comes at a cost of bandwidth increase; this is again in contrast with M-ary ASK, M-ary PSK and M-ary QAM, where the bandwidth decreases with M. Hence, M-ary FSK is power-efficient but not bandwidth-efficient.

8.3.4.3   PSD of Coherent Orthogonal M-ary FSK

A coherent orthogonal M-ary FSK signal given by (8.120) may be rewritten as

$$s_i(t) = \sqrt{\frac{2E}{T}} \cos w_i t \, \Pi(t/T), \quad i = 1, 2, \cdots, M$$

$$(8.129)$$

where the rectangular function is defined by $\Pi(t/T) = 1$ for $0 < t \le T$ and 0 otherwise. Using the similarity with (8.30), the PSD of (8.129) may be directly written from (8.33) as

$$G_{s_i}(f) = \frac{E}{2} \, \mathrm{sinc}^2[(f - f_i)T]$$
$$+ \frac{E}{2} \mathrm{sinc}^2[(f + f_i)T] \quad (W/Hz)$$
$$(8.130)$$

where $T = T_b \log_2 M$ denotes the symbol duration. The PSD of M-ary FSK signal is then given by the sum of the spectra of M signals:

$$G_s(f) = \frac{E}{2} \sum_{i=1}^{M} [\mathrm{sinc}^2\{(f + f_i)T\}$$
$$+ \mathrm{sinc}^2\{(f - f_i)T\}] \quad (W/Hz)$$
$$(8.131)$$

Note that the spectrum defined by (8.131) is not totally occupied during a symbol, but only the spectrum of the i-th symbol is used and the rest

is empty during that symbol. The PSD of coherent 4-ary FSK is plotted in Figure 8.38 at the baseband for $\Delta f = 1/(2T)$ and $\Delta f = 1/T$. To maintain orthogonality of coherently detected M-ary FSK signals, adjacent signals need to be separated from each other by a frequency difference of $\Delta f = 1/(2 T)$. Hence, the bandwidth required to transmit orthogonal coherent M-ary FSK signals is approximately given by

$$B_M \cong M \, \Delta f = \frac{M}{2T} = \frac{M}{2T_b \log_2 M} = \frac{M}{2 \log_2 M} B_2 \ (Hz)$$
$$(8.132)$$

where $B_2 = 2/(2T_b) = R_b$ When the frequency tones are separated by $1/T$, the bandwidth may be approximated by $B \approx M/T$, which is twice that given by (8.132) (see Figure 8.38). Using (8.132), the spectrum efficiency of orthogonal coherent M-ary FSK signals is found as follows:

$$\eta = \frac{R_b}{B_M} = \frac{2 \log_2 M}{M} \ (bps/Hz) \qquad (8.133)$$

As shown in Table 8.6, the spectrum efficiency of M-ary FSK decreases with increasing values of M. This is in contrast with M-PSK, M-ASK and M-QAM, for which, the bandwidth efficiency is improved with increasing values of M. However, in contrast with M-PSK, M-ASK and M-QAM, average symbol energy required by M-FSK for a given bit error probability level does not increase with M.

Example   8.13 Spectrum   Efficiency   of Coherent Orthogonal M-ary FSK.
Consider a coherent M-ary FSK signaling system operating at a bit rate of $R_b = 1$ Mbps. Assume that $f_i T$ is an integer. The minimum frequency spacing required for coherent 8-ary and 16-ary FSK systems with a bit rate of 1 Mbps is given by

$$\Delta f = \frac{1}{2T} = \frac{R_s}{2} = \frac{R_b}{2 \log_2 M} = \begin{cases} 166.67 \ kHz & M = 8 \\ 125 \ kHz & M = 16 \end{cases}$$
$$(8.134)$$

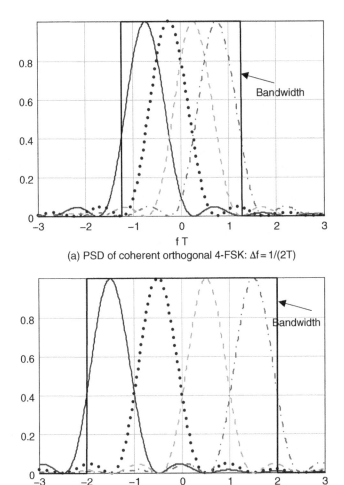

(a) PSD of coherent orthogonal 4-FSK: Δf = 1/(2T)

(b) PSD of coherent orthogonal 4-FSK: Δf = 1/T

Figure 8.38 PSD of 4-ary Orthogonal FSK.

Table 8.6 Spectral Efficiency of Orthogonal Coherent M-ary FSK Signals Versus M When Frequency Tones Are Separated by $\Delta f = 1/(2T)$.

M	2	4	8	16	32	64
η (bps/Hz)	1	1	0.75	0.5	0.31	0.19

The transmission bandwidth is given by

$$B = M \ \Delta f = \frac{M}{2\log_2 M} R_b = \begin{cases} 1.33MHz & M = 8 \\ 2MHz & M = 16 \end{cases}$$

$$(8.135)$$

It is clear from (8.135) that the transmission bandwidth required to support a bit rate of 1 Mbps is 1.33 MHz for 8-ary FSK, and 2 MHz for 16-ary FSK. Hence, 8-ary FSK has a higher spectral efficiency compared to 16-ary FSK.

We now determine whether 8-ary FSK or 16-ary FSK has higher symbol error probability when the two have the same E_b/N_0. The symbol error probability is known to be inversely proportional to the minimum Euclidean distance between the signal points. It is evident from (8.128) that the minimum Euclidean distance is larger for 16-FSK; thus, 16-FSK has an

SNR advantage of $\log_2 16/\log_2 8 = 1.33$ (1.25 dB) (see also Figure 8.37).

Example 8.14 Choice of Spectrum-Efficient Modulation.

A passband digital communication system transmits at $R_b = 15$ Mb/s in a $B = 10$ MHz transmission bandwidth by using a raised-cosine transmit filter with a roll-off factor $\alpha = 0.25$. Assuming that a coherent receiver with a carrier-to-noise ratio of 13 dB is used, we determine the modulation technique that meets the above requirements. [2]

Since $R_b/B = 1.5$ and $R_b/B \leq 1$ is for FSK modulation (see Table 8.6), a bandwidth efficient modulation, such as M-PSK, M-ASK or M-QAM, is needed. On the other hand, transmission bandwidth of a raised cosine filter with $\alpha = 0.5$ should satisfy

$$B \geq R_s(1+\alpha) = \frac{R_b}{\log_2 M}(1+\alpha) \Rightarrow$$

$$\log_2 M \geq \frac{R_b}{B}(1+\alpha) = \frac{15}{10}(1+0.25) \quad (8.136)$$

$$= 1.875 \Rightarrow M = 4$$

Therefore $M \geq 4$ should be chosen. Assuming a coherent receiver and a carrier-to-noise power ratio of 13 dB is available, we firstly compute E_b/N_0 from C/N:

$$\frac{C}{N} = \frac{E_b R_b}{N_0 B} \Rightarrow$$

$$\frac{E_b}{N_0} = \frac{C}{N}\frac{B}{R_b} = 20\frac{10}{15} \quad (8.137)$$

$$= 13.33 \ (11.25 \ dB)$$

The bit error probability for QPSK (same for 4-QAM) for $E_b/N_0 = 11.25$ dB:

$$P_2 = Q\left(\sqrt{2E_b/N_0}\right) - 0.5Q\left(\sqrt{2E_b/N_0}\right)^2$$

$$= 1.21 \times 10^{-7} \quad (8.138)$$

From (8.27), the bit error probability for 4-ASK for $E_b/N_0 = 11.25$ dB:

$$P_2 = \frac{M-1}{M}Q\left(\sqrt{\frac{6\log_2 M}{(M^2-1)}\frac{E_{b,av}}{N_0}}\right) = 4 \times 10^{-4} \quad (8.139)$$

QPSK is the obvious choice since it provides lower bit error probability for a given average E_b/N_0. To achieve the same bit error probability as QPSK, 4-ASK neeeds a 3.89 dB increase in E_b/N_0, hence $E_b/N_0 = 15.14$ dB.

Example 8.15 Power-Efficiency Comparison of Modulation Schemes.

Modulation techniques may be compared in terms of their effective use of the spectrum and the transmit power. Spectrum efficiency of a modulation techniques is determined by (8.6). We know from (8.35) and (8.82) that the spectrum efficiencies of M-ary ASK, M-ary PSK and M-ary QAM are the same. However, as given by (8.133), M-ary FSK uses the spectrum less efficiently than the other modulation techniques.

One of the easiest ways of comparing the power efficiencies of the modulation techniques is to compare their minimum Euclidean distances. The Euclidean distances of the modulation techniques considered so-far are given by (8.29) for M-ary ASK, (8.76) for M-ary PSK, (8.91) for M-ary QAM and (8.128) for M-ary FSK:

$$d_{min} = \begin{cases} 2\sqrt{\dfrac{3\log_2 M}{M^2-1}E_{b,av}} & M-ASK \\[2ex] 2\sqrt{\log_2 M \ E_b}\ \sin(\pi/M) & M-PSK \\[2ex] 2\sqrt{\dfrac{3\log_2(LT)}{L^2+T^2-2}E_{b,av}} & M-QAM \\[2ex] \sqrt{2\log_2 M \ E_b} & M-FSK \end{cases} \quad (8.140)$$

The minimum Euclidean distances normalized by $\sqrt{E_b}$ of the considered modulation techniques are compared in Table 8.7, where the square of the listed values indicate the relative SNR advantage of the modulation of interest. For example, 64-FSK has an E_b/N_0 advantage of $20\log(3.464/0.756) = 13.2$ dB over 64-QAM.

Table 8.7 Comparison of $d_{\min}/\sqrt{E_b}$, the Minimum Euclidean Distance Normalized by $\sqrt{E_b}$, For Several Modulation Techniques.

Modulation	$M = 2$	$M = 4$	$M = 8$	$M = 16$	$M = 32$	$M = 64$
M-ary ASK	2	1.265	0.736	0.434	0.242	0.133
M-ary PSK	2	2	1.326	0.78	0.438	0.24
M-ary QAM	2	2	1.414	1.265	0.877	0.756
M-ary FSK	1.414	2	2.449	2.828	3.162	3.464

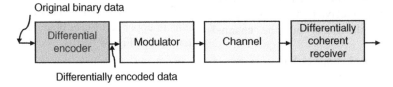

Original binary data

Differentially encoded data

Figure 8.39 Block Diagram of the DPSK Transceiver.

Similarly, 32-QAM has an E_b/N_0 advantage of $20\log(0.877/0.438) = 6.02$ dB over 32-PSK, which is the same as shown in Table 8.5. The SNR advantage of QPSK over 4-ASK is $20\log(2/1.265) = 3.98$ dB, which can be compared with the exact result of 3.88 dB, as found in Example 8.14.

8.4 Noncoherently Detected Passband Modulations

In coherent detection, the receiver is assumed to be perfectly synchronized to the transmitter, that is, the receiver can perfectly estimate the frequency, time and phase of the carrier. The only channel impairment is assumed to be AWGN.

Noncoherent or differentially coherent detection is employed in cases where the receiver cannot follow the channel-induced phase of the received signals, or it is uneconomical to use phase-recovery circuits at the receiver. Systems using noncoherent detection can detect signals with no knowledge of the channel-induced phase, albeit at the expense of some degradation in the bit error probability performance. However, perfect recovery of the carrier frequency and symbol timing of the received signals is still assumed.

In differentially coherent detection, the transmitted phase, which carries the bit information,

is detected by looking at the phase difference between two successive symbols, based on the assumption that the channel-induced phase does not change appreciably during two symbol intervals.

8.4.1 Differential Phase Shift Keying (DPSK)

8.4.1.1 Binary DPSK

Differential phase-shift keying (DPSK) is an orthogonal modulation and uses differential encoding at the transmitter and differentially coherent detection at the receiver. As shown in Figure 8.39, the input binary stream is first differentially encoded. If the input bit sequence is denoted by $\{b_k\}$, then the differentially encoded bit d_k can be written as $d_k = \overline{b_k \oplus d_{k-1}}$. To encode the first bit b_1, a reference bit d_0 is needed. Hence, to send $b_k = 0$, the phase of the current signal waveform is advanced by 180 degrees with respect to the phase of the previous signal. To send $b_k = 1$, the phase of the current signal waveform is not changed. Differentially encoded bit sequence is PSK modulated before transmission.

The receiver uses differentially coherent demodulation/detection for the received noisy signals. Noncoherent demodulation in DPSK does not require the estimation of the channel-induced phase of the received signal, hence no need for a coherent local oscillator. Instead, the receiver is equipped with a two-bit buffer so as to measure the phase difference between signals received during two successive bit intervals. The channel-induced phase of the received signal may be assumed to be constant over two bit intervals, if it varies slowly enough with time. Then, the phase difference between signals received in two successive bit intervals will cancel out the channel-induced phase.

The transmitted binary bit sequence is described by the signal

$$\sqrt{\frac{2E_b}{T_b}}\cos(2\pi f_c t + \phi_k), \quad \begin{matrix} 0 \le t \le T_b \\ k = 1, 2 \end{matrix} \quad (8.141)$$

where the signal phases $\phi_k = 0$, π stand for binary 1 and binary 0, respectively; hence, antipodal signaling is assumed for the input bit sequence. The binary DPSK signal at the receiver input may be written as

$$s_k(t) = \sqrt{2E_b/T_b}\cos(w_c t + \phi_k - \theta_k)$$

$$= \sqrt{2E_b/T_b}\cos(w_c t)\cos(\phi_k - \theta_k)$$

$$- \sqrt{2E_b/T_b}\sin(w_c t)\sin(\phi_k - \theta_k)$$

$$= x_k \psi_1(t) - y_k \psi_2(t)$$

$$(8.142)$$

where the basis function are defined by (8.46) and

$$x_k = \sqrt{E_b}\ \cos(\phi_k - \theta_k)$$
$$y_k = \sqrt{E_b}\ \sin(\phi_k - \theta_k) \quad (8.143)$$

Here, θ_k shows slowly varying channel phase, during k-th bit, which is not predicted by the receiver but is assumed to remain essentially constant during two-bits intervals, $\theta_k \approx \theta_{k-1}$.

The signal components at the correlator outputs in upper and lower branches may be written as (see Figure 8.40)

$$\int_0^{T_b} [s_k(t) + w_k(t)]\ \psi_1(t)\ dt = x_k + w_{k,1}$$

$$\int_0^{T_b} [s_k(t) + w_k(t)]\ \psi_2(t)\ dt = -y_k + w_{k,2}$$

$$w_{k,i} = \int_0^{T_b} w_k(t)\ \psi_i(t)\ dt, \quad i = 1, 2$$

$$(8.144)$$

It is important to note that $w_{k,i}$ is a Gaussian random variable with zero-mean and variance $N_0/2$ and $w_{k,1}$ and $w_{k,2}$ are orthogonal to each other (see (8.12)). The decision variable then becomes

$$z = x_k x_{k-1} + y_k y_{k-1} + x_k w_{k-1,1} + x_{k-1} w_{k,1}$$
$$- y_k w_{k-1,2} - y_{k-1} w_{k,2}$$

$$(8.145)$$

Assuming $f_c T_b$ to be an integer and the channel phase does not change in two successive bit

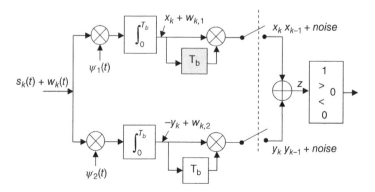

Figure 8.40   Block Diagram of a Binary DPSK Receiver.

durations $\theta_k \approx \theta_{k-1}$, the signal component of the decision variable reduces to

$$x_k x_{k-1} + y_k y_{k-1} = E_b \cos(\phi_k - \phi_{k-1} - \theta_k + \theta_{k-1})$$

$$\approx E_b \cos(\phi_k - \phi_{k-1}) = \begin{cases} + E_b & \phi_k = \phi_{k-1} \\ - E_b & \phi_k \neq \phi_{k-1} \end{cases}$$

(8.146)

Then, the observation vector (decision variable) may be decomposed into its inphase and quadrature components:

$$z_1 = E_b \cos(\phi_k - \phi_{k-1}) + \sqrt{E_b}[\cos(\phi_k - \theta_k)w_{k-1,1}$$

$$+ \cos(\phi_{k-1} - \theta_{k-1})w_{k,1}]$$

$$z_2 = -\sqrt{E_b}[\sin(\phi_k - \theta_k)w_{k-1,2} + \sin(\phi_{k-1} - \theta_{k-1})w_{k,2}]$$

(8.147)

Since θ_k is a random variable, the terms with sine and cosine may be absorbed into the noise terms, each with a variance

$$E\left[\left|w'_{k-1,1}\right|^2\right] = E\left[|\cos(\phi_k - \theta_k)w_{k-1,1}|^2\right]$$

$$= E\left[|\cos(\phi_k - \theta_k)|^2\right] E\left[|w_{k-1,1}|^2\right] = N_0/4$$

(8.148)

Hence, the decision variables y_1 and y_2, which yield identical results, may be written in terms of in-phase and quadrature components as

$$y_1 = z_1/\sqrt{E_b} = \pm\sqrt{E_b} + \left(w'_{k-1,1} + w'_{k,1}\right)$$

$$= \underbrace{\pm\sqrt{E_b} + w_{1r}}_{N(\pm\sqrt{E_b},\sigma^2)} + j \underbrace{w_{1i}}_{N(0,\sigma^2)}$$

$$y_2 = z_2/\sqrt{E_b} = \sqrt{E_b}\left(w'_{k-1,2} + w'_{k,2}\right)$$

$$= \underbrace{w_{2r}}_{N(0,\sigma^2)} + j \underbrace{w_{1i}}_{N(0,\sigma^2)}$$

(8.149)

where $\sigma^2 = N_0/4$. The + sign above corresponds to $\phi_k = \phi_{k-1}$, while − sign to $\phi_k \neq \phi_{k-1}$. In-phase and quadrature components of y_1 and y_2 are uncorrelated Gaussian random variables with variance σ^2. The pdf of the envelope

$r_1 = |y_1|$ of the in-phase component is given by (F.61):

$$f(r_1) = \frac{r_1}{\sigma^2}\exp\left(-\frac{r_1^2 + E_b}{2\sigma^2}\right) I_0\left(\frac{\sqrt{E_b}r_1}{\sigma^2}\right)$$

(8.150)

where $I_0(x) = (1/2\pi)\int_0^{2\pi} \exp(x\cos\varphi)\,d\varphi$ denotes the modified Bessel function of the first kind and zeroth order. The pdf of r_2, the envelope of y_2, is obtained by simply inserting $E_b = 0$ into (8.150):

$$f(r_2) = \frac{r_2}{\sigma^2}\exp\left(-\frac{r_2^2}{2\sigma^2}\right)$$

(8.151)

Irrespective of whether binary 1 or 0 is transmitted, the transmitted bits are correctly detected if $r_1 > r_2$. The probability of bit error for binary DPSK may then be written as

$$P_2 = \text{Prob}(r_2 > r_1)$$

$$= \int_0^\infty \text{Prob}(r_2 > r_1 | r_1 = x) f_{r_1}(x)dx$$

$$= \int_0^\infty \exp\left(-\frac{x^2}{2\sigma^2}\right) f_{r_1}(x)dx$$

$$= \int_0^\infty \frac{x}{\sigma^2}\exp\left(-\frac{2x^2 + E_b}{2\sigma^2}\right) I_0\left(\frac{\sqrt{E_b}\,x}{\sigma^2}\right) dx$$

$$= \frac{1}{2}\exp\left(-\frac{E_b}{N_0}\right)$$

(8.152)

where the last expression is obtained using (D.57).

Here, it may be worth to reconsider the fact that binary DPSK is an orthogonal signalling over two bit intervals. If we have a binary symbol 1 at the transmitter input for $T_b \leq t < 2T_b$, then transmission of symbol 1 leaves carrier phase unchanged for $0 \leq t < T_b$:

$$S_1(t) = \begin{cases} \sqrt{2E_b/T_b}\cos(2\pi f_c t) & 0 \leq t < T_b \\ \sqrt{2E_b/T_b}\cos(2\pi f_c t) & T_b \leq t < 2T_b \end{cases}$$

(8.153)

Table 8.8 Comparison of Coherent, Differentially Encoded and Differential PSK ($\gamma_b = E_b/N_0$).

Modulation	Encoding	Demodulation	BER
PSK	Absolute	Coherent	$P_2 = Q(\sqrt{2\gamma_b})$
DEPSK	Differential	Coherent	$P_2 = 2Q(\sqrt{2\gamma_b})$
DPSK	Differential	Noncoherent	$P_2 = \frac{1}{2}\exp(-\gamma_b)$

On the other hand, if we have a binary symbol 0 for $T_b \leq t < 2T_b$, then the carrier phase will be changed by 180 degrees with respect to that in the $0 \leq t < T_b$ interval:

$$S_2(t) = \begin{cases} \sqrt{2E_b/T_b}\cos(2\pi f_c t) & 0 \leq t \leq T_b \\ \sqrt{2E_b/T_b}\cos(2\pi f_c t + \pi) & T_b \leq t \leq 2T_b \end{cases}$$

$$(8.154)$$

Noting that the signals $S_1(t)$ and $S_2(t)$ may be represented in the two-bit interval as [5]

$$\mathbf{S}_1 = \begin{bmatrix} \sqrt{E_b} & \sqrt{E_b} \end{bmatrix}, \quad \mathbf{S}_2 = \begin{bmatrix} \sqrt{E_b} & -\sqrt{E_b} \end{bmatrix}$$

$$(8.155)$$

$\mathbf{S}_1$ and $\mathbf{S}_2$ are orthogonal over the two-bit interval $0 \leq t < 2T$, and the Euclidean distance between them is $\sqrt{2}$ times smaller than that for antipodal signals (see Figure 6.22). Therefore, DPSK performs 3 dB worse than coherent BPSK at small SNR values but is only slightly (less than 1 dB) inferior to coherent BPSK at large SNR. Since DPSK does not require the estimation of the carrier phase, it is often used in digital communication systems. Table 8.8 presents a comparison of binary PSK, DEPSK and DPSK.

8.4.1.2 Differential Quarternary PSK

In differential QPSK, the relative phase shifts between successive intervals are 0^0, 90^0, 180^0, and -90^0, corresponding to the information bits 00, 01, 11, and 10, respectively. The

probability of bit error for four-phase DPSK with Gray coding is given by: [5]

$$P_b = Q_1(a,b) - \frac{1}{2}I_0\left(\sqrt{2}\,\gamma_b\right)\exp[-2\gamma_b]$$

$$(8.156)$$

where the parameters a and b are defined as:

$$a = \sqrt{2\gamma_b\left(1 - 1/\sqrt{2}\right)}, \quad b = \sqrt{2\gamma_b\left(1 + 1/\sqrt{2}\right)}$$

$$(8.157)$$

Marcum Q function $Q_1(a,b)$ is defined by (see (B.16))

$$Q_1(a,b) = \int_b^\infty x\exp\left(-\frac{x^2 + a^2}{2}\right)I_0(ax)\,dx$$

$$(8.158)$$

where $I_0(x)$ denotes the modified Bessel function of the first kind and order zero.

Figure 8.41 presents a comparison of various versions of binary and quarternary PSK modulations. Four-phase DPSK is approximately 2.3 dB poorer in bit error performance than QPSK at high SNR values.

8.4.2 Noncoherent Orthogonal Frequency Shift Keying (FSK)

M-ary noncoherent FSK signal may be written as

$$s_i(t) = \sqrt{E}\,\Phi_i(t), \quad i = 1, \cdots, M \qquad (8.159)$$

where the basis functions are defined by (8.111). The signal at the receiver input,

$$x(t) = \sqrt{\frac{2E}{T}}\cos(w_i t + \theta(t)) + w(t), \quad \begin{cases} 0 \leq t \leq T \\ i = 1, \cdots, M \end{cases}$$

$$(8.160)$$

undergoes a slowly varying random phase shift $\theta(t)$ induced by the channel with uniform pdf:

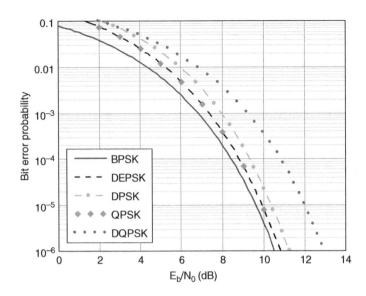

Figure 8.41 Comparison Between Various Binary and Quarternary PSK Modulation Techniques.

$$f_\Theta(\theta) = \begin{cases} 1/2\pi & -\pi < \theta \leq \pi \\ 0 & otherwise \end{cases} \qquad (8.161)$$

In noncoherent FSK, the receiver is expected to estimate the transmitted symbol, using the received signal $x(t)$ with a random carrier phase $\theta(t)$, which is unknown by the receiver.

8.4.2.1 Noncoherent Orthogonal Binary FSK (NCFSK)

Let us first consider the noncoherent reception of binary FSK signals. The receiver has the task of determining whether the information carried by the received noisy signal (8.160) is a 1 or a 0. Since the receiver has no provision for estimating the channel phase $\theta(t)$, it uses correlation receivers to determine in-phase and quadrature components of the received signal for both frequencies, then takes the sum of the squares of in-phase and quadrature components and takes the largest of the two branches as the estimate of the transmitted symbol (see Figure 8.42).

Assuming that the symbol s_1 is sent, the branch outputs following the integrators may be written as

$$x_{11} = \int_0^{T_b} x(t)\,\Phi_1(t)\,dt = +\sqrt{E_b}\cos\theta(t) + w_{11}$$

$$x_{12} = \int_0^{T_b} x(t)\,\Theta_1(t)\,dt = -\sqrt{E_b}\sin\theta(t) + w_{12}$$

$$x_{21} = \int_0^{T_b} x(t)\,\Phi_2(t)\,dt = w_{21}$$

$$x_{22} = \int_0^{T_b} x(t)\,\Theta_2(t)\,dt = w_{22}$$

$$(8.162)$$

where $\Phi_i(t)$, $i = 1, 2$ is defined by (8.111) and

$$\Theta_i(t) = \sqrt{2/T}\,\sin w_i t, \quad i = 1, 2, \cdots, M \quad (8.163)$$

It is important to note that in order to maintain orthogonality of the basis functions, given by (8.111) and (8.163), with the received noisy signal (8.160), the minimum frequency spacing between carriers should be chosen as $\Delta f = 1/T_b$ (see (8.108)). The noise terms, defined by

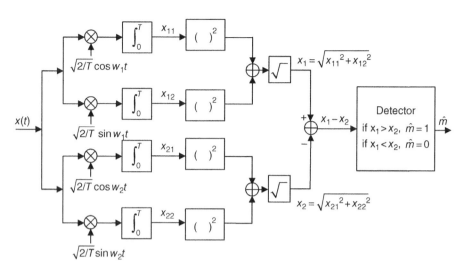

Figure 8.42   Noncoherent Detection of Binary Orthogonal FSK.

$$w_{i1} = \int_0^{T_b} w(t)\, \Phi_i(t)\, dt, \quad i = 1, 2$$
$$w_{i2} = \int_0^{T_b} w(t)\, \Theta_i(t)\, dt, \quad i = 1, 2$$
(8.164)

have zero means and equal variances $N_0/2$ (see (8.12)).

The conditional pdf of $x_1 = (x_{11}^2 + x_{12}^2)^{1/2}$ when s_1 is sent is given by (F.61):

$$f(x_1|s_1) = \frac{x_1}{\sigma^2}\exp\left(-\frac{x_1^2 + E_b}{2\sigma^2}\right) I_0\left(\frac{\sqrt{E_b}\,x_1}{\sigma^2}\right)$$
(8.165)

where $I_0(x) = (1/2\pi)\int_0^{2\pi} \exp(x\cos\varphi)\, d\varphi$ denotes the modified Bessel function of the first kind and order zero. The conditional pdf of $x_2 = (x_{21}^2 + x_{22}^2)^{1/2}$ may be obtained directly by inserting $E_b = 0$ into (8.165):

$$f(x_2|s_1) = \frac{x_2}{\sigma^2}\exp\left(-\frac{x_2^2}{2\sigma^2}\right)$$
(8.166)

The probability of bit error for equiprobable signals may be determined as follows:

$$P_2 = \frac{1}{2}\Pr(x_1 > x_2|s_2) + \frac{1}{2}\Pr(x_2 > x_1|s_1)$$

$$= \Pr(x_2 > x_1|s_1)$$

$$= \int_0^\infty \left[\int_{x_1}^\infty f(x_2|s_1)\, dx_2\right] f(x_1|s_1)\, dx_1$$

$$= \int_0^\infty \frac{x_1}{\sigma^2} e^{-\frac{2x_1^2 + E_b}{2\sigma^2}} I_0\left(\frac{\sqrt{E_b}}{\sigma^2}x_1\right) dx_1$$
(8.167)

Using (D.57) and $\sqrt{\sigma^2} = N_0/2$ one gets

$$P_2 = \frac{1}{2}\exp\left(-\frac{E_b}{2N_0}\right)$$
(8.168)

Similar to orthogonal BFSK, DPSK is a noncoherent orthogonal modulation with $T = 2T_b$ and $E = 2E_b$. The bit error probability for binary DPSK modulation, (8.152), may also be directly obtained from that for binary orthogonal noncoherent FSK if E_b is replaced by $2E_b$ in (8.168). Hence DPSK outperforms noncoherent FSK by 3 dB for the same value of E_b/N_0.

8.4.2.2 Noncoherent Orthogonal M-ary FSK

Let us now consider the transmission of M-ary orthogonal equal-energy signals over an AWGN channel. Assume that the M signals are equally probable and that the signal $s_1(t)$ is transmitted in the symbol interval $0 \leq t \leq T$. The noncoherent receiver for M-ary FSK is similar to that used for binary FSK, which is shown in Figure 8.142. with the exception that the receiver for M-ary orthogonal FSK has M branches instead of 2.

Based on the assumption that the signal $s_1(t)$ is transmitted, the pdf of x_1 is given by (8.165). The pdf's of the decision variables x_m, m = 2,3,...,M are given by (8.166):

$$f(x_m|s_1) = \frac{x_m}{\sigma^2}\exp\left(-\frac{x_m^2}{2\sigma^2}\right), \quad m=2,3,\cdots,M$$

(8.169)

The probability of correct decision is simply the probability that $x_1 > x_m$, m = 2, 3,..., M:

$$P_c(s_1) = P(x_2 < x_1, x_3 < x_1, \cdots, x_M < x_1)$$

$$= \int_0^\infty P(x_2 < x_1, x_3 < x_1, \cdots, x_M < x_1 | x_1 = z)f(z|s_1)dz$$

(8.170)

where $f(z|s_1)$ is given by (8.165). Since the random variables x_m, m = 2, 3,..., M are statistically independent and identically distributed (i.i.d.), the above simplifies to

$$P_c(s_1) = \int_0^\infty [P(x_2 < x_1|x_1 = z)]^{M-1} f(z|s_1) \, dz$$

(8.171)

Using

$$P(x_2 < x_1|x_1 = z) = \int_0^z f_{x_2}(x_2)dx_2 = 1 - \exp\left(-\frac{z^2}{2\sigma^2}\right)$$

(8.172)

and the following binomial expansion (see (D.11))

$$\left(1-e^{-z^2/2\sigma^2}\right)^{M-1} = \sum_{n=0}^{M-1}(-1)^n\binom{M-1}{n}e^{-nz^2/2\sigma^2}$$

(8.173)

the probability of correct decision given by (8.171) is found using (D.57):

$$P_c = \frac{1}{M}\sum_{m=1}^{M}P_c(s_m) = P_c(s_1)$$

$$= \sum_{n=0}^{M-1}(-1)^n\binom{M-1}{n}\frac{1}{n+1}\exp\left(\frac{-nE}{(n+1)N_0}\right)$$

(8.174)

The probability of symbol error for M-ary noncoherent FSK is found to be

$$P_M = 1 - P_c = \frac{1}{M}\sum_{n=1}^{M-1}(-1)^{n+1}\binom{M}{n+1}$$

$$\exp\left(\frac{-n\log_2 M E_b}{(n+1)N_0}\right)$$

(8.175)

which reduces to (8.168) for binary noncoherent FSK, as expected. Figure 8.43 compares the symbol error probability of M-ary noncoherent and coherent FSK with E_b/N_0 for various values of M. Similar to M-ary coherent FSK, the E_b/N_0 required to achieve a given symbol error probability decreases with increasing values of M. For large values of M, at a symbol error probability of 10^{-5}, noncoherent M-ary FSK has approximately 0.6 dB worse performance compared to coherent M-ary FSK.

The PSD of noncoherent orthogonal M-ary FSK signals is similar to that for coherent orthogonal M-ary FSK (see Figure 8.38b) and can easily be found by following the procedure outlined in (8.129)–(8.131).

In continuous-phase FSK, the phase of the transmitted tone should be the same as the phase of the tone of the preceeding symbol. This implies that the frequency synthesizer at the transmitter should synthesize tones at the

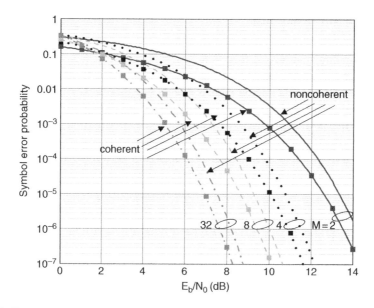

Figure 8.43   Symbol Error Probability For Coherent and Noncoherent Orthogonal M-ary FSK.

start of each symbol interval with a phase so as to ensure phase continuity between symbols at switching times. However, irrespective of whether the transmitted tones have the same phase or not, each transmitted tone undergoes phase changes induced by the channel. Hence, the tones at the receiver input may suffer different phase changes.

In coherent reception of M-ary FSK signals, the phase of each tone should be estimated at the receiver by a carrier recovery circuit (see Figure 8.4) which drives the local oscillator in the receiver (see Figure 8.5 and Figure 8.36). If these phases can be perfectly estimated at the receiver, then coherent FSK offers a superior performance compared to noncoherent FSK. However, inaccurate estimation of the phase would increase symbol errors and decrease the received mean SNR.

In noncoherent reception of the FSK signals, the receiver does not bother for estimating the phases of the received tones and employs noncoherent detection (see Figure 8.42), albeit at the cost of degraded symbol error performance. Ignoring the carrier recovery circuit, a noncoherent M-ary FSK

receiver has higher complexity compared to the coherent reception.

8.5   Comparison of Modulation Techniques

The study of modulation techniques justifies asking the following fundamental question: What are the most important features of a modulation technique and which modulation technique should be chosen for a certain application?

Modulation techniques may be classified depending on whether they have a constant envelope or not. In constant envelope signaling, the information is transmitted by modulating either the phase or the frequency of the carrier as in PSK and FSK. Therefore, envelope fluctuations induced by the channel do not affect the receiver performance. Constant envelope modulations are more suitable for wireless channels, which induce short- and/or long-term fades in the signal envelope due to fading and shadowing. In systems using non-constant envelope modulations, such as ASK and QAM, the receiver may not be able to predict whether the envelope changes are due to modulation or due to the channel. This increases bit errors. Non-constant envelope signaling is more suitable for wireless channels with free-space propagation conditions so that fading and shadowing effects are minimal. Similarly, they can be used for wired channels with stable and predictable path-loss and free of multipath effects. [1][2][3][5][6]

In addition to the signal envelope, the phase and the frequency of signals also undergo random fluctuations as they propagate through a channel. Besides, the receiver may not be able to predict the time of arrival of a signal with sufficient accuracy, due to either problems related to synchronization or random fluctuations in the propagation delay. Consequently, as shown in Section 8.1, the lack of precise information or inability of the receiver to predict random fluctuations in frequency, phase and time due to channel degrades the receiver performance.

Coherent modulation techniques require the receiver to predict the fluctuations in time, frequency and phase due to channel with sufficient accuracy. This is usually accomplished by using some special circuits at the receiver. Noncoherent modulation techniques may be used to demodulate and detect received signals without using the information of some or all of the three parameters, albeit at the cost of some performance degradation compared to coherent signaling.

Modulation techniques may also be compared on the basis of their efficiency in using the channel bandwidth to support a given bit rate for a E_b/N_0 level required for a certain bit error probability. Noting that the main resources provided by the system are the channel bandwidth and the transmit power, modulation techniques may be compared in terms of their spectrum- and power- efficiencies. Under free-space propagation conditions, the E_b/N_0 at the receiver is related to the transmitter power, receiver and antenna noise, the bit rate R_b and the total loss L (including propagation and implementation

losses) between transmitter and the receiver as follows:

$$\frac{E_b}{N_0} = \frac{P_r}{k(T_A + T_R) \, R_b} = \frac{P_t G_t G_r}{k(T_A + T_R) \, R_b \, L}$$

(8.176)

where $k = 1.38 \times 10^{-23}$ $(Joule/K)$ is the Boltzmann constant, P_t and P_r denote transmit and receive powers, respectively. G_t and G_r denote transmit and receive antenna gains, respectively. T_A and T_R, denote the noise temperatures of the receive antenna and the receiver, respectively, and provide a measure of the receiver system noise. Knowing that the bit/symbol error probability is uniquely determined by E_b/N_0, a power efficient modulation requires lower E_b/N_0 to achieve a given bit/symbol error probability. Therefore, power efficiencies of modulation techniques can be compared in terms of the E_b/N_0's required to achieve a certain bit/symbol error probability. This comparison leads to meaningful results if the chosen bit/symbol error probability levels are sufficiently low. A more heuristic and simple way of power-efficiency comparison of modulation techniques is presented in Table 8.7 based on the minimum Euclidean distances of various modulation techniques as given by (8.140). The results obtained by this approach agree closely with the results based on exact analytical expressions for the bit/symbol error probability.

A particular modulation technique requires a certain bandwidth to support a given bit rate and alphabet size M. In that sense, it is reasonable to use the bandwidth (spectrum) efficiency, defined by (8.6), as a measure of the efficiency in using the available bandwidth by a particular modulation technique and a given value of E_b/N_0.

The spectral occupancy of modulated signals is determined by the pulse shape, the line code, the symbol duration T and the statistics of the input bit stream. For a random input data stream, the spectral occupancy is determined by the pulse shape, the line code and the symbol duration T. In M-ary signaling, a signal is transmitted for each group of $\log_2 M$ bits during $T = T_b(\log_2 M)$ (T_b denotes the bit duration). Therefore, transmission (symbol) rate is proportional to $1/\log_2 M$, implying that the transmission bandwidth decreases with M, except for M-ary FSK. Note that there is no unique definition of the transmission bandwidth. The bandwidth of a rectangular NRZ pulse of duration T is approximately given by $1/T$. Since the symbol duration may be written as $T = (\log_2 M)T_b$, the bandwidth efficiency is the same as for M-ary PSK, QAM or ASK (see (8.35), (8.82)):

$$\eta = \frac{R_b}{B_M} = \frac{R_b}{1/T} = \log_2 M \quad (bps/Hz) \quad (8.177)$$

The bandwidth efficiency increases with M, since the signaling rate, proportional to $1/T = R_b/\log_2 M$, decreases with M. However, the E_b/N_0 required to achieve this bandwidth efficiency is not the same for these three types of modulations. Tables 8.2, 8.3, 8.5 and 8.7 clearly show that M-ary ASK requires the highest E_b/N_0 value (has lowest power-efficiency) among the three to achieve a certain bit error probability since it uses the signal-space in only one dimension. M-ary QAM is most power-efficient since it requires the lowest E_b/N_0 to achieve a given bit/symbol error probability, and it uses the signal-space in two dimensions

more economically compared to M-ary ASK and M-ary PSK.

However, in M-ary FSK, the transmission bandwidth occupied by M frequency tones, each carrying $\log_2 M$ bits, increases with M. In noncoherent orthogonal signaling, neighboring carriers are separated by $1/T$ and the bandwidth efficiency is given by

$$\eta = \frac{R_b}{B_M} = \frac{R_b}{M/T} = \frac{\log_2 M}{M} \quad (8.178)$$

In coherent orthogonal signaling, neighboring carriers are separated by $1/(2T)$ and the corresponding bandwidth efficiency is given by (8.133)

$$\eta = \frac{R_b}{B_M} = \frac{R_b}{M/(2T)} = \frac{2\log_2 M}{M} \quad \left(\frac{bps}{Hz}\right)$$
$$(8.179)$$

The bandwidth efficiency is increased by a factor of two compared to the case when neighboring carriers are separated by $1/T$.

The bandwidth required for biorthogonal signaling is one-half of that for orthogonal signals since the number of orthognal carriers is $M/2$; hence the bandwidth efficiency increases by a factor of two.

Table 8.9 presents a comparison of various coherently and noncoherently detected modulation techniques for a symbol error probability

Table 8.9 Comparison of Bandwidth Efficiency and E_b/N_0 (dB) Required to Achieve a Symbol Error Probability of 10^{-5} For Coherently and Noncoherently Detected Modulation Schemes. [3][5]

		M-ary ASK, M-ary PSK, M-ary QAM			Coherent M-FSK		Noncoherent M-FSK	
		M-ASK	M-PSK	M-QAM				
M	η	E_b/N_0 (dB)	E_b/N_0 (dB)	E_b/N_0 (dB)	η	E_b/N_0 dB	η	E_b/N_0 (dB)
2	1	9.585	9.585		1	12.596	0.5	13.35
4	2	13.745	9.89	9.89	1	10.06	0.5	10.775
8	3	18.286	13.463	12.995	0.75	8.635	0.375	9.29
16	4	23.135	18.065	14.04	0.5	7.665	0.25	8.286
32	5	28.214	23.075	17.253	0.31	6.948	0.156	7.54
64	6	33.445	28.293	18.574	0.19	6.39	0.094	6.95

of 10^{-5}. The channel bandwidth in orthogonal M-ary FSK increases but the bandwidth efficiency decreases as M gets higher. However, the decrease in bandwidth efficiency is compensated by the decrease in E_b/N_0 required to achieve a given symbol error probability. Hence, M-ary FSK is not bandwidth-efficient but is power-efficient. On the other hand, ASK, PSK and QAM are bandwidth-efficient since the required channel bandwidth decreases with increasing values of M at the expense of higher E_b/N_0 required to achieve a given symbol error probability. Therefore, ASK, PSK and QAM modulations are bandwidth-efficient but not power-efficient.

Because of their high bandwidth-efficiencies, PAM, PSK and QAM are appropriate for bandwidth-limited communication channels. They provide spectrum efficiencies $\eta > 1$ but also require higher SNRs for higher values of M (narrower bandwidths). They are preferred in bandwidth-limited channels. However, M-ary orthogonal signals have lower spectrum efficiency $\eta \leq 1$, but require smaller E_b/N_0 to achieve a given bit/symbol error probability. As M increases, η decreases due to an increase in the required channel bandwidth but the E_b/N_0 required to achieve a given error probability also decreases. Consequently, M-ary orthogonal signals are appropriate for power-limited channels with sufficiently large bandwidths.

A more meaningful and quantitative comparison of the modulation techniques may be related to the lowest E_b/N_0 needed to achieve a data rate R_b in a given bandwidth and for a given bit error probability Shannon's capacity theorem states that the capacity C in bits/s of an AWGN channel band-limited to B is given by

$$\frac{C}{B} = \log_2(1 + SNR) = \log_2\left(1 + \frac{E_b R_b}{N_0 B}\right)$$

(8.180)

The capacity $C(\geq R_b)$ in bits/s represents the highest achievable bit rate in this channel. Hence, it serves as the upper bound on the bandwidth efficiency of any type of modulation. As B goes to infinity, the error probability can be made as small as desired, provided that $E_b/N_0 > 0.693$ (−1.6 dB):

$$\frac{C}{B} \leq \log_2\left(1 + \frac{E_b}{N_0}\frac{C}{B}\right) \Rightarrow \frac{E_b}{N_0} \geq \frac{2^{C/B} - 1}{C/B} \Rightarrow$$

$$\lim_{B \to \infty} \frac{E_b}{N_0} \geq \ln 2 = 0.693 \ (-1.6 \ dB)$$

(8.181)

This is the minimum E_b/N_0 required to achieve reliable transmission (bit error probability approaching zero) in the limit as the channel bandwidth B goes to infinity and the corresponding spectrum efficiency C/B approaches zero. The realistic values of E_b/N_0 required to achieve a certain bit/symbol error probability are much higher than −1.6 dB. A modulation technique is therefore as good as it is close to the Shannon limit. Figure 8.44 shows the variation of the spectrum efficiency versus E_b/N_0 for a symbol error probability of 10^{-5}. The Shannon limit C/B denotes the maximum value of the spectrum efficiency for a given value of E_b/N_0 and no practical system can have a spectrum efficiency exceeding the Shannen limit. In agreement with (8.181), the E_b/N_0 required to achieve arbitrarily low bit error probabilities approaches to −1.6 dB, as the spectrum efficiency vanishes. Using Figure 8.44, one can compare the spectrum efficiencies of modulation schemes for a given value of E_b/N_0. Modulations with higher spectrum efficiencies are located closer to the Shannon limit. The region determined by $R_b/B \geq 1$ is the so-called bandwidth-limited region and includes M-ASK, M-PSK, M-QAM and their derivatives. On the other hand, the region determined by $R_b/B < 1$ is the so-called power-limited region and includes M-FSK.

The gap between the Shannon limit and the practical operating values of E_b/N_0 can be decreased by using some advanced techniques including eror correcting coding. Their use

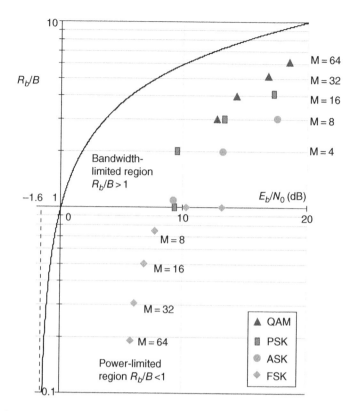

Figure 8.44   Comparison of Different Modulation Techniques for a Symbol Error Probability of 10^{-5}.

decreases the required E_b/N_0 to achieve a given probability of symbol error usually at the expense of an increase in the transmission bandwidth.

Example 8.16   Shannon's Capacity Versus Realistic Data Rate.

Consider a telecommunication system using 16-QAM at a symbol error rate of 10^{-5}. The E_b/N_0 required to achieve a symbol error probability of 10^{-5} for 16-QAM is found from Table 8.9 as 14.04 dB. The spectrum efficiency predicted by the Shannon capacity theorem is found from Figure 8.44 as 7.59. However, using (8.177), the spectrum efficiency for 16-QAM is found to be 4, almost half the Shannon limit. On the other hand, M-ary PSK and M-ary ASK have the same spectrum efficiency of 4 as 16-QAM. However, this is achieved at E_b/N_0

values of 18.07 dB for 16-PSK and 23.14 dB for 16-ASK (see Table 8.9) (see Figure 8.44). Consequently, for example in a 1 MHz bandwidth, the above considered modulations can support 4 Mbps compared to 7.59 Mbps predicted by the Shannon capacity limit. However, required E_b/N_0 values by these modulations are not the same.

References

[1] S. Haykin, *Communication Systems* (3rd ed.), John Wiley: New York, 1994.

[2] A. B. Carlson, P. B. Crilly, and J. C. Rutledge, *Communication Systems* (4th ed.), McGraw Hill, Boston, 2002.

[3] B. Sklar, *Digital communications, Fundamentals and Applications* (2nd ed.), Prentice Hall: New Jersey, 2003.

[4] J. Lodewijk, *An even better atomic clock*, IEEE Spectrum, October 2014, pp.42-64.

[5] J. G. Proakis and M. Salahi, *Communication Systems Engineering* (2nd ed.), Prentice Hall: New Jersey, 2002.
[6] I. A. Glover, and P. M. Grant, *Digital Communications* (2nd ed.), Pearson Prentice Hall: Harlow, 2004.

Problems

1. Consider a bandpass signal $s(t) + w(t)$,

$$s(t) = \begin{cases} Ag(t)\cos(2\pi f_c t) & 0 \le t \le T \\ 0 & otherwise \end{cases}.$$

where T denotes the signalling period such that $f_c T = $ integer and $w(t)$ is the AWGN with single-sided PSD N_0. We consider the following alternative pulses:

$$g_1(t) = \Pi\left(\frac{t - T/2}{T}\right), \quad 0 \le t < T$$

$$g_2(t) = \Lambda\left(\frac{t - T/2}{T/2}\right), \quad 0 \le t < T$$

$$g_3(t) = t/T, \quad 0 \le t < T$$

$$g_4(t) = \text{sinc}(2W(t - T/2))$$

a. Determine the impulse response of the filter matched to these signals.
b. Determine the matched filter output at $t = T$.
c. Assuming that a correlation receiver is used for the signal $s(t)$, determine the receiver output at $t = T$ and compare it with the result you found in part (b).
d. What conclusions can you draw concerning the use of different pulses?

2. Consider matched filter detection of equally likely BPSK signals $s_{1,2}(t) = \pm \sqrt{2E_b/T}\cos w_c t$. Assume that the matched filter output is $x_1 = \pm\sqrt{E_b} + w(T)$, where $w(T)$ is a random variable accounting for the channel noise, which is described by the Laplacian pdf:

$$f(n) = \frac{1}{\sqrt{2}\sigma_w}\exp\left(-\frac{\sqrt{2}|n|}{\sigma_w}\right)$$

with $\sigma_w^2 = N_0/2$. The received E_b/N_0 is 6 dB.

a. Determine the optimum threshold level and the minimum probability of bit error P_b.
b. Determine P_b if the decision threshold is $\lambda = 0.1\sqrt{E_b}$.
c. The threshold of $\lambda = 0.1\sqrt{E_b}$ is optimum for a particular set of a priory probabilities $P(s_1)$ and $P(s_2)$. Compute these probabilities.

3. The Euclidean distance-squared between two signal-space points is given by (7.10)

$$d^2 = \int_0^T [s_1(t) - s_2(t)]^2 dt = \|s_1 - s_2\|^2$$
$$= \|s_1\|^2 + \|s_2\|^2 - 2(s_1 \cdot s_2) = 2E_b(1 - \rho)$$

where $\quad E_b = \|s_i\|^2 = \int_0^{T_b} |s_i(t)|^2 dt, \ i = 1, 2$

from (7.9). The correlation coefficient between $s_1(t)$ and $s_2(t)$ is defined by (6.57):

$$\rho = \frac{1}{E_b}\int_0^{T_b} s_1(t)s_2(t)dt = \frac{s_1}{\sqrt{E_b}} \cdot \frac{s_2}{\sqrt{E_b}}.$$

The average probability of bit error may be expressed as, in terms of the minimum Euclidean distance as $P_2 = Q(d/\sqrt{2N_0}) = Q(\sqrt{E_b(1-\rho)/N_0})$ from (6.62) where $(1-\rho)E_b/N_0$ denotes the SNR at the matched filter output for correlated signals. Assuming that $f_c T_b$ is an integer and $f_c T_b \gg 1$,

a. Determine d, ρ and the average probability of bit error for binary PSK where

$$s_1(t) = -s_2(t)$$
$$= \sqrt{2E_b/T_b}\cos(2\pi f_c t), \quad 0 \le t \le T_b$$

b. Repeat (a) for binary orthogonal FSK where

$$s_{1,2}(t)=\sqrt{2E_b/T_b}\cos[2\pi(f_c\pm\Delta f/2)t],\ 0\le t\le T_b$$

c. Determine the required increase in E_b/N_0 in coherent binary FSK so that it has the same bit error probability performance as a coherent PSK system.

4. Consider a coherently detected binary PSK modulation where

$$s(t)=\begin{cases} A\cos(2\pi f_c t) & for\ m_1 \\ -B\cos(2\pi f_c t) & for\ m_0 \end{cases}$$

where $A, B>0$. A priori probabilities of symbol m_1 and m_0 are given by $p(m_1)$ and $p(m_0)$, respectively. The correlation receiver shown in Figure 8.14 is used with a threshold level λ.

a. Determine the average E_b/N_0. How can you minimize the average E_b/N_0? Which symbol is more vulnerable to errors?
b. Write down the pdf of $\mathbf{x}$ when m_1 and m_0 is transmitted.
c. Determine the probability $P(\hat{m}=m_1|m_0)$ of receiving m_1 when m_0 was transmitted and the probability $P(\hat{m}=m_0|m_1)$ of receiving m_0 when m_1 was transmitted. Determine the average probability of bit error.
d. Determine the optimum threshold level λ_{opt}.
e. Compare the average bit error probability and the average power for the following cases:

$A=B$	$\lambda=0$
$A=B$	$\lambda=\lambda_{opt}$
$A\ne B$	$\lambda=0$ (you can change the relative values of A and B so that $P(\hat{m}=m_0|m_1)p(m_1)=P(\hat{m}=m_1|m_0)p(m_0))$

f. Repeat (e) for $p(m_1)=p(m_0)$.

5. Consider a BPSK signal with pilot carrier added for synchronization purposes

$$s_1(t)=+A\cos w_c t+\alpha A\cos(w_c t+\theta)$$
$$s_2(t)=-A\cos w_c t+\alpha A\cos(w_c t+\theta)$$
$$0\le t\le T_b$$

in an AWGN channel with $w(t)$ of zero mean and PSD $N_0/2$. The arbitrary phase θ is assumed to be zero.

a. Determine the energy E_i of $s_i(t)$, $i=1,2$, the average bit energy and the correlation coefficient between $s_1(t)$ and $s_2(t)$, which is defined as

$$\rho=\frac{1}{\sqrt{E_1 E_2}}\int_0^T s_1(t)\ s_2(t)\ dt$$

b. Draw the block diagram of an optimum receiver for this signaling system when there is no pilot signal. Write down the bit error probability.
c. Draw the block diagram of an optimum receiver in the presence of the pilot signal. Write the analytical expressions for the variables at appropriate places. Determine the noise variance at the receiver output.
d. Determine the mean $\gamma_b=E_b/N_0$ at the receiver output and show that the bit error probability for the optimum coherent receiver operating in an AWGN channel is given by

$$P_e=Q\left(\sqrt{\frac{2\gamma_b}{1+\alpha^2}}\right)$$

e. Assuming $\gamma_b=8.35dB$, determine the degradation in SNR and increase in the bit error probability for $\alpha=0.3$, that is, when 9% of the transmit power is allocated to the pilot signal.

6. A 2 Mbps bit stream is to be transmitted using QPSK by multiplexing even- and odd-numbered bits with two quadrature carriers in an AWGN channel.

a. Determine the symbol rate.

b. Determine the required channel bandwidth for NRZ polar signalling if the output signal is filtered so as to pass only the main lobe of the signal spectrum.

c. Determine the symbol error probability if the received carrier power is 100 μW and $N_0 = 5 \times 10^{-12}$ W/Hz?

d. If one of the quadrature carriers is desired to carry 10 Mbps while the other still supports 1 Mbps, determine the relative amplitudes of the signals for the two carriers so that the E_b/N_0 for the two channels are the same.

e. Determine the four phases representing the four dibits in case (d). Are the phase separations the same as when the quadrature carriers support the same data rate?

7. Consider the constellation shown below. The ratio of radii of the two rings is assumed to be $r_1/r_2 = 1/2$ and $r_1 = \sqrt{E_0}$ where E_0 denotes the minimum symbol energy.

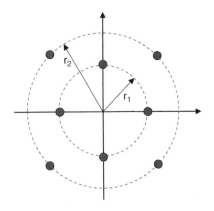

a. Determine the average symbol energy, $E_{s,av}$ in terms of E_0.

b. Determine peak-to-average power ratio (PAPR).

c. Using the Euclidean distances between signal points, determine an upper bound for the symbol error probability in terms of the average $E_{b,av}/N_0$.

d. Bearing in mind the Euclidean distances between the symbols, use Gray code to assign three bits to each of the signal points.

e. Using the result of parts (a)–(d), how would you improve the performance of the above constellation. Determine the improvement that your suggestion provides on the required $E_{b,av}/N_0$ to achieve a given symbol error probability.

8. Consider the eight-point constellations shown below. Assuming that these constellations have the same symbol error probabilities,

a. Compare their peak to average power ratios (PAPRs).

b. Compare the average powers required by these constellations. Which constellation is more power efficient?

c. Compare E_b/N_0 ratios required by these constellations to achieve the same symbol error probability.

9. Assuming that the symbol error performances of the modulation schemes 64-PSK and 64-QAM are determined mainly by the arguments of the Q function, the factors in front of the Q functions may be ignored.

a. For the same E_b/N_0 ratio, which modulation scheme shows smaller symbol error probability, that is, which shows higher power efficiency.

b. For the same bit error probability, which modulation scheme shows higher E_s/N_0 and higher spectral efficiency.

10. Consider an unequalized passband channel whose baseband equivalent is shown in Figure 6.31, where $g(t)$ denotes the

transmit pulse, $h(t)$ is the impulse response of the channel and $c(t)$ is the receive filter which is matched to $g(t)$. Assume that a square-root raised cosine filter with roll-off factor $\alpha = 0.3$ is used for pulse shaping. The chanel is assumed to have a flat frequency transfer function around a sufficiently high carrier frequency

$$H(f) = \begin{cases} 1 & f_c - 0.3 \; MHz \leq f \leq f_c + 0.3 \; MHz \\ 0 & otherwise \end{cases}$$

a. Determine the maximum bit rate that coherent 8-PSK can support with no ISI.
b. Repeat (a) for coherent orthogonal 8-FSK.
c. Compare the results in parts (a) and (b) and explain the reason for their difference.

11. Passband data transmission at 19.2 kbps is desired in a 4 kHz bandwidth channel.

a. Select a symbol rate and a constellation size for a power efficient modulation.
b. If a raised cosine pulse is used for transmssion, select the roll-off factor.
c. If a symbol error probability of 10^{-6} is desired, what is the required E_b/N_0? Determine the average E_b and the received carrier power if $N_0 = 10^{-9}$ W/Hz.

12. Binary data is to be transmitted at $R_b = 1$ Mbps in a wireless channel with 400 kHz bandwidth.

a. Determine the modulation technique that minimizes signal energy needed to achieve $P_b \leq 10^{-6}$ and determine the required E_b/N_0 in dB.
b. Repeat part (a) if we choose a constant envelope signal. What is the penalty for using a constant-envelope signal?

13. Determine the bit rate that can be transmitted through a 1 kHz bandpass channel using

a. binary PSK, QPSK and 8-QAM using a transmit pulse with a raised-cosine spectrum with $\alpha = 0.3$.
b. coherently-detected orthogonal M-ary FSK with M = 2, 4 and 8.
c. noncoherently-detected orthogonal M-ary FSK with M = 2, 4 and 8.

14. Consider a binary communication system with $P_t = 200$ mW transmit power operating at a carrier frequency of 900 MHz over a range of $d = 3$ km under free-space propagation conditions. Transmit and receive antenna gains are assumed to be $G_t = 12$ dBi and $G_r = 0$ dBi, respectively. Assuming that the noise spectral density at the receiver input is $N_0 = 10^{-15}$ W/Hz and the probability of bit error satisfies $P_2 \leq 10^{-4}$, compare the maximum allowable bit rate for noncoherent FSK, coherent FSK, DPSK, and coherent PSK.

15. Prove that the joint cdf in (8.124) can be written as the product of marginal cdf's. Show that for $i \neq j$, $w_j - w_1$ and $w_i - w_1$ are uncorrelated.

16. The considered frequency tones for equally likely coherently detected binary FSK signals with amplitudes A = 1mV and bit duration 0.1 ms are $f_1 = 98$ kHz and $f_2 = 102$ kHz; $f_1 = 97.5$ kHz and $f_2 = 102.5$ kHz; and $f_1 = 97$ kHz and $f_2 = 103$ kHz. The power spectral density of the noise is $N_0 = 5 \times 10^{-12}$ W/Hz.

a. Determine the correlation coefficients for the three considered frequency assignments. Are these two tones orthogonal?
b. Determine the corresponding bit error probabilities. Which case provides the lowest bit error probability? Why?

17. Consider a binary continuous-phase FSK (CPFSK) signaling scheme where the symbols are represented by

$$s_i(t) = \sqrt{\frac{2E_b}{T_b}} \cos(2\pi f_i t + \theta(0)), \quad \begin{array}{c} i=1,2 \\ 0 \le t \le T_b \end{array}$$

where $\theta(0)$ denotes the value of the phase at $t = 0$. In classical binary FSK, $\theta(0)$ is a sample value of uniformly distributed random variable in the interval $[0, 2\pi]$ and is not the same for the two symbols; this stems from the fact that the local oscillator at the transmitter can not have the same phase at the start of each symbol interval. The above equation may also be written as

$$s_i(t) = \sqrt{\frac{2E_b}{T_b}} \cos(2\pi f_c t + \theta_i(t)), \quad \begin{array}{c} i=1,2 \\ 0 \le t \le T_b \end{array}$$

where

$$\theta_i(t) = \theta(0) + b_i \frac{\pi h}{T_b} t, \quad i = 1, 2, \quad b_1 = -b_2 = 1$$

Here b_1 and b_2 denote binary symbols. Since the phase $\theta_i(t)$ is a continuous function of time, $s_i(t)$ is also continuous at all times.

a. Express the carrier frequency and the parameter h, called as deviation ratio, in terms of f_1 and f_2 and find the value of $\theta_i(T_b) - \theta(0)$.
b. Determine the minimum value of h and the corresponding value of the frequency spacing $\Delta = |f_2 - f_1|$ which makes $s_1(t)$ and $s_2(t)$ orthogonal to each other.
c. CPFSK is called as minimum shift keying (MSK) for $h = 1/2$, which corresponds to minimum tone spacing that preserves orthogonality between $s_1(t)$ and $s_2(t)$. Assuming $\theta(0) = 0$, show that MSK is an example for orthogonal signaling:

$$s_i(t) = \sqrt{\frac{2E_b}{T_b}} \cos\left[\frac{\pi t}{2T_b}\right] \cos(2\pi f_c t)$$

$$\mp \sqrt{\frac{2E_b}{T_b}} \sin\left[\frac{\pi t}{2T_b}\right] \sin(2\pi f_c t)$$

Determine the PSD of MSK signal.

18. Consider the following 4-ary signaling:

$$s_n(t) = \sqrt{\frac{2E_s}{T}} \cos\left(\frac{2\pi n t}{T}\right) \quad 0 \le t \le T,$$

$$n = 1, 2, 3, 4$$

where E_s denotes the symbol energy and n is an arbitrary integer. Assume that $s_n(t)$, $n = 1, 2, 3, 4$ have equal a priori probabilities,

a. Classify this signaling and determine the dimensionality of the signal-space.
b. Sketch the signal space and decision boundaries.
c. Determine the transmission bandwidth and the spectrum efficiency.

19. Consider a sinusoidal signal of amplitude A, frequency f_c and a random phase θ, $x(t) = A\cos(2\pi f_c t + \theta) + w(t)$ where $w(t)$ denotes AWGN with zero mean and variance $N_0/2$.

a. The figure below shows a noncoherent correlation receiver followed by an envelope detector. Determine x_1, x_2 and the output y.

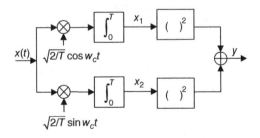

b. For the same input signal, determine the output of a receiver (basically a spectrum analyzer) which consists of a correlator followed by a Fourier transformer as shown below. Compare the value of the PSD of $x(t)$ at $f = f_c$ with the mean value of y from part (a).

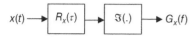

$$x(t) \longrightarrow \boxed{R_x(\tau)} \longrightarrow \boxed{\Im(.)} \longrightarrow G_x(f)$$

20. The PSD of passband signals given by (8.4), (8.33), and (8.80) are applicable for a NRZ polar rectangular pulse, which has a PSD of $T \, \text{sinc}^2(fT)$, where T denotes the symbol duration. Determine the corresponding PSDs for a raised-cosine pulse and a Nyquist pulse.

21. Consider M-ary PSK transmission in 1 MHz bandwidth. In order to avoid adjacent–channel interference, the spectral envelope outside the bandwidth is required to be lower than 30 dB below the maximum. Determine the maximum data rate R_b to achieve this objective for M = 2, 4, 8 and 16?

22. Consider a communication system operating at 6 GHz. This system uses 64-QAM and is required to operate at bit error probabilities less than 10^{-6}.

   a. Determine the required mean E_s/N_0 including 2 dB implementation losses at the receiver.
   b. Determine the PSD of receiver noise if the antenna noise temperature is 290° K and a receiver noise figure of 4 dB.
   c. Determine the average received symbol and bit energies.
   d. Determine the average received carrier power under free space propagation conditions for $P_t = 100$ mW, $G_t = G_r = 20$ dBi and 30 km range.
   e. Determine the data rate that can be supported by this system.

23. A noncoherent binary FSK signal with $R_b = 100$ kbps is transmitted over an AWGN channel. The receiver has a noise figure of 6 dB an antenna noise temperature of $T_0 = 290$ K. The antenna sees a 50 Ω receiver input impedance. If the received signal level at the receiver input is 1μV, then determine the bit error probability.

24. A wireless system using BPSK modulation operates with a bit error probability 10^{-6}. The contribution of the atmospheric noise temperature to a BS antenna noise is about 70–100 K at 900 MHz, and man-made noise contributes another 10–120 K depending on the area (rural or urban). The BS antenna noise temperature T_A is thus in the range of 100–200 K. The receiver noise figure is assumed to be 10 dB.

   a. Assuming a BS antenna noise temperature of 200 K, determine the required received bit energy E_b in dB. How this result would be affected if the antenna noise temperature were 100 K?
   b. Assuming that the maximum required bit rate is 171 kbps and a 3 dB implementation loss, determine the minimum power level required by a receiving BS antenna in dBm.
   c. Assuming that the gains of MS and BS antennas are 2.15 dBi and 15 dBi and the range is 20 km, determine the minimum required transmit power by a MS so as to establish a link between MS and BS in a wireless system operating at 900 MHz. *Hint:* Assume free-space propagation conditions.

25. Consider an Earth-space-Earth link between a spacecraft and the Earth. Since the generation of electric power onboard the spacecraft is more difficult, space-Earth links are more vulnerable due to their lower transmit powers. This can be compensated by using Earth-based antennas with higher gains (diameters). Such a choice can also provide flexibility for the spacecraft receiver to use lower gain (smaller diameter) antennas. In

such systems, downlink frequencies are chosen to be higher than uplink frequencies to enable the spacecraft to achieve higher antenna gains with physically smaller antennas. Consider a narrowband Earth-space link with free-space propagation conditions where uplink and downlink frequencies are very close to each other, implying that free-space losses in up-and down-links are practically the same. On the other hand, noise temperature of the Earth-based receive antenna is expected to be much higher than that of the spacecraft in the 'cold' space environment.

a. If the E_b/N_0's at Earth- and space-based receivers are required to be the same, determine the relation between the bit rates in up- and down-links. Which link can support higher data rates?

b. If system noise temperature of the Earth-based receiver is 4 times higher than that of the spacecraft receiver, and the received power level by the spacecraft receiver is 20 dB higher than the power received by the Earth-based receiver, determine the relative bit rates in up- and down-links. State whether up-link or the down-link forms the bottleneck in this communication system.

26. A spacecraft that travels to the planet Mercury sends data to Earth through a distance of 1.6 10^{11} meters (changes between 0.77 10^{11}m–2.22 10^{11} m). Transmit antenna gain is 27 dBi, carrier frequency is $f = 2.3$ GHz and the transmit power is 17W. The Earth station receive antenna has 64-meter diameter and aperture efficiency 0.65. Effective noise temperature

of the receiver is $T_r = 15$ K and the antenna noise temperature is assumed to be $T_A = 4$ K.

a. If the desired E_b/N_0 is 6.8 dB (bit error probability is 10^{-3} for BPSK modulation), determine the data rate that can be supported in the spacecraft-Earth link (down-link).

b. Noting that data rate and the transmission bandwidth in down- and up-links are the same it is reasonable to choose up- and down-link E_b/N_0 to be the same. Noting that the overall C/N is given by

$$\frac{C}{N} = \frac{1}{(C/N)_u^{-1} + (C/N)_d^{-1}}$$

determine the overall E_b/N_0.

c. If P_{eu} and P_{ed} denote the probability of bit error in up- and down-links, respectively, then show that the overall probability of bit error in the link is given by

$$P_e = P_{eu}(1 - P_{ed}) + P_{ed}(1 - P_{eu})$$
$$= P_{eu} + P_{ed} - 2P_{ed} P_{eu}$$

Explain whether P_{eu} or P_{ed} dominates the overall bit error probability.

27. Consider transmission at 1 Mbps in a 150 kHz channel bandwidth.

a. Calculate the bandwidth efficiency.

b. Assuming $E_b/N_0 = 10$ dB in an AWGN channel, determine the capacity.

c. Determine the required E_b/N_0 that will enable a 150 kHz bandwidth to support 1 Mbps?

9

Error Control Coding

A communication system should provide information transmission in a cost-effective way, at a rate, level of reliability and quality that are acceptable to a user. The key parameters available to the system designer are transmitted signal power, channel bandwidth and the receiver noise, which determine the energy per bit to noise PSD ratio E_b/N_0 required to achieve a given error probability. E_b/N_0 uniquely determines the bit error probability for a particular modulation system. Practical considerations, for example, signal attenuation, fading, shadowing, distortion and multi-user interference, usually limit the level of the E_b/N_0. Accordingly, for a modulation scheme of interest, it may not be possible to achieve acceptable bit error probability performance at the desired data rate. For a fixed value of the E_b/N_0, error-control coding is one of the available options for improving the reliability of data transmission, that is, with less bit errors. Error control coding may be exercised mainly as *source coding* that aims to remove the redundancy in the signal to be transmitted and to minimize the transmission rate with minimum loss in its information content; and *channel coding* that

aims to minimize the number of errors in the data delivered to the user due to signal attenuation, fading, shadowing, noise and distortion due to channel. In this chapter, we will consider only channel coding.

9.1 Introduction to Channel Coding

Figure 9.1 shows the block diagram of a communication system with source and channel coding. The *channel encoder* inserts redundant (parity) bits according to a prescribed rule to the binary symbols generated by the discrete source, or at the output of the source encoder at the transmitter. For real-time transmission, the encoded data is required to be transmitted at the same time as the data with no redundant bits. Consequently, the encoded data is transmitted at a rate higher than that without coding; hence the transmission rate and the channel bandwidth are increased. A *channel decoder* at the receiver exploits the redundancy provided by the parity bits to detect (error detection coding) and/or correct (error correction coding)

Digital Communications, First Edition. Mehmet Şafak.
© 2017 John Wiley & Sons Ltd. Published 2017 by John Wiley & Sons Ltd.
Companion website: www.wiley.com/go/safak/Digital_Communications

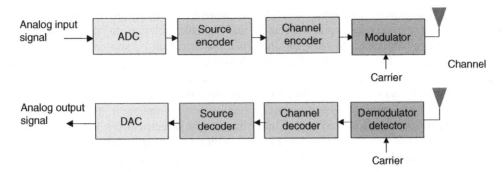

Figure 9.1 Block Diagram of a Digital Communication System with Source and Channel Coding.

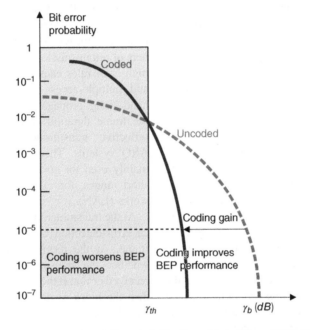

Figure 9.2 Comparison of Coded and Uncoded Bit Error Probability (BEP) Performance and the Coding Gain.

errors in the regenerated message digits. Parity bits are discarded following the detection/ correction process. Each duplex communication system using coding is equipped with a *codec*, which is the generic name for the encoder-decoder combination.

The use of redundant (parity) bits in error control coding increases transmission bandwidth and brings added complexity to the system, especially for the implementation of

decoding operations at the receiver. Therefore, the design of an error-control coding requires a trade-off between improved error performance and increased transmission bandwidth and system complexity.

Figure 9.2 shows the typical variation of the bit error probability with and without coding. Note that the duration of coded bits is shorter than the uncoded bit duration, since increased number of bits must be transmitted per unit time with coding. Consequently, bit error probability increases at the decoder input since the coded bit energy becomes lower than the uncoded bit energy. However, when E_b/N_0 is higher than a threshold level (see Figure 9.2), the decoder can correct some of these bit errors and hence the bit error probability performance is improved by coding. However, below the threshold level, the use of the parity bits decreases the coded bit energy to such levels that coding can not mitigate the increased bit errors.

For a given bit error probability level, the *coding gain* G_c is defined as the reduction in the $\gamma_b = E_b/N_0$ that can be realized by coding:

$$G_c(dB) = \gamma_{b,uncoded}(dB) - \gamma_{b,coded}(dB) \quad (9.1)$$

Coding gain is achieved at the expense of increase in transmission bandwidth. Note that Figure 9.2 may also be interpreted in terms of the decrease in the error probability achieved by coding for a given value of the γ_b.

Error control by coding may be exercised in mainly two forms: *Error-detection coding* aims to only detect some (bu not all) errors due to channel but no attempt is made for correcting them. These codes require smaller numbers of redundant bits (narrower transmission bandwidths and lower transmission rates) and their capability is limited to detection only of the bit errors. *Error-correction coding* (also called forward error correction coding, FEC) is designed to detect, locate and correct the erroneous bits in a codeword; therefore, they require higher percentage of redundant bits (increased transmission bandwidth and higher transmission rates). In hybrid automatic repeat request (H-ARQ) systems, error control coding is used for simultaneous error detection and correction. Then, the code may be designed to both detect and correct a few bit errors and only detect a larger number of bit errors.

Error detection codes are used mainly in ARQ systems, which are employed when the channel conditions are good and the observed bit error probabilities are very low. Then,

error-detection coding provides an economical and reliable solution to data transmission because it requires lower transmission rates/bandwidths. The ARQ systems necessitate the use of a return transmission path (feedback channel) with low bit errors and additional hardware to implement retransmission of frames with detected errors. Forward transmission bit rate must make allowance for retransmissions since the effective transmission rate would be decreased due to retransmissions.

The use of ARQ protocols are more appropriate for wired channels where transmission impairments like fading and shadowing are mostly non-existent. However, it may be unsuitable for wireless links where large numbers of retransmissions may be required due to high error rates caused by fading, shadowing and multiple-access interference. On the other hand, long propagation delays compared to the frame duration decreases the throughput (effective transmission rate) achieved by ARQ systems. Therefore, ARQ systems are mainly used for good channel conditions and short ranges, for example, in local area networks (LAN).

At the transmitter, data bits are encoded with an error-detection code and transmitted as a frame. At the receiver, the decoder decides whether the transmitted frame is correctly received or not. If the received frame is in error, then the block of data with error is discarded and the receiver requests the transmitter via the feed-back link to retransmit the erroneous frame. The transmitter retransmits that frame and until the receiver detects no error. Note that retransmission is requested when the data bits and/or the parity check bits are received in error, since the errors can not be located in the frame. If a frame is received correctly, then it is forwarded to the end user.

On the other hand, FEC codes are desirable when a reverse link is not available, the delay with ARQ is excessively long, retransmission strategy is not conveniently implemented and/or expected number of errors without corrections would require excessive number of retransmissions. Since error correction requires more parity check bits than for error detection, FEC codes require wider transmission bandwidths than error-detection codes. In channels where the SNR is relatively high (bit errors are very infrequent), then FEC coding wastes the channel resources. Hybrid ARQ (H-ARQ) systems provide a compromise between error-detection and FEC codes since they may correct a limited number of bit errors but detect more bit errors; hence they provide a tradeoff between the decreased number of retransmissions and increased transmission bandwidth.

FEC codes are mainly classified as *block codes*, which are used to encode a block of bits, and *convolutional codes* that treat several bits at a time at the encoder but decoding takes place at a longer time span. A (n, k) block FEC encoder maps each successive k-bit blocks of message into n-bit codewords, with the addition of $n-k$ $(n > k)$ redundant bits that are algebraically related to the appropriate k-bit message words. The n-bit codeword is said to have a block length of n. At the receiver, the received n-bit sequence is demodulated and passed through an FEC decoder. The FEC decoder compares the received n-bit sequence with all possible codewords that might have been transmitted and estimates the transmitted codeword. The following possibilities may arise at the decoder output: decoder detects no errors, detects and corrects bit errors, detects but cannot correct bit errors and detects no bit errors though errors are present.

The encoding operation in convolutional coding may be viewed as the discrete-time convolution of the input bit sequence with the encoder impulse response. The duration of the impulse response equals the memory of the encoder. The encoder accepts message bits as a continuous sequence and thereby generates a continuous sequence of encoded bits at a higher rate.

Figure 9.3 provides a simple classification of the error-control codes. Cyclic-redundancy check (CRC) highly capable of detecting bit errors though it requires a few tens of parity bits for thousands of data bits. As we will see in the following sections, error-correcting codes can be classified as random-error correcting and burst-error correcting codes. Random errors occur due to AWGN. However, in some channels with fading and shadowing, the errors may occur in bursts. For example, when a signal undergoes a fade for a given time duration, the receiver may observe a burst of errors during a fade and almost no errors for long durations of time. FEC coding strategies differ for mitigating random and burst errors. Bose-Chaudhuri and Hocquenhem (BCH) and convolutional codes are some notable examples for random error correcting codes. Reed-Solomon (RS) and low-density parity check (LDPC) codes are powerful in correcting burst errors. In practice, communication systems undergoing both random and burst errors use random- and burst-error correcting codes

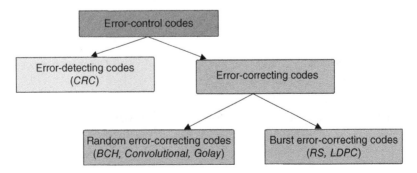

Figure 9.3 Classification of Error-Control Codes.

together; this is referred to as concatenated coding.

An effective method of implementing FEC coding is to combine coding with modulation. In this approach, coding is redefined as a process of imposing certain patterns on the transmitted signal. Trellis-coded modulation (TCM) is an example of coded-modulation techniques.

9.2 Maximum Likelihood Decoding (MLD) with Hard and Soft Decisions

Consider a baseband communication system, as shown in Figure 9.4, where a message word $\mathbf{m}_i$ with k equally-likely bits,

$$\mathbf{m} = [\mathbf{m}_0 \ \mathbf{m}_1 \ \cdots \ \mathbf{m}_i \ \cdots \ \mathbf{m}_{2^k-1}]$$
$$\mathbf{m}_i = [m_{i0} \ m_{i1} \ \cdots \ m_{ik-1}], \ m_{ik} \in \{0,1\},$$
$$i = 0,1,\cdots,2^k-1$$

(9.2)

is encoded by an encoder to obtain one of the 2^k n-bit codewords:

$$\mathbf{c} = [\mathbf{c}_0 \ \mathbf{c}_1 \ \cdots \ \mathbf{c}_i \ \cdots \ \mathbf{c}_{2^k-1}]$$
$$\mathbf{c}_i = [c_{i0} \ c_{i1} \ \cdots \ c_{in-1}], \ c_{ik} \in \{0,1\},$$
$$i = 0,1,\cdots,2^k-1$$

(9.3)

where $n-k$ parity bits in (9.3) are determined according to a prescribed rule so as to mitigate bit errors in the channel. For BPSK baseband modulation we assume that $\sqrt{R_c E_b}$ is transmitted for $c_{ij}=1$ and $-\sqrt{R_c E_b}$ for $c_{ij}=0$. The transmitted symbol vector may be written as

$$\mathbf{c}_i^t = \begin{bmatrix} c_{i0}^t & c_{i1}^t & \cdots & c_{ij}^t & \cdots & c_{in-1}^t \end{bmatrix}$$
$$c_{ij}^t = \sqrt{R_c E_b}\left(-1 + 2c_{ij}\right)$$

(9.4)

where $R_c E_b$ denotes the energy of a coded bit. Since the dimensionality of the BPSK signal is 1, the components of the observation vector at the matched-filter demodulator output consist of single elements for each signaling interval:

$$\mathbf{x} = [x_0 \ x_1 \ \cdots \ x_{n-1}]$$
$$x_j = c_{ij}^t + w_j, \ j = 0,1,\cdots n-1$$

(9.5)

where w_j denotes j-th noise sample at the matched-filter demodulator output. The observation vector given by (9.5) may differ from the transmitted code vector $\mathbf{c}_i^t$ due to channel impairments. Given the observation vector $\mathbf{x}$, the detector/decoder is required to make an estimate $\hat{\mathbf{m}}$ of the transmitted message vector by simultaneous use of all elements of the observation vector. Because of the one-to-one correspondence between the message vector $\mathbf{m}_i$ and the code vector $\mathbf{c}_i$, the decoder first estimates the code vector $\hat{\mathbf{c}} = \mathbf{c}_i$ and then decides as $\hat{\mathbf{m}} = \mathbf{m}_i$. Otherwise, a decoding error is made at the receiver. The decoding rule for choosing the estimate $\hat{\mathbf{c}}$, given the received vector $\mathbf{x}$, is said

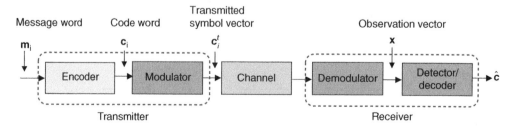

Figure 9.4 Encoding and Modulation of a Message Word.

to be optimum when the probability of decoding error is minimized, assuming that all transmitted codewords are equally probable. For equiprobable messages, the probability of decoding error is minimized if the estimate $\hat{m}$ is chosen to maximize the likelihood function $f(x|c_i)$, which denotes the conditional probability of receiving observation vector x, given that c_i was sent. The transmitted code vector c_i^t and the observation vector $\mathbf{x}$ differ from each other because of errors due to channel impairments. In a memoryless channel, the demodulator output in a given time interval depends only on the signal transmitted in that interval. The likelihood function $f(\mathbf{x}|c_i)$ may therefore be written as the product of the conditional likelihood functions of its n components:

$$f(\mathbf{x}\,|\,\mathbf{c}_i) = \prod_{j=0}^{n-1} f\left(x_j\,|\,c_{i,j}\right), \quad i=0,1,\cdots 2^k-1$$

$$(9.6)$$

For maximum likelihood decoding (MLD) in memoryless channels, likelihood functions are determined for all code words and the codeword with maximum likelihood is estimated as the transmitted codeword:

$$\hat{\mathbf{c}} = \max_{i=0,1,\cdots 2^k-1} f(\mathbf{x}\,|\,\mathbf{c}_i) \qquad (9.7)$$

Decoding is usually carried out by quantizing the components x_j, $j = 0,\ 1,\dots,n-1$ of the observation vector with one bit (**1** or **0**) (hard-decision decoding) or with more than one bit (soft-decision decoding). In *hard-decision MLD*, each element of the observation vector is quantized with **1** or **0** by using a hard-limiter. The received vector $\mathbf{x}$ quantized with 1-bit is compared with each possible transmitted code vector c_i, that is, Hamming distances between 1-bit quantized observation vector and all allowed codewords are calculated. The codeword with the smallest Hamming distance to the 1-bit quantized observation vector is estimated as the transmitted codeword.

A discrete memoryless channel (DMC) is defined as a channel with discrete input symbols and with no memory, that is, with symbol-by-symbol demodulation. The DMC shown in Figure 9.5 represents the combination of the modulator, the channel, and the demodulator. A DMC is completely described by the set of transition probabilities p_{ji}, which denotes the probability of receiving symbol j at the demodulator output, given that symbol i at the modulator input was sent. The simplest

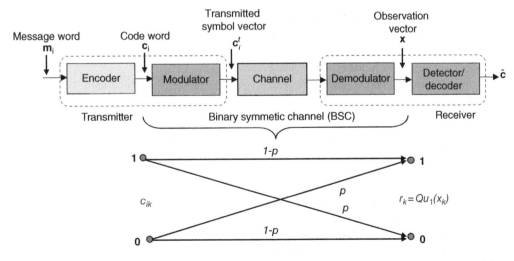

Figure 9.5 Binary Symmetric Channel (BSC) Where the Observation Vector Is Quantized by 1-Bit.

DMC results from the use of binary input and binary output symbols, the so-called binary symmetric channel (BSC). A BSC relates the transmitted bits to the received bits at the demodulator output. Therefore, the modulator input and demodulator output have only the bits **0** and **1**. Since the observation vector is quantized with a single bit, a hard decision is made at the demodulator output. Channel is symmetric since the transition probability from **1** to **0** and from **0** to **1** is assumed to be the same.

In an AWGN channel, the BSC is completely described by the transition probability p:

$$f(r_j|c_{i,j}) = \begin{cases} p & \text{if } r_j \neq c_{i,j} \quad j=0,1,\cdots,n-1 \\ 1-p & \text{if } r_j = c_{i,j} \end{cases}, \quad i=0,1,\cdots,2^k-1$$

(9.8)

where **r** denotes the 1-bit quantized version of the observation vector **x**:

$$\mathbf{r} = Qu_1(\mathbf{x}) \qquad (9.9)$$

For BPSK, the transition probability p is given by

$$p = Q\left(\sqrt{2E_b/N_0}\right) \leq 0.5 \qquad (9.10)$$

Suppose also that the received vector **r** differs from the transmitted code vector $\mathbf{c}_i$ in exactly d_i positions, that is, d_i is the Hamming distance between vectors **x** and $\mathbf{c}_i$. Then, using (9.8), the likelihood function given by (9.6) may simply be written as

$$f(\mathbf{r}|\mathbf{c}_i) = p^{d_i}(1-p)^{n-d_i} = \left(\frac{p}{1-p}\right)^{d_i}(1-p)^n$$

(9.11)

Noting that $p < 1/2$, $p/(1-p) < 1$ and the factor $(1-p)^n$ is constant for all codewords, $f(\mathbf{r}|\mathbf{c}_i)$ is maximized when Hamming distance d_i between **r** and $\mathbf{c}_i$ is minimum. Then, according to (9.7), MLD rule for BSC consists of chosing

the estimate $\hat{\mathbf{c}}$ that minimizes the Hamming distance between the 1-bit quantized version **r** of the observation vector for all possible transmitted vectors $\mathbf{c}_i$. Hard-decision decoding is simpler to implement than soft-decision decoding but causes an irreversible loss of information at the decoder because of quantization with a single bit.

In *soft-decision MLD decoding*, elements of the observation vector are quantized with more than one bit. This is achieved by a multiple-bit analog-to digital convertor (ADC) at the decoder input. For example in Figure 9.6, a 2-bit ADC is used, where the first bit denotes the sign $\pm$ and the second bit shows the reliability of the received bit. Information carried by the extra quantization bits showing the reliability of the matched-filter demodulator output is therefore passed to the decoder. In MLD with soft decisions, the likelihood function (9.16) for i-th codeword may be written as

$$f(\mathbf{x}|\mathbf{c}_i) = \prod_{j=0}^{n-1} f(x_j|c_{i,j})$$

$$= \prod_{j=0}^{n-1} \frac{1}{\sqrt{\pi N_0}} \exp\left(-\frac{|x_j - c_{i,j}^t|^2}{N_0}\right)$$

$$= \frac{1}{(\pi N_0)^{n/2}} \exp\left(-\frac{d_{E,i}^2}{N_0}\right)$$

$$d_{E,i}^2 = \|\mathbf{x} - \mathbf{c}_i^t\|^2 = \|\mathbf{x}\|^2 + \|\mathbf{c}_i^t\|^2 - 2\mathbf{x}.\mathbf{c}_i^t$$

(9.12)

Note that the likelihood function is maximized when the Euclidean distance $d_{E,i}$ is minimized. Since $\|\mathbf{c}_i^t\|^2$ is the same for all codewords and $\|\mathbf{x}\|^2$ is independent of the codeword, MLD corresponds to the minimum Euclidean distance and to the maximum the metric U_i, $i=0,1,\ldots,2^k-1$, which denotes the projection of the observation vector **x** onto the i-th codeword $\mathbf{c}_i^t$.

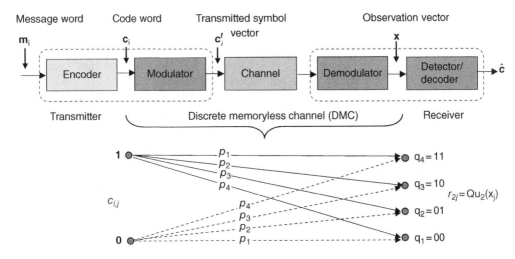

Figure 9.6 Channel Transition Diagram For Binary Input 4-ary Output (4-Level Quantizer: 2 Bits/Sample) Discrete Memoryless Channel (DMC).

$$\hat{\mathbf{c}} = \max_{i=0,1,\cdots 2^k-1} f(\mathbf{x}\,|\mathbf{c}_i) = \min_{i=0,1,\cdots 2^k-1} d_{E,i}$$

$$= \max_{i=0,1,\cdots 2^k-1} U_i$$

$$U_i = \mathbf{x}.\mathbf{c}_i^t / \sqrt{R_c E_b} = \sum_{r=0}^{n-1} x_r \, c_{i,r}^t$$

$$(9.13)$$

In an AWGN channel, the likelihood function is maximized when the corresponding Euclidean distance is minimum. The MLD for soft-decision consists of choosing the codeword, with shortest Euclidean distance to the observation vector $\mathbf{x}$. In summary, in the AWGN channel, a ML decoder with soft-decision chooses the codeword with minimum Euclidean distance to the received sequence:

$$\hat{\mathbf{c}} = \mathbf{c}_j \ \text{ if } \ U_j = \max_{i=0,1,\cdots 2^k-1} U_i \qquad (9.14)$$

The Euclidean distances between the observation vector, quantized with more than 1-bit, and all allowed codewords are calculated. The codeword with the smallest Euclidean distance is estimated as the transmitted codeword. Soft-decision decoding offers (approximately 2 dB) performance improvement over hard-decision decoding, but makes the implementation of the decoder more complex.

Example 9.1 Relationship Between Hamming and Euclidean Distances.
Euclidean distance between two vectors $\mathbf{x}$ and $\mathbf{y}$, each of dimension n,

$$\mathbf{x} = [x_0 \ x_1 \ \cdots \ x_{n-1}], \ x_k \in \{-1, \ +1\}$$
$$\mathbf{y} = [y_0 \ y_1 \ \cdots \ y_{n-1}], \ y_k \in \{-1, \ +1\}$$
$$(9.15)$$

may be written as

$$d_E^2(\mathbf{x},\mathbf{y}) = \|\mathbf{x}-\mathbf{y}\|^2 = \sum_{i=0}^{n-1}(x_i-y_i)^2 \qquad (9.16)$$

The Euclidean distance between x_i and y_i, each with energy E, equals

$$d_E(x_i,y_i) = |x_i - y_i| = \begin{cases} 0 & x_i = y_i \\ v\sqrt{E} & otherwise \end{cases}$$
$$(9.17)$$

where $v = 2$ if $\mathbf{x}$ and $\mathbf{y}$ are antipodal and $v = \sqrt{2}$ if they are orthogonal. If the Hamming distance is given by d_H between the two vectors, from (9.16) and (9.17), Euclidean and Hamming distances are interrelated by:

$$d_E^2 = d_H \left(\nu \sqrt{E} \right)^2 = \nu^2 E \; d = \begin{cases} 4 \, E \, d & antipodal \\ 2 \, E \, d & orthogonal \end{cases}$$
(9.18)

Example 9.2   Decode-and-Forward Relay (or Bent-Pipe (Non-Generative) Satellite Relay) Channel.

In some satellite communication channels, uplinks and downlinks are isolated from each other. In such channels, uplink signals are demodulated, frequency converted and retransmitted. Similar applications also exist in so-called decode-and-forward relay channels. These channels can be represented by two cascaded BSCs, as shown in Figure 9.7, when they use hard-decision decoding.

From Figure 9.7, the probability of correct reception and probability of error for binary symbols may be written as

$$P_e = (1-p_1)p_2 + (1-p_2)p_1$$
$$P_c = 1 - P_e = p_1 \, p_2 + (1-p_1)(1-p_2)$$
(9.19)

Note that probability of correct reception implies correct reception or error in both links at the same time. However, probability of error consists of two terms accounting for correct reception in the uplink though downlink is in error and error in the uplink while the downlink signal is correctly received.

Example 9.3   Metrics For Hard- and Soft-Decision Decoding.

Consider BPSK modulated codewords defined by (9.3) and (9.4) where $R_c E_b = 1$ is assumed. The observation vector $\mathbf{x}$ differs from $\mathbf{c}_i'$ by the additive noise term in an AWGN channel. Quantizing the observation vector (9.5), using a 2-bit ADC (soft-decision decoding), one gets

$$\mathbf{r}_2 = Qu_2(\mathbf{x}) = [r_{20} \; r_{21} \; \cdots \; r_{2j} \; \cdots \; r_{2,n-1}],$$
$$r_{2j} \in \{00, \, 01, \, 10, \, 11\}$$
(9.20)

For quantizing the observation vector with 2 bits, four decision regions are designated: $x_j > 0.5, 0.5 \geq x_j > 0, 0 \geq x_j > -0.5$ and $x_j \leq -0.5$. Each component r_{2j} of the observation vector may then be represented by one of the four representation levels: -0.75 V for $q_1 = 00$, -0.25 V for $q_2 = 01$, 0.25 V for $q_3 = 10$, and 0.75 V for $q_4 = 11$. As a symmetric discrete memoryless channel (DMC), the transition probabilities shown in Figure 9.6 satisfy $\sum_{r=1}^{4} p_r = 1$ where p_r denotes the conditional probability $p_r = P(r_{2j} = q_r | c_{i,j} = 0)$, $r = 1, 2, 3, 4$.

In soft-decision decoding, the path metrics U_i for the AWGN channel are defined by (9.13) as the inner product of the received word and the codewords. Maximization of U_i is equivalent to finding the codeword $\mathbf{c}_i'$ that is

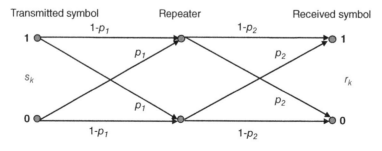

Figure 9.7   Two Binary Symmetric Channels (BSC) in Cascade.

closest to **x** in terms of the Euclidean distance, that is, with minimum Euclidean distance between c_i^t and **x**. However, instead of using the Euclidean distance or U_i, it may be more convenient to work with normalized bit metrics defined as [1]

$$M(p_i) = \frac{\log_2(p_i) - \log_2(p_4)}{\log_2(p_1) - \log_2(p_4)}, \quad i = 1, 2, 3, 4$$

(9.21)

This definition limits the values of the bit metrics between 0 and 1 so that $M(p_1) = 1 > M(p_2) > M(p_3) > M(p_4) = 0$. In this case, the highest value of the bit metric corresponds to the most-likely transmitted bit. Alternatively, one may also define the following normalized bit metrics

$$M(p_i) = \frac{\log_2(p_i) - \log_2(p_1)}{\log_2(p_4) - \log_2(p_1)}, \quad i = 1, 2, 3, 4$$

(9.22)

with $M(p_1) = 0 < M(p_2) < M(p_3) < M(p_4) = 1$ so that the lowest value of the metric corresponds to the most-likely transmitted bit, as in hard-decision decoding. Hereafter we will use the metric (22).

For example, assume that the conditional (transition) probabilities are as given in Table 9.1 in an AWGN channel. Using the values of p_i shown in Table 9.1 and (9.22), the values of the bit metrics corresponding to

conditional log-likelihood functions are shown in Table 9.2 and Figure 9.8.

Let us now consider a coding scheme, the so-called repetition coding, where bit 1 is encoded into 111 and bit 0 into 000. Any transmission error in a received codeword changes a 1 to 0, or vice versa. The codewords transmitted using BPSK modulation with $R_c E_b = 1$ are as follows:

$$c_1^t = (\, +1 \quad +1 \quad +1\,)$$
$$c_2^t = (-1 \quad -1 \quad 1\,)$$

(9.23)

The observation vector (sampled outputs of the matched filter demodulator) is assumed to be given by

Table 9.1 Transition Probabilities For Two-Bit Quantized Soft-Decision Decoding in a Symmetric DMC.

	$q_1 = 00$	$q_2 = 01$	$q_3 = 10$	$q_4 = 11$
$c_{1j} = 0$	$p_1 = 0.6$	$p_2 = 0.3$	$p_3 = 0.08$	$p_4 = 0.02$
$c_{2j} = 1$	$p_4 = 0.02$	$p_3 = 0.08$	$p_2 = 0.3$	$p_1 = 0.6$

Table 9.2 Normalized Bit Metrics $M(p_i)$, $i = 1, 2, 3, 4$, For 2-Bit Quantized Soft-Decision Decoding in a Symmetric DMC Based on Table 9.1 and (9.22).

	$q_1 = 00$	$q_2 = 01$	$q_3 = 10$	$q_4 = 11$
$c_{1j} = 0$	0	0.204	0.592	1
$c_{2j} = 1$	1	0.592	0.204	0

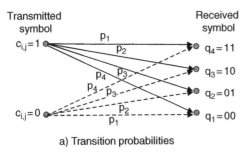

a) Transition probabilities

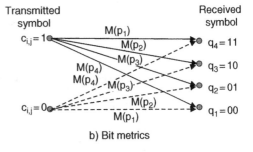

b) Bit metrics

Figure 9.8 Channel Transition Diagram and Bit Metrics For Binary-Input 4-ary Output DMC.

$$\mathbf{x} = (0.45 \quad 0.6 \quad -0.3) \qquad (9.24)$$

In hard-decision decoding, the observation vector is quantized by one bit, that is, each sample is component-wise detected by a hard-limiter as 1 or 0, at the decoder input. Hence, 1-bit quantized observation vector at the decoder input may be written as

$$\mathbf{r} = Q u_1(\mathbf{x}) = (1 \quad 1 \quad 0) \qquad (9.25)$$

The decoder uses the Hamming distance between $\mathbf{r}$ and 111 and 000 as the metric and decides that 1(0) is transmitted if $d(\mathbf{r}, 111)$ is smaller (larger) than $d(\mathbf{r}, 000)$. Since

$$d(\mathbf{r},\ 111) = 1 \qquad d(\mathbf{r},\ 000) = 2 \qquad (9.26)$$

the decoder decides that 1 is transmitted.

Now let us consider soft decision decoding with 2-bit quantization, where the representation levels are ± 0.25 and ± 0.75. Therefore, when the observation vector given by (9.24) is quantized with 2 bits, one gets

$$\mathbf{r}_2 = (10 \quad 11 \quad 01) \qquad (9.27)$$

Noting that 10 corresponds to the quantized sample value of 0.25 V, 11 to 0.75 V and 01 to -0.25 V, Euclidean distances are calculated between the received soft information $\mathbf{r}_2$ and the valid codewords:

$$d_E^2(\mathbf{r}_2, 111) = (1-0.25)^2 + (1-0.75)^2$$
$$+ (1+0.25)^2 = 2.2$$
$$d_E^2(\mathbf{r}_2, 000) = (-1-0.25)^2 + (-1-0.75)^2$$
$$+ (-1+0.25)^2 = 5.2$$
$$(9.28)$$

Since $d_E^2(\mathbf{r}, 111) < d_E^2(\mathbf{r}, 000)$ a soft-decision decoder decides that 1 is transmitted.

Using directly U_i defined by (9.13), one gets U_0 for the codeword 000 and U_1 for the codeword 111

$$000: \ U_0 = \sum_{r=0}^{2} r_{2,r}\, c_{0,r}^t = -r_{2,0} - r_{2,1} - r_{2,2} = -0.75$$

$$111: \ U_1 = \sum_{r=0}^{2} r_{2,r}\, c_{1,r}^t = +r_{2,0} + r_{2,1} + r_{2,2} = +0.75$$
$$(9.29)$$

Since $U_1 > U_0$, the soft-decision decoder decides that the codeword 111 is transmitted. Alternatively, using the bit metric definition by (9.22), soft-decision decoding of (9.27) is based on the calculation of the metrics for the codewords 111 and 000 instead of direct calculation of the Euclidean distances. The path metrics for the codewords 000 and 111 are found as the sum of bit metrics corresponding to (9.27):

$$000: \ M(p_3) + M(p_4) + M(p_2) = 0.592 + 1$$
$$+ 0.204 = 1.796$$
$$111: \ M(p_3) + M(p_4) + M(p_2) = 0.204 + 0$$
$$+ 0.592 = 0.796$$
$$(9.30)$$

The decoder decides that the codeword 111 is transmitted, since it has the lowest path metric.

9.3   Linear Block Codes

A block encoder outputs n bits for every k bits at its input (see Figure 9.9). A block encoder

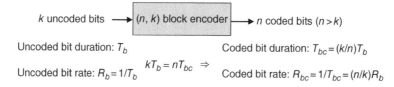

Figure 9.9   Input-Output Relationship in a (n, k) Block Encoder.

stores a block of k uncoded bits in a buffer, encodes them into n coded bits with the addition of $n-k$ parity check bits and transmits them. The requirement for real-time transmission dictates that the transmision time of n coded bits remains the same as for k uncoded bits:

$$kT_b = nT_c \quad \Rightarrow \quad R_{bc} = \frac{1}{T_c} = \frac{n}{k}R_b = \frac{R_b}{R_c} \geq R_b \tag{9.31}$$

where T_b and T_c denote respectively uncoded and coded bit durations. Similarly, R_b and R_{bc} are uncoded and coded bit rates. Since $n \geq k$, the code rate

$$R_c = \frac{k}{n} \leq 1 \tag{9.32}$$

is always less than or equal to unity and $R_c = 1$ implies uncoded transmission. The so-called redundancy is defined as the ratio of the number of redundant (parity) bits to the number of data bits:

$$\frac{n-k}{k} = \frac{1}{R_c} - 1 \tag{9.33}$$

Note that the increase in the channel data (transmission) rate, $R_{bc} = (n/k)R_b \geq R_b$ increases the channel bandwidth by a factor of $n/k = 1/R_c$. In summary, coding mitigates channel bit errors

at the expense of increased transmission rate (bandwidth).

A linear block code is a k-dimensional subspace of an n-dimensional space. The encoding maps each of the 2^k message sequences of k-bit long to one of the 2^n sequences of n-bit long (see Figure 9.10). Since there can be only 2^k codewords for 2^k message sequences, the choice of 2^k codewords among 2^n sequences is a critical issue. This choice should lead to a good code, that is, with best performance.

Here we consider codes where the mapping transformation between 2^k message sequences and 2^n sequences is linear. Linearity follows from the closure property: modulo-2 sum of two message words m_i and m_j represents a new message word and this implies the modulo-2 sum of the corresponding codewords (c_i and c_j):

$$If \quad m_i \Rightarrow c_i \quad and \quad m_j \Rightarrow c_j$$
$$then \quad m_i \oplus m_j \ (also \ a \ message \ word)$$
$$\Rightarrow c_i \oplus c_j \ (also \ a \ codeword) \tag{9.34}$$

In linear codes, all zero sequence is always a codeword since $c_i \oplus c_i \equiv 0$. The full set of codewords is generated by letting the message vector $\mathbf{m}$ range through the set of all 2^k binary k-tuples ($1 \times k$ vectors).

Example 9.4 (4, 2) Binary Linear Block Code. Assume a (4, 2) binary linear block code. 2^k message sequences of length $k = 2$ are given as follows:

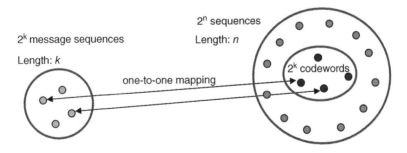

Figure 9.10 (n, k) Block Coding.

$$\mathbf{m} = \{00 \quad 01 \quad 10 \quad 11\} \qquad (9.35)$$

2^n sequences of length $n = 4$ are given by

0000 0001 0010 0011 0100 0101 0110 0111
1000 1001 1010 1011 1100 1101 1110 1111
$$\qquad (9.36)$$

Let us choose four codewords of length n with different mappings to (9.35), as follows:

$$\mathbf{c} = \{0000 \quad 0101 \quad 1010 \quad 1111\}$$
$$\mathbf{C} = \{0101 \quad 0000 \quad 1010 \quad 1111\} \qquad (9.37)$$

Linearity of the code depends not only on the codewords, also on the mapping. We observe that the code $\mathbf{c}$ is linear, since it satisfies (9.34). However, $\mathbf{C}$ does not satisfy (9.34); hence it is not linear.

9.3.1 Generator and Parity Check Matrices

Let $\mathbf{m}$ constitute a block of k arbitrary message bits

$$\mathbf{m} = [m_0 \quad m_1 \quad \cdots \quad m_{k-1}], \quad m_i \in \{0,1\} \quad (9.38)$$

Since each element of $\mathbf{m}$ can be either 0 or 1, there can be 2^k distinct message words. A (n, k) linear block code is called systematic if k bits of the n-bits codeword are always identical to the message bit sequence. Systematic block codes simplify implementation of the decoder. A (n, k) linear block systematic encoder produces an n-bit codeword whose elements are denoted by

$$\mathbf{c} = [c_0 \quad c_1 \quad \cdots \quad c_{n-1}]$$
$$c_i = \begin{cases} b_i & i = 0, \ 1, \dots, n-k-1 \\ m_{i+k-n} & i = n-k, \ n-k+1, \dots, n-1 \end{cases}$$
$$\qquad (9.39)$$

where the $n-k$ parity-check bits b_i are computed from the message bits according to a prescribed encoding rule that determines the mathematical structure of the code. The systematic code vector $\mathbf{c}$ may then be expressed as a partitioned row vector in terms of the parity-check vector $\mathbf{b}$ and the message vector $\mathbf{m}$:

$$\mathbf{c} = \begin{bmatrix} \mathbf{b} & \vdots & \mathbf{m} \end{bmatrix}$$
$$= \begin{bmatrix} \underbrace{b_0 \ b_1 \ \cdots \ b_{n-k-1}}_{n-k \ parity \ bits} & \underbrace{m_0 \ m_1 \ \cdots \ m_{k-1}}_{k \ message \ bits} \end{bmatrix}$$
$$\mathbf{b} = [b_0 \ b_1 \ \cdots \ b_{n-k-1}]$$
$$\qquad (9.40)$$

The $(n-k)$ parity bits are linear sums of the k message bits:

$$b_i = P_{i0} m_0 + P_{i1} m_1 + \dots + P_{i,k-1} m_{k-1},$$
$$i = 0, 1, \cdots n-k-1$$
$$\qquad (9.41)$$

where the coefficients are defined as follows:

$$P_{ij} = \begin{cases} 1 & if \ b_i \ depends \ on \ m_j \\ 0 & otherwise \end{cases} \qquad (9.42)$$

The coefficients P_{ij} are chosen in such a way that the rows of the generator matrix become linearly independent, hence parity equations are unique. This leads to a code that can better mitigate bit errors. Using (9.41), the parity bits for a systematic code may be written in matrix form:

$$\mathbf{b} = [b_0 \ b_1 \ \cdots \ b_{n-k-1}] = \mathbf{mP}$$
$$\mathbf{P} = \begin{bmatrix} P_{00} & P_{01} & \cdots & P_{0,n-k-1} \\ P_{10} & P_{11} & \cdots & P_{1,n-k-1} \\ \vdots & \vdots & \ddots & \vdots \\ P_{k-1,0} & P_{k-1,1} & \cdots & P_{k-1,n-k-1} \end{bmatrix}$$
$$\qquad (9.43)$$

where $\mathbf{P}$ of size $k \times (n-k)$ is called as the parity matrix. Using (9.38) and (9.43), the code vector given by (9.40) may be rewritten as

$$\mathbf{c} = \begin{bmatrix} \mathbf{b} & \vdots & \mathbf{m} \end{bmatrix} = \begin{bmatrix} \mathbf{mP} & \vdots & \mathbf{m} \end{bmatrix} = \mathbf{mG}$$

$$\mathbf{G} = \begin{bmatrix} \mathbf{P} & \vdots & \mathbf{I}_k \end{bmatrix}$$

(9.44)

where $\mathbf{I}_k$ is the $k \times k$ identity matrix and $\mathbf{G}$ is defined as the $k \times n$ generator matrix. The full set of codewords may be generated via $\mathbf{c} = \mathbf{mG}$ by letting the message vector $\mathbf{m}$ span through the set of all 2^k binary k-tuples ($1 \times k$ vectors). The generator equation, $\mathbf{c} = \mathbf{mG}$, uniquely provides the description and operation of a linear block code. The k rows of the generator matrix $\mathbf{G}$ are linearly independent, that is, it is not possible to express any row of the matrix $\mathbf{G}$ as a linear combination of the remaining rows.

The so-called $(n-k) \times n$ parity check matrix $\mathbf{H}$ is defined by

$$\mathbf{H} = \begin{bmatrix} \mathbf{I}_{n-k} & \vdots & \mathbf{P}^T \end{bmatrix}$$

(9.45)

where $(n-k) \times k$ matrix $\mathbf{P}^T$ denotes the transpose of $\mathbf{P}$, and $\mathbf{I}_{n-k}$ is the $(n-k) \times (n-k)$ identity matrix. From (9.44) and (9.45), one may show that

$$\mathbf{cH}^T = \mathbf{mGH}^T \equiv \mathbf{0}$$

$$\mathbf{GH}^T = \begin{bmatrix} \mathbf{P} & \vdots & \mathbf{I}_k \end{bmatrix} \begin{bmatrix} \mathbf{I}_{n-k} \\ \cdots \\ \mathbf{P}^T \end{bmatrix}$$

(9.46)

$$= \mathbf{P}\,\mathbf{I}_{n-k} + \mathbf{I}_k\,\mathbf{P}^T$$

$$= \mathbf{P}^T + \mathbf{P}^T = \mathbf{0}$$

In view of (9.46), the decoding operation may easily be carried out at the receiver. A received code word $\mathbf{r}$, single-bit quantized version of the observation vector (see (9.9), may differ from the transmitted codeword $\mathbf{c}$ if there is transition error in one or more of the n coded bits. Then, in view of (9.46), one may state that $\mathbf{r}$ is a valid codeword if $\mathbf{s} = \mathbf{r}\,\mathbf{H}^T = \mathbf{0}$ but transmitted codeword $\mathbf{c}$ is received in error if $\mathbf{s} = \mathbf{r}\,\mathbf{H}^T \neq \mathbf{0}$ (see Figure 9.11), where $\mathbf{s}$ is called as the syndrome.

From (9.46), one may write $\mathbf{GH}^T = \mathbf{HG}^T = \mathbf{0}$, which suggests that every (n, k) linear block code with generator matrix $\mathbf{G}$ and parity-check matrix $\mathbf{H}$ has a dual code $(n, n-k)$, with generator matrix $\mathbf{H}$ and parity check matrix $\mathbf{G}$.

Note that the closure property given by (9.34) may be rewritten in terms of the generator matrix as follows:

$$\mathbf{c}_i \oplus \mathbf{c}_j = \mathbf{m}_i \mathbf{G} \oplus \mathbf{m}_j \mathbf{G} = (\mathbf{m}_i \oplus \mathbf{m}_j)\mathbf{G}$$

(9.47)

This implies that the modulo-2 sum of any two code words ($\mathbf{c}_i$ and $\mathbf{c}_j$) is another code word. Similarly, modulo-2 sum of any two message words ($\mathbf{m}_i$ and $\mathbf{m}_j$) represents a new message word. All-zero code vector is always a code word since $\mathbf{c}_i \oplus \mathbf{c}_j \equiv \mathbf{0}$.

Example 9.5 Error Detection by Parity-Checking.

Parity checking is the simplest error detection $(k+1, k)$ block code. A parity-check bit is added at the end of a message word, to make the number of **1**'s in the codeword thus formed an even number (even parity) or an odd number (odd parity). At the decoder output, if the number of decoded bits has an even number of **1**'s, then the received message is assumed to be

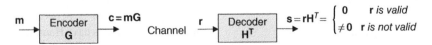

Encoder G: $m \rightarrow$ Encoder G $\xrightarrow{c=mG}$ Channel $\xrightarrow{r}$ Decoder $\mathbf{H}^T$ $\xrightarrow{s=rH^T=}$ $\begin{cases} 0 & r \text{ is valid} \\ \neq 0 & r \text{ is not valid} \end{cases}$

Figure 9.11 Encoding and Syndrome Decoding For Block Codes.

correctly received. Otherwise, a transmission error is assumed. A parity-checking decoder can detect only an odd number of erroneously received bits. An even number of errors preserves valid parity and are undetected. Code rate is $R_c = k/(k + 1)$.

Consider, for example, $k = 3$ bit message words

$$\mathbf{m} = [000 \ 001 \ 010 \ 011 \ 100 \ 101 \ 110 \ 111] \tag{9.48}$$

The parity check bit is calculated as the sum of the message bits: $b_0 = m_0 \oplus m_1 \oplus m_2$ (even parity) and the corresponding generator matrix is given by

$$\mathbf{G} = \left[\mathbf{P} \ \vdots \ \mathbf{I}_k\right] = \begin{bmatrix} 1 & 1 & 0 & 0 \\ 1 & 0 & 1 & 0 \\ 1 & 0 & 0 & 1 \end{bmatrix} \tag{9.49}$$

Codewords of this (4,3) even-parity code are obtained using (9.48) and (9.49) as follows:

$$\mathbf{c} = [0000 \ 1001 \ 1010 \ 0011 \ 1100 \ 0101 \ 0110 \ 1111] \tag{9.50}$$

This code can not correct any bit error since the location of the erroneously received bits within the codeword can not be determined. For example, assume that the codeword 0000 is transmitted and 0100 is received at the demodulator output. Noting that the codewords 0000, 0101, 0110 and 1100 differ from the received word 0100 by only a single bit, one of these four codewords might have been transmitted with equal likelihood.

This code can detect all single and triple errors. However, it can not detect two- and four-bit errors since parity is preserved. Hence the probability of undetected errors in a 4-bit codeword is given by the sum of the probabilities of two- and four-bit errors:

$$P_u(4) = P(2, 4) + P(4, 4)$$
$$= 6p^2 - 12p^3 + 7p^4 \approx 6p^2 \tag{9.51}$$

where p denotes the transition probability for coded bits and $P(i, n)$ shows the probability of i errors in a n-bit codeword:

$$P(i, n) = \binom{n}{i} p^i (1-p)^{n-i} \qquad \binom{n}{i} = \frac{n!}{i!(n-i)!} \tag{9.52}$$

The probability of correct reception of a 4-bit codeword implies no bit errors:

$$P_c(4) = P(0, 4) = (1-p)^4 \tag{9.53}$$

The probability of detecting an error in a 4-bit codeword is then given by

$$P_d(4) = 1 - P_c(4) - P_u(4) = P(1, 4) + P(3, 4) \cong 4p \tag{9.54}$$

For $p = 10^{-3}$, the undetected error probability, probability of correct reception and the probability of detecting an error are found to be $P_u(4) = 6 \times 10^{-6}$, $P_c(4) = 0.996$ and $P_d(4) = 4 \times 10^{-3}$.

Example 9.6 Repetition Coding.

Repetition coding is a brute-force approach to binary communications. Each message is encoded into a block of n identical bits, producing an $(n, 1)$ block code, for example, 11....1 for 1 and 00....0 for 0. Any transmission error alters the transmitted codeword by changing a 1 to 0, or vice versa. Received sequence is decoded as a 0 or 1 if more than half of received bits are 0's or 1's, respectively. Let us consider a (3, 1) repetition coding, which has two codewords: An all-zero codeword 000 and an all-one codeword 111. A total of eight 3-bit words may be located at the vertices of a cube as shown in Figure 9.12, as a function of the Hamming distance between them. The 3-bit words 111 and 000 (shown as squares) are chosen as the codewords since they have a maximum Hamming distance of 3 between them. The points shown in circles have Hamming distance of 2 with 111 but a Hamming distance of 1 with 000. Similarly, the points shown in triangles

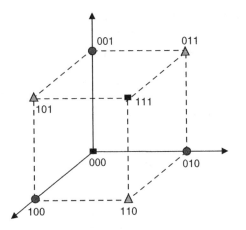

Figure 9.12 Representation of a (3,1) Repetition Code By the Vertices of a Cube.

Table 9.3 Likelihoods of the Received Words When 111 is Transmitted in a (3,1) Repetition Code.

Received bit sequence	Prob. of reception	Decoded sequence (majority voting)
111	$(1-p)^3$	1
110	$(1-p)^2 p$	1
101	$(1-p)^2 p$	1
100	$(1-p)p^2$	0
001	$(1-p)p^2$	0
010	$(1-p)p^2$	0
011	$(1-p)^2 p$	1
000	p^3	0

have Hamming distance of 2 with 000 but a Hamming distance of 1 with 111.

The message vector in the (3,1) repetion code consists of a single binary symbol, that is, $\mathbf{m} = 0$, or $\mathbf{m} = 1$. The generator matrix may be written as

$$\mathbf{G} = \begin{bmatrix} \mathbf{P} & \vdots & \mathbf{I}_k \end{bmatrix} = \begin{bmatrix} 1 & 1 & \vdots & 1 \end{bmatrix} \qquad (9.55)$$

The parity-check matrix is given by

$$\mathbf{H} = \begin{bmatrix} \mathbf{I}_{n-k} & \vdots & \mathbf{P}^T \end{bmatrix} = \begin{bmatrix} 1 & 0 & 1 \\ 0 & 1 & 1 \end{bmatrix} \qquad (9.56)$$

which satisfies $\mathbf{HG}^T = \mathbf{GH}^T \equiv \mathbf{0}$, as expected.

We transmit 111 for the bit 1 and 000 for the bit 0. If p denotes the transition probability, that is, we receive 1(0) when 0(1) is transmitted. Received bit sequences and probability of reception are tabulated in Table 9.3.

This code can correct single bit errors because the decision is made by majority voting for estimating the transmitted codeword. For example, when 111 is transmitted, the received words 110, 101, 011 are decoded as 111. Probability of correct reception may therefore be written as the sum of the probability of

no error $P(0, 3)$ and the probability of single error $P(1, 3)$:

$$P_c(3) = P(0, 3) + P(1, 3)$$
$$= \underbrace{(1-p)^3}_{\text{no error}} + \underbrace{3p(1-p)^2}_{1-\text{bit errors}} = 1 - 3p^2 + 2p^3$$

$$(9.57)$$

Since an error occurs when more than half of the bits in a codeword are received in error, the probability of error is given by

$$P_e(3) = 1 - P_c = P(2, 3) + P(3, 3)$$
$$= \underbrace{3p^2(1-p)}_{2-\text{bit errors}} + \underbrace{p^3}_{3-\text{bit errors}} = 3p^2 - 2p^3$$

$$(9.58)$$

Consider that uncoded bit error probability is given by 10^{-6} for BPSK transmission. Hence the uncoded mean SNR per bit is given by

$$p_u = Q\left(\sqrt{2\gamma}\right) = 10^{-6} \Rightarrow \gamma = 11.28 \ (10.53 \ dB)$$

$$(9.59)$$

With repetition coding, coded SNR per bit is decreased by $n = 3$ times, that is, the coded $R_c E_b/N_0 = \gamma/3$ is 5.76 dB. Consequently, the transition probability with coding at the decoder input increases from 10^{-6} to

$$p = Q\left(\sqrt{2\gamma/3}\right) = 3.1 \times 10^{-3} \qquad (9.60)$$

However, by correcting single bit errors, the probability of receiving a codeword in error is found to be

$$P_e(3) \cong 3p^2 = 2.8 \times 10^{-5} \qquad (9.61)$$

The average bit error probability for the coded case is equal to 1/3 of the value given in (9.61), namely 9.3×10^{-6}. This shows an increase in the bit error probability compared with 10^{-6} for the uncoded transmission, in spite of three-fold increase in the channel transmission rate and bandwidth. (see Figure 9.2).

9.3.2   Error Detection and Correction Capability of a Block Code

The Hamming distance between two binary codewords is equal to the number of differing bits in the two codewords. For example, Hamming distance between $c_1 = 011011$ and $c_2 = 110001$ is $d(c_1, c_2) = 3$. The Hamming weight $w(c_i)$ of a codeword c_i denotes the number of 1's in this codeword. Equivalently, the Hamming weight of a codeword is the distance between that codeword and the all-zero codeword, $w(c_i) = d(c_i, 0)$. The minimum distance d_{min} of a linear block code represents the smallest Hamming distance between any pairs of codewords in the code:

$$
\begin{aligned}
d_{\min} &= \min_{i \neq j, i,j = 0,1,\cdots 2^k - 1} d(c_i, c_j) \\
&= \min_{i \neq j, i,j = 0,1,\cdots 2^k - 1} d(c_i + c_j, 0) \\
&= \min_{i \neq j, i,j = 0,1,\cdots 2^k - 1} w(c_i + c_j) \\
&= \min_{m = 0,1,\cdots 2^k - 1} w(c_m)
\end{aligned}
\qquad (9.62)
$$

One may determine the minimum distance d_{min} of a codeword in various ways. If the generator matrix is known, then one may obtain all the codewords using (9.44) and determine the minimum distance using the codewords. A second approach is to use the parity-check matrix since the minimum distance d_{min} is related to the structure of the parity-check matrix. Let the matrix H be expressed in terms of its columns as follows:

$$H = [h_0 \ h_1 \ \cdots \ h_{n-1}] \qquad (9.63)$$

If a code vector c has a minimum distance of d_{min}, then sum of d_{min} columns of H must be equal to the zero vector 0 so as to satisfy the condition $cH^T = 0$. Conversely, if d_{min} columns of H vector sum up to zero, then the related codeword c has a Hamming weight d_{min}.

Let a codeword $c_i = (c_0 \ c_1 \ c_2 \ ... c_{n-1})$ has a Hamming weight of d_{min}. Since the Hamming weight of a codeword is equal to the number of 1's in that codeword, let us assume that the first d_{min} elements of c_i are 1 and the rest are zero, that is, $c_i = (c_0 \ c_1 \ c_2 \ c_{d_{min}-1} \ 0 \ 0 \ 0...0)$, where $c_0 = c_1 = ... = c_{d_{min}-1} = 1$. Then

$$
c_i H^T = [c_0 \ c_1 \ \cdots \ c_{d_{min}-1} \ 0 \ \cdots \ 0]
\begin{bmatrix}
h_0^T \\
h_1^T \\
\vdots \\
h_{n-1}^T
\end{bmatrix}
$$

$$
= c_0 h_0^T + c_1 h_1^T + c_2 h_2^T + \cdots + c_{d_{min}-1} h_{d_{min}-1}^T
$$

$$
= h_0 + h_1 + h_2 + \cdots + h_{d_{min}-1} \equiv 0
$$

$$(9.64)$$

Since the Hamming weight of a linear block code equals the minimum distance of the code except for the all-zero codeword, the minimum distance of a linear block code is defined as the minimum number of columns of the matrix H whose sum is equal to the zero vector.

The minimum distance d_{min} of a (n, k) linear block code determines the error-correcting capability of that code. Let a codeword c_i is transmitted and the observation vector at the decoder input (demodulator output) is quantized (hard-decision) by 1-bit. Consequently, the decoder input is characterized by the received codeword r, which differs from the transmitted codeword c_i by an error vector e, defined as follows:

$$r = c_i \oplus e$$

$$
\begin{aligned}
r &= [r_0 \ r_1 \ \cdots \ r_{n-1}], \ r_i \in \{0,1\} \\
c_i &= [c_{i0} \ c_{i1} \ \cdots \ c_{in-1}], \ c_{ij} \in \{0,1\} \\
e &= [e_0 \ e_1 \ \cdots \ e_{n-1}], \ e_i \in \{0,1\}
\end{aligned}
$$
(9.65)

Note that c_i, r and e consist of entries which are either 1 or 0. Elements of the error vector are defined by

$$
e_j =
\begin{cases}
1 & r_j \neq c_{ij} \\
0 & r_j = c_{ij}
\end{cases}
$$
$$
=
\begin{cases}
1 & \text{if an error has occured in the } j^{th} \text{ location} \\
0 & \text{otherwise}
\end{cases}
$$
(9.66)

where $j = 0,1,\ldots n-1$. If the 2^k codewords of a code are transmitted with equal probability, then the best strategy for the MLD decoder in a BSC is to select the codeword closest, that is, the one with minimum Hamming distance $d(c_i, r)$, to the received vector r as the transmitted codeword (see (9.11)). Hence, the decision in a BSC is in favor of c_i if

$$\hat{c} = c_i \quad if \quad d(r,c_i) = \min_{j=0,1,\cdots,2^k-1} d(r,c_j) \quad (9.67)$$

Consider for example two codewords c_i and c_j, $i \neq j$, $i,j = 0,1,\ldots,2^k-1$, which are separated from each other with a Hamming distance of d as shown in Figure 9.13. As shown in Figure 9.13(a), the codewords c_i and c_j, with a Hamming distance of $d(c_i, c_j) = 5$, can correct a maximum of 2 bit errors. In Figure 9.13a, the received codeword r is closer to c_i with a Hamming distance of 2, but its distance to c_j is 3; r can not be at equal Hamming distances to c_i and c_j at the same time. However, the code shown in Figure 9.13(b) can correct only single bit errors. Since $d(c_i, c_j) = 4$, the received codeword can not be correctly decoded if r is at equal Hamming distances of 2 to both codewords. Similarly, the code shown in Figure 9.13(c) is characterized by $d(c_i, c_j) = 3$ and can correct only single bit errors.

A decoder is able to detect and correct all error patterns of Hamming weight $d(r, c_i) = w(e) \leq t$, provided that $d_{min} \geq 2t+1$. In other words, if a code vector c_i is transmitted and $r = c_i + e$ is received, the decoder output will be $\hat{c} = c_i$ whenever the error pattern e has a $w(e) \leq t$ provided that $d_{min} \geq 2t+1$. In conclusion, a (n, k) linear block code can correct all error patterns of weight t or less if and only if,

$$d(c_i, c_j) = w(c_i) \geq 2t+1 \quad i \neq j, \ i,j = 0,1,\ldots,2^k-1$$
(9.68)

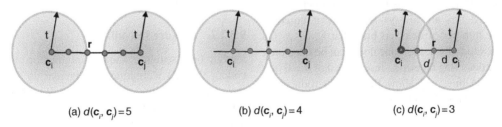

(a) $d(c_i, c_j) = 5$ (b) $d(c_i, c_j) = 4$ (c) $d(c_i, c_j) = 3$

Figure 9.13 Hamming Distance and Error Correction Capability of Block Codes. a) Hamming distance d $(c_i, c_j) \geq 2t+1$ where $t = 2$. Two bit errors are correctable. b) Hamming distance $d(c_i, c_j) = 2t$, where $t = 2$. Single bit error is correctable c) Hamming distance $d(c_i, c_j) < 2t$, where $t = 2$. Single bit error is correctable.

In summary, a (n, k) linear block code of minimum distance d_{min} can correct up to t errors if and only if

$$d_{min} \geq 2t + 1 \implies t \leq \left\lfloor \frac{d_{min} - 1}{2} \right\rfloor \qquad (9.69)$$

where $\lfloor . \rfloor$ denotes the largest integer less than or equal to the enclosed quantity.

9.3.2.1 Error Detection Capability of a Block Code

A code can be used to detect errors prior to, or instead of, correcting them. A block code with d_{min} guarantees the detection of all error patterns of

$$e \leq d_{min} - 1 \qquad (9.70)$$

It is also capable of detecting a large fraction of patterns with d_{min} or more errors.

It is possible to trade correction capability for the maximum guaranteed errors, with the ability to simultaneously detect a number of errors. A code can be used for simultaneous correction of α errors and detection of β errors. Inserting $\alpha + \beta = 2t$ into (9.69), error detection capability of a code may be written as[2]

$$d_{min} - 1 \geq 2t = \alpha + \beta \qquad (9.71)$$

Since error correction requires first the detection of the erroneous bits, the number of corrected bits are always less than or equal to the number of detected bits, that is, $\alpha \leq \beta$. On the other hand, the same code can be used only for detecting bit errors. Insert $\beta = 0$ or $\alpha = 0$ into (9.71) one may show that (9.71) reduces to (9.69) and (9.70) respectively:

$$\alpha = 0 \implies e = \beta \leq d_{min} - 1$$
$$\beta = 0 \implies t = \left\lfloor \frac{d_{min} - 1}{2} \right\rfloor \qquad (9.72)$$

Other combinations of α and β imply simultaneous detection and correction of bit errors. For example, consider a code with $d_{min} = 5$. This code can potentially be used for simultaneous detection and correction of the following combinations of α and β:

$$\begin{array}{cccc} \beta: & 2 & 3 & e = 4 \\ \alpha: & t = 2 & 1 & 0 \end{array} \qquad (9.73)$$

The combination $\alpha = \beta = 2$ implies a FEC code that can detect and correct 2 bit errors. The combination $\alpha = 1$, $\beta = 3$ implies that this code can detect 3 bit errors but can correct only one of the detected bit errors. The last combination $\alpha = 0$, $\beta = 4$ corresponds to an error detecting code that can detect 4 bit errors but can not correct any of the detected errors. Therefore, this code can be used as an error detection code with a capability of detecting $e = 4$ bit errors, as a FEC code with $t = 2$ bit error correction capability or for simultaneous detection of 3 bits and correction of single bit errors.

In fact, a (n, k) linear code is capable of detecting exactly $2^n - 2^k$ error patterns of length n (see Figure 9.10). There are $2^k - 1$ error patterns that are identical to the $2^k - 1$ nonzero codewords. If any of these $2^k - 1$ error patterns occurs, it alters the transmitted codeword into another codeword of length n. This implies that there are $2^k - 1$ undetectable error patterns. Since $2^k - 1 << 2^n$ for large n, only a small fraction of error patterns pass through the decoder without being detected.

9.3.2.2 Erasure Decoding

It might be wise to use adaptive modulation and coding (AMC) techniques for increasing the throughput in fading channels by making best use of the randomly changing channel conditions. FEC coding mitigates bit errors due to channel impairments. However, lower protection levels against bit errors than provided by FEC encoding may be sufficient when the instantaneous SNR is usually high. Then, the code rate may be increased by puncturing (erasing, deleting) some of the parity bits generated by a mother encoder. This implies the transmission of less parity bits and more data bits. At the receiver side, a soft-decision decoder uses erasures to indicate the reception of a signal whose corresponding symbol value is doubtful and estimates them. The so-called erasure decoding does not provide significant additional gain for AWGN channels, but it provides substantial improvement over fading and bursty channels. Some block codes provide very efficient means for erasure decoding.

Consider a received block code with ρ erased coordinates, that is, the number ρ and the locations of the erased bits are known. Any pattern of ρ or fewer erasures can be corrected if

$$d_{\min} \geq \rho + 1 \qquad (9.74)$$

Given ρ erased coordinates, the code will have an effective minimum distance of at least $d_{min} - \rho$ over the unerased coordinates. This code can correct $t = \lfloor d_{\min} - \rho - 1 \rfloor / 2$ errors in the unerased coordinates of the received codeword. In other words, t errors and ρ erasures can be corrected as long as $d_{\min} \geq 2t + \rho + 1$. Note that the code can correct twice as many erasures as it can correct errors. This may be intuitively explained by noting that the locations of the erasures are known, but no information is available as to the location of the bit errors.

In binary erasure decoding, the following simple algorithm may be used: Given a received word $\mathbf{r}$, place zeros in all erased coordinates and decode normally to estimate the transmitted codeword. Now place ones in all erased coordinates and decode normally to estimate a second codeword. Compare the two and select the codeword that is closest in Hamming distance to $\mathbf{r}$ as the final decoded output.

Example 9.7   Erasure Decoding.
Consider the following $(6, 3)$ code with $d_{min} = 3$:

$$\mathbf{c} = \begin{bmatrix} 000000 & 110100 & 011010 & 101110 \\ 101001 & 011101 & 110011 & 000111 \end{bmatrix}$$

$$(9.75)$$

Noting first that $d_{min} = 3$, this code can correct $\rho = 2$ erasures. Suppose that the codeword 110011 was transmitted and that the two leftmost digits were declared by the receiver to be erasures. Comparing the rightmost digits of xx0011 with each of the codewords in (9.75). The codeword that is closest in Hamming distance to the sequence with erasure (hence 110011) is declared to be actually transmitted.

9.3.3   Syndrome Decoding of Linear Block Codes

As shown in Figure 9.11, $k \times n$ generator matrix $\mathbf{G}$ and $(n-k) \times n$ parity-check matrix $\mathbf{H}$ are used in the encoding and decoding operations, respectively. The receiver has the task of decoding the code vector $\mathbf{c}_i$ using the one bit quantized $1 \times n$ observation vector $\mathbf{r}$. Given $\mathbf{r}$, decoding starts with the computation of a $1 \times (n-k)$ vector called the error-syndrome vector or simply the syndrome:

$$\mathbf{s} = \mathbf{r}\mathbf{H}^T = (\mathbf{c}_i + \mathbf{e})\mathbf{H}^T = \mathbf{c}_i\mathbf{H}^T + \mathbf{e}\mathbf{H}^T = \mathbf{e}\mathbf{H}^T$$

$$= \begin{bmatrix} e_0 & e_1 & \cdots & e_{n-1} \end{bmatrix}\mathbf{H}^T = \begin{bmatrix} s_0 & s_1 & \cdots & s_{n-k-1} \end{bmatrix}$$

$$(9.76)$$

One may easily observe from (9.76) that the syndrome depends only on the error pattern but not on the transmitted code word, that is, the syndrome is independent of $\mathbf{c}_i$. Having a dimension of $1 \times (n-k)$, the syndrome vector can have $2^{n-k} - 1$ non-zero patterns. The all-zero pattern implies that the codeword is correctly received. Therefore a (n, k) block code can correct up to $2^{n-k} - 1$ error patterns.

For k message bits, there are 2^k distinct code vectors denoted as $\mathbf{c}_i$, $i = 0, 1, \ldots, 2^k - 1$. For any error pattern $\mathbf{e}_j$, $j = 0, 1, \ldots, 2^{n-k} - 1$, we define the following 2^k distinct vectors as

Table 9.4   Standard Array For an (n, k) Block Code.

c_0	c_1	c_2	....	c_i	.....	c_{2^k-1}
e_1	$c_1 + e_1$	$c_2 + e_1$	....	$c_i + e_1$	....	$c_{2^k-1} + e_1$
e_2	$c_1 + e_2$	$c_2 + e_2$	....	$c_i + e_2$	....	$c_{2^k-1} + e_2$
$\vdots$	$\vdots$	$\vdots$	$\vdots$	$\vdots$	$\vdots$	$\vdots$
e_j	$c_1 + e_j$	$c_2 + e_j$	....	$c_i + e_j$	....	$c_{2^k-1} + e_j$
$\vdots$	$\vdots$	$\vdots$	$\vdots$	$\vdots$	$\vdots$	$\vdots$
$e_{2^{n-k}-1}$	$c_1 + e_{2^{n-k}-1}$	$c_2 + e_{2^{n-k}-1}$	....	$c_i + e_{2^{n-k}-1}$	....	$c_{2^k-1} + e_{2^{n-k}-1}$

$$c_i + e_j, \quad i = 0, 1, \cdots, \ 2^k - 1, \ j = 0, 1, \cdots, \ 2^{n-k} - 1$$

$$\text{(9.77)}$$

Syndrome decoding may be better understood by using the so-called standard array shown in Table 9.4. The standard array is formed by letting i and j vary as defined by (9.77). For a given value of j, the set of vectors $\{c_i + e_j, \ i = 0, 1, \ldots, 2^k - 1\}$ is called a coset of the code. A coset has exactly 2^k elements that differ by a code vector. Thus, an (n,k) linear block code has 2^{n-k} possible cosets, each with a coset leader e_j which corresponds to a correctable error pattern. The coset leader (error pattern) $e_0 = 0$, implies no error; therefore, $e_0 = 0$ is not shown in the first row of the standard array in Table 9.4. Hence, each coset is characterized by a unique syndrome and a correctable unique error pattern. Therefore, a (n, k) linear block code is capable of correcting $2^{n-k} - 1$ error patterns. For a given channel, the probability of decoding error is minimized when the most likely error patterns (i.e., those with the largest probability of occurrence) are chosen as the coset leaders. In a BSC, the smaller the Hamming weight of an error pattern, the more likely it is to occur. Accordingly, the standard array is constructed with coset leaders with minimum Hamming weights.

It might be useful to note that 2^n sequences of length n in the standard array, are the same as shown on the right-hand-side of Figure 9.10 and the first row of the standard array are identical to those shown as 2^k codewords in Figure 9.10.

Table 9.5   Shortest Code $(n, 8)$ with t Error Correction Capability.

t	n	$R_c = k/n$	t/n
1	12	0.67	0.083
2	15	0.53	0.133
3	18	0.44	0.167
4	21	0.38	0.190

The number of cosets is related to the error-correcting capability of a linear block code by the Hamming bound:

$$2^{n-k} \geq \underbrace{\binom{n}{0}}_{no \ error} + \underbrace{\binom{n}{1}}_{1 \ error} + \underbrace{\binom{n}{2}}_{2 \ errors} + \cdots + \underbrace{\binom{n}{t}}_{t \ errors} = \sum_{i=0}^{t} \binom{n}{i}$$

$$\text{(9.78)}$$

The codes for which the equality holds are called as perfect codes. Perfect codes can correct an integer number of bit errors and do not provide partial correction.

Example 9.8   Hamming Bound

Consider a $(n, 8)$ block code which is required to correct t bit errors. Using the Hamming bound (9.78), the value of n of the shortest code for $k = 8$ and $t = 1, 2, 3$ and 4 is shown in Table 9.5. The ratio t/n, of the number of correctable errors to the code length, is a measure of the effectiveness of the code in error correction. One may observe that the code effectiveness increases with n, at the expense of increasing the transmission bandwidth, which is proportional to $1/R_c$.

Example 9.9 Generator and Parity Check Matrices and Standard Array for a (4, 2) Block Code.

Consider the (4, 2) linear block code with the following message and codewords as in Example 9.4:

$$\mathbf{m} = \{00 \ \ 01 \ \ 10 \ \ 11\}$$
$$\mathbf{c} = \{0000 \ \ 0101 \ \ 1010 \ \ 1111\} \tag{9.79}$$

Since any codeword consists of the repetition of the message word, parity, generator and parity-check matrices may be written as

$$\mathbf{c} = \mathbf{m}\left[\mathbf{P} \ \vdots \ \mathbf{I}_2\right] = [m_1 \ m_2 \ m_1 \ m_2] \ \Rightarrow \ \mathbf{P} = \begin{bmatrix} 1 & 0 \\ 0 & 1 \end{bmatrix}$$

$$\mathbf{G} = \left[\mathbf{P} \ \vdots \ \mathbf{I}_k\right] = \begin{bmatrix} 1 & 0 & 1 & 0 \\ 0 & 1 & 0 & 1 \end{bmatrix}$$

$$\mathbf{H} = \left[\mathbf{I}_{n-k} \ \vdots \ \mathbf{P}^T\right] = \begin{bmatrix} 1 & 0 & 1 & 0 \\ 0 & 1 & 0 & 1 \end{bmatrix} \tag{9.80}$$

This code has a minimum distance $d_{min} = 2$, which implies $t = 0$ and $e = 1$. Therefore, this code can not correct any bit error but can detect single bit errors. The standard array of this code has $2^k = 4$ columns and $2^{n-k} = 4$ rows as shown in Table 9.6. Only some of the patterns with single bit error are chosen as the coset leaders. The error pattern $e_4 = 0001$ is not chosen since 4 rows can not accomodate all-zero pattern and four patterns with single-bit errors. Note that e_1 and e_3 lead to similar words. For example,

Table 9.6 Standard Array For (4,2) Linear Block Code.

$c_0 = e_0 = 0000$	$c_1 = 0101$	$c_2 = 1010$	$c_3 = 1111$
$e_1 = 1000$	1101	0010	0111
$e_2 = 0100$	0001	1110	1011
$e_3 = 0010$	0111	1000	1101

$\mathbf{r} = 1101$ may be decoded as $\mathbf{c}_1$ or $\mathbf{c}_3$ with equal likelihood since $d(\mathbf{r}, \ \mathbf{c}_1) = d(\mathbf{r}, \ \mathbf{c}_3) = 1$. The choice of $\mathbf{e}_4 = 0001$ as a coset leader does not provide any improvement; this code can not correct any bit error.

Syndrome patterns are found as follows:

$$\mathbf{s}_i = \mathbf{e}_i \ \mathbf{H}^T \ \Rightarrow \ \begin{matrix} e_0 = 0000 & s_0 = 00 \\ e_1 = 1000 & s_1 = 10 \\ e_2 = 0100 & s_2 = 01 \\ e_3 = 0010 & s_3 = 10 \end{matrix} \tag{9.81}$$

Syndrome for the received word $\mathbf{r} = 1101$ and the corresponding error patterns are given by

$$\mathbf{r} \ \mathbf{H}^T = [1 \ 1 \ 0 \ 1] \ \mathbf{H}^T = [1 \ 0] \ \Rightarrow \mathbf{e}_1 \text{ or } \mathbf{e}_3 \tag{9.82}$$

Example 9.10 (6, 3) Linear Block Code.

Consider a (6, 3) block code with the generator matrix:

$$\mathbf{G} = \begin{bmatrix} 1 & 1 & 0 & 1 & 0 & 0 \\ 0 & 1 & 1 & 0 & 1 & 0 \\ 1 & 0 & 1 & 0 & 0 & 1 \end{bmatrix} \tag{9.83}$$

Since the generator matrix is already in the systematic form, the parity check matrix is readily found using $\mathbf{G}$ as follows:

$$\mathbf{H} = \begin{bmatrix} 1 & 0 & 0 & 1 & 0 & 1 \\ 0 & 1 & 0 & 1 & 1 & 0 \\ 0 & 0 & 1 & 0 & 1 & 1 \end{bmatrix} \tag{9.84}$$

The standard array for this (6,3) code, which has $2^k = 8$ columns and $2^{n-k} = 8$ rows, is shown in Table 9.7. This code has a minimum distance $d_{min} = 3$. This can be observed from the first row of the standard array or from the parity-check matrix (9.84) where the sum of first, second and fourth colums is a zero vector. This code can therefore correct all single-bit errors. Eight rows of the standard array correspond

Table 9.7   Standard Array For (6, 3) Block Code.

000000	110100	011010	101110	101001	011101	110011	000111
000001	110101	011011	101111	101000	011100	110010	000110
000010	110110	011000	101100	101011	011111	110001	000101
000100	110000	011110	101010	101101	011001	110111	000011
001000	111100	010010	100110	100001	010101	111011	001111
010000	100100	001010	111110	111001	001101	100011	010111
100000	010100	111010	001110	001001	111101	010011	100111
010001	100101	001011	111111	111000	001100	100010	010110

Table 9.8   Syndrome Table For (6, 3) Code.

$s = rH^T = eH^T$	Error pattern, e
000	000000
101	000001
011	000010
110	000100
001	001000
010	010000
100	100000
111	010001

to no-error pattern in the first row, six single-bit error patterns and a single two-bit error pattern, namely, 010001, in the last row. The choice of the two-bit error pattern 010001 ensures that the standard array contains all $(2^n = 64)$ combination of 6-bit sequences; any received 6-bit sequence is directly mapped to a single error pattern and a valid codeword. Hence this code can correct 48 error patterns with single bit errors and 8 error patterns with two bit errors.

The syndrome and the corresponding bit error patterns are found using (9.76) and shown in Table 9.8.

Now assume a received sequence of $r = 010010$. Using (9.76) the syndrome corresponding to r is found to be $s = 001$. From Table 9.8, the error pattern corresponding to this syndrome pattern is estimated to be $\hat{e} = 001000$. The correct codeword is then $\hat{c} = r \oplus \hat{e} = 010010 \oplus 001000 = 011010$. This result may also be obtained by inspection from the standard array shown in Table 9.7.

9.3.4   Bit Error Probability of Block Codes with Hard-Decision Decoding

Consider a digital communication system using coherent BPSK modulation and a (n, k) block code, which is capable of correcting t bit errors, in an AWGN channel. The transition probability p^u, the probability of correct reception P_c^u and probability of error P_e^u of uncoded k-bit message word are given by

$$p^u = Q\left(\sqrt{2\gamma}\right)$$
$$P_e^u = 1 - P_c^u = 1 - (1 - p^u)^k \tag{9.85}$$

The transition probability, probability of correct reception and the probability of error of a codeword, of n-coded bits, may be written as

$$p = Q\left(\sqrt{2R_c\gamma}\right)$$
$$P_c = \sum_{i=0}^{t} \binom{n}{i} p^i (1-p)^{n-i} \tag{9.86}$$
$$P_e = 1 - P_c = \sum_{i=t+1}^{n} \binom{n}{i} p^i (1-p)^{n-i}$$

where $R_c = k/n$ denotes the code rate. The corresponding bit error probability is given by

$$P_b = \sum_{i=t+1}^{n} \frac{i}{n} \binom{n}{i} p^i (1-p)^{n-i}$$
$$= p - p(1-p)^{n-1} \sum_{i=0}^{t-1} \binom{n-1}{i} \left(\frac{p}{1-p}\right)^i \tag{9.87}$$

where the last expression follows from (D.35). It might be useful to note that nP_b shows the average number of errors in a codeword of length n.

Example 9.11 Coding Gain of a (63, 39) BCH Code.

Assume that we transmit data over an AWGN channel using BPSK modulation using (63, 39) BCH code which is capable of correcting $t = 4$ bit errors. For the uncoded case, the SNR required for a bit error probability of 10^{-7} is found from (9.85) to be $\gamma = 13.51$ (11.3 dB). We now determine the $\gamma = E_b/N_0$ required for $P_b = 10^{-7}$ with coding. Inserting $P_b = 10^{-7}$, $t = 4$ and $n = 63$ into (9.87) for the coded bit error probability, one gets $p = 0.0029$. Using the expression for the transition probability for BPSK with coding $p = Q(\sqrt{2R_c\gamma}) = 0.0029$, one gets $\gamma = 6.15$ (7.9 dB) to achieve $P_b = 10^{-7}$. This implies a coding gain of 11.3–7.9 = 3.4 dB.

Figure 9.14 shows the bit error probability of a (63,39) block (BCH) code which can correct up to four-bit errors. Note that the transition probability p, given by (9.86), at the decoder input is much higher than the uncoded bit error probability p^u given by (9.85); it requires an SNR of 13.37 dB compared to 11.30 dB for the uncoded case at 10^{-7} bit error probability level. As the decoder corrects up to four bit errors, the bit error probability at the decoder output is much lower. A coding gain of 3.4 dB is achieved at 10^{-7} bit error probability level at the expense of $1/R_c = 63/39 = 1.61$-fold increase in the transmission bandwidth. Alternatively, at an SNR value of 8 dB, coding decreases the bit error probability by a factor of $(1.91 \times 10^{-4}/5.82 \times 10^{-8}) = 3278$.

9.3.5 Bit Error Probability of Block Codes with Soft-Decision Decoding

Let $\mathbf{x}$ represent the n sampled matched filter output (observation vector) assuming that the i-th codeword $\mathbf{c}_i$ of a (n, k) block code is transmitted. For coherent BPSK signaling in AWGN channel, the sampled observation vector may be written in terms of the transmitted codeword $\mathbf{c}_i'$ as (see (9.4) and (9.5))

$$\mathbf{x} = \mathbf{c}_i' + \mathbf{w}$$

$$x_j = c_{ij}' + w_j = (2c_{ij} - 1)\sqrt{R_c E_b} + w_j, \qquad (9.88)$$

$$j = 0, 1, \cdots n-1, \quad i = 0, 1, \cdots 2^k - 1$$

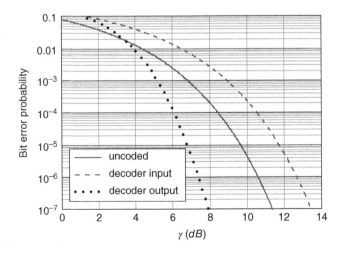

Figure 9.14 Bit Error Probability of the (63, 39) BCH Code with $t = 4$ Bit Error Correction Capability.

where $c_{ij} \in \{0,1\}$ denotes the j-th bit of the i-th codeword; the weighting factor $2c_{ij}-1 = 1$ for $c_{ij} = 1$, and it is equal to -1 for $c_{ij} = 0$. The variable w_j represents the j-th noise sample with zero mean and variance $N_0/2$. Using the knowledge of 2^k codewords, the ML decoder formes 2^k decision variables by correlating the sampled matched filter output with 2^k codewords (see (9.13)):

$$U_m = \sum_{j=0}^{n-1} (2c_{mj}-1)\, x_j, \quad m = 0,1,\cdots,2^k-1$$

$$= \sum_{j=0}^{n-1} \left[(2c_{mj}-1)(2c_{ij}-1)\sqrt{R_c E_b} + (2c_{mj}-1)w_j \right],$$

$$m = 0,1,\cdots,2^k-1$$

(9.89)

One may observe from (9.89) that U_m represents the projection of the observation vector $\mathbf{x}$ onto $\mathbf{c}_m^t$. All decision variables are Gaussian-distributed with noise variance $nN_0/2$. The mean value of the i-th decision variable U_i corresponding to the transmitted codeword, will be $n\sqrt{R_c E_b}$ while the other 2^k-1 decision variables will have smaller mean values.

In determining the probability of error for linear block codes with optimum soft-decision decoding, we assume that the transmission probability of all the codewords is the same. For the sake of simplicity, we assume that the all-zero codeword $\mathbf{c}_0$ is transmitted, that is, $c_{ij} \equiv 0$. For correct decoding of $\mathbf{c}_0$, the decision variable U_0 must exceed all the other 2^k-1 decision variables $U_1, U_2, \cdots, U_{2^k-1}$. The pair-wise error probability between the metrics U_0 and U_m denotes the probability that the soft-decision decoder estimates U_m as being transmitted when U_0 is transmitted. The codewords $\mathbf{c}_0$ and $\mathbf{c}_m$ are assumed to have a Hamming distance of d between them, which is identical to the Hamming weight w_m of the m-th codeword $\mathbf{c}_m$. The pairwise error probability may be written as

$$P_2(d) = Prob(U_m \geq U_0) = Prob(U_m - U_0 \geq 0),$$

$$m = 1,2,\cdots 2^k-1$$

(9.90)

Since the coded bits in the two decision metrics are identical except for in d positions, inserting (9.89) into (9.90) and retaining d non-zero terms, the pair-wise error probability may be rewritten in a simpler form as

$$P_2(d) = Prob\left(2\sum_{\ell=0}^{d-1} (-\sqrt{R_c E_b} + w_\ell) \geq 0 \right)$$

$$= Prob\left(\sum_{\ell=0}^{d-1} w_\ell \geq d\sqrt{R_c E_b} \right)$$

$$= Q\left(\sqrt{2R_c \gamma d} \right)$$

(9.91)

where $\gamma = E_b/N_0$. The last expression is obtained by noting that the sum of d AWGN terms is characterized by zero mean and variance $d\,N_0/2$.

The average probability of a codeword error is upper bounded by the sum of the pair-wise error probabilities, when $\mathbf{c}_0$ is sent:

$$P_e \leq \sum_{d=d_{min}}^{n} A_d P_2(d) = \sum_{d=d_{min}}^{n} A_d\, Q\left(\sqrt{2R_c \gamma d} \right)$$

(9.92)

where $\{A_d\}$ denotes the weight distribution of the code, that is, the number of codewords with weight d. Note that, except for the all-zero codeword, which is assumed to be transmitted, $A_d = 0$ for $d < d_{min}$. Therefore, (9.92) accounts for the sum of pair-wise error probabilities due to codewords with Hamming weights larger than d_{min}. The computation of the average symbol error probability (9.92) for soft-decision decoding requires the knowledge of the weight distribution of the code. On the other hand (9.92) is an upper bound since pair-wise error probabilities are not necessarily mutually exclusive.

If N_d denotes the number of 1's in the m-th message word of k-bit long, there will be N_d bit errors if the corresponding codeword $\mathbf{c}_m$ is received in error. Therefore, the resulting average bit error probability is found by multiplying each term in (9.92) by the corresponding number N_d of 1's in the message word codeword:

$$P_b \le \sum_{d=d_{min}}^{n} A_d \, N_d \, Q\left(\sqrt{2R_c\gamma d}\right) \qquad (9.93)$$

The fact that the argument of the Q-function in (9.93) has an additional factor $R_c \, d \ge R_c \, d_{min} > 1$ compared to the uncoded BPSK (see for example (9.85), the coding improves the bit error probability performance by a factor of $R_c \, d_{min}$, which also determines the coding gain.

9.3.6   Channel Coding Theorem

It was shown in Chapter 8 that the capacity C (bps) in an AWGN channel, normalized by the transmission bandwidth B, is a logarithmic function of the SNR. Hence it represents the highest achievable data rate in an AWGN channel of bandwidth B and also serves as an upper bound to the bandwidth efficiency. As $B \to \infty$, the error probability can be made as small as desired,

provided that $E_b/N_0 > 0.693$ (−1.6 dB) (see (8.181)). This, the so-called the Shannon limit, is the minimum E_b/N_0's required to achieve reliable transmission as the channel bandwidth $B \to \infty$ ($C/B \to 0$). The gap between the Shannon limit and the practical operating points (see Figure 8.44) may be decreased by using FEC codes, which lower E_b/N_0's required to achieve a given probability of symbol error. However, this decrease occurs at the expense of increase in the transmission bandwidth.

In the design of coded communication systems, we aim to minimize the E_b/N_0' required for a given error probability; this is equivalent to maximizing the coding gain. The cost of this improvement is increased transmission rate, hence larger transmission bandwidth. Shannon's coding theorem provides a theoretical limit on the minimum E_b/N_0 required for a code rate R_c to achieve communication with arbitrarily small error probability. If E_b/N_0 exceeds the Shannon limit, then there exists a (perhaps complex) code which achieves error-free communications.

Figure 9.15 shows the minimum E_b/N_0, required for error-free communications, for a

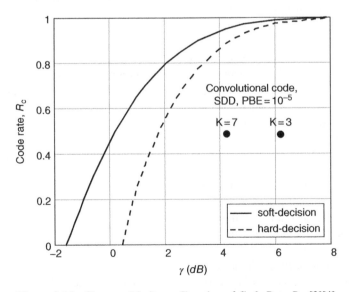

Figure 9.15   Shannon Limit as a Function of Code Rate R_c. [3][4]

given code rate using hard- and soft-decision decoding (see Appendix 9A). [3][4] One may observe from Figure 9.15 that the minimum $\gamma = E_b/N_0$ required to achieve the capacity is -1.6 dB and 0.37 dB for soft- and hard-decision decoding, respectively, as R_c goes to zero (transmission bandwidth approaching infinity). Hence, Shannon limit for soft-decision decoding shows 1.97 dB higher coding gain compared for hard-decision decoding but this difference becomes smaller for higher code rates, for example, ~1.5 dB for $R_c = 0.8$. Figure 9.15 shows that arbitrarily small bit error probabilities may be achieved for code rate-½ at $\gamma = 0.188$ dB for soft-decision decoding and 1.77 dB for hard decision decoding. Compared with uncoded BPSK transmission that requires $\gamma = 9.6$ dB at 10^{-5} bit error probability, this implies a maximum achievable coding gain of 9.4 dB and 7.8 dB for a rate-½ code with soft- and hard-decision decoding, respectively.

Now let us consider using a rate- ½ convolutional code with constraint length 3 with soft-decision decoding. This code achieves 10^{-5} bit error probability at $E_b/N_0 = 6.17$ dB. This implies a coding gain of $9.6 - 6.17 = 3.4$ dB but it is still 5.98 dB away from the Shannon limit. If we use a more power convolutional code of rate-½ with constraint length 7, then an $E_b/N_0 = 4.15$ dB is required to achieve the same bit error probability. Then, the coding gain increases to $9.6 - 4.15 = 5.44$ dB, its distance to the Shannon limit is decreased to 3.96 dB (see Figure 9.15). Coding gain of codes can be improved by using longer and more powerful codes and efficient decoding algorithms. For example, long LDPC codes can operate a fraction of a dB away from the Shannon limit.

Channel coding theorem states that if a DMC has capacity C and a source generates information at a rate less than C, then there exists a coding technique such that the output of the source may be transmitted over the channel with an arbitrarily small probability of symbol error. The channel coding theorem for the special case of a binary symmetric channel (BSC) may be stated as follows: If the data rate is less than the channel capacity C, then it is possible to find a code that achieves error-free transmission over the channel. Conversely, it is not possible to find such a code if the data rate is greater than the channel capacity C.

The channel capacity C provides a fundamental limit on the rate at which messages can be transmitted reliably (error-free) over a DMC using good codes. The issue that matters is not the signal-to-noise ratio, so long as it is sufficiently large. The real issue is to determine good codes whose existence is already asserted by the theorem. Therefore, we still need to find codes that ensure reliable information transmission over the channel.

9.3.7 Hamming Codes

Hamming codes is a family of (n,k) linear block codes with

Block length:	$n = 2^m - 1, m \geq 3$
Number of message bits:	$k = n - m = 2^m - m - 1$
Minimum distance:	$d_{min} = 3$

Hamming codes are perfect codes (see (9.78)) and can correct only single bit errors.

Example 9.12 (7, 4) Hamming Code. Consider the following generator matrix for the (7, 4) Hamming code:

$$\mathbf{G} = \begin{bmatrix} 1 & 1 & 0 & \vdots & 1 & 0 & 0 & 0 \\ 0 & 1 & 1 & \vdots & 0 & 1 & 0 & 0 \\ 1 & 1 & 1 & \vdots & 0 & 0 & 1 & 0 \\ 1 & 0 & 1 & \vdots & 0 & 0 & 0 & 1 \end{bmatrix} \tag{9.94}$$

From (9.94), the code and the corresponding parity equations may be written as

Table 9.9 Weight Distribution of (7, 4) Hamming Code.

Message vector, **m**	Code vector, **c**	Weight	Message vector, **m**	Code vector, **c**	Weight
0000	0000000	0	1000	1101000	3
0001	1010001	3	1001	0111001	4
0010	1110010	4	1010	0011010	3
0011	0100011	3	1011	1001011	4
0100	0110100	3	1100	1011100	4
0101	1100101	4	1101	0001101	3
0110	1000110	3	1110	0101110	4
0111	0010111	4	1111	1111111	7

$$\mathbf{c} = \begin{bmatrix} b_0 & b_1 & b_2 & \vdots & m_0 & m_1 & m_2 & m_3 \end{bmatrix}$$

$$b_0 = m_0 + m_2 + m_3 \tag{9.95}$$

$$b_1 = m_0 + m_1 + m_2$$

$$b_2 = m_1 + m_2 + m_3$$

Note that the parity equations are chosen so that a message bit is used at least twice in determining $\{b_0, b_1, b_2\}$. The corresponding parity-check matrix is obtained from (9.94) as

$$\mathbf{H} = \begin{bmatrix} \mathbf{I}_{n-k} \vdots \mathbf{P}^T \end{bmatrix} = \begin{bmatrix} 1 & 0 & 0 & \vdots & 1 & 0 & 1 & 1 \\ 0 & 1 & 0 & \vdots & 1 & 1 & 1 & 0 \\ 0 & 0 & 1 & \vdots & 0 & 1 & 1 & 1 \end{bmatrix} \tag{9.96}$$

The $2^k = 16$ distinct message words, codewords and their weights are tabulated in Table 9.9.

As one may observe from Table 9.9, the weight distribution of the message vector is $N_0 = 1$, $N_1 = 4$, $N_2 = 6$, $N_3 = 4$, $N_4 = 1$ (see (9.93)). The (7, 4) Hamming code has a weight distribution of $A_0 = 1$, $A_1 = A_2 = 0$, $A_3 = A_4 = 7$, $A_5 = A_6 = 0$, $A_7 = 1$. In addition to a single all-zero and all-1 codewords, there are 7 codewords with weight three and seven codewords with weight four. The minimum Hamming weight for the nonzero codewords is 3; hence, the minimum distance of the code is $d_{min} = 3$. This also implies that the smallest number of

Table 9.10 Syndrome Table for (7, 4) Hamming Code.

Syndrome, **s**	Error pattern, **e**
000	0000000
100	1000000
010	0100000
001	0010000
110	0001000
011	0000100
111	0000010
101	0000001

columns in **H** that sums to zero vector is 3 (see (9.96)), confirming $d_{min} = 3$.

Seven coset leaders (single-error patterns) and corresponding syndromes are shown in Table 9.10. Suppose that the codeword [1010001] is sent, and the received vector is $\mathbf{r} = [1110001]$ with an error in the second bit. The syndrome is found to be:

$$\mathbf{s} = \mathbf{r}\mathbf{H}^T = \begin{bmatrix} 0 & 1 & 0 \end{bmatrix} \tag{9.97}$$

Using Table 9.10, the corresponding coset leader (i.e., error pattern with the highest probability of occurrence) is found to be $\hat{\mathbf{e}} = [0100000]$, indicating that the second bit of the received vector is erroneous. Thus, adding this error pattern to the received vector yields the correct codeword actually sent: $\hat{\mathbf{c}} = \mathbf{r} \oplus \hat{\mathbf{e}} = 1110001 \oplus 0100000 = 1010001$.

Uncoded and coded bit error probabilities for (7, 4), (15, 11) and (31, 26) Hamming codes all with $t = 1$ are determined using (9.85) and (9.87), and compared in Figure 9.16. At 10^{-7} bit error probability, coding gain is 0.59 dB, 1.44 dB and 1.83 dB for (7, 4), (15, 11) and (31, 26) Hamming codes, respectively. The respective transmission bandwidths are increased due to coding by factors of $n/k = 7/4 = 1.75$, $n/k = 15/11 = 1.27$ and $n/k = 31/26 = 1.19$, respectively.

Figure 9.17 shows the bit error probability of the (7, 4) Hamming code with hard- and soft-decision decoding using (9.87) and (9.93) and Table 9.9 for the values of A_d and N_d. Soft decision decoding has a coding gain of 1.6 dB at

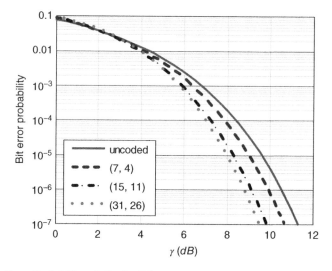

Figure 9.16 Bit Error Probability of (7, 4), (15, 11) and (31, 26) Hamming Codes for Hard-Decision Decoding.

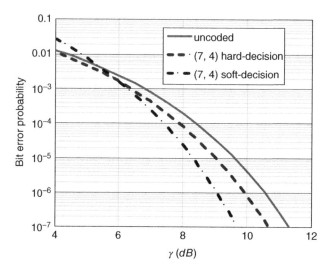

Figure 9.17 Bit Error Probability for (7, 4) Hamming Code with Hard- and Soft-Decision Decoding.

10^{-7} bit error probability, which is approximately 1 dB higher than the coding gain for hard-decision decoding.

9.4 Cyclic Codes

Cyclic codes form an important subclass of linear block codes. They possess a well-defined mathematical structure and are easy to encode and decode. They are used to implement for long block codes with many codewords. Most of the commonly used linear block codes are cyclic because of the availability of very efficient decoding algorithms.

A binary code is said to be cyclic if any cyclic shift of a codeword is also a codeword. This implies that a linear (n, k) block code

$$\mathbf{c} = (c_0, c_1, \cdots, c_{n-1}) \qquad (9.98)$$

is cyclic if the following n-tuples are also the codewords:

$$(c_0, c_1, \cdots, c_{n-1})$$
$$(c_{n-1}, c_0, c_1, \cdots, c_{n-2})$$
$$(c_{n-2}, c_{n-1}, c_0 \cdots, c_{n-3}) \qquad (9.99)$$
$$\vdots$$
$$(c_1, c_2, \cdots, c_{n-1}, c_0)$$

To develop the algebraic properties of cyclic codes, we define the code polynomial

$$c(X) = c_0 + c_1 X + c_2 X^2 + \cdots + c_{n-2} X^{n-2} + c_{n-1} X^{n-1} \qquad (9.100)$$

where each power of X represents a one-bit shift in time. Multiplication of the polynomial $c(X)$ by X may be viewed as a one-bit shift in time:

$$X c(X) = c_{n-1} + c_0 X + c_1 X^2 + c_2 X^3 + \cdots + c_{n-2} X^{n-1} \qquad (9.101)$$

where $X^n = 1$ is assumed.

Consider that the code polynomial $c(X)$ is multiplied by X^i

$$X^i c(X) = \left[c_0 X^i + c_1 X^{i+1} + c_2 X^{i+2} + \cdots + c_{n-i-1} X^{n-1} \right]$$
$$+ \left[c_{n-i} X^n + \cdots + c_{n-1} X^{n+i-1} \right]$$
$$= \left[c_0 X^i + c_1 X^{i+1} + c_2 X^{i+2} + \cdots + c_{n-i-1} X^{n-1} \right]$$
$$+ \left[c_{n-i} + \cdots\cdots + c_{n-1} X^{i-1} \right] X^n \qquad (9.102)$$

Since $c_j \oplus c_j \equiv 0$ in modulo-2 addition, we may reorganize (9.102) as follows:

$$X^i c(X) = \left[c_{n-i} + \cdots + c_{n-1} X^{i-1} \right]$$
$$+ \left[c_0 X^i + c_1 X^{i+1} + c_2 X^{i+2} + \cdots + c_{n-i-1} X^{n-1} \right]$$
$$+ \underbrace{\left[c_{n-i} + \cdots + c_{n-1} X^{i-1} \right]}_{q(X)} (X^n + 1)$$
$$= q(X)(X^n + 1) + c^{(i)}(X) \qquad (9.103)$$

where the polynomial $c^{(i)}(X)$, which is the i-th cyclic shift of $c(X)$, is also a code polynomial

$$c^{(i)}(X) = c_{n-i} + \cdots + c_{n-1} X^{i-1} + c_0 X^i + c_1 X^{i+1}$$
$$+ c_2 X^{i+2} + \cdots + c_{n-i-1} X^{n-1} \qquad (9.104)$$

Noting that $c^{(i)}(X)$ is the remainder that results from dividing $X^i c(X)$ by $(X^n + 1)$, one may express $c^{(i)}(X)$ as:

$$c^{(i)}(X) = X^i c(X) \mod(X^n + 1) \qquad (9.105)$$

Modulo $(X^n + 1)$ implies that the multiplication is subject to the constraint $X^n = 1$; the application of this constraint restores the polynomial $X^i c(X)$ to the order $n-1$ for all $i < n$.

9.4.1 Generator Polynomial and Encoding of Cyclic Codes

The polynomial $X^n + 1$ and its factors play a major role in the generation of cyclic codes.

Let $g(X)$ be a generator polynomial of degree $n-k$ and is a factor of $X^n + 1$. Then $g(X)$ is a polynomial of least degree in the code and may be expressed as follows:

$$g(X) = 1 + \sum_{i=1}^{n-k-1} g_i X^i + X^{n-k}, \quad g_i \in \{0,1\}$$

$$(9.106)$$

where the coefficients $\{g_i\}$ are equal to 0 or 1. A cyclic code is uniquely determined by its generator polynomial $g(X)$ since each code polynomial in the code may be expressed in the form of a polynomial product as follows:

$$\underbrace{c(X)}_{n-1} = \underbrace{a(X)}_{k-1} \underbrace{g(X)}_{n-k} \qquad (9.107)$$

where $a(X)$ is a polynomial in X with degree $k-1$. The $c(X)$ so formed is of degree $n-1$ and satisfies (9.105) since $g(X)$ is a factor of $X^n + 1$.

Given the generator polynomial $g(X)$, 2^k codewords a (n,k) cyclic code may be constructed in non-systematic form if $a(X)$ is replaced by the message polynomial of degree $k-1$:

$$m(X) = m_0 + m_1 X + m_2 X^2 + \cdots + m_{k-1} X^{k-1}$$

$$(9.108)$$

which corresponds to the message word of k bits-long. A simple approach in constructing the code is to replace $a(X)$ by $1, X,\ldots, X^{k-1}$ in (9.107). Note from (9.108) that $m(X) = X^i$ implies a message word whose i-th bit is 1 and all others are zero. Then, the code (generator matrix G) may be constructed since k polynomials $g(X), Xg(X),\ldots, X^{k-1}g(X)$ span the code. Hence, n-tuples corresponding to these polynomials are used as rows of the $k \times n$ generator matrix $\mathbf{G}$. The generator matrix thus formed is in non-systematic form, but it can be put into systematic form using well-known matrix operations.

Now consider that we are given the generator polynomial $g(X)$ and we want to encode the message sequence $\mathbf{m} = (m_0,\ m_1,\ldots,\ m_{k-1})$ into an (n, k) systematic cyclic code:

$$\mathbf{c} = \begin{pmatrix} b_0 & b_1 & \cdots & b_{n-k-1} \vdots & m_0 & m_1 & \cdots & m_{k-1} \end{pmatrix} \Rightarrow$$

$$c(X) = b_0 + b_1 X + \cdots + b_{n-k-1} X^{n-k-1}$$

$$+ m_0 X^{n-k} + m_1 X^{n-k+1} + \cdots m_{k-1} X^{n-1}$$

$$(9.109)$$

The message polynomial of degree $k-1$ is given by (9.108), while the parity-check polynomial of degree $n-k-1$ is defined as

$$b(X) = b_0 + b_1 X + b_2 X^2 + \cdots + b_{n-k-1} X^{n-k-1}$$

$$(9.110)$$

Since we want the code polynomial to be in the systematic form as in (9.109), we shift the message polynomial by $n-k$ bits so that it follows the $n-k$ bits of the parity-check polynomial. Therefore the code polynomial (9.109) may be rewritten as

$$c(X) = b(X) + X^{n-k} m(X) \qquad (9.111)$$

Using (9.107) and modulo-2 addition property, (9.111) may also be expressed as

$$\frac{X^{n-k} m(X)}{g(X)} = a(X) + \frac{b(X)}{g(X)} \qquad (9.112)$$

This implies that the parity-check polynomial $b(X)$ is the remainder left over after dividing $X^{n-k} m(X)$ by $g(X)$. It also offers a simple approach for encoding cyclic codes in systematic form: Divide $X^{n-k} m(X)$ by $g(X)$; find the remainder which is the parity-check polynomial $b(X)$; and use (9.111) to obtain the corresponding code polynomial.

Example 9.13 (7, 4) Cyclic Hamming Code in Systematic Form with Generator Polynomial $g(X) = 1 + X + X^3$.

To construct a (7, 4) cyclic Hamming code, we start by factorizing $X^7 + 1$ into three irreducible polynomials:

$$X^7 + 1 = (1 + X)(1 + X^2 + X^3)(1 + X + X^3)$$
(9.113)

An irreducible polynomial can not be factored further. An irreducible polynomial of degree m is said to be primitive if the smallest positive integer n for which the polynomial divides $X^n + 1$ is $n = 2^m - 1$. The two polynomials $(1 + X^2 + X^3)$ and $(1 + X + X^3)$ are therefore primitive. The generator polynomial for a (7,4) code must be of degree $n - k = 3$. Therefore, both of these primitive polynomials can be used as the generator polynomial for this code. For comparison with the code in Example 9.12, the generator polynomial $g(X)$ is defined by

$$g(X) = 1 + X + X^3$$
(9.114)

In this example, we will determine the codewords of this (7, 4) cyclic code and compare with those listed in Table 9.9. Let us first determine the codeword for the message $m_1 = (1000)$ for which the message polynomial is $m_1(X) = 1$ (see (9.108)). Dividing $X^{n-k}m_1(X) = X^3$ by $g(X)$, one gets

$$X^3 = \underbrace{1}_{\text{quotient } a(X)} \underbrace{(1 + X + X^3)}_{g(X)} + \underbrace{(1 + X)}_{\text{remainder } b(X)}$$
(9.115)

Then using (9.111), one gets the following code polynomial:

$$c_1(X) = b(X) + X^3 m_1(X) = 1 + X + X^3$$
$$c_1 = \underbrace{(1\ 1\ 0)}_{\text{parity bits}}\ \underbrace{(1\ 0\ 0\ 0)}_{\text{message bits}}$$
(9.116)

which is identical to the codeword for the message 1000 listed in Table 9.9 and in the first row

of the generator matrix (9.94). Similarly, we now use the message polynomial $m_2(X) = X^2$ for the message $m_2 = (0010)$. Dividing $X^{n-k}m_2(X) = X^5$ by $g(X)$, the remainer is found to be $b(X) = 1 + X + X^2$. The corresponding code polynomial and codeword are given by

$$c_2(X) = b(X) + X^3 m_2(X) = 1 + X + X^2 + X^5$$
$$c_2 = \underbrace{(1\ 1\ 1)}_{\text{parity bits}}\ \underbrace{(0\ 0\ 1\ 0)}_{\text{message bits}}$$

(9.117)

Next we select the message $m_3 = (1010)$, which may be written as $m_3 = m_1 \oplus m_2$. The corresponding message polynomial and the remainder are respectively given by $m_3(X) = 1 + X^2$ and $b(X) = X^2$. Then, the corresponding code polynomial and codeword are found to be

$$c_3(X) = b(X) + X^3 m_3(X) = X^2 + X^3 + X^5$$
$$c_3 = \underbrace{(0\ 0\ 1)}_{\text{parity bits}}\ \underbrace{(1\ 0\ 1\ 0)}_{\text{message bits}}$$

(9.118)

Note that the linearity principle holds since $m_3 = m_1 \oplus m_2 \Rightarrow c_3 = c_1 \oplus c_2$ (see Table 9.9 and (9.47)).

The encoding procedure for an (n,k) cyclic code in systematic form may be summarized as follows:

Multiply the message polynomial $m(X)$ by X^{n-k}.

Divide $X^{n-k} m(X)$ by the generator polynomial $g(X)$, and obtain the remainder $b(X)$.

Add $b(X)$ to $X^{n-k} m(X)$, to obtain the code polynomial.

Obtain other codewords from the code polynomial found above

The steps outlined above may be implemented by means of an encoder consisting of linear feedback shift registers of $n-k$ stages as shown in Figure 9.18. Note that the ith value g_i of the generator polynomial may be either 1 or 0, which implies respectively the presence or absence of a physical connection to the corresponding modulo-2 adder. The coefficients g_0 and g_{n-k} are always unity as defined by (9.106).

Encoding operation by the encoder in Figure 9.18 proceeds as follows:

As the gate is switched on, the k message bits are fed into the encoder channel and are transmitted at the same time. As soon as the k message bits have entered the encoder, the gate is switched off and the contents of the encoder (resulting $n-k$ parity bits in the encoder) are fed into the channel. Note that the $n-k$ parity bits in the encoder are the same as the coefficients of the remainder $b(X)$. Hence, the encoder performs the division operation defined in (9.112).

9.4.2 Parity-Check Polynomial

A (n, k) cyclic code is uniquely specified by its generator polynomial $g(X)$ of order $n-k$. Such a code may also be uniquely specified by its parity-check polynomial $h(X)$ of degree k,

$$h(X) = 1 + \sum_{i=1}^{k-1} h_i X^i + X^k, \quad h_i \in \{0,1\} \quad (9.119)$$

which has a similar form as the generator polynomial. The generator polynomial $g(X)$ is equivalent to the generator matrix $\mathbf{G}$ for characterizing a block code. Correspondingly, the parity-check polynomial $h(X)$ is an equivalent representation of the parity-check matrix $\mathbf{H}$. Consequently, the matrix relation $\mathbf{GH}^T = 0$ corresponds to the following:

$$g(X)\, h(X) \bmod (X^n + 1) = 0 \qquad (9.120)$$

The generator polynomial $g(X)$ and the parity-check polynomial $h(X)$ are factors of the polynomial $X^n + 1$:

$$g(X)\, h(X) = X^n + 1 \qquad (9.121)$$

However, the construction of the parity-check matrix $\mathbf{H}$ of a cyclic code using the parity-check polynomial requires special attention. Multiplying both sides of (9.121) by $a(X)$ and then using $c(X) = a(X)\, g(X)$ (see (9.107)), one gets

$$\underbrace{c(X)}_{degree \le n-1}\ \underbrace{h(X)}_{degree \le k} = \underbrace{a(X)}_{degree \le (k-1)} + \underbrace{X^n a(X)}_{n \le degree \le (k+n-1)}$$
$$(9.122)$$

The product $c(X)\, h(X)$ contains terms with powers extending up to $n + k - 1$, while the polynomial $a(X)$ has degree $k-1$ or less; thus $X^n a(X)$ contain terms $X^n,...., X^{k+n-1}$. This implies that the powers of $X^k, X^{k+1},...., X^{n-1}$ do not appear on the right-hand side of (9.123). Therefore, we set the coefficients of $X^k, X^{k+1},...., X^{n-1}$ in the expansion of the product $c(X)\, h(X)$

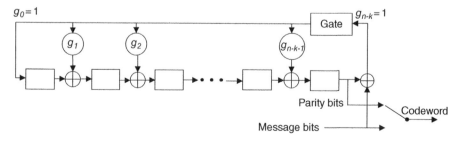

$g_0 = 1$

Gate $g_{n-k} = 1$

g_1 g_2 g_{n-k-1}

Parity bits

Codeword

Message bits

Figure 9.18 Encoder For an (n, k) Cyclic Code.

equal to zero, and obtain the following set of $n-k$ equations:

$$\sum_{i=j}^{j+k} c_i \, h_{k+j-i} = 0 \ \ for \ \ 0 \le j \le n-k-1 \quad (9.123)$$

Comparing (9.123) with $\mathbf{c}\mathbf{H}^T = \mathbf{0}$, one observes that the coefficients of the parity-check polynomial $h(X)$ in the above equation are arranged in reversed order with respect to the coefficients of the parity-check matrix $\mathbf{H}$ in $\mathbf{c}\mathbf{H}^T = \mathbf{0}$. This observation suggests that we define the reciprocal of the parity-check polynomial as follows:

$$X^k h\left(X^{-1}\right) = X^k \left(1 + \sum_{i=1}^{k-1} h_i X^{-i} + X^{-k}\right)$$

$$= 1 + \sum_{i=1}^{k-1} h_{k-i} X^i + X^k$$

$$(9.124)$$

The n-tuples pertaining to the $n-k$ polynomials $X^k \ h(X^{-1})$, $X^{k+1}h(X^{-1}),\ldots,X^{n-1}h(X^{-1})$ may therefore be used as rows of the $(n-k) \times n$ parity-check matrix $\mathbf{H}$.

9.4.3   Syndrome Decoding of Cyclic Codes

Suppose a codeword defined by the code polynomial (9.100) is transmitted over an AWGN channel. The received polynomial at the decoder input, that is, the 1-bit quantized version of the observation vector, may be written as

$$r(X) = r_0 + r_1 X + r_2 X^2 + \cdots + r_{n-1} X^{n-1} \quad (9.125)$$

In view of (9.65), the received word may be written as

$$r(X) = c(X) + e(X) \quad (9.126)$$

where $c(X)$ is defined by (9.100) and the error polynomial defined by

$$e(X) = e_0 + e_1 X + e_2 X^2 + \cdots + e_{n-1} X^{n-1} \quad (9.127)$$

Similar to (9.66), the coefficients of the error polynomial $e(X)$ are defined as follows:

$$e_i = \begin{cases} 1 & r_i \ne c_i \\ 0 & r_i = c_i \end{cases}, \quad i = 0, 1, \cdots, n-1 \quad (9.128)$$

The received polynomial $r(X)$ may be also be expressed as follows:

$$r(X) = q(X) \, g(X) + s(X) \quad (9.129)$$

where the so-called syndrome polynomial $s(X)$ is of degree $n-k-1$ or less, similar to the syndrome vector in (9.76). One may write (9.129) in a similar form as (9.112); therefore, $s(X)$ is the remainder when the received polynomial r (X) is divided by $g(X)$. Since a valid code polynomial is divisible by the generator polynomial without a remainder (see (9.107)), $s(X)$ is zero if the received code polynomial $r(X)$ is a valid codeword, that is, if $r(X) = c(X)$, then $e(X) = 0$, $s(X) = 0$ and $q(X) = a(X)$. Consequently, (9.129) may be used to determine whether a received polynomial is in error. The first step in decoding of a linear block code is therefore to divide the received polynomial $r(X)$ by $g(X)$; the remainder is the syndrome polynomial. If the syndrome polynomial is zero, then there are no transmission errors in the received word. If the syndrome is nonzero, the received word contains transmission errors to be corrected.

Using the syndrome polynomial thus found, the location of error is estimated by $\hat{e}(X)$ and the received polynomial is corrected to get an estimate $\hat{c}(X)$ of the transmitted codeword:

$$\hat{c}(X) = r(X) + \hat{e}(X) \quad (9.130)$$

The syndrome polynomial $s(X)$ has the following useful properties.

a. The syndrome of a received codeword polynomial is also the syndrome of the corresponding error polynomial:

$$r(X) = c(X) + e(X) = q(X)g(X) + s(X) \implies$$
$$e(X) = r(X) + \underbrace{c(X)}_{a(X)g(X)} = [\underbrace{a(X) + q(X)}_{u(X)}]g(X) + s(X)$$

$$(9.131)$$

(9.131) implies that the syndrome polynomial is the remainder when the error polynomial or the received polynomial is divided by the generator polynomial. The syndrome polynomial depends only on the error polynomial and is hence independent of the code polynomial.

b. If $s(X)$ is the syndrome polynomial of a received word polynomial $r(X)$ as given by (9.129), then

$$X^i r(X) = X^i q(X)g(X) + X^i s(X) \quad (9.132)$$

the syndrome of $X^i r(X)$, i^{th} cyclic shift of $r(X)$, is given by $X^i s(X)$.

c. If the errors are confined to the $n-k$ parity-check bits of the received code polynomial, then the error polynomial $e(X)$ is of degree $n-k-1$ or less. Then the syndrome polynomial $s(X)$ becomes identical to the error polynomial $e(X)$. Because $g(X)$ is of degree $n-k$, the term $u(X)g(X)$ is of degree $\geq n-k$. In order that $e(X) \leq n-k-1$, one must have $u(X)g(X) \equiv 0$ in (9.131), implying $e(X) = s(X)$.

The syndrome calculator shown in Figure 9.19 is identical to the encoder in Figure 9.18 except for the fact that the received bits

are fed into the $(n-k)$ stages of the feedback shift register from the left. As soon as all the received bits have been shifted into the shift register, the gate is opened and the contents of the $(n-k)$ shift registers define the syndrome.

Example 9.14 Correspondance Between Generator and Parity-Check Polynomials and Generator and Parity-Check Matrices For (7, 4) Hamming Code.

Consider the (7, 4) cyclic Hamming code considered in Examples 9.12 and 9.13. The codeword corresponding to $m(X) = 1$, is given by $c(X) = m(X)g(X) = g(X)$. Noting that $X^i c(X)$ is also codeword, we can form the generator matrix in non-systematic form using the generator polynomial $g(X)$ and its three cyclic shifts:

$$G = \begin{bmatrix} g(X) \\ Xg(X) \\ X^2 g(X) \\ X^3 g(X) \end{bmatrix} = \begin{bmatrix} 1+X+X^3 \\ X+X^2+X^4 \\ X^2+X^3+X^5 \\ X^3+X^4+X^6 \end{bmatrix}$$

$$(9.133)$$

$$= \begin{bmatrix} 1 & 1 & 0 & 1 & 0 & 0 & 0 \\ 0 & 1 & 1 & 0 & 1 & 0 & 0 \\ 0 & 0 & 1 & 1 & 0 & 1 & 0 \\ 0 & 0 & 0 & 1 & 1 & 0 & 1 \end{bmatrix}$$

By adding the first row to the third row, and adding the sum of the first two rows to the fourth row, one obtains the generator matrix **G** in systematic form given by (9.94).

One may construct the 3×7 parity-check matrix **H** from the parity-check polynomial $h(X)$ using (9.124). We first take $X^4 h(X^{-1})$ and two cyclic shifts of it to get

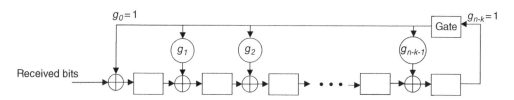

Figure 9.19 Syndrome Calculator For a (n, k) Cyclic Code.

$$\mathbf{H} = \begin{bmatrix} X^4 h(X^{-1}) \\ X^5 h(X^{-1}) \\ X^6 h(X^{-1}) \end{bmatrix} = \begin{bmatrix} 1 + X^2 + X^3 + X^4 \\ X + X^3 + X^4 + X^5 \\ X^2 + X^4 + X^5 + X^6 \end{bmatrix}$$

$$= \begin{bmatrix} 1 & 0 & 1 & 1 & 1 & 0 & 0 \\ 0 & 1 & 0 & 1 & 1 & 1 & 0 \\ 0 & 0 & 1 & 0 & 1 & 1 & 1 \end{bmatrix}$$

$$(9.134)$$

The matrix **H** in systematic form as given by (9.96) may be obtained by adding the third row to the first row in (9.134).

Figure 9.20 shows the block diagram of the encoder for the (7, 4) code with generator polynomial $g(X) = 1 + X + X^3$. Table 9.11 shows the contents of the shift register in the encoder for this code for the input message sequence 1001. Figure 9.21 shows the block diagram of the syndrome calculator for this (7, 4) block code. Table 9.12 shows the operation of syndrome calculator for this code for the received codeword $\mathbf{r} = 0110001$, where the last bit enters

Table 9.11 Operation of Encoder For (7, 4) Code For the Input Bit Sequence 1001.

Shift	Input	Register contents
		000 (inital state)
1	1	110
2	0	011
3	0	111
4	1	011 (parity bits)

Table 9.12 Contents of the Shift Registers of the Syndrome Calculator Shown in Figure 9.21.

Shift	Input bit	Contents of shift register
		000 (initial state)
1	1	100
2	0	010
3	0	001
4	0	110
5	1	111
6	1	001
7	0	110 (syndrome)

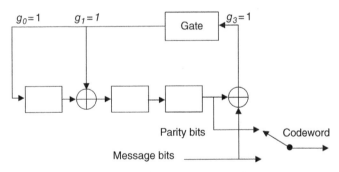

Figure 9.20 Encoder For the (7, 4) Cyclic Code Generated By $g(X) = 1 + X + X^3$.

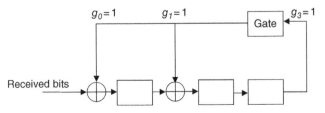

Figure 9.21 Syndrome Calculator For the (7,4) Cyclic Code Generated By the Polynomial $g(X) = 1 + X + X^3$.

first the syndrome calculator. Using Table 9.10, the error pattern corresponding to the syndrome $s = 110$ is estimated to be $\hat{e} = 0001000$. The codeword is then estimated as $\hat{c} = r \oplus \hat{e} = 0111001$.

9.4.4 Cyclic Block Codes

9.4.4.1 Maximum–Length Codes

Maximum-length codes are defined by $(n, k) = (2^m - 1, m)$ for any positive integer $m \geq 3$, where m denotes the number of shift registers in the encoder. These codes, which have a minimum distance $d_{min} = 2^{m-1}$, are generated by polynomials satisfying (9.120), where $g(X)$ is any primitive polynomial of degree m. The encoder consists of m-stage shift registers with feedback. Since any cyclic code generated by a primitive polynomial is a Hamming code of minimum distance 3, maximum-length codes are the dual of Hamming codes.

Consider a $(7, 3)$ maximum-length code with the following generator and parity-check polynomials:

$$g(X) = 1 + X + X^3$$
$$h(X) = (X^7 + 1)/g(X) = 1 + X + X^2 + X^4$$
$$(9.135)$$

The encoder for the generator polynomial (9.135) is shown in Figure 9.22 with the initial state 100, which is determined by the source. The contents of the shift registers are shifted one stage at a time to the right at each clock pulse of duration T_c for a total of $n = 2^m - 1$

shifts, hence, a systematic code of length $n = 2^m - 1$ and duration nT_c is generated.

Figure 9.23 shows the evolution of the states and the output sequence (codeword) starting from the initial state 100. Figure 9.23 clearly shows that the maximum length codes are cyclic since the encoder returns to its initial state after $2^m - 1 = 7$ shifts by the clock. The 000 initial state is never used, since it forces the encoder to remain always in this state. Therefore, the encoder can have one of $2^m - 1$ initial

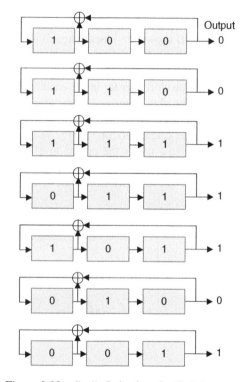

Figure 9.23 Cyclic Behavior of a $(7, 3)$ Maximal-Length Code.

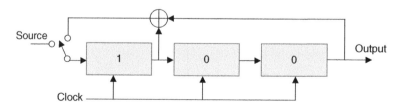

Figure 9.22 Encoder For the $(7, 3)$ Maximal-Length Code with the Initial State 100.

Table 9.13 Seven States For the $m = 3$ Maximum-Length Shift Register and the Corresponding Codewords.

Initial state	Codewords
000	0000000
001	0011101
010	0100111
011	0111010
100	1001110
101	1010011
110	1101001
111	1110100

states and can generate $2^m - 1$ different codewords each of length $2^m - 1$ (see Table 9.13).

As one may also observe from Table 9.13, all codewords have identical weights $d_{min} = 2^{m-1}$; each codeword contains 2^{m-1} ones and $2^{m-1} - 1$ zeros. Therefore, maximum-length codes satisfy the balance property, that is, in each period nT_c of the sequence, the difference between the number of 1's and 0's is always unity. Bearing in mind that a *run* is defined as a sequence of a single type of binary digits, the run lengths appear random in the codewords. Consequently, maximal-length codes appear as random binary sequences for sufficiently large values of n.

Binary sequences generated by maximum-length codes are periodic with a period $nT_c = (2^m - 1)T_c$. Therefore, they have a sharp (impulse-like) autocorrelation function, which is also periodic with a period nT_c. The autocorrelation function of a discrete periodic signal c(t) of period T_b may be written as

$$R_c[k] = \frac{1}{n}\sum_{i=0}^{n-1} c[i]\, c[i-k], \quad k = 0, 1, \cdots n-1$$

(9.136)

When a binary m-sequence is correlated by a shifted version of itself, the difference between the number of agreements and the number of disagreements is always equal to -1. For example, the m-sequences 0011101 and

1001110 (one shifted version of itself) have 3 agreements and 4 disagreements. Consequently, the autocorrelation function $R_c(i) = -1/n$ for $1 \le i \le n-1$. The value of the autocorrelation function between $R_c(0) = 1$ and $R_c(\pm 1) = -1/n$ behaves as the autocorrelation of a single rectangular pulse of duration T_c, which is triangular (see Example 1.6 and Example 1.13). On the other hand, the periodicity of the m-sequence implies $R_c(in) = 1$, $i = 0, 1, 2, \ldots$ Therefore, the autocorrelation function and the corresponding PSD a maximum-length sequence may be written as

$$R_c(\tau) = \begin{cases} 1 - \dfrac{n+1}{nT_c}|\tau| & |\tau| \le T_c \\ \\ -1/n & \text{otherwise} \end{cases}$$

$$S_c(f) = \frac{1}{n^2}\delta(f) + \frac{1+n}{n^2}\sum_{\substack{i=-\infty \\ i \ne 0}}^{\infty} sinc^2\left(\frac{i}{n}\right) \delta\left(f - \frac{i}{nT_c}\right)$$

(9.137)

For large values of n, $1/n$ approaches zero; then the autocorrelation function given by (9.137) approximates closely the ideal autocorrelation function $R_c[k] = \delta[k]$, $k = 0, 1, \cdots n-1$. As the shift-register length m, and thus n, of the maximum-length sequence increase, the maximum-length sequence becomes increasingly similar to a PN sequence. In the limit, the two sequences become identical when n is made infinitely large. However, large n implies an infinite bandwidth and increasing storage requirements for the generation of the sequence.

Figure 9.24 shows the variation of the autocorrelation function and its PSD. As the shift-register length m (hence n) increases, the maximum-length sequence becomes increasingly similar to a random binary sequence. In the limit, the autocorrelation function looks more like a delta function. Assume a maximum-length codes with a period $T = nT_c$, where T denotes the uncoded symbol duration and T_c is the chip (coded symbol) duration, that

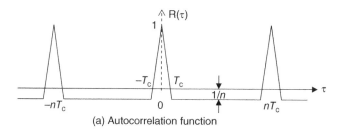

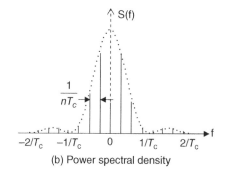

(b) Power spectral density

Figure 9.24   Autocorrelation Function and the Corresponding PSD For Maximal-Length Codes.

is, $1/n$ of the symbol duration. Then, the null-to-null bandwidth $2/T_c$ of the PN sequence is $n = 2^m - 1$ times larger than $2/T$, null-to-null bandwidth of the uncoded data. This implies that encoding by the maximum-length sequence spreads the bandwidth of the uncoded data signal by a factor of n. As $n = 2^m - 1$ increases, the autocorrelation function becomes sharper and the PSD becomes more spread.

Due to their noise-like appearance, the maximal-length codes are also called as PN (pseudo-noise) sequences since they show the properties of PN codes used in spread-spectrum applications and in data scrambling. Nevertheless, Gold sequences constructed by the XOR'-ing two m-sequences with the same clocking or Kasami sequences are more robust against jamming and interception. If maximum-length codes are used in CDMA applications, each of the $2^m - 1$ different encoder states can be allocated to serve one of $2^m - 1$ different users. Orthogonality of codes, and the number of users that can be served are evidently increased by increasing m, the number of shift registers.

Table 9.14   Connection Diagram for Maximal-Length Codes. [3][5]

m	Feedback taps
2	(2,1)
3	(3,1)
4	(4,1)
5	(5,2) (5,4,3,2), (5,4,2,1)
6	(6,1), (6,5,2,1), (6,5,3,2)

The properties of maximum-length codes are determined by their generator polynomials. Optimum feedback connections (generator polynomial) for a given number of shift registers in the encoder is determined by using extensive computer simulations. Note that for a given set of feedback connections, there is an image set that generates an identical maximum-length code, reversed in time sequence. The optimum generator polynomials for some small values of m are given in Table 9.14. [3][5]

PN sequences may also be used as data scramblers. A scrambler prevents the transmission of long strings of 0 s and 1 s that may appear in raw data or in other circumstances. Long sequences of 0 s or 1 s may cause difficulties in synchronization/tracking circuits and can lead to non-even distribution of the PSD of the transmitted signal; this may potentially cause interference to other signals in adjacent channels. Scramblers can also be used for data encryption in the most simple form, if the initial state of the PN sequence generator is known only to the transmitter and the receiver.

A scrambler may be implemented by modulo-2 adding the output of a PN sequence generator with the data. At the receiver, the incoming scrambled data is modulo-2 added with the output of a de-scrambler, which is synchronized with the scrambler at the transmitter. Synchronization implies that that scrambler and de-scrambler should be initialized to the same state at the same point in the data sequence. Scrambler, which changes the data, should not be confused with interleaver, which does not change the data but simply alters the order of transmission.

9.4.4.2 Walsh-Hadamard Codes

Walsh-Hadamard codes are perfectly orthogonal codes and the number of codewords is limited by the code length. Perfect orthogonality between codewords distinguishes these codes to be widely used in radar applications and in CDMA-based cellular radio systems. In cellular radio applications, these codes of length n can be used to serve n users per cell without multi-user interference (MUI). Nevertheless, the orthogonality which exists between the user signals sharing the same channel in a fully synchronized communication system may still be destroyed in a multipath channel.

These codes are generated by the rows of the Hadamard matrices. Since the orders of the Hadamard matrices are restricted to the powers of 2, the number of codewords is limited by the matrix size. A $n \times n$ Hadamard matrix is

constructed recursively by lower order Hadamard matrices as

$$\mathbf{H}_1 = 1, \quad \mathbf{H}_2 = \begin{bmatrix} 1 & 1 \\ 1 & 0 \end{bmatrix}, \quad \mathbf{H}_{2n} = \begin{bmatrix} \mathbf{H}_n & \mathbf{H}_n \\ \mathbf{H}_n & \mathbf{H}_n^c \end{bmatrix}$$

(9.138)

where $\mathbf{H}_n^c$ is the complement of the $\mathbf{H}_n$ matrix, that is, each element of $\mathbf{H}_n$ is replaced by its complement. This means that Walsh-Hadamard codes can have code lengths $n = 2$, 4, 8, 16, 32,... For example, Hadamard matrices $\mathbf{H}_4$ and $\mathbf{H}_8$ are given by

$$\mathbf{H}_4 = \begin{bmatrix} 1 & 1 & 1 & 1 \\ 1 & 0 & 1 & 0 \\ 1 & 1 & 0 & 0 \\ 1 & 0 & 0 & 1 \end{bmatrix}, \quad \mathbf{H}_8 = \begin{bmatrix} 1 & 1 & 1 & 1 & 1 & 1 & 1 & 1 \\ 1 & 0 & 1 & 0 & 1 & 0 & 1 & 0 \\ 1 & 1 & 0 & 0 & 1 & 1 & 0 & 0 \\ 1 & 0 & 0 & 1 & 1 & 0 & 0 & 1 \\ 1 & 1 & 1 & 1 & 0 & 0 & 0 & 0 \\ 1 & 0 & 1 & 0 & 0 & 1 & 0 & 1 \\ 1 & 1 & 0 & 0 & 0 & 0 & 1 & 1 \\ 1 & 0 & 0 & 1 & 0 & 1 & 1 & 0 \end{bmatrix}$$

(9.139)

The orthogonality of Walsh-Hadamard codes implies that $\mathbf{H}_n \mathbf{H}_n^T = n\mathbf{I}$, One may also note $d_{\min} = n$ for $\mathbf{H}_{2n}$; this implies that Walsh-Hadamard codewords for $\mathbf{H}_4$ can detect one bit error but can correct no errors. However, Walsh-Hadamard codes for $\mathbf{H}_8$ have $d_{\min} = 4$; hence they can detect three bit errors or correct single bit errors (see (9.72)) [6].

Example 9.15 Variable-Rate Spreading With Walsh-Hadamard Codes.

Consider a wideband CDMA system with two users. User A transmits at bit rate R_b and user B transmits at bit rate $2R_b$. The codes for users A and B are given by the following Walsh-Hadamard codes, where bit 1 is represented by +1 and bit 0 by −1

$$c'_A = [+1 \quad +1 \quad +1 \quad +1 \quad -1 \quad -1 \quad -1 \quad -1]$$
$$c'_B = [+1 \quad -1 \quad +1 \quad -1]$$

$$(9.140)$$

so that they occupy the same spread spectrum bandwidth. User A transmit bit **1** at bit duration $T_b = 1/R_b$. However, user B, transmitting twice as fast as user A, transmits the bit sequence **1 0** during T_b.

The transmitted coded sequences for users A and B and the transmitted signal at the channel output (i.e., at the receiver input) may then be written as:

$$s_A = [+1 \quad +1 \quad +1 \quad +1 \quad -1 \quad -1 \quad -1 \quad -1]$$
$$s_B = [+1 \quad -1 \quad +1 \quad -1 \quad -1 \quad +1 \quad -1 \quad +1]$$
$$s_A + s_B = [+2 \quad 0 \quad +2 \quad 0 \quad -2 \quad 0 \quad -2 \quad 0]$$

$$(9.141)$$

Assuming that no bit errors are induced by the channel, the signal at the channel output (i.e., at the receiver input) is the same as the total transmitted signal, $r = s_A + s_B$, ignoring the noise. The receiver communicating with user A correlates the received signal r with c'_A to decode the desired signal:

$$r = [+2 \quad 0 \quad +2 \quad 0 \quad -2 \quad 0 \quad -2 \quad 0]$$
$$c'_A = [+1 \quad +1 \quad +1 \quad +1 \quad -1 \quad -1 \quad -1 \quad -1]$$
$$\langle r.c'_A \rangle = \underbrace{+2+0+2+0+2+0+2+0 = 8}_{\Rightarrow +1}$$

$$(9.142)$$

The receiver communicating with user B correlates the received signal r with c'_B over $2T_b$ to decode the desired signal:

$$r = \left[+2 \quad 0 \quad +2 \quad 0 \vdots -2 \quad 0 \quad -2 \quad 0\right]$$
$$c'_B = \left[+1 \quad -1 \quad +1 \quad -1 \vdots +1 \quad -1 \quad +1 \quad -1\right]$$
$$\langle r.c'_B \rangle = \underbrace{+2+0+2+0 = 4}_{\Rightarrow +1} \quad \underbrace{-2+0-2+0 = -4}_{\Rightarrow -1}$$

$$(9.143)$$

9.4.4.3 Bose-Chaudhuri-Hocquenghem (BCH) Codes

BCH codes, a class of random error-correcting cyclic codes, were invented in 1959 independently by Bose, Chaudhuri and Hocquenghem. They are generalization of Hamming codes to allow multiple error correction and commonly employ binary alphabet. BCH codes are highly flexible and have acceptable error thresholds. Hence BCH codes are multilevel, cyclic, random error-correcting, variable-length codes. They are used to correct multiple random error patterns. At block lengths of a few hundred, BCH codes outperform all other block codes with comparable block lengths and code rates. BCH codes can be easily decoded via syndrome decoding.

BCH codes are characterized by the following parameters:

The codeword length:	$n = 2^m - 1$, $m = 3, 4,...$
The number of correctable errors:	$t \geq (n-k)/m$
The minimum distance:	$d_{min} \geq 2t+1$

Figure 9.25 shows the bit error probability performance of $(63, k)$ BCH codes for $n = 63$ ($m = 6$). The number k of message bits at the encoder input is selected so that the code has $t = 1, 2, 3$ and 4 bit error correcting capability. The corresponding values of k are found as 57, 51, 45 and 39. The coding gain was observed to saturate for $t \geq 3$ at sufficiently high SNR values. Table 9.15 shows tradeoff between the error correction capability of the code (measured in terms of t/n and coding gain) and the increase in the transmission bandwidth.

9.4.4.4 Golay Codes

Golay code is a $(23,12)$ perfect cyclic code with the generator polynomial, $g(X) = X^{11} + X^9 + X^7 + X^6 + X^5 + X + 1$. It has a minimum distance of $d_{min} = 7$ and hence the number of correctable errors is $t = 3$. The extended Golay code (24, 12

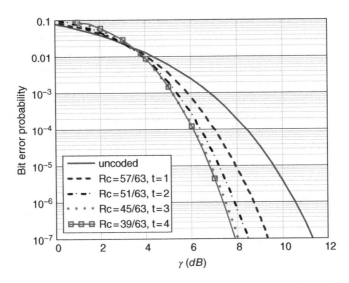

Figure 9.25   Bit Error Probability Performance of BCH Codes with $n = 63$ ($m = 6$) and $t = 1,2,3,4$.

Table 9.15   Error Correcting Capability Versus Required Bandwidth Increase in Some $(63, k)$ and $(127, k)$ BCH codes. [2][4]

n	k	t	R_c	Generator Polynomial (octal)[*]	Bandwidth increase, $1/R_c$	Coding effectiveness, t/n	Coding gain, G_c (dB) @ 10^{-5}
63	63	0	1	1	1	0	0
63	57	1	0.905	103	1.105	0.016	1.7
63	51	2	0.810	12471	1.235	0.032	2.5
63	45	3	0.714	1701317	1.4	0.048	2.7
63	39	4	0.619	166623567	1.615	0.064	2.8
127	120	1	0.945	211	1.058	0.00787	1.7
127	113	2	0.89	41567	1.124	0.0157	2.6
127	106	3	0.835	11554743	1.198	0.0236	3.1

*For (63, 57) BCH code generator polynomial in octal form 103 may be written as 001 000 011 in binary form. The generator polynomial is then $x^6 + x + 1$.

is obtained by adding an additional parity bit to the perfect (23, 12) Golay code. (24, 12) code is easier to implement because of the code rate $R_c = 1/2$ and has a minimum distance $d_{min} = 8$. This code is guaranteed to correct all triple errors and can be designed to correct some but not all four-error patterns. The weight distributions of perfect (23, 12) and (24, 12) extended Golay codes are shown in Table 9.16.

Probability of symbol and bit errors are respectively given by (9.86) and (9.87) for hard-decision decoding and by (9.92) and (9.93) for soft-decision decoding. Figure 9.26 shows the symbol error probability of the (23, 12) Golay code for hard- and soft-decision decoding. For comparison, the error probability of an uncoded symbol of 12-bits is also shown. Note the 1.9 dB performance advantage for soft-decision decoding over hard-decision decoding at 10^{-7} symbol error probability.

Figure 9.27 shows the bit error probability of some of the FEC codes already considered; (15, 11) Hamming code with $t = 1$, (23, 12) Golay and (31, 16) BCH codes with $t = 3$ and (63, 39) BCH code with $t = 4$. It might be useful to compare the performances of the two $t = 3$ bit error correcting block codes with rates close to each other. (23, 12) Golay code has $R_c = 12/23 = 0.522$, $t/n = 3/23 = 0.13$ and $G_c = 2.6$ dB at 10^{-5} bit error probability while (31, 16) BCH code has $R_c = 16/31 = 0.516$, $t/n = 3/31 = 0.097$ and $G_c = 1.3$ dB at 10^{-5} bit error probability. Thus, Golay code outperforms the similar BCH code. Nevertheless, the performance difference becomes smaller at high SNR

values. The (63, 39) BCH code has the following performance parameters: $R_c = 39/63 = 0.62$, $t/n = 4/63 = 0.063$ and the coding gain of $G_c = 2.79$ dB at 10^{-5} bit error probability, which is only 0.19 dB higher than that for (23, 12) Golay code with $t = 3$. Table 9.17 shows, the error correction capabilities of some FEC block codes. The interested reader may refer to [4] for more detailed tables.

9.4.4.5 Cyclic Redundancy Check (CRC) Codes

CRC codes are one of the best cyclic codes which are widely used for error detection in ARQ systems. The generator polynomials are standardized for interoperability purposes (see Table 9.18). A k-bit input message sequence may be encoded, as described by (9.107), to generate a n-bit codeword, which is exactly divisible by the generator polynomial. The decoder divides the received n-bit sequence by the generator polynomial. The received sequence is accepted as error-free if there is no remainder. Otherwise, it is assumed to be

Table 9.16 Weight Distribution of (23, 12) Golay and (24, 12) Extended Golay Codes. [7]

Weight	(23, 12)	(24, 12)
0	1	1
7	253	0
8	506	759
11	1288	0
12	1288	2576
15	506	0
16	253	759
23	1	0
24	0	1

Table 9.17 Comparison some block codes [4]

Type	n	k	R_c	t
Hamming	7	4	0.57	1
	15	11	0.73	1
	31	26	0.84	1
BCH	15	7	0.46	2
	31	16	0.516	3
	63	39	0.62	4
	63	51	0.81	2
	63	45	0.71	3
	127	120	0.945	1
	127	113	0.89	2
	127	106	0.835	3
	127	85	0.67	6
	255	239	0.94	2
	255	207	0.81	6
	511	457	0.89	6
	1023	963	0.90	6
Golay	23	12	0.522	3

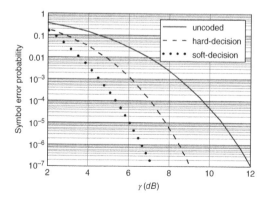

Figure 9.26 Symbol Error Probability of (23,12) Golay Code with Hard- and Soft-Decision Decoding.

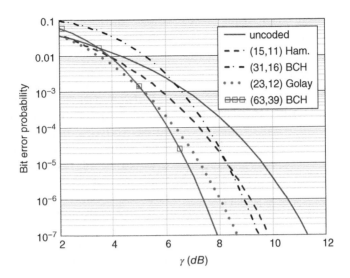

Figure 9.27   Bit Error Probability Performance of Some Codes. Note that $t = 1$ for (15, 11) Hamming code, $t = 3$ for (23, 12) Golay code and (31, 16) BCH code, and $t = 4$ for (63, 39) BCH code.

Table 9.18   Standard Generator Polynomials $g(X)$ For Various CRC Codes.

CRC code	Generator polynomial, $g(X)$
CRC 4	$X^4 + X^3 + X^2 + X + 1$
CRC 7	$X^7 + X^6 + X^4 + 1$
CRC 12	$X^{12} + X^{11} + X^3 + X^2 + X + 1$
CRC 16-ANSI	$X^{16} + X^{15} + X^2 + 1$
CRC 16-CCITT	$X^{16} + X^{12} + X^5 + 1$
CRC 24	$X^{24} + X^{23} + X^{14} + X^{12} + X^8 + 1$
CRC 32	$X^{32} + X^{26} + X^{23} + X^{22} + X^{16} +$ $X^{12} + X^{11} + X^{10} + X^8 + X^7 + X^5 +$ $X^4 + X^2 + X + 1$

in error and a request is made to the transmitter for retransmission.

CRC codes typically detect a large class of bit errors, hence provide a vast improvement over simple parity checking at low cost (see Example 9.5). For example, consider that we send packets of 188 encoded bytes of 8 bits each using an error detecting code. If we use parity-checking code with one parity bit per byte, 188 parity bits are needed; this implies $1/8 = 12.5$ % increase in bandwidth. Parity-checking can detect only 1, 3, 5 and 7 bit errors. However, even number of

bit errors can not detected since they do not violate the parity (see Example 9.5). If we use 24-bit CRC, only 24 CRC bits are needed for all the bits to be transmitted. This implies only $24/(188 \times 8 - 24) = 1.6\%$ increase in transmission bandwidth for detecting many more errors than parity-checking. Hence, CRC is more effective in error detection and has higher bandwidth efficiency.

9.5   Burst Error Correction

In some channels, an entire block of bits can be wiped out occasionaly by some bursty disturbances, such as shadowing, fading, a stroke of lightning or a man-made electrical disturbance. Then, bit errors appear in bursts but not randomly distributed in time. Burst length, b, is defined as a sequence of bits in which the first and the bth bits are in error, with the b-2 bits in between are either in error or received correctly. For example, an error vector $e = \{0010010100\}$ has a burst length of $b = 6$. If parity checking is used for detecting burst errors of length b or less in a block code of length n, then b parity check bits are necessary and sufficient. For example, consider the following message word **m** of length $k = 20$, and the codeword **c** of length $n = 24$:

$$\mathbf{m} = \left[1011\vdots0101\vdots1101\vdots1000\vdots1100\right]$$

$$\mathbf{c} = \left[\mathbf{b}\vdots\mathbf{m}\right] = \left[0111\vdots1011\vdots0101\vdots1101\vdots1000\vdots1100\right]$$

$$(9.144)$$

where the use of even parity-check bits **b** of length 4 enables this code to detect bursts of length $b = 4$. If a bit sequence of length b or less is in error, parity will be violated. The number of parity-check bits, b, is determined so as to detect the burst errors with most likely burst lengths and is independent of k. This is useful in packet switching, where k may vary depending on channel conditions. In slowly fading channels with large coherence time, k may be chosen to be high so as to minimize the signalling overheads. A higher percentage of burst lengths longer than b may be detected. In order to correct burst errors of length b or less, a linear block code must have larger number of parity-check bits.

9.5.1 Interleaving

It is generally more difficult for a FEC code to correct burst errors compared to random bit errors. Interleaving is an effective means of avoiding burst errors at the decoder input. The encoded data is interleaved before transmission; hence neighboring bits propagate in the channel at sufficiently separated time intervals (larger than channel coherence time) so that they do not suffer simultaneous fading. At the receiver a deinterleaver at the decoder input performs the inverse operation and deinterleaved bits with randomized bit errors are fed to the decoder. Then the decoder can correct them more easily.

A *block interleaver* of depth $n \times m$ acts as a memory buffer. It accepts a block of $n \times m$ bits and transmits the bits at the interleaver output in a different order (see Figure 9.28). An interleaver produces a delay of $n \times m \times T$ (*T*: symbol duration) in the transmission of the message signal. Data is written in an $n \times m$ rectangular array from the channel encoder in row-wise. Once the array is filled, the contents are transmitted column-wise; the neighboring bits are then separated from each other by n bits in the channel. At the receiver, the inverse operation (deinterleaving) is performed; the contents of the array in the receiver are written column-wise, and once the array is filled, it is read out row-wise into the decoder. Then consecutive symbols undergoing a burst in the channel will be separated from each other by n symbols following the de-interleaving

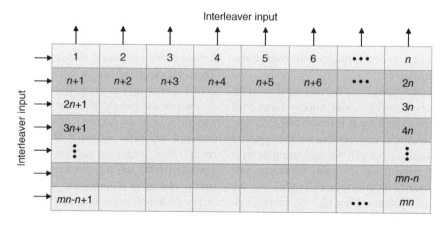

Figure 9.28 Block Interleaver.

process. If the value of nT is chosen to be larger than the burst length, then the probability that two consecutive symbols will fade at the same time will be very low. The erroneous symbols due to a bursty channel are thus separated by $n-1$ correct symbols when the deinterleaver output is read into the decoder.

Figure 9.29 presents a pictorial view of interleaving and deinterleaving processes. At the transmitter, interleaving changes the order of transmission of the bits. Therefore, in case of an error burst in the channel, the bits affected by the burst will appear as if they suffer random bit errors at the deinterleaver output. Then, a burst-error or a random-error correcting code can more easily correct deinterleaved random bit errors.

Example 9.16 Interleaving Depth in a Fading Channel.
A mobile radio operating at $f_c = 900$ MHz transmits data at 13.4 kbps and is required to serve users with vehicle speeds up to 125 km/hour. The maximum Doppler shift for a user speed of 125 km/hr at 900 MHz carrier frequency is found as

$$f_{d,\max} = \frac{v}{c} f_c = \frac{125 \ km/hr}{3 \times 10^8 \ m/s} 900 \ MHz = 104 \ Hz$$

$$(9.145)$$

Coherence time T_{coh} of the channel denotes the time duration over which the channel remains the same

$$T_{coh} = \frac{1}{2f_{d,\max}} = \frac{\lambda/2}{v} = 4.8 \ ms \qquad (9.146)$$

Note from (9.146) that the coherence time may also be interpreted as the time required to displace a distance of half-wavelength at speed v. The coherence time may also be interpreted as the mean burst duration. The burst length, which corresponds to the number of bits suffering fading burst, at a data rate of 13.4 kbps is therefore given by

$$b = R_b \ T_{coh} = 13.4 \ kbits/s \times 4.8 \ ms = 64 \ bits$$

$$(9.147)$$

On the average, only half of these bits are expected to be in error, since bit error probability is less than or equal to 0.5. Nevertheless, a burst error correcting code should either be capable of correcting $b = 64$ consecutive bits or use interleaving.

In *the convolutional interleaver* shown in Figure 9.30, the encoded bits to be interleaved are arranged in blocks of M symbols. For each block, the encoded symbols are sequentially shifted into and out of a bank of I registers by means of two synchronized input and output

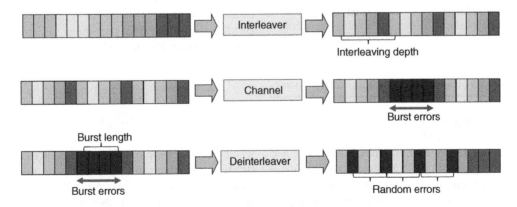

Figure 9.29   Interleaving Randomizes Burst Errors.

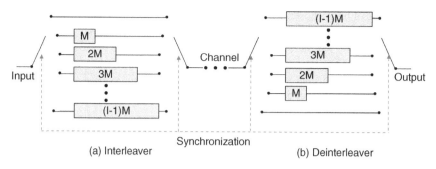

Figure 9.30   Convolutional Interleaver and Deinterleaver. [8]

commutators. Consecutive symbols are separated from each other by $M \times I$ symbols, hence an interleaving depth of $M \times I$. The memory requirement is $(I-1) \times M/2$ in both the interleaver and deinterleaver. The total end-to-end delay in convolutional interleaving is $(I-1) \times M$ symbols, which is one-half of the corresponding value in a block interleaver/deinterleaver for a similar level of interleaving. Hence, convolutional interleaving is usually preferred over block interleaving.

9.5.2   Reed-Solomon (RS) Codes

RS codes form a subclass of nonbinary BCH codes and they are suitable for burst error correction. "Nonbinary" implies that the data is processed in symbols of m bits. An (n, k) RS code has the following parameters:

Symbol length:	m bits per symbol
Block length:	$n = 2^m - 1$ symbols
Data length:	k symbols
Number of correctable errors:	$t = (n-k)/2$ symbols
Minimum distance:	$d_{min} = 2t + 1$ symbols

For example, if $m = 8$ and $t = 16$, then $(n, k) = (255, 223)$ and $R_c = 223/255 \approx 7/8 = 0.875$. This code, which has a block length of $n = 255$ symbols $= 255 \times 8$ bits $= 2040$ bits, can correct up to $16 \times 8 = 128$ consecutive bit errors. Then there must be an error-free region of $255 - 16 = 239$ symbols $= 239 \times 8 = 1912$

bits. If the errors were random, and there were at most one error per symbol, then this RS code could correct only 16 bit errors in 255 symbols. Therefore, RS codes are very effective for correcting burst errors, but not for random errors.

The Hamming distance between nonbinary codewords is defined as the number of symbols in which the codewords differ. The minimum distance d_{min} of the code determines the maximum number of symbol errors that can be corrected; symbol errors are corrected without considering the bit errors in the symbols. The error correction capability of the code increases with block length. The RS codes achieve the largest possible d_{min} of any linear codes with the same encoder input and output lengths. For RS codes, there are efficient hard-decision decoding algorithms, which enable the implementation of long codes in many practical applications.

To determine the error probability of RS codes, consider an RS-encoded system using BPSK modulation. The probability of error in a symbol of m RS encoded bits may be written as

$$P_s = 1 - (1-p)^m$$
$$p = Q(\sqrt{2R_c\gamma})$$

$$(9.148)$$

where p denotes the transition probability for BPSK modulation, and R_c is the code rate. Since a codeword consists of $n = 2^m - 1$ symbols, the corresponding symbol error probability with RS encoding is given by (see (9.87))

$$P_s = \frac{1}{(2^m-1)} \sum_{i=t+1}^{2^m-1} i \binom{2^m-1}{i} P_s^i (1-P_s)^{2^m-1-i}$$

(9.149)

where t denotes the number of correctable burst of symbol errors. For binary coding of the symbols, bit error probability P_b may be written in terms of the symbol error probability as

$$P_b = \frac{2^{m-1}}{2^m-1} P_s$$

(9.150)

Figure 9.31 shows the bit error probability for RS codes with code rate $R_c \approx 1/2$. The coding gain at 10^{-7} bit error probability level is 2.19 dB for $R_c = 7/15$, 3.35 dB for $R_c = 15/31$, 4.13 dB for $R_c = 31/63$ and 4.63 dB for $R_c = 63/127$. It might be meaningful to observe the relationship between output and input bit error probabilities of a RS decoder. As shown in Figure 9.32, RS decoder provides a significant reduction of the bit error probability, especially for input bit error probabilities lower than 10^{-3}.

RS code is particularly suitable for the transmission of M-ary modulation techniques, if a RS symbol has the same length as that of one of the $M = 2^m$ possible modulation symbols. Specifically, in M-ary FSK a code consists of $n = 2^m - 1$ orthogonal signals, where each signal is selected from the set of $M = 2^m$ possible orthogonal M-FSK signals of length m. For example, each of the 256 symbols in a 256-FSK consists of $\log_2 256 = 8$ bits. Each 256-FSK symbol may be transmitted as a RS symbol of length $m = 8$. Depending on the desired error correction capability, the message block length k may be determined as a function of t.

Example 9.17 Energy Dispersion, RS Encoding and Convolutional Interleaving in DVB-T and DVB-S systems. [8]
Figure 9.33 shows the block diagram of the first stages of DVB-S and DVB-T transmitters. The MPEG-2 encoded signals are first randomized to ensure the data sequence is balanced between 1's and 0's; RS encoder is used for mitigating burst errors; convolutional interleaving serves the purpose of randomizing burst errors so as to help the RS decoder at the receive side for mitigating the burst errors; and finally the convolutional encoder is used for the correction of random bit errors.

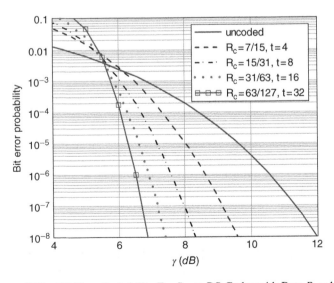

Figure 9.31 Bit Error Probability For Some RS Codes with Rate $R_c \approx 1/2$.

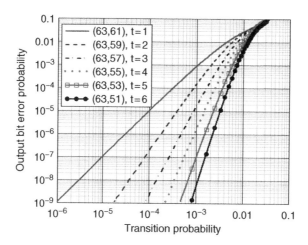

Figure 9.32   Output Versus Input Bit Error Probability For (63, k) Code with $t = 1,2,...,6$.

Figure 9.33   Energy Dispersion, RS Encoding and Convolutional Interleaving in DVB-T and DVB-S Systems.

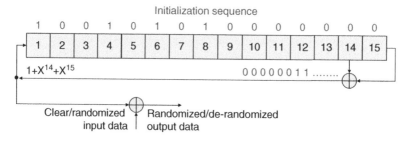

Figure 9.34   Schematic Diagram of Scramber/Descrambler. [8]

The total packet length of the MPEG-2 transport multiplex packet (consisting of data for video, audio and service information) is 188 bytes, that is, 1 synchronization byte ($47_{\text{HEX}} = 01000111$) and 187 data bytes. The data of the MPEG-2 multiplexer output is randomized using the scrambler/descrambler shown in Figure 9.34. In addition to scrambling the transmitted signals, the randomizer also ensures adequate binary transitions, that is, even distribution of the signal PSD over the bandwidth. The generator polynomial of the pseudo-random binary sequence (PRBS) generator is given by

$$g(X) = 1 + X^{14} + X^{15} \qquad (9.151)$$

The initialization sequence 100101010000000 is loaded into the PRBS registers at the start of every eight transport packets. The first bit at the output of the PRBS generator is applied to the first bit of the first byte following the synchronization byte. To provide an initialization signal for the descrambler, MPEG-2 synchronization bytes ($47_{HEX} = 01000111$) of the first transport packet in a group of eight packets is bit-wise inverted ($B8_{HEX} = 10111000$); this process is referred as the 'Transport Multiplex Adaptation'. To aid other synchronization functions, during the MPEG-2 synchronization bytes of the subsequent 7 transport packets, the PBRS generation continues, but its output is disabled, leaving these bytes unrandomized. Thus, the period of the PRBS sequence is $188 \times 8 = 1504$ bytes. The randomization remains active even when the modulator input bit-stream is non-existant, or when it is non-compliant with the MPEG-2 transport stream format (1 synchronization byte and 187 packet bytes). Total transmission time of one code block is $204 \times 17 = 3468$ byte periods, which corresponds to approximately 3.5 ms at a data rate of 8 Mbps.

The DVB-T system employs an outer (204, 188) RS code, shortened from the original (255, 239) RS code, This code is applied to each randomized 188-byte transport packet where each byte (symbol) is 8 bits long. This implies the addition of 239 $-188 = 51$ empty (all set to zero) bytes at the RS encoder input and $255-204 = 51$ output empty bytes are discarded at the RS encoder output. This code has a minimum distance of up to about 10 and can correct a burst of $t = (204-188)/2 = 8$ bytes, hence a burst of 64 bits. Modulations used range from QPSK to 64-QAM.

The convolutional interleaver, placed at the output of a (204,188) RS encoder, is composed of $I = 12$ branches, cyclically connected to the input byte stream by the input switch. Each branch is equipped with first-input first-output (FIFO) shift registers with $M = 204/I = 17$ bytes; thus the output of the shortened RS encoder (204,188) is interleaved. The maximum delay through the interleaver is $12 \times 17 = 204$ symbols, that is, one code block of the shortened (204,188) RS (outer) code. For synchronization purposes, the synchronization bytes are always routed in the branch corresponding to a null delay of the interleaver. The deinterleaver is similar to the interleaver, but the branch indexes are reversed, that is, the first branch corresponds to the largest delay. The deinterleaver synchronization is carried out by routing the first synchronization byte in the top branch. Consequently, the synchronization bytes always arrive first to the receiver, as they should.

9.5.3 Low-Density Parity Check (LDPC) Codes

LDPC codes, which are linear block codes specialized in burst error correction, are designed by R. Gallager in the early 1960s. However, these codes are ignored until late 1990s because of the required computation complexity, that was not available at that time. They are rediscovered by several publications in the mid 1990s. LDPC codes are powerful codes with large block lengths and low error floor. Their performance approaches the theoretical limit for Shannon's channel capacity within a fraction of a dB.

LDPC codes are defined by their parity check matrix **H** rather than the generator matrix **G**, so as to facilitate the decoding process. Dimensions of the parity-check matrix **H** (block size of the code) are very high. However, they are sparse, that is, the ratio of the number of 1's to the total number of entries is very low in each row and column, hence low density. Simple algorithms exist for constructing good parity-check matrices. However, the size of the generator matrix is very large and is not generally sparse, hence the encoding process requires large number of operations. As a block code, an LDPC encoder may use algebraic or geometric methods; encoding basically consists of expressing **H** as in (9.45) and then determining the **G** matrix as given by (9.44). Such structured codes may be encoded with shift register circuits. Therefore, encoding is not a serious issue.

In regular LDPC codes, parity check matrix **H** contains exactly W_c 1's per column and exactly W_r 1's per row, where $W_r \ll n$. $W_c \geq 3$ is necessary for good codes. In irregular LDPC codes, the number of 1's per column or row is not constant. Usually irregular LDPC codes outperform regular LDPC codes. If all rows are linearly independent, then the resulting code rate is $1 - W_c/W_r$. LDPC codes have large minimum distances, which are proportional to the code length, and good burst error correcting performance.

The code length of a linear block code should be sufficiently large in order to approach the Shannon's capacity limit, and large block lengths cause an exponential increase in the computational complexity of a ML decoder. Sparcity of the parity-check matrix allows decoding LDPC codes with reasonable complexity. LDPC codes have efficient iterative decoding algorithms which can be split into a number of manageable steps with linear decoding complexity in time and are suitable for parallel implementation.

Figure 9.35 shows the bit error probability for a (65520, 61425) LDPC code with a soft-decision near-ML detection algorithm. The Shannon limit for this code of rate $R_c = 61425/65520 = 15/16 = 0.9375$ is 3.91 dB (see Figure 9.15). At a bit error probability of 10^{-5}, this code performs approximately 0.5 dB worse than the Shannon limit.[4]

9.6 Convolutional Coding

In block coding, an encoder maps a k-bit message block at its input into a n-bit codeword at its output. An entire message block is buffered by the encoder before generating the associated codeword, a block of n encoded bits. In convolutional coding, an encoder operates on the incoming message sequence continuously in a serial manner rather than in large blocks; hence a convolutional encoder does not need a large buffer. Coded bits result from the convolution of the incoming bits with the impulse response of the encoder. Convolutional codes are commonly used for random error correction and in concatenation with RS codes to mitigate both random and burst errors. In contrast with block

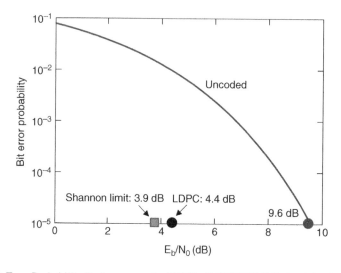

Figure 9.35 Bit Error Probability Performance of a (65520, 61425) LDPC Code with a Soft-Decision Near-ML Detection Algorithm Compared to Shannon Limit, for 10^{-5} Bit Error Probability. [4]

coding, the use of non-systematic convolutional codes is preferred over systematic codes.

The encoder of a binary convolutional code may be viewed as a finite-state machine that consists of a number of shift registers with prescribed connections to n modulo-2 adders and a multiplexer that serializes the adder outputs. As shown in Figure 9.36, k bits are fed into the encoder at a time and n encoded bits are obtained at the output; k and n are generally small. Connections between the k shift registers of each of the K stages characterizes the encoder output. Since a k-bit message sequence affects following $K-1$ blocks of k-bits, the convolutional encoder is said to have memory, which is characterized by the so-called constraint length. The constraint length K of a convolutional encoder is defined as the maximum number of bits in a single output stream that can be affected by any input bit. Alternatively, it denotes the maximum number of shifts over which a single message bit can influence the encoder output. Therefore, in a (n, k, K) convolutional encoder, a n-bit codeword depends on current block of k input bits and previous $K-1$ blocks of k input bits. A bit error due to the channel impairments would therefore affect the decoding of the following Kk bits; this information provides a valuable tool in correcting the bit errors, albeit at the expense of increased decoding complexity. A convolutional encoder always starts encoding with all the shift registers at zero state. Similarly, when the data bits are finished, the so-called tail (all

zero) bits are sent so as to leave the encoder again at zero state; the tail bits are assumed to yield n_{tail} output bits. Therefore, a k-bit message sequence produces a coded output sequence of $n + n_{tail}$ bits. The code rate is then defined as

$$R_c = \frac{k}{n + n_{tail}} \qquad (9.152)$$

Since $n \gg n_{tail}$ in most cases, the code rate simplifies to

$$R_c \cong k/n \qquad (9.153)$$

9.6.1   A Rate-½ Convolutional Encoder

In this section, we will consider one of the simplest convolutional encoders of rate ½ and constraint length $K = 3$, as shown in Figure 9.37. The encoder processes the incoming data bit and the contents of the two rightmost shift-registers, that is, the encoder state. Hence the encoder can assume one the following four states: **a**: 00, **b**: 10, **c**: 01 or **d**: 11. When a bit (0 or 1) enters the encoder, it continues to affect the output coded bits for the duration of three shifts, hence a constraint length of 3. For each input bit, one gets two coded bits at the encoder output; therefore, the duration of a coded bit is ½ of the input data bit. This implies that the rate of the encoded data is twice

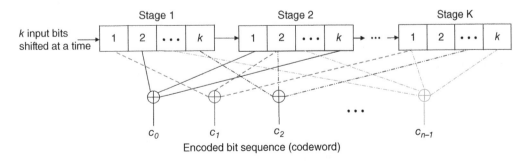

Figure 9.36   Block Diagram of a (n, k) Convolutional Encoder with Constraint length K.

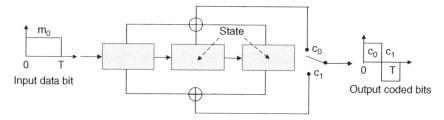

Figure 9.37   Rate ½, and Constraint Length 3 Convolutional Encoder ($k = 1$, $n = 2$, $K = 3$).

Table 9.19   States of the 1/2–Rate Convolutional Encoder Shown in Figure 9.37.

Current state	Input	Next state	Output
a: 00	0	a	00
	1	b	11
b: 10	0	c	10
	1	d	01
c: 01	0	a	11
	1	b	00
d: 11	0	c	01
	1	d	10

Table 9.20   Impulse Response of the 1/2–Rate Convolutional Encoder Shown in Figure 9.37.

Input	Current state	Next state	Output
1	a	B	11
0	b	C	10
0	c	A	11

the input data rate and that the transmission bandwidth is doubled.

The output of the encoder shown in Figure 9.37 is determined by the input bit and the encoder state. When a new bit arrives, it shifts the bits in the encoder to the right; this leads to a change in the encoder state and the output codeword. Therefore, the input bit and the current state determines the next state and the output codeword. Table 9.19 shows the encoder state and the output codeword for all combinations of the three bits (input bit and the two bits forming the current state of the encoder).

9.6.2   Impulse Response Representation of Convolutional Codes

Each path connecting one of n coded bits to the input of a convolutional encoder may be characterized in terms of its impulse response. Impulse response of a path is defined as the response of that path to a symbol 1 applied to

its input, when each register in the encoder is initially set in the zero state. Hence, impulse response refers to the response of a convolutional encoder to a single **1** bit that goes through it.

As an example, let us determine the impulse response of the encoder shown in Figure 9.37; that is for an input bit sequence of 100 where 1 is assumed to be the first bit to arrive the encoder. The following 00 are the flushing bits to reset the encoder at zero state. The output sequence and the evolution of states are shown in Table 9.20. Using the linearity property of the convolutional codes, one may use the impulse response to determine the output sequence for an arbitrary input sequence. For example, the response of the above-considered convolutional encoder for the input sequence 1101 is shown in Table 9.21.

9.6.3   Generator Polynomial Representation of Convolutional Codes

In generator polynomial approach, a convolutional encoder is described by a set of generator polynomials

Table 9.21 Response of the 1/2–Rate Convolutional Encoder For the Input Bit Sequence 1101.

Input bit	Output
1	11 10 11
1	11 10 11
0	00 00 00
1	11 10 11
Modulo-2 sum	11 01 01 00 10 11

$$\{g_0(D),\ g_1(D),\ \cdots,\ g_{n-1}(D)\} \qquad (9.154)$$

for each codeword, that is, $g_i(D)$ characterizes the ith path (coded bit), and D denotes the unit-delay variable. The generator polynomial of the ith path is defined by

$$g_i(D) = g_{i0} + g_{i1}D + g_{i2}D^2 + \cdots + g_{iM}D^M,$$
$$g_{ik} \in \{0,1\}$$

$$(9.155)$$

where M denotes the number of shift registers in the path. The coefficient $g_{ik} \in \{0,1\}$ shows whether k^{th} shift register of the ith path is connected or not. For the convolutional encoder of rate ½ and constraint length $K = 3$ shown in Figure 9.37, the generator polynomial for the upper and lower paths are given by

$$g_0(D) = 1 + D + D^2$$
$$g_1(D) = 1 + D^2 \qquad (9.156)$$

Consider the message sequence 1101, used in Table 9.21, which may be represented by the following message polynomial $m(D) = 1 + D + D^3$. Using (9.156) and $m(D)$, coded bit sequences for upper and lower branches in Figure 9.37 may be written as

$$m(D)g_0(D) = 1 + 0.D + 0.D^2 + 0.D^3 + 1.D^4 + 1.D^5$$
$$m(D)g_1(D) = 1 + 1.D + 1.D^2 + 0.D^3 + 0.D^4 + 1.D^5$$
$$c(D) = (1,1) + (0,1)D + (0,1)D^2 + (0,0)D^3 + (1,0)D^4 + (1,1)D^5$$
$$\mathbf{c} = \ \ 11 \ \ \ \ 01 \ \ \ \ 01 \ \ \ \ 00 \ \ \ \ 10 \ \ \ \ 11$$

$$(9.157)$$

where $c(D)$, obtained by interleaving the outputs of the two branches, is the same as found in Table 9.21.

9.6.4 State and Trellis Diagram Representation of a Convolutional codes

State and trellis diagrams provide a compact representation of the structural properties of convolutional encoders. Figure 9.38 shows the state and trellis diagrams for the ½ rate $K = 3$ convolutional encoder shown in Figure 9.37. State diagram is basically a pictorial representation of the state table shown in Table 9.19. Figure 9.38 shows that for each shift (a new input bit arrives) two incoming and two outgoing paths exist for each state(node). A transition from one state to another in response to input 0 is represented by a solid branch and input 1 is represented by a dashed branch. Trellis diagram, which has a treelike structure with emerging and converging branches, is simply an extension of the state diagram and shows the encoding process as a function of time (see Figure 9.38b).

The state/trellis diagrams are useful for determining the encoder output for any incoming message sequence. Encoding always starts at state **a**, the all-zero initial state, and walk through the state/trellis diagram in accordance with the message sequence. A solid branch is followed if the input bit is a 0 and a dashed branch if it is a 1. As each branch is traversed, the output is shown as the binary label on the branch. After transmitting the message bits, encoding ends again with state **a** through the use of tail bits.

We now consider encoding of a 3-bit message sequence

$$\mathbf{m} = (m_0 \ \ m_1 \ \ m_2 \ \ 0 \ \ 0), \ m_i \in \{0,1\} \quad (9.158)$$

with convolutional encoder shown in Figure 9.37. Encoding of three bits makes full use of the encoder constraint length $K = 3$. The last

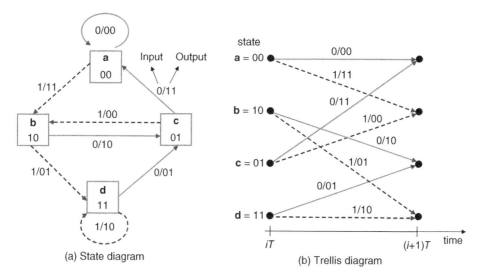

Figure 9.38   State and Trellis Diagrams For the ½ rate $K = 3$ Convolutional Encoder Shown in Figure 9.33.

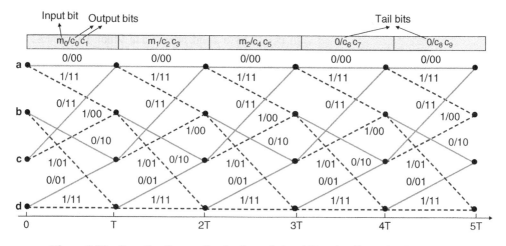

Figure 9.39   Encoding Process By the Convolutional Encoder Shown in Figure 9.37.

two tail bits in (9.158) are used to reset the encoder state to the initial state **a**. The corresponding trellis diagram is shown in Figure 9.39. Note that there are only eight paths in the trellis starting and ending at state **a** to uniquely represent the eight possible combinations of the three input bits m_0, m_1 and m_2 (see Table 9.22). Therefore, every 3-bit block is uniquely encoded to yield eight valid codewords of length $n = 10$ (including tail bits).

For example, the input bit sequence 010 00 is encoded as {00 11 10 11 00} via the path {**a a b c a a**} in the trellis. Table 9.22 shows that this code looks like a (10, 5) linear block code with a minimum distance of $d_{min} = 5$, hence with a two-bit error correction capability.

To generalize, a rate k/n constraint length K convolutional code is characterized by 2^k branches entering and 2^k branches leaving each

Table 9.22 Trellis Paths For Encoding Eight Possible Combinations of the Input Sequence of Three Bits. Hamming distances between $\mathbf{r} = 10\ 01\ 10\ 11\ 00$ and the codewords are also shown.

Input bit sequence	Path	Codeword	Hammimg weight	Hamming distance to $r = 1001101100$
000 00	a a a a a a	00 00 00 00 00	0	5
001 00	a a a b c a	00 00 11 10 11	5	6
010 00	a a b c a a	00 11 10 11 00	5	2 (minimum)
011 00	a a b d c a	00 11 01 01 11	6	7
100 00	a b c a a a	11 10 11 00 00	5	6
101 00	a b c b c a	11 10 00 10 11	6	7
110 00	a b d c a a	11 01 01 11 00	6	3
111 00	d d d d c a	11 01 10 01 11	7	4

state (node). The trellis/state diagrams each have $2^{k(K-1)}$ possible state [3].

9.6.5 Decoding of Convolutional Codes

There are various algorithms for decoding convolutional codes such as sequential decoding by Wozencraft and Fano, feedback decoding by Massey and maximum likelihood (ML) decoding by Viterbi. ML decoding, which is already studied in Section 9.2, is optimum in AWGN channels. For ML decoding in a BSC (hard-decision decoding), the decoder chooses the estimate $\hat{c}$ that minimizes the Hamming distance between the received vector $\mathbf{r}$ and all possible transmitted vectors. For soft-decision decoding, Euclidean distance is used as a metric instead of the Hamming distance. The equivalence between ML decoding and minimum distance decoding implies "decoding a convolutional code by choosing the path in the trellis diagram (hence a codeword) with minimum Euclidean distance between the received sequence". Viterbi decoding is feasible for constraint lengths shorter than ~10. Sequential decoding is preferred for higher constraint lengths for which the complexity of Viterbi decoding increases.

For the message sequence given by (9.158), the decoder input is assumed to be $\mathbf{r} = 10\ 01\ 10$

11 00, including some errors. For hard-decision decoding using the trellis diagram, the decoder estimates the output $\hat{c}$, that is, the most likely transmitted code, and the corresponding message bit sequence $\hat{m}$, by comparing the received word with all valid codewords and selecting the one with minimum Hamming distance. For this purpose, we can use Table 9.22 which tabulates all possible eight paths through the trellis and the corresponding 10-bit codewords for eight combinations of the three message bits. The Hamming distance between $\mathbf{r}$ and the valid codewords are also shown. We estimate the transmitted code as $\hat{c} = (00\ 11\ 10\ 11\ 00)$ since it has the minimum Hamming distance (of 2) with $\mathbf{r}$. The corresponding input message bits are found from Table 9.22 as $\hat{m} = (01000)$. Hence two coded bit errors are corrected.

9.6.5.1 Viterbi Algorithm

In decoding convolutional codes, the number of paths likely to be followed in the trellis exponentially increases with the number of transmitted bits. In such cases, the Viterbi algorithm provides a systematic approach for the minimum distance (optimum ML) decoding in AWGN channels. We observe from Figures 9.38 and 9.39 that at any level, there are two incoming and two outgoing paths for each of the four nodes (states) in the trellis. Starting from and ending at the all-zero state, the Viterbi algorithm considers the likely paths to be followed through the trellis depending on the received coded bits. At any node, there may be two incoming paths (in this particular example) and the Viterbi decoder may make a decision at that node as to which of these two paths to discard or to retain. This sequence of decisions is exactly what the Viterbi algorithm does as it walks through the trellis.

The algorithm for hard-decision decoding operates by computing Hamming distance (as a metric) for every possible paths in the trellis. The metric for a particular path is defined as the Hamming distance between the coded sequence represented by that path and the received sequence. For each node (state), the algorithm compares the two paths entering the node, and the path with the lower metric (smaller Hamming distance) is retained and the other path is discarded. If metrics are identical, a coin tossing is used. In summary, the Viterbi decoding algorithm compares then received sequence **r** with all possible transmitted sequences, and it chooses the path whose coded sequence differs from the received sequence in the fewest number of places. Once a valid path is selected as the correct path, the decoder can estimate the input data bits from the output coded bits of the selected path.

The paths that are retained by the algorithm are called survivor or active paths. This computation is repeated for every level j of the trellis in the range $(K-1) \le j \le L$, where K denotes the constraint length and L denotes the length of the incoming message sequence. A convolutional code of constraint length K, needs a buffer to store up to 2^{K-1} survivor paths and their metrics before decision as to the transmitted message. When the received sequence is very long, the storage requirement of the Viterbi algorithm becomes excessively high. Then, a decoding window of length ℓ is specified, and the algorithm operates on a corresponding frame of the received sequence, always stopping after ℓ steps. The decoding window of length ℓ is on the order of $5\,K$. In practice, sequential decoding

by Wozencraft and Fano is preferred for decoding convolutional codes with long constraint lengths.

The Viterbi decoding algorithm consists of the following steps:

Initialization: Start with the all-zero state.
Computation step $j, j = 1, 2, \ldots$ Assume that all survivor paths are identified and their metric are stored at step j for each state of the trellis. Compute the metric for all paths entering each state of the trellis by adding the metric of the incoming branches to the metric of the connecting survivor path from level j. For each state, identify the path with the lowest metric as the survivor of step $j + 1$.
Final step: Continue the computation until the algorithm completes its forward search through the trellis and therefore reaches the termination node (i.e., all-zero state). Then, the sequence of symbols associated with the ML path are estimated (decoded) as the transmitted message signal.

Example 9.18 Hard-Decision Decoding with Viterbi Algorithm.
The steps outlined above of the Viterbi algorithm for hard-decision decoding may be observed from Figure 9.40 for the received coded bit sequence **r** = 10 01 10 11 00.
Step $j = 1$: The decoding starts at state **a**. The path metric, expressed in Hamming distance and shown in circles, is initialized to zero.
Step $j = 2$: Two outgoing branches from state **a** at step $j = 1$ go to states to **a** and **b** for input bit 0 and 1, respectively. Therefore, at step $j = 2$, the encoder state can be either **a** or **b**. Hamming distances between 10 of the received sequence and the branch outputs (00 for input bit 0 and 11 for input bit 1) are shown as metrics of the respective branches. The new branch metrics, shown in circles at $j = 2$, are found by adding the current Hamming distances (1 for both) to the initial branch metric at $j = 1$ (which is 0).

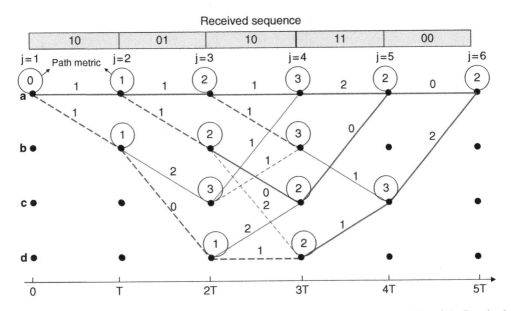

Figure 9.40 Pictorial Representation of Viterbi Algorithm For Hard-Decision Decoding of the Received Sequence **r** = 10 01 10 11 00.

Step j = 3: The outgoing branches from the states **a** and **b** at step j = 2 go to states **a**, **b**, **c** and **d** (see Figure 9.40). Hamming distances between the coded bits 01 of the received sequence and the branch outputs are shown at the relevant branches. Path metrics are found by adding the current Hamming distances to the path metrics at j = 2 and are shown as new metrics in circles at step j = 3. At this level observe four likely paths; path metrics for **aaa** and **aab** are both 2 while the path metric is 3 for the path **abc** and 1 for **abd**.

Step j = 4: Hamming distance between 10 of the receive sequence and the branch outputs are shown at the relevant branches. Two paths, **aaaa** (with metric 3) and **abca** (with metric 4), diverge at state **a**. The path **aaaa** with a lower metric is the survivor; hence, the path **abca** is discarded. Similarly, the paths **abcb**, **abdc** and **aabd** ending respectively at nodes **b**, **c** and **d** are discarded. Starting from the step j = 4 onwards only the survivor paths are considered for decoding since they have lower metrics and are more likely to be followed.

New path metrics of the survivor paths are calculated by adding the Hamming distance of each survivor path to the path metric of the departure node and shown at the nodes at j = 4. The new branch metrics for the states **a**, **b**, **c** and **d** are found to be 3,3,2 and 2, respectively.

Step j = 5: Note that the decoding terminates at state **a** and the state **a** can be reached only from states **a** and **c**. Therefore, we need to consider the survivor paths leading to states **a** and **c** only at step j = 5; these are **aaaaa**, **aabca**, **aaabc** and **abddc**. We firstly determine the Hammimg distance between 11 of the received sequence and the relevant branch outputs. Secondly we find the path metric as 5 for **aaaaa** and 2 for **aabca**; hence **aabca** is the survivor path at state **a** at j = 5. Similarly, the path **abddc** with metric 3 is survivor at state **c** at j = 5; we discard the path **aaabc** with metric 4. The relevant total path metrics are shown with circles.

Step j = 6: Here we calculate the Hamming distance between 00 of the received sequence and the output of the branch **aa** and the branch

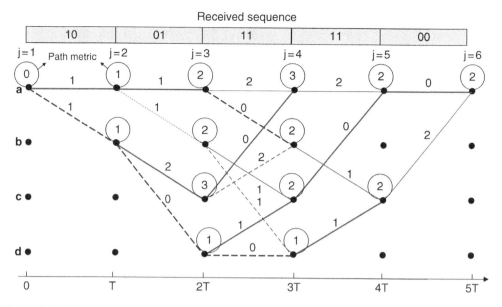

Figure 9.41   Pictorial Representation of Viterbi Algorithm For Hard-Decision Decoding of the Received Sequence **r** = 10 01 11 11 00.

ca, which are 0 and 2, respectively. Then, the metrics of the two paths **aabcaa** and **abddca** are 2 and 5 respectively. Therefore, the ML decoder chooses the path **aabcaa** as the most likely path. The estimated message word and the codeword are then given by $\hat{\mathbf{m}} = 01000$ and $\hat{\mathbf{c}} = 00\ 11\ 10\ 11\ 00$, respectively (see Figure 9.39). Comparing $\hat{\mathbf{c}}$ with **r** = 10 01 10 11 00, we observe that first and third coded bits of **r** were received in error and are corrected by this code (see also Table 9.22).

Now assume that we send the codeword **c** = 00 11 10 11 00 corresponding to the message word **m** = 010 00 as before and the received sequence **r** = 10 01 11 11 00 at the decoder input contains three coded bit errors. Following the steps shown in Figure 9.41, similar to those in Figure 9.40, the Viterbi algorithm predicts the transmitted message and the corresponding codeword as $\hat{\mathbf{m}} = 110\ 00$ and $\hat{\mathbf{c}} = 11\ 01\ 01\ 11\ 00$. Obviously this code with $d_{\min} = 5$ (see Table 9.22) cannot correct three coded bit errors. This is reasonable since the Hamming distance between **r** and **c** is 3 while

the Hamming distance between **r** and $\hat{\mathbf{c}}$ is equal to 2 (see Table 9.22 and Figure 9.13).

Example 9.19   Soft Decision Decoding with Viterbi Algorithm.
We consider soft-decision decoding of the following observation vector

$$\mathbf{x} = [0.3 \quad -0.4 \quad -0.1 \quad 0.6 \quad 0.2 \quad -0.15$$
$$0.6 \quad 0.1 \quad -0.6 \quad -0.4] \qquad (9.159)$$

One may observe that (9.159) reduces to the received sequence **r** = 10 01 10 11 00, the same as considered in Example 9.18, when it is hard-decision decoded.

Soft-decision decoding is based on finding the codeword with shortest Euclidean distance to the observation vector **x** (see (9.13) and (9.14)). In practical applications of soft-decision decoding, bit metrics similar to those given by (9.22) and Table 9.2 are used. However, for the sake convenience, we will here use the metric defined by (9.13). The projection of the observation vector **x** on each of the eight codewords listed in Table 9.22 (after replacing

0's in the codewords by -1), leads to the following the decision metric vector:

$$U = x.c^t = [-0.15 \ -0.85 \ 2.65 \ -1.25$$
$$-0.45 \ -1.35 \ 1.95 \ -0.55] \quad (9.160)$$

where $U_i = x.c_i^t$, $i = 0, 1, \cdots, 7$. For this particular observation vector, the largest metric is $U_2 = 2.65$; the soft-decision decoder will therefore estimate $\hat{c} = c_2 = 00 \ 11 \ 10 \ 11 \ 00$ as the transmitted codeword, as in Example 9.18. One may observe from Table 9.22 that U_2 corresponds to a Hamming distance of 2 while the next largest metric $U_6 = 1.95$ corresponds to a Hamming distance of 3.

Now assume that the second element x_1 of the observation vector x is changed from -0.4 to -0.005. In this case, the new observation vector and the corresponding decision metric vector will be

$$x = [0.3 \ -0.005 \ -0.1 \ 0.6 \ 0.2 \ -0.15$$
$$0.6 \ 0.1 \ -0.6 \ -0.4]$$
$$U = x.c^t = [-0.55 \ -1.25 \ 2.26 \ -1.66$$
$$-0.06 \ -0.96 \ 2.35 \ -0.16] \quad (9.161)$$

In this case, $U_6 = 2.35$ is the largest metric and the soft-decision decoder estimates that $\hat{c} = c_6 = 11 \ 01 \ 01 \ 11 \ 00$ is the transmitted codeword.

9.6.6 Transfer Function and Free Distance

The performance of a convolutional code depends not only on the employed decoding algorithm but also on the distance properties of the code. Bearing in mind that decoding starts and ends at node **a**, the shortest path in the trellis corresponds to the all-zero codeword, which is assumed to be transmitted. Then, any error in the received sequence **r** corresponds to a detour from the all-zero path in the trellis; the decoder follows paths in the trellis which diverge from

and converge again to the all-zero path. The most likely path, other than the all-zero path, followed by the decoder corresponds to the most likely error pattern in the code. It should be intuitively obvious that a path has a lower likelihood of being followed by the decoder as its Hamming distance with the all-zero path increases. This in turn implies that the error correction capability of a convolutional code is determined mainly by the path with minimum Hamming distance to the all-zero path, which is called as the free distance of the code. The free distance d_{free} of a convolutional code is defined as the minimum Hamming distance between any two code words in the code. A convolutional code with free distance d_{free} can correct t errors if and only if $d_{free} > 2t$. The free distance and some other performance indices of a convolutional code can be directly obtained from the so-called transfer function.

For determining the transfer function of the rate ½ convolutonal code considered above, we modify its state diagram in Figure 9.38b as a signal-flow graph with a single input and a single output by splitting the zero state into starting (**a**) and ending (**e**) nodes (see Figure 9.42). Hence the all-zero path is eliminated and any other path between starting and ending nodes corresponds to a detour from the all-zero path due to a certain error pattern in the received sequence **r**. Each branch in the graph is labelled by the following indices:

D: Hamming weight (number of 1's) of the encoder output (codeword) corresponding to that branch

L: A counter to indicate the number of branches in any given path from state $a = 00$ to state $e = 00$.

N: A factor showing a branch transition is caused by the input bit 1. Thus, as each branch is traversed, the cumulative exponent on N increases by 1, only if that branch transition is due to an input bit 1.

For determining the transfer function, we write the state equations using the dummy variables $X_a, X_b, \ldots, X_e$ as follows:

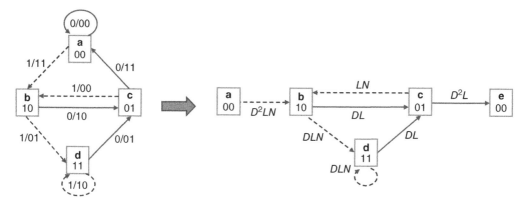

Figure 9.42 Transfer Function of the Rate-½ Convolutional Code in Figure 9.37.

$$X_b = D^2 LNX_a + LNX_c$$

$$X_c = DLX_b + DLX_d$$

$$X_d = DLNX_b + DLNX_d \qquad (9.162)$$

$$X_e = D^2 LX_c$$

Solving the above equations, we find the transfer function $T(D, L, N)$ and its two modified versions as follows:

$$T(D,L,N) = \frac{X_e}{X_a} = \frac{D^5 L^3 N}{1 - DL(1+L)N}$$

$$= D^5 L^3 N \sum_{i=0}^{\infty} [DL(1+L)N]^i$$

$$T_2(D,N) = T(D,1,N) = D^5 N + 2D^6 N^2$$

$$+ 4D^7 N^3 + 8D^8 N^4 + 16D^9 N^5$$

$$+ 32D^{10} N^6 + \cdots$$

$$T_1(D) = T(D,1,1) = D^5 + 2D^6 + 4D^7 + 8D^8$$

$$+ 16D^9 + 32D^{10} + \cdots$$

$$(9.163)$$

where the binomial expansion (D.8) is used in obtaining the first expression.

Figure 9.42 and (9.163) show that the three paths in the trellis with the smallest Hamming distance to the all-zero path are given by

a. The path **abce** with weight 5, length 3 and data weight 1, (has a Hamming distance 5 with all-zero path)

b. The path **abdce** with weight 6, length 4 and data weight 2, (has a Hamming distance 6 with all-zero path) and

c. The path **abcbce** with weight 6, length 5 and data weight 2 (has a Hamming distance 6 with all-zero path).

The most likely detours (error patterns) from the all-zero path for this code is shown in Figure 9.43 foll all message words listed in Table 9.22. The three paths shown in bold in Figure 9.43, namely **abceee, aabcee** and **aaabce,** all have a Hamming distance 5 to the all-zero pattern. These three paths correspond to the transmission of the message words 100, 010 and 001 as shown in Table 9.22. It is therefore clear that the transfer function formulation (9.163) commonly used in the litterature ignores the transmission of 010 and 001 message words. This conclusion is based on the observation that the transfer function in (9.163) shows only a single path with Hamming distance 5. Consequently, the errors due to these message words are ignored. This code with $d_{free} = 5 > 2\,t$ is capable of correcting up to two coded bit errors in the received sequence; two or fewer errors will cause the received sequence to be at most at a Hamming distance of 2 from the transmitted sequence but at least at a Hamming distance

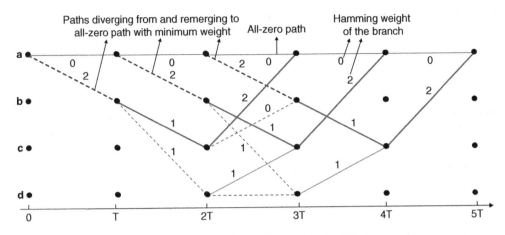

Figure 9.43   Free Distance in Convolutional Codes.

of 3 from any other code sequence in the code (see Figure 9.13).

For a convolutional code, the transfer function may be written as:

$$T(D,L,N) = \sum_{d=d_{free}}^{\infty} a_d \, D^d \, L^{c_d} \, N^{n_d} \quad (9.164)$$

where a_d represents the number of paths with the Hamming distance d, that is, the number of codewords with Hamming distance d from the all-zero codeword; n_d denotes the number of incorrectly decoded message (input) bits for the corresponding incorrect path; and c_d acts as a counter for the number of branches crossed for a path. Here, it may be useful to observe the correspondance between a_d defined by (9.164) for convolutional codes and the weight distribution A_d for linear block codes (see (9.92)). Similarly, note the analogy between n_d in (9.164) and N_d in (9.93).

The distance transfer function $T_1(D)$ gives the number of code words that are a given distance apart. Using (9.164) two versions of the distance transfer function may be expressed as

$$T_1(D) = T(D,1,1) = \sum_{d=d_{free}}^{\infty} a_d \, D^d$$

$$T_2(D,N) = T(D,1,N) = \sum_{d=d_{free}}^{\infty} a_d \, D^d \, N^{n_d}$$

$$(9.165)$$

It is assumed that the power series $T_1(D)$ is convergent, that is, its sum has a finite value. If $T_1(D)$ is nonconvergent, an infinite number of decoding errors are caused by a finite number of transmission errors; convolutional code is then called a catastrophic code. A systematic convolutional code can not be catastrophic but has smaller free distances compared to non-systematic convolutional codes.

9.6.7   Error Probability of Convolutional Codes

An exact formulation for determining the error probability of convolutional codes is not available in the literature. Instead, usually upper bounds and simulation results are used. The usual approach for determining an upper bound on the symbol error probability consists of first calculating the probability, $Prob(\hat{\mathbf{c}} \neq \mathbf{c}_i | \mathbf{c}_i)$, $i = 0, 1, \cdots 2^k - 1$, that the decoder estimates an incorrect codeword when a given codeword is transmitted, and average this probability over all codewords (see (7.64)). This is an upper bound since error events may not be mutually exclusive. Noting that each codeword corresponds to a distinct path in the trellis, the error probability is the same as the probability at the decoder follows a trellis path other than the all-zero path. In order to avoid

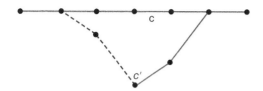

Figure 9.44 Pair-Wise Error Probability $P(\mathbf{c}, \mathbf{c}')$ Between Paths $\mathbf{c}$ and $\mathbf{c}'$.

repeating the cumbersome calculations for each codeword, we assume that only the all-zero codeword is transmitted; this is valid assumption because of the linearity of the code. We then evaluate pair-wise error (error event) probability (probability that the decoder chooses the path $\mathbf{c}'$ instead of the correct path $\mathbf{c}$ (all-zero path) (see Figure 9.44). An error event occurs at a time instant in the trellis if a non-zero path leaves the all-zero path and remerges to it at a later time.

9.6.7.1 Error Probability for ML Hard-Decision Decoding

Pair-wise error probability $P_2(d)$ between the two paths (codewords) $\mathbf{c}$ and $\mathbf{c}'$, which have a Hamming distance d between them, is defined as the probability that the received sequence $\mathbf{r}$ is closer in Hamming distance to $\mathbf{c}'$ than to $\mathbf{c}$, which is assumed to be transmitted: [3]

$$P_2(d) \equiv P(\mathbf{c}, \mathbf{c}')$$
$$= \begin{cases} \displaystyle\sum_{i=(d+1)/2}^{d} \binom{d}{i} p^i (1-p)^{d-i} & d \text{ odd} \\[2em] \displaystyle\sum_{i=d/2+1}^{d} \binom{d}{i} p^i (1-p)^{d-i} + \frac{1}{2}\binom{d}{d/2}[p(1-p)]^{d/2} & d \text{ even} \end{cases}$$

$$(9.166)$$

There are a_d paths with Hamming distance $d \ge d_{\min}$ that diverge from and remerge to the correct path, the first-event probability may be found by the weighted sum of the pair-wise error probabilities of all such paths:

$$P_e \le \sum_{d=d_{free}}^{\infty} a_d \, P_2(d) \qquad (9.167)$$

Since n_d bit errors occur in a path with Hamming distance d from the all-zero path, the bit error probability is found by multiplying each term in (9.167) by the number n_d of input bits in the corresponding message word:

$$P_b \le \frac{1}{k} \sum_{d=d_{free}}^{\infty} a_d \, n_d \, P_2(d) \qquad (9.168)$$

where k denotes the number of message bits that enter the encoder at a time.

An upper bound to the pairwise error probability is commonly used:

$$P_2(d) < p^{d/2}(1-p)^{d/2} \sum_{i=(d+1)/2}^{d} \binom{d}{i} < \left[2\sqrt{p(1-p)}\right]^d$$

$$\sum_{i=(d+1)/2}^{d} \binom{d}{i} \cong 2^{d-1} < \sum_{i=0}^{d} \binom{d}{i} = 2^d$$

$$(9.169)$$

Inserting (9.169) into (9.167), one gets a loose bound on the first-event error probability, which may also be expressed in terms of the transfer function $T_1(D)$:

$$P_e \le \sum_{d=d_{free}}^{\infty} a_d \, P_2(d) < \sum_{d=d_{free}}^{\infty} a_d \left[2\sqrt{p(1-p)}\right]^d \Rightarrow$$

$$P_e \le T_1(D)\Big|_{D=2\sqrt{p(1-p)}}$$

$$(9.170)$$

where the last expression follows from (9.165). Bit error probability is found by multiplying each term in (9.170) by the number n_d of 1's (corresponding to the number of bits in error) in the corresponding message word. Since n_d is given by the exponent of N in $T_2(D, N)$ in

(9.165), an upper bound on the bit error probability is obtained by taking the derivative of $T_2(D, N)$ with respect to N:

$$P_b \leq \frac{1}{k} \sum_{d=d_{free}}^{\infty} a_d \, n_d \, P_2(d) \leq \frac{1}{k} \frac{\partial T_2(D,N)}{\partial N} \bigg|_{\substack{N=1 \\ D=2\sqrt{p(1-p)}}}$$

(9.171)

9.6.7.2 Error Probability For ML Soft-Decision Decoding

We now consider the performance of convolutional codes with soft-decision MLD for BPSK in an AWGN channel. The Euclidean distance between paths **c** and **c'**, which are at a Hamming distance d from each other, are related by (9.18) for antipodal signaling:

$$d_E = \sqrt{4 \, d \, R_c \, E_b}$$

(9.172)

where R_c denotes the code rate and d shows the Hamming distance between **c** and **c'**. The sampled antipodal signal at the output of the matched-filter demodulator is characterized by the Gaussian distribution with means $\pm \sqrt{R_c E_b}$ and noise variance $N_0/2$ (see (6.37)). Therefore, the pair-wise error probability, $P_2(d)$, is defined as the probability that the decoder chooses the path **c'** instead of the correct path **c** (see (7.66)):

$$P_2(d) = Q\left(\frac{d_E}{\sqrt{2N_0}}\right) = Q\left(\sqrt{2R_c \gamma d}\right)$$

(9.173)

where $\gamma = E_b/N_0$. Since error events are not mutually exlusive, a union bound on the first error probability is given by

$$P_e \leq \sum_{d=d_{free}}^{\infty} a_d \, P_2(d)$$

(9.174)

An upper bound on the bit error probability for BPSK signaling in the AWGN channel is given by

$$P_b \leq \frac{1}{k} \sum_{d=d_{free}}^{\infty} a_d \, n_d \, Q\left(\sqrt{2R_c \gamma d}\right)$$

(9.175)

Similar to for hard-decision decoding, a loose upper bound for the probabilities of codeword and bit errors may be obtained by using the asymptotic approximation of the Q function and the transfer function of the code:

$$Q(\sqrt{x}) < \frac{1}{2} e^{-x/2}$$

$$P_e \leq \sum_{d=d_{free}}^{\infty} a_d \, P_2(d) = \frac{1}{2} T_1(D) \big|_{D=\exp(-R_c \gamma)}$$

$$P_b \leq \frac{1}{k} \sum_{d=d_{free}}^{\infty} a_d \, n_d \, Q\left(\sqrt{2R_c \gamma d}\right)$$

$$\leq \frac{1}{2k} \frac{\partial T_2(D,N)}{\partial N} \bigg|_{\substack{N=1 \\ D=\exp(-R_c \gamma)}}$$

(9.176)

Figure 9.45 shows the probability of bit error of the rate-½ convolutional code with constraint length $K = 3$ for hard- and soft-decision decoding. For hard-decision decoding, (9.166) is inserted into (9.168), where a_d and n_d given by (9.163) for $d \leq 15$. It is important to note that $a_5 = 1$ in (9.163) but the trellis diagram in Figure 9.43 suggests that the correct value is $a_5 = 3$. This difference stems from the fact that the message words 010 and 001 are not taken into consideration in determining the transfer function by (9.163) (see Table 9.22). The curve for hard-decision decoding includes this correction. On the other hand, Table 9.22 clearly shows that this code may also be considered as a (10, 5) block code with $d_{min} = 5$ ($t = 2$). Therefore, the bit error probability expression (9.87) for hard-decision decoding of this (10, 5) block code is also shown. In view of the asymptotic and upper-bound nature of

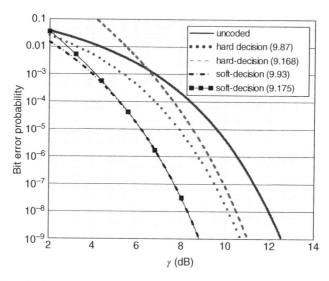

Figure 9.45 Bit Error Probability For ½, K = 3 Convolutional Code with Hard-Decision and Soft-Decision Decoding.

(9.168), the expression (9.87) for a (10,5) linear block code with $d_{min} = 5$ seems to describe the bit error probability of this code more accurately.

For soft-decision decoding, the bit error probability expressions (9.93) and (9.175) for block and convolutional codes respectively have similar forms. The values of a_d and n_d (except for the correction $a_5 = 3$) in (9.175) are given by (9.163) while N_d and A_d in (9.93) are directly found from Table 9.22 ($A_0 = A_7 = 1$, $A_5 = A_6 = 3$, $N_0 = N_3 = 1$, $N_1 = N_2 = 3$); one may observe that the parameters a_d (even if a_1 is corrected as $a_1 = 5$) given by (9.163) for convolutional codes is not the same as A_d for block codes. Similarly, n_d and N_d both denote the number of 1's in the message word corresponding to a codeword with a Hamming distance d from the all-zero codeword.

The binomial expansion (9.163) of the transfer function for the convolutional codes accounts for paths with infinitely large Hamming distances in the trellis. However, the weight distribution is limited by 7 based on the block code assumption and in Table 9.22. Nevertheless the close agreement between the curves based on formulations for convolutional and block codes suggests that, at least for short codes, the bit error probability formulation for block codes can also be used for the convolutional codes. In addition, this also suggests that, in calculating the bit error probability of convolutional codes, one may not need to consider paths in the trellis longer than the constraint length, since path lengths longer than or equal to the constraint length carries all the characteristic behavior of the code.

The coding gain for hard-decision decoding at 10^{-9} bit error probability is found to be 1.58 dB and 1.86 dB by (9.168) and (9.87), respectively. The coding gain for soft-decision decoding is 3.75 dB in both cases. Hence, soft-decision decoding performs 2.2 dB and 1.89 dB better than hard-decision decoding estimated by (9.168) and (9.87), respectively.

Example 9.20   2/3 Convolutional Code with $K = 2$.

We now consider a rate-2/3 convolutional encoder with constraint length $K = 2$ as shown in Figure 9.46. This encoder accepts 2 input bits at a time and outputs three encoded bits, hence the code rate is 2/3. Two input bits will form the encoder state when they are shifted once.

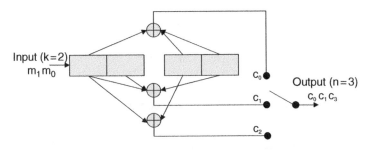

Figure 9.46   Block Diagram of Rate-2/3 Convolutional Encoder with $K = 2$.

Table 9.23   Codewords For the 2/3 Convolutional Code with Constraint length $K = 2$.

Input bit sequence	Path	Codeword	Hammimg weight
0000 00	a a a a	000 000 000	0
0001 00	a a b a	000 010 110	3
0010 00	a a c a	000 111 101	5
0011 00	a a d a	000 101 011	4
0100 00	a b a a	010 110 000	3
0101 00	a b b a	010 100 110	4
0110 00	a b c a	010 001 101	4
0111 00	a b d a	010 011 011	5
1000 00	a c a a	111 101 000	5
1001 00	a c b a	111 111 110	8
1010 00	a c c a	111 010 101	6
1011 00	a c d a	111 000 011	5
1100 00	a d a a	101 011 000	4
1101 00	a d b a	101 001 110	5
1110 00	a d c a	101 100 101	5
1111 00	a d d a	101 110 011	6

The constraint length is equal to two, since the input dibit will affect the encoder output twice. At each of the four states, four different combinations of dibits, namely 00, 01, 10 and 11, will cause four different codewords at the output. Hence, there will be four branches merging into and four branches leaving each node (state). The state table of this encoder is shown in Table 9.23, where the four message bits (two input bits and two bits forming the encoder state) can have 16 different combinations of message word of 6 uncoded bits (two input bits, two bits forming the encoder state and two tail

bits). Consequently, there will be 16 codewords of length 9. A close look at Table 9.23 clearly shows that this convolutional code behaves as a (9,6) block code with $d_{min} = 3$, hence a $t = 1$ bit error correcting capability. The weight distribution of this code is obtained from Table 9.23 as follows:

$A_0 = 1$, $A_3 = 2$, $A_4 = 4$, $A_5 = 6$, $A_6 = 2$, $A_8 = 1$.

The corresponding distribution of the input bit sequence is $N_0 = 1$, $N_1 = 4$, $N_2 = 6$, $N_3 = 4$, $N_4 = 1$. Each of the 16 codewords listed in Table 9.23 corresponds to a unique path in the state and trellis diagrams in Figures 9.47 and 9.48.

We have already shown in Figure 9.45 that the bit error probability of convolutional codes may be determined by ML decoding approach used for block codes. Therefore, we used (9.87) and (9.93) (with parameters N_d and A_d listed in Table 9.23) to plot the bit error probability curves in Figure 9.49 for respectively hard- and soft-decision decoding of this rate-2/3, $K = 2$ convolutional code. Uncoded bit error probability is also shown for comparison purposes. The coding gain at 10^{-9} bit error probability is found to be 1.9 dB and 3.5 dB for hard-decision and soft-decision decoding, respectively.

9.6.8   Coding Gain of Convolutional Codes

From (9.175) which gives the bit error probability of convolutional codes with soft-

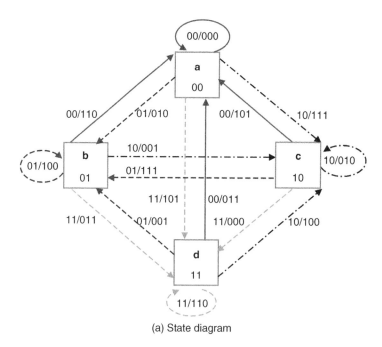

(a) State diagram

Figure 9.47   State Diagram of the 2/3 Convolutional Code with Constraint Length $K = 2$.

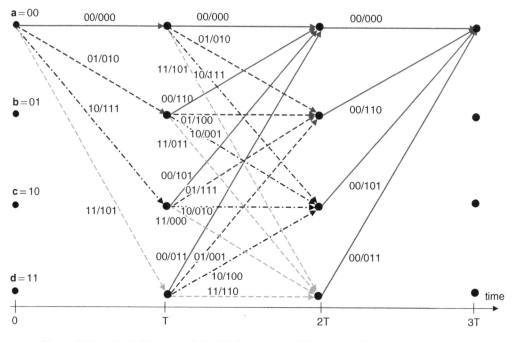

Figure 9.48   Trellis Diagram of the 2/3 Convolutional Code with Constraint Length $K = 2$.

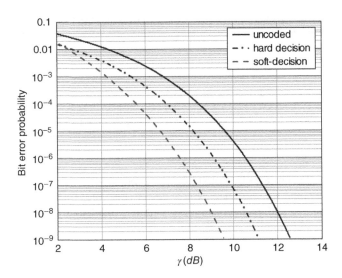

Figure 9.49   Bit Error Probability For 2/3 K = 2 Convolutional Code with Hard- and Soft-Decision Decoding.

Table 9.24   Rate-½ and Rate−1/3 Maximum Free Distance for Non-Systematic Convolutional Codes. [2][9][10]

Rate	Constraint length K	Generators			Free distance d_{free}
1/2	3	111	101		5
1/2	4	1111	1011		6
1/2	5	10111	11001		7
1/2	6	101111	110101		8
1/2	7	1001111	1101101		10
1/2	8	10011111	11100101		10
1/2	9	110101111	100011101		12
1/3	3	111	111	101	8
1/3	4	1111	1011	1101	10
1/3	5	11111	11011	10101	12
1/3	6	10111	110101	11101	13
1/3	7	1001111	1010111	1101101	15
1/3	8	11101111	10011011	10101001	16

decision decoding, one may observe that the coding gain for a convolutional code with soft-decision decoding for BPSK/QPSK modulations is given by

$$G_c \leq 10 \log_{10}\left(d_{free} R_c\right) \quad (dB) \qquad (9.177)$$

Coding gain for hard-decision decoding is approximately 2 dB lower in the AWGN channel as code rate approaches zero (see Figure 9.15). As shown in Table 9.24, the free distance d_{free} can be increased either by decreasing the code rate R_c and/or by increasing the constraint length K. Hence, there is a tradeoff between d_{free} and the code rate R_c. The use of codes with higher constraint lengths requires higher processing power for encoding and decoding. Table 9.24 gives a list of some short

Table 9.25 Coding Gain (dB) For Soft-Decision Viterbi Decoding. [3][11]

		$R_c = 1/3$		$R_c = 1/2$			$R_c = 2/3$		$R_c = 3/4$	
P_b	γ (dB)	$K=7$	$K=8$	$K=5$	$K=6$	$K=7$	$K=6$	$K=8$	$K=6$	$K=9$
10^{-3}	6.8	4.2	4.4	3.3	3.5	3.8	2.9	3.1	2.6	2.6
10^{-5}	9.6	5.7	5.9	4.3	4.6	5.1	4.2	4.6	3.6	4.2
10^{-7}	11.3	6.2	6.5	4.9	5.3	5.8	4.7	5.2	3.9	4.8

constraint length rate-½ and rate–1/3 convolutional codes with maximal free distance. Note that, for a given code rate and given constraint length, the generator polynomials, that is, the connection pattern of shift registers, are determined so as to maximize the free distance. This may generally require exhaustive computer simulations. Free distances achieved by systematic convolutional codes are generally lower than those for non-systematic codes listed in Table 9.24. Table 9.25 gives coding gains for several convolutional codes with soft-decision Viterbi decoding. Note that coding gain increases with increasing values of γ and constraint length K.

Example 9.21 Convolutional Coding and Interleaving in a Fading Channel
Let us reconsider Example 9.16 in a fading channel where burst errors of length $b = 64$ bits are expected. One should use either a burst error correcting code, which is capable of correcting $b = 64$ consecutive bits, or interleaving so as to shorten the burst length to no more than a random-error correcting FEC code can correct or concatenated coding with both burst- and random-error correcting codes.

For example, a nonsystematic convolutional code with constraint length $K = 7$ code has a free distance of $d_{free} = 10$. In view of $d_{free} > 2t$, such a code can be expected to handle a maximum of only four consecutive bit errors. Therefore it is not capable of correcting a burst of 64 consecutive bit errors. Unless we use a powerful burst error correcting code which can correct burst lengths up to 64 bits, interleaving is necessary. Since the increased capability

of a burst-error correcting code also implies lower code rates (increased transmission bandwidths), interleaving is generally a preferred option.

9.7 Concatenated Coding

FEC codes can not be effective against both random and burst errors. Therefore, it is useful to use concatenated coding in channels suffering both burst and random errors (see Example 9.17). In serial concatenation of two codes, the output of the first (outer) encoder, which is usually specialized in burst errors, is fed into the input of the second (inner) encoder designed to mitigate random errors, usually after an interleaver. At the receiver, the second (inner) code is decoded first and its output is subsequently decoded by the outer code for burst error correction. Traditionally, convolutional and RS codes are serially concatenated to mitigate random and burst errors, respectively, for example, in deep space missions, first-generation digital video broadcasting (DVB) and satellite communications. New generation systems employ concatenation of LDPC codes for mitigating burst errors and BCH codes for random errors.

As we already considered in Example 9.17, each code is usually followed by an interleaver. Interleavers aim to randomize/shorten the burst errors and alleviate error propagation at the receiver so that decoders can successfully correct the burst/random errors. Interleaving depth is usually dictated by the channel coherence time and the data rate.

Figure 9.50 Simplified Block Diagram of a Second Generation DVB System.

Example 9.22 Concatenated Coding in Second Generation DVB Systems. [12][13][14] A simplified block diagram of a second generation DVB (T2 for terrestrial, S2 for satellite and C2 for cable) transmission system is shown in Figure 9.50. An MPEG encoder output is characterized by R_u bps. A (48600, 48408) BCH encoder uses 192 parity bits to correct Gaussian bit errors. The output of the low-density parity-check (LDPC) encoder of rate ¾ has 64800 bits. Following the concatenated FEC encoder (BCH + LDPC), the QPSK (or higher order QAM)-modulated symbols are transmitted via OFDM with 2048 subcarriers in 8 MHz bandwidth (except for S2 systems). The duration T_G of the cyclic prefix in the OFDM is 1/32 of the useful symbol duration T (see Chapter 10).

The frequency separation between neighbouring subcarriers is given by

$$\Delta f = \frac{8\ MHz}{2048} = 3.906\ kHz \qquad (9.178)$$

The useful symbol duration and that including the cyclic prefix are given by

$$T = 1/\Delta f = 0.256\ ms$$
$$T + T_G = T(1 + T_G/T) = T(1 + 1/32) = 0.264\ ms \qquad (9.179)$$

The orthogonality between subcarriers is assumed to be preserved if the maximum Doppler shift does not exceed 1% of Δf. Then, this system operating at 300 MHz can successfully serve user terminals with velocities up to

$$f_{d,max} = f_c \frac{v}{c} \leq \frac{\Delta f}{100} \quad \Rightarrow \quad v \leq \frac{c\ \Delta f}{100 f_c}$$
$$\Rightarrow \quad v = 140.6 \left(\frac{km}{hr}\right)$$
$$(9.180)$$

The symbol rate R_s (symbols/s) at the OFDM output may be expressed in terms of the raw bit rate R_u (bits/s) at the output of the MPEG encoder as follows:

$$R_s = R_u \frac{48600}{48408} \frac{64800}{64800} \frac{1}{2} \frac{T}{T+T_G} = 0.649\ R_u$$
$$(9.181)$$

where $T/(T+T_G) = 32/33$. The factor ½ accounts for the reduction in transmission rate due to the use of QPSK. Higher order modulations can be accounted for by replacing 2 with $\log_2 M$ for M-ary QAM.

The maximum bit rate that can be supported by this system is given by

$$R_b = \frac{2048\,(subcarriers) \times 2\,(bits/subcarrier)}{T + T_G}$$
$$= \frac{2048 \times 2}{0.264\ (ms)} = 15.5\ (Mbps)$$
$$(9.182)$$

Assuming that a SDTV channel requires approximately 4 Mbps, at least three digital TV channels can be multiplexed in an analog SDTV channel of 8 MHz bandwidth. The corresponding maximum raw bit rate R_u (bits/s) is found from (9.181) and (9.182) as follows:

$$R_u = \frac{R_s}{0.649} = \frac{R_b/2}{0.649} = 11.17\ (Mbps) \quad (9.183)$$

The use of concatenated powerful codes and interleaving leads to drastic improvements in the system bit error probability performance. Figure 9.51 shows the significant performance improvement provided by the use of BCH and LDPC concatenation in second generation DVB systems compared to the concatenated RS and convolutional coding in 1st generation

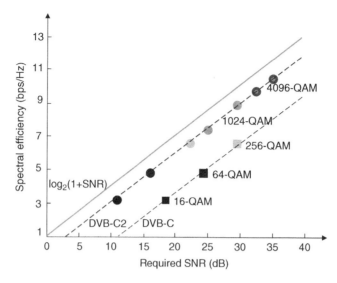

Figure 9.51 Spectral Efficiency of DVB-C2. Note that the horizontal axis shows the SNR = $(E_bR_b)/(N_0B)$ required for achieving the spectral efficiency in bps/Hz. [15]

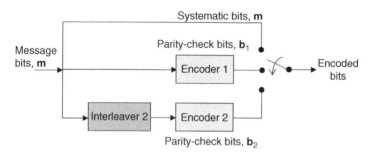

Figure 9.52   Block Diagram of Turbo Encoder.

DVB systems. The use of concatenated BCH and LDPC codes approaches the Shannon limit where the horizontal axis shows the required SNR required for achieving the spectral efficiency in bps/Hz. [15]

9.8   Turbo Codes

Turbo encoders use two constituent systematic convolutional encoders which are concatenated in a parallel fashion and separated from each other by an interleaver (see Figure 9.52). Apart from the systematic bits, the input data stream is applied directly to encoder 1, and the pseudo-randomly reordered (by a random interleaver) version of the same data stream is applied to encoder 2. Although the constituent codes are convolutional, turbo codes are linear block codes with the block size being determined by the size of the interleaver. The resulting code rate is 1/3.

The block code behavior of turbo codes presents some practical problems about starting and finishing of a block of input bits. As in convolutional codes, the encoder is initialized to the all-zero state and then the data is encoded. After encoding a block of data bits, a number of tail bits are added so as to reset the encoder to the all-zero state. There are basically two approaches for terminating the turbo codes. The first approach consists of terminating Encoder 1 and leaving Encoder 2 unterminated. In this case, the bits at the end of a block are more vulnerable to noise than the other bits due to unterminated Encoder 2. In the second approach, both Encoder 1 and Encoder 2 are terminated in a symmetric manner. Through the combined use of a good interleaver and dual termination, the error floor can be reduced significantly.

The convolutional codes previously considered are non-systematic, where data and parity bits are not separated, and they do not have feedback paths. At high SNR values, non-systematic convolutional codes perform better while, at low SNR values, systematic convolutional codes have superior performance. Therefore, non-systematic convolutional codes are not suitable for parallel concatenation and systematic codes are required.

Recursive systematic convolutional (RSC) codes are used as constituent encoders in turbo codes. RSC codes can be generated from non-recursive codes by connecting some of the encoder outputs directly to the input (see Figure 9.53), which implies that the state of the encoder depends on past outputs as well as past inputs. This affects the behavior of the error patterns and leads to improved performance. Due to the use of RSC codes and pseudo-random interleaver, the turbo code appears essentially random to the channel, yet it possesses sufficient structure to enable decoding.

An interleaver is an input-output mapping device that permutes the ordering of a sequence of symbols from a fixed alphabet in a completely deterministic manner; it takes the symbols at the input and produces identical symbols at the output but in a different temporal order. The use of pseudo-random interleaving in turbo codes serves the purpose of tying together errors that are easily made in one half of the turbo encoder to errors that are unlikely to occur in the other half. This is one of the main reasons why turbo codes perform better than the traditional codes. They provide robust performance with respect to mismatched decoding, which is a problem that arises when the channel statistics are not known or have been incorrectly specified.

In receivers with soft-decision decoding, demodulators produce soft-decisions, that is, the observation vector is quantized with more than 1 bit (see Figure 9.6). However, the decoding process usually results in hard-decisions, that is, in bits, by using hard-limiters; hence, a soft-input/hard-output process. However, decoding in turbo coding involves feeding

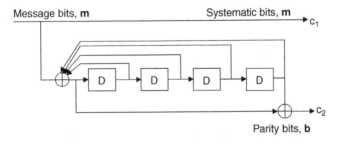

Figure 9.53 Recursive Systematic Convolutional (RSC) Encoder with Generator Polynomial $G(D) = (1, b(D)/m(D))$, $m(D) = 1 + D + D^2 + D^3 + D^4$ and $b(D) = 1 + D^4$.

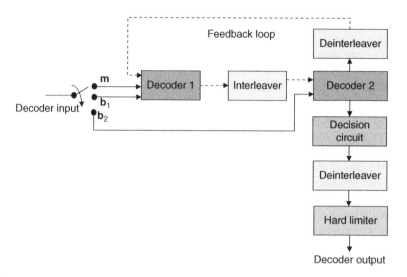

Figure 9.54   Turbo (Iterative) Decoder (Dashed Lines Show Extrinsic Information, Continuous Lines Show Intrinsic Information). [2]

outputs from one decoder to the input of the other decoder in an iterative fashion. Therefore, a hard-output decoder is not suitable because hard decision input to a decoder leads to an irrecoverable loss of information. Soft-input soft-output decoders are used in turbo coding.

Figure 9.54 shows the basic operation of a turbo decoder. At the input, the received systematic and parity bits (intrinsic information) are demultiplexed. Systematic bits and parity bits b_1 are fed to the input of the decoder 1. In the first iteration, the feedback loop is open that is, only 0's are fed to Decoder 1. Parity bits b_2 are put in a buffer until the interleaver output is fed to Decoder 2, which processes these input bits and yields a soft-output. The soft output of Decoder 2, is deinterleaved and fed-back to Decoder 1; this is called as the extrinsic information. The iteration continues as many times as required until it is interrupted by the decision circuit. Then, the soft output is deinterleaved and a hard decision is made to yield the decoded output through a hard-limiter. The extrinsic (soft) information, shown in dashed lines, is represented by the log-likelihood ratio for each bit; the sign corresponds to the hard-

decision 1 or 0 while its magnitude provides confidence/reliability information. The Decoders 1 and 2 may use the BCJR (Bahl, Cocke, Jelink, Raviv) MAP algorithm, which works with soft-input soft-output. [2] Note that the Viterbi algorithm is originally a soft-input hard-output decoding algorithm. However, a soft-output version of it (SOVA) is also developed. [16] For practical implementations, decoders must operate much faster than the data rate so as to be able to make a decision before the consecutive data arrives. Alternatively, a single iterating decoder may be replaced by series of interconnected decoders. In either case, decoding process is stopped after a preset number of iterations and the combination of intrinsic and extrinsic information is used to determine the decoded output.

Figure 9.55 shows the error performance based on Monte-Carlo simulations of a ½ rate, $K = 5$ turbo encoder with two RSC encoders shown in Figure 9.53 for binary transmission over an AWGN channel. An interleaver size of $256 \times 256 = 65536$ was used. The modified Bahl algorithm was employed for decoding. At small values of E_b/N_0, the bit error

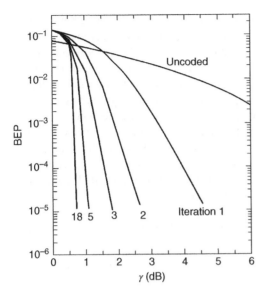

Figure 9.55 Evolution of Bit Error Probability with the Number of Iterations of a $R_c = 1/2$ Rate Turbo Code, Uncoded BPSK Transmission in AWGN Channel. [17]

probability for turbo-coding is higher than for uncoded transmission. The bit error probability drops very rapidly after several iterations. After 18 iterations, the bit error probability was observed to be less than 10^{-5} at $E_b/N_0 = 0.7$ dB. Note that the Shannon limit is -1.6 dB as the required system bandwidth approaches infinity and the capacity (code rate) approaches zero. This impressive performance is attained at the expense of large interleaver (long block length) large number of iterations (long decoder delay) and speed restrictions on the decoder hardware.

Turbo codes may also be used for adaptive coding purposes. One example, when the channel conditions are favorable, the rate of the turbo code shown in Figure 9.52 can be increased from 1/3 to ½ by simply puncturing one of the two parity bits according to a predetermined pattern so that the receiver knows the punctured bits and can decode accordingly. Other code rates can also be obtained by using different puncturing schemes. At

the decoder, the punctured bits are replaced by inserting dummy bits.

9.9 Automatic Repeat-Request (ARQ)

In some applications, for example, wired communications at short ranges, error probability may be very low. Then the use of FEC codes is not justified, because of the required higher transmission rates/bandwidths. In these cases, ARQ becomes a feasible approach for correcting occasional bit errors. Error detecting codes, for example, cyclic redundancy check (CRC) codes, are generally used for detecting bit errors in ARQ systems. The requirements for transmission rates/bandwidths by ARQ systems are not as high as in FEC codes, since error detecting codes require less redundancy (have higher code rates). In addition, they require much simpler decoding equipment. However, a duplex channel is needed to inform the transmitter as to the correct or incorrect reception of the the packets.

A transmitter using ARQ protocol breaks up the data into packets, encodes them with an error detecting code, for example, CRC, stores them in a buffer and transmits them sequentially. The packet size is chosen to be shorter than the channel coherence time, since longer packets are more vulnerable to fading in the channel. The receiver firstly checks whether a packet is decoded as correct. If a packet is decoded as correct, then it is forwarded to the end user. Depending on the adopted protocol, the transmitter may wait for a positive acknowledgment (ACK) signal from the receiver via a feedback link before transmitting the next packet. When a packet is decoded as incorrect, that is, an error is detected, the receiver sends a negative acknowledgment (NAK) signal to request the transmitter to retransmit the erroneous packet. If the time interval between the two transmissions is longer than the channel coherence time, then ARQ introduces time diversity into the system, since it is highly unlikely that retransmitted packet will suffer again from noise, interference and/or fades. ARQ systems are adaptive in the sense that a packet is retransmitted only when it is incorrectly received. The transmitter will then retransmit the same packet until the packet is correctly decoded at the receiver or a maximum number of retransmissions is reached. The receiver may either discard the erroneous packet or can combine with the retransmitted ones for a more reliable decision.

The performance of an ARQ system is usually measured in terms of average probability of error, transmission (throughput) efficiency, and the average delay with which a packet is received. Throughput efficiency is defined as the ratio of average number of information bits accepted at receiver per unit time without ARQ to the average number of information bits accepted at the receiver per unit time with ARQ. It may as well be defined as the ratio of effective data rate to the data transmission rate.

A packet may be declared as 'correctly received' (accepted) by the decoder if it is correctly received or the decoder can not detect error(s) existing in the received packet (undetected error). When an error is detected, the receiver requests the transmitter to retransmit the erroneously received packet. Since error-detection and retransmission probabilities are the same in ARQ systems, it is more meaningful to use undetected error probability for characterizing the error probability performance of ARQ protocols. If the undetected error probability is sufficiently low, then the use of FEC becomes unfeasible. However, when undetected error probability is too high, the ARQ system has a poor performance because of the long delays caused by retransmissions. Then, it might be feasible to use FEC coding.

Error control mechanisms in ARQ typically include the following:

a. Error detection: Receiver detects errors and discards the damaged packets (or saves them for future use)

b. Positive acknowledgement (ACK): Receiver acknowledges the transmitter via a feedback channel that the packet is received with no detected error. Output controller and buffer assemble the output bit stream from accepted packets and forward to the end-user.

c. Negative acknowledgement (NAK): Receiver informs the transmitter via a feedback channel, with negligibly small error probability, to retransmit the packet in error.

d. Retransmission after timeout: The transmitter retransmits the (supposedly lost) packet when no ACK or NAK signal is received within a predetermined duration of time.

Consider an ARQ system where k data bits are encoded with an error detecting code of e-error-detection capability to form a n-bit packet. The probability of error of i bits in a packet of n bits $P(i, n)$ is given by (9.52), where p denotes the transition probability. For coherent BPSK it may be written as $p = Q(\sqrt{2R_c\gamma})$ where R_c denotes the code rate and γ is the uncoded mean E_b/N_0. Probability of correct reception of an n-bit codeword implies that no bit is in error:

$$P_c = P(0,n) = (1-p)^n \qquad (9.184)$$

An ARQ receiver accepts a packet as correct when it is either correctly received or errors are undetected. An upper bound for the probability of receiving a codeword with undetected error(s) P_u may be written as

$$P_u = \sum_{i=e+1}^{n} \binom{n}{i} p^i (1-p)^{n-i} \qquad (9.185)$$

Exact expression for the undetected bit error probability may be expressed in terms of the weight distribution $\{A_i\}$ of the employed code as

$$P_u = \sum_{i=e+1}^{n} A_i \, p^i (1-p)^{n-i} \qquad (9.186)$$

The probability of acceptance may be written as

$$P_a = 1 - P_{ret} = P_c + P_u \qquad (9.187)$$

Codewords with detected errors are retransmitted as many times as necessary, so that the only output errors appear in words with undetected errors. In practice, the total number of transmissions may be limited by a predetermined number N_{max}. The total probability of receiving a n-bit message in error is given by the sum of the probabilities of message error on the first, second, third,... transmissions:

$$P_B = P_u + P_u \, P_{ret} + P_u \, P_{ret}^2 + \cdots + P_u \, P_{ret}^{N_{max}-1}$$

$$= P_u \frac{1 - P_{ret}^{N_{max}}}{1 - P_{ret}}$$

$$(9.188)$$

where (D.12) is used to obtain the last expression. Note that the sum of the probability of correct transmission P_c, the probability of detectable errors (retransmission probability) P_{ret}, and the probability of undetectable errors P_u equals unity (see (9.187)). For sufficiently large values of N_{max}, and/or small values of P_{ret}, using (9.187) and (9.188), the total probability of receiving an n-bit message in error may be expressed as

$$P_B \cong \frac{P_u}{1 - P_{ret}} = \frac{P_u}{P_c + P_u} \qquad (9.189)$$

9.9.1 Undetected Error Probability

Consider a hybrid ARQ system with a (n,k) block code and coherent BPSK modulation in an AWGN channel. A code with minimum Hamming distance d_{min}, designed for both error detection and correction, can correct up to $t = \lfloor (d_{min}-1)/2 \rfloor$ errors per received word and can detect $d_{min}-t-1$ error patterns (see (9.70)). When an uncorrectable error pattern is detected, a retransmission request is generated. A decoder (undetected) error can also occur if the received word is incorrectly

accepted as a valid code word. This happens when the received word is within a Hamming distance t of an invalid codeword, that is, a codeword other than the transmitted one.

Let the weight distribution of this code is denoted by $\{A_0, A_1, \ldots, A_n\}$, where A_i denotes the number of codewords of weight i in this (n,k) linear block code. If this code is used for both error detection and correction in a BSC, then the probability that the decoder will fail to detect the presence of errors can be computed from its weight distribution. An undetected error occurs only when the error pattern is identical to a nonzero codeword of this code. Assuming that the all-zero code word was transmitted, the probability that a received word has a Hamming distance d with a weight-j binary code word is given by [1]

$$P_d^j = \sum_{r=0}^{d} \binom{j}{d-r}\binom{n-j}{r} p^{j-d+2r}(1-p)^{n-j+d-2r}$$

$$(9.190)$$

where $p = Q(\sqrt{2R_c\gamma})$. The undetected error probability, that is, the probability that a code word is incorrectly accepted, is then

$$P_u = \sum_{j=d_{min}}^{n} A_j \sum_{d=0}^{t} P_d^j \qquad (9.191)$$

Inserting P_0^j from (9.190) into (9.191), one may easily show that the undetected error probability for $t=0$ (for error detection only) reduces to (9.186).

Consider now a $(n,k) = (31,21)$ BCH code with $d_{min} = 5$. This code can be used as a double-error-correcting ($t=2$) FEC code, for correcting single bit errors and detecting three bit errors or as a four-error-detecting ($e=4$) code (see (9.73)). Generator polynomial of this code is given by: [1]

$$g(X) = X^{10} + X^9 + X^8 + X^6 + X^5 + X^3 + 1$$

$$(9.192)$$

The weight distribution of this code is given by [1]

$A(X) = 1 + 186X^5 + 806X^6 + 2635X^7$

$\quad + 7905X^8 + 18910X^9 + 41602X^{10}$

$\quad + 85560X^{11} + 142600X^{12} + 195300X^{13}$

$\quad + 251100X^{14} + 301971X^{15}$

$\quad + 301971X^{16} + 251100X^{17} + 195300X^{18}$

$\quad + 142600X^{19} + 85560X^{20}$

$\quad + 41602X^{21} + 18910X^{22} + 7905X^{23}$

$\quad + 2635X^{24} + 806X^{25} + 186X^{26} + X^{31}$

$$\hfill (9.193)$$

Figure 9.56 shows undetected and detected error probabilities in a hybrid ARQ system for this code used for detecting four bit errors but no error correction ($e = 4$, $t = 0$). Figure 9.56 clearly shows that one should be cautious in using the upper bound (9.185) for undetected error probability, since it is very loose compared to the exact undetected error probability given by (9.186). However, the exact expression (9.186) requires a knowledge

of the weight distribution of the code. For comparison purposes, Figure 9.56 also shows the uncoded error probability (the probability that k uncoded bit sequence is received in error with one or more bit errors)

$$P_{u,u} = 1 - \left[1 - Q\left(\sqrt{2\gamma} \right) \right]^k \qquad (9.194)$$

where $\gamma = E_b / N_0$ denotes the mean SNR in the AWGN channel.

For a given error probability, ARQ systems can operate with lower values of γ compared to uncoded transmission with no ARQ, since some of the resulting additional bit errors can be mitigated by retransmissions. Therefore, similar to FEC systems, ARQ systems can achieve power savings compared to uncoded transmission at the expense of decrease in effective data rate. In Figure 9.56, for a undetected symbol error probability of 10^{-5}, the error-detection coding provides a 7.4 dB improvement in γ at the expense of a $31/21 = 1.48$ fold increase in the transmission rate/ bandwidth.

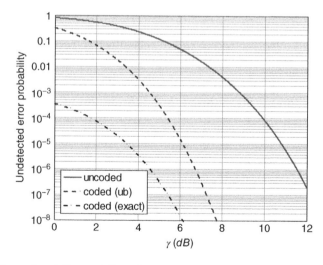

Figure 9.56 Undetected and Detected Error Probability For (31, 21) BCH Code Used For Detecting Four-Bit Errors($e = 4$). Uncoded error probability, coded upper bound and coded exact error probability are given by (9.194), (9.185) and (9.186), respectively.

9.9.2   Basic ARQ Protocols [6]

There are basically three ARQ systems, stop-and-wait (SW), Go-back-N (GBN) and select-ive repeat (SR). These three ARQ protocols are in increasing order of complexity and at the same time in increasing quality of performance.

Stop-and-wait (SW) ARQ:

In SW ARQ, an encoded n-bit packet of dur-ation T_f is transmitted and an ACK signal is expected before transmitting the next packet (see Figure 9.57). When a NAK signal is received, the last transmitted packet is retrans-mitted. If neither ACK nor NAK is received within a preset duration of time, the transmitter retransmits the last packet assuming that it is lost. The time needed by the transmitter between the end of the previous packet and the transmission of the next packet is given by

$$T_I = 2T_{prop} + T_f + 2T_p + T_{ack} \qquad (9.195)$$

where T_{prop} denotes the propagation time between transmitter and receiver, T_p is the pro-cessing time at the receiver, T_{ack} denotes the duration of the acknowledgment frame and T_f denotes the packet duration.

Average number of transmissions required for acceptance of a packet is given by

$$N_{SW} = \sum_{i=1}^{\infty} \underbrace{(No\ of\ transmissions)}_{i} \underbrace{(\ Prob.\ of\ i\ transmissions)}_{(1-P_a)^{i-1}P_a}$$

$$= \sum_{i=1}^{\infty} i P_a (1-P_a)^{i-1} = \frac{1}{P_a}$$

$$(9.196)$$

where (D.15) is used to evaluate the series. The total time devoted to a single attempt to get the receiver to accept a packet is given by the sum $T_f + T_I$. Hence, the average time required to transmit one packet is given by

$$D_{SW} = (T_f + T_I)N_{SW} = T_f \frac{(1 + T_I/T_f)}{P_a} \qquad (9.197)$$

Since the time needed to transmit k uncoded bits with no ARQ is simply given by $R_c T_f$ (R_c is the code rate), the throughput efficiency is sim-ply given by

$$\eta_{SW} = \frac{R_c T_f}{(T_f + T_I)/P_a} = \frac{R_c P_a}{1 + T_I/T_f} \qquad (9.198)$$

Go-Back-N (GBN) ARQ:

In GBN ARQ protocol, the transmitter con-tinuously transmits encoded packets and the receiver informs transmitter via a feedback chan-nel only when a received packet is in error. Hence, this ARQ algorithm does not employ ACK signals and the receiver sends a NAK sig-nal for the erroneously received packets. When a transmitter receives this message, it goes back N packets and retransmits the packets starting from the erroneous one. Here N denotes the number of transmitted packets after the erroneously received packet. Hence, if a packet is received in error, the last $N + 1$ packets are retransmitted. The receiver replaces the erroneous and the fol-lowing N packets with the retransmitted ones (see Figure 9.58). The value of N is given by

$$N = \lfloor T_I/T_f \rfloor + 1 \qquad (9.199)$$

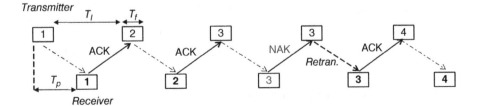

Figure 9.57   Stop and Wait ARQ.

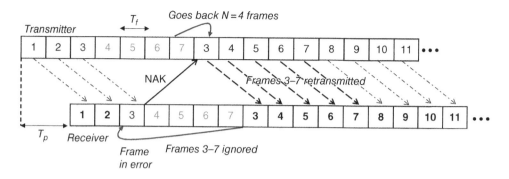

Figure 9.58   Go-Back N ARQ.

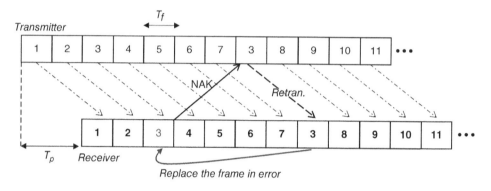

Figure 9.59   Selective Repeat (SR) ARQ.

Average number of codeword transmissions required for acceptance of a single packet is found using (D.14) and (D.15):

$$N_{GBN} = 1.P_a + (N+1).P_a(1-P_a)$$

$$+ (2N+1).P_a(1-P_a)^2 + \dots \quad (9.200)$$

$$= 1 + N(1-P_a)/P_a$$

As expected, (9.200) reduces to unity for as P_a goes to unity. The average time needed to transmit one packet is given by

$$D_{GBN} = T_f N_{GBN} = T_f [1 + N(1-P_a)/P_a] \quad (9.201)$$

Since the time needed to transmit k uncoded bits with no ARQ is equal to $R_c T_f$, the throughput efficiency is given by

$$\eta_{GBN} = \frac{R_c T_f}{T_f N_{GBN}} = \frac{R_c}{1 + N(1-P_a)/P_a} \quad (9.202)$$

Selective Repeat (SR) ARQ:

In SR ARQ protocol, the receiver sends a NAK signal for an erroneously received packet, as in GBN ARQ protocol. However, in contrast with GBN ARQ protocol, only the packet received in error is retransmitted (see Figure 9.59). The receiver, which is required to have a capability of reordering the received packets, replaces the erroneously received packet with the retransmitted packet when it is correctly decoded. Therefore, the average time needed to transmit one packet is given by

$$D_{SR} = D_{SW}|_{T_l=0} = T_f/P_a \quad (9.203)$$

On the other hand, the time needed to transmit k uncoded bits with no ARQ is equal to

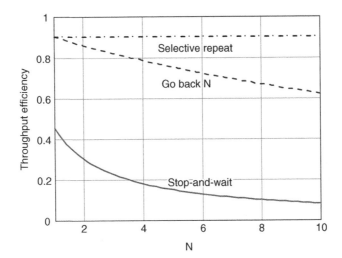

Figure 9.60   Comparison of the Throughput Efficiencies of SW, GBN and SC ARQ Protocols For $P_a = 0.99$ and $R_c = 0.95$.

$R_c T_f$. Therefore, the throughput efficiency may be written as

$$\eta_{SR} = \frac{R_c T_f}{T_f / P_a} = R_c P_a \qquad (9.204)$$

Figure 9.60 shows a comparison of the throughput efficiencies of SW, GBN and SR ARQ as a function of N for $P_a = 0.99$ and $R_c = 0.95$. The value of N denotes the round-trip time between transmitter and receiver relative to the packet duration T_f (see (9.199). The throughput efficiency of the SW protocol is very low because of inefficient use of the time. On the other hand, the throughput efficiencies of SW and GBN protocols are adversely affected by increased values of N (hence as range increases). The throughput efficiency of SR ARQ would not be affected by the range, except for the propagation delay, since time is economically used by retransmitting only erroneous packets. Nevertheless, ARQ protocols generally perform much better for short ranges as in local area networks (LANs). For example, consider a LAN network which can support an uncoded bit rate of 10 Mbps within

a maximum range of 3 km; this corresponds to a propagation delay of 10 µs. Assuming a packet size of 1000 uncoded bits, the packet duration is $T_f = 1$ ms. Hence, the propagation time in LAN using ARQ systems does not seriously degrade the system performance.

The actual (effective) data rate may be expressed in terms of the throughput efficiency as

$$R_{effective} = \eta R \qquad (9.205)$$

where η shows the throughput efficiency for any of the above considered ARQ protocols. For example, when packets are transmitted at a rate of 100 kbps in an ARQ system with a throughput efficiency of 0.9, then the effective data rate will be 90 kbps due to retransmissions.

Figure 9.61 provides a comparison of the throughput efficiencies of the above ARQ schemes as a function of the transition probability p when BCH (31,21) code with $d_{min} = 5$ is used for a 4 bit error detecting code (see (9.184), (9.186) and (9.187)). For small values of the transition probability, that is, low bit errors, the throughput efficiency is relatively constant but rapidly decreases with increasing

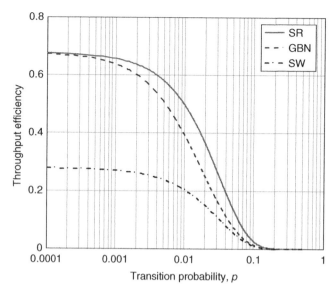

Figure 9.61   Comparison of the Throughput Efficiencies of SW, GBN and SR ARQ Schemes.

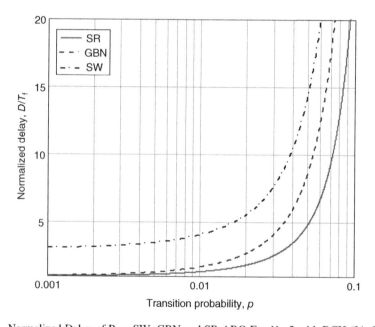

Figure 9.62   Normalized Delay of Pure SW, GBN and SR ARQ For $N = 2$ with BCH (31, 21) ($d_{min} = 5$).

values of p. Poor performance of SW ARQ is due to inefficient use of the time. SR ARQ performs slightly better than the GBN ARQ, due to its capability in reordering packets when packets in error are retransmitted. At high values of p, all these schemes have similar performances since P_a becomes very low and more frequent retransmissions are required (see Figure 9.62). Note that the delay due to re-transmissions are negligible when the channel operates at high SNRs but becomes inacceptably long for p exceeds 0.01. Figures 9.61–9.64 clearly show that pure ARQ protocols are suitable when the channel conditions are very good (low transition probability, high γ) and at short ranges (due to constraints on delay and signal attenuation), for example, for LAN applications.

Example 9.23 Comparison SW, GBN and SR ARQ Schemes.

Consider an ARQ system using a $(n, k) = (8, 7)$ simple parity check code with the capability of detecting odd numbers of bit errors. This ARQ system is required to operate at effective data rates of 50 kbps. The separation between transmitter and repeater is assumed to be 30 km. If the uncoded data rate does not exceed $R_b = 100$ kbps, we will determine whether GBN and SW ARQ schemes would be acceptable.

Since the packets coded with (8,7) parity-check code are transmitted at 100 kbps, the packet duration T_f, round-trip time T_I and the ratio T_I/T_f are given by, for $P_a = 0.99$ and $N = 3$,

$$T_f = k/R_b = 7/10^5 = 0.07 \ ms$$

$$T_I = 2d/c = 0.2 \ ms \tag{9.206}$$

$$T_I/T_f = 0.2/0.07 = 2.86$$

The corresponding values of the throughput efficiencies are given by (for $N = 3$ and $P_a = 0.99$)

$$\eta_{SR} = \frac{k}{n}P_a = 0.87$$

$$\eta_{GBN} = \frac{k}{n}\frac{P_a}{nP_a + N(1 - P_a)} = 0.85 \tag{9.207}$$

$$\eta_{SW} = \frac{k}{n}\frac{P_a}{1 + T_I/T_f} = 0.22$$

Since the throughput efficiency is required to be higher than $50 \ kbps/100 \ kbps = 0.5$ (see (9.205)), the throughput efficiency of SW ARQ scheme is below the operational requirement of 0.5; hence it is not a feasible option.

9.9.3   Hybrid ARQ Protocols

In some channels, the bit error probability is not very low; hence more frequent retransmissions may be needed. This in turn leads to longer delays and lower throughput efficiencies. Therefore it might be feasible to combine ARQ with some limited FEC capability. Such schemes are called as hybrid ARQ. In this approach, error control coding with error correction and detection capability is required. For example, when single- and/or double-bit errors are corrected at the receiver, there will be no need for retransmissions for such errors. Only the received packets with more bit errors than that can be corrected by FEC will be retransmitted. Hence, there is a trade-off between the number of corrected bit errors (increased transmission bandwidth) and reduction in the number of retransmissions (delay). Hybrid ARQ provides reliable and adaptive communications especially over wireless channels. Hybrid ARQ self-optimizes and adjusts automatically to channel conditions without requiring frequent SNR measurements. It adds redundancy only when needed, the receiver may save failed (re)transmissions to help future decoding, and each retransmission helps to increase the packet acceptance probability.

Some of the commonly used hybrid ARQ protocols are summarized below:

In *Hybrid ARQ Type I with Packet (Chase) Combining*, which is the simplest Hybrid ARQ method, the packets received in error are stored in a buffer. Each retransmission is self decodable. The corresponding soft values at bit level of all received codewords are combined according to the weights of the received SNRs.

In *Hybrid ARQ Type-II (Full Incremental Redundancy, IR)*, when a packet is received in error, only additional redundancy bits are retransmitted. The packets received in error are stored in buffer and the code rate is gradually decreased in each retransmission with the newly arrived redundancies.

Corresponding bits of the retransmitted packets and previously received packets are combined.

In *Hybrid ARQ Type III (Partial IR)*, the code rate is decreased by sending additional redundancy bits while maintaining self-decodability in each retransmission. The retransmitted packet can be Chase-combined with previous packets to increase the diversity gain.

Incremental Redundancy techniques often make use of Rate Compatible Punctured Convolutional (RCPC) Codes or Rate Compatible Punctured Turbo (RCPT) Codes. Rate compatibility is achieved by choosing puncturing matrices in such a way that the coded bits of higher puncturing belong to the coded bits of lower puncturing rates. Table 9.26 shows typical puncturing patterns and

Table 9.26 Puncturing Patterns and Transmitted Sequences After Parallel-To-Series Conversion For Some Code Rates Obtained From the Mother Code of Rate$-1/2$ with Constraint Length 7. The first convolutionally encoded bit of a symbol always corresponds to X_1. Here **1** denotes the transmitted bit and **0** shows the non-transmitted bit. The generator polynomials of the rate ½ mother code are $G_1 = 171_{OCT} = (1111001)$ for X output and $G_2 = 133_{OCT} = (1011011)$ for Y output.[18]

Code rate R_c	Puncturing pattern	Transmitted sequence (after parallel to series conversion)
1/2	X: 1 Y: 1	$X_1\ Y_1$
2/3	X: 1 0 Y: 1 1	$X_1\ Y_1\ Y_2$
3/4	X: 1 0 1 Y: 1 1 0	$X_1\ Y_1\ Y_2\ X_3$
5/6	X: 1 0 1 0 1 Y: 1 1 0 1 0	$X_1\ Y_1\ Y_2\ X_3\ Y_4\ X_5$
7/8	X: 1 0 0 0 1 0 1 Y: 1 1 1 1 0 1 0	$X_1\ Y_1\ Y_2\ Y_3\ Y_4\ X_5\ Y_6\ X_7$

transmitted sequences after parallel-to-series conversion for some code rates obtained from the mother code of rate$-1/2$ and constraint length 7.[18] Adaptive hybrid ARQ shows significant improvements over link adaptation alone; it can compensate for link adaptation errors and provides a finer granularity of code rate thus giving a better throughput performance. However, it is useful to note that the punctered codes are not as powerful as the mother code.

Consider a hybrid ARQ system using a *n*-bit codeword that corrects *t* bit errors and detects $e > t$ bit errors. This code can correct up to *t* bit errors, can detect but not correct *i* bit errors if $t + 1 \le i \le e$, and can not detect bit errors if $e < i \le n$. Since the system can correct up to *t* bit errors, the probability of correct reception may be written as

$$P_c = \sum_{i=0}^{t} P(i,n) = \sum_{i=0}^{t} \binom{n}{i} p^i (1-p)^{n-i}$$

$$(9.208)$$

where $P(i,n)$ is defined by (9.52). In hybrid ARQ, undetected error probability is given by (9.190) and (9.191) for a code with minimum Hamming distance d_{min} and the error correction capability *t*. Then the acceptance probability may be determined using (9.187). Figure 9.63 shows the throughput efficiency of Type I Hybrid SRe ARQ protocol using (31,21) BCH code with $d_{min} = 5$. Note that $t = 0$ corresponds to pure SRe ARQ, while $t = 1$ implies one of the three detected errors are corrected (see (9.73)). The throughput efficiency is approximately equal to the code rate $R_c = 0.68$ at small values of *p*, as expected (since $P_a \to 1$). The hybrid SRe ARQ (curve for $t = 1$) provides a trade-off between error detection ($t = 0$) and error correction ($t = 2$). At small values of *p*, the throughput efficiency is higher for hybrid ARQ. Similar arguments are valid for the retransmission probability of (31,21) BCH code, shown in Figure 9.64. For $p = 10^{-3}$, hybrid ARQ reduces the retransmission

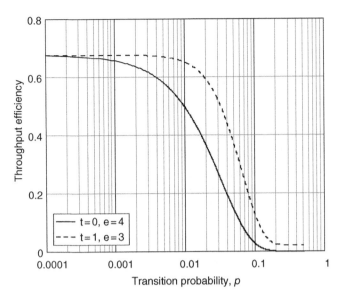

Figure 9.63   Throughput efficiency for hybrid SRe ARQ using BCH code (31,21) with $d_{min} = 5$ for $t = 0$ (pure ARQ) and $t = 1$ (hybrid ARQ).

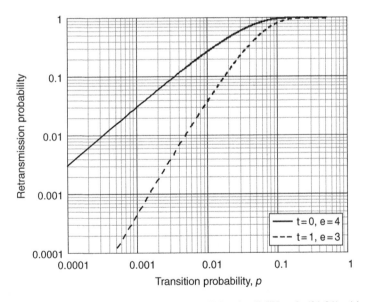

Figure 9.64   Retransmission probability for hybrid SRe ARQ using BCH code (31,21) with $d_{min} = 5$ for $t = 0$ (pure ARQ) and $t = 1$ (hybrid ARQ).

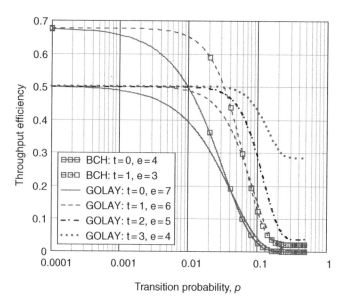

Figure 9.65   Comparison of throughput efficiencies of BCH (31,21) code with $d_{min} = 5$ and extended Golay (24,12) code with $d_{min} = 8$.

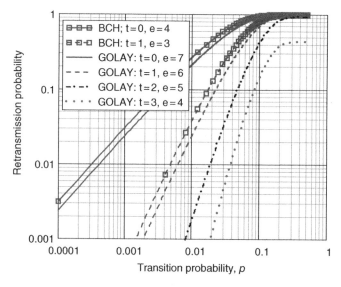

Figure 9.66   Comparison of Retransmission Probability For Hybrid SR ARQ For Golay Code (24,12) with $d_{min} = 8$ and BCH Code (31,21) with $d_{min} = 5$.

probability by a factor of 67, compared to pure ARQ.

Similar arguments may be repeated for the performances for Type I Hybrid SRe ARQ of (24,12) extended Golay code (with $d_{min} = 8$) and BCH code (31,21) with $d_{min} = 5$, compared in Figures 9.65–9.67. Using the weight distribution given by Table 9.16, the throughput

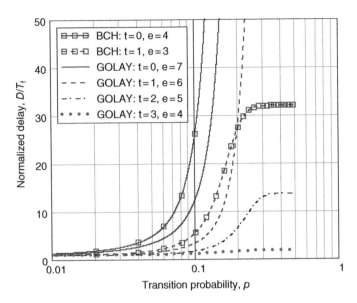

Figure 9.67 Comparison of normalized delay of BCH (31,21) code with $d_{min} = 5$ and extended Golay (24,12) code with $d_{min} = 8$.

efficiency is shown in Figure 9.65 for various values of t. For small values of p, the efficiency is approximately equal to the code rate $R_c = 0.5$ for (24, 12) Golay code and to $R_c = 0.68$ for (31, 21) BCH code. The BCH code has a clear advantage for throughput efficiency for small values of p, for which small code rates are not justified. Figure 9.66 and 9.67 show that delay and retransmission probabilities both decrease as the error correction capability of the code increases. The delay curve in Figure 9.67 for Golay code with $t = 3$ is almost flat because the code can correct three bit errors and there is almost no need for retransmissions.

Appendix 9A Shannon Limit For Hard-Decision and Soft-Decision Decoding

This Appendix aims to provide insight for Figure 9.15, which shows the Shannon limit as a function of the code rate in AWGN

channels. For *hard-decision decoding* (BSC), the channel capacity in bits per code symbol is given by [3]

$$I = 1 + p\log_2 p + (1-p)\log_2(1-p) \quad (9.209)$$

where $p = Q(\sqrt{2R_c\gamma})$ denotes the coded BEP for coherent BPSK in AWGN channel. Figure 9.68 shows the variation of I and p versus γ in dB for the code rate $R_c = 1/2$. As the uncoded bit SNR $\gamma = E_b/N_0$ goes to zero, one gets $p = 1/2$, and the channel capacity approaches zero. On the other hand, as γ goes to infinity, the error probability p vanishes monotonically, whereas I approaches unity.

Letting $I = R_c$ in (9.209), the value of γ that satisfies (9.209) gives the curve in Figure 9.15, showing the capacity versus γ for *hard-decision decoding*. Approximating p for small values of R_c as (see (B.11))

$$p = Q\left(\sqrt{2R_c\gamma}\right) \approx \frac{1}{2} - \sqrt{R_c\gamma/\pi}, \quad R_c \to 0 \quad (9.210)$$

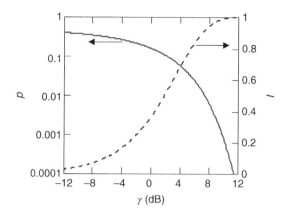

Figure 9.68   Variation of I, Given by (9.209), and Error Probability p Versus γ for a BSC for $R_c = 0.5$.

and using $\ln(1+x) \approx x - x^2/2$ from (D.2), the channel capacity reduces to [3]

$$I = \frac{2}{\pi \ln 2} R_c \gamma \qquad (9.211)$$

Letting $I = R_c$ in (9.211), one gets the minimum value of γ required to satisfy the capacity equation in the limit as R_c approaches zero (see Figure 9.15):

$$\gamma = \pi \ln 2 / 2 \quad (0.37\ dB) \qquad (9.212)$$

For *soft-decision decoding* in the binary-input AWGN channel, the capacity in bits per code symbol is given by [3]

$$I = \int_{-\infty}^{\infty} \left[f(y|m_0)p(m_0)\log_2\left(\frac{f(y|m_0)}{f(y)}\right) \right.$$

$$\left. + f(y|m_1)p(m_1)\log_2\left(\frac{f(y|m_1)}{f(y)}\right) \right] dy$$

$$(9.213)$$

where $p(m_0)$ and $p(m_1)$ denote a priori probabilities of binary symbols m_0 and m_1, respectively, and

$$f(y) = f(y|m_0)p(m_0) + f(y|m_1)p(m_1) \quad (9.214)$$

The conditional probabilities for m_0 and m_1 for binary antipodal signaling (BPSK) are given by

$$f(y|m_1) = \frac{1}{\sqrt{2\pi}\sigma} e^{-\frac{(y - \sqrt{R_c E_b})^2}{2\sigma^2}}$$

$$(9.215)$$

$$f(y|m_0) = \frac{1}{\sqrt{2\pi}\sigma} e^{-\frac{(y + \sqrt{R_c E_b})^2}{2\sigma^2}}$$

where $\sigma^2 = N_0/2$. Assuming equally likely bits, $p(m_0) = p(m_1) = 1/2$, (9.213) may be rewritten as

$$I = \frac{1}{2\sqrt{2\pi}\ln 2} \int_{-\infty}^{\infty} \left[e^{-\frac{\left(z - \sqrt{2R_c \gamma}\right)^2}{2}} \ln\left(\frac{2}{1 + e^{-2z\sqrt{2R_c \gamma}}}\right) + e^{-\frac{\left(z + \sqrt{2R_c \gamma}\right)^2}{2}} \ln\left(\frac{2}{1 + e^{2z\sqrt{2R_c \gamma}}}\right) \right] dz \quad (9.216)$$

The curve in Figure 9.15 for soft-decision decoding is plotted by letting $I = R_c$ in (9.216). As the code rate R_c approaches zero, we may use $\ln(2/(1 + e^x)) = -\ln(1 + x/2 + x^2/4) \cong -x/2 - x^2/8$ to approximate the capacity as [3]

$$I \simeq \frac{R_c \gamma}{\ln 2}, \quad R_c \to 0 \qquad (9.217)$$

Inserting $I = R_c$ into (9.217), we obtain the well-known Shannon capacity limit given by (8.181) (see Figure 9.15):

$$\gamma = \ln 2 \ (-1.61 \ dB), \quad R_c \to 0 \qquad (9.218)$$

Note, however, that (9.218) is valid for codes with arbitrarily small rates and does not give any information about how to construct such codes. The interested reader may refer to [3], [4] for further details.

References

[1] S. B. Wickers, *Error Control Systems for Digital Communication and Storage*, Prentice Hall: New Jersey, 1995.

[2] B. Sklar, *Digital Communications: Fundamentals and Applications* (2nd ed.), Prentice Hall: New Jersey, 2001.

[3] J. G. Proakis, *Digital Communications* (3rd ed.), McGraw Hill: New York, 1995.

[4] S. Lin and D. J. Castello, Jr., *Error Control Coding: Fundamentals and Applications* (2nd ed.), Pearson Prentice Hall: New Jersey, 2004.

[5] S. Haykin, *Communication Systems* (3rd ed.), John Wiley:New York, 1994.

[6] H. Taub, and D. L. Schilling, *Principles of Communication Systems* (2nd ed.), McGraw Hill: New York, 1986.

[7] W. W. Peterson and E. J. Weldon Jr., *Error Correcting Codes* (2nd ed.), MIT Press Cambridge, Mass., 1972.

[8] EN 300 421 V1.1.2 (1997-08) Digital Video Broadcasting (DVB): Framing structure, channel coding and modulation for 11/12 GHz satellite services.

[9] J. P. Odenwalder, *Error control coding Handbook*, Linkabit Corp., San Diego, Cal., July 1976.

[10] K. J. Larsen, Short convolutional codes with maximal free distance for rates ½, 1/3, and ¼, *IEEE Trans. Information Theory*, vol. **IT-19**, pp.371–372, May 1973.

[11] I. M. Jacobs, Practical applications of coding, *IEEE Trans. Information Theory*, vol. **IT-20**, pp. 305–310, May 1974.

[12] ETSI EN 302 755 V1.1.1 (2009-09), Digital Video Broadcasting (DVB); Frame structure channel coding and modulation for a second generation digital terrestrial television broadcasting system (DVB-T2).

[13] ETSI EN 302 307 V1.3.1 (2013-03), Digital Video Broadcasting (DVB); Second generation framing structure, channel coding and modulation systems for Broadcasting, Interactive Services, News Gathering and other broadband satellite applications (DVB-S2).

[14] ETSI EN 302 769 V1.2.1 (2011-04), Digital Video Broadcasting (DVB); Frame structure channel coding and modulation for a second generation digital transmission system for cable systems (DVB-C2).

[15] DVB-C2 Fact sheet, July 2012, www.dvb.org

[16] J. Hagenauer, Source controlled channel decoding, *IEEE Trans. Commun.*, vol. **43**, no.9, pp. 2449–2457, Sept 1995.

[17] C. Berrou, A. Glavieux, and P. Thitimajshima, Near Shannon limit error-correcting coding and decoding: Turbo codes, IEEE Proc. Int., Conf. Communications, Geneva, Switzerland, May 1993 (ICC'93), pp. 1064–1070.

[18] ETSI EN 300 744 V1.6.1 (2009-01), Digital Video Broadcasting (DVB); framing structure, channel coding and modulation for digital terrestrial television.

Problems

1. Consider a $(3,1)$ repetition code where $(-1 -1 -1)$ corresponds to binary 0 while $(+1 + 1 + 1)$ corresponds to binary 1. Assuming that the received signal is $(-0.1 \ 0.7 \ -0.2)$ when the transmitted codeword is $(+1 + 1 + 1)$.

   a. Estimate codeword at the decoder output using hard-decision decoding.

   b. Estimate codeword at the decoder output using soft-decision decoding.

   c. Determine the probability of error for hard and soft-decision decoding in terms of uncoded E_b/N_0 and compare the results for $E_b/N_0 = 6.8$ dB. Determine the coding gain for soft-decoding compared to uncoded transmission.

2. Construct the standard array of the (7, 4) Hamming code in Example 9.12 and show that the coset leaders are the same as the error patterns in Table 9.10.

3. In Example 9.13 obtain all the codewords and show that they are identical to those listed in Table 9.9.

4. The figure below shows a m-sequence generator consisting of a four-stage shift register with [1, 4] configuration. The initial state is 0100 and it is driven by a 10 MHz clock.

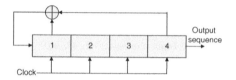

a. Determine the output sequence
b. What is the length and the period of the PN sequence?
c. What is the chip rate (rate of the output sequence)?

5. Weight distribution of an error correcting code C is defined as the $n+1$ integers $W(C) = \{A_i, \ 0 \le i \le n\}$, such that there are A_i codewords of Hamming weight i in C. BPSK modulation is assumed.

a. Calculate the probability that the error vector e has weight i in a BSC.
b. For an undetected error to occur, the error vector e must be a nonzero codeword. The probability of an undetected error P_u is therefore defined as the probability that the received word differs from the codeword but the syndrome equals zero. Write the exact expression for undetected error probability.
c. For most codes of practical interest, the weight distribution $W(C)$ is unknown. For such codes, the following upper bound is used:

$$P_u \le \sum_{i=d_{min}}^{n} \binom{n}{i} p^i (1-p)^{n-i}$$

For a binary linear (4, 2) code with $d_{min} = 2$, the weight distribution is given by $W(C) = \{1,0,1,2,0\}$. Calculate and compare exact and upper bound values for undetected error probability for a transition probability of $p = 10^{-3}$.

6. Given a code with the parity check matrix:

$$\mathbf{H} = \begin{bmatrix} 1 & 0 & 0 & \vdots & 1 & 1 & 0 & 1 \\ 0 & 1 & 0 & \vdots & 1 & 1 & 1 & 0 \\ 0 & 0 & 1 & \vdots & 0 & 1 & 1 & 1 \end{bmatrix}$$

a. Write down the generator matrix $\mathbf{G}$.
b. Obtain all the codewords and determine weight structure and the Hamming distance. Determine the minimum Hamming distance. How many errors can this code detect and/or correct? Can this code be used for simultaneous detection and correction?
c. Determine the minimum Hamming distance using only the $\mathbf{H}$ matrix.
d. Construct the syndrome table for this code.
e. Decode the received sequence 0111010.

7. Construct a (6, 3) systematic linear code and determine its minimum Hamming distance. Can this code be described by a generator polynomial? Can you describe a code generated by a generator polynomial as a linear systematic code?

8. Minimum distance of a (n, k) Hamming code is $d_{min} = 3$. Hence, these codes can correct single-bit errors. Assuming $k = 11$, determine the minimum value of n.

9. Consider that a block code with $k = 6$ is required to have a single bit error correction capability.

   a. Determine the minimum value of n.
   b. Construct a generator matrix for this code.

10. In one of the operation modes in DVB-T2, a BCH code is desired to have codeword length of 7200 bits for each 7032 bit at BCH encoder input. This code is concatenated with a LDPC code of rate 4/9 so that the overall codeword length becomes 16200 bits.

    a. Determine the minimum distance and the error correcting capability of this code.
    b. Determine m.

11. Consider a (n, k) CRC encoder using 4, 7, 12, 16, 24 or 32 parity bits for error detection.

    a. Determine the probability of an undetected error.
    b. Determine the probability of detecting an error.
    c. Calculate detected and undetected error probabilities for 4, 7, 12, 16, 24 or 32 parity bits and compare them.

12. Following BCH codes and BPSK modulation are considered for operation in an AWGN channel so as to achieve average bit error probability of 10^{-4} with minimum transmitted power.

(n, k)	$R_c = k/n$	t
(63,57)	0.905	1
(63,51)	0.81	2
(63,45)	0.714	3
(63,39)	0.619	4

Which code should be chosen? Calculate the power saving in dB by each code compared to uncoded transmission and the corresponding increase in the transmission bandwidth.

13. Consider data transmission at a bit rate 10 kbps using a raised-cosine pulse with $\alpha = 1$. The SNR at the receiver is $E_b/N_0 = 10$ dB. Compare the probability of bit errors and the transmission bandwidths for

    a. Uncoded QPSK.
    b. (24, 12) Golay encoding $(d_{min} = 8)$ followed by interleaving and QPSK modulation.
    c. (24, 12) Golay encoding $(d_{min} = 8)$ followed by interleaving and 16-PSK modulation.

14. You are required to provide a real-time communication system to support 3.2 Mbps with a bit-error probability not exceeding 10^{-5} within an available bandwidth of 1 MHz. The E_b/N_0 available at the detector input is 15.5 dB. Choose a suitable modulation scheme and a BCH code from Table 9.15 that meets the above performance requirements. Justify and verify your design choices.

15. DVB-S2 system is the second generation system for broadcasting of digital TV signals by geostationary satellites at 12 GHz carrier frequency. The satellite transmitter, at 42000 km away from the earth surface, is assumed to have an EIRP = 51 dBW. Nyquist filtered and ¾ encoded symbol rate of QPSK signals is 30.9 Msymbols/s. The ground station receiver has a diameter of 0.6 m and 65 % aperture efficiency.

    a. Determine the raw (useful) bit rate at the satellite. If standard TV (SDTV) signals are broadcasted at approximately 4

Mbps, how many SDTV channels can be multiplexed in this system?

b. This system requires a C/N power ratio of 5.1 dB at the Earth-based satellite receiver in a 27.5 MHz bandwidth. Determine the maximum allowed noise temperature of the receiver system.

c. If the antenna noise temperature is 147 K, then determine the noise figure of the receiver system.

d. Now assume we use a low-noise amplifier (LNA) just after the receiver antenna with 20 dB gain. Determine the noise figure of this LNA if the rest of the receiver system has an equivalent noise figure of 10 dB.

e. How the link budget is altered when there is a rain, with physical temperature of 290 K, causing 3 dB signal loss? How can you overcome this problem?

16. Consider the convolutional interleaver/deinterleaver considered in Figure 9.30. Assuming that $M = 2$ and $I = 5$, show that the interleaver output is given by s_1, 0, 0, 0, 0, s_6, 0, 0, 0, 0, s_{11}, s_2, 0, 0, 0, s_{16}, s_7, 0, 0, 0, s_{21}, s_{12}, s_3, 0, 0, s_{26}, s_{17}, s_8, 0, 0, s_{31}, s_{22}, s_{13}, s_4, 0, s_{36}, s_{27}, s_{18}, s_9,0, s_{41}, s_{32}, s_{23}, s_{14}, s_5, s_{46}, s_{37}, s_{28}, s_{19}, s_{10}, s_{51}, ... for the input sequence s_1, s_2, s_3, ... Show that the distance between consecutive symbols are separated by MI symbols from each other at the interleaver output. Show that deinterleaving restores the input symbol sequence.

17. Using the initialization sequence 100101010000000 in Figure 9.34 derive the first ten bits of the PRBS sequence.

18. A channel transmitting at 9600 bps often undegoes fades lasting 30 ms with a repetition interval of at least 30 ms.

a. Using a (255, 207) BCH code with $t = 6$ error correction capability, design an interleaving system to eliminate the effects of these error bursts. Determine

the maximum delay between interleaving and deinterleaving? Is this delay acceptable for real-time video transmission?

b. Repeat part (a) for (1023, 963) BCH code also with $t = 6$. Which code you would choose? Explain.

19. Prove (9.169).

20. Consider a convolutional encoder, which is described by the generating polynomials.

$$g_1(D) = 1$$
$$g_2(D) = D^2 + D^3$$
$$g_3(D) = 1 + D + D^3$$

a. Draw the block diagram of this encoder.

b. What is the code rate and the constraint length?

c. Is this a systematic or nonsystematic encoder?

d. Determine the coded bit sequence for an input bit sequence $m = \{1011011100010000...\}$.

21. Consider a rate–1/2, constraint length-3 convolutional encoder defined by the generator polynomials $1 + D + D^2$ and $1 + D^2$. The all-zero sequence is transmitted in a BSC and the received sequence is 10 00 10 00 0.... Using Viterbi algorithm, compute the decoded sequence and show the path metric, the branch metric and the surviving/deleted paths.

22. Use the transfer function of the rate–1/2 constraint length-3 convolutional encoder, defined by the generator polynomials $1 + D + D^2$ and $1 + D^2$, to compute the probability of bit error for BPSK in an AWGN channel with hard- and soft-decision decoding. Compare the probability of bit error performances for $E_b/N_0 = 7$ dB.

23. An upper bound on the bit error probability tighter than (9.176) may be found as

follows: Replace d by the expression $d = d_{free} + (d - d_{free})$ in (9.173) and use the upper bound for Q-function $Q(\sqrt{x+y}) \le Q(\sqrt{x})e^{-y/2}$ to show that (see (B.12))

$$Q(\sqrt{2R_c \gamma_b\, d}) \le Q(\sqrt{2R_c \gamma_b\, d_{free}})$$
$$e^{R_c \gamma_b\, d_{free}} e^{-R_c \gamma_b\, d}$$

$$\le Q(\sqrt{2R_c \gamma_b\, d_{free}})$$
$$e^{R_c \gamma_b\, d_{free}} \frac{dT_2(D,N)}{dN} \bigg|_{\substack{N=1 \\ D=e^{-R_c \gamma_b}}}$$

Plot the bit error probability using the upper bound found above and (9.176) versus E_b/N_0 and compare them for the rate-1/2 constraint length-3 convolutional code with the transfer function (9.163).

24. Consider the (3,1) nonsystematic convolutional encoder with the following connection vectors

$$g_1 = (1\ 1\ 1)$$
$$g_2 = (1\ 1\ 0)$$
$$g_3 = (1\ 0\ 1)$$

a. What is the constraint length of this encoder?
b. Draw the block diagram.
c. Draw the trellis diagram.
d. Find the codeword and the code polynomial corresponding to the information sequence $\{10011\}$.

25. Find the lowest-rate cyclic code whose generator polynomial is $g(D) = D^8 + D^7 + D^6 + D^4 + 1$. Determine the rate, the minimum distance and the coding gain of this code.

26. The generator polynomial for the recursive systematic code (RSC) shown in Figure 9.53 may be written as

$$g(D) = [1,\ b(D)/m(D)]$$

where $m(D)$ denote the transform of the message sequence m_i, $i = 1, 2, \cdots, k$ and $b(D)$ denotes the transform of the parity sequence b_i, $i = 1, 2, \cdots, n-k$

$$\frac{b(D)}{m(D)} = \frac{1 + D^4}{1 + D + D^2 + D^3 + D^4}$$

a. Cross-multiply the above relation and invert into the time-domain to obtain the parity-check equation, which is satisfied by the RSC encoder at each time step. Show that the encoder output depends on past inputs as well as past outputs.
b. Determine the code rate and the constraint length.
c. Assuming zero initial states and an input bit sequence $m = \{110100101...\}$, determine the systematic bits, the parity bits and the encoder output in Figure 9.53.

27. Consider the following recursive systematic convolutional code transfer function with the generator polynomial

$$G(D) = \left[1 \quad \frac{1 + D + D^2}{1 + D^2} \right]$$

a. Draw the block diagram of this encoder.
b. Draw the state diagram.
c. Determine the impulse response.
d. Is this encoder systematic or nonsystematic?
e. Determine the code rate and the constraint length.
f. Write-down the parity-check equation.
g. Determine the encoder output for the input sequence $m(D) = 1 + D^2 + D^3 + D^6 + D^8$.
h. Can an all-zero sequence bring the encoder state back to 00?

28. A hybrid selective-repeat ARQ system uses an (n, k) block code with $d_{min} = 2t + 2$ to correct up to t errors and detect $t + 1$ errors per codeword.

a. Assuming that the transition probability
$$p = Q\left(\sqrt{2R_cE_b/N_0}\right) \ll 1, \quad \text{determine}$$
probability of correct reception, undetected codeword error probability, retransmission probability and probability of acceptance.

b. Evaluate p and the above probabilities for a (24, 12) extended Golay code with $d_{min} = 8$ and for an undetected bit error probability $P_b = 10^{-6}$.

c. Compare the undetected bit error probability that you determined in part (b) with the average bit error probability for uncoded transmission in AWGN channel for the same E_b/N_0 as in part (b). Discuss the advantages of using hybrid selective-repeat ARQ.

29. Obtain (9.217) from (9.216).

30. Obtain (9.211) from (9.209).

10

Broadband Transmission Techniques

Ever increasing demands for higher and differing data rates resulted in broadband transmission of multimedia signals. Broadband transmission has to cope with frequency-selective fading and other issues related to, amongst others, multiple access and multiplexing. In addition, associated time-synchronization problems and complex equalization techniques may be required. Spread spectrum (SS) and orthogonal frequency-division multiplexing (OFDM) represent the two main technologies for transmission of broadband signals in such adverse conditions.

Historically, direct sequence (DS) and frequency hopping (FH) SS systems are developed and used for military applications, for example, for anti-jamming (AJ) and low-probability of intercept (LPI). SS is a transmission technique in which a data sequence is transmitted in a bandwidth much in excess of the minimum bandwidth necessary to send it, with a correspondingly reduced PSD. The spectrum spreading is accomplished before transmission through the use of a code that is independent of the data sequence. The same code is used at the receiver, operating in synchronism with the transmitter, to despread the received signal for recovering the original data sequence. Spreading the user signals accross a wide spreading bandwidth improves the anti-jamming (AJ) capability of these signals. Since SS signals appear like random noise, they are difficult to detect and/or decode by receivers other than the intended ones, hence low probability of interception (LPI). Therefore, SS provides the opportunity for secure communications in a hostile environment.

The interference between SS signals sharing the same spread spectrum bandwidth is negligible if their codes are orthogonal. The code-division multiple access (CDMA) systems use this property to enable multiple users to share the same spread spectrum bandwidth with minimum multi-user interference (MUI). The spreading codes for user signals with different data rates are chosen so that all signals occupy the same spread bandwidth. The quasi-orthogonality of PN codes also provides a unique opportunity for accurate ranging, accurate timing and range rate (velocity) measurements. Modern navigations systems such as global positioning satellites (GPS), Glonass and Galileo systems exploit these properties

Digital Communications, First Edition. Mehmet Şafak.
© 2017 John Wiley & Sons Ltd. Published 2017 by John Wiley & Sons Ltd.
Companion website: www.wiley.com/go/safak/Digital_Communications

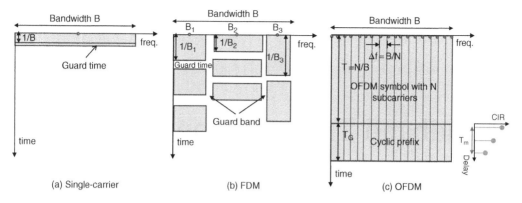

Figure 10.1 Spectrum Sharing in Single Carrier, FDM and OFDM Systems.

of PN codes to disseminate accurate positioning and timing information. The multipath rejection capability of SS signals can also be used to reduce self-interference due to multipath propagation in fading environments. In some applications, spectrum spreading via PN sequences is used to reduce the PSD's of signals in order to meet specified thresholds on PSD, for example, for health concerns, interference reduction purposes, and so on. Spread spectrum signals suffer frequency selective fading and Rake-receiver is generally used to equalize and maximal-ratio combining the signal replicas with different delays. This requires accurate channel estimation. SS signals may also suffer synchronization problems under adverse conditions.

OFDM is a bandwidth-efficient multi-carrier transmission technique. In OFDM, a high speed input data is divided into blocks that modulate each of the large numbers of subcarriers at low-data rates. The total bandwidth B allocated to the OFDM system is shared by N orthogonal subcarriers, hence the frequency separation between the neighbouring subcarriers is given by $\Delta f = B/N$ (see Figure 10.1). The subcarriers overlap but are still orthogonal to each other if $\Delta f = 1/T$ where T denotes the OFDM symbol duration (see (8.108)). For practical values of N, OFDM symbol duration T is N times longer than that

of the original broadband signal. Therefore, T is much longer than the maximum channel excess delay, hence multi-path fading problem is alleviated. On the other hand, since narrowband subcarriers suffer flat-fading, simple one-tap equalization for each subcarrier becomes sufficient. The number of subcarriers are usually chosen as an exponent of 2, $N = 2^n$, for FFT and IFFT operations required at modulation and demodulation stages.

Figure 10.1 shows a comparison of single-carrier FDM and OFDM systems. A single-carrier system, transmitting high-speed data at a bandwidth of B, has a relatively short symbol duration proportional to $1/B$. The corresponding guard time, which is chosen as a fraction of the symbol duration, can not mitigate ISI since the CIR is usually longer than the guard time. In FDM, several user signals of differing bandwidths share the transmission bandwidth B with symbol durations inversely proportional to their bandwidth. Sufficiently long guard time and wide guard bands are needed to separate these signals in time- and frequency-domains; this makes FDM relatively inefficient. On the other hand, OFDM is a bandwidth-efficient multi-carrier transmission technique where high-speed data is transmitted by dividing between multiple overlapping but still orthogonal subcarriers. Since the transmission bandwidth B is divided by N subcarriers, each subcarrier is assigned a transmission bandwidth $\Delta f = B/N$. This N-fold decrease in transmission bandwidth implies an N-fold increase in symbol duration. The signals transmitted by subcarriers hence undergo flat-fading in contrast with frequency selective fading suffered by single-carrier signals. Equalization in OFDM signals is therefore not a serious issue. However, depending on the number of subcarriers used, the ICI due to Doppler frequency shift and ISI due to delay spread may cause problems. The ICI is usually mitigated by Doppler correction. On the other hand, the guard-time (cyclic prefix) is chosen, as a fraction of the N-times increased symbol duration, to be longer than the maximum excess delay T_m of the channel in order to mitigate the ISI with minimum penalty in transmission rate. Time- and frequency-synchronization problems in OFDM are solved by using pilot subcarriers. Multiple-access in OFDM systems is accomplished either by time-division, that is, a user is allowed to use all sub-carriers during a limited-number of symbol durations, or by the so-called OFDMA (orthogonal frequency-division multiple-access). In OFDMA, users are allowed to transmit via non-overlapping resource blocks (RBs), consisting of a number of subcarriers and symbol durations. OFDM is adopted for applications like ADSL, DVB and 4G cellular radio.

This chapter will present the fundamentals and some applications of DS- and FH-SS systems and OFDM.

10.1 Spread Spectrum

In SS systems, two stages of modulation are used. In DSSS, a wide-band pseudo-random code spreads the incoming narrow-band data sequence into a noise-like (pseudo-random) wide-band signal. The resulting wide-band signal undergoes a second modulation, for example, M-PSK or M-QAM. In FHSS systems, the frequency of the usually M-FSK modulated carrier hops according to a pseudo-

random code across a large hopping bandwidth. DSSS uses coherent demodulation and it uses the whole spreading bandwidth at any instant. On the other hand, FHSS employs noncoherent M-FSK with a narrow instantaneous bandwidth, which is equal to that of the M-FSK. The performance of both of these techniques is closely related to the so-called pseudo-random or pseudo-noise (PN) spreading codes.

10.1.1 PN Sequences

CDMA systems require a sufficiently large set of spreading codes which allow as many users as possible to share the spread spectrum bandwidth with minimum mutual interference. Here, the number of codes and their auto/cross correlation properties become the most important performance metric. In military applications, the priority is in spreading codes with best low-probability of intercept (LPI) and anti-jamming (AJ) protection performances. The protection against and decoding by the adversary then become the most important priority in the code design.

A pseudonoise (PN) sequence is a periodic binary sequence with a noislike waveform that is usually generated by means of a feedback shift register. A feedback shift register consists of m flip-flops (two-state memory stages) and a logic circuit that is interconnected to form a multiloop feedback circuit. The shifts of the flip-flops is controled by a single timing clock. At each pulse (tick) of the clock, the state of each flip-flop is shifted to the next one. With each clock pulse the logic circuit computes a Boolean function of the states of the flip-flops. The result is fed-back as the input to the first flip-flop (see Figure 10.2).

In Chapter 9, we studied maximal-length and Walsh-Hadamard sequences that are widely used as PN sequences in SS systems. The maximal-length sequence is determined by the length m of the shift register, its initial state, and the feedback logic. A feedback shift

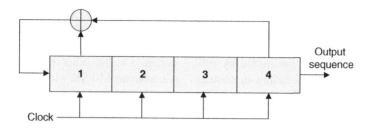

Figure 10.2   A PN Generator with $m = 4$ and the Feedback Logic. [1][4]

register is said to be linear when the feedback logic consists entirely of modulo-2 adders. With a total number of m flip-flops, the number of possible states of the shift register is 2^m. Since the contents of shift register are shifted one bit at a time for a total of 2^m-1 shifts, a periodic sequence of output length (period) less than or equal to $n = 2^m-1$ is generated. The zero-state (the state for which all the flip-flops are in state 0) is not permitted. If the zero-state would be permitted, the shift register would always remain in the zero state. Consequently, the period of a sequence produced by a linear feedback shift register with m flip-flops cannot exceed $n = 2^m-1$. When the period is exactly $n = 2^m-1$ (an odd number), the sequence is called a maximum-length-sequence or simply m-sequence.

PN sequences employed in practice are mostly derived from maximum-length sequences. Maximum-length sequences have many of the properties possessed by a truly random binary sequence, for example, each codeword contains 2^{m-1} ones and $2^{m-1}-1$ zeros, and all codewords have identical weights $w = d_{min} = 2^{m-1}$. Though maximum-length sequences are not completely random, they possess the properties of a truely random sequence as the sequence length n increases. Some properties of maximum-length sequences are as follows:

Balance property: In each period of a maximum-length sequence, the number of 1 s is always one more than the number of 0 s.

Run property: A *run* is defined as the number of consecutive bits of the same polarity, that is,

1 or 0. Among the runs of 1 s and 0 s in each period of a maximum-length sequence, 1/2 of the runs of each kind are of length one, 1/4 of the runs of each kind are of length two, 1/8 of the runs of each kind are of length three, and so on as long as these fractions represent meaningful numbers of runs.

Correlation property: The autocorrelation function of a maximum-length sequence is periodic and binary-valued. The normalized autocorrelation function of the random sequence x of length n may be written as

$$R_x(k) = E\left[\frac{1}{n}\sum_{i=0}^{n-1} x_i x_{i+k,\text{mod}(n)}\right]$$

$$= \begin{cases} \dfrac{1}{n}\sum_{i=0}^{n-1} E[x_i] E[x_{i+k,\text{mod}(n)}] = 0 & k \neq 0 \\[3mm] \dfrac{1}{n}\sum_{i=0}^{n-1} E\left[|x_i|^2\right] = 1 & k = 0 \end{cases}$$

(10.1)

where $i + k,\text{mod}(n)$ reflects the cyclic nature of the correlation. Autocorrelation function of a PN sequence is important since DS SS receivers synchronize to the incident SS signals by correlating the locally generated PN sequence with the PN sequence of the received signal. The autocorrelation function of a maximum-length sequence is shown in Figure 9.24 when the period of the maximum-length sequence is $T_b = nT_c$ where $n = 2^m-1$ and T_c is the chip duration. Note that the autocorrelation function given by (9.136) and (9.137) approaches that of a true random sequence with decreasing pulse duration T_c[1]

$$R(\tau) = \begin{cases} 1 - \dfrac{n+1}{nT_c}|\tau| & |\tau| \le T_c \\ 0 & otherwise \end{cases} \qquad (10.2)$$

that is, as n becomes very large. One may easily observe that the autocorrelation function of a true random sequence is identical to that of a single pulse of duration T_c.

The cross-correlation properties of PN codes are also important. The normalized cross-correlation function of two random binary sequences $\mathbf{x}$ and $\mathbf{y}$ is defined by

$$R_{\mathbf{xy}}(k) = \frac{1}{n}\sum_{i=0}^{n-1} x_i y_{i+k,\bmod(n)}$$

$$E[R_{\mathbf{xy}}(k)] = 0$$

$$E\left[|R_{\mathbf{xy}}|^2\right] = E\left[\frac{1}{n^2}\sum_{i=0}^{n-1}\sum_{j=0}^{n-1} x_i x_j y_{i+k,\bmod(n)}\, y_{j+k,(\bmod(n))}\right]$$

$$= \frac{1}{n^2}\sum_{i=0}^{n-1} E\left[x_i^2\right] E\left[y_{i+k,\bmod(n)}^2\right] = \frac{1}{n}$$

$$(10.3)$$

The PN codes assigned to CDMA users are desired to be mutually orthogonal in order to mitigate multi-user interference (MUI). However, they exhibit some cross-correlation, which limits the multiple-access capability of CDMA systems. Cross-correlation function between any pair of m-sequences of the same period has a peak value, which is a large percentage of the maximum value of the autocorrelation function. Since the number of m-sequences, hence the number of potential users in a CDMA system, increases rapidly with m, CDMA systems employing m-sequences suffer serious MUI. A small subset of the m-sequences, which have lower cross-correlation peaks, may not be sufficient for serving the ever-increasing numbers of CDMA users.

One needs to determine the optimum connection diagram for the feedback shift register to obtain a robust PN sequence (see Table 9.14). In military SS applications, m-sequences can not potentially provide desired AJ and LPI

performance. For example, a careful look at Figure 9.24(b) shows that one may easily determine T_c and n by measuring the first null and the frequency separation between frequency components in the PSD. Due to the linearity of the m-sequences, a jammer can determine the feedback connections by observing $2m$ chips of the PN sequence. The number m of flip-flops must be sufficiently large so as to force a potential jammer to spend long time to try all combinations of feedback connections of the PN generator. Then, a computer search for $n = 2^m$ possible codes and correlating them with the received signal would be required in order to determine the code. In military applications, nonlinear sequences, for example, Gold sequences, are preferred since they are more difficult to be determined. Further protection may be obtained by changing the feedback connections and/or the number of stages according to some predetermined plan known only by the transmitter and the intended receiver.

Gold sequences are constructed from modula-2 addition of two m-sequences with the same length and synchronized clock rate (see Figure 10.3). The period of Gold codes is the same as that of the constituent m-sequences. Gold sequences have the same three peaks in their cross correlation and autocorrelation functions. These peaks normalized with respect to the autocorrelation peak, $2^m - 1$, are given by [2][3]

$$\frac{-1}{2^m - 1}, \frac{-t(m)}{2^m - 1}, \frac{t(m)-2}{2^m - 1},$$

$$t(m) = \begin{cases} 2^{(m+1)/2} + 1 & m \ odd \\ 2^{(m+2)/2} + 1 & m \ even \end{cases} \qquad (10.4)$$

The peaks tend to become smaller as the code length increases. For example for $m = 5$, the peaks are $-1/31$, $-9/31$ and $7/31$. The autocorrelation properties of Gold sequences are not as good as those of m-sequences. However, they exhibit better cross-correlation properties.

Two m-sequences of length n with cross-correlation peaks given by (10.4) are called

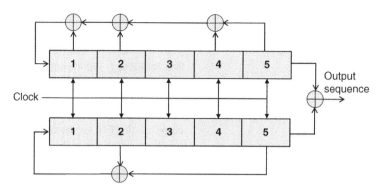

Figure 10.3 Construction of a Gold Sequence Using the Preferred Sequences with Generator Polynomials $g_1(D) = D^5 + D^2 + 1$ and $g_2(D) = D^5 + D^4 + D^2 + D + 1$. [2]

preferred sequences. In addition to two preferred sequences c_1 and c_2 of length n, n other Gold sequences of period $n = 2^m - 1$ can be constructed by taking the modula-2 sum of c_1 with n cyclic shifts of c_2; hence $n + 2$ Gold sequences may be obtained. [1] Gold sequences are commonly used in CDMA systems since they are robust, supply a large number of codes by using only a simple circuitry and provide more security compared to m-sequences.

Kasami sequences are also widely employed in CDMA and military applications. They are of length $2^m - 1$ where m (an even number) denotes the degree of the primitive polynomial to generate Kasami sequences. The normalized maximum value of the correlation function is $t(m)/(2^m - 1)$, which is the same as that of the Gold code for even m in (10.4).

Walsh-Hadamard codes, which are of limited number, are also used in coherent CDMA systems especially in eliminating transmissions from different cells/services. For example, 64 orthogonal Walsh-Hadamard codes are used for spreading in the IS-95 system. The Walsh-Hadamard codes do not have good autocorrelation characteristics. Since they have more than one peak, a receiver can not detect the beginning of the codeword without an external synchronization scheme. The cross-correlation can also be non-zero for a number of time-shifts; therefore unsynchronized users can interfere with

each other. This is why Walsh-Hadamard codes can only be used in synchronous CDMA systems. On the other hand, the spreading behavior of these codes is worse than PN sequences because their PSD is concentrated in small numbers of discrete frequencies.

Walsh-Hadamard codes are fixed-length orthogonal codes and also form the basis for the so-called codes with *orthogonal variable spreading factors (OVSF)*. These codes are useful in applications where users with different data rates share the same SS bandwidth. For example, users may have different data rate requirements depending on whether they use voice, data or image transmission. By assigning different spreading factors (SFs), these users can spread their data accross the same spreading bandwidth.

The generation algorithm of these codes is similar to that for the Walsh-Hadamard codes. Figure 10.4 shows the generation of OVSF codes in a binary tree structure using Walsh sequences. New levels in the code are generated by using a root codeword and a replica of itself. Note that all the codes in the tree are not orthogonal to each other. However, they become orthogonal only if they satisfy the following rules:

a. The codes in the same layer constitute the Walsh functions, and they are orthogonal.

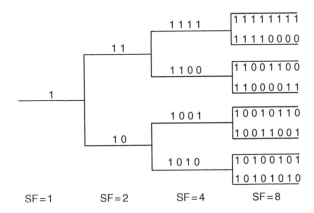

Figure 10.4 Walsh sequences form Orthogonal Variable Spreading Factor (OVSF) codes.

b. Any two codes of different layers are also orthogonal except for the condition that one code is the mother code of the other. [3]

OVSF codes may be used to support a variety of data services from low to very high bit rates. Since the SS bandwith is the same for all users, multiple-rate transmissions require the use of OVSFs. For example if the symbol rate is R_s and the spreading factor is SF so that the chip rate $R_c = R_s SF$, the chip rate (spreading bandwidth) would be unchanged if the symbol rate is multiplied by K and the SF is divided by K. For example, for a chip rate of 3.84 Mchips/s, one can use $SF = 4$ and $R_s = 960$ ksps for uncoded transmissions while for transmissions at $R_s = 480$ ksps, the spreading factor should be chosen as $SF = 8$.

Wideband CDMA (WCDMA) systems limit the channel bandwidth to 5 MHz. They allow multi-rate transmissions; hence the system can support multiple simultaneous services from multiple users. Spreading user signals with two different codes is also commonly used in CDMA-based applications. The OVSFs are used as channelization code (short code) to distinguish signals of different cells or services. The users in the same cell or using the same services are distinguished by the so-called

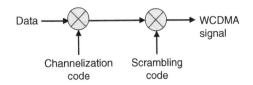

Figure 10.5 Spreading by Channelization and Scrambling Codes in WCDMA.

scrambling codes, usually Gold or Kasami sequences (see Figure 10.5). [4]

10.1.2 Direct Sequence Spread Spectrum

Let $\{b_k\}$ and $\{c_k\}$ denote a binary data sequence, and a PN sequence (spreading code), respectively. Let information-bearing signal $b(t)$ and the PN signal $c(t)$ denote their respective polar NRZ representations, namely, ± 1. The signal $m(t)$ which is obtained by spreading the data sequence $b(t)$ by the code sequence $c(t)$ may be written as

$$m(t) = b(t)\, c(t) \tag{10.5}$$

Noting that the bandwidths of $b(t)$ and $m(t)$ are proportional to $1/T_b$ and $1/T_c$ respectively, the ratio of the bandwidths is given by

$$\frac{W_{ss}}{W} = \frac{1/T_c}{1/T_b} = \frac{T_b}{T_c} = n \tag{10.6}$$

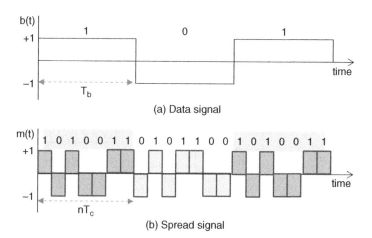

(a) Data signal

(b) Spread signal

Figure 10.6   Spreading the Data Signal 101 with the m-Sequence 1010011 with $n = 7$, as Listed in Table 9.13.

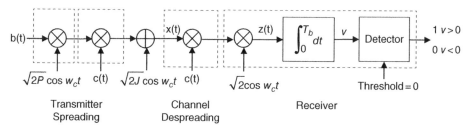

Figure 10.7   Block Diagram of a Spread-Spectrum Transceiver For BPSK Modulation.

Note that the spreading period nT_c is taken equal to the duration T_b of one bit. When long codes are used, as in AJ applications, the spreading period may be chosen to be much longer than one bit.

Figure 10.6 shows the spreading of a data sequence $b(t) = [1 \ -1 \ 1]$ by a code sequence $c(t) = [1 \ -1 \ 1 \ -1 \ -1 \ 1 \ 1]$. Since $T_b = 7T_c$, the bandwidth of $b(t)$ is spread by a factor of $n = 7$. The signal m(t) is said to be despread when it is multiplied by the same code c(t) and the original data signal b(t) is recovered since $m(t)c(t) = b(t) \, c^2(t) = b(t)$. Here it is important to note that for successful despreading the signal $m(t)$ must have perfect time synchronization with the code to be multiplied.

Figure 10.7 shows the block diagram of a DSSS system, where the data sequence b(t) is first spread by the PN code c(t) and then

up-converted by the carrier. Due to the linearity of these operations, the order of spreading and up-converting is reversed in Figure 10.7. Hence, one may observe that a DSSS system is the same as a coherent BPSK system when spreading and despreading operations are removed. Then, the block diagram in Figure 10.7 reduces to that for BPSK signalling.

Assuming that the jamming/interference signal j(t) is much stronger than AWGN, the DSSS signal at the receiver input may be written as

$$x(t) = s(t) + j(t)$$
$$s(t) = \sqrt{2P}\, b(t)\, c(t) \cos(w_c t + \varphi) \qquad (10.7)$$
$$j(t) = \sqrt{2J} \cos(w_c t + \varphi')$$

The jammer signal is assumed to be a sinusoid of frequency f_c and mean power J.

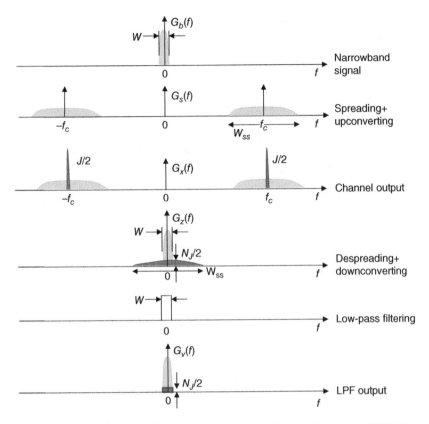

Figure 10.8 PSD of the DSSS Signal and the Jammer at Various Stages of a DSSS System.

The phases φ and φ' denote channel-induced phase shifts suffered by user and jammer signals, respectively. Since locations of user and jammer transmitters are not the same, these random phase shifts differ from each other. As the message signal $b(t)$ is narrowband and the PN signal $c(t)$ is wideband, the spread signal $s(t)$ will have a spectrum wider than that of $b(t)$ (see Figure 10.6).

The PSD of $x(t)$ is given by (10.135) (see Appendix 10A and Figure 10.8)

$$G_x(f) = G_s(f) + G_J(f) \qquad (10.8)$$

where the autocorrelation function and the corresponding PSD for $s(t)$ and $j(t)$ are given by (10.134), (10.136) and (10.138) (see Figure 10.8):

$$R_s(\tau) = P\ R_b(\tau)\ R_c(\tau)\ \cos(w_c\tau)$$

$$\Leftrightarrow\ G_s(f) = \frac{PT_c}{2}\left[\operatorname{sinc}^2\{(f-f_c)T_c\}\right.$$
$$\left. + \operatorname{sinc}^2\{(f+f_c)T_c\}\right]$$

$$R_J(\tau) = J\ \cos(w_c\tau)\ \Leftrightarrow$$
$$G_J(f) = \frac{J}{2}[\delta(f-f_c) + \delta(f+f_c)] \qquad (10.9)$$

To recover the original message signal $b(t)$, the received signal $x(t)$ is down-converted using the local oscillator of carrier frequency f_c and correlated with the locally generated code, which is assumed to be in perfect synchronism with the transmit code. This is followed by an integrator, and a detector (see Figure 10.7). Following down-conversion of the received signal $x(t)$ using a local oscillator with unity power and correlation with the locally generated code, we obtain

$$z(t) = \sqrt{2}\, x(t)\, c(t) \cos(w_c t + \hat{\varphi}) \qquad (10.10)$$

where $\hat{\varphi}$ denotes the estimation for the phase of the user signal. The PSD of $z(t)$ at the integrator input is given by (10.141) (see Figure 10.8).

$$G_z(f) = PT_b\ \mathrm{sinc}^2(fT_b)\, E\left[\cos^2(\varphi - \hat{\varphi})\right]$$
$$+\ JT_c\, \mathrm{sinc}^2(fT_c)\, E\ \left[\cos^2(\varphi' - \hat{\varphi})\right]$$
$$(10.11)$$

where $E[\cos^2(\varphi - \hat{\varphi})] = 1$ is assumed based on the perfect estimation of the channel-induced phase. However, $E[\cos^2(\varphi' - \hat{\varphi})] = 1/2$ is assumed since the jammer is mostly unable to predict the channel phase. Because the chip rate of the spreading code is much higher than the data rate ($T_b \gg T_c$), the PSD of the jammer signal may be assumed to be constant over the receiver bandwidth. If $z(t)$ is passed through an ideal IF filter of bandwidth $1/T_b \ll 1/T_c$, the SNR at the receiver output is given by (10.145):

$$SNR = \frac{P}{J/G_p} = G_p\, SNR_{\mathrm{input}} \qquad (10.12)$$

where $SNR_{\mathrm{input}} = P/J$ denotes the SNR at the input of the DSSS receiver. Hence, a DSSS receiver reduces the effectiveness of a jammer signal by a factor of G_p. In other words, the SNR performance of a DSSS system is enhanced by the processing gain G_p, which is given by (10.143):

$$G_p = 2n = 2T_b/T_c \qquad (10.13)$$

The above is based on the assumption that the jammer can not estimate the channel-induced phase. In this case, the jammer is forced to divide its power into two quadrature channels, thus wasting half of its power in jamming the channel orthogonal to the user signal. Otherwise, it can match its carrier phase to that of the user to increase its effectiveness by 3 dB.

Single-sided PSD of the jammer signal over the SS bandwidth is given by (10.144)

$$N_J = JT_c\ E\left[\cos^2(\varphi' - \hat{\varphi})\right]$$
$$= \begin{cases} JT_c & \hat{\varphi} = \varphi' \\ JT_c/2 & otherwise \end{cases} \quad (W/Hz) \qquad (10.14)$$

Hence, (10.14) shows that the jammer/interference signal is spread in the SS bandwidth with a single-sided PSD N_J. If jammer can correctly estimate the chanel-induced phase, then the jammer PSD would be increased by 3 dB. Noting that the PSD of the jammer signal looks "flat" in the receiver bandwidth $(0, 1/T_b)$ for large n, it is usually assumed to be AWGN.

Applying the correlator output to a low-pass filter (integrator) with a bandwidth of $1/T_b$ ensures the recovery of the data signal $b(t)$ by filtering out the spectral components of the spread jammer signal $c(t)\, j(t)$ outside the bandwidth $(0, 1/T_b)$. Low-pass filtering is actually performed by the integrator that evaluates the area under the signal produced at the multiplier output in $0 \le t \le T_b$. Since $1/T_b \ll W_{ss}$, only the fraction $1/G_p$ of the jammer/interference power within this band appears at the integrator output. Therefore, spreading the jammer signal by the PN code over W_{ss} and integration over the bit duration $[0, T_b]$ (or alternatively low-pass filtering over $[0, 1/T_b]$) results in a reduction of the effective jammer power by a factor of G_p. In other words, the performance of the DS SS system is enhanced by the processing gain, G_p.

The integration in the bit interval $0 \le t \le T_b$ provides the sampled output $v(T_b)$ (see Figure 10.7). This output is the same as that of the matched filter for a coherent BPSK signal. Assuming perfect synchronization and $b(t) = +1$ for bit 1 and $b(t) = -1$ for bit 0, the normalized decision variable is given by (10.154) in Appendix 10B:

$$v(T_b)/\sqrt{E_b} = \begin{cases} +\sqrt{E_b} + w(T_b) & 1 \\ -\sqrt{E_b} + w(T_b) & 0 \end{cases} \qquad (10.15)$$

where $w(T_b)$ has zero mean and a single-sided PSD N_J (see (10.14) and (10.153). The probability of bit error is found using (10.12):

$$P_e = Q\left(\sqrt{SNR}\right) = Q\left(\sqrt{\frac{G_p P}{J}}\right)$$

$$= Q\left(\sqrt{\frac{G_p E_b / T_b}{(N_J/2)/T_c}}\right) = Q\left(\sqrt{\frac{2E_b}{N_J}}\right)$$

$$(10.16)$$

which is identical to (10.155). In AWGN channels, one has $N_J = N_0$; then, (10.16) reduces to the well-known bit error probability expression for BPSK modulation. Since AWGN is white (already spread) and not affected by (de-) spreading, SS does not provide any performance advantage in AWGN channels. However, SS is effective when spreading/despreading is exploited, for example, in interference/jamming channels. It is clear from (10.12) that a DSSS receiver suffering narrowband jamming improves the SNR at the output by a factor of G_p compared to the input SNR.

We demonstrated above the basic principles of DSSS with coherent binary PSK signals. For QPSK/DSSS signaling, following serial-parallel conversion (S/P), the data modulates the orthogonal carriers (see Figure 10.9). Modulation on orthogonal carriers is spread by codes c_1 and c_2. In this case, spreading codes c_1 and c_2 do not have to be orthogonal since quadrature carriers $\cos(w_c t)$ and $\sin(w_c t)$ already provide the desired orthogonality.

Example 10.1   Link Budget For a Wi-Fi Link. A BPSK/DSSS Wi-Fi system operates at 2.4 GHz carrier frequency in an AWGN channel at a range d. Transmit and receive antenna gains are 0 dBi. The receiver noise figure is equal to $F = 5$ (7 dB) and the receive antenna noise temperature is $T_A = T_0 = 290$ K. 10 dB shadowing margin is assumed. Channel model is given by (3.92) with $d_0 = 1$ meter and path-loss exponent is 3.

From (4.62), the noise PSD is given by $N_0 = kFT_0$ and the E_b/N_0 at the output of a SS receiver is the same as that of a BPSK receiver:

$$\frac{E_b}{N_0} = \frac{P_t G_t G_r}{N_0 R_b L_{fs}} = \frac{P_t G_t G_r}{kFT_0 R_b}\left(\frac{\lambda}{4\pi d_0}\right)^2 \left(\frac{d_0}{d}\right)^3$$

$$(10.17)$$

where $k = 1.38 \times 10^{-23}$ (J/K). The bit error probability of a SS system encoded with a (127,113) BCH code with $t = 2$ bit error correction capability (see (9.87) and Table 9.17) is shown in Figure 10.10 as a function of the range d for several values of the transmit power level. One needs transmit power levels higher than 10 dBm for a coverage radius of 200 m at 10^{-6} bit error probability level.

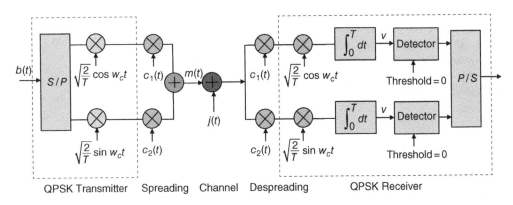

QPSK Transmitter   Spreading   Channel   Despreading   QPSK Receiver

Figure 10.9   Block Diagram of a QPSK/DSSS System.

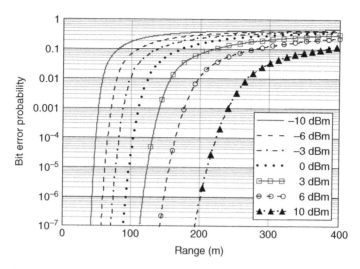

Figure 10.10 Bit Error Probability of a DSSS System as a Function of Range For Several Values of the Transmit Power Level. Frequency 2.4 GHz, antenna noise temperature $T_A = 290$ K, receiver noise figure $F = 5$ and (127,113) BCH code with $t = 2$. Transmit and receive antenna gains are 0 dBi. 10 dB shadowing margin is assumed. Path loss model is given by (3.92) with $d_0 = 1$ m and path-loss exponent 3.

Example 10.2 Use of DSSS to Reduce the PSD of SATCOM Signals.

A satellite operating at 7 GHz is used to transmit signals over a region on earth. The satellite transmit dish antenna has a half-power beamwidth of is 1 degrees and 65% antenna efficiency. The satellite has a transmit power of $P_t = 50$ W over a bandwidth W. The overall losses in the system are specified as $L_{dB} = 5$ dB. The maximum power density on the surface of earth which is $d = 40000$ km away from the satellite is limited to -120 dBW/m^2 over W Hz.

We will first determine the PSD of the satellite signals on the earth's surface. From (2.84), an aperture antenna with 1 degrees half-power beamwidth and 65% aperture efficiency has a diameter of $D_d/\lambda = 72$. The corresponding antenna gain is found from (2.101) as

$$G = 10 \log \left\{ \eta (\pi D_a/\lambda)^2 \right\} = 44.87 \ dBi \quad (10.18)$$

The power density of signals on the earth surface over W Hz is given by

$$Pr_{dB} = 10 \log \left\{ \frac{P_t \, G}{L 4\pi d^2} \right\}$$

$$= 10 \log \left\{ \frac{50 \ 10^{4.487}}{10^{0.5} 4\pi \left(4 \ 10^7 \right)^2} \right\} \quad (10.19)$$

$$= -106.17 \ \left(\frac{dBW}{m^2} \right)$$

Since the received level does not meet the requirement -120 dBW/m^2 over the bandwidth W, the bandwidth of the transmitted signal needs to be spread by a factor of

$$n = 10^{-0.1 \times 106.17} / 10^{-0.1 \times 120} = 10^{1.383} \cong 24 \quad (10.20)$$

Hence, the satellite signals may be spread 24 times by using DSSS, that is, $W_{ss} = 24$ W.

10.1.2.1 Synchronization in DSSS Systems

DSSS systems are coherent in the sense that the code timing and the carrier phase/frequency of the input signal need to be estimated at the receiver. In case of perfect synchronization, $E[\cos^2(\varphi-\hat{\varphi})]=1$ in (10.10); then the degradation due to phase offsets vanishes. The degradation due to time offset between the incoming and locally generated PN sequences as well as the phase estimation errors are already studied in Section 8.3.

The synchronization in DSSS systems is usually carried out in a two-dimensional plane formed by frequency (phase) shift and delay. In the first step, the receiver sweeps the frequency/phase of its local oscillator to observe a signal maximum at the output. This compensates for the frequency shift, for example, Doppler frequency shifts, suffered by the received signal, the phase shift induced the channel and the differences in the carrier frequencies generated the local oscillators of the transmitter and the receiver.

The next step in the synchronization is to align the locally-generated PN sequence with that of the received sequence. This process may be divided into acquisition (coarse synchronization) and tracking (fine synchronization) phases. In acquisition, the two PN codes are aligned to within a fraction of the chip as fast as possible. PN acquisition proceeds in two steps: The received code which is transmitted for synchronization purposes, is downconverted and then multiplied with the locally generated code. The resulting product signal is integrated over a number of chips ($\kappa \gg 1$) to produce a measure of correlation between

them. An appropriate decision-rule and search strategy is used to process the measure of correlation between the two codes. In the serial acquisition algorithm shown in Figure 10.11, the output of the correlator is compared to a preset threshold level. If the correlator output exceeds the threshold, the acquision is declared. Otherwise, the search control circuit advances/delays the locally generated code by a fraction (typically one-half) of a chip duration and search continues until acquisition is achieved. The maximum time required for searching an uncertainty region of N_c chips long in half-chip increments is therefore [1]

$$\left(T_{acq}\right)_{\max} = 2N_c\kappa T_c \qquad (10.21)$$

Once the incoming PN code has been acquired, tracking process starts by using phase-lock techniques. The aim of tracking is to avoid the loss of synchronization against random changes in the channel characteristics in terms of time, delay and carrier frequency offsets.

10.1.2.2　Code Division Multiple Access (CDMA)

CDMA allows multiple users, occupying the same SS bandwidth to communicate simultaneously with tolerable interference with each other. Each of the K users is assigned a separate code $c_i(t)$, $i = 1,2,\ldots K$, of length n, that is, $0 \le t \le nT_c$, with low cross-correlation properties in code period T. If the code period T coincides

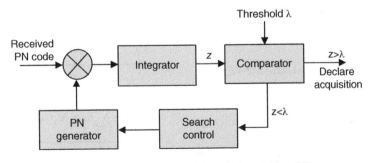

Figure 10.11　DSSS Serial-Search Acquisition. [1]

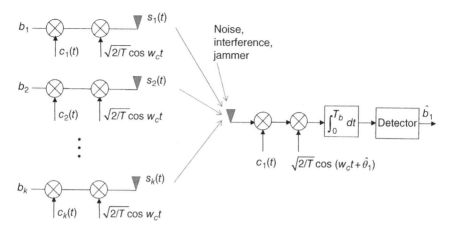

Figure 10.12 Block Diagram of a CDMA System with K Users.

with the symbol duration T_s, then there are n chips of duration $T_c = T_s/n$ in each code. The SS bandwidth is therefore given by n/T_s, implying that the bandwidth $1/T_s$ of transmitted symbols is spread by n times. Users of a CDMA system can share the full SS bandwidth synchronously or asynchronously. In asynchronous operation, symbol transition times of different users do not have to coincide with each other. Hence, there is no need for coordination between users albeit at the expense of some performance degradation compared to synchronous operation.

Consider a CDMA system as shown in Figure 10.12 where each of the K users is transmitting data $\{b_k\}$ using his own particular PN code $c_k(t)$ at a common carrier frequency f_c. Each code is kept secret and its use is restricted to the community of authorized users. The signal at the receiver input may therefore be written as the sum of the signals of K uncoordinated users:

$$r(t) = \sqrt{\frac{2E}{T}} \sum_{k=1}^{K} h_k b_k c_k(t) \cos(w_c t + \theta_k)$$

$$+ w(t), \quad 0 \le t \le T$$

$$\text{(10.22)}$$

where $w(t)$ denotes the AWGN noise with variance $N_0/2$, E is the mean symbol energy,

and θ_k denotes the carrier phase of the k-th user. As users may be located at different distances from the BS receiver, the complex channel gain $h_k = \alpha_k e^{j\phi_k}$ accounts for the differences between the received user signal levels at the BS unless automatic power control is used.

The receiver wanting to receive the data b_1 correlates the received signal with

$$\sqrt{\frac{2}{T}} c_1(t) \cos(w_c t + \hat{\theta}_1) \qquad \text{(10.23)}$$

where $\hat{\theta}_1$ denotes the estimate of the channel-induced phase of the wanted signal. The received signal may then be written as:

$$y(T) = \int_0^T r(t) \sqrt{\frac{2}{T}} c_1(t) \cos(w_c t + \hat{\theta}_1) \, dt$$

$$= \underbrace{\sqrt{E} h_1 b_1 \cos(\theta_1 - \hat{\theta}_1)}_{\text{wanted signal}}$$

$$+ \underbrace{\sqrt{E} \sum_{k=2}^{K} h_k b_k R_{c_1 c_k} \cos(\theta_k - \hat{\theta}_1)}_{MUI} + n(T)$$

$$\text{(10.24)}$$

The noise term is characterized by zero-mean and variance $N_0/2$:

$$n(T) = \sqrt{\frac{2}{T}} \int_0^T c_1(t) \cos(w_c t + \hat{\theta}_1) \, w(t) dt$$

$$E[n(T)] = 0$$

$$E\left[|n(T)|^2\right] = \frac{2}{T} E\left[\int_0^T \int_0^T c_1(t) c_1(t') \cos(w_c t + \hat{\theta}_1)\right.$$

$$\left.\underbrace{\cos(w_c t' + \hat{\theta}_1) w(t) w(t')}_{N_0/2\delta(t-t')} \, dt dt'\right] = N_0/2$$

$$(10.25)$$

The cross-correlation function between $c_1(t)$ and $c_k(t)$ is defined by (10.3). The multi-user interference (MUI) terms are due to non-zero correlation between $c_1(t)$ and $c_k(t)$, $k = 2, 3, ..,$ K. The choice of PN codes with low cross-correlation is therefore critical in CDMA. In view of the central-limit theorem, the MUI term in (10.24) may be approximated by AWGN for large values of K with a variance[5]

$$\sigma_{MUI}^2 = E\left[\left|\sqrt{E} \sum_{k=2}^K h_k b_k R_{c_1 c_k} \cos(\theta_k - \hat{\theta}_1)\right|^2\right]$$

$$= E \sum_{k=2}^K \sum_{j=2}^K h_k h_j^* \underbrace{E[b_k b_j^*]}_{\delta(k-j)} \underbrace{E[R_{c_1 c_k} R_{c_1 c_j}]}_{1/n}$$

$$\underbrace{E[\cos(\theta_k - \hat{\theta}_1) \cos(\theta_j - \hat{\theta}_1)]}_{1/2}$$

$$= \frac{1}{n} E \sum_{k=2}^K \alpha_k^2$$

$$= \frac{K-1}{n} \alpha_1^2 E\left(\frac{1}{K-1} \sum_{k=2}^K \frac{\alpha_k^2}{\alpha_1^2}\right) \quad (10.26)$$

where (10.3) is used to determine the expectation of the square of the cross-correlation function. Assuming that the PSD's of noise and MUI are additive and the intended receiver can perfectly estimate the channel phase, that is, $\hat{\theta}_1 = \theta_1$, the SINR is found from (10.24) as

$$SINR = \frac{E\left[|\sqrt{E} h_1 b_1|^2\right]}{N_0/2 + \sigma_{MUI}^2} = 2 \frac{\alpha_1^2 E/N_0}{1 + \frac{2}{N_0} \sigma_{MUI}^2}$$

$$(10.27)$$

The following special cases apply in relation to *MUI* and the *SINR*:

a. When K users with different channel gains share the same SS bandwidth, the SINR suffers a degration with respect to a single-user case by the factor:

$$D_{MUI} = 1 + \frac{2}{N_0} \sigma_{MUI}^2$$

$$= 1 + \frac{(K-1)}{n} \frac{\alpha_1^2 E}{N_0} \left(\frac{1}{K-1} \sum_{k=2}^K \left(\frac{\alpha_k}{\alpha_1}\right)^2\right)$$

$$(10.28)$$

b. For a single user ($K = 1$), $D_{MUI} = 1$ and (10.28) reduces to $2\alpha_1^2 E/N_0$, as expected.

c. When the users can perfectly control their transmit powers, the signal power levels received from all users are the same at the CDMA receiver. Then, the degradation caused by the *MUI* reduces to

$$D_{MUI} = 1 + \frac{K-1}{n} \frac{E}{N_0}, \quad \alpha_k = 1, \quad k = 1, 2, ..., K$$

$$(10.29)$$

d. When automatic power control is not implemented in the CDMA system, the term

$$D_{PI} = \frac{1}{K-1} \sum_{k=2}^K \left(\frac{\alpha_k}{\alpha_1}\right)^2 \quad (10.30)$$

which accounts for the effect of the power imbalance (PI) between received user signals, will differ from unity. If all user signals are not received at the same power level, bit errors observed by a user will be higher than those of the other users that are located closer to the receiver. This is

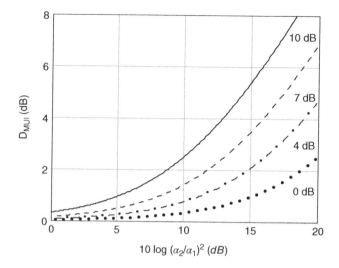

Figure 10.13 Degradation Suffered By the First User For Several Values of the E/N_0 of the First User Due to Power Imbalance Between Two Users For $n = 127$, $K = 2$ and $\alpha_1 = 1$. α_2/α_1 denotes the ratio of the received signal level of the second user to that of the first user. [5]

the so-called *near-far effect* which occurs when CDMA users, at different distances from the same receiver, transmit with the same transmit power (no power control). CDMA signals at the receiver with higher powers cause intolerable interference to signals with lower-power. The near-far effect can be mitigated by adjusting each user's transmit power so that the received signal levels are at the same minimum acceptable level; this will evidently minimize the mutual interference between users. Therefore, automatic power control is a must in CDMA systems. Figure 10.13 shows the degradation (10.28) suffered by the first user at the receiver as a function of the ratio of the SNRs of second user to the first user for various values of the SNR of the first user in a $K = 2$ user CDMA system. [5] For example, if the SNR of the first user is assumed to be 10 dB in a CDMA system with $n = 127$, the presence of a second user with SNR $= 15$ dB suppresses the SNR of the first user by 1 dB; hence, the effective SNR of the first user becomes 10 dB -1 dB $= 9$ dB.

Transmit power control (TPC) is essential in CDMA systems in order to solve the near-far problem and to increase the system capacity. Power control may be implemented by measuring the power received at the receiver and then relaying to the transmitter so that it adjusts its transmit power. Two implementation issues related to power control need serious consideration. Accurate estimation of amount of increase/decrease of the transmit power, and the difference between the time the power is estimated and the time the correction is applied. Inaccurate estimation and delayed update of the transmit power would offset the desired benefit. There is a tradeoff between the two. To minimize processing delay, simple processing may be performed on the received signal with minimal averaging at the expense of estimation accuracy. In open loop TPC of WCDMA systems, the receiver estimates the transmission channel's path loss, and transmitter power is calculated on the basis of path loss. In SIR-based closed loop TPC, the receiver SIR is measured in every power control cycle (typically

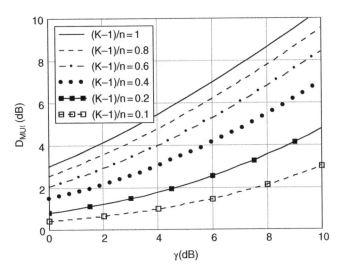

Figure 10.14 Degradation Due to MUI in a CDMA System with Perfect Power Control as a Function of $\gamma = E/N_0$ (dB) For Various Values of $(K-1)/n$.

0.625 ms). Then, a TPC-down or a TPC-up command is sent depending on whether the measured value is higher or lower than the target SIR value.

Figure 10.14 shows the variation of (10.29), the degradation in E/N_0 due to MUI in a CDMA system with perfect power control, as a function of E/N_0 for various values of $(K-1)/n$. The degradation evidently increases with increasing values of $(K-1)/n$ and E/N_0. For example, in a CDMA system with $n = 1000$, a user with $E/N_0 = 8$ dB suffers a MUI of 2 dB when the system serves $K-1 = 0.1n = 100$ users.

Figure 10.15 shows the SNR suppression suffered by the first user in a K user CDMA system, where the relative signal levels are defined by three different scenarios for $n = 127$. The small-signal suppression is therefore a serious issue and may hinder the capability of reliable communication of weaker signals. Therefore automatic power control is essential in CDMA systems.

For perfect power control, that is, $\alpha_k = 1$, $k = 1, 2, \ldots, K$, the variance of the MUI given by (10.26), reduces to

$$\sigma_{MUI}^2 = \frac{K-1}{2n} E = \frac{N_{MUI}}{2}$$

$$N_{MUI} = \frac{K-1}{n} E = (K-1)PT_c = (K-1)E_c$$

$$(10.31)$$

where N_{MUI} denotes the two-sided PSD of the MUI and $E_c = E/n$ is the energy per chip.

Inserting (10.31) into (10.27), the bit error probability of a BPSK/CDMA system with K asynchronous users using perfect power control is given by,

$$P_b = Q\left(\sqrt{\frac{2E_b}{N_{MUI} + N_0}}\right) = Q\left(\sqrt{\frac{2E_b/N_0}{\frac{E_b}{N_0}\frac{K-1}{n} + 1}}\right)$$

$$\approx Q\left(\sqrt{\frac{2n}{K-1}}\right)$$

$$(10.32)$$

where $E_b = PT_b$ and T_c is the chip duration. Figure 10.16 shows the effect of the MUI on the bit error probability, where K shows the number of users accessing the system and

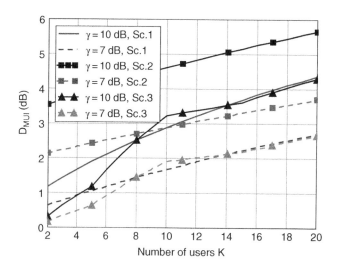

Figure 10.15 Degradation Suffered By User 1 Due to Power Imbalance in a K user CDMA System For Scenario 1: $\alpha_1 = 1$, $\alpha_2 = 2$, $\alpha_k = 1$, $3 \le k \le 20$; Scenario 2: $\alpha_1 = 1$, $\alpha_2 = 4$, $\alpha_k = 1$, $3 \le k \le 20$; and Scenario 3: $\alpha_k = 1$, $1 \le k \le 5$, and $15 \le k \le 20$; $\alpha_k = \sqrt{2}$, $k = 6,7,\ldots,10$ $\alpha_k = 1/\sqrt{2}$, $11 \le k \le 15$.

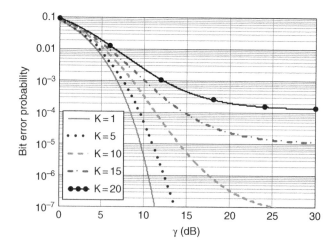

Figure 10.16 Effect of the Number of Users on the Performance of a CDMA System for $n = T_b/T_c = 127$.

$n = 127$. Note that there is a graceful degradation in the bit error probability performance as the number of users increases. Unlike TDMA based systems, there is no hard limit for the number of users that can be served by the system. As the number of users increases, the error floor gets higher and the system degrades

gradually but it still continues to serve, though with lower quality of service.

Example 10.3 Capacity Limitation Due to Multi-User Interference in CDMA Systems. A BPSK/CDMA system with $n = 400$ (26 dB) operates in an AWGN channel. The bit error probability in the multi-user environment is required to be lower than 10^{-6}. Noting that the bit error probability level of 10^{-6} corresponds to $E_b/N_0 = 11.29$ for BPSK modulation, the number of users that can be supported by a CDMA system is found using (10.32)

$$\frac{E_b/N_0}{\dfrac{E_b}{N_0}\dfrac{K-1}{n}+1} = 11.29$$

$$\Rightarrow K = 1 + n\left(\frac{1}{11.29} - \frac{1}{E_b/N_0}\right)$$
$$(10.33)$$

If the received $E_b/N_0 \to \infty$, that is, when the noise is neglected, then the system can serve $K = 36$ users. Assuming $E_b/N_0 = 17$ dB when there is a single user in the CDMA system, the number of equal power users is found from (10.33) as 28. On the other hand, if each user's transmit power is reduced by 3 dB, then $E_b/N_0 = 25$ (14 dB) and the system can serve $K = 20$ users.

10.1.2.3   AJ and LPI

Spread spectrum was historically designed in order to provide protection against against jamming and interception for military communication systems in which high data rate transmissions is not the primary requirement. A jammer aims to deny communications to his adversary at minimum cost. Knowing that the complete protection is not possible, a communicator aims to develop a jam-resistant system under the following assumptions: a) jammer has a-priori knowledge of system parameters, such as frequency band, timing, and so on, but no a-priori knowledge of PN spreading codes; b) jammer should not gain any appreciable jamming advantage by choosing a jammer waveform and strategy other than wideband Gaussian noise, that is, being clever should provide no additional advantage for jammer; c) make as costly as possible for the

jammer to be successful in jamming the user system. [1][6]

There is a variety of jamming strategies depending on the system to be jammed, the resources for jamming and the knowledge available about the user waveform. Jamming waveforms usually exploit time-frequency resources, for example, the frequency band (full-band, partial band) and the time domain (continuous, pulsed, swept, stepped) in many different ways by using noise or frequency tones as jamming signals (see Figure 10.17). In jamming FH systems, a jammer may use noise for broadband noise jamming, if its power is sufficiently high. If not, it can maximize the degradation caused to user signals by partial-band jamming. The fraction of the jammed bandwidth increases with the jammer power J. The optimum strategy for jamming DSSS signals is to use pulsed tone jamming in order to be least affected by spectrum spreading. Pulsing the jamming signals may disturb the synchronization circuits of DSSS systems and helps achieving higher peak powers for jamming. Intelligent jamming strategies aim to attack the user systems through their weakest points, for example, synchronisation, automatic gain control, and so on.

Protection against jamming and interception is provided by purposely making the user signal occupy a bandwidth far in excess of the minimum bandwidth necessary to transmit it. The use of spreading code at the transmitter produces a wideband signal that appears noise-like to an unintended receiver that has no knowledge of the spreading code (see Figure 10.8). The longer the period of the spreading code, the wider the spread spectrum bandwidth, the lower the user PSD, the closer will the user signal be to noise, and the harder it is to detect by an unintended receiver. The transmitted signal may then propagate through the channel undetected by the unintended receivers who do not know the spreading code; hence low-probability of interception (LPI). A jammer becomes less effective with this approach by being forced to jam the wide SS bandwidth

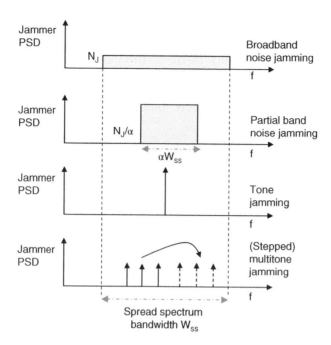

Figure 10.17 Typical Strategies For Jamming the Spread Spectrum Bandwidth. Jammer PSD for broadband noise jamming is $N_J = JT_c/2$.

instead of the much narrower data bandwidth. The user receiver, knowing the spreading code, is not affected by spreading and thus obtains a significant advantage over an interceptor/jammer.

In DSSS systems, a PN sequence is used to modulate a phase-shift-keyed signal to achieve instantaneous spreading of the data bandwidth. The ability of such a system to mitigate jamming is determined by its processing gain. The SNR at the input and the output of a DSSS system are related by (10.12), which shows that the SNR at the output of a DSSS receiver is G_p times higher than the input SNR. However, this advantage is available to the intended receiver who knows the PN code of the incoming signal. Therefore, a jammer who does not know the PN code has an SNR disadvantage by a factor of G_p with respect to the intended receiver. This provides AJ and LPI to a DSSS user in proportion with the processing gain G_p. The processing gain can be made larger by

employing longer PN sequences with narrower chip durations which, in turn, lead to wider SS bandwidths and transmission of more chips per bit. However, generation of PN sequences with narrower chip durations imposes a practical limit on the achievable processing gain and AJ performance.

Since the signal energy per bit is related to the average received signal power P by $E_b = PT_b$, one may express E_b/N_J as

$$\frac{E_b}{N_J} = \frac{PT_b}{JT_c/2} \Rightarrow \frac{J}{P} = \frac{2n}{E_b/N_J} \qquad (10.34)$$

where AWGN is neglected. The J/P ratio is called as the jamming margin,

$$10\log_{10}(J/P) = 10\log_{10}(2n) - 10\log_{10}(E_b/N_J)_{min} \qquad (10.35)$$

and denotes the maximum jammer-to-signal power ratio that can be supported by a DSSS system operating at a bit error probability level corresponding to $(E_b/N_J)_{min}$. For example, if we let $T_b = 1.024$ ms and $T_c = 1$ µs, then n = $T_b/T_c = 1024$ (30.1 dB). The period of the PN sequence is $2^m - 1 = 1024$, where $m = 10$ shows the number of flip-flops. For $E_b/N_0 = 10.53$, which corresponds to $P_b = 10^{-6}$ for coherent BPSK, the jamming margin is found from (10.35) as $33.1 - 10.53 = 22.57$ dB. This implies that a DSSS system with a processing gain $G_p = 2n = 33.1$ dB can operate reliably, that is, with a bit error probability less than 10^{-6}, as long as J/P ratio at the receiver input does not exceed 22.57 dB. In other words, in order to disable this DSSS, a jammer must realize a power at the receiver input at least 181 times (22.57 dB) higher than that of the user power. Such high EIRP levels usually require the use of higher transmit powers and/or antenna gains, hence higher initial and running costs.

A tone jammer may be most effective against DSSS systems by continuous or pulsed jamming in the time domain using a tone at the same carrier frequency as the user signal. In pulsed jamming, a jammer increases its peak power during the pulsing periods to be more effective in increasing the bit error rate and disturbing synchronization and automatic gain control (AGC) circuits. In pulsed jamming a BPSK/DSSS system, the jammer with an average power J jamms a DSSS system only a fraction α $(0 < \alpha \leq 1)$ of the time. The bit error probability may then be written as (see Figure 10.17)

$$P_b = \alpha\, Q\left(\sqrt{\frac{2E_b}{N_0 + N_J/\alpha}}\right) + (1-\alpha)Q\left(\sqrt{\frac{2E_b}{N_0}}\right)$$

$$(10.36)$$

Assuming that noise PSD is negligible compared to that of the jammer $(N_0 \ll N_J)$, the worst-case pulsed jamming occurs for the value of the duty cycle α, which is found differentiating (10.36) with respect to α and equating to zero:

$$\alpha_{wc} = \begin{cases} 0.71/(E_b/N_J) & E_b/N_J \geq 0.71 \\ 1 & E_b/N_J < 0.71 \end{cases}$$

$$(10.37)$$

The corresponding worst-case error probability is given by

$$P_{b,wc} = \begin{cases} 0.083/(E_b/N_J) & E_b/N_J \geq 0.71 \\ Q\left(\sqrt{\dfrac{2E_b}{N_J}}\right) & E_b/N_J < 0.71 \end{cases}$$

$$(10.38)$$

Figure 10.18 shows the variation of the bit error probability (10.36) for pulsed tone jamming for various values of the duty cycle α. If a jammer can predict the E_b/N_J of the user receiver, then it can adopt an clever strategy by choosing the worst-case duty cycle (10.37) corresponding to the operating value of the E_b/N_J. The user would then suffer worst-case pulsed jamming and its bit error probability would be given by (10.38).

Example 10.4 Worst-Case Pulsed Tone Jamming the Uplinks of Military SATCOM Systems.
Consider a military satellite communication (SATCOM) system as shown in Figure 10.19, operating at 9 GHz band with a satellite at the geostationary orbit. The system uses DSSS with n = 400. The user satellite ground terminal has a 2 m diameter antenna with 65% aperture efficiency and has a range of 40000 km to the satellite. Fading effects, atmospheric losses and implementation losses are ignored.

We will consider the performance of this system under worst-case pulsed tone jamming. Under these circumstances, the user system is assumed to sustain communications using the lowest modulation alphabet, that is, BPSK, at a bit error probability level of 10^{-3}. Then, (10.38) implies that a minimum E_b/N_J ratio of 83 (19.2 dB) is required for achieving a bit error probability less than 10^{-3}. For $E_b/N_J = 83$, the

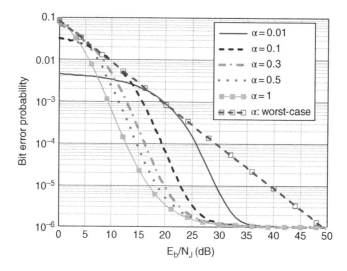

Figure 10.18   Bit Error Probability For Pulsed Jamming of DSSS System Using BPSK. Unjammed bit error probability is 10^{-6} ($E_b/N_0 = 10.53$ dB).

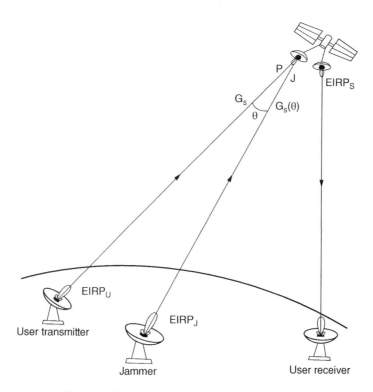

Figure 10.19   Uplink Jamming of SATCOM Links.

jammer to signal power ratio at the receiver input is given by (10.34):

$$\frac{J}{P} = \frac{2n}{E_b/N_J} = \frac{800}{83} \cong 9.64 \qquad (10.39)$$

Assuming that the user and the jammer on the ground have LOS with the geostationary satellite, the received jammer to signal power ratio J/P can be related to the transmit powers and antenna gains of the user and the jammer as follows:

$$\frac{J}{P} = \frac{EIRP_J\, G_S(\theta)/Lfs_J}{EIRP_U\, G_S/Lfs_U} = \frac{P_J G_J G_S(\theta)/Lfs_J}{P_U G_U G_S/Lfs_U}$$

$$(10.40)$$

where $EIRP_U$ and $EIRP_J$ denote respectively user and jammer EIRPs. Similarly, P_J and G_J (P_U and G_U) denote respectively the transmit power and the antenna gain of the jammer (user). Lfs_J and Lfs_U denote the corresponding free space losses. G_S and $G_S(\theta)$ denote the gain of the satellite receive antenna in the directions of the user and the jammer transmitters, respectively. Assuming that the distance of the jammer to the satellite is approximately equal to the distance of the user, and $G_S(\theta) \cong G_S$, (10.40) may be simplified, using (2.101), as

$$\frac{J}{P} = \frac{P_J G_J}{P_U G_U} = \frac{P_J}{P_U}\left(\frac{D_{aJ}}{D_{aU}}\right)^2 = 9.64 \qquad (10.41)$$

If the diameters of the user and jammer antennas are the same, that is, $D_{aJ} = D_{aU}$, the jammer must have transmit power which is 9.64 times (9.8 dB) higher than that of the user. In case when the transmit powers of the user and the jammer are the same, that is, $P_J = P_U$, then jammer antenna must have a 6.2 m diameter in comparison with the 2 m diameter antenna of the user:

$$D_{aJ} = D_{aU}: \quad \frac{J}{P} = \frac{P_J}{P_U} = 9.64 \Rightarrow P_J = 9.64\, P_U$$

$$P_J = P_U: \quad \frac{J}{P} = \left(\frac{D_{aJ}}{D_{aU}}\right)^2 = 9.64 \Rightarrow$$

$$D_{aJ} = \sqrt{9.64}\, D_{aU} = 6.2\ m$$

$$(10.42)$$

Example 10.5 BPSK/DSSS System Performance Under Broadband Jamming.
A coherent BPSK/DSSS system is transmitting at a data rate of $R_b = 10$ kbps in the presence of a broadband jammer. Propagation losses are assumed to be the same for the communicator and the jammer. The probability of bit error for BPSK in the presence of AWGN and a broadband jammer is obtained inserting $\alpha = 1$ and $N_J = (J\ T_c\ E_b)/(2P\ T_b)$ into, where effective SNR may be rewritten as

$$\frac{2E_b}{N_0 + N_J} = \frac{2E_b/N_0}{1 + N_J/N_0}$$

$$(10.43)$$

$$= \frac{2E_b/N_0}{1 + (E_b/N_0)(J/P)(1/2n)}$$

The bit error probability may then be expresed in terms of E_b/N_0, J/P and the processing gain as follows:

$$P_b = Q\left(\sqrt{\frac{2E_b/N_0}{1 + (E_b/N_0)(J/P)(1/2n)}}\right)$$

$$\cong Q\left(\sqrt{\frac{4n}{J/P}}\right)$$

$$(10.44)$$

The last expression in (10.44) is valid for $(E_b/N_0)(J/P)(1/2n) \gg 1$. Note the similarity between the bit error probability expression (10.44) against jamming with (10.32) in a multi-user interference environment, where $J/P = K-1$.

Figures 10.20 and 10.21 show the bit error probability as a function of E_b/N_0 for $n = 200$ and $J/P = 100$, respectively. Note that increasing the E_b/N_0 ratio alone does not cause a significant improvement in the bit error probability performance unless sufficient protection is provided by the processing gain for high J/P ratios. If user and jammer transmit power levels are 10 kW and 60 kW

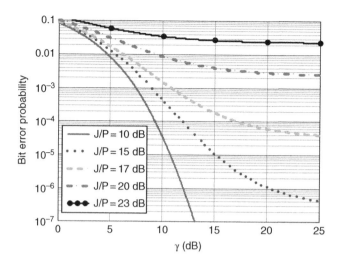

Figure 10.20   Bit Error Rate Versus $\gamma = E_b/N_0$ For Several Values of J/P ratio For $n = 200$ (23 dB).

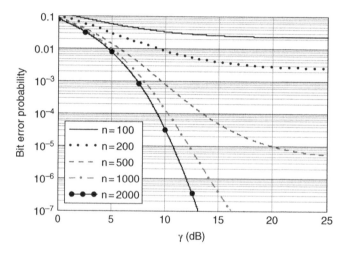

Figure 10.21   Bit Error Probability for $J/P = 100$ (20 dB).

respectively, the required SS bandwidth for the data rate of 10 kbps to achieve a bit error probability of $P_b = 10^{-5}$ is given by

$$P_b = Q\left(\sqrt{\frac{4n}{J/P}}\right) = 10^{-5} \Rightarrow$$

$$n = \frac{1}{4}\frac{J}{P}4.26^2 = \frac{160\ kW}{4\,10\ kW}4.26^2 = 27.2$$

$$W_{ss} = R_b n = 4.54 R_b \frac{J}{P} = 272\,kHz$$

$$\text{(10.45)}$$

10.1.2.4   Multipath Rejection and Rake Receiver

Consider a BPSK/DSSS system, as shown in Figure 10.22, that transmits, during $0 \leq t \leq T_b$, the bit b which is spread by the code $c(t)$ of length n. The bit and the code both consist of elements $\{+1,-1\}$. The transmitted signal is assumed to arrive the receiver via two paths, namely a path of zero delay and a second path with a relative delay of τ. The received signal may then be written as

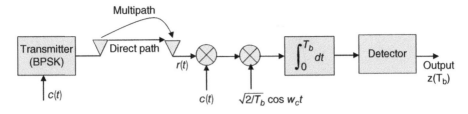

Figure 10.22   A Direct Sequence BPSK System Operating Over a Multipath Channel. [7]

$$r(t) = \sqrt{\frac{2E_b}{T_b}} \, bc(t)\cos(w_c t)$$

$$+ \alpha \sqrt{\frac{2E_b}{T_b}} \, bc(t-\tau)\cos(w_c t + \theta) + w(t),$$

$$0 \le t \le T_b$$

$$\text{(10.46)}$$

where $w(t)$ denotes the additive white Gaussian noise, τ is the differential time delay between the two paths, θ is the phase induced by the channel including $-w_c\tau$, and α shows the relative attenuation of the signal with delay τ.

Assuming that the DSSS receiver is matched to the direct path, the correlator output may be written as

$$z(T_b) = \sqrt{2/T_b} \int_0^{T_b} r(t) \, c(t) \, \cos(w_c t) \, dt$$

$$= \sqrt{E_b} \, b + \sqrt{E_b} \, \alpha b \cos\theta \, R(\tau) + w_0(t)$$

$$w_0(t) = \sqrt{2/T_b} \int_0^{T_b} w(t) c(t)\cos(w_c t) \, dt$$

$$\text{(10.47)}$$

where $w_0(t)$ denotes the zero-mean Gaussian random variable with the same variance as $w(t)$. The autocorrelation function $R(\tau)$ of the PN sequence $c(t)$ is equal to $-1/n$ for maximal-length sequences (see (9.137)), and vanishes for completely random sequences (see (10.2)) for $\tau > T_c$. Therefore, if the multipath excess delay T_m is longer than the chip duration T_c, then the multipath components with delays larger than T_c will not contribute

to the signal at the receiver output. Therefore (10.47) clearly shows that the DSSS eliminates multipath components with delays larger than T_c. For example, a DSSS system with a chip rate of $R_c = 10$ Mchips/s can successfully eliminate multipath signal components with delays longer than $T_c = 0.1$ µs longer than $T_c = 0.1$ microseconds, that is, signals with path lengths 30m longer than the direct path.

Above we observed that a DSSS receiver synchronized to the direct path ignores the multipath components with delays longer than the chip duration for detecting a symbol. COST 207 path models in Chapter 3 show that the maximum excess delay in wireless channels is generally between 2 µs and 20 µs.

In CDMA systems, replicas of a transmitted chip arrive to the CDMA receiver with delays in excess of chip duration, thus causing significant ISI to the symbols following the symbol of interest. The CDMA systems operating in ISI channels commonly use *Rake receiver* structure in order to combine multipath components for obtaining higher received signal levels. The RAKE receiver achieves improved performance by adding the SNRs of multipath components (maximal ratio combining, MRC) compared to simply recovering the dominant component and ignoring the other components as noise. A Rake receiver may consist of $L + 1$ fingers with T_c-spaced elements (see Figure 11.51). The number of channel taps is often chosen to combine only the strongest multipath components. A Rake receiver samples the incoming signal at the chip rate and

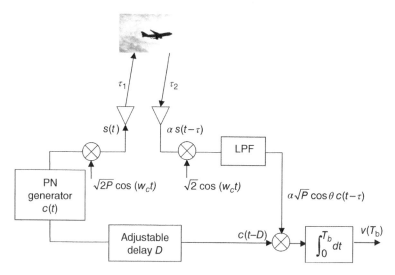

Figure 10.23   Block Diagram of a Ranging System Using DSSS. [7]

then passes them through T_c-spaced delay elements. At each delay, they are multiplied by the corresponding tap weight which consists of the estimate of the complex channel gain. The channel gain also accounts for the Doppler frequency shift, if the channel is time-variant. The multipath interference is similar in form to the multiple-access interference as in (10.24). For sufficiently large values of the processing gain, the multipath interference is negligible. [5] The channel gains usually vary slowly in time; therefore they are updated by the channel estimator circuits at time-intervals shorter than the channel coherence time. In most applications, the channel taps are modelled as independent Rayleigh-fading random variables, which are constant during a symbol interval. Assuming that the channel gains at $L+1$ taps are perfectly estimated in the multipath channel, the Rake receiver provides an MRC output signal for each bit (see (11.213)). Hence, an appropriately designed Rake receiver can turn the multipath propagation in the wide SS bandwidth into an advantage by obtaining an L-th order MRC diversity gain. The resulting bit error probability for BPSK modulation is given by (11.221)

and shown in Figure 11.55. The Rake receiver will be studied in Chapter 11.

10.1.2.5   Ranging

Consider that we want to predict the range R of a target in the far-field of a user transceiver by transmitting an unmodulated DSSS signal of power P:

$$s(t) = \sqrt{2P}\, c(t) \cos(w_c t) \qquad (10.48)$$

where $c(t)$ denotes a PN sequence with chip duration T_c (see Figure 10.23).

The signal scattered back by the target is received with a delay $\tau = \tau_1 + \tau_2$ seconds:

$$r(t) = \alpha s(t-\tau) = \alpha \sqrt{2P}\, c(t-\tau) \cos(w_c t + \theta) \qquad (10.49)$$

where α denotes the signal attenuation in the propagation path and θ is the random phase shift induced by the channel, including the term $-w_c \tau$ related to the propagation delay, and the term due to the Doppler frequency shift if the target is nonstationary. The received PN sequence $c(t-\tau)$ is correlated with the PN sequence $c(t-D)$ where the adjustable delay D is believed to correspond to the total channel delay. The integrator output at T_b is proportional to the autocorrelation function $R(D-\tau)$ of the PN sequence:

$$v(T_b) = a\sqrt{P} \cos \theta \ R(D-\tau) \qquad (10.50)$$

The output $v(T_b)$ is maximized when $D = \tau$ and becomes negligibly small if the magnitude of the delay difference $D-\tau$ exceeds the chip duration. Therefore, the range is estimated by adjusting D to maximize $v(T_b)$. In practice, the delay can be measured from the triangular autocorrelation function (10.2) with an accuracy of $|D-\tau| \leq \varepsilon T_c$, where $\varepsilon \leq 1$. Assuming LOS with the target and $\tau_1 = \tau_2$, the range R of the target is estimated as

$$R = \frac{c}{2}\tau = \frac{c}{2}(D \pm \varepsilon T_c) \qquad (10.51)$$

where c denotes the velocity of light. It is clear from (10.51) that the measurement accuracy can be improved by decreasing εT_c relative to D. For example, consider a DSSS system with $T_c = 100$ ns (chip rate 10 Mchip/s). If the correlator output is maximized for $D = 100$ µs, then the target range is estimated as

$$R = \frac{c}{2}(D \pm \varepsilon T_c) = \frac{3 \ 10^8}{2}(100\mu s \pm \varepsilon 0.1\mu s)$$
$$= 15 \ km \pm \varepsilon 15 \ m$$

$$(10.52)$$

with a measurement accuracy of 7.5 meters for $\varepsilon = 0.5$. In practice, DSSS is widely employed for accurate ranging applications.

10.1.2.6 Satellite Navigation Systems

DSSS systems are widely used in satellite navigation systems for positioning and time dissemination. The United States NAVigation Satellite Timing And Ranging (NAVSTAR) system's

global positioning system (GPS), the Russian GLONASS and the European GALILEO are the major satellite navigation systems. The operational characteristics of these systems are quite similar to each other.

The *GPS system* enables users to determine their position on or above the earth's surface by measuring their range to four to six GPS satellites whose positions are accurately known. The GPS system design is based on 24 GPS satellites in circular orbits 20000 km above the earth. The orbital period of these satellites is 12 hours. There are eight satellites in each of three orbits that are inclined 55 degrees with respect to the equator and are separated in longitude by 120 degrees. Each of the 24 satellites on the orbit maintains a highly accurate clock. The clocks are synchronized and orbital positions are updated through a control station near Colorado Springs, USA.

As shown in Figure 10.20, two DSSS signals L1 (1.57542 GHz) and L2 (1.2276 GHz) are transmitted from each satellite towards the earth. A GPS user anywhere on or above the surface of the earth receives signals from at least four satellites at any time. The L1 signal, which is an unbalanced QPSK/DSSS signal, uses a short Gold code of length 1023 chips for spreading in the I-channel. This so-called the clear/acquisition (C/A) code has a chip rate of 1.023 Mchips/s, and thus a code period of 1 ms. A different Gold code used by each satellite enables a GPS receiver to distinguish them. The so-called precise (P) code used by the L2 BPSK/DSSS signal provides more precise range information because of its 10-fold higher chip rate of 10.23 Mchips/s. The P code is a long nonlinear code with a period of seven days and is available for only military users. Each satellite uses a different phase of the long code in the L2 signal. The P- and C/A-channel signals are both modulated by the same 1500-bit data message of 50 bps and repeated cyclically. This 1500-bit message contains satellite position information as well as data used to correct for errors in propagation time, satellite clock bias, and so on.

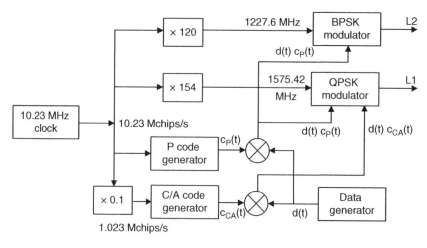

Figure 10.24   Block Diagram of a GPS Transmitter. [8]

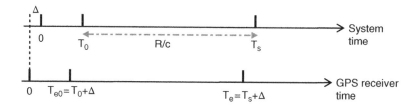

Figure 10.25   System Time, Which Is Assumed To Be the Same For All Satellites, and the Time At the GPS Receiver.

A GPS receiver may determine its range to a satellite using either the C/A or the P code by correlating the received GPS signal with the locally generated Gold code and determining the correlation peak for each of the satellites. The receiver then records the time at which, say, the C/A code epoch occurs. Note that the time reference is the receiver's own clock, which is assumed to be offset from the system clock by Δ (see Figure 10.25). The time differences between the correlation peaks are measures of the relative path differences to the satellites of interest. Then the arrival times for the C/A code epoch from all four satellites are recorded. The data transmitted from each satellite gives the exact position of the satellite and the system time when the clock epoch event occurred. Receiver records the time of the received epoch, corrects by time offset and relates to the range.

Let T_0 denote the system time at which the signals are transmitted by a GPS satellite and T_S denotes the system time at which the signals are received by the GPS receiver on the earth. Because of the time offset between satellite and GPS receiver clocks, the receiver time for the reception of the signal is given by $T_e = T_S + \Delta$. Consequently, the so-called pseudo-range R and the range with reference to the GPS receiver clock Re may be written as

$$\text{Re} = (T_e - T_0)c = R + \Delta c$$
$$R = (T_S - T_0)c \tag{10.53}$$

Let (x, y, z) and (X_i, Y_i, Z_i) denote respectively the coordinates of the user receiver and the ith satellite with respect to the origin, located at the center of the earth. Solving the following $N_{sat} \geq 4$ range equations

$$R_i = \left[(X_i - x)^2 + (Y_i - y)^2 + (Z_i - z)^2 \right]^{1/2}$$
$$- \Delta c, \quad i = 1, 2, \cdots, N_{sat}$$
$$(10.54)$$

the receiver can accurately determine its position (x, y, z) and time.

GPS provides Precise Positioning Services (PPS) and Standard Positioning Services (SPS). SPS code is intentionally degraded through Selective Availability (SA). Published specifications without enhancement are given as follows: The PPS provide 17.8 meter horizontal accuracy, 27.7 meter vertical accuracy, and 100 nanosecond time accuracy. The SPS (with SA on) provides 100 meter horizontal accuracy, 156 meter vertical accuracy, and 167 nanosecond time accuracy.

There are various ways to improve the positioning accuracy. For example, differential GPS (DGPS) improves the positioning accuracy to about 10 cm. A DGPS network of fixed, ground-based reference stations broadcast on short ranges the difference between the measured satellite pseudo-ranges and the computed pseudo-ranges to help GPS receivers to correct their pseudo-ranges. [5] Currently, the rms signal-in-space user range error, which is defined as the difference between a GPS satellite's navigation data (position and clock) and the truth, projected on the LOS of the user, is reported to be less than 1 m. [9]

Global Navigation Satellite System (GLONASS) represents the Russian counterpart of the GPS. The GLONASS constellation comprises 24 Glonass-M satellites that are uniformly deployed in three roughly circular orbital planes at an inclination of 64.8° to the equator; each orbit comprises eight equally spaced satellites. The orbital planes are separated by 120 degrees. The altitude of the orbit is 19,100 km; hence, the orbital period of a satellite is approximately 11 hours, 16 minutes.

GLONASS uses both frequency division multiple access (FDMA) and code division multiple access (CDMA) techniques. It accommodates L1 (~1601 MHz), L2 (~1246 MHz) and L3 (~1202 MHz) signals. The GLONASS used alone is slightly less accurate than GPS. However, on high latitudes, GLONASS is more accurate than GPS due to the orbital position of the satellites. Some receivers use GLONASS and GPS signals together, providing improved coverage in indoor, urban canyon or mountainous areas.

The European Satellite Navigation System (*Galileo*) is the European counterpart of the GPS. The nominal Galileo constellation comprises a total of 27 operational satellites, which are evenly distributed in three orbital planes inclined at 56 degrees relative to the equator; hence, nine evenly distributed operational satellites in each of the three orbital slots. The Galileo satellites are placed in circular Earth orbits with a radius of about 30,000 km and an approximate revolution period of 14 hours.

Galileo satellites transmit radio-navigation signals in four different operating frequency bands: E1 (1559 ~ 1594 MHz), E6 (1260 ~ 1300 MHz), E5a (1164 ~ 1188 MHz) and E5b (1195 ~ 1219 MHz). The Galileo E1 band comprises two signals that can be used alone or in combination with signals in other frequency bands, depending on the performance required by the application. The signals are provided for the open, and the public regulated services. Moreover, an integrity message for the safety-of-life service is included in the open service signal. The Galileo E6 signal comprises signals for commercial and public regulated services. Both signals include a navigation message and encrypted ranging codes. The wideband Galileo E5 signal comprises E5a and E5b signals. The Galileo E5a signal provides a second signal (dual frequency reception) for the open service and safety-of-life services, both including navigation data messages. The Galileo E5b signal provides a safety-of-life service, including a navigation message with an integrity information message.

Table 10.1   Galileo Open-Service Target Positioning, Navigation and Timing Performance Specifications.
[9][10]

Performance specification	Single-frequency open service user (E1)	Dual-frequency open service user (E1-E5b)
Horizontal accuracy (95%)	15 m	4 m
Vertical accuracy (95%)	35 m	8 m
Timing accuracy (95%)	–	
Galileo open service availability (averaged over system lifetime)	99.5%	99.5%

The search-and-rescue downlink signal is also transmitted by the Galileo satellites in the frequency range between 1544 and 1545 MHz.

The European Geostationary Navigation Overlay Service (EGNOS) provides an augmentation signal to the GPS standard positioning service. The EGNOS signal is transmitted in the same frequency band and modulation as the GPS L1 (1575.42 MHz) C/A signal. While the GPS consists of positioning and timing signals generated from spacecraft orbiting the Earth, thus providing a global service, EGNOS provides correction and integrity information intended to improve positioning navigation services over Europe. The EGNOS space segment consists of three navigation transponders onboard three geostationary satellites and broadcast corrections and integrity information for GPS satellites in the L1 frequency band (1575.42 MHz). The performance specifications of the Galileo system is listed in Table 10.1.

10.1.3   Frequency Hopping Spread Spectrum

In frequency-hopping (FH) spread spectrum (FHSS) systems, the data-modulated carrier randomly hops from one frequency to the next in a wide hopping bandwidth. The hopping bandwidth will be called as spread spectrum bandwidth W_{ss} for comparison purposes with DSSS systems. In contrast with DSSS systems, the instantaneous bandwidth of FHSS systems is equal to the data bandwidth, which is a small

fraction of the hopping bandwidth. Therefore, the rate at which the carrier frequency hops (hop rate) becomes a critical performance parameter. FHSS systems are usually categorized as slow- and fast-hopping. In slow-FH systems, the symbol rate R_s is an integer multiple of the hop rate R_h. Therefore, several symbols may be transmitted during each frequency hop. In fast-FH, the hop rate R_h is an integer multiple of the symbol rate R_s, which implies that the carrier frequency will hop several times during the transmission of one symbol.

Since the data-modulated carrier randomly hops in a very wide hopping bandwidth, frequency synthesizers need very short settling times for generating a new carrier for each hop with low phase noise and minimum spurious radiation. As the hop rate increases, the synthesizers will be forced to generate the carrier signal in shorter time intervals. Spectral purity and settling times of the synthesized carrier tones may create serious problems in fast-FH systems. Another issue is the difficulty of maintaining coherence (carrier phase) between carriers generated for each hop in such short times. Therefore, most FHSS systems use noncoherent M-ary modulation schemes. Coherent detection is possible only within each hop since a frequency synthesizer is unable to maintain phase coherence over successive hops. M-ary noncoherent frequency-shift keying (M-FSK) is the most commonly used modulation in FHSS systems. When a carrier hops to a new frequency, the data is transmitted by one of the M frequencies around the newly hopped carrier. We know from (8.108) that the frequency separation between tones in noncoherent orthogonal M-ary FSK should be $\Delta f = 1/T_s$, where T_s denotes the symbol duration. Therefore the bandwidth occupied by a noncoherent orthogonal M-ary FSK signal is equal to

$$W = M\Delta f = M/T_s = MR_b/\log_2 M \quad (10.55)$$

Figure 10.26 shows the block diagram of a FHSS transmitter and receiver using M-FSK modulation. Basically, the transmitter consists of a M-FSK transmitter and a FH mixer unit which changes the carrier frequency randomly from one hop to the other. The FH unit has a PN generator consisting of m flip flops as already

studied in Section 10.1.1. The 2^m combinations of the flip-flop states dictates the frequency synthesizer to generate one of the 2^m hopping frequencies during each hop. Therefore, the number N of frequencies to be hopped in a hopping bandwidth of W_{ss} is chosen as

$$N = 2^m, \quad m = \lfloor \log_2(W_{ss}/W) \rfloor \quad (10.56)$$

The modulation frequency f_0 of the M-FSK signal is up-converted by the frequency f_i generated by the frequency synthesizer and the resulting MFSK/FH signal is transmitted after a band-pass filter which eliminates either the sum or the difference frequency, $f_i \pm f_0$. At the receiver side, the FH mixer unit down-converts the FH signal to the frequency f_0 and filters out

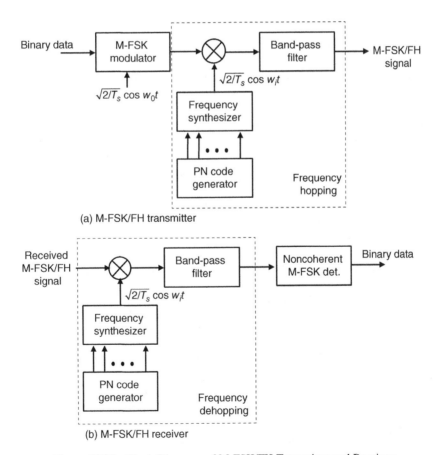

(a) M-FSK/FH transmitter

(b) M-FSK/FH receiver

Figure 10.26   Block Diagrams of M-FSK/FH Transmitter and Receiver.

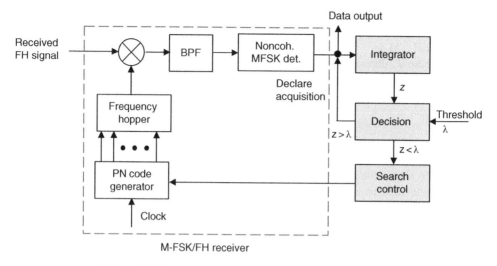

Figure 10.27   Serial Search Acquision of FHSS Signals. [1]

the out-of-band radiations via a band-pass filter. The signal at mixer output is an ordinary
M-FSK signal and demodulated and decoded
using well-known techniques for noncoherent
M-FSK signalling. For successful operation
of a FHSS system, the frequency synthesizer
at the receiver must be synchronized with the
one at the transmitter. This requires an accurate
estimation of the propagation delay and the
channel phase.

With present-day technology, FH bandwidths
on the order of several GHz are achievable. This
is an order of magnitude larger than W_{ss} for
DSSS. On the other hand, FH systems can also
operate by hopping over noncontiguous bands.
For this purpose, generation of the frequencies
corresponding to the forbidden bands may be
inhibited by programming the PN generator.

10.1.3.1   Synchronization of FHSS Signals

As an example to various techniques for the synchronization of FHSS systems, Figure 10.27
shows a serial acquisition scheme for the acquisition of the hopping pattern. The hopping frequency is searched serially by sliding the code
timing of the PN code generator, which controls

the hopping pattern. The averaging provided by
the integrator following the M-FSK detector
minimizes false alarms for acquisition due to
temporary exceedance of the threshold by
instantaneous noise peaks and/or interferences.
Acquisition is declared when the local hopping
pattern is aligned with that of the received FH
signal. Serial search of the hopping frequency
is time consuming. However, search time may
be shortened at the expense of complexity and
cost, by using multiple correlators operating
parallel and searching in non-overlapping time
slots. [2]

10.1.3.2   Slow Frequency Hopping

Consider a M-FSK modulator with $M = 2^k$
orthogonal signals, where k denotes the number
of bits transmitted by each of the M frequency
tones. A slow FH/M-FSK signal is characterized by having multiple symbols transmitted
per hop, that is, the hop duration $T_h = 1/R_h$
(R_h: hop rate) is longer than the symbol duration $T_s = 1/R_s$ (R_s: symbol rate),

$$T_h = rT_s \quad \Rightarrow \quad R_s = \frac{R_b}{\log_2 M} = rR_h \quad (10.57)$$

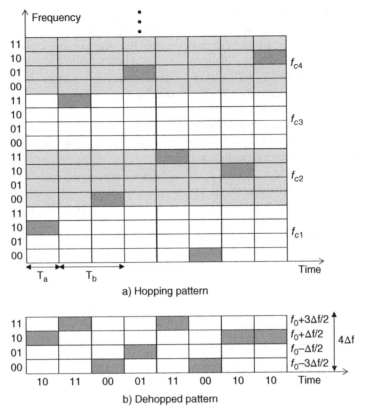

Figure 10.28 (a) Frequency Hopping Pattern, (b) Dehopped Pattern For 4-FSK.

where $r \geq 1$ is an integer and denotes the number of symbols sent during each hop. For the special case of $T_h = T_s$, hop and symbol rates are identical and a single symbol is sent during each hop. At each hop, the M-FSK tones are separated in frequency by an integer multiple of the symbol rate R_s, to ensure orthogonality between them (see (10.55)). This implies that any transmitted symbol will not produce any interference in the other M-1 noncoherent matched filter outputs. In AWGN channel, the performance of slow FH/M-FSK is the same as that for the unhopped M-FSK. Depending on whether coherent or noncoherent detection is used at the receiver, the symbol error probability is given by (8.126) or (8.175), respectively, and repeated below for convenience

$$P_M = 1 - \left[1 - Q\left(\sqrt{E/N_0}\right)\right]^{M-1}$$

$$\leq (M-1)Q\left(\sqrt{E/N_0}\right), \quad coh.$$

$$P_M = \sum_{n=1}^{M-1} \binom{M-1}{n} \frac{(-1)^{n+1}}{n+1} \qquad (10.58)$$

$$\exp\left(-\frac{n}{n+1}\frac{E}{N_0}\right), \quad noncoh.$$

where E denotes the symbol energy.

Figure 10.28(a) shows a typical hopping pattern of a 4-FSK/FHSS with time. The dehopped pattern at the input of the 4-FSK detector (see Figure 10.28b) is identical to the input of a classical 4-FSK receiver. For $T_h = T_a$, a symbol is sent during each hop, that is $T_h = T_s = T_a$. However, $T_h = T_b$ implies slow-FH where two symbols are sent during each hop, that is $T_h = 2T_s = T_b$. On the other hand, if $T_s = T_b$, then one has fast-FH where a symbol is sent during two hops, that is $T_s = T_b = 2T_h$.

On a single hop, the bandwidth, W, of the transmitted signal is the same as that resulting from the use of a convential MFSK with an alphabet of $M = 2^k$ orthogonal signals. Consider for example a slow-FH system using noncoherent orthogonal 4-FSK at a bit rate of 10 kbps. Frequency separation between the tones is 5 kHz and the instantaneous bandwidth is equal to 20 kHz. This implies that this 4-FSK/FHSS system with a hopping bandwidth of 500 MHz can potentially hop to 500 MHz/ 20 kHz = 25000 carrier frequencies. However, in virtue of (10.56), only $N = 2^{14} = 16384$ hopping frequencies will be used.

Example 10.6 Hopping Bandwidth and Processing Gain in FH Systems.
A 8-FSK/FHSS system communicates at 100 kbps in a hopping bandwidth $W_{ss} = 300$ MHz. The instantaneous bandwidth of the 8-FSK is found as

$$W = 8\,\Delta f = \frac{8}{T_s} = \frac{8}{3T_b} = \frac{8R_b}{3} = 266.67\,kHz$$

$$(10.59)$$

The minimum number of flip flops in the PN generator and the number of hop frequencies in W_{ss} are given by

$$m = \lfloor \log_2(W_{ss}/W) \rfloor = 10 \quad \Rightarrow \quad N = 2^{10} = 1024$$

$$(10.60)$$

This implies an effective spread spectrum bandwidth of

$$W_{ss,\mathit{eff}} = N \times \Delta f = 273.1\,MHz \qquad (10.61)$$

Assuming that a broadband noise jammer spreads its power J over the entire FH bandwidth W_{ss}, the jammer's effect is equivalent to an AWGN with PSD of $N_J = J/W_{ss}$. The spread-spectrum system is thus characterized by

$$\frac{E}{N_J} = \frac{P/R_s}{J/W_{ss}} = \frac{P/J}{R_s/W_{ss}} = \frac{G_p}{J/P} \qquad (10.62)$$

where the processing gain is defined by

$$G_p = W_{ss}/R_s \qquad (10.63)$$

Depending on its power, a jammer may jam the whole SS bandwidth or resort to partial band jamming. However, a jammer may be more effective by jamming the band with power levels barely sufficient to cause untolerable bit errors. In partial band jamming, the processing gain realized by the receiver may not be described by (10.63).

10.1.3.3 Fast Frequency Hopping

In fast M-FSK/FH, multiple hops are used to transmit a M-ary symbol:

$$T_s = rT_h \quad \Rightarrow \quad R_h = rR_s \qquad (10.64)$$

where $r \geq 1$ is an integer and denotes the number of hops used to sent a symbol. For example, in Figure 10.28(a), $T_a = T_h$ and $T_b = T_s = 2T_h$ corresponds to a fast-FH system where a symbol is transmitted during two hopping periods. Therefore, the decision as to which symbol is transmitted is given by noncoherent combining of the symbol components received over a period of two hops.

To determine the bit error probability for FHSS signals in AWGN channels, we assume binary orthogonal FSK with noncoherent detection in a AWGN channel with single-sided PSD N_0. Bit error probability for slow-FH (1 hop/bit) is given by

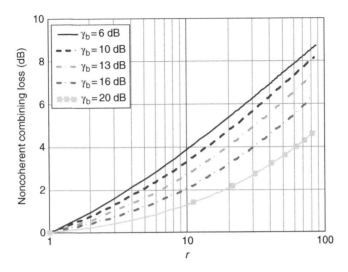

Figure 10.29   Losses for Square-Law Combining of NonCoherent Binary FSK Signals.

$$P_b = \frac{1}{2}\exp\left(-\frac{\gamma_b}{2}\right) \qquad (10.65)$$

where $\gamma_b = E_b/N_0$.

The detection procedure in fast-FH is quite different from slow-FH. In fast-FH (multiple hops/bit), the bit duration consists of r subintervals (hop durations); each of the chips received in r subintervals are combined to estimate whether the transmitted bit is a 1 or a 0. In *hard-decision decoding* (HDD), a separate decision is made for each BFSK/FH symbol based on the r frequency-hop chips received, and the decoder uses majority voting to estimate the transmitted BFSK symbol. The decoder decides in favor of a 1 if the majority of signals received in r subintervals have the frequency corresponding 1 or vice versa. The probability of bit error for majority voting is therefore given by

$$P_b = \sum_{i=(r+1)/2}^{r} \binom{r}{i} p^i (1-p)^{r-i}$$

$$p = \frac{1}{2}\exp\left(-\frac{\gamma_b}{2r}\right) \qquad (10.66)$$

A second approach consists of *square-law combining* (SLC) r chips of each of the two orthogonal matched-filters for 1 and 0 in order to determine the signal power levels corresponding to symbols 1 and 0. A decision is made in favor of 1 if the square-law combined signal for 1 is stronger than that for 0 or vice versa. The bit error probability of the fast-FH BFSK signal with square-law combining is given by:[1], [2]

$$P_b = \frac{1}{2}\exp\left(-\frac{\gamma_b}{2}\right) L$$

$$L = \frac{1}{2^{2r-2}}\sum_{n=0}^{r-1} K_n \left(\frac{\gamma_b}{2}\right)^n, \quad K_n = \frac{1}{n!}\sum_{k=0}^{r-1-n}\binom{2r-1}{k}$$

$$(10.67)$$

where the so-called noncoherent combining loss $L > 1$ denotes the increase in bit error probability due to noncoherent combining. Equating P_b in (10.67) to $0.5\exp(-0.5\gamma_{eq})$ where $\gamma_{eq} < \gamma_b$ accounts for the combining losses, we determine degradation in SNR due to combining losses as

$$\Delta = \frac{\gamma_b}{\gamma_{eq}} = \left(1 - \frac{2}{\gamma_b}\ln(L)\right)^{-1} > 1 \qquad (10.68)$$

Hence noncoherent combining degrades the SNR required for a given probability of bit error by Δ times. Figure 10.29 shows the variation of

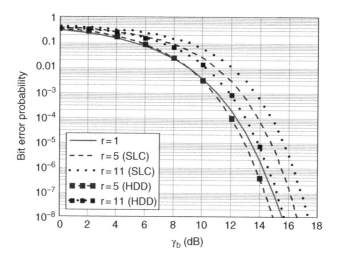

Figure 10.30 Bit errror probability for fast FH with binary orthogonal FSK with majority voting (HDD) and square-law combining (SLC). [2]

Δ in dB as a function of the number of combined hops. One may observe that the noncoherent combining loss increases with r but decreases with increasing values of γ_b.

Figure 10.30 shows the bit error probability for binary orthogonal FSK signals with fast FH, where one bit is transmitted in r hops. Note that $r = 1$ corresponds to the case where one bit is transmitted per hop. One may observe that HDD yields lower probability of bit error compared to noncoherent combining of r chips. The bit error probability for HDD of fast FH signals was observed to be lower than that for slow-FH for $r \le 7$.

10.1.3.4 Bit Error Probability For Multi-User FH Systems

Consider BFSK/FH with $R_h = R_b$ where carrier frequencies of K users hop over N carrier frequencies independently from each other. If only a single user is using a frequency slot during a hop, then the probability of error is the same as BFSK in an AWGN channel, as given by (10.65). However, if two or more users hop simultaneously over the same carrier frequency, then the bit error probability is $P_b =$

1/2 due to the collision. Thus, the overall probability of bit error may be written as

$$P_b = \frac{1}{2}p_{collision} + \frac{1}{2}\exp\left(-\frac{\gamma_b}{2}\right)(1 - p_{collision})$$

(10.69)

where $p_{collision}$ denotes the probability of collision.

Assume a synchronous (slotted) BFSK/FH system, where the users are allowed to transmit so that their signals arrive to the receiver at only the beginning of each hop duration. This implies that the duration of a potential collision will be equal to the bit (hop) duration. Since there are N carrier frequencies to hop, the probability that another user signal will be present in the desired user's slot is $1/N$. Then the collision probability by at least one of the $K-1$ interfering users is given by

$$P_{collision} = 1 - \left(1 - \frac{1}{N}\right)^{K-1} \approx \frac{K-1}{N} \quad \text{for large } N$$

(10.70)

In the special case of a single-user ($K = 1$), (10.69) reduces to (10.65) as expected. For high γ_b values, (10.69) shows an

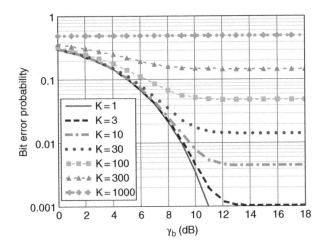

Figure 10.31   Bit Error Probability in a Synchronous Multi-User BFSK/FH System for $N = 1024$.

irreducible bit error probability due to multiple-access interference (MAI):

$$\lim_{\gamma_b \to \infty} P_b \simeq \frac{1}{2} P_{collision} \qquad (10.71)$$

If a (n,k) block code with t error correction capability is used for FEC, then the probability of block error may be written as

$$P_{block} = \sum_{i=t+1}^{n} \binom{n}{i} P^i (1-P)^{n-i}$$

$$P = \frac{1}{2} p_{collision} + \frac{1}{2} \exp\left(-\frac{1}{2} R_c \gamma_b\right) (1 - p_{collision})$$

$$(10.72)$$

Probability of decoded bit error is given by

$$P_b = \sum_{i=t+1}^{n} \frac{i}{n} \binom{n}{i} P^i (1-P)^{n-i} \qquad (10.73)$$

Since hops are not synchronized in asynchronous (unslotted) FHSS systems, the hop period of one user may intersect two hop periods of other users. Depending on the degree of overlap between the collided symbols, the

collision probability will be higher than the synchronous (unslotted) case.

Figure 10.31 shows the uncoded bit error probability in a synchronous multi-user BFSK/FH system for $N = 1024$. The error floor (irreducible error probability) clearly hinders the system's operation as the number of users increases and shows the need for the use of FEC coding. Figure 10.32 shows that the use of (23,12) Golay code with $t = 3$ error correction capability provides significant performance improvement. However, coding can not provide sufficient performance improvement as the number of users becomes intolerably high. In this case, collision probability reaches very high levels and the (23,12) Golay code not not correct the resulting bit errors. For example, in this slotted BFSK/FH system with $N = 1024$ hopping frequencies, the collision probability is equal to 12.7% for $K = 300$ and 22.2% for $K = 600$.

Example 10.7   Performance of Multi-User FH/BFSK Systems.
A fast FH/BFSK system is required to operate over a continuous 40 MHz spectrum. A single user is assumed to operate at $\gamma_b = 20$ dB. The bit rate is 20 kbps and each bit is transmitted during two hops ($T_b = 2T_h$). The hop rate if

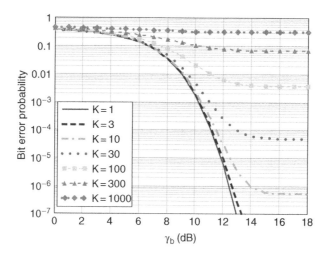

Figure 10.32   Bit Error Probability in a Synchronous Multi-User FH/BFSK System For $N = 1024$ with (23,12) Golay Coding.

each user transmits at 20 kbps will be $R_h = 2$ hops/bit × 20 kbits/s = 40 khops/s. Using Figure 10.29 and (10.68), we find non-coherent combining losses to be negligiblly small for $r = 2$ and $\gamma_b = 20$ dB.

Assuming an AWGN channel, the probability of error for a single-user is given by

$$P_b = \frac{1}{2}\exp\left(-\frac{\gamma_b}{2}\right) = \frac{1}{2}\exp\left(-\frac{100}{2}\right) \quad (10.74)$$
$$= 9.64 \times 10^{-23} \approx 0$$

We find the data bandwidth using (10.55) as $W = 40$ kHz. The number of FH channels is thus found to be

$$N = \frac{W_{ss}}{W} = \frac{40\ MHz}{40\ kHz} = 1000 \quad (10.75)$$

The probability of error for a user operating at $\gamma_b = 20$ dB in the presence of $K-1$ other independently hopping BFSK/FH users is found using (10.69)

$$P_b \cong \frac{1}{2}P_{collision} = \frac{1}{2}\left[1 - \left(1 - \frac{1}{N}\right)^{K-1}\right]$$
$$= \begin{cases} 9.4\ 10^{-3} & K = 20 \\ 4.7\ 10^{-2} & K = 100 \end{cases}$$
$$(10.76)$$

Now let us make a simple comparison of the multiuser capabilities of CDMA and FH systems for $n = N = 1000\ W_{ss}$. For a fair comparison, we assumed a bit error probability level of 10^{-6} for a single-user scenario. This implies $E_b/N_0 = 14.2$ dB for noncoherent BFSK/FH and $E_b/N_0 = 10.53$ dB for coherent BPSK/CDMA. Now assume that the bit error probability in a multi-user environment is 10^{-3} for both systems. Using (10.69), we see that the FH system can support $K = 3$ users, while the BPSK/CDMA system can serve $K = 122$ simultaneous users (see (10.32)). This clearly shows the relative advantage of CDMA over FH as a multiple-access technique.

10.1.3.5   Strategies for Jamming FHSS Systems

In FHSS systems, the carrier frequency of a narrow-bandwidth user signal hops over a very wide hopping bandwidth. Therefore, broadband jamming may not be effective against FHSS systems as the jamming power has a low PSD N_J over the hopping bandwidth. However, jamming effectiveness can be increased by partial-band jamming, that is, by distributing the jammer power J across a fraction α of the hopping bandwidth and selecting the optimim value of α as a function of J (see Figure 10.17). In partial-band tone jamming, the jammer power J is divided between tones so that each jamming tone has a power barely sufficient to cause a bit error probability 0.5 when it hits user signal.

The probability of bit error for a noncoherently detected binary orthogonal FSK/FH system under partial-band jamming may be written as

$$P_b = \frac{\alpha}{2}\exp\left(-\frac{E_b}{2(N_0+N_J/\alpha)}\right)$$
$$+\frac{(1-\alpha)}{2}\exp\left(-\frac{E_b}{2N_0}\right) \quad (10.77)$$

where $N_J \gg N_0$ is assumed. Differentiating (10.77) with respect to α and equating to zero, one gets the worst-case partial band factor as

$$\alpha_{wc} = \begin{cases} 2/(E_b/N_J) & E_b/N_J > 2 \\ 1 & E_b/N_J \leq 2 \end{cases} \quad (10.78)$$

The worst-case bit error probability for partial-band noise jamming is found by inserting (10.78) into (10.77):[1]

$$P_{b,wc} = \begin{cases} e^{-1}/(E_b/N_J) & E_b/N_J > 2 \\ \frac{1}{2}\exp\left(-\frac{E_b}{2N_J}\right) & E_b/N_J \leq 2 \end{cases} \quad (10.79)$$

Comparison of (10.79) with (10.38) shows that bit error probability performances of worst-case partial-band noise jamming of FHSS and worst-case pulsed tone jamming DSSS systems are similar. However, BFSK/FH under worst-case partial band noise jamming performs $1/(0.083e) = 4.44$ (6.47 dB) worse than the BPSK/DS against worst-case pulsed tone jamming for a given bit error probability. Figure 10.33 shows the bit error probability for a noncoherent orthogonal BFSK/FH system against partial-band noise jamming, as given by (10.77) and (10.79). Note that partial band noise jamming for a given value of α is most effective for only a specific value of the E_b/N_J. The worst-case partial band jamming results in a $(E_b/N_J)^{-1}$ type behavior of the error probability (see (10.79)), which is evidently much worse than the bit error probability performance in AWGN channels.

Some other intelligent jamming strategies may also be used if some knowledge of the user signals is available to the jammer. For example,

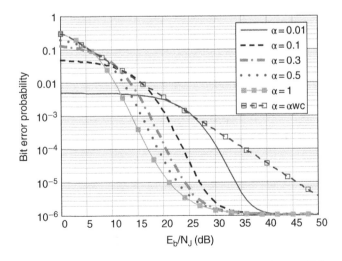

Figure 10.33 Partial-Band Noise Jamming of a Noncoherent Orthogonal BFSK/FH System. Unjammed bit error probability is 10^{-6}, that is, $E_b/N_0 = 14.18$ dB.

a jammer can exploit the vulnerability of the synthesizers, if any, by forcing the FHSS system to lose synchronization. Repeat-back jamming (RBJ) is generally the optimum strategy for slow-hopping FH signals for deceiving the user receiver. RBJ strategy involves the measurement of the spectral content of the hopping bandwidth, identifying the instantaneous frequency and returning some version of the intercepted signal in that portion of the frequency band during the same symbol interval. A repeat-back-jammer must therefore be sufficiently close to the LOS path so that it has sufficient time to process, modify and retransmit the intercepted signals to ensure that they arrive to the user receiver within the duration of the jammed symbol. This implies that RBJ is feasible for slow-FH systems and for jammers closely located to the propagation path. Fast-FH mitigates RBJ by hopping to a

new frequency before the jammer is able to complete the processes cited above.

Example 10.8 Downlink Jamming SATCOM Links using MFSK/FHSS.
Consider a FH SATCOM link under worst-case partial-band downlink jamming at 9 GHz band as shown in Figure 10.34. This implies that the jammer can estimate the worst-case partial-band factor to maximize its effectiveness against the user system. As a countermeasure, the user system reverts to noncoherent binary FSK to counter the worst-case partial-band jamming scenario. Then, (10.79) implies that the user system needs a minimum $E_b/N_J =368$ (25.7 dB) to sustain communications at a bit error probability level 10^{-3}. The worst-case partial band factor is found from (10.78) as $2/368 = 0.54\%$.

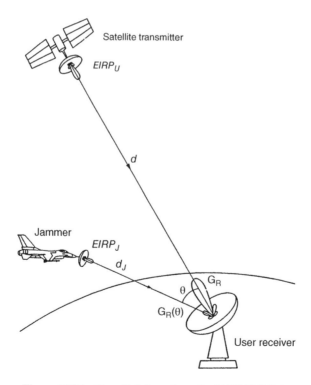

Figure 10.34 Downlink Jamming of a SATCOM Link.

Assuming that the LOS exists for satellite and jammer transmitter terminals with ground-based user receiver, the energy per bit of the user signal is given by

$$E_b = \frac{EIRP_U\, G_R}{R_b\, Lfs_U} = \frac{EIRP_U\, G_R}{R_b\, (4\pi d/\lambda)^2} \qquad (10.80)$$

where the satellite range is assumed to be $d = 40000$ km. On the other hand, a jammer located at a distance d_J from the ground-based user receiver and looking at the user receiver antenna with θ degrees offset from its boresight has a PSD as

$$N_J = \frac{J}{W_{ss}} = \frac{EIRP_J\, G_R(\theta)}{W_{ss}\, Lfs_J} = \frac{EIRP_J\, G_R(\theta)}{W_{ss}\, (4\pi d_J/\lambda)^2} \qquad (10.81)$$

From (10.80) and (10.81), the E_b/N_J ratio may be written as follows:

$$\frac{E_b}{N_J} = \frac{W_{ss}}{R_b}\left(\frac{d_J}{d}\right)^2 \frac{EIRP_U}{EIRP_J}\frac{G_R}{G_R(\theta)} \qquad (10.82)$$

For a hopping bandwidth of $W_{ss} = 500$ MHz and user bit rate $R_b = 100$ kbps, the processing gain is approximately $W_{ss}/R_b = 5000$ $(37\,dB)$. Assuming that the jammer looks at the user terminal with an offset of half-power beamwidth, one has $G_R/G_R(\theta) = 2$. Then, the following must satisfy in order to to achieve a $E_b/N_J = 368$:

$$\frac{EIRP_J}{EIRP_U} = \frac{W_{ss}}{R_b}\left(\frac{d_J}{d}\right)^2 \frac{1}{E_b/N_J}\frac{G_r}{G_r(\theta)} = 27.17\left(\frac{d_J}{d}\right)^2 \qquad (10.83)$$

If the jammer (air-borne) terminal is located at a distance of $d_J = 40$ km from the user receiver, then EIRP of the jammer terminal could be 45.66 dB lower than that of the user terminal:

$$\frac{EIRP_J}{EIRP_U} = 27.17\left(\frac{d_J}{d}\right)^2 = 27.17 \times 10^{-6} \;\; (-45.66\ dB) \qquad (10.84)$$

Assuming that the antenna diameter of an airborne jammer is 1/5 of the satellite transmit antenna, which is reasonable assumption, then the ratio of transmit powers is found as

$$\frac{EIRP_J}{EIRP_U} = \frac{P_J\, G_J}{P_U\, G_U} = \frac{P_J}{P_U}\left(\frac{D_J}{D_U}\right)^2 = 27.17\left(\frac{d_J}{d}\right)^2$$

$$\Rightarrow \frac{P_J}{P_U} = 27.17\left(\frac{d_J\, D_U}{d\, D_J}\right)^2 = 27.17\left(\frac{40}{40000}\frac{5}{1}\right)^2$$

$$= 6.8 \times 10^{-4} \;\; (-31.68\ dB) \qquad (10.85)$$

10.2 Orthogonal Frequency Division Multiplexing (OFDM)

OFDM is a multi-carrier broadband transmission technique in contrast with DSSS which uses a single carrier to transmit broadband data. In OFDM, an input data sequence is split into N lower-rate data sequences each of which modulates a different subcarrier. Hence, each subcarrier transmits at a rate, which is N times slower than the input data rate. Because N times longer symbol durations enable collecting more multipath components, OFDM systems reduce the time dispersion due to multipath propagation. Multicarrier structure of OFDM also offers a significant advantage in equalization in frequency selective propagation environments. Due to N-fold increase in the symbol duration, each of the N subcarriers undergo frequency-flat fading but not frequency-selective fading. Increased symbol durations also allow the use of longer guard intervals between symbols in order to eliminate ISI due to multipath fading. The use of narrowband orthogonal subcarriers leads to a channel that is roughly constant over each subband, which makes frequency-domain equalization much simpler at the receiver. A subcarrier with N times lower data-rate can be equalized with a single-tap;

this is much easier compared to the equalization of wideband single-carrier DSSS signals.

OFDM subcarriers are mutually orthogonal though they overlap with each other. This implies that, at the peak spectral response of a subcarrier, all other subcarrier spectral responses vanish; hence there is no intercarrier interference (ICI) between them. This also leads to improved spectral efficiency compared to conventional FDM, where adjacent frequency channels do not overlap and are separated by guard bands. Nevertheless, the OFDM is sensitive to time-selective fading and frequency offsets (Doppler shifts, phase noise) which may destroy the orthogonality between subcarriers and cause ICI.

The long symbol duration in OFDM also facilitates the design of single frequency networks (SFNs), where the signals from several neighboring transmitters sending the same signal simultaneously at the same frequency may be combined constructively, rather than interfering destructively as would typically occur in traditional single-carrier systems. This is especially a desired and useful feature for single-frequency networking of DVB-T transmitters with overlapping coverage areas. In addition, compared with single-carrier systems, the OFDM can cope better with undesired channel conditions, for example, frequency-dependent attenuation in cables and narrow

band interference. Despite all these benefits, OFDM has also some drawbacks, for example, sensitivity to Doppler shift, synchronization problems, and inefficient power consumption due to high peak-to-average power ratio (PAPR). OFDM is widely used in modern telecommunication systems such as ADSL, IEEE 802.11 (Wi-Fi), DVB-T and LTE.

10.2.1   OFDM Transmitter

The block diagram of an OFDM transceiver is shown in Figure 10.35. The high-speed binary data of rate R is FEC-encoded and interleaved against burst errors. The data rate at the encoder/interleaver output is R/R_c where R_c denotes the code rate. As a coherent transmission technique, OFDM uses coherent M-PSK or M-QAM. Assuming that all the subcarriers use the same alphabet size, the encoded data is mapped into symbols consisting of $\log_2 M$ encoded bits. Hence the symbol rate at the modulator output is given by $R/(R_c \log_2 M)$. Then, the encoded and modulated data is serial-to-parallel converted to modulate N orthogonal subcarriers. Assuming all subcarriers are used and carry the same number of symbols, the symbol rate carried by each of the N subcarriers is given by $R/(N R_c \log_2 M)$. Hence, the duration of an OFDM symbol is given by

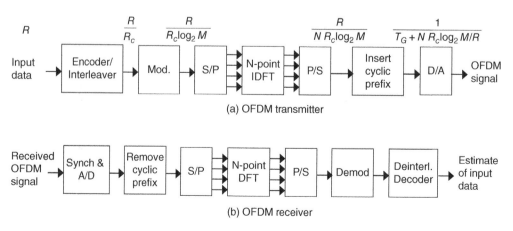

(a) OFDM transmitter

(b) OFDM receiver

Figure 10.35   Block Diagram of an OFDM Transceiver.

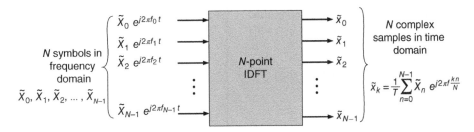

Figure 10.36 Block Diagram of IDFT Operation. IDFT converts the parallel input signals in the frequency domain into N complex time-samples.

$$T = \frac{NR_c\log_2 M}{R} \qquad (10.86)$$

The serial-to-parallel conversion of the data evidently decreases the symbol rate of each subcarrier by a factor of N, which leads to a N-fold increase in the symbol duration. For uncoded binary transmission, the OFDM symbol duration T is therefore N times the symbol duration of input encoded binary data. As a result of increased symbol duration, the data carried by the subcarriers suffers flat fading but not frequency-selective fading. The inverse discrete Fourier transform (IDFT) converts the parallel input signals consisting of N modulated subcarriers in the frequency domain into N complex discrete time-samples. Hence, the IDFT acts as a multiplexer, where the symbols carried by N sub-carriers are transmitted as a single discrete signal with N samples per OFDM symbol duration T (see Figure 10.36). As given by (8.108), the subcarriers become orthogonal with each other if the subcarriers are separated by multiples of $\Delta f = 1/T$. The use of overlapping of mutually orthogonal subcarriers makes OFDM a spectrally efficient transmission technique.

If the maximum excess delay of the channel is longer than the symbol duration, multipath components will create intersymbol interference (ISI) between neighboring symbols. In order to eliminate ISI, the duration (number of samples) of the cyclic prefix (CP), which is added at the beginning of each OFDM symbol, should exceed the maximum excess delay of the channel. Following the insertion of the cyclic prefix of duration T_G, the channel symbol duration T_s and the corresponding symbol rate R_s may be written as follows:

$$R_s = \frac{1}{T_s} = \frac{1}{T+T_G} = \frac{1}{T}\frac{1}{1+T_G/T} \qquad (10.87)$$

The cyclic prefix (CP), which is a realization of the guard interval, is simply a repetition of the last part of an OFDM symbol at the beginning of the same symbol.

Now we consider the IDFT operation shown in Figure 10.36 in order to gain more insight about the OFDM transmission. A bandpas signal $x(t)$ bandlimited to B can be expressed in terms of its complex envelope $\tilde{x}(t)$ as follows:

$$x(t) = \mathrm{Re}\left[\tilde{x}(t)\,e^{jw_c t}\right] \qquad (10.88)$$

The complex envelope signal $\tilde{x}(t)$ is defined by the following inverse Fourier transform:

$$\tilde{x}(t) = \frac{1}{B}\int_B \tilde{X}(f)\,e^{j2\pi f t}df \qquad (10.89)$$

over the OFDM bandwidth $B = N\,\Delta f$. We can express $\tilde{x}(t)$ as the inverse DFT of modulated subcarriers:

$$\tilde{x}(t) = \frac{1}{N}\sum_{n=0}^{N-1}\tilde{X}_n\,e^{j2\pi f_n t},\quad 0 \le t < T \qquad (10.90)$$

$$f_n = f_0 + \Delta f\,n,\quad n = 0,1,\cdots N-1$$

where $\tilde{X}_n = \tilde{X}(f_n)$ and $\Delta f = 1/T$. Discrete-time representation of (10.90) with $\tilde{x}_k = \tilde{x}(t_k)$ yields

$$\tilde{x}_k = \frac{1}{N}\sum_{n=0}^{N-1}\tilde{X}_n e^{j2\pi\frac{k\cdot n}{N}}, \quad k=0,1,\dots,N-1$$

$$t_k = \frac{k}{N}T, \quad k=0,1,\dots N-1$$

$$(10.91)$$

where the common term $\exp(j2\pi f_0 t)$ is ignored. Note that f_0 may be assumed to be zero or to be equal to $f_0 = -\Delta f(N-1)/2$. The latter value makes the spectrum of the baseband OFDM signal to be symmetrical around dc. The corresponding inverse Fourier transform expression may be written as

$$\tilde{X}_n = \sum_{k=0}^{N-1}\tilde{x}_k e^{-j2\pi\frac{k\cdot n}{N}}, \quad n=0,1,\dots,N-1 \quad (10.92)$$

Note that each term in (10.92) represents a band-pass signal centered at frequency f_n and the aggregate of all the bandpass signals represents the overall signal to be transmitted. From (10.90) and (D.36) we observe that the subcarriers are orthogonal with each other

$$\frac{1}{T}\int_0^T e^{j2\pi f_n t} e^{-j2\pi f_m t}\, dt = \frac{1}{T}\int_0^T e^{j2\pi(f_n-f_m)t}dt = \delta(n-m)$$

$$(10.93)$$

Square-windowed sinusoid in time domain as in (10.90) implies a sinc shaped subchannel spectrum response (transfer function):

$$e^{j2\pi f_n t}, \ 0<t<T \ \Leftrightarrow \ T\,\frac{\sin(\pi(f-f_n)T)}{\pi(f-f_n)T}$$

$$(10.94)$$

where the phase term is ignored for the sake of convenience. Note that the sinc function above is unity at $f=f_n$ but vanishes for $f=f_m$, $m \neq n$,

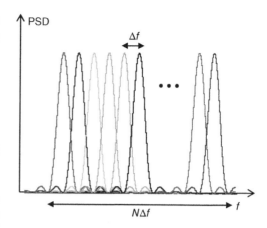

Figure 10.37 PSD of an OFDM Signal with N Orthogonal Subcarriers.

(n and m are integers). Therefore, the subcarriers separated from each other by Δf in the frequency domain are orthogonal and do not cause ICI unless the orthogonality is destroyed by Doppler shift and/or oscillator phase noise. The OFDM signal, which uses the frequency spectrum efficiently, suffers no interference between subcarriers. Figure 10.37 shows the PSD of an OFDM signal with N orthogonal subcarriers.

10.2.1.1 Cyclic Prefix

Guard intervals are inserted between the OFDM symbols for mitigating the ISI. Instead of blanking out the signal during the guard intervals, a copy of the symbol tail is inserted at the beginning of each OFDM symbol. Specifically, the cyclic extension of an OFDM symbol is simply the periodic extension of the DFT output:

$$\tilde{x}_{-i} = \tilde{x}_{N-i}, \quad i=1,2,\dots v \quad (10.95)$$

where v last samples of an OFDM symbol are appended at the beginning of the same symbol (see Figure 10.38). Here N denotes the number

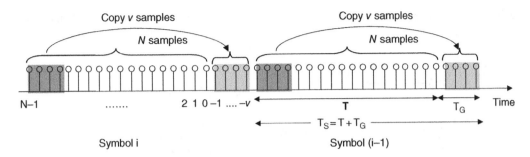

Figure 10.38 Insertion of the Cyclic Prefix.

of subcarriers in the OFDM signal or alternatively the number of samples at the IFFT output. Note that the duration of the v samples of cyclic prefix should exceed the maximum excess delay T_m of the wireless channel, that is, $(T/N)v \geq T_m$ for mitigating the ISI. The use of cyclic prefix converts the discrete-time linear convolution between the transmitted signal and the channel impulse response into a discrete-time circular convolution. Hence, the received signal can be modeled as a circular convolution between the channel impulse response and the transmitted data block. In the frequency domain, this corresponds to a point-wise multiplication of the DFT frequency components. [11]

Due to the extension of the symbol duration from T to $T + T_G$ (see (10.87)), the use of cyclic prefix requires more transmit energy and decreases the transmission rate by the factor $1/(1 + T_G/T)$. If each subcarrier transmits b bits in a symbol duration, then the overall bit rate in an OFDM system with N subcarriers will be $Nb/(T + T_G)$ bps with cyclic prefix as compared to the bit rate of Nb/T bps with no cyclic prefix. The loss in SNR (or data rate) can be decreased by choosing the symbol period T much longer than the length of the cyclic prefix T_G as long as the condition $T_G > T_m$ is satisfied. For example, in IEEE 802.11a and g (Wi-Fi), the duration of the cyclic prefix is T_G $= 0.8$ ms while the symbol duration is $T = 3.2$ ms; hence $T_G/T = 1/4$. IEEE 802.16a (Wi-

Max) and DVB systems use different symbol lengths and T_G choices, for example, $T_G/T =$ 1/4, 1/8, 1/16 or 1/32.

10.2.2 OFDM Receiver

An OFDM receiver reverses the operations carried out at the transmitter following the down-conversion and analog-to-digital (A/D) conversion stages (see Figure 10.35). Various techniques are available for estimating symbol timing and carrier frequency and subsequently synchronization with the received signals. The cyclic prefixes are removed before the S/P conversion takes place. The use of a sufficiently long cyclic prefix removes the ISI by separating consecutive symbols from each other. Then the output of the S/P converter is in the correct form for FFT. The FFT output is P/S converted and subsequently demodulated, deinterleaved and decoded. To demodulate the subcarriers using M-PSK or M-QAM modulations, reference phase and amplitude of each subcarrier are estimated using pilot subcarriers. To overcome the unknown phase and amplitude ambiguities, coherent and differential detection techniques are used. [12]

Let us consider a multipath channel with an impulse response (CIR) $h(\tau, t)$, $0 < \tau < T_m$, where T_m denotes the maximum excess delay with L multipath components with channel gains $\{\alpha_\ell(t)\}$, delays $\{\tau_\ell(t)\}$ and Doppler shifts $\{\theta_\ell(t)\}$:

$$h(\tau,t) = \sum_{\ell=0}^{L-1} \underbrace{\alpha_\ell(t) e^{-j\theta_\ell(t)}}_{h_\ell(t)} \delta(\tau - \tau_\ell(t))$$

$$h_k = \sum_{\ell=0}^{L-1} h_{k,\ell} \, \delta(\tau - \tau_{k,\ell}(t)) \qquad (10.96)$$

where $h_{k,\ell} = h_\ell(t_k) = \alpha_\ell(t_k) e^{-j\theta_\ell(t_k)}$ denotes complex CIR of the ℓ-th multipath component at time t_k. The channel transfer function at the nth subcarrier frequency at time t_k is found by the Fourier transform of the first expression in (10.96) with respect to τ:

$$H(f,t) = \mathfrak{F}_\tau[h(\tau,t)] = \sum_{\ell=0}^{L-1} h_\ell(t) e^{-j2\pi f \tau_\ell}$$

$$H_{n,k} = \sum_{\ell=0}^{L-1} h_{k,\ell} e^{-j\frac{2\pi n \ell}{N}}$$

$$(10.97)$$

where (10.90) and (10.91) are used to obtain the last expression and $\tau_\ell(t)$ is assumed to be time-invariant.

If the duration of the cyclic prefix T_G is chosen to be larger than T_m, then the delayed multipath components do not induce ISI. The samples $\tilde{r}_k$ of the complex envelope of the received OFDM symbol may be obtained from the convoluton of the CIR and $\tilde{x}_k$ over the symbol interval $(0,T)$:

$$\tilde{r}_k = \sum_{\ell=0}^{L-1} \tilde{x}_{k-\ell} h_{k,\ell} + w_k$$

$$= \frac{1}{N} \sum_{n=0}^{N-1} X_n H_{n,k} \, e^{j\frac{2\pi n k}{N}} + w_k, \quad k = 0, 1, \cdots, N-1$$

$$(10.98)$$

The last expression is obtained by inserting (10.91) for $\tilde{x}_k$ into (10.98) and then using (10.97).

If when the duration of the cyclic prefix is longer than the channel excess delay T_m, that

is, $L-1 \le v$, then using (10.98), the channel frequency response at the nth subcarrier may be written as

$$\tilde{R}_n = \sum_{k=0}^{N-1} \tilde{r}_k e^{-j2\pi n k/N}$$

$$= \frac{1}{N} \sum_{k=0}^{N-1} \sum_{r=0}^{N-1} \tilde{X}_r H_{r,k} e^{j\frac{2\pi k(r-n)}{N}} + W_n$$

$$= \underbrace{\frac{1}{N} \sum_{k=0}^{N-1} \tilde{X}_n H_{n,k}}_{desired\ signal}$$

$$+ \underbrace{\frac{1}{N} \sum_{k=0}^{N-1} \sum_{r=0, r \neq n}^{N-1} \tilde{X}_r H_{r,k} e^{j\frac{2\pi k(r-n)}{N}}}_{ICI} + \underbrace{W_n}_{noise}$$

$$(10.99)$$

For a *time invariant CIR*, that is, the channel gain $\alpha_\ell(t)$, the channel-induced phase $\theta_\ell(t)$ and the delay $\tau_\ell(t)$ do not vary over an OFDM symbol period (no Doppler shift). Then, in view of (10.97), the channel transfer function is also time-invariant, that is, $H_{n,k} \triangleq H_n$. In this case, (10.99) simplifies to

$$\tilde{R}_n = \tilde{X}_n H_n + W_n \qquad (10.100)$$

Since, using (D.13), the ICI term in (10.99) may be shown to vanish:

$$\frac{1}{N} \sum_{r=0, r \neq n}^{N-1} \tilde{X}_r H_r \sum_{k=0}^{N-1} e^{j\frac{2\pi k(r-n)}{N}}$$

$$= \frac{1}{N} \sum_{r=0, r \neq n}^{N-1} \tilde{X}_r H_r \frac{\sin(\pi(r-n))}{\sin(\pi(r-n)/N)} = 0$$

$$(10.101)$$

Note however that (10.100), which states that OFDM transmission is achieved via N orthogonal transmission paths, is valid if $L-1 \le v$, that is, the cyclic prefix is longer than the channel excess delay. In this case, the

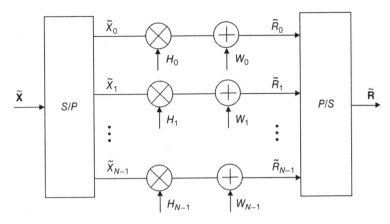

Figure 10.39   Representation of an OFDM System with Cyclic Prefix Exceeding CIR Length, the Channel Being Constant During the Transmission of One OFDM Symbol, and with No ICI.

convolution in the first line of (10.98) becomes cyclic and the convolution process is confined within the symbol of interest. Otherwise, the channel excess delay exceeds the duration of the cyclic prefix and convolution above extends to the previous symbol, hence ISI occurs.

The term W_n in (10.99) is defined by the FFT of the sampled noise terms

$$W_n = \sum_{k=0}^{N-1} w_k \, e^{-j2\pi nk/N} \qquad (10.102)$$

and has a variance which is equal to N times the variance of w_k.

As given by (10.100) and shown in Figure 10.39, the received symbol in a time-invariant channel is proportional to the transmitted symbol $\tilde{X}_n$ with a proportionality factor H_n, which denotes the value of the channel transfer function at n^{th} subcarrier. The OFDM transmission in a time-invariant channel can therefore be modelled as N interference-free parallel sub-channels. Each received subcarrier carries the transmitted symbol scaled by the value of the channel transfer function at that sub-carrier frequency and corrupted by the additive noise. The subchannels may be equalized in the frequency domain using parallel

one-tap equalizers by estimating $\hat{H}_n$ of the channel transfer function at that subcarrier. Then, the receiver detects the transmitted symbol at n^{th} subcarrier by dividing (10.100) with the estimate $\hat{H}_n$ of the channel transfer function. Such simple channel equalization justifies and motivates the use of a cyclic prefix and the OFDM technique itself.

The SNR of the nth subcarrier is found using (10.100) and (10.102):

$$SNR = \frac{E\left[\left|H_n\tilde{X}_n\right|^2\right]}{E\left[\left|W_n\right|^2\right]} = |H_n|^2 \frac{E\left[\left|\tilde{X}_n\right|^2\right]}{N N_0/2}$$

$$= |H_n|^2 \frac{T}{T+T_G} \frac{2E_b}{N_0}$$

$$E\left[\left|\tilde{X}_n\right|^2\right] = \frac{T}{T+T_G} N E_b$$

$$(10.103)$$

Discarding the cyclic prefix decreases the bit energy by a factor of $T/(T+T_G)$. On the other hand, since the OFDM symbol duration is N-times the duration of the input symbol, energy of an OFDM symbol is N times the energy E_b of the input symbol. Consequently, except for the degradation caused by the cyclic prefix, the SNR and the error probability of an

OFDM symbol remains the same as that of the input symbol. The degradation caused by the longest cyclic prefix of $T_G = T/4$ used in practice is 4/5, which leads to a SNR loss of maximum 1 dB.

Example 10.9   Fast Fourier Transform (FFT) and Inverse FFT (IFFT).

Using (10.91) and (10.92), the discrete Fourier transform (DFT) and its inversion may also be written as

$$\tilde{x}_k = \frac{1}{N}\sum_{n=0}^{N-1}\tilde{X}_n\, e^{j\frac{2\pi}{N}kn} = \frac{1}{N}\sum_{n=0}^{N-1}\tilde{X}_n\, W_N^{kn}, \quad k=0,1,\ldots,N-1$$

$$\tilde{X}_n = \sum_{k=0}^{N-1}\tilde{x}_k\, e^{-j\frac{2\pi}{N}kn} = \sum_{k=0}^{N-1}\tilde{x}_k\, W_N^{-kn}, \qquad n=0,1,\ldots,N-1$$

$$W_N \triangleq e^{j2\pi/N}$$

$$(10.104)$$

Since $(N-1)$ multiplications are required for each value of n in (10.104), the DFT requires $(N-1)^2$ complex multiplications and $N(N-1)$ complex additions. However, if $\tilde{x}_k$ is split into two sequences with even and odd subscripts, then

$$\tilde{X}_n = \sum_{k=0}^{N-1}\tilde{x}_k\, W_N^{-kn}$$

$$= \sum_{k=0}^{N/2-1}\tilde{x}_{2k}\, W_N^{-2kn} + \sum_{k=0}^{N/2-1}\tilde{x}_{2k+1}\, W_N^{-(2k+1)n}$$

$$= \sum_{k=0}^{N/2-1}\tilde{x}_{2k}\, W_{N/2}^{-kn} + W_N^{-n}\sum_{k=0}^{N/2-1}\tilde{x}_{2k+1}\, W_{N/2}^{-kn},$$

$$n=0,1,\ldots,N-1$$

$$(10.105)$$

The total number of multiplications required in (10.105) is given by $2(N/2-1)^2$, which is approximately ½ of that for (10.104). In the Cooley-Tukey FFT algorithm, one chooses N as $N=2^m$, which can be divided by 2 continuously. The DFT can then be implemented as FFT by continuing to split each term into its even and odd terms. Thus the FFT is very efficient and fast since it requires only $(N/2)\log_2 N$ complex multiplications and $N\log_2 N$ complex additions. For example, for an OFDM system with $N=2^{10}=1024$ subcarriers, the FFT provides drasic time savings since it requires $(N/2)\log_2 N = 5120$ multiplications while the DFT requires $(N-1)^2 = 1046529$ multiplications. [13]

Example 10.10   Why and How Cyclic Prefix Removes ICI?

Based on (10.96), the CIR of a time-invariant multipath channel with L multipath components may be written in vector form as

$$\mathbf{h} = [h_0 \quad h_1 \quad \cdots \quad h_{L-1}] \tag{10.106}$$

As one may observe from (10.95), the number v of samples in the cyclic prefix is not necessarily equal to the length of the CIR. To study the effect of unequal lengths of the CIR and the cyclic prefix, we define Γ as the length of the CIR exceeding the length v of the cyclic prefix:

$$\Gamma = L - 1 - v \tag{10.107}$$

We can then rewrite the convolution operation in (10.98) in discrete form for the ith symbol as follows:

$$\tilde{r}_i[k] = \sum_{\ell=0}^{L-1}\tilde{x}_i[k-\ell]\, h_\ell, \quad k=0,1,\ldots,N-1 \tag{10.108}$$

Now let us determine the first term in (10.108):

$$\tilde{r}_i[0] = \sum_{\ell=0}^{L-1}\tilde{x}_i[-\ell]\, h_\ell$$

$$= h_0\tilde{x}_i[0] + h_1\tilde{x}_i[-1] + \cdots + h_v\tilde{x}_i[-v]$$

$$+ \underbrace{h_{v+1}\tilde{x}_i[-v-1]}_{\tilde{x}_{i-1}[N-1]} + \cdots + \underbrace{h_{L-1}\tilde{x}_i[-L+1]}_{\tilde{x}_{i-1}[N-\Gamma]}$$

$$(10.109)$$

It is clear from (10.95) and (10.109) that, when the length of the CIR exceeds the length of the cyclic prefix, the terms corresponding to (i-1)-th symbol cause ISI to the i-th symbol (see also Figure 10.38). Using the v cyclic prefixes for the i-th symbol, as defined by (10.95), the convolution in (10.108) may be written in matrix form as follows:

$$\tilde{r}_i[0] = h_0 \tilde{x}_i[0] + h_1 \underbrace{\tilde{x}_i[-1]}_{\tilde{x}_i[N-1]}$$

$$+ h_2 \underbrace{\tilde{x}_i[-2]}_{\tilde{x}_i[N-2]} + \cdots + h_{v-1} \underbrace{\tilde{x}_i[-v+1]}_{h_{L-2} \quad \tilde{x}_i[N-v+1]}$$

$$+ \underbrace{h_v \quad \tilde{x}_i[-v]}_{h_{L-1} \quad \tilde{x}_i[N-v]}$$

(10.111)

$$\begin{pmatrix} \tilde{r}_i[0] \\ \tilde{r}_i[1] \\ \vdots \\ \tilde{r}_i[N-1] \end{pmatrix} = \begin{pmatrix} h_{L-1} & \cdots & h_v & \cdots & h_0 & 0 & \cdots & 0 \\ 0 & \ddots & & \ddots & & \ddots & & \vdots \\ \vdots & & \ddots & & \ddots & & \ddots & 0 \\ 0 & \cdots & \cdots & h_{L-1} & \cdots & h_v & \cdots & h_0 \end{pmatrix} \begin{pmatrix} \tilde{x}_{i-1}[N-\Gamma] \\ \vdots \\ \tilde{x}_{i-1}[N-1] \\ \tilde{x}_i[-v] \\ \vdots \\ \tilde{x}_i[0] \\ \vdots \\ \tilde{x}_i[N-1] \end{pmatrix} + w_i$$

(10.110)

The terms in CIR with indices higher than the length v of the cyclic prefix are convolved with (i-1)-th symbol but not with the i-th symbol; this evidently leads to ISI.

Now let us consider the case where the length of the cyclic prefix is greater than or equal to the length of the CIR, that is, $v \geq L-1$. Assuming $\Gamma = 0$, hence $v = L-1$, and using (10.95), (10.109) may be rewritten as

The convolution in (10.108) may then be written in matrix form in (10.112):

The use of the cyclic prefix longer than or equal to the CIR, makes the $N \times N$ channel matrix in (10.112) a circulant matrix, that is, its rows are circularly shifted versions of each other. This makes the linear convolution defined by (10.108) to be a circular convolution. Consequently, the OFDM output can be

$$\begin{pmatrix} \tilde{r}_i[0] \\ \tilde{r}_i[1] \\ \vdots \\ \tilde{r}_i[N-1] \end{pmatrix} = \begin{pmatrix} h_0 & 0 & \cdots & 0 & h_{L-1} & \cdots & h_1 \\ \vdots & \ddots & \ddots & & \ddots & \ddots & \vdots \\ \vdots & & \ddots & \ddots & & \ddots & h_{L-1} \\ h_{L-1} & & & \ddots & \ddots & & 0 \\ 0 & \ddots & & & \ddots & \ddots & \vdots \\ \vdots & \ddots & \ddots & & & \ddots & 0 \\ 0 & \cdots & 0 & h_{L-1} & \cdots & \cdots & h_0 \end{pmatrix} \begin{pmatrix} \tilde{x}_i[0] \\ \tilde{x}_i[1] \\ \vdots \\ \vdots \\ \vdots \\ \tilde{x}_i[N-2] \\ \tilde{x}_i[N-1] \end{pmatrix} + w_i$$

(10.112)

written as in (10.100), where the ISI is cancelled as the receiver takes FFT after the removal of the CP. Nevertheless, the circulant nature of the channel matrix may be destroyed in time-varying channels, because the CIR coefficients in a row (corresponding to a sample of the OFDM symbol) may be potentially different than those in some other row. [14]

Example 10.11   OFDM System Design.
Consider a UMTS system operating at 2.2 GHz carrier frequency over a $B = 5$ MHz bandwidth with $N = 2^9 = 512$ subcarriers. The system accomodates mobile speeds up to 120 km/hr ($f_{d,max} = 244$ Hz), and a maximum excess delay of $T_m = 10$ μs. Then the subcarrier spacing must be $\Delta f = B/N = 9.77\,kHz$ and the corresponding symbol duration is $T = 1/\Delta f = 102.4\ \mu s$. For avoiding the ISI, the duration of cyclic prefix must be larger than maximum delay spread $T_G > T_m$. For example, the choice of $T_G = T/8 = 12.8$ μs leads to a 11% reduction in the data rate:

$$E_{loss} = \frac{T_G}{T + T_G} = \frac{T/8}{T + T/8} = \frac{1}{9} = 0.11$$

The E_b/N_0 will be decreased by a factor of (see (10.103))

$$1 - E_{loss} = \frac{T}{T + T_G} = \frac{T}{T + T/8} = \frac{8}{9} = 0.89 \quad (0.51\ dB)$$
$$(10.113)$$

If all subcarriers are modulated with a QPSK symbol, then each subcarrier transmits 2 bits, which results in an overall raw data rate of

$$\frac{512\ (subcarriers) \times 2\ (bits/subcarrier)}{102.4 + 12.8\ (\mu s)}$$
$$= 8.9\ (Mbps)$$

Note, however, that some of the subcarriers are used as pilots for channel estimation and some subcarriers at the edge are not modulated for reducing adjacent channel interference.

Assuming that 80% of the subcarriers are used for data transmission, the data rate would be 7.1 Mbps for QPSK.

Example 10.12   Single-Frequency Networks (SFNs) in DVB-T Systems.
In analog terrestrial TV broadcasting, the neighboring TV base stations use different carrier frequencies for avoiding interference in TV receivers that can receive both base stations. However, because of the long symbol durations, OFDM offers the possibility of SFNs for digital video broadcasting (DVB). [15] This means that all terrestrial TV transmitters can use the same carrier frequency for DVB without interfering with each other. When the OFDM symbol duration is chosen to be sufficiently long, signals of neighboring DVB base stations arrive to a TV receiver with delays within a fraction of the long symbol duration; hence these signals, which are not resolvable, are treated as a single signal at a digital TV receiver.

Consider a DVB-T system which allows mobile operation at user velocities up to 120 km/h. A cyclic prefix duration of 200 μs is chosen in order to avoid ISI due to multipaths in the SFN. To provide acceptable spectral efficiencies, the cyclic prefix must not be longer than one-quarter of the useful symbol duration T. Hence the symbol duration should satisfy $T \geq 4T_G = 800$ μs. The corresponding frequency spacing between subcarriers is then $\Delta f \leq 1/T = 1.25$ kHz, hence $N = B/\Delta f = 4000$ subcarriers for B = 5 MHz. For compatibility with the MPEG-2 coded video standard, a useful data rate of about 10 Mbps is assumed to be required when operating with QPSK (modulation orders up to 64-QAM can be used) with Gray-coded constellation mapping. When all subcarriers are used for data transmission, this system can support a maximum data rate.

$$\frac{4000\ (subcarriers) \times 2\ (bits/subcarrier)}{800 + 200\ (\mu s)}$$
$$= 8\ (Mbps)$$
$$(10.114)$$

Noting that $2^{12} = 4096$, we use the next larger power of 2, that is, $2^{13} = 8192$ subcarriers, in order to have data rates in excess of 10 Mbps and to allow for additional overheads, for example, the use of some subcarriers for channel estimation. Therefore the so-called 8 k mode is usually chosen for SFNs. The actual choice of T is 896 µs, with a cyclic prefix duration 224 µs. In view of the COST 207 model, which is characterized by a maximum excess delay of 20 µs, this is sufficient to overcome the worst case multipath. For verifying the feasibility of SFN, consider a TV receiver, which is tuned to nearby a DVB-T broadcasting station, also receives TV signals from another station $d = 30$ km away and broadcasting at the same carrier frequency. Since the delay $d/c = 100\mu s$ is much shorter than the 224 µs cyclic prefix duration, the signal with 100 µs delay is not resolvable and is received by the TV receiver as a single signal with that from the nearest broadcasting station.

10.2.3 Intercarrier Interference (ICI) in OFDM Systems

If the subcarriers remain orthogonal at the receiver input, then they can be easily separated by the receiver with no ICI, even though they overlap in the frequency domain. Orthogonality between subcarriers may be destroyed due to Doppler frequency shift in the channel, when the clocks are not properly synchronized, and/ or when the phase noise of synthesizers leads to non-negligible random frequency shifts around subcarrier frequencies. Frequency selectivity of the channel also contributes to frequency shifts of the subcarriers. Consequently, as the orthogonality between subcarriers is destroyed, then the ICI begins to degrade the system performance.

If the duration T_G of cyclic prefix is chosen to be larger than T_m, then the delayed multipath components do not induce ISI, hence have no affect on the orthogonality between the subcarriers. However, in a time-varying

channel where the channel gain and phase change over an OFDM symbol period due to, for example, Doppler shift, then the channel transfer function changes not only with frequency but also with time. In this case, the ICI may become a major source of degradation in the system performance. Assuming that the channel is WSS and ignoring the noise, average value of the signal-to-ICI ratio (SIR) is obtained by the ratio of the powers of the desired and ICI signal components of (10.99) as follows (see Appendix 10C for the details of the derivation). [16][17][18]

$$SIR_i = \frac{N + 2\sum_{p=1}^{N-1}(N-p)J_0\left(\dfrac{\pi p \beta}{N}\right)}{\sum_{\substack{k=0, \\ k \neq i}}^{N-1}\left\{N + 2\sum_{p=1}^{N-1}(N-p)J_0\left(\dfrac{\pi p \beta}{N}\right)\cos\left(\dfrac{2\pi p(i-k)}{N}\right)\right\}}$$

$$SIR = \frac{1}{N}\sum_{i=0}^{N-1}SIR_i$$

$$(10.115)$$

where β is defined as the ratio of the OFDM symbol duration to the channel coherence time, which is inversely related to the Doppler spread $2f_{d,\max}$:

$$\beta \triangleq T / \Delta t_c = 2f_{d,\max}T \qquad (10.116)$$

In a time-invariant channel, β approaches zero and the SIR given by (10.115) goes to infinity; then, the channel does not suffer any Doppler frequency shift and ICI vanishes. The SIR rapidly degrades with increasing Doppler shift (time selectivity of the channel) as orthogonality is lost between the subcarriers. This clearly shows the sensitivity of OFDM signals to Doppler shifts and hence the need for Doppler correction. In OFDM systems, measures are taken for the (pre)correction of the frequency offsets.

Figure 10.40 shows the variation of the SIR as a function of β, the ratio of OFDM symbol duration to the channel coherence time Δt_c. The degradation in SIR (increase in ICI) is

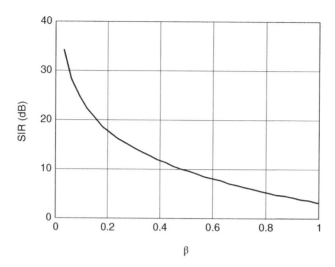

Figure 10.40   SIR Versus β, the Ratio of the Symbol Duration To the Channel Coherence Time. The SIR was observed to be insensitive to N.

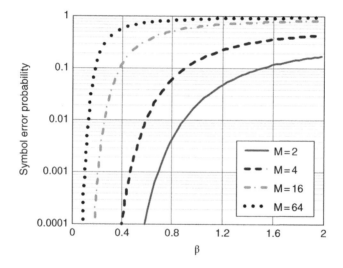

Figure 10.41   Symbol Error Probability For M-ary QAM for $N = 64$.

obvious with decreasing channel coherence time. The SIR was observed to be insensitive to N. To have a better feeling about the sensitivity of OFDM systems to Doppler frequency shift, consider the IEEE 802.11a system operating at the 2.4 GHz industrial, scientific and medical (ISM) band with $T = 3.2$ ms and $T_G = 0.8$ ms. A mobile user of 10 km/hr velocity suffers a maximum Doppler shift $f_{d,\max} = v/\lambda = 22.2\,Hz$. Using (10.116), the value of β is found to be $\beta = 0.142$ for which the value of SIR is acceptable. However, for a user velocity of 100 km/hr $\beta = 1.42$; hence IEEE 802.11a system can not serve at these user velocities. Figure 10.41 shows the symbol error performance of M-ary QAM for $N = 64$. The symbol error probability increses with M and β but is insensitive to N, as expected.

10.2.4 Channel Estimation by Pilot Subcarriers

For a time-invariant channel with the cyclic prefix longer than the CIR, the received symbol from a subcarrier is the same as the transmitted symbol except for a multiplier, which is equal to the value of the channel transfer function at that subcarrier frequency (see (10.100)). Therefore, an OFDM receiver must estimate H_n, $n = 0, 1, \cdots, N-1$ for correctly decoding the received signals. This is usually achieved by using some of the subcarriers as pilots. Pilot subcarrier patterns are designed so as to track the changes of the channel transfer function in time, frequency and phase. The H_n's of the pilots are used with some form of interpolation to estimate the H_n's of the other subcarriers. Hence, the magnitude and phase shift offsets of the received symbols are corrected (equalized) for coherent demodulation of MPSK/MQAM. The pilots can also be used for frame/frequency/time synchronization, channel estimation, transmission mode identification and to follow phase noise. The use of differential demodulation does not require the use of pilots. The loss due to differential demodulation may be partially offset by the reduction in overhead. [19]

For an error-free interpolation, the density of the pilots is chosen so as to satisfy the Shannon sampling theorem in time and frequency, as a function of coherence time and coherence bandwidth. This implies that the pilots must be used more frequently in time (frequency) as the channel becomes more time (frequency)-selective. As will be discussed in Chapter 11 in more detail, the frequency separation between the subcarriers must not exceed the so-called coherence bandwidth $\Delta f_c \propto 1/T_m$ of the channel, where T_m denotes the maximum excess delay of the channel (or in simple terms the length of the CIR). This implies that the pilot symbols separated from each other by frequency intervals not exceeding Δf_c allow accurate estimation of the frequency, phase and delay of the subcarriers in between.

In a *time-selective channel*, the pilot subcarriers should be updated at time intervals not exceeding the channel coherence time $\Delta t_c \propto 1/B_d$ where $B_d = 2f_{d,\max}$ denotes the Doppler spread of the channel. In *time- and frequency-selective channels*, efficient pilot schemes, as used in DVB systems, make use of interpolation both in frequency and time domains. Figure 10.42 shows a typical pilot pattern where subcarriers are separated from each other by no more than Δf_c in the frequency-domain and Δt_c in the time-domain. Hence, the H_n's of the other sub-carriers may be obtained by two-dimensional interpolation in time- and frequency-domains.

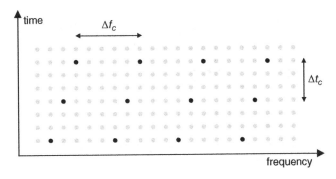

Figure 10.42 A Typical Pilot Pattern In an OFDM System. An efficient pilot scheme makes use of interpolation both in frequency and time domain.

10.2.5 Synchronization of OFDM Systems

Carrier recovery by pilots is already discussed in the previous section. As for the frame/symbol synchronization, it is usually achieved by using a sliding window, as shown in Figure 10.43, which exploits the redundancy provided by cyclic prefix. Samples of cyclic prefix of a duration T_G at the beginning of a symbol are correlated with the samples at the end of the same symbol, which are separated from each other by the symbol duration T. Since they are identical, the correlation is maximized when they are aligned, but fluctuates with small amplitudes when they are not. Starting time of symbol is determined by observing the correlation peak. [20]

10.2.6 Peak-to-Average Power Ratio (PAPR) in OFDM

An OFDM signal consists of the phasor sum of N modulated subcarriers, evenly distributed across the OFDM bandwidth. Therefore, peak and average signal power levels differ

significantly, resulting in a high peak-to-average power ratio (PAPR). PAPR is a measure of the envelope fluctuations and is widely employed in the design of multicarrier systems. However, the cubic metric (CM) is recently suggested by 3GPP to predict HPA power derating in multicarrier systems. [21] Here, we will be interested in the PAPR approach. PAPR, which is used to characterize envelope fluctuations of OFDM signals, is defined as follows:

$$PAPR = \frac{\max_{0 \le k \le LN-1} |\tilde{x}_k|^2}{E\left[|\tilde{x}_k|^2\right]} = \frac{\max_{0 \le k \le LN-1} |\tilde{x}_k|^2}{\frac{1}{LN} \sum_{k=0}^{LN-1} |\tilde{x}_k|^2}$$

(10.117)

where $\tilde{x}_k$, which is given by (10.91), is a complex Gaussian signal whose in-phase and quadrature components are modeled as Gaussian random variables with zero-mean and variance ½. Bearing in mind that an analog signal can not be obtained precisely by D/A conversion of Nyquist rate sampled signals, L-times oversampled signal samples

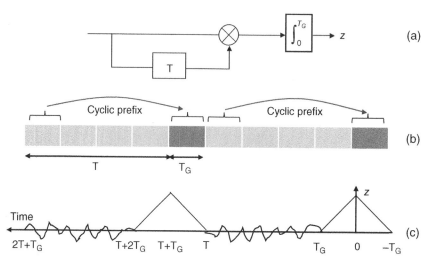

Figure 10.43 Symbol Synchronization in OFDM Using a Sliding Window Correlator. (a) Block diagram of the correlator, (b) Description of the OFDM symbol and the cyclic prefix, (c) Time variation of the correlator output.

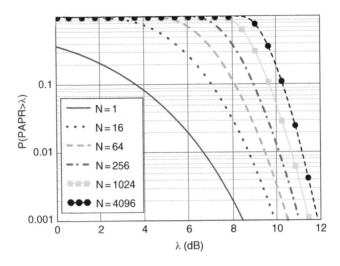

Figure 10.44 Complementary CDF of the PAPR Based on (10.128).

$\tilde{x}_k$, $k = 0, 1, \ldots, LN - 1$ are generally used. Oversampled time-domain samples are obtained by appending $(L-1)N$ zeros in (10.91). The so-called oversampling ratio L is an integer larger than or equal to 1 and $L = 1$ corresponds to Nyquist sampling.

Phasor sum of N modulated complex subcarriers of an OFDM signal can add up constructively or destructively. Noting that the signal peaks occur from the constructive interference of the subcarrier signals, the PAPR may mitigated by preventing the occurrence of constructive interference. This approach is distortionless and very effective, but it is not always easy to implement for large values of N. The peak signal level reaches N and its power N^2. As the number N of subcarriers is usually very large, central-limit theorem states that the distribution of in-phase and quadrature-components of an OFDM signal are Gaussian random variables. Since the magnitude of each of these complex Gaussian signals is characterized by a Rayleigh pdf, the statistics of the OFDM signal may easily be derived based on the assumption that they are i.i.d. However, this assumption is not realistic since the subcarriers, which are closely separated from each other in the frequency band, are not independent from each other. Nevertheless, this assumption leads to useful insight about the PAPR. The probability that the PAPR, normalized to its average power, will be higher than a threshold level Λ is given by (see (F.133))

$$P(PAPR > \Lambda) = 1 - F_{PAPR}(\Lambda) = 1 - \left(1 - e^{-\Lambda}\right)^N$$

(10.118)

where the threshold power level Λ is also normalized with respect to the average power level. Figure 10.44 shows the complementary cdf of the PAPR as a function of the normalized threshold level. For example, for $\Lambda = 4$ (peak power is 6 dB higher than the average power), the PAPR of a single carrier will be higher than its average value for 1.9% of the time, compared with 99.1% for $N = 256$. Hence, as one may easily observe from Figure 10.44, for the practical values of N, the peak power level will be significantly higher than the average power level.

Signals with high PAPR require highly linear high power amplifiers (HPA's), which allow the amplification of signals up to a maximum preset transmit power level without intermodulation distortion. However, the HPA's are usually nonlinear since the use of linear amplifiers are neither practical nor cost-effective. On the other hand, the use of non-linear transmit amplifiers destroys the orthogonality between subcarriers and leads to in-band and out-of band emissions, for example, intermodulation products. Therefore, an OFDM system with nonlinear HPA wastes the transmit power (reducing the power efficiency), jams itself via in-band radiations (increasing the bit error probability) and causes interference to the neighboring channels (decreasing the system capacity). The PAPR may be mitigated by predistorting a nonlinear power amplifier, called linearization of an amplifier. However, the usual approach is to back-off the output power level of the nonlinear HPA for forcing to operate in the linear range of its characteristics. This leads to low power efficiency, higher cost and a heavy burden on the battery of mobile terminals.

A variety of techniques are developed and adopted as standard for mitigating the harmful effects of PAPR. There is no unique best technique for PAPR reduction; each technique has its advantages and drawbacks. The cost of reducing PAPR may be high computational complexity, higher system cost, loss in data rate, inefficient use of the transmit power or the degradation of the bit error probability performance.

Block *coding* and its variants accross subcarriers is used for PAPR reduction by selecting the codewords which minimize the PAPR. For example, using a code rate of $(N-k)/N$ in an OFDM system with N subcarriers, $N-k$ data bits modulate $N-k$ subcarriers and k redundancy bits are transmitted by the remaining k subcarriers. By a suitable choice of the code and the code rate, the PAPR can be reduced to reasonable levels and one may even obtain some modest coding gains. However, the lower the PAPR level, the lower the achievable code rate. Complementary block coding (CBC)[22] uses complementary redundancy bits to reduce to peak value of the OFDM signal by avoiding code sequences with some patterns, for example,

all bits equal and alternating bits. CBC can also provide some error correction capability, since the receiver can choose the stronger of the information bit or its complementary bit for decoding. For large N, high PAPR can still be observed by using modified CBC (MCBC), which consists of dividing a long bit sequence into several subblocks, and encoding each subblock with CBC. Interleaving of short codes in the frequency domain may make the system more robust against burst errors. Complementary Golay codes and generalized Reed-Muller codes are suitable for M-PSK OFDM systems. Coding leads to no ICI but reduces the throughput depending on the code rate. For example, 3 dB PAPR reduction is reported for $R_c > (N-2)/N$ by using CBC with large frame sizes and using MCBC with code rate ¾ for any frame size. Complementary Golay codes are reported to provide significant PAPR reduction (~2 dB) and good error correction capability. [22][23]

Example 10.13 Coding for PAPR Reduction. [23]
Consider an BPSK/OFDM system with $N = 4$ subcarriers. Oversampling is used for improving the accuracy of the D/A conversion at the transmitter. The L-times oversampled signal $\tilde{x}_k$, $k = 1, 2, \ldots LN - 1$ is obtained as

$$\tilde{x}_k = \frac{1}{N} \sum_{n=0}^{N-1} \tilde{X}_n e^{j\frac{2\pi k\, n}{LN}}, \quad k = 0, 1, \ldots, LN - 1$$

(10.119)

by using N frequency-domain samples $\tilde{X}_n$, $n = 0, 1, \ldots, N-1$. Note that (10.119) may also be interpreted as IDFT of $\{\tilde{X}_n\}_{n=0}^{LN-1}$ with $(L-1)N$ zero-padding, that is, $\tilde{X}_n = 0$, $n = N, N+1, \ldots, LN-1$. This implies that only NL equidistant samples of $\tilde{x}_k$ will be considered where the oversampling ratio L is an integer larger than or equal to unity. Since the samples of $\tilde{x}_k$ will then be closer to each other in the time domain, D/A output will allow a more accurate replica of the original signal. However, the time samples will be more correlated with each other, which in turn requires more processing for obtaining the CDF of the PAPR. The resulting PAPR is computed from the L-times oversampled time-domain signal in (10.117).

Table 10.2 shows the PAPR for all possible data blocks for an BPSK/OFDM signal with $N = 4$ subcarriers and an oversampling ratio $L = 4$. To gain more insight about the PAPR values presented in Table 10.2, we use (10.119) to determine the 4-times oversampled signals for the data block $\{1\,1\,-1\,-1\}$ as follows:

$$\{\tilde{X}_n\} = \{1\quad 1\quad -1\quad -1\}$$

$$\{|x_k|^2\} = \{0\quad 0.14\quad 0.43\quad 0.59\quad 0.5\quad 0.26\quad 0.07$$

$$0\quad 0\quad 0\quad 0.07\quad 0.26\quad 0.5\quad 0.59\quad 0.43\quad 0.14\}$$
$$(10.120)$$

Table 10.2 PAPR Values of Eight Possible 4-Bit Data Blocks For an OFDM Signal with $N = L = 4$ and BPSK Modulation. [23] In view of (10.119), the PAPR values for the other eight combinations, which are given by $-\tilde{\mathbf{X}} = \begin{bmatrix} -\tilde{X}_0 & -\tilde{X}_1 & -\tilde{X}_2 & -\tilde{X}_3 \end{bmatrix}$, are the same as for $\tilde{\mathbf{X}}$.

$\tilde{X}_0$	$\tilde{X}_1$	$\tilde{X}_2$	$\tilde{X}_3$	PAPR (dB)
1	1	1	1	6.0
1	1	1	-1	2.3
1	1	-1	1	2.3
1	1	-1	-1	3.7
1	-1	1	1	2.3
1	-1	1	-1	6.0
1	-1	-1	1	3.7
1	-1	-1	-1	2.3

Inserting $|x_k|^2$ from (10.120) into (10.117), one gets a PAPR value of 3.7 dB, as also shown in Table 10.2. Table 10.2 shows that four data blocks result in a 6 dB PAPR, another four data blocks lead to a PAPR of 3.7 dB and eight data block have a PAPR of 2.3 dB. One may thus conclude that rate-¾ block coding of 3-bit data-words by appropriately chosing the parity bit, as shown in Table 10.2, limits the PAPR to 2.3 dB; hence an 3.7 dB improvement compared to an uncoded system. For example, for the 3-bit data block (1 1 1), we choose the parity bit -1 to obtain (1 1 1 -1), which has a PAPR $= 2.3$ dB. However, this approach requires an exhaustive search of the best codewords that yield the lowest PAPR and large look-up tables for encoding and decoding, especially for large values of N. Moreover, the issue of error correction is not addressed.

In *selective mapping* (SLM), a set of alternative OFDM signals are generated and the one with lowest PAPR is selected. As shown in Figure 10.45, N subcarriers of an OFDM symbol are multiplied by U distinct fixed vectors $\mathbf{B}^{(u)}$, $u = 1, 2, \ldots, U$ in the frequency domain,

$$\tilde{\mathbf{X}}^{(u)} = \mathbf{B}^{(u)}\tilde{\mathbf{X}} = \begin{bmatrix} \tilde{X}_0^{(u)} & \tilde{X}_1^{(u)} & \cdots & \tilde{X}_{N-1}^{(u)} \end{bmatrix}, \quad u = 1, 2, \cdots, U$$
$$(10.121)$$

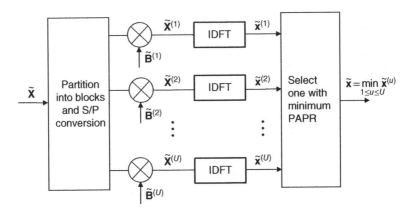

Figure 10.45 A Block Diagram of the SLM Technique. [23][24][25]

where $\tilde{X}_i^{(u)} = B_i^{(u)} \tilde{X}_i$. The resulting U OFDM symbols are transformed into the time domain by IDFT and the one with the lowest PAPR is transmitted together with the side information (SI) about the chosen fixed vector. This method is very effective in reducing the PAPR but has high computational complexity.

Example 10.14 Selective Mapping Technique. [23]

We will demonstrate the basic principles of the SLM technique in an OFDM system with $N = 8$ subcarriers. The data block to be transmitted is assumed to be

$$\tilde{X} = [1 \; -1 \; 1 \; 1 \; 1 \; -1 \; 1 \; -1] \quad (10.122)$$

We select the following four phase sequences as follows:

$$\mathbf{B}^{(1)} = [1 \; 1 \; 1 \; 1 \; 1 \; 1 \; 1 \; 1]$$
$$\mathbf{B}^{(2)} = [1 \; -1 \; 1 \; 1 \; 1 \; 1 \; -1 \; -1]$$
$$\mathbf{B}^{(3)} = [1 \; -1 \; 1 \; -1 \; 1 \; 1 \; -1 \; -1]$$
$$\mathbf{B}^{(4)} = [-1 \; -1 \; 1 \; 1 \; 1 \; 1 \; -1 \; -1]$$
$$\quad (10.123)$$

The modified data $\tilde{X}^{(u)}$, $u = 1, 2, \ldots, 4$ are obtained by component-wise multiplication of $\tilde{X}$ and $\mathbf{B}^{(u)}$ as shown in Figure 10.45. Using an oversampling ratio of $L = 4$, the PAPR values corresponding to $\tilde{X}^{(1)}$, $\tilde{X}^{(2)}$, $\tilde{X}^{(3)}$, $\tilde{X}^{(4)}$ were computed to be 5.34 dB, 3.01 dB, 2.18 dB and 3.73 dB, respectively. Hence, $\tilde{X}^{(3)}$ is selected for transmission. Noting that $\tilde{X}^{(1)}$ represents the original data, the PAPR is reduced from 5.34 dB to 2.18 dB, hence a PAPR reduction of 3.2 dB at the cost of 4 IDFT operations and the transmission of 2 bits as SI for informing the receiver about the chosen phase sequence.

In the method called as *partial-transmit sequences*, an input block of N symbols is partioned into U non-overlapping arbitrary sub-blocks (see Figure 10.46). Total number of subcarriers included in any of these U sub-blocks may be arbitrary. N-point IDFT is applied to each sub-block by zero-padding subcarriers which are not within the sub-block. Using the linearity of IDFT operation, U resulting OFDM signals could be added to each other to form the desired OFDM signal. However, before adding, subcarriers in each partial transmit sequence are weighted by a complex phase factor with unity amplitude, for example, $\Psi_u \in \{1, e^{j\pi/2}, e^{j\pi}, e^{j3\pi/2}\}$. U phase factors are chosen so as to minimize the PAPR.

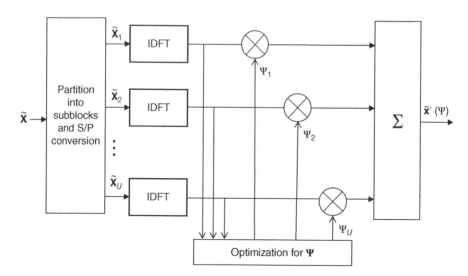

Figure 10.46 A Block Diagram of PTS Technique. [23][25]

For example, consider the following sequence for an OFDM system with $N = 8$ subcarriers:

$$\tilde{\mathbf{X}} = [1 \;\; -1 \;\; 1 \;\; -1 \;\; 1 \;\; -1 \;\; -1 \;\; -1]$$

(10.124)

which is partioned into $U = 4$ sub-blocks as shown in Figure 10.47. Taking the IFFT of each of the sub-blocks and taking a weighted sum implies

$$\tilde{\mathbf{x}}'(\mathbf{\Psi}) = \sum_{u=1}^{U} \Psi_u \, \tilde{\mathbf{x}}_u$$

(10.125)

where

$$\mathbf{\Psi} = [\Psi_1 \;\; \Psi_2 \;\; \cdots \;\; \Psi_U]$$

(10.126)

If we assign only the phases of 0 and π, that is, $\Psi_u = \pm 1$, then the vector $\mathbf{\Psi}$ has 16 combinations. However, it is sufficient to consider only eight combinations shown in Table 10.3 since the other eight combinations are simply the complements of the considered ones and have the same PAPR values. The PAPR values determined using (10.125) are tabulated for the corresponding phase factors. The phase factor $\mathbf{\Psi}_8 = (1 -1 -1 -1)$ was observed to yield the lowest PAPR value of 2.18 dB, which is 3.15 dB lower than the 5.33 dB of the original sequence. The cost is 4 IDFT operation and 3 bit side information to indicate the preferred phase factor between the $2^4 = 16$ possible combinations and thus enable the receiver to estimate $\tilde{\mathbf{X}}$. In addition to phase rotation, cyclic shifting, time reversing and conjugation may also be used to increase the number of alternatives. This technique has relatively low overhead and creates no ICI.

Perhaps the most intuitive idea for PAPR reduction is to *clip* the subcarrier amplitudes which exceed a preset threshold level λ before transmission. The clipped output signal may therefore be written as

Table 10.3 PAPR Values By PTS For an BPSK/OFDM Signal with $N = 8$ and $L = 4$.

$\tilde{\mathbf{X}}$	1 −1	1 −1	1 −1	−1 −1	PAPR (dB)
Ψ_1	1	1	1	1	5.33
Ψ_2	1	1	1	−1	6.25
Ψ_3	1	1	−1	1	4.65
Ψ_4	1	1	−1	−1	2.78
Ψ_5	1	−1	1	1	6.65
Ψ_6	1	−1	1	−1	5.58
Ψ_7	1	−1	−1	1	4.65
Ψ_8	1	−1	−1	−1	2.18

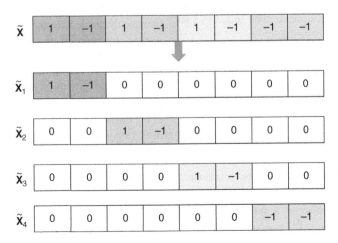

Figure 10.47 An Example of Adjacent Subblock Partioning in PTS with $L = 4$. [23]

$$\tilde{x}_c(\lambda) = \begin{cases} \tilde{x} & |\tilde{x}| \leq \lambda \\ \lambda e^{j\phi(\tilde{x})} & |\tilde{x}| > \lambda \end{cases} \qquad (10.127)$$

Clipping with a rectangular window widens the signal spectrum; this results in out-of-band emissions and ICI (hence increased bit error probability). Filtering after clipping (using better shaped window functions) decreases the out-of-band emissions but may also cause some peak regrowth. Out-of-band emissions may be decreased if all peaks exceeding a certain threshold level are cancelled by the subtraction of appropriately scaled and time-shifted *reference signals* (RS), for example, a sinc function multiplied by a raised cosine window. The RS is subtracted from the OFDM signal at the transmitter after cyclic prefix insertion, windowing and P/S conversion. The SI about the RS are transmitted so that the receiver can recover the original OFDM signal. As in clipping, the disadvantage of this method is the increased error probability. The more peaks are cancelled, the more the original signal is distorted.

In the *tone reservation method*, a data-block dependent time-domain correction signal $\tilde{c}$ is added to the original signal $\tilde{x}$ at the transmitter. The added correction signal, which is removed at the receiver, should be smooth so as not to cause out-of-band interference. A small subset of subcarriers are reserved for transmitting the correction signal. The objective is then to determine a time domain signal $\tilde{c}$ to be added to $\tilde{x}$ so that the following is satisfied:

$$\tilde{x} + \tilde{c} = IFFT\left(\tilde{X} + \tilde{C}\right) \qquad (10.128)$$

Since different subcarriers are used for sending $\tilde{X}$ and $\tilde{C}$, the time domain signal $\tilde{c}$ is related to the frequency-domain signal $\tilde{C}$ by IFFT. The optimal value of the vector $\tilde{C}$ is found by linear programming. In wireline systems, subcarriers with low SNRs are used for sending $\tilde{C}$ while in wireless systems reserve a number of subcarriers for sending $\tilde{C}$, regardless of their

SNRs. Though very efficient, this method requires substantial amount of computation for determining the correction symbols. The correction signal increases the bit error probability since it acts as a noise and wastes the transmit power. [26][27]

In the *tone injection* method, the location of a symbol in the original constellation is mapped into one of several equivalent points in an expanded constellation. These extra degrees of freedom is exploited for PAPR reduction. Hence, the original constellation $\tilde{X}_n$ of n-th subcarrier is shifted in in-phase and quadrature axes as $\tilde{X}_n + pD + jqD$ where p and q are integers and D is a positive real number known by the receiver. In other terms, a tone of magnitude $pD + jqD$ is injected at the subcarrier frequency used for sending $\tilde{X}_n$. For example, for M-QAM with minimum Euclidean distance d between the constellation points, the real and imaginary parts of $\tilde{X}_n$ can take values of $\{\pm d/2, \pm 3d/2, \ldots, \pm \sqrt{M}d/2\}$. If the expansion of the original constellation can be estimated correctly at the receiver, then the expanded constellation point can be used to reduce the PAPR. In order to maintain the minimum Euclidean distance d between the symbols of the extended constellation, D must satisfy $D = \rho d\sqrt{M}, \rho \geq 1$. [23] If a signal peak is detected at the IFFT output, the constellations of the input symbols with opposite phase to the signal peak are expanded in order to compensate the peak. The algorithm used for determining the appropriate expanded constellation points requires iterative computations and additional IFFT operations. However, at the receiver side, only the modulo-D shifts are performed on the received symbols until they fall in the decision regions of the original constellation. Depending on the value of ρ, this method reduces the PAPR by increasing the average transmit power instead of decreasing the peak power.

The method of *active constellation* extension is a special case of the tone injection method where only the outer constellation points are

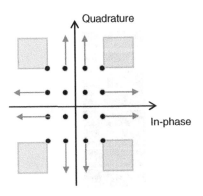

Figure 10.48 The Active Constellation Extension (ACE) Technique For 16-QAM.

expanded towards the outside of the original constellation. This method is specifically used for expanding the signal positions of M-ary QAM constellations in one or two dimensions depending on their position on the constellation (see Figure 10.48). Consequently, the Euclidean distance between the constellation points will increase leading to lower bit error probability at the expense of higher average transmit power. This technique is more meaningful for large constellation sizes having larger numbers of outer constellation points.

10.2.7 Multiple Access in OFDM Systems

Orthogonal frequency division multiple access (OFDMA) is a multiple access scheme based on OFDM where multiple users share the OFDM resources in mainly two ways. In the first approach, only a single user is allowed to transmit using the assigned time slots (comprising one or more symbol durations) using all OFDM subcarriers dedicated for data transmission. In this TDMA scheme, users access the OFDM system by time sharing as shown in Figure 10.49(a). In the second approach shown in Figure 10.49 (b), the available time-frequency resource is divided in basic units called resource blocks (RBs). Users are allowed to transmit simultaneously over non-overlapping time-

frequency RBs assigned to them. In this approach, the assignment of subcarriers to multiple users may be either localized or distributed, as shown Figure 10.50. Assuming that all terminals transmit L symbols per block, the system can handle $Q = N/L$ simultaneous transmissions by Q users. In the localized mode, L subcarriers assigned to a single user are contiguous and the rest of the subcarriers are loaded with zeros. In the distributed mode, L subcarriers assigned to each user are interleaved with a separation of Q subcarriers between them. Unused subcarriers by the user of interest are loaded with zeros so that they can be used by other users.

The 3GPP LTE downlink resource block (RB) consists of 12 subcarriers with a frequency separation of $\Delta f = 15\ kHz$, hence occupying a $12\Delta f = 180\ kHz$ bandwidth, and a transmit time interval (TTI) of 7 symbol duration, that is, $7T = 7/\Delta f \cong 0.5 ms$. A user may be assigned multiple RBs depending on its requirements and the availability of the OFDM resources. The modulation format and hence the transmission rate is adopted so as to match the current channel conditions for each user terminal. Allocation of the resource blocks to multiple users at a given time interval is determined by the adopted scheduling algorithm. [28]

OFDMA downlinks allow for the operation of low-performance receivers, the frequency-domain scheduling, the use of high order M-QAM and improved spectral efficiency and capacity. Orthogonality of the subcarriers can be maintained and sophisticated signal processing operations can be performed at the downlink transmitters in order to reduce ICI and PAPR.

10.2.7.1 Uplink Design and SC-FDMA

Frequency offset between the clocks of the uncoordinated mobile user terminals, that transmit simultaneously, destroys orthogonality between the subcarriers in the OFDM uplinks. On the other hand, high PAPR quickly drains the battery of mobile terminals. As a remedy, single carrier frequency divison

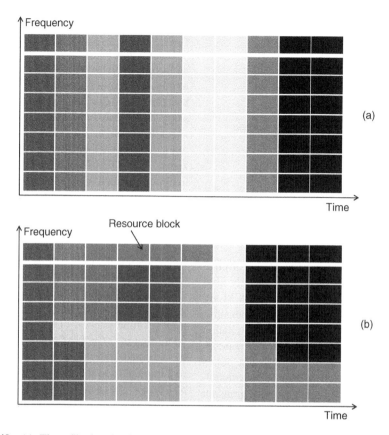

Figure 10.49   (a) Time Sharing in OFDMA, and (b) Resource Blocks in OFDMA with Localized Subcarriers.

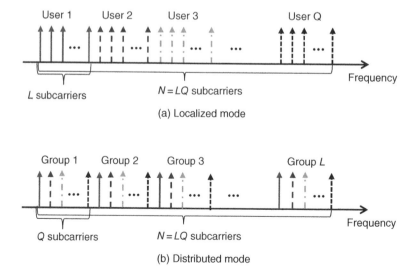

Figure 10.50   Allocation of Subcarriers to Users in Distributed and Localized Modes.

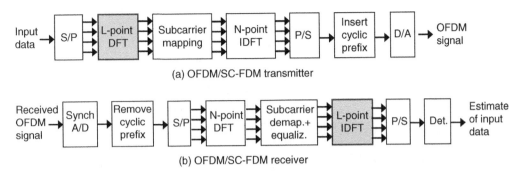

(a) OFDM/SC-FDM transmitter

(b) OFDM/SC-FDM receiver

Figure 10.51 Transmitter and Receiver Block Diagrams of OFDMA and SC-FDMA Systems. In OFDMA systems L-point FFT and IFFT does not exist. [11]

multiple access (SC-FDMA), which is a precoded version of OFDMA, is used in OFDM uplinks mainly because of its improved PAPR performance, hence lower battery-power consumption. SC-FDMA is accepted as access scheme for the uplinks in the 3GPP LTE system. Otherwise, OFDMA and SC-FDMA both use orthogonal subcarriers, and have similar performances as far as throughput, multipath mitigation and low-complexity equalization are concerned.

Figure 10.51 shows the block diagram of a SC-FDMA in comparison with that of the OFDMA. The main difference between OFDMA and SC-FDMA is that the SC-FDMA performs a DFT prior to the IDFT operation. In SC-FDMA, the user data are first grouped into blocks of L symbols, which are spread by the L-point DFT over L subcarriers allocated to that user. Hence, L-point DFT of the input symbols $\tilde{x}_k$, $k = 0, 1, ..., L-1$ leads to L frequency-domain signals $\tilde{X}_\ell$, $\ell = 0, 1, ..., L-1$. Because of the L-point DFT operation, each of the $\{\tilde{X}_\ell\}_{\ell=0}^{L-1}$ depends on all $\{\tilde{x}_k\}_{k=0}^{L-1}$ instead of a single $\tilde{x}_k$. However in OFDMA, each subcarrier only carries information related to one specific symbol. This property, which leads to lower PAPR than OFDMA, makes SC-FDM attractive for uplink transmissions. The signals $\tilde{X}_\ell$, $\ell = 0, 1, ..., L-1$ of one of users modulate the selected $L < N$ of the SC-FDMA subcarriers. L subcarriers of a user can be mapped into N-subcarriers in either

localized or distributed mode, as shown in Figure 10.50. The resulting signal undergoes N-point IFFT as in OFDM. Thus, the signals of Q users, each using L non-overlapping subcarriers, in distributed or localized mode, arrive the uplink receiver as a composite OFDM signal.

Figure 10.52 shows the formation of a SC-FDMA signal for $N = 12$ subcarriers and $L = 4$ subcarriers per user. Each of the signals $\tilde{X}_0, \tilde{X}_1, \tilde{X}_2, \tilde{X}_3$ obtained by 4-point DFT carry information about each of the time domain signals $\tilde{x}_0, \tilde{x}_1, \tilde{x}_2, \tilde{x}_3$. The assignment of $L = 4$ signals to 12 subcarriers may be achieved either in distributed or localized mode as shown in Figure 10.52. Figure 10.53 shows PAPR performance results for SC-FDMA with $N = 8192$ subcarriers, of which only 6000 are used. [29] Each user is assumed to use at most 600 of the 6000 subcarriers. Assuming that a RB comprises 15 subcarriers, 40 RBs of the 400 available are assigned to a user. For localized allocation, SC-FDM has almost 2 dB better PAPR performance compared to OFDMA. However, the PAPR advantage reduces to 0.3 dB for the distributed allocation. Therefore, if the RBs are randomly distributed across the subcarriers, the advantage of SC-FDM with respect to OFDMA disappears. As for OFDMA, the distributed allocation exhibits approximately 1 dB higher PAPR than for the localized allocation, while for SC-FDMA it is 2.7 dB.

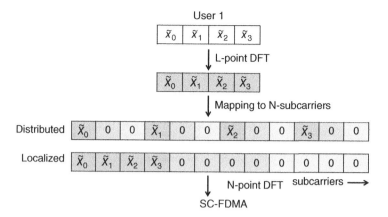

Figure 10.52 Distributed and Localized Mapping in SC-FDMA with $N = 12$ Subcarriers, $L = 4$ Subcarriers Per User and $Q = 3$ Users.

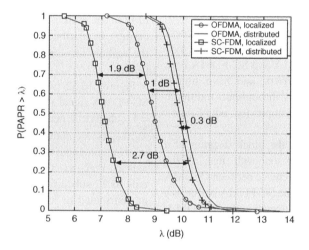

Figure 10.53 Comparison of the PAPR Performances of Localized and Distributed SC-FDMA with OFDMA for N = 8192, 16 QAM and 10% BW Usage. [29]

OFDMA and SC-FDMA may be compared based on two performance metrics, namely PAPR and throughput. Distributed SC-FDMA has higher PAPR than localized SC-FDMA and both of them perform better than OFDMA in both distributed and localized modes. Since configurations with the lowest PAPR tend to have lower throughput, a tradeoff should be sought to meet the specific needs. Nevertheless, distributed SC-FDMA is preferable when many users transmit at moderate data rates, whereas localized SC-FDMA performs better in a system with several high data rate users.

Since each subcarrier contains information about all symbols, distributed SC-FDMA provides frequency diversity in a frequency-selective channel. Even if some of the subcarriers suffer deep fade, the symbols can still be decoded using other subcarriers experiencing better channel conditions. The localized SC-FDMA, despite its low PAPR performance, is more sensitive to resource allocation since the subcarriers assigned to a single mobile terminal occupy only a small fraction of the system bandwidth. Therefore, it can potentially provide higher throughput in a frequency- selective channel by adaptive scheduling, that is, by assigning a user to subcarriers with higher channel gains. Localized SC-FDMA also reduces the need for linearity in power amplifiers of MS's at the cost of complex signal processing at the BS. [11]

In analogy with CDMA systems where the multiple access capability depends on the orthogonality between the codes, the multiple access capability of OFDMA is closely related to the orthogonality between the subcarriers. A BPSK/CDMA system with n = 400 supports 28 users at 10^{-6} bit error probability (see Example 10.3), when the E_b/N_0 of a single user is 17 dB. The required spread spectrum bandwidth is given by $W_{ss} = n R_b$ where R_b denotes the bit rate supported by each user. Now consider an BPSK/OFDMA downlink with $N = 400$ subcarriers and each of the subcarriers is assigned to a separate user. Assuming an OFDM symbol duration of $T = 1/R_b$, the OFDMA bandwidth will be $B = N \Delta f = N/T = NR_b$. The data rates supported by OFDM and CDMA systems in the same bandwidth are NR_b (N = 400) and KR_b (K=28), respectively. Hence, for the same transmission bandwidth $B = W_{ss}$, OFDMA seems to have higher capacity than CDMA. Note however that some of the subcarriers need to be used as pilots for channel estimation and equalization and the presence Doppler shift will have a negative impact on the OFDMA performance; SIR will then be decreased drastically calling for Doppler correction mechanisms and/or increase in the transmit power.

10.2.8 Vulnerability of OFDM Systems to Impulsive Channel

OFDM is well known for its advantages for broadband communications, such as simple equalization, ISI mitigation and higher data rates. However, OFDM systems are vulnerable to channels disturbed by impulsive noise. Because of the very high sampling rates in OFDM systems, the noise samples may be correlated and impulse noise might affect many subchannels at the same time.

Before studying the OFDM performance in impulsive channels, let us first consider Middleton's class A pdf for characterizing impulsive noise:

$$f(n) = \sum_{k=0}^{\infty} e^{-A} \frac{A^k}{k!} \frac{1}{\sqrt{2\pi}\sigma_k} e^{-n^2/(2\sigma_k^2)} \quad (10.129)$$

where

$$\sigma_k^2 = \sigma_G^2 \frac{k/A + \Gamma}{\Gamma} = \begin{cases} \sigma_G^2 & \sigma_I^2 \to 0 \\ k\dfrac{\sigma_I^2}{A} & \sigma_G^2 \to 0 \end{cases}.$$

$$\Gamma = \sigma_G^2/\sigma_I^2$$

$$(10.130)$$

Here Γ provides a measure of the impulsiveness of the channel. The variances of Gaussian and impulsive noise components are denoted by σ_G^2 and σ_I^2, respectively. Hence, the impulsiveness of the class A pdf can be controlled by the parameters Γ and A. Here, A denotes the number of impulsies observed in a given time interval (see (F.153)). Note that $\sigma_I^2 \to 0$ ($\Gamma \to \infty$) corresponds to as purely Gaussian noise channel, while $\sigma_G^2 \to 0$ ($\Gamma \to \infty$) represents a purely impulsive noise channel with variance equal to the product of the random number k of impulses (k is a specific value of the Poisson distributed random variable with parameter A) and the impulsive noise variance per impulse, σ_I^2/A (see Example F.17).

The probability of bit error for BPSK, observed by a subcarrier, in an impulsive noise channel may be written as the weighted sum of the error probabilities for the Poisson distributed number of impulses:

$$P_b(\gamma) = \sum_{k=0}^{\infty} e^{-A} \frac{A^k}{k!} Q\left(\sqrt{\frac{2\Gamma\gamma}{\Gamma + k/A}}\right) \quad (10.131)$$

where $\gamma = E_b/(2\sigma_G^2) = E_b/N_0$

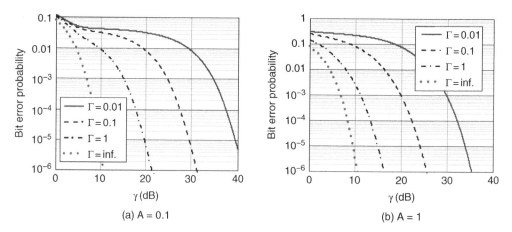

Figure 10.54 Bit Error Probability For BPSK in An Impulsive Noise Channel Characterized By class A Distribution.

$$\frac{2\Gamma\gamma}{\Gamma+k/A} = \begin{cases} E_b/\left(2\sigma_G^2\right) & \sigma_I^2 \to 0 \\ E_b/\left(k\sigma_I^2/A\right) & \sigma_G^2 \to 0 \end{cases} \quad (10.132)$$

Figure 10.54 shows the bit error probability for BPSK in an impulsive noise channel described by (10.129) for $\Gamma = 0.1$ and $\Gamma = 1$. Note the worsening bit error probability performance for $A = 0.1$ since the channel is more impulsive than for $A = 1$. The channel behaves as a AWGN channel as Γ approaches infinity. Therefore, Figure 10.54 may also be used for determining the degradation caused by the channel as its impulsive behavior becomes more pronounced. For $A = 0.1$ and a bit error probability level of 10^{-6}, the required SNR levels are 10.5 dB, 21.3 dB, 31.1 and 41.1 dB for Γ = infinity, 1, 0.1 and 0.01, respectively. This implies a performance degradation of 10.8 dB, 21.5 dB and 31.5 dB, respectively. Hence, there is a strong requirement for FEC coding. Also note that the impulsive noise degrades the performance of multiple subcarriers simultaneously, because an impulsive signal has a very wide spectrum.

10.2.9 Adaptive Modulation and Coding in OFDM

Until now we assumed that each subcarrier has the same modulation and hence carries the same number of bits. For example, in using Q-PSK modulation, two bits are assigned to each subcarrier, irrespective of whether the SNR of that subcarrier is sufficiently high to provide the desired bit error probability. If the subcarrier SNR is not sufficiently high to carry two bits, then these two bits would be decoded with high bit errors. On the other hand, the SNR of some of the subcarriers may be higher so that they can support more than two bits. In that case, the power of those subcarriers will be wasted by transmitting only two bits with much lower error probabilities. Therefore, a bit-loading algorithm may be used to assign differing number of bits to each subcarrier depending on their channel gain in order to increase the throughput and to ensure data transmission at an acceptable level of error probability. The achievable throughput will evidently depend on the chosen threshold level for the error probability.

Design of a *bit loading algorithm* requires that the transmitter knows the channel state information (CSI) about all subcarriers. Since uplink and downlink channels operate at the same frequency in the time-division duplex (TDD) mode, these algorithms work better for systems operating in TDD mode in slowly-fading channels. In the FDD mode of operation, downlink and uplink frequencies are different. Therefore, the CSI estimated with sufficient accuracy by the receiver must be rescaled in frequency and send to the transmitter with shortest delay so that it does not become out-dated when it is received by the transmitter.

The *adaptive modulation and coding* schemes take into consideration of the maximum and minimum values of the SNR that can be observed in the channel. The SNR interval is chosen to correspond the preset minimum and maximum error probabilities for a given modulation. If the SNR falls below a minimum threshold level, then either no bits is assigned to that subcarrier, or the power level of that subcarrier may be raised so as to support BPSK, the modulation requiring the lowest SNR. The size and the location of the training sequences for channel equalization should be carefully chosen depending on the channel coherence time and coherence bandwidth. For slowly fading channels, the update rate may be slower. On the other hand, smaller packet sizes are preferred in channels with higher Doppler shifts (shorter coherence time). In fast-changing channels, the update rate of the bit-loading (adaptive modulation) algorithm may not be able to follow the changes in the channel, leading to decrease in the spectral efficiency and increase in the error probability. Combined use of adaptive modulation and FEC coding or hybrid-ARQ leads to higher spectral efficiencies. However, such approaches require the transmission of some considerable CSI, as a function of the up-date rate, about the chosen modulation and coding for each or a group of subcarriers. [30] This topic will be treated in the next chapter in more detail.

Appendix 10A Frequency Domain Analysis of DSSS Signals

We consider the block diagram of a spread-spectrum system for BPSK modulation shown in Figure 10.7. Assuming that the channel dominated by a sinusoidal jamming/interference signal at carrier frequency f_c and power J, the signal at the input of a spread spectrum receiver may be written as

$$x(t) = s(t) + \sqrt{2J}\cos(w_c t + \varphi')$$
$$s(t) = \sqrt{2P}\, b(t)\, c(t)\cos(w_c t + \varphi) \tag{10.133}$$

where $P = E_b/T_b$ denotes the power the desired (user) signal, $b(t)$ is the user signal of duration T_b and $c(t)$ denotes the spreading code with chip duration $T_c = T_b/n$. Here, $n = T_b/T_c >> 1$ denotes the number of chips per bit. The phases φ and φ' denote channel-induced phase shifts suffered by user and jamming signals, respectively. Since locations of the user transmitter and the jammer are not the same, these random phases differ from each other.

Using (10.133), the autocorrelation function of $x(t)$ may be written as

$$R_x(\tau) = R_s(\tau) + J\,\cos(w_c\tau)$$
$$R_s(\tau) = P\,R_b(\tau)\,R_c(\tau)\cos(w_c\tau) \tag{10.134}$$

The PSD corresponding to $R_x(\tau)$ is given by

$$G_x(f) = G_s(f) + G_J(f) \tag{10.135}$$

where $G_J(f)$ denotes the PSD of the jammer signal:

$$J\,\cos(w_c\tau)\;\Leftrightarrow\;G_J(f) = \frac{J}{2}[\delta(f - f_c) + \delta(f + f_c)] \tag{10.136}$$

Since the multiplication of signals in the time domain leads to the convolution of their Fourier transforms in the frequency domain, $G_s(f)$ is given by the convolution of the PSDs of $b(t)$, $c(t)$ and $\cos w_c t$. PSD's of $b(t)$ and $c(t)$, assuming that they are rectangular NRZ pulses, are given by (1.104)

$$G_b(f) = T_b\,\text{sinc}^2(fT_b)$$
$$G_c(f) = T_c\,\text{sinc}^2(fT_c) \tag{10.137}$$

Using (10.134), (10.136) and (10.137), we find the PSD of $s(t)$ as follows: [31]

$$G_s(f) = P G_c(f) \otimes G_b(f) \otimes \frac{1}{2}[\delta(f-f_c) + \delta(f+f_c)]$$

$$= \frac{PT_b}{2} G_c(f) \otimes \left[\sin c^2\{(f-f_c)T_b\} + \sin c^2\{(f+f_c)T_b\}\right]$$

$$= \frac{PT_bT_c}{2} \int_{-\infty}^{\infty} \sin c^2\{(f'-f_c)T_b\} \underbrace{\sin c^2\{(f-f')T_c\}}_{\approx \sin c^2\{(f-f_c)T_c\}} df'$$

$$+ \frac{PT_bT_c}{2} \int_{-\infty}^{\infty} \sin c^2\{(f'+f_c)T_b\} \underbrace{\sin c^2\{(f-f')T_c\}}_{\approx \sin c^2\{(f+f_c)T_c\}} df'$$

$$= \frac{PT_bT_c}{2} \sin c^2\{(f-f_c)T_c\} \underbrace{\int_{-\infty}^{\infty} \sin c^2\{(f'-f_c)T_b\} df'}_{=1/T_b}$$

$$+ \frac{PT_bT_c}{2} \sin c^2\{(f+f_c)T_c\} \underbrace{\int_{-\infty}^{\infty} \sin c^2\{(f'+f_c)T_b\} df'}_{=1/T_b}$$

$$= \frac{PT_c}{2}\left[\sin c^2\{(f-f_c)T_c\} + \sin c^2\{(f+f_c)T_c\}\right]$$

$$\tag{10.138}$$

where (D.41) is used for the evaluation of the integrals. As the message signal $b(t)$ is narrowband and the PN code $c(t)$ is wideband, the PSD of $b(t)c(t)$ resembles to that of $c(t)$. Inserting (10.136) and (10.138) into (10.135), one gets the PSD of the signal at the input of the DSSS receiver:

$$G_x(f) = \frac{PT_c}{2}\left[\text{sinc}^2\{(f-f_c)T_c\}\right.$$
$$\left. + \text{sinc}^2\{(f+f_c)T_c\}\right] + \frac{J}{2}[\delta(f-f_c) + \delta(f+f_c)]$$

$$\tag{10.139}$$

The DSSS receiver correlates the received signal with an exact replica of the code $c(t)$ and locally generated carrier, followed by an integrator and a decision device. The correlator is supplied with locally generated PN code $c(t)$, that is desired to be in perfect synchronism with the code generated at the transmitter. The output of the despreading mixer is given by (see Figure 10.7)

$$z(t) = x(t)c(t)\sqrt{2}\,\cos(w_ct + \hat{\varphi})$$

$$= 2\sqrt{P}\,b(t)\underbrace{c(t)c(t)}_{=1}\,\cos(w_ct + \varphi)\,\cos(w_ct + \hat{\varphi})$$

$$+ 2\sqrt{J}\,c(t)\cos(w_ct + \varphi')\,\cos(w_ct + \hat{\varphi})$$

$$= \sqrt{P}\,b(t)\,[\cos(\varphi - \hat{\varphi}) + \cos(2w_ct + \varphi + \hat{\varphi})]$$

$$+ \sqrt{J}\,c(t)[\cos(\varphi' - \hat{\varphi}) + \cos(2w_ct + \varphi' +)\hat{\varphi}]$$

$$\tag{10.140}$$

where $\hat{\varphi}$ denotes the estimation of φ by the receiver. Integration of the mixer output at DSSS receiver, shown in Figure 10.7, filters out the frequency components at $f = 2f_c$. Therefore, we ignore the high frequency terms, to determine the autocorrelation function and the PSD of (10.140)

$$R_z(\tau) = P\,R_b(\tau)\,E\left[\cos^2(\varphi - \hat{\varphi})\right]$$
$$+ J\,R_c(\tau)\,E\left[\cos^2(\varphi' - \hat{\varphi})\right]$$

$$G_z(f) = P\,T_b\,\text{sinc}^2(fT_b)\,E\left[\cos^2(\varphi - \hat{\varphi})\right]$$
$$+ J T_c\,\text{sinc}^2(fT_c)\,E\left[\cos^2(\varphi' - \hat{\varphi})\right]$$

$$\tag{10.141}$$

If $z(t)$ is passed through an ideal IF filter of bandwidth $1/T_b \ll 1/T_c$, the effective received power is given by

$$P_{\text{out}} = \int_0^{1/T_b} G_z(f)\,df$$

$$= PT_b\,E\left[\cos^2(\varphi - \hat{\varphi})\right]\int_0^{1/T_b} \text{sinc}^2(fT_b)\,df$$

$$+ JT_c\,E\left[\cos^2(\varphi' - \hat{\varphi})\right]\underbrace{\int_0^{1/T_c} \text{sinc}^2(fT_c)}_{\approx 1}\,df$$

$$= P\,E\left[\cos^2(\varphi - \hat{\varphi})\right]\int_0^1 \text{sinc}^2 x\,dx$$

$$+ \frac{J}{n}E\left[\cos^2(\varphi' - \hat{\varphi})\right]$$

$$\approx P + \frac{J}{G_p}$$

$$\tag{10.142}$$

where $\text{sinc}^2(fT_c) \approx 1$ for $0 \le f \le 1/T_b$, the PSD of the jammer signal can be assumed to be constant over the bandwidth of the receiver. In a coherent DSSS receiver, phase recovery circuit of the receiver is assumed to accurately estimate the channel phase, hence, $E[\cos^2(\varphi - \hat{\varphi})] = 1$. The integral of $\text{sinc}^2(x)$ over its first lobe is approximately equal to 0.897.

One may observe from (10.142) that the jammer's contribution is decreased by the so-called processing gain G_p, which is defined as

$$G_p = \frac{n}{E[\cos^2(\varphi' - \hat{\varphi})]} = \begin{cases} n & \hat{\varphi} = \varphi' \\ 2n & otherwise \end{cases}$$

(10.143)

Since a jammer is unable to accurately estimate the channel-induced phase, the expected value of the phase term is equal to $E[\cos^2(\varphi' - \hat{\varphi})] = 1/2$; then, the processing gain is $G_p = 2n$. PSD of the jammer signal over the SS bandwidth is found from (10.141) as

$$N_J = JT_c\, E\left[\cos^2(\varphi' - \hat{\varphi})\right]$$

$$= \begin{cases} JT_c & \hat{\varphi} = \varphi' \\ JT_c/2 & otherwise \end{cases} \quad 0 \le f \le 1/T_c$$

(10.144)

If a jammer can estimate the phase of the user carrier correctly, then it can match its carrier phase to that of the user to increase its effectiveness by 3 dB. Then, the jammer PSD would be increased by 3 dB. Signal-to-noise power ratio at the DSSS receiver output is given (10.142):

$$SNR = G_p \frac{P}{J} = G_p\, SNR_{input}$$

(10.145)

where $SNR_{input} = P/J$ denotes the SNR at the input of the DSSS receiver. Hence, a DSSS receiver reduces the effectiveness of a jammer signal by a factor of G_p. In other words, the SNR performance of a DSSS system is enhanced by the processing gain G_p.

Appendix 10B Time Domain Analysis of DSSS Signals

The signal at the input of a DSSS receiver is given by

$$x(t) = s(t) + j(t)$$

$$s(t) = \sqrt{2P}\, b(t)c(t)\, \cos(w_c t + \varphi) \quad (10.146)$$

$$j(t) = \sqrt{2J}\cos(w_c t + \varphi')$$

The interference signal is assumed to be a tone of frequency f_c with mean power J and phase θ_J. To recover the original message signal $b(t)$, the received signal $r(t)$ is first down-converted using the local oscillator of carrier frequency f_c and a phase offset θ with the incoming carrier. Then it is correlated with the locally generated code followed by an integrator, and a detector. The correlator is supplied with a locally generated PN code that is desired to be in perfect synchronism with the code used in the transmitter. The signal and interference components at the integrator input may be written as

$$v_1(T_b) = \int_0^{T_b} x(t)c(t)\, \sqrt{\frac{2}{T_b}}\cos(w_c t + \hat{\varphi})\, dt$$

$$= x_1(T_b) + w_1(T_b)$$

(10.147)

where, in contrast with (10.10) and Figure 10.7, the locally generated carrier is assumed to have unit energy for the sake of convenience. The signal component at the integrator output is found to be

$$x_1(T_b) = \sqrt{\frac{2}{T_b}} \int_0^{T_b} s(t)c(t)\, \cos(w_c t + \hat{\varphi})\, dt$$

$$= 2\sqrt{\frac{P}{T_b}}b(t)$$

$$\underbrace{\int_0^{T_b} c(t)c(t)}_{=1}\, \cos(w_c t + \varphi)\cos(w_c t + \hat{\varphi})\, dt$$

$$= \sqrt{E_b}\, b(t)\cos(\varphi - \hat{\varphi})$$

(10.148)

In case where the PN code generated at the receiver is not perfectly synchronized but has a time offset τ with the received code, then (10.2) will also appear in (10.148). Here, we assumed BPSK modulation with $b(t) = 1$ for binary 1 and -1 for binary 0. The interference term $w_1(T_b)$ at the detector input has a zero-mean:

$$w_1(T_b) = \sqrt{\frac{2}{T_b}} \int_0^{T_b} j(t)c(t)\,\cos(w_c t + \hat{\varphi})dt \tag{10.149}$$

In order to determine its variance, we express the jammer waveform $j(t)$ in terms of basis functions Φ_k and $\tilde{\Phi}_k$, which are orthonormal over each chip duration:

$$
\begin{aligned}
j(t) &= \sqrt{2J}\cos(w_c t + \varphi') \\
&= \sqrt{2J}\left[\cos\varphi'\cos(w_c t) - \sin\varphi'\sin(w_c t)\right] \\
&= \sqrt{JT_c}\left[\cos\varphi'\sum_{k=1}^{n-1}\Phi_k - \sin\varphi'\sum_{k=1}^{n-1}\tilde{\Phi}_k\right] \\
&= \sum_{k=1}^{n-1} j_k\Phi_k - \sum_{k=1}^{n-1}\tilde{j}_k\tilde{\Phi}_k
\end{aligned}
\tag{10.150}
$$

where

$$
\Phi_k = \sqrt{\frac{2}{T_c}}\cos(w_c t), \quad kT_c \le t \le (k+1)T_c
$$

$$
\tilde{\Phi}_k = \sqrt{\frac{2}{T_c}}\sin(w_c t), \quad kT_c \le t \le (k+1)T_c
\tag{10.151}
$$

$$
j_k = \sqrt{JT_c}\cos\varphi'
\tag{10.152}
$$

$$
\tilde{j}_k = \sqrt{JT_c}\sin\varphi'
$$

Using (10.150)–(10.152), one can rewrite the interference term given by (10.149) as follows:

$$
\begin{aligned}
w_1(T_b) &= \sqrt{\frac{2}{T_b}}\int_0^{T_b} j(t)c(t)\,\cos(w_c t + \hat{\varphi})dt \\
&= \sqrt{\frac{2}{T_b}}\sum_{k=0}^{n-1} c_k \int_{kT_c}^{(k+1)T_c} j(t)\cos(w_c t + \hat{\varphi})dt \\
&= \sqrt{\frac{1}{n}}\sum_{k=0}^{n-1} c_k \left[\cos\hat{\varphi}\underbrace{\int_{kT_c}^{(k+1)T_c} j(t)\,\Phi_k(t)\,dt}_{j_k} - \sin\hat{\varphi}\underbrace{\int_{kT_c}^{(k+1)T_c} j(t)\,\tilde{\Phi}_k(t)\,dt}_{\tilde{j}_k}\right] \\
&= \sqrt{\frac{1}{n}}\sum_{k=0}^{n-1} c_k \left[\cos\hat{\varphi}\,j_k - \sin\hat{\varphi}\,\tilde{j}_k\right] \\
&= \sqrt{\frac{JT_c}{n}}\cos(\hat{\varphi}+\varphi')\sum_{k=0}^{n-1} c_k
\end{aligned}
\tag{10.153}
$$

$$
\sigma_{w_1}^2 = E\left[|w(T_b)|^2\right]
$$

$$
= \frac{JT_c}{n}\underbrace{E\left[\cos^2(\hat{\varphi}+\varphi')\right]}_{1/2}\underbrace{\sum_{k=0}^{n-1}|c_k|^2}_{n} = \frac{JT_c}{2}
$$

If the jammer can estimate the phase of the user carrier correctly, then it can match its carrier phase to that of the user, that is, $\varphi' = -\hat{\varphi}$, to increase the jamming effectiveness; then one would have $E[\cos^2(\hat{\varphi}+\varphi')] = 1$. However, this is highly unlikely.

The integration in the bit interval $0 \le t \le T_b$ provides the sampled output $v_1(T_b)$. This output is the same as that of the matched filter for a coherent BPSK signal, where the single-sided PSD of the noise term is given by N_J. Assuming perfect synchronization, and $b(t) = +1$ for bit 1 and $b(t) = -1$ for bit 0, the decision variable $v_I(T_b)$ is given by (10.147), (10.148) and (10.153):

$$
v_1(T_b) = \begin{cases}
+\sqrt{E_b}\,\cos(\varphi-\hat{\varphi}) + w_1(T_b) & 1 \\
-\sqrt{E_b}\,\cos(\varphi-\hat{\varphi}) + w_1(T_b) & 0
\end{cases}
\tag{10.154}
$$

where $w_1(T_b)$ has zero mean and a single-sided PSD N_J. Assuming perfect synchronization, the average probability of error for large values of the processing gain (for AWGN assumption) is found from (10.154) as

$$
\begin{aligned}
P_e &= \frac{1}{2}P\left(-\sqrt{E_b} + w_1(T_b) > 0 | 0\right) \\
&\quad + \frac{1}{2}P\left(\sqrt{E_b} + w_1(T_b) < 0 | 1\right) \\
&= Q\left(\sqrt{\frac{2E_b}{N_J}}\right)
\end{aligned}
\tag{10.155}
$$

Appendix 10C SIR in OFDM systems

The channel transfer function is defined by (10.97) as

$$
H_{n,k} = \sum_{\ell=0}^{L-1} h_{k,\ell}\,e^{-j\frac{2\pi n\ell}{N}} = \sum_{\ell=0}^{L-1} \alpha_{k,\ell}\,e^{-j\theta_{k,\ell}}\,e^{-j\frac{2\pi n\ell}{N}}
\tag{10.156}
$$

where $\alpha_{k,\ell}$ and $\theta_{k,\ell}$ denote respectively the channel gain and channel phase of the ℓ-th multipath component at time t_k. The channel phase is assumed to be induced only by the Doppler shift in the channel:

$$
\theta_{k,\ell} = 2\pi\frac{v_\ell}{\lambda}\cos\phi_\ell t_k = 2\pi f_{d,\max,\ell}\cos\phi_\ell\, t_k
\tag{10.157}
$$

Here v_ℓ shows the relative velocity of the terminal and the angle ϕ_ℓ denotes the angle of incidence of the scattered signal arriving to the receiver. The maximum Doppler shift for the ℓ-th multipath component is given by $f_{d,\max,\ell} = v_\ell/\lambda$. When the multipath components undergo the same Doppler shift due to temporal changes in the channel, the autocorrelation function of the channel impulse response is assumed to the same for all multipath components:

$$E\left[h_{k,\ell} h^*_{k+p,\ell}\right] = E\left[\alpha_{k,\ell}\alpha^*_{k+p,\ell}\right] E\left[e^{-j\theta_{k,\ell}} e^{j\theta_{k+p,\ell}}\right]$$

$$= R_\alpha\left(\frac{pT}{N}\right) J_0\left(\frac{\pi\beta p}{N}\right)$$

$$E\left[e^{-j\theta_{k,\ell}} e^{j\theta_{k+p,\ell}}\right] = E\left[e^{j2\pi f_{d,\max,\cos\phi}\left(t_{k+p}-t_k\right)}\right]$$

$$= \int_0^{2\pi} e^{j\frac{\pi\beta p}{N}\cos\phi_\ell} f_\phi(\phi)\, d\phi$$

$$= \frac{1}{2\pi}\int_0^{2\pi} e^{j\frac{\pi\beta p}{N}\cos\phi}\, d\phi = J_0\left(\frac{\pi\beta p}{N}\right)$$

$$\text{(10.158)}$$

where $t_k = kT/N$ from (10.91) and the direction of arrival of multipath components is assumed to be uniformly distributed in $[0,2\pi]$. Here, β is defined as the ratio of the symbol duration to the channel coherence time Δt_c, which is related to the Doppler spread as follows:

$$\beta = \frac{T}{\Delta t_c} = 2f_{d,\max} T \qquad (10.159)$$

The received signal at the n-th subcarrier frequency is rewritten from (10.99) as follows:

$$\tilde{R}_n = \frac{1}{N}\tilde{X}_n \sum_{k=0}^{N-1} H_{n,k}$$

$$+ \frac{1}{N}\sum_{\substack{r=0,\\ r\neq n}}^{N-1} \tilde{X}_r \sum_{k=0}^{N-1} H_{r,k}\, e^{j\frac{2\pi k(r-n)}{N}} + W_n$$

$$\text{(10.160)}$$

Using (10.160), the signal power is found to be

$$S_n = E\left[\left|\frac{1}{N}\tilde{X}_n \sum_{k=0}^{N-1} H_{n,k}\right|^2\right] = E\left[\frac{1}{N^2}|\tilde{X}_n|^2 \sum_{k=0}^{N-1}\sum_{k'=0}^{N-1} H_{n,k} H^*_{n,k'}\right]$$

$$= \frac{1}{N^2} P_n\, E\left[\sum_{k=0}^{N-1}\sum_{k'=0}^{N-1} H_{n,k} H^*_{n,k'}\right]$$

$$= \frac{1}{N^2} P_n E\left[\sum_{k=0}^{N-1}\sum_{k'=0}^{N-1}\sum_{\ell=0}^{L-1}\sum_{\ell'=0}^{L-1} \underbrace{h_{k,\ell} h^*_{k',\ell'}}_{\delta(\ell-\ell')}\, e^{-j2\pi n(\ell-\ell')/N}\right]$$

$$= \frac{1}{N^2} P_n E\left[\sum_{\ell=0}^{L-1}\sum_{k=0}^{N-1}\sum_{k'=0}^{N-1} h_{k,\ell} h^*_{k',\ell}\right] \qquad (10.161)$$

$$= \frac{1}{N^2} P_n E\left[\sum_{\ell=0}^{L-1}\sum_{k=0}^{N-1}|h_{k,\ell}|^2 + \sum_{\ell=0}^{L-1}\sum_{k=0}^{N-1}\sum_{\substack{k'=0,\\ k'\neq k}}^{N-1} h_{k,\ell} h^*_{k',\ell}\right]$$

$$= \frac{1}{N^2} P_n E\left[LN + \sum_{\ell=0}^{L-1}2\sum_{p=1}^{N-1}(N-p)R_\alpha(pT/N)J_0(\pi\beta p/N)\right]$$

$$= \frac{L}{N^2} P_n\left[N + 2\sum_{p=1}^{N-1}(N-p)R_\alpha(pT/N)J_0(\pi\beta p/N)\right]$$

where

$$
E\left[\sum_{k=0}^{N-1}\sum_{\substack{k'=0,\\k'\neq k}}^{N-1} h_{k,\ell}h^*_{k',\ell}\right] = E\left[\sum_{\substack{p=1-N\\p\neq 0}}^{N-1} h_{k,\ell}h^*_{k+p,\ell}\right] = E\left[\sum_{\substack{p=1-N\\p\neq 0}}^{N-1} \alpha_{k,\ell}\alpha^*_{k+p,\ell}\,e^{-j2\pi\frac{v_\ell}{\lambda}\cos\phi_\ell pT}\right]
$$

$$
= 2\sum_{p=1}^{N-1}(N-p)R_{\alpha_\ell}(pT/N)\,J_0(\pi\beta p/N) \tag{10.162}
$$

$$
E\left[|h_{k,\ell}|^2\right]=1 \;\Rightarrow\; E\left[\sum_{\ell=0}^{L-1}\sum_{k=0}^{N-1}|h_{k,\ell}|^2\right]=LN
$$

Using (10.160), the ICI power is foud as follows:

$$
I = E\left[\left|\frac{1}{N}\sum_{\substack{r=0,\\r\neq n}}^{N-1}\tilde{X}_r\sum_{k=0}^{N-1}H_{r,k}e^{j\frac{2\pi k(r-n)}{N}}\right|^2\right]
$$

$$
= E\left[\frac{1}{N^2}\sum_{\substack{r=0,\\r\neq n}}^{N-1}\sum_{\substack{r'=0,\\r'\neq n}}^{N-1}\underbrace{\tilde{X}_r\tilde{X}^*_{r'}}_{|\tilde{X}_r|^2\delta(r-r')}\sum_{k=0}^{N-1}\sum_{k'=0}^{N-1}H_{r,k}H^*_{r',k'}e^{j\frac{2\pi k(r-n)}{N}}e^{-j\frac{2\pi k'(r'-n)}{N}}\right]
$$

$$
= E\left[\frac{1}{N^2}\sum_{\substack{r=0,\\r\neq n}}^{N-1}|\tilde{X}_r|^2\sum_{k=0}^{N-1}\sum_{k'=0}^{N-1}\sum_{\ell=0}^{L-1}\sum_{\ell'=0}^{L-1}\underbrace{h_{k,\ell}h^*_{k',\ell'}\,e^{-j2\pi n(\ell-\ell')/N}}_{\delta(\ell-\ell')}\,e^{j\frac{2\pi(r-n)(k-k')}{N}}\right]
$$

$$
= E\left[\frac{1}{N^2}\sum_{\ell=0}^{L-1}\sum_{\substack{r=0,\\r\neq n}}^{N-1}|\tilde{X}_r|^2\sum_{k=0}^{N-1}\sum_{k'=0}^{N-1}h_{k,\ell}h^*_{k',\ell}e^{j\frac{2\pi(r-n)(k-k')}{N}}\right] \tag{10.163}
$$

$$
= E\left[\frac{1}{N^2}\sum_{\ell=0}^{L-1}\sum_{\substack{r=0,\\r\neq n}}^{N-1}|\tilde{X}_r|^2\left(\sum_{k=0}^{N-1}|h_{k,\ell}|^2+\underbrace{\sum_{k=0}^{N-1}\sum_{\substack{k'=0,\\k'\neq k}}^{N-1}h_{k,\ell}h^*_{k',\ell}e^{j\frac{2\pi(r-n)(k-k')}{N}}}_{I_1}\right)\right]
$$

$$
= \frac{1}{N^2}\sum_{\ell=0}^{L-1}\sum_{\substack{r=0,\\r\neq n}}^{N-1}P_r\left[N+2\sum_{p=1}^{N-1}(N-p)R_\alpha\left(\frac{pT}{N}\right)J_0\left(\frac{\pi\beta p}{N}\right)\cos\left(\frac{2\pi(r-n)p}{N}\right)\right]
$$

where

$$E[I_1] = E\left[\sum_{k=0}^{N-1} \sum_{\substack{k'=0, \\ k' \neq k}}^{N-1} h_{k,\ell} h_{k',\ell}^* e^{j\frac{2\pi(r-n)(k-k')}{N}}\right]$$

$$= E\left[\sum_{\substack{p=1-N \\ p \neq 0}}^{N-1} h_{k,\ell} h_{k+p,\ell}^* e^{j\frac{2\pi(r-n)p}{N}}\right]$$

$$= E\left[\sum_{\substack{p=1-N \\ p \neq 0}}^{N-1} \alpha_{k,\ell} \alpha_{k+p,\ell}^* e^{-j2\pi\frac{v_\ell}{\lambda}\cos\phi_\ell pT} e^{j\frac{2\pi(r-n)p}{N}}\right]$$

$$= \sum_{\substack{p=1-N \\ p \neq 0}}^{N-1} (N-|p|) R_{\alpha_\ell}\left(\frac{pT}{N}\right) J_0\left(\frac{\pi\beta_\ell p}{N}\right) e^{j\frac{2\pi(r-n)p}{N}}$$

$$= 2\sum_{p=1}^{N-1} (N-p) R_{\alpha_\ell}\left(\frac{pT}{N}\right) J_0\left(\frac{\pi\beta_\ell p}{N}\right) \cos\left(\frac{2\pi(r-n)p}{N}\right)$$

$$(10.164)$$

Based on the assumption that the channel gain is constant over an OFDM symbol duration, the autocorrelation function of the channel gain in a non-fading channel is taken to be equal to unity $R_\alpha(pT/N) = 1$. Then, the S_n/I ratio given by (10.115) is obtained by taking the ratio of (10.161) to (10.163).

References

[1] B. Sklar, *Digital Communications* (2nd ed.), Prentice Hall: New Jersey, 2001.

[2] J. G. Proakis, *Digital Communications* (4th ed.), McGraw Hill: New York, 2001.

[3] A. Mittra, On pseudo-random and orthogonal binary spreading codes, World Academy of Science, *Engineering and Technology*, vol. 2, pp. 12–28, 2008.

[4] E. H. Dinan and B. Jabbari, Spreading codes for direct sequence CDMA and wideband CDMA cellular networks, IEEE Communications Magazine, September 1998.

[5] S. Haykin and M. Moher, *Modern Wireless Communications*, Pearson/Prentice Hall: New Jersey, 2005.

[6] D. J. Torrieri, *Principles of Secure Communication Systems*, Artech House: Norwood, 1985.

[7] H. Taub and D. L. Schilling, *Principles of Communication Systems* (2nd ed.), McGraw Hill: New York, 1986.

[8] P. Moore and P. Crossley, GPS applications in power systems Part 1 Introduction to GPS, Power Engineering Journal, February 1999, pp. 33–39.

[9] United Nations Office for Outer Space Affairs, Current and Planned Global and Regional Navigation Satellite Systems and Satellite-based Augmentations Systems, *International Committee on Global Navigation Satellite Systems Provider's Forum, 2010.*

[10] http://en.wikipedia.org/wiki/Satellite_navigation

[11] H. G. Myung, J. Lim, and D. J. Goodman, Single carrier FDMA for uplink wireless transmission, IEEE Vehicular Technology Magazine, pp. 30–38, September 2006.

[12] V. Chakravarthy, A. S. Nunez, J. P. Stephens, A. K. Shaw, and M. A. Temple, TDCS, OFDM, and MC-CDMA: A Brief Tutorial, *IEEE Communications Magazine*, Volume: 43, Issue: 9, Sept. 2005, pp. S11–16, September 2005.

[13] S. Haykin, *Communication Systems* (3rd ed.), John Wiley: New York, 1994.

[14] M. D. Nisar, W. Utschick, H. Nottensteiner and T. Hindelang, On channel estimation and equalization of OFDM systems with insufficient cyclic prefix, VTC April 2007.

[15] ETSI TR 101 190 V1.3.1 (2008-10), Technical Report, Digital Video Broadcasting (DVB); Implementation guidlines for DVB terrestrial services; Transmission Aspects.

[16] Ö. Karabacak and M. Şafak, Performance of an OFDM system in frequency non-selective, fast fading channels, Proc. 7th International OFDM-Workshop (InOWo'02), pp. 241–244, September 10-11, 2002, Hamburg, Germany.

[17] Ö. Karabacak, Performance of an OFDM system in frequency non-selective, fast fading channels, M.Sc. thesis, Hacettepe University, Dept. Electrical and Electronics Eng., Ankara, Turkey, 2002.

[18] V. D. Nguyen, T. H. Nguyen, D. T. Ha and T. Kaiser, Performances analysis of Vehicle-to-X communication systems, IEEE Int. Symposium on Signal Processing and Information Technology (ISSPIT), 2012, pp. 326–330.

[19] M. K. Özdemir, H. Arslan, Channel estimation for wireless OFDM systems, *IEEE Communications Surveys & Tutorials*, vol. 9, no. 2, pp. 18–48, 2nd quarter 2007.

[20] M. Morelli, C.-C. Jay Kuo, and M.-O. Pun, Synchronization techniques for orthogonal frequency division multiple access (OFDMA): A tutorial review, *Proc. IEEE*, vol. 95, no. 7, pp. 1394–1427, July 2007.

[21] M. Deumal, A. Behravan, and J. L. Pijoan, On Cubic Metric Reduction in OFDM Systems by Tone Reservation, *IEEE Trans. Communications*, vol. 59, no. 6, pp. 1612–1620, June 2011.

[22] T. Jiang, and G. Zhu, Complement block coding for reduction in peak-to-average power ratio of OFDM signals, *IEEE Communications Magazine*, vol. **43**, no. 9, Sept. 2005, pp. S17–S22.

[23] S. H. Han and J. H. Lee, An overview of peak-to-average power ratio reduction techniques for multicarrier transmission, IEEE Wireless Communications, pp. 56–65, April 2005.

[24] S.H. Müller and J. B. Hüber, A comparison of peak power reduction schemes for OFDM, Proc. IEEE GLOBECOM 97, Phoenix, AZ, Nov. 1997, pp. 1–5.

[25] S.H. Müller and J. B. Hüber, OFDM with reduced peak to-average power ratio by optimum combination of partial transmit sequences, *Electronics Letters*, vol. **33**, no. 5, Feb. 1997, pp. 368–369.

[26] A. F. Molisch, *Wireless Communications*, John Wiley: Chichester, 2005.

[27] H. Bogucka, Directions and recent advances in PAPR reduction methods, 2006 IEEE Int. Symposium in Signal Processing and Information Technology, pp. 821–827.

[28] http://www.3gpp.org

[29] G. Berardinelli, L.A. M. R. De Temino, S. Frattasi, M. I. Rahman and P. Mogensen OFDMA vs SC-FDMA: performance comparison in local area IMT-A scenarios, IEEE Wireless Communications, pp. 64–72, October 2008.

[30] M. Çakır, Adaptive modulation schemes for orthogonal frequency division multiplexing systems, M.Sc. thesis, Hacettepe University, Dept. Electrical and Electronics Eng., Ankara, Turkey, June 2005.

[31] R. E. Ziemer and R. L. Peterson, *Introduction to Digital Communications* (2nd ed.), Prentice Hall: New York, 2001.

Problems

1. The figure below shows a pseudo-noise (PN) sequence generator consisting of a five-stage shift register. The initial state is 01000 and it is driven by a 1 MHz clock.

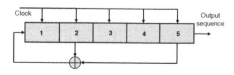

   a. Determine output sequence.
   b. Determine the length and the period of the PN sequence.

   c. Determine the chip rate (rate of the output sequence)
   d. Plot the autocorrelation function produced by PN generator.

2. Determine the output sequence of the gold code shown in Figure 10.3 for the same initial state 10110 for both shift registers.

3. Determine and plot the cross-correlation between the outputs of the two shift registers used in the gold code in Figure 10.3 using the same initial state 10110 for both shift registers.

4. Despread the signal $m(t)$ shown in Figure 10.6 to obtain the original data sequence.

5. Derive (10.26).

6. Consider a CDMA system with the following four spreading sequences:

$$c_1 : (1 \quad 1 \quad 1 \quad 1)$$
$$c_2 : (1 \quad 1 \quad -1 \quad -1)$$
$$c_3 : (1 \quad -1 \quad 1 \quad -1)$$
$$c_4 : (1 \quad -1 \quad -1 \quad 1)$$

The system, operating at frequency f_c, employs BPSK where −1 V and +1 V are transmitted for message bits 0 and 1, respectively. Consider that the user i, $i =$ 1, 2, 3, 4 uses the spreading sequence c_i to transmit the message sequence m_i:

$$m_1 : 00, \quad m_2 : 01, \quad m_3 : 11, \quad m_4 : 10$$

   a. Draw the block diagram of this CDMA system.
   b. Write down the spread signals transmitted by each of the four users and the signal at the input of a CDMA receiver, ignoring the AWGN.
   c. Obtain each of the transmitted data sequences by correlating the input signal with the appropriate spreading sequence.

7. Consider a BPSK/CDMA system with a bit rate of 100 kbps and a maximum length shift register of length $m = 10$ stages.

 a. What is the chip rate, the processing gain and the null-to-null bandwidth of this spread-spectrum signal?
 b. Assuming that $E_b/N_0 = 6.8$ dB, determine the bit error probability and the number of users that can be supported by this system if the spreading codes are perfectly orthogonal to each other.
 c. For the same bit error probability performance for BPSK modulation as in part (b), determine the number of users that can be supported by this system in the presence of synchronous users with long spreading sequences so that multi-user interference dominates the AWGN.

8. 50 MHz bandwidth is allocated to deploy a BPSK/CDMA system with 10-kbps for each simplex voice channel and $E_b/N_0 = 10$ dB. Bit error probability is required to be less than or equal to 10^{-3} for BPSK modulation under multi-user interference and AWGN. Determine the system capacity in number of channels per cell.

9. Assuming that the locations of four GPS satellites are known, write down the steps of a software program to determine the position (x,y,z) and time for a ground based GPS receiver using (10.54).

10. The block diagram of a baseband BPSK/DSSS transmitter is shown below.

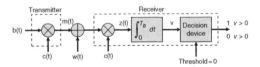

 Here $b(t)$ denotes the input bit sequence, with a bit duration T_b, and $c(t)$ denotes the code, with a chip duration of

$T_c = T_b/N$, where N denotes the number of chips per bit. Assume that binary 1's are encoded by +1 V and binary 0's are encoded as −1 V. Similarly, assume that the bit sequence is given by $b(t) = \{101\}$, the chip sequence is $c(t) = \{10110\}$; hence $N = 5$.

 a. Draw the waveforms $b(t)$, $m(t)$ and $z(t)$, ignoring the AWGN $w(t)$.
 b. Compare the bandwidths of $b(t)$, $m(t)$ and $z(t)$.
 c. Determine the E_c/N_0 in terms of E_b/N_0, where E_c denotes the chip energy.
 d. Assuming perfect synchronization, determine the bit error probability of this signalling scheme.
 e. Discuss the implications of increasing N. How can be used for anti-jamming (AJ) purposes, assuming that the code $c(t)$ is known only by the transmitter and the receiver? What are the advantages and disadvantages of this transmission technique?

11. Consider a CDMA system with processing gain of G_p.

 a. Using (10.32) and (10.34), show

 $$\frac{J}{P} = G_p\left(\frac{1}{\text{SINR}} - \frac{1}{E_b/N_0}\right)$$

 where SINR is defined as

 $$\text{SINR} = \frac{E_b}{N_0 + N_{MUI}} = \frac{1}{\frac{1}{E_b/N_0} + \frac{1}{E_b/N_{MUI}}}$$

 b. Determine the number equal-power CDMA users that can share the SS bandwidth, if $SINR = 10$ dB is required, $G_p = 30$ dB and $E_b/N_0 = \infty$ (E_b/N_0 very large), 16 dB, 13 dB and 10 dB. How can you explain the variation of the number of users with E_b/N_0?

12. The frequency reuse factor of a CDMA system is equal to unity, that is, all users

in all cells employ the same frequency spectrum. The users from other cells are distinguished by their codes, which are desired to be orthogonal to those of the cell of interest. However, this can not be easily achieved. The interference power from other-cell users may be approximated by an *other-cell relative interference factor* γ_{oc}. Thus, the multiuser interference power is multiplied by $1 + \gamma_{oc}$, where $\gamma_{oc} = 0.6$ is usually assumed.

For further interference mitigation, CDMA systems monitor users' activity such that users switch off their transmitters or reduce their transmit power during the periods of no/lower user activity. Therefore, interference power can, in principle, be reduced by the factor γ_v, which is called the *voice activity factor*. In a two-way telephone conversation, $\gamma_v \approx 8/3$. Similarly, assuming that the user population is uniformly distributed in a cell, employing sectored antennas reduce the interference power by the *antenna sectoring factor* γ_s. For a 120^0 sectoring, this factor is estimated as $\gamma_s \approx 2.4$.

Hence, considering other-cell interference, voice activity and antenna sectoring, total interference PSD given by (10.31) may be rewritten as

$$N_{MUI} = (K-1)PT_c \frac{1 + \gamma_{oc}}{\gamma_s \, \gamma_v}$$

where P denotes the average user power. In WCDMA, each channel occupies 5 MHz of 60 MHz bandwidth allocated in each direction.

Assume that the user rate 12 kb/s, the received signal power $P = 5 \times 10^{-12}$ W, the one-sided AWGN PSD is $N_0 = 10^{-17}$ W/Hz, and the other-cell relative interference factor $\gamma_{oc} = 0.6$. If the bit error probability is required to be less than 10^{-5}, determine the channel capacity and the total capacity of a CDMA system assuming:

a. No other cell interference, omnidirectional base station antennas and no voice activity detection

b. Other cell interference with $\gamma_{oc} = 0.6$, omnidirectional base station antennas and no voice activity detection.

c. Other cell interference, 120^0 sectoring at the base station with $\gamma_s = 2.4$ and voice activity detection with $\gamma_v = 2.6$.

13. Orthogonal Frequency Division Multiplexing (OFDM) is a multi-carrier transmission technique. Consider a single-carrier system transmitting at a data rate of R, hence a symbol duration of $1/R$. However, each of the N subcarriers of an equivalent OFDM system (having the same bandwith) transmits at a rate of R/N; thus has an increased symbol duration of $T = N/R$. A subcarrier in an OFDM system may be written as

$$\phi_i(t) = \sqrt{\frac{2}{T}}\cos(2\pi f_i t), \quad 0 \leq t \leq T$$
$$f_i = f_0 + (i-1)\Delta f, \quad i = 1, 2, \cdots, N$$

Here $f_0 = k/T$ where k is a large integer and $\Delta f = 1/T$. Note that the difference between consecutive subcarrier frequencies are separated from each other by an amount of Δf, that is, $f_{i+1} - f_i = \Delta f$.

a. Show that the subcarriers are orthonormal.

b. Determine the Fourier transform, $\Phi_i(f)$, of the subcarrier $\phi_i(t)$ and plot $\Phi_i(f)$, $\Phi_{i+1}(f)$ and $\Phi_{i+2}(f)$. Show that the spectra of the subcarriers overlap.

c. If the subcarriers remain orthogonal at the receiver, then they can be easily separated at the receiver without interfering to each other, even though they overlap in the frequency domain. Discuss how can the orthogonality of the subcarriers be destroyed by the channel and/or the transmitter/receiver? Discuss how the orthogonality is affected when two adjacent subcarriers are separated from each other by $(1+k)\,\Delta\,f$, where k is a number smaller than a few percent.

14. Show from (10.115) that the ICI vanishes for $\beta=0$.

15. Assuming that i-th and (i + 1)-th OFDM symbols of 16-bit long are given by 1101 0011 0101 1001 and 1111 1100 1011 1010, redraw Figure 10.43(c) assuming that length of the cyclic prefix is 4.

16. Consider a SC-FDMA system with $N=12$ subcarriers, $Q=3$ users and $L=4$ subcarriers per user. Assuming that the data $\tilde{x}$ of three users are given by 1 1 1 1, −1 1 −1 1 and −1 −1 −1 1, determine the PAPR for distributed and localized mapping and compare their performances.

17. A simplified block diagram of a DVB-T (digital video broadcasting-terrestrial) system is shown below:

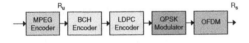

An MPEG encoder output is characterized by R_u bps. 48408 bits at the input of the BCH encoder become 48600 bits with the addition of parity bits to correct random bit errors. The output of the low-density parity-check (LDPC) encoder of rate ¾ has 64800 bits. Following the concatenated FEC encoder (BCH + LDPC), the QPSK-modulated symbols are transmitted via OFDM with 2048 subcarriers in a 8 MHz bandwidth allocated for an analog TV channel. The cyclic prefix (guard interval) in the OFDM is 1/32 of the useful symbol duration.

a. Calculate the frequency separation between neighbouring subcarriers and the useful symbol duration and that including the cyclic prefix.

b. The orthogonality between subcarriers may be assumed to be preserved if the maximum Doppler shift does not

exceed 1% of the frequency separation between the subcarriers. Then, determine the maximum velocity of the user terminal that can be served by this system operating at 300 MHz.

c. Determine the symbol rate R_s (symbols/s) at the output of the OFDM in terms of the raw bit rate R_u (bits/s) at the output of the MPEG encoder.

d. Determine the maximum bit rate that can be supported by this system. Assuming that a standard definition TV (SDTV) channel requires approximately 4 Mbps, how many SDTV digital TV channels can be multiplexed in an analog TV channel of 8 MHz bandwidth.

18. Digital terrestrial TV broadcasting systems use OFDM for TV transmissions. Consider such a system operating in a multipath environment with a maximum excess delay of 10 μs. The system specification is such that the duration T_G of the cyclic prefix is equal to or greater than 1/16 of the OFDM symbol duration T.

a. Determine the OFDM symbol duration and the frequency spacing between nearest sub-carriers.

b. If a terrestrial digital TV broadcasting channel is allocated a bandwidth of 4 MHz, determine the number N of subcarriers in this OFDM system. Note that the number of subcarriers should be expressed as $N=2^n$ where n is an integer so as to allow FFT and IFFT needed for the OFDM transceiver. Using the value of N thus found, recalculate the corresponding values of T and T_G. Are the requirements stated above all satisfied? If no, what is your suggestion for designing this system?

c. Assuming that, including coding and overhead bits, a data rate of 4 Mbps is sufficient for the transmission of a TV channel, what would be the alphabet size M of a M-QAM modulation used for broadcasting.

19. To ensure the orthogonality between subcarriers, the channel should not change appreciably during an OFDM symbol. This implies that OFDM symbol duration should be shorter than the channel coherence time. Similarly, the guard time should be larger than the maximum excess delay of the channel for avoiding ISI. Maximum Doppler shift is required not to exceed 2.5% of the frequency spacing between the subcarriers of the OFDM system with 5 MHz bandwidth, 2 GHz carrier, and 10 µs maximum excess delay.

   a. Determine the number of subcarriers for user velocities no higher than 120 km/hr. Determine the loss in E_b/N_0 in dB due to the use of cyclic prefix.
   b. Repeat part (a) for a carrier frequency of 4 GHz.
   c. Discuss how you can minimize the E_b/N_0 loss as a function of coherence time, maximum excess delay and carrier frequency.

   d. Determine the carrier frequency for $T_G<T/32$, a maximum user velocity of 120 km/hr and maximum excess delay of 10 µs.

20. Consider an OFDM system, with 1024 subcarriers, operating over $B = 10$ MHz bandwidth at 1 GHz carrier frequency. This system is allowed to take the ratio of cyclic prefix duration T_G to that of the symbol duration, T, as $T_G/T = 1/4, 1/8, 1/16$ or $1/32$.

   a. Assume first that this system operates in outdoor environment with a maximum excess delay of 20 µs. What length cyclic prefix is needed for this system? What is the throughput efficiency?
   b. Repeat part (a) for an indoor environment with a maximum excess delay of 1 µs.
   c. If the number of OFDM tones is fixed, derive the throughput efficiency in terms of N, T_G and B. How does the throughput efficiency change as a function of the bandwidth? How does the efficiency change as a function of the maximum excess delay?

11

Fading Channels

In Chapter 4, we considered the propagation of electromagnetic waves mainly in VHF and UHF bands, where the wavelength varies in the range 10-0.1 m. In lower parts of the VHF band where the wavelength is longer than 10 m, larger obstacles, located in the first Fresnel zone, are needed for the signal transmission via reflection, diffraction and scattering. The number of such obstacles contributing the signal transmission is not generally large. The propagation paths between transmitter and receiver may then be determined by ray tracing via sufficiently large obstacles which reflect, diffract or scatter electromagnetic waves. Reflection occurs when a signal encounters a smooth surface that is large compared to the wavelength. The strength of the reflected signal is determined by the reflection coefficients for vertical and horizontal polarizations. Diffraction occurs at sharp edges of conductive obstacles that are large compared to the wavelength. Scattering is caused by irregularities smaller than the wavelength of obstacles with rough surfaces. In this band, small variations in distance do not result in rapid fluctuations in the received signals.

With increasing frequencies starting from the upper end of the UHF band, the wavelength becomes smaller than 10 cm. Consequently, the number of obstacles contributing to multi-path propagation increases as even physically small obstacles have significantly large electrical sizes. Eventually, ray-tracing becomes practically impossible not only because of increased number of paths but also due to variations in electrical characteristics (conductivity, dielectric constant) and sizes of the obstacles in addition to the angle of arrival fluctuations of waves. Then, channel modeling requires statistical approaches since deterministic techniques become unmanageable and very costly. The received signal does not any more have the form e^{-jkr}/r of an electromagnetic wave as in free-space propagation, but is rather characterized by the channel (path) gain $\alpha_k(t) < 1$, phase shift $\phi_k(t)$, and delay $\tau_k(t)$ of the path via the kth obstacle, where $k = 1,2,\dots,L$, (L denotes the number of multipaths). As $\alpha_k(t), \phi_k(t)$ and $\tau_k(t)$ vary randomly, the signal received from L paths, in this so-called fading channel, can be described in terms of its probability density function (pdf) and

Digital Communications, First Edition. Mehmet Şafak.
© 2017 John Wiley & Sons Ltd. Published 2017 by John Wiley & Sons Ltd.
Companion website: www.wiley.com/go/safak/Digital_Communications

cumulative distribution function (cdf). This results in a linear time-varying multipath fading channel. The received signal, which consists of the phasor addition of L multipath signals, undergoes random variations in time within a large dynamic range. Severe fluctuations in the received signal level due to fades lead to drastic decreases in the mean E_b/N_0 ratio of the received signal. Fading channels are encountered in ionospheric propagation, cellular radio, LOS microwave radio, tropospheric scattering, satellite communications, and so on.

Other contributions to the propagation loss include the path loss, which is caused by square-law spreading, absorption losses due to atmospheric gases, and the precipitation losses due to rain, snow, fog and mist. Atmospheric absorption and precipitation losses are frequency dependent and become significant at frequencies above several GHz. In addition, distortion of received signals in time- and frequency-domains results in intersymbol interference (ISI). Similarly, one suffers co-channel interference in cellular radio and Doppler frequency shift in time-varying channels mainly due to user mobility.

The phasor addition of multipath signals results in rapid changes (fast/short-term fading, over distances in the order of $\lambda/2$) in the received signal (see Figure 11.1). Short-term

fading is usually characterized by Rayleigh, Rician or Nakagami-m pdf. In addition to multipath effects, the presence of large obstacles like large buildings, hills and tunnels decreases the local-mean power level (usually averaged over few tens of wavelengths) of the received signal; these changes are much slower (over distances in the order of 10s of λ) compared to the multipath fading. This phenomenon, which is called shadowing, is usually characterized by log-normal pdf.

Countermeasures against fading may include diversity (transmitting the same signal from different channels, for example, frequency, time, or space (antennas)), interleaving (efficient when a fade affects many bits/symbols at a time), frequency hopping, error control coding, multi-user diversity and time-scheduling, which mainly consists of transmitting to a user only when its channel gain is sufficiently high.

11.1   Introduction

A transmitted bandpass signal may be written, in terms of its low-pass equivalent $s_\ell(t)$ and its carrier frequency f_c, as

$$s(t) = \mathrm{Re}[s_\ell(t)e^{jw_c t}], \quad 0 \le t \le T \qquad (11.1)$$

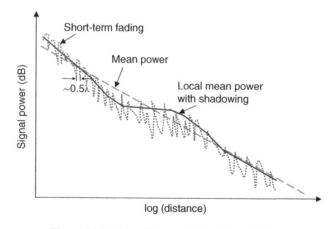

Figure 11.1   Long-Term and Short-Term Fading.

where T denotes the symbol duration. Similarly, the received bandpass signal is given by

$$r(t) = \text{Re}[r_\ell(t)e^{jw_c t}], \quad 0 \le t \le T \quad (11.2)$$

where no Doppler frequency shift is assumed in the channel. For the sake of simplicity, we will hereafter be concerned only with low-pass equivalent signals propagating through low-pass equivalent (baseband) channels and use (11.1) and (11.2) to obtain the corresponding bandpass signals.

In an AWGN channel, the received signal may be expressed as the sum of the transmitted signal and the complex-valued AWGN $w(t)$ with zero mean and double-sided PSD $N_0/2$:

$$r_\ell(t) = s_\ell(t) + w(t) \quad (11.3)$$

In a multipath channel, the transmitted signal arrives the receiver via L multipaths with different channel gains $\alpha_k(t)$, phase shifts $\phi_k(t)$ and delays $\tau_k(t)$, which may vary with time:

$$r_\ell(t) = \sum_{k=0}^{L-1} \alpha_k(t)e^{-j\phi_k(t)}s_\ell(t - \tau_k(t)) + w(t)$$

$$(11.4)$$

In an AWGN channel, the signals arrive the receiver via a single path ($L = 1$); this may easily be observed comparing (11.4) with (11.3), where the channel gain and the phase are absorbed into the noise term. Multipath propagation results in dispersion of the signal over time at the channel output. When dispersion becomes significantly large so that the symbol detection is affected by preceeding symbols, this leads to intersymbol interference (ISI).

When the channel gains $\alpha_k(t)$, phase shifts $\phi_k(t)$, delays $\tau_k(t)$ and the number of significant multipath components L do not vary with time, the channel is said to be linear time-invariant (LTI). Then, signal propagation through the channel is not a function of time. If the channel changes in time more slowly than the symbol duration, then the channel is said to be slowly

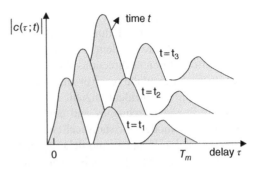

Figure 11.2 Impulse Response of a Time-Invariant Dispersive Channel.

fading and the channel parameters may be assumed to remain constant during a symbol period. In the so-called block fading, the channel parameters may be assumed constant during a block of symbols. Figure 11.2 shows a pictorial view of the impulse response $c(\tau; t)$ of a time invariant dispersive channel with continuous multipath. The channel is time invariant because it does not change with time, and is said to be dispersive if the channel response exceeds the symbol duration, that is, $T_m > T$, where T_m denotes the maximum excess delay and T is the symbol duration.

11.2 Characterisation of Multipath Fading Channels

Multipath propagation results in spreading of the signal in mainly three different dimensions. *Doppler spread* (time-selective fading) is caused by relative motion of transmitter, receiver and/or scatterers. The channel impulse response will change with time due to time variations in the structure of the medium. The time rate of change of the channel impulse response is characterized mainly by the Doppler spread of the received carrier frequency. *Delay spread* (frequency-selective fading) results from the superposition of time-shifted and scaled replicas of the transmitted signal. Signal replicas are received with different time delays due to multipath scattering of the transmitted signal from obstacles at different locations. *Angle spread* (space-selective fading) occurs when multipath signals are incident at the receiving antenna from different directions due to scatterers located around the receiver and/or transmitter. When an incident signal is spread randomly in angle, frequency and delay, one needs realistic models for statistical characterization of angle-, Doppler- and delay spreads. In summary, the received signal is not any more received at a single carrier frequency but spread in a frequency band determined by the maximum Doppler shift, its arrival is not any more confined to the symbol duration but is spread in delay, and does not arrive at the receive antenna from a single direction, but its angle of arrival is spread. Therefore, the receiver design should be optimized so as to collect all/maximum signal power spread in delay, frequency and space.

In the presence of delay and Doppler spread, the received bandpass signal may be written from (11.2) and (11.4) as

$$r(t) = \text{Re}\left[\begin{array}{c} \sum_{k=0}^{L-1} \alpha_k(t) s_\ell[t - \tau_k(t)] \\ e^{j\left[2\pi\left(f_c + f_{t,k} + f_{r,k}\right)(t - \tau_k(t)) - \phi_k(t)\right]} \end{array} + w(t) \right]$$

(11.5)

where $f_{t,k}$ $(f_{r,k})$ shows the Doppler frequency shift due to transmitter (receiver) velocity on kth multipath signal. The Doppler frequency shift from kth multipath due to relative movement of the transmitter is given by

$$f_{t,k} = f_c \frac{v_t}{c} \cos \varphi_{t,k} = \frac{v_t}{\lambda} \cos \varphi_{t,k}$$

(11.6)

where v_t denotes the velocity of the transmitter, $\varphi_{t,k}$ is the angle between the direction of displacement of the transmitter and the direction of arrival of the kth multipath, c is velocity of light, λ is the wavelength and f_c is the carrier frequency.

The instantaneous frequency of the kth received multipath signal is found from (11.5) as

$$f_{i,k} = f_c + f_{t,k} + f_{r,k} - \frac{1}{2\pi} \frac{d\phi_k(t)}{dt}$$

(11.7)

This shows that each multipath component may have a different instantaneous frequency. Using (11.2) and (11.5), the received low-pass equivalent signal may be written as

$$r_\ell(t) = \sum_{k=0}^{L-1} \alpha_k(t) e^{-j\theta_k(t)} s_\ell[t - \tau_k(t)] + w(t)$$

$$= c(\tau;t) \otimes s_\ell(t) + w(t)$$

(11.8)

which is simply the convolution of the low-pass equivalent transmitted signal and the time-variant impulse response of the equivalent low-pass channel:

$$c(\tau;t) = \sum_{k=0}^{L-1} \alpha_k(t) e^{-j\theta_k(t)} \delta[\tau - \tau_k(t)]$$

(11.9)

From (11.5) the phase term in (11.8) and (11.9) is given by

$$\theta_k(t) = 2\pi f_c \tau_k(t) - 2\pi \left(f_{t,k} + f_{r,k}\right)(t - \tau_k(t)) + \phi_k(t)$$

(11.10)

Figure 11.3 shows a pictorial view of a multipath channel and the channel impulse response with $L = 3$ paths. First replica of an impulse transmitted at time $t = 0$ is received at the received at $t = \tau_0$. In free-space propagation conditions $\tau_0 = d/c$, where d denotes the distance between transmitter and receiver. The second and third replicas arrive at times $t = \tau_1$ and $t = \tau_2$. Relative delays of these replicas are $\tau = \tau_1 - \tau_0$ and $\tau = \tau_2 - \tau_0$ in the delay coordinates. The maximum excess delay T_m, which is defined as delay difference between first- and last-arriving multipath components, is given by $T_m = \tau_2 - \tau_0$. It is important to note that $\alpha_k(t)$, $\theta_k(t)$ and $\tau_k(t)$ vary with time in a time-varying channel.

Large dynamic changes are required for $\alpha_k(t)$ to change sufficiently to cause a significant

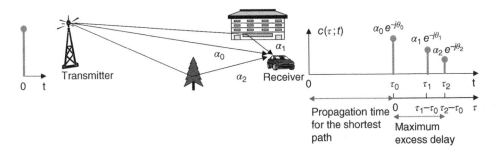

Figure 11.3   Multipath Propagation and Channel Impulse Response.

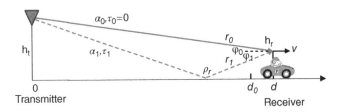

Figure 11.4   Two-Path Channel with Smooth Earth-Reflected Ray.

change in $r_\ell(t)$. However, noting from (11.10) that $\theta_k(t)$, which is dominated by the term $2\pi f_c \tau_k$, will change by 2π radian whenever τ_k changes by $1/f_c$. Therefore, $\theta_k(t)$ can change by 2π radian with relatively small variations in the medium at sufficiently high frequencies, for example, 1.1 ns delay variation ($\approx 1/3$ m path length) at 900 MHz. Since delays τ_k associated with different paths change randomly at different rates, the resulting phase and the Doppler frequency shift (being proportional to the time rate of change of the phase) will also change randomly.

In some channels, for example, the tropospheric channel, the received signal may be viewed as consisting of a continuum of multipath components. Then, the low-pas equivalent received signal given by (11.8) reduces to

$$r_\ell(t) = c(\tau;t) \otimes s_\ell(t) + w(t)$$

$$= \int_{-\infty}^{\infty} \alpha(\tau;t)e^{-j\theta(\tau;t)}s_\ell(t-\tau(t))d\tau + w(t)$$

$$(11.11)$$

Low-pass equivalent time-variant impulse response of a continuous multipath channel is then given by

$$c(\tau;t) = \alpha(\tau;t)e^{-j\theta(\tau;t)} \qquad (11.12)$$

Example 11.1   Two-Path Channel with Earth-Reflection.

Consider a two-path channel which consists of direct and Earth-reflected rays, where ρ_r denotes the reflection coefficient over the smooth Earth surface. The heights of transmit and receive antennas are denoted by h_t and h_r, respectively. Transmit and receive antennas are separated by a distance d_0 at time $t=0$ as shown in Figure 11.4. We assume that the mobile receiver starts to move at $t=0$ with a constant velocity v in the horizontal direction. Therefore, the distance between transmit and receive antennas at time t may be written as $d(t) = d_0 + vt$. The path-lengths are given by

$$r_0(t) = \sqrt{d(t)^2 + (h_r - h_t)^2}$$

$$r_1(t) = \sqrt{d(t)^2 + (h_r + h_t)^2} \qquad (11.13)$$

The channel impulse response may be written, in terms of direct and reflected rays, as

$$c(\tau; t) = \alpha_0(t) e^{-j\theta_0(t)} \delta(\tau) + \alpha_1(t) e^{-j\theta_1(t)} \delta(\tau - \tau_1(t))$$

$$\tau_1(t) = \frac{r_1(t) - r_0(t)}{c}$$

$$(11.14)$$

The path gains are simply given by Friis transmission formula (2.114):

$$\alpha_0(t) = \sqrt{G_t G_r} \left(\frac{\lambda}{4\pi r_0(t)} \right)$$

$$\alpha_1(t) = |\rho_r| \sqrt{G_t G_r} \left(\frac{\lambda}{4\pi r_1(t)} \right) \qquad (11.15)$$

The channel phases and Doppler frequency shifts are given by

$$\theta_0(t) = k r_0 + 2\pi f_{d_0} t$$

$$\theta_1(t) = \operatorname{sgn}(\rho_r) k r_1 + 2\pi f_{d_1} t$$

$$f_{d_i}(t) = \frac{v}{\lambda} \cos \varphi_i(t), i = 0, 1 \qquad (11.16)$$

$$\varphi_0(t) = -\tan^{-1}[(h_r - h_t)/d(t)]$$

$$\varphi_1(t) = -\tan^{-1}[(h_r + h_t)/d(t)]$$

where f_{d_i} denotes the Doppler frequency shift of the i-th path.

11.2.1   Delay Spread

Assuming that it is wide-sense stationary (WSS), the autocorrelation function of the low-pass time-variant channel impulse reponse $c(\tau;t)$ may be written as

$$R_c(\tau_1, \tau_2; \Delta t) = E[c^*(\tau_1; t) c(\tau_2; t + \Delta t)]$$

$$(11.17)$$

Since different delays imply uncorrelated scattering from different obstacles, that is, the attenuation and phase shifts associated with path delays τ_1 and τ_2 are uncorrelated, (11.17) reduces to

$$R_c(\tau_1, \tau_2; \Delta t) = R_c(\tau_1, \tau_1; \Delta t) \delta(\tau_1 - \tau_2) \quad (11.18)$$

Here, $R_c(\tau, \tau; \Delta t)$ shows the variation of the average power at the channel output for an input impulse as a function of the time delay and the difference Δt in the observation time. This may be measured by transmitting two very narrow pulses with a time interval Δt and cross-correlating the two received signals. On the other hand,

$$R_c(\tau) = R_c(\tau, \tau; 0) = E\left[|c(\tau; t)|^2 \right] \quad (11.19)$$

is called as multipath intensity profile (MIP) (also called as delay power spectrum) and shows the variation of the average received power versus delay at the channel output due to an impulse at the channel input. In other words, MIP provides a measure of the delay disperion suffered by the signal in propagating through a channel.

The MIP for continuous and discrete multipaths are obtained from (11.9), (11.12) and (11.19) as

$$R_c(\tau) = \begin{cases} E\left[|\alpha(\tau; t)|^2 \right] & \text{continuous multipath} \\ \sum_{k=0}^{L-1} E\left[|\alpha_k(t)|^2 \delta(\tau - \tau_k(t)) \right] & \text{discrete multipath} \end{cases}$$

$$(11.20)$$

where $E[|\alpha_k(t)|^2]$ denotes the average power of the kth multipath signal.

Figure 11.5 shows a typical MIP curve, hence the delay power spectrum of continuous multipath signals at the channel output. Noting that the area under the MIP curve is not equal to unity, the mean τ_m, and the variance σ_τ^2, which are used to characterise the MIP, are given by

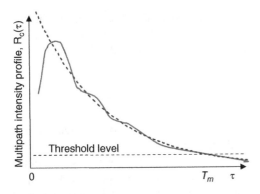

Figure 11.5 A Typical MIP and the Exponential Model. T_m denotes the maximum excess delay.

$$\tau_m = \frac{\int_0^\infty \tau R_c(\tau)d\tau}{\int_0^\infty R_c(\tau)d\tau}$$

$$\sigma_\tau^2 = \frac{\int_0^\infty (\tau-\tau_m)^2 R_c(\tau)d\tau}{\int_0^\infty R_c(\tau)d\tau} \qquad (11.21)$$

As also shown in Figure 11.5, a simple exponential model is usually used to approximate the MIPs encountered in practice. Normalized MIP expression for the exponential model may be written as

$$R_c(\tau) = \frac{1}{\tau_0}\exp\left(-\frac{\tau}{\tau_0}\right), \quad \tau>0 \qquad (11.22)$$

The mean value and the standard deviation of (11.22) are given by (see (F.104))

$$m_\tau = \sigma_\tau = \tau_0 \qquad (11.23)$$

In practice, multipath components with powers less than 5-10 % of the total power are considered to be insignificant and are generally ignored. In some applications, it may be useful to characterize the channel delay by the so-called maximum excess delay T_m, which

is defined as the delay difference between the first- and last-arriving significant multipath components.

Delay spread limits the transmission rate over the channel. If the symbol duration T is shorter than T_m, then the multipath components with delays in excess of the symbol duration will interfere with the consecutive symbols, hence causing intersymbol interference (ISI). In order to avoid ISI, the channel forces the communicator to use symbol durations longer than T_m, hence to reduce the symbol rate. In summary, the channel MIP dictates a maximum symbol rate to the communication system. Signaling at higher symbol rates will cause ISI and higher bit error probabilities.

The limitation imposed by the channel delay on the symbol rate may be studied in the frequency domain as well. The channel transfer function $C(f;t)$ which is given by the Fourier transform of the impulse response

$$C(f;t) = \Im_\tau[c(\tau;t)] = \int_0^\infty c(\tau;t)e^{-j2\pi f\tau}d\tau$$
$$(11.24)$$

may be interpreted as the low-pass frequency response of the multipath fading channel for a transmitted impulse at time t.

The so-called spaced-frequency, spaced-time correlation function of a low-pass equivalent channel is defined by

$$R_C(\Delta f;\Delta t) = E[C(f+\Delta f;t+\Delta t)C^*(f;t)]$$
$$(11.25)$$

and may be measured by transmitting a pair of sinusoids separated by Δf and cross-correlating the two separately received signals with a relative time difference Δt. Inserting (11.24) into (11.25) one gets the Fourier transform relation between the spaced-frequency, spaced-time correlation function and the autocorrelation function of the channel impulse response:

$R_C(\Delta f; \Delta t)$

$$= E\left[\int_0^\infty \int_0^\infty c^*(\tau_1;t)c(\tau_2;t+\Delta t)e^{j2\pi f(\tau_1-\tau_2)}e^{-j2\pi\Delta f\tau_2}d\tau_1 d\tau_2\right]$$

$$= \int_0^\infty \int_0^\infty R_c(\tau_1,\tau_2;\Delta t)e^{j2\pi f(\tau_1-\tau_2)}e^{-j2\pi\Delta f\tau_2}d\tau_1 d\tau_2$$

$$= \int_0^\infty \int_0^\infty R_c(\tau_1;\Delta t)\delta(\tau_1-\tau_2)e^{j2\pi f(\tau_1-\tau_2)}e^{-j2\pi\Delta f\tau_2}d\tau_1 d\tau_2$$

$$= \int_0^\infty R_c(\tau;\Delta t)e^{-j2\pi\Delta f\tau}d\tau$$

$$\text{(11.26)}$$

The so-called scattering function is defined by the inverse Fourier transform of the spaced-frequency, spaced-time correlation function:

$$S_C(\tau,\lambda) \triangleq$$

$$\int\int_{\Delta f \, \Delta t} R_C(\Delta f; \Delta t)e^{j2\pi\lambda\Delta t}e^{j2\pi\Delta f\tau}d(\Delta t)d(\Delta f)$$

$$\text{(11.27)}$$

The scattering function shows delay and Doppler spreads of the received signal in orthogonal coordinates. Each multipath component may be characterized by a different delay and Doppler spread, hence by a different scattering function.

It is evident from (11.26) that the Fourier transform of the MIP $R_c(\tau)$ gives the spaced-frequency correlation function $R_C(\Delta f) = R_C(\Delta f; 0)$, that is, the correlation between two frequency tones separated by Δf, as they go through the channel. For an exponential MIP, one gets

$$R_C(\Delta f) = R_C(\Delta f, 0) = \int_0^\infty R_c(\tau)e^{-j2\pi\Delta f\tau}d\tau$$

$$= \int_0^\infty \frac{1}{\tau_0}e^{-\tau/\tau_0}e^{-j2\pi\Delta f\tau}d\tau = \frac{1}{1+j2\pi\tau_0\Delta f}$$

$$\text{(11.28)}$$

Since $R_C(\Delta f)$ represents the autocorrelation of the channel transfer function, the magnitude

of (11.28) is equated to $1/\sqrt{2}$ to determine the so-called 3-dB coherence bandwidth of the channel for the exponential MIP model

$$\Delta f_c = 2\Delta f = \frac{1}{\pi\tau_0} = \frac{1}{\pi\sigma_\tau} \qquad \text{(11.29)}$$

where $\sigma_\tau = \tau_0$ from (11.23). Coherence bandwidth Δf_c provides a measure of the frequency band over which the channel passes transmitted spectral components with approximately equal gain and linear phase (flat fading), that is, a measure of bandwidth that will fade in a correlated fashion at any instant in time. A commonly used approximation for the coherence bandwidth is given by

$$\Delta f_c \approx \frac{1}{5\sigma_\tau} \qquad \text{(11.30)}$$

The coherence bandwidth is also expressed in terms of the maximum excess delay T_m:

$$\Delta f_c \approx 1/T_m \qquad \text{(11.31)}$$

Note from (11.29)-(11.31) and Figure 11.6 that the channel coherence bandwidth decreases with increasing values of the rms delay spread (or T_m). In the limiting case where the T_m goes to infinity (a flat delay spectrum), the channel coherence bandwidth narrows down to an impulse at $f = 0$; hence it becomes practically impossible to communicate through this channel. On the other hand, as T_m becomes shorter, the coherence bandwidth gradually increases and signals undergo flat-fading in a perfectly correlated manner in the channel.

11.2.1.1 Flat Fading Versus Frequency-Selective Fading

The bandwidth of a bandpass signal with Nyquist pulse shaping is given by (6.106): $W = R_s = 1/T$ where R_s and T denote the symbol rate and the symbol duration, respectively.

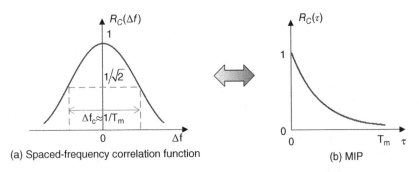

(a) Spaced-frequency correlation function

(b) MIP

Figure 11.6   Relationship Between Coherence Bandwidth and Multipath Intensity Profile.

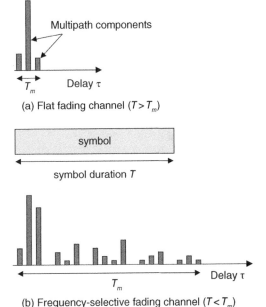

(a) Flat fading channel $(T > T_m)$

(b) Frequency-selective fading channel $(T < T_m)$

Figure 11.7   Flat Fading and Frequency-Selective Fading Channels.

A signal with a transmission bandwidth $W = 1/T$, propagating through a channel of coherence bandwidth $\Delta f_c \approx 1/T_m$ is said to suffer *frequency non-selective* (flat) fading if $W < \Delta f_c$ $(T > T_m)$ (see Figure 11.7). Signal (multipath) components may then be assumed to be equally affected by the channel, that is, they will suffer correlated fading. The maximum excess delay being much shorter than the symbol duration,

the system cannot resolve the multipath components and considers them to arrive with the same delay. Therefore, one-tap is sufficient for channel equalization. The resulting ISI is negligible since delayed components do not cause significant interference to the neighboring symbols. This is based on the fact that the decisions, which are taken in the middle of symbol durations, are not affected by the delayed multipath components of the preceeding symbols. The system performance in such channels may be improved via diversity, for example, combining signal replicas received by different antennas, which are sufficiently separated from each other.

A signal is said to undergo *frequency selective fading* if $W > \Delta f_c$; then, signal components with a frequency difference larger than Δf_c will fade independently in the channel. In such channels, the symbol duration is shorter than the maximum excess delay of the channel $(T < T_m)$ (see Figure 11.7). Delays, gains and phases of the multipath components, which are resolvable by the system, play a significant role in the system performance. The resulting ISI can cause significant degradation in the bit error probability which has a floor approaching ½ and may not be improved by simply increasing the E_b/N_0. A frequency-selective fading channel may be modelled by a multi-tap FIR filter. The resulting ISI need to be removed before decision making by adaptive equalizers.

11.2.1.2   Commonly Used MIP Models

The propagation environment for mobile communications is usually classified as urban, suburban, rural and hilly areas and different MIP models are adopted for each of these environments. Such a classification accounts for different density and heights of buildings, hills and vegetation. The maximum excess delay of the MIP model for urban areas is limited to 7 μs, which corresponds to a maximum excess path length of 2.1 km. However, the maximum excess delay is limited to 10 μs in suburban areas where signals may be received from scatterers as far as 3 km due to lower building heights. In areas with hilly terrain, the first-multipath component has a maximum excess delay of 2 μs, due to scattering from nearby objects, but the second multi-path component accounts for scattering by a hill as distant as 6 km. Figure 11.8 shows the MIP models commonly used for these areas. The analytical expressions and the corresponding delay spreads for these MIP models are shown in Table 11.1. In some applications, discrete models are used instead of continuous models

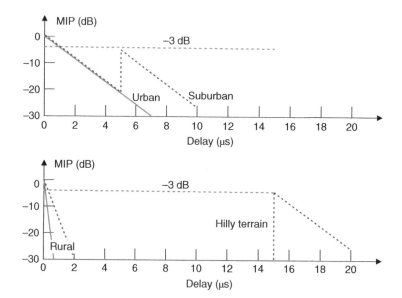

Figure 11.8   Cost 207 Typical MIPs For Rural, Urban, Suburban and Hilly Terrain Areas. Note that 1 μs excess delay corresponds to a path difference of $c \times 1\,\mu s = 3 \times 10^8 \times 1\,\mu s = 300$ m.

Table 11.1   MIP Models Considered in Figure 11.8.

Propagation area	MIP (not normalized)	Delay spread
Rural area	$\exp(-9.2\tau)\ \ 0\le\tau\le0.7\mu s$	0.11 μs
Typical urban	$\exp(-\tau)\qquad 0\le\tau\le7\mu s$	1.00 μs
Suburban	$\exp(-\tau)\qquad 0\le\tau\le5\mu s$	2.53 μs
Hilly terrain	$0.5\exp(5-\tau)\ \ 5\mu s\le\tau\le10\mu s$ $\exp(-3.5\tau)\qquad 0\le\tau\le2\mu s$ $0.5\exp(15-\tau)\ \ 15\mu s\le\tau\le20\mu s$	6.88 μs

Table 11.2 A Discrete MIP Model For Typical Urban and Suburban Regions.

Path number	Typical urban		Suburban	
	Delay (μs)	Fractional power	Delay (μs)	Fractional power
1	0	0.189 (−7.24 dB)	0	0.164 (−7.85 dB)
2	0.2	0.379 (−4.21 dB)	0.3	0.293 (−5.33 dB)
3	0.5	0.239 (−6.22 dB)	1.0	0.147 (−8.33 dB)
4	1.6	0.095 (−10.22 dB)	1.6	0.094 (−10.27 dB)
5	2.3	0.061 (−12.15 dB)	5.0	0.185 (−7.33 dB)
6	5.0	0.037 (−14.32 dB)	6.6	0.117 (−9.32 dB)

for the MIP. Table 11.2 shows discrete MIP models for typical urban and suburban areas.

The ITU-R Recommendation P.1407 applies for LOS propagation in an urban high-rise environment at carrier frequencies around 2.5 GHz and for ranges d of 50–400 meters. The delay spread σ_τ at a distance of d meters follows a normal distribution with the following mean and standard deviation:[1]

$$E[\sigma_\tau] = 99.35 \log(d + 13.24) \, (ns)$$
$$\left(E\left[(\sigma_\tau - E[\sigma_\tau])^2\right]\right)^{1/2} = 39.90 \log(d) - 27.67 \, (ns)$$

$$(11.32)$$

The MIP is modeled by

$$R_c(\tau) = P_{peak} + 50\left(e^{-\tau/\tau_0} - 1\right) \quad (dB)$$
$$\tau_0 = 4\sigma_\tau + 266 \quad\quad\quad\quad (ns)$$

$$(11.33)$$

where P_{peak} denotes the peak power in dB and τ is in ns. The energy arriving in the first 40 ns has a Rician distribution with K-factor about 6-9 dB, while the energy arriving later has a Rayleigh or Rician distribution with a K-factor of up to about 3 dB.

Example 11.2 Mean Delay and Delay Spread of the Typical Urban and Suburban Channels.
Consider the multipath fading environments for typical and bad urban regions described by Table 11.2. Mean delay and delay spread for the typical urban model are found to be

$$\tau_m = \sum_{k=0}^{5} \tau_k P_k = 0.67 \, \mu s$$
$$\sigma_\tau = \left(\sum_{k=0}^{5} (\tau_k - \tau_m)^2 P_k\right)^{1/2} = 1 \, \mu s$$

$$(11.34)$$

The corresponding values for the suburban are as follows:

$$\tau_m = \sum_{k=0}^{5} \tau_k P_k = 2.1 \, \mu s$$
$$\sigma_\tau = \left(\sum_{k=0}^{5} (\tau_k - \tau_m)^2 P_k\right)^{1/2} = 2.4 \, \mu s$$

$$(11.35)$$

The maximum excess delay for typical and suburban models are $T_m = 5$ μs and 6.6 μs respectively. Note that the suburban areas have larger mean delays and the rms delay spreads compared to typical urban areas, as expected.

Example 11.3 Symbol Duration Versus Delay Spread.
We assume that the channel equalisation is not required in a flat-fading channel, since the transmission bandwidth is narrower than the channel coherence bandwidth. Using the definition by (11.30), the symbol rate is limited by the delay spread as follows:

$$W < \Delta f_c \;\Rightarrow\; T > 5\sigma_\tau \;\Rightarrow\; R_s < 1/(5\sigma_\tau)$$

$$(11.36)$$

Using (11.36), the maximum symbol rate that can be supported without equalization in the considered channels is presented in Table 11.3. Table 11.4 shows whether fading is flat- or frequency-selective at some data rates that are typically used in mobile communications. In frequency-selective fading channels, higher order modulations, for example M-ary QAM, may be used to increase the symbol duration. Very narrow subcarriers bandwidths (very large symbol durations) in OFDM leads to flat fading.

Table 11.3 Maximum Symbol Rate That Can Be Supported in Various Channels Without Equalization.

Channel	Delay spread σ_τ (μs)	Max. symbol rate (ksymbols/s)
Rural area	0.11	1820
Urban area	1	190
Suburban area	2.53	83
Hilly terrain	6.88	29

Table 11.4 Effect of the Data Rate on the Frequency-Selectivity of a Channel.

Bit rate R_b	Bit duration, T_b	$\sigma_\tau <$ $T_b/5$	Channel
10 kbps	100 μs	20 μs	Flat-fading
271 kbps (GSM max.)	3.7 μs	0.74 μs	Flat fading in rural area. Frequency-selective fading in urban, suburban and hilly areas
3.84 Mbps (WCDMA)	0.26 μs	0.052 μs	Frequency-selective fading
54 Mbps (WLAN)	18.5 ns	3.7 ns	Frequency-selective fading

Example 11.4 ISI Due to MIP.

Assume that the transmitted bit stream over a channel is represented as a sequence of delta functions:

$$s(t) = \sum_{n=-\infty}^{\infty} \delta(t - nT_b) \qquad (11.37)$$

where T_b denotes the duration between impulses. Based on the assumption that the bit duration is arbitrarily short, this representation simplifies the analysis and provides valuable insight for visualising the effect of the delay-spread caused by the channel. It represents a worst-case ISI scenario since all impulses have the same polarity, that is, transmitted bits will be either all 1's or all 0's. Therefore, the results to be obtained may be considered to be qualitative.

The delta function transmitted at $t = 0$ is assumed to undergo a delay spread described by an exponential MIP, given by (11.22). The sample of the MIP at time $t = 0$ corresponds to the received signal, while the samples at $t = kT_b$, $k > 0$ represent the ISI. Therefore, mean signal to interference power ratio (SIR) may be written using (11.22) and (D.12) as follows:

$$SIR = \frac{R_c(0)}{\sum_{k=1}^{L-1} R_c(kT_b)} = \frac{R_c(0)}{\sum_{k=1}^{L-1} e^{-kT_b/\tau_0}} \qquad (11.38)$$

$$= \frac{1-z}{z-z^L} \approx \exp(T_b/\tau_0) - 1.$$

where $z = \exp(-T_b/\tau_0)$ and $L = \lfloor T_m/T_b \rfloor + 1$. The term $z^L \ll z$ is ignored to obtain the last expression.

Assuming that the *ISI* power is much higher than that of the AWGN, the E_b/N_{ISI} may be expressed in terms of the $SIR = S/ISI$ as follows:

$$SIR = \frac{S}{ISI} = \frac{E_b}{N_{ISI}} \frac{R_b}{B} \qquad (11.39)$$

where N_{ISI} denotes the PSD of the ISI. Assuming $B \approx R_b$, and the mean $\bar{\gamma}_b = E_b/N_{ISI}$ of the channel may be equated, apart from a constant factor, to the mean SIR. The average bit error probability for DPSK modulation in a Rayleigh fading ISI channel, described by an exponential pdf (see (F.103)), may be written as

$$P_2 = \frac{1}{2}\int_0^\infty e^{-\gamma}\frac{1}{\gamma}e^{-\gamma/\bar{\gamma}_b}d\gamma = \frac{1}{2(1+\bar{\gamma}_b)} \Rightarrow$$

$$\bar{\gamma}_b = \frac{1}{2P_2}-1 = e^{T_b/\tau_0}-1$$

$$(11.40)$$

Using (11.40), one may determine the maximum bit rate that can be supported in this ISI channel:

$$R_b = \frac{1}{T_b} = \frac{1}{\tau_0 \ln[1+\bar{\gamma}_b]} = -\frac{1}{\sigma_\tau \ln(2P_2)}$$

$$= -\frac{5}{\ln(2P_2)}\Delta f_c$$

$$(11.41)$$

Note from (11.41) that R_b is directly proportional to the coherence bandwidth, hence inversely proportional to the delay spread. The bit rate that can be supported in an ISI channel decreases as the required bit error probability becomes lower or, alternatively, as the required mean SIR $\bar{\gamma}_b$ increases. The ISI causes an irreducible error floor and the delay spread imposes a limit on the maximum bit rate in a multipath environment. [2]

Now assume that bit error probability for DPSK is required to be lower than 10^{-3} in a

ISI channel in a typical urban area with $\sigma_\tau = 1\,\mu s$. From (11.40) this corresponds to a SIR level of approximately $\bar{\gamma}_b = 27dB$. The maximum bit rate that can be supported by this channel is found from (11.41) as 161 kbps, whereas the prediction by (11.36) is 200 kbps (190 kbps is predicted in Table 11.3). For a bit error probability of 10^{-5}, the maximum supportable bit rate reduces to 92 kbps. The corresponding bit rates for BPSK modulation are 181 kbps and 99 kbps for bit error probabilities 10^{-3} and 10^{-5}, respectively.

11.2.2 Doppler Spread

Consider a mobile terminal at position **x**, with respect to the origin moves with a velocity $v = dx/dt$, where $x = |\mathbf{x}|$, (see Figure 11.9). A signal is incident at the mobile terminal making an angle φ with the direction of movement and is described by the following wave vector:

$$\mathbf{k} = -k\hat{r} = -\frac{2\pi}{\lambda}\hat{r} \qquad (11.42)$$

The received signal may be written as

$$r(t) = \mathrm{Re}\left[s_\ell(t)e^{j(w_ct-\mathbf{k}.\mathbf{x})}\right]$$

$$= \mathrm{Re}\left[s_\ell(t)e^{j(w_ct+kx\cos\varphi)}\right]$$

$$(11.43)$$

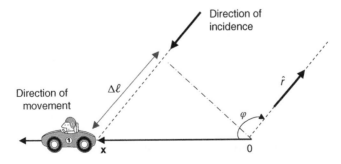

Figure 11.9 Geometry for Movement of a Mobile Terminal with Respect to the Origin at $x=0$.

As the position of the mobile terminal changes with time, the instantaneous frequency of the received signal is found from (11.43) as

$$f = \frac{1}{2\pi}\frac{d}{dt}(w_c t + kx\cos\varphi) = f_c\left(1 + \frac{v}{c}\cos\varphi\right)$$
$$= f_c + f_d$$

$$(11.44)$$

where v and φ are assumed to be time-invariant. The instantaneous frequency of the received signal is hence shifted due to the relative motion of the receiving mobile terminal. The so-called Doppler frequency shift f_d, which is given by

$$f_d = f_{d,max}\cos\varphi, \quad f_{d,max} = f_c\frac{v}{c} = \frac{v}{\lambda} \quad (11.45)$$

varies between $-f_{d,max}$ and $f_{d,max}$ depending on the values of φ. This results in a Doppler spread of $B_d = 2f_{d,max}$.

Doppler frequency shift may also be explained intuitively in terms of the path and phase differences of the ray at a distance x and the origin (see Figure 11.9):

$$\Delta\phi = k\Delta\ell = kx\cos\varphi = \frac{2\pi}{\lambda}x\cos\varphi$$

$$(11.46)$$

$$f_d = \frac{1}{2\pi}\frac{d}{dt}(\Delta\phi) = f_{d,max}\cos\varphi$$

The velocity v and the direction of incidence φ will often vary randomly with time depending

on the movement of the mobile terminal. Doppler spread B_d is a measure of the frequency band over which the instantaneous received carrier frequency would fluctuate randomly due to the relative movements of the transmitter, receiver and/or the scatterer. Therefore, it is modeled statistically and characterizes the random variations in the instantaneous carrier frequency.

We will first express the Doppler spectrum in terms of the pdf of the angle of incidence φ, for a constant value of v. Bearing in mind that f_d changes in $[-f_{d,max}, f_{d,max}]$ while φ changes randomly in the interval $[0,2\pi]$, using (11.45) and (F.53), the pdf of f_d may be related to that of the angle of incidence φ as follows:

$$f_{f_d}(\lambda) = \left[f_\varphi(\varphi) + f_\varphi(-\varphi)\right]\left|\frac{d\varphi}{df_d}\right|_{\varphi = \cos^{-1}(\lambda/f_{d,max})}$$

$$= \frac{2f_\varphi(\varphi)}{|f_{d,max}\sin\varphi|} = \frac{2f_\varphi(\varphi)}{\sqrt{f_{d,max}^2 - \lambda^2}}$$

$$(11.47)$$

In the special case where multipath signals arrive to the receiver uniformly from all directions, φ is uniformly distributed in $[0,2\pi]$ (see Figure 11.10a):

$$f_\varphi(\varphi) = 1/(2\pi), \quad 0 \le \varphi \le 2\pi \qquad (11.48)$$

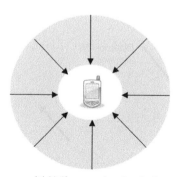

(a) Uniform angle-of-arrival

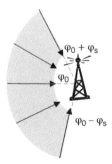

(b) Nonuniform angle-of-arrival

Figure 11.10 Angle of Arrival Distribution of Signals Incident on a Terminal.

Inserting (11.48) into (11.47) one gets the so-called Jake's Doppler spectrum:

$$f_{f_d}(\lambda) = \frac{1}{\pi\sqrt{f_{d,\max}^2 - \lambda^2}}, \quad -f_{d,\max} < \lambda < f_{d,\max}$$

(11.49)

According to the Jakes's Doppler spectrum shown in Figure 11.11, the received instantaneous frequencies are more likely to be concentrated at $f_c \pm f_{d,\max}$. The so-called channel coherence time is usually defined as the time interval between the points where the time-correlation function of the channel impulse response $R_c(\Delta t)$ becomes equal to $1/\sqrt{2}$, which corresponds to half-power points.

Assuming that a mobile terminal moves in a direction characterized by an angle of arrival φ, and the path gain does not change appreciably at the considered distances, the time-variant channel impulse response may be written from (11.12) and (11.43) as follows:

$$c(\tau;t) = e^{jkx(t)\cos\varphi}$$

(11.50)

Autocorrelation function given by (11.17) of the channel impulse response for a time difference of Δt is directly related to the time correlation function and is simply the Fourier transform of the Doppler spectrum:

$$R_c(\Delta t) = E[c^*(\tau;t)c(\tau;t+\Delta t)]$$

$$= E\left[e^{jk\cos\varphi\{x(t+\Delta t)-x(t)\}}\right]$$

$$= E\left[e^{j2\pi f_d \Delta t}\right]$$

$$= \int_{-f_{d,\max}}^{f_{d,\max}} e^{j2\pi f_d \Delta t} f_{f_d}(f_d) df_d$$

(11.51)

$$= \int_0^{2\pi} e^{j2\pi f_{d,\max}\Delta t\cos\varphi} f_\varphi(\varphi) d\varphi$$

$$= \frac{1}{2\pi}\int_0^{2\pi} e^{j2\pi f_{d,\max}\Delta t\cos\varphi} d\varphi$$

$$= J_0(2\pi f_{d,\max}\Delta t)$$

where we used (D.92) and the coordinate transformation between f_d and φ using (11.45). One should bear in mind that (11.49) and (11.51) are Fourier transform pairs and (11.51) shows the time selectivity of the channel according to Jake's Doppler spectrum. The Bessel function $J_0(x)$ of the first kind and order zero behaves as a damped sinusoid with increasing values of x and $J_0(0) = 1$ (see (D.94) and Figure D.1). Noting that $J_0^2(1.126) = 0.5$, the coherence time corresponding to Jake's Doppler spectrum is given by

$$\Delta t_c = 2\Delta t = \frac{1.126}{\pi f_{d,\max}} \cong \frac{0.72}{B_d}$$

(11.52)

(11.52) shows that the channel coherence time is inversely proportional to the Doppler spread

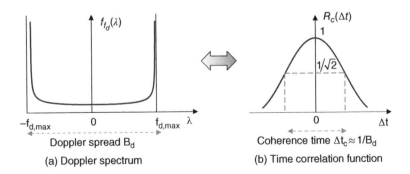

(a) Doppler spectrum

(b) Time correlation function

Figure 11.11   Jake's Doppler Spectrum and the Corresponding Time Correlation Function.

and correlation between the received signal samples decreases with increasing values of the velocity, v, and/or the time interval, Δt, between the time-samples (see (11.51)). Coherence time Δt_c is a measure of the expected time duration over which the channel's response is essentially invariant and fading appears to be highly correlated. Coherence time Δt_c may also be interpreted as the time to traverse a distance $\lambda/2$ when travelling at a constant velocity v:

$$\Delta t_c \approx \frac{1}{B_d} = \frac{1}{2f_{d,\max}} = \frac{\lambda/2}{v} \qquad (11.53)$$

If a mobile user is driving his car along the x-axis with a constant velocity v as shown in Figure 11.9 and the angle of arrival φ of incident signals with respect to the direction of movement is uniformly distributed in $[0,2\pi]$, then the time correlation function is given by (11.51). The assumption about the uniform distribution of the angle-of-arrival is generally valid for mobile terminals since scattering around them contribute uniformly to the received signal. However, signals usually arrive to base stations with a restricted angle-of-arrival (see Figure 11.10b). Therefore, as shown in Figure 11.12, the signals arriving to base stations from a restricted range of angle-of-arrival are more correlated, since

they are more likely to be scattered by the same scatterers.

If the channel is time-invariant, that is, $f_{d,\max} = 0$ and $R_c(\Delta t) = 1$, the received signal components are perfectly correlated with each other, irrespective of the differences in their transmission times. Then, from the Fourier transform relationship, one has $f_{f_d}(\lambda) = \delta(\lambda)$, which implies no spectral (Doppler) broadening at the receiver when a pure frequency tone is transmitted. On the other hand, as the velocity of the mobile terminal or the carrier frequency increases, the Doppler spread increases, which in turn leads to shorter coherence time. This evidently has a negative impact on the synchronization performance of the receiver.

Example 11.5   Effect of Random User Velocity on the Spaced-Time Correlation Function. Assume that the velocity of the mobile user considered in Figure 11.9 changes uniformly between v_1 and v_2. Hence the pdf of the velocity is given by

$$f_v(v) = \frac{1}{v_2 - v_1}, \quad v_1 < v < v_2 \qquad (11.54)$$

The corresponding time correlation function for uniformly distributed angle-of-arrival is found using (11.51) and (D.64)

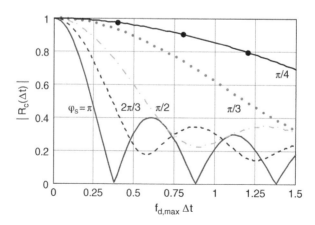

Figure 11.12   Time Autocorrelation Function $R_c(\Delta t) = 1/(2\varphi_s) \int_{-\varphi_s}^{\varphi_s} e^{j2\pi f_{d,\max} \Delta t \cos\varphi} d\varphi$ of the Impulse Response For a Uniformly Distributed Angle of Arrival in $[-\varphi_s, \varphi_s]$.

$$E_v[R_c(\Delta t)] = \int_{v_1}^{v_2} J_0(kv\Delta t) f_v(v) dv$$

$$= \frac{1}{kv_2\Delta t - kv_1\Delta t} \int_{kv_1\Delta t}^{kv_2\Delta t} J_0(z) dz$$

$$= \frac{2}{kv_2\Delta t - kv_1\Delta t}$$

$$\sum_{k=0}^{\infty} [J_{2k+1}(kv_2\Delta t) - J_{2k+1}(kv_1\Delta t)]$$

(11.55)

Figure 11.13 shows the variation of (11.55) for various values of v_2/v_1. As v_2-v_1 approaches zero, the correlation coefficient given by (11.55) reduces to $J_0(kv\Delta t)$, which implies that the correlation coefficient is not much influenced by the user velocity as long as it is uniformly distributed in a very narrow range. Figure 11.13 also shows that the correlation function decays faster with increasing spreads of the user velocity. In other words, the assumption of constant user velocity implies slower decay of time correlations between the received signal samples with Δt intervals.

11.2.2.1 Slow Versus Fast Fading

The fading is said to be slow if $T < \Delta t_c$ $(W > B_d)$. A slowly fading channel has a large coherence time Δt_c, or equivalently, a small Doppler spread. However, a channel is said to suffer fast fading whenever $T > \Delta t_c$ $(W < B_d)$ which implies smaller coherence time Δt_c and higher Doppler spreads due to faster user velocities.

Example 11.6 Efect of Coherence Time on the Minimum Data Rate.
Consider a mobile user travelling at a velocity of v km/h receives a 900 MHz signal. The variation of the coherence time from (11.53) and the corresponding transmission rate $R_s = 1/T > 1/\Delta t_c$ in a slowly fading channel is given Table 11.5.

The coherence time, which is inversely proportional to the user velocity (see (11.53)), decreases as the users move faster. The symbol duration should also decrease if slow-fading conditions are desired. Thus, the symbol duration must be shorter than the coherence time; then transmitted symbols are least affected by the

Table 11.5 Effect of User Velocity on the Transmission Rate.

Velocity v (km/h)	Coherence time Δt_c (ms)	Transmission rate (symbols/s)
5	120	8.33
50	12	83.3
100	6	166.7
150	4	250
200	3	333.3

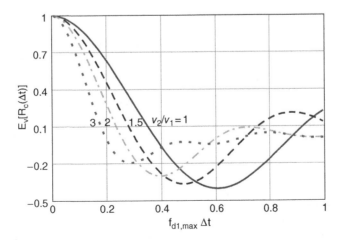

Figure 11.13 Time Correlation Function For User Velocity Changing Uniformly Between v_1 and v_2.

changes in the channel. The implication is that Doppler spread is usually not a serious issue for high-speed wireless communications. However, the Doppler spread causes an irreducible error floor when differential detection is used due to the decorrelation of the reference signal. The error floor decreases with increasing data rates for a given value of the Doppler spread (coherence time), as expected. [3] On the other hand, increasing user velocities impose larger transmission rates, which tend to make the channel frequency-selective. As the symbol duration decreases, the condition $T > T_m$ may be violated in a flat fading channel.

Example 11.7 Power Control in Mobile Radio Systems.

Power control is a serious issue in both TDMA and CDMA-based cellular radio systems. In the uplink, the transmit power of the MS must be sufficiently high so as to achieve the desired E_b/N_0 at the BS. On the other hand, it must be kept as low as possible so as to reduce co-channel interference to nearby BSs, to alleviate health concerns and to increase battery life. In CDMA systems, automatic power control is essential to ensure reception with equal-powers of multiple users' signals so as to avoid small-signal suppression (see Figure 10.13 and Figure 10.15). To achieve the desired E_b/N_0 at the BS, transmit powers of the MSs in the close vicinity of the BS do not need to be as high as those near the cell edges. Otherwise the received power levels at the BS would differ several tens of dB's.

Consider now that two MSs with transmit powers P_{t1} and P_{t2} are transmitting to the same BS from distances r_1 and r_2, respectively. Assuming a path-loss exponent n, the ratio of the received powers from the two MSs located at r_1 and r_2 is given by (3.92):

$$\frac{P_r(r_1)}{P_r(r_2)} = \frac{\dfrac{P_{t1}G_tG_r}{(4\pi r_0/\lambda)^2}(r_0/r_1)^n}{\dfrac{P_{t2}G_tG_r}{(4\pi r_0/\lambda)^2}(r_0/r_2)^n} = \frac{P_{t1}}{P_{t2}}\left(\frac{r_2}{r_1}\right)^n$$

(11.56)

If the transmit powers of the two MSs are the same $(P_{t1} = P_{t2})$, then the ratio of received powers at the BS is 30 dB for $n = 3$ and $r_2/r_1 = 10$. Alternatively, if the powers received by the BS are required to be the same, then P_{t1} may be 30 dB below P_{t2}. This implies significant saving in battery power for the user closer to the BS.

Hence, a MS can adjust its transmit power as it moves in the cell. In open-loop power control, the MS looks at the received signal level in the downlink, and adjusts its uplink transmit power by frequency scaling. In closed-loop power control, the BS monitors the received signal power and informs the MS to adjust its transmit power for proper reception at the BS. In a fading channel, the update interval in the transmit power level should be shorter than the channel coherence time so as to be able to follow the time changes in the channel. Consider that power control system of a cellular radio system, operating at 900 MHz, is required to track MSs velocities up to 250 km/h. The corresponding channel coherence time is $\Delta t_c = 2.4\,\text{ms}$ (see (11.53)). The time interval between two updates must therefore be shorter than the channel coherence time. In other words, the power control to be effective, the update rate of the power control system should be faster than the Doppler rate:

$$\Delta t_{update} < \Delta t_c = 2.4ms \Rightarrow$$
$$f_{update} = 1/\Delta t_{update} > 417Hz$$

(11.57)

This implies that the power control system is required to update the transmit power in every 2.4 ms, that is, 417 times per second.

Example   11.8 Spaced-Time   Spaced-Frequency and Scattering Function Characterization of a Two-Path Channel.

Two-path channel impulse response, defined by

$$c(\tau;t) = \alpha_1 e^{-j\theta_1}\delta(\tau-\tau_1) + \alpha_2 e^{-j\theta_2}\delta(\tau-\tau_2)$$

(11.58)

is frequently used in wireless communication systems. Multipath delays τ_1 and τ_2, phases θ_1

and θ_2, and channel gains α_1 and α_2 may all vary randomly as a function of time t. The model assumes a propagation environment with no LOS and the presence of two different clusters, each of which consists of a large number of scatterers. The propagation environment is assumed to undergo Rayleigh fading; θ_1 and θ_2 are independent and uniformly distributed in $[0,2\pi]$ while α_1 and α_2 are uncorrelated complex Gaussian circularly symmetric random variables with zero mean and variances $E\left[|\alpha_i|^2\right] = 2\sigma_i^2, \quad i=1,2.$

Multipath intensity profile of the channel is given by (11.20):

$$R_c(\tau) = E\left[|c(\tau;t)|^2\right]$$

$$= E\left[\left(\alpha_1 e^{-j\theta_1}\delta(\tau-\tau_1) + \alpha_2 e^{-j\theta_2}\delta(\tau-\tau_2)\right)\right.$$

$$\left.\left(\alpha_1 e^{j\theta_1}\delta(\tau-\tau_1) + \alpha_2 e^{j\theta_2}\delta(\tau-\tau_2)\right)\right]$$

$$= E\left[\alpha_1^2\delta(\tau-\tau_1) + \alpha_2^2\delta(\tau-\tau_2)\right.$$

$$\left. + \alpha_1\alpha_2\cos(\theta_1-\theta_2)\delta(\tau-\tau_1)\delta(\tau-\tau_2)\right]$$

$$= 2\sigma_1^2\delta(\tau-\tau_1) + 2\sigma_2^2\delta(\tau-\tau_2)$$

(11.59)

where the assumption $\tau_1 \neq \tau_2$ implies uncorrelated scattering of the two multipath components from separate clusters of obstacles. The channel transfer function, which is defined by the Fourier transform of the impulse response

$$C(f;t) = \mathfrak{F}_\tau[c(\tau;t)]$$

$$= \alpha_1 e^{-j(w\tau_1+\theta_1)} + \alpha_2 e^{-j(w\tau_2+\theta_2)}$$

(11.60)

also varies randomly with time to the randomness of α_i, τ_i and θ_i.

For the sake of simplicity, let $\tau_1 = 0$ and $\theta_1 = 0$ and the second multipath component is caused by a mobile obstacle with a constant velocity v_2, that is, $\theta_2 = kv_2\cos\varphi t = 2\pi f_d t$, where $f_d = f_{d,max}\cos\varphi$ with $f_{d,max} = v_2/\lambda$. Here φ denotes the angle between the direction of movement and the direction of incidence of the considered multipath signal. The spaced-frequency, spaced-time correlation function of the channel is determined using (11.60):

$$R_C(\Delta f, \Delta t) \triangleq E[C(f+\Delta f;t+\Delta t)C^*(f;t)]$$

$$= E\left[\left\{\alpha_1 + \alpha_2 e^{-jkv_2\cos\varphi(t+\Delta t)}e^{-j2\pi(f+\Delta f)\tau_2}\right\}\right.$$

$$\left. \times \left\{\alpha_1 + \alpha_2 e^{jkv_2\cos\varphi t}e^{j2\pi f\tau_2}\right\}\right]$$

$$= 2\sigma_1^2 + 2\sigma_2^2 e^{-j2\pi(f_d\Delta t + \Delta f\tau_2)}$$

(11.61)

Figure 11.14 shows the variation of the normalized spaced-frequency correlation function $R_C(\Delta f)$ with $\Delta f\tau_2$ and the normalized spaced-time correlation function $R_C(\Delta t)$ as a function of $f_d\Delta t$. Frequency and time selectivity of the channel depends on the ratio σ_2^2/σ_1^2. The channel becomes more selective as the mean powers of the two multipath components become comparable to each other. One may also observe that the channel becomes more frequency (time) selective for increasing values of τ_2 (f_d).

The 3-dB coherence bandwidth of the channel is defined as the separation between frequencies at where $|R_C(\Delta f,0)| = 1/\sqrt{2}$ of its maximum value (or between 3-dB points in Figure 11.14):

$$\frac{|R_C(\Delta f,0)|}{|R_C(\Delta f,0)|_{max}} = \frac{1}{\sqrt{2}} \Rightarrow \Delta f_c = 2\Delta f = \frac{\xi(\beta)}{\tau_2}$$

$$\xi(\beta) = \frac{1}{\pi}\cos^{-1}\left(\frac{-(1-\beta)^2}{4\beta}\right)$$

(11.62)

where $\beta = \sigma_2^2/\sigma_1^2$. One may easily observe from Figure 11.15 that the factor $\xi(\beta)$ approaches its minimum value of 0.5 as the mean powers of the two multipath components become equal to each other. Then, the coherence bandwidth reaches its minimum value. For example, for $\beta = 0.4$, the channel coherence bandwidth is given by $\Delta f_c = 0.572/\tau_2$ where τ_2 denotes the maximum excess delay in this example. A maximum excess delay of $\tau_2 = 7\mu s$ in a typical urban area allows distortionless signal transmission over 82 kHz coherence bandwidth.

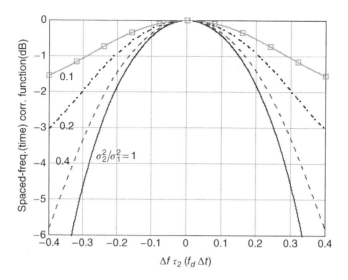

Figure 11.14   Spaced-Frequency ($\Delta t = 0$) and Spaced-Time ($\Delta f = 0$) Correlation Functions.

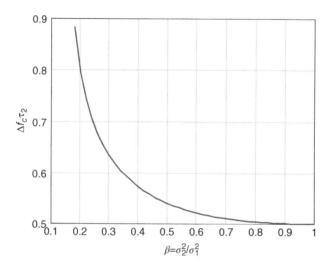

Figure 11.15   The Variation of the Coherence Bandwidth and Coherence Time with the Ratio of Powers σ_2^2/σ_1^2 of Channel Gains of a Two-Path Wireless Channel.

Similarly, the channel coherence time is found from (11.61) as

$$\frac{|R_C(0,\Delta t)|}{|R_C(0,\Delta t)|_{max}} = \frac{1}{\sqrt{2}} \Rightarrow \Delta t_c = 2\Delta t = \frac{1}{f_d}\xi(\beta)$$

(11.63)

where f_d denotes the Doppler shift. For example, for $\beta = 0.4$, the channel coherence time is given by $\Delta t_c = 0.572/f_d$. A mobile station (MS) operating at 900 MHz and with 120 km/h velocity will experience a maximum Doppler shift of $f_{d,\,max} = 100$ Hz; using (11.63), the resulting the coherence time will be

5.72 ms. If the MS has a transmission rate of 100 ksymbols/s (10 µs symbol duration), the required interleaving depth will be $5.72\,\text{ms}/10\,\mu\text{s} = 572$ symbols for 120 km/h mobile velocity.

Using (11.27), (11.61) and (D.36), the scattering function of the channel is found as

$$S_C(\tau,\lambda)$$

$$\triangleq \iint_{\Delta f\,\Delta t} R_C(\Delta f,\Delta t)e^{j2\pi\lambda\Delta t}e^{j2\pi\Delta f\tau}d(\Delta t)d(\Delta f)$$

$$= 2\sigma_1^2\delta(\lambda)\delta(\tau) + 2\sigma_2^2\delta(\lambda-f_d)\delta(\tau-\tau_2)$$

$$\tag{11.64}$$

The scattering function given by (11.64) shows the presence of two scatterers. The first scattered signal has an average power of $2\sigma_1^2$, does not cause any Doppler shift and arrives to the receiver with no delay. The signals scattered from the second obstacle arrive to the receiver with an average power of $2\sigma_2^2$, with a relative delay of τ_2 and undergo a Doppler frequency shift of f_d. The scattering function is shown in Figure 11.16 for a constant value of v_2. The first ray with an average power σ_1^2 is located at $(\tau,\lambda) = (0,0)$ whereas the second ray of average power σ_2^2 is located at $(\tau,\lambda) = (\tau_2, f_d)$. The angle φ is assumed to have a constant value. When φ is uniformly distributed in $[0,2\pi]$, the Doppler frequency shift at $\tau = \tau_2$

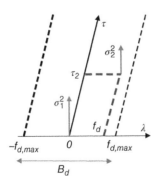

$$-f_{d,max} \quad 0 \quad f_{d,max}$$
$$B_d$$

Figure 11.16   Scattering Function Given by (11.64).

has is characterized by Jakes Doppler spread (see (11.49)).

The expressions for the received signal in baseband and passband are given by

$$r_\ell(t) = c(\tau;t) \otimes s_\ell(t)$$

$$= \alpha_1 s_\ell(\tau) + \alpha_2 e^{-j2\pi f_d t}s_\ell(\tau-\tau_2)$$

$$r(t) = \text{Re}\left[r_\ell(t)e^{jw_c t}\right] \tag{11.65}$$

$$= \alpha_1 s_\ell(\tau)\cos(2\pi f_c t)$$

$$+ \alpha_2 s_\ell(\tau-\tau_2)\cos(2\pi(f_c-f_d)t)$$

Example 11.9   Scattering Function. Consider a wideband channel which is characterized by the following scattering function:

$$S_c(\tau,\lambda) =$$

$$\begin{cases} \dfrac{1}{\tau_0}e^{-\tau/\tau_0}\dfrac{1}{\sqrt{2\pi}\sigma}e^{-\dfrac{(\lambda-f_c)^2}{2\sigma^2}} & -\infty \leq \lambda \leq \infty,\; 0 \leq \tau \leq \infty \\ \\ 0 & else \end{cases}$$

$$\tag{11.66}$$

Spaced-time spaced-frequency correlation function is given by the Fourier transform of (11.66):

$$R_C(\Delta f,\Delta t)$$

$$\triangleq \int_{-\infty}^{\infty}\int_0^{\infty} S_C(\tau,\lambda)e^{-j2\pi\Delta f\tau}e^{-j2\pi\lambda\Delta t}d\tau d\lambda$$

$$= R_C(\Delta f)R_C(\Delta t)$$

$$\tag{11.67}$$

The magnitude of the spaced-frequency correlation function, given by (11.28), may be written as

$$|R_C(\Delta f)| = \left|\int_0^{\infty}\frac{1}{\tau_0}e^{-\tau/\tau_0}e^{-j2\pi\Delta f\tau}d\tau\right|$$

$$= \frac{1}{\sqrt{1+(2\Delta f/\Delta f_c)^2}} \tag{11.68}$$

where the channel coherence bandwidth is defined in terms of the frequency separation at which the spaced-frequency coherence function drops to $1/\sqrt{2}$ of its maximum value, hence $\Delta f_c = 2\Delta f = 1/(\pi\tau_0) = 1/(\pi\sigma_\tau)$ (see (11.29)).

Spaced-time correlation function is found using (11.66), (11.67) and (D.55):

$$R_C(\Delta t) = \frac{1}{\sqrt{2\pi}\sigma} \int_{-\infty}^{\infty} e^{-\frac{(\lambda - f_c)^2}{2\sigma^2}} e^{-j2\pi\lambda\Delta t} d\lambda$$

$$= \frac{1}{\sqrt{2\pi}\sigma} e^{-j2\pi f_c \Delta t} \int_{-\infty}^{\infty} e^{-\frac{z^2}{2\sigma^2}} e^{-j2\pi z\Delta t} dz$$

$$= \frac{2}{\sqrt{2\pi}\sigma} e^{-j2\pi f_c \Delta t} \int_{0}^{\infty} e^{-\frac{z^2}{2\sigma^2}} \cos(2\pi z\Delta t) dz$$

$$= e^{-j2\pi f_c \Delta t} e^{-2\pi^2 \sigma^2 \Delta t^2}$$

$$(11.69)$$

The coherence time is obtained by equating $|R_C(\Delta t)| = 1/\sqrt{2}$:

$$\Delta t_c = 2\Delta t = \frac{\sqrt{\ln 2}}{\pi\sigma} = \frac{1}{3.77\sigma} \quad (11.70)$$

For example if the standard deviation of the Doppler spread is $\sigma = 30$ Hz, then the channel coherence time is 8.8 ms. Figure 11.17 shows the magnitudes of the spaced-time and spaced-frequency correlation functions. Note that the spaced-time correlation function decays faster than the spaced-frequency correlation function.

11.2.3 The Effect of Signal Characteristics on the Choice of a Channel Model

Received low-pass equivalent signal, which is the convolution of the low-pass equivalent transmitted signal and the channel impulse response, may be written as (see (1.46))

$$r_\ell(t) = c(\tau;t) \otimes s_\ell(t) + w(t)$$

$$= \int_{-\infty}^{\infty} S_\ell(f)C(f;t)e^{j2\pi ft} df + w(t)$$

$$(11.71)$$

The channel transfer function $C(f;t)$ is defined by (11.24) as the Fourier transform of the impulse response. In an ideal low-pass

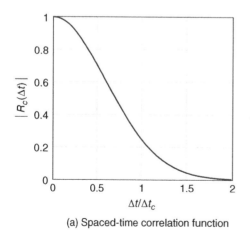

(a) Spaced-time correlation function

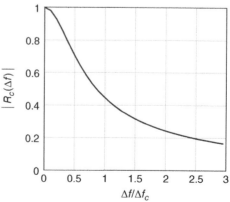

(b) Spaced-frequency correlation function

Figure 11.17 Spaced-Time and Spaced-Frequency Correlation Functions Corresponding to the Scattering Function Given By (11.66).

channel $r_\ell(t)$ is desired to be the same as $s_\ell(t)$. This implies $C(f;t) = 1$, (or equivalently $c(\tau;t) = \delta(\tau)$); hence, a channel with no multipath and no delay.

Now consider digital information transmission over a channel by modulating $s_\ell(t)$ at rate $W = 1/T$, which is also equal to the passband transmission bandwidth for the Nyquist pulse shape (see (6.106)). Since $S_\ell(f)$ is then bandlimited to $W/2$, one may rewrite (11.71) as

$$r_\ell(t) = \int_{-W/2}^{W/2} S_\ell(f)C(f;t)e^{j2\pi ft}\,df + w(t)$$

$$(11.72)$$

This implies that the received signal is determined by the behavior of $C(f;t)$ over $[-W/2, W/2]$ only. In other words, transmission of a signal $s_\ell(t)$ of bandwidth W over a channel is closely related to the coherence bandwidth Δf_c of a multipath fading channel. If the transmission bandwidth, W ($\approx 1/T$) is narrower than the coherence bandwidth ($\Delta f_c \approx 1/T_m$) of the

channel, that is, $W < \Delta f_c$, then the channel is said to undergo *flat fading*. In a flat-fading channel, all frequency components of the signal fade in a highly correlated manner, hence the channel transfer function $C(f;t)$ does not change appreciably over the bandwidth W. In a flat fading channel, (11.72) may be approximated by

$$r_\ell(t) \cong C(0;t) \int_{-W/2}^{W/2} S_\ell(f)e^{j2\pi ft}\,df + w(t)$$

$$= \alpha(t)e^{-j\phi(t)}s_\ell(t) + w(t), \quad 0 < t < T$$

$$(11.73)$$

where the channel transfer function at the mid-band, described by $C(f,t) \cong C(0,t) = \alpha(t)e^{-j\phi(t)}$, is a complex Gaussian process (see Figure 11.18). It is clear from (11.73) that the frequency nonselective (flat) fading results in multiplicative distortion of the transmitted signal. The flat fading condition $W < \Delta f_c$ implies that the symbol duration is longer than the maximum excess delay $T > T_m$ (see

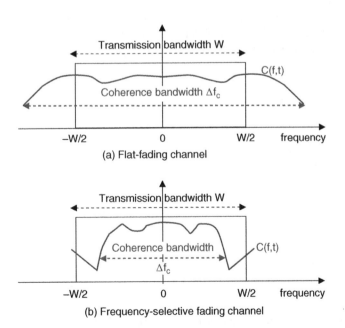

Figure 11.18 Coherence Bandwidth Versus Transmission Bandwidth in Flat- and Frequency-Selective Fading Channels.

Figure 11.7). The received multipath components are not resolvable and the receiver accepts these components as if they arrive via a single fading path. Consequently, flat fading channels may be equalized by a single tap FIR filter with a tap coefficient inversely proportional to $C(0,t)$.

If the transmission bandwidth W is larger than the coherence bandwidth Δf_c of the channel, then the channel suffers *frequency-selective fading*. In such channels, frequency components of a transmitted signal, which are separated more than Δf_c from each other, will fade almost independently from each other, that is, will have different channel gains and phase shifts over the transmission bandwidth (see Figure 11.18). $W > \Delta f_c$ implies that the maximum excess delay will be longer than the symbol duration $T_m > T$ (see Figure 11.7). Then the received multipath components are resolvable and some multipath components arrive to the receiver during consecutive symbol durations, thus causing intersymbol interference (ISI). Sophisticated equalization techniques are needed to mitigate ISI.

On the other hand, the relation between the channel coherence time Δt_c and the symbol duration T determines whether fading is fast or slow.

If $T > \Delta t_c$, then the channel is *fast-fading*, but the channel fades slowly for $T < \Delta t_c$. *Slow fading* is usually caused by shadowing by large obstructions between transmitter and receiver, but fast fading is due to multipath effects. Since the symbol duration is shorter than the channel coherence time in a *slow fading channel*, the multiplicative process, described by (11.73), may be regarded as constant during a signaling interval

$$\alpha(t) \cong \alpha, \quad \phi(t) \cong \phi, \quad 0 < t < T \qquad (11.74)$$

Assuming that the signal undergoes Rayleigh fading, the envelope α of the received signal in (11.74) has a Rayleigh pdf while the channel phase ϕ is uniformly distributed. In a so-called *block-fading channel*, $\alpha(t)$ and $\phi(t)$ are assumed to be constant over a block of bits.

In summary, a fading channel may be classified as slow or fast fading depending on whether the symbol duration is shorter or longer than the channel coherence time. Similarly, the channel is said to be flat-fading if $W < \Delta f_c$, otherwise it is frequency-selective. Hence, channels may be divided into four categories, based on coherence time and frequency in comparison with the signal transmission bandwidth (see Figure 11.19). In a *flat and slow fading*

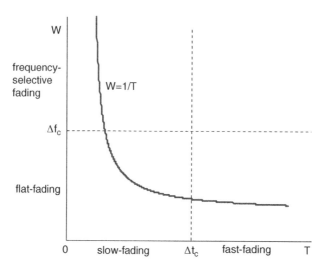

Figure 11.19   Channel Classification.

($W \ll \Delta f_c$ and $T \ll \Delta t_c$) environment, diversity overcomes the fading effects. However, in a frequency-selective and slow fading ($W \gg \Delta f_c$ and $T \ll \Delta t_c$) environment, one may subdivide the transmission channel into a number of sub-channels with bandwidths narrower than the coherence bandwidth (to operate in flat-fading); hence use multi-carrier transmission as in OFDM. Rake receiver is a well-known technique for diversity reception of broadband signals, such as CDMA, in frequency-selective fading environments. In a Rake receiver, SNRs of the multipath components (chips) of a received signal are combined. As will be studied in the next chapter, this leads to considerable improvement in the output SNR and the bit error probability performance. The usual approach in fast fading channels is to increase the transmission rate so as to force the symbol duration to be smaller than the coherence time and the signal to undergo slow-fading.

Example 11.10   Coherence Time and Coherence Bandwidth of the GSM900 System.
The GSM900 system comprises 124 frequency channels each with $W = 200$ kHz bandwidth in the downlink and the uplink. Hence 25 MHz bandwidth is allocated to each of uplink and downlink. Uplink and downlink frequency channels are separated by 45 MHz from each other for frequency isolation purposes. Transmission in each frequency channel is organized in frames of 4.6155 ms duration. A frame is divided into 8 time slots, each with $T_{slot} = 4.6155$ ms/8 = 0.577 ms. Each frame is shared by 8 users in a TDMA format.

Now consider a pedestrian user with a speed of $v = 5$ km/hr (1.39 m/s) and a user on board a high-speed train at 250 km/hr ($\cong 70$ m/s) The coherence time is given by

$$\Delta t_c = \frac{\lambda/2}{V} = \begin{cases} 120 \ ms & v = 5 km/h \\ 2.38 \ ms & v = 250 km/h \end{cases} \quad (11.75)$$

Since $\Delta t_c \gg T_{slot}$, the GSM900 system fades slowly for all user speeds of practical interest.

As a rule of thumb, an interleaving depth of $10\Delta t_c$ is reasonable to distribute block errors due to fades so that random error correcting codes can correct them. However, an interleaving depth of $10\Delta t_c = 1.2$ s for $v = 5$ km/h is not tolerable in real time transmissions such as voice. However, interleaving depths on the order of $10\Delta t_c = 23.8$ ms are feasible for real-time applications for high-speed users.

In a suburban environment, which is characterized by a delay spread of $\sigma_\tau = 2.53$ µs, the coherence bandwidth is found to be

$$\Delta f_c = 1/(5\sigma_\tau) = 79 kHz \quad (11.76)$$

Since $W > \Delta f_c$, the GSM900 system undergoes frequency-selective fading irrespective of the terminal speed. Therefore, the GSM900 system requires equalization, which is achieved using a training sequence of 26 bits (of 96 µs duration)) in each time slot. This corresponds to nearly 17% of the slot duration.

Example 11.11   Effect of Coherence Time and Coherence Bandwidth in the OFDM System Design.
Inter-carrier interference (ICI) and inter-symbol interference (ISI) limit the performance of OFDM systems. The ISI may be reduced by suitable choices of symbol and cyclic prefix durations. Cyclix prefix provides a guard interval T_G between two consecutive symbols. If the channel delay requirements permit, losses in SNR and in the data rate caused by the use of cyclic prefix may be reduced by choosing symbol period $T \gg T_G$. ISI requirement dictate the OFDM symbol duration to be longer than the maximum excess delay T_m of the channel:

$$T = \frac{N}{N\Delta f} \gg T_m \Rightarrow N \gg BT_m \quad (11.77)$$

where N denotes the number of subcarriers, separated from each other by Δf, in the OFDM bandwidth $B = N\Delta f = N/T$.

On the other hand, if the symbol duration T is chosen too long, then the frequency separation

Δf between orthogonal subcarriers would become narrower. Then, the orthogonality between subcarriers would be more susceptible to Doppler frequency shifts; that would increase ICI which degrades the system performance. Therefore, one needs to choose $\Delta f = 1/T$ to be much larger than the maximum Doppler shift $f_{d,max}$ in order to reduce the ICI:

$$\Delta f = \frac{N\Delta f}{N} >> f_{d,\max} \Rightarrow N << \frac{B}{f_{d,\max}} \quad (11.78)$$

The above condition also implies $f_{d,max}T$ $<<1$, hence an underspread channel.

The above two constraints on N may be combined to show the limitations imposed by delay- and Doppler-spreads on the OFDM system design:

$$BT_m << N << B/f_{d,\max}$$
$$T_m << T << 1/f_{d,\max} \quad (11.79)$$

The ISI may be assumed to be negligibly small if we choose the OFDM symbol duration T for $T_G/T = 1/4$ as:

$$T > 5T_m \quad (11.80)$$

Similarly, the ICI may be neglected if the maximum Doppler shift is less than 3% of the frequency separation between subcarriers Δf:

$$f_{d,\max} < 0.03\Delta f \Rightarrow T < 0.03/f_{d,\max} \quad (11.81)$$

Using (11.79)–(11.81), OFDM symbol duration is bounded by

$$5T_m < T < 0.03/f_{d,\max}$$
$$\Rightarrow 5BT_m < N < 0.03B/f_{d,\max} \quad (11.82)$$

For $B = 5$ MHz, $T_m = 20\,\mu s$ (hilly terrain), and $f_{d,max} = 250$ Hz (120 km/h at 2.2 GHz), the number of OFDM subcarriers should satisfy $250 < N < 600$. A reasonable choice would be $N = 2^9 = 512$.

11.3 Modeling Fading and Shadowing

The mathematical background for probability, random variables and stochastic processes is already introduced in Appendix F. Therefore, this section will only provide a brief review of the statistical tools for characterizing channels suffering fading and/or shadowing. We will study Rayleigh, Rician and Nakagami-m fading, log-normal shadowing and composite fading and shadowing. The emphasis will be on the physical interpretations rather than the mathematical details.

11.3.1 Rayleigh Fading

Consider the transmission of a low-pass equivalent signal $s_\ell(t)$ in a slow flat-fading discrete multipath channel over a signaling interval $[0,T]$. Assuming that the maximum excess delay is shorter than the symbol duration, that is, $\tau_k << T$, $k = 0, 1,.., L-1$, the received equivalent low-pass signal given by (11.8) may be written as the phasor sum of the L time-variant multipath components:

$$r_l(t) \cong s_\ell(t) \sum_{k=0}^{L-1} \alpha_k(t) e^{-j\theta_k(t)} + w(t)$$

$$= s_\ell(t)[x(t) - jy(t)] + w(t), \quad 0 \le t < T$$

$$x(t) = \sum_{k=0}^{L-1} \alpha_k(t)\cos\theta_k(t) \quad y(t) = \sum_{k=0}^{L-1} \alpha_k(t)\sin\theta_k(t)$$

$$(11.83)$$

Path gains $\alpha_k(t)$ and phases $\theta_k(t)$ may change differently with time, hence comprise different Doppler shifts. Although each individual multipath component may be small, their sum may become considerably high. For large number L of multipaths, the central limit theorem states that $x(t)$ and $y(t)$ are independent Gaussian random processes in time with zero mean and variance σ^2. Therefore, $x(t)$, $y(t) \sim N(0, \sigma^2)$ and are described by a Doppler spectrum.

The received signal in (11.83) may be rewritten as

$$r_\ell(t) = a(t)e^{-j\phi(t)}s_\ell(t) + w(t) \quad 0 \le t < T$$

$$a(t) = \sqrt{x(t)^2 + y(t)^2} \quad \phi(t) = \tan^{-1}(y(t)/x(t))$$

$$(11.84)$$

In slow fading ($T < \Delta t_c$), $a(t)$ and $\phi(t)$ may be assumed constant during the signaling interval, that is, $a(t) \approx a$ and $\phi(t) \approx \phi$. The pdfs of a and ϕ may be derived from the joint pdf of x and y, which are independent from each other:

$$f_{X,Y}(x,y) = \frac{1}{\sqrt{2\pi}\sigma}e^{-\frac{x^2}{2\sigma^2}} \frac{1}{\sqrt{2\pi}\sigma}e^{-\frac{y^2}{2\sigma^2}} = \frac{1}{2\pi\sigma^2}e^{\frac{x^2+y^2}{2\sigma^2}}$$

$$(11.85)$$

We obtain joint and marginal pdf's of a and ϕ by using the variable transformation with $x = a\cos\phi$, $y = a\sin\phi$ (see (F.63)):

$$f_{a,\Phi}(a,\phi) = a f_{X,Y}(a\cos\phi, a\sin\phi)$$

$$= \frac{a}{2\pi\sigma^2}\exp\left(-\frac{a^2}{2\sigma^2}\right)$$

$$f_a(a) = \int_0^{2\pi} f_{a,\Phi}(a,\phi)d\phi = \frac{a}{\sigma^2}\exp\left(-\frac{a^2}{2\sigma^2}\right), \quad a \ge 0$$

$$f_\Phi(\phi) = \int_0^\infty f_{a,\Phi}(a,\phi)da = \frac{1}{2\pi}, 0 \le \phi \le 2\pi$$

$$(11.86)$$

The channel gain a, which is assumed to be constant over a signalling interval T, is Rayleigh distributed with the following moments (see (F.136)):

$$E[a^k] = \int_0^\infty a^k f_a(a)da = (2\sigma^2)^{k/2}\Gamma(1+k/2)$$

$$E[a] = \sqrt{\pi/2}\sigma$$

$$E[a^2] = 2\sigma^2$$

$$(11.87)$$

Figure 11.20 shows the pdf of a for various values of σ^2, which also affects the mean value of the channel gain (see (11.87)). Figure 11.21 shows a typical realization of the channel gain a of a slowly fading channel for 200 iterations and $\sigma^2 = 1/2$. The channel undergoes slow-fading as long as a does not change appreciably over the symbol duration and is said to undergo block-fading if a does not change appreciably during a block of symbols. Fast fading implies that a may not be assumed to be constant during a symbol period.

The pdf of $p = a^2$ may be found via a change of variable:

$$f_P(p) = f_a(a)\left|\frac{da}{dp}\right|\bigg|_{a=\sqrt{p}} = \frac{1}{P_0}\exp\left(-\frac{p}{P_0}\right), \quad p \ge 0$$

$$(11.88)$$

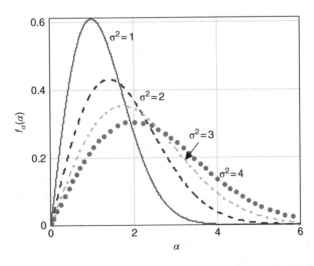

Figure 11.20 The Pdf of the Envelope of the Received Signal Experiencing Rayleigh Fading For Various Values of the Rariance σ^2.

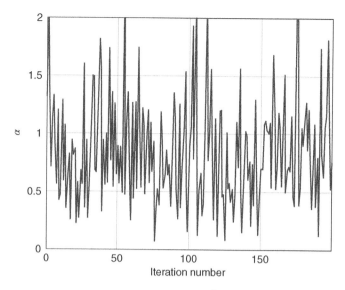

Figure 11.21 Typical Variation of α with Time For $\sigma^2 = 1/2$ in a Rayleigh Fading Channel.

where $P_0 = E[P] = E[\alpha^2] = 2\sigma^2$. It is important to note that the envelope of a Rayleigh fading signal is described by Rayleigh distribution (11.86), but the power of a Rayleigh fading signal has exponential distribution (11.88). The outage probability, that is, the probability that P is below a threshold level $P \leq P_{th}$, is found from (11.88) as

$$Pr(P \leq P_{th}) = \int_0^{P_{th}} f_P(p)\,dp = 1 - \exp(-P_{th}/P_0)$$

(11.89)

The moment generating function of the power of Rayleigh-faded signals is given by

$$M_P(s) = E[e^{-sP}] = \frac{1}{1 + sP_0}$$

(11.90)

11.3.2 Rician Fading

In addition to randomly moving scatterers, if the received signal has a dominant component A due to, for example, fixed scatterers/reflectors or a LOS, then the received low-pass

equivalent signal may be written as the sum of the contributions from diffuse and specular components:

$$r_\ell(t) = s_\ell(t)\left(A + \sum_{k=0}^{L-1} \alpha_k(t)e^{-j\theta_k(t)}\right) + w(t)$$

$$= s_\ell(t)[A + x(t) - jy(t)] + w(t)$$

(11.91)

where $x(t)$ and $y(t)$ are defined by (11.83). For a large number of multipaths, the central limit theorem states that the envelope of $r_\ell(t)$ undergo Rayleigh fading in the absence of the specular component A. In case of a specular component A as in (11.91), apart from the noise term, $r_\ell(t)$ consists of the sum of an in-phase Gaussian component with mean A, and a quadrature component with zero-mean (see Figure 11.22). Note that in case where A is complex, in-phase and quadrature components $x(t)$ and $y(t)$ in (11.91) will both have non-zero means (refer to Example F.7 for a more general treatment). The joint pdf of $x' = A + x$ and y, which are independent from each other, may be written as

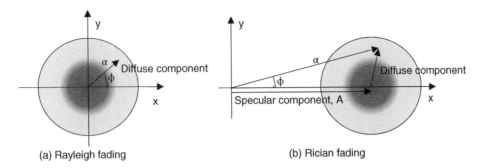

Figure 11.22 Rayleigh and Rician Distributed Random Variables in Complex Plane.

$$f_{X',Y}(x',y) = \frac{1}{2\pi\sigma^2}\exp\left(-\frac{(x'-A)^2+y^2}{2\sigma^2}\right)$$

(11.92)

When $A = 0$, the complex Gaussian distribution is centered at the origin of the complex plane and its magnitude is Rayleigh distributed.. However, for $A \neq 0$, the complex Gaussian distribution is centered around the specular component A and its magnitude is Rician distributed. As one may easily observe from Figure 11.22, the probability of a deep fade in Rician fading is much smaller than in the Rayleigh case, since the received signal power is dominated by the specular component. The received signal is hence composed of the sum of the diffused components with power $2\sigma^2$ and a specular component with power A^2.

Using the variable transformation with $x' = \alpha\cos\phi, y = \alpha\sin\phi$ in (11.92), one gets the pdf of the envelope and the phase of $r_e(t)$ as follows (see Example F.7):

$$f_{\alpha,\Phi}(\alpha,\phi) = \alpha f_{X',Y}(\alpha\cos\phi,\alpha\sin\phi)$$

$$= \frac{\alpha}{2\pi\sigma^2}\exp\left(-\frac{\alpha^2+A^2-2\alpha A\cos\phi}{2\sigma^2}\right)$$

$$f_\alpha(\alpha) = \int_0^{2\pi} f_{\alpha,\Phi}(\alpha,\phi)d\phi$$

$$= \frac{\alpha}{\sigma^2}\exp\left(-\frac{\alpha^2+A^2}{2\sigma^2}\right)I_0\left(\frac{\alpha A}{\sigma^2}\right), \alpha \geq 0$$

$$f_\Phi(\phi) = \int_0^\infty f_{\alpha,\Phi}(\alpha,\phi)d\phi = \frac{1}{2\pi}\exp\left(-\frac{\rho^2}{2}\right)$$

$$\left[1 + \sqrt{2\pi}\rho\cos\phi e^{\rho^2\cos^2\phi/2}\{1-Q(\rho\cos\phi)\}\right]$$

$$\rho = A/\sigma$$

(11.93)

where $Q(x)$ denotes the Gaussian Q function (see Appendix B). For $A = 0$, the pdf of the phase reduces to uniform distribution given by (11.86), as expected, but, for larger values $\rho = A/\sigma$, it peaks around $\phi = 0$ with decreasing variance (see Figures 8.22 and 11.22). The modified Bessel function $I_n(x)$ of the second kind of order n is shown in Figure 11.23. Its asymptotic behavior for large and small values of the argument is given by (D.99):

$$I_n(x) \approx \begin{cases} \dfrac{1}{n!}\left(\dfrac{x}{2}\right)^n & x^2 \ll 1 \\[2mm] \dfrac{e^x}{\sqrt{2\pi x}} & x \gg 1 \end{cases}$$

(11.94)

Moments of α, given by

$$E(\alpha) = \sqrt{\frac{\pi}{2}}\sigma e^{-K/2}[(1+K)I_0(K/2)+KI_1(K/2)]$$

$$E(\alpha^2) = A^2 + 2\sigma^2 = 2\sigma^2\left(1+\frac{A^2}{2\sigma^2}\right) = 2\sigma^2(1+K)$$

(11.95)

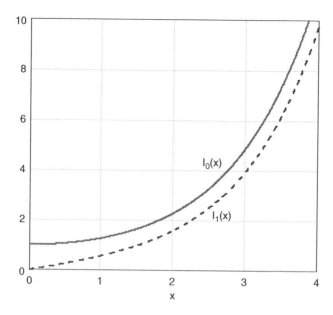

Figure 11.23   Modified Bessel Function $I_n(x)$ of Second Kind and Order n.

The Rice factor K is defined as the ratio of the power of the specular component to the diffuse component of a Rician fading signal:

$$K = \frac{A^2}{2\sigma^2} \qquad (11.96)$$

For the special case where K, hence A, approaches zero, the specular component will be negligibly small compared to the diffuse component and $I_0(x)$ approaches unity. Then, the Rician pdf given by (11.93) reduces to Rayleigh pdf described by (11.86). Similarly the moments given by (11.95) reduce to those given by (11.87) for Rayleigh fading.

For large values of K and A, the average power of the specular component will be much stronger than that of the diffuse component and the signal will look more like Gaussian (see also Figure 11.24). This may easily be observed inserting the asymptotic expression given by (11.94) for large values of the argument of $I_0(x)$ into (11.93):

$$f_\alpha(\alpha) \approx \lim_{K \to \infty} \frac{\sqrt{\alpha/A}}{\sqrt{2\pi\sigma}} \exp\left(-\frac{\{\alpha-A\}^2}{2\sigma^2}\right) \to \delta(\alpha-A)$$

$$(11.97)$$

which looks like a Gaussian pdf for $\alpha \approx A$. Indeed, inserting the asymptotic expression for $I_n(x)$ from (11.94) into (11.95) that mean and the variance of α for $A \to \infty$ reduce to

$$\begin{aligned} E[\alpha] &\cong A \\ E[\alpha^2] - \{E[\alpha]\}^2 &\cong 2\sigma^2 \end{aligned} \qquad as\ A \to \infty \quad (11.98)$$

Noting that the average signal power is given by the sum of those specular and diffuse components, $P_0 = E[\alpha^2] = A^2 + 2\sigma^2$, we defined the normalized power of a Rician-faded signal as follows:

$$\gamma = \frac{\alpha^2}{P_0} = \frac{\alpha^2}{2\sigma^2 + A^2} = \frac{\alpha^2}{2\sigma^2(1+K)} \qquad (11.99)$$

The pdf of γ is obtained from (11.93) as

$$\begin{aligned} f(\gamma) &= f(\alpha) \left|\frac{d\alpha}{d\gamma}\right|_{\alpha = \sqrt{2\sigma^2(1+K)\gamma}} \\ &= (1+K)e^{-K-(1+K)\gamma} I_0\left(2\sqrt{K(1+K)\gamma}\right) \end{aligned}$$

$$(11.100)$$

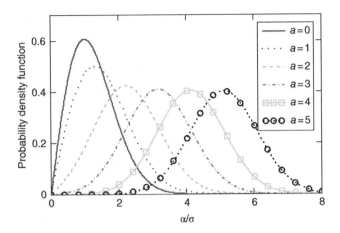

Figure 11.24 The Rician Pdf as a Function of α/σ For Several Values of $a = A/\sigma$.

Since $E[\gamma] = 1$, (11.100) with $K = 0$ reduces to (11.88) with $P_0 = 1$. The cdf of Rician distribution is given by

$$F(R) = Pr(\alpha \le R) = \int_0^R f_\alpha(\alpha)\,d\alpha = 1 - Q_1\left(\frac{A}{\sigma},\frac{R}{\sigma}\right)$$

$$= 1 - Q_1\left(\sqrt{2K},\sqrt{2(K+1)\rho^2}\right)$$

$$\rho^2 = \frac{R^2}{2\sigma^2(1+K)}$$

$$(11.101)$$

where the so-called Marcum Q function is defined by

$$Q_1(0,b) = \int_b^\infty x\exp\left(-\frac{x^2+a^2}{2}\right)I_0(ax)\,dx$$

$$(11.102)$$

The reader may refer to Appendix B for further details. Figure 11.25 shows the cdf $F(R)$, given by (11.101), that the received signal envelope, experiencing Rician fading, is less than R. For convenience the horizontal axis is shown as ρ^2, the threshold level normalized to the mean received power $P_0 = (1+K)2\sigma^2$. The probability that the envelope remains

below the threshold level rapidly decreases as the Rice factor K increases, since the specular component dominates the received signal.

The moment generating function of a Rician faded signal is found using (D.60) and (11.100):

$$M_\gamma(s) = E\left[e^{-sP_0\gamma}\right]$$

$$= \frac{1+K}{1+K+P_0s}\exp\left(\frac{-KP_0s}{1+K+P_0s}\right)$$

$$(11.103)$$

11.3.3   Nakagami-m Fading

Nakagami-m distribution provides a flexible fading model for signals propagating through multipath fading channels. The pdf of the signal envelope is characterized by the so-called fading parameter m and its average power P_0.

$$f_\alpha(\alpha) = \frac{2}{\Gamma(m)}\left(\frac{m\alpha^2}{P_0}\right)^m\frac{1}{\alpha}\exp\left(-\frac{m\alpha^2}{P_0}\right),\quad \alpha \ge 0$$

$$P_0 = E[\alpha^2]$$

$$m = \frac{P_0^2}{E\left[(\alpha^2 - P_0)^2\right]} \ge \frac{1}{2}$$

$$(11.104)$$

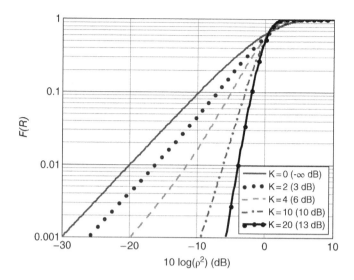

Figure 11.25 Probability That the Received Signal Envelope, Experiencing Rician Fading, is Less Than a Threshold Level ρ^2, Which is Related to R By (11.101).

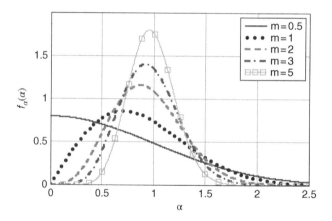

Figure 11.26 Pdf of Nakagami-m Distribution For $P_0 = 1$.

Figure 11.26 shows the Nakagami-m pdf for various values of the fading parameter m. Nakagami-m pdf reduces to one-sided Gaussian pdf for $m = 1/2$, Rayleigh pdf for $m = 1$ (see (11.86)) and deterministic channel (no fading) as $m \to \infty$. Hence, the fading parameter m controls the severity of fading in the channel; increased values of m corresponds to fading channels with higher mean values and smaller variances.

The moments given by

$$E\left[\alpha^k\right] = \frac{\Gamma(m+k/2)}{\Gamma(m)}\left(\frac{P_0}{m}\right)^{k/2} \qquad (11.105)$$

may be shown to reduce to (11.87) for $m = 1$ and $P_0 = 2\sigma^2$, as expected.

The cdf of the signal envelope is given by

$$F_\alpha(R) = \int_0^R f_\alpha(\alpha)d\alpha = \frac{\gamma(m, m\rho^2)}{\Gamma(m)}, \quad \rho = R/\sqrt{P_0}$$

$$(11.106)$$

where $\gamma(\alpha, z)$ denotes the incomplete Gamma function (see (D.106) and (D.111)):

$$\gamma(\alpha, z) \triangleq \int_0^z t^{\alpha-1} e^{-t} dt$$

$$= \Gamma(\alpha)\left[1 - e^{-z}\sum_{k=0}^{\alpha-1}\frac{z^k}{k!}\right], \quad \alpha = 1, 2, \ldots$$

$$(11.107)$$

which is related to the well-known Gamma function:

$$\Gamma(\alpha) = \gamma(\alpha, \infty) = \int_0^\infty t^{\alpha-1} e^{-t} dt \qquad (11.108)$$

The pdf of the signal power $p = \alpha^2$ may easily be found from (11.104) by variable transformation:

$$f_P(p) = f_\alpha(\alpha)\left|\frac{d\alpha}{dp}\right|_{\alpha=\sqrt{p}}$$

$$= \frac{1}{\Gamma(m)}\left(\frac{mp}{P_0}\right)^m\frac{\exp(-mp/P_0)}{p}, \quad p \geq 0$$

$$(11.109)$$

Using (D.49) and (D.111), the cdf of P is found to be

$$F_P(P_{th}) = \int_0^{P_{th}} f_P(u)du$$

$$(11.110)$$

$$= 1 - \frac{1}{\Gamma(m)}\gamma\left(m, \frac{mP_{th}}{P_0}\right)$$

The moment generating function corresponding to (11.109) is obtained using (D.50):

$$M_P(s) = E\left[e^{-sP}\right] = (1 + sP_0/m)^{-m} \quad (11.111)$$

Example 11.12 Outage Probability Versus Bit Error Probability in Nakagami-m Fading. The power outage probability, widely used as a performance measure for wireless communication systems, is defined as the probability that the received power level falls below some threshold value P_{th}:

$$P_{out} = P(P < P_{th}) = F_P(P_{th}) \qquad (11.112)$$

where $F_P(P_{th})$ denotes the cdf of the power. Note that the outage probability may also be expressed in terms of the SNR, instead of the received power. Outage probability is typically applied to slowly fading signals due to shadowing, since once the signal power fades below some threshold level P_{th}, it is likely to stay there for relatively long time durations. Then, the correction of the large number of corrupted bits may not be easy by using FEC codes and/or the use of long interleaving depths may not be fesible for real-time communications. The outage threshold is generally determined by the 'receiver sensitivity' which corresponds to the minimum required power (SNR) for acceptable performance. Therefore, a slowly-fading channel becomes basically disconnected when the received power falls below the given threshold level. For example, quasi-stationary pedestrian users stepping into a deep fade would experience long outages. Outage probability is not generally applied to multipath (short-term) fading channels since a fade lasts for the duration of several bits for a mobile user at vehicle velocities. Such errors may easily be corrected by FEC codes.

In this example we compute the outage probability in a Nakagami-m fading channel using (11.110). Figure 11.27 shows the variation of the outage probability as a function of P_{th}/P_0, where the received signal envelope has a Nakagami-m fading distribution with average power P_0 and P_{th} is the threshold power level. The outage probability increases by $10\,m$ times per 10 dB increase in P_{th}/P_0, hence a significant improvement with the fading parameter m. For example, for $P_{th}/P_0 = -10$ dB, the outage

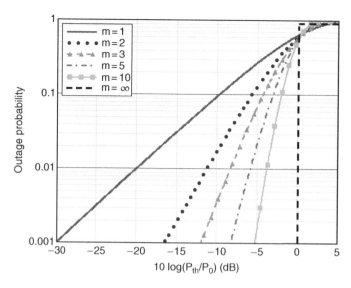

Figure 11.27   Outage Probability For Nakagami-m Fading.

probability is $P_{out} = 0.095$ for $m = 1$ and $P_{out} = 0.018$ for $m = 2$. Figure 11.27 also shows that the outage probability may be represented by the unit step function $U(P_{th}/P_0 - 1)$ as m approaches infinity (implying a deterministic channel). If an application requires a power outage probability of 0.001 in a Rayleigh fading channel, then we have $P_{th}/P_0 = -30$ dB (see the curve for $m = 1$ in Figure 11.27). This implies that the receiver should be designed to continue to operate even when the received power level drops up to 30 dB below the average power level; hence the required receiver sensitivity is 30 dB below the average received power level.

Bit error probability provides another metric for the system performance evaluation. When the received power level falls below P_{th}, the bit error probability is usually unacceptable for the desired application and the service provided by the system often becomes interrupted. However, bit errors gradually increase as the received power level decreases and approaches P_{th}. Hence, the bit error probability provides an average performance measure as compared to the outage probability which is a kind of 'hard'

performance measure, indicating whether the system provides services or not.

Example 11.13   Amount of Fading (AF).
Amount of fading (AF), which is used to quantify the severity of fading, is defined as follows in terms of the instantaneous channel gain α or the power $\gamma = \alpha^2$:

$$AF_\alpha = \frac{Var[\alpha^2]}{(E[\alpha^2])^2} = \frac{E[\alpha^4] - (E[\alpha^2])^2}{(E[\alpha^2])^2} = \frac{E[\alpha^4]}{(E[\alpha^2])^2} - 1$$

$$AF_\gamma = \frac{Var[\gamma]}{(E[\gamma])^2} = \frac{E[\gamma^2] - (E[\gamma])^2}{(E[\gamma])^2} = \frac{E[\gamma^2]}{(E[\gamma])^2} - 1$$

(11.113)

AF is hence defined as the ratio of the variance to the square mean of the instantaneous power γ and provides a measure of the fluctuation of the received power level around its mean value. Note that $AF = 0$ corresponds to an ideal (Gaussian) channel and $AF = \infty$ to severe fading. For Nakagami-m fading, we find $AF = 1/m$ using (11.105). Hence, $AF = 0$ as $m \to \infty$ and $AF = 1$ for a Rayleigh fading channel ($m = 1$). [4][5]

11.3.4 Log-Normal Shadowing

In applications such as cellular radio, TV/radio broadcasting and UHF satellite communications, the received signals, may differ vastly from the average value predicted by the channel models due to changes in the surrounding environmental clutter. Clutter refers to obstacles which lead to unwanted signals that interfere with the desired signals. Log-normal pdf is usually used to account for shadowing effects due to large obstacles, such as large buildings, hills and trees, obstructing the LOS path between transmitter and receiver.

In order to gain more insight about log-normal distribution, let us consider electromagnetic wave propagation through an obstacle which shadows a LOS path. Such an obstacle may be characterized as a lossy propagation medium. In a lossy propagation environment with an attenuation coefficient α, the amplitude of an electric field attenuates with distance r as follows:

$$E = E_0 \frac{e^{-(jwt + \alpha r)}}{r} \qquad (11.114)$$

where E_0 denotes a parameter accounting for directional property, transmit power and frequency dependence of the transmitted signal, as explained in Chapter 2. If the electromagnetic wave propagates through a lossy medium with an attenuation constant α_1 and thickness d_1, then the ratio of the electric field intensity at the output (at distance r_2) to that at the input (at distance r_1) of the medium may be written as

$$\frac{E_2}{E_1} = \frac{E_0 e^{-\alpha_1 r_2} / r_2}{E_0 e^{-\alpha_1 r_1} / r_1} \approx e^{-\alpha_1 d_1} \qquad (11.115)$$

where $d_1 = r_2 - r_1$ shows the thickness of the lossy medium. If the propagation medium is layered with arbitrary attenuation constants and thicknesses, then the normalized electric field at the output of the Nth layer may be written as

$$X = E_N / E_1 = \prod_{i=1}^{N} e^{-\alpha_i d_i} = e^{-\Sigma_{i=1}^{N} \alpha_i d_i} = e^{W'} = 10^W$$

$$W = (\log_{10} e) W' = -(\log_{10} e) \sum_{i=1}^{N} \alpha_i d_i$$

$$(11.116)$$

Based on the central limit theorem, W will have a normal distribution as $N \to \infty$ whereas X will be log-normally distributed. If we let

$$Z = 10 \log_{10} X^2 = 20 \log_{10} X \quad (dB) \qquad (11.117)$$

then the pdf of the random variable Z (representing signal power or SNR expressed in dB) is normally distributed

$$f_Z(z) = \frac{1}{\sqrt{2\pi}\sigma_{dB}} \exp\left(-\frac{(z - \mu_{dB})^2}{2\sigma_{dB}^2}\right), \quad -\infty < z < \infty$$

$$(11.118)$$

Here μ_{dB} and σ_{dB} denote respectively the mean and the standard deviation of Z:

$$\mu_{dB} = E[Z] = E[20 \log_{10}(X)] \quad (dB)$$
$$\sigma_{dB}^2 = Var[Z] = Var[20 \log_{10}(X)] \quad (dB)$$

$$(11.119)$$

Note that the probability density function of X, which is log-normally distributed, is directly obtained from (11.118) as

$$f_X(x) =$$

$$\frac{\xi}{x\sigma_{dB}\sqrt{2\pi}} \exp\left(-\frac{(20 \log(x) - \mu_{dB})^2}{2\sigma_{dB}^2}\right), \quad x > 0$$

$$(11.120)$$

where $\xi = 20/\ln(10) = 20 \log_{10} e = 8.868$ (see (D.76)). The moments of the log-normal random variable X are readily found as

$$E[X^k] = \exp\left(\frac{k\mu_{dB}}{\xi} + \frac{k^2 \sigma_{dB}^2}{2\xi^2}\right) \qquad (11.121)$$

The mean value of X may be expressed in dB as follows:

$$20\log(E[X]) = \mu_{dB} + \frac{\sigma_{dB}^2}{2\xi}(dB) \qquad (11.122)$$

It should be noted that (11.121) denotes the linear mean of X while μ_{dB} in (11.119) represents the average of the dB values of X. They are evidently not identical. In practice, μ_{dB} and σ_{dB} are used to characterize log-normal shadowing rather than the moments of X.

Measurements show that the received signal power $P(r)$(dB) at a particular distance r has a normal distribution about the local mean power level, $P_m(r)$(dB) at that point:

$$P(r) = P_m(r) + P(dB)$$
$$P_m(r) = P(r_0) + 10n \log(r_0/r)(dB) \qquad (11.123)$$

where $P_m(r)$ is given by (3.92) for the dual slope path-loss model. Here P (dB), which denotes a zero-mean log-normally distributed random variable with standard deviation σ_{dB} (dB), represents the contribution of the environmental clutter to the path-loss. The pdf of the received shadowed signal power in dB may then be written as

$$f_{P(r)}(u)$$
$$= \frac{1}{\sqrt{2\pi}\sigma_{dB}} \exp\left(-\frac{(u - P_m(r))^2}{2\,\sigma_{dB}^2}\right), -\infty < u < \infty \qquad (11.124)$$

The outage probability due to shadowing, that is, the probability that the received signal power level will be less than a treshold level P_{th}, is simply given by

$$Prob[P(r) < P_{th}] = \int_{-\infty}^{P_{th}} f_{P(r)}(u)\,du \qquad (11.125)$$
$$= Q\left(\frac{P_m(r) - P_{th}}{\sigma_{dB}}\right)$$

Figure 11.28 shows the cdf of the received signal power level as a function of the normalized threshold level $P_{th} - P_m(r)$ (dB) for various values of the shadowing variance. It is clear that the outage performance deteriorates

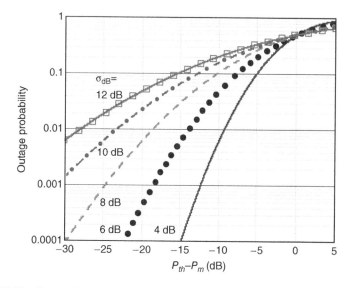

Figure 11.28   Outage Probability For Log-Normal Shadowing For Various Values of σ_{dB}.

quickly with increasing values of the shadowing variance.

Example 11.14 Log-Normal Shadowing
Consider a communication receiver with a noise power level of $N = -120$ dBm operating in a log-normal shadowing environment with $\sigma_{dB} = 8$ dB and path-loss exponent $n = 4$. We determine the required average signal power $P_m(r)$ and the corresponding distance r if an SNR $= 20$ dB is required for at least 95% of the time. This implies that the required received signal power level should be higher than $P_{th} =$ SNR $+ N = -100$ dBm for at least 95% of the time.

Using (11.125), this requirement may be expressed as

$$\text{Prob}(P(r) > P_{th}) = 1 - Q\left(\frac{P_m(r) - P_{th}}{\sigma_{dB}}\right) > 0.95$$

(11.126)

where the threshold level P_{th}, the instantaneous received power level $P(r)$, the mean signal level $P_m(r)$, and the shadowing standard deviation σ_{dB} are all expressed in dB. Since (11.126) suggests that P_{th} is lower than the mean received power level $P_m(r)$, the mean signal power received at distance r is given by (see Table B.1)

$$\frac{P_m(r) - P_{th}}{\sigma_{dB}} = 1.64 \Rightarrow$$
$$P_m(r) = 1.64\sigma_{dB} + P_{th} = -86.9 \ dBm$$

(11.127)

Now consider a cellular radio system operating at 900 MHz with BS transmit power level of 10 W, BS antenna gain of 10 dBi and a MS antenna gain 0 dBi. The average received power at $r_0 = 1$ km, assuming free-space propagation conditions, is

$$P(r_0) = P_t G_t G_r \left(\frac{\lambda}{4\pi r_0}\right)^2$$
$$= 7 \times 10^{-5} mW(-41.5 \ dBm)$$

(11.128)

The distance corresponding to the average received power level -86.7 dBm is found as

$$P_m(r) = P(r_0)(r_0/r)^n \Rightarrow$$
$$r = r_0 \left[\frac{P(r_0)}{P_m(r)}\right]^{1/n} = 1km \left[\frac{10^{-4.15}}{10^{-8.69}}\right]^{1/4} = 13.65 \, km$$

(11.129)

Example 11.15 Useful Coverage Area.
Consider a cell with a circular coverage area of radius R as shown in Figure 11.29. Due to local variations in the clutter, some locations within the coverage area will undergo shadowing and the received signal power will be below a particular threshold level P_{th}. We will determine the surface of useful coverage area normalized that of the coverage area of radius R, where the received signal power level is higher than the threshold level P_{th}.

Using (11.123), the mean received signal power at an arbitrary distance r in the cell may be related to the mean received signal power at the cell border $r = R$ as

$$\left.\begin{array}{l} P_m(r) = P(r_0)_{dB} + 10n \log(r_0/r) \\ P_m(R) = P(r_0)_{dB} + 10n \log(r_0/R) \end{array}\right\}$$
$$\Rightarrow P_m(r) = P_m(R) - 10n \log(r/R) \quad (11.130)$$

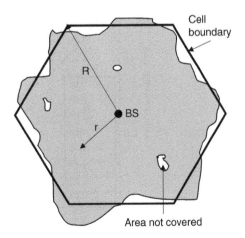

Figure 11.29 Useful Coverage Area. The received signal power level in the unshaded regions is assumed to be below the threshold level.

Using (11.126) and (11.130), the probability that the received power in dB at a point r exceeds a threshold level P_{th} (dB) is found to be

$$\mathrm{Prob}[P(r) > P_{th}] = Q\left(\frac{P_{th} - P_m(r)}{\sigma_{dB}}\right)$$

$$= Q\left(\frac{P_{th} - P_m(R) + 10n\log(r/R)}{\sigma_{dB}}\right)$$

$$= Q\left(a + b\ln\frac{r}{R}\right)$$

(11.131)

where the property $Q(-z) = 1 - Q(z)$ of the Gaussian Q-function is used and

$$a = \frac{P_{th} - P_m(R)}{\sigma_{dB}}$$

$$b = \frac{n}{\sigma_{dB}}10\log_{10}e = 4.343\frac{n}{\sigma_{dB}}$$

(11.132)

The useful service area normalized by πR^2, where the received power level exceeds the threshold power level P_{th}, is given by

$$U(\gamma) = \frac{1}{\pi R^2}\int_0^{2\pi}\int_0^R \mathrm{Prob}[P(r) > P_{th}]r\,dr\,d\phi$$

$$= \frac{2}{R^2}\int_0^R \mathrm{Prob}[P(r) > P_{th}]r\,dr$$

$$= 2\int_0^1 \rho Q(a + b\ln\rho)d\rho$$

$$= Q(a) + \exp\left(\frac{2 - 2ab}{b^2}\right)Q\left(\frac{2 - ab}{b}\right)$$

(11.133)

The last integral is evaluated by integration by parts and using the derivative of the Gaussian-Q function given by (B.13).

Figure 11.30 shows the useful coverage area as a function of σ_{dB}/n for various values of $Q(a)$, which gives the probability that the signal power is above threshold at the cell boundary (see (11.131)). Useful coverage area serves as a metric for the QoS for cellular radio users.

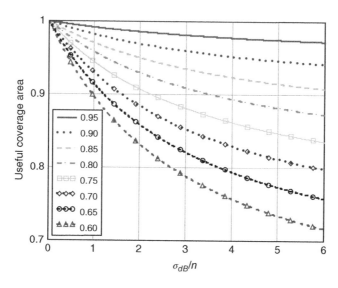

Figure 11.30   Useful Coverage Area $U(\gamma)$, That is, Fraction of Total Area with Signal Above Threshold For Various Values of $Q(a)$, the Probability That the Signal Power is Above Threshold at the Cell Boundary.

If the mean signal level at the cell edge $P_m(R)$ equals the threshold level P_{th} (i.e., $a=0$), then $Q(0)=0.5$ and (11.133) reduces to

$$U(\gamma)=\frac{1}{2}+\exp\left(\frac{2}{b^2}\right)Q\left(\frac{2}{b}\right) \qquad (11.134)$$

Assuming that the average received power at $r_0=1$ km is $P(r_0)=-41.5$ dBm, $P_{th}=-100$ dBm, $\sigma_{dB}=8$ dB and $n=4$ as in Example 11.14, the mean received signal power at the edge of a cell with $R=20$ km radius found using (11.123):

$$P_m(R)=-41.5\,dBm+40\log(1/20)$$
$$=-93.5\,dBm$$
$$\qquad (11.135)$$

The parameters for determining the useful coverage area are found to be

$$a=\frac{P_{th}-P_m(R)}{\sigma_{dB}}=\frac{-100+93.5}{8}=-0.81$$
$$b=4.343\frac{n}{\sigma_{dB}}=2.17$$
$$\qquad (11.136)$$

Noting that $Q(-0.81)=0.79$ (see (B.9) and Table B.1), the useful coverage area is given by

$$U(\gamma)=Q(-0.81)+\exp\left(\frac{2-2\times(-0.81)\times2.17}{2.17^2}\right)$$
$$\times Q\left(\frac{2-(-0.81)\times2.17}{2.17}\right)$$
$$=0.925$$
$$\qquad (11.137)$$

This implies that 92.5% of the cell area is covered, based on the assumption that the received signal power at the cell edge is higher than the threshold level for $Q(a)=79\%$ of the time.

The users located in regions between the two neighboring cells have the flexibility of being covered by two BSs. Such users experience outage only when they lose connection with both BSs at the same time. The coverage probability for such users is approximately given by

$$P_{\text{cov}}=1-[1-Q(a)]^2 \qquad (11.138)$$

For $Q(a)=0.79$, the coverage probability for mobile users located on the borders of a cell is 96%, which is even higher than 92.5% for the users closer to the BS. This percentage decreases to 84% for $Q(a)=0.6$.

Example 11.16   Handover in Cellular Radio Systems.
In cellular radio systems, the mean signal power received by the BS decreases as a MS approaches the cell boundaries, but the signal received by the BS of the adjacent cell tends to increase. Consequently, the MS is more likely being served by the BS of the adjacent cell as it crosses the cell boundary. The process of a MS changing its BS as it crosses the boundaries of its cell is called *handover*. The handover process is not always smooth due to mainly the shadowing effects. This process should be able to distinguish between short term and long-term fades, monitor if an increase in the power level would be sufficient to restore the channel quality, to test the validity of measurements (averaging is necessary), and to inquire whether the cell chosen for handover has available channels.
   Suppose that a MS is moving along a straight line between BS_1 and BS_2, which are separated from each other by $2R$, where R denotes the cell radius. The received power in dB at BS_1 and BS_2 from the MS is modeled by (11.123) as

$$P(r_i)=P_m(r_i)+P_i \;\;(dB)\;\; i=1,2 \qquad (11.139)$$

where r_i denotes the distance between the MS and BS_i and $r_1+r_2=2R$. $P_m(r_i)$ is the mean received power at distance r_i. P_i, $i=1,2$ are independent zero-mean Gaussian random variable with standard deviation σ_{dB} and they serve to model the random variations in the received signal level due to shadowing.

Let P_{min} denote the minimum received signal power (sensitivity) level above which a BS can successfully demodulate signals and $P_{HO} > P_{min}$ is the handover threshold below which handover process is initiated. Assume that the MS is currently connected to BS_1. Handover occurs when the received signal at the BS_1 from the MS drops below threshold P_{HO} and the signal received by the BS_2 becomes greater than the minimum acceptable level P_{min}; then, handover to BS_2 can be initiated. Since these two events are independent of each other, the probability that handover occurs may be written as

$$P_{HO} = Pr(P(r_1) < P_{HO})Pr(P(r_2) > P_{min})$$

(11.140)

where $P(r_i)$ denotes the signal power received at BS_i and

$$Pr(P(r) < P_{HO})$$

$$= \int_{-\infty}^{P_{HO}} \frac{1}{\sqrt{2\pi}\sigma_{dB}} \exp\left[-\frac{(x - P_m(r))^2}{2\sigma_{dB}^2}\right] dx$$

$$= 1 - Q\left(\frac{P_{HO} - P_m(r)}{\sigma_{dB}}\right)$$

$$Pr(P(r) > P_{min})$$

$$= \int_{P_{min}}^{\infty} \frac{1}{\sqrt{2\pi}\sigma_{dB}} \exp\left[-\frac{(x - P_m(r))^2}{2\sigma_{dB}^2}\right] dx$$

$$= Q\left(\frac{P_{min} - P_m(r)}{\sigma_{dB}}\right)$$

(11.141)

The handover fails to occur when the handover procedure is initiated by BS_1 but the MS can not be connected to BS_2 since power level received by BS_2 is below P_{min}: [6]

$$P_{fail} = Pr(P(r_1) < P_{HO}) \ Pr(P(r_2) < P_{min})$$

$$= \left[1 - Q\left(\frac{P_{HO} - P_m(r_1)}{\sigma_{dB}}\right)\right]$$

$$\times \left[1 - Q\left(\frac{P_{min} - P_m(r_2)}{\sigma_{dB}}\right)\right]$$

(11.142)

The outage probability is defined as the probability that the power levels received by BS_1 and BS_2 are both below P_{min}:

$$P_{out} = Pr(P(r_1) < P_{min}) \ Pr(P(r_2) < P_{min})$$

$$= \left[1 - Q\left(\frac{P_{min} - P_m(r_1)}{\sigma_{dB}}\right)\right]$$

$$\times \left[1 - Q\left(\frac{P_{min} - P_m(r_2)}{\sigma_{dB}}\right)\right]$$

(11.143)

This case occurs when the power level received by BS_2 is below P_{min} while the connection to BS_1 is lost.

Figure 11.31 shows the variation of the handover probability, the probability that handover fails and the probability of outage as a MS moves along a straight line connecting BS_1 and BS_2 that are 4 km apart from each other. For example, the handover occurs with 53% and 88% probabilities when the MS is located at distances 2 km and 3 km from the BS_1, respectively.

11.3.5 Composite Fading and Shadowing

In the presence of both fading and shadowing, the pdf of the received signal power P may be determined by

$$f_P(\gamma) = \int_0^{\infty} f_{P|\Omega}(\gamma|\Omega) f_{\Omega}(\Omega) d\Omega \qquad (11.144)$$

where Ω denotes the instantaneous local mean power of the received signal. The pdf of P is usually described by Rayleigh, Rice or Nakagami-m distributions, while shadowing is usually accounted for by the log-normal pdf. The Suzuki model, which is based on Rayleigh fading and log-normal shadowing, is commonly used in modelling signal propagation in urban areas. [7] The so-called Loo model is proposed for modeling composite fading and shadowing in land mobile satellite links. [8] In this model,

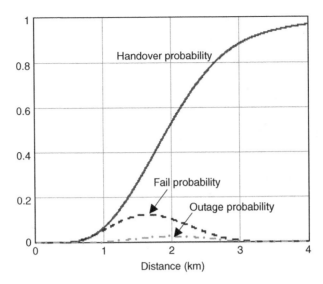

Figure 11.31   Handover Probability, the Probability that Handover Fails and the Probability of Outage For a MS Moving Along a Straight Line Connecting BS$_1$ and BS$_2$ That Are 4 km Away From Each Other. $P(r_0) = -20$ dBm, $n = 3$, $\sigma_{dB} = 6$ dB, $P_{min} = -65$ dBm, $P_{HO} = -57$ dBm and $R = 2$ km.

shadowing accounts for the presence/absence of the LOS in the satellite link while Rayleigh fading takes care of the multipath effects.

However, a closed-form expression for the composite pdf is not available when log-normal pdf is used to model shadowing. Instead, a Gamma distribution is proposed:

$$f_\Omega(X) = \frac{1}{\Gamma(m_s)} \left(\frac{m_s X}{\Omega_s}\right)^{m_s} \frac{\exp(-m_s x/\Omega_s)}{X},$$

$$x \geq 0, m_s \geq 0$$

$$(11.145)$$

where m_s and Ω_s denote respectively the shadow-fading parameter and the mean power level of the shadow fading. When the means and variances of (11.145) are matched to those of the log-normal pdf, given by (11.121), the values of m_s and Ω_s may be expressed in terms of μ_{dB} and σ_{dB} as follows: [9][10]

$$m_s = 1/\left(e^{\sigma_{dB}^2/\xi^2} - 1\right)$$

$$\Omega_s = e^{\mu_{dB}/\xi + \sigma_{dB}^2/(2\xi^2)} = e^{\mu_{dB}/\xi}\sqrt{1 + 1/m_s}$$

$$(11.146)$$

Figure 11.32 shows a comparison of log-normal and Gamma pdf's with the same means and variances for the values of $\mu_{dB} = 0$ dB and $\sigma_{dB} = 4$ dB and 8 dB. The agreement between the two pdf's seem to be better for light shadowing. However, it is reported that none of the two pdf's are superior to the other in describing the empirical results. [11][12][13].

For the sake of flexibility, let us assume Nakagami-m pdf to model multipath fading:

$$f_\Omega(\gamma|\Omega)$$

$$= \frac{1}{\Gamma(m)} \left(\frac{m\gamma}{\Omega}\right)^m \frac{\exp(-m\gamma/\Omega)}{\gamma}, \quad \gamma \geq 0, m \geq 1/2$$

$$(11.147)$$

Inserting (11.145) and (11.147) into (11.144) and using (D.65), one gets a closed form expression for the composite fading channel:

$$f_c(\gamma) = \frac{2}{\Gamma(m)\Gamma(m_s)} \left(\frac{m m_s}{\Omega_s}\right)^{(m+m_s)/2}$$

$$\gamma^{(m+m_s-2)/2} K_{m_s-m}\left(2\sqrt{\frac{m m_s}{\Omega_s}\gamma}\right), \gamma > 0$$

$$(11.148)$$

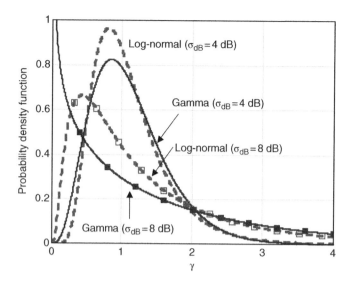

Figure 11.32 Log-Normal and Gamma Pdf's with the Same Means and Variances For $\mu_{dB} = 0$ and Various Values of σ_{dB}.

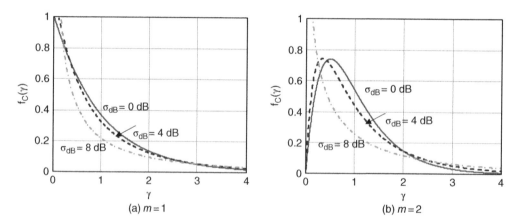

Figure 11.33 Composite Nakagami-m and Gamma Pdf Given By (11.144) with $\mu_{dB} = 0$ in (11.146).

where $K_v(x)$ denotes the modified Bessel function of second kind and order v (see Figure D.3). For $m = 1$ and $m_s \to \infty$ (shadowing variance $\sigma_{dB} \to 0$ dB), (11.148) reduces to Rayleigh distribution. Figure 11.33 clearly shows the increasing likelihood of getting lower values of γ as the shadowing variance increases. One may also note from Figure 11.33b that the pdf curves for all shadowing variances are shifted to the right as the severity of fading decreases (m increases). On the other hand, it

may be useful to note that (11.148) may be used as the pdf of either the received signal power or the SNR, since they differ only by a constant of proportionality.

The moments of γ are found using (11.148) and (D.66):

$$E\left[\gamma^k\right] = \frac{\Gamma(m+k)\Gamma(m_s+k)}{\Gamma(m)\Gamma(m_s)}\left(\frac{\Omega_s}{mm_s}\right)^k$$

(11.149)

The amount of fading (AF) for the composite Nakagami-m faded and Gamma shadowed channel is found using (11.113) as follows:

$$AF_\gamma = \frac{E[\gamma^2]}{\left(E[\gamma]^2\right)} - 1 = \frac{1}{m} + \frac{1}{m_s} + \frac{1}{mm_s}$$

(11.150)

$$= \left(1 + \frac{1}{m}\right) e^{\sigma_{dB}^2/\xi^2} - 1$$

where the last expression is obtained using (11.146). It is clear from (11.150) that AF = $1/m_s$ for $m \to \infty$, while AF = $1/m$ for $\sigma_{dB} \to 0$ ($m_s \to \infty$). One may hence conclude that the AF is dominated by $\min(m, m_s)$. Figure 11.34 shows the effect of shadowing variance on the mean, variance and AF of γ describing the composite fading and shadowing for $m = 1$ and 2.

The outage probability of the composite channel is obtained by (11.148) and (D.68) [10]

$$P_{out}(\gamma_{th}) = \text{Prob}(\gamma \le \gamma_{th}) = \int_0^{\gamma_h} f_c(\gamma) d\gamma$$

$$= \frac{\Gamma(m_s - m)}{\Gamma(m_s)\Gamma(m+1)} \left(\frac{mm_s\gamma_{th}}{\Omega_s}\right)^m$$

$$\times {}_1F_2\left(m; 1-m_s+m, 1+m; \frac{mm_s\gamma_{th}}{\Omega_s}\right)$$

$$+ \frac{\Gamma(m-m_s)}{\Gamma(m)\Gamma(m_s+1)} \left(\frac{mm_s\gamma_{th}}{\Omega_s}\right)^{m_s}$$

$$\times {}_1F_2\left(m_s; 1-m+m_s, 1+m_s; \frac{mm_s\gamma_{th}}{\Omega_s}\right)$$

(11.151)

where ${}_1F_2(a;b,c;z)$ denotes the hypergeometric function (see D.6.3). Figure 11.35 shows the variation of the outage probability versus the threshold level normalized with Ω_s, where the effects of fading only, shadowing only and the composite channel are also shown for the values of $m = 1$ and 2 of the fading parameter. The values of m_s are related to the standard deviation of shadowing σ_{dB} by (11.146). For small values of the argument (γ_{th}/Ω_s), the hypergeometric function approaches to unity. Then the asymptotic behavior of the outage probability may be written as

$$P_{out}(\gamma_{th}) = \frac{\Gamma(m_s - m)}{\Gamma(m_s)\Gamma(m+1)} \left(\frac{mm_s\gamma_{th}}{\Omega_s}\right)^m$$

$$+ \frac{\Gamma(m-m_s)}{\Gamma(m)\Gamma(m_s+1)} \left(\frac{mm_s\gamma_{th}}{\Omega_s}\right)^{m_s}, \frac{mm_s\gamma_{th}}{\Omega_s} \ll 1$$

(11.152)

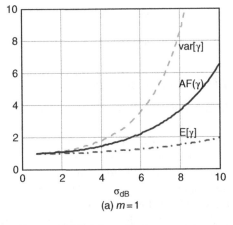

(a) $m = 1$

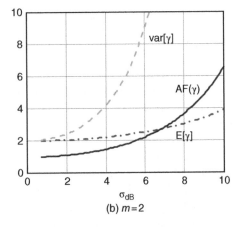

(b) $m = 2$

Figure 11.34 Mean, Variance and Amount of Fading (AF) of the Composite Pdf (11.148).

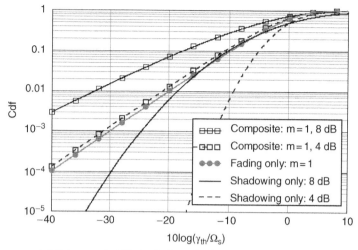

(a) Outage probability for fading, shadowing and composite channels

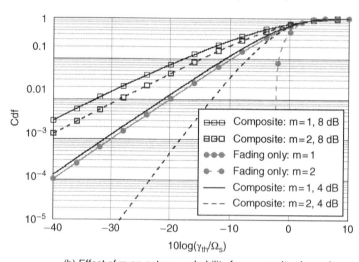

(b) Effect of m on outage probability for composite channels

Figure 11.35 Outage Probability of the Composite Rayleigh Fading and Gamma Shadowing Channel.

Noting the symmetry between m and m_s in (11.151) and (11.152), one may easily observe from Figure 11.35 that the outage probability is dominated by $\min(m, m_s)$. This implies that in light shadowing scenarios $(m_s > m)$ the outage probability becomes proportional to Ω_s^{-m} but is proportional to $\Omega_s^{-m_s}$ for $m_s < m$ $\left(\sigma_{dB} > \sqrt{\ln(1 + 1/m)}\right)$ where fading is not sufficiently severe and the outage probability is dominated by shadowing.

11.3.6 Fade Statistics

In a multipath channel, the received signal level fluctuates randomly around its local mean value. This happens not only because of the changes with time in the scattering environment but also due to the relative motion of transmitter and/or receiver. The rate at which the signal envelope undergoes fades and flares is determined by the channel coherence time. Consequently, the receiver cannot decode the received signals when the signal envelope drops below some threshold level determined by the receiver sensitivity. Then, signal outages are observed and communication services may be disrupted if necessary precautions are not taken.

The *level crossing rate* (LCR) N_R is defined as the rate at which the fading signal crosses of the envelope level R with a positive slope $(\dot{\alpha} = d\alpha/dt > 0)$, that is, in the positive-going direction (see Figure 11.36):

$$N_R = \int_0^\infty \dot{\alpha} f_{\alpha,\dot{\alpha}}(R, \dot{\alpha}) d\dot{\alpha} \qquad (11.153)$$

Determination of N_R requires the joint pdf $f_{\alpha,\dot{\alpha}}(\alpha, \dot{\alpha})$ of the envelope level α and the envelope slope, $\dot{\alpha} = d\alpha/dt$.

We assume that the joint pdf of α and $\dot{\alpha}$ are independent:

$$f_{\alpha,\dot{\alpha}}(\alpha, \dot{\alpha}) = f_\alpha(\alpha) f_{\dot{\alpha}}(\dot{\alpha}) \qquad (11.154)$$

which is generally valid for typical fading environments such as Rician, Rayleigh and Nakagami-m. Inserting (11.154) into (11.153), the level crossing rate reduces to

$$N_R = f_\alpha(R) \int_0^\infty \dot{\alpha} f_{\dot{\alpha}}(\dot{\alpha}) d\dot{\alpha} \qquad (11.155)$$

It is evident from (11.155) that the level crossing rate is directly proportional to the value of the PDF $f_\alpha(R)$ of α at level R. Assuming that the slope of the envelope has a Gaussian distribution with standard deviation $\dot{\sigma}$, the LCR is found as follows:

$$f_{\dot{\alpha}}(\dot{\alpha}) = \frac{1}{\sqrt{2\pi}\dot{\sigma}} \exp\left(-\frac{\dot{\alpha}^2}{2\dot{\sigma}^2}\right) \Rightarrow$$
$$N_R = f_\alpha(R)\dot{\sigma}/\sqrt{2\pi} \qquad (11.156)$$

The standard deviation of the envelope slope $\dot{\sigma}$ is related to the maximum Doppler shift in a Nakagami-m fading channel as follows:

$$\dot{\sigma} = \sqrt{\frac{P_0}{m}} \pi f_{d,\max} \qquad (11.157)$$

Using (11.104), (11.156) and (11.157), the level crossing rate for Nakagami-m fading is found as [14]

$$N_R = \frac{1}{\Gamma(m)} \sqrt{2\pi} \, f_{d,\max} \left(m\rho^2\right)^{m-1/2} e^{-m\rho^2}$$
$$\rho = R/\sqrt{P_0}$$
$$\qquad (11.158)$$

The LCR for Rayleigh fading is found by inserting $m = 1$ into (11.158):

$$N_R = \sqrt{2\pi} f_{d,\max} \rho e^{-\rho^2} \qquad (11.159)$$

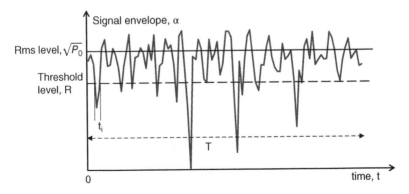

Figure 11.36 Crossing the Envelope Level R and Average Fade Duration.

Level crossing rate for Rician fading is obtained using (11.100): [15]

$$N_R = f_\alpha(R)\frac{\dot\sigma}{\sqrt{2\pi}} = \sqrt{2\pi(K+1)}f_{d,\max}$$

$$\times \rho\, e^{-K-(K+1)\rho^2} I_0\left(2\rho\sqrt{K(K+1)}\right)$$

$$\dot\sigma = \sqrt{2\sigma^2}\pi f_{d,\max}$$

$$\rho = R/\sqrt{2\sigma^2(1+K)}$$

$$(11.160)$$

Figure 11.37 shows the LCR normalized with respect to the maximum Doppler frequency shift versus ρ in dB for various values of the fading parameter m. For all values of ρ, the LCR increases as the fading parameter m becomes smaller, implying more frequent and deeper fades and flares. The LCR approaches its maximum value, which is approximately equal to the maximum Doppler shift, around the mean value $\rho \approx 0$ dB for large m. However for $m = 1$ (Rayleigh fading), the maximum value of the LCR occurs at a level which is 3 dB below the mean value. For $\rho > 0$ dB, the LCR decreases with increasing values of ρ, which corresponds to higher envelope levels and slower fluctuations.

Average envelope fade duration is defined as the average duration that the envelope level remains below a specified level R. Consider a sufficiently long time interval of length T and let t_i be the duration of the ith fade below the envelope level R, as shown in Figure 11.36. The cdf of the envelope, that is, the probability of the envelope level remains below the level R, is given by

$$\Pr(\alpha \le R) = \int_0^R f_\alpha(\alpha)d\alpha = \lim_{T\to\infty}\frac{1}{T}\sum_i t_i$$

$$(11.161)$$

The average envelope fade duration may then be written as

$$\bar{t} = \frac{\Pr(\alpha \le R)}{N_R} \qquad (11.162)$$

For Nakagami-m fading, the average envelope fade duration is found by inserting (11.106) and (11.158) into (11.162):

$$\bar{t} = \frac{\exp(m\rho^2)\gamma(m, m\rho^2)}{\sqrt{2\pi}f_{d,\max}(m\rho^2)^{m-1/2}} \qquad (11.163)$$

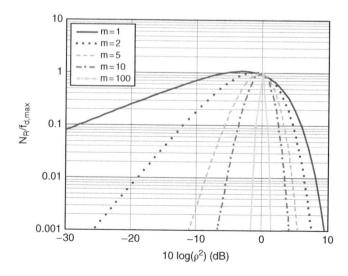

Figure 11.37 Normalized LCR $N_R/f_{d,\max}$ For Nakagami-m Fading.

Average fade duration in Rayleigh fading is found by inserting $m = 1$ into (11.163):

$$\bar{t} = \frac{e^{\rho^2} - 1}{\sqrt{2\pi}\rho f_{d,\max}} \qquad (11.164)$$

Using (11.101) and (11.160), the average envelope fade duration for Rician fading is given by [15]

$$\bar{t} = \frac{\left[1 - Q_1\left(\sqrt{2K}, \sqrt{2(K+1)\rho^2}\right)\right] e^{K + (K+1)\rho^2}}{\sqrt{2\pi(K+1)}\rho f_{d,\max} I_0\left(2\rho\sqrt{K(K+1)}\right)}$$

$$(11.165)$$

Figure 11.38 shows the average normalized fade duration for Nakagami-m fading as a function of the normalized threshold level. When the threshold is below the rms envelope level, with which we are mainly interested in, the average fade duration decreases as m increases. In this region, average fade duration increases with increasing values of the threshold level.

Example 11.17 LCR and Average Fade Duration For Rayleigh Fading.

Consider a mobile terminal communicating onboard a vehicle travelling at 120 km/hr. We will compare average LCR and average fade durations at $f_c = 900\,\text{MHz}$ and $2100\,\text{MHz}$. The maximum Doppler frequency shift is $f_{d,\max} = 100\,\text{Hz}$ at $900\,\text{MHz}$ and $233.3\,\text{Hz}$ at $2100\,\text{MHz}$. Using (11.159) and (11.164), the LCR and mean fade duration for Rayleigh fading are listed in Table 11.6 for various values of the threshold level. The average fade duration monotonically decreases as the envelope level becomes lower. However, LCR reaches a maximum at a level 3 dB below the average power level and decreases as the threshold level is set below or above that level. Since the LCR is directly proportional and the average fade duration is inversely proportional to the maximum Doppler frequency shift, LCR and average fade duration can easily be frequency-scaled.

Example 11.18 Effect of Average Fade Duration on the Choice of the Transmission Rate. Fades cause symbol errors if they last longer than the symbol duration. FEC codes cannot correct all errors in a burst unless they are specifically designed to mitigate burst errors. One

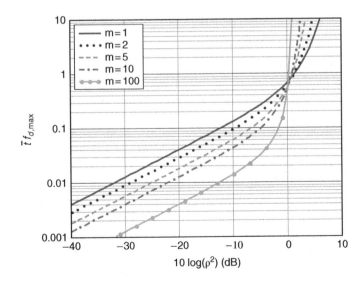

Figure 11.38 Normalized Average Fade Duration For Nakagami-m Fading.

Table 11.6 Average LCR and Fade Duration in Rayleigh Fading For 120 km/hr Vehicle Speed at 900 MHz and 2100 MHz.

Envelope level, ρ	Relative power level (dB)	LCR, N_R (fades/s)		Average fade duration (ms)	
		900 MHz	2100 MHz	900 MHz	2100 MHz
1	0	92.2	214.9	6.86	2.94
$1/\sqrt{2}$	−3	107.5	250.5	3.66	1.57
		(max.)	(max.)		
0.1	−20	24.8	57.8	0.40	0.17
0.01	−40	2.5	5.8	0.04	0.02

way to overcome this problem is to use inter-leaving, that is, change the order of transmission of the symbols in the channel at the transmitter and deinterleave them (restore their original order) at the receiver. Then the consecutive symbol errors caused by fading are randomized at the deinterleaver output so that the FEC codes can correct them more easily.

An alternative approach to alleviate this problem is to limit the coded bit rate. For example, if the FEC code can correct at most t consecutive bit errors, then the coded bit rate should be limited so that the duration of t coded bits should be longer than the average fade duration. Consider now a FEC code that can correct at most t consecutive errors in the coded bit stream transmitted over a Rayleigh fading channel. Coded bit errors are assumed occur in block when the threshold signal power level P_{th} is more than 20 dB below the average value P_{av}. Assume that the considered system operates at 900 MHz carrier frequency and is required to serve mobile users with velocities not exceeding 180 km/hr; hence the maximum Doppler shift of the carrier frequency is $f_{d,max} = 150$ Hz. Inserting $\rho^2 = P_{th}/P_{av} = 0.01(-20\,dB)$ into (11.164), the average fade duration is found to be

$$\bar{t} = \frac{\exp(P_{th}/P_{av})-1}{f_{d,max}\sqrt{2\pi P_{th}/P_{av}}} = \frac{\exp(0.01)-1}{150\sqrt{2\pi 0.01}} = 0.267\, ms \tag{11.166}$$

Assuming, for example, that no more than t coded bits can be transmitted during a fade on the average, the required coded bit duration and the coded bit rate should satisfy the following:

$$T > \bar{t}/t, \quad R < t/\bar{t} \tag{11.167}$$

Assuming a FEC code which can correct $t = 3$ consecutive coded bits, the coded bit rate should not exceed 11.24 kbps. To be able to communicate at higher bit rates in such channels, one needs to use bit interleaving with interleaving depths much larger than average fade duration and/or burst error correcting codes which are capable of correcting long bursts of errors. On the other hand, the average fade duration, which is inversely proportional to the mobile speed (see (11.164)), would be increased by 36 times (9.6 ms) for pedestrian users moving with 5 km/h compared for 180 km/hr. The 3-error correction capability forces the transmission rate to be lower than 11.24 kbps/36 = 312 bps and interleaving depths on the order of ten times the fade duration (~96 ms) are not feasible for real-time communications.

11.4 Bit Error Probability in Frequency-Nonselective Slowly Fading Channels

Frequency-nonselective (flat) fading leads to multiplicative distortion of the transmitted signal as given by (11.73). In a slowly fading channel, the symbol duration T is shorter than the channel coherence time $T \ll \Delta t_c$. Assuming that the signal undergoes Rayleigh fading, the envelope α has a Rayleigh pdf while the phase ϕ is uniformly distributed. Hence, the instantaneous bit SNR γ_b in Rayleigh fading is found from (11.73) as

$$\gamma_b = \frac{\frac{1}{2}\int_0^T |\alpha e^{-j\phi} s_\ell(t)|^2 dt}{E[|w(t)|^2]} = \frac{\alpha^2 \int_0^T |s_\ell(t)|^2 dt}{N_0} = \alpha^2 \frac{E_b}{N_0}$$

(11.168)

where the factor 1/2 accounts for the average power of the sinusoidal carrier. Since α is Rayleigh distributed, the pdf of γ_b is exponentially distributed (see (11.88))

$$f_{\gamma_b}(\gamma_b) = \frac{1}{\bar{\gamma}_b} \exp\left(-\frac{\gamma_b}{\bar{\gamma}_b}\right), \gamma_b \geq 0$$

$$\bar{\gamma}_b = E[\alpha^2] \frac{E_b}{N_0} = 2\sigma^2 \frac{E_b}{N_0} = \frac{E_b}{N_0} \text{ if } \sigma^2 = 0.5$$

(11.169)

Here, $E[\alpha^2] = 1$ implies that variances of in-phase and quadrature components, defined by (11.83), are equal to 1/2.

11.4.1   Bit Error Probability for Binary Signaling

In a channel with no fading, α has a fixed (non-random) value, which is usually assumed to be unity; hence $\gamma_b = E_b/N_0$. Then, the bit error probability (BEP) in an AWGN channel is given by (see (6.62) and (8.118))

$$P_2(\gamma_b) = Q\left(\sqrt{(1-\rho)\gamma_b}\right)$$

(11.170)

where $\rho = -1$ for antipodal signaling (e.g., coherently-detected binary PSK) and $\rho = 0$ for coherent orthogonal signaling (e.g., coherently-detected binary orthogonal FSK).

In fading conditions, channel gain α and channel phase ϕ are random. For coherent detection, the channel phase ϕ is assumed to be sufficiently slow so that it can be estimated perfectly from the received signal. The BEP is then found by averaging $P_2(\gamma_b)$ in (11.170) with the PDF of γ_b given by (11.169):

$$P_2 = \int_0^\infty P_2(\gamma_b) f_{\gamma_b}(\gamma_b) d\gamma_b$$

(11.171)

Using (D.70) one gets

$$P_2 = \frac{1}{2}\left[1 - \sqrt{\frac{(1-\rho)\bar{\gamma}_b}{2 + (1-\rho)\bar{\gamma}_b}}\right]$$

(11.172)

Approximation for large values of $\bar{\gamma}_b$ is obtained by first rewriting P_2 as

$$P_2 = \frac{1}{2}\left(1 - \sqrt{1 - \frac{2}{2 + (1-\rho)\bar{\gamma}_b}}\right)$$

(11.173)

and then using the binomial expansion given by (D.9) to get

$$P_e \cong \frac{1}{2(2 + (1-\rho)\bar{\gamma}_b)} \quad \bar{\gamma}_b \gg 1$$

(11.174)

If fading is not sufficiently slow, then a stable estimate of the channel phase ϕ can be achieved only by averaging over several signaling intervals. Then, DPSK, which does not require an estimate of the channel phase, becomes an alternative to coherent PSK since it is quite robust in the presence of signal fading, and phase stability is required over only two consecutive signaling intervals. However, higher transmitter power is necessary to achieve the same BEP levels as for coherent BPSK.

For a fixed (non-random) value of α, that is, in non-fading conditions, the BEP for noncoherent detection in an AWGN channel is given by (see (8.152) and (8.168))

$$P_2(\gamma_b) = \frac{1}{2}\exp(-\beta\gamma_b)$$

(11.175)

where $\beta = 1$ for binary DPSK and $\beta = 1/2$ for noncoherently detected binary orthogonal FSK. The BEP in a Rayleigh fading channel is determined by averaging $P_2(\gamma_b)$ given by (11.175) with the pdf of γ_b given by (11.169):

$$P_2 = \frac{1}{2(1 + \beta \bar{\gamma}_b)} \qquad (11.176)$$

The expressions for BEP of the considered binary modulations are presented in Table 11.7 for AWGN and Rayleigh fading channels.

Asymptotic behavior for large SNR values are also shown. Figure 11.39 shows the BEP for the considered binary modulations in AWGN and Rayleigh fading channels. For all the considered modulations, the average E_b/N_0 required for achieving the same BEP in a Rayleigh fading channel is much higher compared to an AWGN channel. For example, the average E_b/N_0 required for a BEP level of 10^{-3} for coherent BPSK modulation is 24 dB in a Rayleigh fading channel compared to 6.8 dB in the AWGN channel, hence a 17.2 dB performance

Table 11.7 Bit Error Probabilities of Some Binary Modulations in AWGN and Rayleigh Fading Channels.

Modulation	$P_2(\gamma_b)$	P_2	$P_2(\text{large } \bar{\gamma}_b)$
2-PSK (coherent)	$Q(\sqrt{2\gamma_b})$	$\frac{1}{2}\left(1 - \sqrt{\frac{\bar{\gamma}_b}{1 + \bar{\gamma}_b}}\right)$	$\frac{1}{4\bar{\gamma}_b}$
2-FSK (coherent)	$Q(\sqrt{\gamma_b})$	$\frac{1}{2}\left(1 - \sqrt{\frac{\bar{\gamma}_b}{2 + \bar{\gamma}_b}}\right)$	$\frac{1}{2\bar{\gamma}_b}$
DPSK (differentially coherent)	$\frac{1}{2}\exp(-\gamma_b)$	$\frac{1}{2(\bar{\gamma}_b + 1)}$	$\frac{1}{2\bar{\gamma}_b}$
2-FSK (noncoherent)	$\frac{1}{2}\exp(-\gamma_b/2)$	$\frac{1}{\bar{\gamma}_b + 2}$	$\frac{1}{\bar{\gamma}_b}$

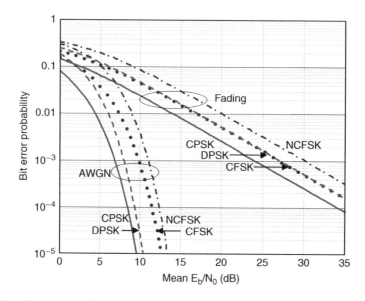

Figure 11.39 Comparison of Some Binary Modulations in AWGN and Rayleigh Fading Channels.

degradation due to fading. Figure 11.39 clearly shows that, for a given value of the BEP, the coherent BPSK requires the lowest average E_b/N_0 in both AWGN and Rayleigh fading channels while noncoherent orthogonal binary FSK requires the highest. However, the BEP performances of DPSK and coherent orthogonal FSK are very close to each other in a Rayleigh fading channel.

To observe the effect of the fading environment on the BEP, we consider the average BEP for coherent BPSK in a Nakagami-m fading channel, which is found using (11.109) and (D.70):

$$P_2 = \int_0^\infty Q\left(\sqrt{2\gamma}\right) f(\gamma) d\gamma$$

$$= \frac{1}{2} - \frac{1}{2}\sqrt{\frac{\bar{\gamma}/m}{1+\bar{\gamma}/m}} \sum_{k=0}^{m-1} \binom{2k}{k} \frac{1}{[4(1+\bar{\gamma}/m)]^k}$$

$$(11.177)$$

The BEP given by (11.177) is shown in Figure 11.40 versus the mean E_b/N_0 for various values of the fading parameter m, which controls the severity of fading. Note that the

BEP performance reduces to that of Rayleigh fading channel for $m = 1$ and AWGN channel for large values of m. Therefore, Nakagami-m pdf provides a flexible model for fading channels.

11.4.2   Moment Generating Function

Moment generating function (MGF) provides a useful tool for determining the moments of random variables, the statistics of the sum of random variables and the BEP. The MGF of a random variable X is defined by

$$M_x(s) \equiv \mathrm{E}\left[e^{-sX}\right] = \int_{-\infty}^{\infty} e^{-sx} f_x(x)\,dx$$

$$= s \int_{-\infty}^{\infty} e^{-sx} F_x(x)\,dx$$

$$(11.178)$$

where $f_X(x)$ and $F_X(x)$ denote respectively the pdf and the cdf of the random variable X. The last expression in (11.178) is obtained by applying integration by parts on the previous one.

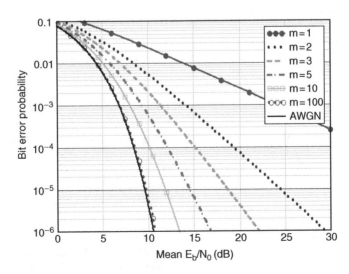

Figure 11.40   BEP For BPSK in a Nakagami-m Fading Channel For Various Values of the Fading Parameter m.

Using MGF the moments of the random variable X may be found as

$$E[X^n] = (-1)^n \frac{d^n}{ds^n} M_x(s)|_{s=0} = \int_{-\infty}^{\infty} x^n f_x(x) \, dx$$

$$(11.179)$$

As an example for the use of the MGF in error probability calculations, consider first the BEP for binary DPSK and NCFSK in a fading channel. The average BEP is found by averaging (11.175) with the pdf of γ and using (11.178):

$$P_2 = \int_0^{\infty} P_2(\gamma) f_\gamma(\gamma) \, d\gamma = \frac{1}{2} \int_0^{\infty} e^{-\beta\gamma} f_\gamma(\gamma) \, d\gamma = \frac{1}{2} M_\gamma(\beta)$$

$$(11.180)$$

where $\beta = 1$ for DPSK and $\beta = 1/2$ for NCFSK. Using the MGF expression given by (11.90) for Rayleigh fading with $P_0 = \bar{\gamma}_b$, one may observe that (11.180) and (11.176) are identical to each other.

It is convenient to use the Craig's definition (see (B.4)) in order to determine the BEP for modulation schemes involving Gaussian Q-functions since the integration limits are finite and independent of the argument. From (B.4) and (B.5), first two powers of the Q function may be written in a compact form as

$$Q^n\left(\sqrt{2\beta\gamma}\right)$$

$$= \frac{1}{\pi} \int_0^{\pi/(2n)} \exp\left(-\frac{\beta\gamma}{\sin^2\phi}\right) d\phi, \quad x > 0, \, n = 1, 2$$

$$(11.181)$$

Using (11.181), the symbol error probability (SEP) of a class of modulation schemes in an AWGN channel may be written as

$$P_s(\gamma) = \frac{\alpha}{\pi} \int_0^{\Gamma} \exp\left(-\frac{\beta\gamma}{\sin^2\phi}\right) d\phi \qquad (11.182)$$

where α, β, and Γ are constants. Note that (11.182) reduces to (11.181) for $\alpha = 1$ and $\Gamma = \pi/(2n)$. Comparing (8.74) with (11.182), the SEP for M-ary PSK may be represented by (11.182) for $\alpha = 2$, $\beta = \log_2 M \, \sin^2(\pi/M)$, $\Gamma = \pi/2$ and $\gamma = E_b/N_0$. For the special case for BPSK, the bit error probability is obtained from (11.182) for $\alpha = \beta = 1$, $\Gamma = \pi/2$ and $\gamma = E_b/N_0$.

Using (11.182), the SEP in a fading channel may be expressed in terms of the MGF as follows:

$$P_s = \int_0^{\infty} P_s(\gamma) f_\gamma(\gamma) \, d\gamma$$

$$= \int_0^{\infty} f_\gamma(\gamma) \, d\gamma \frac{\alpha}{\pi} \int_0^{\Gamma} \exp\left(-\frac{\beta\gamma}{\sin^2\phi}\right) d\phi$$

$$= \frac{\alpha}{\pi} \int_0^{\Gamma} M_\gamma\left(\frac{\beta}{\sin^2\phi}\right) d\phi$$

$$(11.183)$$

where the MGF is given by (11.90), (11.103) and (11.111) for Rayleigh, Rician and Nakagami-m fading channels, respectively. If analytical solution is not available, then the integration in (11.183) may be evaluated more accurately due to finite integration limits.

Example 11.19 The Pdf and the Cdf of the Sum of Nakagami-m Random Variables.

In multipath fading channels, the receivers are usually designed to sum the powers (SNRs) of the multipath components in order to extract the power of a signal suffering delay dispersion (see 11.22). The receiver first estimates the gains and phases of multipaths, multiplies the received signal of a particular path with the complex conjugate of its estimations and finally sums them all.

As will be studied in the next chapter, diversity reception and combining of multipath componens necessitates the determination of the statistics of the sum of usually independent random variables. The pdf of the sum of random variables is obtained by the convolution of the individual pdf's. This tedius calculation procedure may be avoided by determining the moment generating function (MGF) of the independent random variables and taking the inverse Fourier transform of the product of the MGFs. This is similar to the convolutional property of the Fourier transform; time convolution of multiple signals may be obtained by using the Fourier transforms of these signals (see (1.46)).

Consider the so-called maximal ratio combining (MRC) diversity system which takes the sum of E_b/N_0's of L channels:

$$\gamma = \sum_{i=1}^{L} \gamma_i \qquad (11.184)$$

where γ_i denotes the instantaneous E_b/N_0 ratio of the ith multipath.

Assuming that $\gamma_i, i = 1 \ldots L$ are independent and undergo Nakagami-m fading with different mean powers $\bar{\gamma}_i$ and fading parameters, m_i, one may use the MGF of γ_i given by (11.111), to find the MGF of γ:

$$M_\gamma(s) \triangleq E[e^{-s\gamma}] = E\left[\exp\left(-s\sum_{i=1}^{L}\gamma_i\right)\right]$$

$$= \prod_{i=1}^{L} M_{\gamma_i}(s) = \prod_{i=1}^{L}(1 + s\bar{\gamma}_i/m_i)^{-m_i} \qquad (11.185)$$

From (11.180) and (11.185), the bit error probability for binary DPSK or binary NCFSK modulations at the output of a combiner that combines SNR's of L independent channels is given by

$$P_2(L) = \frac{1}{2}M_\gamma(\beta) = \frac{1}{2}\prod_{i=1}^{L}(1 + \beta\bar{\gamma}_i/m_i)^{-m_k} \qquad (11.186)$$

where $\beta = 1$ for DPSK and $\beta = 1/2$ for NCFSK. For example, the average BEP for dual MRC diversity DPSK system with, $m_1 = 2$, and

$m_2 = 1$ and equal mean branch powers $\bar{\gamma}_1 = \bar{\gamma}_2 = \bar{\gamma}$ is simply given by

$$P_2(2) = \frac{1}{2(1+\bar{\gamma}/2)^2(1+\bar{\gamma})} \qquad (11.187)$$

When there is no diversity ($L = 1$), depending on whether the first channel with $m_1 = 2$, or the second channel with $m_2 = 1$ is used, the BEP may be written as

$$P_2(1) = \begin{cases} \dfrac{1}{2(1+\bar{\gamma}/2)^2} & m_1 = 2 \\[4mm] \dfrac{1}{2(1+\bar{\gamma})} & m_2 = 1 \end{cases} \qquad (11.188)$$

Figure 11.41 shows the BEP in a Nakagami-m fading channel for diversity reception ($L = 2$) and reception by a single-branch receiver ($L = 1$) for fading factor $m_1 = 2$ and $m_2 = 1$. The diversity receiver outperforms single-branch receivers, while a single-branch receiver with $m_1 = 2$ has a better BEP performance than that for $m_2 = 1$, because of less severe fading conditions.

The mean SNR's required to achieve $P_2 = 10^{-4}$ for no-diversity ($L = 1$, $m_1 = 2$) and two-branch ($L = 2$) diversity systems are given by

$$P_2(1) = \frac{1}{2(1+\bar{\gamma}/2)^2} = 10^{-4} \Rightarrow \bar{\gamma} = 21.43\,dB$$

$$P_2(2) = \frac{1}{2(1+\bar{\gamma}/2)^2(1+\bar{\gamma})} = 10^{-4}$$

$$\Rightarrow \bar{\gamma} \cong 14.05\,dB \qquad (11.189)$$

For the BEP level of 10^{-4}, the diversity gain, that is, the reduction in the required SNR compared to the no-diversity case, is 21.43-14.05 = 7.38 dB. Hence, an SNR of 14.05 dB is sufficient to realize a BEP of 10^{-4} when two-branch diversity is used, but 21.43 dB SNR would be required to realize the same BEP if a single branch-receiver were used in a Nakagami-m fading propagation environment with $m_1 = 2$.

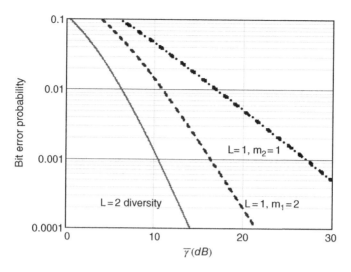

Figure 11.41 Bit Error Probability with No Diversity ($L=1$) and Two-Branch Diversity ($L=2$) in a Nakagami-m Fading Channel with $m_1 = 2$, and $m_2 = 1$ and Equal Mean SNRs $\bar{\gamma}_1 = \bar{\gamma}_2 = \bar{\gamma}$.

11.4.3 Bit Error Probability for M-ary Signalling

In this section, we consider the BEP performance of M-ary PSK, M-ary QAM, as well as coherent and noncoherent orthogonal M-ary FSK modulations in a Rayleigh fading channel. Using (8.74), which is applicable for AWGN channel, the asymptotic SEP for M-ary PSK for high SNRs in a Rayleigh fading channel is found using (D.70):

$$P_M = 2\int_0^\infty Q\left(\sqrt{2\beta\gamma}\right)f_\gamma(\gamma)d\gamma = 1 - \sqrt{\frac{\beta\bar{\gamma}_b}{1+\beta\bar{\gamma}_b}}, \quad M>4$$

(11.190)

where $\beta = \log_2 M\sin^2(\pi/M)$ and $\bar{\gamma}_b$ denotes the average E_b/N_0. The exact BEP expression for $M=2$ is given by (11.172), which is ½ of (11.190). For $M=4$ (QPSK), the exact SEP expression is given by (8.59) in AWGN channel and will be determined as part of the M-ary QAM formulation which will follow. Figure 11.42 shows the SEP for M-ary PSK in a Rayleigh fading environment. Note that SEPs for $M=2$ and $M=4$ are nearly equal to each

other since $\beta = 1$ in both cases. As β gets smaller with increasing values of M, this reduction in β is compensated by an increase in $\bar{\gamma}$. For example, to compensate for the reduction from $\beta = 1$ for $M=2$ to $\beta = 0.152$ for $M=16$, the mean SNR required for 16-PSK is nearly $10\log(1/\beta) = 8.2$ dB (see Figure 11.42). It may be useful to take a close look at the BEP of M-ary PSK, which is presented in Figure 11.43. Because of the factor $1/\log_2 M$ in front of (11.190) for BEP with Gray encoding, QPSK shows an improved BEP performance compared to BPSK for a given $\bar{\gamma}_b$, as expected.

Now consider the SEP performance of M-ary QAM in a Rayleigh fading channel. From (8.90), the SEP for M-ary QAM in AWGN channel is given by

$$P_M(\gamma_b) = \alpha_1 Q\left(\sqrt{2\beta\gamma_b}\right) - \alpha_2 Q^2\left(\sqrt{2\beta\gamma_b}\right)$$

$$\alpha_1 = 2(2-1/L-1/T)$$

$$\alpha_2 = 4(1-1/L)(1-1/T)$$

$$\beta = \frac{3\log_2(LT)}{L^2+T^2-2}$$

(11.191)

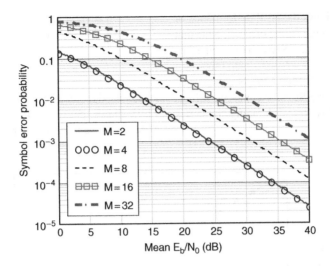

Figure 11.42 Symbol Error Probability For M-ary PSK in a Rayleigh Fading Channel.

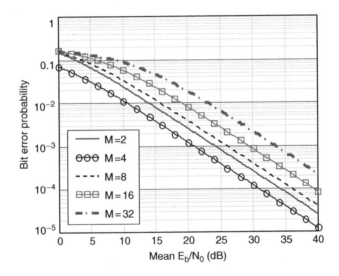

Figure 11.43 Bit Error Probability For M-ary PSK in a Rayleigh Fading Channel.

From (11.183), the SEP in a Rayleigh fading channel may be written as

$$P_M = \int_0^\infty \left[\alpha_1 Q\left(\sqrt{2\beta\gamma}\right) - \alpha_2 Q^2\left(\sqrt{2\beta\gamma}\right)\right] f_{\gamma_b}(\gamma)\, d\gamma$$

$$= \frac{\alpha_1}{\pi} \int_0^{\pi/2} M_\gamma\left(\frac{\beta}{\sin^2\phi}\right) d\phi - \frac{\alpha_2}{\pi}\int_0^{\pi/4} M_\gamma\left(\frac{\beta}{\sin^2\phi}\right) d\phi$$

$$= \frac{\alpha_1}{\pi} \int_0^{\pi/2} \frac{\sin^2\phi}{\sin^2\phi + \beta\bar{\gamma}_b}\, d\phi - \frac{\alpha_2}{\pi}\int_0^{\pi/4} \frac{\sin^2\phi}{\sin^2\phi + \beta\bar{\gamma}_b}\, d\phi$$

$$\tag{11.192}$$

The exact SEP expression in closed-form for M-ary QAM in a Rayleigh fading channel is obtained by evaluating the integrals in (11.192) with the help of (D.42):

$$P_M = \frac{\alpha_1}{2}\left(1 - \sqrt{\frac{\beta\bar{\gamma}_b}{1+\beta\bar{\gamma}_b}}\right)$$

$$- \frac{\alpha_2}{4}\left(1 - \sqrt{\frac{\beta\bar{\gamma}_b}{1+\beta\bar{\gamma}_b}}\frac{4}{\pi}\tan^{-1}\left(\sqrt{\frac{1+\beta\bar{\gamma}_b}{\beta\bar{\gamma}_b}}\right)\right)$$

$$\tag{11.193}$$

Figure 11.44 shows the BEP performance of M-QAM in a Rayleigh fading channel. Note that the performances for $M = 8$ and 16 as well as for $M = 32$ and 64 are quite close to each other (see Figure 8.30 for the SEP in AWGN channel).

Using (8.126), the SEP for coherent orthogonal M-ary FSK in a Rayleigh fading channel is given by

$$P_M = 1 - \int_0^\infty \left[1 - Q\left(\sqrt{\log_2 M \gamma} \right) \right]^{M-1} \frac{e^{-\gamma/\bar{\gamma}_b}}{\bar{\gamma}_b} d\gamma$$

(11.194)

where $\bar{\gamma}_b$ denotes the average E_b/N_0. Figure 11.45 shows that the SEP for different values of M are close to each other.

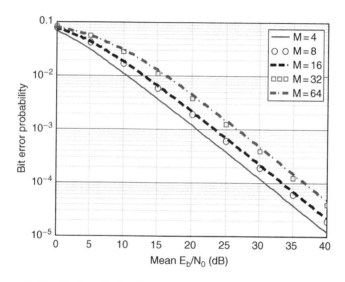

Figure 11.44   Bit Error Probability of M-QAM in a Rayleigh Fading Channel.

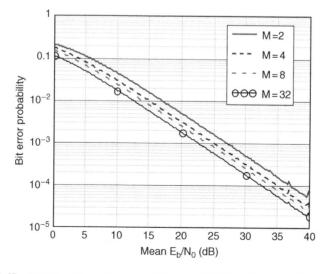

Figure 11.45   BEP For M-ary Coherent Orthogonal FSK in a Rayleigh Fading Channel.

The SEP for noncoherent M-ary FSK in AWGN channel is given by (8.175):

$$P_M(\bar{\gamma}_b) = \frac{1}{M}\sum_{m=1}^{M-1}\binom{M}{m+1}(-1)^{m+1}e^{-\beta\bar{\gamma}_b}$$

$$(11.195)$$

where $\beta = m\log_2 M/(m+1)$. The corresponding SEP performance in a Rayleigh fading channel is found by averaging (11.195) with (11.169) or directly from (11.180):

$$P_M = \frac{1}{M}\sum_{m=1}^{M-1}\binom{M}{m+1}(-1)^{m+1}$$
$$\times \frac{1}{1+m\log_2 M\bar{\gamma}_b/(m+1)}$$

$$(11.196)$$

Figure 11.46 shows the BEP for M-ary noncoherent orthogonal FSK. Comparison with Figure 11.45 clearly demonstrate the superior performance of coherently detected M-ary FSK over noncoherent detection.

11.4.4 Bit Error Probability in Composite Fading and Shadowing Channels

Using (11.148) and (D.69)), the bit error probability for DPSK and noncoherent binary orthogonal FSK in the the composite channel is found as [9]

$$P_e = \int_0^\infty \frac{1}{2}e^{-\beta\gamma}f_C(\gamma)d\gamma$$

$$= \frac{\Gamma(m_s-m)}{2\Gamma(m_s)}\left(\frac{mm_s}{\beta\Omega_s}\right)^m {}_1F_1\left(m;m-m_s+1;\frac{mm_s}{\beta\Omega_s}\right)$$

$$+ \frac{\Gamma(m-m_s)}{2\Gamma(m)}\left(\frac{mm_s}{\beta\Omega_s}\right)^{m_s} {}_1F_1\left(m_s;m_s-m+1;\frac{mm_s}{\beta\Omega_s}\right)$$

$$(11.197)$$

where $\beta = 1$ for binary DPSK modulation and $\beta = 1/2$ for noncoherently detected binary orthogonal FSK. ${}_1F_1(a;b;x)$ denotes the hypergeometric function (see (D.88)). A close look at (11.197) shows that the MGF of (11.148) is equal to $2P_e$ when β is replaced by s.

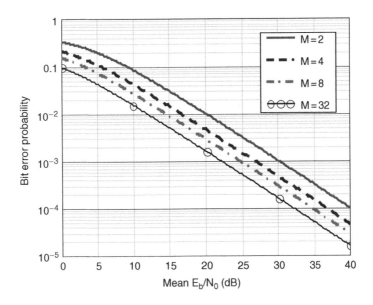

Figure 11.46 BEP For M-ary Noncoherent Orthogonal FSK in a Rayleigh Fading Channel.

For large values of the SNR Ω_s, the hypergeometric functions in (11.197) approaches unity (see (D.88)), and the BEP behaves asymptotically as follows [9]:

$$P_e = \frac{\Gamma(m_s - m)}{2\Gamma(m_s)} \left(\frac{mm_s}{\beta\Omega_s}\right)^m$$
$$+ \frac{\Gamma(m - m_s)}{2\Gamma(m)} \left(\frac{mm_s}{\beta\Omega_s}\right)^{m_s}, \quad \Omega_s \gg 1$$
$$(11.198)$$

Figure 11.47 shows the average BEP for fading only, shadowing only and composite fading and shadowing for $m = 1$ of the Nakagami-m fading parameter and $m_s = 4.233$ and 0.749 of the Gamma pdf, corresponding respectively to $4\,dB$ and $8\,dB$ log-normal shadowing variances. One may easily observe from (11.198) and Figure 1.48 that the slope of the bit error probability curve versus the mean local mean power for the composite channel behaves as $\Omega_s^{-\min(m,m_s)}$. This implies that the BEP performance in a composite fading and shadowing channel is determined by the worser of fading and shadowing parameters.

11.5 Frequency-Selective Slowly-Fading Channels

In a slowly fading frequency-selective channel one has $W > \Delta f_c$ and $T < \Delta t_c$. In such a channel, we consider the transmission of a bandpass signal, $s(t)$, with bandwidth W and symbol duration T as described by (11.1). The corresponding low-pass equivalent signal $s_\ell(t)$ is bandlimited to $W/2$ (see Figure 8.1).

A multipath channel may consist of M physical paths which are not necessarily uniformly spaced. Nevertheless, the channel impulse response is usually sampled by an ADC, at the Nyquist sampling rate W for the low-pass equivalent signal bandlimited to $W/2$, using L complex-valued samples uniformly spaced at sampling interval $1/W$:

$$L = \lfloor T_m / (1/W) \rfloor + 1 = \lfloor WT_m \rfloor + 1 \quad (11.199)$$

With a transmitted signal of bandpass bandwidth $W \gg \Delta f_c$, we achieve a resolution of $1/W$ in the multipath delay profile, limited by the maximum excess delay, T_m. Since the sampling interval $1/W$ in such a channel is much shorter

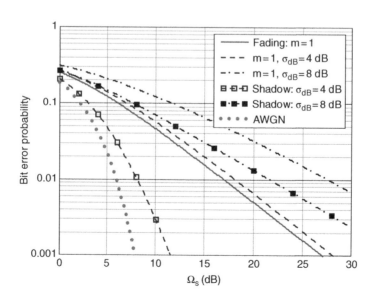

Figure 11.47 Bit Error Probability For the Composite Channel For Binary DPSK Modulation.

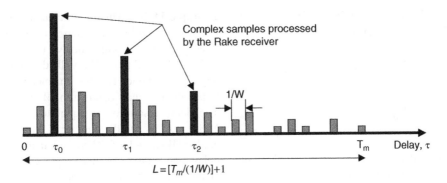

Figure 11.48 Sampled Channel Impulse Response and the Strongest $L=3$ Samples To Be Processed By a Rake Receiver.

than $1/\Delta f_c \approx T_m$, the adjacent samples may be correlated. Therefore, the channel estimation circuit selects the strongest independent samples to be processed in L fingers of the Rake receiver (see Fig. 11.48). Consequently, the channel impulse response, sampled in the delay coordinate, may be written as

$$c(\tau;t) = \sum_{k=0}^{L-1} c_k(t)\delta(\tau-\tau_k)$$

(11.200)

$$c_k(t) = \alpha_k(t)e^{-j\theta_k(t)}$$

Hence, the received signal spread in the delay domain consists of the sum of L independently-fading replicas of the transmitted signal with different delays, channel gains and phase shifts.

11.5.1 Tapped Delay-Line Channel Model

According to the Nyquist sampling theorem, a baseband signal $s_\ell(t)$, which is bandlimited to $W/2$, may be represented by its samples $s_\ell[n/W]$ taken at the sampling frequency, $f_s = W$. Hence $s_\ell(t)$ and its Fourier transform are given by (see (5.5) and (5.6))

$$s_\ell(t) = \sum_{n=-\infty}^{\infty} s_\ell\left[\frac{n}{W}\right] \text{sinc}\left(W\left\{t-\frac{n}{W}\right\}\right)$$

$$S_\ell(f) = \Im[s_\ell(t)]$$

$$= \begin{cases} \displaystyle\sum_{n=-\infty}^{\infty} s_\ell\left[\frac{n}{W}\right] \exp(-j2\pi nf/W) & |f| \le W/2 \\ \\ 0 & |f| > W/2 \end{cases}$$

(11.201)

Inserting $S_\ell(f)$ into (11.72), one gets $r_\ell(t)$

$$r_\ell(t) = \sum_{n=-\infty}^{\infty} s_\ell\left[\frac{n}{W}\right] \int_{-W/2}^{W/2} C(f;t)$$

$$\times \exp\left[j2\pi f\left(\tau-\frac{n}{W}\right)\right] df + w(t)$$

$$\cong \sum_{n=-\infty}^{\infty} s_\ell\left[\frac{n}{W}\right] c\left(\tau-\frac{n}{W};t\right) + w(t)$$

$$= \sum_{n=-\infty}^{\infty} c\left(\frac{n}{W};t\right) s_\ell\left[t-\frac{n}{W}\right] + w(t)$$

(11.202)

Note that the last equation follows from the convolution property. Hence, the low-pass equivalent received signal, propagating through a frequency-selective channel may be written as the convolution of the CIR (11.200) with the input low-pass equivalent signal:

$$r_\ell(t) = \sum_{n=-\infty}^{\infty} c_n(t) s_\ell\left[t - \frac{n}{W}\right] + w(t),$$

$$c_n(t) = c\left(\frac{n}{W}; t\right)$$

$$(11.203)$$

where $c_n(t)$ denotes the nth delay sample of the channel transfer function.

It is clear from (11.203) that a time-variant frequency-selective channel may be modeled as a tapped delay line with L taps, which denotes the maximum number of resolvable multipaths, and tap spacing $1/W$, as shown in Figure 11.49. Tap weights $\{c_n(t)\}$ represent the values of the time-variant channel impulse response at delays n/W and are complex-valued uncorrelated Gaussian random variables. As shown in Figure 11.48, based on this tapped delay line model of a frequency selective channel, L replicas of the transmitted signal at delays n/W, $n = 1,..,L$ are available at the receiver input.

Example 11.20 Three-path channel model
Consider a multipath channel consisting of three paths with relative delays $(\tau_1, \tau_2, \tau_3) = (0.5\ \mu s,\ 1\ \mu s,\ 2\ \mu s)$. We will determine an appropriate model for this channel when the transmitted signal bandwidth W is 2 MHz. The delay resolution is equal to $1/W = 1/(2\ MHz) = 0.5\ \mu s$ and the number of required taps are $L = \lfloor WT_m \rfloor + 1 = 5$. Therefore, in the tapped-delay line model for this frequency-selective channel, 1st, 2nd and 4th taps are connected in order to extract the powers of the multipath components with delays (τ_1, τ_2, τ_3) (see Figure 11.50). The signal at the channel output may be written as

$$r_\ell(t) = \sum_{k=1,2,4}^{L-1} c_k(t) s_\ell[t - k/W] + w(t)$$

$$(11.204)$$

where $\{c_k(t)\}$ denote the value of the channel impulse response at delay k/W.

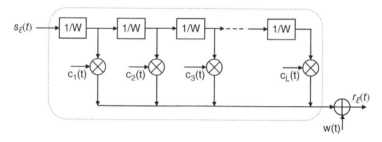

Figure 11.49   Tapped Delay Line Model of a Frequency Selective Channel.

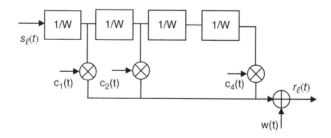

Figure 11.50   Three-Path Frequency-Selective Channel Model with $1/W = 0.5\ \mu s$.

11.5.2 Rake Receiver

The tapped delay line model with statistically independent tap weights provides the receiver with L independent replicas of the transmitted signal with different delays. Rake receiver, that processes these replicas in an optimum manner, acts as a maximal ratio combiner (MRC) where branch SNR's corresponding to the delay bins are added. Thus, the Rake receiver turns the delay dispersion of the multipath channel into an advantage.

For binary signaling, the signal at the output of a frequency-selective channel (at the receiver input) may be written as

$$r_{\ell m}(t) = \sum_{k=0}^{L-1} g_{k,m}(t) + w(t), \quad m = 1,2$$

$$g_{k,m}(t) = c_k s_{\ell m}[t - k/W] = \alpha_k e^{-j\theta_k} s_{\ell m}[t - k/W] \tag{11.205}$$

where slow fading is asumed so that $c_k(t)$ can be estimated perfectly (without noise) and is treated as a constant, c_k, during a signaling interval. Since the input is a discrete-time signal, an optimum filter matched to this signal should also be a discrete-time matched filter (see Example 6.3). An optimum receiver for orthogonal signaling consists of two filters matched to $r_{\ell 1}(t)$ and $r_{\ell 2}(t)$ followed by samplers and a decision device that selects the signal corresponding to the largest output. For antipodal signaling where $s_{\ell 2}(t) = -s_{\ell 1}(t)$,

optimum receiver consists of a single matched filter. The decision device will select either $s_{\ell 1}(t)$ or $s_{\ell 2}(t)$ depending on whether the matched filter output is positive or negative if these two signals are equally likely.

Following the approach described in Example 6.3, we use a discrete-time matched filter receiver with tap weights ξ_k, $k = 0, 1, \ldots$, $L-1$ for optimum reception of the input signal (11.205). Figure 11.51 shows a discrete matched filter realization of a Rake receiver. Note that the Rake receiver should have the ability to estimate the channel gain c_k and the channel phase θ_k so as to determine the tap weights.

The output signal $x_{\ell m}(T_s)$ at the sampling time $T_s = (L-1)/W$, which also coincides with the maximum excess time T_m, may be written as the convolution of the input signal with the impulse response $h(t)$ of the Rake receiver realized as a discrete-time matched filter (see Figure 11.51):

$$x_{\ell m}(T_s) = r_{\ell m}(t) \otimes h(t)|_{t=T_s}$$

$$= \sum_{k=0}^{L-1} \xi_k \, g_{L-1-k,m}(T_s) + \sum_{k=0}^{L-1} \xi_k \, w_{L-1-k}$$

$$g_{k,m}(T_s) = c_k \, s_{\ell m}[T_s - k/W]$$

$$= \alpha_k \, e^{-j\theta_k} s_{\ell m}[(L-1-k)/W] \tag{11.206}$$

where w_k denotes the sample of the AWGN noise $w(t)$ in (11.205) at delay k/W. Assuming

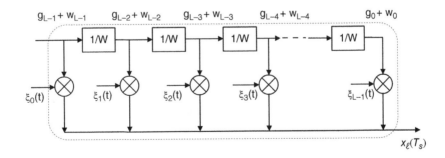

$$g_{L-1} + w_{L-1} \qquad g_{L-2} + w_{L-2} \qquad g_{L-3} + w_{L-3} \qquad g_{L-4} + w_{L-4} \qquad g_0 + w_0$$

Figure 11.51 Discrete Matched Filter Realization of a Rake Receiver.

that the transmitted signal is $s_{\ell 1}(t)$, the optimum tap weights of the discrete-time filter matched to $s_{\ell 1}(t)$ are found to be (see Example 6.3)

$$\xi_k = g_{L-1-k,1}^*(T_s)$$

$$= \alpha_{L-1-k}\, e^{j\theta_{L-1-k}}\, s_{\ell 1}^*[k/W], \qquad (11.207)$$

$$k = 0, 1, \cdots, L-1$$

Inserting (11.207) into (11.206), one gets

$$x_{\ell m}(T_s) = \sum_{k=0}^{L-1} g_{L-1-k,1}^*(T_s)\, g_{L-1-k,m}(T_s) + N_1(T_s)$$

$$= \sum_{k=0}^{L-1} |\alpha_{L-1-k}|^2 s_{\ell m}[k/W]\, s_{\ell 1}^*[k/W] + N_1(T_s)$$

$$= \sum_{k=0}^{L-1} |\alpha_k|^2 s_{\ell m}[(L-1-k)/W]\, s_{\ell 1}^*[(L-1-k)/W] + N_1(T_s)$$

$$N_m(T_s) = \sum_{k=0}^{L-1} g_{L-1-k,m}^*(T_s)\, w_{L-1-k} = \sum_{k=0}^{L-1} g_{k,m}^*(T_s)\, w_k$$

$$= \sum_{k=0}^{L-1} \alpha_k e^{j\theta_k} s_{\ell m}^*[(L-1-k)k/W]\, w_k$$

$$(11.208)$$

where $N_m(T_s)$, $m = 1, 2$ denotes the noise sample at the output of the filter matched to $s_{\ell m}(t)$. The variance $E\left[|N_m(T_s)|^2\right] = (E_b N_0/2)$ $\sum_{k=0}^{L-1} \alpha_k^2$, $m = 1, 2$ is the same for $m = 1$ and 2.

For binary antipodal signaling where $s_{\ell 1}(t) = -s_{\ell 2}(t)$, the decision variables are found from (11.208) as

$$x_{\ell 1}(T_s) = \pm \sum_{k=0}^{L-1} |\alpha_k|^2 |s_{\ell 1}[(L-1-k)/W]|^2 + N_1(T_s)$$

$$(11.209)$$

where $+ (-)$ sign applies when $s_{\ell 1}(t)$ $(s_{\ell 2}(t))$ is sent. Hence, for antipodal signaling, a single filter matched to $s_{\ell 1}(t)$ is sufficient because of $s_{\ell 1}(t) = -s_{\ell 2}(t)$ (see Figure 11.52). The decision rule is given by

$$\begin{cases} s_{\ell 1} & \text{if } x_{\ell 1}(T_s) > 0 \\ s_{\ell 2} & \text{if } x_{\ell 1}(T_s) < 0 \end{cases} \qquad (11.210)$$

For orthogonal signaling, two filters matched to $s_{\ell 1}(t)$ and $s_{\ell 2}(t)$ are required (see Figure 11.53). The decision variables are found from (11.208) as follows:

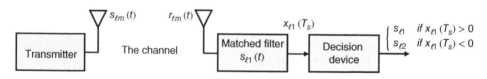

Figure 11.52   Optimum MF Rake Receiver For Binary Antipodal Signaling.

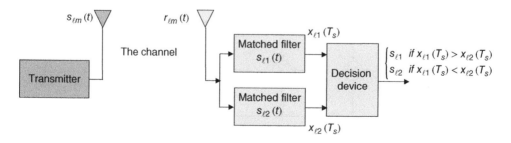

Figure 11.53   Optimum MF Rake Receiver For Binary Orthogonal Signals.

$$x_{\ell 1}(T_s) = \begin{cases} \sum_{k=0}^{L-1} |\alpha_k|^2 |s_{\ell 1}[(L-1-k)/W]|^2 + N_1(T_s) & s_{\ell 1} \text{ sent} \\ \\ N_1(T_s) & s_{\ell 2} \text{ sent} \end{cases}$$

$$x_{\ell 2}(T_s) = \begin{cases} N_2(T_s) & s_{\ell 1} \text{ sent} \\ \\ \sum_{k=0}^{L-1} |\alpha_k|^2 |s_{\ell 2}[(L-1-k)/W]|^2 + N_2(T_s) & s_{\ell 2} \text{ sent} \end{cases}$$

$$(11.211)$$

The decision rule for orthogonal signaling is given by

$$\begin{cases} s_{\ell 1} & \text{if } x_{\ell 1}(T_s) > x_{\ell 2}(T_s) \\ s_{\ell 2} & \text{if } x_{\ell 1}(T_s) < x_{\ell 2}(T_s) \end{cases} \qquad (11.212)$$

If the two signals have the same average bit energies at each delay bin, then the SNR at the matched filter output may be written as the sum of SNR's at each delay:

$$\gamma = \frac{\frac{1}{2} E \left[\left| \sum_{k=0}^{L-1} |\alpha_k|^2 |s_{\ell 1}[(L-1-k)/W]|^2 \right|^2 \right]}{E \left[|N_1(T_s)|^2 \right]}$$

$$= \frac{\left(E_b \sum_{k=0}^{L-1} \alpha_k^2 \right)^2}{N_0 E_b \sum_{k=0}^{L-1} \alpha_k^2} = \sum_{k=0}^{L-1} \gamma_k$$

$$\bar{\gamma}_k = E[\gamma_k] = E[\alpha_k^2] \frac{E_b}{N_0}$$

$$(11.213)$$

Assuming independent fading in L Rayleigh fading channels, γ_k's are statistically independent, and the MGF of γ is found using (11.90) and (11.213):

$$M_\gamma(s) = E \left[\exp \left(s \sum_{k=0}^{L-1} \gamma_k \right) \right] = \prod_{k=0}^{L-1} M_{\gamma_k}(s)$$

$$= \prod_{k=0}^{L-1} \frac{1}{1 + s\bar{\gamma}_k}$$

$$(11.214)$$

The pdf corresponding to (11.214) will be determined for two special cases. For the case of unequal mean SNR's,

$$\bar{\gamma}_k \neq \bar{\gamma}_j, \quad j,k = 0,1, \cdots L-1 \quad \text{and} \quad j \neq k$$

$$(11.215)$$

the pdf of γ may be determined using the partial fraction expansion of (11.214):

$$M_\gamma(s) = \prod_{k=0}^{L-1} \frac{1}{1 + s\bar{\gamma}_k} = \sum_{k=0}^{L-1} \frac{\pi_k}{1 + s\bar{\gamma}_k} \iff$$

$$f(\gamma) = \sum_{k=0}^{L-1} \pi_k \frac{\exp(-\gamma/\bar{\gamma}_k)}{\bar{\gamma}_k},$$

$$\pi_k = \prod_{\substack{i=0 \\ i \neq k}}^{L-1} \frac{\bar{\gamma}_k}{\bar{\gamma}_k - \bar{\gamma}_i}$$

$$(11.216)$$

The pdf given by (11.216) is simply the weighted sum of the pdf's of γ_k, $k = 0,1 \ldots, L-1$. For identical mean SNR's in (11.214),

$$\bar{\gamma}_k = \bar{\gamma}, \quad k = 0,1, \cdots, L-1 \qquad (11.217)$$

one gets the central chi-square distribution for SNR (see (F.115)):

$$f(\gamma) = \frac{1}{(L-1)!} \left(\frac{\gamma}{\bar{\gamma}} \right)^{L-1} \frac{\exp(-\gamma/\bar{\gamma})}{\bar{\gamma}} \qquad (11.218)$$

The corresponding cdf is obtained using (D.49)

$$F_\gamma(x) = \frac{\gamma(L, x/\bar{\gamma})}{\Gamma(L)} = 1 - e^{-x/\bar{\gamma}} \sum_{k=0}^{L-1} \frac{1}{k!} \left(\frac{x}{\bar{\gamma}} \right)^k$$

$$(11.219)$$

which reduces to (11.89) for $L = 1$. The mean value of γ is found using (11.213):

$$E[\gamma] = \sum_{i=0}^{L-1} E[\gamma_i] = \sum_{i=0}^{L-1} \bar{\gamma}_i = L\bar{\gamma}, \quad \text{for } \bar{\gamma}_i = \bar{\gamma}$$

$$(11.220)$$

The pdf and the cdf given by (11.218) and (11.219), respectively, are plotted in

Figure 11.54. The likelihood of the small values of SNR at the Rake receiver output becomes smaller as L increases. Hence, the outage probability at the Rake receiver output becomes lower with increasing values of L.

For coherent signaling, conditional BEP in AWGN channel is given by (11.170). Using (D.70), the BEP in a Rayleigh fading channel for the considered two cases, described by (11.215) and (11.217), is found as follows:

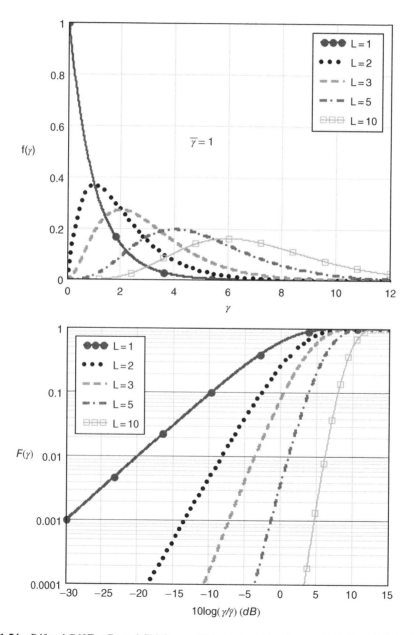

Figure 11.54   Pdf and Cdf For Central Chi-Square Distribution at the Output of a L-Branch Rake Receiver.

$$P_2 = \int_0^\infty Q\left(\sqrt{2(1-\rho)\gamma}\right) f(\gamma) \, d\gamma$$

$$= \begin{cases} \dfrac{1}{2} - \dfrac{1}{2}\sqrt{\dfrac{\bar{\gamma}(1-\rho)}{2+\bar{\gamma}(1-\rho)}} \displaystyle\sum_{k=0}^{L-1} \binom{2k}{k} \dfrac{1}{[2(2+\bar{\gamma}(1-\rho))]^k} & \bar{\gamma}_k = \bar{\gamma}, \ k=0,1,\dots,L-1 \\[4ex] \dfrac{1}{2}\displaystyle\sum_{k=0}^{L-1}\pi_k\left[1-\sqrt{\dfrac{\bar{\gamma}_k(1-\rho)}{2+\bar{\gamma}_k(1-\rho)}}\right] & \bar{\gamma}_k \neq \bar{\gamma}_j, \ k,j=0,1,\dots,L-1 \end{cases} \qquad (11.221)$$

where $\rho = -1, 0$ for antipodal and orthogonal signals, respectively. For large values of SNR per bit, the asymptotic behavior of the BEP is given by

$$P_2 = \begin{cases} \dbinom{2L-1}{L} \dfrac{1}{[2\bar{\gamma}(1-\rho)]^L} & \bar{\gamma}_k = \bar{\gamma}, \ k=0,1,\dots,L-1 \\[4ex] \dbinom{2L-1}{L} \displaystyle\prod_{k=0}^{L-1} \dfrac{1}{2\bar{\gamma}_k(1-\rho)} & \bar{\gamma}_k \neq \bar{\gamma}_j, \ k,j=0,1,\dots,L-1 \end{cases} \qquad (11.222)$$

The fact that the BEP is proportional to $1/\bar{\gamma}^L$ for large values of $\bar{\gamma}$ (see (11.222)) implies a rapid decrease of the BEP with increasing values of L. This may also be observed from Figure 11.55, which shows the effect of the number of Rake receiver fingers on the BEP for BPSK modulation. Note that the BEP performance of a Rake receiver in a Rayleigh fading channel approaches that of AWGN channel as the number L of Rake fingers increases. For 10^{-4} BEP, one needs an average

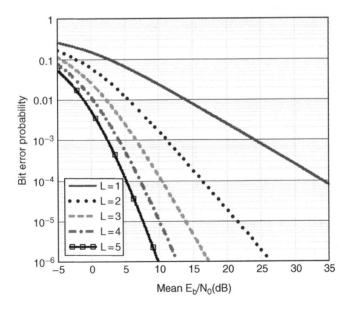

Figure 11.55 BEP For BPSK Modulation For a L-Branch Rake Receiver in a Rayleigh Fading Channel For Identical Mean Branch SNR's.

$E_b/N_0 = 33.95$ dB for $L = 1$, 16.28 dB for $L = 2$ and 10.3 dB for $L = 3$. This implies a diversity gain of $33.95 - 16.28 = 17.67$ dB for $L = 2$, and $33.95 - 10.3 = 23.65$ dB for $L = 3$.

11.6 Resource Allocation in Fading Channels

Fading is usually considered as a nuisance to communication systems, since fading channels cause random fluctuations in the received signals due to Doppler, delay and angular spreads of the received signals. Consequently, a receiver is forced to collect the incoming signal power, which is spread in frequency, delay and angular domains. This makes the reception of faded signals more difficult and lead to performance degradation compared to AWGN channels (see Figure 11.39). However, fading can also be used as a means for improving the system performance. We observed that a Rake receiver turns delay spread into an advantage by combining the SNR's of the signal components with different delays. Diversity reception, adaptive coding and modulation (ACM), scheduling, and multi-user diversity (MUD) may be cited among other means of exploiting fading for the benefit of communication systems.

11.6.1 Adaptive Coding and Modulation

Since the channel conditions vary over time, the system design requires the use of a set of channel statistics to optimize system parameters, such as modulation, coding, signal bandwidth, transmit power, training period, channel estimation and automatic gain control (AGC). The so-called channel state information (CSI), which comprises channel gain, phase and delay of the received signals, changes randomly with time and distance in a fading channel. If a receiver is equipped with a recovery circuit that estimates the CSI, then modulation and coding formats can be adapted to instantaneous variations in the channel. As a typical example of the use of ACM in cellular radio, the users close to the BS are typically assigned higher order modulations with higher code rates (e.g., 64 QAM with $R = 3/4$ turbo codes) since the degradations due to channel are minimal. However, the users close to the cell boundary are assigned lower-order modulations with lower code rates (e.g., QPSK with $R = 1/2$ turbo codes).

As an example for ACM, consider the variation of the E_b/N_0 with time as shown in Figure 11.56. We assume that the adaptive modulation system can follow the channel

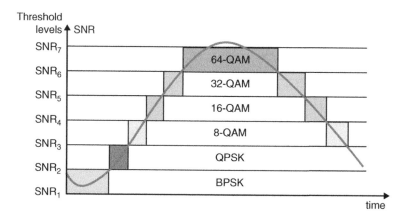

Figure 11.56 Pictorial View of Adaptive Modulation.

fluctuations, that is, the channel is sufficiently slow and does not change appreciably before the receiver reports the CSI to the transmitter. One may designate E_b/N_0 intervals for transmission in desired ranges of BEP levels for different modulation orders. Consequently, the transmitter uses the appropriate modulation order during a time slot where a certain E_b/N_0 is observed. Depending on the channel coherence time (Doppler spread), the packet size as well as the length and the location of a training sequence within the packet greatly affects the system throughput. As shown by Figure 11.57, transmission with adaptive date rate yields higher spectral efficiency at tolerable BEP levels, since the alphabet size is increased during time intervals where higher E_b/N_0 values were observed. The curve for adaptive modulation in Figure 11.57 will shift to the right as the BEP is desired to be lower, and the gap with the channel capacity will increase. If combined with FEC coding, ACM provides a robust transmission strategy in fading channels. An alternative implementation is to use automatic power control whereby the transmit power is adapted to compensate for the random variations in the channel gain, so as to maintain transmissions using a preset coding and modulation.

11.6.2 Scheduling and Multi-User Diversity

Harmful effects of fading may also be alleviated by *scheduling* the transmissions in a fading channel. Scheduling exploits the fact that independent signal paths have a low probability of experiencing deep fades simultaneously. For example, Figure 11.58 shows the received signal power levels of two independently Rayleigh fading channels; they evidently have a low probability of simultaneous occurrence of deep fades. Independently fading channels may be realized by separating the two signals in time, frequency, space, polarization, and so on.

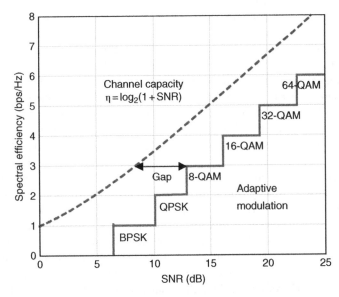

Figure 11.57 Typical Variation of the Spectral Efficiency in An Adaptive Modulation Scheme For a Given BEP Level.

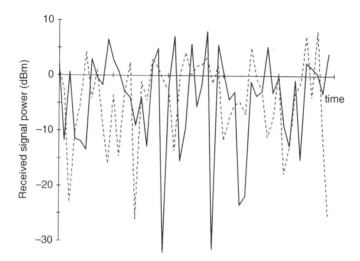

Figure 11.58   Received Powers of Two Independently Rayleigh Fading Signals.

Exploiting the independence of fading of the signals received by each user, the transmission is scheduled to the user with the highest channel gain during a specified time interval. Appropriate scheduling algorithms can be designed for both uplink, and downlink. For example, MS's receiving the downlinks of a mobile radio system feed back their CSI to the BS and the BS selects for transmission the MS with the highest channel gain.

In the uplink, the BS estimates the CSI of all the MSs from their transmissions and instructs the MS with the highest channel gain to transmit. The scheduling aims to provide the appropriate QoS measures, for example, maximum packet delay, packet loss probability and throughput, to individual users and to ensure better use of the resources such as channel bandwidth.

Scheduling algorithms are mainly categorized as round robin (RR), greedy, and proportionally fair (PF). In the *RR scheduling* algorithm, the users are allocated a transmission opportunity one after another, with no priority. In RR, a user may be scheduled to transmit during an interval when its channel undergoes a deep fade. Therefore, the throughput of RR scheduling is rather low. *Greedy scheduler* always selects the user with the highest SNR. A greedy scheduler in a MS calculates the SINR for each BS and selects and transmits to the BS with the best SINR. Alternatively, a MS with the highest SINR is always allowed to transmit. In greedy scheduling, despite the low BEP's achieved, the fairness between the users is a serious issue. The *proportional fair scheduling* (PFS) algorithm, in a given time slot, allows the transmission of packets waiting in the user queue that have the highest ratio of requested transmission rate to the average rate. If a tie occurs, it is broken randomly. The throughput depends not only on the average SINR but also on the SINR standard deviation around the average. In PFS, the individual throughputs of MSs as well as the overall system throughput are improved considerably compared to RR scheduling.

Figure 11.59 shows the behavior of the greedy (top) and proportional-fair (PFS) schedulers (bottom)for users (represented by continuous and dashed-lines) with different average rates. Circles and squares on the curves indicate the scheduled user at each sampling time. In greedy scheduling, the user represented by dashed lines has more access to the system, since it has generally higher SINRs, whereas in PFS the access to the system is balanced between the two users. Only the PFS scheduler

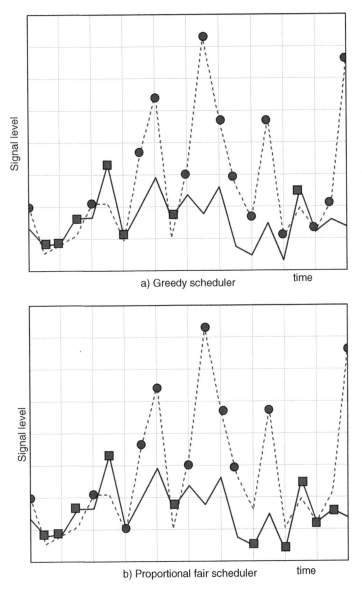

a) Greedy scheduler

b) Proportional fair scheduler

Figure 11.59 Greedy and Proportional-Fair Schedulers With Two Users. Squares and Circles Show the Tranmissions of the Two Users at Each Signaling Interval. [16]

can guarantee long-term fairness and grants access to both users the same number of times. In the presence of more users in the system, the scheduler will have more opportunities to select a user with its SINR well above the average. The benefit offered by the presence of multiple users in such a scheduling system is called as the multi-user diversity (MUD). Figure 11.60 shows the sum rate as a function of the number of active users in Rayleigh and Ricean fading channels in comparison with the AWGN channel. The sum capacity for a Rayleigh fading channel was observed to be higher than for a Ricean fading channel.

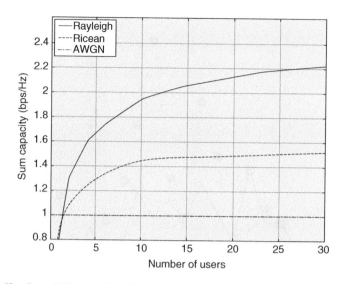

Figure 11.60   Overall Capacity Due To MUD in Rayleigh and Rician Fading Channels. [16]

References

[1] Recommendation ITU-R P.1407-1, Multipath propagation and parameterization of its characteristics, 2003.

[2] J. C.-I. Chuang, "The effects of time delay spread on portable radio communications channels with digital modulation," *IEEE Journal JSAC*, vol. **5**, no. 5, pp. 879–889, June 1987.

[3] M. D. Yacoub, *Foundations of Mobile Radio Engineering*, CRC Press, 1993

[4] B. Holter and G. E. Oien, On the amount of fading in MIMO diversity systems, *IEEE Trans. Wireless Communications*, vol. **4**, no. 5, pp. 2498–2507, Sept. 2005.

[5] M. T. Ivrlac and J. A. Nossek, Diversity and correlation in Rayleigh fading MIMO channels, *VTC 2005 spring IEEE 61st*, vol.**1**, 30 May-1 June 2005, pp. 151–155.

[6] T. S. Rappaport, *Wireless communications, principles and practice* (2nd ed.), Prentice Hall PTR: New Jersey, 2002.

[7] H. Suzuki, A statistical model for urban radio propagation, *IEEE Trans. Commun.*, vol. **25**, no. 7, pp. 673–680, July 1977.

[8] C. Loo, A statistical model for a land mobile satellite link, *IEEE Trans Vehicular Technology*, vol. **34**, no. 3, pp. 122–127, August 1985.

[9] I. M. Kostic, Analytical approach to performance analysis for channel subject to shadowing and fading, *IEE Proc.-Commun.*, vol. **152**, no. 6, pp. 821–827, Dec. 2005.

[10] P. M. Shankar, Outage probabilities in shadowed fading channels using a compound statistical model, *IEE Proc.-Commun.*, vol. **152**, no. 6, pp. 828–832, Dec. 2005.

[11] A. Abdi and M. Kaveh, *On the utility of gamma pdf in modeling shadow fading (slow fading)*, Proc. IEEE Conf. Vehicular Technlogy, Houston, TX, 1999, pp. 2308–2312.

[12] A. Abdi and M. Kaveh, K-distribution: an approximate substitute for Rayleigh-lognormal distribution in fading-shadowing wireless channels, *Electronics Letters*, vol. **34**, no. 9, pp. 851–852, 30th April 1998.

[13] S. Atapattu, C. Tellambura and H. Jiang, Representation of composite fading and shadowing distributions by using mixtures of Gamma distributions, IEEE Wireless Communications and Networking Conference (WCNC), pp. 1–5, 18-21 April 2010.

[14] M.D. Yacoub et al., On higher order statistics of the Nakagami-m distribution, *IEEE Trans. Vehicular Technology*, vol. **48**, no. 3, p. 790–793, May 1999.

[15] G.L. Stuber, *Principles of Mobile Communications*, Kluwer Publications, 1996.

[16] C. Antón-Haro, P. Svedman, M. Bengtsson, A. Alexiou and A. Gameiro, *Cross-layer scheduling for multiuser MIMO systems*, IEEE Communications Magazine, Sept 2006, pp. 39–45.

Problems

1. Consider a multipath fading channel with three rays; a direct ray with power P_0, a Rayleigh distributed ray with average power $P_1 = 2\sigma_1^2$, arriving to the receiver with same delay τ_0 as the direct ray, and a delayed Rayleigh distributed ray of average power

$P_2 = 2\sigma_2{}^2$. The average delay τ_m, relative to that of the direct ray, and the delay spread σ_τ are given by

$$\tau_m = \frac{\displaystyle\sum_{i=0}^{M}(\tau_i - \tau_0)\, P_i}{\displaystyle\sum_{i=0}^{M} P_i} \qquad \sigma_\tau^2 = \frac{\displaystyle\sum_{i=0}^{M}(\tau_i - \tau_m)^2 P_i}{\displaystyle\sum_{i=0}^{M} P_i}$$

where $M = 2$.

a. Assuming that the delay of the direct wave is equal to zero and the delay difference between the direct wave and the delayed wave is denoted by $\Delta\tau$, show that

$$\tau_m = \Delta\tau\, \frac{P_2}{P_0 + P_1 + P_2} \qquad \sigma_\tau^2 + \tau_m^2 = \Delta\tau\,\tau_m$$

b. If $s^2 = (P_0 + P_1 + P_2)/P_0$ and $P_0 = a^2$ then show that

$$\frac{P_1}{P_0} = \frac{2\sigma_1^2}{a^2} = \frac{s^2\,\sigma_\tau^2}{\sigma_\tau^2 + \tau_m^2} - 1 \qquad \frac{P_2}{P_0} = \frac{2\sigma_2^2}{a^2} = \frac{s^2\,\tau_m^2}{\sigma_\tau^2 + \tau_m^2}$$

c. Show that the pdf's of the amplitudes of the direct rays and the delayed ray can be written as

$$f_1(r_1) = \frac{r_1}{\sigma_1^2}\exp\left(-\frac{a^2 + r_1^2}{2\sigma_1^2}\right) I_0\left(\frac{a\,r_1}{\sigma_1^2}\right)$$

$$f_2(r_2) = \frac{r_2}{\sigma_{12}^2}\exp\left(-\frac{r_2^2}{2\sigma_2^2}\right)$$

where $I_0(x)$ denotes the modified Bessel function of the first kind and order zero.

2. The bit error probability for DPSK signalling is given by $P_e = 0.5\, e^{-\alpha^2\gamma_b}$, in AWGN channel with double-sided PSD $N_0/2$. In a fading channel, the channel gain α is modelled by the following pdf:

$$f(\alpha) = \sum_{\ell=0}^{L-1} p_\ell\, \delta(\alpha - \alpha_\ell)$$

where $\sum_{\ell=0}^{L-1} p_\ell = 1$ and $\alpha_0 = 0$, $\alpha_\ell > 0$, $\ell = 1, 2, \ldots, L-1$.

a. Determine the average probability of error for matched-filter reception.
b. Determine the asymptotic value of the error probability as $\gamma_b = E_b/N_0$ approaches infinity. Error floor (minimum bit error probability) is determined by which term?
c. Determine the value of p_0 so as to achieve an asymptotic bit error probability level lower than 10^{-3}.

3. Assuming that the impulse response of a discrete multipath channel is given by

$$c(\tau;t) = \sum_{\ell=0}^{L-1} \alpha_\ell(t)\, e^{-j\theta_\ell(t)}\, \delta(\tau - \tau_\ell(t))$$

a. Determine the multipath intensity profile (MIP), $R_c(\tau)$, for this channel.
b. Assuming that the E_b/N_0 denotes the overall energy per bit to noise power ratio, show that the average energy per bit to noise power ratio of each multipath component is given by

$$\bar{\gamma}_\ell = E\left(\alpha_\ell^2\right) \frac{E_b}{N_0}$$

where

$$\sum_{\ell=0}^{L-1} E\left(\alpha_\ell^2\right) = 1$$

c. Approximating the MIP by an exponential function normalized to unity, that is,

$$R_c(\tau) = \frac{1}{\tau_0}\, e^{-\tau/\tau_0},$$

where τ_0 denotes the rms delay spread and taking, $\tau_n = nT_c$, determine the E_b/N_0 of each multipath delay component and their sum in terms of E_b/N_0, T_c and τ_0.

d. Assuming that these multipath components undergo Rayleigh fading have different average powers, as found in (c), determine the pdf of

$$\gamma_b = \sum_n^{L-1} \gamma_n$$

4. Consider a two-delay channel with the following instantaneous power delay profile

$$P(\tau;t) = \alpha_1{}^2 \, \delta(\tau) + \alpha_2{}^2 \, \delta(\tau - \tau_0)$$

where α_1 and α_2 are independent Rayleigh distributed variables with variances $2\sigma_1{}^2$ and $2\sigma_2{}^2$.

a. Determine the mean power delay profile as defined by $R_c(\tau) = E_t\{P(\tau;t)\}$.
b. Determine the rms delay spread σ_τ, defined by (11.21), in terms of $r = \sigma_1/\sigma_2$. Show that σ_τ is unchanged under the transformation $r \to 1/r$, and $\sigma_\tau = \tau_0/2$ for $r = 1$.

5. Rederive (11.51) if the angle of arrival is uniformly distributed in $[-\varphi_s, +\varphi_s]$ and show that your result reduces to (11.51) for $\varphi_s = \pi$.
6. The MIP of a wireless channel is given by

$$R_c(\tau) = p_0 \, \delta(\tau) + p_1 \, \delta(\tau - 0.1 \ \mu s) + p_2 \, \delta(\tau - 0.4 \ \mu s)$$

where $p_0 = 0$ dB, $p_1 = -3$ dB, and $p_2 = -6$ dB. Determine the maximum symbol rate that this channel can support without using an equalizer.

7. A multipath channel comprises three rays; a direct ray with average power $P_0 = -10$ dB and two rays with average powers $P_1 = -20$ dB and $P_2 = -40$ dB. The delays for the three rays are given by $\tau_0 = 0$ μs, $\tau_1 = 1$ μs and $\tau_2 = 2$ μs.

a. What is the maximum excess delay in this channel?
b. Determine the average delay τ_m and the delay spread σ_τ for the channel.

Compare σ_τ with the maximum excess delay.
c. If the channel needs equalization when the bit duration is less than $10\sigma_\tau$, determine the maximum bit rate that can be supported in this channel without using an equalizer.
d. Determine the coherence time for a mobile user travelling at 30 km/hr and receiving 900 MHz signals. Compute also the coherence time for 10 km/hr user velocity.

8. Doppler power spectrum and the time autocorrelation function according to Jake's model are given by (11.49) and (11.51), respectively.

a. Assuming that the coherence time Δt_c is defined for $R_c(\Delta t_c)=0$ and $J_0(2.4)=0$, determine the coherence time in a channel for $v=120$ km/hr and the carrier frequency of $f_c=900$ MHz.
b. Is this channel slow-fading or fast-fading for transmission rates of 1 kbps and 10 kbps?

9. A TDMA-based mobile radio system operating at 1900 MHz is considered for use on a high-speed train with 250 km/hr velocity. For equalization purposes, each TDMA frame also comprises training bits. The 26 bit-long training sequence is embedded into a total of 156 bits, at least every $\Delta t_c/4$ s, where Δt_c denotes the channel coherence time. Determine the slowest transmission rate that meets these requirements in a slow fading environment.

10. Impulse response of a multipath channel is given by $h(\tau) = \delta(\tau) + \alpha \delta(\tau - \tau_0)$.

a. Determine the maximum excess delay T_m and the delay spread σ_τ.
b. Determine the frequency response and the PSD of the channel.

c. Plot the PSD, normalized to be unity at $f = 0$, versus $f\tau_0$ for $\alpha = 0.1, 0.3, 1$. For $\tau_0 = 2$ ms, determine the maximum bandwidth over which the channel could be considered frequency-flat (e.g., less than 1 dB variation in the frequency response) Determine the constants k_1 and k_2 such that the bandwidth found above is the same as the analytical expressions $\Delta f_c = k_1/T_m = k_2/\sigma_\tau$ for the coherence bandwidth.

d. Discuss the effect of α and τ_0 on the coherence bandwidth.

e. Assuming that a signal pulse with unit amplitude and duration T propagates through this channel, determine the minimum null-to-null bandwidth of the signal relative to τ so that the flat-fading assumption holds.

11. A Rayleigh fading channel has a maximum excess delay of $T_m = 2$ ms and a Doppler spread of $B_d = 20$ Hz. Assuming that the transmission bandwidth is 25 kHz and the coherence time is equal to $1/T_m$,

a. Determine the coherence bandwidth and the coherence time of the channel.

b. Is the channel frequency selective? Slowly fading?

c. Repeat (b) if the Doppler spread is increased to 100 Hz.

d. Repeat (b) if we use an OFDM system with 64 subcarriers in this channel.

12. Multipath intensity profile for a channel is given by

$$R_c(\tau) = 0.5\,\delta(\tau) + 0.35\,\delta(\tau - 1\mu s) + 0.15\,\delta(\tau - 2\mu s)$$

a. Determine the received power.

b. Determine the mean delay and rms delay spread.

c. Determine the coherence bandwidth. Is equalization necessary?

d. Assuming that the channel requires an equalizer if $T < 10\,\sigma_\tau$ where T denotes the symbol duration, determine the maximum symbol rate that can be supported over this channel with minimum ISI without using an equalizer.

e. If a mobile user traveling at 30 km/h receives a signal at 900 MHz carrier frequency, determine the Doppler spread and the channel coherence time. Explain if this is a fast or a slow fading channel. What is the effect of Doppler spread on the receiver bandwidth?

13. A channel is characterized by the scattering function

$$S_C(\tau, \lambda) = \begin{cases} 1 & -25Hz \le \lambda \le 25Hz,\ 0 \le \tau \le 1\mu s \\ 0 & else \end{cases}$$

Signals are assumed to be uncorrelated when the spaced-time spaced frequency correlation function drops to ½ of its maximum value.

a. Determine delay and Doppler spreads, and also coherence time and coherence bandwidth of the channel.

b. Determine the minimum value of the frequency separation Δf between two sinusoidal tones propagating through the channel so that they fade independently of each other. Does this channel undergo flat- or frequency-selective fading if the signal has a transmission bandwidth of 200 kHz, 2 MHz? What is the maximum signal bandwidth for which a flat-fading assumption is valid?

c. Determine the minimum value of the time difference between two sinusoidal tones $\cos\{2\pi ft\}$ and $\cos\{2\pi f(t + \Delta t)\}$ propagating through the channel so that they fade independently of each other. If the channel impulse response is sampled every T seconds, for what value of T will the samples be independent?

14. Suppose you build an OFDM system with a total system bandwidth of 50 MHz operating in a channel with maximum excess delay $T_m = 0.55$ μsec.

   a. Is this an indoor or an outdoor channel?
   b. How many subchannels are needed so that subchannels suffer flat fading? Flat fading assumption implies that $\Delta f_c \geq 10\Delta f$, where $\Delta f = 1/T$ denotes the subchannel bandwidth, which is assumed to be equal to the inverse of the OFDM symbol duration T. Noting that FFT/IFFT requires that the number of subcarriers N should be equal to an exponent of 2, that is, $N = 2^n$ where n is an integer, determine the appropriate value of N that can be used in practice.
   c. Determine the duration T_G of the cyclic prefix to ensure no ISI between FFT blocks? If the duration of a cyclic prefix can assume one of the values $T_G/T = 1/32, 1/16, 1/8$ or $1/4$, determine the appropriate duration of the cyclic prefix.
   d. If the maximum Doppler shift should be less than 1% of the subchannel bandwidth, determine the maximum mobile speeds that this system can support.

15. An OFDM system operates at 1 GHz carrier frequency over a bandwidth of 5 MHz in an outdoor environment in a fading channel with a maximum excess delay of 17 μs.

   a. OFDM systems usually take the ratio of cyclic prefix duration, T_G, to that of the symbol duration, T, as $T_G/T = 1/4, 1/8, 1/16$ or $1/32$. Choose a suitable length for cyclic prefix and symbol duration in order to mitigate ISI.
   b. Noting that the number of subcarriers should be determined as a power of 2, determine the number of subcarriers. Determine the corresponding throughput efficiency.

   c. Repeat parts (a) and (b) for an indoor environment with a maximum delay spread of 2μs.
   d. If the number of subcarriers is fixed, how does the efficiency change as a function of the bandwidth? How does the efficiency change as a function of the maximum excess delay?

16. Consider an OFDM system that operates in a 4 MHz bandwidth channel with a maximum excess delay of 17μs.

   a. Assuming that the bandwidth of each subchannel should be less than or equal to 0.1 of the coherence bandwidth to ensure flat fading in each subchannel, determine how many subchannels are needed?
   b. Can intersymbol interference (ISI) be neglected in this channel? Do you need a cyclic prefix? Explain.
   c. Determine the total required bandwidth if raised cosine pulses with $\alpha = 0.5$ are used. Is the 4 MHz OFDM bandwidth is sufficient? If not what should be the value of α?
   d. Assuming an average E_b/N_0 of 17 dB on each subchannel, determine the alphabet size that maximizes the symbol rate with a bit error probability lower than 10^{-3} using M-ary QAM. Determine the total data rate of the system.

17. In order for subcarriers to remain orthogonal, the channel must be fixed for the duration of an OFDM symbol. Therefore, an OFDM symbol is required to be no longer than the channel coherence time. On the other hand, the maximum Doppler shift be no longer than 2% of the frequency separation between tones. Consider an OFDM system operating at 5 GHz carrier frequency over an RF bandwidth of 75 MHz in a channel with 2 μs maximum excess delay.

a. Design an OFDM system with the maximum number of subcarriers such that tone orthogonality is maintained, assuming that users have a maximum velocity of 30 m/s. Determine the corresponding efficiency.

b. Derive an expression for the maximum possible efficiency of an OFDM system as a function of coherence time and maximum excess delay.

c. Discuss the effect of the carrier frequency on the maximum efficiency. Determine the carrier frequency at which the maximum efficiency is equal to 90 % for a maximum excess delay of 2μs and 30 m/s maximum user velocity.

18. Verify (11.133).

19. The propagation channel is described by the path loss model given by (3.92) with $n = 3.5$ and the average received power of -40 dBm at $r_0 = 1$ km. The signals also suffer log-normal shadowing with 6 dB standard deviation. The receiver has a noise power of -100 dBm. The SNR level is required to be higher than 20 dB for 95% of the time in the coverage area.

a. Determine the required average received signal power at a distance r.

b. Determine the distance corresponding to the average received signal power found in part (a).

c. Suppose that the system has a cell radius of $R = 30$ km. Determine the useful coverage area.

20. The average power received at a MS is 0 dBm at a given distance from the BS under lognormal shadowing conditions.

a. Determine the probability that the received power at a MS at that distance from the BS will exceed 0 dBm.

b. If a received signal power of 10 dBm or higher is required for proper operation, determine the probability that a MS

will have an acceptable signal level, for the log-normal standard deviation $\sigma = 6$ dB, 8 dB and 10 dB. Repeat for an acceptable received signal levels of 7 dBm and 13 dBm.

c. If an outage occurs when the received power level falls below 20 dB less than the mean received power level, determine the outage probability for $\sigma = 6$ dB, 8 dB and 10 dB.

21. The sum of L independent random variables as defined by $\gamma = \sum_{\ell=1}^{L} \gamma_k$ is often encountered in applications in multipath channels where γ_k represents the SNR of a multipath component. This corresponds to a receiver which combines the SNRs of the received multipath components. The pdf $f_\gamma(\gamma)$ of the sum γ is given by (11.218) for Rayleigh fading when γ_k have identical mean values. The bit error performance of such systems for some coherent modulation systems require the evaluation of the following integral:

$$P_2 = \int_0^\infty Q\left(\sqrt{2\beta\gamma}\right) f_\gamma(\gamma) d\gamma$$

$$= \int_0^\infty Q\left(\sqrt{2\beta\gamma}\right) \frac{1}{\Gamma(L)} \left(\frac{\gamma}{\bar{\gamma}}\right)^{L-1} \frac{1}{\bar{\gamma}} e^{-\gamma/\bar{\gamma}} d\gamma$$

where $\beta = 2$ for BPSK and $\beta = 1$ for coherent binary orthogonal FSK. In the litterature, there are mainly three formulas for the integrals above

$$P_2 = \frac{1}{2} - \frac{1}{2\sqrt{\pi}} \sqrt{\frac{\beta\bar{\gamma}}{1+\beta\bar{\gamma}}} \sum_{k=0}^{L-1} \frac{\Gamma(k+1/2)}{\Gamma(k+1)} \frac{1}{(1+\beta\bar{\gamma})^k}$$

$$= \frac{1}{2} - \frac{1}{2} \sqrt{\frac{\beta\bar{\gamma}}{1+\beta\bar{\gamma}}} \sum_{k=0}^{L-1} \binom{2k}{k} \frac{1}{[4(1+\beta\bar{\gamma})]^k}$$

$$= \left[\frac{1-\mu}{2}\right]^L \sum_{k=0}^{L-1} \binom{L-1+k}{k} \left[\frac{1+\mu}{2}\right]^k, \quad \mu = \sqrt{\frac{\beta\bar{\gamma}}{1+\beta\bar{\gamma}}}$$

Show that these three equations are identical. Hint: You may use (D.27).

22. Prove that (11.222) describes the asymptotic behavior of the bit error probability given by (11.221).

23. Using the identity (D.28), show that the bit error probability expressions below are identical to each other:

$$\int_0^\infty Q\left(\sqrt{2\beta\gamma}\right)\frac{1}{\Gamma(L)}\left(\frac{\gamma}{\bar\gamma}\right)^{L-1}\frac{1}{\bar\gamma}e^{-\gamma/\bar\gamma}d\gamma$$

$$= \frac{1}{2} - \frac{1}{2}\sqrt{\frac{\beta\bar\gamma}{1+\beta\bar\gamma}}\sum_{k=0}^{L-1}\binom{2k}{k}\frac{1}{[4(1+\beta\bar\gamma)]^k}$$

$$= \frac{1}{2}\sqrt{\frac{\beta\bar\gamma}{2+\beta\bar\gamma}}\sum_{k=L}^{\infty}\binom{2k}{k}\frac{1}{[4(1+\beta\bar\gamma)]^k}$$

24. For the class of modulation techniques characterized by (11.182), show that the symbol error probability in a fading channel may also be expressed in terms of the cdf of γ as

$$P_s = \int_0^\infty \alpha\, Q\left(\sqrt{2\beta x}\right) f_\gamma(x)\, dx$$

$$= \frac{\alpha}{2}\sqrt{\frac{\beta}{\pi}}\int_0^\infty \frac{e^{-\beta x}}{\sqrt{x}} F_\gamma(x)\, dx$$

Hint: Use (B.13) and apply the following integration by parts:

$$u=\alpha Q(\sqrt{2\beta x}),\ \ dv=f_\gamma(x)\,dx\ \Rightarrow$$

$$du = -\frac{\alpha}{2}\sqrt{\frac{\beta}{\pi}}\frac{e^{-\beta x}}{\sqrt{x}},\ \ v=\int_0^\gamma f_\gamma(x)\,dx=F_\gamma(\gamma)$$

Use the above formula to determine the SEP corresponding to $f(\gamma)=\exp(-\gamma/\bar\gamma)/\bar\gamma$.

25. Consider an uplink multi-user diversity (MUD) system with one BS and multiple users. The number of multiple users is a random variable with Poisson distribution [17]:

$$f(\lambda,k)=e^{-\lambda}\frac{\lambda^k}{k!},\ \ k=0,1,2,\dots$$

where λ denotes the traffic rate. The signal received by the BS from nth user may be written as in a Rayleigh fading channel:

$$r_n = x_n h_n + w_n$$

where w_n is AWGN with zero mean and variance $\sigma^2 = N_0/2$. The channel gain h_n and the transmitted signal x_n both have unity variance.

a. Determine the instantaneous and $\bar\gamma$, the mean E_b/N_0, for each user.

b. Assuming that the BS selects the user with highest instantaneous E_b/N_0, determine the cdf and the pdf of the received E_b/N_0.

c. Compute the outage probability assuming that $\lambda=0.1$ and $\gamma_{th}/\bar\gamma=0.01$.

26. Consider that the baseband representation of a received signal $r(t)$ in a scattering medium is represented by $r(t)= x(t)+jy(t)$. Assume that $x(t)$ and $y(t)$ are two correlated Gaussian processes with zero mean, equal variances σ^2 and a correlation coefficient ρ between them. The joint pdf is given by (F.28)

$$f_{XY}(x,y)=\frac{1}{2\pi\sigma^2(1-\rho^2)}\exp\left[-\frac{x^2-2\rho xy+y^2}{2\sigma^2(1-\rho^2)}\right]$$

a. Show that the pdf of the envelope of r, $z=|r|$, is given by

$$f_Z(z)=\frac{z}{(1-\rho^2)\,\sigma^2}\exp\left[\frac{-z^2}{2\,(1-\rho^2)\,\sigma^2}\right]$$

$$\times I_0\left[\frac{\rho\, z^2}{2(1-\rho^2)\,\sigma^2}\right]$$

Verify that for $\rho=0$, that is, x and y are independent, the envelope of r has a Rayleigh pdf.

b. Determine the pdf of SNR defined by $\gamma = z^2$.

c. Determine the bit error probability for DPSK in this correlated channel and plot it as a function of the mean E_b/N_0 for various values of the correlation coefficient. Discuss the effect of correlation on the bit error probability.

27. Determine

$$\int_0^\infty Q^2\left(\sqrt{2\beta\gamma}\right) \frac{e^{-\gamma/\bar{\gamma}}}{\bar{\gamma}} d\gamma$$

using (D.71) and show that it is identical to the result given by (11.193), which is based on the integration of the MGF.

28. A system employing automatic coding and modulation uses BPSK modulation when the instaneous values of E_b/N_0 changes between 7 dB and 10.5 dB. Assuming that the change in E_b/N_0 is uniformly distributed, determine the lowest, the highest and the average BEP.

29. Consider DPSK signalling in a Rayleigh fading channel.

a. Determine the average probability of error and the degradation in dB caused by fading at 10^{-3} bit error probability.

b. Determine the average bit error probability in a cell, assuming that the average SNR in a cell is uniformly distributed between 3 dB and 10 dB.

30. If γ denotes the sum of the E_b/N_0's of L independent Nakagami fading signals with the same average E_b/N_0 $\bar{\gamma}$ and fading figures m, determine the pdf of γ using MGF given by (11.185). Determine the average E_b/N_0 and the fading figure of the sum signal. Determine the bit error probability for DPSK

signalling. Determine the diversity gain for bit error probability 10^{-5}.

31. In CDMA cellular radio systems, multiuser interference (MUI) reduces the effective SNR of the system. The total instantaneous received bit energy from the desired user-to-noise and MUI PSD ratio is given by

$$R = \frac{P_1 T_b}{N_0 + \frac{1}{G_p}\sum_{i=2}^{K} P_i T_b} = \frac{\gamma_1}{1 + \frac{\gamma_{MUI}}{G_p}}$$

where P_1 denotes the power of the desired signal, T_b (T_c) is the bit (chip) duration, and $G_p = T_b/T_c$ is the processing gain. Similarly, $\gamma_1 = P_1 T_b/N_0$ denotes the instantaneous bit energy to the PSD ratio of the desired signal and

$$\gamma_{MUI} = \sum_{i=2}^{K} \gamma_i = \begin{cases} \sum_{i=2}^{K} P_i T_b/N_0 & P_i \neq P_j \\ (K-1)PT_b/N_0 & P_i = P \end{cases}$$

is instantaneous bit energy to noise PSD ratio of $K-1$ interferers. The second expression refers to the case where $(K-1)$ interfering users of different locations in a cell use automatic power control so as to produce the same received power levels P at the BS receiver.

In a fading channel, numerator and denominator of R are both random. Consequently, the pdf of γ_{MUI} should be firstly determined. The next step would be to determine the pdf of the ratio of two random variables. To simplify the analysis in a frequency nonselective fading channel with sufficiently large number users, one may use the central limit theorem to approximate the MUI by a Gaussian noise whose variance determined by the sums of the mean powers of the $(K-1)$ interfering signals. In other words, the MUI term above may be approximated by

$$\bar{\gamma}_{MUI} = \sum_{i=2}^{K} \bar{\gamma}_i = \begin{cases} \sum_{i=2}^{K} \bar{P}_i T_b/N_0 & \bar{P}_i \neq \bar{P}_j \\ (K-1)\bar{P}T_b/N_0 & \bar{P}_i = \bar{P} \end{cases}$$

Using the last expression for accounting for the MUI,

a. Determine the expected value of R.
b. Assuming that the desired signal suffers Rayleigh fading, determine the pdf of R.
c. Noting that the bit error probability for BPSK for a deterministic value of $R = r$ is given by $P_2(r) = Q(\sqrt{2r})$, calculate the probability of bit error in a Rayleigh fading channel.
d. Assuming $\bar{P}_i = \bar{P}$, $i = 1, 2, \ldots, K$ and average $E_b/N_0 = 10$ dB for all users, determine the probability of bit error and the degradation in dB due to multiple access for $G_p = 20$ dB.

32. Repeat the previous question by starting to determine the pdf of γ_{MUI} and the pdf of R assuming that all SNRs are exponentially distributed with the same mean values. Compare the results with those of the previous question and discuss the validity of the approach described in the previous question.

33. Consider a cellular radio system where the signal and the interferer both undergo Rayleigh fading, that is, the received signal and interference powers are characterized by

$$f_{\gamma_s}(\gamma_s) = \frac{1}{\bar{\gamma}_s} \exp(-\gamma_s/\bar{\gamma}_s), \quad \gamma_s > 0$$

$$f_{\gamma_I}(\gamma_I) = \frac{1}{\bar{\gamma}_I} \exp(-\gamma_I/\bar{\gamma}_I), \quad \gamma_I > 0$$

a. Derive the probability that the signal to interference power ratio (SIR) is

less than a predetermined threshold level of γ_{th}, i. e., $\mathrm{Prob}(SIR < \gamma_{th}) = \mathrm{Prob}(\gamma_s/\gamma_I < \gamma_{th})$.
b. Determine the probability that the SIR is less than a threshold level, which is 10 dB below the mean SIR.
c. Determine the threshold level which corresponds to $\mathrm{Prob}(SIR < \gamma_{th}) = 1/2$.

34. The signal-to-interference power ratio (SIR) in cellular radio systems may be written as

$$SIR = \frac{S}{I} = \frac{1/R^n}{\sum_{i=1}^{N} \chi_i/D^n} = \frac{q^n}{\sum_{i=1}^{N} \chi_i}$$

where $q = D/R$ denotes the normalized frequency reuse distance, n is the path-loss exponent, R is the cell radius and D is the distance between nearest co-channel cells.

The random variable $\chi_i \in \{0, 1\}$ is described by the Bernoulli pdf:

$$f_{\chi_i}(\chi) = p\, \delta(\chi - 1) + (1 - p)\, \delta(\chi)$$

where χ_i takes value 1 with probability p (also called as voice-activity factor) and value 0 with probability $1-p$. Therefore, $f_{\chi_i}(\chi_i = 1) = p$ denotes the probability that the i-th co-channel user is active (communicating) and $f_{\chi_i}(\chi_i = 0) = 1 - p$ shows the probability that the co-channel interferer is not active. Hence, $Y = \sum_{i=1}^{N} \chi_i$ shows the number, out of a total N, of active co-channel interferers with probability p. The outage probability of SIR is defined by

$$P_{out} = P(SIR < SIR_{th}) = P\left(\frac{q^n}{\sum_{i=1}^{N} \chi_i} < SIR_{th} \right)$$

a. Show that

$$E[\chi_i] = p$$

$$Var[\chi_i] = p(1-p)$$

$$E\left[\sum_{i=1}^{N}\chi_i\right] = N\,E[\chi_i] = Np$$

$$Var\left[\sum_{i=1}^{N}\chi_i\right] = N\,Var[\chi_i] = Np(1-p)$$

b. If N is sufficiently large, then the random variable $Y = \sum_{i=1}^{N}\chi_i$ which is described by the binomial pdf

$$f_Y(p,N) = \binom{N}{k}p^k(1-p)^{N-k}$$

can be approximated as a Gaussian random variable with $E[\chi_i] = p$, $var[\chi_i] = p(1-p)$ Based on the Gaussian approximation, derive the outage probability as a function of N, the voice activity factor p, the normalized frequency distance q and the SIR_{th}.

c. Show that the classical SIR formula $SIR = q^n/N$ is valid for $p = 1$, which corresponds to the worst-case scenario where N co-channel interferers are always active.

35. Consider a sampled low-pass channel impulse response $c[n]$ satisfying

$$c[n] = \alpha_0 + \alpha_1 e^{j2\pi\nu n}$$

where v denotes the maximum Doppler frequency shift of the 2nd path. The channel gains α_0 and α_1 are i.i.d. complex Gaussian random variables with zero-mean and variance ½.

a. Determine the autocorrelation of the channel impulse response.

b. Determine the minimum separation between channel samples that guarantees

independence. What is the relation between the minimum separation and the channel coherence time?

c. Determine the optimal value of the symbol separation for user velocities 60 km/h and 120 km/h for binary transmission at a bit rate of $1/T = 100$ kbps and 900 MHz carrier frequency.

36. Power outage probability, that is, the probability that the received signal power falls below some power threshold level P_{th}, is used as a 'hard' performance measure. During signal outages due to occasional fades, the receiver cannot decode the received signals or decodes with unacceptable number of bit errors. Outage probability is typically applicable to slowly varying signals. Such signals stay below the threshold level for a relatively long time, once they undergo a deep fade. Since the outage threshold is generally set to the minimum power required for acceptable performance, a slowly varying channel with received power falling below P_{th} is not usable. Power outages are observed more under shadowing. The outage probability is not meaningful for multipath fading channels for high mobility users, since a fade lasts for the duration of small number of bits. However, pedestrians are more vulnerable to power outages since they remain almost stationary as they communicate. Thus, if a pedestrian suffers a deep fade, his channel may be unusable for duration several hundreds and even thousands of bits. In this problem we compute outage probability for pedestrian users in Nakagami-m fading channels.

a. Using the pdf of the SNR for Nakagami-m fading, determine the probability that $P < P_{th}$, where $P_{th} \le P_{av}$ determines the outage threshold level below the average power level P_{av}.

b. Determine the outage probability for $m = 1$, 2 and outage thresholds 0 dB, 10 dB, and 20 dB below P_{av}.

c. For a power outage probability of P_{out}, determine the ratio P_{th}/P_{av} in terms of the fading parameter m, and outage probability P_{out}. Compute P_{th}/P_{av} in dB for $m = 1$ and $P_{out} = 0.01$.

d. Noting that the outage probability also shows the fraction of time that the channel is not usable, determine the effective data rate for a 1 Mbps system for $m = 1$ and 2 and outage thresholds 0 dB, 10 dB, and 20 dB below the average power level.

e. How can you reduce signal power outages?

37. A flat Rayleigh fading signal at 1 GHz is received by a mobile travelling at 100 km/hr.

a. Determine the number of level crossings below the average power level in the positive direction, which occur during 1 s.

b. Determine the average duration of a fade at 0 dB, 10 dB and 20 dB below the average power level.

38. Determine the value of the normalized threshold level $\rho = \sqrt{P_{th}/P_{av}}$ that maximizes the level crossing rate in a Rayleigh fading channel for a given value of the Doppler frequency. Determine the average fade duration for the threshold level $\rho = 0.707$ when the Doppler frequency shift is 25 Hz. Is Rayleigh fading slow or fast for 50 bps binary transmission? Assuming that a bit error occurs whenever a bit encounters a fade, that is, $\rho < 0.1$, determine the average number of bit errors per second for the given data rate.

39. A mobile station traveling with constant velocity receives 900 MHz transmission.

The local mean power is assumed to remain constant during travel.

a. If the average fade duration at a threshold level 10 dB below the rms signal level is 2 ms, determine the distance travelled by the MS during the 1 s interval. Determine the number of fades experienced by the signal during 1 s interval.

b. Repeat (a) for a threshold level 20 dB below the rms signal level.

40. A user traveling at 80 km/hr decides to send a text message which fails whenever the received signal power falls below half its average value due to Rayleigh fading. How long on the average will the user be unable to send text? If the text data is sent via DPSK modulation at a data rate of 10 kbps, on the average how many consecutive bits will have SNR below half the average SNR? Determine the E_b/N_0 required to to sustain communications at a bit error probability of 10^{-4}.

41. Verify the level crossing rate expression given by (11.160).

42. A mobile radio system operating at 1800 MHz uses automatic power control to update the transmit power with 2 ms intervals. Determine the maximum speed of mobile terminals that can be tracked by the system.

43. Tapped delay-line representation of the equivalent low-pass received signal through a frequency-selective channel is given by (11.202). Assuming a symbol rate of 800 kbps, determine the number of taps L, the tap spacing $1/W$ for raised-cosine pulsing with $\alpha = 0.25$ and tap amplitudes in the tapped delay-line model of a COST 207 typical suburban channel, defined by the MIP function in Table 11.1.

44. Consider a channel with a uniform impulse response in $[0,T]$. Suppose that a transmitter has a transmission bandwidth of $W = 4/T$. Determine the number of taps required by the Rake receiver. What is the order of diversity provided by this Rake receiver? Determine the bit error probability as a function of E_b/N_0.

45. Consider a channel with exponential MIP $R_c(\tau) = (1/\tau_0)\exp(-\tau/\tau_0)$.

    a. Determine the signal to ISI ratio for this system using NRZ pulse shaping. What happens as $\tau_0 \rightarrow \infty$? Discuss how the SIR is affected by the sequence of transmitted symbols. Which sequence minimizes the SIR?.

    b. Consider a Rayleigh fading channel with $\tau_0 = 1$ μs delay spread and AWGN is negligible compared to ISI. Determine the maximum data rate in this ISI-limited channel for coherent BPSK and bit error probability $\leq 10^{-5}$.

46. Consider the operation of a GSM system with 271 kbps maximum bit rate in rural,

typical urban, suburban and hilly terrain areas with multipath intensity profiles (MIPs) listed in Table 11.1.

    a. Determine the number of bits affected by intersymbol interference in the signal transmission in these channels.

    b. Determine the number taps in a RAKE demodulator that would be needed for the above channels. Which fingers of RAKE receiver will contain signal components?

47. Rake receiver is widely used in CDMA systems for maximal-ratio combining the chip powers where Rake fingers correspond to the delay bins used to spread a bit. Consider a Rake receiver which operates in a frequency-selective Rayleigh fading environment with a rectangular MIP, hence identical mean branch SNRs. The system can tolerate BEPs lower than 10^{-4} only 1% of the time when using BPSK in a channel for an average branch SNR of $\bar{\gamma} = 17$ *dB*. Determine the number of Rake fingers in order to meet the above requirements.

12

Diversity and Combining Techniques

In wireless communications, there is an increasing demand for reliable transmission, higher data rates and increased user mobility. These require more efficient transmission technologies, modulation, coding and signal processing techniques with comparable quality of service at similar cost as wireline technologies. However, the wireless channel is in general a non-LOS channel and often suffers time- and frequency-selective fading. Consequently, the received signals may not be successfully detected at all times. Performance degradation may be due to time selectivity (Doppler spread), frequency selectivity (delay spread), angular selectivity (angular spread) of the fading channel, and/or the interference. Use of FEC coding, increasing transmit power and/or antenna gains may not always be economical or sufficient for achieving reliable communications.

Diversity may provide an alternative to these approaches. If several replicas of the message signal can be made to simultaneously reach the receiver over independently fading (frequency, space, time, delay, direction or polarization) channels, then at least one of these signals will not be degraded by fading. This will increase the probability of successful detection of the received signals. Performance of diversity systems also depends largely on how these signals are combined and the correlation between them. Proper combination of received signals greatly reduces the severity of fading and improves the transmission reliability. Therefore, the diversity reception of partially correlated signals and the choice of the proper combining technique characterize this process.

Combining the replicas of a signal received by sufficiently separated multiple receive antennas (antenna/space diversity) or processing multipath components of a signal by a single-antenna receiver, for example, delay diversity by Rake receiver, can provide higher received SNR levels, lower bit error probabilities and higher data rates. For example, data may be transmitted over parallel spatial channels using multiple transmit and receive antennas in order to increase the throughput. Similarly, in space-time coding, data is transmitted over multiple antennas by

Digital Communications, First Edition. Mehmet Şafak.
© 2017 John Wiley & Sons Ltd. Published 2017 by John Wiley & Sons Ltd.
Companion website: www.wiley.com/go/safak/Digital_Communications

coding symbols in time and space. Beamforming capability of multiple antennas may also be used in order to focus beams in desired direction(s) or nulling interference/jamming signals impinging onto the receiver from certain other direction(s). Multiple antenna systems are also widely used in positioning and navigation systems for precise ranging, positioning, direction finding, tracking and time measurements. Therefore, the use of multiple antennas at the transmitter and/or the receiver offers significant performance improvements in communication systems.

Multiple antenna systems in wireless communications may be categorized as shown in Figure 12.1. In the baseline system, called as *single-input single-output* (SISO), transmitter and receiver are both equipped with single antennas. The so-called *single-input multiple-output* (SIMO) systems are characterized by a single transmit antenna and multiple receive antennas. In this scheme, also called as receive diversity, multiple and independently fading replicas of the transmitted signal are combined at the receiver. In *multiple-input single-output* (MISO) systems, also called as transmit diversity systems, transmissions from multiple transmit antennas are received and combined by a single receive antenna. Finally, *multiple-input multiple-output* (MIMO) systems are equipped with multiple antennas at both transmitter and receiver.

The basic building block of multiple antenna systems is the antenna array. An antenna array consists of an assembly of antennas (also called array elements) in a geometrical configuration (linear, circular etc.) and acts as a single antenna. Arrays are usually in the form of uniform linear arrays (ULA), where array elements (antennas) are located along a line (the array axis) uniformly, that is, the distance between neighboring array elements is the same. Elements of a transmit array (MISO system) are fed by dividing the transmit power uniformly between them or with different weights. The weights might be adapted to a time-varying channel to meet certain operational requirements. Receive diversity (SIMO system) has an inherent advantage over transmit diversity since a single transmit antenna uses the full transmit power, and the signals received by the elements of a receive array are created by the channel.

At the receive side of a SIMO system, the outputs of receive array elements may be combined in different ways. Combining might be in the form of selecting the output of an array element with the highest SNR (*selection combining*, SC), summing the signal envelopes at the outputs of array elements after eliminating the phase-differences between them (*equal gain combining*, EGC) or summing SNR's of the element outputs (*maximal ratio combining*, MRC). Several other combining techniques also exist, such as *optimum combining* and *square-law combining*. As we shall study in detail in the following sections, each combining technique has its strengths, weaknesses and application areas. The performance of multiple antenna systems depends not only on the sizes of transmit and receive arrays but also

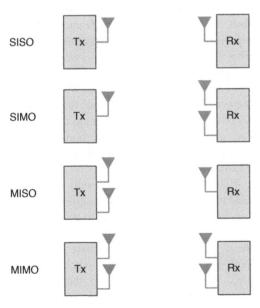

Figure 12.1 Classification of Multi-Antenna Systems.

on the scattering characteristics of the channel. For example, there are $N_t N_r$ propagation paths in a MIMO system with N_t transmit and N_r receive antennas. Based on the assumption that these paths suffer independent fading, this improves the ability of the receiver to detect transmitted signals since at least one of the N_t N_r signals would arrive the receiver with sufficiently high SNR. If the fadings suffered in these paths are uncorrelated from each other, then the probability that $N_t N_r$ signals will fade at the same time will be extremely low and the receiver can decode the transmitted signal with much lower error probabilities. This implies that the number of antennas and the correlation between the received signals determine the performance of multi-antenna systems. The data rate supported (multiplexing gain), the BEP (diversity gain) and the resulting output SNR (array gain) of a multiple antenna system are determined by the number of antennas, the correlation between antennas/channel fading and the propagation channel itself.

Correlation between the signals received by the receive array elements is determined by the distance between antennas at the transmitter, at the receiver as well as the scattering environment in the channel. As antenna spacing at the transmitter and/or the receiver become shorter, the received signals will tend to become more correlated as they will be scattered from the same obstacles in the propagation medium. The correlation between multi-path signals is also affected by the richness of the scattering environment; signals tend to become less correlated as the number of scattering obstacles increases.

We will first briefly review the theory of array antennas in a non-fading channels (for free space propagation), as is usually treated in classical antenna books, and then extend this study to fading channels. The rest of the Chapter will be devoted to the study of diversity and combining techniques for SIMO and MISO systems. MIMO system will be treated in the next Chapter.

12.1   Antenna Arrays in Non-Fading Channels

Consider a ULA consisting of N antennas separated from each other by a distance of d along the z-axis of a Cartesian coordinate system as shown in Figure 12.2. The antennas are assumed to have isotropic patterns in the plane of the page; for example, such a configuration may be obtained by using dipole antennas located perpendicular to the page. The propagation environment is assumed to be free space. The array elements may be fed with different current amplitudes and phases. The rays radiated by array elements interfere with each other in a complex manner in the near-field region but have plane wave front at the input of a receiver, which is assumed to be located in the far-field.

For far-field radiation in angular direction θ, the path difference between neighboring elements is given by $d\cos\theta$ (see Figure 12.2); this implies a phase difference of $kd\cos\theta$ and a delay difference of $(d/c)\cos\theta$, where $k = 2\pi/\lambda$ and c denotes the velocity of light. The array size limits the maximum symbol rate since the symbol duration T should be much longer than the maximum delay difference between signals transmitted by different array elements. This implies $(r_1 - r_N)/c = (N - 1)d\cos\theta/c <<T$. For example, for $N = 10$ and $d = 0.5$ m, one gets $(N-1)d\cos\theta/c = 1.5 \times 10^{-8}\cos\theta$, which limits the symbol rate to 67 Mbps along the array axis (z-axis), but there is no limitation on the data rate along the boresight direction $(\theta \approx \pi/2)$, which represents the direction maximum transmission and reception.

Here, we will be interested only in the far-field radiation where the rays of array elements may be assumed to be parallel to each other, as shown in Figure 12.2. The signal transmitted by the nth antenna undergoes a phase change by $\exp(-jkr_n)$ and an amplitude attenuation by $1/r_n$. The waves incident on the receiving antenna, which are assumed to be in the far-field of the N-antenna array, have plane wave fronts. Then, based on the assumption that

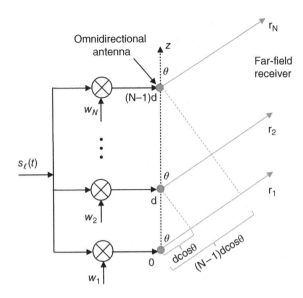

Figure 12.2 Far-Field Geometry of an N-Element Transmitting Uniform Linear Array (ULA) of Isotropic Sources Positioned Along Z-Axis with a Spacing d Between Them.

the rays represented by r_1 to r_N are parallel to each other, one may write

$$r_n = \begin{cases} r_1 - (n-1)d\cos\theta & \text{for phase} \\ r_1 & \text{for amplitude} \end{cases}, \quad n = 1, 2, \cdots N$$

(12.1)

Consider that a lowpass signal $s_\ell(t)$ is transmitted by the nth antenna with a weighting factor w_n, which denotes the relative strength of the transmitted signal (current) level of the nth antenna. In LOS conditions, where amplitude varies as $1/r$ with distance, and ignoring the noise, the received low-pass equivalent signal due to n^{th} array element antenna may be written as

$$r_{\ell_n}(t, \theta) = w_n s_\ell(t) \frac{e^{-jkr_n}}{r_n} \cong w_n r_{\ell_1}(t) e^{jk(n-1)d\cos\theta}$$

$$r_{\ell_1}(t) = s_\ell(t) \frac{e^{-jkr_1}}{r_1}$$

(12.2)

Assuming that the nth array element is a half-wave dipole, the received electric field intensity in the equatorial plane is found using (2.29):

$$E_{\theta,n} = j60I_n \frac{e^{-jkr_n}}{r_n}$$

(12.3)

where I_n denotes the current of the $\lambda/2$ dipole. Comparing (12.2) and (12.3), one gets

$$w_n = I_n/I_1$$

$$s_\ell(t) = j60I_1$$

(12.4)

The far-field received signal is given by the phasor sum of the field intensities of the N elements:

$$r_\ell(t, \theta) = \sum_{n=1}^{N} r_{\ell_n}(t) + n(t)$$

$$= r_{\ell_1}(t) \sum_{n=1}^{N} w_n e^{jk(n-1)d\cos\theta} + n(t) = r_{\ell_1}(t)\mathbf{w}\mathbf{u} + n(t)$$

$$\mathbf{w} = \begin{bmatrix} w_1 & w_2 & \cdots & w_N \end{bmatrix},$$

$$\mathbf{u} = \begin{bmatrix} 1 & e^{jkd\cos\theta} & \cdots & e^{jk(N-1)d\cos\theta} \end{bmatrix}^T$$

$$= \begin{bmatrix} u(\theta)^0 & u(\theta)^1 & \cdots & u(\theta)^{N-1} \end{bmatrix}^T$$

$$u(\theta) = e^{jkd\cos\theta}$$

(12.5)

where T denotes the transpose operator, $n(t)$ is the AWGN and $\mathbf{u}$, the so-called steering vector, accounts for the phase differences between rays transmitted by the array elements. The *array factor* is defined as the ratio of the received field strength from ULA to that of the first (reference) antenna, when AWGN is ignored:

$$f_a(\theta) = r_\ell(t,\theta)/r_{\ell_1}(t) = \mathbf{wu} \qquad (12.6)$$

The array factor represents the effect of N array elements on the transmitted/received signal compared to the case where a single antenna ($N = 1$) is used.

Let us choose the weights in (12.6) as

$$w_n = \frac{1}{\sqrt{N}} e^{-j(n-1)\beta}, \quad \beta = kd\cos\theta_0 \qquad (12.7)$$

where β denotes the phase shift between neighboring array elements in the direction of maximum radiation θ_0. (12.7) shows that the weights of the ULA are normalized:

$$\|\mathbf{w}\|^2 = \sum_{n=1}^{N} |w_n|^2 = 1 \qquad (12.8)$$

This implies that the transmit power of the array is constant independently of N. This choice of the weights in the transmitter also compensates for the phase differences between the N rays received at a far-field receiver, which is directed for maximum reception from direction θ_0 (matched-filter!). Therefore, N equal magnitude signals transmitted by the N-element ULA may be coherently added at the receiver in the direction θ_0; this implies $\mathbf{wu} = \sqrt{N}$. This choice of the antenna weights also suggests that the direction of maximum radiation can be changed electronically by simply changing θ_0. Then, the signals received from each of the N array elements will be equi-phased in the direction θ_0.

For uniform weighting, inserting (12.7) into (12.6) and using (D.13), the array factor may simply be written as

$$f_a(\theta) = \mathbf{wu} = \frac{1}{\sqrt{N}} \sum_{n=1}^{N} e^{j(n-1)\Psi}$$

$$= \frac{1}{\sqrt{N}} \frac{\sin(N\Psi/2)}{\sin(\Psi/2)} e^{j(N-1)\Psi/2} \qquad (12.9)$$

$$\Psi = kd[\cos\theta - \cos\theta_0]$$

The maximum value of the array factor approaches $\sqrt{N}$ as Ψ goes to zero:

$$|f_a(\theta)|_{\max} = |f_a(\theta)|_{\Psi=0} = \sqrt{N} \Rightarrow$$
$$|r_\ell(t,\theta)| = \sqrt{N}|r_{\ell_1}(t)| \ as \ \Psi \to 0 \qquad (12.10)$$

Hence, the use of an N-element ULA increases the transmitted/received power level by a factor N compared to a single antenna. Figure 12.3 shows the array factor of a uniformly illuminated array with 10-elements. The phases of the array elements are adjusted so that the direction of maximum radiation is $\theta_0 = \pi/2$. The maximum value of the array factor is equal to $\sqrt{10}$, as predicted by (12.10). For a given value of N, as antenna spacing (array size) increases, the number of sidelobes increases and beamwidth gets narrower. This is in agreement with the results of Chapter 2 that the beamwidth is inversely proportional to the antenna size. Nevertheless, irrespective of the value of d/λ, the sidelobes' envelope remains the same since it is determined by the current (weight) distribution over the array.

For a receive array, (12.7) and (12.8) suggest that the signal amplitudes received by the elements of a receive array are equally weighted and the phases are adjusted for maximum reception from the direction θ_0. Therefore, one may control the amplitude and phase distributions over an array so as to control its radiation pattern. Similar arguments are valid for a transmit array, since, according to the Lorentz' reciprocity theory in the electromagnetics, a ULA in the receive mode performs the same as in the transmit mode. As shown in Figure 12.4, signals received by a ULA in the far-field region of a transmitter may be written as

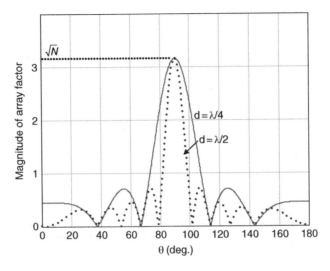

Figure 12.3 The Array Factor For a 10-Element ULA.

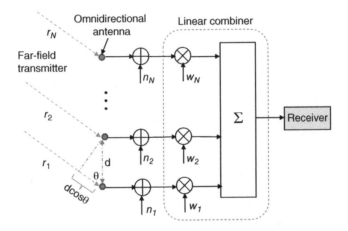

Figure 12.4 A Receiving ULA.

$$r_\ell(t,\theta) = r_{\ell 1}(t)f_a(\theta) + n(t)$$

$$n(t) = \sum_{n=1}^{N} w_n n_n \tag{12.11}$$

where the noise $n(t)$ is simply the weighted sum of the noises received by N array elements.

Since $0 \leq \theta \leq \pi$, using $\Psi = kd\cos\theta$ one gets $-kd \leq \Psi \leq kd$, that is, $-2\pi d/\lambda \leq \Psi \leq 2\pi d/\lambda$. In order that Ψ lies in the range $[-\pi, \pi]$ the condition $d \leq \lambda/2$ should be satisfied. In this case, there is one-to-one correspondence between the values of θ and Ψ without ambiguity. However, if this requirement is not satisfied, then the array pattern will have so-called grating lobes. This requirement may be viewed as the spatial analogy of the sampling theorem; spectrum aliasing is observed if the sampling rate is lower than the Nyquist range (see Figure 5.4). Noting that the time of arrival difference between signals of neighboring array elements is given by (d/c) $\cos\theta$, the array elements receive the samples

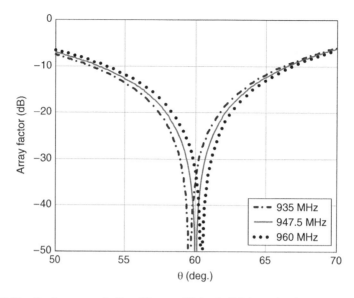

Figure 12.5 Nulling Performance of a Two-Element ULA of a BS Operating in the GSM Band (935–960 MHz). The center frequency is $f_0 = 947.5$ MHz and the direction of maximum radiation is $\theta_0 = 60$ degrees.

of the same signal with a sampling interval of $(d/c)\cos\theta$ (see Figure 12.4). Hence the values of d satisfying $d \leq \lambda/2$ correspond to sampling rates in excess of the Nyquist rate and hence no aliasing and no grating lobes are observed.

Direction of maximum radiation is defined by

$$\Psi = kd(\cos\theta - \cos\theta_0) = 0 \implies \theta = \theta_0 \quad (12.12)$$

In a *broadside array*, the direction of maximum radiation is perpendicular to the array axis, that is, $\theta_0 = \pi/2$ and $\beta = kd\cos\theta_0 = 0$. In an *end-fire array*, the radiation is maximized along the line of the array. This implies $\beta = kd$ for $\theta_0 = 0$ and $\beta = -kd$ for $\theta_0 = \pi$. Hence, by adjusting the phase shift β between the neighbouring array elements, the direction of maximum radiation may be changed electronically. Therefore, instead of mechanically turning the array, the direction of maximum radiation can be scanned electronically by simply adjusting the phase change θ_0 between array elements.

Example 12.1   Antenna Nulling.
In order to demonstrate the interaction between array elements, this example aims to demonstrate the principle of placing a null in direction θ_0 at the carrier frequency f_0 by a 2 element ULA. Using (12.6) the array factor may be written as

$$f_a(\theta) = \left(w_1 + w_2 e^{jkd\cos\theta}\right) \equiv 0 \ \ for \ \theta = \theta_0 \ and \ f = f_0 \implies$$
$$w_2/w_1 = -e^{-jk_0 d\cos\theta_0}$$

$$(12.13)$$

where the distance between the elements is assumed to be $d/\lambda_0 = 1/2$ and $w_1 = 1$. The array factor and its magnitude may then be written as

$$f_a(\theta) = 1 - e^{j(kd\cos\theta - k_0 d\cos\theta_0)}$$

$$|f_a(\theta)| = 2\left|\sin\left(\frac{\pi}{2}\left(\frac{f}{f_0}\cos\theta - \cos\theta_0\right)\right)\right|$$

$$(12.14)$$

Figure 12.5 shows the variation of the array factor of a BS equipped with two antennas designed so as to place a narrow null in the direction of $\theta_0 = 60$ degrees and $f_0 = 947.5$ MHz. The downlink is assumed to operate in the

935-960 MHz band with a center frequency of 947.5 MHz. The nulling system is designed for operation at 947.5 MHz to form a null at $\theta_0 = 60^0$, that is, the separation between antennas is 16 cm. At 947.5 MHz, the null depth is more than -12.74 dB for θ varying in the range $60^0 \pm 5^0$, and -7 dB for $60^0 \pm 10^0$, compared with the signal level induced by a single antenna. The nulling performance deteriorates at the band edges; the location of the null shifts from θ_0 and the array factor does not vanish but has a value of -33.7 dB at $\theta_0 = 60^0$. The same system can be used for forming a maximum at $\theta_0 = 60^0$ by changing the sign of w_2/w_1 in (12.13).

Example 12.2   Non-Uniform Weighting.
Let the normalized weights of a 3-element array be given as

$$w_1 = 1/\sqrt{6}, \quad w_2 = 2e^{-j\beta}/\sqrt{6}, \quad w_3 = e^{-j2\beta}/\sqrt{6}$$
$$(12.15)$$

where β defined by (12.7) implies that the relative phases of the normalized weights are adjusted so that the direction of maximum radiation is θ_0. The array elements at the edge have weights with unity magnitude while the centre element has a weight with a magnitude of 2; hence, they have a triangular amplitude

distribution and satisfy (12.8). The array factor of this antenna may be written from (12.6) as

$$f_a(\theta) = \frac{1}{\sqrt{6}}\left(1 + 2e^{j\Psi} + e^{j2\Psi}\right)$$
$$= \frac{1}{\sqrt{6}}\left(1 + e^{j\Psi}\right)^2 = \frac{4}{\sqrt{6}}e^{j\Psi/2}\cos^2(\Psi/2)$$
$$(12.16)$$

Figure 12.6 shows a comparison of the array factor given by (12.16) with those of uniformly illuminated arrays given by (12.9) with $N = 2$ and 3. Note that the array factor given by (12.16) with triangular current distribution is proportional to the square of the array factor of a 2-element array with unity weights. Maximum value of (12.16) is 0.5 dB lower than the maximum value $\sqrt{3}$ of the array factor for a uniformly illuminated array of 3-elements. As the number of antennas increases, the current (weight) distribution over the array axis can be assumed to be continuous. Consequently, the observations we previously made on the effects of current distribution over linear antennas are also valid for the array antennas. This implies that uniform distribution yields the highest array factor but higher sidelobe levels. As the current distribution over the array becomes tapered, the

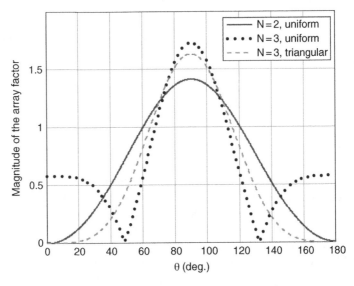

Figure 12.6   Effect of Current Distribution on the Array Factor For $d/\lambda = 0.5$ and $\theta_0 = \pi/2$.

sidelobe levels become lower at the expense of decreased maximum array factor and broadened beamwidth (see Examples 2.9–2.11).

Example 12.3 Mutual Impedance Between Array Elements.

The elements of a ULA are mostly fed by a single power source. The feeding circuit is critical in determining the amplitudes and phases of the currents of the array elements. On the other hand, the input impedance of a ULA element depends not only on its self-current and self-impedance but also on the currents of and mutual impedances between neighboring elements. This dependence is closely related to the inter-element spacing and the lengths of the ULA elements measured in wavelength.

Generalizing the formulation presented in Example 2.22, the total power transmitted by an N-element ULA may be written as

$$P_t = \frac{1}{2} \sum_{n=1}^{N} Z_n I_n^2 \qquad (12.17)$$

where Z_n and I_n denote respectively the input impedance and the input current of nth antenna in the ULA. The input impedance, Z_n, of nth array element is determined not only by its self-impedance Z_{nn} but also by mutual impedances, Z_{nk}, $k = 1, 2,.. N$, between neighboring array elements:

$$Z_n = Z_{nn} + \sum_{\substack{k=1 \\ k \neq n}}^{N} w_k Z_{nk}, \quad n = 1, 2, ..., N \quad (12.18)$$

where $w_k = I_k / I_1$.

The mutual impedance Z_{nk} between n^{th} and k^{th} antennas is a function of the electrical distance between them. As the distance between array elements becomes smaller, the fading of their signals become more correlated and the mutual impedances between array elements increase; then, the input impedance becomes more influenced by the presence of the neighboring antennas.

The input impedance of a dipole antenna may be found using (2.34) and (2.31) which gives an analytical expression for the radiation resistance. The input resistance of a half-wave dipole is equal to its radiation resistance, which is 73.2 Ω. The input reactance of a half-wave dipole antenna is equal to 42.5 Ω. Hence the input impedance of a half-wave dipole is equal to $Z_{nn} = 73.2 + j42.5$ Ω. The sensitivity of the input reactance to dipole length decreases as the diameter of the wire increases. Therefore, thick wires are used for broadband applications (see Figure 2.9).

The mutual resistance R_m and mutual reactance X_m between two half-wave dipole antennas which are side-by-side and separated by a distance d are given by (2.165) and repeated here for convenience: [1]

$$R_m = \frac{\eta}{4\pi} [2 \cdot Ci(kd) - Ci(\mu_1) - Ci(\mu_2)]$$

$$X_m = -\frac{\eta}{4\pi} [2 \cdot Si(kd) - Si(\mu_1) - Si(\mu_2)]$$

$$\mu_1 = k\left(\sqrt{d^2 + \ell^2} + \ell\right), \quad \mu_2 = k\left(\sqrt{d^2 + \ell^2} - \ell\right)$$

$$(12.19)$$

where $\ell = \lambda/2$. Cosine and sine integrals $S_i(x)$ and $C_i(x)$ are defined by (D.119) and (D.120). The variations of mutual resistance and mutual reactance with d/λ are shown in Figure 12.7. One may observe that the mutual impedance reduces to self (input) impedance, that is, $Z_{in} = 73.2 + j42.5$ Ω, as d/λ goes to zero, and behaves as a damped sinusoid with increasing values of d/λ.

Impedance matching between transmitter/receiver and the array is usually implemented by broadband feeder-circuits. As a first-order approximation, assume that the input impedances of all array elements are the same and independent of the ratio of current elements, that is, $Z_1 = Z_n = Z_{nn}$, which implies that the mutual impedances are negligible (see Figure 12.7). If the total power transmitted by

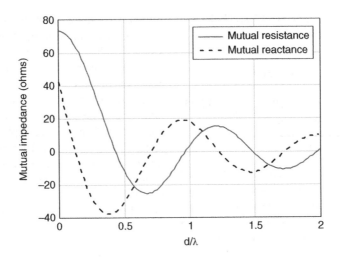

Figure 12.7 Mutual Impedance Between Two Parallel Half-Wave Dipoles Versus the Electrical Separation d/λ Between Them.

a N-element ULA is the same as that of a single antenna, that is, $P_t = Z_1 I_1^2 / 2$, then we observe from (12.7) and (12.8) that the total transmit power is normalized:

$$P_t = \frac{1}{2} Z_1 \sum_{n=1}^{N} I_n^2 = \frac{1}{2} Z_1 I_1^2 \sum_{n=1}^{N} |w_n|^2 \Rightarrow$$

$$\|\mathbf{w}\|^2 = \sum_{n=1}^{N} |w_n|^2 = 1 \tag{12.20}$$

Normalization of the total transmit power in a MISO system allows us to see the effect of the number of antennas in a transmit ULA without changing the total transmitted power. Otherwise, it would be difficult to distinguish the effects of the increase in the transmit power and the number of antennas on the ULA performance.

12.1.1   SNR

Using (12.5), the received instantaneous SNR may be written as

$$\gamma(\theta) = \bar{\gamma}_c |f_a(\theta)|^2 = \bar{\gamma}_c |\mathbf{w}\mathbf{u}|^2 = \bar{\gamma}_c \mathbf{w}\mathbf{u}(\mathbf{w}\mathbf{u})^H = \bar{\gamma}_c \mathbf{w}\mathbf{u}\mathbf{u}^H \mathbf{w}^H \tag{12.21}$$

where the mean SNR is found using (2.114):

$$\bar{\gamma}_c = \frac{E_b}{N_0} = \frac{P_r}{R_b N_0}$$

$$P_r = \frac{r_{\ell_I}^2}{2\eta} A_e = \frac{r_{\ell_I}^2}{2\eta} \frac{\lambda^2}{4\pi} G_r \tag{12.22}$$

Here $\eta = 120\pi$ is the intrinsic impedance of free space, R_b denotes the bit rate, and A_e and G_r are the effective receiving area and the gain of the receiver antenna.

The mean received SNR due to an N element ULA may be written in terms of the channel covariance matrix $\mathbf{R}$ as

$$\bar{\gamma} = E[\gamma(\theta)] = \bar{\gamma}_c \, \mathbf{w} \, \mathbf{R} \, \mathbf{w}^H$$

$$\mathbf{R} = E[\mathbf{u}\mathbf{u}^H] \tag{12.23}$$

When the AOA distribution of the incident signal energy is described by the pdf $f_\theta(\theta)$, over $[\theta_0 - \theta_s, \theta_0 + \theta_s]$. Using (12.5), one may express the covariance matrix of the received signals as

$$\mathbf{R} = E[\mathbf{u}\mathbf{u}^H] = \int_{\theta_0 - \theta_s}^{\theta_0 + \theta_s} f_\theta(\theta) \mathbf{u}\mathbf{u}^H d\theta$$

$$R_{nm} = \int_{\theta_0 - \theta_s}^{\theta_0 + \theta_s} f_\theta(\theta) e^{j(n-m)kd\cos\theta} d\theta, \quad n, m = 1, 2, \ldots, N \tag{12.24}$$

Using the SNR received from direction θ given by (12.21), the mean SNR at the output of a uniformly weighted (see (12.7)) N-element ULA is found as:

$$\bar{\gamma} = E[\gamma(\theta)] = \int_{\theta_0 - \theta_s}^{\theta_0 + \theta_s} f_\theta(\theta)\gamma(\theta)d\theta$$

$$= \bar{\gamma}_c \underbrace{\int_{\theta_0 - \theta_s}^{\theta_0 + \theta_s} f_\theta(\theta) \frac{1}{N} \left| \frac{\sin(N\Psi/2)}{\sin(\Psi/2)} \right|^2 d\theta}_{Array\ gain} \quad (12.25)$$

where the AOA of the incident rays is described by the pdf $f_\theta(\theta)$, where $\theta_0 - \theta_s < \theta < \theta_0 + \theta_s$. For the special case when the signal arrives from only $\theta = \theta_0$ only (no AOA spread), one has $\theta_s = 0$ and $\Psi = 0$, the mean SNR at the ULA output is given by

$$f_\theta(\theta) = \delta(\theta - \theta_0) \Rightarrow$$

$$\lim_{\theta_s \to 0} \bar{\gamma} = \bar{\gamma}_c \int_{\theta_0 - \theta_s}^{\theta_0 + \theta_s} \delta(\theta - \theta_0) \underbrace{\frac{1}{N} \left| \frac{\sin(N\Psi/2)}{\sin(\Psi/2)} \right|^2}_{=N\ as\ \Psi \to 0} d\theta = N\bar{\gamma}_c$$

$$(12.26)$$

The mean SNR given by (12.25) due to an N-element ULA, normalized with respect to mean SNR $\bar{\gamma}_c$ of a single array element, is

shown in Figure 12.8 when the AOA is described by a uniform pdf $f_\theta(\theta)$ in $\theta_0 - \theta_s < \theta < \theta_0 + \theta_s$. When $\theta_s = 0$, the mean SNR is equal to $N\bar{\gamma}_c$ as given by (12.26). However, the mean SNR was observed to decrease with increasing values of θ_s and becomes insensitive to the number of array elements for sufficiently large values of θ_s. This may be explained by the fact that the array is designed for maximum transmission/reception of signals to/from a given direction. Therefore, signals arriving from different directions are not received in-phase by the ULA and this leads to a decrease in the mean SNR. [2][3]

Example 12.4 Correlation Matrix of a 2-element ULA For Uniform and Gaussian AOA Distributions.

For a 2-element ULA, the covariance matrix given by (12.24) reduces to

$$\mathbf{R} = \begin{bmatrix} 1 & \rho(d/\lambda) \\ \rho^*(d/\lambda) & 1 \end{bmatrix}$$

$$(12.27)$$

$$\rho(d/\lambda) = \int_{\theta_0 - \theta_s}^{\theta_0 + \theta_s} e^{jkd\cos\theta} f_\Theta(\theta)d\theta$$

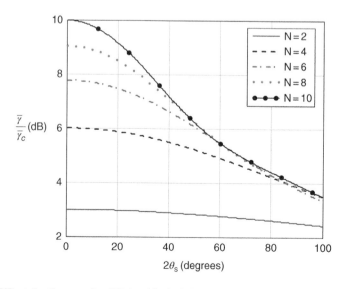

Figure 12.8 SNR at the Output of an ULA with $d = 0.25\lambda$ Versus $2\theta_s$, the Angle-of-Arrival Spread, For $N = 2$, 4 and 8. The pdf of the AOA is $f_\Theta(\theta) = 1/(2\theta_s)$, and $\theta_0 = \pi/2$.

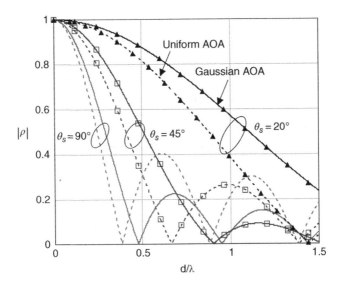

Figure 12.9 The Correlation Coefficient For Uniform and Gaussian AOA Distributions with the Same AOA Variance $\sigma^2 = \theta_s^2/3$ For $\theta_s = 90°$, $45°$, $20°$.

If the AOA of the signals is uniformly distributed in $[\theta_0 - \theta_s, \theta_0 + \theta_s]$, then the correlation coefficient given by (12.27) reduces to

$$\rho(d/\lambda) = \frac{1}{2\theta_s} \int_{\theta_0 - \theta_s}^{\theta_0 + \theta_s} e^{jkd\cos\theta} d\theta$$

$$= J_0(kd), \quad \theta_s = \theta_0 = \pi/2 \qquad (12.28)$$

where $J_0(x)$ denotes the Bessel function of the first kind and order zero (see (D.92)).

To see the effect of the pdf of the AOA on the correlation coefficient, we also consider a Gaussian distribution for AOA centered around θ_0 with a rms spread σ:

$$f_\Theta(\theta) = \frac{1}{\sqrt{2\pi}\sigma C} \exp\left(-\frac{(\theta - \theta_0)^2}{2\sigma^2}\right), \quad -\pi \leq \theta \leq \pi$$

$$(12.29)$$

For a fair comparison between uniformly and Gaussian distributed AOAs, the rms spreads for both distributions in $(\theta_0 - \theta_s, \theta_0 + \theta_s)$ are assumed to be the same: $\sigma = \theta_s/\sqrt{3}$ (see (F.93)). Here, $C = 1 - 2Q(\sqrt{3})$ denotes the normalization constant which normalizes the area under (12.29) in the angular region $\theta_0 - \theta_s < \theta < \theta_0 + \theta_s$.

Figure 12.9 presents a comparison of the magnitude of the complex correlation coefficient for uniform and Gaussian distributions as a function of d/λ for $\theta_0 = \pi/2$ and various values of θ_s. The correlation coefficient approaches unity for $d/\lambda \to 0$, irrespective of the value of θ_s, but it decreases with increasing antenna separation and AOA spread. The dependence of the correlation coefficient on the AOA spread may be explained by the fact that signals impinging on a receiving ULA with a wider AOA spread will be less correlated with each other since they are more likely to be scattered from different obstacles. Then, the array elements may be located closer to each other for an accepted value of the correlation coefficient. For small electrical separations between the antennas, the correlation coefficient for the Gaussian model decays more slowly compared to uniform AOA. However, it has lower peaks and decays more rapidly at larger separations. This clearly shows the need for accurate modeling of the AOA of received signals for a realistic performance evaluation of wireless systems.

Example 12.5 Downlink Versus Uplink Correlations in Cellular Radio.

In this example we consider antenna spacing required for up- and down-links of cellular radio systems in order to keep fading correlations between signals at acceptable levels. The AOA distribution of signals incident at the BS may be significantly different than at the MS, particularly in macro-cells.

The ULA at the BS is usually tilted downwards in order to focus radiations/receptions to/from the cell to be covered. The beamwidth in the elevation plane is chosen to be rather narrow so as to cover the cell but not to cause (suffer) co-channel interference to (from) neighboring cells. At the BS receiver, the AOA spread in the elevation plane becomes narrower as the BS height is increased. Consequently, the received signals in the elevation plane might be highly correlated with each other (see Figure 12.9). Hence, larger antenna spacings are required to keep correlation between received signals at reasonably low levels, especially if array elements are stacked vertically at the BS. However, vertically erected arrays are rarely used at the BSs but the ULA axis is usually chosen parallel to the Earth's surface. In the azimuthal plane, the coverage is either omnidirectional or sectored (usually with sectors of 60 or 120 degrees). Therefore, at the BS, the AOA spread (beamwidth) in the azimuthal plane is larger than in the elevation plane. Antenna spacing required to keep fading correlations at acceptable levels becomes even larger for off-broadside incidence. Generally, space does not impose a serious limitation at the BS. Low transmit powers of the MS may be compensated in the uplinks by using arrays at the BSs. In the down-links, electrical energy at the BS is not as valuable a resource as at the MS.

Signals arrive to MSs from all directions after scattering from nearby objects, hence an omnidirectional AOA. This implies from Figure 12.9 that the received signals may be assumed to be uncorrelated if the distance between array elements satisfy $d = 0.5\lambda$. For 900 MHz operation, it may be difficult to separate array antennas by half-wavelength (≈ 17 cm) at the MS. However, noting that the correlation effects do not become detrimental unless the correlation coefficient is higher than 0.8 or so, a two-element array used at a MS may still allow reasonably uncorrelated reception.

12.2 Antenna Arrays in Fading Channels

Now consider a propagation scenario where transmitter and receiver are not in LOS of each other. Transmitted signals reach the receiver in the far-field region via multi-paths through a rich scattering environment. Consider a propagation scenario as shown in Figure 12.10, where only the receiver is equipped with an N-element ULA, that is, a SIMO system.

Assume that a low-pass equivalent binary signal, $s_{\ell m}(t)$, $m = 1, 2$, arrives to the receiver via N paths with complex path gains

$$h_i(t) = \alpha_i(t)e^{-j\phi_i(t)}, \, i = 1, 2, \ldots, N \quad (12.30)$$

where $\alpha_i(t) > 0$ denotes the magnitude and $\phi_i(t)$ the phase of the channel gain, which are both time-dependent. In contrast with the free-space propagation scenario based on the factor $\{e^{-jkr_i}/r_i\}$ as in (12.2), signal propagation in a multipath channel is characterized by random channel gains $\{h_i(t)\}$, which are complex Gaussian random processes with zero mean and unity variance (see (11.4)). In multi-path propagation scenario, signal transmission takes place as a result of a random number of reflection, refraction, diffraction and scattering. Therefore, the received signal level is described by the random channel gains $\{h_i(t)\}$.

The receiver combines the received multi-path signals with appropriate weighting coefficients $\{w_i\}$. The output of the linear combiner may be written as (see Figure 12.10):

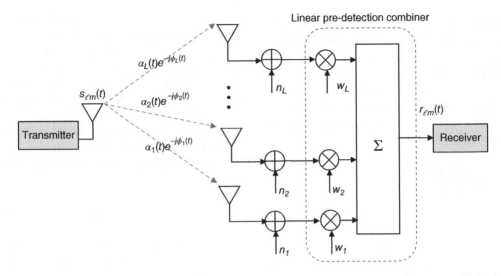

Figure 12.10 Block Diagram of a Linear Pre-Detection Combiner. Receive antennas are sufficiently separated from each other to ensure uncorrelated fading between branches. In a post-detection combiner, each branch has a separate receiver.

$$r_{\ell m}(t) = \sum_{i=1}^{N} w_i(t)\left[h_i(t)s_{\ell m}(t) + n_i(t)\right] = s(t) + n(t),$$

$$0 \le t < T, m = 1,2 \tag{12.31}$$

where T denotes the symbol interval. The noise component is given by

$$n(t) = \sum_{i=1}^{N} w_i(t)n_i(t), \; 0 \le t < T \tag{12.32}$$

The signal component of the combined received signal may be written as:

$$s(t) = s_{\ell_m}(t) \sum_{i=1}^{N} w_i(t)h_i(t) = s_{\ell_m}(t)\mathbf{wh}, \; 0 \le t < T \tag{12.33}$$

where $s_{\ell_m}(t)$ is assumed to be constant during the symbol interval T. It is important to note that (12.33) has the same form as (12.5) with the exception that the deterministic steering vector $\mathbf{u}$ is replaced by the random vector $\mathbf{h}$ of complex channel gains:

$$\mathbf{h} = \begin{bmatrix} h_1 & h_2 & \cdots & h_N \end{bmatrix}^T \tag{12.34}$$

The complex channel gains $\{h_i(t)\}$ will assume the same values if they follow the same propagation path. However, signals radiated by the transmit antenna within a given beamwidth will follow different paths in the propagation environment, hence they undergo different delays and are scattered by different (numbers of) obstacles. Consequently, correlations between the channel gains will depend on the characteristics of the propagation environment. The correlation coefficient ρ between channel gains h_i and h_j, with zero mean and standard deviations σ_i and σ_j of two multiplicative flat fading channels is defined as follows

$$\rho = \frac{E\left[h_i \, h_j^*\right]}{\sigma_i \sigma_j}, \quad \sigma_i^2 = E\left[|h_i|^2\right] = 1 \tag{12.35}$$

Note that (12.35) defines the correlation between field strengths but not between the power levels. The correlation coefficient between the power levels is usually taken as the square of field correlation coefficient.

Similar to (12.21), the instantaneous SNR may be written as

$$\gamma = \bar{\gamma}_c \frac{|\mathbf{wh}|^2}{\mathbf{ww}^H} = \bar{\gamma}_c \frac{\mathbf{whh}^H \mathbf{w}^H}{\mathbf{ww}^H} \tag{12.36}$$

where $\mathbf{w}\mathbf{w}^H = \sum_{i=1}^{N} |w_i|^2$ is used to normalize the weights as in (12.8). Since $\mathbf{h}$ is a random vector whose elements have zero mean and unit variance, $\mathbf{h}\mathbf{h}^T$ is an $N \times N$ random matrix and determines the statistical characteristics of the instantaneous SNR. The mean SNR may be expressed in terms of the channel covariance matrix $\mathbf{R}$ similar to (12.23):

$$\bar{\gamma} = \bar{\gamma}_c \frac{\mathbf{w}\mathbf{R}\mathbf{w}^H}{\mathbf{w}\mathbf{w}^H}$$

$$\mathbf{R} = E\left[\mathbf{h}\mathbf{h}^H\right] = E\left[\begin{pmatrix} |h_1|^2 & h_1 h_2^* & \cdots & h_1 h_N^* \\ h_2 h_1^* & |h_2|^2 & \cdots & h_2 h_N^* \\ \vdots & \vdots & \ddots & \\ h_N h_1^* & h_N h_2^* & \cdots & |h_N|^2 \end{pmatrix}\right]$$

$$\text{(12.37)}$$

The value of the mean SNR is closely related to the behavior of the channel covariance matrix, which provides a measure of the correlations between the channel gains, which are in turn determined by the distances between the elements of the receive array and the scattering processes in the wireless channel. The channel covariance matrix is usually characterized by its eigenvalues which correspond to eigenmodes of this multi-antenna system.

We will consider below three special cases which correspond to various degrees of correlations between the channel gains.

Case 1: Uncorrelated Case

If the signals received by the array elements undergo uncorrelated fading, then the channel gains are uncorrelated with each other:

$$\rho = \frac{E\left[h_i h_j^*\right]}{\sigma_i \sigma_j} = \begin{cases} 0 & i \neq j \\ 1 & i = j \end{cases}, \quad i,j = 1,2,\ldots,N$$

$$\text{(12.38)}$$

In view of (12.38), the covariance matrix reduces to an identity matrix $\mathbf{R} = \mathbf{I}_{N \times N}$, that is, only diagonal elements are nonzero and equal to unity. This corresponds to the case for $d \gg \lambda/2$ in a non-fading channel, where the covariance matrix $\mathbf{R}$ in (12.24) has only non-zero diagonal elements (see (12.8)).

If the receiver has the capability to estimate the magnitudes and phases of the channel gains, then the weighting coefficients may be chosen as

$$\mathbf{w} = \mathbf{h}^* \Rightarrow w_i = h_i^* = \alpha_i e^{j\phi_i} \quad \text{(12.39)}$$

which is referred to as maximal ratio combining (MRC). Inserting (12.39) into (12.36), one gets the combiner output which provides the sums of the SNR's of N branches:

$$\gamma = \bar{\gamma}_c \sum_{i=1}^{N} |\alpha_i|^2 = \sum_{i=1}^{N} \gamma_i \quad \text{(12.40)}$$

where $\bar{\gamma}_c = E_b/N_0$. The mean SNR then reduces to

$$\bar{\gamma} = E[\gamma] = \bar{\gamma}_c \sum_{i=1}^{N} E\left[|\alpha_i|^2\right] = N\bar{\gamma}_c \quad \text{(12.41)}$$

For uncorrelated fading, $\mathbf{R} = \mathbf{I}_{N \times N}$ has rank N and all eigenvalues have equal means $\bar{\gamma}_i = \bar{\gamma}_c$, $i = 1, 2, \ldots, N$. The mean SNR at combiner output is therefore increased by N times compared to that of a single branch.

Noting the similarity between (12.40) and (11.213), the pdf of the sum of N uncorrelated Rayleigh-fading SNRs is given by the chi-squared pdf (11.218):

$$f(\gamma) = \frac{1}{(N-1)!} \frac{\gamma^{N-1}}{\bar{\gamma}_c^N} e^{-\gamma/\bar{\gamma}_c} \quad \text{(12.42)}$$

This pdf is more concentrated around the mean compared to that of a single antenna, that is, better outage performance (see Figure 11.54). Noting that the bit error and outage probabilities are proportional to $1/\bar{\gamma}_c^N$, the diversity order is N.

Case 2: Perfectly Correlated Case

If the two channel gains are perfectly correlated, that is, $h_j = kh_i$ where k is an arbitrary constant, then

$$\rho = \frac{E\left[h_i h_j^*\right]}{\sigma_i \sigma_j} = \frac{E\left[h_i k\, h_i^*\right]}{\sigma_i k \sigma_i} = \frac{E\left[|h_i|^2\right]}{\sigma_i^2} \equiv 1$$

(12.43)

In view of (12.43), all elements of the channel covariance matrix given by (12.37) are equal to unity, that is, $\mathbf{R} = \mathbf{1}_{N \times N}$ where $\mathbf{1}_{N \times N}$ denotes an $N \times N$ matrix with all elements equal to unity. Since such a matrix has unity rank, it has a single non-zero eigenvalue with a mean value of $\lambda_1 = N\bar{\gamma}_c$ while the other eigenvalues vanish $\lambda_n \equiv 0, n \geq 2$. Note that this case corresponds to the covariance matrix in (12.23) with all elements equal to unity, that is, the case for $d \to 0$.

Using (12.36), the instantaneous SNR for the perfectly correlated case may be written as

$$\gamma = \bar{\gamma}_c \frac{|\mathbf{wh}|^2}{\mathbf{ww}^H} = \bar{\gamma}_c \frac{\left|\sum_{i=1}^{N} w_i h_i\right|^2}{\sum_{i=1}^{N} |w_i|^2} = \bar{\gamma}_c \frac{\left(\sum_{i=1}^{N} \alpha_i^2\right)^2}{\sum_{i=1}^{N} \alpha_i^2}$$

$$= \bar{\gamma}_c \frac{(N\alpha^2)^2}{N\alpha^2} = N\bar{\gamma}_c \alpha^2$$

(12.44)

where the weights are chosen as in (12.39). Hence, (12.44) shows that the mean SNR at combiner output is given by $N\bar{\gamma}_c$ (see also (12.41)). The only non-nonzero eigenvalue also represents the mean SNR.

The pdf of γ, given by (12.44), for the perfectly correlated case is that of a Rayleigh faded SNR with mean value $N\bar{\gamma}_c$:

$$f(\gamma) = \frac{1}{N\bar{\gamma}_c} \exp\left(-\frac{\gamma}{N\bar{\gamma}_c}\right)$$

(12.45)

Note that this is also the pdf of a Rayleigh-faded SNR with a mean $N\bar{\gamma}_c$, received by a single antenna. The BEP corresponding to (12.45) is shown in Table 11.7 for different modulation schemes, where the asymptotic behavior is given by $1/(N\bar{\gamma}_c)$.

Case 3: Partially Correlated Case

For partially correlated Rayleigh fading, the pdf of the SNR at the receive array output is determined using the partial fraction expansion of the MGF (see (11.216)):

$$f(\gamma) = \sum_{k=1}^{N} \pi_k f_k(\gamma) = \sum_{k=1}^{N} \pi_k \frac{1}{\lambda_k \bar{\gamma}_c} \exp\left(-\frac{\gamma}{\lambda_k \bar{\gamma}_c}\right)$$

$$\pi_k = \prod_{i=1, i \neq k}^{N} \frac{\lambda_k}{\lambda_k - \lambda_i}$$

(12.46)

where $\{\lambda_k\}$ denote the normalized distinct eigenvalues of the covariance matrix $\mathbf{R}$. In this case, the rank of the covariance matrix, hence the number of non-zero eigenvalues, is less than or equal to N depending on the values of the correlation coefficients. Consequently, eigenvalues correspond to different transmission modes with different mean SNR levels.

Using (12.46), the expected value of γ may be shown to be

$$E[\gamma] = \bar{\gamma}_c \sum_{k=1}^{N} \pi_k \lambda_k = \bar{\gamma}_c \sum_{k=1}^{N} \lambda_k = N\bar{\gamma}_c$$

$$\sum_{k=1}^{N} \lambda_k = tr(\mathbf{R}) = N$$

(12.47)

where $tr(\mathbf{R})$ denotes the trace of the matrix $\mathbf{R}$. Note that the mean SNR at the combiner output for $\rho = 0$, $\rho = 1$ and $0 < \rho < 1$ are all equal to $N\bar{\gamma}_c$ (see (12.41), (12.45) and (12.47)). For $\rho = 0$, all eigenvalues have identical mean values. For $\rho = 1$, there is a single nonzero eigenvalue which captures all the power of the system. For $0 < \rho < 1$, all eigenvalues have distinct mean values. However, the sum of normalized eigenvalues is always equal to N, the number of array elements.

Example 12.6 Partially Correlated Fading. The normalized eigenvalues of a 2×2 covariance matrix $\mathbf{R}$, given by (12.27), are given by

$$|\mathbf{R} - \lambda \mathbf{I}| = \begin{vmatrix} 1-\lambda & \rho \\ \rho^* & 1-\lambda \end{vmatrix} = 0 \Rightarrow \begin{array}{l} \lambda_1 = 1 + |\rho| \\ \lambda_2 = 1 - |\rho| \end{array}$$
(12.48)

For uncorrelated fading ($\rho = 0$, $\mathbf{R} = \mathbf{I}_{2 \times 2}$), the covariance matrix has two distinct eigenvalues of equal mean values, $\lambda_1 = \lambda_2 = 1$. The pdf of the SNR is given by (12.42) with $N = 2$. However, for the perfectly correlated case ($\rho = 1$, $\mathbf{R} = \mathbf{1}_{2 \times 2}$), the covariance matrix has a single non-zero eigenvalue, that is, $\lambda_1 = 2$ and $\lambda_2 = 0$. The corresponding pdf of the SNR is obtained by inserting $N = 2$ into (12.45).

For the partially correlated case ($0 < \rho < 1$), using (12.46) and (12.48), the pdf of the SNR may be written in terms of the two distinct normalized eigenvalues λ_1 and λ_2 of $\mathbf{R}$ as follows

$$f(\gamma) = \frac{1}{(\lambda_1 - \lambda_2)\bar{\gamma}_c} \exp\left(-\frac{\gamma}{\lambda_1 \bar{\gamma}_c}\right)$$

$$+ \frac{1}{(\lambda_2 - \lambda_1)\bar{\gamma}_c} \exp\left(-\frac{\gamma}{\lambda_2 \bar{\gamma}_c}\right)$$

$$= \frac{1}{2|\rho|\bar{\gamma}_c} \left[\exp\left(-\frac{\gamma}{(1+|\rho|)\bar{\gamma}_c}\right) - \exp\left(-\frac{\gamma}{(1-|\rho|)\bar{\gamma}_c}\right) \right]$$
(12.49)

The effect of correlation may be observed on the BEP for DPSK modulation in a correlated fading channel by averaging (11.175) with (12.49):

$$P_2 = \frac{1+|\rho|}{2|\rho|} \frac{1}{2[1+(1+|\rho|)\bar{\gamma}_c]}$$

$$- \frac{1-|\rho|}{2|\rho|} \frac{1}{2[1+(1-|\rho|)\bar{\gamma}_c]}$$
(12.50)

$$= \frac{1}{2\left[1 + 2\bar{\gamma}_c + \left(1 - |\rho|^2\right)\bar{\gamma}_c^2\right]}$$

One may observe from (12.50) that the $BEP \sim 1/\bar{\gamma}_c^2$ and thus has a diversity order

$d = 2$. However, for finite values of $\bar{\gamma}_c$, its asymptotic behavior is described by $1/\bar{\gamma}_c$ as in the fully correlated case ($\rho = 1$).

Consider now the following covariance matrix for a $N = 3$ element SIMO system:

$$\mathbf{R} = E\left[\mathbf{h}\mathbf{h}^H\right] = \begin{bmatrix} 1 & \rho & \rho^2 \\ \rho & 1 & \rho \\ \rho^2 & \rho & 1 \end{bmatrix}$$
(12.51)

Using the same approach as in (12.48), the normalized eigenvalues of this matrix are found to be

$$\lambda_{1,2} = 1 + \frac{\rho^2}{2} \pm \frac{\rho}{2}\sqrt{\rho^2 + 8}, \ \lambda_3 = 1 - \rho^2 \quad (12.52)$$

Figure 12.11 shows the variations of the three normalized eigenvalues as a function of the correlation coefficient. For the uncorrelated case ($\rho = 0$), the three eigenvalues are distinct and have equal means: $\lambda_1 = \lambda_2 = \lambda_3 = 1$. For the perfectly correlated case ($\rho = 1$), there is a single non-zero eigenvalue with a normalized mean value $\lambda_1 = 3$ while the others vanish ($\lambda_2 = \lambda_3 = 0$). However, for any value of the correlation coefficient, the sum of normalized eigenvalues is always equal to N, as predicted by (12.47).

12.3 Correlation Effects in Fading Channels

A receiver operating in a fading channel suffers Doppler, delay and angular spreads and has a challenging task of collecting the incident signal power in frequency, delay and space coordinates. Nevertheless, fading may also be exploited to achieve performance improvements using, amongst others, diversity and combining techniques. Since independent channel gains have a low probability of experiencing deep fades simultaneously, independent fading signals can be achieved by separating them in time, frequency, space, polarization, and so on. Usual approach to achieve independent fading paths

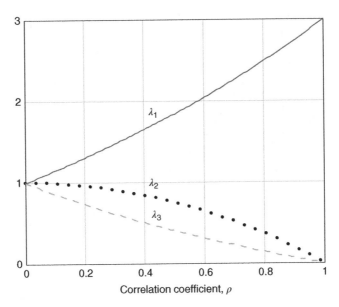

Figure 12.11 Normalized Eigenvalues of a 3×3 Covariance Matrix Versus the Correlation Coefficient ρ.

includes sending the same signal with time intervals larger than the channel coherence time, at frequencies separated larger than the channel coherence bandwidth, or by antennas separated larger than the channel coherence distance. Sending the same signal over independently fading channels and properly combining them reduces outage and bit error probabilities. Before proceeding further, it might be useful to consider fading channels that are not independent.

Now let us consider a SIMO system with $N = 3$ antennas undergoing correlated fading as modeled by the covariance matrix and the eigenvalues given by (12.51) and (12.52). The pdf of the SNR, given by (12.46) for an arbitrary value of ρ in a Rayleigh fading channel, reduces to (12.42) and (12.45) for the special cases of $\rho = 0$ and $\rho = 1$, respectively. Figure 12.12 shows the pdf of the SNR for various values of the correlation coefficient. The probability that the SNR assumes lower values increases with increasing values of the correlation coefficient. In the extreme case of $\rho = 1$, the pdf is the same as that for a single

antenna with a mean SNR three times higher than that of a single antenna (see (12.45)). For small values of the correlation coefficient, the pdf looks similar to that for $\rho = 0$.

The cdf of the SNR for an N antenna system is given by (11.219) for $\rho = 0$ and is obtained using (12.46) for $0 < \rho < 1$ and (12.45) for $\rho = 1$:

$$F(\gamma) = \begin{cases} 1 - \exp\left(-\dfrac{\gamma}{\bar{\gamma}_c}\right) \displaystyle\sum_{k=0}^{N-1} \dfrac{1}{k!} \left(\dfrac{\gamma}{\bar{\gamma}_c}\right)^k & \rho = 0 \\[2ex] \displaystyle\sum_{n=1}^{N} \pi_n \left[1 - \exp\left(-\dfrac{\gamma}{\lambda_n \bar{\gamma}_c}\right)\right] & 0 < \rho < 1 \\[2ex] 1 - \exp\left(-\dfrac{\gamma}{N\bar{\gamma}_c}\right) & \rho = 1 \end{cases}$$

$$(12.53)$$

where π_n is defined by (12.46). Using (12.53), the effect of correlation on the outage probability is shown in Figure 12.13. As a reference, the cdf for $N = 1$ and $N = 2$ are also shown for

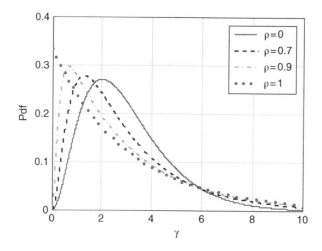

Figure 12.12 The Effect of Correlation on the Pdf of the SNR For a Three Element ULA.

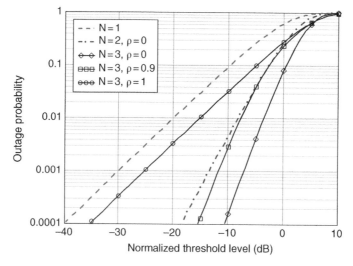

Figure 12.13 Effect of Correlation Between Signals Received By an N-element ULA on the Outage Probability.

uncorrelated fading. Degradation in the outage performance was observed to be insignificant for small values of the correlation coefficient. However, for $N = 3$, $\rho = 0.9$, it is slightly better than that for two uncorrelated antennas ($N = 2$, $\rho = 0$). The outage performance rapidly degrades as ρ becomes higher than 0.8. In the extreme case of $\rho = 1$, the outage curve has the same slope as that for a single antenna ($N = 1$), except for $10 \log(3) = 4.78\, dB$ performance improvement simply because of the 3 times higher mean SNR (see (12.45)).

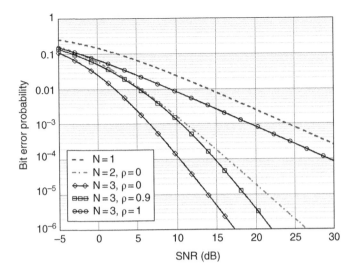

Figure 12.14   The BEP For BPSK For Several Values of N and the Correlation Coefficient.

The BEP for an N-antenna system for coherent BPSK is obtained using (11.221):

$$P_2 = \begin{cases} \dfrac{1}{2} - \dfrac{1}{2}\sqrt{\dfrac{\overline{\gamma}_c}{1+\overline{\gamma}_c}}\sum_{k=0}^{N-1}\binom{2k}{k}\dfrac{1}{[4(1+\overline{\gamma}_c)]^k} & \rho=0 \\[18pt] \dfrac{1}{2}\sum_{n=1}^{N}\pi_k\left[1-\sqrt{\dfrac{\overline{\gamma}_c\lambda_k}{1+\overline{\gamma}_c\lambda_k}}\right] & 0<\rho<1 \\[18pt] \dfrac{1}{2} - \dfrac{1}{2}\sqrt{\dfrac{N\overline{\gamma}_c}{1+N\overline{\gamma}_c}} & \rho=1 \end{cases}$$

(12.54)

where λ_k denotes the kth, $k=1,2,\ldots,N$, eigenvalue of the $N \times N$ covariance matrix.

Figure 12.14 shows the effect of correlation between signals received by an $N=3$ element array on the BEP for BPSK in a Rayleigh fading environment. The results for $N=1$ and 2 are also shown for comparison purposes. The comments for Figure 12.13 also apply for Figure 12.14. The degradation in outage and bit error probabilities is sufficiently small for reasonable values of the correlation coefficient but becomes intolerably large for $\rho > 0.8$.

12.4   Diversity Order, Diversity Gain and Array Gain

We know that fading increases the BEP (see Figure 11.39) and outage probabilities. Nevertheless, fading channels may also be exploited to mitigate the above effects and to increase the mean output SNR level, improve the spectrum efficiency and to decrease BEP and outage probabilities. ACM, scheduling, and diversity are some of the techniques available for this purpose. In the diversity scheme, multiple copies of the same signal are received via preferably independently fading channels and combined according to a given rule for improved performance. Independence of channels may be achieved in space, time, frequency, polarization etc. The performance of the diversity system depends strongly on the fading statistics (pdf), number of channels, fading correlations between channels and mean channel SNR levels.

In the following, some definitions will be given as performance measures for diversity reception in fading channels. *Diversity order* is defined as

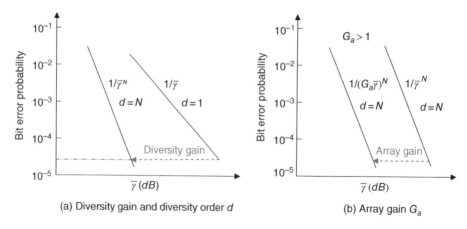

Figure 12.15   Effect of Diversity Order, Diversity Gain and Array Gain on the BEP in Fading Channels with N Independently Fading Diversity Branches.

$$d = - \lim_{\bar{\gamma} \to \infty} \frac{P_e(\bar{\gamma})}{\log \bar{\gamma}} \qquad (12.55)$$

The diversity order denotes the slope of the BEP curve versus SNR $\bar{\gamma}$ in dB as $\bar{\gamma} \to \infty$. Figure 12.15 provides a pictorial description of diversity order, diversity gain and array gain for diversity reception via N independently fading channels. A receiver equipped with N antennas will have a diversity order of N and the BEP curve will decrease as $1/\bar{\gamma}^N$, that is faster than for $N = 1$. For a SISO system in a Rayleigh fading channel, the BEP curves shown in Figures 11.42-46 for various modulation schemes all have unity diversity order, that is, $BEP \sim 1/\bar{\gamma}$. Figure 12.14 shows that the BEP at the output of a N-element receiving ULA achieves a diversity order $d = N$, when its elements are sufficiently apart from each other. Figure 12.13 shows that the diversity order may also be expressed in terms of the slope of the outage probability curve. Note that the diversity order is defined for large values of the mean SNR. Since the slope of the BEP curve at finite SNR values is not the same as its slope at infinite SNR (the diversity order), one needs to be cautious about using the concept of diversity order as a performance measure for SNR levels encountered in practice.

The *diversity gain* of a SIMO system is defined, for a given BEP level, as the decrease in the SNR in dB compared to a SISO system ($N = 1$) (see Figure 12.15a). For a BEP level of 10^{-3} in Figure 12.14, a single antenna receiver requires $\bar{\gamma}_c = 23.96 dB$, but the use of a two element array reduces the required SNR level to $\bar{\gamma}_c = 11.09 dB$, hence a diversity gain of 12.87 dB. The diversity gain becomes higher at lower BEP levels. One may also observe from Figure 12.14 that increased correlation has a negative impact on both the diversity gain and the diversity order at finite SNR values. For example, the curve for $N = 3$, $\rho = 1$ in Figure 12.14 has the same diversity order as for $N = 1$ at finite SNR values and shows only 4.78 dB better performance compared to $N = 1$. However, the diversity gain for $N = 3$, $\rho = 0$ is 17.41 dB.

Outage and BEP curves also depend on the value of the *array gain*, which is defined as the increase in the average channel SNR, usually in dB, with N. An increase in the array gain shifts the BEP curve to smaller SNR values (see Figure 12.15b)). For example, consider two receivers, each equipped with N-element ULAs. When they operate with N independently fading channels with mean SNRs $\bar{\gamma}$ and $G_a \bar{\gamma}$ ($G_a > 1$), the slope (diversity orders) of both BEP curves will be the same $1/\bar{\gamma}^N$ and $1/(G_a \bar{\gamma})^N$, hence no diversity order advantage to each other. However, the shift between the two curves is simply due to the array gain (see Figure 12.15b). For example, Figure 12.14 shows that curve for $N=3$, $\rho=1$ has an array gain $G_a = 3$ in contrast with $G_a = 1$ for $N=1$. However, they both have unity diversity order (see (12.45)). The 4.78 dB shift between them is evidently due to the difference in their array gains. For example, a receiving array with N uncorrelated elements has N identical eigenvalues $\lambda_i = 1, i = 1, 2, \cdots, N$; hence $G_a = 1$, $d = N$ for $\rho = 0$ (see (12.42)). For $\rho = 1$, $\lambda_1 = N, \lambda_i = 0, i = 1, 2, 3, \cdots, N$; hence $G_a = N$ and $d = 1$ (see (12.45)). For $0 < \rho < 1$, the N non-zero eigenvalues will be different from each other with $G_a = 1$, $d = N$ (see (12.46)).

12.4.1 Tradeoff Between the Maximum Eigenvalue and the Diversity Gain.

We already observed from (12.41), (12.44) and (12.47) that the mean SNR at the output of a SIMO system is increased by a factor of N, independently of the correlation between the array elements. For $\rho = 0$, all eigenvalues have equal means $\bar{\gamma}_c$ but, for $\rho = 1$, the only nonzero

eigenvalue has a mean value of $N\bar{\gamma}_c$. For $0 < \rho < 1$, there are N distinct eigenvalues each with a different mean level. However, their sum is the same as for uncorrelated and fully correlated arrays. There are some operational modes, such as MIMO beamforming, which exploit only the largest eigenvalue for minimizing the BEP. The number of antennas and the degree of correlation between them determine the value of the maximum eigenvalue λ_1 in (12.52) for a SIMO system. For example, we observe from Figure 12.11 that λ_1 increases with increasing values of the correlation coefficient. For example, higher values of λ_1 lead to longer ranges and lower BEPs in communication systems. On the contrary, increased correlation decreases the diversity gain and hence degrades the BEP performance for finite values of $\lambda_1\bar{\gamma}_c$ (see Figure 12.14). The tradeoff between these two conflicting consequences of correlation between the signals of a receiving array may be observed in Figure 12.16, where the loss in the diversity gain and the increase in the value of the maximum eigenvalue are shown as a function of the correlation coefficient. The loss in the diversity gain becomes significant for

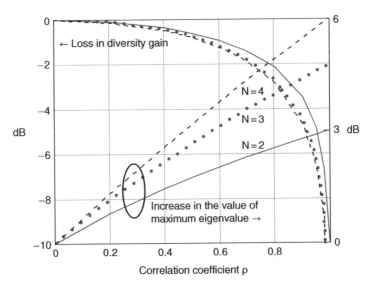

Figure 12.16 The Variation of the Maximum Eigenvalue and the Loss in the Diversity Gain with the Correlation Coefficient at a BEP $= 10^{-4}$ For a Receiving ULA with $N = 2$, 3 and 4 Elements. For $\rho = 0$, SNR $= 16.28$ dB is required to achieve a BEP $= 10^{-4}$ for $N = 2$; SNR $= 10.31$ dB for $N = 3$; and SNR $= 7.15$ dB for $N = 4$ (see Figure 12.14).

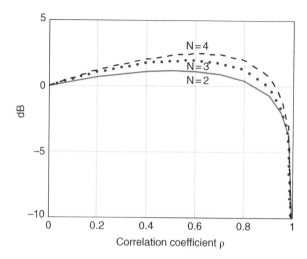

Figure 12.17 The Difference Between the Increase in the Maximum Eigenvalue and the Loss in the Diversity Gain at a BEP level of 10^{-4} For $N = 2, 3, 4$. Note that the increase in the maximum eigenvalue offsets the losses in the diversity gain for $\rho < 0.84$ for $N = 2$; $\rho < 0.9$ for $N = 3$, and $\rho < 0.92$ for $N = 4$.

$\rho > 0.8$ while the maximum eigenvalue shows almost a linear increase with ρ. Figure 12.17 shows the difference between the increase in the maximum eigenvalue and the loss in diversity gain as a function of the correlation coefficient. As long as $\rho < 0.8$, a few dB additional gain can be realized by the system by purposely inserting correlation between the received signals, for example, by controlling the distance between array elements. This also alleviates the restriction on the separation between the array elements, that is, the array elements can be placed closer to each other.

12.5   Ergodic and Outage Capacity in Fading Channels

We already considered the fading effects on the BEP and the outage probability. Another performance measure is related to the maximum or minimum data rate that can be supported by a system in a fading environment, that is, ergodic and outage capacities. Here, we prefer to work with the spectral efficiency, which is defined as the capacity normalized by the transmission bandwidth. The instantaneous spectral

efficiency of a MISO system with N array elements is given by

$$\eta \triangleq C/B = \log_2(1+\gamma), \quad \gamma = \bar{\gamma}_c \frac{\mathbf{w}\mathbf{h}\mathbf{h}^H\mathbf{w}^H}{\mathbf{w}\mathbf{w}^H}$$
(12.56)

where the instantaneous value of γ is given by (12.36).

The ergodic (mean) spectral efficiency may be written as

$$E[\eta] = E[\log_2(1+\gamma)] = \int_0^\infty \log_2(1+\gamma)f(\gamma)d\gamma$$
(12.57)

For the fully correlated case ($\rho = 1$), the mean spectral efficiency is found using (D.73):

$$E[\eta] = \int_0^\infty \log_2(1+\gamma)\frac{1}{N\bar{\gamma}_c}e^{-r/(N\bar{\gamma}_c)}d\gamma$$
(12.58)

$$= (\log_2 e)e^{1/(N\bar{\gamma}_c)}E_1(1/(N\bar{\gamma}_c))$$

where $E_1(x) = \int_x^\infty (e^{-t}/t)dt$ is defined by (D.115). For the uncorrelated case ($\rho = 0$), the ergodic spectral efficiency is found using (D.74): [4]

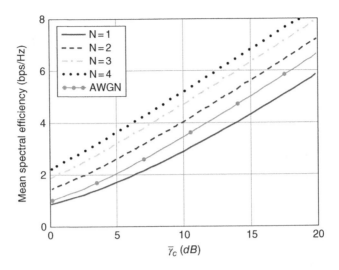

Figure 12.18   Spectral Efficiency of a SIMO System with N Antennas For Identical Mean Branch SNR's in a Rayleigh Fading Channel.

$$E[\eta] = \int_0^\infty \log_2(1+\gamma) \frac{1}{(N-1)!} \frac{\gamma^{N-1}}{\bar{\gamma}_c^N} e^{-\gamma/\bar{\gamma}_c} d\gamma$$

$$= \frac{1}{(N-1)!} \int_0^\infty e^{-x} x^{N-1} \log_2(1+\bar{\gamma}_c x) dx$$

$$= \log_2 e \sum_{k=0}^{N-1} \frac{1}{k!(-\bar{\gamma}_c)^k} \left[e^{1/\bar{\gamma}_c} E_1(1/\bar{\gamma}_c) + \sum_{i=1}^{k} (i-1)!(-\bar{\gamma}_c)^i \right]$$

$$(12.59)$$

For $0 < \rho < 1$, the mean spectral efficiency is found using (D.73) and (12.46):

$$E[\eta] = \int_0^\infty \log_2(1+\gamma) \sum_{k=1}^{N} \pi_k \frac{1}{\lambda_k \bar{\gamma}_c} \exp\left(-\frac{\gamma}{\lambda_k \bar{\gamma}_c}\right) d\gamma$$

$$= (\log_2 e) \sum_{k=1}^{N} \pi_k \int_0^\infty \ln(1+\lambda_k \bar{\gamma}_c x) \exp(-x) dx$$

$$= (\log_2 e) \sum_{k=1}^{N} \pi_k e^{1/(\lambda_k \bar{\gamma}_c)} E_1(1/(\lambda_k \bar{\gamma}_c))$$

$$(12.60)$$

where π_k is defined by (12.46). Figure 12.18 shows that the spectral efficiency in bps/Hz increases with mean SNR and N, as expected. Also note that the spectral efficiency in a fading channel is lower than in the AWGN channel unless diversity reception and combining is used. For finite values of the mean SNR, the slope of the spectral efficiency curve is practically the same for all values of N. However, spectral efficiency increases with N, implying that the use of SIMO systems increases the throughput.

Figure 12.19 shows the variation of the ergodic spectral efficiency with mean SNR for various values of the correlation coefficient for $N = 3$. The results for $N = 1$ and 2 are also shown for comparison purposes. Note that the correlation effects are not as strong as in outage and bit error probabilities. At a given spectral efficiency value, the performance difference for $\rho = 0$ and $\rho = 1$ is less than 2 dB for $N = 3$. In other terms, the spectral efficiency for $\rho = 1$ is approximately 12% less than that for $\rho = 0$ at a given value of the mean SNR. Nevertheless, the ergodic spectral efficiency is still higher than that for $\rho = 0$, $N = 2$.

Figure 12.20 shows the variation of the mean spectral efficiency with the number of antennas for $\rho = 0$ and $\rho = 1$. The increase in the spectral efficiency with N shows similar trends for uncorrelated and perfectly correlated cases.

The cdf of the ergodic capacity, that is, the probability that the ergodic capacity is below a

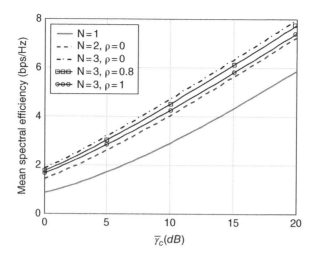

Figure 12.19   Effect of Correlation on the MeanSpectral Efficiency.

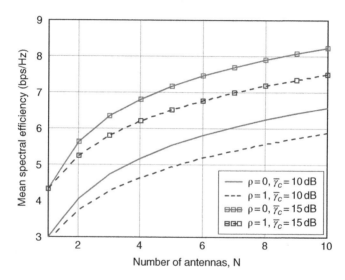

Figure 12.20   The Mean Spectral Efficiency (bps/Hz) as a Function of the Number of Antennas For $\bar{\gamma}_c = 10, 15 (dB)$ and $\rho = 0, 1$.

threshold normalized capacity level $\eta_{th} = C_{th}/B$, may be formulated as follows:

$$\text{Prob}(\eta < \eta_{th}) = \text{Prob}(\log_2(1+\gamma) < \eta_{th})$$
$$= \text{Prob}(\gamma < 2^{\eta_{th}} - 1) = F(2^{\eta_{th}} - 1)$$
$$(12.61)$$

where $F(\gamma)$ denotes the cdf of γ and is given by (12.53). Figure 12.21 shows the correlation

effects on the cdf of the spectrum efficiency. The outage probability decreases with increasing values of the number of antennas, as expected. However, strong correlations between the channel gains may offset this decrease. For example, let $N=2, \bar{\gamma}_c = 10\ dB$ and $\eta_{th} = 2$. For $\rho = 1$, the probability that the spectral efficiency is below η_{th} is found using (12.53) and (12.61) as

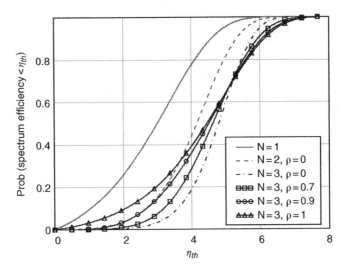

Figure 12.21 Cdf of the Spectral Efficiency For a SIMO System with N Antennas For 10 dB Mean Channel SNR.

$$\text{Prob}(\eta < 2) = 1 - \exp(-3/20) = 0.139 \ (13.9\%) \tag{12.62}$$

while, for $\rho = 0$, it is given by

$$\text{Prob}(\eta < 2) = 1 - e^{-3/10}(1 + 3/10) = 0.037 \ (3.7\%) \tag{12.63}$$

Hence, for $\rho = 1$, the spectrum efficiency will be higher than 2 for 86.1% of the time. However, it is higher than $\eta_{th} = 2$ during 96.3% of the time for $\rho = 0$. In other words, the outage probability for the spectrum efficiency is increased from 3.7% to 13.9% as the correlation coefficient ρ changes from 0 to 1.

Figure 12.22 shows the variation of the probability that the spectrum efficiency is less than $\eta_{th} = 2$ as a function of $\bar{\gamma}_c$. The outage performance of the spectrum efficiency is evidently improved as the mean SNR increases. The curve for $\rho = 1$, $N = 3$ is shifted by 4.78 dB compared to that for $N = 1$ due to the 3-fold increase in the array gain, as we already observed in Figure 12.13 and Figure 12.14.

In some applications, it may be more desirable to use outage capacity rather than the ergodic capacity. Outage capacity is defined as the capacity that remains below a certain threshold capacity level, C_{th}, for a given time percentage p %. In other words, the capacity will higher than C_{th} for $(1-p)$ % of the time. Let us now determine the probability that the capacity is below a threshold level. Figure 12.23 shows the variation of the 2% outage spectral efficiency $(\eta_{th} = C_{th}/B)$ for $N = 1$, 2 and 3. The degradation in the outage spectrum efficiency is observed to be relatively low for $\rho < 0.8$.

12.5.1 Multiplexing Gain

Multiplexing gain is defined as the slope of the ergodic channel capacity (spectrum efficiency) with SNR:

$$r = \lim_{\bar{\gamma} \to \infty} \frac{C(\bar{\gamma})}{\log \bar{\gamma}} \tag{12.64}$$

where $C(\bar{\gamma})$ denotes the ergodic (average) channel capacity at the mean SNR level $\bar{\gamma}$. Multiplexing gain provides a measure of the increase in the maximum data rate that can be transmitted by the system with the number of antennas on transmit and/or receive side. In

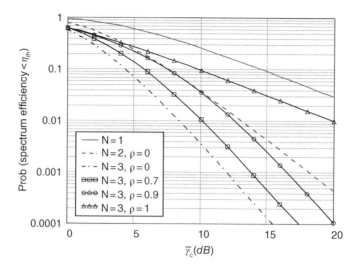

Figure 12.22   Cdf of the Spectral Efficiency For an N-Antenna SIMO System Versus Mean SNR For $\eta_{th} = 2$.

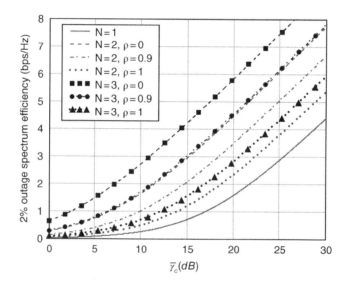

Figure 12.23   The Spectrum Efficiency that Falls Below the Values on the Ordinate For Only 2% of the Time. The spectrum efficiency is higher than the values shown on the ordinate for 98% of the time.

other words, it gives the number of parallel channels that is used for transmission in the presence of multiple antennas at the transmit and/or receive side. One may observe from Figure 12.18 that the multiplexing gain is given by the slope of the bandwidth efficiency as the SNR approaches infinity. The conditions under which the multiplexing gain changes will be discussed in Chapter 13.

12.6   Diversity and Combining

Diversity and combining techniques offer significant performance improvements in fading channels in terms of array, diversity and multiplexing gains. However, they require the availability of a number of transmission paths, which carry the same message, have approximately the same mean power levels, and preferably exhibit independent fading statistics. Since the probability of simultaneous fading of independent channels is much lower than that of a single channel, or correlated multiple channels, the availability of a number of independent diversity channels allows more reliable data transmission with lower error probabilities. In other terms, diversity mitigates fading. Properly combined diversity signals at the receiver lead to mean output SNRs higher than that of any of the individual channels (array gain) and pdf's with lower tails, leading to decreased outage and bit error probabilities (diversity gain) (see Example 11.19).

Diversity may be categorized as macroscopic and microscopic diversity. *Macroscopic diversity* mitigates long-term (in the order of ms) shadow fading, which is caused by hills, mountains, and buildings in the propagation path. It results in random variation of the local mean power of the received signal, which is usually characterized by log-normal pdf whose standard deviation depends on the terrain topology and the propagation distance. Shadowing is mitigated by macroscopic diversity provided by physically separated antennas for transmission and/or reception.

Microscopic diversity, which is used to mitigate the effects of short-term (in the order of μs) fading, is usually implemented as frequency, time or space diversity. *Space diversity* may be realized mainly as antenna diversity, angular diversity or polarization diversity. In *antenna diversity*, multiple antennas are used to receive/transmit signals. Antenna spacings must be adjusted so that the faded signals incident on each receive antenna become independent from each other. Antenna diversity becomes more effective if reception conditions depend on the location, and/or the diversity channels are time variant. In *angle diversity*, the signals received from different angles of arrival are resolved and combined at the receiver. Angle diversity requires a number of directional antennas, each selecting signals arriving from a narrow ranges of angles with low correlations between them.

Polarization diversity is a special case of space (antenna) diversity and is based on the assumption that wireless signals with orthogonal polarizations (vertical/horizontal, RHCP/LHCP) undergo independent fading in wireless channels. Signals simultaneously transmitted from two orthogonal polarizations are resolved and combined at the receiver. If the propagation loss is appreciably different at these polarizations, then the transmit power may be divided accordingly so as to realize equal powers in two diversity branches. This scheme requires the isolation of the two polarizations from each other, since signals suffer depolarization in the wireless channel due to scattering and/or antennas can not isolate them perfectly. Transmit and receive antennas should therefore be designed with low depolarization characteristics. Monopole and patch antennas are commonly used in polarization diversity systems.

Frequency diversity implies the transmission of a signal with different carrier frequencies separated from each other by at least the channel coherence bandwidth. Frequency diversity is feasible when the channel undergoes frequency selective fading. The major drawbacks of the frequency diversity include the requirement for larger transmission bandwidths due to multiple frequency channels, and the generation and combining of signals of different carriers with the same phases (for coherent signalling).

In *time diversity*, the same information bearing signal is transmitted at different time slots that are separated by at least the channel coherence time. Time or delay diversity is feasible when fading is time-selective. In slow fading channels, cyclic delay diversity may be more effective. Delay diversity requires information storage at the transmitter and the receiver. In frequency selective channels, the multipath signals arriving to the receiver with delays longer than the symbol duration, can be treated as diversity signals. As shown in Section 11.5.2, turning multipath propagation into its advantage, a Rake receiver uses MRC to cophase and sum the powers of multipath components in proportion to their respective SNR ratios. Synchronization is an important design factor for a MRC system, which operates coherently and, hence, requires channel state information (CSI) at the receiver.

In the receive diversity (SIMO) systems, the replicas of a signal transmitted by a single transmit antenna are received by an ULA and combined at the receiver (see Figure 12.10). Combining the received signals may be implemented at pre-detection or post-detection stages. In SIMO systems with pre-detection combining, the signals received by the array elements are combined at the receiver front-end and a single receiver processes the combined signals. However, in post-detection combining, separate receivers are required for each channel since signals are combined at the demodulator output. The performance difference between pre-detection and post-detection combining is observed to be negligibly small for ideal coherent detection. However, a slight performance difference is observed for differentially coherent detection.

In transmit diversity (MISO) systems, the transmit power is divided between the array elements. Consequently, signals transmitted by each branch have lower transmit powers, compared to receive diversity systems where the transmit power is allocated to a single transmit antenna. Therefore, the transmit diversity has potentially lower performance compared to the receive diversity. However, if the channel state information (CSI) is made available to the transmitter, then the transmit power may be divided between transmit antennas so as to level the SNR's received by each element of the receiver ULA. The CSI, which includes the gain, the phase and the delay of each channel, may be provided to the transmit side using feedback (closed-loop), training information but no feedback (open-loop), and blind schemes. This approach, which is called as maximal ratio transmission (MRT), provides a performance identical to MRC.

12.6.1   Combining Techniques for SIMO Systems

There are various techniques for combining diversity branches. In *selection combining* (SC), for each symbol duration, the branch with the highest SNR is selected and signals of other branches are ignored. Switched combining and

switch and stay combining (SSC) are special versions of the SC. [5]

As given by (12.31), the output of a linear combiner consists of a weighted sum of signals received by each branch. The weights are used to co-phase the received signals; hence linear combination is a coherent process and requires CSI at the receiver. The receiver can acquire the CSI by using a recovery circuit at its input; hence this process is not as difficult as in MISO systems. *Maximal ratio combiner* (MRC) and equal-gain combiner (EGC) are examples linear-combining techniques. In MRC, signals received by diversity branches are co-phased and are weighted in proportion to their received signal levels; hence the MRC output consists of the sum of branch SNR's as in (11.213) and (12.40). Therefore, MRC requires the estimation of phases and amplitudes of signals received by each diversity branch. For receivers operating in frequency-selective fading channels, for example, Rake receiver, the delays are also required. MRC gives the best possible performance among the diversity combining techniques and provides highest protection against fading.

In *EGC*, received signals in diversity branches are co-phased and added with equal weights. Therefore, the EGC output, which consists of the sum of co-phased signal envelopes, requires the estimation of only the channel phases. In frequency-selective channels, estimation of the signal delays in each diversity branch is also required.

MRC emphasizes signals with higher SNRs by choosing the weights to be complex conjugate of the channel gain (see (12.39)). Therefore, it is not optimum in the presence of strong interference, since higher signal levels are due to interference. *Optimum combining* (OC) is based on minimizing the mean square error between a training sequence and the combiner output, or maximizing signal-to-noise and interference power ratio (SINR). OC reduces to MRC in the absence of interference. *Square-law combining* (SLC) is commonly used in channels where CSI is not available, for example, for noncoherent detection of orthogonal modulations. SLC suffers

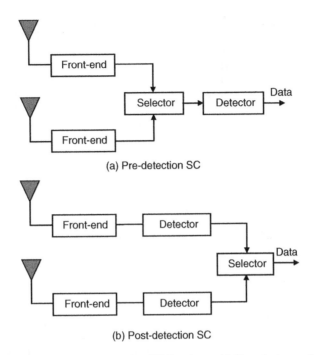

Figure 12.24 Pre-Detection and Post-Detection SC Receiver with Two Antennas For a SIMO System.

noncoherent combining losses, since branch signals and noises are combined together, and performs worse than linear combining techniques.

All these combining techniques have advantages and drawbacks and different areas of applications. Their performances differ in terms of outage and bit error probabilities as well as the output SNR.

12.6.1.1 Selection Combining

SC is often used for macroscopic diversity by BS to increase the coverage area in cellular radio systems against shadowing. For example, a second BS installed in a tunnel may be help user receivers, suffering heavy losses due to shadowing by tunnel, to select between the two signals with the highest SNR. Macroscopic diversity makes use of a number of antennas that are well separated from each other so that the received signals undergo independent shadowing. At each symbol duration, the selection combiner at a receiver selects the received signal with the largest SNR. A single receiver is required

if selection is made at the front-end (pre-detection diversity) but two receivers are necessary when the signal with highest SNR is selected at the demodulator output (post-detection diversity) (see Figure 12.24).

Consider a N-branch SC system with a single transmitter and N receivers. The selected SNR Z is the strongest of the SNRs X_i, $i = 1, 2, ..., N$ of the N diversity branches:

$$Z = \max\{X_1, X_2, \cdots\cdots, X_N\} \qquad (12.65)$$

Since the cdf of Z is defined as the probability that $Z \leq z$ during a symbol period, all branch SNRs should be less than or equal to z during the same symbol period:

$$
\begin{aligned}
F_z(z) &= \mathrm{Prob}(Z \leq z) \\
&= \mathrm{Prob}\{X_1 \leq z, X_2 \leq z, \cdots\cdots, X_N \leq z\}
\end{aligned}
$$
$$(12.66)$$

When the receive antennas are located sufficiently far from each other, then X_i's are independent and the cdf of Z reduces to

$$F_z(z) = \prod_{i=1}^{N} \mathrm{Prob}\{X_i \leq z\} = \prod_{i=1}^{N} F_{X_i}(z) \quad (12.67)$$

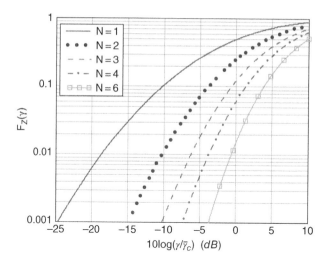

Figure 12.25 The Cdf of the Output SNR of a SIMO System with N-branch SC in Log-Normal Shadowing For 8 dB Shadowing Variance.

The pdf of Z then becomes

$$f_Z(z) = \frac{d}{dz}F_Z(z) = \sum_{i=1}^{N} f_{X_i}(z) \prod_{j=1,j\neq i}^{N} F_{X_j}(z)$$

(12.68)

where $f_{X_i}(z)$ denotes the pdf of X_i.

If the signals undergo log-normal shadowing, then the CDF of the ith branch may be written as

$$F_{X_i}(z) = \int_{-\infty}^{z} f_{X_i}(u)\,du$$

$$= \frac{1}{\sqrt{2\pi}\sigma_i} \int_{-\infty}^{z} \exp\left\{-\frac{(u-\bar{\gamma}_i)^2}{2\,\sigma_i^2}\right\}du = 1 - Q\left(\frac{z-\bar{\gamma}_i}{\sigma_i}\right)$$

(12.69)

where $\bar{\gamma}_i$ and σ_i^2 denote, respectively, the mean SNR and the variance of shadowing, both expressed in dB. Using (12.67), the CDF of output SNR for N branch diversity may be written as

$$F_Z(z) = \prod_{i=1}^{N}\left[1 - Q\left(\frac{z-\bar{\gamma}_i}{\sigma_i}\right)\right]$$

$$= \left[1 - Q\left(\frac{z-\bar{\gamma}_c}{\sigma}\right)\right]^{N}, \quad \begin{matrix}\bar{\gamma}_i = \bar{\gamma}_c, \sigma_i = \sigma \\ i = 1,2,\ldots,N\end{matrix}$$

(12.70)

The outage probability in a log-normal shadowing environment with no diversity

($N=1$) is already depicted in Figure 11.28 for various values of the shadowing variance. Figure 12.25 clearly shows the improvement provided by SC on the outage probability for log-normal shadowing.

The mean SNR value of the correlated SC output in a log-normal shadowing channel is given by: [6]

$$E[Z] = \begin{cases} 2\bar{\gamma}_c\left[1 - Q\left(\sigma\sqrt{(1-\rho)/2}\right)\right] & N=2 \\ 3\bar{\gamma}_c\left[1 - Q\left(\sigma\sqrt{(1-\rho)/2}\right)\right] \\ \quad -6T\left(\sigma\sqrt{(1-\rho)/2}, 1/\sqrt{3}\right) & N=3 \end{cases}$$

$$T(\alpha,\beta) = \frac{1}{2\pi}\int_0^{\beta} \frac{\exp(-0.5\alpha^2(1+x^2))}{1+x^2}dx$$

(12.71)

where shadowing variance and mean branch SNRs are assumed to be identical $\bar{\gamma}_c = \bar{\gamma}_i$, $\sigma = \sigma_i$, $i = 1,2$. Here, $E[Z]/\bar{\gamma}_c$ denotes the increase in the mean SNR due to diversity.

Figure 12.26 shows the normalized output SNR for SC with $N=2$ and 3 branches as a function of the correlation coefficient between branches. The output SNR was observed to become higher for higher shadowing variances and decreases rapidly for $\rho > 0.8$. Figure 12.9 and (12.28) show that the $\rho = 0.8$ corresponds

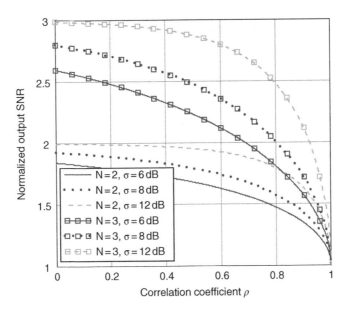

Figure 12.26 Normalized Mean SNR at the Output of a Selection Combiner with $N=2$ and 3 in a Log-Normal Shadowing Environment.

to a distance of 4.8 cm between antennas for uniform AOA with $\theta_s = \pi/2$ at $f = 900$ MHz. In view of the required antenna separation of 4.8 cm, hand-held telephones can potentially use SC between signals of two nearest BSs.

Now let us consider SC in a Rayleigh fading channel. The probability that X_i, denoting the instantaneous E_b/N_0 in i^{th} branch, is less than or equal to z is given by

$$F_{X_i}(z) = \text{Prob}(X_i \le z) = \int_0^z \frac{1}{\bar\gamma_i} e^{-z/\bar\gamma_i} \, dx = 1 - e^{-z/\bar\gamma_i}$$

$$(12.72)$$

where $\bar\gamma_i$ denotes the mean value of E_b/N_0 in i^{th} branch. The cdf of Z for independently Rayleigh-fading branches is obtained from (12.67) as

$$F_z(\gamma) = \prod_{i=1}^N \left(1 - e^{-\gamma/\bar\gamma_i}\right)$$

$$= \left(1 - e^{-\gamma/\bar\gamma_c}\right)^N, \quad \bar\gamma_c = \bar\gamma_i, i = 1,2,...,N$$

$$(12.73)$$

The cdf (12.73) for Rayleigh fading is shown in Figure 12.27. The outage probability performance improves drastically with increasing values of N since the slope of the outage curve, hence the diversity order, is given by γ^N. Assuming that the local mean power $\bar\gamma_c$ is 10 times higher than the normalized threshold level γ, that is, $\gamma/\bar\gamma_c = 0.1 (-10 \, dB)$, Figure 12.27 shows that the probability that at least one of the channels has an instantaneous power higher than the threshold level is 90% for $N=1$ (no diversity), 99% for $N=2$, 99.9% for $N=3$, and 99.99% for $N=4$. Figure 12.27 may also be interpreted for a constant outage level. For example, $F_z(\gamma) = 0.01$ implies 99% probability that at least one of the diversity channels has an instantaneous power greater than the normalized threshold level $\gamma/\bar\gamma_c$. For $N=1$ (no diversity), the required normalized threshold level is -20 dB while it is -10 dB for $N=2$; this implies 10 dB diversity advantage with respect to $N=1$. The threshold level of -5 dB for $N=4$ implies 15 dB diversity advantage with respect to $N=1$.

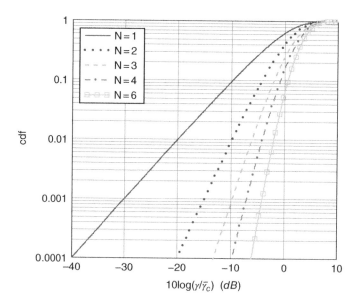

Figure 12.27 The Cdf at the Output of a N-Branch SC in a Rayleigh Fading Channel.

The pdf of Z is determined using (12.73) and (D.11):

$$f_z(z) = \frac{d}{dz}F_z(z) = \frac{N}{\bar{\gamma}_c}e^{-z/\bar{\gamma}_c}\left(1-e^{-z/\bar{\gamma}_c}\right)^{N-1}$$

$$= \frac{N}{\bar{\gamma}_c}\sum_{n=0}^{N-1}(-1)^n\binom{N-1}{n}e^{-(n+1)z/\bar{\gamma}_c}$$

(12.74)

Using (D.50) and (D.24), mean E_b/N_0 value at the SC output is found to be

$$E[Z] = \int_0^\infty zf_z(z)dz$$

$$= \frac{N}{\bar{\gamma}_c}\sum_{n=0}^{N-1}(-1)^n\binom{N-1}{n}\int_0^\infty z\,e^{-(n+1)z/\bar{\gamma}_c}dz$$

$$= N\bar{\gamma}_c\sum_{n=0}^{N-1}(-1)^n\binom{N-1}{n}\frac{1}{(n+1)^2}$$

$$= \bar{\gamma}_c\sum_{n=1}^{N}\binom{N}{n}\frac{(-1)^{n+1}}{n} = \bar{\gamma}_c\underbrace{\sum_{k=1}^{N}\frac{1}{k}}_{array\ gain}$$

(12.75)

The array gain in (12.75) accounts for the increase in the mean output E_b/N_0 due to SC. Figure 12.28 shows the improvement of the pdf of the E_b/N_0 at the SC output with the number of antennas. The advantage provided by SC is clearly visible since the pdf and its mean value, given by (12.75), shift towards higher E_b/N_0 values as N increases.

We now study the effect of SC diversity on the BEP performance for noncoherent BFSK and DPSK modulations in a slowly Rayleigh fading channel. Averaging $P_2(\gamma) = 0.5\,e^{-\beta\gamma}$ ($\beta = 1$ for DPSK and ½ for NC-FSK) given by (11.175) with (12.74), one gets

$$P_2 = \int_0^\infty P_2(\gamma)f_z(\gamma)d\gamma$$

$$= \frac{N}{2}\sum_{n=0}^{N-1}(-1)^n\binom{N-1}{n}\frac{1}{1+n+\beta\bar{\gamma}_c}$$

(12.76)

Figure 12.29 shows (12.76), the BEP for binary DPSK, for various values of the number of diversity branches. SC diversity with 6 branches was observed to eliminate the

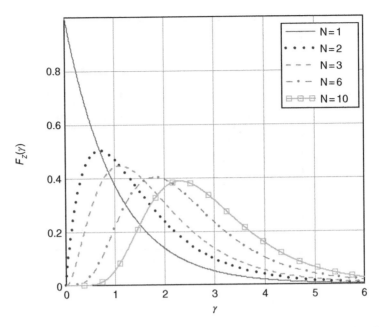

Figure 12.28 Pdf of the SNR When N Independent Signals with $\bar{\gamma}_c = 1$ Are Selection-Combined.

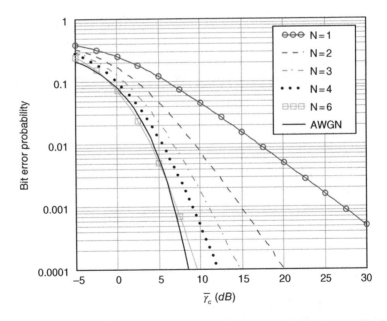

Figure 12.29 BEP For Binary DPSK For SC of N Independent Signals in a Rayleigh Fading Channel.

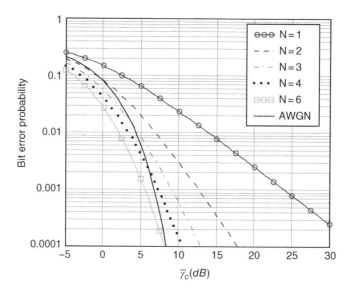

Figure 12.30 BEP For Coherent Binary PSK with SC in a Rayleigh Fading Channel.

degradation caused by Rayleigh fading; then the BEP performance becomes very close to that for AWGN channel.

Averaging $P_2(\gamma) = Q(\sqrt{2\gamma})$ with $f_z(z)$ given by (12.74) one gets the BEP for coherent BPSK in a slowly Rayleigh fading channel as follows:

$$P_2 = \int_0^\infty Q(\sqrt{2z}) f_z(z) dz$$

$$= \frac{1}{2} \sum_{n=0}^{N-1} (-1)^n \binom{N}{n+1} \left(1 - \sqrt{\frac{\bar{\gamma}_c}{n+1+\bar{\gamma}_c}}\right)$$

$$= \frac{1}{2} \sum_{n=0}^{N} (-1)^n \binom{N}{n} \sqrt{\frac{\bar{\gamma}_c}{n+\bar{\gamma}_c}}$$

$$\hspace{6cm} (12.77)$$

where the identity given by (D.23) is used to obtain the last expression. Figure 12.30 shows the variation of (12.77) with $\bar{\gamma}_c(dB)$ for various values of N. The diversity order for SC of N signals is N and the BEP performance for Rayleigh fading becomes better than in AWGN channel for $N \geq 6$.

Example 12.7 Unequal Branch SNRs in SC. Consider a dual-branch SC diversity receiver which selects the largest of the SNR's γ_1 and γ_2 of two independent Rayleigh fading signals with unequal mean values $\bar{\gamma}_1 \neq \bar{\gamma}_2$. The cdf of the SNR at the output of this dual-branch SC is given by (12.67):

$$F_Z(\gamma) = F_{X_1}(\gamma) F_{X_2}(\gamma) \hspace{1.5cm} (12.78)$$

The pdf of the dual-branch SC output is obtained using (12.78):

$$f_Z(\gamma) = \frac{d}{d\gamma} F_Z(\gamma) = f_{X_1}(\gamma) F_{X_2}(\gamma) + F_{X_1}(\gamma) f_{X_2}(\gamma)$$

$$= f_{X_1}(\gamma) + f_{X_2}(\gamma) - \left(\frac{1}{\bar{\gamma}_1} + \frac{1}{\bar{\gamma}_2}\right) \exp\left[-\left(\frac{1}{\bar{\gamma}_1} + \frac{1}{\bar{\gamma}_2}\right)\gamma\right]$$

$$\hspace{6cm} (12.79)$$

The mean value of SNR at the SC combiner output is found from (12.79):

$$E[Z] = \bar{\gamma}_1 + \bar{\gamma}_2 - \frac{1}{1/\bar{\gamma}_1 + 1/\bar{\gamma}_2} = \bar{\gamma}_1 + \bar{\gamma}_2 - \frac{\bar{\gamma}_1 \bar{\gamma}_2}{\bar{\gamma}_1 + \bar{\gamma}_2}$$

$$\hspace{6cm} (12.80)$$

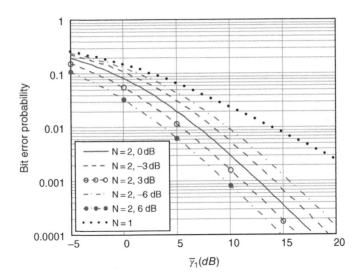

Figure 12.31 BEP For Dual-Branch SC of BPSK Modulated Signals in a Rayleigh Fading Channel. The ratio of the mean SNRs is chosen as $a = \bar{\gamma}_2/\bar{\gamma}_1 = 0, \pm3, \pm6\ dB$.

Averaging $P_2(\gamma) = Q(\sqrt{2\gamma})$ with (12.79), the BEP for BPSK modulation is found using (D.70):

$$P_e = \frac{1}{2}\left(1 - \sqrt{\frac{\bar{\gamma}_1}{1 + \bar{\gamma}_1}}\right) + \frac{1}{2}\left(1 - \sqrt{\frac{a\bar{\gamma}_1}{1 + a\bar{\gamma}_1}}\right)$$
$$- \frac{1}{2}\left(1 - \sqrt{\frac{a\bar{\gamma}_1}{1 + a + a\bar{\gamma}_1}}\right)$$

$$(12.81)$$

where $a = \bar{\gamma}_2/\bar{\gamma}_1$. For the special case $\bar{\gamma}_1 = \bar{\gamma}_2 = \bar{\gamma}_c$, (12.81) reduces to (12.77), as expected. Figure 12.31 shows the effect of unequal mean branch SNRs in a dual-branch SC diversity system. The BEP performance for SISO ($N = 1$) is also shown as a reference. The effectiveness of SC diversity was observed to degrade as the branch SNRs differ from each other; the diversity order of $d = 2$ remains the same but diversity gain decreases due to the decrease in the array gain (see (12.80)). Lower mean SNR values of the second branch compared to the first one forces the SC to choose the first branch more often; this leads to degradation in the BEP performance. In the limiting case where $\bar{\gamma}_2$ approaches zero, the SC acts as a SISO receiver.

Example 12.8 Improvement in Level Crossing Rate (LCR) Due to SC.

Consider a receiver employing Nth order SC diversity in Rayleigh fading conditions. We will derive analytical expressions for the envelope LCR and average fade duration and determine the improvement in LCR provided by SC diversity.

The cdf and the pdf of the envelope of a Rayleigh-faded signal at the output of N-order SC may be written from (11.86):

$$F_\alpha(r) = \text{Prob}(\alpha < r) = \left(1 - \exp\left(-\frac{r^2}{2\sigma^2}\right)\right)^N$$

$$f_\alpha(r) = \frac{dF_\alpha(r)}{dr} = \frac{Nr}{\sigma^2}e^{-\frac{r^2}{2\sigma^2}}\left(1 - \exp\left(-\frac{r^2}{2\sigma^2}\right)\right)^{N-1}$$

$$(12.82)$$

Using (11.156) and (11.157), the LCR of the envelope at $r = R$ is found to be

$$N_R(N) = \sqrt{\pi}\sigma f_{d,\max}f_\alpha(R)$$

$$= \sqrt{2\pi}f_{d,\max}N\rho e^{-\rho^2}\left(1 - \exp(-\rho^2)\right)^{N-1}$$

$$\frac{N_R(N)}{N_R(1)} = N\left(1 - \exp(-\rho^2)\right)^{N-1}$$

$$(12.83)$$

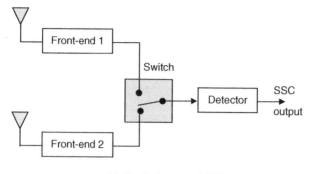

(a) Block diagram of SSC

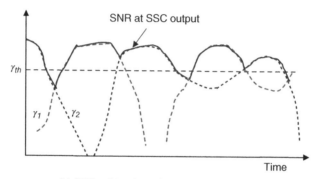

(b) SNRs of two branches and the SSC output

Figure 12.32 Pre-Detection Switch and Stay Combining (SSC).

where $\rho = R/(\sqrt{2}\sigma)$ denotes the normalized threshold level. The pdf of the envelope slope (see (11.156)) is assumed to be unaffected by SC diversity. One may observe from (12.83) that the LCR rapidly decreases with N. The average fade duration is determined using (12.83) and (11.162):

$$\bar{t}(N) = \frac{F_\alpha(R)}{N_R} = \frac{e^{\rho^2}-1}{\sqrt{2\pi}f_d\rho N} \Rightarrow \frac{\bar{t}(N)}{\bar{t}(1)} = \frac{1}{N}$$

(12.84)

Average fade duration is decreased by a factor N with N^{th} order SC.

Switch and Stay Combining (SSC)

In SC, the diversity branches are scanned during each symbol period and the branch with

the highest instantaneous E_b/N_0 is chosen. However, scanning in SSC stops whenever the combiner finds a diversity branch with an E_b/N_0 above a specified threshold γ_{th}. Regardless of the E_b/N_0 levels of other branches, this diversity branch is used until the E_b/N_0 drops below the threshold. Then, another branch is chosen that has an E_b/N_0 exceeding the threshold. Hence, SSC is practical since it switches between the diversity branches less frequently.

Figure 12.32 shows a dual-branch pre-detection SSC, variations of E_b/N_0's in the two branches and the E_b/N_0 at the combiner output. SSC performs worse than SC since it does not always operate with the diversity branch having the highest instantaneous E_b/N_0. However, they have the same performance at the switching threshold. Since SSC offers its best

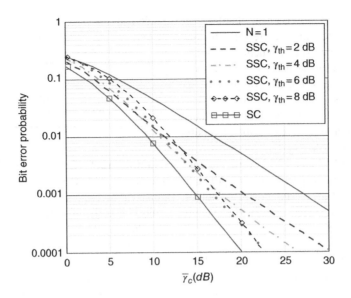

Figure 12.33 BEP For a Dual-Branch Pre-Detection SSC of Binary DPSK Signals in a Rayleigh Fading Channel.

performance above the threshold level, γ_{th}, the threshold level should be chosen just above the minimum acceptable instantaneous E_b/N_0.

BER for binary DPSK for SSC is given by [5]

$$P_2 = \frac{1}{2(1+\bar{\gamma}_c)} \Upsilon(\bar{\gamma}_c, \gamma_{th}) \qquad (12.85)$$

where

$$\Upsilon(\bar{\gamma}_c, \gamma_{th}) = q + (1-q)e^{-\gamma_{th}}$$
$$q = F_\gamma(\gamma_{th}) = \text{Prob}(\gamma \le \gamma_{th}) = 1 - e^{-\gamma_{th}/\bar{\gamma}_c}$$
$$(12.86)$$

Here, q denotes the probability that E_b/N_0 is less than the threshold level γ_{th}. Frequency of switching can be minimized by keeping q (thus γ_{th}) as small as possible. The limiting case $\gamma_{th} = 0$ implies no diversity and no switching. For a given value of $\bar{\gamma}_c$, there is an optimum threshold γ_{th} that minimizes both the factor $\Upsilon(\bar{\gamma}_c, \gamma_{th})$ and the BEP. The factor $\Upsilon(\bar{\gamma}_c, \gamma_{th})$ decreases with increasing values of $\bar{\gamma}_c$. Figure 2.33 shows

the BEP for a dual-branch pre-detection SSC of binary DPSK signals in a Rayleigh fading environment. Reference BEP curves for DPSK for $N = 1, 2$ are obtained using (12.76). Figure 12.33 clearly shows the need for adaptive choice of the threshold level, since the SSC performance at lower and higher values of $\bar{\gamma}_c$ is reversed.

12.6.1.2 Maximal Ratio Combining

In maximal ratio combining (MRC), diversity branches are weighted by complex conjugates of their respective complex channel gains, as given by (12.39) (Figure 12.10). Then, the SNR at the MRC output, given by (12.40), consists of the sum of SNR's of diversity branches. This choice of the weighting coefficients implies that, during every symbol period, the MRC amplifies the diversity branches with higher gains at the disadvantage of those with lower gains. This means that the receiver has the capability for estimating the CSI, that is, the channel gain, phase and delay.

We also observed that full benefits of diversity can be achieved with uncorrelated channel gains, that is, when array elements are sufficiently separated from each other in a rich scattering environment. Then, the components of the channel gain vector, given by (12.34), become uncorrelated with each other. Otherwise, the correlations between the diversity branches degrade the system performance. Figures 12.12, 12.13 and 12.14 show the correlation effects on, respectively, the pdf, cdf and the BEP for BPSK modulation at the output of a three-branch MRC in a correlated Rayleigh fading environment.

One may also show that MRC maximizes the SNR at its output and minimizes the BEP. Using the Schwartz inequality, the SNR at MRC output given by (12.36) may be rewritten as

$$
\gamma = \bar{\gamma}_c \frac{\left| \sum_{i=1}^{N} w_i h_i \right|^2}{\sum_{i=1}^{N} |w_i|^2} \leq \bar{\gamma}_c \frac{\sum_{i=1}^{N} |w_i|^2 \sum_{i=1}^{N} |h_i|^2}{\sum_{i=1}^{N} |w_i|^2} = \bar{\gamma}_c \sum_{i=1}^{N} \alpha_i^2
$$

(12.87)

We know that the equality in (12.87) holds when weights are chosen as in (12.39), that is $w_i = h_i^*$:

$$
\left| \sum_{i=1}^{N} w_i h_i \right|^2 \leq \sum_{i=1}^{N} |w_i|^2 \sum_{i=1}^{N} |h_i|^2
$$
(12.88)

Hence, the signals received by diversity branches are co-phased and their SNR's are summed.

The resulting pdf and the cdf are given by (12.42) and (12.53), respectively. The pdf and the cdf of MRC output are shown in Figure 11.54 for various values of the number of branches. Similarly, the BEP for BPSK is depicted in Figure 11.55 for a Rake receiver, which uses MRC to combine signal replicas with different delays.

Table 12.1   Normalized SNR Levels Required by MRC and SC For 99.8% reliability, That Is, $1 - F_\gamma(z) = 0.998$.

Combiner	$N=1$	$N=2$	$N=3$	$N=4$
MRC	−27	−11.9 dB	−6.1 dB	−2.84 dB
SC	−27	−13.4 dB	−8.7 dB	−6.24 dB

Example 12.9   Comparison of Outage Performances of MRC and SC.
Consider that the design of a SIMO system limits the transmit power level to 20 dB less than that required by a SISO system. Therefore, we desire to use diversity reception for achieving the same outage performance with 20 dB lower transmit power. We are asked to determine the required number of uncorrelated diversity branches in order to obtain a 20 dB gain at a level where the received signal is below the threshold level only 0.2% of the time.

From the outage curve for MRC given by Figure 11.54, 0.2% outage corresponds to a normalized SNR level of −27 dB for a single branch, −11.9 dB for $N = 2$, −6.1 dB for N = 3 and −2.84 dB for $N = 4$. In order to obtain 20 dB gain with respect to −27 dB for a single branch (SISO), the signal level must remain at −7 dB or higher for 99.8% of the time. Hence, a three branch MRC diversity provides the desired gain since it requires a normalized SNR level of −6.1 dB for 0.2% outage.

Table 12.1 presents a comparison of the normalized SNR levels required by SC and MRC for 0.2% outage probability using the curves given by Figure 12.27 and Figure 11.54 respectively for SC and MRC. For dual diversity, MRC outperforms SC approximately 1.5 dB for $N = 2$, 2.6 dB for $N = 3$ and 3.4 dB for $N = 4$.

Example 12.10   BEP Performance of MRC in a Nakagami-m Fading Channel.
The pdf and cdf of the power sum of i.i.d. distributed N Nakagami-m fading signals are given by (11.109) and (11.110), respectively. Using (11.109) and (D.70), the BEP for BPSK in a Nakagami-m fading channel is found to be

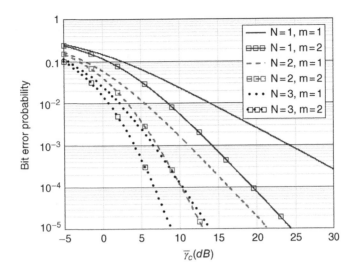

Figure 12.34   BEP For N-Branch MRC of BPSK Signals in a Nakagami-m Fading Channel.

$$P_2 = \int_0^\infty Q\left(\sqrt{2\alpha\gamma}\right)f_\gamma(\gamma)d\gamma$$

$$= \frac{1}{2}\left[1 - \sqrt{\frac{\alpha\bar{\gamma}_c/m}{1+\alpha\bar{\gamma}_c/m}}\sum_{k=0}^{mN-1}\binom{2k}{k}\frac{1}{\{4(1+\alpha\bar{\gamma}_c/m)\}^k}\right]$$

$$(12.89)$$

where $\alpha = 1$ for BPSK and ½ for coherent orthogonal FSK. Here m denotes the fading parameter, $\bar{\gamma}_c$ mean channel SNR and N is the number of diversity branches. For $m = 1$, the above expression reduces to that for Rayleigh fading given by (12.54). Figure 12.34 shows the effect of the fading parameter m on the BEP for BPSK modulation in a Nakagami-m fading channel for $N = 1$, 2, and 3. Note that the product of the fading parameter m and N determine the diversity order $d = mN$. The effect of m on the BEP performance also shows the importance of correct modelling of the fading channel.

Similarly, using (11.109) and (D.50), the BEP for binary DPSK and noncoherent FSK (NCFSK) is found to be

$$P_2 = \int_0^\infty \frac{1}{2}e^{-\beta\gamma}f_\gamma(\gamma)d\gamma = \frac{1}{2(1+\beta\bar{\gamma}_c/m)^{mN}}$$

$$(12.90)$$

where $\beta = 1$ for DPSK and $\beta = 1/2$ for NC-FSK.

Example 12.11   Unequal Branch SNRs in Dual MRC Combining.
Consider a dual branch MRC diversity receiver with unequal mean branch SNR's $\bar{\gamma}_1$ and $\bar{\gamma}_2$. Branch fadings are assumed to uncorrelated with each other. The MGF for a dual-branch MRC system with unequal mean branch SNRs is given by (11.216):

$$M_\gamma(s) = \frac{1}{(1+s\bar{\gamma}_1)(1+s\bar{\gamma}_2)}\qquad(12.91)$$

The BEP for DPSK is found using the above MGF:

$$P_2 = \frac{1}{2}M_\gamma(s=1) = \frac{1}{2(1+\bar{\gamma}_1)(1+\bar{\gamma}_2)}$$

$$= \frac{1}{2\left[1+(1+a)\bar{\gamma}_1+a\,\bar{\gamma}_1^2\right]},\ a=\bar{\gamma}_2/\bar{\gamma}_1$$

$$(12.92)$$

The BEP of DPSK modulation for N-branch MRC with equal mean branch SNRs is obtained by inserting $m=\beta=1$ into (12.90). Figure 12.35 shows that the BEP degrades

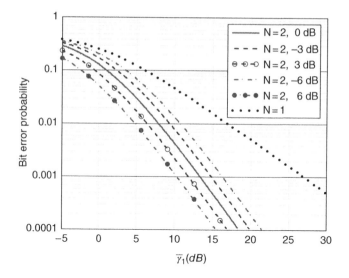

Figure 12.35 Effect of Unequal Mean SNRs on the BEP For DPSK Signals With Dual-Branch MRC. The ratio of the mean SNRs is chosen as $a = \bar{\gamma}_2/\bar{\gamma}_1 = 0, \pm 3, \pm 6dB$.

considerably when the mean SNRs of the two branches are not equal to each other. The results are similar to those for dual-branch SC in a Rayleigh fading channel for BPSK modulation (see Figure 12.31).

12.6.1.3 Equal Gain Combining

It may not always be easy or convenient to have a variable weighting capability as required by the MRC, where the weight coefficients are determined by the estimates of gains and phases of diversity branches (see (12.39)). This is necessary for co-phasing the signals before combining. Otherwise, complex random signals would be combined noncoherently and thus interfere with each other destructively.

EGC is simpler than MRC since it needs the estimation of only the channel phase but not the channel gain. Hence, the weights in the linear combiner shown in Figure 12.10 are chosen as

$$w_i = e^{j\phi_i}, i = 1, 2, ..., N \qquad (12.93)$$

based on the assumption that the channel phase is perfectly estimated (compare with (12.39) for

MRC). Inserting (12.93) into (12.31), one gets the EGC output as

$$r_{\ell m}(t) = \sum_{i=1}^{N} \left(\alpha_i(t) \, s_{\ell m}(t) + n_i(t) \, e^{j\phi_i(t)} \right)$$

$$(12.94)$$

where $\alpha_i(t) \, s_{\ell m}(t)$ denotes the signal envelope received by the ith branch. Hence, the EGC output consists of the coherent sum of signal envelopes. Therefore, EGC is convenient for coherently detected modulation techniques with equal energy symbols, for example, M-PSK. In slow fading, time dependence during a symbol may be suppressed. Assuming slow fading, the instantaneous SNR of the EGC output is found from (12.94):

$$\gamma = \frac{\frac{1}{2}E\left[\left|\sum_{i=1}^{N} \alpha_i s_{\ell m}\right|^2\right]}{E\left[\left|\sum_{i=1}^{N} n_i e^{j\phi_i}\right|^2\right]} = \frac{\frac{1}{2}E\left[|s_{\ell m}|^2\right]}{NN_0}\left|\sum_{i=1}^{N} \alpha_i\right|^2$$

$$= \bar{\gamma}_c \frac{1}{N}\left|\sum_{i=1}^{N} \alpha_i\right|^2$$

$$(12.95)$$

Note from (12.95) that the SNR in the EGC output is proportional the square of the sum of branch envelopes. The AWGN noise samples in diversity branches are independent and have double-sided PSD $= N_0/2$. The cdf and the pdf of γ given by (12.95) do not have closed-form expressions for $N > 2$.

Assuming that α_i are i.i.d. Rayleigh variables, the mean SNR at the EGC output may be written as

$$E[\gamma] = \bar{\gamma}_c \frac{1}{N} E\left[\left| \sum_{i=1}^{N} \alpha_i \right|^2 \right] = \bar{\gamma}_c \frac{1}{N} \sum_{i=1}^{N} \sum_{j=1}^{N} E[\alpha_i \alpha_j]$$

$$= \bar{\gamma}_c \frac{1}{N} \left[\sum_{i=1}^{N} E[\alpha_i^2] + \sum_{i=1}^{N} \sum_{j=1, i \neq j}^{N} E[\alpha_i] E[\alpha_j] \right]$$

$$(12.96)$$

where the moments are defined by (11.87):

$$E[\alpha_i \alpha_j] = \begin{cases} \pi \sigma^2 / 2 & i \neq j \\ 2\sigma^2 & i = j \end{cases} \quad (12.97)$$

Inserting (12.97) into (12.96), the mean SNR at the EGC output is found to be

$$E[\gamma] = \bar{\gamma}_c \frac{1}{N} \left[N 2\sigma^2 + (N^2 - N) \frac{\pi}{2} \sigma^2 \right]$$

$$= \underbrace{\bar{\gamma}_c \left[1 + (N-1) \frac{\pi}{4} \right]}_{Array\ gain} \quad (12.98)$$

where $E[\alpha^2] = 2\sigma^2 = 1$ is based on the assumption that a Rayleigh random variable is formed by two independent zero-mean Gaussian random variables with variances ½. The array gain of EGC is 0.49 dB, 0.67 dB and 0.76 dB poorer than that of MRC (see (12.41)), for $N = 2, 3$, and 4, respectively. However, it 0.76 dB, 1.47 dB and 2.1 dB better than the array gain of SC, given by (12.75), for $N = 2$, 3, and 4, respectively.

The BEP for binary coherent PSK for EGC diversity with $N = 2$ and 3 in a Rayleigh-faded channel is given by [7][8]

$$P_2 = \frac{1}{2} \left(1 - \frac{\sqrt{\bar{\gamma}_c (2 + \bar{\gamma}_c)}}{1 + \bar{\gamma}_c} \right), \quad N = 2$$

$$P_2 = \frac{1}{2} - \frac{1}{2} \sqrt{\frac{\bar{\gamma}_c (2\bar{\gamma}_c + 3)^2}{3(1 + \bar{\gamma}_c)^3}} \, _2F_1\left(-\frac{1}{2}, -\frac{1}{2}; \frac{1}{2}; \frac{\bar{\gamma}_c^2}{(2\bar{\gamma}_c + 3)^2} \right)$$

$$+ \frac{\pi}{4} \sqrt{\frac{\bar{\gamma}_c^3}{27(1 + \bar{\gamma}_c)^3}}, \quad N = 3$$

$$(12.99)$$

where $\bar{\gamma}_c$ denotes the mean E_b/N_0 of diversity branches and $_2F_1(a, b; c; z)$ is the Gauss hypergeometric function (see (D.82)). Figure 12.36 compares the BEP performances of EGC and MRC systems for $N = 2$ and 3. Performance difference between MRC and EGC at 10^{-4} BEP is observed to be 0.62 dB for $N = 2$ and 0.85 dB for $N = 3$.

12.6.1.4 Generalised Selection Combining

In SC, during each signal interval, one of the N receive antennas with the highest SNR is selected and the selected antenna is connected to the receiver. The signals of the other antennas are ignored. However, in MRC and EGC systems all N receive antennas are used simultaneously. In certain applications, especially in post-detection combining systems, use of multiple antennas may be costly because N separate receivers are required. As a solution to this problem, one may use the so-called generalized selection combining (GSC) technique. In GSC, signals of a subset N_r of the N receive antennas with the highest SNRs are selected and selected signals are combined via either MRC or EGC (see Figure 12.37).

Instantaneous SNRs $\gamma_i, i = 1, 2, \cdots, N$ of the diversity branches are assumed to be i.i.d. random variables with the same mean $\bar{\gamma}_c$. The channel is assumed to be Rayleigh fading; hence, the branch SNRs are exponentially distributed. The ordered γ_i's, shown as $\gamma_{(i)}, i = 1, 2, \cdots, N$, represent ordered instantaneous SNR's at the outputs of the N receive antennas:

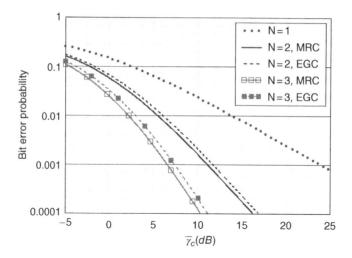

Figure 12.36 The BEP For MRC and EGC of Rayleigh Faded Coherent BPSK Signals with Equal Mean Branch SNRs.

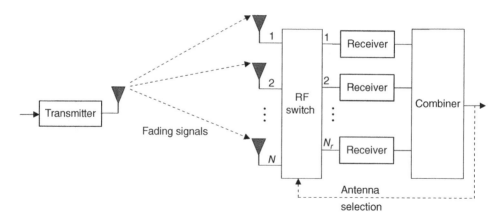

Figure 12.37 Generalized Selection Combining.

$$\gamma_{(1)} > \gamma_{(2)} > \cdots > \gamma_{(N)} > 0 \qquad (12.100)$$

Note that the possibility of at least two equal $\gamma_{(i)}$'s is excluded, since $\gamma_{(i)} \neq \gamma_{(j)}$ almost surely for continuous random variables. The joint pdf of the ordered random variables is given by (see section F.6 of Appendix F) [9]

$$f\left(\gamma_{(1)}, \gamma_{(2)}, \cdots, \gamma_{(N)}\right) = \frac{N}{\bar{\gamma}_c^N} \exp\left(-\frac{1}{\bar{\gamma}_c} \sum_{m=1}^N \gamma_{(m)}\right)$$
$$(12.101)$$

It is important to note that, in contrast with γ_i's, $\gamma_{(i)}$'s are not independent with each other. Therefore, the evaluation of the MGF

$$M_{\gamma_{out}}(s) = E\left[\exp\left(-s \sum_{n=1}^N a_n \gamma_{(n)}\right)\right]$$

$$= \int_0^\infty \int_0^{\gamma_{(1)}} \cdots \int_0^{\gamma_{(N-1)}} \exp\left(-s \sum_{n=1}^N a_n \gamma_{(n)}\right)$$

$$f\left(\gamma_{(1)}, \gamma_{(2)}, \cdots, \gamma_{(N)}\right) d\gamma_{(N)} \cdots d\gamma_{(2)} d\gamma_{(1)}$$

$$(12.102)$$

involves nested N-fold integrals, which is cumbersome and time consuming. Here, $a_i = 1$ if is selected, and $a_i = 0$ if it is not selected to be connected to the receiver. To solve this problem, ordered instantaneous branch SNR's $\gamma_{(i)}$ may be transformed into a new set of virtually independent instantaneous branch SNR's V_n's as follows: [9]

$$\gamma_{(i)} = \sum_{n=i}^{N} \frac{\bar{\gamma}_c}{n} V_n \qquad (12.103)$$

With this transformation, one may express the output SNR in terms of the instantaneous SNR's of the virtual branches as

$$\gamma_{GSC} = \sum_{i=1}^{N} a_i \, \gamma_{(i)} = \sum_{n=1}^{N} b_n V_n \qquad (12.104)$$

where $a_i \in (0,1)$. Inserting (12.103) into (12.104) and using the orthogonality between V_n's, one gets

$$b_n = \frac{\bar{\gamma}_c}{n} \sum_{i=1}^{n} a_i \qquad (12.105)$$

If N_r antennas with highest SNR's are selected among N antennas, then b_n may be written as follows:

$$b_n = \begin{cases} \bar{\gamma}_c & n \le N_r \\ \bar{\gamma}_c \, N_r/n & n > N_r \end{cases} \qquad (12.106)$$

Note that $N_r = 1$ corresponds to SC while $N_r = N$ implies MRC. Using independent virtual branches, the MGF of the output SNR is simply found as

$$M(s) = E\left[\exp\left(-s \sum_{n=1}^{N} b_n V_n\right)\right]$$

$$= \prod_{n=1}^{N} \frac{1}{1 + s b_n}$$

$$= \frac{1}{(1 + s\bar{\gamma}_c)^{N_r}} \prod_{n=N_r+1}^{N} \frac{1}{1 + s\bar{\gamma}_c N_r/n} \qquad (12.107)$$

The pdf and the cdf for GSC, shown in Figures 12.38 and 12.39 for $N = 4$, are found by using the inverse Laplace transform of the partial fraction expansion of (12.107). One

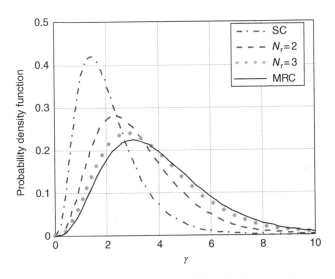

Figure 12.38 Pdf of the SNR For GSC with $N = 4$ and $\bar{\gamma}_c = 1$.

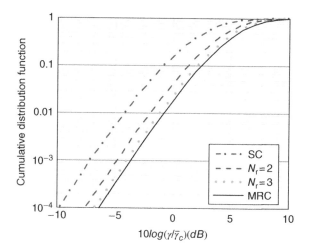

Figure 12.39   Cdf of the SNR For GSC with $N = 4$.

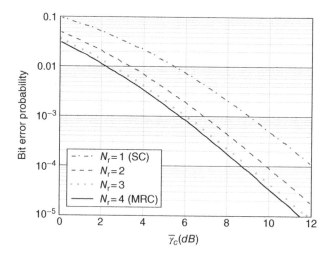

Figure 12.40   The BEP For GSC of DPSK Signals For $N = 4$.

may observe a smooth transition of pdf and cdf between SC ($N_r = 1$) and MRC ($N_r = 4$).

The BEP for DPSK signalling with GSC is obtained using the MGF (12.107):

$$P_b = \frac{1}{2} M(1)$$

$$= \frac{1}{2} \prod_{n=1}^{N} \frac{1}{1 + b_n} \qquad (12.108)$$

$$= \frac{1}{(1 + \bar{\gamma}_c)^{N_r}} \prod_{n = N_r + 1}^{N} \frac{1}{1 + \bar{\gamma}_c N_r / n}$$

Figure 12.40 shows the BEP for $N = 4$ with $N_r = 1, 2, 3, 4$. $N_r = 1$ and 4 correspond respectively to SC and MRC. The curves for $N_r = 2$ and 3 remain between those for SC and MRC, as expected. One may also observe from Figure 12.40 that SC performs 3.41 dB worse than MRC at 10^{-5} BEP.

However, selection of two or three antennas with highest SNR's decreases this performance difference to 1.18 dB and 0.31 dB, respectively. Hence, instead of connecting four

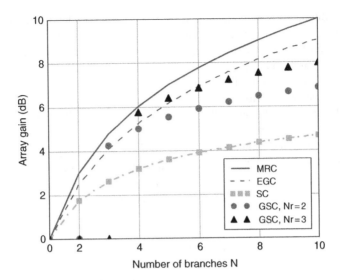

Figure 12.41 Normalized Array Factor in dB For MRC, EGC, SC and GSC Versus the Number of Diversity Branches.

antennas to four RF chains, only 1.18 dB performance degradation is suffered if two RF chains are connected to two antennas with highest SNR's. This evidently leads to significant cost savings compared to MRC with minimum performance degradation.

Using (12.106) and (12.107), the mean SNR for GSC with MRC is found as follows [9]:

$$\bar{\gamma}_{GSC} = E[\gamma_{GSC}]$$

$$= -\frac{dM(s)}{ds}|_{s=0} = \sum_{n=1}^{N} b_n$$

$$= \bar{\gamma}_c N_r \left(1 + \sum_{k=N_r+1}^{N} \frac{1}{k} \right), \bar{\gamma}_{SC} \leq \bar{\gamma}_{GSC} \leq \bar{\gamma}_{MRC}$$

$$(12.109)$$

Noting from (12.109) that that $N_r = 1$ corresponds to SC, while $N_r = N$ implies MRC (see (12.47)), this formulation includes SC and MRC as special cases. The array gains provided by MRC, EGC given by (12.98), SC and GSC are presented in Figure 12.41 as a funtion of the number of antennas. The array gain of EGC is a fraction of dB poorer than MRC but the array

gain of SC is much lower. The array gain of GSC is between SC and MRC and depends on the values of N_r and N. Nevertheless, the diversity order is not influenced negatively with this selection, that is, it is $d = N$ but not N_r (see Figure 12.40).

12.6.1.5 Square-Law Combining

MRC and EGC rely on the ability of the receiver to estimate the CSI (gain, phase and delay) in diversity branches so as to combine them coherently. Often, such a procedure is not practical due to physical separation of diversity branches, channel conditions or hardware limitations. Square-law combining (SLC) is a technique which does not require the CSI. Therefore, unlike MRC and EGC used for coherently demodulated signals, SLC is applicable to orthogonal modulations with noncoherent demodulation, for example, noncoherent orthogonal FSK.

In SLC, the received signals in diversity branches are squared and combined noncoherently, since they cannot be co-phased in the absence of the CSI. Noncoherently combined output suffers additional losses compared to coherent combining. Noncoherent combining losses, depending on the number of branches and branch SNR's, impair the diversity gain. Therefore, SLC is usually not the first choice in combining the diversity signals.

Consider a binary modulation with two orthogonal signals $s_{\ell 0}(t)$ and $s_{\ell 1}(t)$. The received signal $r_{\ell k}(t)$ at the input of the kth diversity branch may be written as

$$r_{\ell k}(t) = \alpha_k e^{-j\phi_k} s_{\ell j}(t) + n_k(t), \quad j = 0, 1$$

$$(1/2) \int_0^T s_{\ell i}(t) s_{\ell j}^*(t)\, dt = E_b\, \delta(i-j), \quad i, j = 0, 1$$

$$(12.110)$$

The optimum receiver for binary orthogonal signals consists of two parallel matched filters matched to $s_{\ell 0}(t)$ and $s_{\ell 1}(t)$. Equivalently, two parallel correlation receivers may be used, as shown in Figure 12.42. Hence, the signals at the output of the two correlators may be written as

$$Z_{ik} = \int_0^T r_{\ell k}(t) s_{\ell i}^*(t)\, dt \quad i = 0, 1 \qquad (12.111)$$

where T denotes the symbol duration. Assuming that $s_{\ell 0}(t)$ is transmitted, the decision variables for the kth branch are found by inserting $r_{\ell k}(t)$ into (12.111):

$$Z_{0k} = 2E_b \alpha_k e^{-j\phi_k} + w_{0k}, \quad \mathrm{var}(Z_{0k}) = 2E_b N_0(1 + \bar{\gamma}_c)$$

$$Z_{1k} = w_{1k}, \qquad\qquad\qquad \mathrm{var}(Z_{1k}) = 2E_b N_0$$

$$w_{ik} = \int_0^T n_k(t) s_{\ell i}^*(t)\, dt, \quad i = 0, 1$$

$$(12.112)$$

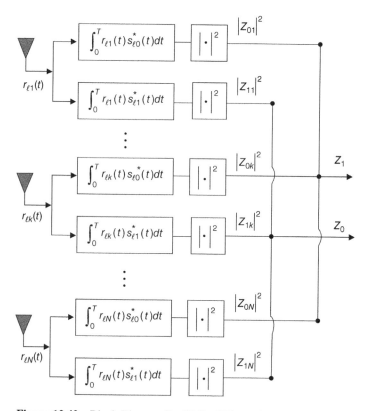

Figure 12.42   Block Diagram For SLC of Binary Orthogonal Signals.

where w_{0k} and w_{1k} are uncorrelated Gaussian random variables with zero mean and variance $E_b N_0$. Based on slow-fading assumption, $\alpha_k(t)$ and $\phi_k(t)$ are assumed to be constant over T.

Mean SNRs in diversity branches are assumed to be identical:

$$\bar{\gamma}_c = E\left[\frac{\alpha_k^2 E_b^2}{E_b N_0}\right] = E(\alpha_k^2)\frac{E_b}{N_0}, \quad k = 1, 2, \ldots, N$$

(12.113)

With receive diversity, the detector outputs of N diversity branches are square-law combined before comparing against the decision variables:

$$Z_0 = \sum_{k=1}^{N} |Z_{0k}|^2, \quad Z_1 = \sum_{k=1}^{N} |Z_{1k}|^2 \quad (12.114)$$

Both Z_0 and Z_1 are the sum of squares of complex zero-mean Gaussian random variables; they both have central chi-square distributions with $2N$ degrees of freedom (see (F.115)). Assuming that $s_{\ell 0}(t)$ was transmitted, the pdf's of SNR's associated with Z_0 and Z_1 (see (12.112)) may be written from (11.218) as

$$f_{\gamma_0}(x) = \frac{1}{(N-1)!(\bar{\gamma}_c + 1)^N} x^{N-1} \exp\left(-\frac{x}{\bar{\gamma}_c + 1}\right)$$

$$f_{\gamma_1}(x) = \frac{1}{(N-1)!} x^{N-1} \exp(-x)$$

(12.115)

Figure 12.43 shows the pdf's of the SNRs at the SLC output. One may observe that the probability of having higher SNRs increases for higher values of N. The pdf of γ_1 shifts to the right with increasing values of $\bar{\gamma}_c$, as expected.

The BEP for noncoherent orthogonal binary modulation for SLC is given by

$$P_2 = \frac{1}{2}\text{Prob}(\gamma_0 < \gamma_1 | s_{\ell 0}) + \frac{1}{2}\text{Prob}(\gamma_1 < \gamma_0 | s_{\ell 1})$$

$$= \text{Prob}(\gamma_0 < \gamma_1 | s_{\ell 0})$$

$$= \int_0^\infty f_{\gamma_0}(\gamma_0) \, d\gamma_0 \int_{\gamma_0}^\infty f_{\gamma_1}(x) \, dx$$

$$= \frac{1}{(N-1)!(\bar{\gamma}_c + 1)^N} \sum_{j=0}^{N-1} \frac{1}{j!} \int_0^\infty x^{j+N-1} e^{-\left(1 + \frac{1}{\bar{\gamma}_c + 1}\right)x} \, dx$$

$$= \frac{1}{(\bar{\gamma}_c + 2)^N} \sum_{j=0}^{N-1} \binom{j+N-1}{j} \left(\frac{\bar{\gamma}_c + 1}{\bar{\gamma}_c + 2}\right)^j$$

(12.116)

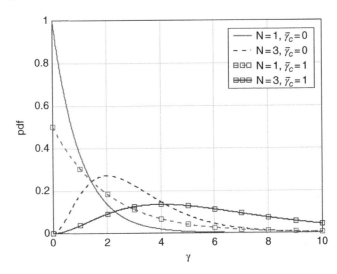

Figure 12.43 The Pdf of the SNR at the Output of a SLC with $N = 1$ and 3 Antennas For Mean Branch SNRs $\bar{\gamma}_c = 0, 1$.

where (D.48) is used to obtain the following:

$$\int_{\gamma_0}^{\infty} f_{\gamma_1}(x)\,dx = \frac{1}{(N-1)!}\int_{\gamma_0}^{\infty} x^{N-1}e^{-x}\,dx$$

$$= e^{-\gamma_0}\sum_{j=0}^{N-1}\frac{\gamma_0^j}{j!}$$

(12.117)

For $N=1$, (12.116) reduces to the BEP of noncoherent orthogonal binary FSK given by (11.176). Now let us determine the asymptotic behavior of (12.116) for large values of $\bar{\gamma}_c$. Using (D.26) and $(\bar{\gamma}_c+1)/(\bar{\gamma}_c+2)\cong 1$ for large values of the SNRs, the asymptotic BEP for square-law combined binary orthogonal signals may be written as

$$P_2 \approx \binom{2N-1}{N}\frac{1}{\bar{\gamma}_c^N}$$

(12.118)

The BEP given by (12.116) and (12.90) for respectively SLC and MRC for binary noncoherent BFSK is shown in Figure 12.44 as function of the mean branch E_b/N_0. The diversity order is N both for SLC and MRC. However,

at BEP $=10^{-4}$ level, MRC performs 0.9 dB, 1.35 and 1.6 dB better than SLC for $N=2$, 3 and 4, respectively.

12.6.2 Transmit Diversity (MISO)

MISO systems incorporate multiple transmit antennas but a single receive antenna. These systems are more suitable for applications where more space, power and processing capability is available at the transmit side compared to the receive side. In these systems, diversity order increases with the number of transmit antennas. However, the diversity gain is offset by the decrease in transmit power per antenna since the transmit power is distributed among transmit antennas. Therefore, transmit diversity systems perform worse than receive diversity systems.

Design and performance of MISO systems depend on the availability of the CSI to the transmitter. If perfect CSI is available at the transmitter, then the transmit power can be optimally allocated to antennas based on their channel gains. Under these conditions, the performance of the transmit diversity becomes

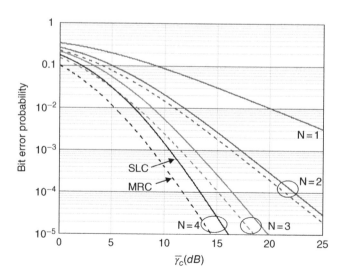

Figure 12.44 BEP For Noncoherent Orthogonal BFSK Modulation with SLC and MRC in a Rayleigh Fading Environment.

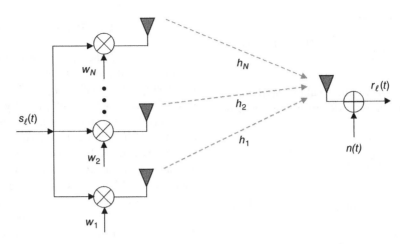

Figure 12.45   Block Diagram of a Multiple-Input Single-Output (MISO) Diversity System.

identical to that of the receive diversity. However, if CSI is not available, then a logical approach would be to distribute the transmit power equally between the transmit antennas. In this case, the power used by an antenna is wasted when that antenna suffers a deep fade. Alternatively, transmit diversity may be accomplished as a combination of space and time diversity, the so-called space-time coding. Alamouti scheme is the best known example of the transmit diversity systems exploiting space-time coding.

It is not straightforward to obtain the CSI at the transmitter. In frequency-division duplex (FDD) systems, the CSI is measured at the receiver (at downlink frequency) and then fed-back to the transmitter at the uplink frequency following a frequency-scaling, albeit with a delay. In time-division duplex (TDD) systems, the CSI can be fed-back to the transmitter more accurately because forward and reverse links are reciprocal, that is, both links operate at the same frequency.

Consider a transmit diversity system where a low-pass equivalent signal $s_\ell(t)$, with energy E_b, is transmitted by N antennas as shown in Figure 12.45. The constant transmit power is divided between the antennas with a

weight distribution defined by a weight vector, with unit norm, as defined by $\|\mathbf{w}\|^2 = \sum_{i=1}^{N} |w_i|^2 = 1$.

The received signal may be written as

$$r_\ell(t) = s_\ell(t) \sum_{i=1}^{N} w_i \alpha_i(t) e^{-j\phi_i(t)} + n(t) \quad (12.119)$$

where the AWGN noise $n(t)$ has a single-sided PSD N_0.

Case 1:   CSI Is Available at the Transmitter

The complex channel gain associated with i^{th} antenna $\alpha_i(t)e^{-j\phi_i(t)}$ is assumed to be known accurately at the transmitter. In the so-called maximal-ratio transmission (MRT), the normalized branch weights at the transmitter may be chosen as in MRC (see (12.39)) so as to maximize the received SNR:

$$w_i = \frac{\alpha_i(t)e^{j\phi_i(t)}}{\|\mathbf{w}\|} = \frac{\alpha_i(t)e^{j\phi_i(t)}}{\sqrt{\sum_{i=1}^{N} |\alpha_i(t)|^2}} \quad (12.120)$$

Inserting (12.120) into (12.119), the received signal reduces to

$$r_\ell(t) = \sqrt{\sum_{i=1}^{N} |\alpha_i(t)|^2 \, s_\ell(t)} + n(t) \qquad (12.121)$$

Using (12.121), the resulting SNR may be written as the sum of SNR's of diversity branches:

$$\gamma = \frac{\dfrac{1}{2} E\left[\left| \sqrt{\sum_{i=1}^{N} |\alpha_i(t)| 2 \, s_\ell(t)} \right|^2 \right]}{E\left[|n(t)|^2 \right]} \qquad (12.122)$$

$$= \bar{\gamma}_c \sum_{i=1}^{N} |\alpha_i(t)|^2 = \sum_{i=1}^{N} \gamma_i$$

where $\bar{\gamma}_c = E_b/N_0$. Comparison between (12.122) and (12.40) shows that transmit and receive diversity have identical performance, if CSI is available at the transmitter [10].

Case 2: No CSI Is Available at the Transmitter

If CSI is not available at the transmitter, the simplest approach may be to equally divide the transmit power between N transmit antennas:

$$w_i = 1/\sqrt{N} \qquad (12.123)$$

Inserting (12.123) into (12.119), one gets the received signal [10]

$$r_{\ell, N \times 1}(t) = \frac{s_\ell(t)}{\sqrt{N}} \sum_{k=1}^{N} h_i + n(t), \quad h_i = \alpha_i e^{-j\phi_i} \qquad (12.124)$$

where $s_\ell(t)$ is assumed to be constant over the bit duration. Since $\sum_{i=1}^{N} h_i/\sqrt{N}$ is a complex Gaussian random variable with zero mean and unity variance, (12.124) has a mean energy of E_b. Therefore, the received signal for this $N \times 1$ MISO channel has the same energy as if we had just used one transmit antenna transmitting the signal $s_\ell(t)$ of bit energy E_b:

$$r_{\ell, 1 \times 1}(t) = h_1 \, s_\ell(t) \qquad (12.125)$$

Assuming that the transmit bit energy, E_b, is equally divided between the branches, the symbol energy per branch is E_b/N. The received symbol energy, when multiplied by an array gain of N, becomes the same as that of a single transmit antenna. Since the transmit power can not be divided intelligently between transmit antennas, transmit diversity with no CSI and equal branch SNRs does not provide an SNR advantage from multiple antennas. However, the diversity order is the same as for the receive diversity, namely $d = N$ (see (12.124)), since N independent multipath signals arrive to the receiver.

12.6.2.1 Alamouti Scheme

The Alamouti transmit diversity scheme [11] is designed to provide both space and time diversity by operation during two signaling intervals. Space diversity is provided with two transmit antennas sufficiently separated from each other. Time diversity is achieved by transmitting each of the two symbols from two different antennas, assuming that the channel gain is constant over two consecutive symbol periods. It is therefore a simple space-time block code with two transmit antennas and a single receive antenna. The simplicity of the receiver is due to the orthogonality of the code. This scheme does not require any bandwith increase or CSI at the transmitter.

The Alamouti scheme, with two transmit and a single receive antenna, may be summarized as follows: The two antennas transmit s_1 and s_2 which are a pair of symbols from either a one- or a two-dimensional signal constellation, each with energy $E_s/2$. Over the first symbol period, the symbols (s_1, s_2) are transmitted simultaneously from antennas 1 and 2, respectively. Over the next symbol period, symbols $\left(-s_2^*, s_1^*\right)$ are transmitted simultaneously from antennas 1 and 2, respectively (see Figure 12.46).

The signal received (r_1, r_2) over the two signaling intervals may be written as

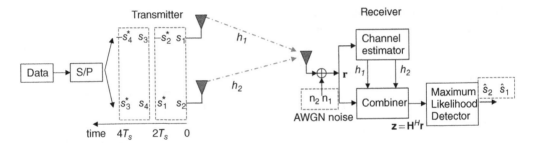

Figure 12.46　Block Diagram of the Alamouti Space-Time Block Coding Scheme.

$$\mathbf{r} = \mathbf{Hs} + \mathbf{n}$$

$$\mathbf{r} = \begin{bmatrix} r_1 \\ r_2^* \end{bmatrix}, \quad \mathbf{H} = \begin{bmatrix} h_1 & h_2 \\ h_2^* & -h_1^* \end{bmatrix}, \quad (12.126)$$

$$\mathbf{s} = \begin{bmatrix} s_1 \\ s_2 \end{bmatrix}, \quad \mathbf{n} = \begin{bmatrix} n_1 \\ n_2^* \end{bmatrix}$$

Here, (h_1, h_2) represent the complex-valued path gains, which are assumed to be zero-mean, independent complex Gaussian random variables with unit variance. In addition, they are assumed to be constant over the two symbol intervals and are assumed to be known to the receiver. The terms (n_1, n_2) represent uncorrelated AWGN terms that have zero-mean and variance N_0.

Assuming that CSI, that is, $\mathbf{H}$ matrix, is known perfectly at the receiver, a new receive vector $\mathbf{z}$ is defined as

$$\mathbf{z} = \begin{bmatrix} z_1 \\ z_2 \end{bmatrix} = \mathbf{H}^H \mathbf{r} = \mathbf{H}^H \mathbf{Hs} + \mathbf{H}^H \mathbf{n}$$

$$= \left(|h_1|^2 + |h_2|^2 \right) \mathbf{s} + \tilde{\mathbf{n}}$$

$$E\left[\tilde{\mathbf{n}} \tilde{\mathbf{n}}^H \right] = E\left[\mathbf{H}^H \mathbf{n} \mathbf{n}^H \mathbf{H} \right] = E\left[\mathbf{H}^H \mathbf{H} \right] N_0$$

$$= \left(|h_1|^2 + |h_2|^2 \right) N_0 \mathbf{I}_2$$

$$(12.127)$$

where $\mathbf{I}_2$ denotes 2×2 identity matrix. One may observe from (12.127) that transmissions of two symbols are decoupled:

$$z_i = \left(|h_1|^2 + |h_2|^2 \right) s_i + \tilde{n}_i, \quad i = 1,2 \quad (12.128)$$

Using (12.127) and (12.128), the received SNRs corresponding to z_1 and z_2 are found to be

$$\gamma_i = \frac{\left(|h_1|^2 + |h_2|^2 \right)^2 E_s / 2}{\left(|h_1|^2 + |h_2|^2 \right) N_0}$$

$$(12.129)$$

$$= \frac{1}{2} \left(|h_1|^2 + |h_2|^2 \right)^2 \frac{E_s}{N_0}, \quad i = 1,2$$

One may observe from (12.129) that the Alamouti scheme has unity array gain, hence no SNR improvement due to the use of two antennas:

$$E[\gamma_i] = \frac{1}{2} E\left[|h_1|^2 + |h_2|^2 \right] \frac{E_s}{N_0} = \frac{E_s}{N_0}, \quad i = 1,2$$

$$(12.130)$$

However, the Alamouti scheme has a diversity order $d = 2$ despite the fact that CSI is not available at the transmitter.

Now consider a 2×2 MIMO system, where the Alamouti scheme is employed at the transmitter and the outputs of the two receive antennas are maximal ratio combined. The SNR at the two signaling intervals may be written as

$$\gamma_i = \frac{1}{2} \left(|h_1|^2 + |h_2|^2 + |h_1'|^2 + |h_2'|^2 \right) \frac{E_s}{N_0}$$

$$= \frac{1}{2} \sum_{k=1}^{4} \gamma_{c,k}, \quad i = 1,2$$

$$(12.131)$$

Here (h_1', h_2') denote the complex path gains between receive antenna 2 and transmit antennas 1 and 2, respectively. Assuming that each of the four paths has the same mean SNR $\bar{\gamma}_{c,k} = \bar{\gamma}_c, k = 1, 2, \ldots, 4$, the pdf of γ is given by

$$f_{\gamma_i}(z) = \frac{1}{(N-1)!} \left(\frac{2z}{\bar{\gamma}_c}\right)^{N-1} \frac{e^{-2z/\bar{\gamma}_c}}{\bar{\gamma}_c/2}, \quad N = 4$$

(12.132)

The corresponding cdf is

$$F_{\gamma_i}(z) = 1 - e^{-2z/\bar{\gamma}_c} \sum_{k=0}^{N-1} \frac{(2z/\bar{\gamma}_c)^k}{k!}, \quad N = 4$$

(12.133)

Note that (12.132) and (12.133) are equally valid for the original Alamouti scheme when $N = 2$ is used. Noting that, except for the ½ factor, (12.129) and (12.131) denote the sum of the channel SNRs. The corresponding BEP expression for the Alamouti scheme may therefore be found from (12.54) for coherent BPSK if $\bar{\gamma}_c$ is replaced by $\bar{\gamma}_c/2$. Hence, Alamouti scheme performs 3 dB worse than the corresponding MRC, due to the non-availability of the CSI at the transmitter.

12.6.2.2 Transmit Antenna Selection

On one hand, the transmit diversity suffers losses by equally dividing the transmit power between transmit antennas, due to the non-

availability of the CSI at the transmitter. On the other hand, the diversity gain increases with increasing numbers of the transmit antennas. Therefore, transmit diversity does not perform as well as receive diversity, where transmit power is used to drive a single transmit antenna. One remedy to this problem may be to select one or a few of the N transmit antennas that provide the highest SNR at the receiver and allocate the total transmit power only to the selected antenna(s). Transmit antenna selection (TAS) is similar to GSC, where only one or a few antennas are selected for connecting to the available receivers. Noting that CSI is available at the receiver, GSC provides significant cost savings with affordable performance degradation. However, since TAS requires CSI at the transmitter, CSI or the information about the selected antenna must be provided to the transmitter via a feed-back channel (see Figure 12.47).

Assuming that the mean SNRs of all channels are $\bar{\gamma}_c$ in a Rayleigh fading channel, the cdf of the highest instantaneous SNR at the receiver due to the transmit antenna selected for transmission is given by (12.73). The total power is allocated to this antenna only. The corresponding pdf and the BEP for BPSK are given by (12.74) and (12.77), respectively.

Figure 12.48 shows a comparison of the BEP for BPSK modulation when MRT, Alamouti scheme and TAS are employed for transmit diversity. MRT has evidently the best

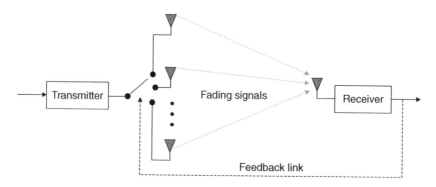

Figure 12.47   Transmit Antenna Selection.

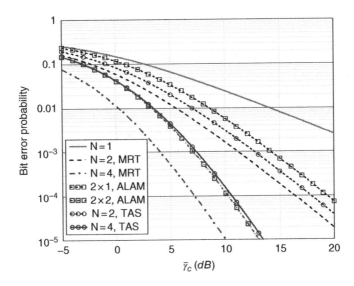

Figure 12.48   Comparison of BEP Between N-branch MRT, TAS Between 2 and 4 Antennas and 2 × 1 and 2 × 2 Alamouti Scheme For BPSK Modulation in a Rayleigh Fading Channel. Total transmit power is assumed to be the same for all values of N. Average SNR at each receive antenna is also the same.

performance, which is the same as that of MRC. On the other hand, 2 × 1 Alamouti scheme performs worse than TAS for $N = 2$ but TAS $N = 4$ antennas performes worse than 2 × 2 Alamouti scheme.

In the so-called hybrid maximal ratio transmission (hybrid MRT), N_t transmit antennas with highest channel gains out of a total of N antennas are selected for transmission. The channel gains of the selected transmit antennas (subject to power constraints) must be such that their superposition at the receiver results in maximal received SNR. Hybrid MRT requires the transmitter to know the N_t transmit antennas with highest SNR's and the relative complex-valued channel gains from each transmit antenna to the receiver. Hence, hybrid MRT needs more feedback than TAS.

References

[1] C. A. Balanis, *Antenna Theory: Analysis and Design*, John Wiley: New York, 1982.

[2] B. Friedlander and S. Scherzer, Beamforming versus transmit diversity in the downlink of a cellular communication system, *IEEE Trans Vehicular Technology*, vol. 53, no. 4, pp. 1023–1034, July 2004.

[3] C. van Rensburg, and B. Friedlander, Transmit diversity for arrays in correlated Rayleigh fading, *IEEE Trans. Vehicular Technology*, vol. 53, pp. 1726–1734, Nov. 2004.

[4] L. S. Gradsteyn and I. M. Ryzhik (7th ed.), *Table of Integrals, Series, and Products, Academic Press*: San Diego, 2007.

[5] G.L. Stuber, *Principles of Mobile Communications*, Kluwer Publications, 1996.

[6] C. Tellambura, Bounds on the distribution of a sum of correlated lognormal r.v.'s and their applications, *IEEE Trans. Comm.*, vol. 56, no. 8, pp. 1241–1248, August 2008.

[7] Q. T. Zhang, Probability of error for equal gain combiners over Rayleigh channels: some closed form solutions, *IEEE Trans. Communications*, vol. 45, no. 3, pp. 270–273, March 1997.

[8] Q. T. Zhang, A simple approach to probability of error for EGc over Rayleigh channels, *IEEE Trans. VT*, vol. 48, no. 4, pp. 1151–1154, July 1999.

[9] M. Z. Win, R. K. Mallik, G. Chrisikios and J. H. Winters, Higher order statistics of the output SNR of hybrid selection/maximal-ratio combining, *Globecomm* 2000, vol. 2, pp. 922–926.

[10] A. Goldsmith, *Wireless Communications*, Cambridge, 2005.

[11] S. M. Alamouti, A simple transmit diversity technique for wireless communications, *IEEE Journal Selected*

Areas Communications, vol. **16**, pp. 1451–1458, October 1998.

[12] L. Fang, G. Bi and A. C. Kot, New method of performance analysis for diversity reception with correlated Rayleigh-fading signals, *IEEE Trans. VT*, vol. **49**, no. 5, Sept. 2000, pp. 1807–1812.

[13] M. K. Simon and M.-S. Alouini, Digital Communication over Fading Channels (2nd ed.) John Wiley, 2005.

[14] M. Schwartz, Information Transmission, Modulation and Noise, Mc Graw Hill: New York, 1970.

Problems

1. Consider a dual MRC diversity system operating in a correlated Rayleigh fading channel. The instantaneous value of the output SNR may be written as the sum of two branch SNRs: $\gamma = \gamma_1 + \gamma_2$. The branch SNRs are assumed to be correlated with the following joint pdf:

$$f_{\gamma_1,\gamma_2}(x_1,x_2) = \frac{1}{\bar{\gamma}_c^2(1-\rho^2)} I_0\left(\frac{2\rho\sqrt{x_1 x_2}}{\bar{\gamma}_c(1-\rho^2)}\right)$$
$$\exp\left(-\frac{x_1+x_2}{\bar{\gamma}_c(1-\rho^2)}\right)$$

where ρ denotes the correlation coefficient and $\bar{\gamma}_c = E[\gamma_1] = E[\gamma_2]$ (see Problem 11.26). The MGF of MRC output SNR may then be written as

$$M_\gamma(s) = E[e^{-s\gamma}]$$
$$= \int_0^\infty \int_0^\infty e^{-s(\gamma_1+\gamma_2)} f_{\gamma_1,\gamma_2}(\gamma_1,\gamma_2)\, d\gamma_1\, d\gamma_2$$

a. Show that the MGF may be written as

$$M_\gamma(s) = \frac{(1-\rho^2)}{a^2-\rho^2}$$
$$= \frac{1}{2\rho\bar{\gamma}_c}\left[\frac{\bar{\gamma}_c(1+\rho)}{1+s\bar{\gamma}_c(1+\rho)} - \frac{\bar{\gamma}_c(1-\rho)}{1+s\bar{\gamma}_c(1-\rho)}\right]$$

where $a = 1 + s\bar{\gamma}_c(1-\rho^2)$.

b. Show that the pdf of the SNR at the output of dual-branch MRC system with correlated branch fadings may be written as (12.49):

$$f_\gamma(x) = \frac{1}{2\rho\bar{\gamma}_c} \exp\left(-\frac{x}{(1+\rho)\bar{\gamma}_c}\right)$$
$$-\frac{1}{2\rho\bar{\gamma}_c} \exp\left(-\frac{x}{(1-\rho)\bar{\gamma}_c}\right)$$

Show that the corresponding cdf is given by

$$F_\gamma(x) = 1 - \frac{1+\rho}{2\rho} \exp\left(-\frac{x}{(1+\rho)\bar{\gamma}_c}\right)$$
$$+ \frac{1-\rho}{2\rho} \exp\left(-\frac{x}{(1-\rho)\bar{\gamma}_c}\right)$$

c. Show that the cdf expression above for correlated signals reduces to that for uncorrelated signals given by (12.53)

$$F_\gamma(x) = 1 - \left(1 + \frac{x}{\bar{\gamma}_c}\right) \exp\left(-\frac{x}{\bar{\gamma}_c}\right), \quad \rho = 0$$

d. Determine the decrease in the diversity gain when the correlation coefficient between the branches is equal to 0.6, relative to the uncorrelated case, for 0.2 percent outage probability.

e. Explain how you could minimize the correlation between the branches?

2. Consider a dual time MRC diversity system with envelope correlation of the two received diversity signals $\rho(\tau) = J_0(2\pi f_d \tau)$ where τ denotes the time difference between the two signals. A MRC diversity receiver sums the the output SNRs of the two branches: $\gamma = \gamma_1 + \gamma_2$. The joint pdf of the two correlated Rayleigh fading SNRs and the pdf of γ is given in Problem 12.1.

a. Determine the average BEP for a non-coherently detected orthogonal FSK system and discuss the effect of correlation.

b. Determine the mean value of γ and the diversity gain.

3. Suppose that two-branch antenna diversity is used with SC. However, the branches have correlated fading so that the maximum diversity gain is not achieved. Let γ_1 and γ_2 be the instantaneous bit energy-to-noise ratio for each diversity branch, and $\bar{\gamma} = E[\gamma_i]$, $i = 1,2$. The joint pdf of γ_1 and γ_2 is given in Problem 12.1.

   a. Derive an expression for the cdf of the output of the selective combiner $\gamma = \max(\gamma_1, \gamma_2)$.

   b. Plot the cdf for various values of ρ and comment on the effect of correlation.

4. [12] Consider a dual-branch diversity system, where the signal received from the kth branch is given by

   $$r_k(t) = (X_k + jY_k)e^{j\phi_m(t)}$$
   $$+ w_k(t), \quad k = 1,2, \quad 0 \le t \le T$$

   where $\phi_m(t)$ denotes the modulation phase. X_k and Y_k are zero-mean Gaussian random variables with variance σ^2 and

   $$E[X_i^2] = E[Y_i^2] = \sigma^2, \quad i = 1, 2$$
   $$E[X_1 X_2] = E[Y_1 Y_2] = \rho\sigma^2$$
   $$E[X_i Y_k] = 0, \quad i,k = 1, 2$$
   $$E[w_1 w_2] = E[w_i X_k]$$
   $$= E[w_i Y_k] = 0, \quad i,k = 1, 2$$
   $$N_0 = E[w_i^2], \quad i = 1, 2$$

   a. Define two new random processes $r_3(t)$ and $r_4(t)$ as

   $$\begin{pmatrix} r_3(t) \\ r_4(t) \end{pmatrix} = \frac{1}{\sqrt{2}} \begin{bmatrix} 1 & 1 \\ -1 & 1 \end{bmatrix} \begin{pmatrix} r_1(t) \\ r_2(t) \end{pmatrix}$$
   $$= \begin{cases} (X_3 + jY_3)e^{j\phi_m(t)} + w_3(t) \\ (X_4 + jY_4)e^{j\phi_m(t)} + w_4(t) \end{cases}$$

Express X_3, X_4, Y_3, Y_4, $w_3(t)$ and $w_4(t)$ in terms of X_1, X_2, Y_1, Y_2, $w_1(t)$ and $w_2(t)$. Show that

   $$E[X_3^2] = E[Y_3^2] = \sigma^2(1+\rho)$$
   $$E[X_4^2] = E[Y_4^2] = \sigma^2(1-\rho)$$
   $$E[X_3 X_4] = E[Y_3 Y_4] = 0$$
   $$E[X_i Y_k] = 0, \quad i,k = 3, 4$$
   $$E[w_i w_k] = E[w_i X_k]$$
   $$= E[w_i Y_k] = 0, \quad i,k = 3,4$$

   Hence, the correlated random processes $r_1(t)$ and $r_2(t)$ are transformed into uncorrelated random processes $r_3(t)$ and $r_4(t)$ with the corresponding instantaneous E_b/N_0's $\gamma_3 = (1+\rho)\gamma$ and $\gamma_4 = (1-\rho)\gamma$ with $\gamma = \sigma^2/N_0$. Consequently, the analysis of a correlated dual-branch diversity combining system can be obtained by uncorrelated branch E_b/N_0's $\gamma_3 = (1+\rho)\gamma$ and $\gamma_4 = (1-\rho)\gamma$.

   b. Using the approach described in (a), derive the pdf of the output E_b/N_0 $\gamma = \max(\gamma_3, \gamma_4)$ of a dual-branch selection diversity combining system with correlated branches in Rayleigh fading.

   c. Using the approach described in (a), derive the pdf of the output E_b/N_0 $\gamma = \gamma_3 + \gamma_4$ of a dual-branch MRC system with correlated branches in Rayleigh fading and show that it is identical to (12.49).

5. An N-branch MRC diversity receiver sums the branch SNRs after weighting them by the corresponding complex conjugated channel gain. Hardware implementation of a MRC receiver is often costly because of the difficulty of estimating the channel gains. In practice, a less complex and cheaper combining technique is preferred where the branch with the highest SNR is selected (SC). A SC receiver checks all the diversity branches during each channel

coherence time and then uses only the signal of the diversity branch with the highest SNR.

a. Consider that N antennas of a SC diversity receiver suffer i.i.d. Rayleigh fading. Determine a high-SNR (as $\bar{\gamma} \to \infty$) approximation to the pdf given by (12.74). Using only the pdf's for SC and MRC, determine how many times the BEP for SC would be higher than MRC.

b. The asymptotic BEP for MRC of N BPSK signals with equal mean SNRs $\bar{\gamma}$ is given by (11.222). Starting with (12.77), determine the asymptotic BEP for SC and the required increase in $\bar{\gamma}$ for SC compared to MRC for N = 2, 3 and 4.

6. A single branch Rayleigh fading signal has a 10% probability of being 10 dB below some mean SNR threshold.

a. Determine the mean SNR of the Rayleigh fading signal as referenced to the threshold.

b. Determine the probability that the SNR of a N-branch selection diversity receiver will be 10 dB below the mean threshold for N = 2, 3, 4.

7. A system makes use of antenna selection diversity in an independently Rayleigh-fading channel. The BEP of the combined signals is required to be less than 10^{-4} for NC BFSK modulation, when the average branch SNR is equal to 10 dB.

a. Determine the number of diversity branches needed to meet this requirement.

b. Determine the BEP for a single branch and the decrease in BEP due to diversity.

c. Determine the average SNR of the combiner output and compare with that of a single branch.

d. Determine the probability that the combiner output achieves an SNR < 3 dB, and compare with the probability that any single branch achieves an SNR < 3 dB?

8. A system uses SC in an independently Rayleigh fading channel with an average branch SNR of 13 dB. The probability of all branches being received with SNR < 3 dB is equal to 10^{-4}.

a. Determine the number of diversity branches N required in the receiver to meet this requirement.

b. Determine the probability that any signal branch achieves an SNR<3 dB.

c. Derive the level crossing rate and the average fade duration for this N-branch selection diversity system at an envelope level r = R and calculate the improvement due to selection diversity.

9. Suppose that dual-branch predetection SC is used. However, the branches are not balanced such that $\bar{\gamma}_1 \neq \bar{\gamma}_2$ where $\bar{\gamma}_i$, i = 1,2 are the average received bit-energy-to-noise ratios for the two branches.

a. Determine the cdf of $\gamma = \max(\gamma_1, \gamma_2)$ and express it in terms of the average received bit-energy-to-noise ratio $\bar{\gamma}_a = (\bar{\gamma}_1 + \bar{\gamma}_2)/2$ and $\xi = \bar{\gamma}_1/\bar{\gamma}_2$.

b. Determine the pdf of γ.

c. Determine the average value of γ.

d. Calculate the effect of dual-branch SC on the BEP for binary DPSK and compare the obtained result with no diversity (single branch) with average received bit-energy-to-noise ratio $\bar{\gamma}_a$.

10. Consider dual-branch SC and MRC diversity receiver in an independent fading environment. The SNR of the first branch γ_1 is uniformly distributed between 6 and

10 (not dB) while γ_2 uniformly distributed between 5 and 20.

a. Determine the pdf and the cdf of the SNR at the combiner output.
b. Determine the average SNR at the combiner output.
c. Determine the outage probability corresponding to 10^{-5} BEP for DPSK modulation.

11. A receive-diversity system uses dual-branch SC in a Rayleigh fading channel with i.i.d. branch SNRs with mean level $\bar{\gamma}$.

a. Determine the probability of outage that occurs when the instantaneous signal-to-noise ratio γ drops 10 dB below $\bar{\gamma}$.
b. Determine the BEP at the SC output for DPSK modulation.
c. Determine how many times the BEP is reduced due to dual branch SC diversity compared with the case of no diversity for $\bar{\gamma} = 13\, dB$.
d. Determine the diversity gain at 10^{-3} BEP level.

12. Using the approach outlined in Problem 12.4a, show that the cdf of the bivariate distribution for correlated SNRs $\gamma_i, i = 1, 2$, in a Nakagami-m fading channel, may be written as

$$F(\gamma_1, \gamma_2) = P\left(m, \frac{m\gamma_1}{\bar{\gamma}_1(1+\rho)}\right) P\left(m, \frac{m\gamma_2}{\bar{\gamma}_2(1-\rho)}\right)$$

where $P(\alpha, x) = \gamma(\alpha, x)/\Gamma(\alpha) = 1 - e^{-x}$ $\sum_{k=0}^{\alpha-1} x^k/k!$ (see (D.111)), ρ denotes the field correlation coefficient and $\bar{\gamma}_i, i = 1, 2$ show the mean SNR's. Show that the above simple formula gives the same result as the following [13]

$$F(\gamma_1, \gamma_2) = \left(1 - \rho^2\right)^m \sum_{k=0}^{\infty} \binom{m-1+k}{k}$$

$$\times \rho^{2k} P\left(k+m, \frac{m\gamma_1}{\bar{\gamma}_1(1-\rho^2)}\right)$$

$$\times P\left(k+m, \frac{m\gamma_2}{\bar{\gamma}_2(1-\rho^2)}\right)$$

which is known to have convergence problems as $\gamma_1, \gamma_2 \to \infty$ and/or $\rho \to 1$.

13. Consider a N-branch SC diversity receiver, designed to operate with a constraint on the outage probability P_{out}, in an i.i.d. Rayleigh fading environment.

a. Derive an expression for the fading margin, that is, the power level below the average power level that can be tolerated by the system, in terms of P_{out} for an arbitrary value of N.
b. Determine the diversity gain for an outage probability of 1% for N = 2 and 3.

14. A receiver with an average $E_b/N_0 = 14$ dB is assumed to operate satisfactorily if the E_b/N_0 is larger than 6 dB in a Rayleigh fading channel.

a. Determine the time availability (complementary cdf) at 6 dB level, that is, the percentage of time that the system has an $E_b/N_0 \geq 6$ dB.
b. Repeat (a) for SC with N uncorrelated antennas. How many branches are required in order to obtain time availability of 99% and 99.99%?

15. Consider a dual SC diversity receiver operating in an i.i.d. Rayleigh fading channel. Assuming that the threshold level is 10 dB below the local-mean power level,

a. Determine the probability that the instantaneous branch power, with no diversity, exceeds the threshold.
b. Determine the probability at least one of the two branches will have an instantaneous power exceeding the threshold.
c. Repeat (a) and (b) when the threshold and the local-mean power levels are equal to each other.

16. Macroscopic diversity is used to mitigate the effects of shadowing. With macroscopic diversity, the signals received by a mobile station (MS) from two or more geographically separated BSs are combined using usually SC. A diversity advantage is obtained with SC if the signals experience some degree of uncorrelated shadowing. In the uplink, signals transmitted by a MS are received by two neighboring BSs with SNR's γ_1 and γ_2, respectively. Since it is usually impractical to do coherent combining across distant BSs, the BS with the largest SNR is selected to demodulate the received signal. Assuming that the channels to both BSs undergo shadowing with the same mean and shadowing variance, determine the BEP for DPSK modulation and the diversity gain at 10^{-3} BEP.

17. Assume that a mobile receiver uses dual branch SC diversity where each branch has i.i.d. fading with linear SNR uniformly distributed between 5 and 15.

a. Determine the average SNR and the average BEP for DPSK.
b. Determine the threshold level where the outage probability is 10^{-3}.

18. (see [14]) Consider the diversity receiver shown below where antipodal signals $\pm V$ are received via two paths. Because of differing attenuation (or fading) along the two paths, the signals are received as $\pm V_1$ and $\pm V_2$ by branches 1 and 2 of the diversity receiver, respectively. Each branch has

AWGN w_1 and w_2 with variances σ_1^2 and σ_2^2, and weights A_1 and A_2, respectively.

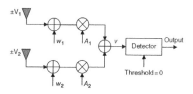

a. Show that the summed output v is a Gaussian variable of mean value $\pm(A_1 V_1 + A_2 V_2)$ and variance $A_1^2 \sigma_1^2 + A_2^2 \sigma_2^2$.
b. Show that the effective SNR is given by $(V_1 + KV_2)^2/(\sigma_1^2 + K^2\sigma_2^2)$, where $K = A_2/A_1$. $K = 0$ and $K = \infty$ correspond to the single-receiver case. Show that the diversity system improves the output SNR and error probability performance compared to a single-receiver.
c. Show that the optimum choice of the gain ratio K is given by $K_{opt} = \left(V_2\sigma_1^2/V_1\sigma_2^2\right)$, which is equivalent to setting $A_1 = kV_1/\sigma_1^2$, $A_2 = kV_2/\sigma_2^2$, where k is an arbitrary constant. The optimum combining system is hence MRC, where each receiver input is weighted by the ratio of the signal voltage to the noise variance measured at that input. Show that the output SNR is given by the sum of the SNR's of two branches, $V_1^2/\sigma_1^2 + V_2^2/\sigma_2^2$.

19. Consider a diversity system with three i.i.d. branch signals operating in a Rayleigh fading channel. At a particular symbol interval, the received signal voltages and the weights of three branches are given by $\mathbf{x}$ = [0.8 1.5 1] and $\mathbf{w}$ = [0.5 0.3 0.7], respectively. Assuming that the average noise power of each branch is equal to $N = 0.1$. Determine the maximum achievable SNR for SC, EGC (with uniform and nonuniform weighting) and MRC.

20. A receiver combines the instantaneous SNRs of N independent and identically distributed (i.i.d.) received signals $\gamma = \sum_{i=1}^{N} \gamma_i$ in a Rayleigh fading channel.

    a. Determine the p.d.f of γ.
    b. Determine the BEP for binary DPSK modulation.
    c. Find the required SNR for BER $= 10^{-6}$ with N $= 1, 2$, and 4 and compare with the SNR required in a non-fading channel.

21. Consider a dual branch MRC diversity receiver with the output SNR $\gamma = \gamma_1 + \gamma_2$. Assume that γ_1 and γ_2 are independent and undergo Rayleigh fading with the pdf's $f_{\gamma_i}(z) = (1/\bar{\gamma}_i) \exp(-z/\bar{\gamma}_i)$, $z \geq 0$, $i = 1, 2$. The average received SNRs in the two branches $\bar{\gamma}_1$ and $\bar{\gamma}_2$ are not equal to each other.

    a. Determine the MGF of γ.
    b. Determine the pdf of γ.
    c. Determine the cdf of γ.
    d. Determine the average value of γ and compare it with the case for $\bar{\gamma}_1 = \bar{\gamma}_2$. Which of the two E[$\gamma$] can be larger for a given constant transmit power?
    e. Determine the BEP for DPSK modulation at the combiner output.
    f. Determine the diversity gain for a BEP of 10^{-3} for $\bar{\gamma}_2 = \bar{\gamma}_1 / 2$ compared to a non-diversity system with average SNR $\bar{\gamma} = (\bar{\gamma}_1 + \bar{\gamma}_2)/2$.
    g. Determine the performance degradation due to unequal average branch powers if $\bar{\gamma}_2 = \bar{\gamma}_1 / 2$ for a BEP of 10^{-3}. Degradation is defined as the difference in dB between the average SNRs for $\bar{\gamma}_1 = \bar{\gamma}_2$ and $(\bar{\gamma}_1 + \bar{\gamma}_2)/2$.

22. Consider a Rayleigh-fading channel with N-branch MRC diversity combining with the same average branch SNRs. Assume

M-ary QAM modulation where the symbol error probability is given by (8.92):

$$P_b = \frac{4}{\log_2 M} \left(1 - \frac{1}{\sqrt{M}}\right) Q\left(\sqrt{\frac{3\log_2 M}{(M-1)}\gamma}\right)$$

where γ denotes the average E_b/N_0. For the sake of simplicity, use the single-term approximation given by (B.7) for the Q function

    a. Determine the average BEP for an arbitrary value of N.
    b. Determine the required average branch SNR for N $= 1, 2, 3$ and 4 corresponding to 10^{-4} average BEP level.
    c. Compute the diversity gain at 10^{-4} average BEP level for N $= 1, 2, 3$ and 4.

23. Consider a wireless channel with three paths that suffer Nakagami-m fading with mean powers proportional to 1, 0.5, 0.25, and fading figures 4, 2 and 1, respectively.

    a. Determine the average BEP for a receiver that uses MRC and DPSK.
    b. Determine the diversity order. Which path is dominant in determining the diversity order?

24. A system using BPSK modulation achieves 10^{-5} BEP in a Rayleigh fading channel without diversity. Estimate the order of MRC diversity required to achieve the same BEP if the transmit power level is required to be 30 dB lower than the non-diversity system. Assume that the transmit and receive antenna gains, the frequency of operation and the range are the same for both cases.

25. [5] Consider N-branch diversity reception of binary DPSK signals in an i.i.d. Rayleigh flat-fading channel. The mutually uncorrelated Gaussian noise in diversity branches has one-sided PSD N_0 (W/Hz).

a. Assume that a separate differential detector is used on each diversity branch to obtain N independent estimates for each transmitted bit. Then, the majority logic is used to combine the N estimates to decode the transmitted bit. Determine the BEP in terms of N and the average received bit energy-to-noise ratio $\bar{\gamma}$ per branch.

b. Determine the BEP if the receiver uses N-branch post-detection MRC and compare the obtained result with the result in part (a).

c. For $N = 3$ and $\bar{\gamma} = 17$ dB, compare the BEP for parts (a) and (b).

26. Consider a diversity reception system which employs MRC combining the SNR's of N receivers. Each of the N receivers is equipped with M antennas and employs selection combining between the M antennas. The channel is assumed to undergo Rayleigh fading with the same mean SNR $\bar{\gamma}$ for each channel.

a. Derive the cdf and the pdf of SNR at the output of a receiver with M antennas. Hint: you may use binomial expansion.

b. Derive the MGF of the output of each receiver for an arbitrary value of M.

c. Derive the BEP at the output of the MRC combiner for DPSK modulation for arbitrary values of M and N.

d. Derive the SNR at the MRC output for arbitrary values of M and N.

e. Plot the BEP for $N = M = 1$, $N = M = 2$, $M = 1, N = 4$ and $M = 4, N = 1$. Compare the BEP performances for $\bar{\gamma} = 10$ dB.

f. Determine the diversity gain for $N = M = 2$ and compare with that for $M = 1$, $N = 4$ and $M = 4, N = 1$ at 10^{-3} BEP.

27. Consider a N-branch MRC diversity system with i.i.d. binary coherent PSK signals with 13 dB mean SNR level in a Rayleigh fading channel.

a. Determine the BEP for transmission with no diversity.

b. Determine the minimum number of diversity branches required to obtain a BEP lower than 10^{-5}.

c. Calculate the BEP in an AWGN channel for the same mean SNR.

d. Determine the increase in SNR in dB for a dual MRC system compared to that for an AWGN channel.

e. Outage and BEP may be related to each other by assuming that BEP = 1/2 in outage and BEP = 0 when the system is available. Based on this approximation, express BEP in terms of the outage probability and compare the obtained result for a threshold level of 13 dB below the mean SNR with the exact BEP value.

28. Consider a dual-branch MRC system, employing DPSK modulation, with two uncorrelated branches in a Nakagami-m fading channel with fading figures m_1 and m_2.

a. Derive an analytic expression for the average BEP as a function of the mean branch SNRs $\bar{\gamma}_1$ and $\bar{\gamma}_2$ and fading figures m_1 and m_2.

b. Assuming that $\bar{\gamma}_1 = \bar{\gamma}_2 = 13$, $m_1 = 1$ and $m_2 = 3$, determine the BEP and compare with that for AWGN channel.

29. Consider an N-branch MRC diversity combining system with equal mean branch SNRs $\bar{\gamma}$. The modulation is assumed to be coherent BPSK.

a. Express the BEP P_b analytically.

b. Analytical determination of the integral involving the Q function may not often be easy and instead approximations are used. Use the following approximation

$$Q\left(\sqrt{2\gamma}\right) = \frac{w_1}{2}e^{-b\gamma} + \frac{w_2}{2}e^{-2b\gamma}$$

where $w_1 = 0.3017$, $w_2 = 0.4389$, and $b = 1.0510$ (see (B.7)) to determine the BEP analytically. Determine the diversity order and the diversity gain in dB for $N=2$ and $P_b=10^{-3}$.

30. If γ denotes the sum of SNRs of N Nakagami fading signals with the same average SNRs $\bar{\gamma}$ and fading figures m, show that the pdf of γ is again Nakagami distributed. Determine the average SNR and the fading figure of the combined signal. Determine the BEP for binary DPSK. Determine the diversity gain for a BER level of 10^{-5}, $N=5$ and $m=1$.

31. Derive the pdf and the cdf corresponding to the MGF given by (12.107), using partial fraction expansion.

32. Derive the marginal pdf of SNR for GSC for $N=3$, $N_r=2$ using (12.107) and (F.88). Show that they are identical.

33. Derive (12.109) using (12.107) and (F.88). Show that the results are identical.

34. GSC may be used instead of MRC in order to reduce the cost and complexity in MRC. Consider a GSC system using DPSK modulation with four antennas. Determine the gain in mean SNR over SC and the loss compared to MRC for $N_r=2$ and 3 at 10^{-4} BEP.

35. Voltage levels received by each branch of a dual-branch diversity receiver for seven symbol durations are given below:

	T	2 T	3 T	4 T	5 T	6 T	7 T
Branch 1	0.9	−0.93	−0.83	0.89	0.89	−0.88	0.92
Branch 2	0.8	−0.81	−0.89	0.93	0.88	−0.95	0.95

a. Determine the combiner output for SC and SSC. The threshold level for SSC is chosen as 0.85 V.
b. Assuming that the above values are co-phased, using (12.39) and (12.93), determine the SNR at the combiner output for EGC and MRC.

36. A communication system employs dual antenna diversity and noncoherent orthogonal BFSK modulation. The received signals at the two antennas is

$$r_i(t)=\alpha_i\,s(t)+w_i(t),\quad i=1,2$$

where α_1 and α_2 are statistically i.i.d. Rayleigh random variables with unity variance and $w_1(t)$ and $w_2(t)$ are statistically independent, zero-mean white Gaussian random processes with PSD $N_0/2$ (W/Hz). The two signals are demodulated, squared and then summed prior to detection.

a. Determine the BEP and compare the diversity order with the case of no diversity.
b. Determine the diversity gain at 10^{-3} BEP and compare with those for SC and MRC.

37. Consider a transmit diversity (MISO) system with N transmit antennas and a single receive antenna in an uncorrelated Rayleigh fading channel. The SNR's of the signals incident on the receive antenna are assumed to have identical mean powers, that is,

$$f_{\gamma_i}(z)=\frac{1}{\bar{\gamma}}e^{-z/\bar{\gamma}},\quad i=1,2,\cdots,N$$

a. Determine the cumulative distribution function (cdf) of each of the Rayleigh-faded signals
b. Now assume that the receives provides the CSI to the receiver so that the transmitter chooses only one of the transmit antennas, with the highest SNR, to transmit a certain number of consecutive symbols. The number of consecutive symbols (block size) to be transmitted by the chosen antenna is determined by the channel coherence time. Determine the cdf and the pdf of the received signal as a result of this selection process.

c. Determine the decrease in outage probability due to this TAS system compared to a SISO system at a threshold level of 10 dB below the average power level. Determine this decrease for N = 2 and 3.
d. Determine the mean SNR at the receiver output for N = 2 and compare this result with the output SNR of a MISO SC diversity system.

38. Consider a transmit diversity system with N antennas in a Rayleigh fading channel. The total transmit power is P.

a. Assume that the CSI is not available at the transmitter and the transmitter equally divides its power P between N transmit antennas. Determine the pdf of the received SNR γ at the output of a

MRC receiver, assuming that N signals undergo i.i.d. fading.
b. The pdf of the SNR of a Nakagami-m fading random variable is given by the Gamma distribution

$$f_P(\gamma) = \frac{1}{\Gamma(m)} \left(\frac{m\gamma}{P_0}\right)^m \frac{\exp(-m\gamma/P_0)}{\gamma}, \quad \gamma \geq 0$$

$$P_0 = E[\gamma]$$

$$m = \frac{P_0^2}{E\left[(\gamma - P_0)^2\right]} \geq \frac{1}{2}$$

where m denotes the fading figure and $\Gamma(m) = (m-1)!$ when m is an integer. Comparing the result that you obtained in part (a) with the above, how can you interpret the Gamma distribution? Comment on the meaning of P_0 and m.

13

MIMO Systems

MIMO systems improve the performance of communication systems by employing multiple antennas at both transmitter and receiver. In MIMO systems the fading is exploited as a channel resource for possible performance improvement. Multiple antenas both at transmit and receive sides provide space (antenna) diversity. The diversity order is defined as the number of paths through which a transmitted signal reaches the receiver, hence is equal to the product of the number of transmit and receive antennas. Increase in the diversity order leads to a significant increase in diversity gain and improvement in the bit error probability performance. On the other hand, the mean SNR at the receiver output also increases; this implies that either the transmit power may be reduced or the range (coverage area) of the telecommunication system may be increased. MIMO systems may also be used in order is to increase the transmission rate since they may also be considered to provide multiple spatial channels for message transmission between transmit and receive antennas. In virtue of the so-called multiplexing gain, the channel capacity (spectral efficiency) may be increased

significantly. Performance improvements in MIMO systems are achieved at the expense of computational complexity while maintaining the primary communication resources (that is, total transmit power and channel bandwidth) fixed. Space-time coding, that is, the data is transmitted over multiple antennas by coding symbols in time and space, and beamforming (adaptive antennas) are widely used applications of MIMO systems.

Transmit antennas may be used either to transmit the same or different symbols during a signaling interval. Transmit power may be divideed between transmit antennas with different weights. Different approaches may be used to detect the transmitted symbols by receive antennas and to combine them.

The performance of MIMO systems is closely related to the channel characteristics and the spacings between antennas. A channel rich in scattering provides more multipath channels which are mostly uncorrelated from each other. Otherwise, the received signals may be correlated with each other. Similarly, channel gains of multipath signals tend to become more correlated with each other as

Digital Communications, First Edition. Mehmet Şafak.
© 2017 John Wiley & Sons Ltd. Published 2017 by John Wiley & Sons Ltd.
Companion website: www.wiley.com/go/safak/Digital_Communications

tansmit and/or receive antennas become more closely spaced. In summary, the correlation between channel gains, which may be due to either antenna spacings or the channel itself, may play a significant role in the MIMO system performance.

13.1   Channel Classification

Channels with multiple antennas at both ends, may be classified as MIMO, multiple-access (MAC), cooperative and broadcast channels, depending on the level of cooperation between antennas at the transmitter and/or the receiver [1].

Let us consider a typical example of a 2×2 MIMO system, as shown in Figure 13.1. In this case, information flow is from two wireless nodes to another two wireless nodes. In the *MIMO channel*, the code applied at the transmit side is simultaneously calculated and applied to the two antennas. Similarly, at the receive side, the decoding can utilize the signals from both receive antennas simultaneously. In the *cooperative channel*, the two transmit nodes do not have prior information about each other's data, and the receive nodes do not know about each other's receive waveform. The information at the transmit or receive sides must be exchanged over a wireless medium, which may be potentially unreliable. The two transmitting nodes are allowed to listen to each

other's transmissions and make decisions based on whatever they might infer (transmit side cooperation). Similarly, the two receiving users are allowed also to exchange information (receive side cooperation). If the link between the two transmit antennas of a standard MIMO channel is removed, then one gets a two-user *MAC channel*, where two single-antenna users communicate with a BS that has two antennas. If we remove the reliable links at both transmit and receive sides then we get a *cooperative channel*. Despite the exchange of information on the wireless links between transmitters and receivers, the users have to share the channel between them. In a *broadcast channel*, there is perfect cooperation between transmit antennas, but no cooperation at the receive side.

The multiplexing gain provided by the MIMO systems is critically dependent on coordination among both transmit and receive antennas. This level of coordination cannot be achieved by the wireless connections available to cooperative communication. For MIMO, MAC and broadcast channels, the multiplexing gain is equal to 2, implying data transmission over two separate channels, while it is equal to unity for the cooperative channel.

In this chapter, we will be concerned only with MIMO channels, where there is a perfect coordination between transmit antennas on the one side, and the receive antennas on the other side.

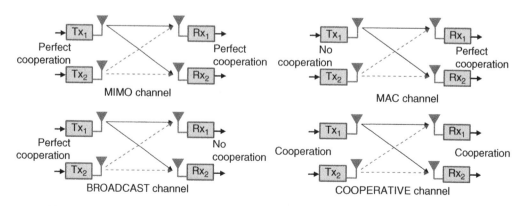

Figure 13.1   Classification of Channels with Multiple Transmit and Receive Antennas. [1]

13.2 MIMO Channels with Arbitrary Number of Transmit and Receive Antennas

Consider a MIMO system with N_t transmit and N_r receive antennas in a Rayleigh fading channel as shown in Figure 13.2. The signal vector **y** received by N_r antennas may be written as

$$\mathbf{y} = \mathbf{Hx} + \mathbf{n}$$

$$\mathbf{H} = \begin{pmatrix} h_{11} & h_{12} & \cdots & h_{1N_t} \\ h_{21} & h_{22} & \cdots & h_{2N_t} \\ \vdots & \vdots & \ddots & \vdots \\ h_{N_r1} & h_{N_r2} & \cdots & h_{N_rN_t} \end{pmatrix} \quad (13.1)$$

$$\mathbf{y} = \begin{pmatrix} y_1 & y_2 & \cdots & y_{N_r} \end{pmatrix}^T$$

$$\mathbf{x} = \begin{pmatrix} x_1 & x_2 & \cdots & x_{N_t} \end{pmatrix}^T$$

$$\mathbf{n} = \begin{pmatrix} n_1 & n_2 & \cdots & n_{N_r} \end{pmatrix}^T$$

where **x** denotes the signals transmitted by N_t transmit antennas and **n** is the AWGN of N_r receiver. The h_{ij} component of the $N_r \times N_t$ channel matrix **H** denotes the channel gain between j^{th} transmit and i^{th} receive antenna and is a complex Gaussian random variable. The magnitude of $h_{ij} = \alpha_{ij} e^{-j\phi_{ij}}$ is Rayleigh distributed with unity variance and its phase is uniformly distributed in $[0, 2\pi]$. Figure 13.3 shows typical temporal variations of the channel gains h_{11}, h_{21} and h_{31} of a 3×3 MIMO system in an uncorrelated Rayleigh fading channel. Since the channel gains are uncorrelated, the their fades are not likely to coincide with each other.

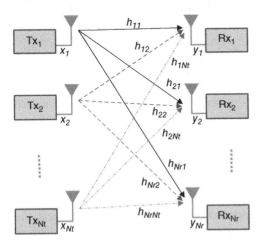

Figure 13.2 A Multiple-Input Multiple-Output (MIMO) System.

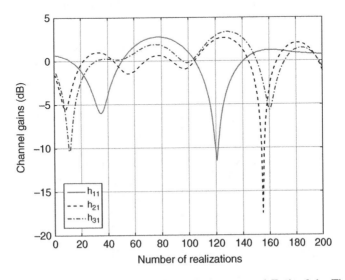

Figure 13.3 Channel Gains Between the First Transmit Antenna and Each of the Three Uncorrelated Receive Antennas in a 3×3 MIMO System in a Rayleigh Fading Channel. [2]

The receiver, having the perfect CSI, multiplies the received vector $\mathbf{y}$ by $\mathbf{H}^H$ in order to obtain the $N_t \times 1$ vector $\mathbf{r}$:

$$\mathbf{r} = \mathbf{H}^H \mathbf{y} = \mathbf{H}^H \mathbf{H} \mathbf{x} + \tilde{\mathbf{n}} \tag{13.2}$$

The $N_t \times 1$ noise vector $\tilde{\mathbf{n}}$ has zero mean and is defined by its covariance matrix:

$$\tilde{\mathbf{n}} = \mathbf{H}^H \mathbf{n}$$

$$E\left[\tilde{\mathbf{n}}\tilde{\mathbf{n}}^H\right] = E\left[\mathbf{H}^H \mathbf{n}\mathbf{n}^H \mathbf{H}\right] = \mathbf{H}^H \mathbf{R}_n \mathbf{H} = \sigma^2 E\left[\mathbf{H}^H \mathbf{H}\right]$$

$$\mathbf{R}_n = E[\mathbf{n}\mathbf{n}^H] = \sigma^2 \, \mathbf{I}_{N_r}$$

$$\tag{13.3}$$

where $\mathbf{I}_{N_r}$ denotes $N_r \times N_r$ identity matrix.

As discussed in Appendix E, the $N_r \times N_t$ channel matrix $\mathbf{H}$ and the $N_t \times N_t$ random Wishart matrix $\mathbf{H}^H \mathbf{H}$ can be expressed in terms of N_r row vectors $\mathbf{t}_i$ of dimension $1 \times N_t$:

$$\mathbf{H} = \begin{pmatrix} h_{11} & h_{12} & \cdots & h_{1N_t} \\ h_{21} & h_{22} & \cdots & h_{2N_t} \\ \vdots & \vdots & \ddots & \vdots \\ h_{N_r 1} & h_{N_r 2} & \cdots & h_{N_r N_t} \end{pmatrix} = \begin{pmatrix} \mathbf{t}_1 \\ \vdots \\ \mathbf{t}_i \\ \vdots \\ \mathbf{t}_{N_r} \end{pmatrix}$$

$$\mathbf{H}^H \mathbf{H} = \sum_{i=1}^{N_r} \mathbf{t}_i^H \mathbf{t}_i$$

$$= \begin{pmatrix} \sum_{k=1}^{N_r} |h_{k1}|^2 & \sum_{k=1}^{N_r} h_{k1}^* h_{k2} & \cdots & \sum_{k=1}^{N_r} h_{k1}^* h_{kN_t} \\ \sum_{k=1}^{N_r} h_{k2}^* h_{k1} & \sum_{k=1}^{N_r} |h_{k2}|^2 & \cdots & \sum_{k=1}^{N_r} h_{k2}^* h_{kN_t} \\ \vdots & \vdots & \ddots & \vdots \\ \sum_{k=1}^{N_r} h_{kN_t}^* h_{k1} & \sum_{k=1}^{N_r} h_{kN_t}^* h_{k2} & \cdots & \sum_{k=1}^{N_r} |h_{kN_t}|^2 \end{pmatrix}$$

$$\mathbf{t}_i = \left(h_{i1} \quad h_{i2} \quad \cdots \quad h_{iN_t} \right), \quad i = 1, 2, \cdots, N_r \tag{13.4}$$

where $\mathbf{t}_i$ denotes the ith row of the channel matrix $\mathbf{H}$. The row vectors $\mathbf{t}_i$ are assumed to be independent of each other but the elements of each row vector may be correlated with each other. The correlation between the elements of $\mathbf{t}_i$ is defined by the $N_t \times N_t$ covariance matrix $\boldsymbol{\Omega}$:

$$\boldsymbol{\Omega} = E\left[\mathbf{t}_i^H \mathbf{t}_i\right], \quad i = 1, 2, \cdots, N_r$$

$$E\left[\mathbf{H}^H \mathbf{H}\right] = E\left[\sum_{i=1}^{N_r} \mathbf{t}_i^H \mathbf{t}_i\right] = N_r \boldsymbol{\Omega} \tag{13.5}$$

The behaviour of the MIMO channel is closely related to the correlation between the channel gains, which are assumed to be complex Gaussian with zero-mean and unity variance. The signals at the transmit side can be correlated if the adjacent transmit antennas are separated from each by less than approximately half wavelength. Hence, at each of the N_r receive antennas, $\boldsymbol{\Omega}$ provides a measure of the correlations between signals transmitted by N_t transmit antennas. In other words, (13.5) accounts for the correlations between transmit antennas. If the transmit antennas are separated sufficiently far from each other, the MIMO channel is uncorrelated and the covariance matrix $\boldsymbol{\Omega}$ becomes an $N_t \times N_t$ identity matrix:

$$E\left[\mathbf{H}^H \mathbf{H}\right] = E\left[\sum_{i=1}^{N_r} \mathbf{t}_j^H \, \mathbf{t}_j\right] \tag{13.6}$$

$$= N_r \, \mathbf{I}_{N_t} \quad uncorrelated \ case$$

In the asymptotic case where N_t and N_r are sufficiently large, the diagonal terms in $\mathbf{H}^H \mathbf{H}$ vanish due to averaging. This implies that (13.6) has rank N_t, that is, N_t identical non-zero eigenvalues. Consequently, in view of (13.2), N_t transmitted signals x_i may be resolved at the receiver:

$$r_i = x_i \sum_{k=1}^{N_r} |h_{ki}|^2 + \tilde{n}_i, \quad i = 1, 2, \cdots, N_t \tag{13.7}$$

The transmitted symbol x_i may then be detected by standard techniques, for example,

ML detection. For the special case of BPSK modulation, the detector will decide for binary symbol 1 for $r_i > 0$ and for binary symbol 0 for $r_i < 0$. The sum of N_r channel power gains in (13.7) will obviously make the detection process more reliable since the use of N_r receive antennas leads to a diversity order of N_r. Using (13.3) and (13.7), the resulting SNR may be written as

$$\gamma_i = \frac{E\left[|x_i|^2\right]\left[\sum_{k=1}^{N_r}|h_{ki}|^2\right]^2}{E\left[|\tilde{n}_i|^2\right]} = \frac{E\left[|x_i|^2\right]\left[\sum_{k=1}^{N_r}|h_{ki}|^2\right]^2}{\sigma^2\sum_{k=1}^{N_r}|h_{ki}|^2}$$

$$= \bar{\gamma}\sum_{k=1}^{N_r}|h_{ki}|^2, \quad i = 1, 2, \cdots, N_t$$

$$\text{(13.8)}$$

where $\bar{\gamma} = E/\sigma^2$. It is important to observe that γ_i in (13.8) represents the MRC of the SNRs of N_r receive antennas when only i^{th} transmit antenna is transmitting. The array gain is therefore given by N_r.

For perfect correlation between channel gains, all elements of $E[\mathbf{H}^H\mathbf{H}]$ will be equal to N_r:

$$E\left[\mathbf{H}^H\mathbf{H}\right] = E\left[\sum_{j=1}^{N_r}\mathbf{t}_j^H\mathbf{t}_j\right]$$

$$= N_r\,\mathbf{1}_{N_t} \quad \text{perfectly correlated case}$$

$$\text{(13.9)}$$

where $\mathbf{1}_{N_t}$ denotes a $N_t \times N_t$ matrix with all elements equal to unity. Hence, $E[\mathbf{H}^H\mathbf{H}]$ will have unity rank and a single non-zero eigenvalue. In this case, the channel behaves like a SISO channel with average SNR equal to N_tN_r.

In some other applications, instantaneous channel correlation matrix $\mathbf{HH}^H$ of dimension $N_r \times N_r$ characterizes the MIMO channel. Then, the channel matrix $\mathbf{H}$ and the Wishart matrix $\mathbf{HH}^H$ can be expressed in terms of the $N_r \times 1$ column vectors denoted by $\mathbf{h}_i$ (see Appendix E):

$$\mathbf{H} = \begin{pmatrix} h_{11} & h_{12} & \cdots & h_{1N_t} \\ h_{21} & h_{22} & \cdots & h_{2N_t} \\ \vdots & \vdots & \ddots & \vdots \\ h_{N_r1} & h_{N_r2} & \cdots & h_{N_rN_t} \end{pmatrix} = \left(\mathbf{h}_1 \cdots \mathbf{h}_j \cdots \mathbf{h}_{N_t}\right)$$

$$\mathbf{HH}^H = \sum_{i=1}^{N_t}\mathbf{h}_i\mathbf{h}_i^H$$

$$= \begin{pmatrix} \sum_{k=1}^{N_t}|h_{1k}|^2 & \sum_{k=1}^{N_t}h_{k1}^*h_{k2} & \cdots & \sum_{k=1}^{N_t}h_{k1}^*h_{kN_t} \\ \sum_{k=1}^{N_t}h_{k2}^*h_{k1} & \sum_{k=1}^{N_t}|h_{2k}|^2 & \cdots & \sum_{k=1}^{N_t}h_{k2}^*h_{kN_t} \\ \vdots & \vdots & \ddots & \vdots \\ \sum_{k=1}^{N_t}h_{kN_t}^*h_{k1} & \sum_{k=1}^{N_t}h_{kN_t}^*h_{k2} & \cdots & \sum_{k=1}^{N_t}|h_{kN_t}|^2 \end{pmatrix}$$

$$\mathbf{h}_j = \left(h_{1j} \quad h_{2j} \quad \cdots \quad h_{N_rj}\right)^T, \quad j = 1, 2, \cdots, N_t$$

$$\text{(13.10)}$$

The column vectors $\mathbf{h}_i$ are assumed to be independent of each other but the elements of each column vector may be correlated with each other. The correlation between the elements of $\mathbf{h}_i$ is defined by the $N_r \times N_r$ covariance matrix:

$$\mathbf{\Sigma} = E\left[\mathbf{h}_i\,\mathbf{h}_i^H\right], \quad i = 1, 2, \cdots, N_t$$

$$E\left[\mathbf{HH}^H\right] = E\left[\sum_{i=1}^{N_t}\mathbf{h}_i\,\mathbf{h}_i^H\right] = N_t\mathbf{\Sigma}$$

$$\text{(13.11)}$$

Hence, $\mathbf{\Sigma}$ measures the correlations between the signals received by N_r antennas. In other words, (13.11) accounts for the correlations at the receive side of the MIMO system when the receive antennas are separated from each other by less than approximately half wavelength. If the receive antennas are separated sufficiently far from each other, then the covariance matrix $\mathbf{\Sigma}$ becomes an identity matrix; then the MIMO channel is said to be uncorrelated.

In contrast with the eigenvalues of Σ and Ω, which are deterministic, $\mathbf{H}^H\mathbf{H}$ and $\mathbf{HH}^H$ are characterized by random eigenvalues for an arbitrary correlation and arbitrary values of N_t and N_r. To account for the cases where N_t may be larger or smaller than N_r, let m and n denote, respectively, the maximum and minimum number of antennas at the two sides of a MIMO channel:

$$m = \max(N_t, N_r)$$
$$n = \min(N_t, N_r) \qquad (13.12)$$

Since the $\mathbf{H}$ matrix is a $N_r \times N_t$ matrix, the matrix $\mathbf{HH}^H$ is $N_r \times N_r$, while $\mathbf{H}^H\mathbf{H}$ is $N_t \times N_t$. In view of (E.3), $\mathbf{HH}^H$ is a $W_{N_r}(N_t, \Sigma)$ central Wishart matrix while $\mathbf{H}^H\mathbf{H}$ is a $W_{N_t}(N_r, \Omega)$ central pseudo Wishart matrix. The N_t eigenvalues of $\mathbf{H}^H\mathbf{H}$ may be used to represent this $N_r \times N_t$ MIMO system. [3] On the other other hand, $N_r \times N_r$ matrix $\mathbf{HH}^H$ has N_r eigenvalues. The n nonzero eigenvalues of $\mathbf{HH}^H$ and $\mathbf{H}^H\mathbf{H}$ are identical to each other. The additional $|N_t - N_r| = m - n$ eigenvalues are identically equal to zero. We therefore can make use of well-known properties of Wishart matrices to characterize the MIMO channels.

Since the Wishart matrices $\mathbf{HH}^H$ and $\mathbf{H}^H\mathbf{H}$ have the same non-zero eigenvalues, one may study the $n \times m$ Wishart matrix $\mathbf{HH}^H$, where the $n \times n$ covariance matrix $\Sigma \neq \mathbf{I}$ defines the correlation between the elements of columns of $\mathbf{H}$, that is, the correlation between signals received by n receive antennas for each of the transmit antennas. The covariance matrix can be expressed as a function of the distance between receive antennas. When transmit antennas are sufficiently close to each other, correlations between signals transmitted by transmit antennas can be expressed by the covariance matrix $\Omega \neq \mathbf{I}$ as a function of distance between them, that is, using the rows of the channel matrix $\mathbf{H}$.

The case for partial correlation between the entries of $\mathbf{H}^H\mathbf{H}$, is usually studied in four different categories: Neither transmit nor receive antennas are correlated among each other, only

the transmit antennas are correlated with each other, only the receive antennas are correlated with each other, and both transmit and receive antennas are correlated among each other. The correlations between the signals of receive antennas may account for the correlations induced by the channel and the correlations between receive antennas. In channels with sufficiently rich scattering, the channel-induced correlations may be ignored. Hereafter, we will focus our attention in the four cases listed below:

a. Uncorrelated case ($\Sigma = \mathbf{I}$, $\Omega = \mathbf{I}$): neither the transmit nor the receive antennas are correlated
b. Correlation on receive side but transmit antennas are uncorrelated ($\Sigma \neq \mathbf{I}$, $\Omega = \mathbf{I}$).
c. Correlation on transmit side but receive antennas are uncorrelated ($\Sigma = \mathbf{I}$, $\Omega \neq \mathbf{I}$)
d. Correlation on both transmit and receive sides ($\Sigma \neq \mathbf{I}$, $\Omega \neq \mathbf{I}$).

Example 13.1 Eigenvalues of a 2×2 MIMO Channel.

The channel matrix $\mathbf{H}$, the Wishart matrix $\mathbf{H}^H\mathbf{H}$ and the corresponding column and row vectors for a 2×2 MIMO system, shown in Figure 13.4, are given by

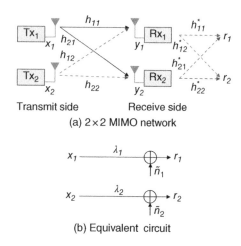

(a) 2×2 MIMO network

(b) Equivalent circuit

Figure 13.4 2×2 MIMO System.

$$\mathbf{H} = \begin{pmatrix} h_{11} & h_{12} \\ h_{21} & h_{22} \end{pmatrix} = (\mathbf{h}_1 \;\; \mathbf{h}_2) = \begin{pmatrix} \mathbf{t}_1 \\ \mathbf{t}_2 \end{pmatrix}$$

$$\mathbf{H}^H\mathbf{H} = \begin{pmatrix} |h_{11}|^2 + |h_{21}|^2 & h_{12}h_{11}^* + h_{22}h_{21}^* \\ h_{12}^*h_{11} + h_{22}^*h_{21} & |h_{12}|^2 + |h_{22}|^2 \end{pmatrix}$$

$$\mathbf{h}_i = (h_{1i} \;\; h_{2i})^T, \quad i = 1,2$$

$$\mathbf{t}_j = (h_{j1} \;\; h_{j2}), \quad j = 1,2$$

(13.13)

From (13.2), we can write the received signal in terms of the ordered eigenvalues $0 \le \lambda_1 \le \lambda_2 \le \infty$ of (13.13) (see (E.27))

$$\begin{pmatrix} r_1 \\ r_2 \end{pmatrix} = \begin{pmatrix} \lambda_1 & 0 \\ 0 & \lambda_2 \end{pmatrix} \begin{pmatrix} x_1 \\ x_2 \end{pmatrix} + \begin{pmatrix} \tilde{n}_1 \\ \tilde{n}_2 \end{pmatrix}$$

(13.14)

As the Wishart matrix has random components, the eigenvalues are also random. Hence, this 2×2 MIMO system has two eigenmodes, each represented by the eigenvalues λ_1 and λ_2. These eigenmodes imply that the link between first transmit and receive antennas is uncoupled from the link between second transmit and second receive antennas (see Figure 13.4b). However, the coupling between the two links is absorbed in the eigenvalues through their dependence on h_{12} and h_{21}.

To clarify the difference between transmit and receive correlations, we write below the covariance matrices Σ and Ω:

$$\Sigma = E\left[\mathbf{h}_i\mathbf{h}_i^H\right] = E\begin{bmatrix} |h_{1i}|^2 & h_{1i}h_{2i}^* \\ h_{2i}h_{1i}^* & |h_{2i}|^2 \end{bmatrix}, \quad i = 1,2$$

$$\Omega = E\left[\mathbf{t}_j^H\mathbf{t}_j\right] = E\begin{bmatrix} |h_{j1}|^2 & h_{j1}^*h_{j2} \\ h_{j2}^*h_{j1} & |h_{j2}|^2 \end{bmatrix}, \quad j = 1,2$$

(13.15)

It is clear from (13.15) and Figure 13.5 that the terms $E[h_{1i}h_{2i}^*]$, $i = 1,2$ in Σ measure the correlation between the channel gains for receive antennas 1 and 2 for the transmissions

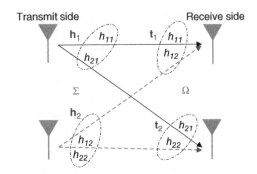

Figure 13.5 Correlations Between Antennas at Transmit and Receive Sides of a 2×2 MIMO System.

of ith transmit antenna. This means that the correlation between these terms may be decreased by increasing the spacing between the two receive antennas. This is based on the assumption that the correlation between the received signals will be less as the two propagation paths diverge from each other. Similarly, the terms $E\left[h_{j1}^*h_{j2}\right]$, $j = 1,2$ in Ω measure the correlation between the channel gains of transmit antennas 1 and 2 for the jth receive antenna. The correlations between received signals can similarly be controlled by adjusting the spacing between transmit antennas.

13.3 Eigenvalues of the Random Wishart Matrix $\mathbf{H}^H\mathbf{H}$

Since the channel gains in (13.4) are zero-mean complex Gaussian random variables with unity variance, $\mathbf{H}^H\mathbf{H}$ is a random Wishart matrix. Consequently, its eigenvalues are also random. Here we will firstly consider the case where channel gains are uncorrelated with each other and secondly the case where transmit and/or receive signals are arbitrarily correlated with each other. One may note that the correlation between received signals may arise from the correlation at the transmitter, in the channel and/or the receiver.

In uncorrelated Rayleigh-fading MIMO channels, the entries of $\mathbf{H}$ are i.i.d. complex Gaussian r.v.'s with zero-mean and unity variance. When receiving antennas are correlated with each other, the columns of $\mathbf{H}$, denoted by $\mathbf{h}_j$, are independent random vectors, but the elements of each column are correlated with each other. For Rayleigh fading, this implies $E[\mathbf{h}_j] = 0$ and the covariance matrix is given by $E[\mathbf{h}_j \mathbf{h}_j^H]$, $j = 1, 2, \ldots, N_t$. (see (13.11)). When transmit antennas are correlated with each other, the rows of $\mathbf{H}$ are independent, but the elements of each row are correlated with each other; the covariance matrix is given by (13.5) [3]–[5].

13.3.1   Uncorrelated Central Wishart Distribution

The joint pdf of m ordered eigenvalues of $\mathbf{H}^H\mathbf{H}$ in an uncorrelated MIMO channel, with zero-mean path gains, is given by (see (E.18)) [3][4][5][6]:

$$f_\mathbf{A}(\lambda_1, \lambda_2, \cdots, \lambda_n)$$

$$= \frac{\exp\left(-\sum_{i=1}^{n}\lambda_i\right) \prod_{i=1}^{n}\lambda_i^{m-n} \prod_{i<j}(\lambda_i-\lambda_j)^2}{\prod_{i=1}^{n}(m-i)! \prod_{i=1}^{n}(n-i)!}$$

$$\mathbf{A} = (\lambda_1 \ \ \lambda_2 \ \ \cdots \ \ \lambda_n) \ \ 0 \le \lambda_1 < \lambda_2 < \cdots < \lambda_n < \infty$$
$$\text{(13.16)}$$

The marginal pdfs are given below for the following special cases:

a. $n = 1, m \ge 1$:

$$f_{\lambda_1}(\lambda) = \frac{1}{(m-1)!}\lambda^{m-1} \ e^{-\lambda} \qquad (13.17)$$

b. $n = 2, m \ge 2$:

$$f(\lambda_1, \lambda_2)$$
$$= \frac{1}{(m-1)!(m-2)!}e^{-\lambda_1-\lambda_2} \ \lambda_1^{m-2}\lambda_2^{m-2} \ (\lambda_1-\lambda_2)^2$$

$$f_{\lambda_2}(\lambda) = \int_0^\lambda f(\lambda_1,\lambda) \ d\lambda_1$$

$$= \frac{1}{(m-1)!(m-2)!}\lambda^{m-2}e^{-\lambda}$$

$$\left[\gamma(m+1,\lambda) - 2\lambda\gamma(m,\lambda) + \lambda^2\gamma(m-1,\lambda)\right]$$

$$f_{\lambda_1}(\lambda) = \int_\lambda^\infty f(\lambda,\lambda_2) \ d\lambda_2$$

$$= \frac{1}{(m-1)!(m-2)!}\lambda^{m-2}e^{-\lambda}$$

$$\left[\Gamma(m+1,\lambda) - 2\lambda\Gamma(m,\lambda) + \lambda^2\Gamma(m-1,\lambda)\right]$$
$$\text{(13.18)}$$

where $\gamma(k, x)$ and $\Gamma(k, x)$ denote respectively lower and upper incomplete Gamma functions:

$$\gamma(k,x) \triangleq \int_0^x t^{k-1}e^{-t}dt = (k-1)!\left[1-e^{-x}\sum_{i=0}^{k-1}\frac{x^i}{i!}\right]$$

$$\Gamma(k,x) \triangleq \int_x^\infty t^{k-1}e^{-t}dt = (k-1)! \ e^{-x}\sum_{i=0}^{k-1}\frac{x^i}{i!}$$
$$\text{(13.19)}$$

c. $n = m = 2$:

$$f_{\lambda_1}(\lambda) = 2e^{-2\lambda}$$
$$f_{\lambda_2}(\lambda) = e^{-\lambda}\left(\lambda^2 - 2\lambda + 2\right) - 2e^{-2\lambda} \qquad (13.20)$$

d. $n = 2, m = 3$:

$$f_{\lambda_1}(\lambda) = e^{-2\lambda}\left(3\lambda + \lambda^2\right)$$
$$f_{\lambda_2}(\lambda) = e^{-\lambda}\left(3\lambda - 2\lambda^2 + \lambda^3/2\right) - e^{-2\lambda}\left(3\lambda + \lambda^2\right)$$
$$\text{(13.21)}$$

e. $n = m = 3$:

$$f_{\lambda_1}(\lambda) = 3 \ e^{-3\lambda}$$
$$f_{\lambda_2}(\lambda) = \frac{1}{2}e^{-2\lambda}\left(12 - 12\lambda + 6\lambda^2 + 2\lambda^3 + \lambda^4\right) - 6e^{-3\lambda}$$
$$f_{\lambda_3}(\lambda) = \frac{1}{4}e^{-\lambda}\left(12 - 24\lambda + 24\lambda^2 - 8\lambda^3 + \lambda^4\right)$$
$$- \frac{1}{2}e^{-2\lambda}\left(12 - 12\lambda + 6\lambda^2 + 2\lambda^3 + \lambda^4\right) + 3e^{-3\lambda}$$
$$\text{(13.22)}$$

Figure 13.6 shows the pdf's of the largest eigenvalue of some MIMO systems based on the random matrix theory, as given by (13.16) – (13.22). Pdf's of all the eigenvalues of 2×2, 2×3 and 3×3 MIMO systems are shown in Figure 13.7.

Tables 13.1 and 13.2 denote respectively the mean values and the variances of the eigenvalues of various MIMO systems. Although the sum of the means of eigenvalues are equal to mn, the largest eigenvalue is dominant, that is, it has the largest mean and variance.

The cdf of the maximum eigenvalue for an uncorrelated $n \times m$ MIMO channel is given by: [8][9][10]

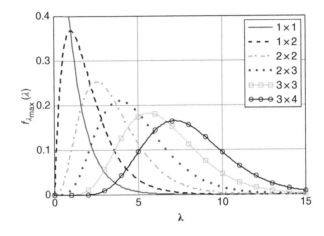

Figure 13.6 The Pdf of the Largest Eigenvalue in Uncorrelated MIMO Channels.

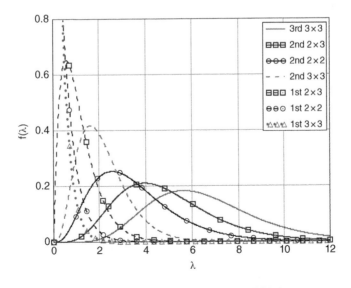

Figure 13.7 Pdf of the Eigenvalues in a MIMO System.

Table 13.1 Mean Values of the Ordered Eigenvalues in a MIMO System. [7]

MIMO	$E[\lambda_4]$	$E[\lambda_3]$	$E[\lambda_2]$	$E[\lambda_1]$	$\sum_{i=1}^{n} E[\lambda_i] = nm$
1×1				1	1
1×2				2	2
2×2			3.5	0.5	4
2×3			4.875	1.125	6
2×4			6.1875	1.8125	8
3×3		6.5208	2.1458	0.3333	9
4×4	9.7723	4.4086	1.5692	0.25	16

Table 13.2 Variances of the Ordered Eigenvalues in a MIMO System. [7]

MIMO System	var $[\lambda_4]$	var $[\lambda_3]$	var $[\lambda_2]$	var $[\lambda_1]$
1×1				1
1×2				2
2×2			3.25	0.25
2×3			4.36	0.61
2×4			5.4023	1.0273
3×3		5.5135	1.1385	0.1111
4×4	7.6392	2.2442	0.5964	0.0625

$$F_{n \times m}(\lambda) = \frac{|\mathbf{A}(\lambda)|}{\prod_{k=1}^{n} \Gamma(m-k+1)\Gamma(n-k+1)}$$

(13.23)

where the elements of the $n \times n$ matrix $\mathbf{A}$ are defined by

$$\mathbf{A}(\lambda)_{i,j} = \gamma(m-n+i+j-1,\lambda), \quad i,j = 1,2,\cdots n$$

(13.24)

The cdf for several combinations of n and m are given by [11]

$$F_{1 \times m}(\lambda) = \gamma(m,\lambda)/(m-1)!$$

$$= 1 - e^{-\lambda} \sum_{k=0}^{m-1} \lambda^k / k!, \quad m = 1,2,\cdots$$

$$F_{2 \times m}(\lambda)$$

$$= \frac{\gamma(m-1,\lambda)\gamma(m+1,\lambda) - \gamma(m,\lambda)^2}{\Gamma(m)\Gamma(m-1)}, \quad m = 1,2,\cdots$$

$$F_{2 \times 2}(\lambda) = 1 - e^{-\lambda}(2 + \lambda^2) + e^{-2\lambda}$$

$$F_{2 \times 3}(\lambda) = 1 - \frac{1}{2} e^{-\lambda}(4 + 4\lambda - \lambda^2 + \lambda^3)$$

$$+ \frac{1}{2} e^{-2\lambda}(2 + 4\lambda + \lambda^2)$$

$$F_{3 \times 3}(\lambda) = 1 - \frac{1}{4} e^{-\lambda}(12 + 12\lambda^2 - 4\lambda^3 + \lambda^4)$$

$$+ \frac{1}{4} e^{-2\lambda}(12 + 12\lambda^2 + 4\lambda^3 + \lambda^4) - e^{-3\lambda}$$

(13.25)

One may observe from (13.17) and (13.25) that the maximum eigenvalue of $1 \times m$ (SIMO and MISO) systems are described by central chi-square distribution (see (F.115)).

Figure 13.8 shows the cdf's of all the eigenvalues of 2×2, 2×3 and 3×3 MIMO systems based on the random matrix theory. One may easily observe that the largest eigenvalue has the highest mean, and the best outage performance.

13.3.2 Correlated Central Wishart Distribution

Correlation effects in MIMO systems are of utmost importance because of their impact

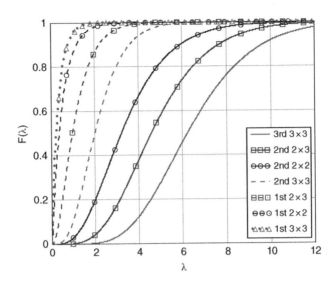

Figure 13.8   Cdf of the Eigenvalues of 2×2, 2×3 and 3×3 MIMO Systems For the Uncorrelated Case.

on the system performance. Correlation may arise either due to close spacing between transceiver antennas and/or the channel itself. For certain applications, it may not be possible and/or practical to locate transceiver antennas sufficiently far from each other. Spacing of transmit and receive antennas are mainly dictated by the size limitations at the transceiver and depend also on the frequency of operation. As the frequency of operation increases, the correlation effects become less significant since, for a given physical antenna spacing, the electrical distance between them increases. On the other hand, the channel may also cause correlation between the propagating signals, depending mainly on the richness of the scattering environment. Signals scattered by the same or closely-spaced scatterers are expected to undergo correlated fading. However, signals propagating through a rich scattering environment undergo uncorrelated fading.

Figure 13.9 shows the effect of correlation on the channel gains between the 2nd receive antenna and each of the 3 correlated transmit antennas of a 3×3 MIMO system operating in a Rayleigh fading channel. The correlation

coefficient between the channel gains is assumed to be 0.8. [2] The diversity aims to exploit channel gains even if some of the channels fade. However, one may observe from Figure 13.9 that the correlated channel gains suffer similar fading effects with time. Such strong correlation between signals prevents achieving the desired diversity gain. The formulation outlined above for the uncorrelated case does not apply for the correlated case.

13.3.2.1   Double-Sided Correlation

Consider a MIMO system where N_t transmit and N_r receive antennas are assumed to be correlated among each other at both transmit and receive sides. In agreement with (13.12) we have $m = \max(N_t, N_r)$ and $n = \min(N_t, N_r)$ and $\tau = m - n$. Hermitian positive definite complex covarance matrices $\Sigma \in C^{m \times m}$ and $\Omega \in C^{n \times n}$ account for the correlation between the group of m and n antennas, respectively. The elements of the covariance matrices Σ and Ω are modelled as $\alpha_{ij} = \alpha^{|i-j|}$ ve $\rho_{ij} = \rho^{|i-j|}$ between i^{th} and j^{th} antennas, where α and

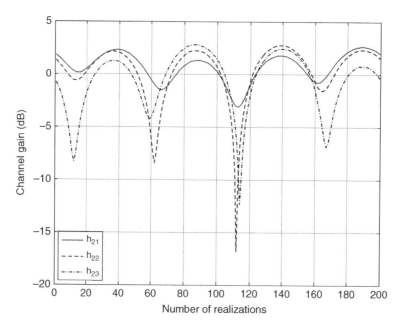

Figure 13.9 Channel Gains Between the Second Receive Antenna and Each of the Three Correlated Transmit Antennas of a 3×3 MIMO System Operating in a Rayleigh Fading Channel. Correlation coefficient between transmit antennas is 0.8. [2]

ρ denote, respectively, the correlation coefficients between the neighboring antennas, and they both vary between 0 and 1:

$$
\Sigma =
\begin{pmatrix}
1 & \alpha & \cdots & \alpha^{m-1} \\
\alpha & 1 & \cdots & \alpha^{m-2} \\
\vdots & \vdots & \ddots & \\
\alpha^{m-1} & \alpha^{m-2} & \cdots & 1
\end{pmatrix},
$$

$$
\Omega =
\begin{pmatrix}
1 & \rho & \cdots & \rho^{n-1} \\
\rho & 1 & \cdots & \rho^{n-2} \\
\vdots & \vdots & \ddots & \\
\rho^{n-1} & \rho^{n-2} & \cdots & 1
\end{pmatrix}
$$

(13.26)

The ordered eigenvalues of the covariance matrices Σ and Ω are denoted by $\sigma_1 < \sigma_2 < \ldots < \sigma_m$ and $\omega_1 < \omega_2 < \ldots < \omega_n$, respectively. For example, for the special case of $m = 3$ and

$n = 2$, the ordered eigenvalues of Σ and Ω are found as follows (see (12.52)):

$$
\sigma_1 = 1 + \frac{\alpha^2}{2} - \frac{\alpha}{2}\sqrt{\alpha^2 + 8}
$$

$$
\sigma_2 = 1 - \alpha^2
$$

$$
\sigma_3 = 1 + \frac{\alpha^2}{2} + \frac{\alpha}{2}\sqrt{\alpha^2 + 8}
$$

(13.27)

$$
w_1 = 1 - \rho
$$

$$
w_2 = 1 + \rho
$$

For the uncorrelated case, that is, when the correlation coefficients α and ρ go to zero, the covariance matrices Σ and Ω will be diagonal and have full rank, that is, all eigenvalues will be identically equal to unity. For the special case of $m = 3$ and $n = 2$, one gets $\sigma_1 = \sigma_2 = \sigma_3 = 1$ and $w_1 = w_2 = 1$ from (13.27), as α and ρ go to zero. On the other hand, as the correlation coefficient α approaches unity, then the corresponding covariance matrix Σ will have unity rank and only a single non-zero eigenvalue; $\sigma_m = m$ and $\sigma_i = 0$, $i = 1, 2, \ldots, m-1$. Similarly, one has $w_n = n$ and $w_i = 0$, $i = 1, 2, \ldots, n-1$ as ρ approaches unity (see (13.27)). Note that the sum of $m(n)$ eigenvalues is always equal to $m(n)$, irrespective of the value of the correlation coefficient.

We will mostly be interested in the maximum eigenvalue since it is generally sufficient to determine the performance of MIMO systems. The marginal cdf of the maximum eigenvalue of a central Wishart matrix with double-sided correlation is given by: [12][13]

$$F_{\lambda_{max}}(x)$$
$$= \frac{(-1)^n \Gamma_n(n) \det(\Omega)^{n-1} \det(\Sigma)^{m-1}}{\Delta_n(\Omega) \Delta_m(\Sigma)(-x)^{n(n-1)/2}}$$
$$\det(\Psi(x)) \tag{13.28}$$

where

$$(\Psi(x))_{i,j}$$
$$= \begin{cases} 1/\sigma_j^{m-i} & i \leq \tau \\ e^{-\frac{x}{w_{i-\tau}\sigma_j}} \frac{1}{(m-1)!} \gamma\left(m; -\frac{x}{w_{i-\tau}\sigma_j}\right) & i > \tau \end{cases}$$

$$\Gamma_m(n) = \prod_{i=1}^m \Gamma(n-i+1)$$

$$\Delta_m(\Sigma) = \prod_{i<j}^m (\sigma_j - \sigma_i)$$

$$\Delta_n(\Omega) = \prod_{i<j}^n (w_j - w_i) \tag{13.29}$$

where $\tau = m - n$ and $\gamma(k, x)$ denotes the lower incomplete Gamma function defined by (13.19).

For the special case for $n = m = 2$, the cdf simplifies to [12]

$$F_{\lambda_{max}}(x) = \frac{w_1 w_2 \sigma_1 \sigma_2}{(\sigma_2 - \sigma_1)(w_2 - w_1)x}$$
$$\sum_{i=2}^2 (-1)^i \prod_{j=1}^2 \left[\exp\left(-\frac{x}{w_{|i-j|+1}\sigma_j}\right) + \frac{x}{w_{|i-j|+1}\sigma_j} - 1 \right] \tag{13.30}$$

where $\sigma_1 = 1-\alpha$, $\sigma_2 = 1+\alpha$, $w_1 = 1-\rho$, $w_2 = 1+\rho$.

The pdf of the maximum eigenvalue of a double-sided correlated MIMO system with with N_t transmit and N_r receive antennas is found by taking the derivative of (13.28): [12]

$$f_{\lambda_{max}}(x) = \frac{(-1)^{n+1} \Gamma_n(n) \det(\Omega)^{n-1} \det(\Sigma)^{m-1}}{\Delta_n(\Omega) \Delta_m(\Sigma)(-x)^{n(n-1)/2}}$$
$$\left\{ \frac{n(n-1)\det(\Psi(x))}{2x} + \sum_{\ell=\tau+1}^m \det(\Psi_\ell(x)) \right\} \tag{13.31}$$

The $m \times m$ matrix $\Psi_\ell(x)$ is defined by

$$(\Psi_\ell(x))_{i,j}$$
$$= \begin{cases} (\Psi(x))_{i,j} & i \neq \ell \\ \frac{e^{-\frac{x}{w_{i-\tau}\sigma_j}}}{w_{i-\tau}\sigma_j \ (m-2)!} \gamma\left(m-1; -\frac{x}{w_{i-\tau}\sigma_j}\right) & i = \ell \end{cases} \tag{13.32}$$

One may observe from (13.31) that, for $n = m$ ($N_t = N_r$), there is a symmetry between correlations at transmit and receive sides, that is, the pdf of the largest eigenvalue is the same when the correlation coefficients α and ρ are interchanged.

13.3.2.2 Correlation at the Side with the Smallest Number of Antennas

Consider a MIMO system where the antennas at the side with the smallest number of antennas are correlated while those at the side with the largest number of antennas are assumed to be uncorrelated. In this case, Σ will be a diagonal matrix with m eigenvalues being equal to unity, $\sigma_i = 1$, $i = 1, 2, \ldots, m$. However, the covariance matrix Ω will have n ordered eigenvalues, $\omega_1 < \omega_2 < \ldots < \omega_n$, which depend on the correlation coefficient ρ.

The marginal pdf of the largest eigenvalue λ_{max} is given by [3][5]

$$f_{\lambda_{max}}(x) = \frac{1}{\Delta_n(\Omega) \prod\limits_{i=1}^{n} \left[(m-i)! \, w_i^m \right]}$$

$$\sum_{k=1}^{n} \sum_{\ell=1}^{n} (-1)^{k+\ell} x^{m-n+k-1} e^{-x/w_\ell} |\Lambda|$$

$$(13.33)$$

which is a special case of (13.31). Here $|\Lambda|$ shows the determinant of the $(n-1) \times (n-1)$ matrix Λ with elements $\Lambda_{i,j}$:

$$\Lambda_{i,j} = \left(w_{r_{j,\ell}} \right)^{m-n+r_{i,k}} \gamma\left(m-n+r_{i,k}, x/w_{r_{j,\ell}} \right),$$

$$i, j = 1, 2, \cdots, n-1$$

$$r_{i,j} \triangleq \begin{cases} i & i < j \\ i+1 & i \geq j \end{cases}$$

$$(13.34)$$

Figure 13.10 shows the pdf of the largest eigenvalue of a 3×3 MIMO channel with single-sided correlation, that is, three antennas are correlated with each other on one side but there is no correlation between the three antennas on the other side. The mean value and the variance of the largest eigenvalue were observed to increase with increasing values of the correlation coefficient. Note that, for the extreme case of $\rho = 1$, the other two eigenvalues disappear and mean value of the maximum eigenvalue becomes equal to $mn = 9$.

Figure 13.11 shows the effect of perfect correlation between only m or n antennas. When n (smallest number of) antennas are correlated with each other, the pdf shifts towards higher SNR values, implying better outage and BEP performance compared to the case where m (largest number of) antennas only are perfectly correlated with each other. In both cases, the MIMO may be considered acting as a SIMO where perfectly correlated antennas behave as a single transmitter yielding a mean SNR value, which is equal to the number of perfectly correlated antennas, whereas the uncorrelated antennas may be considered to perform MRC at the receiver; hence their number determines the diversity order. Therefore, if n antennas are

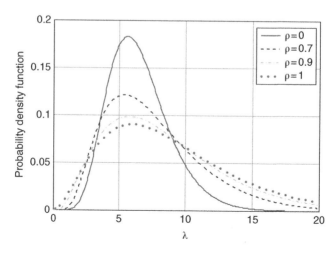

Figure 13.10 The Pdf the Largest Eigenvalue of a Semi-Correlated 3×3 MIMO Channel For Various Values of the Correlation Coefficient.

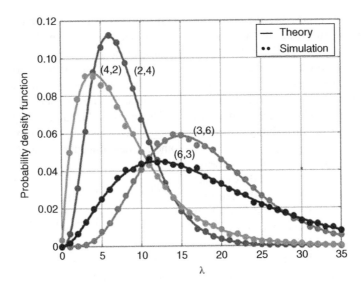

Figure 13.11 Pdf of the Largest Eigenvalue in Semi-Correlated Rayleigh Fading MIMO Channels. (2,4) and (3,6) denotes that antennas at the MIMO side with smallest number of antennas are fully correlated. However, (4,2) and (6,3) denote that antennas at the MIMO side with largest number of antennas are fully correlated. [2]

perfectly correlated with each other, then each of the m uncorrelated receive antennas receives a Rayleigh-faded (n perfectly correlated) signals with a mean value $x_0 = n$ (see (12.45)), and the outputs of the m antennas are maximal-ratio combined hence a mean output SNR of $mx_0 = nm$. The pdf of the power sum of n signals received by each of the m uncorrelated antennas at the receive side, acting as a MISO system with perfect CSI, is described by (12.42) with a diversity order $N = m$ and mean value of $\bar{\gamma}_c = n$; the output SNR is equal to $N\bar{\gamma}_c = nm$ [14]. On the other hand, if m antennas are perfectly correlated with each other, then $x_0 = m$, the diversity order equal to n, and the output SNR $nx_0 = nm$. Figure 13.11 shows that the SNR corresponding to the largest eigenvalue is more likely to take lower values when m antennas are perfectly correlated but n antennas are uncorrelated with each other.

If n antennas are perfectly correlated but m antennas are uncorrelated with each other, inserting $\gamma(k,x) = (k-1)!$ and $w_\ell = 0$, $\ell = 1$,

$2...,n-1$, $w_n = n$ into (13.33), one may easily show that the pdf of the largest eigenvalue reduces to the Gamma pdf with mean value $m_{\lambda_{max}} = mn$ and standard deviation $\sigma_{\lambda_{max}} = n\sqrt{m}$. Similarly, when m antennas are perfectly correlated but n antennas are uncorrelated with each other, the pdf of the largest eigenvalue has the Gamma pdf with mean value $m_{\lambda_{max}} = mn$ and the standard deviation $\sigma_{\lambda_{max}} = m\sqrt{n}$: [13]

$$f_{\lambda_{max}}(\lambda) = \begin{cases} \dfrac{1}{(m-1)!}\dfrac{\lambda^{m-1}}{n^m}e^{-\lambda/n} & \begin{array}{l} n \text{ antennas} \\ perfectly\ correlated \end{array} \\[2em] \dfrac{1}{(n-1)!}\dfrac{\lambda^{n-1}}{m^n}e^{-\lambda/m} & \begin{array}{l} m \text{ antennas} \\ perfectly\ correlated \end{array} \end{cases}$$

$$(13.35)$$

Note that the output SNR for the cases where MIMO is perfectly correlated on one side is the same as the SNR of a $1 \times mn$ MRC system. [7]

It is evident from (13.35) that, the mean output SNRs are identical irrespective of whether m or/and n antennas are perfectly correlated with each other. However, the diversity order of the maximum eigenvalue is higher (is equal to m) when only n antennas are perfectly correlated with each other. This implies improved outage and bit error probability performance. [7]

The fact that the sum of eigenvalues is always equal to $N_tN_r = mn$ may be explained by the fact that the receiver employs MRC and the transmitter uses MRT. This result is identical to that of MRC in a SIMO system since the above result should be divided by N_t when the power transmitted by N_t antennas is the same as that of a single antenna. However, the two cases differ from each other since the mean branch SNRs in MIMO systems are not equal to each other, in contrast with SIMO system, where the channels have the same branch SNRs.

Since the maximum array gain, mn, is already achieved when only one side is fully correlated, correlation at both sides does not help for further increasing the array gain. In addition, the diversity order was observed to be higher when the

correlation is at the side with the smallest number of antennas. This implies that, because of higher diversity order, improved outage and bit error probability performances are achieved when the correlation is "only" at the side with the smallest number of antennas.

The mean value of the largest eigenvalue $m_{\lambda_{max}}$, when n antennas are correlated with each other, is determined by using (13.28):

$$m_{\lambda_{max}} = E[\lambda_{max}] = \int_0^\infty \lambda\, f_{\lambda_{max}}(\lambda)\, d\lambda$$
$$= \int_0^\infty [1 - F_{\lambda_{max}}(\lambda)]\, d\lambda \qquad (13.36)$$

where the last expression is obtained from the first via integration by parts. Expected value of the largest eigenvalue is shown in Figure 13.12 for the case where n antennas are correlated but m antennas are uncorrelated with each other. For $\rho = 0$, array gain is minimum since $w_i = 1$, $i = 1,\ldots, n$. However, as ρ approaches unity, w_i, $i = 1, 2,\ldots, n - 1$ vanish and σ_n reaches its maximum value of n. The array gain increases monotonously with increasing values of ρ. At $\rho = 1$, the mean value

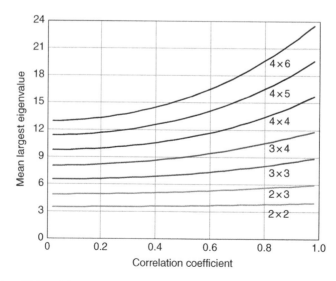

Figure 13.12 Mean Value of the Largest Eigenvalue (Not in dB) Versus Correlation Coefficient Between n Antennas in a $n \times m$ MIMO system. m antennas are assumed to be uncorrelated.

$m_{\lambda_{max}}$ and the standard deviation $\sigma_{\lambda_{max}}$ of the largest eigenvalue of a (m,n) MIMO system is equal to mn and $n\sqrt{m}$, respectively. The increase in array gain at an arbitrary value of ρ is defined as the ratio of the array gain at ρ to its value at $\rho = 0$. Note that the array gain for $\rho = 1$ is equal to mn and the difference between the array gains at $\rho = 0$ and $\rho = 1$ increases as the MIMO size gets larger.

Figure 13.13 shows the variation of the array gain when both sides of a 3×5 MIMO system is correlated. Analytical and simulation results are observed to be in perfect agreement with each other. Increased correlation at either side

causes an increase in the array gain. However, the increase of the correlation coefficient at the side with the highest number of antennas is more effective in increasing the array gain, albeit at the expense of decreased diversity order.

Now consider the special case of a MIMO system with $n = 2$, and $m \ge 2$, where $n = 2$ antennas are correlated with each other while m antennas are uncorrelated. The ordered eigenvalues of the 2×2 covariance matrix are given by $w_1 = 1 - \rho$, $w_2 = 1 + \rho$, while $\sigma_1 = \sigma_2 = \ldots = \sigma_m = 1$ Using (13.33), the pdf of the largest eigenvalue $\lambda_{max} = \lambda_2$ is found as

$$f_{\lambda_{max}}(x) = K_{\Sigma}\, x^{m-2} \left[\begin{array}{l} e^{-x/w_1}\, w_2^m\, \gamma(m, x/w_2) - e^{-x/w_2}\, w_1^m\, \gamma(m, x/w_1) \\ -x\, e^{-x/w_1}\, w_2^{m-1}\, \gamma(m-1, x/w_2) + x\, e^{-x/w_2}\, w_1^{m-1}\, \gamma(m-1, x/w_1) \end{array} \right] \tag{13.37}$$

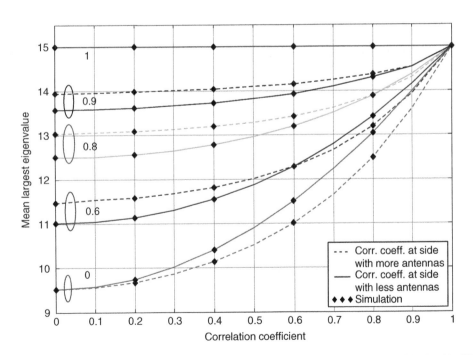

Figure 13.13 Mean Value of the Largest Eigenvalue in a 3×5 MIMO system with Double-Sided Correlation. [2]

where $\gamma(k, x)$ is defined by (13.19) and

$$K_\Sigma = \left[2\rho(m-1)!(m-2)!\left(1-\rho^2\right)^{m-1} \right]^{-1}$$

(13.38)

Figures 13.14 and 13.15 show the effect of correlation on the pdf's of maximum and minimum eigenvalues for $N_t = 2$ and $N_r = 2,4$, where $N_t = 2$ antennas are assumed to be correlated, while N_r receive antennas suffer uncorrelated

Rayleigh fading. Note that increased correlation between transmit antennas forces the eigenvalue (SNR) to assume smaller values.

13.4   A 2×2 MIMO Channel

Consider a 2×2 MIMO system in a Rayleigh fading channel as shown in Figure 13.4. The signals received may be written from (13.1) – (13.3) as follows:

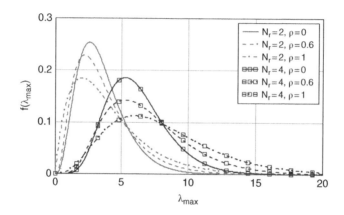

Figure 13.14   The Pdf of the Maximum Eigenvalue For $N_t = 2$. The correlation is on the side with the smallest number of antennas.

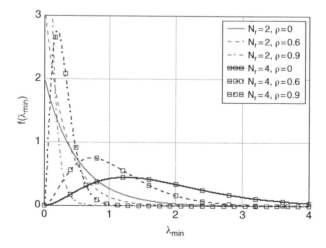

Figure 13.15   The Pdf of the Minimum Eigenvalue For $N_t = 2$. The correlation is on the side with the smallest number of antennas.

$$\mathbf{r} = \mathbf{H}^H \mathbf{y} = \mathbf{H}^H \mathbf{H} \mathbf{x} + \tilde{\mathbf{n}}$$

$$\begin{pmatrix} r_1 \\ r_2 \end{pmatrix} = \begin{pmatrix} |h_{11}|^2 + |h_{21}|^2 & h_{12}h_{11}^* + h_{22}h_{21}^* \\ h_{12}^*h_{11} + h_{22}^*h_{21} & |h_{22}|^2 + |h_{12}|^2 \end{pmatrix}$$

$$\begin{pmatrix} x_1 \\ x_2 \end{pmatrix} + \begin{pmatrix} \tilde{n}_1 \\ \tilde{n}_2 \end{pmatrix}$$

$$(13.39)$$

The zero-mean noise vector $\tilde{\mathbf{n}}$ is defined by:

$$\tilde{\mathbf{n}} = \mathbf{H}^H \mathbf{n} \;\Rightarrow\; \begin{pmatrix} \tilde{n}_1 \\ \tilde{n}_2 \end{pmatrix} = \begin{pmatrix} h_{11}^* & h_{21}^* \\ h_{12}^* & h_{22}^* \end{pmatrix} \begin{pmatrix} n_1 \\ n_2 \end{pmatrix}$$

$$E[\tilde{\mathbf{n}}\tilde{\mathbf{n}}^H] = E[\mathbf{H}^H \mathbf{n}\mathbf{n}^H \mathbf{H}] = \sigma^2 E[\mathbf{H}^H \mathbf{H}]$$

$$(13.40)$$

In this system, there are $N_t N_r = 4$ four paths through which the transmitted signal arrives to the receiver (see Figure 13.4). The eigenvalues of the 2×2 random matrix $\mathbf{H}^H \mathbf{H}$ may be used to characterize this MIMO system.

Case 1: *Uncorrelated Transmit and Receive Signals.*

For 2×2 uncorrelated MIMO channel, all channel gains h_{11}, h_{12}, h_{21} and h_{22} are uncorrelated with each other. Consequently, signals at each receive antenna arrive via two independent paths; the received signals are maximal ratio combined. This implies that four parallel and independent channels characterize this MIMO system, hence a diversity order of 4. The two distinct eigenvalues (3.5 and 0.5) of matrix $\mathbf{H}^H \mathbf{H}$ represent the mean SNRs of the two independent channels. Their sum is equal to 4 (see Table 13.1). The pdf's of the maximum and minimum eigenvalues are given by (13.20).

Case 2: *Perfectly Correlated Transmit and Receive Signals.*

In this case, the channel gains h_{11}, h_{12} and h_{21}, h_{22} are perfectly correlated with each other. In view of (12.43), we insert $h_{11} = h_{12} = h_{21} = h_{22}$ into (13.39) and get

$$\begin{pmatrix} r_1 \\ r_2 \end{pmatrix} = 2|h_{11}|^2 \begin{pmatrix} 1 & 1 \\ 1 & 1 \end{pmatrix} \begin{pmatrix} x_1 \\ x_2 \end{pmatrix} + \begin{pmatrix} \tilde{n}_1 \\ \tilde{n}_2 \end{pmatrix}$$

$$(13.41)$$

This implies perfect correlation between signals at both transmit and receive antennas. The matrix with all unity elements in (13.41) has unity rank and is characterized by the eigenvalues, $\lambda_1 = 0$, $\lambda_2 = 2$. This implies that this system behaves as a SISO system with a mean SNR equal to 4 and has unity a diversity order. The pdf of the non-zero eigenvalue is then given by

$$f_{\lambda_2}(\lambda) = \frac{1}{\bar{\lambda}} \exp\left(-\frac{\lambda}{\bar{\lambda}}\right)$$

$$(13.42)$$

$$\bar{\lambda} = 4E[|h_{11}^2|]$$

Case 3: *Perfectly Correlated Transmit Signals.*

In this case, the signals transmitted by two transmit antennas are perfectly correlated at each receive antenna; hence $h_{12} = h_{11}$, and $h_{21} = h_{22}$. However, the channel gains $\{h_{11}, h_{12}\}$ are uncorrelated with $\{h_{22}, h_{21}\}$. Then the uncorrelated receive antenna outputs may be maximal-ratio combined. In this case (13.39) may be rewritten as

$$\begin{pmatrix} r_1 \\ r_2 \end{pmatrix} = \underbrace{\left(|h_{11}|^2 + |h_{22}|^2\right)}_{h_{11}\ and\ h_{22}\ are\ indep.} \begin{pmatrix} 1 & 1 \\ 1 & 1 \end{pmatrix} \begin{pmatrix} x_1 \\ x_2 \end{pmatrix} + \begin{pmatrix} \tilde{n}_1 \\ \tilde{n}_2 \end{pmatrix}$$

$$(13.43)$$

The matrix with all unity elements in (13.43) has unity rank and is characterized by the eigenvalues, $\lambda_1 = 0$, $\lambda_2 = 2$. In this case, the system behaves as a SIMO system which allocates all its transmit power (of 2) to the second antenna and performs MRC at the receiver. The diversity order is 2 and the mean SNR is 4.

Case 4: *Perfectly Correlated Receive Signals.*

In this case, the signals received by the two receive antennas are perfectly correlated with each other; hence $h_{21} = h_{11}$, and $h_{12} = h_{22}$. However, the signals transmitted by the two transmit antennas are uncorrelated; the channel gains $\{h_{11}, h_{21}\}$ are therefore uncorrelated with the channel gains $\{h_{22}, h_{12}\}$. Then (13.39) may be rewritten as

$$\begin{pmatrix} r_1 \\ r_2 \end{pmatrix} = 2 \begin{pmatrix} |h_{11}|^2 & 0 \\ 0 & |h_{22}|^2 \end{pmatrix} \begin{pmatrix} x_1 \\ x_2 \end{pmatrix} + \begin{pmatrix} \tilde{n}_1 \\ \tilde{n}_2 \end{pmatrix}$$

(13.44)

In this case, (13.44) behaves as a 2×2 uncorrelated MIMO channel where uncorrelated transmit antenna signals experience MRT while the powers of perfectly correlated signals of the receive antennas are coherently added. Hence, the diversity order is equal to 2 and the mean SNR is 4.

For a 2×2 MIMO channel with one-sided correlation treated in *Cases 3* and *4* above, the pdf of the maximum eigenvalue is found by inserting $m = 2$ into (13.37) or directly from (E.25):

$$f_{\lambda_{max}}(\lambda,\rho)$$
$$= \frac{-2}{1-\rho^2} e^{-\frac{2\lambda}{1-\rho^2}} + \frac{(1+\rho)}{2\rho(1-\rho)}\left(1-\frac{\lambda}{1+\rho}\right)e^{-\frac{\lambda}{1-\rho}}$$
$$- \frac{(1-\rho)}{2\rho(1+\rho)}\left(1-\frac{\lambda}{1-\rho}\right)e^{-\frac{\lambda}{1+\rho}}$$

(13.45)

The pdf of the minimum eigenvalue may be written as

$$f_{\lambda_{min}}(\lambda,\rho) = \frac{2}{1-\rho^2} e^{-\frac{2}{1-\rho^2}\lambda}$$

(13.46)

$$\lim_{\rho \to 1} f_{\lambda_{min}}(\lambda,\rho) = \delta(\lambda)$$

The expected values of maximum and minimum eigenvalues are found as follows:

$$E[\lambda_{max}] = \int_0^\infty \lambda f_{\lambda_{max}}(\lambda,\rho)\, d\lambda = -\frac{1}{2}(1-\rho^2) + 4$$

$$E[\lambda_{min}] = \int_0^\infty \lambda f_{\lambda_{min}}(\lambda,\rho)\, d\lambda = \frac{1}{2}(1-\rho^2)$$

(13.47)

As one may also observe from (13.47), the sum of the two eigenvalues is always equal to $N_t N_r = 4$ independently of the correlation coefficient. For the uncorrelated case, normalized mean values of the largest and the smallest eigenvalues are 3.5 and 0.5 respectively. For the special case of perfect correlation ($\rho = 1$), the mean value of the smallest eigenvalue vanishes and that of the largest eigenvalue goes to 4, as expected.

Based on (13.30) and (13.36), Figure 13.16 shows the variation of the mean value of the

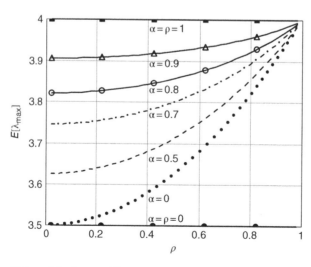

Figure 13.16   Mean Value of the Largest Eigenvalue of a Double-Sided Correlated 2×2 MIMO System Versus ρ For Various Values of α.

maximum eigenvalue for a 2×2 MIMO system with double-sided correlation versus the correlation coefficient ρ and α is given as a parameter. One may observe that the mean value of the maximum eigenvalue of a 2×2 MIMO system is bounded between 3.5 (uncorrelated case) and 4 (both sides perfectly correlated). Increased values of the correlation coefficients α and ρ cause accompanying increases in the mean value of λ_{max}. One may note for comparison purposes that the mean SNR for the 1×4 SIMO system is also equal to 4.

Figure 13.17 shows the effect of correlation on the pdf of a 2×2 double-sided correlated MIMO system. The pdf was observed to shift to the left as the correlation level in the system increases. In the extreme case where both sides are perfectly correlated, the pdf reduces to that of the SISO system described by (13.42). One may observe from (13.31) that the pdf for $\alpha = 0.5$ and $\rho = 0.9$ is the same as for $\alpha = 0.9$ and $\rho = 0.5$. Figures 13.16 clearly shows that mean value of the maximum eigenvalue increases with increasing correlation. However, Figure 13.17 indicates degradation of outage and bit error probability performances with increased correlation between channel gains.

Example 13.2 Performance Comparison For 2×2 MIMO, SISO, MRC, TAS-MRC and Alamouti Techniques.

In this example, we will compare the outage performance of a SISO system, 2×2 MIMO system, 1×4 MRC (SIMO) system, 2×2 TAS-MRC system and 2×2 Alamouti system in an uncorrelated fading environment. A SISO baseline system is characterized by the following pdf, cdf and mean SNR (see (13.17) and (13.25)):

$$f(\lambda) = e^{-\lambda}$$

$$F_\lambda(x) = 1 - e^{-x} \tag{13.48}$$

$$E[\lambda] = 1$$

The pdf and the cdf of the largest eigenvalue of an uncorrelated 2×2 MIMO system are given by (13.20) and (13.25): [15]

$$f_{\lambda_2}(\lambda) = e^{-\lambda}(\lambda^2 - 2\lambda + 2) - 2e^{-2\lambda}$$

$$F_{\lambda_2}(x) = 1 - e^{-x}(x^2 + 2) + e^{-2x} \approx x^4/12 \quad x << 1$$

$$E[\lambda_2] = 3.5$$

$$\tag{13.49}$$

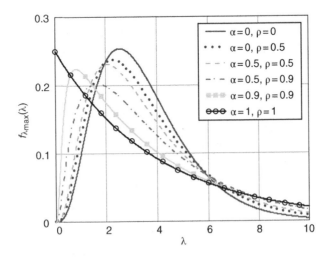

Figure 13.17 The Pdf of the Largest Eigenvalue of a 2×2 - MIMO Channel with Double-Sided Correlation.

The smallest eigenvalue of an uncorrelated 2×2 MIMO system is characterized by pdf (13.20), the cdf and the mean as follows:

$$f_{\lambda_1}(\lambda) = 2e^{-2\lambda}$$
$$F_{\lambda_1}(x) = 1 - e^{-2x} \qquad (13.50)$$
$$E[\lambda_1] = 0.5$$

The pdf, cdf and the mean SNR of a $L = 4$ branch MRC (1×4 SIMO) system are found by inserting $\bar{\gamma}_c = 1$ into (12.41), (12.42) and (12.53): [15]

$$f_{\lambda_{MRC}}(\lambda) = \frac{1}{(L-1)!} \lambda^{L-1} e^{-\lambda}$$

$$F_{\lambda_{MRC}}(x) = 1 - e^{-x} \sum_{k=0}^{L-1} \frac{x^k}{k!} \approx x^4/24 \quad x \ll 1$$

$$E[\lambda_{MRC}] = L$$
$$(13.51)$$

Similarly, the pdf and cdf of 2×2 Alamouti system (L = 4) are found by inserting $\bar{\gamma}_c = 1$ into (12.132) and (12.133), respectively:

$$\gamma_{AL} = \frac{1}{2} \sum_{k=1}^{L} \gamma_k \quad \Rightarrow$$

$$f_{\gamma_{AL}}(\lambda) = \frac{2^L \lambda^{L-1} e^{-2\lambda}}{(L-1)!} \qquad (13.52)$$

$$F_{\lambda_{AL}}(x) = 1 - e^{-2x} \sum_{k=0}^{L-1} \frac{(2x)^k}{k!}$$

$$E[\lambda_{AL}] = L/2$$

When one of the two transmit antennas is selected (TAS) for transmission, the cdf of the output signal SNR γ may be written as

$$F_\gamma(x) = F_{\gamma_1}(x) \, F_{\gamma_2}(x) = [1 - e^{-x}(1+x)]^2$$
$$F_{\gamma_i}(x) = 1 - e^{-x}(1+x), \quad i = 1,2$$
$$(13.53)$$

where $F_{\gamma_i}(x)$ denotes the cdf of the output SNR due to the i-th transmit antenna and is found by

inserting $L = 2$ into (13.51). Note that γ_i is obtained by maximal-ratio combining of the outputs of two receive antennas when i-th transmit antenna is transmitting. The pdf and the mean value of the output SNR are found as follows: [16]

$$f_\gamma(x) = \frac{d}{dx} F_\gamma(x) = 2 \, x \, e^{-x} [1 - e^{-x}(1+x)]$$
$$E[\gamma] = 2.75$$
$$(13.54)$$

where (D.50) is used for determining the mean value.

Figure 13.18 shows the cdf of the smallest and largest eigenvalues of a 2×2 uncorrelated MIMO system in comparison with a SISO system, 2×2 Alamouti scheme, a TAS-MRC system where one of two transmit antennas are selected for transmission and the outputs of two receive antennas are maximal-ratio combined, and finally a 1×4 MRC (SIMO) system, which provides the best performance. Note that 1×4 MRC, 2×2 MIMO (maximum eigenvalue), 2×2 TAS and 2×2 Alamouti systems have the same slope with SNR and hence display the same diversity order of 4. The cdf of the maximum eigenvalue λ_2 of a 2×2 MIMO has a 0.75 dB degraded outage performance compared to a 1×4 MRC system. In other terms, compared to a 1×4 MRC system, it has twice the cumulative probability for the same power level, or a shift of $2^{1/4} = 0.75$ dB in power for a given cumulative probability. 2×2 TAS and 2×2 Alamouti systems perform respectively 1 dB and 2 dB worse than the largest eigenvalue of the 2×2 MIMO system. [15]

13.5   Diversity Order of a MIMO System

Diversity order of a MIMO system may be studied in terms of the diversity order provided by its eigenvalues. The diversity orders of $n = \min(N_t, N_r)$ eigenvalues are given by

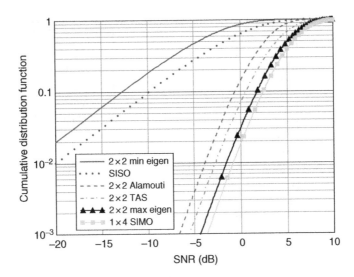

Figure 13.18 The Comparison of the Cdf of Smallest and Largest Eigenvalues of a 2×2 MIMO System with Some Other Transmission Scenarios.

$$d_{n-k+1} = (N_t - k + 1)(N_r - k + 1), \quad k = 1, 2, \cdots, n \tag{13.55}$$

starting from the largest to the smallest. The diversity order of the largest eigenmode, $d_n = N_t N_r$, is same as that for MRC and maximal radio transmission (MRT). The diversity order of the smallest eigenmode is $d_l = m - n + 1$, which becomes equal to unity for $N_t = N_r$.

The marginal pdf of each eigenmode λ_k of a $N_t \times N_r$ MIMO system may be approximated by the Gamma distribution for N_t, $N_r \leq 4$: [6]

$$f_{\lambda_k}(x, d_k, \bar{\gamma}_k) = \frac{x^{d_k - 1} e^{-d_k x / \bar{\gamma}_k}}{\Gamma(d_k)(\bar{\gamma}_k / d_k)^{d_k}}, \quad \bar{\gamma}_k = E[\lambda_k] \tag{13.56}$$

To check the validity of the Gamma pdf approximation, we consider an uncorrelated 3×3 MIMO channel ($N_t = N_r = 3$). The diversity orders of the three eigenvalues are found from (13.55) as $d_3 = 9$, $d_2 = 4$ and $d_1 = 1$. The diversity order 9 of the largest eigenvalue is evidently much larger than those of the remaining eigenvalues. The corresponding values of the mean SNRs are given by Table 13.1 as

$\bar{\gamma}_3 = 6.5208$, $\bar{\gamma}_2 = 2.1458$ and $\bar{\gamma}_1 = 0.333$. The pdf's of the three eigenvalues based on the Wishart matrix approach are given by (13.22). Figure 13.19 clearly shows a close agreement between the pdf's of the three eigenvalues for an uncorrelated 3×3 MIMO channel based on (13.22) and the Gamma approximation given by (13.56).

13.6 Capacity of a MIMO System

Input-output relation of a MIMO channel is described by (13.1). The covariance matrix of the received signal vector $\mathbf{y}$ is given by

$$\mathbf{R_y} = E[\mathbf{y} \, \mathbf{y}^H] = E[(\mathbf{H} \, \mathbf{x} + \mathbf{n}) \, (\mathbf{H} \, \mathbf{x} + \mathbf{n})^H]$$
$$= E[\mathbf{H} \, \mathbf{x}\mathbf{x}^H\mathbf{H}^H + \mathbf{n}\mathbf{n}^H] = \mathbf{H}E[\mathbf{x}\mathbf{x}^H]\mathbf{H}^H + \mathbf{R_n}$$
$$= \mathbf{H}\mathbf{R_x}\mathbf{H}^H + \sigma^2\mathbf{I}$$

$$\mathbf{R_x} = E[\mathbf{x}\mathbf{x}^H] \tag{13.57}$$

where $\mathbf{R_n}$ is defined by (13.3). Assuming that the transmitted signals x_i, $i = 1, 2, \ldots, N_t$ are uncorrelated, the correlation matrix of the transmit signal vector $\mathbf{x}$ is diagonal and the

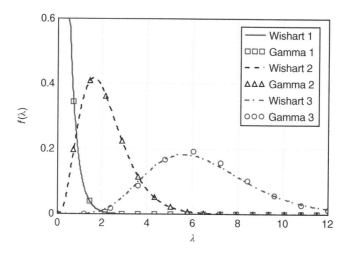

Figure 13.19 Comparison of Wishart Formulation Given by (13.22) and Gamma Approximation (13.56) For the Pdf's of λ_1, λ_2 and λ_3 of an Uncorrelated 3×3 MIMO Channel.

k-th element denotes the power transmitted by the k-th transmit antenna:

$$\mathbf{R_x} = E\left[\mathbf{xx}^H\right] = \{P_k\}, \quad k = 1, 2, \ldots, N_t \quad (13.58)$$

Also assume that the total transmit power, that is, the sum of powers transmitted by all transmit antennas, is constant and is less than or equal to P:

$$\mathrm{tr}[\mathbf{R_x}] = \sum_{k=1}^{N_t} P_k \leq P \quad (13.59)$$

where $\mathrm{tr}[\mathbf{R_x}]$ shows the trace of $\mathbf{R_x}$.

The normalized instantaneous capacity (spectral efficiency) in bps/Hz of this MIMO channel with uncorrelated transmit signals may be written, using (13.57), as

$$\eta = \log_2 \left| \mathbf{I}_{N_r} + \frac{1}{\sigma^2} \mathbf{H} \mathbf{R_x} \mathbf{H}^H \right| \quad (13.60)$$

Noting that $|\mathbf{I} + \mathbf{AB}| = |\mathbf{I} + \mathbf{BA}|$, the normalized instantaneous capacity may be rewritten as

$$\eta = \log_2 \left| \mathbf{I}_{N_t} + \frac{1}{\sigma^2} \mathbf{R_x} \mathbf{H}^H \mathbf{H} \right| \quad (13.61)$$

If no CSI is available at the transmitter, it is reasonable to divide the total transmit power equally between uncorrelated data streams transmitted by each antenna, hence P/N_t:

$$\mathbf{R_x} = (P/N_t)\mathbf{I}_{N_t} \quad (13.62)$$

The normalized instantaneous capacity of a MIMO channel then reduces to

$$\eta = \log_2 \left| \mathbf{I}_{N_t} + \frac{\bar{\gamma}}{N_t} \mathbf{H}^H \mathbf{H} \right| \quad (13.63)$$

where $\bar{\gamma} = P/\sigma^2$ denotes the average SNR and the $N_t \times N_t$ Wishart matrix $\mathbf{H}^H\mathbf{H}$ is given by (13.4). We know that the $N_t \times N_t$ Wishart matrix $\mathbf{H}^H\mathbf{H}$ is characterized by its R_H non-zero eigen-values λ_i, $i = 1, 2, \ldots$, R_H where R_H denotes the rank of $\mathbf{H}^H\mathbf{H}$. $R_H = 1$ if the channel is perfectly correlated at both transmit and receive sides but $R_H = n = \min(N_t, N_r)$ when the channel is uncorrelated; hence $1 \leq R_H \leq n$. Therefore, the normalized instantaneous capacity expression given by (13.63) may be written in terms of its non-zero eigenvalues, since vanishing eigenvalues do not have any contribution:

$$\eta = \log_2 \left| \mathbf{I}_{N_t} + \frac{\bar{\gamma}}{N_t} \mathbf{H}^H \mathbf{H} \right|$$

$$= \log_2 \begin{vmatrix} 1 + \bar{\gamma}\lambda_1/N_t & 0 & \cdots & 0 \\ 0 & 1 + \bar{\gamma}\lambda_2/N_t & 0 & 0 \\ 0 & 0 & \ddots & 0 \\ 0 & 0 & \cdots & 1 + \bar{\gamma}\lambda_{R_H}/N_t \end{vmatrix}$$

$$= \log_2 \prod_{i=1}^{R_H} \left(1 + \frac{\bar{\gamma}}{N_t} \lambda_i \right) = \sum_{i=1}^{R_H} \log_2 \left(1 + \frac{\bar{\gamma}}{N_t} \lambda_i \right)$$

$$(13.64)$$

The normalized capacity of this MIMO channel is clearly the sum of ergodic capacities of R_H eigenmodes. On the other hand, noting that the eigenvalues of the random Wishart matrix $\mathbf{H}^H\mathbf{H}$ are random, the normalized mean capacity is found by the expectation of (13.64).

Case 1: Transmit and Receive Antennas are Perfectly Correlated Among Each Other.

When all transmit and receive antennas are very close to each other, transmitted and receive signals undergo perfectly correlated fading among themselves. Since all channel gains are perfectly correlated with each other, this MIMO channel behaves as a SISO channel (see (13.41) – (13.42)). Then, $\mathbf{H}^H\mathbf{H}$ has a unity rank and is characterized by a single non-zero eigenvalue with the following pdf:

$$f(\lambda) = \frac{1}{N_r N_t} \exp\left(-\frac{\lambda}{N_r N_t} \right) \qquad (13.65)$$

which has a mean value of $N_t N_r$ (see Figures 3.12 and 3.13). The normalized instantaneous capacity is obtained by inserting $R_H = 1$ into (13.64):

$$\eta = \log_2 \left(1 + \frac{\bar{\gamma}}{N_t} \lambda \right) \qquad (13.66)$$

Replacing λ by its mean value $N_t N_r$ in (13.66), one observes an N_r-fold increase in the SNR, that is, $N_r \bar{\gamma}$. Hence, the mean capacity increases logarithmically with the number of receive antennas. This demonstarates the SISO behavior of a perfectly correlated MIMO channel.

The mean capacity is found by averaging (13.66) with the pdf given by (13.65):

$$E[\eta] = \int_0^\infty \log_2\left(1 + \frac{\lambda}{N_t} \right) f(\lambda) \, d\lambda$$

$$= \frac{1}{\ln 2} e^{\frac{1}{N_r \bar{\gamma}}} E_1\left(\frac{1}{N_r \bar{\gamma}} \right) \qquad (13.67)$$

where $E_1(x)$ is defined by (D.116) and (D.73) is used to evaluate the integral. The cdf of the normalized capacity is defined as the probability that the normalized capacity is below a given threshold level, η_{th}:

$$\text{Prob}(\eta < \eta_{th}) = \text{Prob}\left(\log_2\left(1 + \frac{\bar{\gamma}}{N_t}\lambda \right) < \eta_{th} \right)$$

$$= 1 - \exp\left(-\frac{2^{\eta_{th}} - 1}{\bar{\gamma} N_r} \right)$$

$$(13.68)$$

where the last expression is obtained by the cdf of (13.65).

Case 2: Transmit and Receive Antennas are Uncorrelated with Each Other.

In this case, all path-gains are independent from each other. Then the channel matrix is full-rank and $\mathbf{H}^H\mathbf{H}$ has $R_H = n$ distinct eigenvalues whose joint pdf is given by (13.16). It is clear from the normalized instantaneous capacity, (13.64), that the normalized capacity increases linearly with the number of antenna elements. The mean normalized capacity may then be written as

$$E[\eta] = \sum_{i=1}^n E\left[\log_2\left(1 + \frac{\bar{\gamma}}{N_t}\lambda_i \right) \right]$$

$$= \sum_{i=1}^n \int_0^\infty \log_2\left(1 + \frac{\bar{\gamma}}{N_t}\lambda_i \right) f_{\lambda_i}(\lambda_i) \, d\lambda_i$$

$$(13.69)$$

where $f_{\lambda_i}(\lambda_i)$ denotes the marginal pdf of the i-th eigenvalue, which is given by (13.17)–(13.22) for various values of n. The cdf of the normalized capacity is given by

$$
\begin{aligned}
\text{Prob}(\eta<\eta_{th}) &= \text{Prob}\left(\log_2\prod_{i=1}^{n}\left(1+\frac{\bar{\gamma}}{N_t}\lambda_i\right)<\eta_{th}\right) \\
&= \text{Prob}\left(\prod_{i=1}^{n}\left(1+\frac{\bar{\gamma}}{N_t}\lambda_i\right)<2^{\eta_{th}}\right)
\end{aligned}
$$

(13.70)

which requires an n-fold integration.

As one may observe from (13.17) – (13.22) and Tables 13.1 and 13.2 that the eigenvalues have different pdf's, means, variances. However, though each eigenvalue contributes to the capacity, the largest eigenvalue is dominant and is generally employed to determine the capacity of MIMO systems.

Example 13.3   Mean and Outage Capacity of an Uncorrelated 2×2 MIMO System.
The instantaneous capacity of an uncorrelated 2×2 MIMO system is given by (13.64):

$$
\begin{aligned}
\eta &= \log_2\left|\mathbf{I}+\frac{\bar{\gamma}}{2}\mathbf{H}^H\mathbf{H}\right| \\
&= \log_2\left|\begin{pmatrix}1 & 0\\0 & 1\end{pmatrix}+\frac{\bar{\gamma}}{2}\begin{pmatrix}\lambda_{max} & 0\\0 & \lambda_{min}\end{pmatrix}\right| \\
&= \log_2\left(1+\frac{\bar{\gamma}}{2}\lambda_{max}\right) \\
&\quad +\log_2\left(1+\frac{\bar{\gamma}}{2}\lambda_{min}\right)\ (bps/Hz)
\end{aligned}
$$

(13.71)

where $\bar{\gamma}$ denotes the average receive SNR and λ_{max} and λ_{min} denote respectively the instantaneous values of the maximum and minimum eigenvalues. The ergodic capacity is the sum of the ergodic capacities provided by the maximum and the minimum eigenvalues. Using the pdf's of λ_{max} and λ_{min} given by (13.20) and

(D.74), the ergodic capacity of this 2×2 MIMO system is found as the sum of the capacities for maximum and minimum eigenvalues:

$$
\begin{aligned}
E[\eta_{max}] &= \int_0^{\infty}\log_2\left(1+\frac{\bar{\gamma}}{2}\lambda\right)f_{\lambda_{max}}(\lambda)\ d\lambda \\
&= \frac{2}{\ln 2}\left[e^{2/\bar{\gamma}}E_1(2/\bar{\gamma})\left(1+\frac{2}{\bar{\gamma}^2}\right)-\frac{1}{\bar{\gamma}}+\frac{1}{2}\right]-E[\eta_{min}]
\end{aligned}
$$

$$
\begin{aligned}
E[\eta_{min}] &= \int_0^{\infty}\log_2\left(1+\frac{\bar{\gamma}}{2}\lambda\right)f_{\lambda_{min}}(\lambda)\ d\lambda \\
&= \frac{\exp(4/\bar{\gamma})}{\ln 2}E_1(4/\bar{\gamma})
\end{aligned}
$$

(13.72)

Note the similarity between the mean capacity for the minimum eigenvalue in (13.72) and the mean capacity for the perfectly correlated case as given by (13.65); there is simply a shift in the SNR's.

Figure 13.20 shows the variation of the ergodic capacities of the two eigenvalues given by (13.72). The maximum eigenvalue is evidently dominant in determining the ergodic capacity. The ergodic capacity for the perfectly correlated case given by (13.67) is also shown for comparison purposes. The ergodic capacity of a perfectly correlated channel is significantly lower than that for the uncorrelated case, as expected.

The cdf of the normalized capacity is given by the probability that the normalized capacity remains below a given threshold level:

$$
\begin{aligned}
&\text{Prob}(\eta<\eta_{th}) \\
&= \text{Prob}[\log_2[(1+0.5\bar{\gamma}\lambda_{max})(1+0.5\bar{\gamma}\lambda_{min})]<\eta_{th}] \\
&= \int_0^{\infty}\text{Prob}\left[\lambda_{min}<\frac{2}{\bar{\gamma}}\left(\frac{2^{\eta_{th}}}{1+0.5\ \bar{\gamma}\lambda_{max}}-1\right)|\lambda_{max}\right] \\
&\quad f_{\lambda_{max}}(\lambda_{max})\ d\lambda_{max} \\
&= 1-e^{4/\bar{\gamma}}\int_0^{\infty}\exp\left[-\frac{4\times 2^{\eta_{th}}}{(1+0.5\ \bar{\gamma}\lambda)\bar{\gamma}}\right]f_{\lambda_{max}}(\lambda)\ d\lambda
\end{aligned}
$$

(13.73)

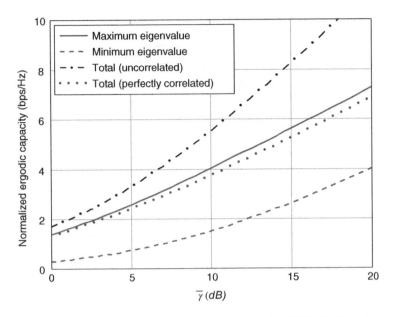

Figure 13.20 Ergodic Capacity For 2×2 Uncorrelated MIMO Channel.

where $0 \leq \lambda_{min} \leq \lambda_{max} \leq \infty$ and the pdf's of maximum and minimum eigenvalues is given by (13.20).

The cdf of normalized capacity for the maximum eigenvalue is given by

$$\text{Prob}(\eta < \eta_{th}) = \text{Prob}[\log_2(1 + 0.5\bar{\gamma}\lambda_{max}) < \eta_{th}]$$

$$= F_{2 \times 2}\left(\frac{2}{\bar{\gamma}}(2^{\eta_{th}} - 1)\right)$$

$$(13.74)$$

where $F_{2 \times 2}(.)$ is given by (13.23). Figure 13.21 shows the cdf of the capacity of an uncorrelated 2×2 MIMO system given by (13.73) as a function of the normalized threshold capacity η_{th} level for various values of the mean SNR $\bar{\gamma}$. The capacity given by (13.73) is also compared with capacity for the maximum eigenvalue given by (13.74). Figure 13.21 clearly shows that the capacity predictions based on the maximum eigenvalue do not seem to be realistic.

Figure 13.22 shows the cdf of the capacity as a function of the mean channel SNR $\bar{\gamma}$ for various values of η_{th}. The results confirm the conclusions drawn from Figure 13.21.

Figure 13.23 shows the increase in 10% outage capacity of an uncorrelated MIMO system with three transmit antennas ($N_t = 3$) as a function of the number of receive antennas. Since the number of transmit antennas is fixed, increase in the number of receive antennas (N_r) will cause an increase in the diversity order and in the outage capacity. However, when we use three receive antennas ($N_r = 3$), as shown in Figure 13.24, the use of more transmit antennas will not cause such a high increase in the outage capacity. With the addition of more antennas to the transmitter, more diversity is incorporated to the system and the capacity is increased. However, because the total transmit power is fixed, the transmit power of each transmit antenna is reduced as their number increases. The reduced transmitted power for each transmit antenna will decrease the channel capacity. Consequently, the advantage in using more

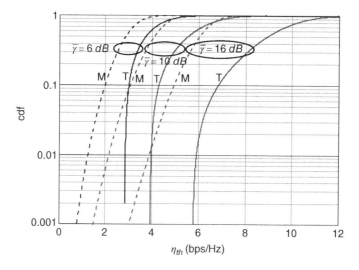

Figure 13.21   Cdf of the Capacity Given by (13.73) as a Function of Normalized Threshold Capacity η_{th} For a 2×2 Uncorrelated MIMO Channel. Outage capacity, (13.74), for the maximum eigenvalue is also shown. T denotes the total capacity and M denotes the capacity due to maximum eigenvalue.

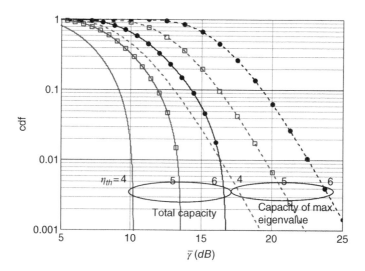

Figure 13.22   Cdf of the Capacity Given by (13.73) For a 2×2 Uncorrelated MIMO Channel Compared with That Given by (13.74) of the Maximum Eigenvalue. T denotes the total capacity and M denotes the capacity due to maximum eigenvalue.

transmit antennas will be offset by the reduced transmit power per transmit antenna. Therefore, increase in the number of transmit antennas is not as effective as the increase in the number of receive antennas.

13.6.1   Water-Filling Algorithm

As one may observe from (13.69), the mean capacity of an uncorrelated MIMO system is the sum of the mean capacities due to non-zero

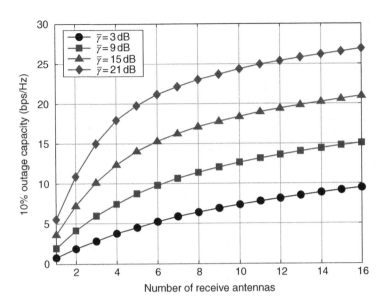

Figure 13.23 The Outage Capacity Increase with the Number of Receive Antennas For $N_t = 3$. Outage probability is 10%. Monte Carlo simulation. [17]

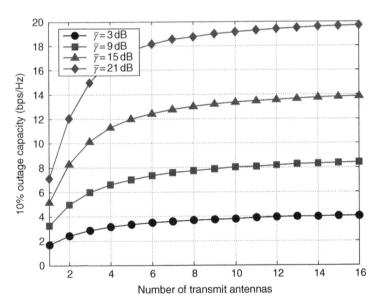

Figure 13.24 The Increase In the 10% Outage Capacity with the Number of Transmit Antennas For Three Receive Antennas ($N_r = 3$) in an Uncorrelated MIMO Channel. Monte Carlo simulation. [17]

eigenvalues. However, as shown by (13.17) – (13.22) and Tables 13.1 and 13.2, the eigenvalues are not identical and the contribution of each eigenvalue to the capacity is different.

Nevertheless, the largest eigenvalue is dominant and is generally employed to determine the capacity of MIMO systems even if the predictions are not sufficiently accurate (see Figures 13.21 and 13.22). Similar arguments also apply for correlated MIMO systems.

When the CSI is not available at the transmitter, it is reasonable to equally divide the transmit power equally between the transmit antennas. However, in case when the CSI is available at the transmitter, the total transmit power P may be assigned adaptively between transmit antennas, P_i, $i = 1, 2,..., N_t$, so as to maximize the ergodic capacity. The so-called water-filling algorithm may be used to determine the transmit powers P_i, $i = 1, 2,..., N_t$ assigned to each transmit antenna as a function of the corresponding eigenvalue.

Assuming that all eigenvalues of $\mathbf{H}^H\mathbf{H}$ are uncorrelated with each other, and all noise powers are the same, the instantaneous capacity may be maximized by assigning different powers, P_i, to each eigenvalue λ_i during each time slot, chosen to be shorter than the channel coherence time. A block of symbols may then be transmitted during a time slot, with an optimum power level in block-fading channels. The duration of a time slot might be as short as a symbol duration in fast-fading channels. Such adaptive time-scheduling can therefore mitigate the degrading effects of random temporal fluctuations in the channel. Consequently, the channel capacity may be maximized with a constraint on the total transmit power P.

Assuming an uncorrelated $N \times N$ MIMO system, the instantaneous normalized capacity may be maximized by Lagrange multipliers approach: [15][18]

$$J = \sum_{i=1}^{N} \log_2\left(1 + \frac{P_i}{\sigma^2}\lambda_i\right) + \alpha\left(P - \sum_{i=1}^{n} P_i\right)$$

(13.75)

We take the derivative of J with respect to P_i and equate to zero:

$$\frac{dJ}{dP_i} = \frac{1}{\ln 2}\frac{\lambda_i}{\sigma^2 + P_i\lambda_i} - \alpha = 0 \implies$$

(13.76)

$$\frac{\sigma^2}{\lambda_i} + P_i = \frac{1}{\alpha \ln 2} = D, \quad i = 1, 2,..., N$$

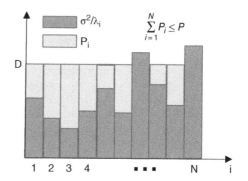

Figure 13.25   Pictorial View of the Water-Filling Algorithm.

The water filling algorithm maximizes the capacity, by filling up each channel to a common level D, which is determined by the total transmit power level allocated to N channels. The channel with the highest SNR, λ_i/σ^2, receives the largest share of the power. In case the level D remains below a certain σ^2/λ_i, then that power is set to zero. Note that water-filling implies that the allocated power level P_i for i-th channel is represented by the difference between the surface level D and the water level σ^2/λ_i as shown in Figure 13.25.

When the transmit power per eigenvalue is determined by the water-filling algorithm, the mean spectrum efficiency is given by

$$\eta = \sum_{i=1}^{N} \log_2\left(1 + \frac{P_i}{\sigma^2}\lambda_i\right)$$

$$= \sum_{i=1}^{N} \log_2\left(\left(\bar{\gamma} + \sum_{\ell=1}^{N}\frac{1}{\lambda_\ell}\right)\frac{\lambda_i}{N}\right)$$

(13.77)

where $\bar{\gamma} = P/\sigma^2$ and the last expression follows from (13.59) and (13.76).

13.7   MIMO Beamforming Systems

In certain applications, a single data stream is transmitted in a MIMO channel in order to benefit the advantages of diversity and array gains. Since the CSI is known at the receiver, the received signals may be combined using MRC. However, the CSI is usually unavailable at the transmitter unless it is provided by the receiver via a fed-back link. If the CSI is unavailable at the transmitter, then the transmit power may be equally divided between the transmit antennas or one or more of the transmit antennas with the highest channel gains may be selected for transmission. If the CSI is available at the transmitter, then transmitted signals may be co-phased with appropriate transmit weights, w_{ti}, $i = 1,2,...,N_t$, so that they are added in phase at the receiver, that is, maximal ratio transmission (MRT) (see Figure 13.26).

In this scheme shown in Figure 13.26, usually referred to as MIMO beamforming, transmitted and received signals are weighted with respectively $\mathbf{w_t}$ and $\mathbf{w_r}$ column vectors. The transmit vector $\mathbf{w_t}$ is normalized, $\|\mathbf{W}_t\| = 1$, to account for the fact that the transmit power is divided among the transmit antennas. Here, a trade-off is required between transmitting using a single antenna with full transmit power and transmitting with N_t transmit antennas, each using only a fraction of the transmit power. Received signal vector in this MIMO system may be written as [16][19][20]

$$y = \mathbf{H} \, \mathbf{w}_t \, s + \mathbf{n} \qquad (13.78)$$

where $\mathbf{H}$ denotes $N_r \times N_t$ channel matrix defined by (13.2), s is the transmitted signal with energy $E = \mathrm{E}[ss^*]$ and

$$\begin{aligned}
\mathbf{y} &= \begin{bmatrix} y_1 & y_2 & \cdots & y_{N_r} \end{bmatrix}^T \\
\mathbf{w}_t &= \begin{bmatrix} w_{t1} & w_{t2} & \cdots & w_{tN_t} \end{bmatrix}^T, \quad \|\mathbf{w}_t\| = 1 \\
\mathbf{n} &= \begin{bmatrix} n_1 & n_2 & \cdots & n_{N_r} \end{bmatrix}^T, \quad E[\mathbf{nn}^H] = N_0 \mathbf{I}_{N_r}
\end{aligned}$$
$$(13.79)$$

The signal vector $\mathbf{y}$ received by N_r receive antennas are combined at the MIMO output with a receive weight vector $\mathbf{w_r}$:

$$r = \mathbf{w}_r \mathbf{y} = \mathbf{w}_r \mathbf{H} \, \mathbf{w}_t \, s + \mathbf{w}_r \mathbf{n} \qquad (13.80)$$

where the receive weight vector has unity norm

$$\mathbf{w}_r = \begin{bmatrix} w_{r1} & w_{r1} & \cdots & w_{rN_r} \end{bmatrix}, \quad \|\mathbf{w}_r\| = 1 \quad (13.81)$$

For MRC combining at the MIMO output, the receive weight vector is chosen as

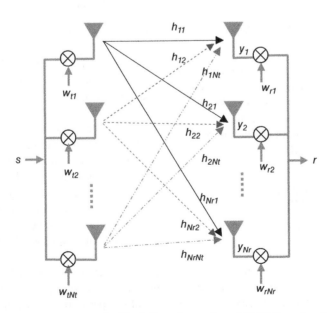

Figure 13.26 Transmission of a Single Data Stream by a MIMO Beamforming System.

$$w_r = (Hw_t)^H = w_t^H H^H \qquad (13.82)$$

Inserting (13.82) into (13.80) we get the MIMO-MRC output

$$r = w_t^H H^H H \; w_t s + w_t^H H^H n \qquad (13.83)$$

The SNR at the MIMO-MRC output is determined as follows:

$$\gamma = \frac{E\left[\left| w_t^H H^H H \; w_t s \right|^2 \right]}{E\left[\left| w_t^H H^H n \right|^2 \right]}$$

$$= E_b \frac{\left| w_t^H H^H H \; w_t \right|^2}{w_t^H H^H E[nn^H] H w_t} = \bar{\gamma} w_t^H H^H H \; w_t$$

$$\qquad (13.84)$$

where $\bar{\gamma} = E_b / \sigma^2$ denotes the average SNR.

We now concentrate on maximizing the SNR given by (13.84). The $N_r \times N_r$ matrix $H^H H$ has a maximum number of R_H eigenvalues, λ_i, $i = 1, 2, \ldots, R_H$ which satisfy

$$H^H H u_i = \lambda_i u_i, \quad i = 1, 2, \cdots, R_H \qquad (13.85)$$

where u_i denotes the normalized eigenvector corresponding to the eigenvalue λ_i. Therefore, the instantaneous SNR is maximized by choosing the transmit weight vector as $w_t = u_{max}$, the eigenvector corresponding to the maximum eigenvalue λ_{max} of $H^H H$:

$$\gamma_{max} = \bar{\gamma} w_t^H \underbrace{H^H H \; w_t}_{= \lambda_{max} u_{max}} \Big|_{w_t = u_{max}}$$

$$= \bar{\gamma} u_{max}^H \lambda_{max} u_{max} = \bar{\gamma} \lambda_{max} \qquad (13.86)$$

It is clear from (13.86) that the pdf of the SNR at the MIMO output is characterized by pdf of the maximum eigenvalue. The expected value of the maximum SNR for a

2×2 MIMO system is found by inserting $E[\lambda_{max}]$ from (13.47) into (13.86) for the one-sided correlated case. For the uncorrelated case, the mean values of the maximum eigenvalue are given by Table 13.1 for arbitrary values of n and m.

13.7.1 Bit Error Probability in MIMO Beamforming Systems

As we already discussed in Chapter 6, the best receiver in AWGN channel is a matched filter or a correlation receiver, that maximizes the output SNR. The probability of error is minimized when the output SNR is maximized. As shown in (13.86), the output SNR is maximized by operating with the maximum eigenvalue. Hence, the average probability of error for BPSK may be expressed in terms of the average SNR per bit $\bar{\gamma} = E_b / \sigma^2$:

$$P_e = \int_0^\infty Q\left(\sqrt{2\bar{\gamma} \lambda} \right) f_{\lambda_{max}}(\lambda) \; d\lambda$$

$$= \frac{1}{2} \sqrt{\frac{\bar{\gamma}}{\pi}} \int_0^\infty \frac{1}{\sqrt{\lambda}} e^{-\bar{\gamma} \lambda} F_{\lambda_{max}}(\lambda) \; d\lambda \qquad (13.87)$$

where the last expression is obtained by applying integration by parts to the first.

Figure 13.27 shows the error probability for BPSK modulation for a 3×3 MIMO system with one-sided correlation versus the branch signal-to-noise ratio (SNR) $\bar{\gamma}$ (in dB) for various values of the correlation coefficient. Figure 13.27 shows that the error probability performance is not much affected by the correlation effects for small values of the correlation coefficient but degrades severely for $\rho > 0.8$. The slope of the bit error probability curve approaches to 2 for the perfectly correlated case ($\rho = 1$) while it is equal to 4 for the uncorrelated case ($\rho = 0$). Note that the diversity order, which is defined as the slope of the bit error probability curve as $\bar{\gamma} \to \infty$, does not provide a realistic performance measure for finite values of $\bar{\gamma}$.

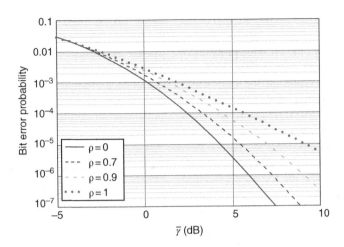

Figure 13.27 Bit Error Probability For BPSK Versus Mean SNR Expressed in dB For Various Values of the Correlation Coefficient in a 3×3 MIMO Channel with One-Sided Correlation.

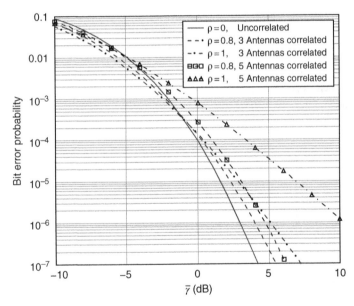

Figure 13.28 BEP For BPSK Versus mean SNR for 3×5 MIMO System When the Smallest or the Largest Number of Antennas is Correlated. The BEP performance is worse when the correlation is at the side with the largest number of antennas. For $\rho = 1$, the diversity order is equal to 5(3) when 3(5) antennas are correlated.

Figure 13.28 shows the bit error probability for a 3×5 MIMO system with double-sided correlation. The bit error probability performance is evidently better when the correlation is at the side with the smallest number of antennas. This is in perfect agreement with (13.35) since the diversity order is five when the correlation is at the side with 3 antennas, but diversity order is three when five antennas are correlated with each other.

Figure 13.29 shows the bit error probability of a 2×2 MIMO system with double-sided correlation with equal values of the correlation coefficients at transmit and receive sides. Uncorrelated 2×2 MIMO system performs approximately 0.75 dB worse than the 1×4 (uncorrelated) MRC system. At $\text{BEP} = 10^{-6}$ level, the degradation in the bit error probability performance is 0.88 dB for $\alpha = \rho = 0.5$ and 2.92 dB for $\alpha = \rho = 0.8$. Note that the diversity order is equal to four for $\alpha = \rho = 0$, but it approaches two for $\alpha = \rho = 1$. As explained in Section 12.4.1 and depicted in Figures 12.16 and 12.17, there is a tradeoff between the mean value of the maximum eigenvalue and the diversity gain as the correlation(s) between N_t transmit and N_r receive antennas change.

Figure 13.30 shows the BEP for BPSK for various sizes of the uncorrelated MIMO beamforming systems. For $\text{BEP} = 10^{-5}$, the average E_b/N_0 required by a SISO system is 33 dB. The relative E_b/N_0 advantage of 2×1, 2×2, 3×2 and 3×3 MIMO systems over the SISO system is 11.66dB, 22.3 dB, 26.07 dB and 28.74 dB, respectively. This clearly shows the great improvement potential provided by MIMO systems in telecommunication links. Note that the transmit power is assumed to be the same irrespetive of the MIMO size. The performance gains come with antenna diversity and signal processing at the transceiver.

13.8 Transmit Antenna Selection (TAS) in MIMO Systems

Consider a MIMO system, with N_t transmit and N_r receive antennas, operating in a flat Rayleigh fading channel. The transmitted and received signals are denoted by x and y, respectively. When the transmitter has no CSI, the usual approach is to divide the transmit power equally between N_t transmit antennas. Alamouti scheme, considered in Chapter 12, uses space-time coding as another solution to this problem if the number of transmit antennas is not large. In this section, we will consider the TAS technique which consists of connecting the RF transmitter to one of N_t transmit antennas with the highest channel gain. Since the

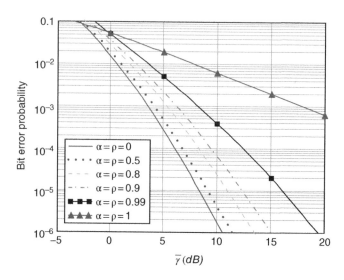

Figure 13.29 Bit Error Probability For BPSK Versus SNR For a 2×2 MIMO System with Double-Sided Correlation with $\alpha = \rho$.

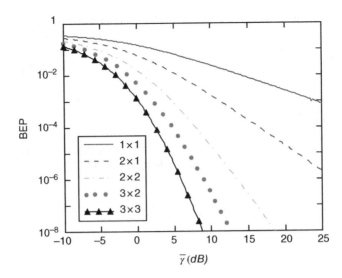

Figure 13.30   Effect of MIMO Size on the BEP For BPSK.

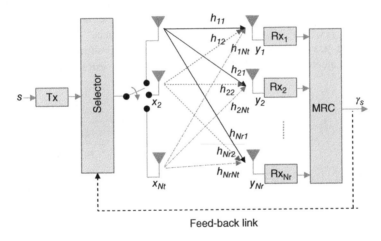

Feed-back link

Figure 13.31   Transmit Antenna Selection (TAS) in a MIMO System.

transmitter does not have the a priori information about the channel gains, this information, required for choosing the transmit antenna, is provided by the receiver via a feed-back link. This technique has the advantage of using a single instead of N_t RF transmitters and the transmit power is used to feed only the chosen transmit antenna. The signals received by N_r receive antennas are maximal ratio combined.

This system, which is depicted in Figure 13.31, will be referred to as $N_t \times N_r$ TAS-MRC system.

If $\bar{\gamma}$ denotes the mean SNR received by any of the N_r receive antennas in Figure 13.31, when only the ith transmit antenna is active, the MRC output γ_i is given by (13.8). Based on the feedback from the receiver MRC output, one of the N_t transmit antennas, which satisfies

$\gamma_s = \max\limits_{1 \le i \le N_t} \{\gamma_i\}$, is selected for transmission. The pdf corresponding to the SNR, given by (13.8), at the MRC output due to ith transmit antenna is already known to be

$$f_{\gamma_i}(\gamma) = \frac{1}{(N_r-1)!}\left(\frac{\gamma}{\bar{\gamma}}\right)^{N_r-1}\frac{\exp(-\gamma/\bar{\gamma})}{\bar{\gamma}},$$

$$i = 1, 2, \cdots, N_t$$

$$(13.88)$$

where $\bar{\gamma} = E[\gamma_i]$, $i = 1,2,...,N_t$ is assumed. The cdf corresponding to (13.88) is given by

$$F_{\gamma_i}(z) = \text{Prob}(\gamma_i < z) = \int_0^z f_{\gamma_i}(\gamma)\, d\gamma$$

$$= 1 - e^{-z/\bar{\gamma}}\sum_{k=0}^{N_r-1}\frac{1}{k!}\left(\frac{z}{\bar{\gamma}}\right)^k, \quad i = 1, 2, ..., N_t$$

$$(13.89)$$

The cdf of $\gamma_s = \max\limits_{1 \le i \le N_t} \{\gamma_i\}$ is defined as the probability that the γ_i's due to N_t transmit antennas are simultaneously (thus SNR γ_s at

MRC output is) less than or equal to a specified threshold value z: [16][19]

$$\text{Prob}(\gamma_s < z) = F_{\gamma_s}(z) = \text{Prob}(\gamma_1 < z, \cdots, \gamma_{N_t} < z)$$

$$= \left[F_{\gamma_i}(z)\right]^{N_t}, \quad E[\gamma_i] = \bar{\gamma}, \quad i = 1, 2,..., N_t$$

$$(13.90)$$

The corresponding pdf is found by taking the derivative of (13.93):

$$f_{\gamma_s}(z) = N_t\, F_{\gamma_i}(z)^{N_t-1}f_{\gamma_i}(z) \qquad (13.91)$$

Figure 13.32 shows the pdf (13.91) of the SNR at the output of N_r maximal-ratio combined receive antennas in a $N_t \times N_r$ MIMO system when one of the N_t transmit antennas is selected for transmission. Figure 13.32 clearly shows that the receive antennas are more effective in increasing the likelihood of the output SNR to assume larger values, compared to transmit antennas.

Figure 13.33 shows the outage performance of TAS-MRC systems (13.90) compared to the

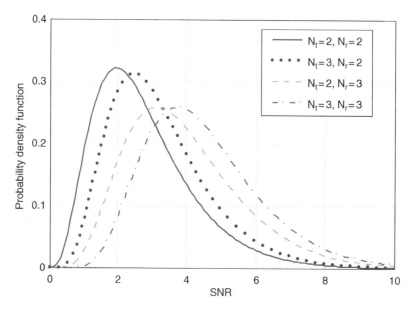

Figure 13.32 The Pdf of the SNR at the MRC Output of N_r Receive Antennas in a $N_t \times N_r$ MIMO System Where One of the N_t Transmit Antennas is Selected.

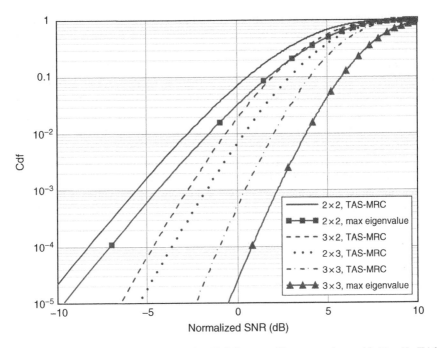

Figure 13.33 The Outage Performance of TAS-MRC Systems. For comparison with $N_t \times N_r$. TAS-MRC system, the outage curves for 2×2 and 3×3 MIMO (maximum eigenvalue) are also shown.

Table 13.3 Outage Performance of TAS-MRC MIMO System with that For SISO and Corresponding MIMO Systems.

Normalized SNR	SISO (13.48)	2×2 TAS-MRC (13.90)	2×2 MIMO, (max. eigen.) (13.49)	1×4 MRC (13.51)	3×2 TAS-MRC (13.90)
0 dB	0.632	0.07	0.032	0.019	0.018
−10 dB	0.095	$2.2 \ 10^{-5}$	$7.5 \ 10^{-6}$	$3.85 \ 10^{-6}$	10^{-7}

MIMO systems (maximum eigenvalue) (see (13.23)). One may observe from Figure 13.33 that the outage performance of a 2×2 TAS-MRC MIMO system is slightly worse than that for the maximum eigenvalue of a 2×2 MIMO system. However, these two curves have the same slope, showing that TAS does not lead to a decrease in the diversity order. It is also clear from Figure 13.33 that the increase in the number of receive antennas is more effective in improving the outage performance.

Example 13.4 Transmit Antenna Selection in a 2×2 MIMO System.
In this example we compare the outage performance of TAS-MRC MIMO system with that for SISO and corresponding MIMO systems at normalized SNR levels of 0 dB and −10 dB (see Figure 13.33). The results are shown in Table 13.3. Table 13.3 and Figure 13.33 show that the outage probability of 2×2 TAS-MRC MIMO system is 2.2 and 2.9 times higher than the 2×2 MIMO system at normalized SNR levels of 0 dB and −10 dB,

respectively. This deterioration in the outage performance is compansated by the use of a single RF transmitter, which is connected to one of the two transmit antennas with the highest channel gain.

The average value of the output SNR, of the $N_t \times N_r$ MIMO system with transmit antenna selection and MRC at the receiver is found by taking the expectation of (13.91). Figure 13.34 shows the variation of the array gain of various TAS-MRC MIMO systems. Note here also that the array gain is more sensitive to the number N_r of receive antennas rather than to N_t.

For the special case for $N_t = 2$, the mean SNR is found as

$$E[\gamma_s] = \int_0^\infty \gamma f_{\gamma_s}(\gamma)\, d\gamma$$

$$= 2N_r \bar{\gamma} \left[1 - \sum_{k=0}^{N_r - 1} \binom{k + N_r}{N_r} 2^{-(k + N_r + 1)} \right]$$

(13.92)

The term $2N_r\bar{\gamma}$ represents the average SNR of a $1 \times 2N_r$ MRC system while the term in paranthesis accounts for the decrease in the average SNR due TAS at the transmitter

compared to that of a $1 \times 2N_r$ MRC system. For $N_r = 2$, the decrease in array gain due to TAS is 1.63 dB and 1 dB compared to 1×4 MRC and 2×2 MIMO systems, respectively.

Figure 13.35 provides a comparison of the TAS-MRC MIMO system with corresponding MIMO and SIMO systems as a function of the number of receive antennas. The array gain of a TAS-MRC $2 \times N_r$ system is approximately 0.8–1 dB lower than that for the corresponding $2 \times N_r$ MIMO system. Slight decrease in array gain may be compansated by an increase in the transmit power. Table 13.4 shows a comparison of the array gains for the considered SIMO, MIMO and TAS systems.

Note that one may extend the transmit antenna selection diversity described above by using generalized selection diversity, which consists of selecting more than a single transmit antenna.

Example 13.5 Bit Error Probability Due to TAS in a 2×2 Uncorrelated MIMO System. In this example we study the degradation in bit error performance due to TAS in a 2×2 uncorrelated MIMO channel. The instantaneous SNR is given by $\gamma = \bar{\gamma}_c \lambda$ where $\bar{\gamma}_c = E_b/N_0$ and λ denote respectively the mean channel

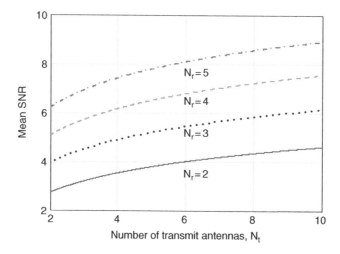

Figure 13.34 Mean SNR at the MRC Output of N_r Receive Antennas in a $N_t \times N_r$ MIMO System Where One of the N_t Transmit Antennas is Selected.

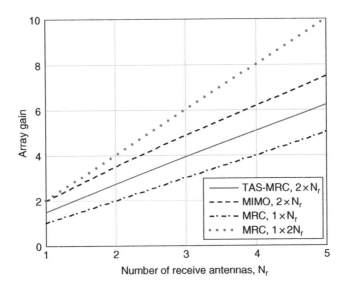

Figure 13.35   Array Gains of $2 \times N_r$ MIMO and $2 \times N_r$ MIMO with TAS at the Transmit Side and MRC at the Receive Side. Array gains for $1 \times N_r$ and $1 \times 2N_r$ MRC systems are also shown.

Table 13.4   Comparison of Output SNR For SIMO and MIMO Systems.

Scenario		SNR at MRC output
1×2 SIMO (MRC)	2	3 dB
2×2 TAS + MRC	2.75	4.37 dB
1×3 SIMO (MRC)	3	4.77 dB
2×2 MIMO (max. eigenvalue)	3.5	5.44 dB
1×4 SIMO (MRC)	4	6 dB

SNR and the maximum eigenvalue. For the considered BPSK modulation, the instantaneous bit error probability for a given value of the SNR is given by $P_b(\gamma) = Q(\sqrt{2\gamma})$.

Using (13.17), the average bit error probability for a SISO system in a Rayleigh fading channel may be written as

$$P_b = \int_0^\infty Q\left(\sqrt{2\bar{\gamma}_c\lambda}\right) e^{-\lambda}\, d\lambda = g(\bar{\gamma}_c,1,1)$$

$$(13.93)$$

where the function $g(c,n,a)$ is defined as follows (see (D.70)):

$$\int_0^\infty e^{-a\lambda}\lambda^{n-1} Q\left(\sqrt{2c\lambda}\right) d\lambda = (n-1)!\ g(c,n,a)$$

$$g(c,n,a) = \frac{1}{2a^n}\left[1 - \sqrt{\frac{c}{c+a}}\sum_{k=0}^{n-1}\binom{2k}{k}\left(\frac{a}{4(c+a)}\right)^k\right]$$

$$(13.94)$$

Average bit error probability for 2×2 MIMO system is obtained as

$$P_b = \int_0^\infty Q\left(\sqrt{2\bar{\gamma}_c\lambda}\right) f_{\lambda_{\max}}(\lambda)\ d\lambda$$

$$= 2\ g(\bar{\gamma}_c,1,1) - 2\ g(\bar{\gamma}_c,1,2)$$

$$- 2\ g(\bar{\gamma}_c,2,1) + 2\ g(\bar{\gamma}_c,3,1)$$

$$(13.95)$$

where the pdf of the maximum eigenvalue is given by (13.20). [21]

Using (13.91), the bit error probability for a 2×2 TAS-MRC system is found to be

$$P_b = \int_0^\infty Q\left(\sqrt{2\bar{\gamma}_c x}\right)[2\ e^{-x}x\{1 - e^{-x}(1+x)\}\ ]dx$$

$$= 2g(\bar{\gamma}_c,2,1) - 2g(\bar{\gamma}_c,2,2) - 4g(\bar{\gamma}_c,3,2)$$

$$(13.96)$$

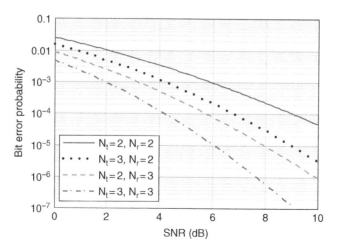

Figure 13.36   Bit Error Probability For BPSK Modulation at the MRC Output of N_r Receive Antennas in a $N_t \times N_r$ MIMO System Where one of the N_t Transmit Antennas is Selected.

The bit error probability for $1 \times n$ MRC (SIMO) system is obtained inserting $N_t = 1$ into (13.91).

$$P_b = \int_0^\infty Q\left(\sqrt{2\bar{\gamma}_c x}\right) \frac{1}{(n-1)!} x^{n-1} e^{-x}\, dx$$

$$= g(\bar{\gamma}_c, n, 1)$$

(13.97)

The corresponding bit error probability for 2×2 Alamouti scheme is obtained by using (12.132) with $n = 4$.

$$P_b = \int_0^\infty Q\left(\sqrt{2\bar{\gamma}_c x}\right) \frac{1}{(n-1)!} (2x)^{n-1} e^{-2x}\, d(2x)$$

$$= g(\bar{\gamma}_c/2, n, 1)$$

(13.98)

(13.98) shows that the Alamouti scheme performs 3-dB worse than the corresponding MRC system described by (13.97).

Figure 13.36 shows the bit error probability of TAS + MRC MIMO systems for various combinations of $N_t \times N_r$. The results are in

agreement with the corresponding pdf's shown in Figure 13.32.

The techniques discussed above are compared in Figure 13.37, where TAS in a 2×2 MIMO system performs 1 dB worse than the 2×2 MIMO system.

13.9   Parasitic MIMO Systems

In classical MIMO systems, the transmit power is divided between transmit antennas. As the number of transmit antennas increases, the improvement in the system performance saturates since the transmit diversity advantage with higher number of antennas is counterbalanced by the decreased transmit power per antenna. However, in parasitic MIMO systems, the transmit power, which feeds only one of the transmit antennas, is used more efficiently than the traditional MIMO systems. Nevertheless, the diversity order is unchanged though all antennas are not fed. The currents induced on the parasitic antennas by the active antennas act as secondary radiators and are used to transmit the same symbol as the active antenna. [22][23][24] Hence, the secondary radiations come with no additional

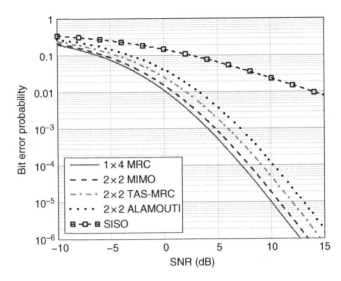

Figure 13.37 BER of a 2×2 MIMO System in Comparison with TAS-MRC, Alamouti and MRC.

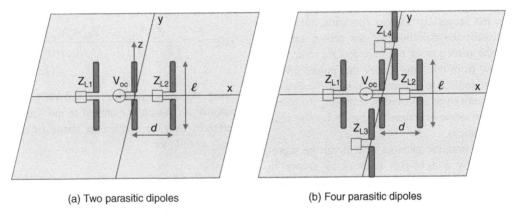

(a) Two parasitic dipoles   (b) Four parasitic dipoles

Figure 13.38 Antenna Arrays Composed of an Active and Two/Four Parasitic Dipoles.

power expenditure. The signals received by the receive antennas are assumed to be uncorrelated and combined using maximal ratio combining (MRC). The system may be designed flexibly by changing the reactive load impedances connected to the parasitic antennas and the distances between antennas. [24] In addition to the efficient use of the transmit power, parasitic MIMO systems are potentially more compact since antennas are located closer to each other which reduces cost and system complexity.

13.9.1   Formulation

Consider an antenna array of three half-wave dipole antennas, as shown in Figure 13.38a. The dipole in the middle is active and connected to a transmitter, with a transmit power P, which feeds the antenna with a voltage V_s in order to support communications using BPSK modulation. The passive antennas are not fed by the transmitter and are loaded with impedances Z_{L1} and Z_{L2}. Electromagnetic

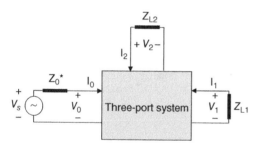

Figure 13.39 A Three-Port Equivalent Circuit For the Parasitic MIMO System with 3-Antennas.

fields radiated by the active antenna cause the parasitic antennas to radiate as secondary sources. The currents induced on the parasitic antennas in the near-field of the active antenna act as secondary sources. It is important to note that the transmit power is economically used by driving only the active antenna and additional advantage may be obtained if the radiations by the secondary sources (parasitic antennas) enhance the radiations of the active antenna in the direction of the intended receiver. We know from the array theory that enhancement of the radiations in the direction of the desired receiver can be achieved by adjusting the phase of the secondary sources, hence by the load impedances.

This parasitic antenna system can be represented by a three-port network as shown in Figure 13.39, where the active antenna is represented by the 0th port, and the parasitic antennas are represented by the 1st and 2nd ports. Here V_s denotes source voltage connected to the terminals of the active antenna. Impedance matching between the active antenna and the source is assumed, that is, the source impedance is matched to the input impedance $Z_0 = V_0/I_0$ of the 0th port. Then the power transferred by the source to the active antenna may be written as

$$P = \frac{V_s^2}{2(Z_0 + Z_0^*)} \qquad (13.99)$$

Using the 3-port system model shown in Figure 13.39, the voltages V_0, V_1, and V_2 across the three antennas may be written in terms of Z_{L1} and Z_{L2} as follows [24]:

$$I_0 Z_0 = V_0 = I_0 Z_{00} + I_1 Z_{01} + I_2 Z_{01}$$
$$-I_1 Z_{L1} = V_1 = I_0 Z_{01} + I_1 Z_{11} + I_2 Z_{12} \quad (13.100)$$
$$-I_2 Z_{L2} = V_2 = I_0 Z_{01} + I_1 Z_{12} + I_2 Z_{22}$$

where $Z_{01} = Z_{02}$ is assumed due to the symmetry in Figure 13.38a. Using the last two equations in (13.100), the ratios α_{01} ve α_{02} of the currents of the parasitic antennas to I_0 of the active antenna may be written as

$$\alpha_{01} = \frac{I_1}{I_0} = \frac{Z_{12} Z_{01} - Z_{01}(Z_{11} + jX_{L2})}{(Z_{11} + jX_{L1})(Z_{11} + jX_{L2}) - Z_{12}^2}$$
$$\alpha_{02} = \frac{I_2}{I_0} = \frac{Z_{12} Z_{01} - Z_{01}(Z_{11} + jX_{L1})}{(Z_{11} + jX_{L1})(Z_{11} + jX_{L2}) - Z_{12}^2}$$
$$(13.101)$$

Using the first equation in (13.100), the impedance Z_0, which is matched to the source impedance, may be written in terms of α_{01} and α_{02} as follows:

$$Z_0 = \frac{V_0}{I_0} = Z_{00} + Z_{01}\alpha_{01} + Z_{02}\alpha_{02} \quad (13.102)$$

Using the reciprocity theorem, we know that

$$Z_{ij} = Z_{ji}^*, \quad i,j = 0,1,2, \quad i \neq j$$
$$Z_{01} = Z_{02} \qquad (13.103)$$

Self impedance of a half-wave $(\ell = \lambda/2)$ dipole is resistive since the reactive part vanishes for $\ell \approx \lambda/2$: [25]

$$Z_{ii} = R_{ii} + jX_{ii}$$

$$R_{ii} = 60 \left\{ \begin{array}{l} C + \ln(k\ell) - Ci(k\ell) + \dfrac{1}{2}\sin(k\ell)[Si(2k\ell) - 2Si(k\ell)] \\[3mm] + \dfrac{1}{2}\cos(k\ell)[C + \ln(k\ell/2) + Ci(2k\ell) - 2Ci(k\ell)] \end{array} \right\}, \quad i = 0,1,2 \qquad (13.104)$$

The mutual impedance between parallel dipoles of length ℓ and separated by a distance d from each other is given by given (12.19) and copied here for the sake of convenience: [25]

$$Z_{ij} = R_{ij} + jX_{ij}, \quad i,j = 0,1,2, \quad i \neq j$$
$$R_{ij} = 30[2 \cdot Ci(kd) - Ci(\mu_1) - Ci(\mu_2)] \quad (13.105)$$
$$X_{ij} = -30[2 \cdot Si(kd) - Si(\mu_1) - Si(\mu_2)]$$

where $C = 0.5772$, $k = 2\pi/\lambda$ (λ is the wavelength), and

$$Ci(x) = -\int_x^\infty \frac{\cos t}{t}\, dt \quad Si(x) = \int_0^x \frac{\sin t}{t}\, dt$$

$$\mu_1 = k\left(\sqrt{d^2 + \ell^2} + \ell\right), \quad \mu_2 = k\left(\sqrt{d^2 + \ell^2} - \ell\right)$$
$$(13.106)$$

Thus, Z_0 can be determined from (13.102) as a function of the load impedances, electrical antenna length and the electrical distance between antennas.

13.9.2 Output SNR

In a 3×3 MIMO beamformer system, all transmit antennas are actively fed, and the maximized output SNR is expressed in terms of the largest eigenvalue λ_{max} of the 3×3 $\mathbf{H}^H\mathbf{H}$ Wishart matrix as $\gamma_{max} = \bar{\gamma}\lambda_{max}$ in (13.86). The pdf of the maximum eigenvalue is given by (13.22) for the uncorrelated case and by (13.33) for one-sided correlation.

For the parasitic MIMO, the transmit weighting vector, which may be expressed as in terms of the ratio of currents,

$$\mathbf{w}_t = [w_0 \ w_1 \ w_2]^T = [1 \ \alpha_{01} \ \alpha_{02}]^T \quad (13.107)$$

does not satisfy the constraint of unity magnitude. As can be observed from (13.102) – (13.105), the weights in (13.107) are determined by the electrical distance between the antennas, electrical antenna lengths, and the load impedances. In order to express the output SNR given by (13.84) in terms the eigenvalues of

$\mathbf{H}^H\mathbf{H}$, we carry out singular value decomposition (SVD) defined by

$$\mathbf{H} = \mathbf{U}\mathbf{\Lambda}\mathbf{V}^H \quad (13.108)$$

where $\mathbf{U}$ is a unitary matrix ($\mathbf{U}\mathbf{U}^H = \mathbf{I}$) and the matrix $\mathbf{V}$ is composed of the eigenvectors corresponding to the ordered eigenvalues of $\mathbf{\Omega}$. Inserting (13.108) into (13.84), the SNR can be rewritten as

$$\begin{aligned}
\gamma &= \bar{\gamma}\mathbf{w}_t^H \mathbf{H}^H \mathbf{H}\mathbf{w}_t \\
&= \bar{\gamma}\underbrace{\mathbf{w}_t^H \mathbf{V}\mathbf{\Lambda}^H}_{\tilde{\mathbf{w}}_t^H} \underbrace{\mathbf{U}^H\mathbf{U}}_{=\mathbf{I}}\mathbf{\Lambda}\underbrace{\mathbf{V}^H\mathbf{w}_t}_{\tilde{\mathbf{w}}_t} \\
&= \bar{\gamma}\tilde{\mathbf{w}}_t^H \mathbf{\Lambda}^H \mathbf{\Lambda}\tilde{\mathbf{w}}_t
\end{aligned} \quad (13.109)$$

where an equivalent weight vector is defined by

$$\tilde{\mathbf{w}}_t = \mathbf{V}^H\mathbf{w}_t \quad (13.110)$$

Inserting

$$\mathbf{\Lambda}^H\mathbf{\Lambda} = \begin{pmatrix} \lambda_0 & 0 & 0 \\ 0 & \lambda_1 & 0 \\ 0 & 0 & \lambda_2 \end{pmatrix} \quad (13.111)$$

into (13.109), one can write the output SNR for a parasitic MIMO system in terms of the eigenvalues of $\mathbf{H}^H\mathbf{H}$ as follows:

$$\gamma = \bar{\gamma}\sum_{i=0}^2 \lambda_i |\tilde{w}_i|^2 = \bar{\gamma}\sum_{i=0}^2 \lambda_i \left| \mathbf{V}^{-H}\mathbf{w}_t \right|_i^2 \quad (13.112)$$

In a correlated MIMO channel, the pdf of the ordered eigenvalues is presented in Appendix E (see (E.13)). The covariance matrix may be written in terms of the correlation coefficients (ρ_{ij}) between the antennas

$$\mathbf{\Omega} = \begin{bmatrix} \rho_{00} & \rho_{01} & \rho_{02} \\ \rho_{10}^* & \rho_{11} & \rho_{12} \\ \rho_{20}^* & \rho_{21}^* & \rho_{22} \end{bmatrix} \quad (13.113)$$

The complex correlation coefficients are defined in terms of the self- and mutual-impedances as follows:

$$\rho_{ij} = Z_{ij}/Z_{ii}, \quad i,j = 0,1,2 \quad (13.114)$$

From the reciprocity theorem and (13.101), we can write $\rho_{ij} = \rho_{ji}^*$, $\rho_{01} = \rho_{02}$. Since the self-impedances of the three antennas are the same since they have the same length and thickness, we have $\rho_{00} = \rho_{11} = \rho_{22} = 1$. On the other hand, in view of (13.103), the correlation coefficients are complex and decrease with distance as expected. The ordered eigenvalues $w_0 > w_1 > w_2$ and the corresponding eigenvectors, forming the vector $\mathbf{V}$,

$$
\begin{aligned}
\mathbf{V} &= (\mathbf{v}_0 \quad \mathbf{v}_1 \quad \mathbf{v}_2) \\
\mathbf{v}_i &= [v_{i1} \quad v_{i2} \quad v_{i2}]^T, \quad i = 0, 1, 2
\end{aligned}
\tag{13.115}
$$

can be found by using standard techniques.

13.9.3 Radiation Pattern

In the equatorial plane, the array factor of this 3×3 array antanna may be written as

$$
|f_a| = \left| w_0 + w_1 e^{-jkd\cos\phi} + w_2 e^{jkd\cos\phi} \right| \tag{13.116}
$$

where the angle ϕ is defined with respect to the direction of the array axis. The radiation pattern, in the equatorial plane, of the considered transmit array is shown in Figure 13.40 for $d = 0.33\,\lambda$ as a function of ϕ. For the uncorrelated case, the parasitic antennas act as director and reflector; hence, the array gain is maximized at $\phi = 0°$. However, for the correlated case, the array gain is maximized in the directions $\phi = 66°$ and $360°–66°$. Nevertheless, the the array factor is still higher than the uncorrelated case in the desired direction $\phi = 0°$.

13.9.4 Bit Error Probability

Here, we compare the performance of the considered 3×3 parasitic MIMO-MRC system with the 3×3 MIMO-MRC beamformer system. Antennas are chosen as half-wave dipoles. The loads Z_{L1} and Z_{L2} connected to the parasitic antennas are chosen so as to maximize the antenna gain and to minimize the bit error probability for each value of the distance d. This implies that the reactive loads are

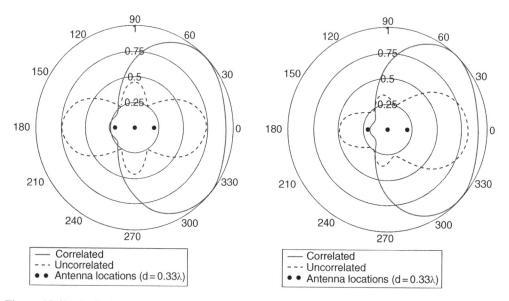

Figure 13.40 Radiation Pattern in the Equatorial Plane of a Parasitic Array with Three Antennas. The left-hand side is for $Z_{L1} = Z_{L2}$ and the right hand side applies for $Z_{L1} \neq Z_{L2}$. [26]

optimized for each value of d. The receiver is assumed to be in the far-field of the transmitter.

For BPSK, the bit error probability in a Rayleigh fading channel may be written as

$$P_b = \int_0^\infty Q\left(\sqrt{2\bar{\gamma}\lambda}\right) f_{\lambda_{max}}(\lambda)\, d\lambda \qquad (13.117)$$

where the pdf of the maximum eigenvalue $f_{\lambda_{max}}(\lambda)$ is given by (13.22) and (13.33) in uncorrelated and correlated MIMO channels, respectively. The Gaussian Q function is defined by (B.1).

Since the induced currents on the parasitic antennas change with distance d, the phase difference between the array elements also change. Consequently, the input impedances, the weighting factors, radiation pattern, array gain and the resulting bit error probability are all influenced by the distance in a complicated manner. Figure 13.41 shows the variation of the bit error probability as a function of d/λ for a mean SNR $\bar{\gamma} = 4\ dB$. The bit error probability was observed to be minimum at $d = 0.33\lambda$. [26][27] Evidently, small values of d/λ

facilitate the interaction between the active and parasitic antennas. This also explains the significant increases in the bit error probability for larger values of d/λ. Parasitic MIMO system evidently outperforms the beamforming system due to the exploitation of the parasitic antennas. Correlation in parasitic MIMO system helps improving the bit error probability performance.

Figure 13.42 shows the comparative performance of the parasitic MIMO-MRC system. In the case of beamforming, the same symbol, which is transmitted from all transmit antennas, are received by the receive antennas and are maximal-ratio combined. However, in the parasitic system, only the active antenna is fed and the signals received by the receive antennas are also maximal-ratio combined. For 10^{-7} bit error probability level, the correlated parasitic MIMO-MRC system performs approximately 3 dB better than the correlated MIMO beamforming system using the maximum eigenvalue. [26][27] If the antennas are uncorrelated, the currents induced on the parasitic antennas and hence the secondary

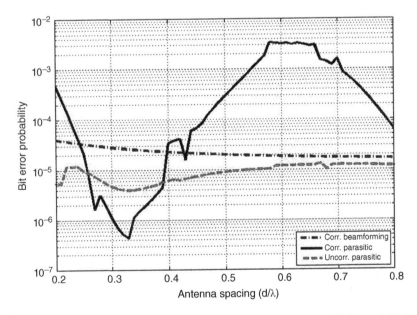

Figure 13.41 Variation of the Bit Error Probability as a Function of d/λ for $\bar{\gamma} = 4 dB$. [26]

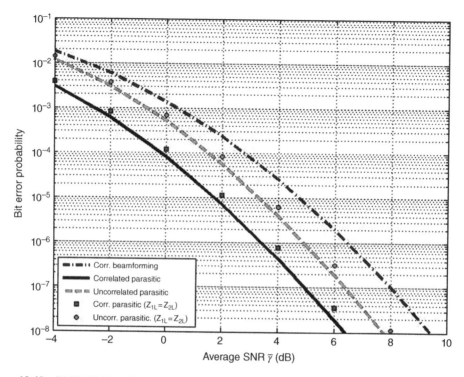

Figure 13.42   BPSK Bit Error Probability For Correlated/Uncorrelated Parasitic 3×3 MIMO-MRC System in Comparison with Conventional MIMO-MRC System For $d = 0.33\lambda$. [26]

radiations are lower. This leads degraded performance compared to correlated parasitic MIMO-MRC system. One may therefore conclude that the correlation between MIMO antennas may be exploited to improve the system performance. On the other hand, by placing MIMO antennas closer to each other one may have the advantage of size, cost and simplification of antenna feeding circuits.

13.9.5 5×5 Parasitic MIMO-MRC

Figure 13.43 shows the variation of the bit error probability as a function of d/λ for a mean SNR $\bar{\gamma} = -8$ dB. The bit error probability was observed to be minimum at $d = 0.4\lambda$. At this value of the distance, the correlated

parasitic MIMO system outperforms correlated beamforming and uncorrelated parasitic systems due to mainly the exploitation of correlation and parasitic antennas. Figure 13.44 shows the bit error probability performance of the parasitic MIMO-MRC system in comparison with correlated beamforming and uncorrelated parasitic MIMO. For a 10^{-7} bit error probability level, the correlated parasitic MIMO-MRC system performs approximately 8 dB better than the correlated MIMO beamforming system. If the parasitic antennas are assumed to be uncorrelated with the active antenna and with each other, the currents induced on the parasitic antennas and hence the secondary radiations will be lower. This leads degraded performance compared to correlated parasitic MIMO-MRC system. [26][27]

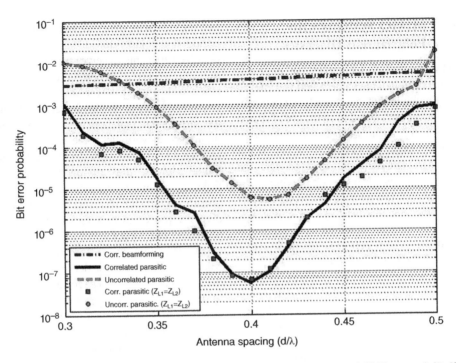

Figure 13.43 BPSK Bit Error Probability For 5×5 MIMO as a Function of d/λ For $\bar{\gamma} = -8\,dB$. [26]

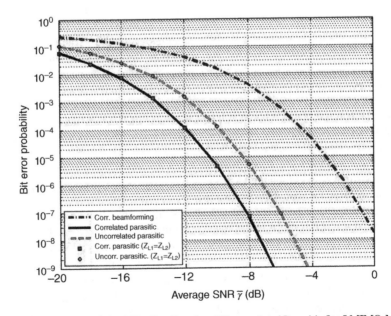

Figure 13.44 BPSK Bit Error Probability For Correlated/Uncorrelated Parasitic 5×5 MIMO-MRC System in Comparison with Conventional MIMO-MRC System For $d = 0.4\lambda$ and $\bar{\gamma} = -8\,dB$. [26]

13.10   MIMO Systems with Polarization Diversity

Spatial diversity is a technique that improves performance in fading channels and is achieved by using multiple antennas. Significant performance gains can be realized from the use of spatial diversity if the separation between antennas is sufficiently large; then, there is a significant degree of statistical independence between the received fading signals from diversity branches. However, when the antennas are located closer to each other, mainly due to space limitations, the statistical independence between signals is lost and the system performance, in terms of diversity and multiplexing gains, degrades.

In addition to the space diversity, MIMO systems can also exploit polarization diversity by using antennas with orthogonal polarizations. The use of dual-polarized antennas provides an additional advantage of removing the space limitations due to the use of multiple antennas. A dual polarized antenna implies a single antenna operating simultaneously at two orthogonal polarizations, that is, the co-location of two antennas with orthogonal polarizations. This minimizes the need for extra space needed for multiple antennas. If, for example, a 2×2 MIMO system is used at both sides, then this channel behaves as a 4×4 MIMO since each antenna operates at two orthogonal polarizations.

The MIMO system may work as a beamformer if the same signal is transmitted at the two orthogonal polarizations; the signals received by orthogonal polarizations are then usually combined by maximal ratio combining (MRC). If such systems are used for spatial multiplexing (SM), then the orthogonally polarized channels carry different data streams. In this case, maximum-likelihood (ML) detection is typically used at the receiver.

In its basic form, a MIMO transmitter with a dual polarized antenna transmits signals on two (linear, circular or elliptical) orthogonal polarizations. Depending on the type of antennas used, transmit and/or receive antennas may not provide a perfect isolation between two orthogonal polarizations. Hence, a vertically polarized transmit (receive) antenna may also transmit (receive) at horizontal polarization, albeit with a much lower antenna gain. With appropriate choice and/or design of antennas, the cross-polar isolation of antennas can be improved. Nevertheless, the scatterers in the propagation channel also depolarise the transmitted signals, thus giving rise to energy leakage into orthogonal polarizations. Consequently, a MIMO receiver with polarization diversity processes two signals at orthogonal polarizations in order to decode the transmitted signal. Therefore, the performance of a polarization diversity system depends strongly on the environment and can be formulated in terms of the cross-polarization discrimination (XPD) and the cross correlations of the signal envelopes between branches. XPD is defined as the ratio of the signal power received in the wanted (transmitted) polarization to that received in the unwanted (orthogonal) polarization expressed in dB. Results show that the XPD is typically between 5–20 dB, and is much higher in suburban than in urban environments. [28][29][30][31][32]. *The cross correlation coefficient* is a measure of the correlation between the signals arriving at the two antennas. The signals can be in the same or in different polarizations. System performance will improve as the cross correlation coefficient between the signals decreases. The uncorrelated signals also have zero cross correlation.

13.10.1   The Channel Model

Assuming that the channel is flat over the frequency-band of interest, the input-output relationship is given by (13.1), $\mathbf{y} = \mathbf{Hx} + \mathbf{n}$, where $\mathbf{x} = [x_1 \ x_2]^T$ is the transmit signal vector whose elements are taken from a finite (complex) constellation with average symbol energy $E_s = E\left[|x_i|^2\right]$, $i = 1, 2$. The receive signal vectors is given by $\mathbf{y} = [y_1 \ y_2]^T$, and $\mathbf{n}$ is complex valued Gaussian noise with $E[\mathbf{n}\mathbf{n}^H] = \sigma_n^2 \mathbf{I}_2$. The channel matrix $\mathbf{H}$, which is also called as the polarization matrix in this case, is given by

$$\mathbf{H} = \begin{bmatrix} h_{11} & h_{12} \\ h_{21} & h_{22} \end{bmatrix} \qquad (13.118)$$

The polarization matrix describes the degree of suppression of individual co-polarized and cross-polarized components, cross correlation, and cross coupling of energy from one polarization to the other. The correlation between the elements of the matrix $\mathbf{H}$ depends on the propagation conditions. The signals x_1 and x_2 are transmitted on two orthogonal polarizations, y_1 and y_2 are the signals received on the corresponding polarizations. The channel is hence a 2×2 MIMO channel, since each polarization mode is treated as an independent physical channel.

Assuming that the channel is Rayleigh fading, that is, the elements of the matrix $\mathbf{H}$ are complex Gaussian random variables with zero mean and unity variance, we can write

$$E\left(|h_{11}|^2\right) = E\left(|h_{22}|^2\right) = 1 \qquad (13.119)$$

The channel gains h_{12} and h_{21} account for leakages from vertical to horizontal and from horizontal to vertical polarizations. Therefore, their variance α is smaller than unity $(0 < \alpha \leq 1)$:

$$E\left(|h_{12}|^2\right) = E\left(|h_{21}|^2\right) = \alpha \qquad (13.120)$$

The case of $\alpha = 1$ implies that two physical antennas on each side of the link employing the same polarization, that is, the classical MIMO channel.

Figure 13.45 shows a schematic description of this dual-polarized channel between transmitter and receiver. Transmitter transmits at vertical (VP) and horizontal polarizations (HP) and receiver receives symbols with VP and HP antenna(s), accordingly.

Cross polarization discrimination, XPD, is defined as the ratio of the power received in wanted polarization (transmitted) to that received in the unwanted (orthogonal) polarization expressed in dB. The XPD for VP is defined as

$$XPD_v = 20 \log_{10} |a_c/b_x|, \quad (dB) \qquad (13.121)$$

Cross polarization isolation, XPI, is defined as the ratio of the power received in wanted (transmitted) polarization to the power of wanted signal with unwanted polarization:

$$XPI_v = 20 \log_{10} |a_c/a_x| \quad (dB) \qquad (13.122)$$

The above definitions, which are valid for wanted signals with VP, can easily be extended for HP. Under certain conditions, XPI and XPI values are assumed to be the same, but this may

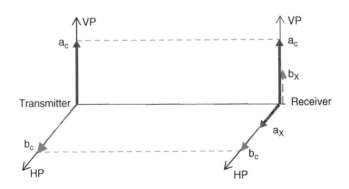

Figure 13.45   Depolarization in a Dual-Polarized (VP and HP) Channel.

not be realistic. Here, we will consider a channel model based on XPD values. In view of (13.119) – (13.121), the XPD may be written as

$$XPD = 10\log_{10}\left(E\left[|h_{11}/h_{12}|^2\right]\right)$$
$$= 10\log_{10}\left(E\left[|h_{22}/h_{21}|^2\right]\right) = 10\log_{10}(1/\alpha)$$
$$(13.123)$$

The correlation coefficient defined by (12.35) may be extended to account for the correlation between orthogonally polarized signals. The so-called *cross correlation coefficient* is defined as

$$\rho = \frac{E[(r_1 - \bar{r}_1)(r_2 - \bar{r}_2)]}{\sqrt{E\left[(r_1 - \bar{r}_1)^2\right]E\left[(r_2 - \bar{r}_2)^2\right]}} \quad (13.124)$$

where r_1 and r_2 are the envelopes of the signals on the diversity branches with orthogonal polarizations. They have zero-means and their variances are given by (13.119) and (13.120). [28] Transmit and receive correlation coefficients are then found to be

$$t = \frac{E\left(h_{11}\,h_{12}^*\right)}{\sqrt{\alpha}} = \frac{E\left(h_{21}\,h_{22}^*\right)}{\sqrt{\alpha}} \quad (13.125)$$

$$r = \frac{E\left(h_{11}\,h_{21}^*\right)}{\sqrt{\alpha}} = \frac{E\left(h_{12}\,h_{22}^*\right)}{\sqrt{\alpha}} \quad (13.126)$$

It is also assumed that

$$E\left(h_{11}\,h_{22}^*\right) = E\left(h_{21}\,h_{12}^*\right) = 0 \quad (13.127)$$

This implies that the channel gains for co-polar and cross-polar channels are uncorrelated. Similarly, channel gains h_{12} and h_{21}, accounting for co-polar to cross-polar and cross-polar to co-polar leakes, are uncorrelated. Measured values of XPD and correlation coefficients are reported in [29][30][31][32][33][34]. In [33] the distribution of the XPD is given for the different measurements for vertically and horizontally polarized antennas. Here XPD, defined by (13.123), is assumed to be a lognormal random variable.

13.10.2   Spatial Multiplexing (SM) with Polarization Diversity

SM exploits multiple antennas at both transmitter and receiver to increase the transmission rate in a wireless radio link with no additional bandwidth consumption. The performance of this technique is highly dependent on the channel statistics and the antenna correlations. In a 2×2 MIMO SM system, as shown in Figure 13.46, the stream of (possibly coded) information symbol $\{b_0, b_1, b_2, b_3, b_4, b_5, b_6,\ldots\}$ is split into two independent substreams $\{b_0, b_2, b_4, b_6,\ldots\}$ and $\{b_1, b_3, b_5,\ldots\}$. These substreams are applied separately to two transmit antennas. At the receiver, each of the two

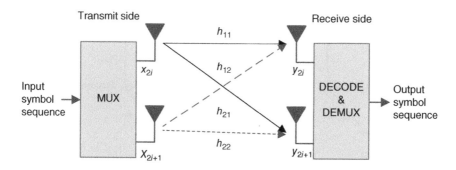

Figure 13.46   SM System Configuration.

substreams is decoded and demultiplexed. A linear SM receiver can be viewed as a bank of superposed spatial weighting filters where every filter aims at extracting one of the multiplexed substreams by spatially nulling the remaining ones. This evidently assumes that the substreams have different signatures.

A $N \times N$ SM system thus allows a transmitter-receiver pair to communicate through N parallel spatial channels hence allowing for a possible N-fold improvement in the link speed. More improvement may actually be obtained by taking into account the diversity gain offered by the multiple antennas and scheduling. The performance parameters are derived based on the assumption that the substreams are independent from each other. In practice, the level of independence between substreams will determine the actual link performance.

Channel fading is assumed Rayleigh and flat. It is also assumed that the channel matrix $\mathbf{H}$ is unknown to the transmitter, but is perfectly known to the receiver to enable the maximum likelihood (ML) decoding. The receiver computes the ML estimate according to

$$\hat{\mathbf{x}} = \underset{\mathbf{x}}{\arg\min} \, \|\mathbf{y} - \mathbf{H}\mathbf{x}\|^2 \qquad (13.128)$$

The minimization is performed over the set of all possible transmitted codevectors, $\mathbf{x}$. Assume that $\mathbf{c}$ denotes a 2×1 transmitted codevector. For a given realization of the channel matrix $\mathbf{H}$, the probability that the receiver decides erroneously in favor of the vector $\mathbf{e}$ is given by the pair-wise error probability [35]

$$P(\mathbf{c} \to \mathbf{e} | \mathbf{H}) = Q\left(\sqrt{\frac{E_s}{2\sigma_n^2} d^2(\mathbf{c}, \mathbf{e} | \mathbf{H})}\right)$$

$$\leq e^{-\frac{E_s}{4\sigma_n^2} d^2(\mathbf{c}, \mathbf{e} | \mathbf{H})}$$

$$(13.129)$$

where the Chernoff bound $Q(x) \leq e^{-x^2/2}$ is used as the upper bound. The squared-Euclidean distance between c and e is given by,

$$d^2(\mathbf{c}, \mathbf{e} | \mathbf{H}) = \|\mathbf{H}(\mathbf{c} - \mathbf{e})\|^2 = \lambda_1 z_1^2 + \lambda_2 z_2^2$$

$$(13.130)$$

is expressed in terms of the eigenvalues λ_1, λ_2 of the Wishart matrix $\mathbf{H}^H\mathbf{H}$ and the 2×1 vector $\mathbf{z} = \mathbf{c} - \mathbf{e}$. The pair-wise error probability averaged over all channel realizations in a Rayleigh fading channel (see (F.135)) is upper-bounded by [35]

$$P(\mathbf{c} \to \mathbf{e}) = E_{\mathbf{H}}[P(\mathbf{c} \to \mathbf{e} | \mathbf{H})]$$

$$\leq \int_0^\infty e^{-\frac{E_s}{4\sigma_n^2}\lambda_1 z_1^2} z_1 e^{-z_1^2/2} dz_1$$

$$\int_0^\infty e^{-\frac{E_s}{4\sigma_n^2}\lambda_2 z_2^2} z_2 e^{-z_2^2/2} dz_2$$

$$= \frac{1}{1 + \lambda_1 \dfrac{E_s}{2\sigma_n^2}} \frac{1}{1 + \lambda_2 \dfrac{E_s}{2\sigma_n^2}}$$

$$(13.131)$$

13.10.3 MIMO Beamforming-MRC System with Polarization Diversity

For the MIMO beamforming system, shown in Figure 13.26, the same signal s is transmitted simultaneously from VP and HP antennas. The output r of the two-branch MRC system given by (13.83) is used by the ML estimator to estimate the transmitted symbol. For PSK signals, the ML estimator at the output of the MRC estimates that the symbol s_i was transmitted if $d^2(r, s_i) \leq d^2(r, s_k)$, $\forall i \neq k$.

13.10.4 Simulation Results

Instead of using the Wishart matrix with double-sided correlation, simulation techniques are employed for generating the channel matrix. This was mainly to account for the log-normally distributed XPD given by (13.123). For this purpose, we firstly generated a 4×1 complex Gaussian random vector $\mathbf{h_W}$ whose elements are i.i.d. with zero mean and unity variance. Then, by using the linear transformation $\mathbf{h} = \mathbf{R}\mathbf{h_w}$, the channel matrix with the desired properties is obtained. Noting that the covariance matrix of $\mathbf{h_w}$ is an identity matrix, the channel covariance matrix is related to $\mathbf{RR}^H$ as follows: $E\left(\mathbf{hh}^H\right) = E\left(\mathbf{Rh_w}\mathbf{h_w}^H\mathbf{R}^H\right) = \mathbf{RR}^H$ where

$$E\left(\mathbf{hh}^H\right) = \begin{bmatrix} 1 & t\sqrt{\alpha} & r\sqrt{\alpha} & 0 \\ t^*\sqrt{\alpha} & 1 & 0 & r\sqrt{\alpha} \\ r^*\sqrt{\alpha} & 0 & 1 & t\sqrt{\alpha} \\ 0 & r^*\sqrt{\alpha} & t^*\sqrt{\alpha} & 1 \end{bmatrix}$$

$$\mathbf{h} = \begin{bmatrix} h_{11} & h_{12} & h_{21} & h_{22} \end{bmatrix}^T$$

$$(13.132)$$

By using Cholesky decomposition, the transformation matrix $\mathbf{R}$ is determined from (13.132). Then, the channel gain vector $\mathbf{h}$ is obtained using $\mathbf{h} = \mathbf{R}\mathbf{h_w}$.

The simulated system, which consists of a dual-polarized transmit and a dual-polarized receive antenna, uses 4-PSK and employs a ML receiver. Figure 13.47 compares the symbol error probability (SEP) of SM and MRC systems in the presence of polarization diversity, obtained by using Monte Carlo simulation for $t = 0.5$, $r = 0.3$, $\alpha = 0.4$ (XPD = 4 dB). [36] The mean signal-to-noise ratio SNR is defined as

$$SNR = 10\log\left(2E_s/\sigma_n^2\right) \qquad (13.133)$$

Performance of a SM system is not as good as for the MRC system. This may be explained by the fact that, in a SM system, two different symbols are simultaneously transmitted from two orthogonally polarized antennas while the same symbol is transmitted from two antennas in a MRC system. Hence the diversity order in MRC system is twice the diversity order of the SM system. However, the transmission rate of the SM system is twice that for the MRC system. Figure 13.48 shows the performance of the MRC system with dual polarized antennas, both at the transmitter and the receiver, when the parameter α is chosen as a constant (deterministic) and a lognormal random variable.

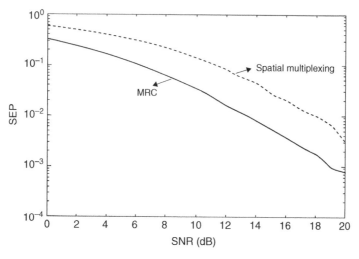

Figure 13.47 Symbol Error Probability of MRC System with Dual Polarized Transmit and Receive Antennas For $t = 0.5$, $r = 0.3$ and $\alpha = 0.4$. [36]

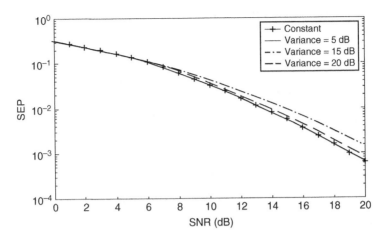

Figure 13.48 Symbol Error Probability of the MRC System with Dual Polarized Antennas Both at Receiver and Transmitter, For $t = 0.5$, $r = 0.3$. The XPD is a lognormal random variable with mean $= 4$ dB ($\alpha = 0.40$) and variance=5, 15 and 20 dB. [36]

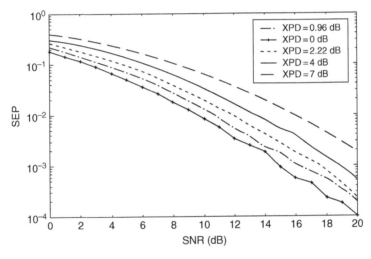

Figure 13.49 Symbol Error Probability of the MRC System with Dual Polarized Antennas Both a Receiver and Transmitter when XPD is Constant (Deterministic) and $t = 0.5$, $r = 0.3$. [36]

[33] Log-normally distributed XPD (dB) has 4 dB mean and variances changing between 5–20 dB. The effect of the variance on the system performance becomes more pronounced at higher SNRs. Figure 13.49 shows the effect of the random variable XPD when its variance is constant and its mean changes. [36]

References

[1] A. Høst-Madsen and A. Nosratinia, "The multiplexing gain of wireless networks," *presented at ISIT'05*, Adelaide Australia, 2005.

[2] M. E. Yanık, "Effects of correlation and antenna polarization on the MIMO system performance", M.Sc. thesis, Hacettepe University, Dept. of Electrical and Electronics Engineering, 2010.

[3] M. Chiani, M.Z. Win and A. Zanella, "On the capacity of spatially correlated MIMO Rayleigh-fading channels", *IEEE Trans. Information Theory*, vol. **49**, no. 10, pp. 2363–2371, October 2003.

[4] M. Chiani, M. Z. Win, A. Zanella, R. K. Mallik, and J. H. Winters, "Bounds and approximations for optimum combining of signals in the presence of multiple co-channel interferers and thermal noise," *IEEE Trans. Commun.*, vol. **51**, pp. 296–307, Feb. 2003.

[5] A. Zanella, M. Chiani, and M. Z. Win, "On the marginal distribution of the eigenvalues of Wishart matrices," *IEEE Trans. Commun.*, vol. **57**, pp. 1050–1060, April 2009, [eq. 41].

[6] P. Xia, H. Niu, J. Oh and C. Ngo, "Diversity analysis of transmit beamforming and the application in IEEE 802.11n systems", *IEEE Trans. Vehicular Technology*, vol. **57**, no. 4, pp. 2638–2642, July 2008.

[7] P. J. Smith, P-H Kuo and L. M. Barth, "Level crossing rates for MIMO channel eigenvalues: implications for adaptive systems", *in Conf. Rec. 2005 IEEE Int. Conf. Commun. (ICC)*, vol. **4**, pp. 2442–2446.

[8] D.–W. Yue, and Q. T. Zhang, Generic approach to the performance analysis of correlated transmit/receive diversity MIMO systems with/without co-channel interference, *IEEE Trans.* Information Theory, vol. **56**, no. 3, pp. 1147–1157, March 2010.

[9] M. Kang and M.-S. Alouini, A comparative study on the performance of MIMO MRC systems with and without cochannel interference, *IEEE Trans. Communications*, vol. **52**, no. 8, pp. 1417–1425, August 2004.

[10] M. Kang and M.-S. Alouini, Largest eigenvalue of complex Wishart matrices and performance of MIMO MRC systems, IEEE J. Sel. Areas Commun. Special Issue on MIMO Syst. Appl. (JSAC-MIMO), vol. **21**, pp. 418–426, April 2003.

[11] A. J. Grant, "Performance analysis of transmit beamforming", *IEEE Trans. Communications*, vol.**53**, no. 4, pp. 738–744, April 2005.

[12] M. R. McKay, A. J. Grant, and I. B. Collings, "Performance analysis of MIMO-MRC in double-correlated Rayleigh environments," *IEEE Trans. Commun.*, vol. **55**, no. 3, pp. 497–507, Mar. 2007.

[13] S. H. Simon, A. L. Moustakas, and L. Marinelli, "Capacity and character expansions: moment-generating function and other exact results for MIMO correlated channels," *IEEE Trans. Information Theory*, vol. **52**, pp. 5336–5351, December 2006.

[14] B. Friedlander and S. Scherzer, "Beamforming versus transmit diversity in the downlink of a cellular communications system," *IEEE Trans. Vehicular Technology*, vol. **53**, pp. 1023–1034, July 2004.

[15] J. B. Anderson, "Array gain and capacity for known random channels with multiple element arrays at the both ends", *IEEE Journal SAC*, vol. **18**, no. 11, pp. 2172–2178, November 2000.

[16] Z. Chen, J. Yuan and B. Vucetic, "Analysis of transmit Antenna Selection/MRC in Rayleigh Fading Channels", *IEEE Trans. Vehicular Technology*, vol. **54**, no. 4, pp. 1312–1321, July 2005.

[17] J. Gong, M. R. Soleymani, and J. F. Hayes, "A rigorous proof of MIMO channel capacity's increase with antenna number", *Wireless Pers Commun.* (2009) vol. **49**, pp. 81–86, DOI 10.1007/s11277-008-9558-2.

[18] S. Haykin, *Communication Systems* (4th ed.), John Wiley.

[19] S. Catreux, L. J. Greenstein, and V. Erceg, "Some results and insights on the performance gains of MIMO systems", *IEEE Journal SAC*, vol. **21**, no. 5, pp. 839–847, June 2003.

[20] P. A. Dighe, R. K. Mallik, and S. S. Jamuar, "Analysis of transmit–receive diversity in Rayleigh fading", *IEEE Trans. Commun.*, vol. **51**, no. 4, pp. 694–703, April 2003.

[21] B. D. Rao and M. Yan, "Performance of maximal ratio transmission with two receive antennas", *IEEE Trans. Communications*, vol. **51**, no. 6, pp. 894–895, June 2003.

[22] A. Mohammadi and F. M. Ghannouchi, Single RF front-end MIMO receivers, *IEEE Commun. Mag.*, pp. 104–109, Dec. 2011.

[23] O. N. Alrabadi, C. B. Papadias, A. Kalis, N. Marchetti, and R. Prasad, "MIMO transmission and reception techniques using three-element ESPAR antennas", *IEEE Communications Letters*, vol. **13**, no. 4, pp. 236–238, April 2009.

[24] L. Petit, L. Dussopt, ans J.-M. Laheurte, "MEMS-Switched Parasitic-Antenna Array for Radiation Pattern Diversity", *IEEE Trans. Antennas and Propagation*, vol. **54**, no. 9, pp. 2624–2631, Sept. 2006.

[25] C. A. Balanis, "Antenna Theory: Analysis and Design" (2nd ed.), pp. 410–418, John Wiley Sons: New York, 1997.

[26] Ö. Haliloğlu, "Performance analysis of parasitic MIMO systems", M.Sc. thesis, Hacettepe University, Dept. of Electrical and Electronics Engineering, 2012.

[27] Ö. Haliloğlu, and M. Şafak, Performance analysis of parasitic MIMO, IEEE 19th Symposium on Signal Processing and Telecommunication Applications (SIU), 20–22 April 2011, Antalya, Turkey.

[28] C.B. Dietrich, K. Dietze, J. R. Nealy, and W. L. Stutzman, Spatial, polarization, and pattern diversity for wireless handheld terminals, *IEEE Trans. Antennas and Propagation*, Vol. **49**, No. 9, pp. 1271–1281, 2001.

[29] J. J. A. Lempiainen, J. K. Laiho-Steffens, and A. F. Wacker, Experimental results of cross polarization discrimination and signal correlation values for a polarization diversity scheme, *IEEE 47th Vehicular Technology Conf.*, Vol. **3**, pp. 1498–1502, 1997.

[30] A.M.D. Turkmani, A. A. Arowojolu, P.A. Jefford, and C. J. Kellett, An experimental evaluation of the performance of two-branch space and polarization

diversity schemes at 1800 MHz, *IEEE Trans. Vehicular Technology*, Vol. **44**, No. 2, pp. 318–326, 1995.

[31] J.F. Lemieux, M. S. El-Tanany, and H.M. Hafez, Experimental evaluation of space/frequency/polarization diversity in the indoor wireless channel, *IEEE Trans. Vehicular Technology*, Vol. 40, No. 3, pp. 569–574, 1991.

[32] P.C.F. Eggers, I.Z. Kovacs, and K. Olesen, Penetration effects on XPD with GSM1800 handset antennas, relevant for BS polarization diversity for indoor coverage, *IEEE 48th Vehicular Technology Conf.*, Ottowa, Canada, pp. 1959–1963, 1993.

[33] R.U. Nabar, V. Erceg, H. Bölcskei, and A. J. Paulraj, Performance of multi-antenna signaling strategies using dual polarized antennas: Measurement results and analysis, Int. Symposium on Wireless Personal Multimedia Communications (WPMC) Aalborg, Denmark, pp. 175–180, 2001.

[34] L. Lukama, A. M. Street, K. Konstantinou, D. J. Edwards, and A. P. Jenkins, Polarization diversity performance for UMTS, IEE 11th Int. Conf. Antennas and Propagation, 2001, pp. 502–507.

[35] H. Bölcskei, R. U. Nabar, V. Erceg, D. Gespert, and A. J. Paulraj, Performance of spatial multiplexing in the presence of polarization diversity, IEEE-ICASSP 2001, Salt Lake City, UT, Vol. **4**, pp. 2437–2440, May 2001.

[36] S. Bozay ve M. Şafak, Performance analysis of spatial multiplexing and maximal ratio combining systems in the presence of polarization diversity, IEEE 11th Symposium on Signal Processing and Telecommunication Applications (SIU 2003), 18-20 June 2003, Istanbul, Turkey, pp. 273–276 (in Turkish).

[37] Z. Chen, J. Yuan and B. Vucetic, Analysis of TAS/MRC in Rayleigh fading channels, *IEEE Trans. VT*, vol. **54**, no. 4, pp. 1312–1321, July 2005.

Problems

1. Using $\gamma'(k,x) = x^{k-1} e^{-x}$, show that (13.25) for $n = 2$, $m = 3$ reduces to (13.21).

2. Derive (13.27) using the covariance matrices given by (13.26).

3. Show that (13.45) can be obtained from (13.37) with some simplifications.

4. Show that the pdf and cdf given by (13.17) and (13.25) for MIMO systems with $n = 1$ and $m \geq 1$ reduce to the pdf and cdf for chi-square distribution for SIMO-MRC

(see (12.42) and (12.53)) and MISO-MRT systems, where CSI is available at the transmitter (see (12.122)).

5. (See [15]) Consider a 2×2 MIMO system in a Rayleigh fading channel. The signals received by the two receive antennas $(y_1\ y_2)^T$ can be written as:

$$\mathbf{y} = \mathbf{Hx} + \mathbf{n} \Rightarrow \begin{pmatrix} y_1 \\ y_2 \end{pmatrix}$$

$$= \begin{pmatrix} h_{11} & h_{12} \\ h_{21} & h_{22} \end{pmatrix} \begin{pmatrix} x_1 \\ x_2 \end{pmatrix} + \begin{pmatrix} n_1 \\ n_2 \end{pmatrix},$$

where $(x_1\ x_2)^T$ denotes the signals transmitted by the two-transmit antennas. Similarly, $(n_1\ n_2)^T$ represents the AWGN at the input of the receive antennas. Here, h_{ij} represents the channel gain between j^{th} transmit and i^{th} receive antennas. The magnitude of the complex Gaussian random variable $h_{ij} = x_{ij} + jy_{ij}$ is Rayleigh distributed and its phase is uniformly distributed:

$$E\left[x_{ij}\right] = E\left[y_{ij}\right] = 0, \quad E\left[x_{ij}^2\right] = E\left[y_{ij}^2\right] = 0.5.$$

Note that there are four channels through which the transmitted signals arrive to the receiver. This is the so-called diversity order which is equal to the product of the number of transmit and receive antennas. By finding the eigenvalues of the channel covariance matrix, $\mathbf{R} = E[\mathbf{H}^H\mathbf{H}]$, one can represent this 2×2 MIMO channel by two independent channels, that is, between 1st transmit antenna and 1st receive antenna and similarly between 2nd transmit and 2nd receive antennas. Thus, the system may be considered to consist of two parallel and independent channels, where the two eigenvalues represent the SNRs of the two channels. If the channels are i.i.d., then the two eigenvalues would be the same. Otherwise, they would differ from each other.

Since the channel correlation matrix is random, the two eigenvalues can be described by their respective pdfs. The pdf of the largest and smallest eigenvalues, normalized with respect to their average values, are given by (13.20).

a. Determine the mean values of the two eigenvalues and show that their sum is equal to 4. Explain why it may be equal to 4? What can you say about the fading statistics of the two eigenvalues?
b. Determine the cdf (outage performance) of the largest eigenvalue, that is, Prob($\lambda_{max} < x$).
c. Determine the outage performance of a receive MRC diversity system with 4 receive antennas.
d. Determine the outage performance of a SISO system as a reference.
e. Compare the performance of this 2×2 MIMO system (two transmit and two receive antennas) with SISO receiver and 1×4 receive MRC diversity system at a threshold level of 10 dB below the average level. Which performs better and how much?

6. The normalized capacity of a $L \times L$ MIMO system is given by (13.61). The signals transmitted by each transmit antenna are assumed to be independent of each other. Assuming that the transmitter has no CSI, the transmit power is divided equally between L antennas. We consider a non-fading channel.

a. Derive the mean channel capacity for the SISO system.
b. Derive the mean channel capacity for the following three cases; 1) all channel gains are perfectly correlated with each other, 2) all channel gains are independent and identically distributed (i.i.d.), and 3) all transmitted signals are perfectly correlated with each other, but there is no correlation between signals at different receive antennas.

c. Compare the normalized capacities for the SISO system and those for the MIMO system considered in (b) for $L = 2$ and the mean channel SNR $\bar{\gamma} = 10$ dB.

7. When the receiver knows the channel perfectly, but no CSI is available at the transmitter, it is optimum to equally divide the transmit power between all transmit antennas, P/L, in a $L \times L$ MIMO system and use uncorrelated data streams. The normalized capacity is given by (13.61). The mean SNR per branch is denoted by $\bar{\gamma}$.

When all path gains h_{ij}, i,j = 1,2,...,L are i.i.d. (independent and identically distributed), the channel matrix is full-rank and has eigenvalues of equal magnitude, $\lambda_1 = \lambda_2 = ... = \lambda_L = \lambda$. This case occurs when the antenna elements are spaced far apart and arranged in a special way. Here, the eigenvalue λ denotes the instantaneous normalized SNR and is a random variable.

a. Show that the normalized capacity is given by $C/B = L \log_2(1 + \gamma\lambda)$ bps/Hz, where the pdf of the eigenvalue λ is given by

$$f_\Lambda(\lambda) = \frac{\lambda^{L-1}}{(L-1)!} e^{-\lambda}, \quad \lambda \geq 0.$$

b. Determine the cdf of the capacity, that is, Prob($C < C_{th}$), where C_{th} denotes the threshold value of the capacity.

8. Consider a $N_t \times N_r$ MIMO system where transmit and receive antennas are all uncorrelated from each other. For $N_t = 3$ and $N_r = 2$, the pdf's of ordered eigenvalues are given by (13.21) and their mean values are listed in Table 13.1. The marginal pdf's of the eigenvalues given above may be approximated by the Gamma distribution given by (13.56) and the diversity order d_k of the k-th eigenvalue is given by (13.55).

a. What are diversity orders of the two eigenvalues for the considered 3×2 MIMO system. Write down the Gamma pdf approximations for the two eigenvalues.

b. Determine the pdf of the received SNR of a SIMO system with L receive antennas. Compare the pdf that you found and compare with the Gamma distribution for the largest eigenvalue in (a). Discuss when the two pdf's give identical results. Also discuss which system has better bit error probability performance.

c. In view of the results obtained in part (b), how can you interpret the pdf of the smallest eigenvalue?

d. Determine analytically the outage probability of the normalized capacity for the largest eigenvalue.

e. Determine bit error probability for BPSK modulation using the Gamma pdf approximations for the largest and the smallest eigenvalues and observe the diversity orders. Hint use the following approximation for Q-function.

$$Q\left(\sqrt{2\gamma}\right) = \frac{w_1}{2}e^{-b\gamma} + \frac{w_2}{2}e^{-2b\gamma}$$

where $w_1 = 0.3017$, $w_2 = 0.4389$, and $b = 1.0510$ (see (B.7)).

9. (See [37]) Consider a MIMO system, with N_t transmit and N_r receive antennas, operating in a flat Rayleigh fading channel. The system chooses one of the N_t transmit antennas to transmit the signal and uses MRC to combine the signals received by the N_r receive antennas.

If $\bar{\gamma}$ denotes the mean SNR received by any of the N_r receiver antennas, the MRC output γ_i from the ith transmit antenna can be written as

$$\gamma_i = \bar{\gamma} \sum_{j=1}^{N_r} |h_{ij}|^2, \quad i = 1, 2, \cdots, N_t$$

Here, h_{ij} denotes the fading coefficient between i^{th} transmitter and j^{th} receiver antennas. They are modelled as independent samples of complex Gaussian random variables with a zero-mean and the variance of 0.5 per dimension.

The transmit antenna is chosen based on the feedback from the receiver MRC output. The single selected transmit antenna is determined by $I = \max\limits_{1 \le i \le L_t} \{\gamma_i\}$.

a. Show that in a Rayleigh fading channel, the pdf of $C_i = \bar{\gamma}|h_{ij}|^2$ is given by

$$f_{C_i}(x) = \frac{1}{\bar{\gamma}} \exp\left(-\frac{x}{\bar{\gamma}}\right)$$

b. Show that the pdf γ_i which is the sum of independent and identically distributed chi-squared variables with $2N_r$ degrees of freedom is given by

$$f_{\gamma_i}(x) = \frac{1}{(N_r-1)!} \left(\frac{x}{\bar{\gamma}}\right)^{N_r-1} \frac{\exp(-x/\bar{\gamma})}{\bar{\gamma}}$$

c. Show that the cdf of γ_I is given by

$$F_{\gamma_i}(x) = 1 - \exp\left(-\frac{x}{\bar{\gamma}}\right) \sum_{i=0}^{N_r-1} \frac{1}{i!} \left(\frac{x}{\bar{\gamma}}\right)^i$$

d. Determine the outage probability, that is Prob($I <$ threshold).

e. Show that the pdf of I is given by

$$f_I(x) = N_t \left[F_{\gamma_i}(x)\right]^{N_t-1} f_{\gamma_i}(x)$$

f. Determine the mean level of I, that is, E(I) and show that it reduces to that for MRC for $N_t = 1$.

g. Write down the expression for the bit error probability for BPSK modulation, for this MIMO system selecting only one of the N_t transmit antennas and maximal ratio combining the outputs of N_r receive antennas.

14

Cooperative Communications

In contrast with traditional networks, cooperative communication networks share resources through distributed transmission and/or processing. [1] In such networks, cooperating users not only transmit/receive their own signals but also serve as (non)regenerative relays for other users' signals to reach their destination (see Figure 14.1). A cooperative network may comprise multiple sources encoding and transmitting their messages in coordination with others. The diversity advantage is achieved through cooperation between users and independence of fading in the links. Cooperative communications provides significant improvements in bit error probability (BEP), multiplexing gain (increase in capacity), and array gain (increase in SNR and in range).

Wireless repeaters used for extending the coverage area of TV and radio broadcasting stations and wired/wireless repeaters used for long-distance communications are classical examples of relaying. On the other hand, nodes in existing computer networks act as relays either in amplify-and-forward (AF) mode, where the received information is amplified and retransmitted, or in detect-and-forward

(DF) mode by detecting, re-modulating and re-transmitting the received information. [2][3] Figure 14.2 shows several cooperation scenarios in cellular radio systems, where the end-user can receive signals from the desired base station (BS) through cooperation with either other users and/or BS's. For example, when a mobile station (MS) is near the edge of the coverage area in a cellular radio, sensor or ad hoc network, the link may not be established without the help of another terminal, which acts as a relay, located closer to the MS. Many other cooperation scenarios can be envisaged, for example, in the absence of close-by BSs in mobile radio systems, in deserted areas or forests. In sensor networks, cooperative sensing may be used for monitoring seismic activities, climate changes, forest fires, pollution, diseases, drug traffic, illegal border crossings and other security applications. [4][5]

As the basic block of cooperative communications, a relay channel consists of a source (S), a relay (R) and a destination (D). In a relay channel, communication is achieved in two time-slots. In time slot I, the source transmits while the destination and the relay receive the

Digital Communications, First Edition. Mehmet Şafak.
© 2017 John Wiley & Sons Ltd. Published 2017 by John Wiley & Sons Ltd.
Companion website: www.wiley.com/go/safak/Digital_Communications

transmitted information. In time-slot II, the relay retransmits the information received in time slot I, using AF, DF or coded-cooperation, and the source may retransmit the same information. The destination uses selection combining (SC) or maximum ratio combining (MRC) to combine the information received from the source and the Relay in the two time slots to have a better estimate of the transmitted signal.

Relaying improves the reliability of communication between the Source and the Destination by following one of the three methods of cooperation. In *AF relaying*, the Relay receives, amplifies and retransmits a noisy and attenuated version of the transmitted signal. Despite the amplification and transmission of the noise at the Relay input, the Destination makes a better estimate of the transmitted signal by combining the information sent by the Source and the Relay via independently fading source-relay-destination (SRD) and source-

destination (SD) channels. Usually, SC or MRC is used to combine the signals from SRD and SD channels, depending the capabilities of the Destination and the system requirements. In *DF relaying*, the Relay attempts to detect the signals transmitted by the Source and retransmits detected and re-modulated signals. If detection is correct, then the Relay suppresses the noise at its input, leading to potentially improved system performance. Source-relay (SR) and relay-destination (RD) links fade independently. The user terminal, acting as a Relay, may be selected using different considerations. *Coded cooperation*, which also incorporates FEC coding, improves the system performance, compared to AF and DF relaying, at the expense of system complexity.

14.1 Dual-Hop Amplify-and-Forward Relaying

Here, we consider dual-hop AF relaying systems in Rayleigh fading channels. The effects of different amplification factors used by the relays are studied. The pdf, cdf, mean SNR, BEP, and mean- and outage-capacity performance of the SRD links with single-relays are considered. The above performance parameters are also determined when the SRD link and direct link are combined with SC and MRC. The results reveal useful information in understanding the performance of cooperative relay networks.

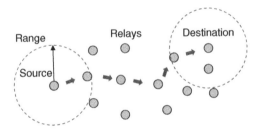

Figure 14.1 Cooperative Communications By Randomly Located Users.

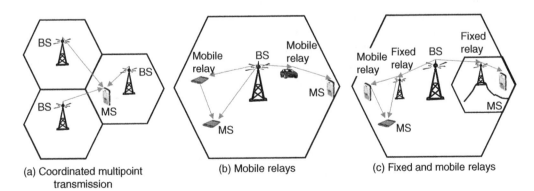

Figure 14.2 Cooperation Scenarios with Fixed and Mobile Relays. [6]

14.1.1 Source-Relay-Destination Link with a Single Relay

Consider a dual-hop AF relaying system, which consists of a Source, a Relay and a Destination as shown in Figure 14.3.

The system is assumed to communicate in time-division multiplexing (TDM) mode in a flat Rayleigh fading channel. This means that in the first half of the transmission period (time slot I), the Source transmits to the Relay, whereas, in time slot II, the Relay retransmits the signal received from the Source after amplifying. Assuming that the Source transmits the signal x, the signal y_1 received by the Relay is given by

$$y_1 = h_1 x + n_1 \qquad (14.1)$$

where $h_1 = \mathbb{CN}(0,1)$ denotes the channel gain of the SR link and $n_1 = \mathbb{CN}(0,\sigma^2)$ is the AWGN. Both h_1 and n_1 are complex Gaussian random variables with zero mean and variances 1 and σ^2, respectively. The Relay amplifies received signal y_1 with an amplification factor A and retransmits to the Destination. The signal y_2 received by the Destination may therefore be written as

$$y_2 = h_2 A(h_1 x + n_1) + n_2 = A h_1 h_2 x + A h_2 n_1 + n_2$$
$$(14.2)$$

Apart from the AWGN n_2 due to the RD link, the additional noise term $A h_2 n_1$ in (14.2) accounts for the noise transmitted by the Relay. Hence, AF relaying leads to accumulation of noise and degradation of the SNR at the Destination output. Due to noise accumulation, the output SNR will be degraded more in multi-hop systems.

14.1.1.1 Equivalent SNR of the Source-Relay-Destination Link

Using (14.2), the instantaneous equivalent SNR of the SRD link may be written as [7][8][9][10][11][12][13][14]

$$\gamma_e = \frac{E\left[|A h_1 h_2 x|^2\right]}{E\left[|A h_2 n_1 + n_2|^2\right]} = \frac{|A|^2 |h_1|^2 |h_2|^2 E_s}{|A|^2 |h_2|^2 \sigma_1^2 + \sigma_2^2}$$
$$(14.3)$$

where $E_s = E\left[|x|^2\right]$ and $\sigma_i^2 = E\left[|n_i|^2\right]$, $i = 1,2$. The above equation may be rewritten as

$$\frac{1}{\gamma_e} = \frac{1}{\gamma_1} + \frac{1}{\gamma_2 |A|^2 |h_1|^2} = \frac{1}{\gamma_1} + \frac{C}{\gamma_1 \gamma_2}$$
$$\Rightarrow \quad \gamma_e = \frac{\gamma_1 \gamma_2}{\gamma_2 + C} \qquad (14.4)$$

where $\gamma_i = |h_i|^2 E_s / \sigma_i^2 = |h_i|^2 \bar{\gamma}_i$, $i = 1,2$ denote the instantaneous SNRs in SR and RD links and $\bar{\gamma}_i = E[\gamma_i] = E_s / \sigma_i^2$, $i = 1, 2$ are the corresponding mean SNRs. The parameter C is defined by

$$\frac{\gamma_1}{C} \triangleq |A|^2 |h_1|^2 = \frac{1}{\bar{\gamma}_1} |A|^2 |h_1|^2 \bar{\gamma}_1 = \frac{1}{\bar{\gamma}_1 / |A|^2} \gamma_1$$

$$C \triangleq \bar{\gamma}_1 / |A|^2$$
$$(14.5)$$

In a Rayleigh fading channel, pdf's, cdf's and MGF's of instantaneous SNRs of SR and RD links are described by

$$f_{\gamma_i}(\gamma) = \frac{1}{\bar{\gamma}_i} \exp(-\gamma/\bar{\gamma}_i)$$

$$F_{\gamma_i}(\gamma) = 1 - \exp(-\gamma/\bar{\gamma}_i), \quad i = 1,2 \qquad (14.6)$$

$$M_{\gamma_i}(s) = E[e^{-s\gamma_i}] = 1/(1 + s\bar{\gamma}_i)$$

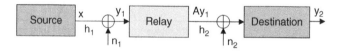

Figure 14.3   Block Diagram of a Dual-Hop Amplify-and-Forward Relaying System.

The amplification factor, A, may be chosen so that the instantaneous output power of the Relay is the same as the transmit power of the Source, that is, $|x|^2 = |Ay_1|^2$:

$$|A|^2 = \frac{E_s}{E_s|h_1|^2 + \sigma_1^2} = \frac{\bar{\gamma}_1}{\gamma_1 + 1}$$ (14.7)

$$\Rightarrow \quad C = \gamma_1 + 1 \quad \Rightarrow \quad \gamma_{eq,0} = \frac{\gamma_1\gamma_2}{\gamma_1 + \gamma_2 + 1}$$

Hence, the effects of signal attenuation and the noise are compensated by the Relay. If the instantaneous channel gain $|h_1|^2$ of the SR link can be estimated at the Relay, then the amplification factor may be chosen so as to compensate the signal attenuation in the SR link, that is, $|A|^2 = 1/|h_1|^2$. Then, (14.4) reduces to

$$C = \frac{\bar{\gamma}_1}{|A|^2} = \frac{\bar{\gamma}_1}{1/|h_1|^2} = \gamma_1 \quad \Rightarrow \quad \gamma_{eq} = \frac{\gamma_1\gamma_2}{\gamma_1 + \gamma_2}$$ (14.8)

Estimation of the instantaneous value of the amplifier gain $|A|^2$ at every transmission interval might be avoided by replacing $|A|^2$ in (14.7) with its average value:

$$C = \frac{\bar{\gamma}_1}{E\left[|A|^2\right]} = \frac{\bar{\gamma}_1}{e^{1/\bar{\gamma}_1}E_1(1/\bar{\gamma}_1)}$$

$$E\left[|A|^2\right] = \int_0^\infty \frac{\bar{\gamma}_1}{1 + \gamma_1\bar{\gamma}_1}e^{-\gamma_1/\bar{\gamma}_1}\, d\gamma_1 = e^{1/\bar{\gamma}_1}E_1(1/\bar{\gamma}_1)$$

$$E_1(z) = \int_z^\infty e^{-t}/t\, dt$$ (14.9)

where the exponential integral $E_1(z)$ is defined by (D.115). In the literature, the average value of C given by (14.7) is also considered:

$$C = \bar{\gamma}_1 + 1 \quad \Rightarrow \quad \gamma_{eq,1} = \frac{\gamma_1\gamma_2}{\gamma_2 + \bar{\gamma}_1 + 1}$$ (14.10)

Alternatively, C defined by (14.8) is replaced by its average value:

$$C = \bar{\gamma}_1 \quad \Rightarrow \quad \gamma_{eq,2} = \frac{\gamma_1\gamma_2}{\gamma_2 + \bar{\gamma}_1}$$ (14.11)

Table 14.1 presents a list of values for C and the corresponding expressions for the equivalent SNR for the link SRD. Definitions by (14.9)–(14.11) are referred to as fixed-gains while (14.7) and (14.8) are referred to as variable-gains. Evidently, variable-gain models provide more accurate representations of the equivalent SNR of the SRD link.

Example 14.1 Amplification Factor and Equivalent SNR For the SRD Link.
Consider the dual hop AF relaying system shown in Figure 14.3. The mean value of the channel gains are assumed to be unity, $E\left[|h_1|^2\right] = E\left[|h_2|^2\right] = 1$. The amplification factor A is chosen so as to make the output powers of the Source and the Relay to be the same. Consequently, the amplification factor and the resulting equivalent SNR of the SRD link are described by (14.7). The noise powers σ_1^2 and σ_2^2 are assumed to be the same; hence $\bar{\gamma}_i = E_s/\sigma_i^2 = 100$, $i = 1, 2$.
Table 14.2 shows the instantaneous SNR's, the amplification factor A and the equivalent SNR for several instantaneous values of the channel gains. Table 14.2 shows that the amplification factor compensates for the lower values of the instantaneous channel gain of the SR

Table 14.1 Different Definitions For C and the Corresponding Expressions For the Equivalent SNR of the SRD Link.

C	
	$\gamma_e = \gamma_1\gamma_2/(\gamma_2 + C)$
$C = \gamma_1 + 1$	$\gamma_{eq,0} = \gamma_1\gamma_2/(\gamma_2 + \gamma_1 + 1)$
$C = \gamma_1$	$\gamma_{eq} = \gamma_1\gamma_2/(\gamma_2 + \gamma_1)$
$C = \bar{\gamma}_1 + 1$	$\gamma_{eq,1} = \gamma_1\gamma_2/(\gamma_2 + \bar{\gamma}_1 + 1)$
$C = \bar{\gamma}_1$	$\gamma_{eq,2} = \gamma_1\gamma_2/(\gamma_2 + \bar{\gamma}_1)$

Table 14.2 Amplification Factor and the Equivalent SNR For Various Values of the Channel Gains For $\bar{\gamma}_i = E_s/\sigma_i^2 = 100$, $i = 1, 2$.

$|h_1|^2$	$|h_2|^2$	γ_1	γ_2	A^2	$\gamma_{eq,0}$
1	1	100	100	1	50
0.1	0.1	10	10	9	5
0.1	0.01	10	1	9	0.83
0.01	0.1	1	10	50	0.83

link. As the channel gain h_1 decreases in a fading channel, the amplification factor is increased. On the other hand, the equivalent SNR of the SRD link is determined by min (γ_1, γ_2). In summary, similar to h_1 and h_2, the amplification factor and the equivalent SNR change randomly in a fading channel.

14.1.1.2 Statistical Characterization of the SRD Link.

The cdf and the pdf of $\gamma_{eq,0} = \gamma_1\gamma_2/(\gamma_1 + \gamma_2 + 1)$, defined by (14.7), are given by (see Appendix 14A for derivations) [10]

$$F_{\gamma_{eq,0}}(\gamma, \alpha, \beta) = \Pr(\gamma_{eq,0} < \gamma)$$

$$= \int_0^\infty \text{Prob}\left(\frac{\gamma_1\gamma_2}{\gamma_1 + \gamma_2 + 1} < \gamma|\gamma_1\right) f_{\gamma_1}(\gamma_1)\, d\gamma_1$$

$$= 1 - \beta \sqrt{\gamma(\gamma+1)}\, \exp(-\alpha\gamma)\, K_1\left(\beta \sqrt{\gamma(\gamma+1)}\right)$$

$$f_{\gamma_{eq,0}}(\gamma, \alpha, \beta) = \beta e^{-\alpha\gamma}\left[\beta\left(\frac{1}{2} + \gamma\right)K_0\left(\beta \sqrt{\gamma(\gamma+1)}\right)\right.$$
$$\left. + \alpha\sqrt{\gamma(\gamma+1)}\, K_1\left(\beta \sqrt{\gamma(\gamma+1)}\right)\right]$$

$$(14.12)$$

where $\alpha = 1/\bar{\gamma}_1 + 1/\bar{\gamma}_2$, $\beta = 2/\sqrt{\bar{\gamma}_1\bar{\gamma}_2}$. The pdf in (14.12) is obtained from the cdf by using (D.105), $K_1'(x) = -K_0(x) - K_1(x)/x$, where $K_0(x)$ and $K_1(x)$ denote the modified Bessel function of the second kind (see Figure D.3).

The cdf of the equivalent SNR of the SRD link, $\gamma_{eq} = \gamma_1\gamma_2/(\gamma_1 + \gamma_2)$ defined by (14.8), is found by inserting $c = 0$ into (14A.2)

$$F_{\gamma_{eq}}(\gamma, \alpha, \beta) = 1 - \beta\,\gamma\,\exp(-\alpha\gamma)\, K_1(\beta\,\gamma)$$

$$(14.13)$$

The pdf corresponding to (14.13) is obtained from (14A.3):

$$f_{\gamma_{eq}}(\gamma, \alpha, \beta) = \beta\gamma\,\exp(-\alpha\gamma)\,[\beta\,K_0(\beta\,\gamma) + \alpha K_1(\beta\,\gamma)]$$

$$(14.14)$$

The moment generating function of γ_{eq} is found from (14.13) or (14.14) using (F.39) and (D.67):

$$M_{\gamma_{eq}}(s, \alpha, \beta) = \int_0^\infty e^{-s\gamma} f_{\gamma_{eq}}(\gamma)\, d\gamma = s \int_0^\infty e^{-s\gamma} F_{\gamma_{eq}}(\gamma)\, d\gamma$$

$$= 1 - s \int_0^\infty e^{-s\gamma}\left[1 - F_{\gamma_{eq}}(\gamma)\right] d\gamma$$

$$= 1 - \frac{2s}{3(s+\alpha)}\, {}_2F_1\left(\frac{1}{2}, 1; \frac{5}{2}; 1 - \frac{\beta^2}{(s+\alpha)^2}\right)$$

$$(14.15)$$

Here, ${}_2F_1(a, b; c; z)$ denotes the hypergeometric function (see (D.82)). [15] The average value of γ_{eq} is found using (F.37):

$$\bar{\gamma}_{eq}(\alpha, \beta) = -\frac{d}{ds}M_{\gamma_{eq}}(s, \alpha, \beta)|_{s=0}$$

$$(14.16)$$

$$= \frac{2}{3\alpha}\, {}_2F_1\left(\frac{1}{2}, 1; \frac{5}{2}; 1 - \frac{\beta^2}{\alpha^2}\right)$$

It is important to note that (14.12)–(14.16) are symmetric with respect to γ_1 and γ_2, that is, the statistical characterization of the SRD channel remains unchanged when γ_1 and γ_2 are interchanged.

The mean equivalent SNR is maximized when the mean SNRs of SR and RD links are balanced, that is, for $\bar{\gamma}_1 = \bar{\gamma}_2$. This implies $\alpha = \beta$ and the hypergeometric function reduces to unity (see (D.82)). Then, the maximum value of the mean SNR of the equivalent channel becomes equal to 1/3 of the mean SNR of a single hop, $\bar{\gamma}_{eq}(\alpha, \alpha) = \bar{\gamma}_1/3$ (see Figure 14.4). Hence, an SRD link with $\bar{\gamma}_1 = \bar{\gamma}_2$ has a mean equivalent SNR 10 $\log(3) = 4.78$ dB worse than a Rayleigh fading single-hop channel with mean SNR $\bar{\gamma}_1$. If the mean SNR's of SR and RD links are not balanced, then the mean equivalent SNR will be lower and the

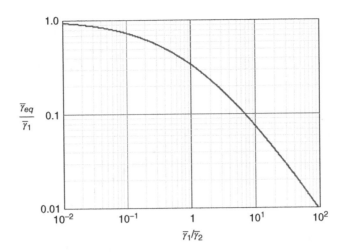

Figure 14.4   Mean Equivalent SNR of the SRD Link Normalized By $\bar{\gamma}_1$ Versus $\bar{\gamma}_1/\bar{\gamma}_2$.

performance of the SRD channel will degrade further. In summary, $\min(\bar{\gamma}_1,\bar{\gamma}_2)$ determines the mean equivalent SNR. For example, mean equivalent SNR is equal to $\bar{\gamma}_{eq}=0.227\bar{\gamma}_1=0.455\bar{\gamma}_2$ for $\bar{\gamma}_1/\bar{\gamma}_2=2$ and $\bar{\gamma}_{eq}=0.455\bar{\gamma}_1$ for $\bar{\gamma}_1/\bar{\gamma}_2=0.5$ (see Figure 14.4). Similarly, the mean equivalent SNR may be written as $\bar{\gamma}_{eq}=0.726\min(\bar{\gamma}_1,\bar{\gamma}_2)$ for $\bar{\gamma}_i/\bar{\gamma}_j=10$, $i,j=1,2$. For the asymptotic case $\bar{\gamma}_i/\bar{\gamma}_j\rightarrow\infty$, $i,j=1,2$, $_2F_1(0.5,1;2.5;1)=3/2$ and the mean equivalent SNR reduces to $\bar{\gamma}_{eq}=\min(\bar{\gamma}_1,\bar{\gamma}_2)$. This implies that the mean equivalent SNR approaches $\bar{\gamma}_{eq}\rightarrow\bar{\gamma}_2$ as $\bar{\gamma}_1\rightarrow\infty$ (the SR link is error-free) and, $\bar{\gamma}_{eq}\rightarrow\bar{\gamma}_1$ as $\bar{\gamma}_2\rightarrow\infty$ (the RD link is error-free).

Rayleigh approximation to the pdf of γ_{eq}, given by (14.14), of the equivalent SRD channel lends itself to convenient and accurate analytical computations:

$$f_{\gamma_{eq,R}}(\gamma,\alpha,\beta)=\frac{1}{\bar{\gamma}_{eq}(\alpha,\beta)}\exp\left(-\frac{\gamma}{\bar{\gamma}_{eq}(\alpha,\beta)}\right)$$

$$F_{\gamma_{eq,R}}(\gamma,\alpha,\beta)=1-\exp\left(-\frac{\gamma}{\bar{\gamma}_{eq}(\alpha,\beta)}\right)$$

$$M_{\gamma_{eq,R}}(s,\alpha,\beta)=1/\left(1+s\,\bar{\gamma}_{eq}(\alpha,\beta)\right)$$

$$(14.17)$$

where the mean value $\bar{\gamma}_{eq}(\alpha,\beta)$ is given by (14.16). This approach is based on the assumption that the SRD channel can be considered as a single-hop channel and be modeled by an exponential pdf with a mean power level of $\bar{\gamma}_{eq}(\alpha,\beta)$ in a Rayleigh fading environment.

The pdf's of $\gamma_{eq,0}$ and γ_{eq}, which are given by (14.12) and (14.14), respectively, are shown in Figure 14.5 in log-scale in order to accentuate their differences in the tail. They agree closely with each other except for a small difference near the origin, due to the omission in (14.8) of the term "1", which appears in the denominator of (14.7). The Rayleigh approximation (14.17) shows small differences compared with those corresponding to (14.12) and (14.14). However, this approximation facilitates mathematical calculations and simulations.

Example 14.2   Outage Performance of the SRD Link.

Consider a SRD link undergoing Rayleigh fading in both hops with mean SNRs $\bar{\gamma}_1$ and $\bar{\gamma}_2$. The outage performance of this link is assumed to be modeled by the cdf given by (14.12). Table 14.3 shows the outage probability for various threshold SNR levels and combinations of mean SNRs $\bar{\gamma}_1,\bar{\gamma}_2$. Because of the cascade

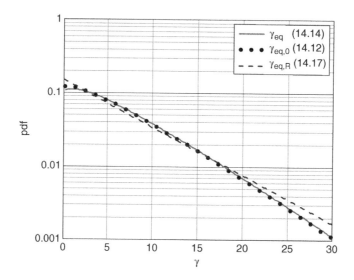

Figure 14.5 Pdf of the SNR of the SRD Link as Given by (14.12) and (14.14) for $\bar{\gamma}_1 = \bar{\gamma}_2 = 13\ dB$. [10] Rayleigh approximation (14.17) for (14.14) is also shown.

Table 14.3 Outage Probability of the SRD Link For Various Combinations of the Mean Link SNR's $\bar{\gamma}_1$, $\bar{\gamma}_2$ in Comparison with a Single-Hop Link.

	Prob($\gamma_{eq,0} \le \gamma_{th}$)			
$\bar{\gamma}_1$, $\bar{\gamma}_2$	$\gamma_{th} = 4$ dB	$\gamma_{th} = 7$ dB	$\gamma_{th} = 10$ dB	$\gamma_{th} = 13$ dB
Single hop ($\bar{\gamma} = 13\ dB$)	0.118	0.221	0.393	0.632
$\bar{\gamma}_1 = 13\,dB, \bar{\gamma}_2 = 13\ dB$	0.285	0.511	0.786	0.964
$\bar{\gamma}_1 = 13\,dB, \bar{\gamma}_2 = 10\ dB$	0.405	0.669	0.906	0.993
Single hop ($\bar{\gamma} = 10\ dB$)	0.221	0.393	0.632	0.865

connection of SR and RD links, the SRD link suffers fading when only one or both of the links undergo fading. Therefore, the outage performance is worse than that of a single hop with the same mean SNR. When the mean SNR levels are not balanced, the outage performance degrades further. The outage performance of a SRD link with unbalanced mean link SNR's is worse than that of a single link with mean SNR $\bar{\gamma} = \min(\bar{\gamma}_1, \bar{\gamma}_2)$.

The average capacity of the SRD link is given by (14B.5)

$$C_{av} = \begin{cases} \dfrac{-1}{2\ \ln 2} + \left(1 + \dfrac{1}{\bar{\gamma}_1}\right) C_m(\bar{\gamma}_1) & \bar{\gamma}_1 = \bar{\gamma}_2 \\[2mm] \dfrac{\bar{\gamma}_1}{\bar{\gamma}_1 - \bar{\gamma}_2} C_m(\bar{\gamma}_2) - \dfrac{\bar{\gamma}_2}{\bar{\gamma}_1 - \bar{\gamma}_2} C_m(\bar{\gamma}_1) & \bar{\gamma}_1 \ne \bar{\gamma}_2 \end{cases}$$

$$(14.18)$$

where $C_m(\bar{\gamma}_i)$ is given by (14B.2):

$$C_m(\bar{\gamma}_i) = \frac{1}{2\ \ln 2}\ e^{1/\bar{\gamma}_i}\ E_1(1/\bar{\gamma}_i), \quad i = 1, 2$$

$$(14.19)$$

The cdf of the capacity of the SRD link is found using (14.13):

$$C_{out,eq}(C_{th}) = \Pr\left(\frac{1}{2}\log_2\left(1+\gamma_{eq}\right) < C_{th}\right)$$

$$= \Pr\left(\gamma_{eq} < 2^{2C_{th}} - 1\right)$$

$$= F_{\gamma_{eq}}\left(2^{2C_{th}} - 1, \ \alpha, \ \beta\right) \quad (14.20)$$

The BEP for BPSK for the SRD link may be expressed in terms of $f_{\gamma_{eq}}(\gamma)$ or $F_{\gamma_{eq}}(\gamma)$ (employing integration by parts in the first expression and using (D.67)): [7]

$$P_{b,eq}(\alpha,\beta) = \int_0^\infty Q\left(\sqrt{2\gamma}\right) f_{\gamma_{eq}}(\gamma) \, d\gamma$$

$$= \frac{1}{\sqrt{2\pi}} \int_0^\infty \frac{1}{\sqrt{2\gamma}} e^{-\gamma} \, F_{\gamma_{eq}}(\gamma) d\gamma$$

$$= \frac{1}{2} - \chi_{eq}(\alpha,\beta)$$

$$\chi_{eq}(\alpha,\beta) = \frac{3\pi}{16\sqrt{2(\alpha+1)}} {}_2F_1\left(\frac{1}{4},\frac{3}{4};2;1-\frac{\beta^2}{(\alpha+1)^2}\right) \quad (14.21)$$

Hereafter we will use γ_{eq}, defined by (14.8), to represent equivalent SNR of the SRD channel, because it lends itself to easy analytical calculations and does not differ appreciably from $\gamma_{eq,0}$. The *variable-gain* expressions, given by (14.7) and (14.8) will be used since they provide better performance compared to *fixed-gain* expressions, given by (14.9)–(14.11). [7][8][11][12]

14.1.2 Combined SRD and Direct Links

Now consider a cooperative communication system, as shown in Figure 14.6, where the SRD link with a single Relay is combined with the direct (SD) link at the Destination. SR, RD and SD links are assumed to undergo Rayleigh fading with mean SNRs $\bar{\gamma}_1$, $\bar{\gamma}_2$, $\bar{\gamma}_3$, respectively.

When the SRD link with a single Relay, characterized by (14.13)–(14.16), and the direct link, with a mean SNR $\bar{\gamma}_3$, are selection-combined, the SNR of the combined link is given by $\gamma_{sc} = \max(\gamma_3, \gamma_{eq})$. In such applications, selection combining may be preferred to MRC because of its reduced complexity.

The resulting cdf, the pdf and the MGF for SC are given by (see Example F.9)

$$F_{\gamma_{sc}}(\gamma) = (1 - \exp(-\gamma/\bar{\gamma}_3)) F_{\gamma_{eq}}(\gamma,\alpha,\beta)$$

$$f_{\gamma_{sc}}(\gamma) = \frac{1}{\bar{\gamma}_3}\exp(-\gamma/\bar{\gamma}_3) + f_{\gamma_{eq}}(\gamma,\alpha,\beta) - f_{\gamma_{eq}}(\gamma,\alpha_1,\beta)$$

$$M_{\gamma_{sc}}(s) = \frac{1}{1+s\bar{\gamma}_3} + M_{\gamma_{eq}}(s,\alpha,\beta) - M_{\gamma_{eq}}(s,\alpha_1,\beta) \quad (14.22)$$

where $\alpha_1 = 1/\bar{\gamma}_1 + 1/\bar{\gamma}_2 + 1/\bar{\gamma}_3$.

When the SRD and SD links are maximum-ratio combined, the pdf of $\gamma_{mrc} = \gamma_3 + \gamma_{eq}$ may be obtained by using the MGF:

$$M_{\gamma_{mrc}}(s) = E\left[e^{-s\left(\gamma_3+\gamma_{eq}\right)}\right] = \frac{1}{1+s\bar{\gamma}_3} M_{\gamma_{eq}}(s,\alpha,\beta) \quad (14.23)$$

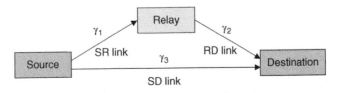

Figure 14.6 Block Diagram of a Cooperative Communication System with SRD and Direct (SD) links.

where $M_{\gamma_{eq}}(s,\alpha,\beta)$ is given by (14.15). Instead of using (14.15) in (14.22) or (14.23) for the SRD link, the MGF $M_{\gamma_{eq,R}}(s,\alpha,\beta)$, defined by (14.17), for the Rayleigh-approximation, may also be used with negligible error (see (14C.12)–(14C.16)).

When direct (SD) and SRD links are selection- or maximal-ratio combined, the mean value of the equivalent SNR is found from (14.22) and (14.23) as follows

$$\bar{\gamma}_{\gamma_{sc}} = \bar{\gamma}_3 + \bar{\gamma}_{eq}(\alpha,\beta) - \bar{\gamma}_{eq}(\alpha_1,\beta)$$
$$\bar{\gamma}_{\gamma_{mrc}} = \bar{\gamma}_3 + \bar{\gamma}_{eq}(\alpha,\beta)$$
(14.24)

One may observe from (14.24) that the mean equivalent SNR for MRC is always higher than that for SC, as expected.

Mean (ergodic) capacity and the cdf of the capacity for the selection combined SRD and SD links are determined by using $F_{\gamma_{sc}}(\gamma)$ in (14.22):

$$C_{av,sc} = \frac{1}{2 \ln 2} \int_0^\infty \ln(1+\gamma) f_{\gamma_{sc}}(\gamma) \, d\gamma$$
$$= \frac{1}{2 \ln 2} \int_0^\infty \frac{1}{1+\gamma} [1 - F_{\gamma_{sc}}(\gamma)] \, d\gamma$$

$$C_{out,sc}(C_{th}) = F_{\gamma_{sc}}(2^{2C_{th}} - 1)$$
(14.25)

The two expressions given for the ergodic capacity are related to each other by integration by parts.

For BPSK modulation, the BEP for the selection-combined direct and SRD links is obtained using the pdf in (14.22): [7]

$$P_{b,sc}(\bar{\gamma}_1,\bar{\gamma}_2,\bar{\gamma}_3) = \int_0^\infty Q\left(\sqrt{2\gamma}\right) f_{\gamma_{sc}}(\gamma) d\gamma$$
$$= I(1,\bar{\gamma}_3) - \chi_{eq}(\alpha,\beta) + \chi_{eq}(\alpha_1,\beta)$$
(14.26)

where $\alpha_1 = 1/\bar{\gamma}_1 + 1/\bar{\gamma}_2 + 1/\bar{\gamma}_3$ and $\chi_{eq}(\alpha,\beta)$ is defined by (14.21). On the other hand,

$I(1,\bar{\gamma}_3) = 0.5\left(1 - \sqrt{\bar{\gamma}_3/(1+\bar{\gamma}_3)}\right)$, defined by (14C.3), denotes the BEP for BPSK for the Rayleigh-fading SD link with a mean SNR $\bar{\gamma}_3$.

14.1.2.1 Selection-Combined Two Direct Links

Let us now consider, as a reference, the BEP performance of two selection-combined direct links, that is, $\gamma_D = \max(\gamma_2,\gamma_3)$, both suffering Rayleigh fading with mean SNRs $\bar{\gamma}_2$ and $\bar{\gamma}_3$. The cdf, the pdf and the MGF of γ_D may be written as (see (12.72), (12.73))

$$F_{\gamma_D}(\gamma) = \left(1 - e^{-\gamma/\bar{\gamma}_3}\right)\left(1 - e^{-\gamma/\bar{\gamma}_2}\right)$$

$$f_{\gamma_D}(\gamma) = \frac{1}{\bar{\gamma}_2} e^{-\gamma/\bar{\gamma}_2} + \frac{1}{\bar{\gamma}_3} e^{-\gamma/\bar{\gamma}_3}$$
$$- \left(\frac{1}{\bar{\gamma}_2} + \frac{1}{\bar{\gamma}_3}\right) e^{-(1/\bar{\gamma}_2 + 1/\bar{\gamma}_3)\gamma}$$

$$M_{\gamma_D}(s) = \frac{1}{1+s\bar{\gamma}_2} + \frac{1}{1+s\bar{\gamma}_3} - \frac{1}{1+s/(1/\bar{\gamma}_2+1/\bar{\gamma}_3)}$$
(14.27)

The mean SNR is easily found from the pdf in (14.27):

$$\bar{\gamma}_D = E[\gamma_D] = \int_0^\infty \gamma f_{\gamma_D}(\gamma) \, d\gamma = \bar{\gamma}_3 + \bar{\gamma}_2 - \frac{\bar{\gamma}_2\bar{\gamma}_3}{\bar{\gamma}_2 + \bar{\gamma}_3}$$
(14.28)

The BEP for BPSK modulation is found to be

$$P_{b,D}(\bar{\gamma}_2,\bar{\gamma}_3) = \int_0^\infty Q\left(\sqrt{2\gamma}\right) f_{\gamma_D}(\gamma) \, d\gamma$$
$$= I(1,\bar{\gamma}_2) + I(1,\bar{\gamma}_3) - I\left(1,(1/\bar{\gamma}_2 + 1/\bar{\gamma}_3)^{-1}\right)$$
(14.29)

where $I(n,x)$ is defined by (14C.3). The mean capacity is found to be

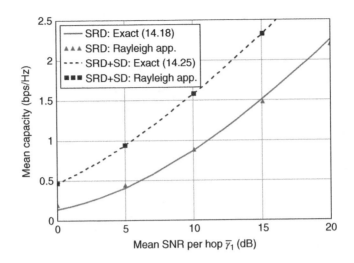

Figure 14.7 The Mean Capacity For the SRD Link and the Selection Combined SRD and SD Links with a Single Relay For $\bar{\gamma}_1 = \bar{\gamma}_2 = \bar{\gamma}_3$. Rayleigh approximation is obtained by replacing $F_{\gamma_{eq}}(\gamma)$ by $F_{\gamma_{eq,R}}(\gamma)$ in (14.18) and (14.25).

$$C_{av,D}(\bar{\gamma}_2, \bar{\gamma}_3) = \frac{1}{2} \int_0^\infty \log_2(1+\gamma) f_{\gamma_D}(\gamma) \, d\gamma$$

$$= C_m(\bar{\gamma}_2) + C_m(\bar{\gamma}_3) - C_m\left(\frac{1}{1/\bar{\gamma}_2 + 1/\bar{\gamma}_3}\right)$$

$$(14.30)$$

where $C_m(\bar{\gamma}_i)$ is given by (14.19). The cdf of the capacity is found as:

$$C_{out,D}(C_{th}) = \Pr\left(\frac{1}{2}\log_2(1+\gamma_D) < C_{th}\right)$$

$$= \Pr\left(\gamma_D < 2^{2C_{th}} - 1\right)$$

$$= F_{\gamma_D}\left(2^{2C_{th}} - 1\right)$$

$$= \int_0^{2^{2C_{th}} - 1} f_{\gamma_D}(\gamma) \, d\gamma$$

$$= 1 - e^{-\left(2^{2C_{th}} - 1\right)/\bar{\gamma}_2} - e^{-\left(2^{2C_{th}} - 1\right)/\bar{\gamma}_3}$$

$$+ e^{-\left(\frac{1}{\bar{\gamma}_2} + \frac{1}{\bar{\gamma}_3}\right)\left(2^{2C_{th}} - 1\right)}$$

$$(14.31)$$

If $\bar{\gamma}_2$ is replaced by $\bar{\gamma}_{eq}(\alpha, \beta)$ in (14.27)–(14.31), then these formulas account for the selection combined direct and Rayleigh-approximated SRD links, described by (14.17) (see also (14C.7)–(14C.11)).

Figure 14.7 shows that the Rayleigh-approximation is very accurate for determining the average capacity of SRD link and the selection-combined SRD and SD links. Based on the chosen mean link SNRs, selection combining SRD and SD links provides significant increase in the mean capacity.

Figure 14.8 shows the outage probability for SRD and selection-combined SRD and SD links. This choice of mean link SNR's lead to significant improvement in the outage performance when SRD and SD links are combined. On the other hand, Rayleigh approximation to the SRD link was observed to be sufficiently accurate.

Figure 14.9 shows the pdf's of SRD and selection combined SRD and SD channels. Rayleigh approximation, which implies that the SRD channel suffers Rayleigh-fading with mean SNR $\bar{\gamma}_{eq}(\alpha, \beta)$, was observed to describe the exact pdf's with sufficient accuracy.

14.2 Relay Selection in Dual-Hop Relaying

Now consider a dual-hop relaying network which consists of a Source, a Destination and

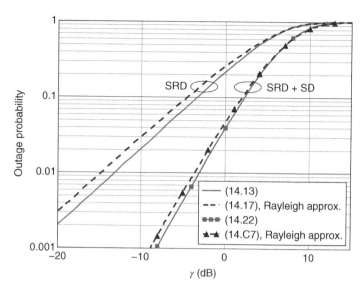

Figure 14.8 Outage Probability For γ_{eq} and $\gamma_{sc} = \max(\gamma_3, \gamma_{eq})$ with Rayleigh Approximations. $\bar{\gamma}_1 = \bar{\gamma}_2 = 2\bar{\gamma}_3 = 10$ is assumed.

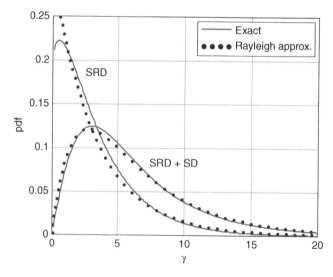

Figure 14.9 Pdf of γ_{eq} and $\gamma_{sc} = \max(\gamma_3, \gamma_{eq})$ with Rayleigh Approximations. $\bar{\gamma}_1 = \bar{\gamma}_2 = 2\bar{\gamma}_3 = 10$ is assumed.

K Relays. The network is assumed to communicate in TDM mode in a flat Rayleigh fading channel. Links between the Source and k^{th} Relay are assumed to have the same mean SNRs $E[\gamma_{1,k}] = \bar{\gamma}_1$, $k = 1, 2..., K$. Similarly, the link between the kth Relay and the Destination has a mean SNR $E[\gamma_{2,k}] = \bar{\gamma}_2$, $k = 1, 2, ..., K$. This implies that all the relays are geometrically located in close vicinity of each other. When $\bar{\gamma}_1 \simeq \bar{\gamma}_2$, the K relays will be in close vicinity of each other near the mid-point between the Source and the Destination. Although all SR links have the same mean SNRs $\bar{\gamma}_1$ and all the RD links have same mean SNR $\bar{\gamma}_2$, the instantaneous SNR levels change randomly due to fading. Consequently, at any signaling interval, a different SRD link may have the highest equivalent SNR, defined by (14.8), among the considered relay links.

14.2.1   Relay Selection Strategies

The relay selection rule may be based on the selection of the RD link with highest SNR. The relay selection rule for fixed- and a variable-gain relays is then given by [7][11][12][13][14]

$$
\gamma_{s0} =
\begin{cases}
\underset{1 \le k \le K}{\max} \left\{ \gamma_1 \gamma_{2,k} / (C + \gamma_{2,k}) \right\} & fixed-gain \\[2ex]
\underset{1 \le k \le K}{\max} \left\{ \gamma_1 \gamma_{2,k} / (\gamma_1 + \gamma_{2,k}) \right\} & variable-gain
\end{cases}
$$

(14.32)

where one of definitions given by (14.9)–(14.11) may be adopted for C. [11] Here, the strategy is based on (14.8) and the selection of the RD link with the highest SNR, regardless of the SNR in the corresponding SR links. Since the best RD link may not correspond to the best SR link, this relay selection scheme limits the system performance.

An alternative relay selection method may be formulated as follows:

$$
\gamma_s = \underset{1 \le k \le K}{\max} \left\{ \gamma_{eq,k} \right\}
$$

(14.33)

where the equivalent instantaneous equivalent SNR of the link between the Source, the kth Relay and the Destination (SR_kD) is given by

$$
\gamma_{eq,k} = \frac{\gamma_{1,k} \gamma_{2,k}}{\gamma_{1,k} + \gamma_{2,k}}
$$

(14.34)

In (14.8), the relays are assumed to be capable of estimating the channel gains (SNRs)

in their respective SR links. Thus, they can adjust the amplification factor so as to compensate for gain variations in the SR_k links. Assuming that the relays and the Destination are capable of estimating the received instantaneous SNRs, the selection based on (14.33) is feasible.

The best relay link, defined by (14.33), may be selection-combined with the direct link with SNR γ_3. The resulting SNR is

$$
\gamma_c = \max\{\gamma_s, \gamma_3\}
$$

(14.35)

14.2.2   Performance Evaluation of Selection-Combined Best SRD and SD Links

In view of (12.65)–(12.68), the cdf of the highest equivalent SNR γ_s of the SRD link is given by

$$
F_{\gamma_s}(\gamma) = \left[F_{\gamma_{eq}}(\gamma) \right]^K
$$

(14.36)

where all K SRD channels are assumed to be independent of each other and all have the same mean SNR's in SR and RD links. Similarly, the cdf of the output SNR $\gamma_c = \max(\gamma_s, \gamma_3)$ of the selection-combined best SRD link and SD link may be written as

$$
F_{\gamma_c}(\gamma) = \left[1 - \exp(-\gamma/\bar{\gamma}_3) \right] \left[F_{\gamma_{eq}}(\gamma) \right]^K
$$

(14.37)

The corresponding pdf's can easily be obtained by taking the derivatives of (14.36) and (14.37). Figure 14.10 shows the effect of relay selection on the pdf of SRD and selection combined SRD + SD links for $\bar{\gamma}_1 = \bar{\gamma}_2 = \bar{\gamma}_3 = 10$. As the Relay with the highest equivalent SNR is selected, the pdf's of γ_s and γ_c are more likely to assume higher values as K increases. The pdf for the SRD + SD link is mainly determined by the relative values of $\bar{\gamma}_s$ and $\bar{\gamma}_3$.

Figure 14.11 shows the variation of the mean equivalent SNR of the SRD link and the selection-combined SRD and SD links with

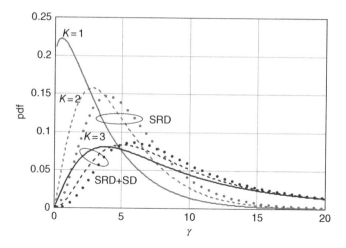

Figure 14.10 The Pdf of $\gamma_s = \max\limits_{1 \le k \le K}(\gamma_{eq,k})$ and $\gamma_c = \max(\gamma_3, \gamma_s)$ For $K = 1,2,3$ and $\bar{\gamma}_1 = \bar{\gamma}_2 = \bar{\gamma}_3 = 10$.

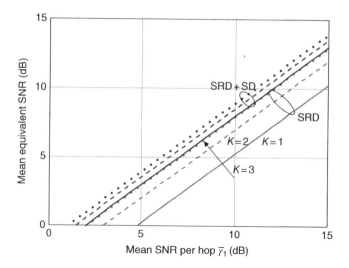

Figure 14.11 Mean Equivalent SNRs γ_s and $\gamma_c = \max(\gamma_3, \gamma_s)$ For $K = 1,2,3$. $\bar{\gamma}_1 = \bar{\gamma}_2 = 2\bar{\gamma}_3$ is assumed.

relay selection. Mean equivalent SNRs are obtained using (14.36) and (14.37) for SRD and selection-combined SRD and SD links, respectively. The value of K is more pronounced in mean equivalent SNR of the SRD link with relay selection compared to the combined link, where the relative values of $\bar{\gamma}_s$ and $\bar{\gamma}_3$ are important.

The BEP for BPSK modulation of the AF relaying system with relay selection may be expressed as

$$P_{b,eq}(\alpha,\beta) = \frac{1}{\sqrt{2\pi}} \int_0^\infty \frac{1}{\sqrt{2\gamma}} e^{-\gamma} F_{\gamma_{eq}}(\gamma)^K \left(1 - e^{-\gamma/\bar{\gamma}_3}\right) d\gamma$$

$$(14.38)$$

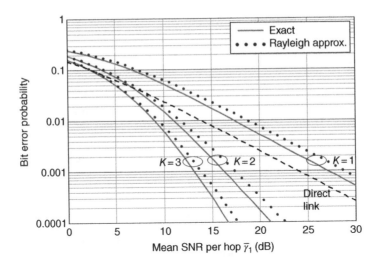

Figure 14.12 The BEP Given by (14.38) Versus Average SNR Per Hop For the SRD Link For $K = 1,2,3$, and $\bar{\gamma}_1 = \bar{\gamma}_2$. The relay selection criterion is based on (14.33).

where the Relay with the highest equivalent SNR is selected. Note that (14.38) applies for the selection combined SRD + SD link. However, it also applies for the SRD link with relay selection if we let $\bar{\gamma}_3 \to \infty$. The BEP expression in (14.38) cannot be evaluated in closed form since the binomial expansion of the Kth power of $F_{\gamma_{eq}}(\gamma)$, given by (14.13), contains powers of the Bessel function $K_1(\gamma)$. However, using the Rayleigh approximation (14.17) for (14.13), the BEP for the SRD link is determined by (14C.2). Similarly, analytical BEP expressions for SRD + SD links with SC and MRC are given by (14C.8) and (14C.13), respectively. Figure 14.12 shows the effect of relay selection on the BEP of the SRD link for $\bar{\gamma}_1 = \bar{\gamma}_2$. The Rayleigh approximation for the SRD equivalent channel leads to 1 dB worse BEP performance compared to (14.38). One may also observe that the BEP performance of the SRD link with a single relay ($K = 1$) is 3.1 dB worse than the BEP curve for the Rayleigh-faded direct link with mean SNR $\bar{\gamma}_1$.

Let us now study the effect of the relay location on the BEP performance of the SRD link. We model the variation of the mean SNR's of the SR and RD links by

$$\bar{\gamma}_i = \bar{\gamma}_0 \frac{1}{d_i^\eta}, \quad i = 1,2 \qquad (14.39)$$

where d_1 and d_2 denote, respectively, the ranges of SR and RD links with $d_1 + d_2 = 1$ and η is the path loss coefficient. Using (14.39), (14.38) and (14C.2), the BEP of the SRD link may be expressed as a function of d_1. As one may easily observe from (14.34), (14.39) and (14C.2), the BEP expression is symmetrical with respect to $\bar{\gamma}_1$ and $\bar{\gamma}_2$ and with respect to $d_1 = 1/2$. Therefore, the BEP is minimized for $\bar{\gamma}_1 = \bar{\gamma}_2 = 2^\eta \bar{\gamma}_0$.

Figure 14.13 shows the variation of BEP versus relay location for $\bar{\gamma}_0 = 10$ and $\eta = 3$. When the relays are located the mid-way between Source and Destination, the mean SNR's of the SR and RD links are 19 dB. Then, for K = 2, the exact value of BEP is found to be 2.65×10^{-4} while it is 4.88×10^{-4} for the Rayleigh approximation. The difference becomes less as the relay location is off the mid-way. Hence, the Rayleigh approximation is believed to be sufficiently accurate. The increased values of K lead to significant reductions in the BEP; for $d = 1/2$, BEP $= 6.5 \times 10^{-3}$, 2.65×10^{-4}, 1.8×10^{-5} for K = 1, 2, 3, respectively.

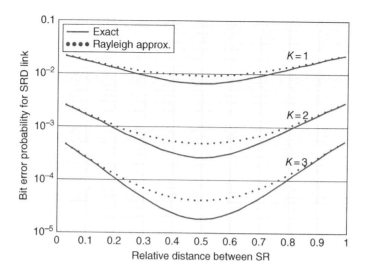

Figure 14.13 BEP Versus Relay Location Between the Source and the Destination, Which are Separated by a Distance Normalized to Unity. The parameters are chosen as $\eta = 3$ and $\bar{\gamma}_0 = 10$.

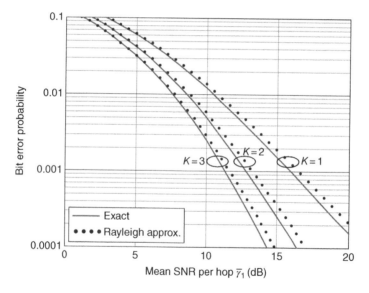

Figure 14.14 BEP Versus Average SNR Per Hop For Selection-Combined SRD and SD Links For $K = 1, 2, 3$, Using the Exact Expression (14.38) and the Rayleigh Approximation (14C.7). $\bar{\gamma}_1 = \bar{\gamma}_2 = 2\bar{\gamma}_3$.

Figure 14.14 shows the variation of BEP with average SNR per hop $\bar{\gamma}$ for the selection-combined SRD and SD links. The average SNRs are assumed to be $\bar{\gamma}_1 = \bar{\gamma}_2 = 2\bar{\gamma}_3$. For $K = 2$, the Rayleigh approximation performs 0.46 dB worse than the exact result at BEP $= 10^{-3}$.

For $K = 2$ and BEP $= 10^{-4}$, selection combining SRD and SD links provides 4.72 dB SNR improvement compared to the SRD link (see Figures 14.12 and 14.14).

Figure 14.15 shows the effect of selection- and maximal ratio combining of SRD and SD

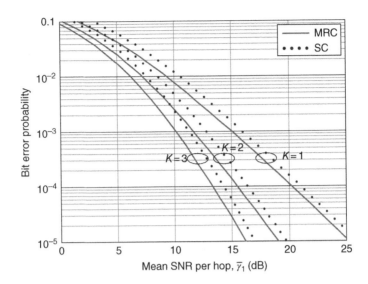

Figure 14.15 BEP Versus Average SNR Per Hop For Selection- and Maximal-Ratio Combined SRD and SD Links For $\bar{\gamma}_1 = \bar{\gamma}_2 = 2\bar{\gamma}_3$. BEP expressions (14C.8) for SC and (14C.13) for MRC are both based on Rayleigh approximation for the pdf of the SNR of the SRD link.

links on the BEP performance based on the assumption that the equivalent SNR of the SRD link is approximated by Rayleigh pdf. The mean link SNR's are assumed to be $\bar{\gamma} = \bar{\gamma}_1 = \bar{\gamma}_2 = 2\bar{\gamma}_3$. For BEP $= 10^{-4}$ and $K = 2$, the MRC requires 0.9 dB lower SNR than SC.

The mean capacity for SRD and selection-combined SRD and SD links, given by (14C.5) and (14C.10), respectively, are shown in Figure 14.16 for $\bar{\gamma} = \bar{\gamma}_1 = \bar{\gamma}_2 = \bar{\gamma}_3$. The results are based on the Rayleigh approximation for the SRD link. For reasonable values of K, the average capacity of the SRD link is upper-bounded by that of the direct link. Similarly, the average capacity of the selection-combined SRD and SD links is upper-bounded by the capacity of the selection-combined two direct links with the same mean hop SNRs.

Based on the Rayleigh approximation for the SRD link, the outage performance of the capacity of the best SRD link is given by (14C.6). For the combined SRD and SD links, it is given by (14C.11) and (14C.16), respectively for SC and MRC. As shown by Figure 14.17, the outage performance of the

capacity of the selection-combined best and SRD link is improved considerably compared to that of the best SRD link. However, for the chosen mean SNR levels, $\bar{\gamma}_1 = \bar{\gamma}_2 = \bar{\gamma}_3 = 10$, this improvement decreases with increasing values of K.

Example 14.3 Effect of Outdated CSI in SRD Links Using AF Protocol and Relay Selection.

The selection of the best Relay in the SRD link with AF protocol is based on (14.33), where perfect channel state information (CSI) is assumed to be available. However, in practice, the CSI used for relay selection may be outdated due to time changes in the channel gain, that is, due to the delay and/or the accuracy of the CSI. Consider a training period during which the Destination measures signals received by from K SRD links in order to determine the link with the highest equivalent SNR to form the CSI. Here, t denotes the time the Destination measures the received signal level from the Relay with the highest equivalent SNR, and $t + \tau$ is the time the message is sent

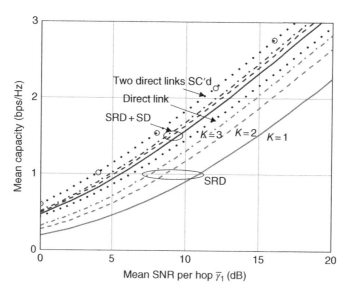

Figure 14.16   The Mean Capacity of the SRD and the Selection-Combined SRD and SD Links For $K = 1,2, 3$. Mean capacity for the direct, and selection-combined two direct links are also shown. All mean hop SNRs are assumed to be the same $\bar{\gamma}_1 = \bar{\gamma}_2 = \bar{\gamma}_3$.

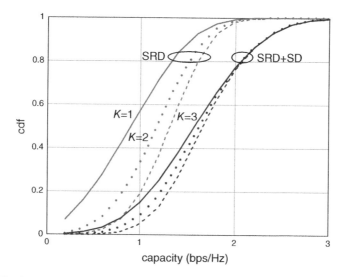

Figure 14.17   The Cdf of the Capacity of the SRD Link with $K = 1, 2, 3$ and the Selection Combined SRD and SD Links For $\bar{\gamma}_1 = \bar{\gamma}_2 = \bar{\gamma}_3 = 10$.

by the Source. The mean value of τ is then $K/2$ times the propagation time for the SRD link. Hence the relay selection based on the CSI at time t may be outdated at the time $t + \tau$ of transmission.

In this example, we study the performance degradation due to outdated CSI in the SRD link in terms of correlation between the values of the equivalent SNR, based on the perfect CSI at time t, and the actual SNR at time $t + \tau$ of

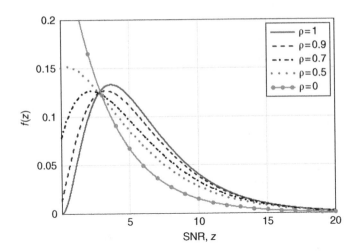

Figure 14.18 Effect of Correlation Between the Outdated CSI and the Actual Equivalent SNR on the Pdf of the SRD Link with Rayleigh Approximation. $K = 3$ and $\bar{\gamma}_1 = \bar{\gamma}_2 = 10$.

transmission. [15] Using Rayleigh approximation for the SRD link, the conditional pdf of the actual SNR z conditioned on a given value of the SNR γ, based on the outdated CSI, is given by

$$f(z|\gamma) = \frac{1}{(1-\rho)\bar{\gamma}_{eq}} \exp\left(-\frac{\rho\gamma+z}{(1-\rho)\bar{\gamma}_{eq}}\right) I_0\left(\frac{2\sqrt{\rho z\gamma}}{(1-\rho)\bar{\gamma}_{eq}}\right)$$

$$(14.40)$$

where ρ denotes power correlation between z and γ and $I_0(x)$ shows the modified Bessel function of the first kind and order zero. Note from (14.40) that, for $\rho = 0$, the pdf of z becomes independent of γ and reduces to (14.17), the Rayleigh approximation to the SRD link. Using (D.60), (14.40) and (14C.1), the pdf of γ corresponding to perfect but outdated CSI, we get the pdf of the actual equivalent SNR as follows:

$$f(z) = \int_0^\infty f(z|\gamma) f(\gamma) d\gamma$$

$$= \sum_{k=1}^{K} \binom{K}{k} (-1)^{k+1} \frac{\exp(-z/\bar{\gamma})}{\bar{\gamma}}$$

$$\bar{\gamma} = (1-\rho+\rho/k)\bar{\gamma}_{eq}$$

$$(14.41)$$

Figure 14.18 shows the effect of outdated CSI on the pdf for $K = 3$ and various values of the correlation coefficient. For $\rho = 1$, (14.41) reduces to (14C.1), the Rayleigh approximation to the SRD link with relay selection. Since this corresponds to the perfect estimation of the CSI (that is, the actual SNR is the same as the estimated SNR based on the CSI), the pdf shifts towards higher SNR values with increasing values of the correlation coefficient. However, for $\rho = 0$ (incorrect CSI), using (D.23), (14.41) can be shown to reduce to (14.17), implying that the SRD link behaves as a single relay channel, independently of the value of K. Then, the system cannot get any benefit from the relay selection.

For the SRD link with perfect CSI and relay selection, the BEP for BPSK is found using (14.41), (D.23), and (D.70) as follows:

$$P_b = \int_0^\infty Q(\sqrt{2z}) f(z) dz$$

$$= \frac{1}{2} \sum_{k=1}^{K} \binom{K}{k} (-1)^{k+1} \left[1 - \sqrt{\frac{\bar{\gamma}}{1+\bar{\gamma}}}\right]$$

$$= \frac{1}{2}\left[1 - \sum_{k=1}^{K} \binom{K}{k} (-1)^{k+1} \sqrt{\frac{\bar{\gamma}}{1+\bar{\gamma}}}\right]$$

$$(14.42)$$

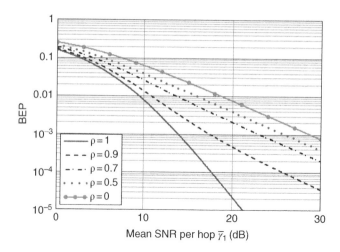

Figure 14.19 Effect of Correlation Between the Outdated CSI and the Actual Equivalent SNR on the BEP of the SRD Link with Rayleigh Approximation. $K = 3$ and $\bar{\gamma}_1 = \bar{\gamma}_2$.

Figure 14.19 shows that the relay selection based on outdated CSI increases the BEP drastically, compared to the curve for $\rho = 1$. For $\rho = 0.9$, that is, the CSI has 90% correlation with the actual SNR values, the BEP performance is degraded by 8 dB at BEP $= 10^{-4}$. For $\rho = 0$, independently of the value of K, the BEP curve reduces to that for Rayleigh approximation to the SRD link with a mean SNR $\bar{\gamma}_{eq}$ (see (14C.3)). Then the relay selection does not bring any performance improvement. This clearly shows the importance of the CSI estimation in relaying, and the difficulty of achieving in practice the performance improvements predicted by the theory.

14.3 Source and Destination with Multiple Antennas in Dual-Hop AF Relaying

In this section, we consider the case where the Source and the Destination are equipped with n and m antennas, respectively. The Relay is assumed to use a single antenna for transmission and reception. The use of multiple antennas at the Source and the Destination may

alleviate the problems that we already observed in relay networks. The Source and the Destination are assumed to employ maximal-ratio transmission (MRT) and MRC, respectively. This means that they both use beamforming with normalized antenna weights, which control the amplitude and phase of the branch signals. Note that the use of MRT requires CSI at the Source. From the equivalence of MRT and MRC when perfect CSI is available at the Source, the results to be obtained by this approach are equally valid for a SRD link with single antennas at the Source and the Destination while the Relay uses MRC with n receive antennas and MRT with m transmit antennas. In the following Sections, we will also study the case where multiple antennas are used at the Source, the Relay and the Destination.

14.3.1 Source-Relay-Destination Link

Consider a dual-hop AF relaying system, which consists of a Source with n antennas, a Relay with a single antenna and a Destination, with m antennas, as shown in Figure 14.20. [9][16][17][18][19][20]

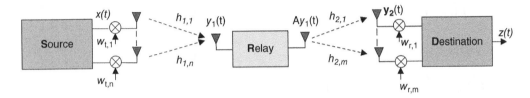

Figure 14.20   Block Diagram of the SRD Link with Multiple Antennas at the Source and the Destination.

The system is assumed to communicate in time-division multiplexing (TDM) mode in a flat Rayleigh fading channel. Assuming that the Source transmits the signal $x(t)$, the signal received by the Relay is given by

$$y_1(t) = \mathbf{w}_t \mathbf{h}_1 x(t) + n_1 \qquad (14.43)$$

where the transmit weight vector and the path gain vector are defined as

$$\begin{aligned} \mathbf{w}_t &= \begin{bmatrix} w_{t,1} & w_{t,2} & \cdots & w_{t,n} \end{bmatrix} \\ \mathbf{h}_1 &= \begin{bmatrix} h_{1,1} & h_{1,2} & \cdots & h_{1,n} \end{bmatrix}^T \end{aligned} \qquad (14.44)$$

Transmit weight coefficients $w_{t,i}$, $i = 1,2,..,n$, are normalized, that is, $\sum_{i=1}^{n} |w_{t,i}|^2 = 1$, so that the weight vector simply controls the distribution of the signal power, which is equal to the transmit power of the Source. The channel gain vector of the SR link consists of complex Gaussian elements with unity variance, $h_{i,j} = \mathbb{CN}(0,1)$, and $n_1 = \mathbb{CN}(0,\sigma_1^2)$ is the zero-mean AWGN with variance σ_1^2. Here, ^{T} denotes the transpose and ^{H} stands for the conjugate transpose.

The Relay amplifies the received signal $y_1(t)$ with an amplification factor of A and retransmits to the Destination. The received signal vector at the Destination may then be written as:

$$\begin{aligned} \mathbf{y}_2(t) &= \mathbf{h}_2 A y_1(t) + \mathbf{n}_2 \\ &= A \mathbf{h}_2 \mathbf{w}_t \mathbf{h}_1 x(t) + A \mathbf{h}_2 n_1 + \mathbf{n}_2 \end{aligned} \qquad (14.45)$$

where

$$\begin{aligned} \mathbf{y}_2(t) &= \begin{bmatrix} y_{2,1}(t) & y_{2,2}(t) & \cdots & y_{2,m}(t) \end{bmatrix}^T \\ \mathbf{h}_2 &= \begin{bmatrix} h_{2,1} & h_{2,2} & \cdots & h_{2,m} \end{bmatrix}^T \\ \mathbf{n}_2 &= \begin{bmatrix} n_{2,1} & n_{2,2} & \cdots & n_{2,m} \end{bmatrix}^T \end{aligned}$$

$$(14.46)$$

Since the Destination knows the path gains of the RD channel, the sufficient statistics for $\mathbf{y}_2(t)$ is obtained by maximum-ratio combining the outputs of the m receive antennas. This implies that the normalized receive weight vector is chosen as $\mathbf{w}_r = \mathbf{h}_2^H / |\mathbf{h}_2|$. The signal $z(t)$ at the output of the Destination may then be written as

$$\begin{aligned} z(t) &= \mathbf{w}_r \mathbf{y}_2(t) = \frac{\mathbf{h}_2^H}{|\mathbf{h}_2|} \mathbf{y}_2(t) \\ &= A |\mathbf{h}_2| \mathbf{w}_t \mathbf{h}_1 x(t) + A |\mathbf{h}_2| n_1 + \frac{\mathbf{h}_2^H}{|\mathbf{h}_2|} \mathbf{n}_2 \end{aligned}$$

$$(14.47)$$

14.3.1.1   Equivalent SNR of the Source-Relay-Destination Link

From (14.47), the instantaneous equivalent (end-to-end) SNR of the SRD link may be written as

$$\gamma_{eq,a} = \frac{|A|^2 |\mathbf{h}_2|^2 |\mathbf{w}_t \mathbf{h}_1|^2 E_s}{|A|^2 |\mathbf{h}_2|^2 \sigma_1^2 + \sigma_2^2} \qquad (14.48)$$

where $E_s = E\left[|x(t)|^2\right]$ denotes the symbol energy and

$$E\left[|n_1|^2\right] = \sigma_1^2$$

$$E\left[|n_{2,i}|^2\right] = \sigma_2^2, \quad i = 1,2 \dots m$$

$$\frac{1}{|\mathbf{h}_2|^2} E\left[|\mathbf{h}_2^H \mathbf{n}_2|^2\right] = \frac{1}{|\mathbf{h}_2|^2} E\left[\mathbf{h}_2^H \mathbf{n}_2 \mathbf{n}_2^H \mathbf{h}_2\right]$$

$$= \frac{1}{|\mathbf{h}_2|^2} E\left[\mathbf{h}_2^H\left(\sigma_2^2 \mathbf{I}\right) \mathbf{h}_2\right] = \sigma_2^2$$

$$(14.49)$$

When the CSI is available at the Source, the normalized transmit weight coefficients may be chosen as $\mathbf{w}_t = \mathbf{h}_1^H/|\mathbf{h}_1|$, which implies that the weight vector $\mathbf{w}_t$ is matched to the channel, that is, the maximal ratio transmission (MRT). Then, the instantaneous equivalent SNR of the SRD link may be rewritten as

$$\gamma_{eq,a} = \frac{|A|^2 |\mathbf{h}_2|^2 |\mathbf{h}_1|^2 E_s}{|A|^2 |\mathbf{h}_2|^2 \sigma_1^2 + \sigma_2^2} \qquad (14.50)$$

Availability of CSI at the Source implies that CSI is already known at the Relay. Therefore, if the amplification factor is chosen so as to compensate for the channel gain as in (14.8), that is, $|A|^2 = 1/|\mathbf{h}_1|^2$, then (14.50) reduces to

$$\gamma_{eq,a} = \frac{\gamma_1 \gamma_2}{\gamma_1 + \gamma_2} \qquad (14.51)$$

where

$$\gamma_i = \frac{E_s}{\sigma_i^2} |\mathbf{h}_i|^2 = \bar{\gamma}_i |\mathbf{h}_i|^2, \quad i = 1,2 \qquad (14.52)$$

Compared with (14.8), in this case, the SNR's γ_1 and γ_2 of SR and RD links are given by the sum of SNRs of n and m paths, respectively. This is simply due to the use of MRT and MRC in SR and RD links, respectively.

The pdf's of γ_1 and γ_2, which are needed in statistical characterization of this relaying system, are given by (12.42), (12.45) and (12.46), and copied below for the sake of convenience:

$$f_{\gamma_1}(x) = \begin{cases} \dfrac{1}{(n-1)!} \dfrac{x^{n-1}}{\bar{\gamma}_1^n} e^{-x/\bar{\gamma}_1} & \rho_s = 0 \\[2ex] \displaystyle\sum_{i=1}^{n} A_i \dfrac{e^{-x/(\bar{\gamma}_1 \lambda_i)}}{\bar{\gamma}_1 \lambda_i} & 0 < \rho_s < 1 \\[2ex] \dfrac{e^{-x/n\bar{\gamma}_1}}{n\bar{\gamma}_1} & \rho_s = 1 \end{cases}$$

$$(14.53)$$

$$f_{\gamma_2}(x) = \begin{cases} \dfrac{1}{(m-1)!} \dfrac{x^{m-1}}{\bar{\gamma}_1^m} e^{-x/\bar{\gamma}_2} & \rho_d = 0 \\[2ex] \displaystyle\sum_{i=1}^{m} B_i \dfrac{e^{-x/(\bar{\gamma}_2 \eta_i)}}{\bar{\gamma}_2 \eta_i} & 0 < \rho_d < 1 \\[2ex] \dfrac{e^{-x/(m\bar{\gamma}_2)}}{m\bar{\gamma}_2} & \rho_d = 1 \end{cases}$$

$$(14.54)$$

where $A_i = \displaystyle\prod_{k \neq i}^{n} \lambda_i/(\lambda_i - \lambda_k)$, $B_i = \displaystyle\prod_{k \neq i}^{m} \eta_i/(\eta_i - \eta_k)$ and the normalized eigenvalues λ_i and η_i are given by (12.48) for $n = m = 2$ and by (12.52) for $n = m = 3$. For $n = m = 1$, one has $\lambda_1 = \eta_1 = 1$, $\lambda_i = \eta_i = 0$, $i \geq 2$ (see (13.17)); thus, $A_1 = B_1 = 1$ and $A_i = B_i = 0$, $i \geq 2$. Here, ρ_s denotes the correlation coefficient between n antennas at the Source, and ρ_d shows the correlation coefficient between m antennas at the Destination.

For the partially correlated case with n antennas at the Source, m antennas at the Destination and a single antenna at the Relay, the cdf of the equivalent SNR $\gamma_{eq,a}$ of the SRD link may be obtained using (14.53) and (14.54) in the derivation of (14A.1):

$$F_{\gamma_{eq,a}}(\gamma) = 1 - \sum_{i=1}^{n} \sum_{j=1}^{m} A_i \, B_j \, \beta_{i,j} \, \gamma$$

$$\times \exp\left(-\alpha_{i,j}\gamma\right) K_1\left(\beta_{i,j} \, \gamma\right)$$

$$(14.55)$$

$$\alpha_{i,j} = \frac{1}{\bar{\gamma}_1 \lambda_i} + \frac{1}{\bar{\gamma}_2 \eta_j}$$

$$\beta_{i,j} = \frac{2}{\sqrt{\bar{\gamma}_1 \lambda_i \, \bar{\gamma}_2 \eta_j}}$$

where $K_1(x)$ denotes the modified Bessel function of the second kind. The corresponding pdf is obtained by the derivative of (14.55):

$$f_{\gamma_{eq,a}}(\gamma) = \sum_{i=1}^{n}\sum_{j=1}^{m} A_i\, B_j\, \beta_{i,j}\gamma\, \exp\left(-\alpha_{i,j}\gamma\right)$$

$$\times \left[\beta_{i,j}\, K_0\left(\beta_{i,j}\,\gamma\right) + \alpha_{i,j} K_1\left(\beta_{i,j}\,\gamma\right)\right]$$

$$(14.56)$$

The corresponding MGF is given by

$$M_{\gamma_{eq,a}}(s) = 1 - \sum_{i=1}^{n}\sum_{j=1}^{m} A_i\, B_j\, \frac{2s}{3\left(s+\alpha_{i,j}\right)}$$

$$\times {}_2F_1\left(\frac{1}{2}, 1; \frac{5}{2}; 1 - \frac{\beta_{i,j}^2}{\left(s+\alpha_{i,j}\right)^2}\right)$$

$$(14.57)$$

For $n = m = 1$, noting that $A_1 = B_1 = 1$ and $A_i = B_i = 0$, $i \geq 2$, (14.55)–(14.57) reduce to (14.13)–(14.15), as expected.

For the special case of $\rho_s = \rho_d = 0$, the cdf of the equivalent SNR is given by (14.D4):

$$F_{\gamma_{eq,a}}(\gamma) = 1 - \frac{2e^{-\alpha\gamma}}{(n-1)!}\left(\frac{\beta\gamma}{2}\right)^n \sum_{j=0}^{m-1}\frac{1}{j!}\left(\frac{\gamma}{\bar{\gamma}_2}\right)^j$$

$$\times \sum_{r=0}^{j+n-1}\binom{j+n-1}{r}\left(\frac{\bar{\gamma}_2}{\bar{\gamma}_1}\right)^{\frac{r}{2}} K_{n-r}(\beta\gamma)$$

$$(14.58)$$

The pdf corresponding to (14.58) is given by (14D.5):

$$f_{\gamma_{eq}}(\gamma,\alpha,\beta) = \frac{2e^{-\alpha\gamma}}{(n-1)!}\left(\frac{\beta\gamma}{2}\right)^n \sum_{j=0}^{m-1}\frac{1}{j!}\left(\frac{\gamma}{\bar{\gamma}_2}\right)^j$$

$$\times \sum_{r=0}^{j+n-1}\binom{j+n-1}{r}\left(\frac{\bar{\gamma}_2}{\bar{\gamma}_1}\right)^{r/2}$$

$$\times \left[\left\{\alpha - \frac{j+r}{\gamma}\right\} K_{n-r}(\beta\gamma) + \beta K_{n-r-1}(\beta\gamma)\right]$$

$$(14.59)$$

The MGF corresponding to (14.59) is given by (14D.6):

$$M_{\gamma_{eq,a}}(s) = 1 - \frac{\sqrt{\pi}\, s}{(n-1)!2^{2n}}\sum_{j=0}^{m-1}\frac{1}{j!(2\bar{\gamma}_2)^j}$$

$$\times \sum_{r=0}^{j+n-1}\binom{j+n-1}{r}\left(\frac{2}{\bar{\gamma}_1}\right)^r \times$$

$$\times \frac{\Gamma(j+r+1)\Gamma(2n-r+j+1)}{(s+\alpha)^{r+j+1}\Gamma(n+j+3/2)}$$

$$\times {}_2F_1\left(\frac{j+r+1}{2}, \frac{j+r+2}{2}; n+j+\frac{3}{2}; 1 - \frac{\beta^2}{(s+\alpha)^2}\right)$$

$$(14.60)$$

For $n = m = 1$, (14.58)–(14.60) reduce to (14.13)–(14.15), as expected.

If the antennas at the Source and the Destination are perfectly correlated among each other ($\rho_s = \rho_d = 1$), one may easily obtain the cdf by inserting the pdf's, given by (14.53) and (14.54), into (14A.1):

$$F_{\gamma_{eq,a}}(\gamma,\alpha_a,\beta_a) = F_{\gamma_{eq}}(\gamma,\alpha_a,\beta_a)$$

$$= 1 - \beta_a\gamma\,\exp(-\alpha_a\gamma)K_1(\beta_a\,\gamma)$$

$$(14.61)$$

where $\alpha_a = 1/(n\bar{\gamma}_1) + 1/(m\bar{\gamma}_2)$, $\beta_a = 2/\sqrt{n\bar{\gamma}_1\, m\bar{\gamma}_2} = \beta/\sqrt{n\, m}$. Since (14.61) has the same form as $F_{\gamma_{eq}}(\gamma,\alpha,\beta)$ given by (14.13), the corresponding pdf, MGF and the mean equivalent SNR may be obtained if $\{\alpha, \beta\}$ are replaced by $\{\alpha_a, \beta_a\}$ in (14.14)–(14.16).

Note that the cdf's given by (14.55), (14.58) and (14.61) must be raised to the power K in order to find the cdf for relay selection in the SRD link, that is, when one of K relays with highest equivalent SNR is selected.

Figure 14.21 shows the effect of correlation between antennas at the Source and at the Destination on the outage probability. The outage performance seems to be more affected from antenna correlations with increasing numbers of antennas. The outage probability in a SRD link with $\bar{\gamma}_1 = \bar{\gamma}_2 = 10$ dB and single antennas at both Source and Destination, the

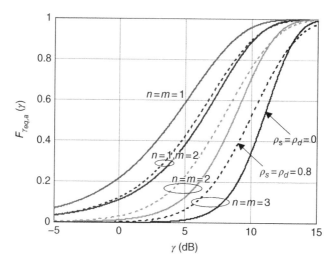

Figure 14.21 Outage Probability For the SRD Link For Various Combinations of $m = n$ and $\rho_s = \rho_d$, the Correlation Coefficient Between Antennas at the Source and the Destination for $\bar{\gamma}_1 = \bar{\gamma}_2 = 10$.

probability that the equivalent SNR of the SRD link remains below 5 dB is 59.3%. For $n = m = 2$, the outage probability reduces to 12.6% and 23.1% for $\rho_s = \rho_d = 0$ and $\rho_s = \rho_d = 0.8$, respectively. As for $n = m = 3$, the outage probability reduces to 1.4% and 6.5% for $\rho_s = \rho_d = 0$ and $\rho_s = \rho_d = 0.8$, respectively. In the latter case, the correlation increases the

outage probability by a factor of $6.5/1.4 = 4.64$. This clearly shows the benefit of using multiple antennas in the SRD link and decreasing the antenna correlations.

The mean equivalent SNR for the uncorrelated and partially and perfectly correlated cases are found using (14.16) and the MGFs given by (14.57) and (14.60):

$$
\bar{\gamma}_{eq,a} = \begin{cases}
\dfrac{\sqrt{\pi}\,2^{-2n}}{(n-1)!}\displaystyle\sum_{j=0}^{m-1}\dfrac{(2\bar{\gamma}_2)^{-j}}{j!}\displaystyle\sum_{r=0}^{j+n-1}\binom{j+n-1}{r}\left(\dfrac{2}{\bar{\gamma}_1}\right)^r\dfrac{\Gamma(j+r+1)}{\alpha^{r+j+1}}\times & \\[2ex]
\qquad\dfrac{\Gamma(2n-r+j+1)}{\Gamma(n+j+3/2)}\,{}_2F_1\left(\dfrac{j+r+1}{2},\dfrac{j+r+2}{2};n+j+\dfrac{3}{2};1-\dfrac{\beta^2}{\alpha^2}\right) & \rho_s=\rho_d=0 \\[3ex]
\displaystyle\sum_{i=1}^{n}\sum_{j=1}^{m}A_i\,B_j\,\dfrac{2}{3\alpha_{i,j}}\,{}_2F_1\left(\dfrac{1}{2},1;\dfrac{5}{2};1-\dfrac{\beta_{i,j}^2}{\alpha_{i,j}^2}\right) & 0<\rho_s,\rho_d<1 \\[3ex]
\dfrac{2}{3\,\alpha_\alpha}\,{}_2F_1\left(\dfrac{1}{2},1;\dfrac{5}{2};1-\dfrac{\beta_\alpha^2}{\alpha_\alpha^2}\right) & \rho_s=\rho_d=1
\end{cases}
\tag{14.62}
$$

Based on (14.62), the effect of correlation between n antennas at the Source and m antennas at the Destination is shown in Figure 14.22. The mean SNR's in SR and RD links are

assumed to be the same. Note that the slope of the curve with the mean hop SNR remains the same but its value increases with increasing values of n and m. The increase in correlation

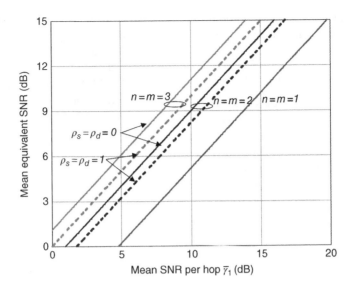

Figure 14.22   Effect of Correlation, Between Antennas at the Source and the Destination, On the Mean Equivalent SNR of a SRD Link with AF Protocol For $\bar{\gamma}_1 = \bar{\gamma}_2$.

coefficient between antennas, which is assumed to be the same at the Source and the Destination, decreases the mean equivalent SNR. For example, for $\bar{\gamma}_1 = \bar{\gamma}_2 = 10$ *dB*, $n = m = 1$ and $\rho_s = \rho_d = 0$, the mean equivalent SNR is 5.23 dB, which is $\bar{\gamma}_1/3$, as shown in Figure 14.4. However, it becomes equal to 9 dB and 11 dB for $n = m = 2$ and $n = m = 3$, respectively. Hence, an SNR advantage of approximately 3.77 dB is achieved by using dual antennas at the Source and the Destination. The use of triple antennas gives an additional 2 dB increase in the mean equivalent SNR. However, for $n = m = 2$ and $n = m = 3$, the mean equivalent SNR decreases from 9 dB to 8.26 dB and from 11 dB to 10 dB, respectively, when, in the worst case, antennas are perfectly correlated with each other. For practical cases of interest, the antenna correlations hardly exceed 0.8, which limits the degradation to less than 0.5 dB.

Depending on the value of the correlation coefficient, we use (14.55), (14.58) or (14.61) in order to determine the average BEP for BPSK modulation for the SRD link with relay selection:

$$P_b = \frac{1}{\sqrt{2\pi}} \int_0^\infty \frac{1}{\sqrt{2\gamma}} e^{-\gamma} F_{\gamma_{eq,a}}(\gamma)^K \, d\gamma \quad (14.63)$$

For relay selection, the cdf of the capacity may be found using (14.55), (14.58) or (14.61), depending on the values of the correlation coefficients between transmit and receive antennas:

$$C_{out,a}(C_{th}) = \text{Prob}\left(\frac{1}{2}\log_2\left(1 + \gamma_{eq,a}\right) < C_{th}\right)$$

$$= F_{\gamma_{eq,a}}^K \left(2^{2C_{th}} - 1\right)$$

$$(14.64)$$

The ergodic capacity is obtained by using (14.55), (14.58) or (14.61) as follows

$$C_{av,a} = \frac{1}{2 \ln 2} \int_0^\infty \frac{1}{1+\gamma} \left[1 - F_{\gamma_{eq,a}}^K(\gamma)\right] d\gamma$$

$$(14.65)$$

The presence of the factor ½ in (14.64) and in front of (14.65) accounts for the use of the first half of the transmission period to the SR link and the second half to the RD link, according to the TDM operation of the relaying protocol.

14.3.2   Source-Destination Link

In the SD link, the signal is transmitted by the Source via n antennas and received by m antennas at the Destination. This link operates as a nxm MIMO beamforming system, described in section 13.7 of Chapter 13. The received signal vector at the Destination, which consists of m components, may simply be written as:

$$\mathbf{y}_d = \mathbf{H}\mathbf{w}x(t) + \mathbf{n}_d \qquad (14.66)$$

where

$$\mathbf{H} = \begin{bmatrix} h_{11} & h_{12} & \cdots & h_{1n} \\ h_{21} & & \cdots & h_{2n} \\ & & \ddots & \\ h_{m1} & & & h_{mn} \end{bmatrix} \qquad (14.67)$$

$$\mathbf{w} = \begin{bmatrix} w_1 & w_2 & \cdots & w_n \end{bmatrix}^T$$
$$\mathbf{y}_d = \begin{bmatrix} y_1 & y_2 & \cdots & y_m \end{bmatrix}^T$$
$$\mathbf{n}_d = \begin{bmatrix} n_1 & n_2 & \cdots & n_m \end{bmatrix}^T$$

As presented in (13.78)–(13.86), the SNR of this MIMO system is maximized by MRT transmission at the Source and MRC reception at the Destination. Therefore, maximal-ratio combining of the directly received signal components yields (see (13.83))

$$r = (\mathbf{H}\mathbf{w})^H \mathbf{y}_d = \mathbf{w}^H \mathbf{H}^H \mathbf{H}\mathbf{w}x(t) + \mathbf{w}^H \mathbf{H}^H \mathbf{n}_d \qquad (14.68)$$

It is shown in (13.86) that the maximum SNR may be expressed in terms of the maximum eigenvalue λ_{max} of the $\mathbf{H}^H\mathbf{H}$ Wishart matrix:

$$\gamma_{max} = \bar{\gamma}\,\lambda_{max} \qquad (14.69)$$

The pdf of the eigenvalues of a Wishart matrix $\mathbf{H}^H\mathbf{H}$ is studied in detail in Chapter 13 for the cases

of arbitrary correlation at the Source and/or the Destination. [21][22][23][24][25] For example, the pdf of the maximum eigenvalue for the uncorrelated case ($\rho_s = \rho_d = 0$) is given by (13.16)–(13.22) for several values of n and m.

The outage capacity due to maximum eigenvalue of a $n \times m$ MIMO system is found as:

$$C_{out, n \times m}(C_{th}) = \text{Prob}\left(\frac{1}{2}\log_2\left(1 + \frac{\bar{\gamma}\lambda_{max}}{n}\right) < C_{th}\right)$$

$$= F_{n \times m}\left(\frac{n}{\bar{\gamma}}\left(2^{2C_{th}} - 1\right)\right) \qquad (14.70)$$

where the cdf $F_{n \times m}(\gamma)$ of the maximum eigenvalue of a $n \times m$ MIMO system is given by (13.23). The mean capacity of the SD ($n \times m$ MIMO) link is obtained from (13.63):

$$C_{m, n \times m} = \frac{1}{2\ln 2} E\left[\ln\left|\mathbf{I} + \frac{\bar{\gamma}}{n}\mathbf{H}^H\mathbf{H}\right|\right] \quad (bps/Hz) \qquad (14.71)$$

For an uncorrelated 2×2 MIMO system ($\rho_s = \rho_d = 0$, $m = n = 2$), operating in TDD mode, the mean capacity of the SD link is given by ½ of the sum of the mean capacities of minimum and maximum eigenvalues in (13.72):

$$C_{m, 2 \times 2, u}(\bar{\gamma}) = 2C_m\left(\frac{\bar{\gamma}}{2}\right)\left(1 + \frac{2}{\bar{\gamma}^2}\right) + \frac{1}{2\ln 2}\left(1 - \frac{2}{\bar{\gamma}}\right) \qquad (14.72)$$

where, the capacity for $n = m = 1$, $C_m(x) = e^{1/x}E_1(1/x)/(2 \ln 2)$ is defined by (14B.2).

The BEP for the BPSK modulation for the SD link may be expressed as

$$P_{b, n \times m} = \int_0^\infty Q\left(\sqrt{2\bar{\gamma}\lambda}\right)f_{\lambda_{max}}(\lambda)\,d\lambda \qquad (14.73)$$

where the pdf of the maximum eigenvalue is given by (13.16)–(13.22).

14.3.3 Selection-Combined SRD and SD Links

The cdf when SRD and SD links are selection combined is given by

$$F_{MIMO}(\gamma) = F_{\gamma_{eq,a}}(\gamma)^K F_{n \times m}(\gamma) \qquad (14.74)$$

where $F_{n \times m}(\gamma)$ denotes the cdf of the SNR for the SD link, which has a $n \times m$ MIMO channel, and is obtained by (13.23)–(13.25) and (14.69) for the uncorrelated case. Note that (14.74) also accounts for the relay selection.

The average SNR of selection combined SRD and SD links may be found using (F.5):

$$E[\gamma_{MIMO}] = \int_0^\infty [1 - F_{MIMO}(\gamma)] \, d\gamma \qquad (14.75)$$

Figure 14.23 and Figure 14.24 show respectively the cdf and the pdf of the equivalent SNR of the SRD link and the selection combined SRD and SD links. Number of antennas at the Source and the Destination are assumed to be the same, $n = m$, and the Relay with the highest equivalent SNR is selected in the

SRD link. Transmitted and received signals are assumed to be uncorrelated with each other. The SNRs in SR and RD links are 10 dB while the SNR of the SD link is assumed to be 7 dB. The mean equivalent SNR of the SRD link is determined by α. For $\bar{\gamma}_1 = \bar{\gamma}_2$, the mean equivalent SNR is $\bar{\gamma}_1/3$ (see (14.4)). On the other hand, bearing in mind that the path length of the SD link is typically twice the path lengths of SR and RD links, and received signal power is proportional to squared-distance, this assumption for $\bar{\gamma}_3$ is reasonable. Figure 14.23 and Figure 14.24 clearly show that increasing the number of antennas is more effective than the relay selection for improving the system performance.

Figure 14.25 shows the mean equivalent SNR of SRD and selection-combined SRD and SD links using (14.75). Increasing the number of antennas increases the equivalent SNR more rapidly than the relay selection. For the chosen mean hop SNRs, $\bar{\gamma}_1 = \bar{\gamma}_2 = 2\bar{\gamma}_3$, the effect of relay selection is even less significant when the SRD and SD links are selection-combined.

The BEP for BPSK is determined using (14.74) and (14.21):

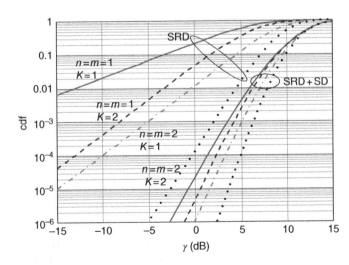

Figure 14.23 The Cdf of the Equivalent SNR of the SRD and Selection-Combined SRD and SD Links For Various Combinations of $n = m$ and K. The antennas are assumed to be uncorrelated and $\bar{\gamma}_1 = \bar{\gamma}_2 = 2\bar{\gamma}_3 = 10$.

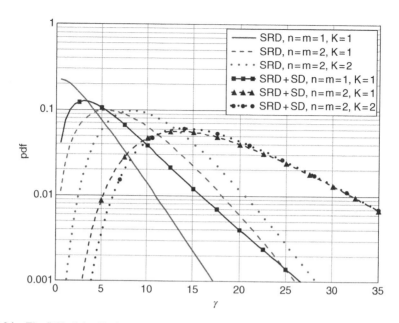

Figure 14.24 The Pdf of the Equivalent SNR of SRD and Selection-Combined SRD and SD Links For Various Combinations of $n = m$ and K. The antennas are assumed to be uncorrelated and $\bar{\gamma}_1 = \bar{\gamma}_2 = 2\bar{\gamma}_3 = 10$.

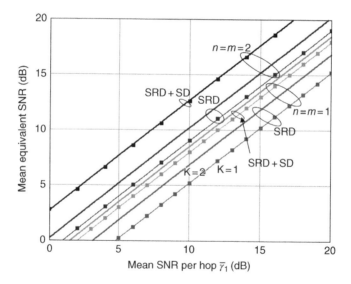

Figure 14.25 Mean Equivalent SNR For the SRD and the Selection-Combined SRD and SD Links For Various Combinations of $n = m$ and K. $\bar{\gamma}_1 = \bar{\gamma}_2 = 2\bar{\gamma}_3$ is assumed.

$$P_{b,MIMO} = \frac{1}{\sqrt{2\pi}} \int_0^\infty \frac{1}{\sqrt{2\gamma}} e^{-\gamma} F_{MIMO}(\gamma) \, d\gamma$$

(14.76)

Figure 14.26 shows the BEP for BPSK modulation for the SRD link as well as selection-combined SRD and SD links for various combinations of $n = m$, the number of antennas, and K, the number of relays for selection.

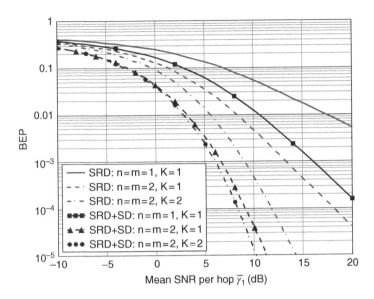

Figure 14.26 BEP For BPSK Modulation For the SRD and the Selection-Combined SRD and SD Links For Various Combinations of $n = m$ and K. $\bar{\gamma}_1 = \bar{\gamma}_2 = 2\bar{\gamma}_3$ is assumed.

The use of multiple antennas was observed to be much more effective than relay selection in decreasing the BEP. The relay selection becomes less effective with increasing number of antennas when SRD and SD links are selection-combined.

The ergodic capacity may be written in terms of (14.74) as follows:

$$C_{MIMO} = \frac{1}{2\,\ln 2}$$

$$\times \int_0^\infty \frac{1}{1+\gamma}[1 - F_{MIMO}(\gamma)]\,d\gamma \quad (bps/Hz)$$

$$(14.77)$$

The cdf of the capacity for selection combined SRD and SD links (due to maximum eigenvalue) of a $n \times m$ MIMO system may be written as

$$C_{out}(C_{th}) = \text{Prob}\left(\frac{1}{2}\log_2(1 + \gamma_{MIMO}) < C_{th}\right)$$

$$= F_{MIMO}\left(2^{2C_{th}} - 1\right)$$

$$(14.78)$$

Figure 14.27 shows the mean (ergodic) capacity for SRD links with relay selection (see (14.65)) and selection-combined SRD and SD links (see (14.77)) as a function of $\bar{\gamma}$ for $\bar{\gamma}_1 = \bar{\gamma}_2 = 2\bar{\gamma}_3$. For the assumed mean SNR levels, increase in the number of antennas and selection-combining SRD and SD links lead to significant improvements in the ergodic capacity. On the other hand, the relay selection seems to be more effective for $n = m = 1$, compared to for $n = m = 2$.

Example 14.4 Design of the SRD Link With AF Relaying Protocol.
Assuming that reliable communications at desired data rates cannot be accomplished between two mobile terminals in a coverage area of $d = 400$ m diameter, we would like to use AF relaying with the cooperation of other users, acting as relays. The frequency of operation is chosen to be $f = 2.4$ GHz in the ISM band. The mobile terminals are assumed to have multiple uncorrelated antennas each with $G_t = G_r = 0$ dBi gain. At 2.4 GHz operating frequency, the antennas are assumed to be uncorrelated as they are separated from each

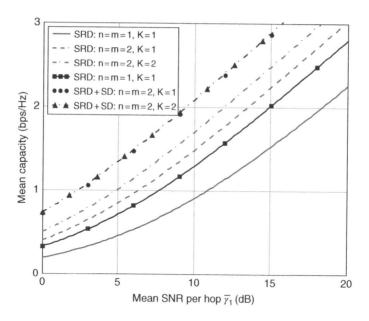

Figure 14.27   Mean Capacity Versus Mean SNR Per Hop For Uncorrelated Antennas at the Source and the Destination For Various Combinations of $n = m$ and K. $\bar{\gamma}_1 = \bar{\gamma}_2 = 2\bar{\gamma}_3$ is Assumed.

other by more than half a wavelength, 6.25 cm. Antenna noise temperature is $T_A = 290$ K and the noise figure $F = 7$ dB. The user density in the coverage area of interest is such that it is highly likely to find at least $K = 2$ users to act as Relay in the vicinity of the mid-way between the Source and the Destination. The outage probability, which is used as a figure of merit for the considered system, is desired to be less than 10^{-3} at a threshold SNR level of 6.8 dB. This threshold level ensures that the BEP at the threshold level remains below than 10^{-3} for BPSK modulation, that is $\text{Prob}(\gamma_{eq,a} \le 6.8\ dB) \le 10^{-3}$. Figure 14.28 shows that this condition is satisfied only if $\bar{\gamma}_1 = \bar{\gamma}_2 = 40$ $(16 dB)$ and $n = m = K = 2$. The mean SNR of the SR link with $n = 2$ antennas at the Source is 2 times higher than the corresponding SISO link (see Table 13.1). This can be intuitively explained as follows: Transmit power is equally shared by two transmit antennas. However, because of MRT, their signals are assumed to be coherently added at the receiver; so, the received power

level is increased by a factor of four. Hence, this corresponds to a 3 dB increase in the mean SNR.

Using the Friis formula (2.114) and (14.69), we determine the relation between the transmit power level P_t and the average bit rate R_b, in free space propagation conditions, as follows:

$$\bar{\gamma}_1 = 2\bar{\gamma}_{1,SISO} = 2\frac{P_t G_t G_r}{R_b (4\pi d_1/\lambda)^2 kT_0 F} = 40$$

$$\Rightarrow\ R_b(Mbps) = 6.4 P_t(mW)$$

where $\bar{\gamma}_{1,SISO}$ denotes the mean SNR of the SR link with single antennas at both the Source and the Relay, which are separated from each other by $d_1 = d/2 = 200$ m. If the transmit power of the mobile terminal is 1 mW, then the SRD link with AF protocol can support 6.4 Mbps. The mean equivalent SNR of the SRD link is given by Figure 14.25 as $\bar{\gamma}_{eq,a} = 16.23\ dB$.

Now let us consider the direct channel between the Source and the Destination, with a maximum range of 400 m, that is, twice the distance between the Source and the Relay.

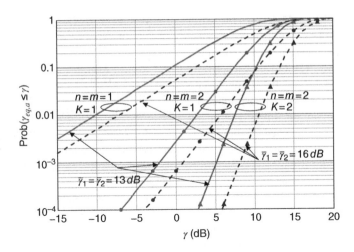

Figure 14.28 Outage Probability For the SRD Link For Dual-Hop AF Relaying with Multiple Uncorrelated Antennas at the Source and the Destination For $\bar{\gamma}_1 = \bar{\gamma}_2 = 13\,dB$, $16\,dB$ and Various Combinations of $n = m$ and K.

This causes a four-fold increase in the free-space propagation loss. On the other hand, the SD link with 2×2 MIMO channel causes a 3.5-fold increase in the mean SNR compared to a SISO system (see Table 13.1). Hence, the mean SNR of the source-destination link with a 2×2 MIMO channel is $3.5/2 = 1.75$ $(2.43\ dB)$ higher than that of the SR link with 2×1 MISO for the same range. Taking into account of the two-fold increase in the range, the mean SNR of the SD link is given by $\bar{\gamma}_{SD} = \bar{\gamma}_1 1.75/4 = 0.438\ \bar{\gamma}_1$. Hence the SD link has a mean SNR of $12.4\,dB$, which is $3.83\,dB$ lower mean SNR compared to the SRD link.

When the SNR's of SRD and SD links are balanced, that is, nearly equal to each other, then SC or MRC of these links lead to higher effective SNRs and hence more reliable system operation. Therefore, SRD and SD links must be designed so as to have comparable mean SNR's. The effective mean SNR for MRC is given by $\bar{\gamma}_{eq,\alpha} + \bar{\gamma}_{SD} = 10^{1.623} + 10^{1.24} = 59.35$ $(17.73\ dB)$, which is 1.5 dB higher than the mean SNR of the SRD link. Assuming SRD and SD links are maximal-ratio combined, the effective SNR $\bar{\gamma}_{eq,\alpha} + \bar{\gamma}_{SD}$ will be

3 dB, 1.76B, and 0.97 dB higher than $\bar{\gamma}_{eq,\alpha}$ for $\bar{\gamma}_{SD}/\bar{\gamma}_{eq,\alpha} = 1$, 0.5 and 0.25, respectively.

14.4 Dual-Hop Detect-and-Forward Relaying

In contrast with AF relaying, where the Relay amplifies and forwards the signal transmitted by the Source, DF relaying is based on detection of the received signal and then re-modulation and forwarding. Consider a dual-hop DF relaying system, which consists of a Source with n antennas, a Relay with a single antenna and a Destination equipped with m antennas, as shown in Figure 14.20.

As in AF relaying operating in the TDM format, the transmission period is divided into two equal-duration time slots. During time slot I, the Source broadcasts an M-ary signal and the Relay detects the signal broadcasted by the Source and performs coherent maximum-likelihood (ML) detection:

$$\hat{s} = \arg \max_{1 \leq j \leq M} |y_1 - h_1 s_j|^2 \qquad (14.79)$$

where $S = \{s_j\}$, $j = 1, 2, \cdots M$ denotes the set of M-ary signal. Here, h_1 and h_2 denote circularly symmetric complex Gaussian gains with unity variance of SR and RD links, respectively. Since the Relay estimates the symbol with the largest argument as being transmitted, the error performance of the SR link is determined by the link SNR γ_1. The detected symbol $\hat{s}$ is remodulated by the Relay and subsequently transmitted during slot II with the same/different average power. If the detection at the Relay is erroneous, then the signal forwarded to the Destination will also be erroneous. In this case, the signal received by the Destination will be correct only if the Destination also makes a detection error. In the contrary, if the Relay correctly detects the signal in the SR link, then the performance of the RD link will be determined only by the SNR γ_2 of the RD link. [26][27]

Let us first consider BPSK signalling with signals $s_1 = \sqrt{E_b}$ and $s_2 = -\sqrt{E_b} = \sqrt{E_b}\, e^{j\pi}$, as shown in Figure 14.29. Assuming that the signal s_1 is broadcasted by the Source, correct detection by the Relay implies $\hat{s} = s_1$. Then, the Relay transmits s_1 with probability $1 - P_e$, where P_e denotes the probability of detection error. In case where the transmitted signal is detected with error, the Relay transmits signal $\hat{s} = s_2$, instead of s_1, with probability P_e. The signal vector received by m antennas at the Destination may then be written as

$$\mathbf{y}_2 = \mathbf{h}_2\left[(1 - P_e(\gamma_1))\, s_1 + P_e(\gamma_1)\, s_2\right] + \mathbf{n}_2 \tag{14.80}$$

The probability of detection error in the SR link is given by $P_e(\gamma_1) = Q(\sqrt{\gamma_1})$ for BPSK modulation. The instantaneous SNRs per bit for SR and RD links are defined as $\gamma_i = |h_i|^2 E_b / \sigma_i^2$, $i = 1, 2$. Similarly, n_1 and $\mathbf{n}_2$ are AWGN noise terms with variances σ_1^2 and σ_2^2, respectively, as defined by (14.49) (see Figure 14.20).

The equivalent SNR per bit of the SRD link for DF relaying may then be written as

$$\gamma_{eq,df} = \frac{E\left[|(1 - P_e)s_1 + P_e\, s_2|^2 |\mathbf{h}_2|^2\right]}{E\left[|\mathbf{n}_2|^2\right]}$$

$$= \frac{|\mathbf{h}_2|^2}{\sigma_2^2} E\left[(1 - P_e)^2 |s_1|^2 + P_e^2 |s_2|^2 + 2(1 - P_e)P_e\, s_1 s_2\right]$$

$$= \gamma_2\, U(\gamma_1) \tag{14.81}$$

where $E_b = E\left[|s_1|^2\right] = E\left[|s_2|^2\right] = -E[s_1 s_2]$ and

$$U(\gamma_1) = (1 - 2P_e(\gamma_1))^2 \tag{14.82}$$

Noting that $P_e(0) = 1/2$ and $P_e(\infty) = 0$, $\gamma_{eq,df}$ varies between zero and γ_2 as γ_1 changes from zero to infinity.

The above analysis can easily be repeated for QPSK modulation, with the constellation shown in Figure 14.29. The complex signals

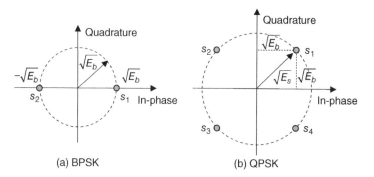

(a) BPSK   (b) QPSK

Figure 14.29 Constellation Diagrams For BPSK and QPSK.

with identical symbol energies, $E_s = E[|s_i|]^2$, $i = 1,2,3,4$, are defined by

$$s_1 = \sqrt{E_s}\, e^{j\pi/4}, \quad s_2 = \sqrt{E_s}\, e^{j3\pi/4},$$
$$s_3 = \sqrt{E_s}\, e^{-j3\pi/4}, \quad s_4 = \sqrt{E_s}\, e^{-j\pi/4} \quad (14.83)$$

Assuming that the symbol s_1 is sent by the Source, the signal received by the Destination may be written as

$$\mathbf{y}_2 = \mathbf{h}_2 \left[(1-P_e)^2\, s_1 + (1-P_e)P_e\, (s_2 + s_4) + P_e^2 s_3 \right] + \mathbf{n}_2$$

$$= \mathbf{h}_2 \left[(1-P_e)^2\, s_1 + P_e^2 s_3 \right] + \mathbf{n}_2$$

$$(14.84)$$

where $s_2 + s_4 \equiv 0$ since s_2 and s_4 are antipodal. $P_e(\gamma_1) = Q\left(\sqrt{\gamma_1/2}\right)$ denotes the probability of error along in-phase or quadrature axis. SNR per symbol in SR and RD links are defined by $\gamma_i = |\mathbf{h}_i|^2 \bar{\gamma}_i = |\mathbf{h}_i|^2 E_s/\sigma_i^2$, $i = 1,2$ (see (7.71)). The first term in the bracket in (14.84) shows that the symbol s_1 is sent by the Relay if the signal is detected correctly both in-phase and quadrature axes, that is, with probability $(1-P_e)^2$. The second term shows that symbols s_2 and s_4 are sent with probability $P_e(1-P_e)$, that is, when an error is made in one axis but no error in the other. Finally, third term indicates that s_3 is sent with probability P_e^2 when the signal is detected erroneously in both in-phase and quadrature axes (see Figure 14.29).

Using (14.84), the equivalent SNR per symbol for the SRD link for QPSK modulation may be written as

$$\gamma_{eq,df} = \frac{|\mathbf{h}_2|^2}{E\left[|\mathbf{n}_2|^2\right]} E\left[|(1-P_e)^2 s_1 + P_e^2 s_3|^2 \right]$$

$$= \frac{|\mathbf{h}_2|^2}{\sigma_2^2} E \left[\begin{array}{l} (1-P_e)^4 |s_1|^2 + P_e^4 |s_3|^2 \\ + (1-P_e)^2 P_e^2\, (s_1 s_3^* + s_1^* s_3) \end{array} \right]$$

$$= \gamma_2\, U(\gamma_1)$$

$$(14.85)$$

where $E[s_1 s_3^* + s_1^* s_3] \equiv -2E_s$ (see (14.83)). Comparison of (14.81) and (14.85) shows that the equivalent instantaneous SNR of the SRD link has the same form, independently of whether the modulation is BPSK or QPSK.

SR and RD links are assumed to fade independently from each other, since they have different propagation channels. For Rayleigh fading, γ_1 and γ_2 are exponentially distributed, if the Source, the Relay and the Destination are all equipped with single antennas. However, when the Source has n antennas and the Destination is equipped with m antennas, as shown in Figure 14.20, pdf's of γ_1 and γ_2 are respectively given by (14.53) and (14.54), expressed in terms of the correlation coefficients ρ_s and ρ_d between the antennas at the Source and the Destination, respectively. When n (m) antennas at the Source (Destination) are separated by more than half wavelength from each other, then one may assume $\rho_s = 0$ $(\rho_d = 0)$.

The mean equivalent SNR for DF relaying system may be found by taking the average value of (14.81). Since γ_1 and γ_2 are i.i.d. with chi-square pdf's, as given by (14.53) and (14.54), the mean value of $\gamma_{eq,df}$ for uncorrelated antennas is given by

$$\bar{\gamma}_{eq,df} = E\left[\gamma_{eq,df}\right] = E[\gamma_2]\, E[U(\gamma_1)] = m\bar{\gamma}_2\, E[U(\gamma_1)]$$

$$(14.86)$$

Using (14.53) for the pdf of γ_1 and (D.71), one gets

$$E[U(\gamma_1)] = \int_0^\infty \left[1 - 2Q\left(\sqrt{\beta\gamma_1}\right) \right]^2 \frac{\gamma_1^{n-1} e^{-\gamma_1/\bar{\gamma}_1}}{(n-1)!\, \bar{\gamma}_1^n}\, d\gamma_1$$

$$= \frac{2\beta\bar{\gamma}_1}{\pi} \sum_{k=1}^{n} \frac{1}{(1+\beta\bar{\gamma}_1)^k}\, {}_2F_1\left(k,1;\frac{3}{2};\frac{\beta\bar{\gamma}_1}{2+2\beta\bar{\gamma}_1}\right)$$

$$(14.87)$$

where $\beta = 1$ for BPSK and $\beta = 1/2$ for QPSK.

In the special case of $n = 1$, using (D.87) and $\tan^{-1}(x) = \pi/2 - \tan^{-1}(1/x)$, $x > 0$, (14.87) reduces to [28]

$$E[U(\gamma_1)] = 2\sqrt{\frac{\beta\bar{\gamma}_1}{2 + \beta\bar{\gamma}_1}} \left[1 - \frac{2}{\pi} \tan^{-1}\left(\sqrt{\frac{2 + \beta\bar{\gamma}_1}{\beta\bar{\gamma}_1}} \right) \right]$$

(14.88)

As $\bar{\gamma}_1 \to \infty$, $\tan^{-1}(1) = \pi/4$ and $E[U(\gamma_1)] \simeq 1$. Therefore, we observe from (14.86) that the mean equivalent SNR approaches $m\bar{\gamma}_2$ as $\bar{\gamma}_1 \to \infty$, since SR link is error-free. Figure 14.30 clearly shows the SNR advantage of the DF protocol over AF relaying for the SRD link with $\bar{\gamma}_1 = \bar{\gamma}_2$. Equivalent SNR for DF relaying is 4.66 dB and 3.98 dB higher than for AF relaying, for $n = 1$ and $n = 2$, respectively. For practical mean hop SNR values of interest, DF relaying has a significant performance advantage over AF relaying. This performance difference becomes less at lower mean SNR per hop, since the

Relay using DF protocol suffers more errors in detecting the signals.

The cdf of $\gamma_{eq,df}$ may be determined as follows:

$$F_{\gamma_{eq,df}}(\gamma) = \Pr(\gamma_{eq,df} < \gamma)$$

$$= \int_0^\infty \Pr[\gamma_2\, U(\gamma_1) < \gamma | \gamma_1] f_{\gamma_1}(\gamma_1)\, d\gamma_1$$

$$= \int_0^\infty F_{\gamma_2}\left(\frac{\gamma}{U(\gamma_1)} \right) f_{\gamma_1}(\gamma_1)\, d\gamma_1$$

(14.89)

Inserting the cdf of γ_2 from (14D.1),

$$F_{\gamma_2}(z) = 1 - \exp\left(\frac{-z}{\bar{\gamma}_2} \right) \sum_{k=0}^{m-1} \frac{1}{k!} \left(\frac{z}{\bar{\gamma}_2} \right)^k \quad (14.90)$$

into (14.89), the cdf and the pdf of $\gamma_{eq,df}$ may be expressed as

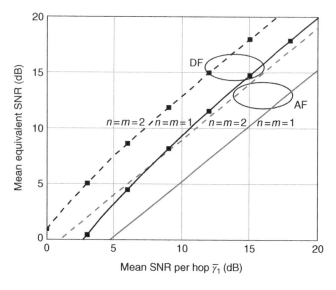

Figure 14.30   Mean Equivalent SNR of the SRD Link with AF and DF Relaying as a Function of $\bar{\gamma}_1 = \bar{\gamma}_2$ For $n = m = 1$, and 2.

$$F_{Y_{eq,df}}(\gamma) = 1 - \int_0^\infty \exp\left(\frac{-\gamma}{\bar{\gamma}_2 U(\gamma_1)}\right)$$

$$\times \sum_{k=0}^{m-1} \frac{1}{k!} \left(\frac{\gamma}{\bar{\gamma}_2 U(\gamma_1)}\right)^k f_{\gamma_1}(\gamma_1) \, d\gamma_1$$

$$f_{Y_{eq,df}}(\gamma) = \int_0^\infty \exp\left(\frac{-\gamma}{\bar{\gamma}_2 U(\gamma_1)}\right) \sum_{k=0}^{m-1} \frac{1}{k!} \left(\frac{\gamma}{\bar{\gamma}_2 U(\gamma_1)}\right)^k$$

$$\times \left(\frac{1}{\bar{\gamma}_2 U(\gamma_1)} - \frac{k}{\gamma}\right) f_{\gamma_1}(\gamma_1) \, d\gamma_1$$

$$(14.91)$$

One may observe from (14.91) that the pdf of the equivalent SNR of the SRD channel cannot be written as the product of the pdf's of γ_1 and γ_2.

Figure 14.31 compares the cdf of the SRD link by DF and AF relaying protocols given by (14.91) and (14.58), respectively. DF relaying was observed to have an improved outage performance compared with AF relaying. Use of multiple antennas at the Source and the Destination provides significant improvement in

the outage performance. Use of multiple antennas is obviously more effective than relay selection. For $n = m = K = 2$, DF protocol was observed to have 2.7 dB performance advantage over AF relaying protocol at 1% outage level.

Figure 14.32 compares the pdf's of the SRD links for AF and DF relaying with n antennas at the Source, m antennas at the Destination and relay selection. The mean SNR's of SR and RD links are assumed to be $\bar{\gamma}_1 = \bar{\gamma}_2 = 10$ dB. The use of two antennas at both Source and Destination offers much better performance than simply selecting one of the two relays with the highest equivalent SNR. The use of multiple antennas at Source and Destination and relay selection provides much improved performance. This is in agreement with the results shown in Figure 14.30 for mean equivalent SNR. On the other hand, as shown in Figure 14.33, Rayleigh approximation to the equivalent SNR of SRD link with DF protocol is sufficiently accurate and might be of interest for ease of computations. Rayleigh approximation given by (14C.1) accurately describes (14.91) for $n = m = 1$ and arbitrary

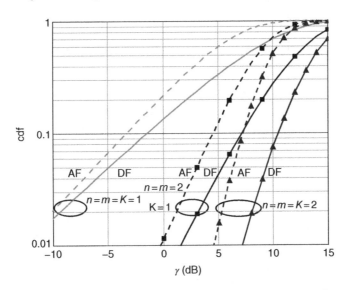

Figure 14.31 The Cdf of the Equivalent SNR of the SRD Link with DF and AF Relaying For $\bar{\gamma}_1 = \bar{\gamma}_2 = 10$ dB.

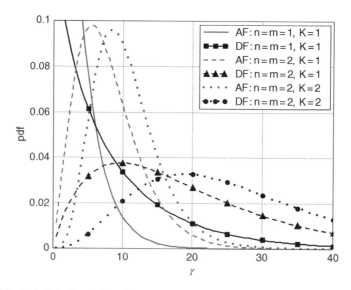

Figure 14.32 The Pdf of the Equivalent SNR of the SRD Link with AF and DF Relaying For $n = m = 1, 2$ and K = 1, 2.

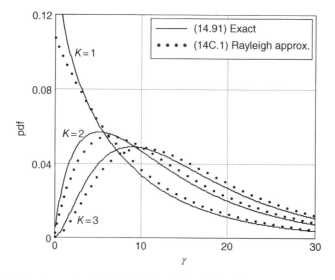

Figure 14.33 Rayleigh Approximation For the Equivalent SNR of the SRD Link with DF Protocol For $n = m = 1$, $K = 1, 2, 3$ and $\bar{\gamma}_1 = \bar{\gamma}_2 = 10$.

values of K. For higher values of n and m, the SRD channel becomes more difficult to be modeled, since SR and RD channels have chi-square distribution (see (14.53) and (14.54)).

The BEP for selection-combined SRD and SD links with multiple antennas at Source and Destination and relay selection may be written as

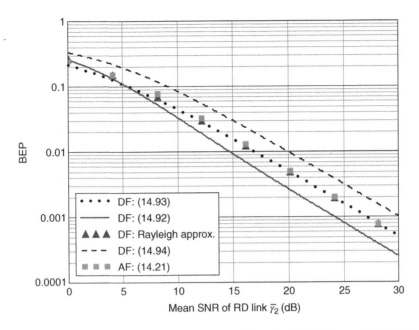

Figure 14.34 BEP Versus the Mean SNR $\bar{\gamma}_2$ of the RD Link with DF Relaying and BPSK Modulation. There is no relay selection and Source, Relay and Destination are all equipped with single antennas, $n = m = K = 1$. We assume $\bar{\gamma}_1 = \bar{\gamma}_2$ except for (14.93).

$$P_{b,df} = \frac{1}{\sqrt{2\pi}} \int_0^\infty \frac{1}{\sqrt{2\gamma}} \exp(-\gamma) F_{\gamma_{eq,df}}(\gamma)^K F_{n \times m}(\gamma) d\gamma$$

(14.92)

where $F_{\gamma_{eq,df}}(\gamma)$ is given by (14.91). $F_{n \times m}(\gamma)$ denotes the cdf of the SD link SNR based on the maximum eigenvalue of the $n \times m$ MIMO system and is obtained by (13.23) and (14.69) for the uncorrelated case.

The BEP for BPSK modulation for the SRD link of a DF relaying system is shown in Figure 14.34 for $\bar{\gamma}_1 = \bar{\gamma}_2$ and single antennas at Source and Destination and a single Relay, that is, $m = n = K = 1$. Since we consider only the SRD link, we take $F_{n \times m}(\gamma) \equiv 1$ and $K = 1$ in (14.92). For the BEP = 10^{-3}, the hop SNRs required by DF relaying based on (14.92) are $\bar{\gamma}_1 = \bar{\gamma}_2 = 24$ dB. In this case, DF protocol requires hop 3 dB lower SNR than the AF protocol (see Figure 14.34 and (14.21)).

However, for the same BEP and $\bar{\gamma}_1 = 10$ dB, $\bar{\gamma}_2 = 26.95$ dB is required for DF relaying in order to compensate for detection errors at the Relay. If the SR link is perfect, that is, for $\bar{\gamma}_1 \to \infty$, then, there are no detection errors in the SR link and the system performance is determined only by the SNR γ_2 of the RD link. For BPSK modulation, the BEP for the RD link with a single antenna at the Destination and no relay selection ($K = 1$), is found from (14C.3):

$$P_b = I(1, \bar{\gamma}_2) = \int_0^\infty Q\left(\sqrt{\beta\gamma}\right) \frac{1}{\bar{\gamma}_2} e^{-\gamma/\bar{\gamma}_2} d\gamma$$

$$= \frac{1}{2} - \frac{1}{2}\sqrt{\frac{\beta\bar{\gamma}_2}{2 + \beta\bar{\gamma}_2}}$$

(14.93)

For small values of $\bar{\gamma}_1 = \bar{\gamma}_2$, (14.92) predicts higher BEP than (14.93), because it also takes into account of the errors in the SR link, whereas $\bar{\gamma}_1 \to \infty$ in (14.93). For larger values of $\bar{\gamma}_1 = \bar{\gamma}_2$ the results predicted by (14.92) are 3 dB better compared to the predictions by (14.93). The Rayleigh approximation shown in Figure 14.34 refers to the use of (14.92) where $F_{\gamma_{eq,df}}(\gamma)$ is replaced by the cdf of Rayleigh distribution with mean (14.86). In other words, the BEP expression is the same as (14.93) with the exception that $\bar{\gamma}_2$ is replaced by $\bar{\gamma}_{eq,df}$, given by (14.86).

For dual-hop DF relaying, the following BEP formulation is proposed by [27] in AWGN and Rayleigh fading channels with single antennas at the Source and the Destination:

$$P_b(\gamma) = [1 - P_e(\gamma_1)] P_e(\gamma_2) + [1 - P_e(\gamma_2)] P_e(\gamma_1)$$
$$P_b = [1 - I(1, \bar{\gamma}_1)] I(1, \bar{\gamma}_2) + [1 - I(1, \bar{\gamma}_2)] I(1, \bar{\gamma}_1)$$
$$(14.94)$$

where $P_e(\gamma) = Q(\sqrt{\gamma})$ denotes the BEP for BPSK in AWGN channel. The BEP expression in (14.94) for BPSK modulation in a Rayleigh fading channel is found by averaging the first expression (14.94) with the pdf's of γ_1 and γ_2 given by (14D.1). It may be useful to note that (14.93) is a special case of (14.94) for $\bar{\gamma}_1 \to \infty$. Then, the SR link is perfect and $I(1, \bar{\gamma}_1)$ goes to zero.

BEP performances for AF and DF relaying for BPSK are compared in Figure 14.35 for the selection-combined SRD and SD links. The mean SNRs of SR, RD and SD links are all assumed to be the same. DF relaying was observed to perform always better than AF relaying for the same values of m, n, and K. Increasing the number of antennas is much more effective than increasing the number of relays for improving the BEP performance. For BEP $= 10^{-3}$ and $n = m = 1$, the required SNR level for DF relaying is 1.4 dB and 1.9 dB less than that for AF relaying for $K = 1$ and 2, respectively. The SNR advantage of DF relaying due to relay selection was

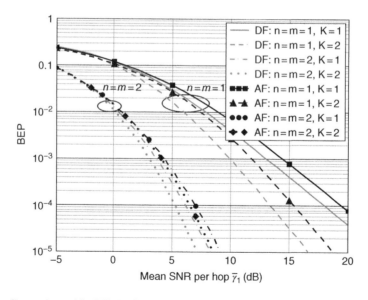

Figure 14.35 Comparison of the BEP Performance of Selection-Combined SRD and SD Links with DF and AF Relaying. BEP for DF and AF relaying are given by (14.92) and (14.76), respectively. The Source and Destination are equipped with n and m antennas, respectively. Relay selection is assumed. Mean SNRs are all the same, $\bar{\gamma}_1 = \bar{\gamma}_2 = \bar{\gamma}_3$ and modulation is BPSK.

observed to decrease for higher values of $n=m$. For $\bar{\gamma}_1 \to \infty$, there are no detection errors in the SR link and the mean SNR $\bar{\gamma}_2$ of the RD link determines the BEP. Then, the equivalent SNR approaches to that of the RD link, $\bar{\gamma}_{eq,df} \to \bar{\gamma}_2$ (see (14.87)); in this case, both AF and DF schemes perform equally well.

When the SRD link, with DF relaying and relay selection, is selection-combined with the SD link, the mean capacity may be expressed as

$$C_{av,df} = \frac{1}{2\ \ln 2} \int_0^\infty \frac{1}{1+\gamma} \left[1 - F_{\gamma_{eq,df}}(\gamma)^K F_{n \times m}(\gamma) \right] d\gamma$$

$$(14.95)$$

where $F_{\gamma_{eq,df}}(\gamma)$ and $F_{n \times m}(\gamma)$ are defined in (14.91) and (13.23), respectively. Figure 14.36 presents the mean capacity for DF relaying in comparison with AF relaying for $\bar{\gamma}_1 = \bar{\gamma}_2 = \bar{\gamma}_3$ and various combinations of $n=m$ and K. The mean capacity for DF relaying is always higher than AF, as expected. For $\bar{\gamma} = 10\, dB$ and $n=m=K=1$, mean capacity is found to be 1.58 bps/Hz and 1.78 bps/Hz for AF and

DF, respectively. Corresponding capacity values for $n=m=K=2$ are 2.51 bps/Hz and 2.63 bps/Hz. Hence, for $n=m=K=2$, DF relaying provides an increase in the mean capacity by a factor of $2.63/1.78 = 1.48$ compared to for $n=m=K=1$. The use of multiple antennas was observed to be more effective than relay selection in improving the mean capacity.

For DF relaying protocol, the cdf of the capacity of the SRD link with multiple antennas and relay selection is given by:

$$C_{out,df}(C_{th}) = F_{\gamma_{eq,df}}\left(2^{2C_{th}} - 1\right)^K F_{n \times m}\left(2^{2C_{th}} - 1\right)$$

$$(14.96)$$

In order to observe the improvement provided by DF relaying, the cdf of the capacity for DF and AF relaying are compared in Figure 14.37 for the SRD link only with $\bar{\gamma}_1 = \bar{\gamma}_2 = 10\, dB$. DF relaying systems were observed to perform better than the corresponding AF relaying systems. However, DF and AF relaying with $n=m=K=1$ performs worse than the direct link, unless multiple antennas

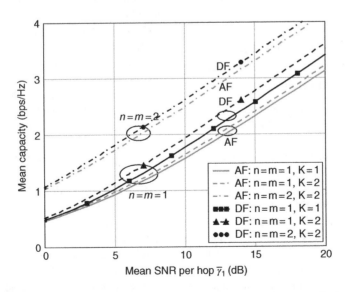

Figure 14.36 Comparison of the Mean Capacity of the SRD Link For AF and DF Relaying For $\bar{\gamma}_1 = \bar{\gamma}_2 = \bar{\gamma}_3$.

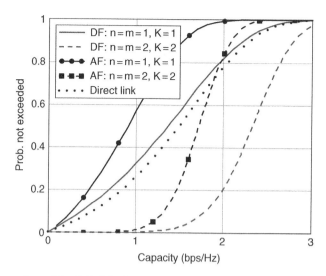

Figure 14.37 Comparison of the Cdf of the Capacity of the SRD Link For AF and DF Relaying. Mean SNR levels are all equal to 10 dB, that is, $\bar{\gamma}_1 = \bar{\gamma}_2 = \bar{\gamma}_3 = 10$.

and/or relay selection is used. Nevertheless, the performance of DF relaying approaches that of the direct link more closely. For DF relaying, the probability that the capacity is lower than 2 bps/Hz is 0.2 for $n = m = K = 2$, while it is 0.8 for $n = m = K = 1$. Relay selection provides additional improvement on the outage capacity.

14.5   Relaying with Multiple Antennas at Source, Relay and Destination

Consider a SRD relay link where the number of antennas used at the Source, the Relay and the Destination is denoted by n, r and m, respectively. Thus, SR and RD links are represented as $n \times r$, and $r \times m$ MIMO channels. The previously considered relaying scenarios with a single antenna at the Relay ($r = 1$), constitute special cases of the MIMO channels.

We know from the Wishart matrix analysis of MIMO channels in Chapter 13 that the eigenvalues of $n \times r$ and $r \times n$ MIMO channels are the same. This symmetry is the result of perfect cooperation between transmitters and receivers in MIMO channels (see

Figure 13.1); transmit antennas use MRT, with perfect CSI provided by the Relay, whereas the receive antennas employ MRC. This may also be observed from the pdf given by (13.17) for both $n \times 1$ SR and $1 \times m$ RD channels for uncorrelated antennas. Also note that (13.17) is identical to (14.53) and (14.54) for the uncorrelated case.

The equivalent SNR for the SRD link with multiple antennas at Source, Relay and Destination can be maximized by applying transmit and receive beamforming (see Section 13.7). The maximum equivalent SNR of the SRD link corresponds to the maximum eigenvalues in SR and SD links (see (13.86)):

$$\gamma_i = \bar{\gamma}_i \lambda_{i,\max}, \quad i = 1, 2 \qquad (14.97)$$

The pdf and the cdf of the maximum eigenvalues for uncorrelated MIMO channels are given by (13.16)–(13.22) and (13.23), respectively. It is important to note that the mean SNRs $\bar{\gamma}_1$ and $\bar{\gamma}_2$ of SR and SD links are determined by the transmit power, antenna gains, distance and the receiver noise of the considered link as well as the frequency of operation and the data rate (see (2.114)).

For DF protocol, the cdf of the equivalent SNR $\gamma_{nrm,df}$ of SRD link with $n \times r$ SR, and $r \times m$ RD MIMO channels may be determined using the pdf and the cdf of the maximum eigenvalues of the corresponding MIMO channels in (14.89):

$$F_{\gamma_{nrm,df}}(\gamma) = \Pr(\gamma_{nrm,df} < \gamma)$$

$$= \int_0^\infty \Pr[\gamma_2 \, U(\gamma_1) < \gamma | \gamma_1] f_{\gamma_1}(\gamma_1) \, d\gamma_1$$

$$= \int_0^\infty F_{\lambda_2, r \times m}\left(\frac{\gamma}{\bar{\gamma}_2 U(\bar{\gamma}_1 \lambda_1)}\right) f_{\lambda_1, n \times r}(\lambda_1) \, d\lambda_1$$

(14.98)

where the pdf $f_{\lambda_1, n \times r}(\lambda_1)$ and the cdf $F_{\lambda_2, r \times m}(\lambda_2)$ of the maximum eigenvalues are given by (13.16)–(13.22) and (13.23), respectively.

Similarly, the cdf of the equivalent SNR for AF relaying with multiple antennas at the Source, the Relay and the Destination may be derived as in (14.D2):

$$F_{\gamma_{nrm,af}}(\gamma) = 1 - \int_\gamma^\infty \left[1 - F_{\gamma_2}\left(\frac{\gamma \gamma_1}{\gamma_1 - \gamma}\right)\right] f_{\gamma_1}(\gamma_1) \, d\gamma_1$$

$$= 1 - \int_{\gamma/\bar{\gamma}_1}^\infty \left[1 - F_{\lambda_2, r \times m}\left(\frac{\gamma \bar{\gamma}_1 \lambda_1}{\bar{\gamma}_2(\bar{\gamma}_1 \lambda_1 - \gamma)}\right)\right] f_{\lambda_1, n \times r}(\lambda_1) \, d\lambda_1$$

(14.99)

Figure 14.38 shows the cdf of the equivalent SNR of the SRD link with DF relaying protocol for various combinations of the number of antennas at the Source, the Relay and the Destination. Based on the equivalence of MRC and MRT, for example, the curve for $n = m = 3$, $r = 1$ also shows the performance of a SRD relay system with single antennas at the Source and Destination ($n = m = 1$), but three antennas at the Relay transmitter and receiver ($r = 3$). In each case, there are three independent propagation paths in both SR and RD links. The outage probability was observed to improve rapidly with increasing values of n and m. In view of the above, it might be of interest to compare the outage performance of the SRD link with $n = m = r = 2$ and with $n = m = 4$,

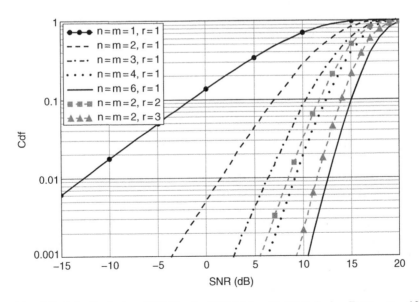

Figure 14.38 Cdf of the Equivalent SNR For the SRD Link with DF Relaying For $\bar{\gamma}_1 = \bar{\gamma}_2 = 10$ *dB*. The number of antennas at the Source, the Destination and the Relay are denoted by n, m, and r, respectively.

$r = 1$. At an outage probability of 10^{-3}, the SRD link with $n = m = r = 2$ performs 0.75 dB worse than the link with $n = m = 4$, $r = 1$ (see also Figure 13.18). Similarly, the SRD link with $n = m = 2$, $r = 3$ performs 1.1 dB worse than the link with $n = m = 6$, $r = 1$. This small reduction in the performance may be attributed to processing at multiple antennas on both sides of the links. These results also show that, depending on the system requirements and cost and space limitations, the number of antennas at the Source, the Relay and the Destination should be chosen adaptively.

14.6   Coded Cooperation

Coded cooperation brings an additional dimension into cooperative communication, since it introduces channel coding into cooperative communications. This consists sending different portions of each user's encoded signal (codeword) via different relays with independently fading links. For example, the data of a user, augmented with CRC for error detection, is encoded into a codeword of length $n = n_1 + n_2$, that is partitioned into two segments, containing n_1 bits and n_2 bits, respectively. Puncturing this codeword down to n_1 bits, we obtain the first segment, which itself is a valid but weaker codeword and the remaining n_2 bits consist of the puncture bits. The transmission period is divided into two time frames of n_1-bit and n_2-bit long, respectively. In the first time frame, each user transmits n_1 bits and checks whether the transmission of its partner is correctly received. If reception is successful (checking with the CRC code), then, in the second frame, it calculates and transmits n_2 bits of its partner. Otherwise, the intended user transmits its own n_2-bit segment. The level of cooperation is hence n_2/n, the ratio of bits for each codeword the user transmits for its partner. The efficiency of cooperation is managed automatically through code design, with no feedback between the users. Such a scheme may be typically realized by using a rate-compatible

punctured convolutional (RCPC) codeword punctured to rate 1/2. This codeword is subsequently repeated by the Relay, resulting in an overall rate of 1/4. Coded cooperation performs better than AF and DF, albeit at the expense of reduced throughput. [29][30][31] Other scenarios for coded cooperation can be considered.

Example 14.5   Network Coding.
Network coding is a method to improve the performance of wireless networks. Consider a relay network as shown in Figure 14.39, where S_1 and S_2 denote the two Sources transmitting $f(x_1)$ and $f(x_2)$ for bits x_1 and x_2, respectively, independently of each other. Let us consider BPSK modulation where $f(x): \{1,0\} \rightarrow \{+\sqrt{E_b}, -\sqrt{E_b}\}$. The transmission interval is divided into three, namely, for receiving transmission of S_1, S_2 and retransmission of combined and encoded messages from S_1 and S_2 to the Destination. The Relay R is assumed to decode the signals from the sources perfectly and retransmits $f(x_1 \oplus x_2)$ to the Destination. Then, the Destination D jointly decodes x_1 and x_2 by using three independent observations $f(x_1) + n_1$, $f(x_2) + n_2$, and $f(x_1 \oplus x_2) + n_3$. The noises n_1, n_2 and n_3 are all assumed to have zero-mean and variance σ^2.

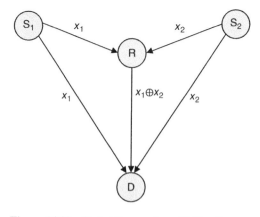

Figure 14.39   Coded Cooperation with Two Sources and a Single Relay.

Such networks generally use iterative decoding using soft-in soft-out decoders. The following provides the basic log-likelihood concepts, which are widely used in soft-in soft-out decoding. The log-likelihood ratio (LLR) of a binary random variable $X = \{0, 1\}$, $L_X(x)$ is defined as

$$L_X(x) = \ln \frac{P(x=0)}{P(x=1)} \qquad (14.100)$$

where $P(x)$ denotes the probability that the random variable X takes the value x. Note that $L_X(x)$ has a soft value; the sign of $L_X(x)$ denotes the hard decision and the magnitude $|L_X(x)|$ shows the reliability of the decision.

Inserting $P(x=1) = 1 - P(x=0)$ into (14.100), one gets

$$P(x=0) = \frac{e^{L_X(x)}}{1 + e^{L_X(x)}}, \quad P(x=1) = \frac{1}{1 + e^{L_X(x)}} \qquad (14.101)$$

The LLR of the binary random variable X conditioned on a different random variable Y, $L_{X|Y}(x|y)$, is obtained using the Bayes' theorem

$$
\begin{aligned}
L_{X|Y}(x|y) &= \ln \frac{P(x=0|y)}{P(x=1|y)} \\
&= \ln \frac{P(y|x=0)P(x=0)/P(y)}{P(y|x=1)P(x=1)/P(y)} \\
&= \ln \frac{P(y|x=0)}{P(y|x=1)} + \ln \frac{P(x=0)}{P(x=1)} \\
&= L_{Y|X}(y|x) + L_X(x)
\end{aligned}
\qquad (14.102)
$$

For $P(x=0) = P(x=1) = 1/2$, we have $L_X(x) = 0$. Assuming that all links are described by AWGN with variance σ^2, Y represents the received noisy signal by the Destination in S_1D or S_2D link, when equiprobable bits X=1 or 0 are transmitted. The pdf of received signals with mean $\pm\sqrt{E_b}$, and variance σ^2 may then be written as:

$$P(y|x=0) = \frac{1}{\sqrt{2\pi}\sigma} \exp\left(-\left(y + \sqrt{E_b}\right)^2 / \left(2\sigma^2\right)\right)$$

$$P(y|x=1) = \frac{1}{\sqrt{2\pi}\sigma} \exp\left(-\left(y - \sqrt{E_b}\right)^2 / \left(2\sigma^2\right)\right)$$

$$\qquad (14.103)$$

Inserting (14.103) into (14.102), one gets

$$L_{X|Y}(x|y) = -\frac{2\sqrt{E_b}}{\sigma^2} y \qquad (14.104)$$

Now consider the RD link, which is also modeled by AWGN with noise variance σ^2. Assuming that x_1 and x_2 are statistically independent random variables, $P(x_1 \oplus x_2 = 0)$ and $P(x_1 \oplus x_2 = 1)$ may be expressed as

$$
\begin{aligned}
P(x_1 \oplus x_2 = 0) &= P(x_1 = 0)P(x_2 = 0) + P(x_1 = 1)P(x_2 = 1) \\
P(x_1 \oplus x_2 = 1) &= P(x_1 = 0)P(x_2 = 1) + P(x_1 = 1)P(x_2 = 0)
\end{aligned}
\qquad (14.105)
$$

Using (14.100) and (14.105), one gets the LLR of $x_1 \oplus x_2$ as follows:

$$
\begin{aligned}
L_{X_1 \oplus X_2}(x_1 \oplus x_2) &= \ln \frac{P(x_1 \oplus x_2 = 0)}{P(x_1 \oplus x_2 = 1)} \\
&= \ln \frac{1 + e^{L_X(x_1) + L_X(x_2)}}{e^{L_X(x_1)} + e^{L_X(x_2)}}
\end{aligned}
\qquad (14.106)
$$

Using (14.106), $P(x_1 \oplus x_2 = 0)$ and $P(x_1 \oplus x_2 = 1)$ may be expresed in terms of the LLR of x_1 and x_2 as follows:

$$P(x_1 \oplus x_2 = 0) = \frac{1 + e^{L(x_1) + L(x_2)}}{(1 + e^{L(x_1)})(1 + e^{L(x_2)})}$$

$$P(x_1 \oplus x_2 = 1) = \frac{e^{L(x_1)} + e^{L(x_2)}}{(1 + e^{L(x_1)})(1 + e^{L(x_2)})}$$

$$\qquad (14.107)$$

Assume that $x_1 = 1$ and $x_2 = 0$ is transmitted by Source 1 and Source 2, respectively, and

Table 14.4   Decoding in the Coded Cooperation.

	S_1D link	S_2D link	RD link
Transmitted bit: x_i	$x_1 = 1$	$x_2 = 0$	$x_3 = x_1 \oplus x_2 = 1$
Transmitted signal: $f(x_i)$	$+1$	-1	$+1$
Noise: n_i	-0.1	0.7	-0.3
Received signal at D: $y = f(x_i) + n_i$	0.9	-0.3	0.7
$L_{x|y}(x|y) = -2y$	-1.8	0.6	-1.4
$P(x_i = 0) = e^{L(x_i)}/\left(1 + e^{L(x_i)}\right)$	0.142	0.65	0.198
$P(x_i = 1) = 1/\left(1 + e^{L(x_i)}\right)$	0.858	0.35	0.802
Decoded bit: $\hat{x}_i$	1	$0?$	1
$L(x_1 \oplus x_3) = L(x_2)$		0.927	
Decoded bit: $\hat{x}_i$ (correction)	1	0	1

$E_b = \sigma^2 = 1$. S_1R and S_2R links are assumed to be perfect, that is, with no error. Table 14.4 shows the steps for decoding the transmitted bits at the Destination.

Based on the LLR, the Destination estimates the transmitted bits as $\hat{x}_1 = \hat{x}_2 = \hat{x}_3 = 1$. Note that the decisions for $\hat{x}_1$ and $\hat{x}_3$ are taken with high confidence but the decision for $\hat{x}_2$ is questionable because of its lower LLR. The next step would be to check whether $\hat{x}_2$, the bit with the lowest LLR, is correct. This can be achieved by determining the LLR of $x_1 \oplus x_3 = x_2$:

$$L_{X_1 \oplus X_3}(x_1 \oplus x_3) = \ln \frac{1 + e^{L_X(x_1) + L_X(x_3)}}{e^{L_X(x_1)} + e^{L_X(x_3)}} = 0.927$$

$$(14.108)$$

Based on (14.101) and (14.108), the Destination decodes $x_1 \oplus x_3 = x_2$ as zero since $P(x_1 \oplus x_3 = 0) = 0.716$ and $P(x_1 \oplus x_3 = 1) = 0.284$. Consequently, with the help of network coding, the Destination decodes as $\hat{x}_2 = 0$ (see the last row of Table 14.4).

Appendix 14A

CDF of γ_{eq} and $\gamma_{eq,0}$

The cdf of γ_{eq}, the equivalent SNR of the SRD link, may be written as

$$F_{\gamma_{eq}}(\gamma) = \Pr(\gamma_{eq} < \gamma)$$

$$= \int_0^\infty \Pr\left(\frac{\gamma_1 \gamma_2}{\gamma_1 + \gamma_2 + c} < \gamma | \gamma_1\right) f_{\gamma_1}(\gamma_1)\, d\gamma_1$$

$$= \int_0^\infty \Pr\left(\gamma_2 < \frac{\gamma(c + \gamma_1)}{\gamma_1 - \gamma} | \gamma_1\right) f_{\gamma_1}(\gamma_1)\, d\gamma_1$$

$$= \underbrace{\int_0^\gamma \Pr\left(\gamma_2 > \frac{\gamma(c + \gamma_1)}{\gamma_1 - \gamma} | \gamma_1\right) f_{\gamma_1}(\gamma_1)\, d\gamma_1}_{\equiv 1}$$

$$+ \int_\gamma^\infty \Pr\left(\gamma_2 < \frac{\gamma(c + \gamma_1)}{\gamma_1 - \gamma} | \gamma_1\right) f_{\gamma_1}(\gamma_1)\, d\gamma_1$$

$$= \int_0^\gamma f_{\gamma_1}(\gamma_1)\, d\gamma_1$$

$$+ \int_\gamma^\infty \left(1 - \exp\left(-\frac{\gamma(c + \gamma_1)}{\bar{\gamma}_2(\gamma_1 - \gamma)}\right)\right) f_{\gamma_1}(\gamma_1)\, d\gamma_1$$

$$= 1 - \int_\gamma^\infty \exp\left(-\frac{\gamma(c + \gamma_1)}{\bar{\gamma}_2(\gamma_1 - \gamma)}\right) \frac{1}{\bar{\gamma}_1} \exp\left(-\frac{\gamma_1}{\bar{\gamma}_1}\right) d\gamma_1$$

$$= 1 - \exp\left(-\left(\frac{1}{\bar{\gamma}_1} + \frac{1}{\bar{\gamma}_2}\right)\gamma\right)$$

$$\times \int_0^\infty \exp\left(-\frac{\lambda}{\bar{\gamma}_1} - \frac{\gamma(c + \gamma)}{\bar{\gamma}_2 \lambda}\right) \frac{1}{\bar{\gamma}_1}\, d\lambda$$

$$(14A.1)$$

where the variable transformation $\lambda = \gamma_1 - \gamma$ is applied in the last equation. Finally, the cdf of γ_{eq} is found using (D.65):

$$F_{\gamma_{eq}}(\gamma, \alpha, \beta) = 1 - \beta \sqrt{\gamma(c + \gamma)}$$

$$\times \exp(-\alpha\gamma) K_1\left(\beta \sqrt{\gamma(c + \gamma)}\right)$$

$$(14A.2)$$

where $K_1(x)$ denotes the modified Bessel function of the second kind, $\alpha = 1/\bar{\gamma}_1 + 1/\bar{\gamma}_2$, $\beta = 2/\sqrt{\bar{\gamma}_1 \bar{\gamma}_2}$. The corresponding pdf is determined by taking the derivative of (14A.2) and using (D.105):

$$f_{\gamma_{eq}}(\gamma, \alpha, \beta) = \beta e^{-\alpha \gamma} \left[\beta \left(\frac{c}{2} + \gamma \right) K_0 \left(\beta \sqrt{\gamma(\gamma + c)} \right) \right.$$
$$\left. + \alpha \sqrt{\gamma(\gamma + c)} \, K_1 \left(\beta \sqrt{\gamma(\gamma + c)} \right) \right]$$

$$(14A.3)$$

The cdf and the pdf for $\gamma_{eq,0}$ are obtained by inserting $c = 1$ into (14A.2) and (14A.3). Similarly, the cdf and the pdf for γ_{eq} are found by inserting $c = 0$ into (14A.2) and (14A.3) (see Table 14.1).

The pdf, cdf and the MGF for $\gamma_{eq,1}$ and $\gamma_{eq,2}$ are found as follows: [11]

$$F_{\gamma_{eq}}(\gamma) = \Pr(\gamma_{eq} < \gamma)$$

$$= \int_0^\infty \Pr\left(\frac{\gamma_1 \gamma_2}{\gamma_2 + C} < \gamma | \gamma_2 \right) f_{\gamma_2}(\gamma_2) \, d\gamma_2$$

$$= \int_0^\infty \Pr(\gamma_1 < \gamma(C/\gamma_2 + 1) | \gamma_2) f_{\gamma_2}(\gamma_2) \, d\gamma_2$$

$$= \int_0^\infty \left[1 - \exp\left(-\frac{\gamma(C/\gamma_2 + 1)}{\bar{\gamma}_1} \right) \right] f_{\gamma_2}(\gamma_2) \, d\gamma_2$$

$$= 1 - \frac{1}{\bar{\gamma}_2} e^{-\gamma/\bar{\gamma}_1} \int_0^\infty e^{-\gamma_2/\bar{\gamma}_2 - \gamma C/(\bar{\gamma}_1 \gamma_2)} \, d\gamma_2$$

$$= 1 - \beta \sqrt{C\gamma} \, e^{-\gamma/\bar{\gamma}_1} \, K_1 \left(\beta \sqrt{C\gamma} \right)$$

$$f_{\gamma_{eq}}(\gamma) = \frac{1}{\bar{\gamma}_1} e^{-\gamma/\bar{\gamma}_1}$$

$$\times \left[\beta \sqrt{C\gamma} \, K_1 \left(\beta \sqrt{C\gamma} \right) + \frac{2C}{\bar{\gamma}_2} K_0 \left(\beta \sqrt{C\gamma} \right) \right]$$

$$M_{\gamma_{eq}}(s) = \frac{1}{1 + s\bar{\gamma}_1} + \frac{sC\bar{\gamma}_1}{\bar{\gamma}_2 (1 + s\bar{\gamma}_1)^2}$$

$$\times \exp\left(\frac{C}{\bar{\gamma}_2 (1 + s\bar{\gamma}_1)} \right) E_1 \left(\frac{C}{\bar{\gamma}_2 (1 + s\bar{\gamma}_1)} \right)$$

$$(14A.4)$$

Note that $C = \bar{\gamma}_1 + 1$ corresponds to $\gamma_{eq,1}$ and $C = \bar{\gamma}_1$ corresponds to $\gamma_{eq,2}$.

Appendix 14B

Average Capacity of $\gamma_{eq,0}$

The derivation of the average (ergodic) capacity for SRD link using the equivalent SNR $\gamma_{eq,0} = \gamma_1 \gamma_2 / (\gamma_1 + \gamma_2 + 1)$ is based on [32]. Here, γ_i, $i = 1, 2$ are assumed to be Rayleigh distributed with mean values $\bar{\gamma}_i$, $i = 1, 2$. The ergodic capacity may be written as follows:

$$C_{av} = \frac{1}{2} E\left[\log_2 (1 + \gamma_{eq,0}) \right]$$

$$= \frac{1}{2 \ln 2} E\left[\ln \left(\frac{(1 + \gamma_1)(1 + \gamma_2)}{1 + \gamma_1 + \gamma_2} \right) \right]$$

$$= \frac{1}{2 \ln 2} \begin{bmatrix} E[\ln(1 + \gamma_1)] + E[\ln(1 + \gamma_2)] \\ - E[\ln(1 + \gamma_1 + \gamma_2)] \end{bmatrix}$$

$$(14B.1)$$

The factor ½ appearing in front of the expectation operator accounts for the division of the transmission period into two. One may easily evaluate the first two integrals using (D.73):

$$C_m(\bar{\gamma}_i) = \frac{1}{2 \ln 2} \int_0^\infty \ln(1 + \gamma_i) \frac{1}{\bar{\gamma}_i} \exp\left(-\frac{\gamma_i}{\bar{\gamma}_i} \right) d\gamma_i$$

$$= \frac{1}{2 \ln 2} e^{1/\bar{\gamma}_i} E_1(1/\bar{\gamma}_i), \quad i = 1, 2$$

$$(14B.2)$$

The random variable $z = \gamma_1 + \gamma_2$, appearing in the last integral, corresponds to the MRC combination of γ_1 and γ_2. The pdf of z is given by (11.216) and (11.218) for distinct and identical mean SNRs, respectively:

$$f_Z(z) = \begin{cases} \dfrac{z}{\bar{\gamma}^2} e^{-z/\bar{\gamma}} & \bar{\gamma}_1 = \bar{\gamma}_2 \\[2mm] \dfrac{1}{\bar{\gamma}_1 - \bar{\gamma}_2} \left(e^{-z/\bar{\gamma}_1} - e^{-z/\bar{\gamma}_2} \right) & \bar{\gamma}_1 \neq \bar{\gamma}_2 \end{cases}$$

$$(14B.3)$$

Therefore, the last term in (14B.1) evaluated as

$$\frac{1}{2\ln2}\int_0^\infty \ln(1+z)f(z)\,dz$$

$$=\begin{cases}\dfrac{1}{2\ln2}+C_m(\bar\gamma_1)\left(1-\dfrac{1}{\bar\gamma_1}\right) & \bar\gamma_1=\bar\gamma_2\\[2mm]\dfrac{\bar\gamma_1}{\bar\gamma_1-\bar\gamma_2}C_m(\bar\gamma_1)-\dfrac{\bar\gamma_2}{\bar\gamma_1-\bar\gamma_2}C_m(\bar\gamma_2) & \bar\gamma_1\neq\bar\gamma_2\end{cases}$$

$$(14B.4)$$

Finally, inserting (14B.2) and (14B.4) into (14B.1), the average capacity of the SRD link with single relay is found to be

$$C_{av}=\begin{cases}\dfrac{-1}{2\ln2}+\left(1+\dfrac{1}{\bar\gamma_1}\right)C_m(\bar\gamma_1) & \bar\gamma_1=\bar\gamma_2\\[2mm]\dfrac{\bar\gamma_1}{\bar\gamma_1-\bar\gamma_2}C_m(\bar\gamma_2)-\dfrac{\bar\gamma_2}{\bar\gamma_1-\bar\gamma_2}C_m(\bar\gamma_1) & \bar\gamma_1\neq\bar\gamma_2\end{cases}$$

$$(14B.5)$$

Appendix 14C

Rayleigh Approximation for Equivalent SNR with Relay Selection

14C.1 SRD Link

Based on the Rayleigh approximation for SRD link with a mean SNR $\bar\gamma_{eq}$, given by (14.16) for AF and (14.86) for DF relaying, one may write the cdf, pdf and MGF of the SNR $\gamma_{s,R}$ when one of the K relays, with highest equivalent SNR γ_{eq}, is selected in the SRD link:

$$F_{\gamma_{s,R}}(\gamma)=\left(1-e^{-\gamma/\bar\gamma_{eq}}\right)^K=\sum_{n=0}^K\binom{K}{n}(-1)^n e^{-n\gamma/\bar\gamma_{eq}}$$

$$f_{\gamma_{s,R}}(\gamma)=\frac{d}{d\gamma}F_\gamma(\gamma)=\sum_{n=1}^K\binom{K}{n}(-1)^{n+1}\frac{n}{\bar\gamma_{eq}}e^{-n\gamma/\bar\gamma_{eq}}$$

$$M_{\gamma_{s,R}}(s)=E[e^{-s\gamma_{eq,R}}]=\sum_{n=1}^K\binom{K}{n}(-1)^{n+1}\frac{1}{1+s\bar\gamma_{eq}/n}$$

$$(14C.1)$$

where the binomial expansion $(1-x)^K=\sum_{n=0}^K\binom{K}{n}(-x)^n$ is used in the first expression.

The BER for BPSK modulation for the SRD link is given by

$$P_b=\int_0^\infty Q\left(\sqrt{2\gamma}\right)f_{\gamma_{eq,R}}(\gamma)\,d\gamma$$

$$(14C.2)$$

$$=\sum_{n=1}^K\binom{K}{n}(-1)^{n+1}I\left(1,\bar\gamma_{eq}/n\right)$$

where $I(L,x)$ is defined by (D.70):

$$I(L,\bar\gamma)=\int_0^\infty Q\left(\sqrt{2\gamma}\right)\frac{1}{(L-1)!}\frac{\gamma^{L-1}}{\bar\gamma^L}e^{-\gamma/\bar\gamma}\,d\gamma$$

$$=\frac{1}{2}-\frac{1}{2}\sqrt{\frac{\bar\gamma}{1+\bar\gamma}}\sum_{k=0}^{L-1}\binom{2k}{k}\frac{1}{[4(1+\bar\gamma)]^k}$$

$$(14C.3)$$

The mean SNR may be obtained from MGF using (F.37) and (D.24):

$$E[\gamma_{eq,R}]=\bar\gamma_{eq,R}=-\frac{d}{ds}M_{\gamma_{eq,R}}(s)\Big|_{s=0}$$

$$=\bar\gamma_{eq}\sum_{n=1}^K\binom{K}{n}\frac{(-1)^{n+1}}{n}=\bar\gamma_{eq}\sum_{n=1}^K\frac{1}{n}$$

$$(14C.4)$$

The average capacity is found as follows:

$$C_{av,R}=\frac{1}{2\ln2}E[\ln(1+\gamma_{eq,R})]$$

$$=\sum_{n=1}^K\binom{K}{n}(-1)^{n+1}\frac{1}{2\ln2}\int_0^\infty\ln(1+\gamma)\frac{n}{\bar\gamma_{eq}}e^{-n\gamma/\bar\gamma_{eq}}\,d\gamma$$

$$=\sum_{n=1}^K\binom{K}{n}(-1)^{n+1}C_m(\bar\gamma_{eq}/n)$$

$$(14C.5)$$

where $C_m(x)$, given by (14B.2), denotes the average capacity of a link with mean SNR, x.

The cdf of the capacity may be written as follows:

$$C_{out,R}(C_{th}) = Prob\left\{\frac{1}{2}\log_2(1+\gamma_{eq,R}) < C_{th}\right\}$$

$$= Prob\left\{\gamma_{eq,R} < 2^{2C_{th}} - 1\right\}$$

$$= F_{\gamma_{eq,R}}\left(2^{2C_{th}} - 1\right)$$

$$(14C.6)$$

14C.2 Selection combined SRD and SD links

The cdf, pdf and MGF of the SNR $\gamma_{c,R} = \max(\gamma_{eq,R}, \gamma_3)$ when one of the K relays, with highest equivalent SNR, is selected in the SRD link and is selection-combined with the SRD link with mean SNR $\bar{\gamma}_3$:

$$F_{\gamma_{c,R}}(\gamma) = \left(1-e^{-\gamma/\bar{\gamma}_{eq}}\right)^K\left(1-e^{-\gamma/\bar{\gamma}_3}\right)$$

$$= \sum_{n=0}^{K}\binom{K}{n}(-1)^n\left[e^{-n\gamma/\bar{\gamma}_{eq}} - e^{-\left(n/\bar{\gamma}_{eq}+1/\bar{\gamma}_3\right)\gamma}\right]$$

$$f_{\gamma_{c,R}}(\gamma) = \frac{1}{\bar{\gamma}_3}e^{-\gamma/\bar{\gamma}_3} + \sum_{n=1}^{K}\binom{K}{n}(-1)^{n+1}$$

$$\times\left[\frac{n}{\bar{\gamma}_{eq}}e^{-n\gamma/\bar{\gamma}_{eq}} - \left(\frac{n}{\bar{\gamma}_{eq}}+\frac{1}{\bar{\gamma}_3}\right)e^{-\left(n/\bar{\gamma}_{eq}+1/\bar{\gamma}_3\right)\gamma}\right]$$

$$M_{\gamma_{c,R}}(s) = \frac{1}{1+s\bar{\gamma}_3} + \sum_{n=1}^{K}\binom{K}{n}(-1)^{n+1}$$

$$\times\left[\frac{1}{1+s\bar{\gamma}_{eq}/n} - \frac{1}{1+s/(n/\bar{\gamma}_{eq}+1/\bar{\gamma}_3)}\right]$$

$$(14C.7)$$

The BER for BPSK modulation for the selection combined SRD and SD links is given by

$$P_b = \int_0^{\infty} Q\left(\sqrt{2\gamma}\right)f_{\gamma_{c,R}}(\gamma)d\gamma$$

$$= I(1,\bar{\gamma}_3) + \sum_{n=1}^{K}\binom{K}{n}(-1)^{n+1}I(1,\bar{\gamma}_{eq}/n)$$

$$-\sum_{n=1}^{K}\binom{K}{n}(-1)^{n+1}I\left(1,\frac{1}{n/\bar{\gamma}_{eq}+1/\bar{\gamma}_3}\right)$$

$$(14C.8)$$

The mean SNR may be written from the MGF using (F.37) and (D.24):

$$E[\gamma_{c,R}] = \bar{\gamma}_{c,R} = \bar{\gamma}_3 + \bar{\gamma}_{eq}\sum_{n=1}^{K}\frac{1}{n}$$

$$-\sum_{n=1}^{K}\binom{K}{n}\frac{(-1)^{n+1}}{(n/\bar{\gamma}_{eq}+1/\bar{\gamma}_3)}$$

$$(14C.9)$$

The average capacity is found as follows:

$$C_{av,c,R} = \frac{1}{2\ln2}E[\ln(1+\gamma_{c,R})]$$

$$= C_m(\bar{\gamma}_3) + \sum_{n=1}^{K}\binom{K}{n}(-1)^{n+1}C_m\left(\frac{\bar{\gamma}_{eq}}{n}\right)$$

$$-\sum_{n=1}^{K}\binom{K}{n}(-1)^{n+1}C_m\left(\frac{1}{n/\bar{\gamma}_{eq}+1/\bar{\gamma}_3}\right)$$

$$(14C.10)$$

where $C_m(x)$ is defined by (B.2).

The cdf of the capacity may be written as follows:

$$C_{out,c,R}(C_{th}) = Prob\left\{\frac{1}{2}\log_2(1+\gamma_{c,R}) < C_{th}\right\}$$

$$= Prob\left\{\gamma_{c,R} < 2^{2C_{th}} - 1\right\}$$

$$= F_{\gamma_{c,R}}\left(2^{2C_{th}} - 1\right)$$

$$(14C.11)$$

14C.3 MRC Combined SRD and SD Links

Now let us consider that the SRD links via K relays are all approximated by the Rayleigh pdf, and the SRD link with the highest equivalent SNR is selected. The equivalent SNR of the selected SRD link, $\gamma_{s,R}$, is MRC combined with the SRD link with SNR $\bar{\gamma}_3$. The cdf, pdf and MGF of the SNR $\gamma_{c,R,MRC} = \gamma_{s,R} + \gamma_3$ are given by

$$M_{\gamma_{c,R,MRC}}(s) = M_{\gamma_{eq,R}}(s)\frac{1}{1+s\bar{\gamma}_3}$$

$$= \sum_{n=1}^{K}\binom{K}{n}\frac{(-1)^{n+1}}{(1+s\bar{\gamma}_{eq}/n)(1+s\bar{\gamma}_3)}$$

$$= \sum_{n=1}^{K}\binom{K}{n}\frac{(-1)^{n+1}}{\bar{\gamma}_{eq}/n-\bar{\gamma}_3}$$

$$\times\left[\frac{\bar{\gamma}_{eq}/n}{(1+s\bar{\gamma}_{eq}/n)}-\frac{\bar{\gamma}_3}{(1+s\bar{\gamma}_3)}\right]$$

$$f_{\gamma_{c,R,MRC}}(\gamma) = \sum_{n=1}^{K}\binom{K}{n}\frac{(-1)^{n+1}}{\bar{\gamma}_{eq}/n-\bar{\gamma}_3}\left[e^{-n\gamma/\bar{\gamma}_{eq}}-e^{-\gamma/\bar{\gamma}_3}\right]$$

$$F_{\gamma_{c,R,MRC}}(\gamma) = 1-\sum_{n=1}^{K}\binom{K}{n}\frac{(-1)^{n+1}}{\bar{\gamma}_{eq}/n-\bar{\gamma}_3}$$

$$\times\left(\frac{\bar{\gamma}_{eq}}{n}e^{-n\gamma/\bar{\gamma}_{eq}}-\bar{\gamma}_3\,e^{-\gamma/\bar{\gamma}_3}\right)$$

$$(14C.12)$$

where (D.23) is used. The BEP for BPSK modulation when the SRD and SD links are maximal-ratio combined is given by

$$P_{c,R,MRC} = \sum_{n=1}^{K}\binom{K}{n}\frac{(-1)^{n+1}}{\bar{\gamma}_{eq}/n-\bar{\gamma}_3}$$

$$\times\left(\frac{\bar{\gamma}_{eq}}{n}I(1,\bar{\gamma}_{eq}/n)-\bar{\gamma}_3\,I(1,\bar{\gamma}_3)\right)$$

$$(14C.13)$$

The mean SNR may be obtained from MGF using (F.37) and (D.24):

$$E[\gamma_{c,R,MRC}] = \bar{\gamma}_{c,R,MRC} = \bar{\gamma}_3 + \bar{\gamma}_{eq}\sum_{n=1}^{K}\frac{1}{n}$$

$$(14C.14)$$

The average capacity is found as follows:

$$C_{av,c,R,MRC} = \frac{1}{2\ \ln2}E\left[\ln\left(1+\gamma_{c,R,MRC}\right)\right]$$

$$= \sum_{n=1}^{K}\binom{K}{n}(-1)^{n+1}\frac{1}{\bar{\gamma}_{eq}/n-\bar{\gamma}_3}$$

$$\times\left[\frac{\bar{\gamma}_{eq}}{n}\,C_m\left(\frac{\bar{\gamma}_{eq}}{n}\right)-\bar{\gamma}_3\,C_m(\bar{\gamma}_3)\right]$$

$$(14C.15)$$

where $C_m(x)$ is defined by (14B.2). The cdf of the capacity may be written as

$$C_{out,c,R,MRC}(C_{th}) = \mathrm{Prob}\left\{\frac{1}{2}\log_2\left(1+\gamma_{c,R,MRC}\right)<C_{th}\right\}$$

$$= \mathrm{Prob}\{\gamma_{c,R,MRC}<2^{2C_{th}}-1\}$$

$$= F_{\gamma_{c,R,MRC}}\left(2^{2C_{th}}-1\right)$$

$$(14C.16)$$

Appendix 14D

CDF of $\gamma_{eq,a}$

Using (F.115), the pdf of the SNR γ_1 of the SR link with mean SNR $\bar{\gamma}_1$ applying MRT with n antennas and the pdf of the SNR γ_2 of the RD link with mean SNR $\bar{\gamma}_2$ applying MRC with m antennas may be written as

$$f_{\gamma_i}(z) = \frac{1}{(k-1)!}\left(\frac{z}{\bar{\gamma}_i}\right)^{k-1}\frac{e^{-z/\bar{\gamma}_i}}{\bar{\gamma}_i}\quad i=1,\ k=n$$

$$F_{\gamma_i}(z) = 1-e^{-z/\bar{\gamma}_i}\sum_{j=0}^{k-1}\frac{1}{j!}(z/\bar{\gamma}_i)^j\quad i=2,\ k=m$$

$$(14D.1)$$

The cdf of $\gamma_{eq,a}$, the equivalent SNR of the SRD link, is given by

$$F_{\gamma_{eq,a}}(\gamma) = P(\gamma_{eq} < \gamma)$$

$$= \int_0^\infty P\left(\frac{\gamma_1\gamma_2}{\gamma_1+\gamma_2} < \gamma \Big| \gamma_1\right) f_{\gamma_1}(\gamma_1)\, d\gamma_1$$

$$= \int_0^\infty P\left(\gamma_2 < \frac{\gamma\gamma_1}{\gamma_1-\gamma} \Big| \gamma_1\right) f_{\gamma_1}(\gamma_1)\, d\gamma_1$$

$$= \int_0^\gamma \underbrace{P\left(\gamma_2 > \frac{\gamma\gamma_1}{\gamma_1-\gamma} \Big| \gamma_1\right)}_{\equiv 1 \text{ since } \gamma_1-\gamma<0} f_{\gamma_1}(\gamma_1)\, d\gamma_1$$

$$+ \int_\gamma^\infty P\left(\gamma_2 < \frac{\gamma\gamma_1}{\gamma_1-\gamma} \Big| \gamma_1\right) f_{\gamma_1}(\gamma_1)\, d\gamma_1$$

$$= 1 - \int_\gamma^\infty P\left(\gamma_2 > \frac{\gamma\gamma_1}{\gamma_1-\gamma} \Big| \gamma_1\right) f_{\gamma_1}(\gamma_1)\, d\gamma_1$$

$$= 1 - \int_\gamma^\infty \left[1 - F_{\gamma_2}\left(\frac{\gamma\gamma_1}{\gamma_1-\gamma} \Big| \gamma_1\right)\right] f_{\gamma_1}(\gamma_1)\, d\gamma_1$$

$$(14D.2)$$

Using (14D.1), (14D.2) may be written as

$$F_{\gamma_{eq,a}}(\gamma) = 1 - \sum_{j=0}^{m-1}\frac{1}{j!}\int_\gamma^\infty \left(\frac{\gamma\gamma_1}{\bar\gamma_2(\gamma_1-\gamma)}\right)^j$$

$$\times \exp\left(-\frac{\gamma\gamma_1}{\bar\gamma_2(\gamma_1-\gamma)}\right)\frac{(\gamma_1/\bar\gamma_1)^{n-1}e^{-\gamma_1/\bar\gamma_1}}{(n-1)!}\frac{d\gamma_1}{\bar\gamma_1}$$

$$= 1 - \frac{e^{-\alpha\gamma}}{(n-1)!\,\bar\gamma_1^n}\sum_{j=0}^{m-1}\frac{1}{j!}\left(\frac{\gamma}{\bar\gamma_2}\right)^j$$

$$\times \int_0^\infty e^{-q/\lambda - p\lambda}\frac{(\lambda+\gamma)^{j+n-1}}{\lambda^j}\, d\lambda \qquad (14D.3)$$

where the variable transformation $\lambda = \gamma_1 - \gamma$ is applied in the last equation. Using the binomial expansion $(\lambda+\gamma)^R = \sum_{r=0}^R \binom{R}{r}\gamma^r\lambda^{R-r}$ above, (14D.3) reduces to

$$F_{\gamma_{eq,a}}(\gamma) = 1 - \frac{e^{-\alpha\gamma}}{(n-1)!\,\bar\gamma_1^n}\sum_{j=0}^{m-1}\frac{1}{j!}\left(\frac{\gamma}{\bar\gamma_2}\right)^j$$

$$\times \sum_{r=0}^{j+n-1}\gamma^r\binom{j+n-1}{r}$$

$$\times \int_0^\infty e^{-q/\lambda - p\lambda}\lambda^{n-r-1}\, d\lambda$$

$$= 1 - \frac{2e^{-\alpha\gamma}}{(n-1)!}\left(\frac{\beta\gamma}{2}\right)^n\sum_{j=0}^{m-1}\frac{1}{j!}\left(\frac{\gamma}{\bar\gamma_2}\right)^j$$

$$\times \sum_{r=0}^{j+n-1}\binom{j+n-1}{r}\left(\frac{\bar\gamma_2}{\bar\gamma_1}\right)^{r/2}K_{n-r}(\beta\gamma)$$

$$(14D.4)$$

For $n = m = 1$, (14D.4) reduces to (14A.2), as expected. Here $K_\nu(x)$ denotes the modified Bessel function of the second kind, $\alpha = 1/\bar\gamma_1 + 1/\bar\gamma_2$, $\beta = 2/\sqrt{\bar\gamma_1\bar\gamma_2}$, $q = \gamma^2/\bar\gamma_2$ and $p = 1/\bar\gamma_1$. The corresponding pdf is determined by taking the derivative of (14D.4) and using (D.105):

$$f_{\gamma_{eq}}(\gamma,\alpha,\beta) = \frac{2e^{-\alpha\gamma}}{(n-1)!}\left(\frac{\beta\gamma}{2}\right)^n\sum_{j=0}^{m-1}\frac{1}{j!}\left(\frac{\gamma}{\bar\gamma_2}\right)^j$$

$$\times \sum_{r=0}^{j+n-1}\binom{j+n-1}{r}\left(\frac{\bar\gamma_2}{\bar\gamma_1}\right)^{r/2}$$

$$\times \left[\left\{\alpha - \frac{j+r}{\gamma}\right\}K_{n-r}(\beta\gamma) + \beta K_{n-r-1}(\beta\gamma)\right]$$

$$(14D.5)$$

For $n = m = 1$, (14D.5) reduces to (14A.3), as expected.

The MGF corresponding to (14D.4) and (14D.5) may easily be found using (F.39) and (D.67):

$$M_{\gamma_{eq,a}}(s) = s \int_0^\infty e^{-sz} F_{\gamma_{eq,a}}(z)\, dz$$

$$= 1 - \frac{s\, 2^{-2n} \sqrt{\pi}}{(n-1)!} \sum_{j=0}^{m-1} \frac{(2\bar{\gamma}_2)^{-jj+n-1}}{j!} \sum_{r=0}^{} \binom{j+n-1}{r}$$

$$\times \left(\frac{2}{\bar{\gamma}_1}\right)^r \frac{\Gamma(j+r+1)\Gamma(2n-r+j+1)}{(s+\alpha)^{r+j+1}\,\Gamma(n+j+3/2)}$$

$$\times \,_2F_1\left(\frac{j+r+1}{2}, \frac{j+r+2}{2}; n+j+\frac{3}{2}; 1 - \frac{\beta^2}{(s+\alpha)^2}\right)$$

$$\hspace{11cm}(14D.6)$$

For $n = m = 1$, the MGF given by (14D.6) reduces to (14.15), as expected.

References

[1] A. Scaglione, D. L. Goeckel, and J. N. Laneman, Cooperative Communications in Mobile Ad Hoc Networks, IEEE Signal Processing Magazine, vol. 18 September 2006.

[2] R. Ahlswede, N. Cai, S.-Y. R. Li, and R. W. Yeung, Network Information Flow, *IEEE Trans. Information Theory*, vol. **46**, no. 4, July 2000, pp. 1204–1216.

[3] X. Tao, X. Xu, and Q. Cui, An overview of cooperative communications, IEEE Communications Magazine, pp. 65–71, June 2012.

[4] K.B. Letaif and W. Zhang, Cooperative communications for cognitive radio networks, *Proc. IEEE*, vol. **97**, no. 5, pp. 878–893, May 2009.

[5] A. Ghasemi, and E. S. Sousa, Spectrum Sensing in Cognitive Radio Networks: Requirements, Challenges and Design Trade-offs, IEEE Communications Magazine, April 2008, pp. 32–39.

[6] C.-X. Wang et al., Cooperative MIMO Channel Models: A Survey, IEEE Communications Magazine, February 2010, pp. 80–87.

[7] N. Yang, M. Elkashlan, and J. Yuan, Impact of opportunistic scheduling on cooperative dual-hop relay networks, *IEEE Trans. Communications*, 2011, vol. **59**, no.3, pp. 689–694.

[8] T. A. Tsiftsis, G. K. Karagiannidis, P. T. Mathiopoulos, and S. A. Kotsopoulos, Nonregenerative dual-hop cooperative links with selection diversity', EURASIP J. Wireless Communications and Networking, 2006, pp. 1–8, DOI 10.1155/WCN/2006/17862.

[9] P. L. Yeoh, M. Elkashlan, and I. B. Collings, Exact and asymptotic SER of distributed TAS/MRC in MIMO relay networks, *IEEE Trans. Wireless Communications*, 2011, vol. **10**, no. 3, pp. 751–756.

[10] D.B. da Costa, and S. Aissa, Amplify-and-forward relaying in channel-noise assisted cooperative networks with relay selection, *IEEE Communications Letters*, 2010, vol. **14**, no. 7, pp. 608–610.

[11] M.O. Hasna, and M-S. Alouini, A performance study of dual-hop transmissions with fixed gain relays, *IEEE Trans. Wireless Communications*, 2004, vol. **3**, no. 6 pp. 1963–1968.

[12] M.O. Hasna, and M-S. Alouini: End-to-end performance of transmission systems with relays over Rayleigh-fading channels, *IEEE Trans. Wireless Communications*, Nov. 2003, vol. **2**, no. 6, pp. 1126–1130.

[13] P. A. Anghel, and M. Kaveh, Exact symbol error probability of a cooperative network in a Rayleigh-fading environment, *IEEE Trans. Wireless Communications*, vol. **3**, no. 5, pp. 1416–1421, September 2004.

[14] P.L. Yeoh, M. Elkashlan, and I. B. Collings, Selection relaying with transmit beamforming: A comparison of fixed and variable gain relaying, *IEEE Trans. Commun.*, vol. **59**, no. 6, pp. 1720–1730, June 2011.

[15] M. Torabi, and D. Haccoun, Capacity analysis of opportunistic relaying in cooperative systems with outdated channel information, *IEEE Communication Letters*, vol. **14**, no. 12, pp. 1137–1139, December 2010.

[16] P. Xin et al., Diversity analysis of transmit beamforming and the application in IEEE 802.11n systems, *IEEE Trans VT*, vol. **57**, no. 4, pp. 2638–2642, July 2008.

[17] P. A. Dighe, R. K. Mallik, and S.S. Jamuar, "Analysis of Transmit-Receive Diversity in Rayleigh Fading," *IEEE Trans. Communications*, vol. **51**, pp. 694–703, April 2003.

[18] S. Prakash and I. McLoughlin, Performance of dual-hop multi-antenna system with fixed-gain AF relay selection, *IEEE Trans. Wireless Communications*, vol. **10**, no. 6, pp. 1709–1714, June 2011.

[19] S. Loyka, and G. Levin, On outage probability and diversity-multiplexing trade-off in MIMO relay channels, *IEEE Trans. Communications*, vol. **59**, no. 6, pp. 1731–1741, June 2011.

[20] M. Torabi, W. Ajib, and D. Haccoun, Performance analysis of amplify-and-forward cooperative networks with relay selection over Rayleigh fading channels, IEEE VTC, Spring 2009.

[21] A. Zanella, M. Chiani, and M. Z. Win, "On the marginal distribution of the eigenvalues of Wishart matrices," *IEEE Trans. Commun.*, vol. **57**, pp. 1050–1060, April 2009.

[22] A. Zanella, M. Chiani, and M. Z. Win, "Performance of MIMO MRC in correlated Rayleigh fading environments," in *Proc. IEEE Vehicular Technology Conference (VTC 2005 Spring)*, Stockholm, Sweden, pp. 1633–1637, vol. **3**, May 2005.

[23] P. J. Smith, P.-H. Kuo and L. M. Barth, Level Crossing Rates for MIMO Channel Eigenvalues: Implications for Adaptive Systems, in Conf. Rec. 2005 *IEEE Int. Conf. Commun. (ICC)*, vol. **4**, pp. 2442–2446.

[24] S. H. Simon, A.L. Moustakas, and L. Marinelli", Capacity and character expansions: moment-generating function and other exact results for MIMO correlated channels," *IEEE Trans. Information Theory*, vol. **52**, pp. 5336–5351, December 2006.

[25] M. R. McKay, A. J. Grant, and I. B. Collings, "Performance analysis of MIMO-MRC in Double-Correlated Rayleigh Environments," *IEEE Trans. Communications*, vol. **55**, pp. 497–507, March 2007.

[26] M. M. Fareed and M. Uysal, On Relay Selection for Decode-and-Forward Relaying, *IEEE Trans. Wireless Communications*, vol. **8**, no. 7, p. 3341–3346, July 2009.

[27] T. Wang, A. Cano, G. B. Giannakis and J. N. Laneman, High-performance cooperative demodulation with decode-and-forward relays, *IEEE Trans. Communications*, vol. **55**, no. 7, pp. 1427–1438, July 2007.

[28] R. M. Radaydeh and M. M. Matalgah, Results for Integrals Involving m-th Power of the Gaussian Q-function Over Rayleigh Fading Channels with Applications, IEEE ICC 2007, pp. 5910–5914.

[29] I. Safak, E. Aktas, and A.O. Yılmaz, Error Rate Analysis of GF(q) Network Coded Detect-and-Forward Wireless Relay Networks Using Equivalent Relay Channel Models, *IEEE Trans. Wireless Communications*, 2013, vol. **12**, no. 8, pp. 3908–3919.

[30] Y. Li, Distributed coding for cooperative wireless networks: an overview and recent advances, IEEE Communications Magazine, pp. 71–77, August 2009.

[31] A. Nosratinia, T. E. Hunter, and A. Hedayat, Cooperative Communication in Wireless Networks, IEEE Communications Magazine, October 2004, pp. 74–80.

[32] L. Fan, X. Lei, and W. Li, Exact closed-form expression for ergodic capacity of amplify-and-forward relaying in channel-noise-assisted cooperative networks with relay selection, *IEEE Communications Letters*, vol. **15**, no. 3, pp. 332–333, March 2011.

[33] H. A. Suraweera, P.J. Smith, A. Nallanathan, and J.S. Thompson, Amplify-and-Forward Relaying with Optimal and Suboptimal Transmit Antenna Selection, *IEEE Trans. Wireless Communications*, vol. **10**, no. 6, pp. 1874–1885, June 2011.

[34] L. S. Gradshteyn, and L. M. Ryzhik, *Table of Integrals, Series and Products (6th ed.)*, Academic Press: San Diego, 2000, p. 17, 0.314.

Problems

1. Prove the following

$$\int_0^\infty \ln(1+\alpha\gamma)f(\gamma)d\gamma = \int_0^\infty \frac{\alpha}{1+\alpha\gamma}[1-F(\gamma)]d\gamma$$

where $f(\gamma)$ and $F(\gamma)$ denote respectively the pdf and the cdf of γ.

2. Using the derivative of the Q function given by (B.13) and integration by parts, prove the following identity [33]

$$P_e = \int_0^\infty Q\left(\sqrt{2\gamma}\right)f_\gamma(\gamma)\,d\gamma$$

$$= \frac{1}{\sqrt{2\pi}}\int_0^\infty \frac{1}{\sqrt{2\gamma}}e^{-\gamma}F_\gamma(\gamma)\,d\gamma$$

3. Derive the pdf in (14.22) by taking the derivative of the corresponding cdf.

4. Plot and compare mean equivalent SNRs, given by (14.24), of the selection-combined and maximal-ratio-combined SRD and SD links.

5. Noting that (14.27)–(14.31) correspond to selection combining of two direct links, that is, $\gamma_D = \max(\gamma_2, \gamma_3)$, derive the expressions corresponding to (14.27)–(14.31) for maximal ratio combining, that is, $\gamma_{MRC} = \gamma_2 + \gamma_3$, where γ_2 and γ_3 are assumed to be independent of each other.

6. The equivalent SNR of the SRD link with Rayleigh approximation and relay selection is given by (14C.1). Using (14C.8) and (14C.13), plot and compare the BEP performances when the SNRs of SRD and SD links are selection- and maximal ratio combined.

7. Show that (14.35) is identical to $\gamma_c = \max\left\{\gamma_{eq,1}, \gamma_{eq,2}, \ldots, \gamma_{eq,K}, \gamma_3\right\}$.

8. The pdf and cdf of a random variable with central chi-square distribution are given below (see (F.115)):

$$f_\gamma(z) = \frac{z^{m-1}}{(m-1)!\bar{\gamma}^m}\exp\left(\frac{-z}{\bar{\gamma}}\right)$$

$$F_\gamma(z) = 1-\exp\left(\frac{-z}{\bar{\gamma}}\right)\sum_{k=0}^{m-1}\frac{1}{k!}\left(\frac{z}{\bar{\gamma}}\right)^k$$

Show that K^{th} power of the cdf, which corresponds to the selection of one of the K relays with the highest equivalent SNR with the chi-square pdf, may be expressed as follows:

$$[F(\gamma)]^K = \sum_{k=0}^{K} \binom{K}{k} (-1)^k$$

$$\times \exp\left(-\frac{kz}{\bar{\gamma}}\right) \sum_{i=0}^{(m-1)k} c_i \left(\frac{z}{\bar{\gamma}}\right)^i$$

Hint: use the following multinomial expansion: [34]

$$\left(\sum_{i=0}^{n-1} a_i x^i\right)^k = \sum_{i=0}^{(n-1)k} c_i x^i$$

$$c_0 = a_0^n$$

$$c_i = \frac{1}{i a_0} \sum_{j=1}^{i} (jk - i + j) a_j c_{i-j},$$

$$i = 1, 2, \ldots (n-1)k$$

9. By using the results of the Problem 14.8, show that the outage probability with fixed channel gain, multiple antennas at source and destination and relay selection, may be expressed analytically as follows:

$$P_{out}(\gamma) = \int_0^\infty \Pr\left(\frac{\gamma_1 \gamma_2}{\gamma_2 + C} < \gamma | \gamma_2\right) f_{\gamma_2}(\gamma_2) d\gamma_2$$

$$= \int_0^\infty F_{\gamma_1}(\gamma(1 + C/\gamma_2)) f_{\gamma_2}(\gamma_2) d\gamma_2$$

$$= 1 - 2 \frac{\bar{\gamma}_2^{-m}}{(m-1)!} \sum_{k=1}^{K} \binom{K}{k} (-1)^{k+1} e^{-k\gamma/\bar{\gamma}_1}$$

$$\times \sum_{i=0}^{(n-1)k} c_i \left(\frac{\gamma}{\bar{\gamma}_1}\right)^i \sum_{r=0}^{i} \binom{i}{r} C^r \left(\frac{k\gamma C \bar{\gamma}_2}{\bar{\gamma}_1}\right)^{\frac{m-r}{2}}$$

$$\times K_{m-r}\left(2\sqrt{\frac{k\gamma C}{\bar{\gamma}_1 \bar{\gamma}_2}}\right)$$

where γ_1 and γ_2 are central chi-square distributed with means $\bar{\gamma}_1$ and $\bar{\gamma}_2$ and degrees of freedom n and m, respectively. Compare the outage probability given above with (11) in [18].

10. Derive and plot the outage probability and the ergodic capacity corresponding to the pdf given by (14.41), and discuss the effect of outdated CSI.

11. The BER of the SRD link with decode-and-forward relaying is expressed as (14.93) [27]

$$P_{eq} = P_b(\gamma_1)[1 - P_b(\gamma_2)] + [1 - P_b(\gamma_1)] P_b(\gamma_2)$$

where $P_b(\gamma_i) = Q(\sqrt{2\gamma_i})$, $i = 1, 2$ for BPSK modulation. Here, γ_1, γ_2 respectively denote the instantaneous SNRs of SR and RD links with mean powers $\bar{\gamma}_1$, $\bar{\gamma}_2$. Assuming the SRD link as a single equivalent link with bit error probability given by $P_{eq} = Q\left(\sqrt{2\gamma_{eq}}\right)$, show that the equivalent instantaneous SNR of the SRD link is given by

$$\gamma_{eq} = \frac{1}{2}\left[Q^{-1}(P_{eq})\right]^2$$

Using the above formulas, propose a method to obtain the pdf of γ_{eq} in terms of the pdf's of γ_1 and γ_2.

When SR and RD links undergo Rayleigh fading and are independent from each other, the average BEP of the SRD link is given by (14.94):

$$P_{eq,av}(\bar{\gamma}_{eq}) = I(1, \bar{\gamma}_1) + I(1, \bar{\gamma}_2) - 2I(1, \bar{\gamma}_1)I(1, \bar{\gamma}_2)$$

where $I(L, x)$ is defined by (14C.3). Assuming that the equivalent SNR γ_{eq} is also Rayleigh fading with a mean $\bar{\gamma}_{eq}$, insert $P_{eq,av}(\bar{\gamma}_{eq}) = I(1, \bar{\gamma}_{eq})$ into the above BER expression and show that

$$\bar{\gamma}_{eq} = \frac{\bar{\gamma}_1 \bar{\gamma}_2}{\bar{\gamma}_1 + \bar{\gamma}_2 + 1}$$

Compare this result with (14.16).

12. Consider a relay network as shown below with two relays which detect the signals at their input, remodulate and forward them after modulation. The SNRs in each hop are denoted by γ_i, i = 1,2,3, and the corresponding error probabilities are p_i, i = 1,2,3. The bit error probability performance of the two relays and the receiver are assumed to be the same.

a. Calculate the error probability of this relay network in terms of p_i, i = 1,2,3.
b. Determine the error probability for p_i = 0.001, i = 1,2,3.
c. Write down the expression for error probability for an N relay network for the special case of $p = p_i$, i = 1,2,..N, show that it reduces to the result found in part (a) for this special case.

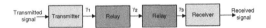

13. Consider that the equivalent SNR is approximated as $\gamma_{eq,m} = \min(\gamma_1,\gamma_2)$. Determine the pdf and the cdf of $\gamma_{eq,m}$ and the BEP for BPSK. Plot them and compare with (14.14), (14.13) and (14.21).

14. Compare SNR levels required to achieve an outage probability of 0.01 for γ_{eq}, $\gamma_{eq,R}$ and $\gamma_{eq,m} = \min(\gamma_1,\gamma_2)$ for the SRD link with $\bar{\gamma}_1 = \bar{\gamma}_2$. Also compare the mean equivalent SNR values.

15. Consider a relay network which consists of a source, a destination and K relays. The SNRs in source-relay and relay-destination links of the k-th relay are denoted by γ_{SR} and γ_{RD}. The mean values of the mean SNRs are denoted by $\bar{\gamma}_{SR_k}$ and $\bar{\gamma}_{RD_k}$. All links are assumed to undergo independent

Rayleigh fading. The equivalent SNR of the k-th relay link is given by

$$\gamma_{eq_k} = \min(\gamma_{SR_k}, \gamma_{RD_k}), \quad k = 1,2,\ldots,K$$

The destination applies selection diversity and selects the strongest equivalent SNR among the K relay links. Hence the output SNR at the destination output is given by

$$\gamma_{SC} = \max(\gamma_{eq_1}, \gamma_{eq_2}, \ldots, \gamma_{eq_K})$$

a. Determine the cdf of equivalent SNR of the k-th link.
b. Determine the cdf at the output of the selection combiner.
c. Assuming that $\bar{\gamma}_{SR_k} = \bar{\gamma}_{RD_k} = 10$ dB, $k = 1,2,\ldots,K$, determine the probability that the output SNR is below -10 dB for K = 1 and K = 2. Calculate the ratio of the outage probability for K = 1 to that for K = 2 and comment on the result that you obtained.

16. Consider a relay network where one of the K relay paths having the highest SNR is selected and selection-combined with the direct path. All relay paths are assumed to undergo Rayleigh fading and have a mean SNR of $\bar{\gamma}_{eq}$. The direct path also undergoes Rayleigh fading and has a mean SNR of $\bar{\gamma}_3$.

a. Determine the outage probability.
b. Determine the BEP in a Rayleigh fading channel for DPSK modulation for $K = 2$.

17. Using (14.98) and (14.99), plot and compare outage and BEP performances for SRD links of AF and DF relays for $n = m = 2$, $r = 1$; $n = m = r = 2$ and $n = m = 2$, r = 3.

18. Derive (14.99).

Appendix A

Vector Calculus in Spherical Coordinates

$$\nabla f = \hat{\mathbf{u}}_r \frac{\partial f}{\partial r} + \hat{\mathbf{u}}_\theta \frac{1}{r}\frac{\partial f}{\partial \theta} + \hat{\mathbf{u}}_\phi \frac{1}{r\sin\theta}\frac{\partial f}{\partial \phi}$$

$$\nabla . \mathbf{F} = \frac{1}{r^2}\frac{\partial}{\partial r}\left(r^2 F_r\right) + \frac{1}{r\sin\theta}\frac{\partial}{\partial \theta}\left(F_\theta \sin\theta\right)$$

$$+ \frac{1}{r\sin\theta}\frac{\partial F_\phi}{\partial \phi}$$

$$(\nabla \times \mathbf{F})_r = \frac{1}{r\sin\theta}\left[\frac{\partial}{\partial \theta}\left(F_\phi \sin\theta\right) - \frac{\partial F_\theta}{\partial \phi}\right]$$

$$(\nabla \times \mathbf{F})_\theta = \frac{1}{r\sin\theta}\frac{\partial F_r}{\partial \phi} - \frac{1}{r}\frac{\partial}{\partial r}\left(r F_\phi\right)$$

$$(\nabla \times \mathbf{F})_\phi = \frac{1}{r}\frac{\partial}{\partial r}\left(r F_\theta\right) - \frac{1}{r}\frac{\partial F_r}{\partial \theta}$$

$$\text{(A.1)}$$

Special case: Calculation of $\nabla \times \mathbf{F}$ in the far-field region ($\mathbf{F}$ is assumed to vary as e^{-jkr}/r with r)

$$(\nabla \times \mathbf{F})_r = \frac{1}{r\sin\theta}\underbrace{\left[\frac{\partial}{\partial \theta}\left(F_\phi \sin\theta\right) - \frac{\partial F_\theta}{\partial \phi}\right]}_{\propto 1/r^2} \cong 0$$

$$(\nabla \times \mathbf{F})_\theta = \underbrace{\frac{1}{r\sin\theta}\frac{\partial F_r}{\partial \phi}}_{\propto 1/r^2} - \frac{1}{r}\frac{\partial}{\partial r}\left(r F_\phi\right) \cong jkF_\phi$$

$$(\nabla \times \mathbf{F})_\phi = \frac{1}{r}\frac{\partial}{\partial r}\left(r F_\theta\right) - \underbrace{\frac{1}{r}\frac{\partial F_r}{\partial \theta}}_{\propto 1/r^2} \cong -jkF_\theta$$

$$\nabla \times \mathbf{F} \cong jk\left(F_\phi \hat{\mathbf{u}}_\theta - F_\theta \hat{\mathbf{u}}_\phi\right) = -jk\,\hat{\mathbf{u}}_r \times \mathbf{F}$$

$$\text{(A.2)}$$

Transformation from Cartesian to spherical coordinates

$$\hat{\mathbf{u}}_r = \sin\theta\cos\phi\,\hat{\mathbf{u}}_x + \sin\theta\sin\phi\,\hat{\mathbf{u}}_y + \cos\theta\,\hat{\mathbf{u}}_z$$

$$\hat{\mathbf{u}}_\theta = \cos\theta\cos\phi\,\hat{\mathbf{u}}_x + \cos\theta\sin\phi\,\hat{\mathbf{u}}_y - \sin\theta\,\hat{\mathbf{u}}_z$$

$$\hat{\mathbf{u}}_\phi = -\sin\phi\,\hat{\mathbf{u}}_x + \cos\phi\,\hat{\mathbf{u}}_y$$

$$\text{(A.3)}$$

Transformation from spherical to Cartesian coordinates

$$\hat{\mathbf{u}}_x = \sin\theta\cos\phi\,\hat{\mathbf{u}}_r + \cos\theta\cos\phi\,\hat{\mathbf{u}}_\theta - \sin\phi\,\hat{\mathbf{u}}_\phi$$

$$\hat{\mathbf{u}}_y = \sin\theta\sin\phi\,\hat{\mathbf{u}}_r + \cos\theta\sin\phi\,\hat{\mathbf{u}}_\theta + \cos\phi\,\hat{\mathbf{u}}_\phi$$

$$\hat{\mathbf{u}}_z = \cos\theta\,\hat{\mathbf{u}}_r - \sin\theta\,\hat{\mathbf{u}}_\theta$$

$$\text{(A.4)}$$

Digital Communications, First Edition. Mehmet Şafak.
© 2017 John Wiley & Sons Ltd. Published 2017 by John Wiley & Sons Ltd.
Companion website: www.wiley.com/go/safak/Digital_Communications

Appendix B

Gaussian Q Function

B.1 Gaussian Q-Function

The Gaussian Q-function, which is defined by

$$Q(x) = \int_x^\infty f_Z(z)\, dz = \frac{1}{\sqrt{2\pi}} \int_x^\infty \exp\left(-z^2/2\right)\, dz \tag{B.1}$$

represents the area under the tail (between x and ∞) of a zero-mean and unit variance Gaussian pdf $f_Z(z)$ (see Figure B.1). Since the area under a pdf is equal to unity, $Q(-\infty) = 1$ and $Q(\infty) = 0$. Owing to the symmetry of the Gaussian pdf with respect to the origin, $Q(0) = 1/2$ and

$$Q(-x) = 1 - Q(x) \tag{B.2}$$

The Craig's definition of the Q-function [1] can easily be derived from (B.1) as follows:

$$Q(x) = \frac{1}{\sqrt{2\pi}} \int_x^\infty \exp\left(-\frac{u^2}{2}\right) du$$

$$= \frac{1}{2\pi} \int_{-\infty}^\infty \int_x^\infty \exp\left(-\frac{u^2 + v^2}{2}\right) du\, dv \tag{B.3}$$

If we make a transformation of Cartesian coordinates to polar coordinates as $u = r \sin\theta$, $v = r \cos\theta$ and inserting $du\, dv = r\, dr\, d\theta$ into (B.3) (see Figure B.2), one gets

$$Q(x) = \frac{1}{2\pi} \int_0^\pi d\theta \int_{x/\sin\theta}^\infty \exp\left(-\frac{r^2}{2}\right) r\, dr$$

$$= \frac{1}{2\pi} \int_0^\pi \exp\left(-\frac{x^2}{2\sin^2\theta}\right) d\theta$$

$$= \frac{1}{\pi} \int_0^{\pi/2} \exp\left(-\frac{x^2}{2\sin^2\theta}\right) d\theta \tag{B.4}$$

Table B.1 gives the values of the Q-function for $0 \le x \le 4$.

It is shown in [1] that the Craig's second formula

$$Q^2(x) = \frac{1}{\pi} \int_0^{\pi/4} \exp\left(-\frac{x^2}{2\sin^2\theta}\right) d\theta \tag{B.5}$$

is a special case of

Digital Communications, First Edition. Mehmet Şafak.
© 2017 John Wiley & Sons Ltd. Published 2017 by John Wiley & Sons Ltd.
Companion website: www.wiley.com/go/safak/Digital_Communications

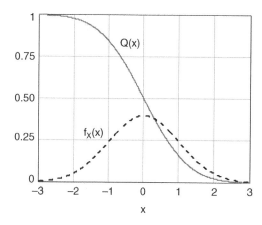

Figure B.1 Gaussian Q-Function and the Gaussian (Normal) Pdf with Zero-Mean and Unity Variance.

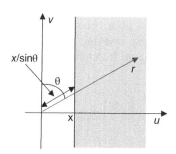

Figure B.2 Coordinate Transformation For Deriving the Craig's Formula For the Gaussian Q Function.

$$Q(x)Q(y) = \frac{1}{2\pi} \int_{\theta=0}^{\theta_0} \exp\left(-\frac{y^2}{2\sin^2\theta}\right) d\theta$$
$$+ \frac{1}{2\pi} \int_{\theta=\theta_0}^{\pi/2} \exp\left(-\frac{x^2}{2\sin^2\theta}\right) d\theta$$

$$(B.6)$$

where $\theta_0 = \tan^{-1}(y/x)$.

It is not always easy to determine the value of the Q-function and even more difficult to determine the integrals involving Gaussian Q-function in fading channels. Therefore, there is a rich list of publications which provide accurate approximations to the Q function. [2][3][4][5] The reader is referred to [6] for a recent review and comparison of these approximations. The authors in [6] also provide an exponential approximation to the Q-function:

$$Q\left(\sqrt{2\gamma}\right) = \frac{1}{2} \sum_{k=1}^{N} w_k e^{-kb\gamma} \qquad (B.7)$$

where the optimized coefficients are given by Table B.2.

Figure B.3 shows close agreement between the exact value of $Q(\sqrt{2\gamma})$ and its approximation by (B.7) with $N = 2$. Table B.3 presents the exact value of $Q(\sqrt{2\gamma})$ for values of 0 dB $\leq \gamma \leq 15$ dB.

B.1.1   Inverse Q-Function

This invertible approximation for $N = 3$ is reported to have an accuracy comparable to the best-known non-invertible exponential-type approximations. [5] For some modulation schemes, error probability P is expressed as $P = Q(\sqrt{2\gamma})$, where γ denotes the SNR. The SNR γ corresponding to a given P is given by the following simple inversion of the Q-function [7]

$$\gamma = \left[\text{erf}^{-1}(1-2P)\right]^2$$

$$\text{erf}^{-1}(x) = \text{sgn}(x)$$

$$\times \sqrt{\sqrt{\left(\frac{2}{\pi a} + \frac{\ln(1-x^2)}{2}\right)^2 - \frac{\ln(1-x^2)}{a}} - \left(\frac{2}{\pi a} + \frac{\ln(1-x^2)}{2}\right)}$$

$$a = 0.147$$

$$(B.8)$$

which is plotted in Figure B.4. The values of γ in dB corresponding to some values of P are listed in Table B.4.

B.1.2   Some Useful Relations

Properties of the Q-function:

$$Q(x) + Q(-x) = 1 \qquad (B.9)$$

Table B.1 Table of Q-Function For $0 \le x \le 4$.

	0.00	0.01	0.02	0.03	0.04	0.05	0.06	0.07	0.08	0.09
0.0	0.5	0,4960106	0,4920217	0,4880335	0,4840466	0,4800612	0,4760778	0,4720968	0,4681186	0,4641436
0.1	0,4601722	0,4562047	0,4522416	0,4482832	0,444330	0,4403823	0,4364405	0,4325051	0,4285763	0,4246546
0.2	0,4207403	0,4168338	0,4129356	0,4090459	0,4051651	0,4012937	0,3974319	0,3935801	0,3897388	0,3859081
0.3	0,3820886	0,3782805	0,3744842	0,370700	0,3669283	0,3631693	0,3594236	0,3556912	0,3519727	0,3482683
0.4	0,3445783	0,3409030	0,3372427	0,3335978	0,3299686	0,3263552	0,3227581	0,3191775	0,3156137	0,3120669
0.5	0,3085375	0,3050257	0,3015318	0,298056	0,2945985	0,2911597	0,2877397	0,2843388	0,2809573	0,2775953
0.6	0,2742531	0,2709309	0,2676289	0,2643473	0,2610863	0,2578461	0,2546269	0,2514289	0,2482522	0,2450971
0.7	0,2419637	0,2388521	0,2357625	0,2326951	0,229650	0,2266274	0,2236273	0,2206499	0,2176954	0,2147639
0.8	0,2118554	0,2089701	0,2061081	0,2032694	0,2004542	0,1976625	0,1948945	0,1921502	0,1894297	0,1867329
0.9	0,1840601	0,1814113	0,1787864	0,1761855	0,1736088	0,1710561	0,1685276	0,1660232	0,1635431	0,1610871
1.0	0,1586553	0,1562476	0,1538642	0,151505	0,1491700	0,146591	0,1445723	0,1423097	0,1400711	0,1378566
1.1	0,1356661	0,1334995	0,1313569	0,1292381	0,1271432	0,1250719	0,1230244	0,1210005	0,1190001	0,1170232
1.2	0,1150697	0,1131394	0,1112324	0,1093486	0,1074877	0,1056498	0,1038347	0,1020423	0,1002726	0,0985253
1.3	0,0968005	0,0950979	0,0934175	0,0917591	0,0901227	0,0885080	0,086915	0,0853435	0,0837933	0,0822644
1.4	0,0807567	0,0792698	0,0778038	0,0763585	0,0749337	0,0735293	0,072145	0,0707809	0,0694366	0,0681121
1.5	0,0668072	0,0655217	0,0642555	0,0630084	0,0617802	0,0605708	0,0593799	0,0582076	0,0570534	0,0559174
1.6	0,0547993	0,0536989	0,0526161	0,0515507	0,0505026	0,0494715	0,0484572	0,0474597	0,0464787	0,045514
1.7	0,0445655	0,0436329	0,0427162	0,0418151	0,0409295	0,0400592	0,0392039	0,0383636	0,037538	0,036727
1.8	0,0359303	0,0351479	0,0343795	0,033625	0,0328841	0,0321568	0,0314428	0,0307419	0,030054	0,029379
1.9	0,0287166	0,0280666	0,0274289	0,0268034	0,0261898	0,0255881	0,0249979	0,0244192	0,0238518	0,0232955
2.0	0,0227501	0,0222156	0,0216917	0,0211783	0,0206752	0,0201822	0,0196993	0,0192262	0,0187628	0,0183089
2.1	0,0178644	0,0174292	0,017003	0,0165858	0,0161774	0,0157776	0,0153863	0,0150034	0,0146287	0,0142621
2.2	0,0139034	0,0135526	0,0132094	0,0128737	0,0125455	0,0122245	0,0119106	0,0116038	0,0113038	0,0110107
2.3	0,0107241	0,0104441	0,0101704	0,0099031	0,0096419	0,0093867	0,0091375	0,008894	0,0086563	0,0084242

(continued overleaf)

Table B.1 (*Continued*)

	0.00	0.01	0.02	0.03	0.04	0.05	0.06	0.07	0.08	0.09
2.4	0,0081975	0,0079763	0,0077603	0,0075494	0,0073436	0,0071428	0,0069469	0,0067557	0,0065691	0,0063872
2.5	0,062097	0,0060366	0,0058677	0,0057031	0,0055426	0,0053861	0,0052336	0,0050849	0,00494	0,0047988
2.6	0,0046612	0,0045271	0,0043965	0,0042692	0,0041453	0,0040246	0,003907	0,0037926	0,0036811	0,0035726
2.7	0,0034670	0,0033642	0,0032641	0,0031667	0,003072	0,0029798	0,0028901	0,0028028	0,0027179	0,0026354
2.8	0,0025551	0,0024771	0,0024012	0,0023274	0,0022557	0,002186	0,0021182	0,0020524	0,0019884	0,0019262
2.9	0,0018658	0,0018071	0,0017502	0,0016948	0,0016411	0,0015889	0,0015382	0,001489	0,0014412	0,0013949
3.0	0,0013499	0,0013062	0,0012639	0,0012228	0,0011829	0,0011442	0,0011067	0,0010703	0,001035	0,0010008
3.1	0,0009676	0,0009354	0,0009043	0,000874	0,0008447	0,0008164	0,0007888	0,0007622	0,0007364	0,0007114
3.2	0,0006871	0,0006637	0,000641	0,000619	0,0005976	0,000577	0,0005571	0,0005377	0,000519	0,0005009
3.3	0,0004834	0,0004665	0,0004501	0,0004342	0,0004189	0,0004041	0,0003897	0,0003758	0,0003624	0,0003495
3.4	0,0003369	0,0003248	0,0003131	0,0003018	0,0002909	0,0002803	0,0002701	0,0002602	0,0002507	0,0002415
3.5	0,0002326	0,0002241	0,0002158	0,0002078	0,0002001	0,0001926	0,0001854	0,0001785	0,0001718	0,0001653
3.6	0,0001591	0,0001531	0,0001473	0,0001417	0,0001363	0,0001311	0,0001261	0,0001213	0,0001166	0,0001121
3.7	0,0001078	0,0001036	9,961E-05	9,574E-05	9,201E-05	8,842E-05	8,496E-05	8,162E-05	7,841E-05	7,532E-05
3.8	7,235E-05	6,948E-05	6,673E-05	6,407E-05	6,152E-05	5,906E-05	5,669E-05	5,442E-05	5,223E-05	5,012E-05
3.9	4,810E-05	4,615E-05	4,427E-05	4,247E-05	4,074E-05	3,908E-05	3,747E-05	3,594E-05	3,446E-05	3,304E-05
4.0	3,167E-05	3,036E-05	2,91E-05	2,789E-05	2,673E-05	2,561E-05	2,454E-05	2,351E-05	2,252E-05	2,157E-05

Table B.2 Optimized Coefficients in (B.7).

Order N	Optimized coefficients
1	$w_1 = 0.4803; \ b = 1.1232$
2	$w_1 = 0.3017; \ w_2 = 0.4389; \ b = 1.0510$
3	$w_1 = 0.3357; \ w_2 = 0.3361; \ w_3 = 0.0305;$ $b = 1.0649$

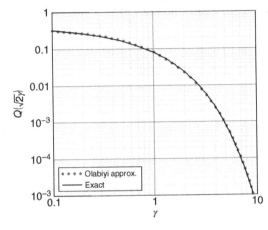

Figure B.3 Gaussian Q-Function $Q(\sqrt{2\gamma})$ and its Approximation Given By (B.7) [6] with $N = 2$.

$$Q(-\infty) = 1$$
$$Q(0) = 1/2 \qquad \text{(B.10)}$$
$$Q(\infty) = 0$$

Asymptotic approximations

$$Q(x) \cong \frac{\exp(-x^2/2)}{\sqrt{2\pi(1+x^2)}}, \quad x \gg 1$$

$$Q(x) \cong \frac{\exp(-x^2/2)}{\sqrt{2\pi}} \frac{1}{0.661 + 0.339\sqrt{x^2 + 5.51}},$$
$$x \gg 1$$

$$Q(x) \cong \frac{1}{2} - \frac{x}{\sqrt{2\pi}}\left(1 - \frac{x^2}{6}\right), \quad x \to 0$$
$$\text{(B.11)}$$

$$Q(\sqrt{x+y}) \le Q(\sqrt{x})e^{-y/2} \qquad \text{(B.12)}$$

The derivative of Gaussian-Q function is found from (B.1) as

$$\frac{d}{dx}Q(x) = \frac{d}{dx}\left(\frac{1}{\sqrt{2\pi}}\int_x^\infty e^{-t^2/2} \, dt\right)$$
$$= -\frac{1}{\sqrt{2\pi}}e^{-x^2/2} \qquad \text{(B.13)}$$

Q function is related to the complementary error function erfc(x) as follows:

$$Q(x) = \frac{1}{2}\text{erfc}\left(\frac{x}{\sqrt{2}}\right) \qquad \text{(B.14)}$$

$$\text{erfc}(x) = 2Q(\sqrt{2}x)$$

B.2 Marcum-Q Function

Marcum-Q function is defined by

$$Q_m(a,b) = \int_b^\infty x\left(\frac{x}{a}\right)^{m-1} \exp\left(-\frac{x^2+a^2}{2}\right)I_{m-1}(ax)dx$$

$$= Q_1(a,b) + \exp\left(\frac{a^2+b^2}{2}\right)\sum_{k=1}^{m-1}\left(\frac{b}{a}\right)^k I_k(ab)$$
$$\text{(B.15)}$$

where $Q_1(a, b)$ denotes the Marcum-Q function of the first order:

$$Q_1(a,b) = \int_b^\infty xe^{-(x^2+a^2)/2} I_0(ax) \, dx$$

$$= e^{-(b^2+a^2)/2}\sum_{k=0}^\infty \frac{1}{k!}\left(\frac{a^2}{2}\right)^k \sum_{i=0}^k \frac{1}{i!}\left(\frac{b^2}{2}\right)^i$$
$$\text{(B.16)}$$

The last expression is obtained by using the series expansion (D.95) of the Bessel function and (D.48) for evaluating the integral. In the special cases for $a = 0$ and $b = 0$, it reduces to

$$Q_1(0,b) = e^{-b^2/2}$$
$$Q_1(a,0) = 1 \qquad \text{(B.17)}$$

Table B.3 The Q-Function $Q(\sqrt{2\gamma})$ For $0\,\text{dB} \le \gamma \le 15\,\text{dB}$.

	0.0	0.1	0.2	0.3	0.4	0.5	0.6	0.7	0.8	0.9
0	0.079	0.076	0.074	0.072	0.069	0.067	0.065	0.063	0.06	0.058
1	.056	0.054	0.052	0.05	0.048	0.046	0.045	0.043	0.041	0.039
2	0.038	0.036	0.034	0.033	0.031	0.03	0.028	0.027	0.025	0.024
3	0.023	0.022	0.02	0.019	0.018	0.017	0.016	0.015	0.014	0.013
4	.013	0.012	0.011	0.01	9,46E-03	8,79E-03	8.16E-03	7.56E-03	6,99E-03	6,46E-03
5	5,95E-03	5,48E-03	5,04E-03	4,62E-03	4,23E-03	3,86E-03	3,52E-03	3,21E-03	2,91E-03	2,64E-03
6	2,39E-03	2,16E-03	1,94E-03	1,75E-03	1,57E-03	1,4E-03	1,25E-03	1,11E-03	9,88E-04	8,75E-04
7	7,73E-04	6,81E-04	5,98E-04	5,24E-04	4,58E-04	3,99E-04	3,46E-04	3,00E-04	2,59E-04	2,23E-04
8	1,91E-04	1,63E-04	1,39E-04	1,18E-04	9,97E-05	8,4E-05	7,05E-05	5,89E-05	4,91E-05	4,07E-05
9	3,36E-05	2,77E-05	2,27E-05	1,85E-05	1,50E-05	1,21E-05	9,74E-06	7,79E-06	6,20E-06	4,91E-06
10	3,87E-06	3,04E-06	2,37E-06	1,84E-06	1,41E-06	1,08E-06	8,26E-07	6,25E-07	4,71E-07	3,52E-07
11	2,61E-07	1,93E-07	1,41E-07	1,03E-07	7,43E-08	5,33E-08	3,79E-08	2,68E-08	1,88E-08	1,31E-08
12	9,01E-09	6,16E-09	4,18E-09	2,81E-09	1,87E-09	1,23E-09	8,06E-10	5,22E-10	3,35E-10	2,12E-10
13	1,33E-10	8,28E-11	5,09E-11	3,10E-11	1,86E-11	1,11E-11	6,49E-12	3,77E-12	2,16E-12	1,22E-12
14	6,81E-13	3,75E-13	2,04E-13	1,09E-13	5,77E-14	3,00E-14	1,54E-14	7,77E-15	3,89E-15	1,89E-15

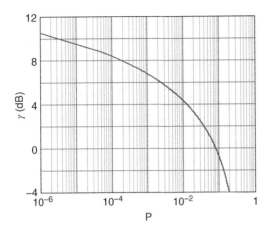

Figure B.4   Inverse Q function.

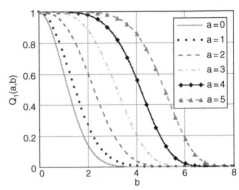

Figure B.5   Marcum-Q function $Q_1(a, b)$.

Table B.4   Values of γ *(dB)* For Various Values of P From (B.8).

P_b	γ *(dB)*	P_b	γ *(dB)*
10^{-1}	−0.854	10^{-6}	10.519
10^{-2}	4.315	10^{-7}	11.302
10^{-3}	6.774	10^{-8}	11.969
10^{-4}	8.382	10^{-9}	12.55
10^{-5}	9.574	10^{-10}	13.064

Table B.5   Optimized Values of the Parameters v and μ For Various Values of a. [8]

a	v	μ
1	−1.1739	2.0921
2	−2.5492	2.7094
3	−4.6291	3.6888
4	−7.1668	4.7779
5	−10.0339	5.9074
6	−13.2014	7.0794

Figure B.5 shows the variation of $Q_1(a, b)$ versus b for several values of a. Similar to the Gaussian Q function, $Q_1(a, b)$ decays exponentially with b, where the value of a determines the shift of the curve along the b-axis. $Q_1(a, b)$ is approximated by [8]

$$Q_1(a,b) \approx \exp\left(-e^{v(a)} b^{\mu(a)}\right) \qquad \text{(B.18)}$$

The accuracy of the approximation by (B-18) is maximized by choosing the non-negative parameters $v(a)$ and $\mu(a)$ so that the error between (B.18) and (B.16) is minimized. Optimized values of $v(a)$ and $\mu(a)$ are listed in Table B.5 for several values of a.

When the parameters $v(a)$ and $\mu(a)$ are chosen as polynomials, the following approximation is proposed for small values of a $(a < 1)$: [8]

$$\mu(a) = 2 + \frac{9a^4}{8(9\pi^2 - 80)}$$

$$v(a) = -\ln 2 - \frac{a^2}{2} + \frac{45\pi^2 + 72 \ln 2 + 36C - 496}{64(9\pi^2 - 80)}a^4$$

$$\text{(B.19)}$$

where $C \approx 0.5772$ denotes the Euler-Mascheroni constant. For large values of a $(a > 1)$, the following approximation is proposed: [8]

$$\mu(a) = 2.174 - 0.592a + 0.593a^2$$
$$\quad -0.092a^3 + 0.005a^4$$
$$v(a) = -0.840 - 0.327a - 0.740a^2$$

$$\text{(B.20)}$$

$$\quad +0.083a^3 - 0.004a^4$$

The approximation by (B.18) for the optimized values listed in Table B.5 is reported to show very close agreement with (B.16). [8]

References

[1] N. C. Beaulieu, Generalization of Craig's second formula, *IEEE Communications Letters*, vol.17, no. 3, pp. 433–434, March 2013.

[2] P. O. Brjesson and C.-E W. Sundberg, Simple approximations of the error function Q(x) for communications applications, *IEEE Trans. Commun.*, vol. 27, pp. 639–643, March 1979.

[3] G. K. Karagiannidis and A. S. Lioumpas, An improved approximation for the Gaussian Q-function, *IEEE Communications Letters*, vol. 11, no. 8, pp. 231, April 2008.

[4] J. S. Dyer and S. A. Dyer, Corrections to and comments on 'An improved approximation for the Gaussian Q-function', *IEEE Communications Letters*, vol. 12, no. 4, pp. 644–646, August 2007.

[5] P. Loskot and N. Beaulieu, Prony and error polynomial approximations for evaluation of the average probability of error over slow-fading channels, *IEEE Trans. Vehicular Technology*, vol. 58, pp. 1269–1280, March 2009.

[6] O. Olabiyi, and A. Annamalai, Invertible exponential-type approximations for the Gaussian probability integral Q(x) with applications, *IEEE Wireless Communications Letters*, vol. 1, no. 5, pp. 544–547, October 2012.

[7] http://en.wikipedia.org/wiki/Error_function

[8] M. Z. Bocus, C. P. Dettmann and J. P. Coon, An approximationof the first order Marcum Q-Function with application to network connectivity analysis, *IEEE Communications Letters*, vol. 17, no. 3, pp. 499–502, March 2013.

Appendix C

Fourier Transforms

Definition of Fourier transform:

$$S(f) = \int_{-\infty}^{\infty} s(t) e^{-jwt} dt$$

$$s(t) = \int_{-\infty}^{\infty} S(f) e^{jwt} df \tag{C.1}$$

Rayleigh's energy theorem:

$$E = \int_{-\infty}^{\infty} |s(t)|^2 dt = \int_{-\infty}^{\infty} |S(f)|^2 df \tag{C.2}$$

Functions used in the Fourier transforms:
Dirac delta function:

$$\delta(x - x_0) = \begin{cases} 1 & x = x_0 \\ 0 & x \neq x_0 \end{cases} \tag{C.3}$$

Sinc function:

$$\text{sinc}(x) = \sin(\pi x)/(\pi x) \tag{C.4}$$

Unit step function:

$$u(x) = \begin{cases} 0 & x < 0 \\ 1/2 & x = 0 \\ 1 & x > 0 \end{cases} \tag{C.5}$$

Signum function:

$$\text{sgn}(x) = \begin{cases} -1 & x < 0 \\ 0 & x = 0 \\ 1 & x > 0 \end{cases} \tag{C.6}$$

Rectangular function:

$$\Pi(t/T) = \begin{cases} 1 & |t| \leq T/2 \\ 0 & |t| > T/2 \end{cases} \tag{C.7}$$

Triangular function:

$$\Lambda(t/T) = \begin{cases} 1 - |t|/T & |t| \leq T \\ 0 & |t| > T \end{cases} \tag{C.8}$$

Digital Communications, First Edition. Mehmet Şafak.
© 2017 John Wiley & Sons Ltd. Published 2017 by John Wiley & Sons Ltd.
Companion website: www.wiley.com/go/safak/Digital_Communications

Properties of Fourier transform:

$s(t)$	$S(f)$
$\sum_i \alpha_i s_i(t)$	$\sum_i \alpha_i S_i(f)$
$s(t \mp \tau)$	$S(f) e^{\mp jw\tau}$
$s(t) e^{\pm jw_c t}$	$S(f \mp f_c)$
$s(\alpha t)$	$|\alpha|^{-1} S(f/\alpha)$
$S(t)$	$s(-f)$
$s^*(\pm t)$	$S^*(\mp f)$
$r(t) s(t)$	$R(f) \otimes S(f)$
$r(t) \otimes s(t)$	$R(f) S(f)$
$\dfrac{d^n s(t)}{dt^n}$	$(j2\pi f)^n S(f)$
$\dfrac{d^n S(f)}{df^n}$	$(-j2\pi t)^n s(t)$
$\displaystyle\int_{-\infty}^{t} s(t')\,dt'$	$\dfrac{1}{2} S(0)\delta(f) + \dfrac{S(f)}{j2\pi f}$
$\displaystyle\sum_{n=-\infty}^{\infty} \delta(t-nT)$	$\dfrac{1}{T}\displaystyle\sum_{n=-\infty}^{\infty}\delta\left(f-\dfrac{n}{T}\right)$
1	$\delta(f)$
$\delta(t)$	1
$\delta(at+b)$	$e^{-jbw/a}$

Fourier transforms:

$\Pi(t/T)$	$T\,\mathrm{sinc}(fT)$
$\Lambda(t/T)$	$T\,\mathrm{sinc}^2(fT)$
$W\,\mathrm{sinc}(Wt)$	$\Pi(f/W)$
$W\,\mathrm{sinc}^2(Wt)$	$\Lambda(f/W)$
$u(t)$	$\dfrac{1}{2}\delta(f) + \dfrac{1}{j2\pi f}$
$\mathrm{sgn}(t)$	$\dfrac{1}{j\pi f}$
$\dfrac{1}{-j\pi t}$	$\mathrm{sgn}(f)$
$t^n \mathrm{sgn} t$	$\dfrac{2n!}{(jw)^{n+1}},\quad n=1,2,\ldots$
$\dfrac{t}{\alpha^2 + t^2}$	$\dfrac{\pi}{a} e^{-\alpha|w|},\quad \mathrm{Re}\,\alpha>0$
$\dfrac{t^{n-1} e^{-\alpha t}}{(n-1)!},\ t>0$	$\dfrac{1}{(\alpha + j2\pi f)^n},\quad n=1,2,\ldots,\ \alpha\geq 0$
$e^{-\alpha t},\ t>0$	$\dfrac{1}{\alpha + j2\pi f},\quad \alpha>0$
$e^{-\alpha|t|},\ \alpha>0$	$\dfrac{2\alpha}{\alpha^2 + w^2}$
$\dfrac{1}{\sqrt{2\pi}\sigma} e^{-t^2/(2\sigma^2)}$	$e^{-\sigma^2 w^2/2}$
$e^{\pm j(w_c t + \phi)}$	$e^{\pm j\phi}\delta(f \mp f_c)$
$\cos(w_c t + \phi)$	$\dfrac{1}{2} e^{-j\phi}\delta(f + f_c) + \dfrac{1}{2} e^{j\phi}\delta(f - f_c)$
$s(t)\cos(w_c t + \phi)$	$\dfrac{1}{2} e^{-j\phi} S(f + f_c) + \dfrac{1}{2} e^{j\phi} S(f - f_c)$

Appendix D

Mathematical Tools

D.1   Trigonometric Identities

$$\cos(x \pm y) = \cos x \cos y \mp \sin x \sin y$$

$$\sin(x \pm y) = \sin x \cos y \pm \cos x \sin y$$

$$2 \cos x \cos y = \cos(x-y) + \cos(x+y)$$

$$2 \sin x \sin y = \cos(x-y) - \cos(x+y)$$

$$2 \sin x \cos y = \sin(x-y) + \sin(x+y)$$

$$\sin x \pm \sin y = 2 \sin\frac{x \pm y}{2} \cos\frac{x \mp y}{2}$$

$$\cos x + \cos y = 2 \cos\frac{x+y}{2} \cos\frac{x-y}{2}$$

$$\cos x - \cos y = 2 \sin\frac{x+y}{2} \sin\frac{x-y}{2}$$

$$\sin^2 x - \sin^2 y = \sin(x+y)\sin(x-y)$$

$$\cos^2 x - \cos^2 y = -\sin(x+y)\sin(x-y)$$

$$\cos^2 x - \sin^2 y = \cos(x+y)\cos(x-y)$$

$$= \cos^2 y - \sin^2 x$$

$$\cos(2x) = \cos^2 x - \sin^2 x = 2\cos^2 x - 1$$

$$= 1 - 2\sin^2 x = \frac{1-\tan^2 x}{1+\tan^2 x}$$

$$\sin(2x) = 2 \sin x \cos x = \frac{2 \tan x}{1+\tan^2 x}$$

$$2\cos^2 x = 1 + \cos(2x)$$

$$2\sin^2 x = 1 - \cos(2x)$$

$$\cos(x \pm \pi/2) = \mp \sin x$$

$$\sin(x \pm \pi/2) = \pm \cos x$$

$$\cos x = \left(e^{jx} + e^{-jx}\right)/2$$

$$\sin x = \left(e^{jx} - e^{-jx}\right)/(j2)$$

$$A \cos x - B \sin x = R \cos(x + \theta)$$

$$R = \sqrt{A^2 + B^2}, \quad \theta = \tan^{-1}(B/A)$$

$$A = R \cos\theta, \quad B = R \sin\theta$$

$$\tan(x \pm y) = \frac{\tan x \pm \tan y}{1 \mp \tan x \tan y}$$

$$\cot(x \pm y) = \frac{\cot x \cot y \mp 1}{\cot x \pm \cot y}$$

$$\tan x \pm \tan y = \frac{\sin(x \pm y)}{\cos x \cos y}$$

$$\cot x \pm \cot y = \frac{\sin(y \pm x)}{\sin x \sin y}$$

Digital Communications, First Edition. Mehmet Şafak.
© 2017 John Wiley & Sons Ltd. Published 2017 by John Wiley & Sons Ltd.
Companion website: www.wiley.com/go/safak/Digital_Communications

$$\tan(2x) = \frac{2\tan x}{1 - \tan^2 x}$$

$$\cot(2x) = \frac{\cot^2 x - 1}{2\cot x} = \frac{1}{2}(\cot x - \tan x)$$

$$\tan\left(\frac{x}{2}\right) = \frac{\sin x}{1 + \cos x} = \frac{1 - \cos x}{\sin x}$$

$$\cot\left(\frac{x}{2}\right) = \frac{\sin x}{1 - \cos x} = \frac{1 + \cos x}{\sin x}$$

$$\sin x = x - \frac{1}{3!}x^3 + \frac{1}{5!}x^5 - \cdots$$

$$\cos x = x - \frac{1}{2!}x^2 + \frac{1}{4!}x^4 - \cdots$$

$$\tan x = x + \frac{1}{3}x^3 + \frac{2}{15}x^5 + \cdots$$

$$\text{sinc}x = 1 - \frac{1}{3!}(\pi x)^2 + \frac{1}{5!}(\pi x)^4 - \cdots$$

$$\sin^{-1}x = x + \frac{1}{2.3}x^3 + \frac{1.3}{2.4.5}x^5 + \frac{1.3.5}{2.4.6.7}x^7 + \cdots$$

$$\cos^{-1}x$$
$$= \frac{\pi}{2} - \left(x + \frac{1}{2.3}x^3 + \frac{1.3}{2.4.5}x^5 + \frac{1.3.5}{2.4.6.7}x^7 + \cdots\right)$$

$$\tan^{-1}x = \begin{cases} x - \frac{1}{3}x^3 + \frac{1}{5}x^5 - \frac{1}{7}x^7 + \cdots & |x| < 1 \\ \frac{\pi}{2} - \frac{1}{x} + \frac{1}{3x^3} - \frac{1}{5x^5} + \cdots & x > 1 \end{cases}$$

$$\frac{d}{dx}\tan^{-1}\left(\frac{x}{a}\right) = \frac{a}{x^2 + a^2}$$

$$\sin^{-1}x = \tan^{-1}\left(\frac{x}{\sqrt{1-x^2}}\right), \quad x^2 < 1$$

$$\tan^{-1}x = \pi/2 - \tan^{-1}(1/x), \quad x > 0$$

D.2   Series

$$e^x = \sum_{k=0}^{\infty} \frac{x^k}{k!} = 1 + x + \frac{x^2}{2!} + \frac{x^3}{3!} + \cdots \qquad \text{(D.1)}$$

$$\ln(1+x) = \sum_{n=1}^{\infty} (-1)^{n-1} \frac{x^n}{n}$$
$$= x - \frac{x^2}{2} + \frac{x^3}{3} - \cdots, \quad x^2 < 1, \quad x = 1 \qquad \text{(D.2)}$$

$$\ln(1-x) \cong -\left(x + \frac{x^2}{2} + \frac{x^3}{3} + \frac{x^4}{4} + \cdots\right)$$
$$x^2 < 1, \quad x = -1 \qquad \text{(D.3)}$$

$$\ln(x) = (x-1) - \frac{(x-1)^2}{2} + \frac{(x-1)^3}{3} - \frac{(x-1)^4}{4}$$
$$+ \cdots, \quad 0 < x \le 2 \qquad \text{(D.4)}$$

$$(a+b+c)^N = \sum_{n=0}^{N} \binom{N}{n} a^{N-n}(b+c)^n$$

$$= \sum_{n=0}^{N} \binom{N}{n} a^{N-n} \sum_{\ell=0}^{n} \binom{n}{\ell} b^\ell c^{n-\ell}$$

$$= \sum_{n=0}^{N} \sum_{\ell=0}^{n} \binom{N}{n} \binom{n}{\ell} a^{N-n} b^\ell c^{n-\ell}$$
$$\text{(D.5)}$$

$$(a+b)^N = \sum_{n=0}^{N} \binom{N}{n} a^n b^{N-n} \qquad \text{(D.6)}$$

$$(1 \pm x)^N = \sum_{n=0}^{N} \binom{N}{n} (\pm x)^n$$

$$= 1 \pm Nx + \frac{N(N-1)}{2!}x^2 \qquad \text{(D.7)}$$
$$\pm \frac{N(N-1)(N-2)}{3!}x^3 + \cdots$$

$$(1+x)^{-1} = 1 - x + x^2 - x^3 + \cdots \qquad \text{(D.8)}$$

$$(1 \pm x)^{1/2} = 1 \pm \frac{1}{2}x - \frac{1.1}{2.4}x^2 \pm \frac{1.1.3}{2.4.6}x^3$$
$$- \frac{1.1.3.5}{2.4.6.8}x^4 \pm \cdots \qquad \text{(D.9)}$$

$$(1 \pm x)^{-1/2} = 1 \mp \frac{1}{2}x + \frac{1.3}{2.4}x^2 \mp \frac{1.3.5}{2.4.6}x^3$$
$$+ \frac{1.3.5.7}{2.4.6.8}x^4 \mp \cdots \qquad \text{(D.10)}$$

$$(1 - e^{-x})^N = \sum_{n=0}^{N} (-1)^n \binom{N}{n} e^{-nx} \qquad \text{(D.11)}$$

D.3   Summations

$$\sum_{n=0}^{N} x^n = 1 + x + x^2 + \cdots + x^N = \frac{1-x^{N+1}}{1-x}$$
(D.12)

$$\sum_{n=0}^{N} e^{jn\phi} = \frac{1-e^{j(N+1)\phi}}{1-e^{j\phi}} = e^{jN\phi/2} \frac{\sin((N+1)\phi/2)}{\sin(\phi/2)}$$
(D.13)

$$\sum_{n=0}^{\infty} x^n = 1 + x + x^2 + \cdots = \frac{1}{1-x}, \quad x^2 < 1 \quad (D.14)$$

$$\sum_{n=1}^{\infty} nx^n = x + 2x^2 + 3x^3 \cdots = \frac{x}{(1-x)^2} \quad (D.15)$$

$$\sum_{n=0}^{\infty} (nN+1)x^n cr = N \sum_{n=0}^{\infty} nx^n + \sum_{n=0}^{\infty} x^n$$
$$= \frac{1+(N-1)x}{(1-x)^2}$$
(D.16)

$$\binom{N}{n} = \binom{N}{N-n} = \frac{N!}{n!(N-n)!}, \quad N \geq n$$
(D.17)

$$\binom{N+1}{n+1} = \sum_{j=n}^{N} \binom{j}{n}, \quad N \geq n \qquad (D.18)$$

$$\binom{N-1}{n-1} = \sum_{j=n}^{N} (-1)^{j-n} \binom{N}{j}, \quad N \geq n \quad (D.19)$$

$$\binom{N+n}{n} = \sum_{j=0}^{n} \binom{N}{j} \binom{n}{j}, \quad N \geq n \quad (D.20)$$

$$\binom{M+N}{N-n} = \sum_{j=0}^{N-n} \binom{N}{n+j} \binom{M}{j}, \quad M \geq n, N \geq n$$
(D.21)

$$\sum_{n=0}^{N} \binom{N}{n} = 2^N \qquad (D.22)$$

$$\sum_{n=0}^{N} (-1)^n \binom{N}{n} = 0 \qquad (D.23)$$

$$\sum_{n=1}^{N} \frac{(-1)^{n+1}}{n} \binom{N}{n} = \sum_{k=1}^{N} \frac{1}{k} \qquad (D.24)$$

$$L \sum_{q=0}^{L-1} \frac{(-1)^q}{q+1} \binom{L-1}{q} = \sum_{\ell=1}^{L} (-1)^{\ell-1} \binom{L}{\ell} = 1$$
(D.25)

$$\frac{1}{2} \binom{2L}{L} = \sum_{j=0}^{L-1} \binom{j+L-1}{j} = \binom{2L-1}{L}$$
$$= \frac{1}{2} \sum_{j=0}^{L} \binom{L}{j}^2$$
(D.26)

$$\binom{2L}{L} \frac{1}{2^{2L}} = \frac{1}{\sqrt{\pi}} \frac{\Gamma(L+1/2)}{\Gamma(L+1)} \qquad (D.27)$$

$$\sum_{k=0}^{\infty} \binom{2k}{k} p^k = \frac{1}{\sqrt{1-4p}}, \quad 0 \leq p < 1/4 \quad (D.28)$$

$$\sum_{n=1}^{N} n = \frac{N(N+1)}{2} \qquad (D.29)$$

$$\sum_{n=1}^{N} n^2 = \frac{N(N+1)(2N+1)}{6} \qquad (D.30)$$

$$\sum_{n=1}^{N} n^3 = \frac{N^2(N+1)^2}{4} \qquad (D.31)$$

$$\sum_{n=1}^{N/2} (2n-1)^2 = \frac{N(N^2-1)}{6} \qquad (D.32)$$

$$\sum_{n=1}^{N} 4^{n-1} = \frac{4^N - 1}{3} \qquad (D.33)$$

$$\lim_{n \to \infty} (1-G/n)^n = e^{-G} \qquad (D.34)$$

$$\sum_{i=t+1}^{n} \frac{i}{n}\binom{n}{i} p^i (1-p)^{n-i}$$

$$= \sum_{i=t+1}^{n} \binom{n-1}{i-1} p^i (1-p)^{n-i}$$

$$= p\sum_{j=t}^{n-1} \binom{n-1}{j} p^j (1-p)^{n-1-j}$$

$$= p\underbrace{\sum_{j=0}^{n-1} \binom{n-1}{j} p^j (1-p)^{n-1-j}}_{\equiv 1}$$

$$-p\sum_{j=0}^{t-1} \binom{n-1}{j} p^j (1-p)^{n-1-j}$$

$$= p - p(1-p)^{n-1}\sum_{j=0}^{t-1} \binom{n-1}{j} \left(\frac{p}{1-p}\right)^j$$

$$\tag{D.35}$$

D.4   Integrals

D.4.1   Integrals Involving Trigonometric Functions

$$\int_{-\infty}^{\infty} e^{j(u-u_0)t}dt = \delta(u-u_0) \tag{D.36}$$

$$\int x\cos x\, dx = \cos x + x\sin x \tag{D.37}$$

$$\int x\sin x\, dx = \sin x - x\cos x \tag{D.38}$$

$$\int x^2\cos x\, dx = 2x\cos x + \left(x^2-2\right)\sin x \tag{D.39}$$

$$\int x^2\sin x\, dx = 2x\sin x - \left(x^2-2\right)\cos x \tag{D.40}$$

$$\int_0^\infty \operatorname{sinc}x\, dx = \int_0^\infty \operatorname{sinc}^2 x\, dx = 1/2 \tag{D.41}$$

$$\int \frac{\sin^2\phi}{\sin^2\phi + a}d\phi = \phi - \sqrt{\frac{a}{a+1}}\tan^{-1}\left(\sqrt{\frac{a+1}{a}}\tan\phi\right)$$

$$\tag{D.42}$$

D.4.2   Integrals Involving Algebraic Functions

$$\int \frac{dx}{a+bx}dx = \frac{1}{b}\ln|a+bx| \tag{D.43}$$

$$\int \frac{dx}{a^2+b^2x^2} = \frac{1}{ab}\tan^{-1}\left(\frac{bx}{a}\right) \tag{D.44}$$

$$\int_0^\infty \frac{x^{m-1}}{1+x^n}dx = \frac{\pi/n}{\sin(m\pi/n)}, \quad n>m>0 \tag{D.45}$$

$$\int \frac{x^2}{a^2+x^2}dx = x - \tan^{-1}\left(\frac{x}{a}\right) \tag{D.46}$$

$$\int \frac{1}{(a^2+x^2)^2}dx = \frac{x}{2a^2(a^2+x^2)} + \frac{1}{2a}\tan^{-1}\left(\frac{x}{a}\right)$$

$$\tag{D.47}$$

D.4.3   Integrals Involving Exponentials

$$\int x^{m-1}e^{-\alpha x}dx = -\frac{\Gamma(m)}{\alpha^m}e^{-\alpha x}\sum_{k=0}^{m-1}\frac{1}{k!}(\alpha x)^k$$

$$\tag{D.48}$$

$$\int_0^z x^{m-1}e^{-\alpha x}dx = \frac{\Gamma(m)}{\alpha^m}\left[1 - e^{-\alpha z}\sum_{k=0}^{m-1}\frac{(\alpha z)^k}{k!}\right]$$

$$\tag{D.49}$$

$$\int_0^\infty x^{m-1}e^{-\alpha x}dx = \frac{\Gamma(m)}{\alpha^m} \tag{D.50}$$

$$\int_0^\infty x^2 e^{-x^2}dx = \sqrt{\pi}/4 \tag{D.51}$$

$$\int_{-\infty}^\infty e^{-a^2x^2+bx}dx = \frac{\sqrt{\pi}}{a}e^{b^2/(4a^2)}, \quad a>0$$

$$\tag{D.52}$$

$$\int_0^\infty e^{-\alpha x}\cos x\, dx = \frac{\alpha}{1+\alpha^2}, \quad \alpha>0 \tag{D.53}$$

$$\int_0^\infty e^{-\alpha x} \sin x\, dx = \frac{1}{1+\alpha^2}, \quad \alpha > 0 \qquad \text{(D.54)}$$

$$\int_0^\infty e^{-\alpha^2 x^2} \cos(\beta x)\, dx = \frac{\sqrt{\pi}}{2\alpha} e^{-\beta^2/(4\alpha^2)} \qquad \text{(D.55)}$$

$$\int_0^\infty \frac{x^3}{e^{ax}-1}\, dx = \frac{\pi^4}{15a^4} \quad a > 0 \qquad \text{(D.56)}$$

D.4.4 Integrals Involving Bessel Functions

$$\int_0^\infty x \exp(-sx^2)\, I_0(ax)\, dx = \frac{1}{2s}\exp\left(\frac{a^2}{4s}\right) \qquad \text{(D.57)}$$

$$\int_0^\infty \exp(-\alpha x)\, I_\nu(\beta x)\, dx = \frac{\left[\alpha - \sqrt{\alpha^2 - \beta^2}\right]^\nu}{\beta^\nu \sqrt{\alpha^2 - \beta^2}},$$
$$\text{Re}\,\alpha > \text{Re}\,\beta, \quad \text{Re}\,\nu > -1 \qquad \text{(D.58)}$$

$$\int J_n(x)\, dx = 2\sum_{k=0}^\infty J_{n+2k+1}(x) \qquad \text{(D.59)}$$

$$\int_0^\infty e^{-sx}\, I_0\left(2\sqrt{bx}\right)\, dx = \frac{e^{b/s}}{s} \qquad \text{(D.60)}$$

$$\int_0^\infty x e^{-sx}\, I_0\left(2\sqrt{bx}\right)\, dx = \frac{e^{b/s}}{s^2}\left(1+\frac{b}{s}\right) \qquad \text{(D.61)}$$

$$\int_0^\infty x\, e^{-(x^2+a^2)/2}\, I_0(ax)\, dx = 1 \qquad \text{(D.62)}$$

$$\int_0^1 J_0(xr)\, r\, dr = J_1(x)/x \qquad \text{(D.63)}$$

$$\int J_\nu(x)\, dx = 2\sum_{k=0}^\infty J_{\nu+2k+1}(x) \qquad \text{(D.64)}$$

$$\int_0^\infty x^{\nu-1} \exp\left(-\alpha x - \frac{\beta}{x}\right)\, dx = 2\left(\frac{\beta}{\alpha}\right)^{\nu/2}$$
$$K_\nu\left(2\sqrt{\alpha\beta}\right), \quad \text{Re}\,\alpha > 0,\ \text{Re}\,\beta > 0 \qquad \text{(D.65)}$$

$$\int_0^\infty x^\mu\, K_\nu(\alpha x)\, dx = 2^{\mu-1}\alpha^{-\mu-1}\Gamma\left(\frac{1+\mu+\nu}{2}\right)$$
$$\Gamma\left(\frac{1+\mu-\nu}{2}\right), \quad \text{Re}(\mu+1\pm\nu) > 0,\ \text{Re}\,\alpha > 0 \qquad \text{(D.66)}$$

$$\int_0^\infty x^{\mu-1} e^{-px} K_\nu(qx)\, dx$$
$$= \frac{p^{\nu-\mu}\sqrt{\pi}\,\Gamma(\mu-\nu)\Gamma(\mu+\nu)}{2^\mu q^\nu\, \Gamma(\mu+1/2)} \times$$
$$_2F_1\left(\frac{\mu-\nu}{2}, \frac{\mu-\nu+1}{2}; \mu+\frac{1}{2}; 1-q^2/p^2\right) \qquad \text{(D.67)}$$

$$\int_0^{\sqrt{\gamma_{th}}} \frac{2c}{\Gamma(m)\Gamma(m_s)}(c\rho)^{\frac{m+m_s-2}{2}}$$
$$K_{m_s-m}\left(2\sqrt{c\rho}\right) d\rho$$
$$= \frac{\Gamma(m-m_s)}{\Gamma(m)\Gamma(m_s+1)}(c\gamma_{th})^{m_s}$$
$$_1F_2(m_s; m_s-m+1, m_s+1; c\gamma_{th})$$
$$+ \frac{\Gamma(m_s-m)}{\Gamma(m_s)\Gamma(m+1)}(c\gamma_{th})^m$$
$$_1F_2(m; m-m_s+1, m+1; c\gamma_{th}) \qquad \text{(D.68)}$$

$$\frac{1}{2}\int_0^\infty e^{-\alpha\gamma} \frac{2c}{\Gamma(m)\Gamma(m_s)}(c\rho)^{\frac{m+m_s-2}{2}} K_{m_s-m}\left(2\sqrt{c\rho}\right) d\rho$$
$$= \frac{1}{2}(c/\alpha)^{\frac{m+m_s-1}{2}} e^{\frac{c}{2a}} M_{-\frac{m+m_s-1}{2}, \frac{m_s-m}{2}}(c/\alpha)$$
$$= \frac{1}{2}(c/\alpha)^{m_s}\frac{\Gamma(m-m_s)}{\Gamma(m)}\, _1F_1(m_s; m_s-m+1; c/\alpha)$$
$$+ \frac{1}{2}(c/\alpha)^m\frac{\Gamma(m_s-m)}{\Gamma(m_s)}\, _1F_1(m; m-m_s+1; c/\alpha) \qquad \text{(D.69)}$$

D.4.5 Integrals Involving Gaussian Q Function [1]

$$P_2(L,\bar\gamma) = \int_0^\infty Q\left(\sqrt{\beta\gamma}\right)\frac{1}{\Gamma(L)}\left(\frac{\gamma}{\bar\gamma}\right)^{L-1}\frac{1}{\bar\gamma}e^{-\gamma/\bar\gamma}d\gamma$$

$$= \frac{1}{2} - \frac{1}{2\sqrt\pi}\sqrt{\frac{\beta\bar\gamma}{2+\beta\bar\gamma}}\sum_{k=0}^{L-1}\frac{\Gamma(k+1/2)}{\Gamma(k+1)}\left(\frac{2}{2+\beta\bar\gamma}\right)^k$$

$$= \frac{1}{2} - \frac{1}{2}\sqrt{\frac{\beta\bar\gamma}{2+\beta\bar\gamma}}\sum_{k=0}^{L-1}\binom{2k}{k}\frac{1}{[2(2+\beta\bar\gamma)]^k}$$

$$= \left[\frac{1-\mu}{2}\right]^L\sum_{k=0}^{L-1}\binom{L-1+k}{k}\left[\frac{1+\mu}{2}\right]^k,$$

$$\mu = \sqrt{\frac{\beta\bar\gamma}{2+\beta\bar\gamma}}$$

$$\text{(D.70)}$$

$$I(n) = \int_0^\infty Q^n\left(\sqrt{\beta\gamma}\right)\frac{1}{\Gamma(L)}\left(\frac{\gamma}{\bar\gamma}\right)^{L-1}\frac{1}{\bar\gamma}e^{-\gamma/\bar\gamma}d\gamma$$

$$= \begin{cases} \dfrac{1}{2} - \dfrac{1}{2}\sqrt{\dfrac{\beta\bar\gamma}{2+\beta\bar\gamma}}\displaystyle\sum_{k=0}^{L-1}\binom{2k}{k}\dfrac{1}{[2(2+\beta\bar\gamma)]^k} & n=1 \\[2ex] I(1) - \dfrac{1}{4} + \dfrac{1}{2\pi}\displaystyle\sum_{k=0}^{L-1}\dfrac{\beta\bar\gamma}{(1+\beta\bar\gamma)^{k+1}} & \\[2ex] \quad {}_2F_1\left(k+1,1;\dfrac{3}{2};\dfrac{\beta\bar\gamma}{2+2\beta\bar\gamma}\right) & n=2 \end{cases}$$

$$\text{(D.71)}$$

D.4.6 Integrals Containing Logarithms [2][3]

$$\int e^{-ax}\ln(1+x)\ dx$$

$$\text{(D.72)}$$

$$= -\frac{1}{a}\left[\ln(1+x)e^{-ax} + e^a E_1(a(x+1))\right]$$

$$\int_0^\infty e^{-x}\ln(1+ax)\ dx = e^{1/a}E_1(1/a) \quad \text{(D.73)}$$

$$\int_0^\infty \ln(1+ax)x^n e^{-x}dx = \sum_{k=0}^n \frac{n!}{k!(-a)^k}$$

$$\left[e^{1/a}E_1(1/a) + \sum_{i=1}^k (i-1)!(-a)^i\right]$$

$$\text{(D.74)}$$

$$\int_0^\infty (1+ax)^m x^n\ e^{-x/b}dx = \frac{n!\ \Gamma(-1-n-m)}{a^{n+1}\ \Gamma(-m)}$$

$${}_1F_1(1+n;\ n+m+2;\ 1/ab)$$

$$+ b^{n+m+1}a^m\ \Gamma(1+n+m)$$

$${}_1F_1(-m;\ -n-m;\ 1/ab)$$

$$\text{Re}\{b\} > 0,\quad n\ge 0,\quad \arg\{a\}\ne\pi \qquad \text{(D.75)}$$

D.5 Useful Relations

D.5.1 Base Changes in Logarithms

$$\ln x = \ln b\ \log_b x$$

$$\ln 2 = 1/\log_2 e = 0.693,\quad \ln 10 = 1/\log_{10} e = 2.3026$$

$$\text{(D.76)}$$

D.5.2 Schwartz's Inequality

$$\left|\int x(t)y(t)dt\right|^2 \le \int |x(t)|^2 dt \int |y(t)|^2 dt$$

$$\left|\sum_i x_i y_i\right|^2 \le \sum_i |x_i|^2 \sum_j |y_j|^2$$

$$\text{(D.77)}$$

Equality holds when $y(t) = kx^*(t)$ where k is a constant.

D.5.3 Poisson's Sum Formula

$$\frac{1}{N}\sum_{n=-\infty}^\infty e^{\pm j2\pi f n/N} = \sum_{n=-\infty}^\infty \delta(f-nN) \quad \text{(D.78)}$$

D.5.4 Stirling's Approximation

$$k! \cong \sqrt{2\pi k}\,k^k e^{-k} \qquad \text{(D.79)}$$

D.6 Functions

D.6.1 Q Function

$$Q(x) = \frac{1}{\sqrt{2\pi}}\int_x^\infty e^{-t^2/2}dt = \frac{1}{\pi}\int_0^{\pi/2}\exp\left(-\frac{x^2}{2\sin^2\theta}\right)d\theta$$

$$\text{(D.80)}$$

D.6.2   Marcum Q Function

$$Q_m(\alpha,\beta) = \int_\beta^\infty x \left(\frac{x}{\alpha}\right)^{m-1} \exp\left(-\frac{x^2+\alpha^2}{2}\right) I_{m-1}(\alpha x) dx$$

$$= Q_1(\alpha,\beta) + \exp\left(\frac{\alpha^2+\beta^2}{2}\right) \sum_{k=1}^{m-1} \left(\frac{\beta}{\alpha}\right)^k I_k(\alpha\beta)$$

$$(D.81)$$

D.6.3   Gauss Hypergeometric Function

$$F(a,b;c;z) = {}_2F_1(a,b;c;z) = \sum_{n=0}^\infty \frac{(a)_n (b)_n}{(c)_n} \frac{z^n}{n!}$$

$$= \frac{\Gamma(c)}{\Gamma(a)\Gamma(b)} \sum_{n=0}^\infty \frac{\Gamma(a+n)\Gamma(b+n)}{\Gamma(c+n)} \frac{z^n}{n!}$$

$$= 1 + \frac{ab}{c}z + \frac{a(a+1)b(b+1)}{c(c+1)} \frac{z^2}{2!} + \cdots, \quad |z| < 1$$

$$(D.82)$$

where a, b, c may assume complex values, $c \neq 0,-1,-2,\ldots$ and

$$(a)_n = \Gamma(a+n)/\Gamma(a) \qquad (D.83)$$

$$F(a,b;c;z) = F(b,a;c;z)$$

$$= (1-z)^{-a} F\left(a,c-b;c;\frac{z}{z-1}\right)$$

$$= (1-z)^{-b} F\left(c-a,b;c;\frac{z}{z-1}\right)$$

$$= (1-z)^{c-a-b} F(c-a,c-b;c;z)$$

$$(D.84)$$

$${}_1F_2(a;b;c;z) = \sum_{n=0}^\infty \frac{(a)_n}{(b)_n(c)_n} \frac{z^n}{n!} = \frac{\Gamma(b)\Gamma(c)}{\Gamma(a)}$$

$$\times \sum_{n=0}^\infty \frac{\Gamma(a+n)}{\Gamma(b+n)\Gamma(c+n)} \frac{z^n}{n!}$$

$$(D.85)$$

$${}_0F_1(a;x) = \sum_{n=0}^\infty \frac{1}{(a)_n} \frac{z^n}{n!}$$

$$= \sum_{n=0}^\infty \frac{\Gamma(a)}{\Gamma(a+n)} \frac{z^n}{n!}$$

$$= 1 + \frac{1}{a}z + \frac{1}{a(a+1)} \frac{z^2}{2!} + \cdots, \quad |z| < 1$$

$$(D.86)$$

$${}_2F_1\left(1,1;\frac{3}{2};z^2\right) = \frac{\sin^{-1}(z)}{\sqrt{z^2(1-z^2)}}$$

$$= \frac{1}{\sqrt{z^2(1-z^2)}} \tan^{-1}\left(\sqrt{\frac{z}{1-z^2}}\right), \quad z^2 < 1$$

$${}_2F_1\left(\frac{1}{2},1;\frac{5}{2};0\right) = 1$$

$${}_2F_1\left(\frac{1}{2},1;\frac{5}{2};1\right) = \frac{3}{2}$$

$$(D.87)$$

D.6.4   Confluent Hypergeometric Function

$${}_1F_1(a;b;z) = \sum_{n=0}^\infty \frac{(a)_n z^n}{(b)_n n!}$$

$$= \frac{\Gamma(b)}{\Gamma(a)} \sum_{n=0}^\infty \frac{\Gamma(a+n)}{\Gamma(b+n)} \frac{z^n}{n!}$$

$$= 1 + \frac{a}{b}z + \frac{a(a+1)}{b(b+1)} \frac{z^2}{2!} + \cdots, \quad |z| < 1$$

$$(D.88)$$

$${}_1F_1\left(1,\frac{1}{2};-a\right) = -e^a \sum_{k=0}^\infty \frac{a^k}{k!(2k-1)} \qquad (D.89)$$

$${}_1F_1(n,n;z) = e^z \qquad (D.90)$$

D.6.5   Bessel Functions

$$J_\alpha(x) = (x/2)^\alpha \sum_{k=0}^\infty (-1)^k \frac{(x/2)^{2k}}{k!\,\Gamma(\alpha+k+1)}, \quad |\arg x| < \pi$$

$$(D.91)$$

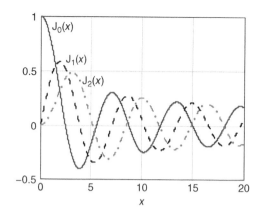

Figure D.1 Bessel Function of the First Kind $J_n(x)$.

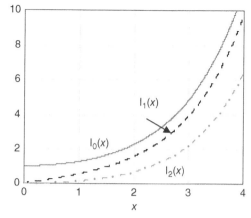

Figure D.2 Modified Bessel Function of the First Kind $I_n(x)$.

$$J_n(x) = \frac{1}{\pi} \int_0^\pi \cos(n\theta - x \sin \theta) \, d\theta$$
$$= \frac{1}{j^n \pi} \int_0^\pi e^{jx\cos\theta} \cos(n\theta) \, d\theta, \quad n=0,1,2,\cdots$$
(D.92)

$$J_{-n}(x) = (-1)^n J_n(x), \quad n=0,1,2,\cdots \quad \text{(D.93)}$$

$$J_n(x) \approx \begin{cases} \dfrac{(x/2)^n}{\Gamma(n+1)}, & n=0,1,2\cdots \ x\ll 1 \\[2mm] \sqrt{\dfrac{2}{\pi x}} \cos\left(x - \dfrac{\pi}{4} - \dfrac{n\pi}{2}\right) & x\gg 1 \end{cases}$$
(D.94)

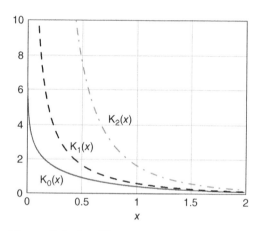

Figure D.3 Modified Bessel Function of the Second Kind $K_n(x)$.

$$I_\alpha(x) = (x/2)^\alpha \sum_{k=0}^\infty \frac{(x/2)^{2k}}{k!\,\Gamma(\alpha+k+1)}, \quad x \geq 0$$
(D.95)

$$I_\alpha(x) = \frac{1}{\pi}\int_0^\pi e^{x\cos\theta}\cos(\alpha\theta)\,d\theta$$
$$\qquad - \frac{\sin(\alpha\pi)}{\pi}\int_0^\infty e^{-\alpha t - x\cosh t}\,dt$$
(D.96)

$$I_n(x) = \frac{(x/2)^n}{\sqrt{\pi}\,\Gamma(n+1/2)}\int_0^\pi e^{\pm x\cos\theta}\sin^{2n}(\alpha\theta)\,d\theta$$
(D.97)

$$I_{-n}(x) = I_n(x) \qquad \text{(D.98)}$$

$$I_n(x) \approx \begin{cases} \dfrac{1}{n!}\left(\dfrac{x}{2}\right)^n & x^2 \ll 1 \\[2mm] \dfrac{e^x}{\sqrt{2\pi x}} & x \gg 1 \end{cases}$$
(D.99)

$$K_\alpha(x) = \frac{\pi}{2}\frac{I_{-\alpha}(x) - I_\alpha(x)}{\sin(\alpha\pi)}$$
$$= \int_0^\infty e^{-x\cosh t}\cosh(\alpha t)\,dt, \quad |\arg x| < \pi/2$$
(D.100)

$$K_{-\alpha}(x) = K_\alpha(x) \qquad \text{(D.101)}$$

$$K_0(x) \sim -\ln x, \quad x \to 0 \qquad \text{(D.102)}$$

$$K_\alpha(x) \sim \frac{1}{2}\Gamma(\alpha)\left(\frac{x}{2}\right)^{-\alpha}, \quad \mathrm{Re}(\alpha) > 0, \ \alpha : \text{fixed},$$

$$x \to 0$$

$$\text{(D.103)}$$

$$K_\alpha(x) \approx \sqrt{\frac{\pi}{2x}}e^{-x}, \quad \mathrm{Re}(\alpha) > 0, \ \alpha : \text{fixed},$$

$$x \to \infty$$

$$\text{(D.104)}$$

$$K'_n(x) = -K_{n-1}(x) - \frac{n}{x}K_n(x) \qquad \text{(D.105)}$$

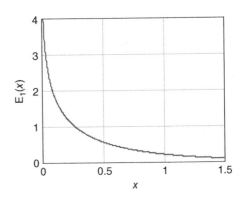

Figure D.4 Exponential Integral $E_1(x)$.

D.6.6 Gamma Function

$$\gamma(\alpha,z) = \int_0^z t^{\alpha-1}e^{-t}\,dt, \quad \mathrm{Re}(\alpha) \ge 0 \qquad \text{(D.106)}$$

$$\Gamma(\alpha,z) = \int_z^\infty t^{\alpha-1}e^{-t}\,dt, \quad \mathrm{Re}(\alpha) \ge 0 \qquad \text{(D.107)}$$

$$\Gamma(\alpha) = \Gamma(\alpha,0) = \int_0^\infty t^{\alpha-1}e^{-t}\,dt, \quad \mathrm{Re}(\alpha) \ge 0$$

$$\text{(D.108)}$$

$$\Gamma(\alpha,z) = \Gamma(\alpha) - \gamma(\alpha,z) \qquad \text{(D.109)}$$

$$\Gamma(\alpha,z) = \sum_{k=0}^\infty \frac{(-1)^k z^{k+\alpha}}{k!(k+\alpha)}, \quad n = 1,2,\cdots$$

$$\text{(D.110)}$$

$$\gamma(n,z) = \Gamma(n)\left[1 - e^{-z}\sum_{k=0}^{n-1}\frac{z^k}{k!}\right], \quad n = 1,2,\cdots$$

$$\text{(D.111)}$$

$$\Gamma(n) = n\,\Gamma(n-1) = (n-1)! \quad n = 1,2,3,\cdots$$

$$\text{(D.112)}$$

$$\Gamma(1/2) = \sqrt{\pi},$$
$$\Gamma(3/2) = \sqrt{\pi}/2, \qquad \text{(D.113)}$$
$$\Gamma(1/3) = 2.6789$$

$$\Gamma(n+1/2) = \frac{1 \cdot 3 \cdot 5 \ldots (2n-1)}{2^n}\sqrt{\pi}, \quad n = 1,2,\cdots$$

$$\text{(D.114)}$$

D.6.7 Exponential Integral

$$E_n(z) = \int_1^\infty \frac{e^{-zt}}{t^n}\,dt = z^{n-1}\int_z^\infty \frac{e^{-t}}{t^n}\,dt, \quad n = 0,1,2,\cdots$$

$$\text{(D.115)}$$

$$E_1(z) = -C - \ln z - \sum_{n=1}^\infty \frac{(-z)^n}{nn!}, \quad |\arg z| < \pi$$

$$C = 0.5772156649$$

$$\text{(D.116)}$$

$$E_n(z) = z^{n-1}\,\Gamma(1-n,z) \qquad \text{(D.117)}$$

$$E_n(z) \sim e^{-z}/z, \quad z \to \infty \qquad \text{(D.118)}$$

D.6.8 Sine and Cosine Integrals

$$\mathrm{Si}(x) = \int_0^x \frac{\sin t}{t}\,dt \qquad \text{(D.119)}$$

$$\mathrm{Ci}(x) = \int_\infty^x \frac{\cos t}{t}\,dt = -\int_x^\infty \frac{\cos t}{t}\,dt$$

$$\text{(D.120)}$$

$$= C + \ln x + \int_0^x \frac{\cos t - 1}{t}\,dt$$

$$C = 0.577215665$$

$$\lim_{x \to \infty} \mathrm{Si}(x) = \pi/2, \quad \lim_{x \to \infty} \mathrm{Ci}(x) = 0 \qquad \text{(D.121)}$$

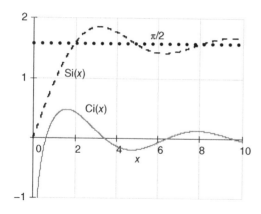

Figure D.5 Sine and Cosine Integrals.

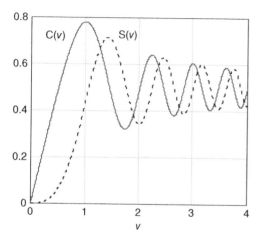

Figure D.6 Fresnel integrals.

D.6.9   Fresnel Integrals

$$F(v) = \int_0^v \exp(-j\pi t^2/2)\,dt = C(v) - jS(v)$$

$$C(v) = \int_0^v \cos(\pi t^2/2)\,dt$$

$$S(v) = \int_0^v \sin(\pi t^2/2)\,dt$$

(D.122)

$$F(0) = 0, \quad F(\infty) = (1-j)/2 \qquad \text{(D.123)}$$

Example D.1   Asymptotic Behavior of Fresnel Integrals.

The Fresnel integrals can be written in terms of the auxiliary functions $f(v)$ and $g(v)$ as follows: [4]

$$C(v) = \frac{1}{2} + f(v)\,\sin\left(\frac{\pi}{2}v^2\right) - g(v)\,\cos\left(\frac{\pi}{2}v^2\right)$$

$$S(v) = \frac{1}{2} - f(v)\,\cos\left(\frac{\pi}{2}v^2\right) - g(v)\,\sin\left(\frac{\pi}{2}v^2\right)$$

(D.124)

The asymptotic behavior of the auxiliary functions is given by: [4]

$$f(v),\ g(v) \sim \frac{1}{\pi v}, \quad v \to \infty \qquad \text{(D.125)}$$

Inserting (D.125) into (D.124), one determines the asymptotic behavior of the Fresnel integrals for $v \to \infty$:

$$C(v) = \frac{1}{2} + \frac{1}{\pi v}\left[\sin\left(\frac{\pi}{2}v^2\right) - \cos\left(\frac{\pi}{2}v^2\right)\right]$$

$$= \frac{1}{2} - \frac{\sqrt{2}}{\pi v}\,\cos\left(\frac{\pi}{2}v^2 + \frac{\pi}{4}\right), \quad v \to \infty$$

$$S(v) = \frac{1}{2} - \frac{1}{\pi v}\left[\cos\left(\frac{\pi}{2}v^2\right) + \sin\left(\frac{\pi}{2}v^2\right)\right]$$

$$= \frac{1}{2} - \frac{\sqrt{2}}{\pi v}\,\cos\left(\frac{\pi}{2}v^2 - \frac{\pi}{4}\right), \quad v \to \infty$$

(D.126)

Consequently, the asymptotic behavior of following expression can be written as

$$\frac{2}{(0.5 - C(v))^2 + (0.5 - S(v))^2} = (\pi v)^2, \quad v \to \infty$$

(D.127)

D.6.10   Sinc Function

$$\text{sinc}(x) = \text{sinc}(-x) = \frac{\sin(\pi x)}{\pi x}, \quad -\infty < x < \infty$$

(D.128)

Table D.1 Sinc(x) Function For the Values of $0 < x < 4.09$.

	0	0.01	0.02	0.03	0.04	0.05	0.06	0.07	0.08	0.09
0.0	1	1	0.999	0.999	0.997	0.996	0.994	0.992	0.99	0.987
0.1	0.984	0.98	0.976	0.972	0.968	0.963	0.958	0.953	0.948	0.942
0.2	0.935	0.929	0.922	0.915	0.908	0.9	0.892	0.884	0.876	0.867
0.3	0.858	0.849	0.84	0.83	0.82	0.81	0.8	0.79	0.779	0.768
0.4	0.757	0.746	0.734	0.722	0.711	0.699	0.687	0.674	0.662	0.649
0.5	0.637	0.624	0.611	0.598	0.585	0.572	0.558	0.545	0.532	0.518
0.6	0.505	0.491	0.477	0.464	0.45	0.436	0.423	0.409	0.395	0.382
0.7	0.368	0.354	0.341	0.327	0.314	0.3	0.287	0.273	0.26	0.247
0.8	0.234	0.221	0.208	0.195	0.183	0.17	0.158	0.145	0.133	0.121
0.9	0.109	0.098	0.086	0.075	0.063	0.052	0.042	0.031	0.02	0.01
1.0	0	-9,90E-3	-0.02	-0.029	-0.038	-0.047	-0.056	-0.065	-0.073	-0.081
1.1	-0.089	-0.097	-0.105	-0.112	-0.119	-0.126	-0.132	-0.138	-0.145	-0.15
1.2	-0.156	-0.161	-0.166	-0.171	-0.176	-0.18	-0.184	-0.188	-0.192	-0.195
1.3	-0.198	-0.201	-0.204	-0.206	-0.208	-0.21	-0.212	-0.213	-0.214	-0.215
1.4	-0.216	-0.217	-0.217	-0.217	-0.217	-0.217	-0.216	-0.216	-0.215	-0.214
1.5	-0.212	-0.211	-0.209	-0.207	-0.205	-0.203	-0.2	-0.198	-0.195	-0.192
1.6	-0.189	-0.186	-0.183	-0.179	-0.176	-0.172	-0.168	-0.164	-0.16	-0.156
1.7	-0.151	-0.147	-0.143	-0.138	-0.133	-0.129	-0.124	-0.119	-0.114	-0.109
1.8	-0.104	-0.099	-0.094	-0.089	-0.083	-0.078	-0.073	-0.068	-0.062	-0.057
1.9	-0.052	-0.046	-0.041	-0.036	-0.031	-0.026	-0.02	-0.015	-0.01	-5,02E-3
2.0	0	4,97E-3	9,89E-3	0.015	0.02	0.024	0.029	0.034	0.038	0.042

(continued overleaf)

Table D.1 (*Continued*)

	0	0.01	0.02	0.03	0.04	0.05	0.06	0.07	0.08	0.09
2.1	0.047	0.051	0.055	0.059	0.063	0.067	0.071	0.075	0.078	0.082
2.2	0.085	0.088	0.091	0.094	0.097	0.1	0.103	0.105	0.108	0.11
2.3	0.112	0.114	0.116	0.118	0.119	0.121	0.122	0.123	0.124	0.125
2.4	0.126	0.127	0.127	0.128	0.128	0.128	0.128	0.128	0.128	0.128
2.5	0.127	0.127	0.126	0.125	0.124	0.123	0.122	0.121	0.119	0.118
2.6	0.116	0.115	0.113	0.111	0.109	0.107	0.105	0.103	0.1	0.098
2.7	0.095	0.093	0.09	0.087	0.085	0.082	0.079	0.076	0.073	0.07
2.8	0.067	0.064	0.06	0.057	0.054	0.051	0.047	0.044	0.041	0.037
2.9	0.034	0.031	0.027	0.024	0.02	0.017	0.013	0.01	6,71E-03	3,34E-03
3.0	0	-3,32E-3	-6,62E-3	-9,89E-3	-0.013	-0.016	-0.019	-0.023	-0.026	-0.029
3.1	-0.032	-0.035	-0.038	-0.04	-0.043	-0.046	-0.049	-0.051	-0.054	-0.056
3.2	-0.058	-0.061	-0.063	-0.065	-0.067	-0.069	-0.071	-0.073	-0.075	-0.076
3.3	-0.078	-0.08	-0.081	-0.082	-0.084	-0.085	-0.086	-0.087	-0.088	-0.088
3.4	-0.089	-0.09	-0.09	-0.091	-0.091	-0.091	-0.091	-0.091	-0.091	-0.091
3.5	-0.091	-0.091	-0.09	-0.09	-0.089	-0.089	-0.088	-0.087	-0.086	-0.085
3.6	-0.084	-0.083	-0.082	-0.08	-0.079	-0.078	-0.076	-0.075	-0.073	-0.071
3.7	-0.07	-0.068	-0.066	-0.064	-0.062	-0.06	-0.058	-0.056	-0.054	-0.051
3.8	-0.049	-0.047	-0.045	-0.042	-0.04	-0.038	-0.035	-0.033	-0.03	-0.028
3.9	-0.025	-0.023	-0.02	-0.018	-0.015	-0.013	-0.01	-7,55E-3	-5,02E-3	-2,51E-3
4.0	0	2,49E-3	4,97E-3	7,43E-3	9,88E-03	0.012	0.015	0.017	0.019	0.022

References

[1] R. M. Radayeh and M. M. Matalgah, Results for integrals involving m-th power of the Gaussian Q-function over Rayleigh fading channels with applications, ICC 2007, pp. 5910–5914.

[2] L.S. Gradshteyn and I.M. Ryzhik, *Table of Integrals, Series and Products*, Academic Press: San Diego, 2000.

[3] W.-C. Jeong, J.-M. Chung, and D. Liu, Characteristic-Function-Based Analysis of MIMO Systems Applying Macroscopic Selection Diversity in Mobile Communications, *ETRI Journal*, vol. **30**, no. 3, pp. 355–364, June 2008.

[4] M. Abramowitz and I. A. Stegun (eds.), *Handbook of Mathematical Tables*, Dover Publications: New York, 1970.

Appendix E

The Wishart Distribution

The Wishart distribution may be considered as a matrix generalization of the chi-squared distribution, or, in the case of non-integer degrees of freedom, of the gamma distribution. It is named in honor of John Wishart, who first formulated the distribution in 1928. It is a family of probability distributions defined over symmetric, positive-definite matrix-valued random variables.

E.1 Introduction to Wishart Distribution

Chi-square distribution is widely encountered in telecommunication applications where the powers/SNRs of multiple Gaussian signals are summed for receive diversity with MRC or transmit diversity with MRT (see Appendix F and Chapter 12). If $x_1, x_2, ..., x_n$ denote real-valued independent Gaussian random variables with identical variances but different means $\mu_i, \ i = 1, 2, ..., n$, then $\sum_{i=1}^{n} x_i^2$ has a chi-square distribution with n degrees of freedom. If $\mu_i = 0, \ i = 1, 2, ..., n$, the chi-square distribution is said to be central, otherwise non-central. The corresponding pdf's are given by (F.111) and (F.120), respectively.

Now consider n random vectors $\mathbf{a}_i, \ i = 1, 2, ..., n$ each with q components. Each component a_{ij} is assumed to be a real-valued Gaussian random variable, described by zero-mean and variance σ^2:

$$\mathbf{a}_i = \begin{pmatrix} a_{1i} & a_{2i} & ... & a_{qi} \end{pmatrix}^T \sim N_q(0, \Sigma), \ i = 1, 2, ..., n \tag{E.1}$$

The $q \times q$ covariance matrix Σ is defined by

$$\Sigma = E\left[\mathbf{a}_i \mathbf{a}_i^H\right], \ i = 1, 2, ..., n \tag{E.2}$$

Suppose $\mathbf{A}$ is a $q \times n$, with $q \leq n$, complex matrix, which consists of q rows and n columns. Each of the n columns $\mathbf{a}_i, \ i = 1, 2, ..., n$ of $\mathbf{A}$ is independently drawn from a q-variate normal distribution with mean vector $\mathbf{\mu}$. This implies that the elements of the column vectors $\mathbf{a}_i$ of length q may be correlated with each other while the n column vectors are mutually independent. The definition of the $q \times q$ positive semidefinite matrix $\mathbf{W} = \mathbf{A}\mathbf{A}^H$ is used to represent the Wishart distribution with covariance matrix $\mathbf{\Sigma}$ and n orders of freedom:

$$\mathbf{W} = \mathbf{A}\mathbf{A}^H = \sum_{i=1}^{n} \mathbf{a}_i \mathbf{a}_i^H$$

$$= \begin{bmatrix} \mathbf{a}_1 & \mathbf{a}_2 & \cdots & \mathbf{a}_n \end{bmatrix} \begin{bmatrix} \mathbf{a}_1^H \\ \mathbf{a}_2^H \\ \vdots \\ \mathbf{a}_n^H \end{bmatrix} \sim W_q(n, \mathbf{\Sigma})$$

(E.3)

If the elements of the matrix $\mathbf{A}$ have zero-means, that is, $\boldsymbol{\mu} = 0$, then the $q \times q$ Hermitian matrix is called *central Wishart* with n orders of freedom. For $\boldsymbol{\mu} \neq 0$, it is called the *noncentral Wishart*. For example, a MIMO channel undergoing Rayleigh fading is characterized by the central Wishart distribution, whereas the same MIMO channel suffering Rician fading is categorized as noncentral Wishart. The covariance matrix $\mathbf{\Sigma} = \sigma^2 \mathbf{I}$ corresponds to *uncorrelated Wishart distribution* while $\mathbf{\Sigma} \neq \sigma^2 \mathbf{I}$ is associated with *correlated Wishart distribution*.

Some important properties of the complex Wishart distribution are listed below: [1][2]

a. The mean value of $\mathbf{W} = \mathbf{A}\mathbf{A}^H$ is given by

$$E[\mathbf{W}] = E\left[\sum_{i=1}^{n} \mathbf{a}_i \mathbf{a}_i^H\right] = n\mathbf{\Sigma} \qquad \text{(E.4)}$$

b. Moment generating function of $\mathbf{W} \sim W_q(n, \mathbf{\Sigma})$, $n \geq q$, $\mathbf{\Sigma} > 0$:

$$\mathbf{M}_\mathbf{W}(\mathbf{S}) = E\left[e^{-tr(\mathbf{S}\mathbf{W})}\right] = |\mathbf{I} + 2\mathbf{S}\mathbf{\Sigma}|^{-n/2} \quad \text{(E.5)}$$

c. If $\mathbf{V}_i$'s are independent and $\mathbf{V}_i \sim W_q(n_i, \mathbf{\Sigma})$, $i = 1, 2, \ldots, k$, then $\sum_{i=1}^{k} \mathbf{V}_i \sim W_q\left(\sum_{i=1}^{k} n_i, \mathbf{\Sigma}\right)$.

d. If $\mathbf{V} \sim W_q(n, \mathbf{\Sigma})$, $n \geq q$, $\mathbf{\Sigma} > 0$, then $|\mathbf{V}| \sim$ $|\mathbf{\Sigma}| \prod_{i=1}^{q} \chi^2_{n-q+i}$, that is, $|\mathbf{V}|/|\mathbf{\Sigma}|$ is distributed as a product of q mutually independent chi-square variables.

e. If $\mathbf{W} = \mathbf{A}\mathbf{A}^H \sim W_q(n, \mathbf{\Sigma})$ and $\mathbf{Q}$ is any $q \times r$ matrix, then $\mathbf{Q}^H \mathbf{W} \mathbf{Q} \sim W_r(n, \mathbf{Q}^H \mathbf{\Sigma} \mathbf{Q})$. As a consequence of this property, one can write $\mathbf{\Sigma}^{-1/2} \mathbf{W} \mathbf{\Sigma}^{-1/2} \sim W_q(n, \mathbf{I})$.

f. If $\mathbf{W} = \mathbf{A}\mathbf{A}^H \sim W_q(n, \mathbf{I})$, then $tr \ \mathbf{W} = tr\mathbf{A}\mathbf{A}^H$

$$\triangleq tr \sum_{j=1}^{n} \mathbf{a}_j \mathbf{a}_j^H = \sum_{j=1}^{n} \mathbf{a}_j^H \mathbf{a}_j$$

$$= \sum_{i=1}^{q} \sum_{j=1}^{n} |a_{ij}|^2 \sim \chi^2_{nq}$$

g. If $\mathbf{W} = \mathbf{A}\mathbf{A}^H \sim W_q(n, \mathbf{\Sigma})$, $n \geq q$, then the joint pdf of the *unordered* eigenvalues of $\mathbf{W}$ is

$$f_\mathbf{W}(\mathbf{W}) = \frac{1}{2^{nq/2} \Gamma_q(n/2) |\mathbf{\Sigma}|^{n/2}} |\mathbf{W}|^{(n-q-1)/2}$$

$$\exp\left[-\frac{1}{2} tr(\mathbf{\Sigma}^{-1} \mathbf{W})\right], \quad \mathbf{W} > 0$$

(E.6)

where

$$\Gamma_q(u) = \pi^{q(q-1)/4} \prod_{i=1}^{q} \Gamma(u - (i-1)/2), \quad u > (q-1)/2$$

(E.7)

To become more familiar with the Wishart distribution, let us first consider the case for $q = 1$, which corresponds to the vector $\mathbf{a}_i$ and the matrices $\mathbf{\Sigma}$ and $\mathbf{W}$ with single elements. For $n = q = 1$, one has

$$\mathbf{a}_1 = (a_{11})$$

$$\mathbf{\Sigma} = E[a_{11}^2] = \sigma^2 \qquad \text{(E.8)}$$

$$\mathbf{W} = \left[|a_{11}|^2\right]$$

The Wishart distribution $W_1(1, \sigma^2)$, corresponding to $n = q = 1$, has a single degree of freedom and is obtained from (E.6):

$$f_W(\gamma) = \frac{1}{\sqrt{\pi \bar{\gamma} \gamma}} \exp\left(-\frac{\gamma}{\bar{\gamma}}\right), \quad \gamma \geq 0$$

$$\bar{\gamma} = 2\sigma^2$$

$$\text{(E.9)}$$

which is identical to (F.107), as expected.

Consider now the special case $q = 1$ and $n \geq 1$, where $E\left[|a_{1i}|^2\right] = \sigma^2$, $i = 1, 2, \ldots, n$. The vector $\mathbf{a}_i$ and the matrices $\boldsymbol{\Sigma}$ and $\mathbf{W}$ for the Wishart distribution $W_1(n, \sigma^2 \mathbf{I})$ with n degrees of freedom has still single elements. Then,

$$\mathbf{W} = \sum_{j=1}^{n} |a_{1j}|^2 \qquad \text{(E.10)}$$

represents the sum of the squares of n independent $N(0, \sigma^2)$ Gaussian random variables. The pdf corresponding to $W_1(n, \sigma^2 \mathbf{I})$ is obtained from (E.6) as

$$f_W(\gamma) = \frac{1}{\Gamma(n/2)} \left(\frac{\gamma}{\bar{\gamma}}\right)^{n/2 - 1} \frac{\exp(-\gamma/\bar{\gamma})}{\bar{\gamma}}, \quad \gamma \geq 0$$

$$\text{(E.11)}$$

which is identical to the chi-square pdf with n degrees of freedom (see (F.111)). The corresponding cdf is given by

$$F_W(\gamma) = 1 - \exp(-\gamma/\bar{\gamma}) \sum_{k=0}^{n/2 - 1} \frac{1}{k!} \left(\frac{\gamma}{\bar{\gamma}}\right)^k, \quad \gamma \geq 0$$

$$\text{(E.12)}$$

One may thus conclude that the chi-square distribution is a special case of the Wishart distribution for $q = 1$. Also note that (E.11) reduces to (E.9) for $n = 1$, as expected. For $q = 1$, $n = 2$, the pdf is exponentially distributed (see (F.103) and (F.133)). Consequently, the analysis based on Wishart distribution is valid for SISO, MISO, SIMO and MIMO systems, that is, for all combinations of $\{n, q\}$. The

Wishart distribution provides a generalized and powerful tool for determining the pdf of the signal power or SNR at the output of communication systems with arbitrary numbers of transmit and receive antennas.

E.2 Full Rank Wishart Matrices

Note that $\mathbf{W}$ given by (E.8) is a scalar quantity and consists of the squared-sum of the real and imaginary parts of a_{11}, a complex Gaussian random variable with variance $\bar{\gamma} = 2\sigma^2$. However, n in (F.109) and (F.111) denotes the number of real-valued Gaussian random variables. Therefore, the sum of the squares of p complex-Gaussian random variables is the same as the sum of the squares of $n = 2p$ real-valued Gaussian random variables.

The joint pdf of the eigenvalues of $W_q(p, \boldsymbol{\Sigma})$ given by (E.6) is valid for real-valued Gaussian matrix elements. In multi-antenna systems, the matrix elements are complex-Gaussian since they represent the amplitude and the phase of channel gain between transmit and receive antennas. Each of the magnitude-squared complex matrix entries consists of the sum of the squares of the real- and imaginary components. Therefore, in the link between q-receive and p-transmit antennas in a MIMO channel, the signal power received by each of the q-receive antennas will consist of the sum of the squares of p complex-Gaussian random variables or equivalently $2p$ real-valued Gaussian random variables. Therefore, the parameter p employed in this section is equivalent to $n/2$ in the previous section, in order to account for the complex-Gaussian matrix elements.

On the other hand, for $q > 1$, determination of the joint and marginal pdf's of the unordered eigenvalues using (E.6) becomes more tedious, especially when the covariance matrix is not diagonal. Recent research efforts on MIMO systems and related topics lead to the expression of the joint pdf of the *ordered* eigenvalues in simpler forms. The joint pdf of the ordered distinct eigenvalues for uncorrelated (central and noncentral) and correlated (central) Wishart matrices is given by [3][4]

$$f_\lambda(\mathbf{x}) = K |\mathbf{\Phi}(\mathbf{x})| \, |\mathbf{\Psi}(\mathbf{x})| \prod_{i=1}^{q} \xi(x_i) \qquad (E.13)$$

where the number of non-zero eigenvalues is equal to q:

$$\mathbf{x} = [x_1 \ x_2 \ \cdots \ x_q]^T$$
$$\boldsymbol{\lambda} = [\lambda_1 \ \lambda_2 \ \cdots \ \lambda_q]^T, \quad \lambda_1 \geq \lambda_2 \geq \cdots \geq \lambda_q$$
$$(E.14)$$

The values of the normalizing constant K, making the area under the pdf equal to unity, $q \times q$ matrix $\mathbf{\Phi}(\mathbf{x})$, $q \times q$ matrix $\mathbf{\Psi}(\mathbf{x})$ and $\xi(\mathbf{x})$ are shown in Table E.1 for the cases of uncorrelated central, uncorrelated noncentral and correlated central Wishart distributions. It is important to note that (E.13) shows the pdf of the normalized eigenvalues, that is, for $\bar{\gamma} = 2\sigma^2 = 1$. This corresponds to channel gains with unity variance, that is, the variances of real- and imaginary-components of the channel gains are both equal to ½. The pdf's of the eigenvalues for arbitrary values of the variance can easily be determined by a simple variable transformation.

Here, $\lambda_1 > \lambda_2 > \cdots > \lambda_q$ are the ordered non-zero eigenvalues of $\mathbf{A}\mathbf{A}^H$ with mean vector $\boldsymbol{\mu} = [\mu_1 \ \mu_2 \ \cdots \ \mu_q]^T$, and $\sigma_1 > \sigma_2 > \cdots > \sigma_q$ denote the ordered eigenvalues of the covariance matrix $\boldsymbol{\Sigma}$, that is, $\boldsymbol{\Sigma} = \mathrm{diag}(\sigma_1 \ \sigma_2 \ \cdots \ \sigma_q)$, with

$\boldsymbol{\sigma} = [\sigma_1 \ \sigma_2 \ \cdots \ \sigma_q]^T$. The $(i, j)^{\text{th}}$ elements of the matrices $\mathbf{F}(\mathbf{x}, \boldsymbol{\mu})$ and $\mathbf{E}(\mathbf{x}, \boldsymbol{\sigma})$ are given by

$$\mathbf{F}(\mathbf{x}, \boldsymbol{\mu}) = \left\{ {}_0F_1\left(p - q + 1; x_i \mu_j\right) \right\}$$
$$\mathbf{E}(\mathbf{x}, \boldsymbol{\sigma}) = \left\{ e^{-x_j/\sigma_i} \right\} \qquad (E.15)$$

The hypergeometric function ${}_0F_1(a; x)$ is defined by (D.86). The Vandermonde matrices $\mathbf{V}_1(\mathbf{x})$ and $\mathbf{V}_2(\boldsymbol{\sigma})$ are given by

$$\mathbf{V}_1(\mathbf{x}) = \begin{bmatrix} 1 & 1 & \cdots & 1 \\ x_1 & x_2 & & x_q \\ x_1^2 & x_2^2 & & x_q^2 \\ \vdots & & \ddots & \vdots \\ x_1^{q-1} & x_2^{q-1} & & x_q^{q-1} \end{bmatrix}$$

$$\mathbf{V}_2(\boldsymbol{\sigma}) = \begin{bmatrix} 1 & 1 & \cdots & 1 \\ -1/\sigma_1 & -1/\sigma_2 & \cdots & -1/\sigma_q \\ -1/\sigma_1^2 & -1/\sigma_2^2 & \cdots & -1/\sigma_q^2 \\ \vdots & \vdots & \ddots & \vdots \\ -1/\sigma_1^{q-1} & -1/\sigma_2^{q-1} & & -1/\sigma_q^{q-1} \end{bmatrix}$$
$$(E.16)$$

If the covariance matrix $\boldsymbol{\Sigma}$ has ℓ coincident eigenvalues, for example, $\sigma_1 = \sigma_2 = \cdots = \sigma_\ell$,

Table E.1 The Constant K, $\mathbf{\Phi}(\mathbf{x})$, $\mathbf{\Psi}(\mathbf{x})$ and $\xi(\mathbf{x})$ in (E.13). [3]

Wishart distribution	K	$\mathbf{\Phi}(\mathbf{x})$	$\mathbf{\Psi}(\mathbf{x})$	$\xi(\mathbf{x})$
uncorrelated central	$K_{uc} = \left[\prod_{i=1}^{q}(p-i)! \prod_{j=1}^{q}(q-j)! \right]^{-1}$	$\mathbf{V}_1(\mathbf{x})$	$\mathbf{V}_1(\mathbf{x})$	$x^{p-q} e^{-x}$
uncorrelated noncentral	$K_{un} = \dfrac{\prod_{i=1}^{q} e^{-\mu_i}}{[(p-q)!]^q |\mathbf{V}_1(\boldsymbol{\mu})|}$	$\mathbf{V}_1(\mathbf{x})$	$\mathbf{F}(\mathbf{x}, \boldsymbol{\mu})$	$x^{p-q} e^{-x}$
correlated central	$K_{cc} = K_{uc} \prod_{i=1}^{q}(i-1)! \, |\boldsymbol{\Sigma}|^{-p}/|\mathbf{V}_2(\boldsymbol{\sigma})|$	$\mathbf{V}_1(\mathbf{x})$	$\mathbf{E}(\mathbf{x}, \boldsymbol{\sigma})$	x^{p-q}

then ℓ columns of $\mathbf{E}(\mathbf{x}, \boldsymbol{\sigma})$ and $\mathbf{V}_2(\boldsymbol{\sigma})$, for correlated central Wishart matrices, are identical and their determinants go to zero. This uncertainty can be removed by taking the limit. [5]

$$\lim_{\sigma_2 = \ldots = \sigma_\ell \to \sigma_1} \frac{\mathbf{E}(\mathbf{x}, \boldsymbol{\sigma})}{\mathbf{V}_2(\boldsymbol{\sigma})} \qquad (E.17)$$

For the uncorrelated case, all the eigenvalues are equal to each other $\sigma_i = 1$, $i = 1, 2, \ldots, q$. Then, the joint pdf of ordered eigenvalues for central correlated can be shown to reduce to that for uncorrelated central Wishart distribution:

$$f_\lambda(\mathbf{x}) = \frac{1}{\prod_{i=1}^{q}(p-i)! \prod_{j=1}^{q}(q-j)!} \prod_{i=1}^{q} x_i^{p-q} e^{-x_i} \prod_{i<j}^{q} (x_i - x_j)^2$$

$$(E.18)$$

Example E.1 Joint and Marginal Pdf For $q = 1, 2$ and Arbitrary Values of p For Uncorrelated Central Wishart Distribution.
For the special case of $q = 1$, the uncorrelated central Wishart distribution $W_1(p, \mathbf{I})$ has a single non-zero eigenvalue (see (E.14)). From (E.18), we find,

$$f_\lambda(x) = \frac{1}{(p-1)!} x^{p-1} e^{-x} \qquad (E.19)$$

which is identical to (E.11) for $p = n/2$ and $\bar{\gamma} = 2\sigma^2 = 1$.
The joint pdf for the uncorrelated central Wishart distribution with $q = 2$ and arbitrary value of p is given by (E.18)

$$f_{\lambda_1, \lambda_2}(x_1, x_2) = K_{uc}(x_1 - x_2)^2 (x_1 x_2)^{p-2}$$
$$\times e^{-x_1 - x_2}, \quad x_1 \geq x_2 \qquad (E.20)$$
$$K_{uc} = [(p-1)!(p-2)!]^{-1}$$

Using (E.20), the marginal pdf's of λ_1 and λ_2 are determined as follows:

$$f_{\lambda_1}(\lambda) = \int_0^\lambda f(\lambda, x_2)\, dx_2$$
$$= K_{uc}\, \lambda^{p-2} e^{-\lambda} [\gamma(p+1, \lambda) - 2\lambda\gamma(p, \lambda) + \lambda^2 \gamma(p-1, \lambda)]$$

$$f_{\lambda_2}(\lambda) = \int_\lambda^\infty f(x_1, \lambda)\, dx_1$$
$$= K_{uc}\, \lambda^{p-2} e^{-\lambda} [\Gamma(p+1, \lambda) - 2\lambda\Gamma(p, \lambda) + \lambda^2 \Gamma(p-1, \lambda)]$$

$$(E.21)$$

Lower and upper incomplete Gamma functions $\gamma(n, x)$ and $\Gamma(n, x)$ are defined by (D.106) and (D.107), respectively:

$$\Gamma(n, z) = \Gamma(n) - \gamma(n, z) = \Gamma(n) e^{-z} \sum_{k=0}^{n-1} \frac{z^k}{k!}$$

$$\Gamma(1, z) = 1 - \gamma(1, z) = e^{-z}$$
$$\Gamma(2, z) = 1 - \gamma(2, z) = e^{-z}(1 + z)$$
$$\Gamma(3, z) = 2 - \gamma(3, z) = 2e^{-z}(1 + z + z^2/2)$$

$$(E.22)$$

Inserting (E.22) into (E.21), one gets the marginal pdf's of the maximum and minimum eigenvalues of the uncorrelated central Wishart distribution for $p = q = 2$:

$$f_{\lambda_1}(\lambda) = -2e^{-2\lambda} + e^{-\lambda}(2 - 2\lambda + \lambda^2)$$
$$f_{\lambda_2}(\lambda) = 2e^{-2\lambda} \qquad (E.23)$$

Example E.2 Joint and Marginal Pdf For $p = q = 2$ For Uncorrelated Noncentral Wishart Distribution.
The joint pdf of the two eigenvalues for uncorrelated noncentral Wishart distribution for $p = q = 2$ may be written from Table E.1 and (D.86) as

$$f_{\lambda_1, \lambda_2}(x_1, x_2) = \frac{x_2 - x_1}{\mu_2 - \mu_1}[G(x_1, \mu_1)G(x_2, \mu_2) - G(x_1, \mu_2)G(x_2, \mu_1)], \quad x_1 \geq x_2$$

$$G(x, \mu) = e^{-x-\mu}\,_0F_1(1, \mu x) = e^{-x-\mu} \sum_{n=0}^\infty \frac{(\mu x)^n}{(n!)^2}$$

$$(E.24)$$

The marginal pdf's of maximum and minimum eigenvalues are obtained by integrating (E.24) with respect to x_2 and x_1, respectively:

$$f_{\lambda_1}(\lambda) = \int_0^{x_1} f_{\lambda_1,\lambda_2}(\lambda, x_2)\,dx_2$$

$$= \frac{1}{\mu_2 - \mu_1}\sum_{n=0}^{\infty}\frac{1}{(n!)^2}$$

$$\times\left[e^{-\mu_2}\mu_2^n G(\lambda,\mu_1) - e^{-\mu_1}\mu_1^n G(\lambda,\mu_2)\right]$$

$$\times\left[\gamma(n+2,\lambda) - \lambda\gamma(n+1,\lambda)\right]$$

$$f_{\lambda_2}(\lambda) = \int_{x_2}^{\infty} f_{\lambda_1,\lambda_2}(x_1,\lambda)\,dx_1$$

$$= \frac{1}{\mu_2 - \mu_1}\sum_{n=0}^{\infty}\frac{1}{(n!)^2}$$

$$\times\left[e^{-\mu_2}\mu_2^n G(\lambda,\mu_1) - e^{-\mu_1}\mu_1^n G(\lambda,\mu_2)\right]$$

$$\times\left[\Gamma(n+2,\lambda) - \lambda\Gamma(n+1,\lambda)\right]$$

$$(E.25)$$

where the Gamma functions are defined by (E.22). Note the symmetry of the pdf's with respect to μ_1 and μ_2, implying that (E.24) and (E.25) remain the same when μ_1 and μ_2 are interchanged. Also note that taking the limit of (E.25) as μ_2 goes to μ_1, one may determine the pdf for equal mean values. Similarly, by letting $\mu_1 = 0$ after taking the limit, one may show that (E.25) reduces to (E.23), which denotes the pdf for uncorrelated central Wishart distribution for $p = q = 2$.

Figure E.1 shows the pdf's of maximum and minimum eigenvalues given by (E.25) for various combinations of the mean values μ_1 and μ_2. Non-zero mean values of the eigenvalues shift the pdf towards higher SNR's, as expected.

Example E.3 Eigenvalues of a Wishart Matrix.
Consider the following Wishart matrix for $p = q = 2$, with complex Gaussian channel gains with unity variance $E\left[|a_{ij}|^2\right] = 1$, $i,j = 1,2$:

$$\mathbf{A}\mathbf{A}^H = \begin{pmatrix} |a_{11}|^2 + |a_{12}|^2 & a_{11}a_{21}^* + a_{12}a_{22}^* \\ a_{11}^* a_{21} + a_{12}^* a_{22} & |a_{21}|^2 + |a_{22}|^2 \end{pmatrix}$$

$$(E.26)$$

Assuming deterministic channel gains, the ordered eigenvalues $\lambda_1 \geq \lambda_2$ are found to be

$$\lambda_{1,2} = \frac{1}{2}\sum_{i=1}^{2}\sum_{j=1}^{2}|a_{ij}|^2$$

$$\pm\frac{1}{2}\sqrt{\left(|a_{11}|^2 + |a_{12}|^2 - |a_{21}|^2 - |a_{22}|^2\right)^2 + 4\left|a_{11}a_{21}^* + a_{12}a_{22}^*\right|}$$

$$(E.27)$$

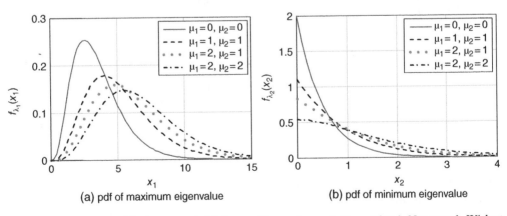

(a) pdf of maximum eigenvalue

(b) pdf of minimum eigenvalue

Figure E.1 Pdf's of Maximum and Minimum Eigenvalues of Uncorrelated Noncentral Wishart Distribution for $p = q = 2$.

Assuming deterministic channel gains and $a_{12} = a_{21} = 0$, that is, no coupling between the two links, the two distinct ordered eigenvalues are given by

$$\lambda_1 = \max\left(|a_{11}|^2, |a_{22}|^2\right)$$
$$\lambda_2 = \min\left(|a_{11}|^2, |a_{22}|^2\right)$$

(E.28)

The two eigenvalues will be identical if the channels described by a_{11} and a_{22} are balanced. Then, the mean values of the eigenvalues in (E.28) will be equal to $E[\lambda_1] = E[\lambda_2] = 1$. However, note that the eigenvalues change randomly, since the channel gains a_{ij}, $i, j = 1, 2$ are complex Gaussian random variables.

The covariance matrix associated with the Wishart matrix in (E.26) is given by

$$\Sigma = E\left[\begin{pmatrix} a_{11} \\ a_{21} \end{pmatrix}(a_{11}^* \quad a_{21}^*)\right]$$

$$= E\begin{bmatrix} |a_{11}|^2 & a_{11}a_{21}^* \\ a_{11}^*a_{21} & |a_{21}|^2 \end{bmatrix} = \begin{pmatrix} 1 & \rho \\ \rho^* & 1 \end{pmatrix}$$

$$= \begin{pmatrix} 1+\rho & 0 \\ 0 & 1-\rho \end{pmatrix}$$

$$\Sigma = E\left[\begin{pmatrix} a_{12} \\ a_{22} \end{pmatrix}(a_{12}^* \quad a_{22}^*)\right]$$

(E.29)

$$= E\begin{bmatrix} |a_{12}|^2 & a_{12}a_{22}^* \\ a_{12}^*a_{22} & |a_{22}|^2 \end{bmatrix} = \begin{pmatrix} 1 & \rho \\ \rho^* & 1 \end{pmatrix}$$

$$= \begin{pmatrix} 1+\rho & 0 \\ 0 & 1-\rho \end{pmatrix}$$

where ρ denotes the correlation coefficient between the zero-mean complex Gaussian elements of the matrix $\mathbf{A}$:

$$\rho = E\left[a_{11}\,a_{21}^*\right] = E\left[a_{12}\,a_{22}^*\right] \qquad \text{(E.30)}$$

The ordered eigenvalues $\sigma_1 \geq \sigma_2$ of the covariance matrix Σ, defined by (E.29), are given by

$$\sigma_1 = 1+\rho$$
$$\sigma_2 = 1-\rho$$

(E.31)

For $\rho = 0$, that is, the uncorrelated case, the pdf's of the two eigenvalues in (E.27) are given by (E.23). If $\rho \neq 0$, that is, the correlated case, then the pdf's of the two eigenvalues will change as a function of the correlation coefficient. The two eigenmodes corresponding to σ_1 and σ_2 imply two uncoupled channels, namely between the first transmit and first receive antennas, and between second transmit and second receive antennas, respectively.

Example E.4 Joint and Marginal Pdf For $q = 2$ and Arbitrary Values of p For Correlated Central Wishart Distribution.

We now consider correlated central Wishart distributions $W_2(p, \Sigma)$ for $q = 2$ and arbitrary values of p. The complex-Gaussian components of the Wishart matrix are assumed to be $E\left[|a_{ij}|^2\right] = 1$, $i, j = 1, 2$.

Using Table E.1, the determinants of the matrices for correlated central Wishart distribution are found to be

$$|\Sigma| = \sigma_1\sigma_2$$

$$|\mathbf{V}_1(\mathbf{x})| = \begin{vmatrix} 1 & 1 \\ x_1 & x_2 \end{vmatrix} = x_2 - x_1$$

$$|\mathbf{V}_2(\boldsymbol{\sigma})| = \begin{vmatrix} 1 & 1 \\ -1/\sigma_1 & -1/\sigma_2 \end{vmatrix}$$

$$= 1/\sigma_1 - 1/\sigma_2 = (\sigma_2 - \sigma_1)/(\sigma_1\sigma_2)$$

$$|\mathbf{E}(\mathbf{x}, \boldsymbol{\sigma})| = \begin{vmatrix} e^{-x_1/\sigma_1} & e^{-x_2/\sigma_1} \\ e^{-x_1/\sigma_2} & e^{-x_2/\sigma_2} \end{vmatrix}$$

$$= e^{-\left(\frac{x_1}{\sigma_1} + \frac{x_2}{\sigma_2}\right)} - e^{-\left(\frac{x_1}{\sigma_2} + \frac{x_2}{\sigma_1}\right)}$$

(E.32)

The joint pdf for the correlated central case with $q = 2$ and arbitrary value of p is given by:

$$f_\lambda(x_1, x_2) = K_{cc}(x_2 - x_1)(x_1 x_2)^{p-2}$$

$$\times \left[e^{-\left(\frac{x_1}{\sigma_1} + \frac{x_2}{\sigma_2}\right)} - e^{-\left(\frac{x_1}{\sigma_2} + \frac{x_2}{\sigma_1}\right)} \right], \quad x_1 \geq x_2$$

$$K_{cc} = \left[(p-1)!(p-2)!(\sigma_1 \sigma_2)^{p-1}(\sigma_2 - \sigma_1) \right]^{-1}$$

$$\text{(E.33)}$$

The marginal pdf's of the eigenvalues are found by the integration of (E.33)

$$f_{\lambda_1}(x) = K_{cc}x^{p-2} \left[\begin{array}{l} -e^{-x/\sigma_1}\sigma_2^p \gamma(p, x/\sigma_2) + e^{-x/\sigma_2}\sigma_1^p \gamma(p, x/\sigma_1) \\ +xe^{-x/\sigma_1}\sigma_2^{p-1}\gamma(p-1, x/\sigma_2) \\ -xe^{-x/\sigma_2}\sigma_1^{p-1}\gamma(p-1, x/\sigma_1) \end{array} \right]$$

$$f_{\lambda_2}(x) = K_{cc}x^{p-2} \left[\begin{array}{l} -e^{-x/\sigma_1}\sigma_2^p \Gamma(p, x/\sigma_2) + e^{-x/\sigma_2}\sigma_1^p \Gamma(p, x/\sigma_1) \\ +xe^{-x/\sigma_1}\sigma_2^{p-1}\Gamma(p-1, x/\sigma_2) \\ -xe^{-x/\sigma_2}\sigma_1^{p-1}\Gamma(p-1, x/\sigma_1) \end{array} \right]$$

$$\text{(E.34)}$$

For the special case of $p = q = 2$, inserting (E.22) into (E.34), one gets

$$f_{\lambda_2}(x) = \left(\frac{1}{\sigma_1} + \frac{1}{\sigma_2} \right) e^{-\left(\frac{1}{\sigma_1} + \frac{1}{\sigma_2}\right)x}$$

$$f_{\lambda_1}(x) = -f_{\lambda_2}(x) + \frac{1}{(\sigma_2 - \sigma_1)\sigma_1\sigma_2}$$

$$\times \left[\sigma_2^2\left(1 - \frac{x}{\sigma_2}\right)e^{-\frac{x}{\sigma_1}} - \sigma_1^2\left(1 - \frac{x}{\sigma_1}\right)e^{-\frac{x}{\sigma_2}} \right]$$

$$\text{(E.35)}$$

Taking the derivative of (E.35) with respect to σ_2 and letting $\sigma_2 \to \sigma_1 = 1$, one can easily show that (E.35) reduces to (E.23).

E.3   Pseudo Wishart Matrices

Some performance parameters of MIMO systems may be studied in terms of the joint pdf of the eigenvalues of the $\mathbf{A}^H\mathbf{A}$, instead of those of $\mathbf{A}\mathbf{A}^H$. In addition, in some cases, for

example, the evaluation of the bit error probability performance of MIMO-MRC systems, one needs to determine the pdf of the largest eigenvalue. In some other applications, the pdf of the smallest eigenvalue or the pdf of the largest two or three eigenvalues are needed. Therefore, determination of the joint pdf of ordered eigenvalues and marginal pdf's of the largest and smallest eigenvalues are desired. The joint pdf of the eigenvalues for the case of correlation among both rows and columns is presented in [8] and [9].

Now consider matrices in the form of $\mathbf{A}^H\mathbf{A}$. The matrix $\mathbf{A}\mathbf{A}^H$ satisfies the condition to be a Wishart matrix, since it can be expressed as in (E.3). However, the matrix $\mathbf{A}^H\mathbf{A}$ cannot be expressed in terms of the column vectors of the matrix $\mathbf{A}$. Instead, it can be expressed in terms of the columun vectors of $\mathbf{A}^H$, in other words, in terms of the row vectors of $\mathbf{A}$:

$$\mathbf{A} = \begin{pmatrix} \mathbf{t}_1 \\ \mathbf{t}_2 \\ \vdots \\ \mathbf{t}_q \end{pmatrix} \tag{E.36}$$

$$\mathbf{t}_i = (a_{i1} \; a_{i2} \; \cdots \; a_{in}), \quad i = 1, 2, \ldots, q$$

Then the matrix $\mathbf{A}^H\mathbf{A}$ can be expressed as

$$\mathbf{A}^H\mathbf{A} = \sum_{j=1}^{q} \mathbf{t}_j^H \mathbf{t}_j = \begin{bmatrix} \mathbf{t}_1^H & \mathbf{t}_2^H & \cdots & \mathbf{t}_q^H \end{bmatrix} \begin{bmatrix} \mathbf{t}_1 \\ \mathbf{t}_2 \\ \vdots \\ \mathbf{t}_q \end{bmatrix}$$

$$\text{(E.37)}$$

The associated covariance matrix is of dimension $p \times p$ and is defined by

$$\mathbf{\Omega} = E\left[\mathbf{t}_j^H \mathbf{t}_j \right], \quad j = 1, 2, \ldots, q \tag{E.38}$$

When the correlation is among the elements of the rows of $\mathbf{A}$ instead of the columns, as in

(E.38), the matrix is referred to as non-full rank Wishart or pseudo Wishart. In that case, the distribution of the eigenvalues can still be written in the form (E.13), but now $\mathbf{\Phi}(x)$ and $\mathbf{\Psi}(x)$ are $p \times p$ matrices. Results applicable for the full rank Wishart matrices can therefore be easily extended to the pseudo Wishart case. [3][6][7] Assuming that $\mathbf{A}$ is a $q \times p$ matrix and $q < p$, $\mathbf{A}\mathbf{A}^H$ is a complex $q \times q$ Wishart matrix and has q eigenvalues. However, $\mathbf{A}^H\mathbf{A}$ is a complex $p \times p$ pseudo Wishart matrix and has p $(p > q)$ eigenvalues; q nonzero eigenvalues of $\mathbf{A}\mathbf{A}^H$ and $\mathbf{A}^H\mathbf{A}$ are identical. The additional $p-q$ eigenvalues are identically equal to zero.

Example E.5 2×3 MIMO Channel with Correlation on the Transmit- or the Receive Side. Consider the following three random vectors each with dimension two, representing a MIMO channel with three transmit antennas and two receive antennas. The 2×3 channel matrix $\mathbf{H}$, which consists of complex Gaussian random variables with zero-mean and unity variance, may be written as

$$\mathbf{H} = [\mathbf{h}_1 \ \mathbf{h}_2 \ \mathbf{h}_3] = \begin{bmatrix} h_{11} & h_{12} & h_{13} \\ h_{21} & h_{22} & h_{23} \end{bmatrix} \quad (E.39)$$

As shown by Figure E.2, there are six propagation paths between three transmit and two receive antennas. If the signals in these paths fade independently of each other, then this MIMO channel has a diversity order of six.

As an example, consider the following realization of the channel gains. The channel matrix $\mathbf{H}$ and $\mathbf{H}\mathbf{H}^H$ may then be written as

$$\mathbf{H} = [\mathbf{h}_1 \ \mathbf{h}_2 \ \mathbf{h}_3]$$

$$= \begin{bmatrix} 1 & 0.9e^{j\pi/2} & 0.7e^{-j\pi/4} \\ 0.6e^{-j\pi/4} & 1.2 & 0.8e^{j\pi/2} \end{bmatrix}$$

$$\mathbf{H}\mathbf{H}^H = \begin{bmatrix} 2.3 & 0.028+j1.108 \\ 0.028-j1.108 & 2.44 \end{bmatrix}$$

$$(E.40)$$

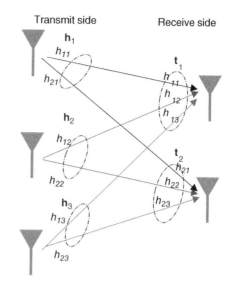

Figure E.2 Correlations Between Antennas at Transmit and Receive Sides of a 2×3 MIMO System.

For this realization of the channel matrix, the two eigenvalues of $\mathbf{H}\mathbf{H}^H$ are found to be 3.48 and 1.26.

Note that the 2×2 covariance matrices, characterizing the correlations on the transmit side, determined by a single realization of $\mathbf{h}_1$, $\mathbf{h}_2$ and $\mathbf{h}_3$ are not the same:

$$\mathbf{h}_1\mathbf{h}_1^H = \begin{bmatrix} h_{11} \\ h_{21} \end{bmatrix} [h_{11}^* \ h_{21}^*]$$

$$= \begin{bmatrix} 1 & 0.6e^{j\pi/4} \\ 0.6e^{-j\pi/4} & 0.36 \end{bmatrix}$$

$$\mathbf{h}_2\mathbf{h}_2^H = \begin{bmatrix} h_{12} \\ h_{22} \end{bmatrix} [h_{12}^* \ h_{22}^*] = \begin{bmatrix} 0.81 & j1.08 \\ -j1.08 & 1.44 \end{bmatrix}$$

$$\mathbf{h}_3\mathbf{h}_3^H = \begin{bmatrix} h_{13} \\ h_{23} \end{bmatrix} [h_{13}^* \ h_{23}^*]$$

$$= \begin{bmatrix} 0.49 & 0.56e^{-j3\pi/4} \\ 0.56e^{j3\pi/4} & 0.64 \end{bmatrix}$$

$$(E.41)$$

One may observe from Figure E.2 that the first covariance matrix provides a measure of the transmit correlation between the signals transmitted by the first transmit antenna towards the first and second receiving antennas. Similarly, the second covariance matrix measures the correlation between the signals transmitted by the second transmit antenna towards the first and second receiving antennas. Finally, the third covariance matrix measures the correlation between signals transmitted by the third transmit antenna towards the first and second receive antennas. Note that, even though these covariance matrices are not the same for a single realization, their expected values, when averaged over sufficiently large number of samples, will be the same:

$$\Sigma = E\left[\mathbf{h}_1 \mathbf{h}_1^H\right] = E\left[\mathbf{h}_2 \mathbf{h}_2^H\right] = E\left[\mathbf{h}_3 \mathbf{h}_3^H\right] \quad (E.42)$$

Now consider the pseudo-Wishart matrix defined by $\mathbf{H}^H \mathbf{H}$:

$$\mathbf{H}^H \mathbf{H} = \begin{bmatrix} 1.36 & 0.509 + j1.409 & 0.156 - j0.156 \\ 0.509 - j1.409 & 2.25 & -0.445 + j0.515 \\ 0.156 + j0.156 & -0.445 - j0.515 & 1.13 \end{bmatrix}$$

$$(E.43)$$

The three eigenvalues of the 3×2 pseudo-Wishart matrix $\mathbf{H}^H \mathbf{H}$ are 3.48, 1.26 and 0. Note that the two non-zero eigenvalues are identical to those of the Wishart matrix $\mathbf{H}\mathbf{H}^H$ and third eigenvalue is equal to zero. The correlations on the transmit side, which affect the signals received by the two receive antennas, may be expressed in terms of the following 3×3 covariance matrices:

$$\mathbf{t}_1^H \mathbf{t}_1 = \begin{bmatrix} h_{11}^* \\ h_{12}^* \\ h_{13}^* \end{bmatrix} \begin{bmatrix} h_{11} & h_{12} & h_{13} \end{bmatrix}$$

$$= \begin{bmatrix} 1 & j0.9 & 0.495 - j0.495 \\ -j0.9 & 0.81 & -0.445 - j0.445 \\ 0.495 + j0.495 & -0.445 + j0.445 & 0.49 \end{bmatrix}$$

$$\mathbf{t}_2^H \mathbf{t}_2 = \begin{bmatrix} h_{21}^* \\ h_{22}^* \\ h_{23}^* \end{bmatrix} \begin{bmatrix} h_{21} & h_{22} & h_{23} \end{bmatrix}$$

$$= \begin{bmatrix} 0.36 & 0.509 + j0.509 & -0.339 + j0.339 \\ 0.509 - j0.509 & 1.44 & j0.96 \\ -0.339 - j0.339 & -j0.96 & 0.64 \end{bmatrix}$$

$$(E.44)$$

The first (second) covariance matrix provides a measure of correlations between the signals transmitted by the three transmit antennas at the first (second) receiver. When averaged over sufficiently large number of realizations, the two covariance matrices will yield the same result:

$$\Omega = E\left[\mathbf{t}_1^H \mathbf{t}_1\right] = E\left[\mathbf{t}_2^H \mathbf{t}_2\right] \quad (E.45)$$

References

[1] M. Bilodeau and D. Brenner, *Theory of Multivariate Statistics*, Springer, 1999.

[2] S. Haykin, *Adaptive Filter Theory*, Prentice Hall: New Jersey, 1996.

[3] A. Zanella, M. Chiani, M. Z. Win, On the marginal distribution of the eigenvalues of Wishart matrices, *IEEE Trans. Communications*, vol. **57**, no. 4, April 2009.

[4] A. Zanella, M. Chiani, M. Z. Win, A general framework for the distribution of the eigenvalues of Wishart matrices, ICC 2008, pp. 1271–1275.

[5] M. Chiani, M. Z. Win and H. Shin, MIMO networks: The effects of interference, *IEEE Trans. Information Theory*, vol. **56**, no. 1, pp. 336–349, January 2010.

[6] M. Chiani, M. Z. Win and A. Zanella, On the capacity of spatially correlated MIMO Rayleigh-fading channels, *IEEE Trans. Information Theory*, vol. **49**, no. 10, pp. 2363–2371, October 2003.

[7] P. Smith, S. Roy, and M. Shafi, Capacity of MIMO systems with semi-correlated flat fading, *IEEE Trans. Inform. Theory*, vol. **49**, no. 10, pp. 2781–2788, Oct. 2003.

[8] S. H. Simon, A. L. Moustakas, and L. Marinelli, Capacity and character expansions: moment-generating function and other exact results for MIMO correlated channels, *IEEE trans. Information Theory*, vol. **52**, no. 12, pp. 5336–5351, Dec. 2006.

[9] M. R. McCay, A. J. Grant, and I. B. Colling, Performance analysis of MIMO-MRC in double-correlated Rayleigh environments, *IEEE Trans. Commun.*, vol. **55**, no. 3, pp. 497–507, March 2007.

Appendix F

Probability and Random Variables

F.1 Random Variable

A random variable (rv) is unknown and unpredictable beforehand, that is, it has a random value, but its value is known completely once it occurs. A rv may be continuous or discrete. For example, the noise voltage generated by an electronic amplifier is random and has a continuous amplitude. On the other hand, coin flipping with outcomes head (H) and tail (T) is discrete. One does not know the outcome before flipping a coin. For each experiment, a value is assigned for each of the possible outcomes in the experiment. For example, the rv may be assumed to be $X(s) = 1$ for the outcome $s = $ H and -1 for $s = $ T. However, once the coin is flipped, the outcome (H or T) and hence $X(s)$ is known. In the sequel we denote the rv simply by X but not as $X(s)$.

A rv is characterized by its probability density function (pdf) or cumulative distribution function (cdf), which are interrelated. The cdf, $F_X(x)$, of a rv X is defined by

$$F_X(x) = P(X \le x), \quad -\infty < x < \infty \tag{F.1}$$

which specifies the probability that the rv X is less than or equal to a real number x. The pdf of a rv X is defined as the derivative of the cdf:

$$f_X(x) = \frac{d}{dx} F_X(x), \quad -\infty < x < \infty \tag{F.2}$$

Conversely, the cdf is defined by the integral of the pdf:

$$F_X(x) = \int_{-\infty}^{x} f_X(u) \, du \tag{F.3}$$

Based on (F.1)–(F.3), a rv has the following properties:

a. $0 \le F_X(x) \le 1$ since $F(-\infty) = \int_{-\infty}^{-\infty} f_X(u) du = 0$
   and $F_X(\infty) = \int_{-\infty}^{\infty} f_X(u) \, du = 1$.

b. The cdf of a continuous rv X is a non-decreasing and smooth function of X.

Digital Communications, First Edition. Mehmet Şafak.
© 2017 John Wiley & Sons Ltd. Published 2017 by John Wiley & Sons Ltd.
Companion website: www.wiley.com/go/safak/Digital_Communications

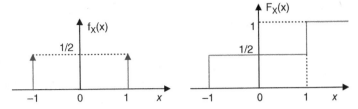

Figure F.1   Pdf and Cdf For Coin Flipping with $P(H) = P(T) = 1/2$ and $X(s) = 1$ For $s = H$ and $X(s) = -1$ For $s = T$.

Therefore, $P(x_1 < X \le x_2) = \displaystyle\int_{x_1}^{x_2} f_X(u)\, du =$

$F_X(x_2) - F_X(x_1) \ge 0$ implies $x_2 \ge x_1$.

c. $f_X(x) \ge 0$ and the area under $f_X(x)$ is always equal to unity: $F_X(\infty) = \displaystyle\int_{-\infty}^{\infty} f_X(u)\, du = 1$.

d. When a rv X is discrete or mixed, its cdf is still a non-decreasing function of X but contains discontinuities.

Figure F.1 shows the pdf and cdf for coin flipping, based on the assumption that the probability of head and tail are equally likely, that is $P(H) = P(T) = 1/2$. The pdf and the cdf may then be written as

$$f_X(x) = P(T)\delta(x+1) + P(H)\delta(x-1)$$
$$F_X(x) = P(T)u(x+1) + P(H)u(x-1)$$
$$\tag{F.4}$$

F.2   Statistical Averages of Random Variables

A rv is characterized by its moments. The n-th moment of a rv X is defined as the expectation of X^n:

$$E[X^n] = \int_{-\infty}^{\infty} x^n f_X(x)\, dx$$
$$= n \int_{-\infty}^{\infty} x^{n-1}[1 - F_X(x)]\, dx \tag{F.5}$$

where integration by parts is used to obtain the last expression. The mean (expected) value m_X of a rv X is given by its first moment:

$$m_X = E[X] = \int_{-\infty}^{\infty} x f_X(x)\, dx \tag{F.6}$$

In some applications we encounter functions of rv's. For example, the mean value of $Y = g(X)$ may be determined as follows:

$$E[Y] = E[g(X)]$$
$$= \int_{-\infty}^{\infty} g(x)\, f_X(x)\, dx \tag{F.7}$$
$$= \int_{-\infty}^{\infty} y\, f_Y(y)\, dy$$

Central moments of X may be determined using $Y = (X - m_x)^n$:

$$E[(X - m_X)^n] = \int_{-\infty}^{\infty} (X - m_X)^n f_X(x)\, dx \tag{F.8}$$

The variance of X is given by

$$\mathrm{var}(X) = \sigma_X^2 = E\left[(X - m_X)^2\right]$$
$$= \int_{-\infty}^{\infty} (X - m_X)^2 f_X(x)\, dx \tag{F.9}$$
$$= E[X^2] - m_X^2$$

When X is discrete or of a mixed type, the pdf contains impulses at the points of discontinuity of $F_X(x)$. In such cases, the discrete part of $f_X(x)$ may be expressed as

$$f_X(x) = \sum_{i=1}^{N} P(X=x_i)\, \delta(x-x_i) \qquad \text{(F.10)}$$

where the rv X is assumed to be discontinuous at N points, $x_1, x_2,\ldots, x_N$. For example, in case of coin flipping, two outcomes may be represented as x_1 for head and x_2 for tail. Then, $P(X=x_1)=p \leq 1$ shows the probability of head, while $P(X=x_2)=1-p$ denotes the probability of tail. Mean value and the variance of a discrete rv is found by inserting (F.10) into (F.6) and (F.9), respectively:

$$m_X = E[X] = \sum_{i=1}^{N} x_i P(X=x_i)$$
$$\sigma_X^2 = \sum_{i=1}^{N} x_i^2 P(X=x_i) - m_X^2 \qquad \text{(F.11)}$$

When the events are equally likely, that is, $P(X=x_i)=1/N$, (F.11) simplifies to

$$m_X = E[X] = \frac{1}{N}\sum_{i=1}^{N} x_i$$
$$\sigma_X^2 = \frac{1}{N}\sum_{i=1}^{N}(x_i-m_X)^2 = \frac{1}{N}\sum_{i=1}^{N} x_i^2 - m_X^2 \qquad \text{(F.12)}$$

The *median* is closely related to the mean value of a rv. In statistics and probability theory, the median is the value of the random number separating the higher half of a data sample from the lower half. The median of N samples can be found by arranging all the observations from lowest value to highest value and picking the middle one. If there is an even number of observations, then there is no single middle value; the median is then usually defined as the mean of the two middle values. The median coincides with the mean value if the rv has a symmetrical pdf.

F.2.1 Statistical Analysis of Multiple Random Variables

Now consider two rv's X and Y, each of which may be continuous, discrete or mixed. The probability that (X,Y) take values in the rectangle shown in Figure F.2 is given by

$$P(x_1 < X \leq x_2, y_1 < Y \leq y_2) = F_{XY}(x_2,y_2)$$
$$- F_{XY}(x_1,y_2) - F_{XY}(x_2,y_1) + F_{XY}(x_1,y_1) \qquad \text{(F.13)}$$

The joint cdf of the rv's X and Y may be obtained by inserting $x_1 = y_1 = -\infty$ and $x_2 = x, y_2 = y$ into (F.13):

$$F_{XY}(x,y) = P(X \leq x, Y \leq y)$$
$$= \int_{-\infty}^{x} \int_{-\infty}^{y} f_{XY}(u_1,u_2)\, du_1\, du_2 \qquad \text{(F.14)}$$

where

$$F_{XY}(\infty,\infty) = 1,\ F_{XY}(x,\infty) = F_X(x),$$
$$F_{XY}(\infty,y) = F_Y(y)$$
$$F_{XY}(-\infty,-\infty) = F_{XY}(x,-\infty)$$
$$= F_{XY}(-\infty,y) = 0 \qquad \text{(F.15)}$$

Joint pdf and joint cdf are related to each other as follows:

$$f_{XY}(x,y) = \frac{\partial^2}{\partial x \partial y} F_{XY}(x,y) \qquad \text{(F.16)}$$

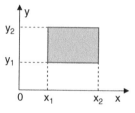

Figure F.2 The Region Defined By (F.13).

The marginal pdf's are found as follows:

$$f_Y(y) = \int_{-\infty}^{\infty} f_{XY}(x,y) \, dx$$

$$f_X(x) = \int_{-\infty}^{\infty} f_{XY}(x,y) \, dy$$

(F.17)

Joint moments of two rv's X and Y may be obtained from their joint pdf $f_{XY}(x,y)$:

$$E[X^k Y^n] = \int_{-\infty}^{\infty} \int_{-\infty}^{\infty} x^k \, y^n \, f_{XY}(x,y) \, dx \, dy$$

(F.18)

Joint central moments are defined as

$$E\left[(X-m_X)^k (Y-m_Y)^n\right]$$

$$= \int_{-\infty}^{\infty} \int_{-\infty}^{\infty} (x-m_X)^k \, (y-m_Y)^n \, f_{XY}(x,y) \, dx \, dy$$

(F.19)

The covariance of X and Y is given by

$$\mu_{XY} = E[(X-m_X)(Y-m_Y)]$$

$$= \int_{-\infty}^{\infty} \int_{-\infty}^{\infty} (x-m_X) \, (y-m_Y) \, f_{XY}(x,y) \, dx \, dy$$

$$= \int_{-\infty}^{\infty} \int_{-\infty}^{\infty} xy \, f_{XY}(x,y) \, dx \, dy - m_X m_Y$$

$$= E[XY] - m_X m_Y$$

(F.20)

Correlation coefficient between X and Y is defined by

$$\rho \triangleq \frac{\mu_{XY}}{\sigma_X \sigma_Y} = \frac{E[(X-m_X)(Y-m_Y)]}{\sigma_X \sigma_Y}$$

$$= \frac{E[XY] - m_X m_Y}{\sigma_X \sigma_Y}$$

(F.21)

If the experiments result in mutually exclusive outcomes, then the probability of an outcome in one experiment is statistically *independent* of the outcome in any other experiment. Then, the joint pdf's (cdf's) may be written as the product of the pdf's (cdf's) corresponding to each outcome:

$$f_{XY}(x,y) = f_X(x) \, f_Y(y)$$

$$F_{XY}(x,y) = F_X(x) \, F_Y(y)$$

(F.22)

Two rv's are said to be *uncorrelated* with each other if the correlation coefficient is identically equal to zero:

$$\rho = \frac{E[XY] - m_X m_Y}{\sigma_X \sigma_Y} = \frac{E[X]E[Y] - m_X m_Y}{\sigma_X \sigma_Y} \equiv 0$$

(F.23)

Two rv's X and Y are perfectly correlated with each other if $Y = a X$ where a is a constant. This implies

$$\rho = \frac{E[XY] - m_X m_Y}{\sigma_X \sigma_Y}$$

$$= \frac{a E[X^2] - a m_X^2}{\sigma_X |a| \sigma_X}$$

$$= \frac{a\left[\sigma_X^2 + m_X^2\right] - a m_X^2}{|a| \sigma_X^2} = \text{sgn}(a)$$

(F.24)

Hence, the correlation coefficient ρ varies between -1 and $+1$. Note that when X and Y are statistically independent, they are also uncorrelated. If X and Y are uncorrelated, then (F.23) holds but they are not necessarily statistically independent.

Two rv's are said to be *orthogonal* when

$$E[XY] = E[X]E[Y] = m_X m_Y = 0 \qquad \text{(F.25)}$$

Hence, X and Y are orthogonal when they are uncorrelated, and m_X and/or m_Y is equal to zero.

Homework F.1 Given $X = \cos\Theta$ and $Y = \sin\Theta$ where Θ is a rv uniformly distributed in $(0, 2\pi)$, show that X and Y are uncorrelated but not independent. Are they orthogonal?

F.2.2   Conditional Probability

The conditional pdf $f_{X|Y}(x|y)$ of the rv X for a given deterministic value y of the rv Y has the following properties:

$$f_{X|Y}(x|y) = \frac{f_{XY}(x,y)}{f_Y(y)} \tag{F.26}$$

$$F_{X|Y}(-\infty|y) = 0, \quad F_{X|Y}(\infty|y) = 1$$

The marginal pdf of X may be determined from its conditional pdf as follows:

$$f_X(x) = \int_{-\infty}^{\infty} f_{XY}(x|y) f_Y(y) \, dy$$
$$= \int_{-\infty}^{\infty} f_{XY}(x,y) \, dy \tag{F.27}$$

Generalization of the above to more than two rv's is straightforward.

Homework F.2   The Pdf of jointly normal rv's X and Y is given by

$$f_{XY}(x,y) = \frac{1}{2\pi\sigma_X\sigma_Y\sqrt{1-\rho^2}}$$

$$\times \exp\left[-\frac{1}{2(1-\rho^2)} \left\{ \frac{(x-m_X)^2}{\sigma_X^2} \right. \right.$$

$$\left. \left. -2\rho\frac{(x-m_X)(y-m_Y)}{\sigma_X\sigma_Y} + \frac{(y-m_Y)^2}{\sigma_Y^2} \right\} \right] \tag{F.28}$$

where ρ denotes the correlation coefficient between X and Y as defined by (F.21). Using (F.27), show that the marginal pdf of X is Gaussian (normal):

$$f_X(x) = \int_{-\infty}^{\infty} f_{XY}(x,y) \, dy$$
$$= \frac{1}{\sqrt{2\pi}\sigma_X} \exp\left[-\frac{(x-m_X)^2}{2\sigma_X^2} \right] \tag{F.29}$$

Show that the conditional pdf is given by

$$f(y|x) = \frac{1}{\sqrt{2\pi}\,\sigma_{Y|X}} \exp\left[-\frac{(y-m_{Y|X})^2}{2\sigma_{Y|X}^2} \right] \tag{F.30}$$

where

$$m_{Y|X} = E[Y|X] = m_Y + \rho\frac{\sigma_Y}{\sigma_X}(x-m_X)$$
$$-1 \le \rho \le 1$$
$$\sigma_{Y|X}^2 = \sigma_Y^2(1-\rho^2) \tag{F.31}$$

Using (F.30) and (F.31) show that

$$E[XY] = E[X\,E(Y|X)] = m_X m_Y + \rho\sigma_X\sigma_Y$$
$$E[X^2 Y^2] = E[X^2\,E(Y^2|X)]$$
$$= \sigma_X^2\sigma_Y^2 + 2\rho^2\sigma_X^2\sigma_Y^2 + \sigma_X^2 m_Y^2$$
$$+ \sigma_Y^2 m_X^2 - 4\rho m_x m_Y\sigma_x\sigma_Y$$
$$E[Y^2|X] = m_{Y|X}^2 + \sigma_{Y|X}^2 \tag{F.32}$$

Example F.1   The Sum of Two Correlated Rv's.
If Z is defined as the sum of two correlated rv's X and Y, then the mean and the variance of Z are given by

$$m_Z = E[X+Y] = m_X + m_Y$$
$$\sigma_Z^2 = E\left[(Z-m_Z)^2\right] = E\left[(X+Y-m_X-m_Y)^2\right]$$
$$= E\left[\{(X-m_X)+(Y-m_Y)\}^2\right]$$
$$= E\left[(X-m_X)^2\right] + E\left[(Y-m_Y)^2\right]$$
$$+ 2E[\{X-m_X\}\{Y-m_Y\}]$$
$$= \sigma_X^2 + \sigma_Y^2 + 2\rho\sigma_X\sigma_Y \tag{F.33}$$

F.3   Moment Generating Function (MGF)

MGF of a rv X, which is defined as

$$M_X(s) \triangleq E[e^{-sX}] = \int_{-\infty}^{\infty} e^{-sx} f_X(x)\, dx$$

$$f_X(x) = \frac{1}{2\pi} \int_{-\infty}^{\infty} e^{sx} M_X(s)\, ds$$

(F.34)

reduces to the Fourier transform of $f_X(x)$ for $s = j2\pi f$. Therefore, the MGF may be used to exploit the advantages of the Fourier analysis.

For example, when the pdf of the sum $Y = \sum_{i=1}^{n} X_i$ of n independent rv's is required, one first determines the MGF of the sum by multiplying the MGF's of the individual rv's by using the linearity of (F.34):

$$M_Y(s) = E\left[e^{-sY}\right] = E\left[\exp\left(-s\sum_{i=1}^{n} X_i\right)\right]$$

$$= E\left[\prod_{i=1}^{n}\left(e^{-sX_i}\right)\right]$$

$$= \int_{-\infty}^{\infty} \cdots \int_{-\infty}^{\infty} \left(\prod_{i=1}^{n} e^{-sX_i}\right)$$

$$\times f_{X_1, X_2, \cdots, X_n}(x_1, x_2, \cdots, x_n)\, dx_1\, dx_2 \cdots dx_n$$

$$= \prod_{i=1}^{n} M_{X_i}(s)$$

$$= [M_X(s)]^n \quad \text{if } X_i \text{ are i.i.d.}$$

(F.35)

where i.i.d. stands for independent and identically distributed. The next step would be to determine the resulting pdf by taking the inverse of the MGF thus found in (F.35). However, it may not always be easy and/or not necessary to take the inverse, since some desired performance parameters, such as bit error probability, can be obtained directly from the MGF.

MGF of $Y = g(X)$ may be obtained as

$$M_{g(X)}(s) \triangleq E\left[e^{-sg(X)}\right] = \int_{-\infty}^{\infty} e^{-sg(x)} f_X(x)\, dx$$

(F.36)

The moments of a rv X can be directly obtained from its MGF:

$$E[X^n] \triangleq \int_{-\infty}^{\infty} x^n f_X(x)\, dx = (-1)^n \frac{d^n M_X(s)}{ds^n}\Big|_{s=0}$$

(F.37)

As an alternative to the MGF, the characteristic function of a rv X is defined by

$$\psi_X(jv) \equiv E\left[e^{-jvX}\right] = \int_{-\infty}^{\infty} e^{-jvx} f_X(x)\, dx \quad \text{(F.38)}$$

One may easily observe that (F.34) and (F.38) are the same if $s = jv$.

Example F.2   Use of the MGF To Determine the Cdf.

Using (F.34) and integration by parts, one obtains the following Fourier transform relation between the cdf and the MGF:

$$M_X(s) = \int_{-\infty}^{\infty} \underbrace{e^{-sx}}_{u} \underbrace{f_X(x)dx}_{dv}$$

$$= \underbrace{e^{-sx} F_X(x)\big|_{-\infty}^{\infty}}_{=0} + s\int_{-\infty}^{\infty} e^{-sx} F_X(x)\, dx$$

$$f_X(x) \iff M_X(s)$$

$$F_X(x) \iff M_X(s)/s$$

(F.39)

Hence, the cdf of a rv may be directly obtained from its MGF.

The pdf and the corresponding MGF of the SNR γ, a rv with exponential distribution in a Rayleigh fading environment, are found using (F.34) as follows:

$$f_\gamma(x) = \frac{1}{\bar{\gamma}} e^{-x/\bar{\gamma}}, \quad x \geq 0$$

$$M_\gamma(s) = E[e^{-s\gamma}] = \frac{1}{\bar{\gamma}} \int_0^\infty e^{-(1/\bar{\gamma}+s)x} dx = \frac{1}{1+s\bar{\gamma}}$$

$$\text{(F.40)}$$

where $\bar{\gamma} = E[\gamma]$ denotes the mean value of the SNR γ. The validity of (F.39) is demonstrated below:

$$F_\gamma(x) = \mathfrak{J}^{-1}\left[\frac{M_\gamma(s)}{s}\right]$$

$$= \mathfrak{J}^{-1}\left[\frac{1}{s} - \frac{\bar{\gamma}}{1+s\bar{\gamma}}\right] = 1 - e^{-x/\bar{\gamma}}, \quad x \geq 0$$

$$s\int_{-\infty}^\infty e^{-sx} F_\gamma(x)\, dx = s\int_0^\infty e^{-sx}\left(1 - e^{-x/\bar{\gamma}}\right) dx$$

$$= s\left[\frac{1}{s} - \frac{1}{s+1/\bar{\gamma}}\right] = M_\gamma(s)$$

$$\text{(F.41)}$$

F.4   Functions of Random Variables

In some applications, one may need to determine the pdf of a rv Y defined as a function of another rv X, that is, $Y = g(X)$. If the mapping $Y = g(X)$ from X to Y is one-to-one, then the determination of $f_Y(y)$ is straightforward. However, when the mapping is not one-to-one, $f_Y(y)$ is determined by using all roots of the function $Y = g(X)$. The mean value of $g(X)$ is given by (F.7).

Example F.3   Linearly Dependent Rv's.
Consider a rv Y which is linearly dependent on another rv X as described by $Y = aX + b$ where $a > 0$ and b are constants. The cdf of Y may be written in terms of the cdf of X as follows:

$$F_Y(y) = P(Y \leq y) = P(aX + b \leq y)$$

$$= P\left(X \leq \frac{y-b}{a}\right) = F_X\left(\frac{y-b}{a}\right) \quad \text{(F.42)}$$

Differentiation with respect to y yields the pdf of Y:

$$f_Y(y) = \frac{1}{a} f_X\left(\frac{y-b}{a}\right) \quad \text{(F.43)}$$

Homework F.3   Consider a Gaussian rv X with zero mean and unity variance. Determine the pdf of $Y = m + \sigma X$, which evidently has a mean m and variance σ^2.

Example F.4   Square-Law Detector $Y = aX^2 + b$, $a > 0$.
The cdf of the signal Y at the output of a square-law detector may be expressed in terms of the cdf of the input signal X as follows:

$$F_Y(y) = P(Y \leq y) = P(aX^2 + b \leq y)$$

$$= P\left(|X| \leq \sqrt{(y-b)/a}\right)$$

$$= P\left(-\sqrt{(y-b)/a} \leq X \leq \sqrt{(y-b)/a}\right)$$

$$= P\left(X \leq \sqrt{(y-b)/a}\right) - P\left(X \leq -\sqrt{(y-b)/a}\right)$$

$$= F_X\left(\sqrt{(y-b)/a}\right) - F_X\left(-\sqrt{(y-b)/a}\right)$$

$$\text{(F.44)}$$

Differentiation with respect to y yields

$$f_Y(y) = \frac{f_X\left(\sqrt{(y-b)/a}\right)}{2a\sqrt{(y-b)/a}} + \frac{f_X\left(-\sqrt{(y-b)/a}\right)}{2a\sqrt{(y-b)/a}}$$

$$\text{(F.45)}$$

If the input to a square-law detector has a normal pdf with zero mean and variance σ_X^2, we use (F.45) to determine the pdf at the output of the square-law detector with $a = 1$ and $b = 0$:

$$f_X(x) = \frac{1}{\sqrt{2\pi}\sigma_X} \exp\left(-\frac{x^2}{2\sigma_X^2}\right), \quad -\infty < x < \infty$$

$$f_Y(y) = \frac{1}{\sqrt{2\pi y}\,\sigma_X} \exp\left(-\frac{y}{2\sigma_X^2}\right), \quad 0 \leq y < \infty$$

$$\text{(F.46)}$$

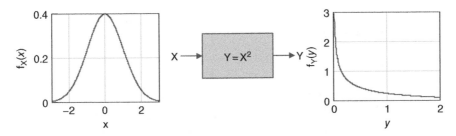

Figure F.3   The Pdf's at the Input and the Output of a Square-Law Detector.

Note that the output of the square-law detector is proportional to the power of the signal X at its input. Therefore, as shown by Figure F.3, the pdf of the output signal Y is concentrated near $y = 0$.

Homework F.4   Determine the pdf of $Y = |X|$, which represents a full-wave rectifier.

Example F.5   Sum of Two Rv's, $Z = X + Y$. Here we determine the pdf and the cdf of a rv Z which is the sum of the rv's X and Y, $Z = X + Y$. The cdf of Z is shown by the shaded area, below the curve $z = x + y$ in Figure F.4, which is defined by $-\infty < x \le z - y$ and $-\infty < y < \infty$: [1]

$$F_Z(z) = P(Z \le z)$$

$$= P(X + Y \le z) = \int_{-\infty}^{\infty} \int_{-\infty}^{z-y} f_{XY}(x,y)dx\, dy$$
$$\text{(F.47)}$$

Differentiation with respect to z yields the pdf:

$$f_Z(z) = \frac{d}{dz} F_Z(z) = \int_{-\infty}^{\infty} f_{XY}(z-y,y)\, dy \quad \text{(F.48)}$$

If X and Y are independent, then their joint pdf may be written as the product of marginal pdf's and (F.48) reduces to the convolution of the marginal pdf's of X and Y:

$$f_Z(z) = \int_{-\infty}^{\infty} f_X(z-y)\, f_Y(y)dy = f_X(x) \otimes f_Y(y)$$
$$\text{(F.49)}$$

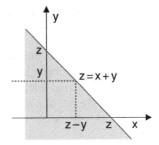

Figure F.4   Sum of Two Rv's, $Z = X + Y$.

Note from (F.35) that, instead of carrying out the convolution in (F.49), the pdf of Z may be found more easily by using the inverse transform of the product of the MGF of X and Y.

Example F.6   A Sinusoidal Carrier with Random Phase. Consider a rv X defined by

$$X = g(\Phi) = A \cos(wt + \Phi) \quad \text{(F.50)}$$

where A, w and t are constants but Φ is a rv with uniform distribution in $[0, 2\pi]$:

$$f_\Phi(\phi) = \frac{1}{2\pi}, \quad 0 \le \phi \le 2\pi \quad \text{(F.51)}$$

The mean value of the rv X may be determined as follows:

$$m_X = E[X] = \int_{-A}^{A} x f_X(x)\, dx$$
$$= \int_{0}^{2\pi} A \cos(wt + \phi) f_\Phi(\phi)\, d\phi = 0$$
$$\text{(F.52)}$$

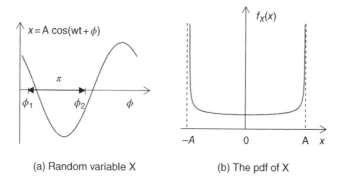

(a) Random variable X   (b) The pdf of X

Figure F.5   The Pdf of the Rv X Defined By (F.50).

Note that the evaluation of the integral on the first line requires the knowledge of the pdf of X, whereas the second integral uses the pdf of Φ (see (F.7)). The pdf of X may be found from the pdf of Φ with a variable transformation:

$$f_X(x) = f_\Phi(\phi)\left|\frac{d\phi}{dx}\right|_{\phi=\phi_1} + f_\Phi(\phi)\left|\frac{d\phi}{dx}\right|_{\phi=\phi_2}$$

$$= \frac{1}{\pi\sqrt{A^2 - x^2}}, \quad -A \le x \le A$$

(F.53)

where $\phi_1 = \phi_2 - \pi = \pi/2 - wt$ are the two roots of the multi-valued function (F.50) (see Figure F.5). Note that the pdf is independent of w and time t, accounting for the time-invariance of the mean. Also note that the pdf peaks at $x = \pm A$.

F.5   Multiple Functions of Multiple Rv's

Let Z and W denote two functions of rv's X and Y:

$$Z = g(X, Y)$$
$$W = h(X, Y)$$

(F.54)

If, for all z and w, $z = g(x,y)$ and $w = h(x,y)$, have a finite number of solutions $\{x_i, y_i\}$, and the determinant of the Jacobian matrix,

$$J(x,y) = \begin{bmatrix} \partial z/\partial x & \partial z/\partial y \\ \partial w/\partial x & \partial w/\partial y \end{bmatrix}$$

(F.55)

is nonzero, then the pdf $f_{Z,W}(w,z)$ is given by

$$f_{ZW}(z,w) = \sum_i \frac{f_{XY}(x_i, y_i)}{|\det J(x_i, y_i)|}$$

(F.56)

Example F.7   Transformation of the Joint Pdf From Cartesian To Polar Coordinates. Transformation of the joint pdf of two independent Gaussian rv's in Cartesian coordinates to polar coordinates is important in many applications. Consider two independent Gaussian rv's X and Y with different means but identical variances:

$$X = N\left(\sqrt{E}\cos\phi' \quad \sigma^2\right)$$
$$Y = N\left(\sqrt{E}\sin\phi' \quad \sigma^2\right)$$

(F.57)

The joint pdf of X and Y may be written as

$$f_{XY}(x,y) = f_X(x)f_Y(y)$$

$$= \frac{1}{2\pi\sigma^2}\exp\left(-\frac{\left(x-\sqrt{E}\cos\phi'\right)^2 + \left(y-\sqrt{E}\sin\phi'\right)^2}{2\sigma^2}\right)$$

$$= \frac{1}{2\pi\sigma^2}\exp\left(-\frac{x^2 + E - 2\sqrt{E}(x\cos\phi' + y\sin\phi')}{2\sigma^2}\right)$$

(F.58)

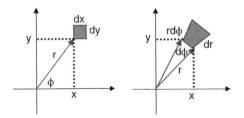

Figure F.6 Transformation Between Rectangular And Polar Coordinates.

The magnitude r and the phase ϕ of a point described by (x, y) in Cartesian coordinates may be written in polar coordinates as follows (see Figure F.6):

$$r = \sqrt{x^2 + y^2} = g(x,y) \qquad x = r\cos\phi$$
$$\phi = \tan^{-1}(y/x) = h(x,y) \implies y = r\sin\phi$$
$$(F.59)$$

Using the determinant of the Jacobian matrix

$$|J(x,y)| = \begin{vmatrix} \partial r/\partial x & \partial r/\partial y \\ \partial\theta/\partial x & \partial\theta/\partial y \end{vmatrix}$$

$$= \begin{vmatrix} \dfrac{x}{\sqrt{x^2+y^2}} & \dfrac{y}{\sqrt{x^2+y^2}} \\ \dfrac{-y}{x^2+y^2} & \dfrac{x}{x^2+y^2} \end{vmatrix} = \frac{1}{\sqrt{x^2+y^2}} = \frac{1}{r}$$

$$(F.60)$$

and based on the fact that (F.59) has a single solution, namely $x = r\cos\phi$ and $y = r\sin\phi$, the joint and marginal pdf's are found using (F.56), (F.58) and (F.60) as follows:

$$f_{R\Phi}(r,\phi) = r\, f_{XY}(r\cos\phi, r\sin\phi)$$
$$= \frac{r}{2\pi\sigma^2}\exp\left(-\frac{r^2 + E - 2\sqrt{E}r\cos(\phi-\phi')}{2\sigma^2}\right)$$

$$f_R(r) = \int_0^{2\pi} f_{R\Phi}(r,\phi)\, d\phi$$
$$= \frac{r}{\sigma^2}\exp\left(-\frac{r^2+E}{2\sigma^2}\right) I_0\left(\frac{\sqrt{E}r}{\sigma^2}\right)$$

$$f_\Phi(\phi) = \int_0^\infty f_{R\Phi}(r,\phi)\, dr$$
$$= \frac{1}{2\pi\sigma^2}\exp\left(-\frac{E\sin^2(\phi-\phi')}{2\sigma^2}\right)$$
$$\times \int_0^\infty r\exp\left(-\frac{[r-\sqrt{E}\cos(\phi-\phi')]^2}{2\sigma^2}\right) dr$$

$$= \frac{1}{2\pi}e^{-\frac{\gamma}{2}}\left[1 + e^{\frac{\gamma}{2}\cos^2(\phi-\phi')}\sqrt{2\pi\gamma}\cos(\phi-\phi')\right.$$

$$\times\left.\left\{1 - Q\left(\sqrt{\gamma}\cos(\phi-\phi')\right)\right\}\right]$$

$$(F.61)$$

where $\gamma = E/\sigma^2$ and $I_0(x) = (1/2\pi)\int_0^{2\pi} e^{x\cos\theta}\, d\theta$ denotes the modified Bessel function of the first kind and order zero. For large values of $\sqrt{\gamma}\cos(\phi-\phi')$, Q function and the term $e^{-\gamma/2}/(2\pi)$ in (F.61) vanish and the pdf of the phase reduces to

$$f_\Phi(\phi) \simeq \sqrt{\frac{\gamma}{2\pi}}e^{-\frac{\gamma}{2}\sin^2(\phi-\phi')}\cos(\phi-\phi')$$
$$\sqrt{\gamma}\cos(\phi-\phi') \gg 1$$
$$(F.62)$$

In view of the fact that (F.59) has a single solution, the joint pdf given by (F.61) may also be determined by using the following variable transformation (see Figure F.6):

$$f_{R\Phi}(r,\phi)|dr\,d\phi| = f_{XY}(x,y)\Big|_{\substack{x=r\cos\phi \\ y=r\sin\phi}} \underbrace{|dx\,dy|}_{r\,dr\,d\phi} \implies$$

$$f_{R\Phi}(r,\phi) = r f_{XY}(r\cos\phi, r\sin\phi)$$
$$(F.63)$$

In case where $E = 0$, then joint and marginal pdf's given by (F.61) reduces to

$$f_{R\Phi}(r,\phi) = \frac{r}{2\pi\sigma^2}e^{-\frac{r^2}{2\sigma^2}}$$

$$f_R(r) = \int_0^{2\pi} f_{R\Phi}(r,\phi)\, d\phi = \frac{r}{\sigma^2}e^{-\frac{r^2}{2\sigma^2}}$$
$$(F.64)$$

$$f_\Phi(\phi) = \int_0^\infty f_{R\Phi}(r,\phi)\, dr$$
$$= \frac{1}{2\pi}\int_0^\infty \frac{r}{\sigma^2}e^{-\frac{r^2}{2\sigma^2}}\, dr = \frac{1}{2\pi}$$

Then, the pdf of Φ is uniformly distributed in $[0,2\pi]$ while R has a Rayleigh distribution.

Example F.8 The Pdf of the Ratio of Two Rv's $Z = X/Y$.

In order to determine the pdf of $Z = X/Y$ using the joint pdf $f_{X,Y}(x,y)$ of X and Y, let $Z = X/Y$ and $W = Y$. Using (F.55), the determinant of the Jacobian matrix is found to be

$$\det J = \det \begin{bmatrix} 1/y & -x/y^2 \\ 0 & 1 \end{bmatrix} = \frac{1}{y} = \frac{1}{w} \quad \text{(F.65)}$$

The joint pdf of Z and W is given by (F.56). The marginal pdf of Z is obtained by integrating the joint pdf of Z and W with respect to w: [1]

$$f_Z(z) = \int_{-\infty}^{\infty} f_{ZW}(z,w) \, dw$$
$$= \int_{-\infty}^{\infty} |w| f_{XY}(zw,w) \, dw \quad \text{(F.66)}$$

Homework F.5 If the correlated rv's X and Y are jointly normal with zero mean, then their joint pdf's can be written as (see (F.28))

$$f_{XY}(x,y) = \frac{1}{2\pi\sigma_X\sigma_Y\sqrt{1-\rho^2}}$$
$$\exp\left[-\frac{1}{2(1-\rho^2)}\left\{\frac{x^2}{\sigma_X^2} - 2\rho\frac{xy}{\sigma_X\sigma_Y} + \frac{y^2}{\sigma_Y^2}\right\}\right] \quad \text{(F.67)}$$

Show that the ratio $Z = X/Y$ has a Cauchy pdf centered at $\rho\sigma_X/\sigma_Y$:

$$f_Z(z) = \frac{1}{\pi} \frac{\frac{\sigma_Y}{\sigma_X\sqrt{1-\rho^2}}}{\left(\frac{\sigma_Y z - \rho\sigma_X}{\sigma_X\sqrt{1-\rho^2}}\right)^2 + 1} \quad \text{(F.68)}$$

Using (D.46) the cdf may be expressed as

$$F_Z(z) = \frac{1}{2} + \frac{1}{\pi}\tan^{-1}\left(\frac{\sigma_Y z - \rho\sigma_X}{\sigma_X\sqrt{1-\rho^2}}\right) \quad \text{(F.69)}$$

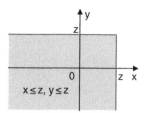

Figure F.7 $Z = \max(X,Y)$.

Example F.9 The Pdf of $Z = \max(X, Y)$. The cdf of Z is defined by: [1]

$$F_Z(z) = P(Z \leq z) = P(\max(X,Y) \leq z)$$
$$= P(X \leq z \text{ and } Y \leq z) = F_{XY}(z,z) \quad \text{(F.70)}$$

The last expression is obtained by inserting $x_1 = y_1 = -\infty$ and $x_2 = y_2 = z$ into (F.13) (see Figure F.2).

The pdf of Z may be written as

$$f_Z(z) = \frac{\partial}{\partial x}F_{XY}(z,z) + \frac{\partial}{\partial y}F_{XY}(z,z)$$
$$= \int_{-\infty}^{z} f_{XY}(z,y) \, dy + \int_{-\infty}^{z} f_{XY}(x,z) \, dx \quad \text{(F.71)}$$

If X and Y are independent, then

$$F_Z(z) = F_X(z)F_Y(z)$$
$$f_Z(z) = f_X(z)F_Y(z) + F_X(z)f_Y(z) \quad \text{(F.72)}$$

Example F.10 The Pdf of $Z = \min(X, Y)$. $Z = \min(X, Y)$ defines the region of the xy-plane where $\min(x, y) \leq z$, that is, $x \leq z$ or $y \leq z$. (see Figure F.8). Hence, the cdf of Z may be written as [1]

$$F_Z(z) = P(Z \leq z) = P(\min(X,Y) \leq z)$$
$$= 1 - P(\min(X,Y) > z)$$
$$= 1 - P(X > z, Y > z) \quad \text{(F.73)}$$
$$= F_X(z) + F_Y(z) - F_{XY}(z,z)$$

The last expression, which corresponds to the shaded area in Figure F.8, may also be obtained

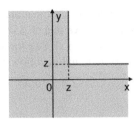

Figure F.8 $Z = \min(X, Y)$.

by inserting $x_1 = y_1 = z$ and $x_2 = y_2 = \infty$ into (F.13) (see also Figure F.2).

The pdf of Z may be obtained from (F.71) and (F.73):

$$f_Z(z) = f_X(z) + f_Y(z) - \int_{-\infty}^{z} f_{XY}(z, y)\, dy$$

$$- \int_{-\infty}^{z} f_{XY}(x, z)\, dx \tag{F.74}$$

If X and Y are independent, then

$$F_Z(z) = F_X(z) + F_Y(z) - F_X(z) F_Y(z)$$

$$f_Z(z) = f_X(z)(1 - F_Y(z)) + f_Y(z)(1 - F_X(z)) \tag{F.75}$$

Example   F.11   Multivariate   Normal Distribution.

Consider the following transformation between the random vectors $\mathbf{z}$ and $\mathbf{x}$, each of dimension n:

$$\mathbf{x} = \boldsymbol{\mu} + \mathbf{H}\mathbf{z} \tag{F.76}$$

where

$$\mathbf{x} = \begin{bmatrix} x_1 \\ x_2 \\ \vdots \\ x_n \end{bmatrix} \quad \boldsymbol{\mu} = \begin{bmatrix} \mu_1 \\ \mu_2 \\ \vdots \\ \mu_n \end{bmatrix} \quad \mathbf{z} = \begin{bmatrix} z_1 \\ z_2 \\ \vdots \\ z_n \end{bmatrix} \tag{F.77}$$

$$\mathbf{H} = \begin{bmatrix} h_{11} & h_{12} & \cdots & h_{1n} \\ h_{21} & h_{22} & \cdots & h_{2n} \\ \vdots & \vdots & \ddots & \vdots \\ h_{n1} & h_{n2} & \cdots & h_{nn} \end{bmatrix}$$

The pdf of $\mathbf{z}$, whose elements z_i are zero-mean and unity variance Gaussian rv's:

$$f_{\mathbf{z}}(\mathbf{z}) = \prod_{i=1}^{n} \frac{1}{\sqrt{2\pi}} \exp\left(-\frac{z_i^2}{2}\right)$$

$$= \frac{1}{(2\pi)^{n/2}} \exp\left(-\frac{\mathbf{z}^H \mathbf{z}}{2}\right) \tag{F.78}$$

Noting that $\mathbf{z} = \mathbf{H}^{-1}(\mathbf{x} - \boldsymbol{\mu})$, the pdf of $\mathbf{x}$:

$$f(\mathbf{x}) = \frac{1}{(2\pi)^{n/2} |\mathbf{H}|}$$

$$\exp\left(-\frac{1}{2}(\mathbf{x} - \boldsymbol{\mu})^H \left(\mathbf{H}\mathbf{H}^H\right)^{-1} (\mathbf{x} - \boldsymbol{\mu})\right)$$

$$|\mathbf{H}| = |\det(\mathbf{H})| = \sqrt{\det\left(\mathbf{H}\mathbf{H}^H\right)} \tag{F.79}$$

Moments of $\mathbf{x} = \boldsymbol{\mu} + \mathbf{H}\mathbf{z}$:

$$E[\mathbf{x}] = E[\boldsymbol{\mu} + \mathbf{H}\mathbf{z}] = \boldsymbol{\mu} + \mathbf{H}\underbrace{E[\mathbf{z}]}_{=0} = \boldsymbol{\mu}$$

$$\operatorname{covar}[\mathbf{x}] = E\left[\mathbf{H}\mathbf{z}(\mathbf{H}\mathbf{z})^H\right] = \mathbf{H}\underbrace{E\left[\mathbf{z}\mathbf{z}^H\right]}_{\operatorname{covar}(\mathbf{z}) = \mathbf{I}} \mathbf{H}^H = \mathbf{H}\mathbf{H}^H \tag{F.80}$$

Moment generating function of $\mathbf{x}$:

$$M(\mathbf{s}) = \int_{-\infty}^{\infty} e^{-\mathbf{s}^T \mathbf{x}} f(\mathbf{x}) d\mathbf{x}$$

$$= \exp\left(-\mathbf{s}^T \boldsymbol{\mu} + \frac{1}{2}\mathbf{s}^T \mathbf{H}\mathbf{H}^H \mathbf{s}\right) \tag{F.81}$$

Homework F.6   Show that (F.79) reduces to (F.28) for $n = 2$ when

$$\mathbf{H}\mathbf{H}^H = \begin{bmatrix} \sigma_1^2 & \rho\sigma_1\sigma_2 \\ \rho\sigma_1\sigma_2 & \sigma_2^2 \end{bmatrix}$$

F.6 Ordered Statistics

Consider N rv's X_k, $k = 1,2,\ldots, N$, each with cdf $F(x)$. These rv's are ordered as follows:

$$X_{(1)} < X_{(2)} < \cdots < X_{(N)} \qquad \text{(F.82)}$$

The cdf of $X_{(k)}$ is denoted by $F_{(k)}(x)$. The cdf of the largest rv:

$$F_{(N)}(x) = P\big[X_{(N)} \leq x\big]$$

$$= P\big[X_{(1)} \leq x,\, X_{(2)} \leq x, \cdots, X_{(N)} \leq x\big]$$

$$= [F(x)]^N \qquad \text{(F.83)}$$

The cdf of the smallest rv:

$$F_{(1)}(x) = P\big[X_{(1)} \leq x\big] = 1 - P\big[X_{(1)} > x\big]$$

$$= 1 - P\big[X_{(1)} > x,\, X_{(2)} > x, \cdots, X_{(N)} > x\big]$$

$$= 1 - [1 - F(x)]^N \qquad \text{(F.84)}$$

The cdf of the kth rv:

$$F_{(k)}(x) = P\big[X_{(k)} \leq x\big]$$

$$= P\big[\text{at least } k\ X_{(i)} \leq x\big]$$

$$= \sum_{i=k}^{N} \binom{N}{i} [F(x)]^i\, [1 - F(x)]^{N-i}$$

$$= 1 - \sum_{i=0}^{k-1} \binom{N}{i} [F(x)]^i\, [1 - F(x)]^{N-i} \qquad \text{(F.85)}$$

where the last expression is obtained by inserting $a = 1 - b = F(x)$ into (D.6). The binomial coefficient

$$\binom{n}{k} = \frac{n!}{k!(n-k)!} \qquad \text{(F.86)}$$

denotes the number of combinations of k in n. One may easily observe that (F.85) reduces to (F.83) for $k = N$ and to (F.84) for $k = 1$. The pdf of the kth rv is found as follows:

$$f_{(k)}(x) = \frac{d}{dx} F_{(k)}(x)$$

$$= \sum_{i=k}^{N} \binom{N}{i} \left\{ \begin{array}{l} i[F(x)]^{i-1}\, [1 - F(x)]^{N-i} f(x) \\[4pt] -[F(x)]^i (N-i)\, [1 - F(x)]^{N-i-1} f(x) \end{array} \right\}$$

$$= k \binom{N}{k} [F(x)]^{k-1}\, [1 - F(x)]^{N-k} f(x)$$

$$+ \sum_{i=k+1}^{N} \binom{N}{i} i[F(x)]^{i-1}\, [1 - F(x)]^{N-i} f(x)$$

$$- \sum_{i=k}^{N} \binom{N}{i} (N-i)[F(x)]^i\, [1 - F(x)]^{N-i-1} f(x) \qquad \text{(F.87)}$$

Since last two terms cancel each other, the pdf of the kth rv reduces to

$$f_{(k)}(x) = k \binom{N}{k} [F(x)]^{k-1}\, [1 - F(x)]^{N-k} f(x) \qquad \text{(F.88)}$$

The pdf of the largest rv ($k = N$):

$$f_{(N)}(x) = N[F(x)]^{N-1} f(x) \qquad \text{(F.89)}$$

The pdf of the smallest rv ($k = 1$):

$$f_{(1)}(x) = N\, [1 - F(x)]^{N-1} f(x) \qquad \text{(F.90)}$$

The pdf's $f_{(N)}(x)$ and $f_{(1)}(x)$ reduce to (F.72) and (F.75), respectively, for $N = 2$ when X and Y are i.i.d. Figure F.9 shows the pdf and the cdf of kth rv for $N = 5$.

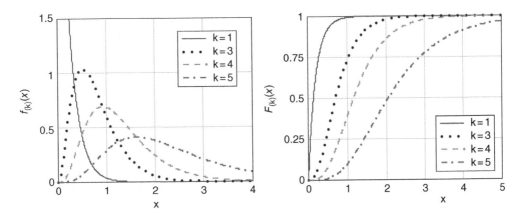

Figure F.9 Pdf and Cdf of the Ordered Rv's for $N=5$ and $k=1,3$ and 5. Exponential Distribution with Unity Mean, $f_X(x)=e^{-x}$, $x \geq 0$, Is Assumed.

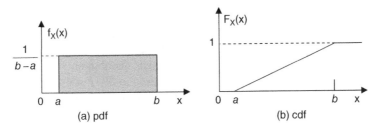

Figure F.10 Pdf and cdf of the uniform distribution.

F.7 Probability Distribution Functions

Uniform distribution:

A continuous rv X is said to be uniformly distributed in [a,b] if it is equally likely in [a,b]:

$$f_X(x) = \frac{1}{b-a}, \quad -\infty < a \leq x \leq b < \infty \quad \text{(F.91)}$$

The cdf is (see Figure F.10)

$$F_X(x) = \begin{cases} (x-a)/(b-a) & a \leq x < b \\ 1 & x \geq b \end{cases} \quad \text{(F.92)}$$

The mean and the variance of a uniformly distributed rv is given by

$$\begin{aligned} m_X &= (a+b)/2 \\ \sigma_X^2 &= (b-a)^2/12 \end{aligned} \quad \text{(F.93)}$$

The MGF is given by

$$M_X(s) = \frac{e^{-sa} - e^{-sb}}{s(b-a)} \quad \text{(F.94)}$$

Gaussian (normal) pdf

The pdf of Gaussian (normal) distribution is defined by

$$f_X(x) = \frac{1}{\sqrt{2\pi}\sigma_X} \exp\left(-\frac{(x-m_X)^2}{2\sigma_X^2}\right), \quad -\infty < x < \infty$$

$$\text{(F.95)}$$

where $m_X = E[X]$ and $\sigma_X^2 = E\left[(X-m_X)^2\right]$ denote respectively the mean (expected) value and the variance of the rv X. The cdf is found to be

$$F_X(x) = \int_{-\infty}^{x} f_X(u)\ du = 1 - \int_{x}^{\infty} f_X(u)\ du$$

$$= 1 - Q\left(\frac{x - m_X}{\sigma_X}\right)$$

(F.96)

The Gaussian $Q(x)$ function

$$Q(x) = \frac{1}{\sqrt{2\pi}} \int_{x}^{\infty} e^{-t^2/2}\ dt \qquad \text{(F.97)}$$

represents the area under the tail of a zero-mean and unity variance Gaussian pdf (see (B.1)). Hence, it is a monotonically decreasing function of x.

The higher order moments of a Gaussian rv are given by

$$\mu_k = E\left[(X - m_X)^k\right]$$

$$= \begin{cases} 1.3 \ \cdots (k-1)\sigma_X^k & k \text{ even} \\ 0 & k \text{ odd} \end{cases} \qquad \text{(F.98)}$$

$$E\left[X^k\right] = \sum_{i=0}^{k} \binom{k}{i} m_X^i \mu_{k-i}$$

where the binomial coefficient is defined by (F.86).

The MGF of X is

$$M_X(s) = \int_{-\infty}^{\infty} e^{-sX} f_X(x)\ dx$$

$$= \exp\left(-sm_X + \frac{1}{2}s^2\sigma_X^2\right)$$

(F.99)

Example F.12 Sum of Gaussian Rv's.
The sum of n statistically independent Gaussian rv's X_i, that is, $Z = \sum_{i=1}^{n} X_i$, with distinct means and variances, is found using (F.35) and (F.99):

$$M_Z(s) = \prod_{i=1}^{n} M_{X_i}(s) = \prod_{i=1}^{n} e^{\left(-sm_i + s\sigma_i^2/2\right)}$$

$$= e^{-sm_Z + s^2\sigma_Z^2/2}$$

$$m_Z = \sum_{i=1}^{n} m_i, \quad \sigma_Z^2 = \sum_{i=1}^{n} \sigma_i^2$$

(F.100)

It is clear from (F.100) that the pdf of the sum of n independent Gaussian rv's is again a Gaussian rv with mean m_Z and variance σ_Z^2.

Log-normal distribution

In many areas of telecommunications, the signal levels are measured in dB. Consequently, the mean and the variance of signals are also expressed in dB. Log-normal distribution is a Gaussian distribution for the random signals expressed in dB. For example, in a shadowing environment, the received signal power level at a distance r from the transmitter may be written as (see (11.123))

$$P(r) = P_m(r) + \chi \quad (dB) \qquad \text{(F.101)}$$

where $P_m(r)$ denotes the mean received power level in dB at distance r and χ in dB denotes the shadow fading with a zero dB mean and a standard deviation σ in dB. Then, the pdf of the received signal at a given distance r has a normal distribution:

$$f_{P(r)}(x) = \frac{1}{\sqrt{2\pi}\sigma} \exp\left(-\frac{(x - P_m(r))^2}{2\sigma^2}\right), \quad -\infty < x < \infty$$

(F.102)

As long as $P(r)$, $P_m(r)$ and σ are in dB, (F.102) obeys all the rules for a normal distribution given by (F.95)–(F.99). Note that the above formula is valid for a given value of the distance r since (F.101) describes a random process in r (see Chapter 1, Section 1.3.2).

Exponential distribution

The pdf and the cdf of an exponentially distributed rv X are given by

$$f_X(x) = \frac{1}{\bar{\gamma}} e^{-x/\bar{\gamma}} \quad , \quad \bar{\gamma} > 0, \ x \geq 0 \qquad (F.103)$$

$$F_X(x) = 1 - e^{-x/\bar{\gamma}}$$

The mean and the standard deviation of X are

$$m_X = \sigma_X = \bar{\gamma} \qquad (F.104)$$

Using (F.34), the MGF is found to be

$$M_X(s) = \frac{1}{1 + s\bar{\gamma}} \qquad (F.105)$$

Homework F.7 Consider a channel which undergoes fading where the received signal power has exponential distibution. Determine the probability that the received signal power level is 10 dB below the average level.

Chi-square distribution

Central chi-square distribution:

If X is a Gaussian rv with $(0, \sigma_X^2)$, then the pdf of

$$Y = X^2 \qquad (F.106)$$

is given by (F.46) and is said to have a *central chi-square distribution*:

$$f_Y(y) = \frac{1}{\sqrt{2\pi y}\sigma_X} e^{-\frac{y}{2\sigma_X^2}}, \ y \geq 0 \qquad (F.107)$$

The MGF corresponding to (F.107) is found from (F.34) as

$$M_Y(s) = \int_0^{\infty} e^{-sy} f_Y(y) \ dy = \frac{1}{\sqrt{1 + s2\sigma_X^2}} \qquad (F.108)$$

The statistics of the power sums of i.i.d. Gaussian rv's with zero-mean and variance σ_X^2

$$Z = \sum_{i=1}^{n} X_i^2 \qquad (F.109)$$

is found using the MGF approach:

$$M_Z(s) = \frac{1}{\left(1 + s2\sigma_X^2\right)^{n/2}} \qquad (F.110)$$

The pdf corresponding to (F.110) is called as the central chi-square pdf with n degrees of freedom and is determined using the Fourier transforms in Appendix C:

$$f_Z(z) = \frac{1}{(2\sigma_X^2)^{n/2}\Gamma(n/2)} z^{n/2-1} e^{-z/2\sigma_X^2}, \ z \geq 0 \qquad (F.111)$$

The Gamma function is defined by (D.108):

$$\Gamma(\alpha) = \int_0^{\infty} t^{\alpha-1} e^{-t} dt, \ \ \text{Re}(\alpha) \geq 0 \qquad (F.112)$$

$$\Gamma(n) = (n-1)! \ , \ \ n = 1, 2, \cdots$$

The moments:

$$m_Z = E[Z] = n\sigma_X^2$$
$$\sigma_Z^2 = E[Z^2] - m_Z^2 = 2n\sigma_X^4 \qquad (F.113)$$

The cdf corresponding to (F.111) is given by

$$F_Z(y) = \int_0^z f_Z(u) \ du$$

$$= \frac{1}{(2\sigma_X^2)^{n/2}\Gamma(n/2)} \int_0^z u^{n/2-1} e^{-u/2\sigma_X^2} du \qquad (F.114)$$

In wireless communications, one often encounters the sum of the powers of m rv's with central chi-square distribution. The power P of a complex rv, $X = X_r + jX_i$, is given by

$P = |X|^2 = |X_r|^2 + |X_i|^2$, where X_r and X_i denote, respectively, the real and the imaginary parts of X. Therefore, the pdf and cdf of the sum are obtained by inserting $m = n/2$ into (F.111) and analytical evaluation of (F.114) using (D.49):

$$f_Z(z) = \frac{1}{(2\,\sigma_X^2)^m \Gamma(m)} z^{m-1} e^{-z/2\sigma_X^2}, \quad z \geq 0$$

$$F_Z(z) = 1 - \frac{1}{\Gamma(m)} \gamma\left(m, \frac{z}{2\,\sigma_X^2}\right)$$

$$= 1 - e^{-z/2\sigma_X^2} \sum_{k=0}^{m-1} \frac{1}{k!} \left(\frac{z}{2\,\sigma_X^2}\right)^k, \quad z \geq 0$$

$$\text{(F.115)}$$

Here, $\gamma(n, x)$ denotes the incomplete Gamma function (see (D.106) and (D.108)):

$$\gamma(\alpha, z) = \int_0^z t^{\alpha-1} e^{-t} dt, \quad \text{Re}(\alpha) > 0$$

$$\text{(F.116)}$$

$$\Gamma(\alpha) = \gamma(\alpha, \infty)$$

The pdf and the cdf given by (F.115) are shown in Figure F.11 for various values of n. The pdf and the cdf were observed to shift towards higher values of z as n increases, implying the increased likelihood of observing higher values of z.

Non-central chi-square distribution:

When the Gaussian rv X has non-zero mean with (m_X, σ_X^2), $Y = X^2$ has a *non-central chi-square distribution* with the pdf obtained by inserting $a = 1$ and $b = 0$ into (F.45):

$$f_X(x) = \frac{1}{\sqrt{2\pi}\sigma_X} \exp\left(-\frac{(x-m_X)^2}{2\,\sigma_X^2}\right)$$

$$f_Y(y) = \frac{1}{\sqrt{2\pi y}\sigma_X} \exp\left(-\frac{y + m_X^2}{2\,\sigma_X^2}\right) \cosh\left(\frac{\sqrt{y}m_X}{\sigma_X^2}\right)$$

$$\text{(F.117)}$$

The MGF is found to be:

$$M_Y(s) = \int_0^\infty e^{-sy} f_Y(y)\, dy$$

$$= \frac{1}{\sqrt{1 + s2\,\sigma_X^2}} \exp\left(\frac{-s\,m_X^2}{1 + s2\,\sigma_X^2}\right)$$

$$\text{(F.118)}$$

The MGF of the power sums, $Z = \sum_{i=1}^n X_i^2$, as defined by (F.109), of n i.i.d. Gaussian rv's with non-zero means m_i, $i = 1, 2, \cdots, n$, and identical variances σ_X^2 is found from (F.118) as follows:

$$M_Z(s) = \frac{1}{(1 + s2\,\sigma_X^2)^{n/2}} \exp\left[\frac{-s\chi^2}{1 + s2\,\sigma_X^2}\right]$$

$$\chi^2 = \sum_{i=1}^n m_i^2$$

$$\text{(F.119)}$$

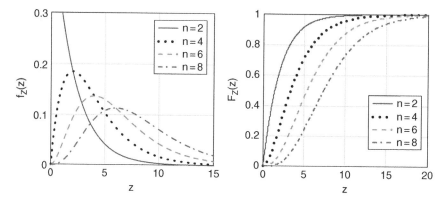

Figure F.11   Pdf and Cdf of the Chi-Square Distribution ($\sigma_X^2 = 1$).

where χ denotes the noncentrality parameter. The pdf of the non-central chi-square distribution with n degrees of freedom) is found using (F.119): [2]

$$f_Z(z) = \frac{1}{2\sigma_X^2}\left(\frac{z}{\chi^2}\right)^{\frac{n-2}{4}}$$

$$\exp\left(-\frac{z+\chi^2}{2\sigma_X^2}\right) I_{\frac{n}{2}-1}\left(\sqrt{z}\frac{\chi}{\sigma_X^2}\right), \quad z \geq 0$$

(F.120)

where $I_\alpha(x)$ is defined by (D.95). The moments are given by

$$m_Z = E[Z] = n\sigma_X^2 + \chi^2$$
$$\sigma_Z^2 = E[Z^2] - m_Z^2 = 2n\sigma_X^4 + 4\chi^2\sigma_X^2$$

(F.121)

When $m = n/2$ is an integer, the cdf may be expressed in terms of the Marcum's Q function as follows:

$$F_Z(z) = \int_0^z f_Z(u)\, du = 1 - Q_m\left(\frac{\chi}{\sigma_X}, \frac{\sqrt{z}}{\sigma_X}\right)$$

(F.122)

The Marcum's Q function is defined by (B.15):

$$Q_m(\alpha,\beta) = \int_\beta^\infty x\left(\frac{x}{\alpha}\right)^{m-1}$$

$$\exp\left(-\frac{x^2 + \alpha^2}{2}\right) I_{m-1}(\alpha x)dx$$

(F.123)

The PDF of the envelope R of the power sums can be obtained from (F.120) via a variable transformation, $z = r^2$:

$$f_R(r) = f_Z(z)\left|\frac{dz}{dr}\right|_{z=r^2}$$

$$= \frac{r^{n/2}}{\sigma_X^2 \chi^{(n-2)/2}} \exp\left(-\frac{r^2 + \chi^2}{2\sigma_X^2}\right) I_{n/2-1}\left(\frac{r\chi}{\sigma_X^2}\right), \quad r \geq 0$$

(F.124)

The moments are given by

$$E[R^k] = \frac{\Gamma((n+k)/2)}{\Gamma(n/2)}(2\sigma_X^2)^{k/2}$$

$$\exp\left(-\frac{\chi^2}{2\sigma_X^2}\right) {}_1F_1\left(\frac{n+k}{2}; \frac{n}{2}; \frac{\chi^2}{2\sigma_X^2}\right), \quad k \geq 0$$

(F.125)

where ${}_1F_1(a; b; z)$ denotes the confluent hypergeometric function defined by (D.88):

$${}_1F_1(a;b;x) = \frac{\Gamma(b)}{\Gamma(a)}\sum_{n=0}^\infty \frac{\Gamma(a+n)}{\Gamma(b+n)}\frac{x^n}{n!} \quad (\text{F.126})$$

The cdf of the envelope is determined from (F.122) using the variable transformation, $z = r^2$:

$$F_R(r) = P(R \leq r) = P\left(\sqrt{Z} \leq r\right) = F_Z(r^2)$$

(F.127)

If $m = n/2$ is an integer

$$F_R(r) = 1 - Q_m\left(\frac{\chi}{\sigma_X}, \frac{r}{\sigma_X}\right), \quad r \geq 0 \quad (\text{F.128})$$

Gamma distribution

The Gamma distribution is similar to central chi-square distribution and is characterized by the following pdf:

$$f_X(x) = \frac{1}{\Gamma(\alpha)\lambda^\alpha}x^{\alpha-1}e^{-x/\lambda}, \quad \alpha, \lambda > 0, x \geq 0$$

(F.129)

where α does not have to be an integer. The cdf may be expressed in terms of the incomplete Gamma function (see (F.115)):

$$F_X(x) = \frac{\gamma(\alpha, x/\lambda)}{\Gamma(\alpha)}, \quad x \geq 0 \quad (\text{F.130})$$

where $\Gamma(\alpha)$ and $\gamma(\alpha, z)$ denote respectively the Gamma function and the incomplete Gamma function (see (F.112) and (F.116)). The mean and the variance of the Gamma distribution are given by

$$m_X = \alpha\lambda, \quad \sigma_X^2 = \alpha\lambda^2 \qquad \text{(F.131)}$$

The MGF is found from (F.34) to be

$$M_X(s) = \frac{1}{(1+s\lambda)^\alpha} \qquad \text{(F.132)}$$

If α is a positive integer, then Gamma distribution reduces to Erlang distribution. For $\alpha = 1$ it reduces to exponential distribution. If $\alpha = m$ for $m = 1, 2, \ldots$ and $\lambda = 2\sigma_X^2$ the Gamma distribution reduces to central chi-square distribution (see (F.115)) [3].

Rayleigh distribution

Rayleigh distribution is frequently used to characterize signals propagating through multipath fading channels. It is a special case of the central chi-square pdf with two degrees of freedom, $Z = X_1^2 + X_2^2$, where X_1 and X_2 are zero-mean statistically independent Gaussian rv's, each with variance σ^2. Z is often used to represent the power of a complex Gaussian rv, $X_1 + jX_2$. The pdf and the cdf of Z are found by inserting $m = 1$ into (F.115):

$$f_Z(z) = \frac{1}{2\sigma^2} e^{-z/2\sigma^2}, \quad z \geq 0 \qquad \text{(F.133)}$$
$$F_Z(z) = 1 - e^{-z/2\sigma^2}$$

We define a new rv R, which denotes the envelope of this complex-Gaussian rv:

$$R = \sqrt{Z} = \sqrt{X_1^2 + X_2^2} \qquad \text{(F.134)}$$

The envelope R is characterized by the Rayleigh pdf:

$$f_R(r) = f_Z(z)\left|\frac{dz}{dr}\right|_{z=r^2} = \frac{r}{\sigma^2} e^{-r^2/2\sigma^2}, \quad r \geq 0 \qquad \text{(F.135)}$$

The moments of the envelope are given by

$$E[R^k] = (2\sigma^2)^{k/2}\Gamma(1 + k/2) \qquad \text{(F.136)}$$
$$\sigma_R^2 = E[R^2] - (E[R])^2 = (2 - \pi/2)\sigma^2$$

The cdf of the envelope is found to be

$$F_R(r) = \frac{1}{\sigma^2}\int_0^r u \, e^{-u^2/2\sigma^2} \, du = 1 - e^{-r^2/2\sigma^2}, \quad r \geq 0 \qquad \text{(F.137)}$$

The variation of the pdf and the cdf given by (F.135) and (F.137) are plotted in Figure F.12 for $\sigma^2 = 1$.

Homework F.8 The rv's X and Y are independent with Rayleigh pdf

$$f_X(x) = \frac{x}{\sigma_X^2}\exp\left(-\frac{x^2}{2\sigma_X^2}\right), \quad x > 0$$
$$\qquad\qquad\qquad\qquad\qquad \text{(F.138)}$$
$$f_Y(y) = \frac{y}{\sigma_Y^2}\exp\left(-\frac{y^2}{2\sigma_Y^2}\right), \quad y > 0$$

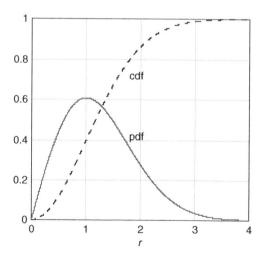

Figure F.12 Pdf and Cdf of the Rayleigh Distribution For $\sigma^2 = 1$.

Show that the pdf and the cdf of $Z = X/Y$ are given by

$$f_Z(z) = \frac{2\,\sigma_X^2}{\sigma_Y^2}\,\frac{z}{(z^2 + \sigma_X^2/\sigma_Y^2)^2}, \quad z \geq 0$$

(F.139)

$$F_Z(z) = \frac{z^2}{z^2 + \sigma_X^2/\sigma_Y^2}$$

Rice distribution

The Rician pdf characterizes the statistics of the envelope of a narrow-band signal with non-zero mean corrupted by additive narrowband Gaussian noise. Rician distribution is a special case of the non-central chi-square pdf with $n = 2$ degrees of freedom, that is, $Z = X_1^2 + X_2^2$. If X_1 and X_2 denote statistically independent Gaussian rv's with $E[X_1] = m_1$, $E[X_2] = m_2$, $\mathrm{var}[X_1] = \mathrm{var}[X_2] = \sigma^2$, then the pdf is a non-central chi-square pdf with two degrees of freedom (see (F.120) and (F.122)):

$$f_Z(z) = \frac{1}{2\sigma^2}\exp\left(-\frac{z+s^2}{2\sigma^2}\right)I_0\left(\sqrt{z}\,\frac{s}{\sigma^2}\right), \quad z \geq 0$$

$$F_Z(z) = 1 - Q_1\left(\frac{s}{\sigma}, \frac{\sqrt{z}}{\sigma}\right)$$

$$s^2 = m_1^2 + m_2^2$$

(F.140)

The pdf of the envelope R is Rician and is given by (F.124) with $n = 2$:

$$f_R(r) = f_Z(z)\left.\left|\frac{dz}{dr}\right|\right|_{z=r^2}$$

$$= \frac{r}{\sigma^2}\exp\left(-\frac{r^2+s^2}{2\sigma^2}\right)I_0\left(\frac{rs}{\sigma^2}\right), \quad r \geq 0$$

(F.141)

where $I_n(x)$ denotes the modified Bessel function of the first kind of order n (see Figure D.2). Note that $I_0(0) = 1$, $I_n(0) = 0$ for $n > 0$ and $I_n(x)$ is a monotonically increasing function of x.

The moments of R are given by

$$E[R] = \sqrt{\frac{\pi}{2}}\,\sigma\,e^{-K/2}[(1+K)I_0(K/2) + K\,I_1(K/2)]$$

$$\sigma_R^2 = 2\sigma^2 + s^2 - (E[R])^2 = 2\sigma^2(1+K) - (E[R])^2$$

$$K = s^2/(2\sigma^2) = (m_1^2 + m_2^2)/(2\sigma^2)$$

(F.142)

where the *Rice factor* K denotes the ratio of the signal power $m_1^2 + m_2^2$ to the noise power, $2\sigma^2$. In a fading environment, the Rice factor shows the ratio of the signal power received from the line-of-sight (LOS) to the received signal power due to diffused scattering.

The cdf of R is obtained by inserting $m = 1$ into (F.128):

$$F_R(r) = 1 - Q_1\left(\frac{s}{\sigma}, \frac{r}{\sigma}\right), \quad r \geq 0 \qquad \text{(F.143)}$$

where the Marcum-Q function is defined by (F.123) and (B.15).

Nakagami-m distribution

Nakagami-m distribution is a flexible pdf which is widely used to characterize signals propagating through multipath fading channels. The pdf and the cdf of the envelope of a signal with Nakagami-m distribution are given by

$$f_R(r) = \frac{2}{\Gamma(m)}\left(\frac{m}{P_0}\right)^m r^{2m-1}\exp\left(-\frac{mr^2}{P_0}\right), \quad r \geq 0$$

$$F_R(r) = 1 - \frac{1}{\Gamma(m)}\gamma\left(m, \frac{mr^2}{P_0}\right)$$

$$= 1 - \exp\left(-\frac{mr^2}{P_0}\right)\sum_{k=0}^{m-1}\frac{1}{k!}\left(\frac{mr^2}{P_0}\right)^k, \quad r \geq 0$$

$$P_0 = E[R^2], \quad m = \frac{P_0^2}{E\left[(R^2 - P_0)^2\right]} \geq \frac{1}{2}$$

(F.144)

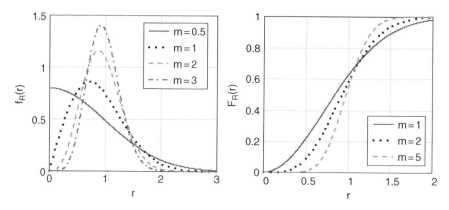

Figure F.13 Pdf and Cdf of a Nakagami-m Distributed Rv with $P_0 = 1$ For Various Values of the Fading Figure m. Note that the curves for $m = 1$ corresponds to Rayleigh fading.

where m denotes the so-called fading figure (parameter). Figure F.13 shows the variation of the pdf and cdf of the Nakagami-m distribution for various values of the fading parameter m. The Nakagami-m distribution reduces to one-sided Gaussian distribution for $m = 1/2$ and Rayleigh distribution for $m = 1$. In the limiting case as m $\to \infty$, (F.144) reduces to a delta function, hence representing a deterministic signal:

$$m = 1/2 \quad f_R(r) = \frac{2}{\sqrt{2\pi P_0}} \exp\left(-\frac{r^2}{2P_0}\right), \quad r \geq 0$$

$$m = 1 \quad f_R(r) = \frac{2r}{P_0} \exp\left(-\frac{r^2}{P_0}\right), \quad r \geq 0$$

$$m \to \infty \quad f_R(r) = \delta(r - \sqrt{P_0}), \quad r \geq 0$$

$$(F.145)$$

The moments of the envelope are found to be

$$E[R^n] = \frac{\Gamma(m + n/2)}{\Gamma(m)} \left(\frac{P_0}{m}\right)^{n/2} \quad (F.146)$$

The PDF of $Y = R^2$ is obtained by a variable transformation:

$$f_Y(y) = f_R(r) \left|\frac{dr}{dy}\right|_{r = \sqrt{y}}$$

$$= \frac{1}{\Gamma(m)} \left(\frac{my}{P_0}\right)^m \frac{\exp(-my/P_0)}{y}, \quad y \geq 0$$

$$(F.147)$$

The cdf of Y is found using (D.49):

$$F_Y(y) = \int_0^y f_Y(u) \, du = 1 - \frac{1}{\Gamma(m)} \gamma\left(m, \frac{my}{P_0}\right)$$

$$= 1 - \exp\left(-\frac{my}{P_0}\right) \sum_{k=0}^{m-1} \frac{1}{k!} \left(\frac{my}{P_0}\right)^k$$

$$(F.148)$$

Comparing with (F.115), one may observe that the power of a Nakagami-m distributed rv has a central chi-square distribution. The MGF is given by

$$M_Y(s) = E[e^{-sY}] = (1 + sP_0/m)^{-m} \quad (F.149)$$

Example F.13 The Statistics of the Power Sums of L Nakagami-Distributed rv's. Consider the power sums of L Nakagami distributed rv's:

$$\gamma = \sum_{n=1}^{L} \gamma_n \quad (F.150)$$

where the pdf of the rv γ_n is defined by (F.147). Assuming that they have identical means, $E[\gamma_n] = \bar{\gamma}$, $n = 1, 2, \cdots, L$, the MGF and the pdf of γ are given by:

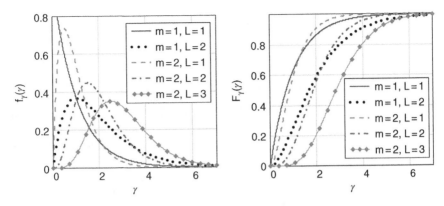

Figure F.14 The Pdf (F.151) and Cdf (F.152) For $\bar{\gamma} = 1$.

$$M_\gamma(s) = (1 + s\bar{\gamma}/m)^{-mL} \Rightarrow$$

$$f_\gamma(\gamma) = \frac{1}{\Gamma(mL)(\bar{\gamma}/m)^{mL}} \gamma^{mL-1} \exp\left(-\frac{\gamma}{\bar{\gamma}/m}\right)$$

(F.151)

In view of (F.115), the above pdf may also correspond to the power sums of mL rv's with mean powers $\bar{\gamma}/m$.

The cdf corresponding to (F.151) is given by

$$F_\gamma(\gamma) = 1 - \frac{1}{\Gamma(mL)} \gamma\left(mL, \frac{m\gamma}{\bar{\gamma}}\right)$$

(F.152)

$$= 1 - e^{-m\gamma/\bar{\gamma}} \sum_{k=0}^{mL-1} \frac{1}{k!} \left(\frac{m\gamma}{\bar{\gamma}}\right)^k$$

One may also observe from Figure F.14 that (F.151) with $m = 2$, $L = 1$ and mean branch power $\bar{\gamma}/2$ has the same form as for $m = 1$, $L = 2$ and mean branch power $\bar{\gamma}$ but they are not the same. Comparing the curves with $m = 1$, $L = 2$ and $m = 2$, $L = 1$ in Figure F.14, it is clear that the pdf with higher values of L shifts the pdf towards right implying higher probabilities for higher power levels.

Poisson distribution

Let the rv k denote the number of events during a time interval $\tau = t_2 - t_1$ where t_1 and t_2 are arbitrary

times with $t_2 \geq t_1$. The events may represent telephone calls, e-mails, accidents, earthquakes, departure/arrival of airplanes, deposits/withdraws from an account, births/deaths, arrivals of customers in a bank/supermarket etc. The events are assumed to be independent. The average number of events per unit time is defined by the arrival rate λ in arrivals/s.

The probability of k events during a time interval τ is given by

$$P_k(\tau) = \frac{(\lambda\tau)^k}{k!} \exp(-\lambda\tau), \quad k = 0, 1, 2, \cdots$$

(F.153)

For example, the probability that no customers arrive to a bank during $\tau = 1$ minutes is given by $P_0(\tau) = \exp(-\lambda\tau) = 0.905$ if one customer arrives on the average per 10 minutes, that is, $\lambda = 0.1$ arrivals/minute. Also note $\sum_{k=0}^{\infty} P_k(\tau) = 1$.

The pdf of the number of events during τ is given by

$$f_k(x) = \sum_{k=0}^{\infty} P_k(\tau)\delta(x-k)$$

(F.154)

The corresponding cdf is found, by integrating (F.154):

$$F_k(x) = e^{-\lambda\tau} \sum_{k=0}^{\infty} \frac{(\lambda\tau)^k}{k!} u(x-k) \qquad \text{(F.155)}$$

If the time intervals of any two events do not overlap, then the corresponding rv's are independent. [3][4] The mean value and the variance of k in [0, t] is then given by

$$E[k] = \sum_{k=0}^{\infty} k \frac{(\lambda t)^k}{k!} e^{-\lambda t} = \lambda t e^{-\lambda t} \sum_{k=1}^{\infty} \frac{(\lambda t)^{k-1}}{(k-1)!} = \lambda t$$

$$E[k^2] = e^{-\lambda t} \sum_{k=0}^{\infty} k^2 \frac{(\lambda t)^k}{k!} = \lambda t e^{-\lambda t} \sum_{k=1}^{\infty} k \frac{(\lambda t)^{k-1}}{(k-1)!}$$

$$= \lambda t e^{-\lambda t} \sum_{k'=0}^{\infty} (k'+1) \frac{(\lambda t)^{k'}}{k'!} = \lambda t [1+\lambda t]$$

$$\sigma_k^2 = E[k^2] - \{E[k]\}^2 = \lambda t$$

$$\text{(F.156)}$$

Using (F.34), the MGF is found to be

$$M_X(s) = e^{-\lambda t(1-e^{-s})} \qquad \text{(F.157)}$$

Example F.14 Satellite Hardening Against Solar Flares.

Solar flares are important for the satellite design, since especially semiconductor devices, including solar panels, need to be protected against nuclear radiations from sun and other stars. Otherwise, the solar radiation shortens the lifetime of a satellite by decreasing the efficiency of solar panels in generating the electrical power. A satellite undergoes intense nuclear radiation due to solar flares with a period of approximately 11 years. In this period, the sun becomes more active during 2–3 years with the occurrence of approximately 3 anomalously large (AL) solar flares. Satellites are hardened so as to withstand a certain number of AL solar flares during their lifetime, since solar flares are the principal contributors of the total dose of nuclear radiation.

The average arrival (occurrence) rate of solar flares is assumed to be

$$\lambda = 3/11 \quad (AL\ solar\ flares/year) \qquad \text{(F.158)}$$

Now assume 15 years lifetime for a satellite with 99.99% protection against solar flares, that is, the probability that the satellite has 15 years of guaranteed lifetime with a confidence level of 99.99% or higher. The probability of observing less than or equal to n solar flares during $\tau = 15$ years with probability 99.99% is given by (F.155):

$$e^{-\lambda\tau} \sum_{k=0}^{n} \frac{(\lambda\tau)^k}{k!} \geq 0.9999 \qquad \text{(F.159)}$$

Insertion of (F.158) into (F.159) shows that the number of AL solar flares should be less than $n = 12$. Therefore, the satellite should be hardened against at least 12 AL solar flares for a guaranteed time of 15 years with 99.99% probability. For a confidence level of 99%, (F.159) implies that the satellite should be hardened against $n = 9$ AL solar flares.

Example F.15 Interarrival Time

The concept of interarrival time is closely linked with the Poisson distribution, which describes the occurrence of a random number of events during a given time interval. The interarrival time, τ, denotes the random time interval between two consecutive arrivals (of calls, packets, e-mails, accidents, earthquakes etc.). The probability that there is no arrival (event) during [0, t] and there is a single arrival during $[t, t + \Delta t]$ is found using (F.153) [4]:

$$f_\tau(t)\Delta t = P_0(t) P_1(\Delta t) = \frac{1}{0!} e^{-\lambda t} \frac{\lambda \Delta t}{1!} e^{-\lambda \Delta t}$$

$$\text{(F.160)}$$

The pdf of the interarrival time $f_\tau(t)$ is determined by taking the limit in (F.160):

$$f_\tau(t) = \lim_{\Delta t \to 0} e^{-\lambda t} \lambda e^{-\lambda \Delta t} = \lambda \exp(-\lambda t) \quad \text{(F.161)}$$

Figure F.15 aims to clarify these concepts. [4] Poisson distribution given by (F.154) characterizes the random number arrivals for a fixed

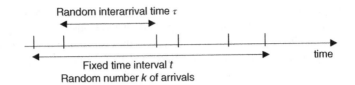

Figure F.15   Random Inter-Arrival Time and the Random Number of Arrivals.

time interval, while the exponential distribution given by (F.161) characterizes the random inter-arrival time. Note that both of these distributions describe the same process but from two different perspectives. For example, if an e-mail user receives 40 e-mails per day with an average $\lambda = 40/24$ (e-mails per hour), then (F.161) implies that the probability of receiving an e-mail is 11% and 72.4% for the next 15 minutes and the next half-hour, respectively.

Example F.16   Football Game.

Consider a 90 minute football game between teams A and B. Assume that the team A scores according to a Poisson process of mean intensity λ_A goals per hour and team B with an intensity of λ_B goals per hour. The probability that team A scores m goals and team B scores n goals during $\tau = 1.5$ hours is given by

$$\frac{(\lambda_A \tau)^m}{m!} e^{-\lambda_A \tau} \frac{(\lambda_B \tau)^n}{n!} e^{-\lambda_B \tau} \qquad \text{(F.162)}$$

For $\lambda_A = 0.5$ and $\lambda_B = 1.5$, the probabilities of 0–0 draw and 0–2 are found to be

$$\frac{(0.5 \times 1.5)^0}{0!} e^{-0.5 \times 1.5} \times \frac{(1.5 \times 1.5)^0}{0!} e^{-1.5 \times 1.5}$$
$$\cong 0.05 \ (5\%)$$
$$\frac{(0.5 \times 1.5)^0}{0!} e^{-0.5 \times 1.5} \times \frac{(1.5 \times 1.5)^2}{2!} e^{-1.5 \times 1.5}$$
$$\cong 0.126 \ (12.6\%)$$

$$\text{(F.163)}$$

The probability that the team A scores first is given by the probability that the scoring time of team A is less that the scoring time of team B. We get using (F.161):

$$P(t_B > t_A) = \int_0^\infty f_{t_A}(t) P(t_B > t_A | t_A = t) dt$$

$$P(t_B > t_A | t_A = t) = \int_t^\infty \lambda_B e^{-\lambda_B x} dx = e^{-\lambda_B t}$$

$$P(t_B > t_A) = \int_0^\infty f_{t_A}(t) e^{-\lambda_B t} dt$$

$$= \int_0^\infty \lambda_A e^{-(\lambda_A + \lambda_B)t} dt$$

$$= \frac{\lambda_A}{\lambda_A + \lambda_B} = 0.25 \ (25\%)$$

$$\text{(F.164)}$$

From (F.164), the probability that the team B scores first is given by $\lambda_B/(\lambda_A + \lambda_B) = 0.75$. Now we determine the probability that four goals are scored during the game, based on the assumption that the two processes are independent from each other. When two independent Poisson processes merge, then the mean intensity of the new Poisson process is equal to the sum of the two intensities. Therefore, the probability of scoring four goals is given by

$$\lambda = \lambda_A + \lambda_B = 2 \ goals/hr$$

$$\frac{(\lambda t)^4}{4!} e^{-\lambda t} = \frac{(2 \times 1.5)^4}{4!} e^{-2 \times 1.5} \cong 0.168 \ (16.8\%)$$

$$\text{(F.165)}$$

Homework F.9   If a geographic region suffers two earthquakes with strength higher than 7 on the Richter scale every 150 years, determine the probability of having an earthquake within the next 25 years.

Example F.17 Impulsive Noise
Some communication and transmission systems, for example, power line communications and OFDM, are highly vulnerable to impulsive noise. The impulsive noise is usually modelled as Middleton's class A interference, which characterizes noise signals for which large amplitudes (impulses) occur with high probability. [5][6][7][8] Middleton's class A pdf may be expressed as

$$f(n) = \int_{-\infty}^{+\infty} f_{N|X}(n|x) f_k(x)\, dx$$

$$= \sum_{k=0}^{\infty} e^{-A} \frac{A^k}{k!} \frac{1}{\sqrt{2\pi}\sigma_k} e^{-n^2/(2\sigma_k^2)}$$

(F.166)

where $f_k(x)$ is defined by the Poisson process with $A = \lambda t$ in (F.154). The impulsive index A characterizes the impulse traffic. Here, λ denotes the arrival rate of the impulses, that is, the number of impulses per unit time, and t denotes the observation period. The class A rv may therefore be modelled as a weighted sum of Gaussian rv with zero mean and variance σ_k^2: [5][6][7][8]

$$f_{N|X}(n|x) = \frac{1}{\sqrt{2\pi}\sigma_X} e^{-n^2/(2\sigma_X^2)}$$

(F.167)

and

$$\sigma_k^2 = P\frac{k/A + \Gamma}{1 + \Gamma} = \sigma_G^2 \frac{k/A + \Gamma}{\Gamma}$$

$$= \begin{cases} \sigma_G^2 & \sigma_I^2 \to 0 \\ k\dfrac{\sigma_I^2}{A} & \sigma_G^2 \to 0 \end{cases}$$

(F.168)

where Γ provides a measure of the impulsiveness of the channel and P stands for the total mean power of the channel noise:

$$P = \sigma_G^2 + \sigma_I^2$$
$$\Gamma = \sigma_G^2 / \sigma_I^2$$

(F.169)

where σ_G^2 and σ_I^2 denote respectively the mean powers of the Gaussian and impulsive components of the channel noise. Hence, the impulsiveness and the power of the class A pdf can be controlled by the parameters Γ and P. Using the identity $e^A = \sum_{k=0}^{\infty} A/k!$, one can show that (F.166) reduces to Gaussian pdf for $\sigma_k^2 = \sigma_G^2$. Note that $\sigma_I^2 \to 0$ ($\Gamma \to \infty$) corresponds to as purely Gaussian noise channel, while $\sigma_G^2 \to 0$ ($\Gamma \to 0$) represents a purely impulsive noise channel with variance equal to the product of the random number k of impulses and the impulsive noise variance per impulse, σ_I^2/A (see (F.168)).

Figure F.16 shows the pdf of a class A impulsive noise for $A = 0.5$ and $A = 1$. One may observe from (F.168) and Figure F.16 that the increased values of Γ decrease the impulsiveness of the class A pdf.

Binomial distribution

The probability that k out of n events occur with a probability p and $n-k$ events occur with a probability $1-p$ is given by

$$P_k(p) = \binom{n}{k} p^k (1-p)^{n-k}, \quad k = 0, 1, 2, \cdots, n$$

(F.170)

where $0 < p < 1$. The binomial coefficient, defined by (F.86), denotes the number of combinations of k events within n. For example, in an experiment, we flip a coin $n = 3$ times where p denotes the probability of head (H) and $1-p$ is the probability of tail (T). Heads can occur $k = 0,1,2$ or 3 times in $n = 3$ trials. For $k = 2$, one has $\binom{3}{2} = 3$ combinations of heads, namely, HHT, HTH, THH.

The binomial pdf is defined by

$$f_X(x) = \sum_{k=0}^{n} \binom{n}{k} p^k (1-p)^{n-k} \delta(x-k) \quad \text{(F.171)}$$

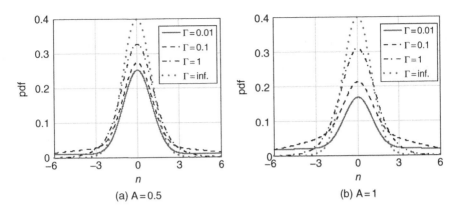

Figure F.16   Pdf of the Class A Distribution For $\sigma_G^2 = 1$ and $A = 0.5, 1$. Note that $\Gamma = \infty$ corresponds to Gaussian pdf.

Integrating (F.171), one gets the cdf of the binomial distribution

$$F_X(x) = \sum_{k=0}^{n} \binom{n}{k} p^k (1-p)^{n-k} u(x-k)$$

$$\text{(F.172)}$$

where $u(x)$ denotes the unit step function. The mean the variance are given by [3]

$$m_X = np$$
$$\sigma_X^2 = np(1-p)$$

$$\text{(F.173)}$$

The characteristic function is

$$M_X(s) = (1 - p + pe^{-s})^n \qquad \text{(F.174)}$$

Example F.18 Selection of Radiation-Hardened Components. [4]
A certain component should be chosen so as to meet the challenging requirements, for example, hardening, for operation in a nuclear radiation environment. We observe that only $p = 0.1\%$ of the available components meet this requirement. The selection must be made among how may components so as to find at

least one component meeting the desired requirement with a probability of 99%.

The probability P that at least one of n components meets the requirement is given by

$$P \geq 1 - P_0(p) = 1 - (1-p)^n \qquad \text{(F.175)}$$

Here, $P_0(p)$ denotes that none of the n components meets this requirement (see (F.170)). Inserting $P = 0.99$ and $p = 0.001$ into (F.175), one gets $n \geq \ln(1-P)/(1-p) = 4603$. This shows that the selection must be made between at least 4603 components.

References

[1] A. Papoulis and S. U. Pillai, *Probability, Random Variables and Stochastic Processes* (4th ed.), McGraw Hill: Boston, 2002.

[2] J. G. Proakis, *Digital Communications* (3rd ed.), McGraw Hill: New York, 1995.

[3] P. Z. Peebles, Jr., *Probability, Random Variables, and Random Signal Principles* (3rd ed.), McGraw Hill: New York, 1993.

[4] P. Beckmann, *Probability in Communication Engineering*, Harcourt, Brace and World, Inc.: New York, 1967.

[5] David Middleton, Statistical-Physical Models of Electromagnetic Interference, *IEEE Trans. Electromagnetic Compatibility*, vol. 19, no. 3, pp. 106–127, August 1977.

[6] Leslie A. Berry, Understanding Middleton's Canonical Formula for Class A Noise, *IEEE Trans.*

Electromagnetic Compatibility, vol. **23**, no. 4, pp. 337–344, November 1981.

[7] G. Pay and M. Şafak, Performance of DMT systems in dispersive channels with correlated impulsive noise, *VTC 2001 Spring*, 6–9 May 2001, vol. **1**, pp. 697–701.

[8] Y. Matsumoto and K. Wiklundh, A simple expression for bit error probability of convolutional codes under class-A interference, *IEEE EMC Conf.*, Kyoto, 22P1-1, 2009. www.ieice.org/proceedings/EMC09/pdf/22P1-1.pdf.

Index

Note: Page numbers followed by "f" refer to figures, page numbers followed by "t" refer to tables and page numbers in "**bold**" refer to first comprehensive discussion of the topic.

Digital Communications, First Edition. Mehmet Şafak.
© 2017 John Wiley & Sons Ltd. Published 2017 by John Wiley & Sons Ltd.
Companion website: www.wiley.com/go/safak/Digital_Communications

Printed and bound by CPI Group (UK) Ltd, Croydon, CR0 4YY

16/04/2025

14658564-0001